# PARASITOLOG

# PARASITOLOGY

*William C. Marquardt*
DEPARTMENT OF ZOOLOGY AND ENTOMOLOGY
COLORADO STATE UNIVERSITY

*Richard S. Demaree, Jr.*
DEPARTMENT OF BIOLOGICAL SCIENCE
CALIFORNIA STATE UNIVERSITY AT CHICO

MACMILLAN PUBLISHING COMPANY
New York
COLLIER MACMILLAN PUBLISHERS
London

Printed in the United States of America



Macmillan Publishing Company
866 Third Avenue, New York, New York 10022

Collier Macmillan Canada, Inc.

**Library of Congress Cataloging in Publication Data**

Marquardt, William C.
Parasitology.

Bibliography: p.
Includes index.
1. Parasitology. I. Demaree, Richard S.
II. Title. Title. [DNLM: 1. Parasites. 2. Parasitic diseases.
QX 4 M357p]
QL757.M33 1985 591.5'249 84-5688
ISBN 0-02-376250-0

Printing: 1 2 3 4 5 6 7 8 Year: 5 6 7 8 9 0 1 2

ISBN 0-02-376250-0

For Barb, Kay, Bill, and Joan

W.C.M.

To Nancy for sharing the time
that was rightfully hers

R.S.D.

# *Preface*

Almost every biology department has an undergraduate parasitology course, generally taught by those of us who find parasites an unending source of interest. Each of us teaches the subject from his or her own viewpoint and emphasizes those organisms we feel to be important or to which we have some attachment, for whatever reason. As such, this book is a personal statement of what we feel to be important in parasitology.

Parasites in the context of this book (protozoa, helminths, arthropods) have a direct impact on humans or on domesticated animals that is almost beyond comprehension. Five of the six diseases considered to be most devastating to human health by the World Health Organization (WHO) are caused by parasites. These are global problems that concern us all and that have an effect on the majority of people on earth. As human populations increase, the need for food and fiber increases, and parasitic diseases of livestock reduce the amount of food available to people everywhere. In a complex, changing world, parasites remain with us and are a part of the ecosystem with which we must contend. Our feeling is that parasitology is an aspect of applied biology, and because of the enormous impact of parasites on human activities we consider the subject of controlling parasitic diseases as the principal focus of our courses.

The question then is, "What do students need to know about principles of control?" A part of the answer is simple—they need to know what organism is causing the problem and to know about certain aspects of its biology. But control is not carried out in a vacuum; not only the parasite but all human institutions, culture, politics, economics, food habits, and religious beliefs influence a program's success or failure. Thus, we attempt to overlay structure and biology of an organism with the so-called real world. We will leave the assessment of our success to others, but our nonrandom, biased survey of selected students indicates that it is well received.

Probably all of us hope that students will complete our courses and go away with certain principles tucked tidily in their minds. The question is how to achieve this exalted goal. We conclude that students need a good grasp of what a particular parasite is all about before they can assess its impact and thus determine what kind of control program may be needed. Examples come from parasitic disease of humans, domesticated animals, and wildlife of interest to us. One should approach a control program differently, depending on what is considered (e.g., human health or economic return from a herd of cattle). We have tried to point out the similarities and differences in what is desirable and possible in various circumstances.

As the late Asa C. Chandler pointed out in his address entitled *The Making of a Parasitologist*, the wide range of knowledge required to be effective in parasitology is one of the truly challenging aspects of the field. Those of us who have worked on parasites for many years may approach adequate knowledge in a number of areas, but students do not usually have a grasp of vertebrate and invertebrate structure and physiology, let alone principles of pathology and pharmacology. We have tried to address these subjects with terminology that is as simple as possible, so that students will not be impeded by arcane words but will be able to see what the phenomenon is without difficulty.

Writing this book has sometimes been interesting, frequently enjoyable, but always educational. We hope that students will find equal dosages of interest, enjoyment, and education in reading and using the book.

## ACKNOWLEDGMENTS

Writing a book of this size becomes such a long and tortuous journey that one sometimes gets lost in the wilds of syntax, organization, and necessary simplifications. We were fortunate that many persons gave freely of their time and experience to guide us back to the main road and the completion of the odyssey. Gregory W. Payne and Eileen Schlesinger of Macmillan were patient and helpful in ensuring that the final product attained high standards. Reviewers of the manuscript in interim stages were Gerald W. Esch, John S. Mackiewicz, Théodore Lund, and William S. Romoser. Norman D. Levine and Donald Heyneman reviewed the completed manuscript and offered crucial insights that improved the final product greatly. We cannot help mentioning that Norman, in his inimitable way, edited the whole manuscript in detail, not only clarifying inept phrasing but pointing out errors of which we were unaware. We are grateful to all of the reviewers for their helpful criticisms and constructive suggestions. Whatever the shortcomings and errors of this book, we remain responsible for them.

Nancy Heisler typed most of the manuscript and eliminated many first-draft glitches; she was encouraging in her own way by admitting that even though the subject matter is repugnant, it is interesting.

W.C.M.
R.S.D.

# Contents

## PART 4

### *Acanthocephala, Nematomorpha, Mollusca, Annelida, and Pentastomida*



## PART 5

### *Arthropods*

# PARASITOLOGY

# 1

# *Introduction*

Nearly everyone has had some experience with parasites and knows in general terms what they are. If we have a dog or cat, it will require worming at some time or need to be treated for mange mites. Horse owners know that treatment for intestinal worms is needed periodically. We nearly all have had chiggers or ticks feed on us after a picnic in a rural area. Travelers to tropical areas take their chloroquine faithfully to prevent malaria. We will be concerned in this book primarily with parasitic worms, protozoa, and arthropods that are in some way important to humans. These organisms comprise only a small fraction of the total number that inhabit various parasitic niches throughout the biotic world.

In this introduction, we will answer four questions:

What is parasitism?

How does the parasitic relationship work?

How do the host and parasite respond to each other?

How are parasitic diseases controlled?

The themes developed here will be reiterated and reexamined throughout the book. It is likely that some of what is set forth here will make more sense as the student becomes better acquainted with the various organisms discussed. We encourage students to return to this introduction periodically with the hope that the basis of the definitions and concepts will become clearer as knowledge is accumulated.

## WHAT IS PARASITISM?

Organisms of the same and different species interact with one another in a myriad of ways. When organisms of different species have an intimate association, it is termed *symbiosis* (Read, 1970; Whitfield, 1979). DeBary coined this term in 1879 to encompass associations involving close physical contact between partners that are not of the same species. Students who have taken a beginning biology course usually know of various symbiotic relationships such as lichens, which are composed of an alga and a fungus, ants that keep aphids as a source of food, crabs that keep a sea anemone on their backs as camouflage, tapeworms that live in the intestines of mammals, and mosquitoes that need a blood meal in order to lay eggs. What these symbiotes have in common is a close association; neither benefit nor harm is implied in the term symbiosis.

The subject of this book is parasites, and parasitism is a particular subset of symbiosis.

**Parasitism:** an association between the populations of two species in which the smaller (parasite) is physiologically dependent on the larger (host), the prevalence of the parasites and the intensity of infection in the host population are nonrandom, and the parasite species has a higher reproductive potential than the host species; the parasite has the potential of harming the host.

Definitions of parasites and parasitism have changed

over time, and through the influence of Read (1970) and Crofton (1971a & b), some of the inconsistencies have been weeded out and the ideas reworked by a number of authors; among them is Whitfield (1979), who wrote a small, readable book about parasitism in general. The definition of parasitism we have given takes ideas from these authors and simplifies some of the concepts for use by incipient parasitologists. Before beginning a discussion of parasitism, we need two more definitions:

**Commensalism:** symbiosis in which there is no discernible damage to the host.
**Mutualism:** symbiosis which benefits both partners.

Consider the major thrust of these three definitions. Note that parasitism implies harm, or at least potential harm, commensalism implies no harm, and mutualism involves a relationship in which both partners benefit. Categorizing a symbiote, and having it remain in that category, is among the more difficult tasks. An example or two will point up the problem. *Entamoeba histolytica* is the cause of amebic dysentery in humans; it can become invasive and cause severe disease, but it often remains as a mild-mannered commensal in the gut. *E. histolytica* is therefore a commensal that can become a parasite. *Trypanosoma musculi* is a flagellated protozoan that lives in the blood of mice and does not usually cause damage to the host. Careful studies have shown that mice actually benefit from nutrients produced by the protozoa (Lincicome, 1971). By definition, then, *T. musculi* moves from a commensal to a mutualistic relationship with its host. Thus, organisms may move into parasitism under some conditions, or their status may change as we gain more knowledge.

In another vein, we know that many parasites damage the host, but we have no way of measuring the damage. Take, for example, the intracellular protozoa called the *coccidia*; the coccidia rupture the host cell at the end of a phase of development and then enter uninfected cells for further development. We know that cells have been destroyed, but in most coccidial infections there is no way to measure the damage. Another example comes from large ruminants in whom extraordinary volumes of tapeworm may be found without any measurable production losses. Nutrient that would have gone into meat, milk, or fiber is now in tapeworm tissue, but we have no way of measuring the economic loss because the host seems to compensate somehow for the presence of the worms.

We hope that the student will recognize that definitions are an essential part of precise communication in shorthand, but no definition fits all the situations for which it was designed. In addition, we are sometimes inconsistent in our use of particular terms. Organisms that traditionally have been called parasites are now covered by the term *symbiote*, but those of us who work in the field still call ourselves parasitologists, not symbiotologists, nor do we often distinguish between parasites and commensals in our research or teaching. Be that as it may, there are important implications in the preceding definition and in those given a little later.

Although this textbook concerns only certain eukaryotic parasites, there are a myriad of parasites in other taxonomic groups of protists, plants, and animals. Consider the range of some parasites:

Viroids 50,000 molecular weight
Viruses $5 \times 10^6$ molecular weight
Rickettsia 0.5 – 2 µm
Bacteria 1 – 10 µm
Yeasts 5 – 10 µm
Protozoa 1 – 150 µm
Nematodes 300 µm to 30 cm (or more)
Platyhelminths 1 mm to 10 m
Mollusks 0.5 – 10 mm
Ticks and mites 0.1 – 15 mm
Insects 0.1 mm to 2 cm
Horsehair worms up to 15 cm
Mesozoa up to 100 µm
Leeches 1 – 5 cm

This list is only a superficial approach to showing the widespread nature and diversity in parasitism.

Parasites pervade our world and our lives, and the organisms that have established themselves as parasites have arisen in nearly all groups of livings things. Parasitism is a subculture that goes almost unrecognized by most people, and few appreciate the impact of parasites in the biotic world. If we consider only a few human parasitic diseases, the World Health Organization (WHO) has stated:

> Furthermore, blood flukes infect 200 million people in 73 countries, and in some areas where dams have been built, the prevalence of infection has risen from 10% to nearly 100%. Filarial nematodes cause gross deformities or blindness in several hundred million people and there are still no satisfactory drugs to treat them. Chagas' disease, which is caused by a flagellated protozoan, affects 24 million people in Latin America, but there is still neither satisfactory treatment nor a vaccine to prevent infection. (*Tropical Disease Research*, No. 18, May 1982)

Parasitic diseases also limit livestock production. There are 160 million cattle and 286 million sheep and goats in Africa, for example, mostly raised by subsistence farmers and pastoralists. It is estimated that twice this number could be supported if two parasitic diseases, trypanosomiasis and East Coast fever, could be eliminated (Lewin, 1982).

In the political arena, the worst pejorative that can be used against individuals or groups is to call them "parasites." It is as if only seed eaters, herbivores, and predators are admirable organisms, and those that live inside them are behavioral degenerates worthy only of disdain. But, like most of us, parasites seek only to find a place in the world where they can live and propagate. If certain structures or sense organs are lost in parasites, it might be better to look upon this evolutionary change as specialization rather than degeneracy. A tapeworm is a marvel of specialization; it maintains intimate contact with the host in order to obtain nutrients through its surface, and it has retained only those structures necessary to live in the intestine of a vertebrate. Surely it is commendable to be economical with one's resources and not to produce eyes where there is no light, ears where there is no sound, or other structures that have no use.

Most of the organisms we deal with are *obligate parasites*, those that require a host for the completion of the life cycle. There are a few *facultative parasites* that can remain as free-living organisms but can invade a host if the proper conditions are presented. Parasites may live on the surface of a host (*ectoparasites*), or they may live in the internal organs (*endoparasites*). Those that come to a host occasionally to feed are sometimes called *temporary parasites*, and those that remain in a host are called *permanent parasites*, but this terminology is not often used.

Some parasites have life cycles in which there is both sexual and asexual development. In this case, it is necessary to distinguish between the kinds of hosts. The host in which the parasite is sexually mature is the *definitive host*, and the one in which the organism is sexually immature or reproduces asexually is the *intermediate host*. Sometimes a parasite is passed from one host to the next, with no development occurring; such a host is called a *paratenic* or *transport host*.

Nearly all parasites that are infectious for humans develop in other hosts as well. Those organisms are called *zoonotic agents* and they cause *zoonoses* (singular: zoonosis). Some authors subdivide this term into *zoonosis*, a disease humans acquire from other animals, and an *anthroponosis*, a disease other animals acquire from humans, but this is an unnecessary complication in terminology. As a general rule, the host from which humans acquire infection, or in which the agent normally cycles, is termed the *reservoir host*. This term is also used in livestock in cases where a domestic animal acquires a disease from a host in which infection is inapparent or the disease is of no importance. In fact, the connotation of the term *reservoir* is that the infection occurs in a host that we do not care much about.

Surveys are often done to determine whether and at what level disease agents occur in a population. The *prevalence of infection* is determined by taking a sample from a population and determining the proportion of the population infected. Note that this term is used when we dip into a population and pull out a sample at a particular time. In contrast, when we determine how many individuals become infected each year, or some other unit of time, we speak of the *incidence of infection*. The terms are also used for the occurrence of disease as well as infection. Figures are often expressed as percentages, but in human public health figures are given as the number of infections per hundred thousand population.

Over the years some looseness of terminology has developed in referring to infections and states of disease. The terms *infection* and *infestation*, for example, have often been used for quite similar conditions. We will use the term *infection* for nearly all parasitic associations, regardless of whether it is outside or inside or whether the organism reproduces in the host or not. Also, it is not easy to distinguish between the suffixes *iasis* and *osis*. We can speak of *helminthiasis*, which means infection with helminths. *Helminthosis*, on the other hand, means to be clinically affected by helminths. Unfortunately, terms such as *trypanosomiasis* refer to a disease state. We have generally adhered to commonly accepted terminology except where it may be necessary to distinguish between mere infection and clinical disease.

Additional kinds of hosts that we must consider are vectors and intermediate hosts. A *vector* is an arthropod, mollusk, or other invertebrate that transmits a parasitic agent to a vertebrate host. Vectors come in two models: *mechanical*, in which no development takes place, and *biological*, in which development or replication occurs. We have already discussed intermediate hosts in a general context, but we wish to point out similarities and differences between intermediate hosts and vectors. Note that a vector is an invertebrate of some sort. Thus, we rule out cattle as a vector of the beef tapeworm of humans, but cattle are intermediate hosts because the worm is immature in them. Mosquitoes are vectors of various viruses, but mosquitoes are definitive hosts as well as vectors of malaria, because the parasite is sexually mature in them. Snails are intermediate hosts and vectors of the blood flukes of humans; the worms are sexually immature in the snails. Tsetse are vectors of African sleeping sickness, but they are neither intermediate nor definitive hosts because the parasite does not reproduce sexually.

Diseases occur both in an explosive form, running through a population in a short period of time, and at a level involving only a small proportion of the population at any one time. When the course through the population is rapid, we speak of its being *epidemic* in humans or *epizootic* in other animals. When there are occasional but continuous cases, the terms are *endemic* and *enzootic* in humans and other animals, respectively. The terms used to cover both humans and animals can also be *epidemic* and *endemic*.

## HOW DOES THE PARASITIC RELATIONSHIP WORK?

With the exception of some kinds of interactions between the parasite and host, parasitism is not a unique relationship compared with free-living organisms. The same general ecological principles apply to both free-living and parasitic organisms, because both need a physical place in the ecosystem, adequate food, and the ability to reproduce.

The life cycles of some parasites seem rather bizarre at first, but many parasites follow a course of development similar to that of their free-living counterparts. A parasite may have a *direct life cycle*, one in which there is only a single species of host and transmission takes place directly from one host to the next by close contact or through development to the infective stage in the free-living environment. Most protozoa that live in the intestines of their hosts are transmitted through a cyst stage from one host to the next; *Entamoeba coli* is a commensal found in the large intestine of humans and it is transmitted by means of a resistant cyst; many free-living amebas also form cysts and use them to bide their time until they find proper conditions again for growth and multiplication. All roundworms or nematodes develop through five larval stages separated by four molts; the same stages are seen in both the free-living and the parasitic nematodes. Many parasitic nematodes develop to the third-stage larva as free-living forms and then wait until they reach a proper host before continuing their development. The parasitic cycle is simply a modification of the free-living one with the addition of a host in which the worm becomes sexually mature.

In other instances, a parasite may have a *complex* or *indirect life cycle*; the completion of the life cycle requires more than one host, most often of different species. A well-known example of an indirect life cycle is that of the organism that causes malaria in humans: transmission takes place through mosquitoes, and without the mosquito the organism would disappear. In some organisms such as the nematodes, the indirect cycle can be readily derived from the cycle of free-living forms; the cycle stops and starts on two or more occasions, but the same stages are always seen. In some instances, such as the tapeworms and digenetic flukes, the free-living precursors are so shrouded in evolutionary mystery that deriving the cycles of the present-day forms requires considerable conjecture.

The large reproductive potential of parasites is sometimes pointed to as a characteristic of parasites, and the numbers are truly impressive. For example, the large ascarid of humans and hogs produces half a million eggs per day, and apparently can do so for many months; the common liver fluke of sheep and cattle, *Fasciola hepatica*, lays about 5,000 eggs a day and can live in a sheep for ten years; one mosquito can infect 50 or more persons and bring about a small malaria epidemic all by itself. The question is whether this reproduction is different from that of free-living organisms. It has been calculated that one pair of house flies could produce $1.9 \times 10^{20}$ offspring from April to August and cover the earth 47 ft deep (Herms & James,

1961). A single oyster produces tens of thousands of eggs in a breeding season, and bacteria often have generation times of 20 min. Each organism develops its own strategy for ensuring the survival of its offspring, and some parasites are highly reproductive. Parasites are opportunistic, and given the chance, they will increase their numbers dramatically in a short period of time.

Most parasites are not randomly distributed through the population of available hosts. Crofton (1971a, b) has referred to this type of distribution as being overdispersed, but we have used the term *nonrandom* in our definition of parasitism. The concept of overdispersion or nonrandomness, together with other mathematical models of population ecology, is being used to better describe host–parasite population dynamics as well as to predict the occurrence of epidemics and the likelihood of controlling them.

Two useful concepts are the *r*- and *K*-strategies (Esch et al., 1977) and the equations expressing the dynamics of parasite populations (Anderson, 1976). In the former, the term *strategy* refers to an inherited characteristic that is shaped by the pressures of natural selection. Parasites that are *r*-selected put their energy into producing large numbers of offspring, usually within a short period. In this type of strategy, there are essentially no density-dependent effects and little or no competition. At the opposite end of the population are *K*-selected species, which produce only a few extremely fit offspring because density effects are maximal and the environment is nearly saturated with organisms (Pianka, 1970).

The equation for the *K*-strategy is

$$\frac{dN}{dt} = rN\left(\frac{K-N}{K}\right)$$

where $dN/dt$ = rate of population increase
$r$ = maximal growth rate
$N$ = number of individuals in the population
$K - N/K$ = realization of the maximal growth rate

The equation for *r*-strategy is

$$\frac{dn}{dt} = rN$$

Most parasites fall into the *r*-selected category because parasites usually are opportunists. Examples of *r*-selected parasites are mosquitoes, which reproduce rapidly when conditions are suitable (Chap. 55), and the dysentery ameba (Chap. 9), which has a short generation time in the host. At the far end of the spectrum we have some *K*-selected parasites such as tsetse, which serve as vectors of African trypanosomes; tsetse females produce one larva at a time and average only about nine offspring during a lifetime. Further complicating the situation are those parasites that have different strategies at different phases of the life cycle; examples are most taeniid tapeworms (Chap. 29) and Monogenea (Chap. 20). Last, reproduction in parasites is influenced by substances produced by the host such as antibody, hormones, and specific nutrients.

Population dynamics for parasites have been reviewed by Anderson (1976) for those that do not mature in the host and for those that multiply in the host. For the former, the expression is

$$\frac{dNt}{dt} = \lambda - \mu Nt$$

If the parasites multiply in the host, the expression is

$$\frac{dNt}{dt} = \lambda - (a + \mu)Nt$$

where $Nt$ = the number of parasites in the system
$t$ = time
$\lambda$ = the immigration rate
$\mu$ = the death rate
$a$ = constant birth rate per parasite per unit time

A more extended discussion is in Anderson (1976) and Hirsch (1977).

One common aspect of reproduction in parasites is alternation of generation. An asexual cycle is alternated with a sexual cycle. Although such a mode of development is not unknown in free-living organisms (ferns and aphids are examples), it is a theme seen in a number of taxa of parasites. The Apicomplexa in the protozoa alternate sexual and asexual development; some nematodes such as *Strongyloides* alternate parthenogenesis with sexual development; the digenetic trematodes have asexual development in the snail intermediate host and sexual reproduction in the vertebrate; some tapeworms have evolved means of asexual reproduction, the most striking one being the replication of the sexual organs in each segment, but a few reproduce also as larvae.

We discussed the supposed degeneracy of parasites

earlier, and the life-cycle patterns of parasites present an opportunity to consider the subject further. An organism such as a digenetic fluke may have three hosts in the life cycle and five or even more stages. Each of these stages requires control of development and has its own niche in the ecosystem. Some stages pass through the external environment for development or exist as free-swimming forms; others have an intimate association with at least two hosts and sometimes three. Fairbairn (1970) has considered the issue of whether parasites so exquisitely dependent on a host that they cannot survive elsewhere have lost genetic capacity. He concludes, in general, that parasites have the genome to survive and develop in other niches but do not use it. He takes his examples principally from helminths with complex life cycles and shows that they have the capacity for development and survival in a wide spectrum of niches as they pass through the life cycle. Further evidence can be drawn from *Plasmodium*, the cause of malaria. We generally consider this organism to be an obligate intracellular parasite, but the oocyst stage in the mosquito is extracellular. The organism has maintained the ability to develop extracellularly, but uses this capacity only in the vector. Thus, we see that parasites are not degenerate genetically; rather, they maintain a genome only a portion of which is used at any specific stage of the life cycle, and they are similar to other eukaryotes.

Nearly all parasites need to contend with the external environment at some stage of their life cycles. Many nematodes with direct life cycles feed, grow, and develop as free-living organisms before entering a host to become sexually mature. The type of environment that the free-living stages require limits the distribution of the parasites. A parasite's ability in its external stages to survive harsh conditions such as freezing and drying usually means that it will have a broader distribution than those susceptible to such conditions. Studies of the limits and optima for development and survival of a particular parasite represent the kind of basic biological study required for developing rational control programs. The parameters that are applied in defining the ecological and physiological requirements of free-living organisms pertain likewise to parasites during this phase of their lives.

It is when a parasite comes into contact with a host that it reaches the stage of physiological dependence that defines the unique nature of parasitism. The pattern of development for a eukaryotic (and some prokaryotic) parasite is to proceed with development in a particular environment (host or external) to the infective stage and then to stop and wait. When the potential host is contacted, development then proceeds. We speak of the organism being "triggered" by the environment in the host, and usually specific signals of temperature, pH, redox potential, and organic chemicals are sensed by the parasite and it is stimulated to proceed to its next stage of the life cycle. In many cases we know what the stimuli are, but we do not know the exact mechanism except in a few instances.

## HOW DO HOSTS RESPOND TO PARASITES?

When two organisms live in close, usually intimate association, each responds to the presence and activities of the other. This is especially true of the host–parasite association. First, we will discuss how the vertebrate hosts respond to the presence of parasites and then how parasites respond to the hosts.

Vertebrate hosts respond to the presence of nonself material, that is, parasites, in two ways. The first is called *nonspecific responses*. The host is able to differentiate "nonself" from "self," but these responses are not dependent upon *specific* recognition of a nonself molecule. The second is called *specific responses*, which depend upon exposure to, and specific recognition of, foreign or nonself molecules.

### *Nonspecific Responses*

One of the first lines of defense against small foreign invaders, be they bacteria or parasitic protozoa, is *endocytosis*, the process of ingestion of material. Endocytosis is often referred to as *phagocytosis*, literally cell eating, or *pinocytosis*, literally cell drinking; it is often impossible to separate the processes of ingestion of fluids and solids; thus, the term *endocytosis* is preferred to describe the overall process. Endocytosis is accomplished by several cell types, including monocytes and polymorphonuclear white blood cells, histiocytes in tissues, and the sinus-lining reticuloendothelial (RE) cells, especially those in the liver and spleen. The function of endocytosis is to engulf foreign material and digest it in lysosomes. Endocytosis can occur independently of specific responses but is greatly facilitated by certain specific responses such as *opsonizing antibodies* and serum proteins called *complement*. Detailed discussion of these topics is beyond the scope of this text (Barriga, 1981).

If the foreign invader is small, it may be surrounded by endocytotic cells and immobilized by a deposition of collagen around it. If the invader is large, a second type of nonspecific response occurs—the *inflammatory response*. This response is often divided into three stages. The first, *acute inflammation*, lasts from days 1 through 3. It is characterized by capillary dilation leading to fluid accumulation or *edema*, and an accumulation of polymorphonuclear white blood cells in the tissues, probably attracted by chemotaxis. The second phase, *subacute inflammation*, lasts from 3 days to over a week. This phase is characterized by the presence of mononuclear cells (monocytes and lymphocytes) in the perivascular space and fibroblasts that secrete collagen. Collagen secretion leads to the production of a fibrous capsule commonly referred to as a *scar*. The third phase, *chronic inflammation*, involves the same cells as subacute inflammation plus *plasma cells*. Often in chronic inflammation, such as that caused by a trapped schistosome egg, a *granuloma* will form. This is an ovoid arrangement of mononuclear cells surrounded by the fibrous connective tissue and cells that secrete it.

The third nonspecific response to parasites is sometimes called *abnormal growth responses*. These include *hyperplasia*, in which the activities of the parasite stimulate the host to produce an increased number of cells. A good example of this is the liver fluke, *Fasciola hepatica*, which induces the host to enlarge greatly the bile ducts, which the parasite then eats. Another abnormal growth response induced by parasites is *neoplasia* or cancer formation. A number of parasites are suspected of inducing neoplasia. The tapeworm larva of *Taenia taeniaformis* in the liver of rats, and the adults of the nematode *Spirocerca lupi*, in the esophagus of dogs, are associated with the formation of sarcomas, connective tissue cancers (Schwabe, 1955). The exact mechanism of parasite-induced sarcomas is unknown.

## *Specific Responses*

Parasite surfaces have characteristic macromolecules such as proteins and polysaccharides, which the host can recognize as nonself. The surface material, or a substance secreted by the parasite, is called an *antigen*. These antigens trigger the specific *immune response* when immunoglobulins (Ig), serum proteins secreted by plasma cells, attach to antigens using *specific molecular recognition*. This attachment usually triggers further host responses. There are five classes of immunoglobulins in humans—IgA, IgD, IgE, IgG, and IgM—which are distinguished by polypeptide differences. The basic Ig molecule is a Y-shaped structure (Fig. 1.1), with the antigen-binding site at the ends of the arms of the Y. Production and secretion of immunoglobulins into the serum is called the *humoral response* and is manifested in a number of different ways.

Each of the Ig molecules has particular functions. IgE responses are often elevated in helminth infections. IgE antibodies bind to mast cells and basophils; this binding of antigen to cell-bound IgE induces these cells to release vasoactive substances such as histamines, which increase capillary permeability.

In addition, IgM and IgG are very important in protozoan infections because they can activate the *complement system*. This system consists of nine protein complexes that can combine with many different antigen–IgM or antigen–IgG complexes. The first component of the complement system binds to antigen-bound IgM or IgG, initiating a sequential series of reactions involving up to nine components of the

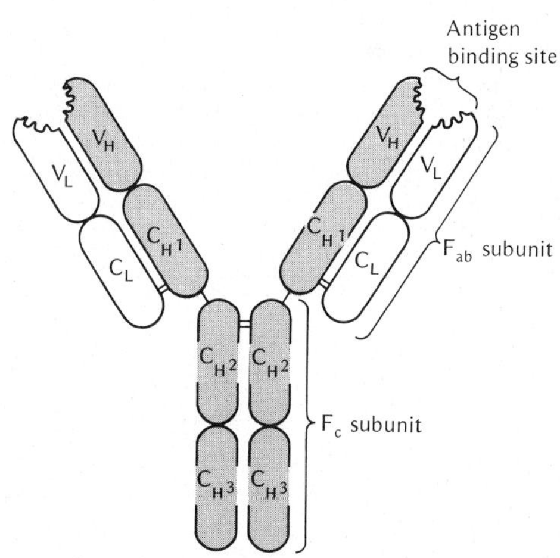


**FIG. 1.1**
Structure of immunologulin G (IgG). [Redrawn from Whitfield, 1979. *The Biology of Parasitism: An Introduction to the Study of Associating Organisms*. University Park Press, Baltimore.]

complement system. This series of reactions triggers a number of biological activities; one of the most important in parasitic infections is the damaging of cell membranes, which often leads to cell lysis and death. A serological test for the presence of activated complement, with its ability to lyse cells, is the *complement fixation test.*

The other half of the immune response involves not secreted immunoglobulins, but another kind of lymphocyte, the T-lymphocyte, which releases proteinaceous substances when bound to antigens. These substances, called *lymphokines*, cause nonspecific reactions on other cells, often leading to inflammation. This part of the immune response is referred to as *cell-mediated immunity (CMI)*. Lymphokines most relevant to parasitic infections include (1) *migration-inhibitory factors*, which prevent the migration of white blood cells; (2) *macrophage-stimulating factors*, which enhance macrophage activity against cells with the target antigen; (3) *chemotactic factors*, which attract inflammatory cells to the site of release; (4) *mitogenic factors*, which stimulate the division of lymphocytes; and (5) *cytostatic factors*, which delay or stop cell proliferation. The exact functions of most of these substances in parasitic infections are poorly understood and are the subject of many studies.

Both Ig-producing lymphocytes (B-lymphocytes, or plasma cells) and lymphokine-producing lymphocytes (T-lymphocytes) originate as stem cells in the bone marrow. The processing of these cells in the thymus (T-cells) or in an organ equivalent, the bursa in birds (B-cells), determines which response will be elicited. An outline of both processes is presented in Figure 1.2.

In general terms, macrophages (endocytotic cells) ingest and process parasite antigens. These "processed" antigens are then distributed on the surface of the macrophage. Both T- and B-lymphocytes contact the surface of the macrophage or parasite directly and are primed against the specific parasite antigen. Both types of lymphocytes are stimulated to divide and then produce either humoral antibodies (Ig) or lymphokines. Antigen combines with IgE bound to mast cells and basophils, causing these cells to release histamines. The subsequent increase in capillary permeability allows white cells to reach the parasite. T-cells combine with parasite surface materials (antigens) and release lymphokines, which attract more endocytotic cells and keep them there, releasing their cytotoxic products. The final result is tissue injury by the cytotoxic products and nonspecific inflammation which changes the parasite's immediate environment, usually to its disadvantage.

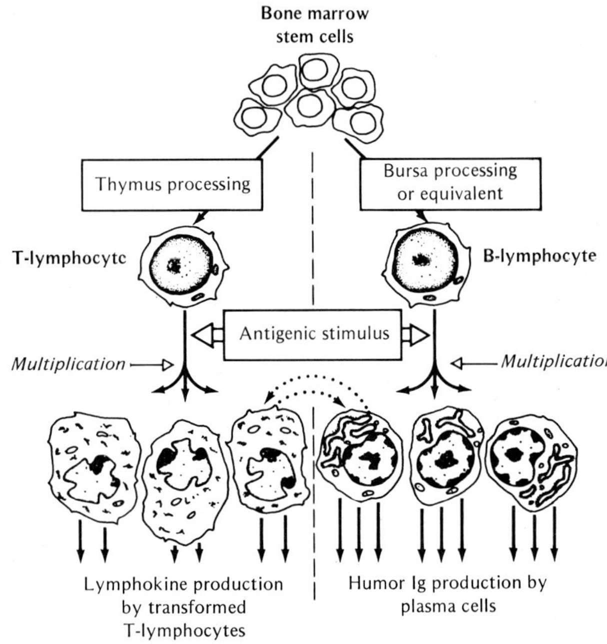


**FIG. 1.2**
The origin, processing, and activities of T- and B-lymphocytes. [Redrawn from Whitfield, 1979. *The Biology of Parasitism: An Introduction to the Study of Associating Organisms.* University Park Press, Baltimore.]

## *Parasite Countermeasures to Host Responses*

Hosts respond to the presence of parasites and take certain measures to eliminate them. Parasites can often react to these responses, and in some cases have evolved ingenious methods of escaping the host's responses. Many of these methods will be discussed with each particular parasite in the text, so only a couple of examples will be mentioned here. For additional information, see Bloom (1979), Whitfield (1979), and Barriga (1981).

One unique way to escape the host response is used by the blood-dwelling protozoan, *Trypanosoma brucei*, which causes African sleeping sickness (Chap. 3). These protozoa undergo *antigenic variation*. As the host recognizes their surface material and attacks it, the remaining parasites change the composition of their surfaces, and the original immune response is not effective against these new organisms. This change in surface can happen many times, leading to parasite population explosions followed by population crashes as the host's immune system responds to the new surfaces, resulting in another explosion, and so on.

Another countermeasure used by some tissue-dwelling parasites is to live inside host cells. Once the foreigner is inside the host cell, the host cannot recognize and attack it. Many protozoa such as *Toxoplasma gondii* and *Trypanosoma cruzi* exploit this method. Some carry this countermeasure one step further by living in macrophages. Macrophages ingest the parasites but are unable to destroy them with internal digestive enzymes. *Leishmania* lives in functionally active macrophages, but the manner in which it escapes the digestive enzymes is incompletely known (Bloom, 1979).

The human blood flukes of the genus *Schistosoma* have evolved *antigenic mimicry*. Adult worms in host blood vessels coat themselves with host-produced molecules and thus appear to the host as self (Damian, 1964).

## HOW ARE PARASITIC DISEASES CONTROLLED?

Parasites cause diseases in humans, in domesticated and companion animals, and in wild animals. Each of these four kinds of hosts poses different problems that require different approaches to alleviating their effects. Although we will approach each agent or group of agents individually in the rest of the book, we can set forth some general principles here that apply to most of them.

As indicated earlier, parasites are constant companions of all groups of animals, and they have the potential of harming the host. We are still rather ignorant of the reasons parasites cause disease in some instances but remain innocuous in others; however, we can list some of the factors that contribute to disease:

1. Overpopulation
2. Manipulation of the environment
3. Monoculture and inbreeding
4. Migration and movement of humans and livestock
5. Individual susceptibility

Parasites are among the many factors that modulate populations of organisms. Cycles in which numbers of animals rise and fall, sometimes with regularity, are familiar. Food, space, social interactions, and predation all contribute to the number that can be supported at any one time, but infectious diseases (in the broad sense) are also a factor. A part of the definition of parasitism is the parasite's higher reproductive potential than that of the host, and this is one of the keys to the inception of disease and modulation of populations. We need merely to tip the scales slightly for the number of parasites in the ecosystem, and therefore exposure to them, to increase.

One way to increase exposure to the parasite is to provide a large number of potential hosts in close association with one another. We have all had the experience of head colds sweeping through a school until every other person seems to have a cold or is trying to get over one. Infectious agents normally in the population are more likely to reach domesticated animals that are crowded together on pasture or in buildings. We do not subscribe to the notion that diseases were nonexistent in primeval conditions, but it is clear that the probability of transmission increases with crowding, and an epidemic may ensue.

Humankind has had an enormous impact on the biosphere, and through manipulation of the environment for our own ends we have brought about disease problems in ourselves, domesticated animals, and wild animal populations. Some of the problems are a result of overpopulation, as discussed in the previous paragraph, but many are a direct result of other environmental changes. An example that readily comes to mind is irrigation, which may increase habitat for vectors of disease agents such as mosquitoes and blackflies. Roads also may interrupt normal migration patterns of wild-life so that animals tend to frequent certain areas on a year-round basis instead of moving to habitats where they have not been for perhaps a year. The ability to move domesticated animals by rail or truck has made markets accessible to those who raise animals in remote areas, but the shipping may precipitate various kinds of microbial and parasitic diseases, because the animals are upset by the noise, crowding, and lack of proper food and water. When animals were trailed to market on foot, the time required was long, and adequate food and water were needed to keep the stock in condition; the more leisurely pace prior to high-speed transport reduced the probability of an epizootic.

As we have gained more control over our environment, we have tended to stick to those things that have worked well and provided a measure of security. Animals and plants are selected for high production in a particular environment, and genetic diversity or heterosis is bred out of a population. The result is that the mutation or importation of a parasitic agent may cause a wholesale die-off of the desirable plant or animal. The related aspect, monoculture, produces the same end result: an area where there is a single crop or species of domestic animal is particularly susceptible to an epizootic.

Related to monoculture and lack of heterosis is the migration of humans and livestock. If a parasitic agent is brought into an area that has had no previous exposure to it, the population is likely to be completely susceptible to it. A few examples will suffice. Smallpox was introduced into one of the Northern Plains Indian tribes, the Mandan, and within a few years, the tribe was completely gone. Tropical Africa was known as the "white man's grave" largely because of complete susceptibility to various diseases such as malaria. Syphilis was probably introduced into central Europe by soldiers who had returned from campaigns in the Middle East and Africa late in the fifteenth century; the disease was so virulent that persons usually died within six months of the first signs of infection. European cattle introduced into sub-Saharan Africa die from the protozoan disease nagana or trypanosomosis, while the local breeds are relatively resistant to the infection.

Although all of these factors leading to disease are interrelated to an extent, individual susceptibility and the current physiological state of a potential host are always factors. The definition of parasitism includes the statement that prevalence is nonrandom. A number of factors, including behavior, age, sex, reproductive state, and so-called stress factors, influence whether a potential host may become infected or diseased as a result of infection. At any rate, for all of these reasons, only a portion of the population harbors a sufficient number of parasites to show the effect. A corollary is that infection is not synonymous with disease, a theme which we reiterate frequently. The question of what precipitates a disease has an answer of extraordinary complexity and, more important, obscurity.

The question then becomes, "What means do we have at hand for reducing the impact of various disease agents?" If we were to make a list of the general techniques, it would include the following:

1. Administer drugs or chemicals to prevent or cure the infection.
2. Reduce the population of reservoir hosts.
3. Reduce the population of vectors.
4. Improve the immunity of the potential hosts either naturally or artificially by vaccination.
5. Alter the physical environment to reduce the probability of transmission; sewage disposal might be included here.
6. Avoid areas of high risk.
7. Protect the individual from infection by some sort of barrier, either physical or spatial.
8. Control the parasites biologically through predators or infectious agents that will reduce their number.
9. Separate by age class.
10. Test and slaughter.

All of these general methods except the last two may be applied equally to parasitic infections of humans and other animals. We will be considering all of these methods in detail throughout the rest of the book, so we will not give examples at this point.

Portions of control programs require dissemination of information about a particular disease complex; we put this in the general category of "education." Education is part of an overall control program, but should not be considered as a control method in itself. Education is unimportant until it brings about a change; the change may be an altered conception of how the world works, or it may be a change in behavior. It is the latter that is the control method. Education can bring about a change in behavior, and thereby prevent disease, in a number of ways: cooking shellfish to prevent paragoniamisis, avoiding cats during pregnancy to prevent congenital toxoplasmosis, not going into water that may have cercariae of blood flukes.

In many instances, students feel that eradication of a particular pathogenic agent is the desired objective. This is true in some instances, but it is seldom feasible, and may not be desirable in the long run. Eradication of a parasitic agent is feasible only when there is a single species of host and survival outside the host is so short as to be negligible. The best-known example of an agent that has been eradicated is smallpox virus; humans are the only host and transmission takes place directly from person to person with short survival outside the host under most conditions. Texas cattle fever was eradicated from the United States by a program of tick control, quarantine, and inspection; the ticks are principally on cattle, and the causative agent develops poorly in other hosts. But note that we speak only of the United States; the disease still occurs in Mexico and other parts of the Western Hemisphere, so the risk of reintroduction is still present. Where such a risk exists, eradication, or even a high level of control, may be inadvisable because a large segment of the population will lose its immunity and reintroduction may result in a catastrophic epidemic.

Finally, when speaking of control we need to consider the physician's maxim: first, do no harm. Another way to put this is, think before you act. We now know of long-term effects of some chemicals such as DDT, but even if this had been known back in the 1940s, the chemical would have been used anyway; perhaps the use would have been different, but the problems needed solution, and DDT was the answer. On the other hand, there are situations that require some deliberation before acting. Stable malaria is an example; in this epidemiologic form of the disease, transmission takes place continuously, or nearly so, and the population maintains a fairly high level of immunity. If transmission is interrupted, immunity in the population wanes. Then, if there is reintroduction, a high proportion of the population may be disabled for an extended period. The same is true in gastrointestinal roundworms of sheep and cattle. If the population of worms is depressed sufficiently, young animals will fail to develop a natural immunity and, when exposed, will be almost completely susceptible. It is essential to control various diseases, but we need to reduce the impact of the diseases as much as possible without bringing about some other, more severe effect. The dilemmas faced by those in disease control are both procedural and moral. There are risks in administering drugs, because no drug lacks side effects. There are long-term effects of controlling certain diseases because of the possibility of overpopulation when one of the modulating factors has been removed. Perhaps the way all parts of the ecosystem interlock is illustrated as well as anywhere in the epidemiology of disease. A change in the system one way or another is likely to have unanticipated ramifications. But this is no reason not to try, which is the reason for the rest of this book.

## REFERENCES AND FURTHER READINGS


Anderson, R. M. 1976. Dynamic aspects of parasite population ecology. In C. R. Kennedy (ed.), *Ecological Aspects of Parasitology*. North Holland, Amsterdam, pp. 431–462.

Barriga, O. O. 1981. *The Immunology of Parasitic Infections*. University Park Press, Baltimore.

Bloom, B. R. 1979. Games parasites play: How parasites evade immune surveillance. *Nature* **279**:21–26.

Crofton, H. D. 1971a. A quantitative approach to parasitism. *Parasitol.* **62**:179–193.

Crofton, H. D. 1971b. A model of host–parasite relationships. *Parasitol.* **63**:343–364.

Damian, R. 1964. Molecular mimicry: Antigen sharing by parasite and host and its consequences. *Am. Naturalist* **98**:129–149.

Esch, G. W., T. C. Hazen, and J. M. Aho. 1977. Parasitism and r- and K-selection. In G. W. Esch (ed.), *Regulation of parasite populations*. Academic Press, New York, pp. 9–62.

Fairbairn, D. 1970. Biochemical adapation and loss of genetic capacity in helminth parasites. *Biol. Rev.* **45**:29–72.

Herms, W. B., and M. T. James. 1961. *Medical Entomology*, 4th ed. Macmillan Publishing Co., New York.

Hirsch, R. P. 1977. Use of mathematical models in parasitology. In G. W. Esch (ed.), *Regulation of Parasite Populations*. Academic Press, New York, pp. 169–207.

Lewin, R. 1982. Nairobi laboratory fights more than disease. *Science* **216**:500–503.

Lincicome, D. R. 1971. The goodness of parasitism: A new hypothesis. In T. C. Cheng (ed.), *Aspects of the Biology of Symbiosis*. University Park Press, Baltimore, pp. 139–227.

Pianka, E. R. 1970. On r- and K-selection. *Am. Naturalist* **104**:592–597.

Read, C. P. 1970. *Parasitism and Symbiology*. Ronald Press Co., New York.

Schwabe, C. W. 1955. Helminth parasites and neoplasias. *Am. J. Vet. Res.* **16**:485–488.

Whitfield, P.J. 1979. *The Biology of Parasitism*. University Park Press, Baltimore.

# PART 1

# *Protozoa*

# 2

# *Introduction to the Protozoa*

The protozoa are single-celled animals. They are generally small, with a size range from 2 to about 2000 μm; most of them are between 5 and 100 μm. Only some of the larger ones can be seen with the naked eye, so it was not until the time of the Dutch microscopist Anton van Leeuwenhoek (1632–1723) that the protozoa were known. Van Leeuwenhoek wrote a series of papers published in the *Philosophical Transactions of the Royal Society* and the *Memoirs of the Paris Academy of Science* between the years 1674 and 1716. He described a number of microorganisms and is considered to be the first person to have seen and described both free-living and parasitic protozoa. Among the parasitic protozoa he saw were oocysts of *Eimeria steidai* and trophozoites of *Giardia duodenalis*.

Protozoa are generally considered to be composed of a nucleus, cytoplasm, and a single limiting outer membrane (Fig 2.1A). By and large, this is true, but many protozoa are multinucleate (ciliates and amebas) (Fig. 2.1B), and some have more than one cell (colonial phytoflagellates and ciliates) (Fig. 2.1C). Clearly, some of them are transitional between the protozoa and the metazoa. Any definition of the protozoa is rather nebulous because they fit no strict category. Kudo (1966) defines protozoa as "unicellular animals. The body of a protozoan is morphologically a single cell and manifests all characteristics common to the living thing."

The protozoa represent a unique line of evolution. Instead of tissues and organ systems, which have evolved in the metazoa, the protozoa have developed *organelles*. Organelles are cellular elaborations that perform the same functions as tissues and organs in the so-called higher animals. The organelles that most biology students know of are *cilia, flagella*, and *pseudopodia*, which occur in the protozoa and protophyta studied in general biology courses. These organelles function in locomotion and food gathering, depending on the species. Some other organelles are the *pulsatory vesicle* (contractile vacuole), which serves to maintain a proper osmotic concentration in the cytoplasm of the organism, the *infraciliature*, which lies under the pellicle of ciliated protozoa and probably serves as a coordinating mechanism for the cilia (Fig. 2.2) and *rhoptries*, which are found in members of the Apicomplexa and are a portion of the complex of organelles that probably aid in penetration of host cells. Additional organelles will be mentioned during discussions of various groups and species.

Protozoa are strictly aquatic organisms. They must have an aqueous environment for feeding, locomotion, and reproduction. Many protozoa form cysts that may be protective or may function in some phases of the reproductive cycle. But the fact remains that the bulk of protozoan activities occur in a fluid. Within this one limitation, the protozoa have been a huge success. Members of the group have invaded every ecologic niche imaginable. The free-living forms are found in all marine and fresh waters as well as in soils.

When we look at the parasitic protozoa, we find that

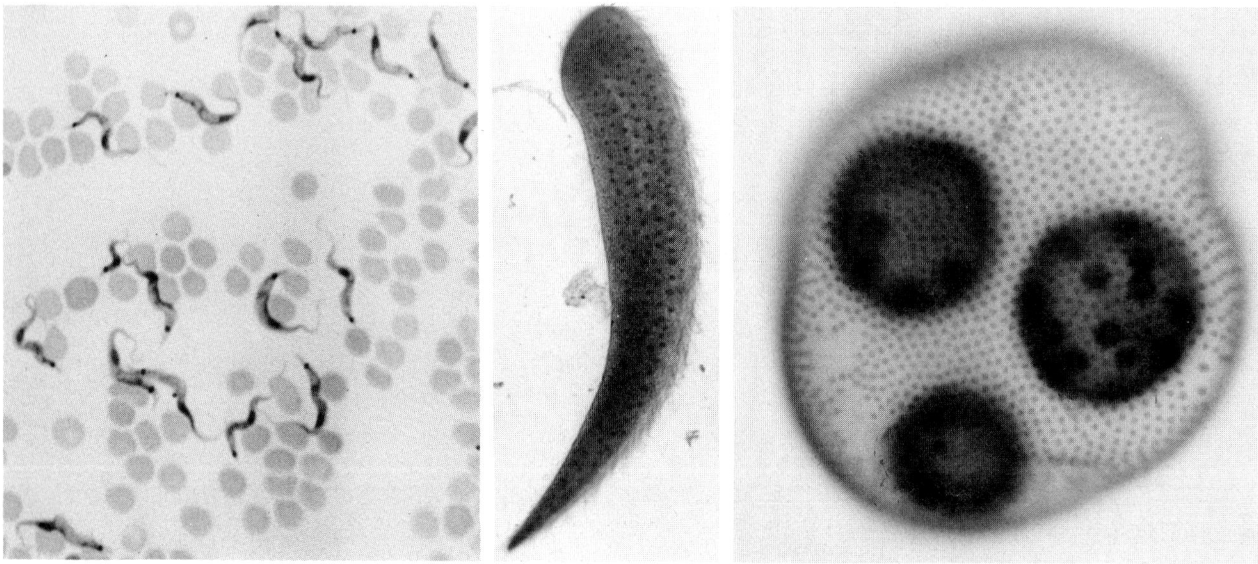

A B C

**FIG. 2.1**
(*A*) Uninucleate *Trypanosoma lewisi*; (*B*) multinucleate *Opalina* sp.; (*C*) colonial *Volvox* sp.

nearly every species of metazoan has a complement of protozoa living in it. Among the members of the nonvertebrate phyla there is almost no species lacking protozoan parasites, and vertebrate species are likely to be veritable zoos.

Within the body of an animal, protozoan parasites are often found in every organ system and sometimes, such as in the intestinal tract, a dozen or more related species may live side by side. To illustrate the tissues and organs that may serve as places of residence and multiplication for parasitic protozoa of humans, the following table gives a single example of protozoa from each of a variety of tissues and organs.

| TISSUE OR ORGAN | PROTOZOAN |
|---|---|
| Skin | *Leishmania tropica* |
| Eye | *Toxoplasma gondii* |
| Brain | *Naegleria fowleri* |
| Mouth | Trichomonas sp. |
| Small intestine, anterior | *Giardia duodenalis* |
| Small intestine, middle | *Pentatrichomonas hominis* |
| Large intestine | *Entamoeba histolytica* |
| Lungs | *Trichomonas* sp. |
| Reproductive tract | *Trichomonas vaginalis* |
| Liver | *Plasmodium vivax* |
| Spleen | *Leishmania donovani* |
| Skeletal muscle | *Sarcocystis* spp. |
| Red blood cells | *Plasmodium falciparum* |
| Blood serum | *Trypanosoma gambiense* |
| Cardiac muscle | *Trypanosoma cruzi* |
| Kidney | *Nosema* sp. |

Of these organisms, all but two or three are medically important—and bear in mind that the list is not exhaustive.

Given that there are a lot of parasitic protozoa, the question we need to consider is what their importance to humans is in causing disease in human populations or bringing about economic losses in domesticated animals or harvestable wild populations. Losses may come about through death, debility, production losses, reproductive failures, or condemnation of food because of disease hazard or esthetic considerations. As emphasized elsewhere, infection and disease are not synonymous; a host infected with a particular pathogen may or may not be clinically ill.

Two examples from human infections, malaria and African sleeping sickness, give some conception of losses from human and animal infections. In the case of malaria, Jones (1967) summarized materials from a variety of sources and states that before intensive efforts

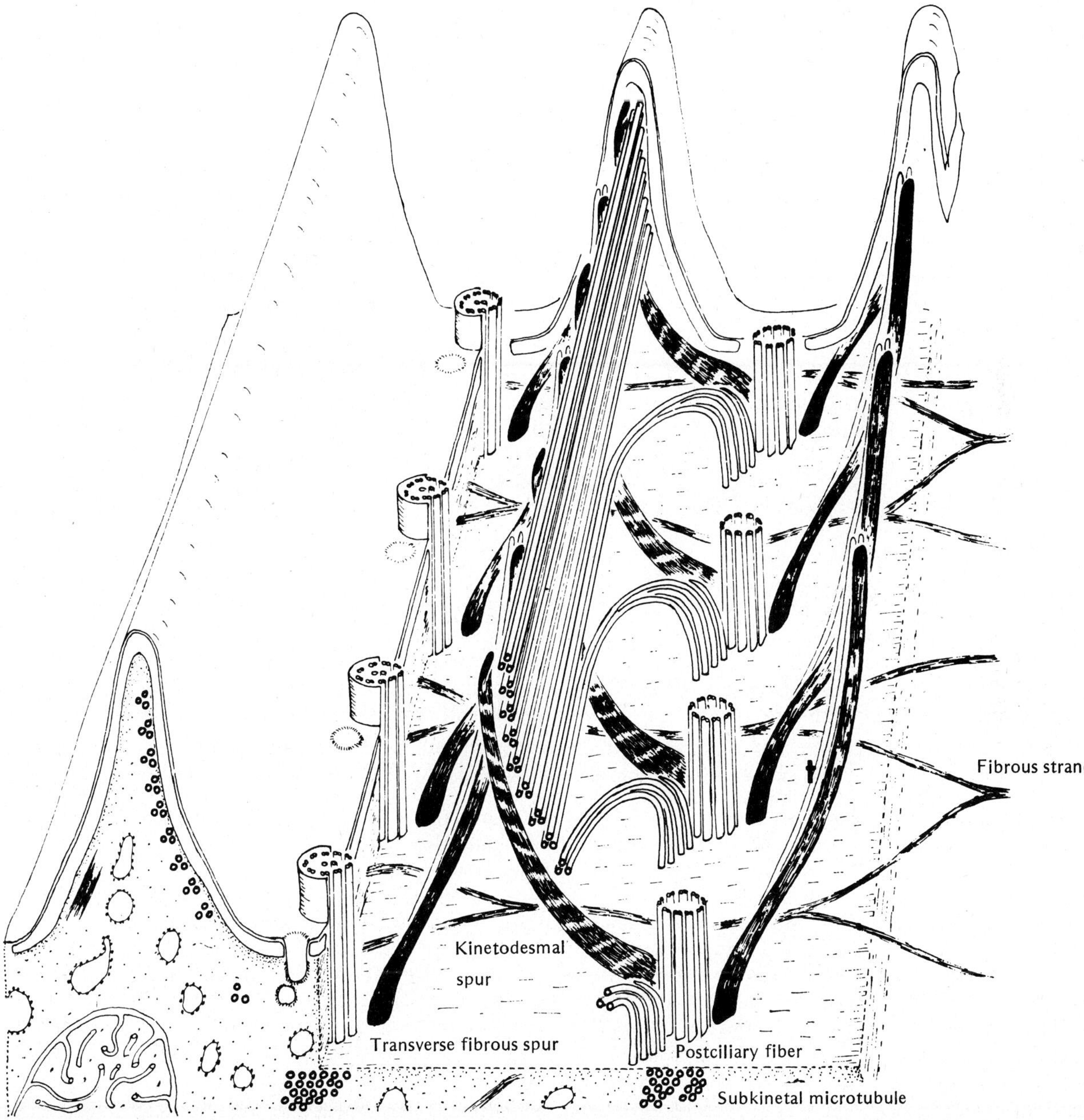


**FIG. 2.2**
Diagram of the infraciliature of *Brooklynella.* [From Lom and Corliss, 1971.]

at control in the late 1940s, there were an estimated 350 million cases of malaria annually worldwide. Of these, about 3 million died. Now, after 30 years of control and eradication programs, there are about 100 million cases of malaria annually.

African sleeping sickness has an impact on both human and animal health. The causative agent, *Trypanosoma brucei*, has subspecies that infect either humans or large herbivores (see Chap. 3). The disease complex is distributed in sub-Saharan Africa over an area of 4 million square miles, an area approaching that of the 48 contiguous United States. Historically, the disease complex has limited human colonization of that huge area, affecting human ability to survive or, alternatively, to raise livestock. Although progress in control has been achieved during the last 25 years, efforts have by no means been completely successful.

In discussing the various diseases in this section, an attempt will be made to assess the impact of each of them on human welfare. Malaria and African trypanosomiasis are only two of many that are important to humankind.

Protozoan classification is based on organelles of locomotion (kind and number), mode of obtaining and storing food, structure at both light and electron microscope levels, and life cycles. Because knowledge of the protozoa is continually unfolding, classifications both at higher and lower taxa continue to be changed. Concepts of classification presented in Kudo (1966) and Levine (1973) differ from each other and from the scheme presented at the end of this chapter.

In looking at the classification, certain features stand out. Grouped together in the first phylum, the Sarcomastigophora, are the organisms that move either by flagella or by forming pseudopodia. At first it would appear that the grouping is unnatural; however, it takes into account that there are organisms that can have either flagella or pseudopodia. Among parasitic protozoa, in this ambivalent position are *Histomonas meleagridis*, the cause of blackhead in turkeys, and *Naegleria fowleri*, a facultative parasite, both of which show ameboid and flagellated phases. Despite the taxonomic association of the flagellates and sarcodines, all but a small percentage fall clearly into one group or another and maintain their particular characteristics under normal conditions.

Members of the subphylum Opalinata (Chap. 11) bear special mention because they were classified with the ciliates in times past. The opalinids are all parasites that live in the intestinal tracts of frogs and toads; a few have been reported from salamanders, reptiles, and fish. They are covered with short flagella and superficially resemble holotrichous ciliates. But there are fundamental differences between opalinids and ciliates in infraciliature, nuclear structure, and asexual and sexual reproduction. They are clearly not ciliates, and the best judgment is that they are closely related to the flagellates (Corliss, 1979).

Within the subphylum Mastigophora there are two classes, Phyto- and Zoomastigophorea—the plantlike and the animal-like flagellates, respectively. The phytoflagellates lie on the borderline between plants and animals; some are clearly plants, obtaining energy from sunlight, and others are clearly animals, obtaining energy by eating particulate food. But there are those that do both and are clearly related, as in *Euglena*, which has chloroplasts, and *Astasia*, which is quite similar but lacks chloroplasts. One is left unable to draw the line between plants and animals. Such anomalies merely confirm that classifications are mental constructs and any man-made model of the world is, at best, an approximation.

With few exceptions the phytoflagellates are free living. Only a few euglenoids are parasitic, and we will not consider the group further. The alga Chlorella sometimes infects mammals (Kaplan et al., 1983).

The Zoomastigophorea have no members that gain energy directly from sunlight, and most of the species are parasitic. The group includes a wide array of members that are medically or economically important, but most of the species normally cause little damage to their hosts or seem to be of no consequence to humans.

The phylum Apicomplexa includes those organisms that, until the late 1960s, were included in the Sporozoa. The definition of the Sporozoa was an embarrassment to protozoologists since their only common feature was that all were parasitic. In the late 1950s, as techniques of fixation and thin-sectioning for the electron microscope developed, patterns of structure and development began to emerge that had only been suspected through studies at the light microscope level. The structures seen in the malarias, the coccidia, and the gregarines made up what came to be called the *apical complex* (see Fig. 12.1). At least some of these organelles were present in certain organisms, and Levine and Scholtyseck independently presented about the same revised classification at the second International Congress of Parasitology in 1970.

The Apicomplexan subclass Piroplasmia has members that do not form spores are are tick transmitted. The subject of sexual reproduction in the piroplasms has been debated for generations by workers, but evidence for sexual reproduction has now been presented for both *Theileria* and *Babesia.*

The Myxozoa and Microspora both have polar filaments—coiled, threadlike organelles that are explosively released when the spores are ingested by a host. The two phyla were formerly placed in a single class, the Cnidosporidia (Kudo, 1966), but there are clear differences between them, and unique characteristics, which justify phylum status for each. For example, the polar filament of the Myxozoa (see Fig. 18.1) is solid and the spore is formed by a number of cells. In contrast, the polar filament of the Microspora is hollow, the organism escapes from the spore through the tube (see Fig. 19.1), and the spore is unicellular in origin.

The ciliates have long been recognized as justifying subphylum status because of their unique structures and modes of reproduction (Chap. 11). Much of the integrative work on the systematics of ciliates was done by students and colleagues of Fauré-Fremiet in France and Corliss in the United States. Their schemes of classification based on infraciliature and modes of development (Corliss, 1979) have gained wide acceptance. Corliss (1979) proposed that the ciliates are sufficiently different from other protozoa to justify phylum status, and this was incorporated into the 1980 classification presented at the end of this chapter.

The parasitic protozoa have characteristics that ally them in some ways more with microbial agents (viruses, bacteria, rickettsiae) than with other eukaryotic parasites (helminths, arthropods). One characteristic is multiplication within the host. Also, they are transmitted from one host to another nearly always as a resistant form or through a biological vector in which there is multiplication.

## KINGDOM PROTISTA

### *Subkingdom Protozoa*

The protozoa range in size from 1 μm to 500 mm or more, most being between 5 and 250 μm. Nearly all are holozoic or saprozoic, but a few are holophytic. The principal line of evolution in the group has been through subcellular specializations or organelles, which function in feeding, locomotion, osmoregulation, and reproduction. Most protozoa have a single vesicular nucleus, but some are multinucleate and the Ciliophora are heterokaryotic. Sexual and asexual reproduction are present, depending on the group. Some protozoa are colonial; many live in loricae, and a number are stalked at some time in their life cycle. They are both free-living and parasitic.

**Phylum 1. Sarcomastigophora Honigberg & Balamuth, 1963**
Single type of nucleus, except in heterokaryotic Foraminiferida: sexuality, when present, essentially syngamy; flagella, pseudopodia, or both types of locomotor organelles.

**Subphylum 1. Mastigophora Diesing, 1866**
One or more flagella typically present in trophozoites; asexual reproduction basically by intrakinetal (symmetrogenic) binary fission; sexual reproduction known in some groups.

**Class 1. Phytomastigophorea Calkins, 1909**
Typically with chloroplasts; if chloroplasts are lacking, relationship to pigmented forms clearly evident; mostly free-living. Ten orders.
*Cryptomonas, Ceratium, Astasia, Euglena, Peranema, Pandorina, Volvox*

**Class 2. Zoomastigophorea Calkins, 1909**
Chloroplasts absent; one to many flagella; ameboid forms, with or without flagella, in some groups; sexuality known in few groups; a polyphyletic group.

**Order 1. Choanoflagellida Kent, 1880**
One flagellum, inserted apically, with proximal part surrounded by ring of tentacles (collar); with membranous sheath or basketlike lorica composed of siliceous costae; stalked or free-swimming; free-living.
*Monosiga*

**Order 2. Kinetoplastida Honigberg, 1963 emend. Vickerman, 1976**
One or two flagella arising from depression; flagella typically with paraxial rod in addition to axoneme; single mitochondrion (nonfunctional in some forms) extending length of body as single tube, hoop, or network of branching tubes, usually containing conspicuous Feulgen-positive (DNA-containing) kinetoplast (nucleoid) located near flagellar kinetosomes; Golgi apparatus typically in region of

flagellar depression, not connected to kinetosomes and flagella; parasitic (majority of known species) and free-living.
*Trypanosoma, Leishmania, Leptomonas, Bodo*

**Order 3. Proteromonadida Grassé, 1952 emend. Vickerman, 1976**
One or two pairs of heterodynamic flagella without paraxial rods; single mitochondrion, distant from kinetosomes, curving around nucleus, not extending length of body, without Feulgen-positive kinetoplast; Golgi apparatus encircling band-shaped rhizoplast passing from kinetosomes near surface of nucleus to mitochondrion; cysts present; parasitic.
*Karotomorpha, Proteromonas*

**Order 4. Retortamonadida Grassé, 1952**
Two to four flagella, one turned posteriorly and associated with ventrally located cytostomal area bordered by fibril; mitochondria and Golgi apparatus absent; intranuclear division spindle; "semi-open" mitosis reported from one genus; cysts present; parasitic.
*Chilomastix, Retortamonas*

**Order 5. Diplomonadida Wenyon, 1926 emend. Brugerolle, 1975**
One or two karyomastigonts; genera with two karyomastigonts with twofold rotational symmetry or, in one genus, primarily mirror symmetry; individual mastigonts with one to four flagella, typically one of them recurrent and associated with cytostome, or, in more advanced genera, with organelles forming cell axis; mitochondria and Golgi apparatus absent; intranuclear division spindle; "semiopen" mitosis reported in one genus; cysts present; free-living or parasitic.
*Giardia, Hexamita*

**Order 6. Oxymonadida Grassé, 1952**
One or more karyomastigonts, each containing four flagella typically arranged in two pairs in motile stages; one or more flagella may be recurrent, adhering to body surface for greater or lesser distance; kinetosomes of flagellar pairs connected by paracrystalline structure ("preaxostyle") in which are embedded anterior ends of axostylar microtubules; one to many axostyles per organism, contractile in many genera; mitochondria and Golgi apparatus (including Janicki-type parabasal apparatus) absent; division spindle intranuclear; cysts in some species; sexuality in some species; parasitic.
*Monocercomonoides, Oxymonas, Pyrsonympha*

**Superorder 1. Parabasalidea Honigberg, 1973**
Typically at least some kinetosomes, bearing flagella or barren, arranged in pattern and associated with rootlet filaments characteristic of trichomonads; one or more argentophilic Janicki-type parabasal apparatuses (Golgi complexes applied to filaments with Type A periodicity) present; mitochondria absent; division spindle extranuclear.

**Order 7. Trichomonadida Kirby, 1947 emend. Honigberg, in Camp, Mattern, & Honigberg, 1974.**
Typically karyomastigonts with four to six flagella, but with only one flagellum in one genus and no flagella in another; karyomastigonts and akaryomastigonts in one family with permanent polymonad organization; in mastigont(s) of typical genera one flagellum recurrent, free or with proximal or entire length adherent to body surface; undulating membrane, if present, associated with adherent segment of recurrent flagellum; pelta and noncontractile axostyle in each mastigont, except for one genus; hydrogenosomes present; true cysts infrequent, known in very few species; all or nearly all parasitic.
*Dientamoeba, Histomonas, Monocercomonas, Trichomonas, Devescovina*

**Order 8. Hypermastigida (Grassi & Foas, 1911)**
Mastigont system with numerous flagella and multiple Janicki-type parabasal apparatuses; "privileged" barren kinetosomes, resembling in arrangement kinetosomes of trichomonads and associated with rootlet filaments characteristic of these flagellates, present in many genera; flagella-bearing kinetosomes distributed in complete or partial circle, in plate or plates, or in longitudinal or spiral rows meeting in a centralized structure; one nucleus per cell; cysts in some species; sexuality in some species; all parasitic.
*Lophomonas, Microjoenia, Barbulanympha, Spirotrichonympha, Trichonympha*

**Subphylum II. Opalinata Corliss & Balamuth, 1963**
Numerous cilia in oblique rows over entire body surface; cytostome absent; nuclear division acentric; binary fission generally interkinetal (symmetrogenic); known life cycles involve syngamy with anisogamous flagellated gametes; all parasitic.

**Class 1. Opalinatea Wenyon, 1926**
With characteristics of the subphylum.

**Order 1. Opalinida Poche, 1913**
With characteristics of the class.
*Opalina*

**Subphylum III. Sarcodina Schmarda, 1871**
Pseudopodia or locomotive protoplasmic flow without discrete pseudopodia; flagella, when pre-

sent, usually restricted to developmental or other temporary stages; body naked or with external or internal test or skeleton; asexual reproduction by fission; sexuality, if present, associated with flagellate or, more rarely, ameboid gametes; most species free-living.

**Superclass 1. Rhizopoda von Siebold, 1845**
Locomotion by lobopodia, filopodia, or reticulopodia, or by protoplasmic flow without production of discrete pseudopodia.

**Class 1. Lobosea Carpenter, 1861**
Pseudopodia lobose or more or less filiform but produced from broader hyaline lobe; usually uninucleate; multinucleate forms not flattened or much-branched plasmodia; no sorocarps, sporangia, or similar fruiting bodies.

**Subclass 1. Gymnamoebia Haeckel, 1862**
Without test.

**Order 1. Amoebida Ehrenberg, 1930**
Typically uninucleate; mitochondria typically present; no flagellate state. Five suborders.
*Amoeba, Entamoeba, Thecamoeba, Flabellula, Mayorella, Acanthamoeba, Echinamoeba*

**Order 2. Schizopyrenida Singh, 1952**
Body with shape of monopodial cylinder, usually moving with more or less eruptive, hyaline, hemispherical bulges; typically uninucleate, nuclear division promitotic; temporary flagellate states in most species.
*Naegleria, Tetramitus, Vahlkampfia*

**Order 3. Pelobiontida Page, 1976**
Body with shape of thick cylinder; monopodial, with true bidirectional fountain flow of cytoplasm common; typically multinucleate; lacking mitochondria but with symbiotic bacteria; in microaerobic habitats; no flagellate stage known, but numerous nonmotile cils visible at fine-structural level, with variations of usual microtubular pattern.
*Pelomyxa*

**Subclass 2. Testacealobosia De Saedeleer, 1934**
Body enclosed by test, tectum, or other complex membrane external to plasma membrane and glycocalyx.

**Phylum II. Labyrinthomorpha Levine et al., 1980**
Trophic stage, ectoplasmic network with spindle-shaped or spherical, nonameboid cells; in some genera ameboid cells move within network by gliding; with sagenogenetosome, unique cell-surface organelle, associated with ectoplasmic network; heterokont zoospores produced by most species; saprobic and parasitic on algae, mostly in marine and estuarine waters.

**Phylum III. Apicomplexa Levine, 1970**
Apical complex (visible with electron microscope), generally consisting of polar ring(s), rhoptries, micronemes, conoid and subpellicular microtubules present at some state; micropore(s) generally present at some stage; cilia absent; sexuality by syngamy, all species parasitic.

**Class 1. Perkinsea Levine, 1978**
Conoid forming incomplete cone: "zoospores" (sporozoites?) flagellated, with anterior vacuole; no sexual reproduction; homoxenous.

**Order 1. Perkinsida Levine, 1978**
With characteristics of the class.
*Perkinsus*

**Class 2. Sporozoea Leuckart, 1879**
Conoid, if present, forming complete cone; reproduction generally both sexual and asexual; oocysts generally containing infective sporozoites, which result from sporogony; locomotion of mature organisms by body flexion, gliding, or undulation of longitudinal ridges; flagella present only in microgametes of some groups; pseudopods ordinarily absent and, if present, used for feeding, not locomotion; homoxenous or heteroxenous.

**Subclass 1. Gregarinia Dufour, 1828**
Mature gamonts large, extracellular; mucron or epimerite in mature organism; mucron formed from conoid; generally syzygy of gamonts; gametes usually similar (isogamous) or nearly so, with similar numbers of male and female gametes produced by gamonts; zygotes forming oocysts within gametocytes; life cycle characteristically consisting of gametogony and sporogony; in digestive tract or body cavity of invertebrates or lower chordates; generally homoxenous.

**Order 1. Archigregarinida Grassé, 1953**
Life cycle apparently primitive, characteristically with merogony, gametogony, sporogony; gamonts (trophozoites) aseptate; in annelids, sipunculids, hemichordates, or ascidians.
*Exoschizon, Selenidioides*

**Order 2. Eugregarinida Léger, 1900**
Merogony absent; gametogony and sporogony present; locomotion progressive, by gliding or undulation of longitudinal ridges, or nonprogressive; typically parasites of annelids and arthropods, but some species in other invertebrates.

**Suborder 1. Blastogregarinina Chatton & Villeneuve, 1936**
Gametogony by gamonts while still attached to intestine, with anisogamous gametes budding off of gamonts; no syzygy; gametocysts absent; oocysts with 10–16 naked sporozoites; gamont composed of single compartment with mucron, without definite protomerite and deutomerite; in marine polychaetes.
*Siedleckia*

**Suborder 2. Aseptatina Chakravarty, 1960**
Gametocysts present; gamont composed of single compartment, without definite protomerite and deutomerite, but with mucron (epimerite?) in some species; syzygy present.
*Lecudina, Monocystis, Selenidium, Urospora*

**Suborder 3. Septatina Lankester, 1885**
Gametocysts present; gamont divided into protomerite and deutomerite by septum; with epimerite; in alimentary canal of invertebrates, especially arthropods.
*Actinocephalus, Cephaloidophora, Gregarina, Stenophora*

**Order 3. Neogregarinida Grassé, 1953**
Merogony, presumably acquired secondarily; in Malpighian tubules, intestine, hemocoel, or fat tissues of insects.
*Caulleryella, Gigaductus, Mattesia*

**Subclass 2. Coccidia Leuckart, 1879**
Gamonts ordinarily present; mature gamonts small, typically intracellular, without mucron or epimerite; syzygy generally absent, if present, involves markedly anisogamous gametes; life cycle characteristically consisting of merogony, gametogony, and sporogony; most species in vertebrates.

**Order 1. Agamococcidiida Levine, 1979**
Merogony and gametogony absent.
*Rhytidocystis*

**Order 2. Protococcidiida Kheisin, 1956**
Merogony absent; in invertebrates.
*Eleutheroschizon, Grellia*

**Order 3. Eucoccidiida Léger & Duboscq, 1910**
Merogony present; in vertebrates and/or invertebrates.

**Suborder 1. Adeleina Léger, 1911**
Macrogamete and microgamont usually associated in syzygy during development; microgamont producing one to four microgametes; sporozoites enclosed in envelope; no endodyogeny; homoxenous or heteroxenous.
*Adelea, Haemogregarina, Klossiella, Hepatozoon*

**Suborder 2. Eimeriina Léger, 1911**
Macrogamete and microgamont developing independently; no syzygy; microgamont typically producing many microgametes; zygote not motile; sporozoites typically enclosed in sporocyst within oocyst; homoxenous or heteroxenous.
*Aggregata, Eimeria, Isospora, Sarcocystis, Toxoplasma*

**Suborder 3. Haemosporina Danilewsky, 1885**
Macrogamete and microgamont developing independently; no syzygy; conoid ordinarily absent; microgamont producing eight flagellated microgametes; zygote motile (oökinete); sporozoites naked, with three-membraned wall; heteroxenous, with merogony in vertebrates and sporogony in invertebrates; transmitted by blood-sucking insects.
*Haemoproteus, Leucocytozoon, Plasmodium*

**Subclass 3. Piroplasmia Levine, 1961**
Piriform, round, rod-shaped, or ameboid; conoid absent; no oocysts, spores, and pseudocysts; flagella absent, usually without subpellicular microtubules, with polar ring and rhoptries; locomotion by body flexion, gliding, or, in sexual stages (in Babesiidae and Theileriidae, at least), by large axopodiumlike *Strahlen*; asexual and probably sexual reproduction; parasitic in erythrocytes and sometimes also in other circulating and fixed cells; heteroxenous, with merogony in vertebrates and sporogony in invertebrates; sporozoites with single-membraned wall; so far as is known, vectors are ticks, but vectors of dactylosomatids are unknown.

**Order 1. Piroplasmida Wenyon, 1926**
With characteristics of the subclass.
*Babesia, Dactylosoma, Theileria*

**Phylum IV. Microspora Sprague, 1977**
Unicellular spores, each with imperforate wall, containing one uninucleate or dinucleate sporoplasm and simple or complex extrusion apparatus always with polar tube and polar cap; without mitochondria; often, if not usually, dimorphic in sporulation sequence; obligatory intracellular parasites in nearly all major animal groups.

**Class 1. Rudimicrosporea Sprague, 1977**
Spore with simple (rudimentary) extrusion apparatus consisting of polar cap and thick (manubroid) polar tube extending backward from cap, bending laterally and terminating in funnel (infundibulum); polaroplast and posterior vacuole absent; spore spherical or subspherical; sporulation sequence with

dimorphism, occurring either in parasitophorous vacuole or in thick-walled cyst; hyperparasites of gregarines in annelids.

**Order 1. Metchnikovellida Vivier, 1975**
With characteristics of the class.
*Amphiacantha, Metchnikovella*

**Class 2. Microsporea Delphy, 1963**
Spore with complex extrusion apparatus of Golgi origin, often including polaroplast and posterior vacuole in addition to polar tube and polar cap; polar tube typically filamentous, extending backward from polar cap and coiling around inside of spore wall; spore shape various, depending largely on structure of extrusion apparatus; spore wall with three layers; proteinaceous exospore, chitinous endospore, and membranous inner layer; outer two layers varying considerably in structure; sporocyst present or absent; often dimorphic in sporulation sequence.

**Order 1. Minisporida Sprague, 1972**
General tendency toward minimum development of accessory spore organelles (components of extrusion apparatus and spore wall) and accompanying tendency toward maximum development of sporocysts; spore without well-developed polaroplast; usually with relatively short polar filament, with little or no endospore; shape spherical or slightly ovoid; merogony present or absent (?); sporulation stages usually separated from host cell cytoplasm by intracellular sporocyst; some genera dimorphic in the sporulation sequence.
*Burkea, Chytridiopsis, Hessea*

**Order 2. Microsporida Balbiani, 1882**
General tendency toward maximum development and varied specialization of accessory spore organelles (components of extrusion apparatus and spore wall), with accompanying reduction of sporocysts; sporocysts inside host cell present or absent; cysts of other kinds sometimes formed from host cell membrane or other host material; merogony present; spore shape variable; often dimorphic in sporulation sequence.

**Suborder 1. Pansporoblastina Tuzet et al., 1971**
Sporulation sequence occurring within more or less persistent intracellular (in host cell) sporocyst (pansporoblastic membrane); often diiorphic, with another sporulation sequence not involving such membrane; sporoblasts and spores usually uninucleate when membrane present, dinucleate when membrane absent.
*Amblyospora, Pleistophora, Thelohania*

**Suborder 2. Apansporoblastina Tuzet et al., 1971**
Pansporoblastic membrane usually absent, vestigial when present, never persisting as sporophorous vesicle; sporoblast most often dinucleate.
*Encephalitozoon, Glugea, Nosema*

**Phylum V. Ascetospora Sprague, 1978**
Spore multicellular (or unicellular); with one or more sporoplasms; without polar capsules or polar filaments; all parasitic.

**Class 1. Stellatosporea Sprague, 1978**
Haplosporosomes present; spore with one or more sporoplasms.

**Order 1. Occlusosporida Perkins, in Sprague, 1978**
Spore with more than one sporoplasm; sporulation involving series of endogenous buddings that produce sporoplasm(s) within sporoplasm(s); spore wall entire.
*Marteilia*

**Order 2. Balanosporida Sprague, 1978**
Spore with one sporoplasm; spore wall interrupted anteriorly by orifice; orifice covered externally by operculum or internally by diaphragm.
*Haplosporidium, Minchinia, Urosporidium*

**Class 2. Paramyxea Levine, in Sprague, 1979**
Spore bicellular, consisting of parietal cell and one sporoplasm; spore without orifice.

**Order 1. Paramyxida Chatton, 1911**
With characters of the class.
*Paramyxa*

**Phylum VI. Myxozoa Grassé, 1970**
Spores of multicellular origin, with one or more polar capsules and sporoplasms; with one, two, or three (rarely more) valves; all species parasitic.

**Class 1. Myxosporea Bütschli, 1881**
Spore with one or two sporoplasms and one to six (typically two) polar capsules, each capsule with coiled polar filament; filament function probably anchoring; spore membrane generally with two, occasionally up to six valves; trophozoite stage well developed, main site of proliferation; coelozoic or histozoic in cold-blooded vertebrates.

**Order 1. Bivalvulida Shulman, 1959**
Spore wall with two valves.

**Suborder 1. Bipolarina Tripathi, 1948**
Spores with polar capsules at opposite ends of spore, or with widely divergent polar capsules located in sutural plane or sutural zone.
*Myxidium, Sinuolinea, Sphaeromyxa*

**Suborder 2. Eurysporina Kudo, 1920 emend. Shulman, 1962**
Spores with two to four polar capsules at one pole in plane perpendicular to sutural plane.
*Ceratomyxa, Chloromyxum, Sphaerospora*

**Suborder 3. Platysporina Kudo, 1920 emend. Shulman, 1962**
Spores with two polar capsules at one pole in sutural plane; spores bilaterally symmetrical, unless with single polar capsule.
*Henneguya, Myxobolus, Thelohanellus*

**Order 2. Multivalvulida Shulman, 1959**
Spore wall with three or more valves.
*Hexacapsula, Kudoa, Trilospora*

**Class 2. Actinosporea Noble, 1980**
Spores with three polar capsules, each enclosing coiled polar filament; membrane with three valves; several to many sporoplasms; trophozoite stage reduced, proliferation mainly during sporogenesis; in invertebrates, especially annelids.

**Subclass 1. Actinomyxia Štole, 1899**
Spores with three polar capsules, each enclosing polar filament; membrane with three valves; several to many sporoplasms; in invertebrates, especially annelids.

**Order 1. Actinomyxida Lom, 1980**
With characteristics of the subclass.
*Triactinomyxon*

**Phylum VII. Ciliophora Doflein, 1901**
Simple cilia or compound ciliary organelles typical in at least one stage of life cycle; with subpellicular infraciliature present even when cilia absent; two types of nuclei, with rare exception; binary fission transverse, basically homothetogenic and generally parakinetal, but budding and multiple fission also occur; sexuality involving conjugation, autogamy, and cytogamy; nutrition heterotrophic; contractile vacuole typically present; most species free-living, but many commensal, some truly parasitic, and large number found as symphorionts on variety of "hosts."

**Class 1. Kinetofragminophorea de Puytorac et al., 1974**
Oral infraciliature only slightly distinct from somatic infraciliature and differentiated from anterior parts, or other segments, of all or some of somatic kineties; stomatogenesis generally telokinetal; cytostome often apical (or subapical) or midventral, on surface of body or at bottom of atrium or vestibulum; cytopharyngeal apparatus commonly prominent; compound ciliature, oral or somatic, typically absent. Four subclasses, 14 orders.
*Balantidium, Entodinium*

**Class 2. Oligohymenophorea de Puytorac et al., 1974**
Oral apparatus, at least partially in buccal cavity, generally well defined, although absent in one group; oral ciliature, clearly distinct from somatic ciliature, consisting of paroral membrane (stichodyad) on right side and small number of compound organelles (membranelles, peniculi, or polykineties) on left side; stomatogenesis parakinetal or buccokinetal; cytostome usually ventral and/or near anterior end, present at bottom of buccal or infundibular cavity; cysts not uncommon; various species loricate; colony formation common in some groups. Two subclasses, three orders.
*Ichthyophthirius*

**Class 3. Polymenophorea Jankowski. 1967**
Dominated by well-developed, conspicuous adoral zone (AZM) of numerous buccal or peristomial organelles (para- or heteromembranelles), often extending out onto body surface; on right side, one or several lines of "paroral" ciliature (mono- or diplo- or polystichomonads or pairs of kinetosomes); stomatogenesis parakinetal or apokinetal; somatic ciliature complete or reduced, or appearing as cirri; cytostome at bottom of buccal cavity or infundibulum; somatic infraciliature rarely including kinetodesmata; postciliodesmata common and prominent; cytoproct often absent; cysts, and especially loricae, very common in some groups; often large and commonly free-living, free-swimming forms in great variety of habitats.
One subclass, four orders.
*Nyctotherus*

## REFERENCES AND FURTHER READINGS

Aikawa, M., and C. R. Sterling. 1974. *Intracellular Parasitic Protozoa*. Academic Press, New York.

Chen, T.-T. (ed.). 1967–1970. *Research in Protozoology*. 4 Vols. Pergamon Press, New York.

Corliss, J. O. 1961. *The Ciliated Protozoa: Characterization, Classification, and Guide to the Literature*. Pergamon Press, New York.

Corliss, J. O. 1979. *The Ciliated Protozoa: Characterization, Classification and Guide to the Literature*, 2nd ed. Pergamon Press, New York.

Dobell, C. 1932. *Anthony van Leeuwenhoek and His "Little Animals."* Harcourt, Brace, New York.

Farmer, J. N. 1980. *The Protozoa: Introduction to Protozoology.* C. V. Mosby Co., St. Louis.

Grassé, P.-P. (ed.). 1952-1953. *Traité de Zoologie.* Vol. I, Fasc. 1 & 2. Masson, Paris.

Grelle, K. G. 1973. *Protozoology,* 1st English ed. Springer-Verlag, New York.

Hall, R. P. 1953. *Protozoology.* Prentice-Hall, New York.

Hyman, L. H. 1940. *The Invertebrates.* Vol. 1. *Protozoa Through Ctenophora.* McGraw-Hill, New York.

Jahn, T. L., E. C. Bovee, and F. F. Jahn. 1979. *How to Know the Protozoa.* W. C. Brown Co., Dubuque.

Jones, A. W. 1967. *Introduction to Parasitology.* Addison-Wesley, Reading, Mass.

Kaplan, W., F. W. Chandler, C. Choudary, and P. K. Ramachandran. 1983. Disseminated unicellular green algal infection in two sheep. *Am. J. Trop. Med. Hyg.* 32:405–411.

Kudo, R. R. 1966. *Protozoology,* 5th ed. Charles C Thomas, Springfield, Ill.

Levine, N. D. 1973. *The Parasitic Protozoa of Domestic Animals and Man,* 2nd ed. Burgess, Minneapolis.

Levine, N.D., et al. 1980. A newly revised classification of the protozoa. *J. Protozool.* 27:37–58.

Lom, J., and J. O. Corliss. 1971. Morphogenesis and cortical ultrastructure of *Brooklynella hostilis,* a dysteriid ciliate ectoparasitic on marine fishes. *J. Protozool.* 18:261–281.

Manwell, R. D. 1961. *Introduction to Protozoology.* St. Martins Press, New York.

Scholtyseck, E. 1979. *Fine Structure of Parasitic Protozoa.* Springer-Verlag, New York.

# 3

# *Trypanosomes*

The Zoomastigophorea, or animal-like flagellates, include organisms with one to many flagella, none of which gain their energy directly from sunlight and nearly all of which are parasitic. In this and subsequent chapters we will be dealing with members of the following orders in some detail: Kinetoplastida, Trichomonadida, Diplomonadida, and Hypermastigida. Members of three additional orders, Proteromonadida, Retortamonadida, and Oxymonadida, will be mentioned briefly in relation to other agents.

The order Kinetoplastida Honigberg, 1963 contains those flagellated protozoa that have a kinetoplast. The kinetoplast is a self-replicating organelle. It contains deoxyribonucleic acid (DNA) and gives rise to one or more mitochondria which remain attached to it. Various members have one to four flagella, and only a few are free-living.

The family Trypanosomatidae is defined as having a single nucleus, a single flagellum, and a leaflike or rounded body. The family contains nine genera, two of which are of medical and economic importance. Members cause African sleeping sickness, Chagas' disease, oriental sore, and dourine among other diseases of humans and domesticated animals.

Some trypanosomatids, members of the genus *Phytomonas*, are parasites of plants. Those best known are in latex-bearing plants such as milkweed; *P. elmassiani* is transmitted by *Oncopeltus* (Hemiptera, Lygaeidae). Two apparently pathogenic *Phytomonas* spp. are known, *P. leptovasorum* of coffee plants and an unidentified species that infects coconut palms in Surinam, South America (Parthasarathy et al., 1976). A few have been found as intranuclear parasites of other protozoa (Wille et al., 1981).

Trypanosomatids range in length from about 3 to 80 µm, with the bulk of them between 5 and 30 µm. The forms in which trypanosomatids may be found are illustrated in Figure 3.1. It has been suggested that the term *amastigote* be replaced by *micromastigote* since these intracellular stages do, in fact, have a flagellum; whether this terminology will be widely used is moot at this time.

Although trypanosomatids may shift back and forth between forms during the life cycle, the basic structure remains much the same. The trypomastigote form is common in the blood of infected vertebrates and it will serve as a type for reference. Trypomastigotes range in size from about 15 to 80 µm. The larger ones are found in some fish; one from a ray reaches 130 µm in length. Most of the trypomastigotes in the blood of mammals, particularly the pathogenic ones, range from 15 to 30 µm. These are the species on which a great deal of morphologic work has been done, and we will use one as a type trypomastigote. *Trypanosoma brucei* is 15 to 30 µm long by 6 to 8 µm wide (Fig. 3.2). The body is sinuous and has a single flagellum, which arises from the basal body lying in a flagellar pocket. Near the flagellar basal body and oriented at right angles to it is a so-called barren basal body. The flagellar pocket is significant for our purposes because surface antigens seem to be

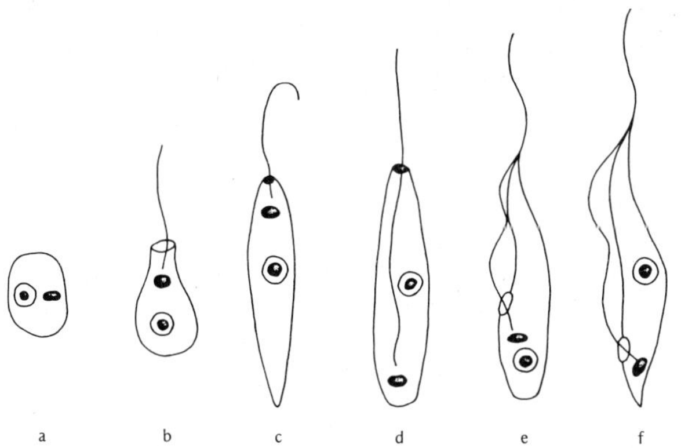


**FIG. 3.1**
Stages of trypanosomatid flagellates. (*a*) Amastigote; (*b*) choanomastigote; (*c*) promastigote; (*d*) opisthomastigote; (*e*) epimastigote; (*f*) trypomastigote.

produced there and move out over the surface of the organism.

In the slender bloodstream forms, the flagellum extends anteriorly beyond the body as a free flagellum. The flagellum has a typical "9 + 2" arrangement. A paraxial rod (Fig. 3.3) is associated with it, and both are enclosed by a sheath that represents the undulating membrane. Subpellicular tubules lie just beneath the outer limiting membrane and extend lengthwise through the organism; their number is variable. There is a single, centrally located nucleus. Near the basal body of the flagellum, but not attached to it, is the kinetoplast. Early studies had indicated that the basal body or blepharoplast of the flagellum was the same as the kinetoplast, but electron microscope (EM) studies have shown them to be separate structures. The basal body and the kinetoplast do retain an association even during transformation to another form, but they are not attached. The kinetoplast is less than 1 μm in length, perhaps twice as long as it is wide, and surrounded by a double membrane. At one or both corners of the kinetoplast arises the mitochondrion of the trypanosome. In the vertebrate bloodstream, the surface area of the mitochondrion is relatively low, indicating that it is not highly functional. Within the kinetoplast are numerous strands of DNA, which extend across the short dimension. Under appropriate treatment, these strands can be seen to be circles. Other cytoplasmic

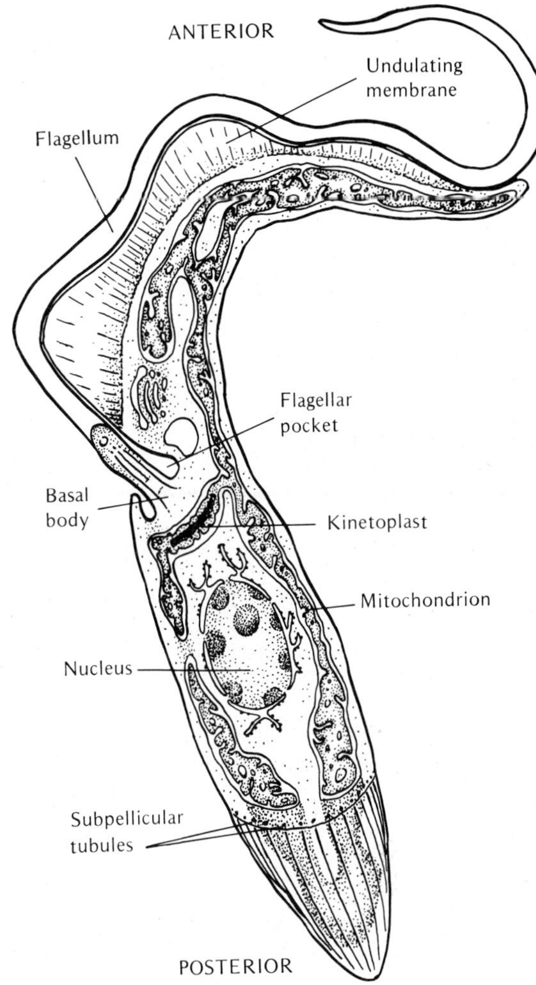


**FIG. 3.2**
Diagram of *Trypanosoma brucei rhodesiense.*

organelles or inclusions are Golgi, endoplasmic reticulum, lysosomes, and volutin granules.

Division in trypanosomatids is essentially like that of other flagellates such as euglenoids—it is longitudinal (symmetrogenic). The nucleus divides by an amitotic process in which the nuclear membrane is retained. The kinetoplast-mitochondrion replicates just prior to nuclear division. Other organelles are replicated during division so that the daughter cells are complete by the time plasmotomy occurs.

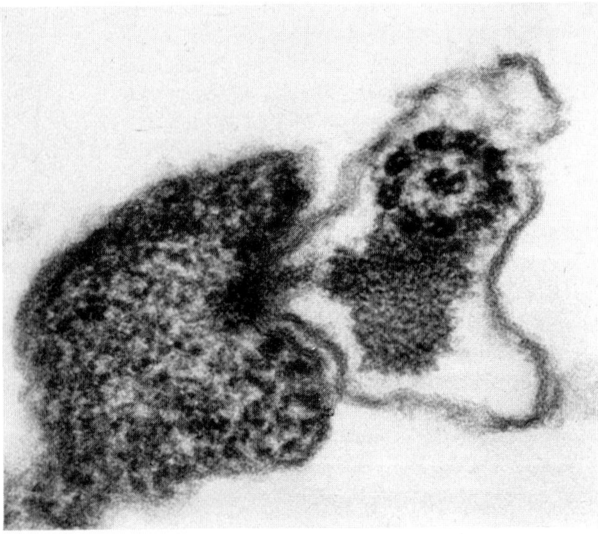

**FIG. 3.3**
Flagellum and paraxial rod of *Trypanosoma avium.*

## GENUS *TRYPANOSOMA* – GENERAL

Members of the genus *Trypanosoma* are widespread as trypomastigotes in the blood of vertebrates. Most remain in tissue fluids, but a few (*T. cruzi*) replicate within cells. All but a few members of the genus are transmitted from one vertebrate to another through biological vectors. Transmission takes place to the vector through blood sucking and to the vertebrate through one of two routes: insect fecal contamination or injection during blood sucking. At this time, there is no scheme for classifying all of the trypanosomes of vertebrates, but one system developed by Hoare (see Hoare, 1967; McGhee, 1968; Lumsden, 1965 & 1970) for trypanosomes of mammals shows patterns in the genus and makes its organization both rational and manageable (Table 3.1).

First, and importantly, the trypanosomes make sense as mammalian parasites if we assume that they were originally parasites of insects. We know, for example, that there are kinetoplastids in a variety of true bugs (Insecta, Hemimetabola, Hemiptera). Since the Hemiptera have sucking mouth parts, it is logical that they would themselves have become infected when preying on other insects. Some of the Hemiptera have become blood-feeders, and the step of injecting the kinetoplastid into the skin of a mammal follows reasonably. Once in the blood, the kinetoplastid replicates readily in such a highly nutritive environment.

The genus *Trypanosoma* has been split into two sections, Stercoraria and Salivaria (Table 3.1). The root *sterc-* refers to feces and transmission from the invertebrate vector to the vertebrate host is through fecal contamination; these organisms are spoken of as developing in the "posterior station." In opposition, the Salivaria include trypanosomes that develop in the "anterior station" and are transmitted from the vector to the vertebrate host by injection during the act of blood sucking.

This classification of trypanosomes is useful in showing the patterns of development, structure, and metabolism within the genus, but it is of little help in identifying an organism in a stained blood film taken from a vertebrate. Trypomastogotes have characteristics that are fairly consistent, and an identification or description can usually be made if there are sufficient organisms in the blood.

The characteristics used in identification are as follows:

1. Size—length and width in micrometers
2. Location of nucleus—central or toward one end
3. Shape of body—straight, sinuous, C-shaped
4. Kinetoplast—present or absent, size, terminal or subterminal
5. Flagellum—free (extending beyond anterior of body) or not
6. Undulating membrane—high or low, number of loops
7. Staining quality of cytoplasm—stains lightly or darkly, has granular inclusions or not

Representative trypomastigotes are illustrated in Figure 3.4.

The greater part of this chapter will concern African sleeping sickness in humans, with some attention paid to nagana in domestic cattle. Both disease complexes are caused by *Trypanosoma* spp. It should be noted that a number of species of *Trypanosoma* in subsaharan

**TABLE 3.1**
Classification of the Genus *Trypanosoma* [a]

| STERCORARIA | |
|---|---|
| Kinetoplast large, not terminal; posterior end of body pointed; typically nonpathogenic; development in vector in posterior station, transmission contaminative | |
| Subgenus *Megatrypanum* | Kinetoplast typically near nucleus, far from posterior end of body; includes *T.(M.) theileri, tragelaphi, ingens, melophagium* and others |
| Subgenus *Herpetosoma* | Kinetoplast subterminal; includes *T.(H.) lewisi, duttoni, nabiasi*, and others |
| Subgenus *Schizotrypanum* | Typically curved; kinetoplast close to posterior end of body; includes *T.(S.) cruzi, vespertilionis, pipistrelli* and others |
| Subgenus *Endotrypanum* | Endoglobular forms; includes only *T.(E.) schaudinni* |
| SALIVARIA | |
| Kinetoplast terminal or subterminal; posterior end of body usually blunt; typically pathogenic; development in vector in anterior station and transmission inoculative. | |
| Subgenus *Duttonella* | Posterior end of body rounded; kinetoplast large, terminal; includes *T.(D.) vivax, uniforme* |
| Subgenus *Nannomonas* | Kinetoplast typically marginal; includes *T.(N.) congolense, dimorphon, simiae* |
| Subgenus *Pycnomonas* | Kinetoplast small, subterminal; includes *T.(P.) suis* |
| Subgenus *Trypanozoon* | Small subterminal kinetoplast; includes some aberrant species transmitted noncyclically or by coitus; includes *T.(T.) brucei, gambiense, rhodesiense, evansi, equinum, equiperdum* |

[a] The type species of each subgenus is cited first.

Africa are tsetse-borne and some species are not. The latter extend into additional areas of the Eastern Hemisphere and some have been introduced into the Western Hemisphere (Table 3.2).

*Trypanosoma evansi* is transmitted by blood-sucking flies (Subord. Brachycera) in the family Tabanidae. Transmission is mechanical in that there is no development inside the fly. If a fly is disturbed while feeding on an infected animal, it may transmit *T. evansi* to an uninfected animal during the next few minutes through its contaminated mouth parts. The disease, called *surra*, occurs north of the tsetse belt in Africa, but has also been spread widely through the Eastern Hemisphere.

Dourine is a disease of horses caused by *T. equiperdum* and transmitted by sexual intercourse. It has been spread worldwide through trade, exploration, and military action. Dourine used to be a serious scourge of horses in the United States, but was eradicated through a test-and-slaughter program during the 1920s and 1930s.

Both *T. evansi* and *T. equiperdum* have been derived from the tsetse-borne trypanosomes of subsaharan Africa. They are structurally similar to *T. brucei*, and the diseases they cause are also similar to nagana. The principal difference is that these species have taken a step or two away from dependence on tsetse as vectors.

NAME OF ORGANISM AND DISEASE CAUSED. *Trypanosoma brucei* causes African sleeping sickness, nagana, trypanosomiasis, and trypanosomosis. *T. brucei*, its subspecies, and some related species form a complex of organisms that cause the African trypanosomoses. Most of them are transmitted by tsetse, members of the genus *Glossina* (Diptera, Muscidae). The term *tsetse* means "fly destructive to cattle" in the Sechuana language of Botswana (Bechuanaland); thus, the term *tsetse flies* is redundant. The term is also both singular and plural.

DISTRIBUTION AND IMPORTANCE. The importance of African trypanosomosis can hardly be overstated. The tsetse-borne trypanosomes extend over tropical Africa south of the Sahara Desert from 15° N latitude to 25° S latitude. The area encompassed is more than 4 million square miles—larger than the continental United States. Until recent years, the three subspecies of

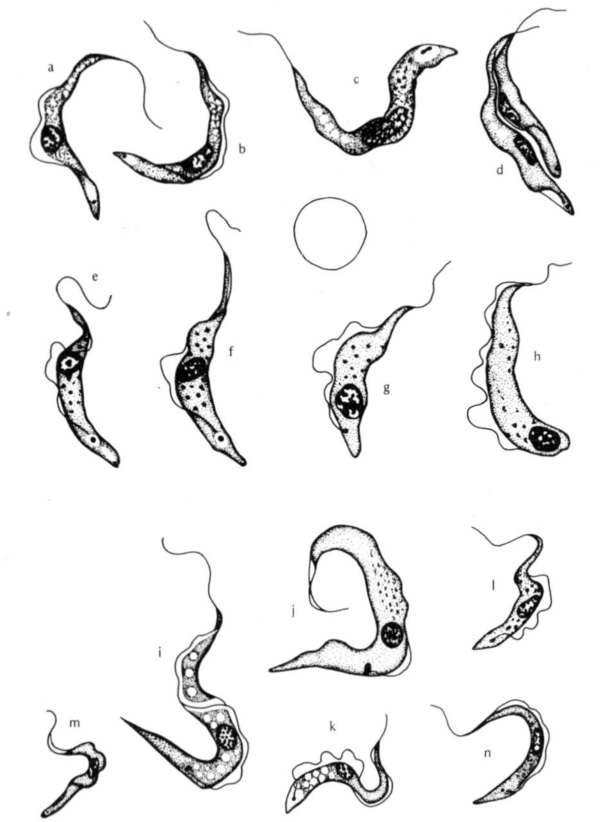


**FIG. 3.4**
Trypomastigotes of various *Trypanosoma* spp. (*a-d*) *T. congolense;* (*e-h*) *T. brucei;* (*i*) *T. theileri;* (*j*) *T. melophagium;* (*k*) *T. evansi;* (*l*) *T. equinum;* (*m*) *T. equiperdum;* (*n*) *T. lewisi.*

*T. brucei*—*T. b. brucei, T. b. gambiense,* and *T. b. rhodesiense*—and *T. congolense* effectively denied human use of this huge area by infecting either humans or their domesticated animals (Fig. 3.5).

In areas where *T. b. brucei* and *T. congolense*, the causes of nagana in cattle, were endemic, it was not possible to raise large domesticated mammals. The effect was to prevent human use of cattle, horses, and swine. The impact of not having oxen or horses was severe. Farming tasks performed for thousands of years in other parts of the world by oxen were still done by hand in large parts of Africa. The soil was tilled by hand. Communication, travel, and war were all dependent on the horse in Asia, Europe, and North Africa. But in tropical Africa, all had to be done on foot. Arab invasions southward were prevented because of the probability of epidemic deaths in their horses, and the Boers in South Africa in more recent times were prevented from migrating northward because of livestock deaths. Thus, trypanosomosis left people in central Africa with a risk of only sporadic incursion but also with primitive farming methods and no beasts of burden or means of rapid transport, except by river.

In western Africa, sleeping sickness in humans was originally present as an endemic focus with some epidemics occurring, probably at fairly regular intervals. Despite infections, it was a disease to which humans had adjusted insofar as the population was able to survive and generally avoid infection. During the late 1800s European efforts at colonization of Africa increased, resulting in greater movement of peoples and introduction of Western technologies. Steamers plied the Congo River and its tributaries, and persons living in endemic areas near the mouth of the Congo were transported hundreds of miles upstream as laborers or soldiers. By the turn of the century, it was clear that a disaster was in progress. In the Congo basin alone, it is estimated that half a million persons died of sleeping sickness in the decade between 1896 and 1906.

The disease spread not only through western Africa but was carried over into east Africa, probably by infected soldiers in Emin Pasha's army. Sleeping sickness broke out on the shores of Lake Victoria in 1896, and by 1908 the original population of about 300,000 had been reduced to about 100,000.

The result of these and other epidemics was that European countries with economic interests in Africa instigated research and control programs that ultimately reduced the prevalence of the disease (Fig. 3.5).

Even today, after extensive control programs, raising livestock is difficult or even impossible in areas that would otherwise be eminently suitable for it. Ready sources of animal protein are often limited to poultry. Cattle, on which people rely for milk, meat, and fiber, and swine, which can scavenge in marginal environments, are not available. Consequently, many peoples in tropical Africa, especially children, suffer from protein deficiencies. As has been well publicized in recent years, protein is essential for proper development of children's minds and bodies during their early years of rapid growth. A striking result of protein deficiency is kwashiorkor, a disease in which children become

**TABLE 3.2.**
Species of *Trypanosoma* Originating in Africa and Infecting Humans and Animals

| TRYPANOSOME | DISEASE | RANGE | MAMMALIAN HOSTS |
|---|---|---|---|
| | TRANSMISSION BY TSETSE | | |
| *T. brucei gambiense* | Chronic sleeping sickness | Central Africa | Human |
| *T. brucei rhodesiense* | Acute sleeping sickness | East Africa | Human, bovids, antelopes |
| *T. brucei brucei* | Nagana | Tropical Africa | All domestic mammals, antelopes |
| *T. congolense* | Nagana | Tropical Africa | Ruminants, pigs, dogs, equids |
| *T. simiae* | Acute or chronic disease | Tropical Africa | Monkeys, pigs, warthogs, possible bovids, equids, camels |
| *T. vivax* | Souma | Central and South America | Ruminants, equids, dogs |
| *T. uniforme* | Relapsing fever, occasionally fatal | East and central Africa; Angola | Ruminants |
| | MECHANICAL TRANSMISSION | | |
| *T. evansi* | Surra | North and N.W. Africa, Sudan, Somalia—all continents | Ruminants, equids, dogs, monkeys, elephants, etc. |
| *T. equiperdum* | Dourine | South America; north, south, and S.W. Africa | Equids[a] |

[a]Transmission by coitus.

pot-bellied and emaciated and their hair becomes pale. But protein levels considerably above this disastrous level are still likely to cause suboptimal development in children.

For those interested in the development of knowledge and the campaigns to control trypanosomosis, two highly readable histories are recommended. McKelvey's (1973) *Man Against Tsetse. Struggle for Africa* sets forth in broad terms the scope of the trypanosomosis problem and details some of the events and people that have increased the problem of trypanosomosis and those who have developed the knowledge and techniques to help control the diseases. The book is a microcosm of human ecology showing the kinds of forces that have impinged on a particular set of circumstances.

Duggan's chapter in Mulligan and Potts (1970), called "An Historical Perspective," is more detailed than McKelvey's book in some ways. He gives more factual material, but does not bring to life the persons who were influential in various aspects of trypanosomosis. Still it is both readable and informative.

The three subspecies of *T. brucei* differ only in host preferences and the clinical course of the diseases they cause. *T. b. gambiense* is the cause of Gambian fever in humans, the domiciliated form of sleeping sickness. *T. b. gambiense* is limited to humans under natural conditions and the disease is usually contracted at home. *T. b. rhodesiense*, the cause of Rhodesian fever, is more virulent than *T. b. gambiense* and is usually contracted when a person leaves home and travels to another area for hunting or fishing. Gambian fever is the more common and widespread form of sleeping sickness.

MORPHOLOGY. All members of *T. brucei* are structurally alike. In the slender form they average 29 μm in length and as stumpy forms they average 18 μm; the total size range is 12 to 42 μm (Fig. 3.4).

The difference among the subspecies is based mainly on the hosts they infect. *T. B. gambiense* is a parasite of humans; the organism can be transmitted experimentally to other animals including cattle, but there is no proven, natural reservoir. *T. b. rhodesiense* is a

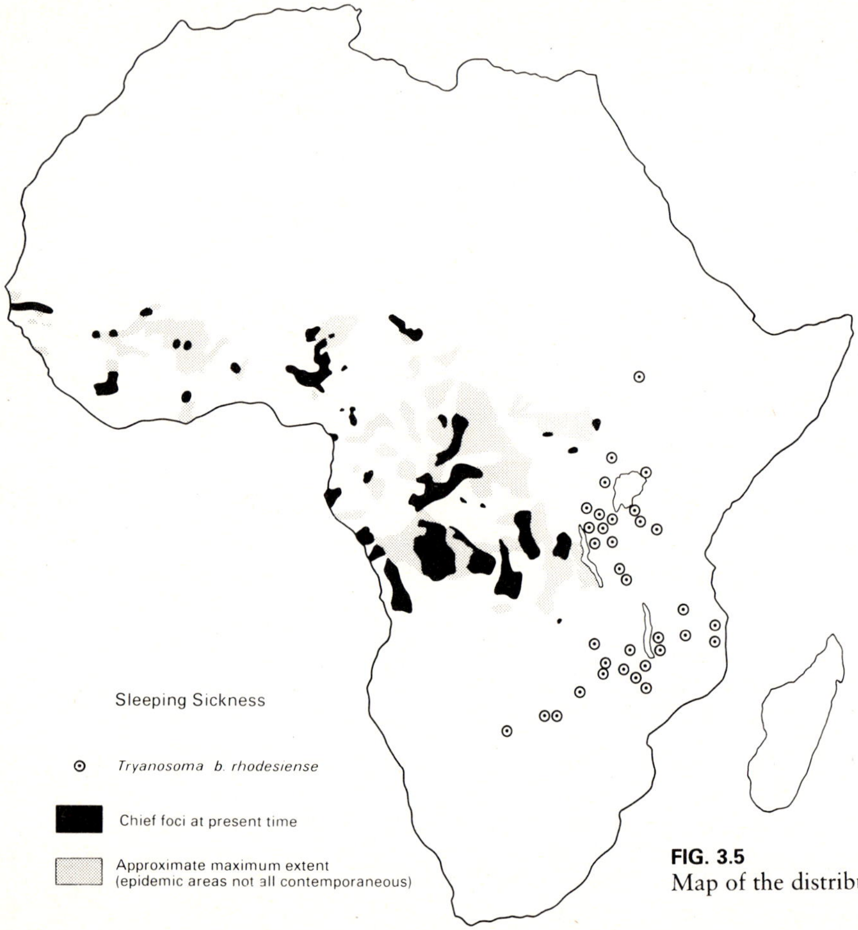


**FIG. 3.5**
Map of the distribution of sleeping sickness in humans.

parasite of wild herbivores, normally, but it has the capability of multiplying in humans if given the opportunity. *T. b. rhodesiense* was isolated from a bushbuck, *Tragelaphus scriptus*, in Kenya and successfully transmitted to a human volunteer (Heisch et al., 1958), but it is understandable why there is only fragmentary information on experimental human infections. *T. b. brucei* has a broad host range among domesticated animals and has a reservoir in wild ruminants, primarily antelope, but does not infect humans.

LIFE CYCLE. Trypomastigotes circulate in human blood and are taken up by *Glossina* (tsetse) during feeding. They multiply as somewhat modified trypomastigotes in the midgut of the fly for about 10 days (Fig. 3.6) and then migrate toward the anterior portion of the gut. Upon reaching the salivary glands a few days later, the organisms transform into epimastigotes, attach to tissue by their flagella, and multiply. About 20 days after infection, the epimastigotes begin to transform back to trypomastigotes. These are called *metacyclic* trypomastigotes and in this form they are infective for humans again. Prior to the appearance of metacyclic trypomastigotes, a tsetse is incapable of transmitting the organism. A mechanical type of transmission may take place when a fly feeding on an infected person is disturbed and within minutes feeds again on a person, but this is not an important type of transmission of Gambian fever.

Upon feeding, a fly carrying metacyclic trypomastigotes injects them with the salivary fluid into the skin of the person. The trypomastigotes remain in the serum and lymph, where they multiply by binary fission. Transmission back to the tsetse can then take place.

This brief description of the life cycle of *T. b. gambiense* raises two questions. What stimulates the organisms to undergo the changes in form? Why is it that only the metacyclic trypomastigotes can reinfect the mammal? Some answers to these questions have been worked out through a variety of investigations (see Vickerman, 1971, for review).

The change in form from trypomastigote to epimastigote and back to trypomastigote in the vector seems due, in part, to mitochondrial changes. In the mammal, slender bloodstream forms are seen early in infections; the slender forms are the dividing forms. Later, stumpy forms predominate (Fig. 3.6); they do not divide. The slender forms have what appear to be nonfunctional mitochondria (Fig. 3.6), and their low level of activity is borne out by the fact that they have aerobic glycolysis, a type of respiration in which metabolism of glucose is incomplete. In the case of the slender trypomastigotes, glucose is broken down to pyruvate and no further; the rest of the Krebs cycle is lacking. The organisms are cyanide-insensitive, indicating that cytochromes are lacking.

In culture forms, which correspond to epimastigotes in the vector, there is a well-developed mitochondrion extending posteriorly as well as anteriorly from the kinetoplast (Fig. 3.6). Epimastigotes have a complete Krebs cycle and oxidize glucose completely. They are also cyanide-sensitive.

Two forms are intermediate between slender trypomastigotes and epimastigotes—stumpy trypomastigotes in the mammal and metacyclic trypomastigotes in the vector. The mitochondrion of the stumpy forms develop tubular cristae seemingly in preparation for being picked up by the tsetse. The metacyclic trypomastigote has a mitochondrion that is smaller and has less internal surface. The implication is that the metacyclic forms are preparing to reenter the mammal.

The change in the mitochondrion is a response to the environment in which the organism finds itself. In the mammal, the trypomastigotes have an almost inexhaustible source of energy and other nutrients. Blood is a rich, complex medium and its constituents remain fairly constant. In a sense, the trypomastigotes have no need to be frugal with carbohydrate. They can maintain only a relatively simple system that allows them to split glucose to pyruvate and then return the pyruvate to the bloodstream, where it will be utilized by host cells or recycled back to glucose. In the fly, on the other hand, the ecosystem is a more limited one. There is a need to be more frugal with energy sources, and complete oxidation of glucose is necessary. Many of the metabolic and antigenic changes have been demonstrated in culture with *T. brucei* (Brown et al., 1973).

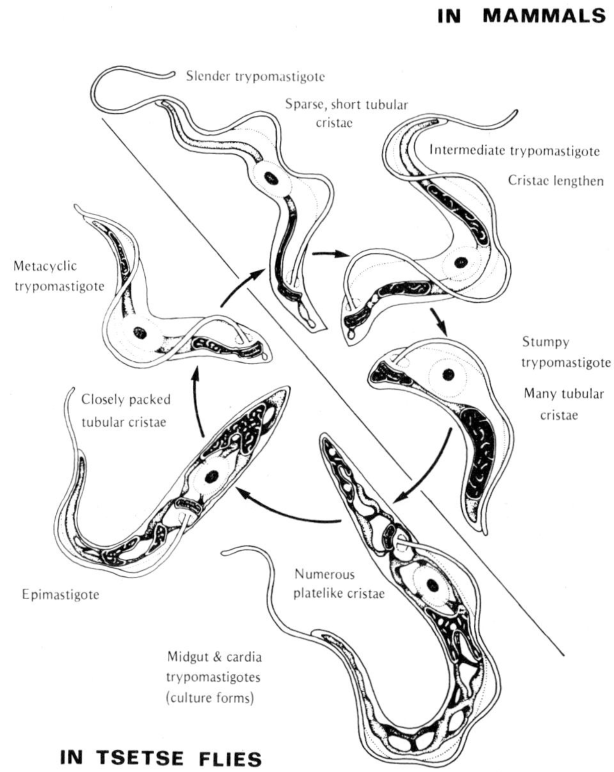


**FIG. 3.6**
Change in body form of *Trypanosoma brucei* during its life cycle. [From Vickerman, K. 1971. Morphological and physiological considerations of extracellular blood protozoa. In A. M. Fallis (ed.), *Ecology and Physiology of Parasites.* University of Toronto Press, Toronto.]

The second question has to do with the ability of only the metacyclic trypomastigotes to infect a mammal. The epimastigotes in the gut of the tsetse are incapable of establishing themselves in mammals. Significantly, the stumpy bloodstream forms are also incapable of being passed by syringe from mammal to mammal as the slender forms are. In the case of the metacyclic trypomastigotes in the vector, the crux of survival in the mammal is the production of a protective coat. The coat can be clearly seen as a layer 15-nm thick covering both the body and the flagellar sheath of the organism in the bloodstream forms (Fig. 3.7). The culture forms, on the other hand, are completely free of such a covering. (Fig. 3.7). There seems to be a sequence of development in which the epimastigotes multiply and then begin to form nonmultiplying, metacyclic trypomastigotes which prepare to reenter a mammal by producing a protective protein coat.

Although we have discussed metabolic capabilities and the protective coat separately, the two are integrated in the same developmental pattern of shifting from the vector to the mammal and back. Both work together and both are essential, but we know little about the stimuli, whether internal or external, that cause the organisms to prepare for the next environment in which they must survive and reproduce.

DIAGNOSIS. Diagnosis of trypanosomosis is based on clinical signs and symptoms, medical history, and finding the organism. In classroom exercises, students usually see large numbers of trypomastigotes in blood films, and are likely to conclude that finding the organisms in a stained film of peripheral blood is a simple, reliable means of determining whether a person is infected. Early in the disease, the trypomastigotes are readily found in the peripheral blood, but later in the

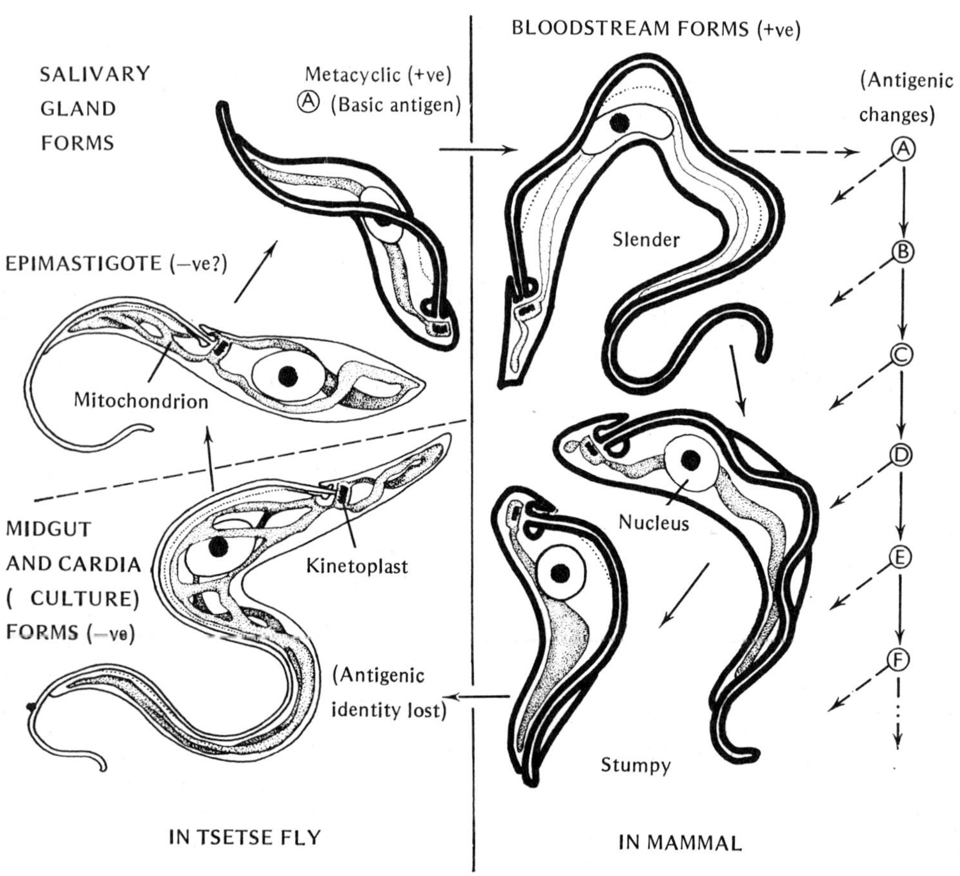


**FIG. 3.7**
Diagram of the life cycle of *Trypanosoma brucei* showing cyclical changes in surface coat, surface antigens, and mitochondrion. [From Vickerman, 1969.]

disease, it may be necessary to examine lymph nodes, bone marrow, or cerebrospinal fluid.

Indirect means of finding the organisms involve animal inoculation and cultivation. Laboratory rats are susceptible to infection, but other laboratory animals may also be used. Culture media usually used are NNN or Weinman's, which are basically agar-based bacteriologic media.

Once one has found an organism in a stained film or culture, it is important to identify it. Because certain trypanosomes have broad host ranges, it should not be assumed that finding one in the blood of a person or animal means that it is a normal inhabitant of the host. Some serological tests have been developed but are limited in usefulness.

HOST–PARASITE INTERACTIONS. Gambian sleeping sickness is almost invariably fatal within 2 to about 7 years after infection if untreated. The disease progresses intermittently in such a way that the affected person lapses into longer and longer periods of drowsiness or coma.

The first signs of infection are swelling at the site of the injection by the tsetse, fever, and swollen lymph nodes; this phase of the disease may be absent in native Africans. These early signs, which may be seen starting a few days to several weeks after infection, are followed by a period of 1 to 2 years in which there is intermittent fever, headache, and swollen lymph nodes (Fig. 3.8), spleen, and liver. As time passes, muscular weakness and emaciation become evident. At 2 to 3 years after infection, drowsiness begins and there ensue longer and longer periods when the person is half awake, asleep, or actually unconscious. Irreversible coma and death ensue. During the later stages of the disease, there may be various kinds of psychopathologies and convulsions.

In Rhodesian sleeping sickness (*T. b. rhodesiense*), the reservoirs of infection are wild ruminants such as the bush buck, and the organism is highly pathogenic for humans. Roughly the same sequence of events takes place as in Gambian sleeping sickness, but the disease is acute rather than chronic; the person will usually die within 6 months to a year. It is generally agreed that *T. b. gambiense* is a domiciliated form that was derived from *T. b. rhodesiense* and that it has become less pathogenic through its association with humans. Although it still causes the death of its host, *T. b. gambiense* is thought to be on its way toward developing a commensal relationship with its human host.

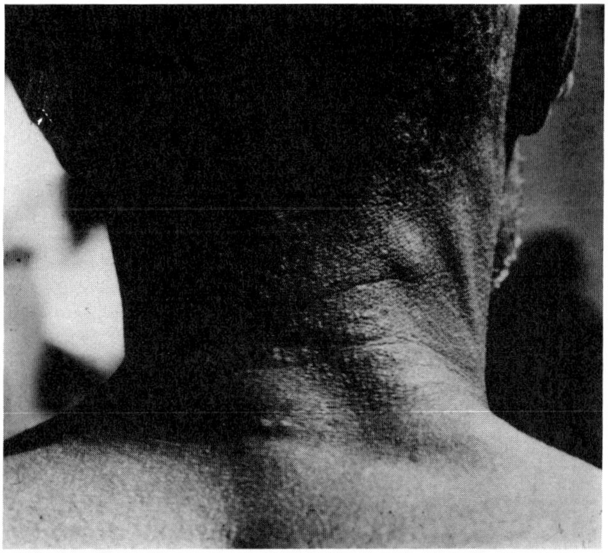

**FIG. 3.8**
Winterbottom's sign. Note the enlarged posterior cervical lymph nodes. [Courtesy U.S. Armed Forces Institute of Pathology. Negative No. 74-8337.]

*T. b. brucei* and *T. congolense* cause nagana in cattle, and the pathogenesis is much the same as that caused by the related organisms in humans. Once animals begin to show clinical signs, the disease progresses rapidly and the animals will usually die within a few weeks. Large ruminants that become seriously ill and lie down must be gotten up and made to move around or the pooling of blood on the lower side of the animal will cause collapse of the circulatory system within a day or so. Cattle therefore will not suffer the long course of disease seen in humans, because once they are down for a day or two, death will quickly follow.

The trypanosomes remain in the skin of the person for the early part of the infection and reproduce there. Within a few weeks they reach the circulating blood and the lymph nodes. In the later stages, corresponding with neurologic involvement, the organisms are found in the cerebrospinal fluid and other tissue spaces and between the cells of the brain. The signs and symptoms of infection as well as the tissue changes seen in the disease have been well known for many years, but the mechanisms by which the trypanosomes damage the host are just now coming to light. Early in the disease most of the changes are seen in the lymph system. When the trypanosomes invade the central nervous system there is evidence of inflammation of the brain. How these changes are brought about has been the subject of considerable study, but the theories have not held up well under scrutiny.

Three theories have been proposed, one based on high potassium levels in the blood of infected persons and the second on high glucose consumption by the trypanosomes. In the former, it has been observed that there is a high blood potassium level, but this is most likely a result, not a cause, of the damage to the host because it comes from lysed red blood cells. The second theory is based on the fact that trypomastigotes consume enormous quantities of glucose to produce pyruvate. Administration of glucose to an animal will prolong its life, and administration of insulin will shorten its life, so it was concluded that low blood glucose is the cause of damage to the host. However, damage to the host occurs when there are few trypomastigotes in the blood or tissues, and the pyruvate that is returned to the bloodstream can be readily used by the host. Both theories, therefore, seem not to be true; however, there is some evidence that pyruvate itself may cause damage in abnormally high blood levels.

The third theory currently accepted suggests that the mechanism of damage in the *T. brucei* complex relates to the immune system. As is true in some other parasitic infections such as filariasis and falciparum malaria, the host defense reaction itself is the source of damage. Kinins are produced in some infections where antibodies are also produced; kinins also control blood pressure and vascular permeability. In the case of Gambian sleeping sickness, kinins seem to give rise to the most severe aspects of the infection. Early in this century, Arthus showed that anaphylaxis could be produced in rabbits by first sensitizing them with small subcutaneous injections of horse serum followed by a large injection of horse serum. Death occurred in about 3 weeks, during which time the rabbits showed wasting, anemia, edema, and scabby patches on the skin. The same syndrome was observed in rabbits infected with *T. brucei* and corresponds to the signs of sleeping sickness in humans. As indicated earlier, many of the tissue changes are related to the congestion and permeability in the vascular system. Kinins bring about these kinds of changes, and the vascular permeability and venous congestion lead to a fall in blood pressure and eventually to cardiovascular collapse. A more complete description of the mechanism involved will be developed as knowledge progresses, but it is important to note that infection, the immune system, and cardiovascular problems are interrelated.

The immune reaction of the vertebrate host against the trypanosome is of considerable importance (1) in pathogenesis, (2) in preparing the parasite to survive in the vertebrate, and (3) in the constant struggle of the host and parasite to survive.

As noted in the discussion of the life cycle, the metacyclic trypomastigote is the only form that is infective when a tsetse bites a susceptible mammal. The metacyclic trypomastigote has a protein coat that protects it from the immune reaction of the host. The development of the protein coat is a preparation for entrance into a new environment—the mammalian blood.

The parasite requires the protein coat for entrance and early survival in the mammal, but it has developed additional mechanisms for continued survival and reproduction in the long term. In trypanosomosis, typically the number of organisms in the blood of the host regularly increases and decreases. Each time there is a resurgence in the number of organisms, there is a new antigenic variant, that is, the protein coat on the trypomastigotes is different (Fig. 3.7). It can be shown by electron microscopy that these antigens originate in the flagellar pocket and flow out over the surface of the organism. Of significance are the observations that there is a predetermined sequence of antigenic types that can be measured. In Figure 3.7, the sequence A→B→C→D→E, and so on, indicates that upon introduction of metacyclic trypomastigotes by a tsetse the sequence of antigenic variants can be predicted. Passing the organism through a tsetse resets the sequence to type A, but organisms passed by syringe lose the capacity to return to type A. Somehow the contact with the environment inside the tsetse causes the organism to return to the starting point of a developmental sequence. An interesting review of unique forms of parasitic immunity including antigenic variation in trypanosomes is presented by Bloom (1979).

There is no way yet to immunize against members of the *T. brucei* complex, largely because of the continual shifting of the antigens. The technical problems of producing an antigenic product capable of stimulating a response to a sufficient number of the antigenic variants are still too great.

TREATMENT. Shortly after David Bruce's discovery in 1896 of *T. b. brucei* as the cause of nagana in cattle, attempts were made to treat both infected livestock and people. Inorganic arsenicals such as sodium arsenite were given, with beneficial results in some instances but catastrophic results in others. Blindness and death often followed treatment and arsenic was abandoned for use in humans, although it was still used for livestock.

Robert Koch was the first (1905) to take an organic arsenical, atoxyl, to the field for studies on its curative powers in human infections. The dosages Koch used, 1 g per week, were such that some patients were blinded and others were in such pain that they refused further treatment. Atoxyl is basically a sodium salt of arsenious acid attached to a phenol ring (arsanilic acid) and was a partially successful attempt to reduce the toxicity of the inorganic arsenic.

In 1917, tryparsamide, a variation on atoxyl, was synthesized and used against sleeping sickness. Its toxicity was lower than that of atoxyl and its efficacy was greater, so that effective treatment could be given with a tolerable risk of side effects. During the intervening 60 years, 11 additional drugs have been introduced and used widely against human and animal infections.

Of the drugs in use, many give a high cure rate in human infections, especially early in the course of the disease. Some, such as the phenanthridines and berenil, can be given prophylactically to prevent the development of infections if a person is bitten by an infected tsetse. Elimination of infections late in the course of the disease, when the central nervous system has been invaded, has been a severe problem. A number of drugs can cure even these kinds of infection, but the brain damage resulting from the infection may not be entirely reversible.

The trypanosomoses of animals are susceptible to the same drugs used in humans, for the most part; however, in domesticated animals, there are seven species of *Trypanosoma* that inhabit at least five species of animals. The drugs not only differ in efficacy, depending on the trypanosome, but also differ in toxicity in different mammals.

Looking broadly at the drugs now available, the problems lie in toxicity, efficacy, drug resistance, and interference with development of immunity or at least with its manifestations. All the drugs must be given by injection, and in some cases, the simple modes of injection, subcutaneous or intramuscular, are not feasible because of severe localized reactions. These drugs must then be given intravenously. There is no one-shot drug that can be administered with the assurance that the person or animal will be cleared of trypanosomes. Protozoa have become resistant to the trypanosomicides and this is a continuing, widespread problem. In some instances, resistance to one drug confers resistance to all members of that drug group.

The perfect drug for use against the African trypanosomoses has not yet been developed, but within the limits of those available, ingenuity of the highest order has been shown. Switching of drugs and doubling up on drugs have reduced the impact of resistance. Organic chemists have produced variations on a theme to reduce toxicity and yet maintain or increase efficacy. For further information see Williamson (in Mulligan & Potts, 1970).

EPIDEMIOLOGY AND CONTROL. The tsetse-borne trypanosomoses can be discussed as a group whether one is concerned with human or animal infections. There are some differences between methods used in human or animal infections, but many of the principles of control are the same.

Control programs are based on

1. Tsetse control
2. Reservoir host control
3. Chemotherapy and chemoprophylaxis
4. Behavioral changes in the vertebrate hosts

Bruce's report in 1895 showed the link between trypanosomes and tsetse through animal experimentation. The transmission experiments Bruce performed in that year showed that tsetse transmitted trypanosomes to domesticated animals and laid the basis for rational control of trypanosomosis. Since that time the bulk of the effort to control or eradicate the disease has been focused on reducing or eliminating populations of tsetse.

Tsetse (*Glossina* spp.) are insects of the order Diptera, family Muscidae, the same family to which house flies, stable flies, and horn flies belong. Tsetse are limited to Africa from south of the Sahara Desert to about 25° S. latitude. The exception, *G. tachinoides*, was discovered in the Arabian peninsula in the early 1900s, but it is not known whether it is still there.

Within the genus *Glossina* there are 22 species and 15 subspecies. The species range from 7.5 to 14 mm in length, most being about 10 mm long. The adults are mottled gray-brown, often with darker bands on the abdomen. The wings have a characteristic discal or "hatchet" cell, which is of diagnostic value.

The life cycle of *Glossina* is basically that of other holometabolic insects (Chap. 50): egg, larva, pupa, adult. It differs from most other muscids in that the females give birth not to eggs, but rather to third-instar larvae. The female nourishes one larva at a time within her body by feeding it with secretions of special glands in the uterus. During the time when a female is nourishing a developing larva, three blood meals (at intervals of several days to a week) are required for its growth.

Tsetse are not among the adventurous dipterans; they do not fly often and they migrate slowly. Any one adult may fly for only about 5 min a day and rarely more than 15 min. In mark and recapture studies, flies were found to migrate about 200 m a week. The flies seem to have a home range from which they move out to feed and then return to about the same resting sites. Infections in flies have sometimes been found to be highly localized, that is, within a single clump of trees and brush. Their limited flight at any one time makes it possible to establish brush-free barriers so that a control program can be established in a particular area in which there will be neither immigration nor emigration.

Since there are more than 20 species of *Glossina* in subsaharan Africa, it would be logical for there to be differences in the habitat, hosts, and biologies of the species (Table 3.3). In any one area usually one, sometimes two, species are the principal vectors, and if one goes to another ecological habitat, a different vector will be used. For example, *Glossina morsitans* is a riverine species that is the important vector of Gambian sleeping sickness in West Africa. It is found along the banks of streams and lakes, and finds its resting sites in brush fairly close to the water. *G. swynnertoni* is a savannah species and is found more diffusely in drier areas than *G. morsitans*. The approaches to control are quite different with the two species.

Since control of trypanosomiasis is largely directed against the vectors, a thorough knowledge of the biology of the vector(s) in the specific locality is essential before a program is started.

The elements that must be coordinated are

1. Selective clearing of trees and brush
2. Application of insecticide
3. Control of land use, that is, farming and grazing practices

Because tsetse do not migrate rapidly or fly far, an area of 2 to 4 km (1–2 miles) is cleared of all trees and brush so that the flies can be contained. Then, if a riverine species such as *G. morsitans* were the vector, an area would be cleared of brush and low-growing plants for about 10 m from the water's edge. This allows the ground to dry out so that pupae die and eliminates resting sites for the adults. Insecticides such as DDT can then be sprayed from a back pack sprayer to kill the remaining flies.

Land use is an important aspect of control in that once an area has been cleared of tsetse habitat, it should not be allowed to return to its former state. It should be kept in a crop or grazed periodically by livestock so that it remains in grass and small forbs. Where all of these steps are taken, and the area is monitored periodically to make sure flies have not returned, the risk of infection to humans or domestic animals is minimal.

When there is risk of infection, the human or livestock population must be monitored to detect new clinical cases and to institute treatment immediately. During the time of continued transmission, it may be necessary to administer drugs either to the human or the animal population on a prophylactic basis. Antrycide, ethidium, and pentamidine have been used either in humans or livestock for protection against the establishment of infection. If an area is made relatively safe, the major tasks then are to maintain monitoring and reporting systems and to exclude immigration of vectors and infected mammals.

The question is whether control has been successful. The answer is a qualified yes. Best current guesses are that perhaps 200 persons die annually from sleeping sickness in Africa. Many areas have been made risk-free for human habitation. As far as livestock are concerned, the risks and losses are still significant, but areas can now be utilized by cattle that were formerly impossible to occupy; however, eternal vigilance is the price of survival, and surveillance is carried on continuously.

Finally, the question must be raised about the probability and even the desirability of eradicating tsetse from the African continent. Clearly, for the tsetse-borne trypanosomoses, no tsetse means no disease. But is complete eradication desirable? Ormerod (1976) proposed that control of trypanosomosis in

**TABLE 3.3**
Major *Glossina* Vectors of African Trypanosomoses

| VECTOR | HABITAT | MAIN HOSTS | TRANSMITS |
|---|---|---|---|
| | | FUSCA GROUP | |
| *G. brevipalpis* | River and stream, thickets | Mammals, reptiles | Not a significant vector |
| | | PALPALIS GROUP | |
| G. palpalis | Forest, river thickets savanna to an extent | Primates, reptiles, bovids | *T.b. gambiense*<br>*T.b. rhodesiense* |
| *G. tachinoides* | Savanna, intermittent river areas | Primates, bovids, reptiles, suids | *T.b. gambiense*<br>*T.b. brucei*<br>*T. congolense, T. vivax* |
| | | MORSITANS GROUP | |
| *G. mortisans* | Savanna, grassland | Suids, bovids, primates | *T.b. rhodesiense*<br>*T.b. brucei* |
| *G. pallidipes* | Rivers, thickets | Bovids, suids | *T.b. rhodesiense*<br>*T.b. brucei* |
| *G. swynnertoni* | Dry, open areas, savanna | Suids, bovids | *T.b. brucei*<br>*T.b. rhodesiense* |

cattle has allowed cattle populations to increase to the point where the Sahel of western Africa is overgrazed and has contributed to drought conditions that have been catastrophic for millions of Africans. Duggan's final few paragraphs on the history of the African trypanosomoses put into perspective perhaps better than any other statement the dilemmas and ethics of disease control (Mulligan & Potts, 1970):

> It is clear that the problem of trypanosomiasis in domesticated animals is now of far greater concern than human infection to the health and prosperity of Africa. Its biological complexity is vast, its physical extent is only vaguely defined and its significance is almost immeasurable. The burden of sleeping sickness in the 1930s was but thistledown in comparison to the leaden weight of this gigantic load.
>
> It may seem necessary to change the face of rural Africa before the scourge is eliminated, yet as Barzun reminds us, there is no use struggling to save a world in which one would find it hell to live when saved. An Africa without game, tsetse and trypanosomes could turn a noble ambition to dust, for they are the guardians of the land. Over-zealous guardians they may be, but their too peremptory dismissal from the scene could permit improvidence and exploitation on a ruinous scale.
>
> The biological complex of trypanosomiasis is in deep and mortal relationship with the life of Africa; to sever this relationship would be to change the bionomics of half a continent. It seems more commendable to leave the relationship alone, and simply to thwart the undesirable effects of its progression by methods which anticipate its own beneficent destiny.

## REFERENCES AND FURTHER READINGS

Baker, J. R. 1963. Speculations on the evolution of the family Trypanosomatidae. *Exp. Parasitol.* 13:219–233.

Bardsley, J. E., and R. Harmsen. 1973. The trypanosomes of anura. *Adv. Parasitol.* 11:1–73.

Bloom, B. R. 1979. Games parasites play: How parasites evade immune surveillance. *Nature* 279:21–26.

Brown, R. C., D. A. Evans, and K. Vickerman. 1973. Changes in the oxidative metabolism and ultrastructure accompanying differentiation of the mitochondrion in *Trypanosoma brucei*. *Intern. J. Parasitol.* 3:691–704.

Cunningham, I., and B. M. Honigberg. 1977. Infectivity reacquisition by *Trypanosoma brucei brucei* cultivated with tsetse salivary glands. *Science* **197**:1279–1282.

Ford, J. 1971. *The Role of Trypanosomiasis in African Ecology*. Clarendon Press, Oxford.

Heisch, R. B., M. P. McMahon, and P. E. C. Manson-Bahr. 1958. The isolation of *Trypanosoma rhodesiense* from a bush-buck. *Brit. Med. J.* **1958** (Nov.):1203–1204.

Hirumi, J., J. Doyle, and K. Hirumi. 1977. African trypanosomes: Cultivation of animal-infective *Trypanosoma brucei* in vitro. *Science* **196**:992–994.

Hoare, C. A. 1964. Morphological and taxonomic studies on mammalian trypanosomes. X. Revision of systematics. *J. Protozool.* **11**:200–207.

Hoare, C. A. 1967. Evolutionary trends in mammalian trypanosomes. *Adv. Parasitol.* 5:47–92.

Hoare, C. A. 1972. *The Trypanosomes of Mammals: A Zoological Monograph*. Blackwell, Oxford.

Honigberg, B. M. 1963. A contribution to the systematics of non-pigmented flagellates. *Proc. 1st Intern. Congr. Protozool.*, Prague, August 1961.

Jahnke, H. E. 1976. *Tsetse Flies and Livestock Development in East Africa. A Study in Environmental Economics*. Humanities Press, New York.

Kudo, R. R. 1966. *Protozoology*. 5th ed. Charles C Thomas, Springfield, Illinois.

Losos, G., and A. Chouinard (eds.). 1979. *Pathogenicity of Trypanosomes*. Workshop Proceedings, Nairobi, Kenya, 20–23 November 1978. International Development Research Center, Ottawa, Canada.

Lumsden, W. H. R. 1965. Biological aspects of trypanosomiasis research. *Adv. Parasitol.* 3:1–58.

Lumsden, W. H. R. 1970. Biological aspects of trypanosomiasis research, 1965, a retrospect, 1969. *Adv. Parasitol.* 8:227–250.

Mackenzie, A. R., and P. F. L. Boreham. 1974. Autoimmunity of trypanosome infections. I. Tissue antibodies in *Trypanosoma (Trypanozoon) brucei* infections of the rabbit. *Immunology* **26**:225–238.

McGhee, R. B. 1968. General characteristics of the blood protists. II. Development and reproduction (vertebrate and arthropod host). In D. Weinman and M. Ristic (eds), *Infectious Blood Diseases of Man and Animals*, Vol. 1. Academic Press, New York, pp. 307–341.

McKelvey, J. J., Jr. 1973. *Man Against Tsetse. Struggle for Africa*. Cornell University Press, Ithaca, N. Y.

Mulligan, H. W., and W. H. Potts. (eds). 1970. *The African Trypanosomiases*. Wiley Interscience, New York.

Ormerod, W. E. 1976. Ecological effect of control of Africa trypanosomiasis. *Science.* **19**:815–821.

Parthasarathy, M. V., W. G. van Slobbe, and C. Coudant. 1976. Trypanosomatid flagellate in the phloem of diseased coconut palms. *Science* **192**:1346–1348.

Talliaferro, W. H., and L. A. Stauber. 1969. Immunology of protozoan infections. In T.- T. Chen (ed.), *Research in Protozoology*, Vol. 3. Pergamon Press, New York, pp. 505–564.

Vickerman, K. 1969. On the surface coat and flagellar adhesion in trypanosomes. *J. Cell Sci.* **5**:163–169.

Vickerman, K. 1971. Morphological and physiological considerations of extracellular blood protozoa. In A. M. Fallis (ed.), *Ecology and Physiology of Parasites*. University of Toronto Press, Toronto, pp. 58–91.

Wille, J. J., E. Weidner, and W. L. Steffens. 1981. Intranuclear parasitism of the ciliate *Euplotes* by a trypanosomatid flagellate. *J. Protozool.* **28**:223–227.

# 4

# *Chagas' Disease*

Chagas' disease, or American trypanosomosis, is a zoonosis that affects some 24 million people in the Western Hemisphere. Although Chagas' disease has not had as great a controlling influence on the people of South America as sleeping sickness has on the people of Africa, it nevertheless has adversely affected the health of millions of persons, and about 65 million are at risk of infection.

NAME OF ORGANISM AND DISEASE CAUSED. *Trypanosoma cruzi* Chagas, 1909 (=*Schizotrypanum cruzi* and others) causes Chagas' disease or American trypanosomiasis or trypanosomosis.

HOSTS AND LOCATION. The organism infects the vertebrate host's blood, somatic and cardiac muscle, and a wide array of other tissues. Normal hosts include many species of wild mammals. The armadillo (*Dasypus*) is the most important reservoir in some areas, and in other localities rodents such as the wood rat (*Neotoma*) and carnivores such as dogs, cats, and raccoons serve as the important reservoirs. The broad host range includes domesticated animals. Humans have also become an important host. Bugs (Insecta, Hemiptera, Reduviidae, Triatominae) of the genera *Triatoma*, *Rhodnius*, and *Panstrongylus* are the vectors. These insects are also known as *triatomins, kissing bugs*, and *cone-nosed bugs*. *T. cruzi* generally remains in the gut of the host insect.

DISTRIBUTION AND IMPORTANCE. *T. cruzi* is limited to the Western Hemisphere, and has generally been considered to range from 31° S latitude (roughly the southern border of the United States) through about three quarters of the length of South America; however, *T. cruzi* has been found as far north as Maryland (39° N) in raccoons, and is endemic in woodrats in the Los Angeles, California, area (Fig. 4.1).

The major centers of human infection are Brazil, Argentina, and Chile, but foci of infection are scattered through northern South America and Central America, and a few cases have been reported from the southwestern United States. A survey of 500 people in the lower Rio Grande Valley in Texas showed a prevalence of infection of 2.4% by serology, indicating that infection may be more common than realized (Burkholder et al., 1980).

In humans, Chagas' disease is a severe, debilitating disease that may take either an acute or chronic form. The acute form is seen mostly in children, and the mortality in infected children may be 10% or higher. The chronic form, seen mostly in adults, affects the heart muscle in about 50% of cases, but also has detrimental effects on the intestinal tract; it may take some years before significant signs and symptoms develop. Adults may become cardiac invalids within a few years after infection and over a period of time become capable of less and less physical activity.

Charles Darwin (1809–1882) may have been a victim of Chagas' disease, contracted perhaps at Iquique,

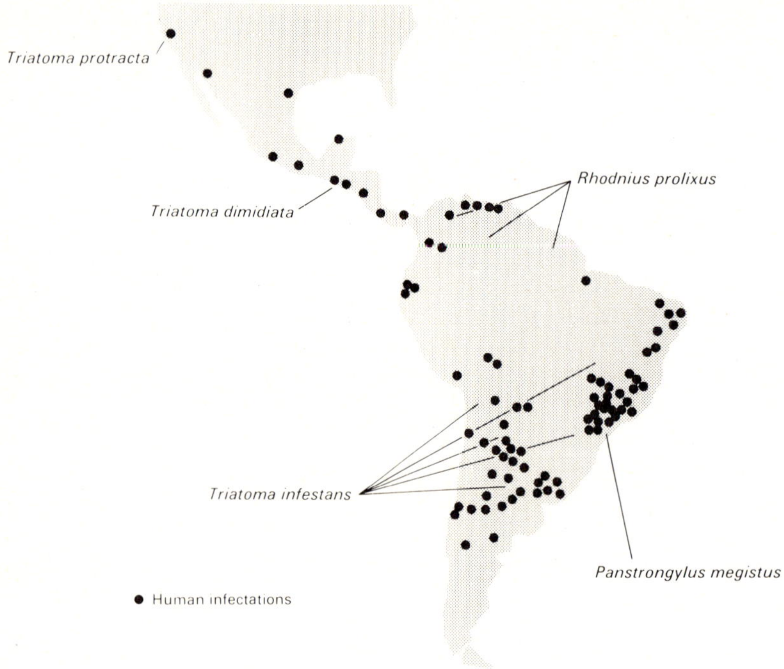


**FIG. 4.1**
The distribution of Chagas' disease.

Chile, during the voyage of the *Beagle* (1831–1836). As a young man, Darwin was sociable, robust, and active outdoors. During the voyage he worked hard both aboard ship and in collecting trips at various places where the *Beagle* stopped. It seems that his health remained good until he returned to England, when he noted that he was "occasionally unwell" during the 2 years he spent at Cambridge working over his notes and collections. By 1839 Darwin was unable to put in a full day's work, to his displeasure, and in 1842, he moved with his family to an estate at Downe, southeast of London. Comparing the portrait painted of him as a young man and a photograph of him late in life, it is easy to see the typical facial changes associated with Chagas' disease. The pattern of physical and mental changes that Darwin exhibited is characteristic of individuals with chronic Chagas' disease. This contention has been disputed by Colp (1977), who did a complete study and concluded that Darwin's illness was a psychological manifestation.

MORPHOLOGY. Trypomastigotes from the blood are 16 to 20 μm in length and generally are C-shaped in a dried blood film; the kinetoplast is large and subterminal (Fig. 4.2). The amastigotes are from 1.5 to 4 μm, but generally are at the upper end of the size range; they have a large nucleus and kinetoplast, but little else can be seen under the light microscope (Fig. 4.3).

LIFE CYCLE. Small numbers of trypomastigotes circulate in the blood of the mammalian host. The organisms are picked up by kissing bugs (Fig. 52.3) during the act of bloodsucking. In the midgut and hindgut of the insect, the organisms multiply as epimastigote and amastigote forms. At 1 week to 2 weeks, depending on temperature, metacyclic trypomastigotes appear in the hindgut and the infection can then be transmitted back to another mammal (Fig. 4.4). *T. cruzi* is a stercorarian trypanosome and as such is transmitted through fecal contamination. While taking a blood meal, kissing bugs swell dramatically and tend to defecate; a species that is a good vector of *T. cruzi* will often defecate while still feeding. The fecal droplet of an infected bug may contain thousands of metacyclic trypomastigotes. The trypomastigotes gain entry into the mammal when it scratches or rubs them into the skin, or they enter the mucous membranes of the eye, mouth,

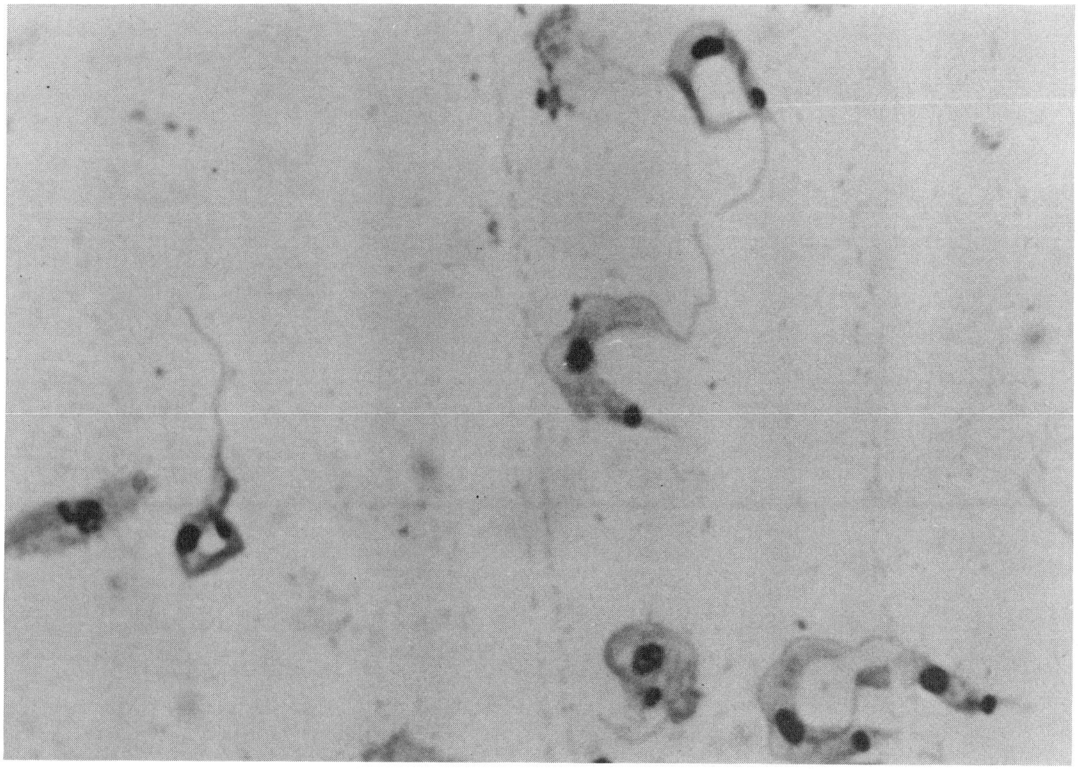

**FIG. 4.2**
Trypomastigote of *Trypanosoma cruzi.*

or nose. Studies on laboratory animals indicate that the organisms cannot enter the unbroken skin, but they readily establish infections when applied to the mucous membranes.

Once regaining a mammalian host, the trypomastigotes may remain in the bloodstream for a time, but they do not multiply there. They invade cells of the reticuloendothelial system (RES or MPS) and skeletal and cardiac muscle cells, among other tissues, and multiply as intracellular amastigotes. Multiplication is by binary fission, and as host cells break down, some organisms are released into the bloodstream. These organisms transform into trypomastigotes and are available to be picked up by a kissing bug; trypomastigotes released from cells do not multiply in the bloodstream.

DIAGNOSIS. Although trypomastigotes circulate in the blood of the vertebrate host, whether human or another animal, they exist only in small numbers except in the acute phase of the disease. The probability of finding *T. cruzi* by examining a stained blood film is quite low during the chronic phase in humans and in most reservoir hosts.

Other direct means of finding organisms are to examine the lesion at the site of entry, to examine the juice of lymph nodes, or to make a sternal biopsy and examine the bone marrow for amastigotes.

Indirect means of determining whether a host is infected involve cultivation, animal inoculation, serology, and clinical signs. *T. cruzi* grows in a variety of media and in tissue culture. Blood or serum may be inoculated into a medium such as NNN (Novy MacNeal–Nicolle) and examined for trypomastigotes after several days' incubation.

A method having good probability of determining whether a vertebrate is infected is xenodiagnosis. In this technique, laboratory-raised, uninfected triatomins are

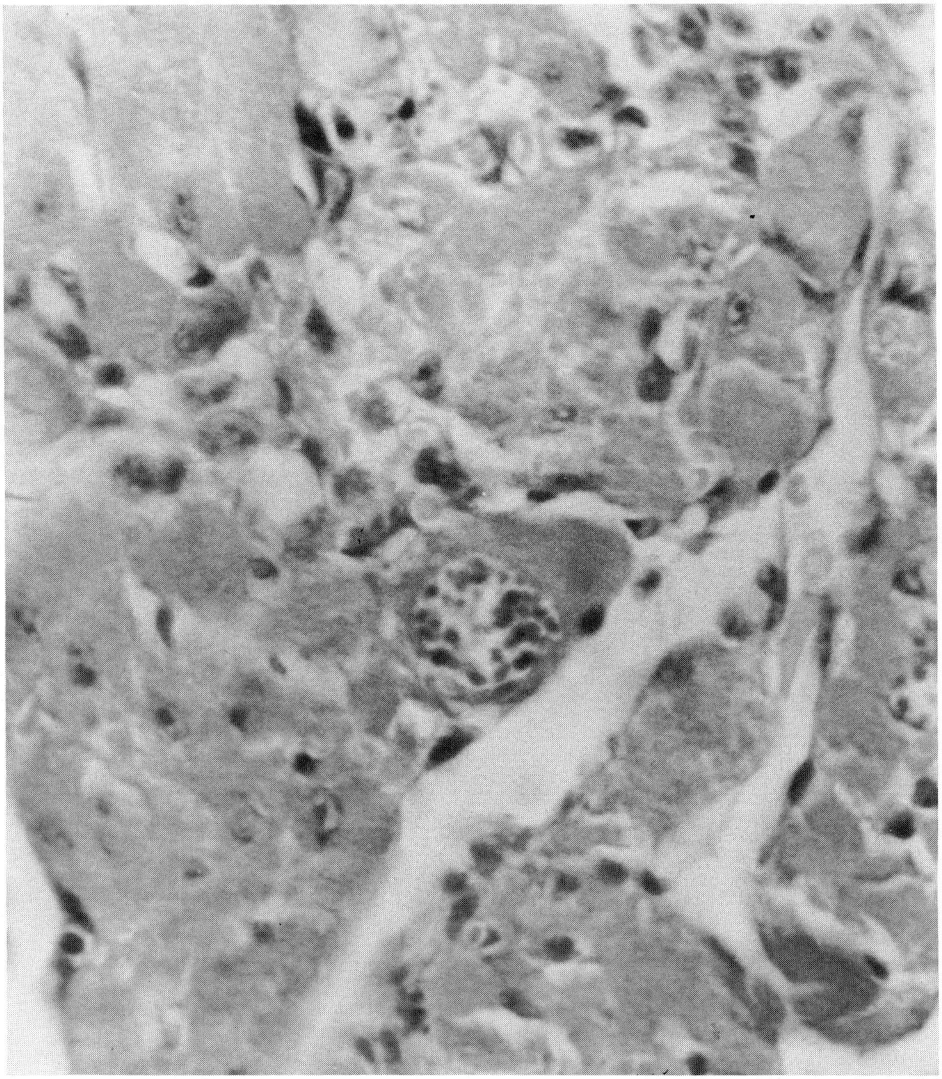

**FIG. 4.3**
Amastigotes of *Trypanosoma cruzi* in a heart section.

allowed to feed on a person or animal suspected of being infected. After 10 days or so, the bugs are killed and the hindgut is examined for trypomastigotes of *T. cruzi*. Where large numbers of bugs are used to feed on a person, the probability of finding *T. cruzi* in one or more bugs is about 70%

Animal inoculation merely uses a living organism as a culture medium, as does xenodiagnosis. Laboratory rodents such as mice and guinea pigs are susceptible to infection; puppies and kittens can also be used, but the cost of uninfected laboratory-raised animals is likely to be high.

Serologic tests indicate by the presence of antibodies whether an animal has been infected. As a general rule, *T. cruzi* elicits good production of antibodies that can be measured through a variety of tests. In some tests, up to 90% accuracy is achieved in determining whether infection has taken place. Although the antibody titer is likely to be higher in the acute phase of the disease, or shortly thereafter, it is not always possible to differenti-

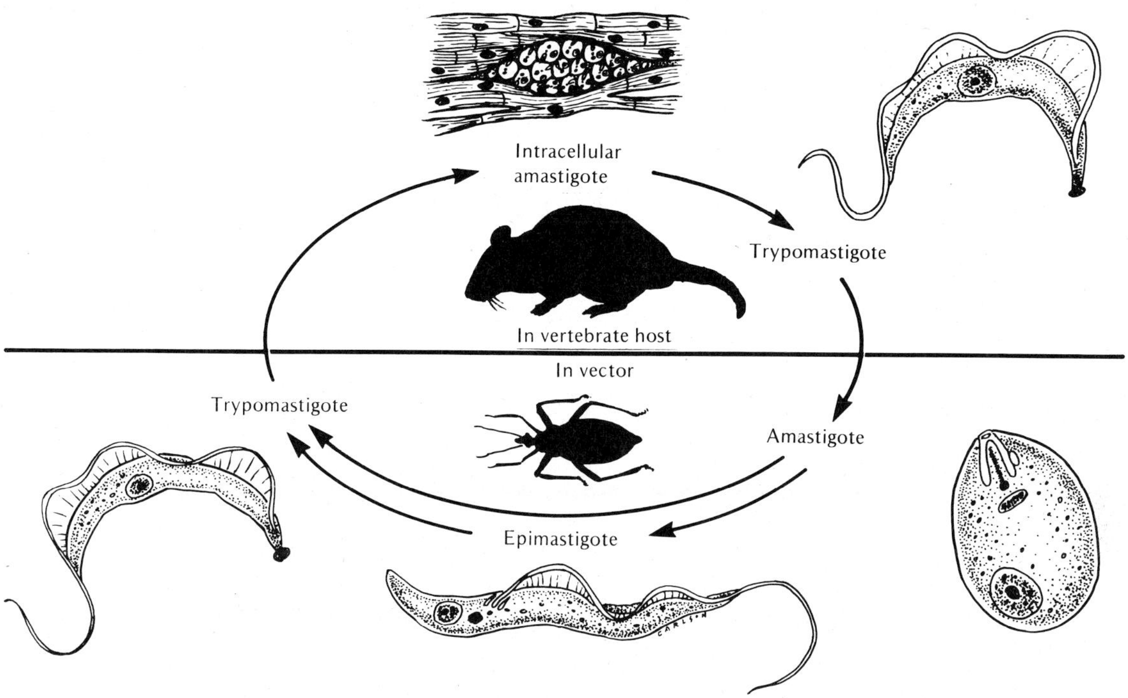


**FIG. 4.4**
Diagram of the life cycle of *Trypanosoma cruzi.*

ate between a recent infection and one of long standing. Interpretation of positive tests also requires keeping in mind that other organisms such as *Leishmania* spp., *T. rangeli,* and *T. minasense* may cross-react with the *T. cruzi* antigens.

A characteristic sign of early infection is swelling around the eye, called Romaña's sign (Fig. 4.5). When using any diagnostic procedure, it is best to remain wary when one finds an organism resembling what one is looking for. As indicated, there can be serological cross-reactions among related organisms. Likewise, organisms may show up on direct or indirect tests which may not be *T. cruzi. T. rangeli* and *T. minasense* may be found in humans and other animals and *T. theileri* in cattle. Also, Chagas' disease and leishmaniasis may occur in some of the same parts of South America. In these instances, the structure of the organism found must be carefully documented. Both quantitative and qualitative characteristics should be tabulated and analyzed before giving an identification.

HOST–PARASITE INTERACTIONS. The damage to the vertebrate host in infections with *T. cruzi* varies with the species and the age of the host and the virulence of the particular organisms. The organism seems to cause no significant damage to the reservoir hosts in the wild. Among domesticated animals, dogs and cats, but not other animals, may be clinically affected. In humans, the disease may be either acute or chronic, and the former may develop into a chronic disease if the individual survives.

Acute disease in humans is usually seen in children. The organisms initially multiply in fat cells under the skin, and then invade muscle cells followed by a variety of white cells. As cells become distended with amastigotes, they break down and release organisms that are carried by the circulatory system to other parts of the body where additional cells are invaded. A wide spectrum of cells is invaded: skeletal and cardiac muscle, lymphoid tissue, bone marrow, nervous tissue, intestine, and endocrine glands, among others. Death

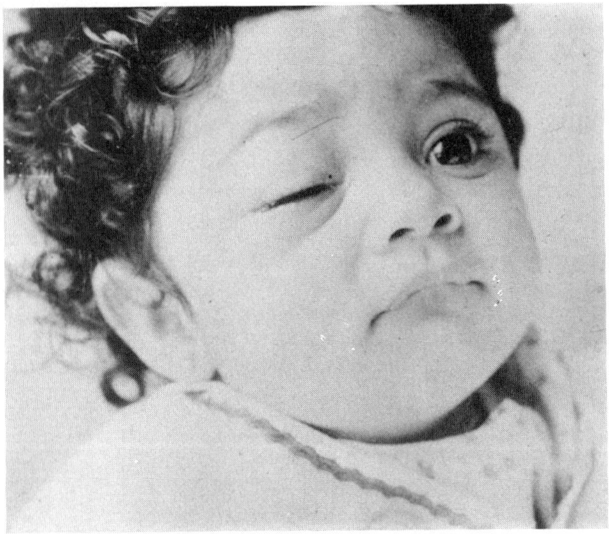

FIG. 4.5
Romaña's sign. [Courtesy of U.S. Armed Forces Institute of Pathology. Negative No. 62-3934-6.]

may occur within a few weeks or months after the onset of symptoms.

If the individual survives the acute stage, or if the person is an adult, the chronic stage of the disease ensues. There is a long, continuous decline in the individual's physical capacities. There may also be psychological disturbances. The main effect is on cardiac muscle in adults and the heart has progressively less and less capacity to work. Another effect of long-standing infections is an enlargement of the esophagus or the colon, or both. The structure becomes flaccid and fails to have normal peristalsis. The colon retains a large quantity of feces because it has lost the capacity to expel its content. These conditions are referred to as *megaesophagus* and *megacolon*, respectively.

The damage to the host tissues in Chagas' disease seems to be caused by a toxin produced by the *T. cruzi*. There is evidence that it is a neurotoxin that acts only in the region in which it is produced. The effect on the gut is denervation of the smooth muscle. The stimulus for the muscle cells to contract is lacking and the organ becomes progressively enlarged.

Persons infected with *T. cruzi* develop antibodies against the agent. The immunity prevents the continued rapid development of the organisms, but it is insufficient to arrest the disease or to clear the host completely of infection. Probably the only time that the host's defenses can act against *T. cruzi* is during the transfer from one host cell to another; there is likely to be little or no immune activity while the organisms are within cells. As indicated in the section on diagnosis, the number of trypomastigotes in the circulating blood is rather low, and it would appear that the best the host can do is to slow down the course of the disease.

Chagas' disease is associated with poverty and substandard living conditions. Crowding and poor housing provide the milieu in which the vectors and the disease agent can become domiciliated (Fig. 4.6).

Before the colonization of the Western Hemisphere by Europeans, Chagas' disease was unrecognized. In fact, it was not until Chagas isolated the causative agent from *Panstrongylus megistus* in 1909 in Brazil that even the suspicion of a problem was raised. Parenthetically, this was the first instance in which a disease agent was isolated from a vector before it was found in the vertebrate animal. Chagas' discovery was greeted with some skepticism by his colleagues, and 20 years passed before the disease that honors his name was accepted by the medical community.

What, then, is the pattern of transmission? Originally, the disease cycled in wild animals such as armadillo, rodents, carnivores, and primates other than humans. In most of these animals, there is no clinical effect and the parasite could be passed from mammal to mammal through the bug with no apparent harm to the participants. With the colonization of the New World by Europeans and Africans, humans invaded and established habitations where the populations of people had probably been relatively sparse. Humankind was inserting itself into the localities where *T. cruzi* was silently cycling. Houses of thatch or adobe were no doubt built in areas close to populations of wild animals harboring *T. cruzi*. Humans provided a ready source of food for the vectors and they moved in with the people. Those triatomins that had been infected by feeding on rodents, for example, passed the infection to persons sleeping on mats or close to the ground, and the cycle began to run in humans.

TREATMENT. The lack of any effective drug for use against *T. cruzi* infections has been catastrophic for the many millions of persons infected. Spontaneous recov-

**FIG. 4.6**
A Meso-American house typical of rural areas. The thatched roof and poorly plastered walls provide hiding places for triatomins.

ery is rare and the only measures available to physicians have been to provide symptomatic relief, that is, to reduce the symptoms of the disease. Further, there is no effective immunization, so there is no way to prevent the infection, such as is done for smallpox, polio, and diphtheria.

The drugs that have been developed for use against the African trypanosomes are completely ineffective against *T. cruzi*. All have been tried and found wanting. The inference is that the metabolic pathways of the African species are quite different from those of *T. cruzi* and its other stercorarian relatives as well. The mitochondrial changes discussed in connection with *T. brucei* are not seen in *T. cruzi* and there is some biochemical evidence that the organisms differ in their oxidative enzymes.

The Bayer Company has developed Lampit for use against *T. cruzi* infections. Although the drug is effective in acute infections, it is ineffective in chronic infections and has rather high toxicity.

EPIDEMIOLOGY AND CONTROL. The lack of any effective drug or means or artificial immunization places the whole emphasis of control of Chagas' disease on prevention. The major effort in control programs is directed toward vector control.

Triatomins (*Triatoma*, *Rhodnius*, *Panstrongylus*) are blood-sucking bugs that live primarily underground with burrowing rodents, but some are also associated with tree-dwelling vertebrates such as birds, bats, and small primates. Given the opportunity, kissing bugs will move in with humans if suitable hiding places and a ready source of food are available. These conditions are met in adobe and thatch houses and where people sleep on or close to the ground.

Kissing bugs are hemimetabolic insects and as such have a gradual metamorphosis in developing toward the winged adults. There are five nymphal stages (instars), all of which feed on blood and require several months to develop to the adult. The nymphs must walk to their source of food, but the adults can fly.

Although all Hemiptera have sucking mouthparts, some are predaceous on other small invertebrates and others are blood feeders. The triatomins seem to be well adapted to blood feeding since they produce little or no pain when feeding. This is an interesting adaptation for an animal that might otherwise find itself squashed or mutilated as soon as it penetrates the skin of its vertebrate host.

Typically, kissing bugs remain in cracks and crevices in the house during the daytime and come out at night to feed. Since they have little preference for one species of host over another, they may feed on both household animals and people. They are attracted to hosts by warmth and by $CO_2$, and this may be the reason they feed on the skin near the mouth (the source of their intriguing name). If undisturbed, the bugs require perhaps 15 min to feed. They are ready to feed again in about 10 days. Before it feeds, a bug is quite flat, but afterward its abdomen is greatly enlarged. *Rhodius prolixus* may consume more than 0.5 ml of blood during its nymphal instars. Adult female bugs often die after egg laying. Whether any one species of triatomin serves as a good vector of *T. cruzi* depends largely on whether it defecates during or soon after feeding. If the bug tends to defecate after it has returned to its hiding place, the probability of its transmitting *T. cruzi* is low. Transmission is likely only if the fecal droplet containing metacyclic trypomastigotes is deposited on the host. Once infected, bugs retain their infection for life. Usually they are not affected by harboring *T. cruzi* unless the organisms escape from the gut and enter the hemocoel, in which case the bug may be killed.

A relatively high percentage of bugs may be infected with *T. cruzi* in endemic areas. A third to two thirds of bugs have been found to be infected in parts of South America; most likely this is because of the relatively low fecundity and prolonged life cycle of the bugs.

Since Chagas' disease is associated with poverty and poor living conditions, preventing infections poses a difficult problem. Over 8500 bugs were found in a single mud hut. Where villages are established, the population of wild animals may be reduced and thereby encourage the domiciliation of the vectors of Chagas' disease (Marsden, 1971). The dwellings themselves, if poorly constructed, provide cracks and crannies into which the bugs can crawl during the daytime to be near a source of food at night (Fig. 4.6). Even the thatch roofs of dwellings provide places for bugs to live and sometimes animals to feed on. These bugs may never come into contact with people, but it is suspected that their contaminated feces may drop on people and thereby infect them. Dogs and cats, both good hosts for *T. cruzi*, are likely to be associated with many households and can substitute for the wild animal reservoirs of *T. cruzi*.

The problem becomes one of indirect bug control through upgraded housing, direct bug control through use of insecticides, prevention of contact with bugs, and control of the reservoirs of infection.

In houses, the first step is to plaster the inside walls to plug up places where bugs can hide. This may be fairly simple in adobe houses but more difficult in other types of construction. Screening of doors and windows keeps the immigration of bugs to a minimum. Personal protection can be accomplished by sleeping under an insect net, but recommendations are that the netting have a solid top so that bug feces cannot drop through. In a warm, humid climate, compliance with this recommendation might be rare because the netting would reduce the flow of air, making sleeping uncomfortable.

Strict control of dogs and cats is recommended by some authors. If one were attempting to control Chagas' disease in a holoendemic area, one would logically eliminate all dogs and cats to remove at least two sources of infection; however, elimination of these reservoirs may be neither feasible nor advisable. Cats, for example, are a low-cost method of rodent control in poor communities.

Control of triatomins through insecticides is effective. DDT has only low activity against bugs, but dieldrin and lindane are effective and have long residual action. Dieldrin has been sprayed three times a year to give a concentration of 1 g per square meter per year. Both insecticides have successfully eradicated triatomins in some locales. Resistance to lindane has been reported.

In southern California, Tarshis (1963) reported good control of triatomins in crawl spaces of houses through the use of Dri-Die 67, a silicone dust. The effect of this compound is to dehydrate the bugs through scratching of the waxy layer on the cuticle.

## REFERENCES AND FURTHER READINGS

Burkholder, J. E., T. C. Allison, and V. P. Kelly. 1980. *Trypanosoma cruzi* (Chagas) (Protozoa:Kinetoplastida) in invertebrate, reservoir, and human hosts of the lower Rio Grande Valley of Texas. *J. Parasitol.* **66**:305–311.

Colp, R., Jr. 1977. *The Illness of Charles Darwin.* University of Chicago Press, Chicago.

Gonzales-Angulo, W., and R. E. Rykman. 1967. Epizootiology of *T. cruzi* in southwestern North America. IX. An investigation to determine the incidence of *T. cruzi* infections in Triatominae and man on the Yucatan Peninsula of Mexico. *J. Med. Ent.* **4**:44–67.

Herman, C. M., and J. J. Bruce, Jr. 1962. Occurrence of *T. cruzi* in Maryland. *Proc. Helm. Soc. Wash.* **29**:55–58.

Kagan, I. G., L. Norman, and D. Allain. 1966. Studies on *T. cruzi* isolated in the U.S.: A review. *Revista Biol. Trop.* **14**:55–73.

Kloetzel, K., and J. C. P. Dias. 1968. Mortality in Chagas' disease: Life table for the period 1949–67 in an unselected population. *Revista Inst. Med. Trop. S. Paulo* **10**:5–8.

Köberle, F. 1963. Chagas' disease and Chagas' syndrome: The pathology of American trypanosomiasis. *Adv. Parasitol.* **6**:63–116.

Laranja, F. S., E. Dias, G. Nobrega, and A. Nuranda. 1956. Chagas' disease. A clinical, epidemiologic and pathologic study. *Circulation* **14**:1035–1060.

Marsden, P. D. 1971. South American trypanosomiasis (Chagas' disease). *Intern. Rev. Trop. Med.* **4**:97–121.

Moseley, V., and H. Miller. 1945. South American trypanosomiasis (Chagas' disease). *Arch. Intern. Med.* **76**:219–229.

Olivier, M. C., L. J. Olivier, and D. B. Segal. 1972. A bibliography on Chagas' disease (1909–1969). *Index Cat. Med. Vet. Zool.*, Special Publ. No. 2, U.S. Department of Agriculture, Washington, D.C.

Pipkin, A. C., Sr. 1969. Transmission of *Trypanosoma cruzi* by arthropod vectors: Anterior versus posterior route infection. *Intern. Rev. Trop. Med.* **3**:1–48.

Puigbó, J. J., J. R. Nava Rhode, H. Garcia Barrios, and C. Gil Yépez. 1968. A 4-year follow-up study of a rural community with endemic Chagas' disease. *Bull. World Health Org.* **39**:341–348.

Talia, R. V. 1949. The Chagas' disease of Uruguay. *Trans. Roy. Soc. Trop. Med. Hyg.* **43**:107–109.

Tarshis, I. B. 1963. Preliminary laboratory and field studies on the utilization of Dri-Die 67 and two percent Dibrom-Dri-Die 67 for the control of the western cone-nosed bug, *Triatoma protracta* (Uhler). *Am. J. Trop. Med. Hyg.* **12**:96–102.

World Health Organization. 1960. Chagas' disease, report of a study group. WHO Tech. Rept. Ser. **202**:1–21.

# 5

# *Leishmania* and *Leishmaniasis*

The *Leishmania* spp. are kinetoplastids that develop intracellularly as amastigotes (Fig. 5.1) in many species of mammals; *Leishmania* of reptiles are mostly extracellular parasites of the gut (Williams & de V. Coelho, 1978). They occur as both amastigotes (in the vertebrate) and promastigotes (in the vector), and amastigotes are among the smallest protozoa. The size range is from 1 to 5 μm, but they usually are about 3 μm. Transmission takes place by blood-sucking insects in the dipteran family Psychodidae, so-called sand flies (see Fig. 54.9).

*Leishmania* spp. cause a complex of infections ranging from localized skin infections of relatively minor importance to generalized systemic disease with a high probability of mortality. Human infections occur worldwide, generally in warm-temperate, subtropical, and tropical areas. Locations where leishmanial infections are important in humans are the Mediterranean, Middle East, Far East, tropical Africa, and Central and South America (Fig. 5.4). A few infections have been found in persons living in the United States.

About 22 species or subspecies of *Leishmania* have been described, half from mammals and half from reptiles. We will be concerned here with the organisms found in mammals, especially those that have been implicated as disease agents of humans. *L. enriettii* Nuniz and Medina was found in the guinea pig, but it is not infective for humans; it is therefore useful as a laboratory model, and has been used to screen drugs for activity against *Leishmania*.

Useful reviews include Adler (1964); Hoogstraal and Heyneman (1969); Lainson and Shaw (1971); Bray (1974); Williams and de V. Coelho (1978); and *Ecologie des leishmanioses* (1977). Heyneman et al. (1980) have produced an exhaustive bibliography that is the single best entry to the vast literature on leishmaniasis (the bibliography itself is almost 700 pages long).

It is interesting to consider the development of knowledge on the *Leishmania* to understand why certain species are currently accepted as valid. At the turn of the century, many infectious diseases in humans and their causative agents were being described; among those found were some protozoa causing both visceral (spleen and other organs) infections and cutaneous infections. These were named *Leishmania donovani* (Laveran & Mesnil, 1903) Ross, 1903 and *L. tropica* (Wright, 1903) Luehe, 1906, respectively. As knowledge progressed, it became clear that the visceral and cutaneous infections intergraded with one another and that there were perhaps 11 clinically distinguishable cutaneous and mucocutaneous infections as well as a number of epidemiologically distinct forms of visceral leishmaniasis. There was a tendency to give a new name to each organism that caused a distinct clinical or epidemiological entity; however, in a number of instances the clinical manifestations intergraded with one another. With the development of increasingly sophisticated and precise serological methods, it was shown that there were

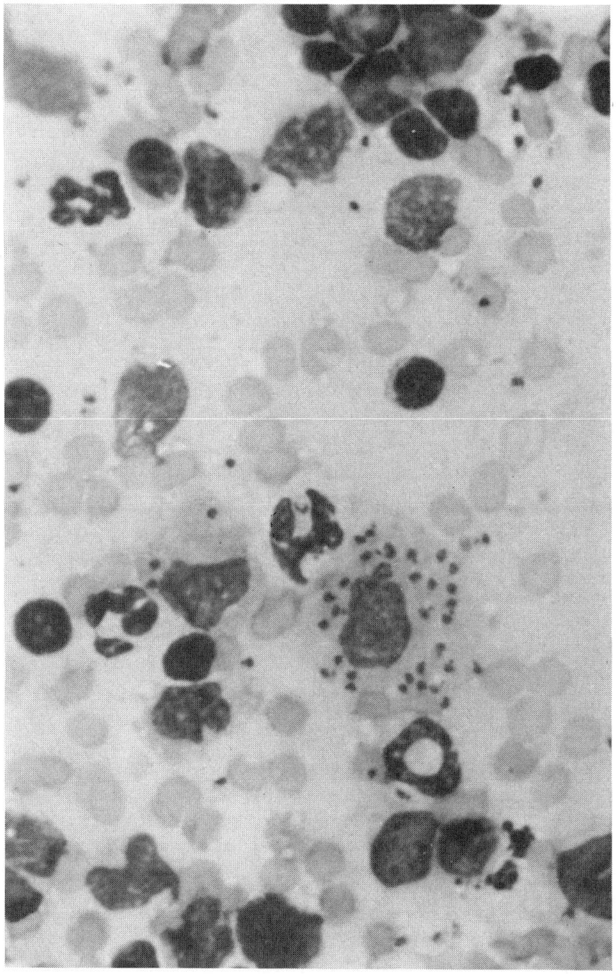

**FIG. 5.1**
Amastigotes of *Leishmania donovani* in a spleen cell.

similarities among *Leishmania* spp. but differences among geographic isolates of what seemed to be the same species.

For a while it seemed best to return to the original two species, but most workers now accept that there are four species that infect humans (Table 5.1). The alternative is to accept 13 or 14 species, as Bray (1974) has suggested (column 2 of Table 5.1). For our purposes, it will be easier to discuss these organisms as four species, recognizing that some of the generalizations may not be entirely accurate.

## VISCERAL LEISHMANIASIS

NAME OF ORGANISM AND DISEASE CAUSED. *Leishmania donovani* (Laveran & Mesnil, 1903) Ross, 1903 is the causative agent of kala azar, visceral leishmaniosis, or visceral leishmaniasis; some other names are Dum-Dum fever, death fever, tropical splenomegaly, and ponos (in Greece).

MORPHOLOGY. Amastigotes are intracellular, ellipsoid bodies 3.3 μm long by 2 μm wide. In the light microscope (LM) in Giemsa-stained material, the nucleus and kinetoplast are red and the cytoplasm is pale blue; the cell membrane can usually be seen (Fig. 5.1). The intracellular rudimentary flagellum can be seen in some preparations. In electron micrographs additional structures can be seen (Fig. 5.2), such as the flagellar pocket, details of the kinetoplast, which is 0.3 by 0.7 μm, subpellicular tubules, and the mitochondrion.

Promastigotes are found in the vector or in culture. They average 20 by 2.5 μm, with a single anterior flagellum that extends about 20 μm beyond the body (Fig. 5.3).

HOSTS AND LOCATION. The host range of *L. donovani* is quite broad, including many mammals if both natural and experimental infections are included. With any of the geographic or epidemiologic forms in which natural transmission takes place, there is usually one or a few reservoir hosts from which humans acquire the infection by means of phlebotomines. Canids (dog, fox, jackal) and sometimes felids (serval) are the principal reservoirs of infection.

In the vertebrate, the organisms develop within cells of the mononuclear-phagocytic system (MPS), also called the *reticuloendothelial system (RES)*, and histiocytes. The infection in the vector remains in the gut.

*Phlebotomus perniciosus*, *P. major*, and *P. longicuspis* in the Mediterranean; *P. chinensis* and *P. sergenti* in China; *P. orientalis* and *P. martini* in East Africa; and *Lutzomyia longipalpis* in South America serve as the vectors.

DISTRIBUTION AND IMPORTANCE. Visceral leishmaniasis is widely distributed (Fig. 5.4), but is usually sharply delimited. Within a geographic area, there is often patchy distribution delimited by the occurrence of proper vector habitat and suitable reservoir hosts.

**TABLE 5.1**
*Leishmania* Species That Infect Mammals

| COMPLEX OR SPECIES | SPECIES OR SUBSPECIES | HEMISPHERE | GROWTH RATE IN CULTURE | DNA (G/ML) NUCLEUS | DNA (G/ML) KINETOPLAST |
|---|---|---|---|---|---|
| *donovani* | | | NA | | |
| | *donovani* | Eastern & Western | | 1.719 | 1.701–1.707 |
| | *infantum* | Eastern | | 1.719 | 1.703–1.704 |
| | *chagasi* | Western | | 1.719 | 1.702–1.704 |
| *tropica* | | | NA | | |
| | *tropica* | Eastern | | 1.719 | 1.703–1.707 |
| | *major* | Eastern | | 1.718 | 1.703 |
| | *aethioipica* | Eastern | | 1.719 | 1.706 |
| *mexicana* | | | Rapid | | |
| | *enriettii*[a] | Western | | 1.718 | 1.701 |
| | *mexicana* | Western | | 1.718 | 1.700 |
| | *amazonensis* | Western | | 1.718 | 1.679–1.699 |
| *braziliensis* | | | Slow | | |
| | *peruviana* | Western | | — | — |
| | *panamaensis* | Western | | 1.717 | 1.694 |
| | *guyanensis* | Western | | 1.716 | 1.692 |
| | *braziliensis* | Western | | 1.716 | 1.691 |
| Unplaced | | | | | |
| | *hertigi*[b] | Western | | 1.714 | 1.700 |

[a] Guinea pig only.
[b] Porcupine only.

The World Health Organization (WHO) considers leishmaniasis to be among the six most important infectious diseases of humans worldwide. Despite that contention, which is probably true, the medical importance of leishmaniasis is difficult to assess. In some areas, the disease occurs only sporadically or epidemically in humans, whereas in others it is endemic. Many of the foci of infection are in remote or underdeveloped areas where public health surveillance may be poor and medical services minimal. In the rural endemic foci (Asia, continental Europe, and perhaps eastern China) there are estimated to be several hundred cases annually. In urban endemic foci (the Mediterranean, central Asia, Russia, and eastern China), there are thousands of cases each year. Hundreds of thousands of cases are estimated to occur annually in India. Visceral leishmaniasis is a significant problem in Central and South America, but data are fragmentary. For example, in Brazil 0.1% of 47,000 persons examined postmortem in connection with a yellow fever study were found to be positive for *L. donovani.*

Some observations in the U.S. Navy's NAMRU-3 final report on a 5-year study in Sudan (Hoogstraal & Heyneman, 1969) are instructive. In 1956 a tribe of 3000 in southern Sudan was reduced by more than half; the tribal unit was destroyed and the survivors scattered. When the high dam at Aswan began filling, persons were relocated from their village on the Egyptian border to a location about 500 km farther south; more than 50 of the 300 persons died because they were settled in an

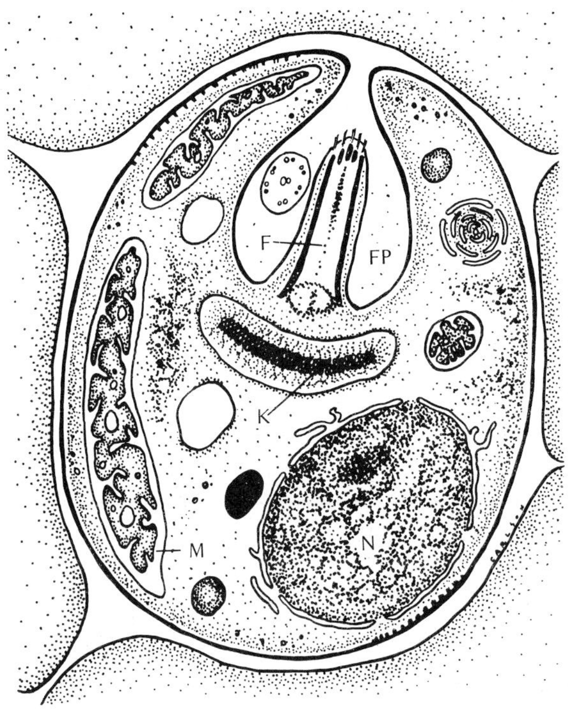


**FIG. 5.2**
Diagram of a *Leishmania* amastigote. F = flagellum; FP = flagellar pocket; K = kinetoplast; M = mitochondrion; N = nucleus.

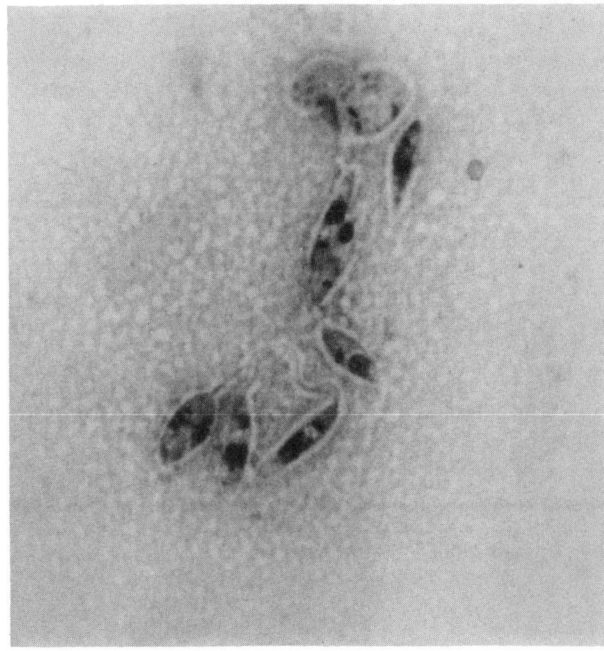

**FIG. 5.3**
Promastigotes of *Leishmania donovani.*

area where kala azar is prevalent. In the area of the NAMRU-3 study, about 600 km south of Khartoum, a medical clinic treated 1000 to 3000 patients each month when there were outbreaks of kala azar.

Most persons who are infected fail to show clinical signs and symptoms, but among those who do, more than 90% die if untreated. Among persons who suffer from visceral leishmaniasis and undergo treatment, about 15% are likely to die.

LIFE CYCLE. All of the leishmaniases of humans are vector-borne diseases in which the vectors are sand flies (Psychodidae, Phlebotominae) in the genera *Phlebotomus* and *Sergentomyia* in the Old World and *Lutzomyia* in the New World. A fly becomes infected by taking a blood meal from an infected animal or, less often, a person who has the amastigotes in circulating white blood cells. Once in the midgut of the fly, the host cell breaks down, releasing the amastigotes that then transform into promastigotes (Fig. 5.3). In a study of *L. donovani* in *Lutzomyia*, multiplication of the Panamanian isolate took place in the stomach of the fly for the first 3 days and then began to move anteriorly. By the fifth day the cardiac stomach was packed with promastigotes and they were seen anterior to the proventricular valve. As numbers increased beyond the third day, promastigotes were seen in the hindgut as well as the proventriculus.

Species that serve as good vectors are those in which the promastigotes multiply well and in which the organisms move to the anterior part of the gut. The esophagus becomes partially blocked; when the fly takes another blood meal, it attempts to clear the obstruction and the promastigotes are injected into the tissue of the new host. Although there is cyclic development, as occurs in *Trypanosoma brucei* or *T. cruzi*, all promastigotes are infective for a vertebrate host; there is no mandatory sequence of developmental stages that must be gone through before infectivity is reached.

When the promastigotes are phagocytized by macrophages, they resume the amastigote form and begin to multiply by binary fission. Development first takes

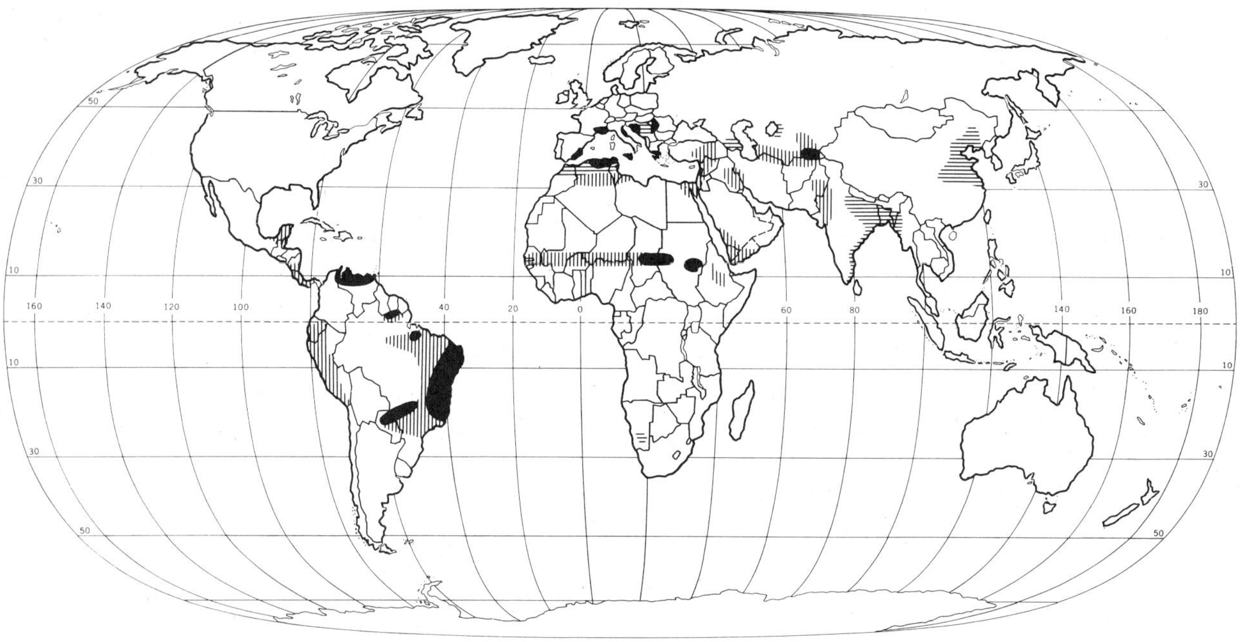

**FIG. 5.4**
World distribution of *Leishmania* spp. of humans. Black areas are visceral forms and hatched areas are cutaneous or mucocutaneous forms.

place at the site where the vector fed, and usually a small lesion develops at that site. In some instances the lesion becomes fairly large and may appear to be infected with *L. tropica*. Within a matter of days the organisms are carried to many parts of the body, where the amastigotes are ingested by fixed phagocytic cells (histiocytes) and further multiplication takes place. Some of the amastigotes infect circulating phagocytic cells or those in the superficial layers of the skin, allowing them to be picked up by another sand fly during feeding.

Although the evidence is not unequivocal, crushing an infected sand fly on the skin may allow promastigotes released on the skin to invade through abrasions (Williams & de V. Coelho, 1978). Feeding by sand flies is painful, and persons almost invariably slap and scratch at the site where a fly is feeding; some persons scratch the skin until it is at least mildly inflamed.

DIAGNOSIS. Determining whether an individual is infected with *Leishmania* spp. is based on the following:

1. Clinical signs and symptoms
2. Visiting or living in an endemic area
3. Serological tests
4. Finding the causative agent

Visceral leishmaniasis may have features of a number of other diseases such as malaria, typhoid fever, or brucellosis, among others. Thus, it is important to determine, by either serology or finding the organism, whether the person has been infected.

Serological tests revealing the presence of antibodies are useful in coming closer to a positive diagnosis, but they do not yet have a high level of reliability. There are a number of antibody-based tests that are useful:

1. High IgG level
2. Fluorescent antibody (FA) test
3. Enzyme-linked immunosorbent assay (ELISA) test

Most important in diagnosis is finding the amastigotes (Fig. 5.1) directly in biopsied material taken from bone marrow or spleen, or sometimes the liver or lymph node. A tiny portion of the material is pressed between two glass slides, allowed to dry, and then stained with a Romanowsky stain such as Giemsa. The remainder is processed for histologic examination. In either event, one searches for the amastigotes (Leishman–Donovan or LD bodies) under an oil immersion objective of a compound microscope. Inoculation of biopsy material into hamsters is a reliable method, but one that may be difficult to do unless a well-equipped diagnostic laboratory is located nearby. If biopsied material is placed in a medium such as NNN (Novy–MacNeal–Nicolle), promastigotes may develop to detectable numbers within a few days when incubated at 22° C; if the number of organisms is small, nearly a month of incubation may be required before they are present in sufficient numbers to be found.

HOST–PARASITE INTERACTIONS. The cells in which the amastigotes multiply are the fixed phagocytes of the MPS, which are found in spleen, lymph nodes, bone marrow, liver, and the lining of smaller blood vessels throughout the body. Leishmaniasis is actually a disease of a part of the immune system.

The way in which the parasites survive and reproduce in the MPS has been a matter of conjecture, but studies using the electron microscope have begun to clarify the relationship (Berman et al., 1981; Chang & Dwyer, 1976). When an amastigote is ingested by a macrophage, the organism first lies in an envelope of host cell membrane, a phagosome. Within hours there is fusion of the phagosome with a lysosome. Normally the ingested organism would be quickly killed and broken down by the multiplicity of enzymes released by the lysosome, but in the case of *Leishmania*, the organism escapes the enzymes and multiplies in the macrophage cytoplasm. It has been suggested that the enzyme acid phosphatase, which occurs on the surface of the amastigotes, may inhibit the action of lysosomal enzymes (Gottlieb & Dwyer, 1981). The mechanism of resistance remains one of the interesting and significant problems of intracellular parasitism (Chang & Dwyer, 1976; Bloom, 1979; Berman et al., 1981). One of the assumptions made about intracellular parasitism is that it occurs at a site where the parasite cannot be reached by the immune system, such as phagocytes, but here we have a parasite that has not retreated from, but rather has coopted, the immune system.

Reproduction of *Leishmania donovani* in the MPS has significance in some aspects of the pathogenesis of the infection. One of the first effects is an overproduction of macrophages and a consequent reduction in the number of polymorphonuclear granulocytes and red blood cells (RBCs). Bone marrow is the principal source of both erythrocytes and leukocytes, and both kinds of cells arise from the same stem cells. If the bone marrow is looked upon as a system whose total output has an upper limit, the effect of overstimulation of the MPS can be appreciated. Normally the stem cells differentiate to produce RBCs or white blood cells (WBCs); each is produced in response to the needs of the body. Further differentiation of the WBCs is into granulocytes (neutrophils, eosinophils, and basophils), mononuclears (histiocytes, macrophages, Kupffer cells), and probably lymphocytes. Now, as the infection progresses, macrophages are increasingly involved in phagocytizing amastigotes and there is a consequent increased production of them from bone marrow. The demand is such that the RBCs and granulocytes decrease in number, because the bone marrow cannot keep up with all of the demands on it. The result for the individual infected with *L. donovani* is the development of anemia and the inability to resist infection by other agents.

The course of visceral leishmaniasis in humans can range from spontaneous cure early in the infection (probably the most likely outcome), to chronic infection lasting many years, to an acute infection that may kill the person within months. Persons who develop signs and symptoms of visceral leishmaniasis stand a greater than 90% chance of dying if they are not treated.

In persons who are treated and recover, a particularly distressing manifestation, post-kala azar dermal leishmanoid (DL), occurs in a small percentage. Areas of the skin become depigmented and nodules develop. In most instances, the swellings later are resolved, but the person may have a seriously deformed face for a period of time.

TREATMENT. The effective drugs are organic antimonials and diamidines, all of which require multiple doses. Even with a complete course of treatment, about 15% of persons retain the infection.

EPIDEMIOLOGY AND CONTROL. Leishmaniasis was originally an infection of wild canids with transmission by sand flies (Fig. 5.5*a*). Human infections were probably rare. When humans, together with the dog,

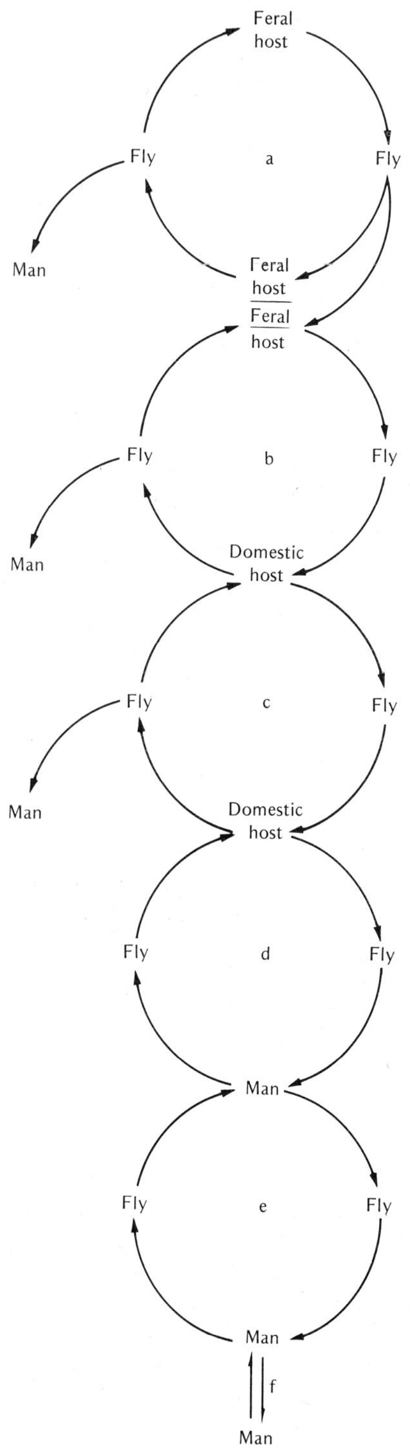


entered an enzootic area, the probability of human infection increased, and the likelihood of dogs carrying the infection was also present. The sequence of events has been theorized to have had four steps:

1. Enzootic foci in wild canids (Fig. 5.5*a*). It is reasoned that the primeval condition was one in which hosts such as the jackal in central Asia maintained the infection and human infections were rare.
2. Rural endemic foci (Fig. 5.5*b*). As humans colonized enzootic areas, infection began to include domestic canids, but with maintenance of the enzootic foci in wild canids. Cases in humans are common. This type of cycle is seen in part of Asia, Soviet Russia, the People's Republic of China, and the area ringing the Mediterranean Sea.
3. Urban endemic foci (Fig. 5.5*c*). With the expansion of human populations and the development of large, permanent towns, it was possible for the wild animal portion of the cycle to drop out (at least locally). Dogs become the reservoir, and transmission to humans was frequent. This cycle occurs in central Asia and the eastern part of the People's Republic of China.
4. Endemic (sometimes epidemic) foci (Fig. 5.5*d* and *e*). In this epidemiologic form, the reservoir may drop out and transmission can take place from person to person by means of sand flies (Fig. 5.5*e*). Foci of kala azar in India are of this type, and probably more human cases occur than in any other type.

**FIG. 5.5**
Modes of transmission of *Leishmania* spp. and their evolution. (*a*) A feral cycle with humans as only a rare and dead-end host. (*b*) A cycle involving a feral host and an animal associated with humans such as the dog; humans are an occasional and dead-end host. (*c*) A domiciliated cycle often with a domiciliated vector as well; humans are an occasional and dead-end host. (*d*) A domiciliated cycle with animals such as the dog as the principal reservoir; humans are integrated into the cycle. (*e*) A cycle in which the reservoirs have dropped out and transmission takes place from person to person through the phlebotomine. (*f*) Person-to-person transmission through direct contact or by means of body fluids. The vector is no longer required.

Each of these four types requires a different approach to control, and all have been quite refractory to such efforts.

Control efforts are directed principally at the following:

1. Control of vectors
2. Protecting the human population against phlebotomines
3. Treatment of affected individuals
4. Elimination of reservoir hosts

Problems of control lie in the following:

1. Finding and reducing populations of reservoirs
2. Finding and identifying the vectors
3. Developing feasible control methods for the vectors
4. Protecting the human population from the vectors
5. Inadequate chemotherapy
6. Lack of a vaccine

In considering the evolution of leishmaniasis from a purely feral cycle through a number of steps to a domiciliated cycle, it can be seen that the hazard to human health increases and that there is a greater probability of being able to manipulate the system. If humans seldom penetrate the nidus of infection, control is not required and, for that matter, is not feasible. If they do interact with the wild animal nidus, attempts to reduce the population of reservoir hosts can be carried out. In some areas of the USSR and China, reduction of human infections has been observed following eradication of rodents in the vicinity of human habitations. Since burrowing rodents are the principal reservoirs in arid and semiarid foci, steps must be taken to destroy burrows and to prevent their recolonization.

As the infection moves into domesticated animals such as the dog, it might be possible to control them by, say, a test-and-slaughter program, but the realities of the kinds of localities where leishmaniasis occurs militate against licensing of dogs and establishment of the administrative structure that it would entail. Dogs are often visibly affected, showing roughened skin, lesions on the muzzle, and hair loss; it would generally be within the powers of public health personnel to remove such animals from a town or city.

In the endemic focus in India, transmission from person to person takes place through phlebotomines and there is no reservoir. The flies have sites of larval development near houses, in garbage and other decaying organic materials. It is therefore feasible to reduce the sites where females oviposit. Phlebotomines have short, hopping flight, and do not usually move more than about 200 m from the site of larval development. Unlike some other flies, such as some species of mosquitoes and blackflies that may fly or be carried long distances from their sites of larval development, phlebotomines can be attacked near houses even if their sites of development are not readily found. Sentinel animals can be used to determine when infected flies are present and where they are most prevalent (Beumer, 1973). Flies can be controlled by source reduction and insecticidal treatment of sites of larval development. Since flies tend to rest in houses, periodic spraying of dwellings with DDT has been effective in reducing the numbers of vectors and transmission of the agent. Where houses have been sprayed for control of vectors of malaria, phlebotomine populations have also been reduced, as well as the transmission of *Leishmania* spp. In epidemiologic types other than the endemic one, breeding sites of flies are diffuse and practical control measures have not generally been successful.

Personal protection is an important aspect of control both for the individual and for the population. Because the flies are small, typical fly screen is inadequate; fly screen has openings of from about 1.3 to 1.5 mm, and phlebotomines are not kept out unless the openings are 0.9 mm or smaller. Organochlorine insecticides with long residual activity can be painted on fly screens and will give protection for a while. Houses can also be sprayed with a residual or a slow-release insecticide. Keeping plants, especially low-growing ones, away from houses also reduces the resting sites of adults. When it is necessary to be out in areas where flies are abundant, insect repellents provide a modicum of protection.

Treatment of persons with leishmaniasis is important not only for the individual, but also to help prevent the spread of the disease among the population. Cases of leishmaniasis in humans begin to appear about 3 months after the fly and rainy seasons begin. Since the signs and symptoms of the disease appear seasonally, control methods can be employed strategically with the most economical use of resources.

## OLD WORLD CUTANEOUS LEISHMANIASIS

As many as three species of *Leishmania* may be associated with the cutaneous lesions of this disease complex (Table 5.1). It appears that the enzootic form causing wet lesions in humans has evolved into the dry endemic form, and for our purposes it is most straightforward to deal with the different clinical forms as a single species.

NAME OF ORGANISM AND DISEASE CAUSED. *Leishmania tropica* (Wright, 1903) Luehe, 1906, causes cutaneous leishmaniasis, oriental sore (Fig. 5.6), Aleppo button, and Delhi boil.

HOSTS AND LOCATION. Humans are the principal host in the dry form and dogs are reservoirs in some localities. In the wet form, reservoirs are various burrowing rodents such as the gerbil *(Rhombomys optimus)* in dry climates of Asia; humans are occasionally the host. The organisms remain in the skin of the vertebrate, but may occasionally be found in human lymph nodes.

Vectors are *Phlebotomus papatasi* and *P. sergenti*, among others.

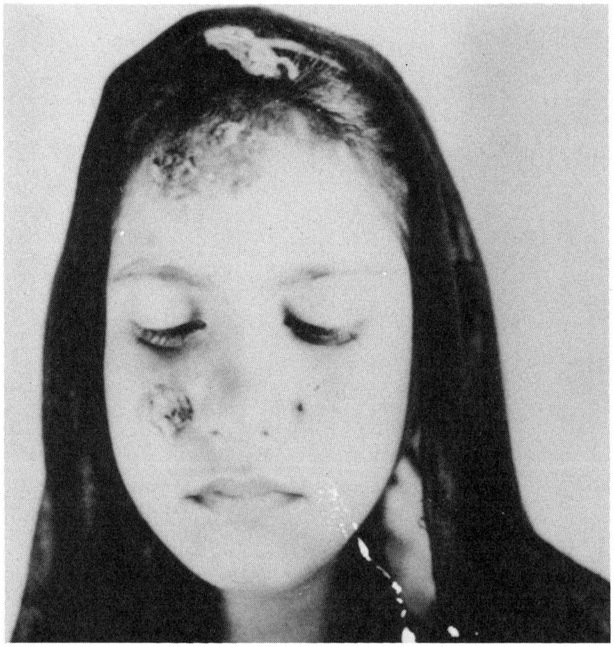

**FIG. 5.6**
Oriental sores on face and forehead. [Courtesy of U.S. Armed Forces Institute of Pathology. Negative No. 80,717.]

DISTRIBUTION AND IMPORTANCE. Cutaneous leishmaniasis is distributed from the high deserts of central Asia through the Middle East and the area surrounding the Mediterranean Sea to India and West Africa. In some instances such as in the USSR, the wet and dry forms may occur in the same area; in other instances the forms are geographically separated.

As a general rule, the lesions caused by this organism are limited to the skin and they heal as immunity develops. There may be some facial disfigurement, especially with the wet form.

MORPHOLOGY. Like most species, *L. tropica* is about 3.5 by 2 µm in the amastigote form.

LIFE CYCLE. The life cycle is the same as that of *L. donovani*. Approximately a week is required for promastigotes to move to the anterior portion of the gut of the fly. In experimental infections in *P. longipes*, more than a quarter (27.2%) of the flies became infected after feeding on patients with lesions (Foster, 1982); thus, despite the small quantity of blood taken by phlebotomines, the likelihood of their becoming infected is high.

DIAGNOSIS. The lesions of cutaneous leishmaniasis are depressed, scaly, and may itch; in an enzootic or endemic area, persistence and growth of such a lesion is reason to strongly suspect cutaneous leishmaniasis. A few instances have been seen in which visceral leishmaniasis has given rise to cutaneous lesions, so one should not leap to a conclusion. Finding the organisms in a lesion is good evidence for cutaneous leishmaniasis, but the material must be taken from the inside margin of the lesion, not a superficial scraping (Garcia & Ash, 1979).

CULTIVATION. See the section on *L. donovani*.

HOST–PARASITE INTERACTIONS. Like *L. donovani*, *L. tropica* was originally an infection that was transmitted among feral animals by phlebotomines

(Fig. 5.5*a*). The original mammalian hosts most likely were rodents, and humans became infected only upon entering the nidus. Much the same evolution as seen with *L. donovani* undoubtedly occurred with *L. tropica* in moving from a feral infection to an occasional zoonosis to a domiciliated zoonosis and, finally, to an endemic disease of humans (Fig. 5.5*a*, *b*, *c*). A further step in the evolutionary process pertains to *L. tropica*, but not to *L. donovani*, namely, that the vector can drop out. It is possible for transmission to take place from an infected person to an uninfected person through personal contact (Fig. 5.5*f*). In some endemic areas, parents may inoculate children on the trunk or buttocks with material taken from an active lesion on another person, thereby avoiding potential disfigurement of the face and also developing immunity against later infection.

The wet lesions seen in enzootic localities develop rapidly and are likely to leave a somewhat disfiguring scar. Wet lesions heal within 12 months and the individual is then immune to not only the wet form but the dry form as well. The dry lesions develop slowly, but usually heal within 12 to 24 months, leaving the individual immune to the dry but not the wet form.

TREATMENT. Same as for *L. donovani*.

EPIDEMIOLOGY AND CONTROL. The same general approaches are taken with cutaneous leishmaniasis as are used against visceral leishmaniasis. Eliminating sites of larval development, personal protection, treatment of active cases, and elimination of infected dogs, where they are reservoirs, are the major steps to be taken.

As indicated, children are sometimes inoculated with material taken from an active lesion of another person so that they may escape later infection and scarring. Artificial immunization has not been successful as yet.

## NEW WORLD CUTANEOUS AND MUCOCUTANEOUS LEISHMANIASIS

The various epidemiologic and clinical forms of cutaneous and mucocutaneous leishmaniasis in the Western Hemisphere have made this complex of diseases exceedingly difficult to study and categorize. Some workers suggest that each distinct entity, by using clinical, geographical, and epidemiologic criteria, be given either a specific or subspecific name (Table 5.1, column 2; Table 5.2). A substantial number of workers use the two species *L. braziliensis* and *L. mexicana* (Table 5.1, column 1; Table 5.2).

NAME OF ORGANISM AND DISEASE CAUSED. *Leishmania mexicana* causes two distinct clinical entities: Chiclero ulcer, in which there may be long-term effects on the pinna of the ear, and a form in the Amazonian basin in which there are skin lesions only (Table 5.2). The organisms in the group have similar growth characteristics and their DNAs are similar (Table 5.1); *L. enriettii* is clearly related to *L. mexicana*, but it infects only the guinea pig and is included in the table only for the sake of completeness.

HOSTS AND LOCATION. As is true with most leishmaniases, humans are accidental hosts in that they play little role in maintaining these infections. Reservoirs are considered to be rodents such as *Oryzomys capito*, *Ototylomys phyllotis*, and *Proechimys guyanensis*, with the possibility of marsupials such as the opossum playing a minor role.

These parasites remain in the skin, with the exception of those cases in which there is involvement of the pinna of the ear.

DISTRIBUTION AND IMPORTANCE. Chiclero ulcer is found in Meso-American forests, and the forms that involve the skin only are found in the Amazon drainage. In general, the lesions heal with some scarring after many months of active infection. Cutaneous leishmaniasis has not been considered to occur in the United States, but two human cases have been found in Texas in persons who probably contracted the agent there (Shaw et al., 1976), and a focus of infection has been found in dogs in Oklahoma (Anderson et al., 1980), but this organism is structurally like *L. donovani* (Kocan et al., 1983).

MORPHOLOGY, LIFE CYCLE, DIAGNOSIS, CULTIVATION. Same as for *L. donovani*.

HOST–PARASITE INTERACTIONS. The initial lesions in cutaneous leishmaniasis occur at the site of feeding by the vector. The ear involvement in Chiclero ulcer results from further invasion of the tissue at the site of the lesion.

Lesions are formed through the presence of large numbers of infected histiocytes and through interfer-

**TABLE 5.2**
New World Cutaneous Leishmaniases

| COMPLEX OR SPECIES[a] | SPECIES OR SUBSPECIES[b] | DISEASE NAME | CLINICAL SIGNS | LOCALITY |
|---|---|---|---|---|
| *mexicana* | *mexicana* | Chiclero ulcer, bay sore | Small skin sore followed by disfiguring, ear involvement | Mexico, Honduras, Guatemala, forest |
| | *amazonensis* | Amazonian leishmaniasis | Skin lesions | Amazon basin, forest |
| *braziliensis* | *peruviana* | Uta | Skin lesions | Peru, high elevations in Andes mountains |
| | *panamaensis* | Panamanian leishmaniasis | Skin lesions | Panama forest |
| | *guyanensis* | Amer. forest leishmaniasis pian bois, buba | Skin lesions 5% nasal mucous membrane involvement, 10% lymph node involvement | Panama, Costa Rica, Guyana, S. American forest |
| | *braziliensis* | Espundia, mucocutaneous leishmaniasis | Chronic skin lesions, mucous membrane involvement, cartilage of the nose | Brazil, & tropical S. American forest |

[a] Names used here.
[b] Specific and/or subspecific names used by some workers.

ence with local circulation of the tissue. Some of the amastigotes multiply in cells lining the smaller blood vessels and the inflammatory response causes reduced blood flow to the skin (or later to the cartilage of the pinna) and ulceration occurs. Because the skin is broken, there is often local infection with bacteria, but it seldom develops into a serious infection.

TREATMENT. Same as for *L. donovani.*

EPIDEMIOLOGY AND CONTROL. These infections are zoonoses and are especially difficult to control. Men going into the forest to collect chicle or to cut wood are likely to be infected. Thus, the best preventive measures under this circumstance are personal protection by repellents, insect netting, and screening of dwellings where there may be contact with the reservoirs.

NAME OF ORGANISM AND DISEASE CAUSED. *Leishmania braziliensis* causes at least four clinically distinct syndromes—uta, Panamanian and American forest leishmaniasis, and espundia (Tables 5.1 and 5.2).

HOSTS AND LOCATION. Uta is the only one of these infections that may lack wild reservoirs; it occurs in high mountain areas of Peru and the dog seems to be a domestic reservoir. In the three other types of infections, reservoirs are in wild animals; sometimes the reservoirs are arboreal, such as in the form in Panama, and in the other forms the main reservoirs seem to be forest rodents. The vectors include *Lutzomyia longipalpis*, also a vector of New World visceral leishmaniasis (*L. d. chagasi* of Tables 5.1 and 5.2), and probably large number of other species of *Lutzomyia* (Johnson et al., 1963).

Uta and Panamanian leishmaniasis remain as relatively superficial skin infections, but American forest leishmaniasis and espundia involve the cartilaginous structures of the face.

DISTRIBUTION AND IMPORTANCE. In some areas on the western slope of the Andes, nearly every child was formerly affected with uta. With successful control, infections now seem to be uncommon. The risk of the other three infections, obtained in forested areas, is significant. For example, in a military exercise in Panama, there was 1 infection for every 1300 man-days in the field; this would mean about one infection per day in a combat battalion. In espundia, the initial skin infection clears up in a few months, but sometimes, years later, there is involvement of the cartilage of the face and an inexorable destruction of those tissues. When such extensive destruction occurs, the person is likely to die from secondary infection.

LIFE CYCLE. See *L. donovani.*

DIAGNOSIS. Same as for *L. mexicana.*

HOST–PARASITE INTERACTIONS. The mucocutaneous forms of leishmaniasis are the most important as far as the long-term health of the infected person is concerned. In Chiclero ulcer, the destruction of the pinna of the ear is a result of the extension of the local infection. In espundia, on the other hand, metastasis (infection at a site other than the original one) occurs months or years after the initial superficial infection. The continued infection represents an inadequate immune response, unlike most of the other forms of cutaneous leishmaniasis in either the Eastern or Western Hemispheres.

CULTIVATION AND TREATMENT. See *L. donovani.*

EPIDEMIOLOGY AND CONTROL. Except for uta, which is a domiciliated infection in Peru, control of the New World leishmaniases has been only sporadically successful. Uta has been largely controlled through elimination of infected dogs, treatment of infected persons, and protection from the vectors. In the less well defined foci in forests, control must rely on individual protection, detection of infections, and treatment of infected persons. In considering all of the leishmaniases, it is clear that there is need not only for better drugs, but also for the development of an efficient vaccine. The only vaccination now used, by medical practitioners in the USSR, is the purposeful inoculation of children with material from an active lesion on another person. Obviously, the only forms that can be used in this sort of purposeful immunization are those caused by *L. tropica* in the Old World and other superficial infections such as Panamanian cutaneous leishmaniasis in the New World. The other infections, visceral leishmaniasis and the mucocutaneous forms, present too much risk for living, virulent agents to be used. In addition, some persons fail to develop immunity at all (Simpson et al., 1968). Some of the difficulties in production of a vaccine lie in the lack of understanding of the basis of immunity in leishmaniasis. It has a strong cellular component, but antibodies are also involved. There are many serotypes in the genus, and although the main groups have antigens in common, any vaccine would probably need to be produced for a particular locale.

## REFERENCES AND FURTHER READINGS

Adler, S. 1964. *Leishmania. Adv. Parasitol.* **2**:35–96.

Anderson, D. C., R. G. Buckner, B. L. Glenn, and D. W. MacVean. 1980. Endemic canine leishmaniasis. *Vet. Pathol.* **17**:94–96.

Berman, J. D., T. B. Fiorevetti, and D. M. Dwyer. 1981. In vivo and in vitro localization of *Leishmania* within macrophage phagolysosomes: Use of colloidal gold as a lysosomal label. *J. Protozool.* **28**:239–242.

Beumer, R. J. 1973. Detection of leishmanial activity in nature by means of sentinel animals. *Trans. R. Soc. Trop. Med. Hyg.* **67**:870–879.

Bloom, B. R. 1979. Games parasites play: How parasites evade immune surveillance. *Nature* **297**:21–26.

Bray, R. S. 1974. *Leishmania. Ann. Rev. Microbiol.* **28**:189–217.

Chang, K. P., and D. M. Dwyer. 1976. Multiplication of a human parasite (*Leishmania donovani*) in phagolysosomes of hamster macrophages in vitro. *Science* **193**:678–680.

*Ecologie des leishmanioses.* 1977. *Colloques Internationaux du Centre National de la Recherche Scientifique No. 239.* Editions du Centre National de la Recherche Scientifique, Paris.

Foster, W. A. 1972. Studies on leishmaniasis in Ethiopia. V: Distribution, infections and assessment of vector potential of *Phlebotomus longipes* (Diptera: Psychodidae). *Ann. Trop. Med. Parasitol.* **66**:445–455.

Garcia, L. S., and L. R. Ash. 1979. *Diagnostic Parasitology*, 2nd ed. C. V. Mosby Co., St. Louis.

Gottleib, M., and D. M. Dwyer. 1981. Protozoan parasite of humans: Surface membrane with externally disposed acid phosphatase. *Science* **212**:939–940.

Heyneman, D., H. Hoogstraal, and A. Djigounian. 1980. Bibliography of *Leishmania* and leishmanial diseases. Spec. Publn. NAMRU-3, Cairo, Egypt. Vols. 1 and 2. (Available from U.S. Dept of Commerce, Nat. Tech. Info. Serv., Springfield, VA 22161.)

Hoogstraal, H., and D. Heyneman. 1969. Leishmaniasis in the Sudan Republic. *Am. J. Trop. Med. Hyg.* **18**:1090–1210.

Johnson, P. T., E. McConnell, and H. Marshall. 1963. Natural infections of leptomonad flagellates in Panamanian *Phlebotomus* sand flies. *Exp. Parasitol.* **14**:107–122.

Kocan, K. M., D. W. MacVean, and J. C. Fox. 1983. Ultrastructure of a *Leishmania* sp. isolated from dogs in an endemic focus in Oklahoma. *J. Parasitol.* **69**:624–626.

Lainson, R., and J. J. Shaw. 1971. Epidemiological considerations of the leishmanias with particular reference to the New World. In A. M. Fallis (ed.), *Ecology and Physiology of Parasites*. University of Toronto Press, Toronto, pp. 21–57.

Scholtyseck, E. 1979. *Fine Structure of Parasitic Protozoa*. Springer-Verlag, New York.

Shaw, P. K., L. T. Quigg, D. S. Allain, D. D. Juranek, and G. R. Healy. 1976. Autochthonous dermal leishmaniasis in Texas. *Am. J. Trop. Med. Hyg.* **25**:788–796.

Simpson, M. H., J. F. Mullins, and O. J. Stone. 1968. Disseminated anergic cutaneous leishmaniasis. *Arch. Dermatol.* **97**:301–303.

Williams, P., and M. de V. Coelho. 1978. Taxonomy and transmission of *Leishmania*. *Adv. Parasitol.* **16**:1– 42.

# 6
# *Trichomonads*

Members of the order Trichomonadida are all parasitic except for a few free-living forms that have been reported infrequently. The organisms are small, generally less than 25 μm in length, but structurally complex flagellates that reproduce asexually and exist only as trophozoites.

The trichomonads are widely spread as parasites of vertebrates and nonvertebrates alike. In vertebrates they are generally found in the digestive tract but also occur normally in the reproductive system, urinary tract, and nasal passages. A number of trichomonads have been described from the gut of termites; *Trichomonas limacis* occurs in the gut and liver of slugs; trichomonads also occur in the intestine of the large hog ascarid *Ascaris suum*. See Kudo (1966) or Levine (1973) for many species not covered here.

Only a few species of trichomonads are pathogenic. Most live in the intestines of their hosts, where they may be readily found in microscopic examination of gut content. Under some conditions, such as the host having enteritis or eating a high-carbohydrate diet, trichomonads may be found in large numbers. At various times, investigators have seen large numbers of trichomonads, or other protozoa, associated with a pathological condition and have concluded that the protozoan caused the disease. Such guilt by association is understandable but nevertheless to be deplored. To produce a causative link between an organism and a disease, the organism should be isolated in pure culture, reintroduced into a susceptible host in which it causes the disease, and then reisolated in pure culture. These straightforward, logical steps are referred to as *Koch's postulates*. They were first set forth in 1840 by one of Koch's teachers, Henle, but established as a working principle through Koch's pioneering work in the late 1800s and early 1900s.

A basic structure underlies all trichomonads, but there is considerable variation on the theme. If we take as an example a generalized trichomonad that might be found in nearly any vertebrate, the organisms are usually longer than wide and have three or more anterior flagella, a recurrent flagellum usually attached to the body by an undulating membrane, an axostyle, a parabasal body, a costa, sometimes a pelta, and a single nucleus (Fig. 6.1). All of these structures except the nucleus are referred to as the *mastigont apparatus*, and the variations in the components plus associated granules are the basis for differentiating among genera and species.

It is beyond the scope and purpose of this book to discuss systematics and structure of trichomonads at length. Honigberg's (1963) revision of the order Trichomonadida provides details for those interested. Two additional revisions have been made within the family Trichomonadidae; the subfamily Protrichomonadinae was established for *Histomonas*, *Parahistomonas*, and *Protrichomonas* (Honigberg & Kuldova, 1969), and the subfamily Dientamoebinae was established for *Dientamoeba fragilis* (Camp et al., 1974).

NAME OF ORGANISM AND DISEASE CAUSED. *Trichomonas foetus* (Riedmuller, 1928) Wenrich & Emmerson, 1933 causes trichomonad abortion, bovine trichomoniasis, and bovine trichomonosis.

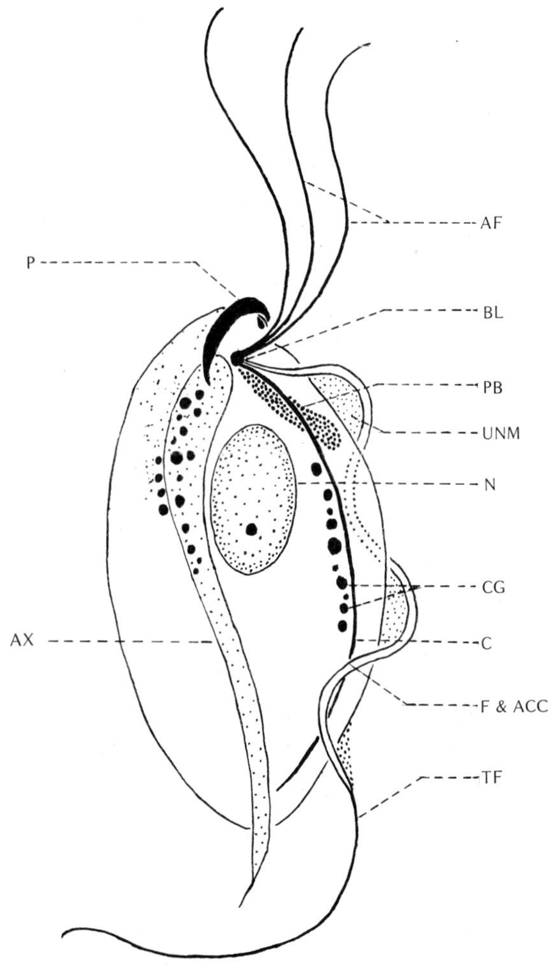


**FIG. 6.1**
A generalized trichomonad showing structures used in identification and descriptions of species. P = pelta, AX = axostyle, AF = anterior flagella, BL = basal body, PB = parabasal body, UNM = undulating membrane, N = nucleus, CG = costal granules, C = costa, F and ACC = flagellum and accessory filament, and TF = recurrent flagellum.

HOSTS AND LOCATION. The principal host of *T. foetus* is the domestic ox (cattle), and at one time, data supported the view that cattle were the only hosts. The clear-cut status of *T. foetus* was changed with the finding that, although the principal site of infection is the reproductive tracts of cows and bulls, *T. foetus*, or a closely related organism, also infects horses and swine. In horses, trichomonads have been found in association with abortion and infections of the reproductive tract. In swine, organisms nearly indistinguishable from *T. foetus* have been found in the nasal passages and intestinal tract, and some of these organisms have established infections in cows and bulls. The organism from swine is usually referred to as *Tritrichomonas suis*, but the division between it and *T. foetus* is not as clear as was once thought (Hibler et al., 1960). Despite the occurrence of *T. foetus* in other hosts, the principal host and the one in which significant economic losses occur is domestic cattle.

DISTRIBUTION AND IMPORTANCE. *T. foetus* is a cosmopolitan parasite. Its importance to the cattle industry has been assessed at about $8 million annually in the United States.

MORPHOLOGY. *T. foetus* is 10 to 25 by 3 to 15 μm, has three anterior flagella, a recurrent flagellum attached to the body by an undulating membrane and continuing free posteriorly (Fig. 6.2).

LIFE CYCLE. Multiplication is by asexual binary fission. Transmission takes place during copulation.

DIAGNOSIS. In the bull, the organisms are found in the prepuce or sheath of the penis. A variety of techniques and instruments have been used for obtaining material for diagnosis. Commonly, a plastic pipet about 10 mm in diameter and perhaps 50 cm long fitted with a rubber bulb is used with a small amount of physiological salt solution to aspirate material from the prepuce. The material may be examined directly under a microscope for motile trichomonads, or it may be placed in a suitable culture medium to encourage growth of the organisms, which may be sought 2 days later. As a general rule, a bull is considered to be uninfected if microscopic examinations and cultures are negative on six occasions.

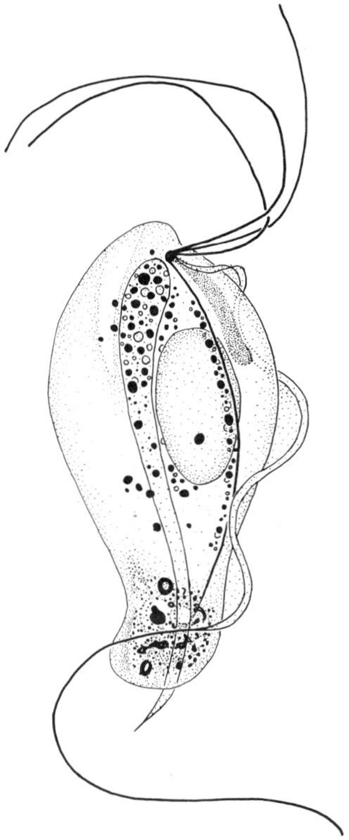

FIG. 6.2
*Trichomonas foetus,* a parasite of the reproductive tract of cattle.

In the cow, material may be aspirated from the vagina or uterus and examined or cultivated as with bulls. Usually, there are sufficient organisms so that they are more easily found on direct examination than in bulls.

When a cow has aborted, the fetus may also be examined for trichomonads. The organisms may be found in the fetal membranes, amniotic fluid, digestive tract, or placenta.

A variety of other infectious agents may also cause abortion in cattle, *Brucella abortus* and *Vibrio foetus* being among the bacteria of importance. *Sarcocystis* sp. may also be involved in abortion (Chap. 15). When a breeding problem is recognized in a herd, it is usually important to consider a number of possible causes for it.

HOST–PARASITE INTERACTIONS. As a general rule, infections in bulls remain benign, but as far as is known, do not clear up spontaneously. The organisms typically remain in the prepuce but on occasion may infect the internal genitalia such as the seminal vesicles and the testicles.

Upon introduction into the reproductive tract of a cow, the trichomonads reproduce in the vagina, where they may cause vaginitis. If the animal is pregnant, the organisms may invade the uterus and infect the developing fetus. Usually when this happens, the cow will abort during the first 16 weeks of pregnancy (gestation in cows takes about 37 weeks). If abortion takes place, the cow will usually be free of infection after two estrus cycles. The infection is eliminated from the cow through the development of immunity.

Sometimes, the fetus may be killed or a portion of the fetal membranes retained. Uterine infections may ensue, which result in permanent sterility.

Although we refer to this disease as *trichomonad abortion*, what the cattle raiser usually perceives is an infertility problem. Some other infectious agents cause abortion late in pregnancy, and in those instances the owner is aware when an animal has aborted. In the case of *T. foetus* infections, the calf is still quite small and the signs of abortion may be missed.

TREATMENT. Until recent years, treatment of bulls for *T. foetus* infection was unreliable with respect to the probability of cure and was laborious to administer. The animal was restrained, locally anesthetized to relax the penile muscles, and then treated topically with a salve containing acriflavine or a related compound.

Dimetridazole (1,2–dimethyl–5–nitroimadazole) was introduced in the early 1960s and approaches 100% effectiveness when administered in the feed, by capsule, or intravenously (McLoughlin, 1963, 1965, 1968, 1970). This drug allows treatment of an animal that might otherwise be sent to slaughter. Cows are not treated because they usually eliminate the infection spontaneously, but dimetridazole can be used in them also.

EPIDEMIOLOGY AND CONTROL. *T. foetus* is transmitted from cow to bull and back through sexual intercourse (coitus). Because trichomonads exist only as trophozoites, the probability of their being transmitted through contaminated bedding or other species of animals is remote. Therefore, control programs revolve around eliminating the organism from a cattle herd and ensuring that it is not reintroduced.

Once an outbreak of trichomonosis has been diagnosed in the herd, the first step is to determine which animals are infected and then to treat, isolate, or dispose of them.

Cows usually eliminate the organism spontaneously after two estrus cycles or heat periods (about 42 days), but they should be examined to ensure that they are free of infection before being bred.

Bulls present a particularly difficult problem. The development of dimetridazole provided an almost 100% guarantee that a bull can be freed of infection. Previously, treatments were laborious and very possibly not completely effective. It was usually recommended that the bull be removed from the herd unless it was a particularly valuable breeding animal. Now the animal can be left in the herd with only a small risk that infection will remain.

In a clean herd it is essential that any bull being introduced into it be examined for trichomonad infection. As a general rule, bulls to be purchased are examined for fertility and for venereally transmitted infectious agents such as *Vibrio foetus* and *T. foetus*. So prevention of bovine trichomonosis should be recognized as part of a good disease prevention program.

Artificial insemination is almost universal in the North American dairy industry and is being used more and more by beef cattle raisers. The question is, does artificial insemination (AI) reduce the risk of bovine trichomonosis? The answer is a qualified yes. In a properly run AI center, no infected bull will be allowed in. Second, semen that has been collected from an animal is diluted and then stored for a matter of days in a refrigerator or for an indefinite time in liquid nitrogen. If trichomonads are present in the collection of semen, they will be reduced in number with the dilution of the semen, thereby reducing the probability of transmission. But the conditions of storage are such that many of them will probably survive. Conditions under which bull sperm is subjected to freezing and then stored are almost identical to the conditions under which maximal numbers of *T. foetus* will survive.

NAME OF ORGANISM AND DISEASE CAUSED. *Trichomonas vaginalis*, Donné, 1836, is the cause of trichomonad vaginitis in women.

HOSTS AND LOCATION. The only normal host of *T. vaginalis* is humans, in whom it is found in the urogenital system. In women it is normally found in the vagina, and in men it lives in the urethra and prostate gland. Some animals such as monkeys, hamsters, and guinea pigs can be infected experimentally.

DISTRIBUTION AND IMPORTANCE. *T. vaginalis* is a cosmopolitan parasite. Most surveys indicate a prevalence of 10 to 25%; although these figures pertain mostly to women, they are no doubt comparable in men (Neva, 1974). In some populations, prevalence has been found to approach 90%.

MORPHOLOGY. The mean size of *T. vaginalis* is about 10 by 7 μm, but it has been reported from 4 to 32 by 2 to 14 μm as extremes in the size range. It has four anterior flagella, the attached flagellum terminates near the posterior of the body, and there is no trailing flagellum. The axostyle is slender and extends beyond the posterior of the body. The parabasal body is single or V–shaped and has a filament associated with it (Fig. 6.3).

LIFE CYCLE. *T. vaginalis* reproduces by binary fission and exists only as a trophozoite. It is transmitted from one host to the next by sexual intercourse. There is some evidence that when hygienic conditions are lacking, transmission may take place through soiled clothing or wash cloths, but such transfer is probably rare because trichomonads survive only a short time when exposed to oxygen.

DIAGNOSIS. Finding the living organisms in material taken from a person is the usual means of determining whether an infection with *T. vaginalis* is present. In women, vaginal swabs are most commonly examined. Trichomonads are sometimes an incidental finding in urine sediments and in Pap smears, but these should not be considered diagnostic methods. Vaginal discomfort and excessive production of mucus are reasons to suspect trichomonad vaginitis, but such conditions may be caused by other infectious agents as well as noninfectious processes.

In men, urethral swabs or secretions obtained through prostatic massage are examined microscopically to find the organisms.

HOST–PARASITE INTERACTIONS. Although both men and women are infected with *T. vaginalis*, it causes disease principally in women. The organism usually is

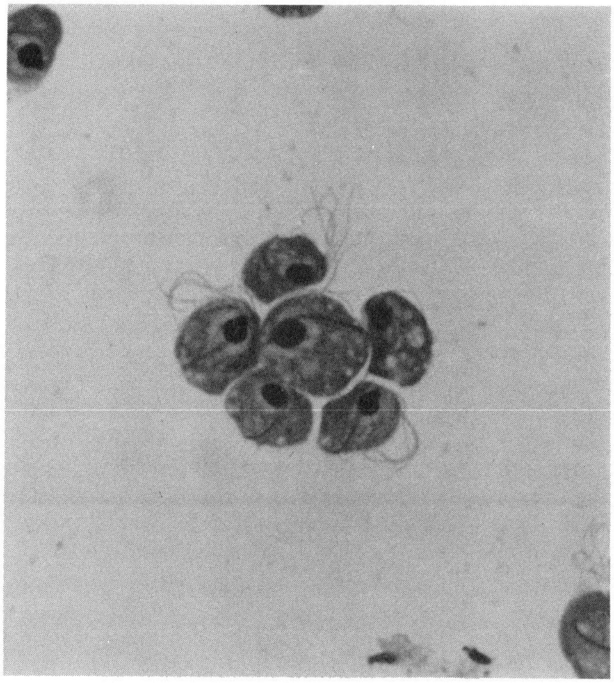

**FIG. 6.3**
*Trichomonas vaginalis*, a parasite of the reproductive tract of men and women.

confined to the vagina, where it multiplies and causes an inflammation of the epithelium. The result of inflammation is tenderness, pain, itching, and excessive production of mucus. Intercourse may be painful.

Although it happens infrequently, *T. vaginalis* infects parts of the urogenital system other than the vagina. There are cases on record of the organism being found in the bladder, ureters, and even the kidneys. For example, there is a case on record of a woman having a large tumorlike swelling in the region of the kidney; the swelling was found to have been caused by trichomonads which had migrated to the kidney, broken out, and colonized the perirenal area (Suriyanon et al., 1975).

Men usually remain asymptomatic, but sometimes there is inflammation of the urethra and the prostate gland.

Vaginal infections with *T. vaginalis* may take place prior to puberty, but they are only transient until the individual has reached sexual maturity. At puberty, the vagina changes in a number of ways so that it secretes more mucus and carbohydrate. The bacterial flora changes to one in which lactobacilli bloom; they consume carbohydrate and produce lactic acid, increasing the hydrogen ion concentration to about pH 4.5. All these changes lead to conditions in the vagina that allow *T. vaginalis* to flourish.

No immunity per se has been demonstrated in infected women, but the course of the infection allows the inference that some does develop. Over a period of time, the severity of the clinical signs and symptoms decreases.

TREATMENT. A variety of drugs has been used more or less successfully in the treatment of both men and women. By and large the drugs have been administered topically – either vaginally in women or into the urethra in men. Some of the drugs that have been used are quinolines, arsonic acid, and tetracyclines. Metronidazole (Flagyl) is highly effective and may be given by mouth. Metronidazole is an imidazole derivative and the imidazole moiety is known to have some carcinogenicity, but the risk is considered to be so low that there is no hazard to the health of a person receiving the drug.

EPIDEMIOLOGY AND CONTROL. Among venereal diseases, trichomonad vaginitis is not the public health problem of gonorrhea, syphilis, or herpes. Although the consequence of infection may be considerable discomfort, there are usually, no catastrophic results for the infected individual. Consequently, trichomonad vaginitis is not a reportable disease and no effort is made by public health personnel to find and treat all sexual partners who may have been exposed during a given period, as is done when gonorrhea is diagnosed.

Control of infection is nearly always limited to individual treatment of women who have shown clinical signs and symptoms of infection. Since men nearly always remain asymptomatic, and many women, too, a relatively high level of infection can be achieved in a population in which sexual partners may be frequently changed. When it has been determined that a woman is infected, it is essential to treat both her and her partner(s) or nothing will have been accomplished.

NAME AND DISEASE CAUSED. *Trichomonas gallinae* (Rivolta, 1878) Stabler, 1938 causes avian trichomonosis and canker.

HOSTS AND LOCATION IN THE HOST. *T. gallinae* has a broad host range, being found in columbiform and falconiform birds. It has been transmitted experimentally to a number of birds in different orders. The organism is a problem principally in doves and pigeons, and secondarily in hawks and falcons, which prey on them. Disease problems have also been reported in domestic chickens. The organisms are most often limited to the upper portion of the digestive tract such as the esophagus, crop, and mouth; however, organisms may spread to liver, lungs, and other internal organs.

DISTRIBUTION AND IMPORTANCE. Distribution is cosmopolitan. Economic losses are principally in doves and pigeons raised either for food or as a hobby. In 1954 annual losses were estimated to be $47,000 in the United States. Trichomonosis may be a factor that controls populations of mourning and ring-neck doves; such reduced populations would be a loss for those who hunt birds.

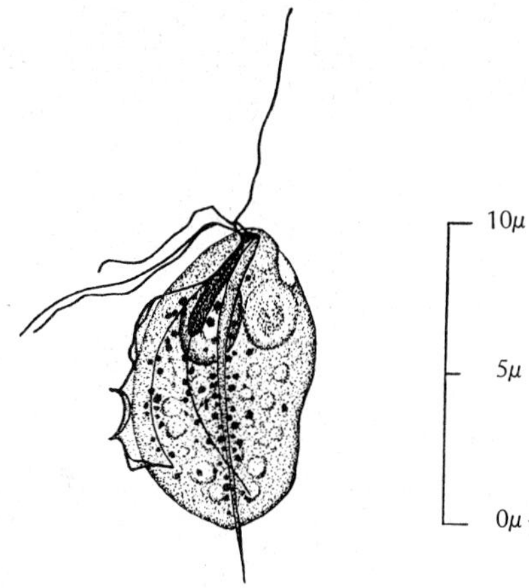


**FIG. 6.4**
*Trichomonas gallinae,* a parasite of the digestive and respiratory tracts of pigeons and doves as well as some other birds. [From Stabler, 1947.]

MORPHOLOGY. The body of *T. gallinae* is about 12 by 6 μm, with four free anterior flagella about 11 μm long. The attached flagellum terminates one half to two thirds of the way down the body, and there is no trailing portion. There is a round parabasal body with two filaments, one of which extends into the posterior portion of the body. The nucleus is closely associated with the anterior somewhat inflated part of the axostyle (Fig. 6.4).

LIFE CYCLE. *T. gallinae* is transmitted from one bird to another through close contact. Young columbiform birds are infected when their parents feed them; in pigeons and doves, parents feed their young with a secretion of the crop called *pigeon's milk*, and in an infected parent this material is laden with trichomonads. Transfer among other birds is by contamination of drinking water; fecal contamination seems not to be a means of transmission. Falcons and hawks are infected through eating infected prey.

DIAGNOSIS. The principal and most effective method of diagnosis is finding the organisms in infected tissues. This may be done by direct examination or by swabbing the throat of birds and cultivating the suspect material to encourage the growth of the trichomonads. The yellow, caseous lesions in the tissues of the gut, lung, and liver are helpful, but other microorganisms may cause similar lesions.

HOST–PARASITE INTERACTIONS. Infection with *T. gallinae* may range from a benign one in which there is uneventful recovery to an acute, rapidly fatal disease, depending on the virulence of the particular isolate of *T. gallinae* and the age of the bird. In young pigeons or doves the mortality from *T. gallinae* infections may approach 90%.

Immunity develops in birds that survive the infection. The immunity is not a sterile one and birds will carry a low-grade infection for an indefinite period.

TREATMENT. The best drug currently available is enheptin (2–amino–5–nitrothiazole). It can be administered either in the drinking water or by mouth individually. It has a high level of efficacy, but some birds may escape receiving a therapeutic level of the drug, especially when it is given in the drinking water.

EPIDEMIOLOGY AND CONTROL. Trichomonosis is a difficult problem for those who raise pigeons or doves. Death losses in squab are likely to be such that few young survive. Enheptin can be used to treat adult birds with the hope that the infection will be eliminated and the squab not infected; however, even a low level of infection among the birds may later precipitate an epizootic through contaminated drinking water.

Further complicating the problem of control is the immunity that surviving birds develop. Birds become immune but retain an infection, which they are fully capable of transmitting. Worse yet, infection with an isolate of low virulence confers immunity to that strain and to virulent strains as well. A bird may become infected with a strain of low virulence, show little clinical effect, and then be infected with organisms that are highly lethal. This means that immune adults can pass the virulent organisms to their offspring with devastating results (Stabler, 1948a). Based on these observations, Stabler (1948b) theorized that the extinction of the passenger pigeon in the early part of this century was due not to human depredations, but rather to these kinds of double infections coupled with nonsterile immunity, which quickly reduced the fecundity of the pigeons and led to their extinction.

A complete control program should be based on finding infected birds, treating or eliminating them, and ensuring that the organism is not reintroduced into the cote with new birds or from wild birds who may be casual visitors for feed and water.

When domestic chickens have been infected, it is probably best to slaughter the birds, clean up the premises, and start with new, clean stock. At that point it is essential that *T. gallinae* not be reintroduced inadvertently by wild pigeons feeding with the chickens.

NAME OF ORGANISM AND DISEASE CAUSED. *Histomonas meleagridis* (Smith, 1895) Tyzzer, 1920, causes blackhead, histomoniasis, infectious enterohepatitis, and histomonosis.

HOSTS AND LOCATION. The principal host in which economic losses are seen is the turkey, *Meleagris gallopavo*, but the organism has a fairly broad host range among gallinaceous birds including the domestic chicken, pheasant, grouse, and quail. The organism is found principally in the digestive tract of the vertebrate host, but also invades the liver.

DISTRIBUTION AND IMPORTANCE. Distribution is cosmopolitan. The U.S. Department of Agriculture (USDA) estimates of losses in turkeys and chickens in the United States total $9.3 million annually.

MORPHOLOGY. The size ranges from 8 to 21 μm; there is a single nucleus and from one to four flagella. Aflagellate stages occur in the liver. The organism often moves by ameboid movement (Fig. 6.5).

*H. meleagridis* was originally classified in the Rhizomastigida, an order containing species that seemed to be transitional between the flagellates and the amebas. In the usual hematoxylin stain of organisms in intestinal content or culture, they show little structure, but silver staining and electron microscopy have allowed investigators to see the affinities of *H. meleagridis* to the trichomonads. These studies reveal the presence of basal bodies, a parabasal body, and an associated filament and an axostyle that continues anteriorly as the pelta (Fig. 6.6) (Schuster, 1968). Serological studies have added further evidence of the relation of *Histomonas* to other trichomonads (Dwyer, 1974).

LIFE CYCLE. *H. meleagridis* reproduces by simple binary fission, as do other trichomonads. Transmission from one bird to another is through the cecal nematode of gallinaceous birds, *Heterakis gallinarum* serving as a biological vector of *H. meleagridis*. The worm itself becomes infected in the cecum of the host bird by ingesting the trophozoites. The trophozoites later infect the reproductive tract of the worm and the protozoa become incorporated into the eggs as they are formed. When a bird ingests an infective egg of *H. gallinarum* containing *H. meleagridis*, it becomes infected with both parasites (Fig. 6.7).

DIAGNOSIS. Determining whether a bird is suffering from histomonosis is usually based on finding the characteristic lesions in the ceca and the liver. Finding the organisms themselves is possible, but not usual.

The cecal lesions, which may be seen early in the infection, are small ulcers in the epithelial tissue. Later, they become larger and tissue necrosis may be seen. The lumen of the ceca also become filled with a yellowish fluid, which is also seen as sulfur-colored droppings. When the histomonads reach the liver, they form a characteristic circular, yellowish, depressed lesion.

Histomonads may be found in the cecum, but if an organism is found, it is rather difficult to identify. One

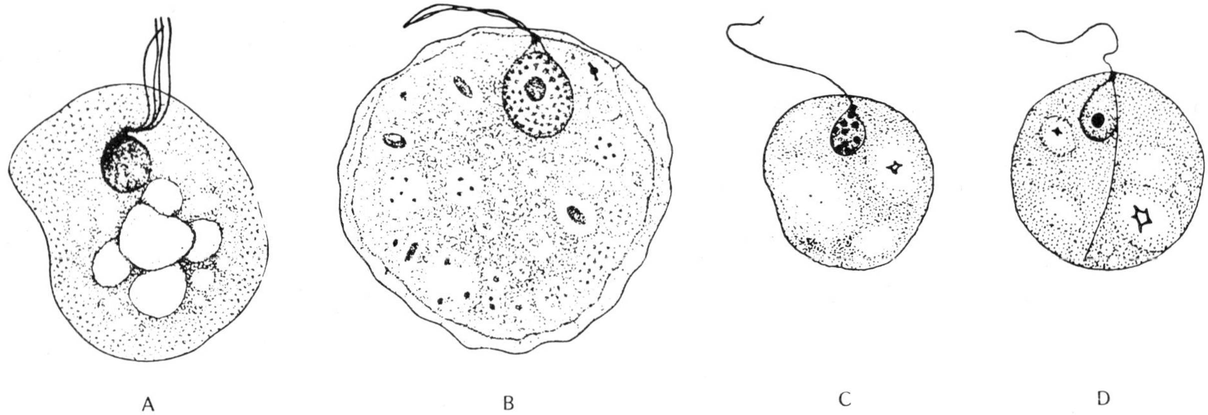


**FIG. 6.5**
*Histomonas meleagridis,* a pathogen of turkeys and other gallinaceous birds. (*A*) From a chicken; (*B*) from a pheasant; (*C* and *D*) from a culture. [From Weinrich, 1943.]

complicating factor is that *Parahistomonas wenrichi,* a nonpathogenic histomonad, may be present, and it cannot be differentiated with certainty from *H. meleagridis* without special staining.

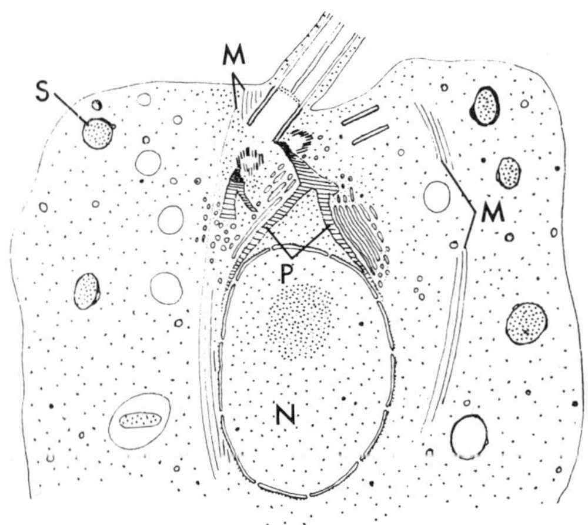


**FIG. 6.6**
Fine structure of *Histomonas meleagridis* showing affinities to other trichomonads. [From Schuster, 1968.]
P = parabasal fibers, M = microtubular elements,
N = nucleus, S = sytoplasm-filled sacs.

HOST–PARASITE INTERACTIONS. Transmission of *H. meleagridis* is one of the few examples in which a parasitic helminth serves as a biological vector for another parasitic agent (see also *Dientamoeba fragilis* and *Nanophyetus salmincola*). Considering the low probability of infecting birds with naked histomonads and their intimate association with the worms, it seems likely that the protozoa were first parasites of the nematodes and later parasites of the avian host.

Since the report in 1920 by Smith and Graybill that *Histomonas* is transmitted by *Heterakis*, a number of studies have been done which have confirmed their report. It has been amply confirmed that the protozoa are inside egg capsules and that only embryonated eggs are capable of transmitting the infection to susceptible birds. Only eggs containing infective second-stage larva hatch on reaching the intestine of the bird; if unembryonated, they cannot respond to the stimuli and merely pass out with the egg capsules intact. It has been only in relatively recent years that there has been cytological demonstration of histomonads in the worms and their eggs (see Lee, 1971, for review).

The requirement that the histomonads be in the eggs of *Heterakis* has been demonstrated through studies showing that even large numbers of histomonads fail to pass through the gizzard of birds under normal conditions. The gizzard of a bird is usually acidic (pH 3)

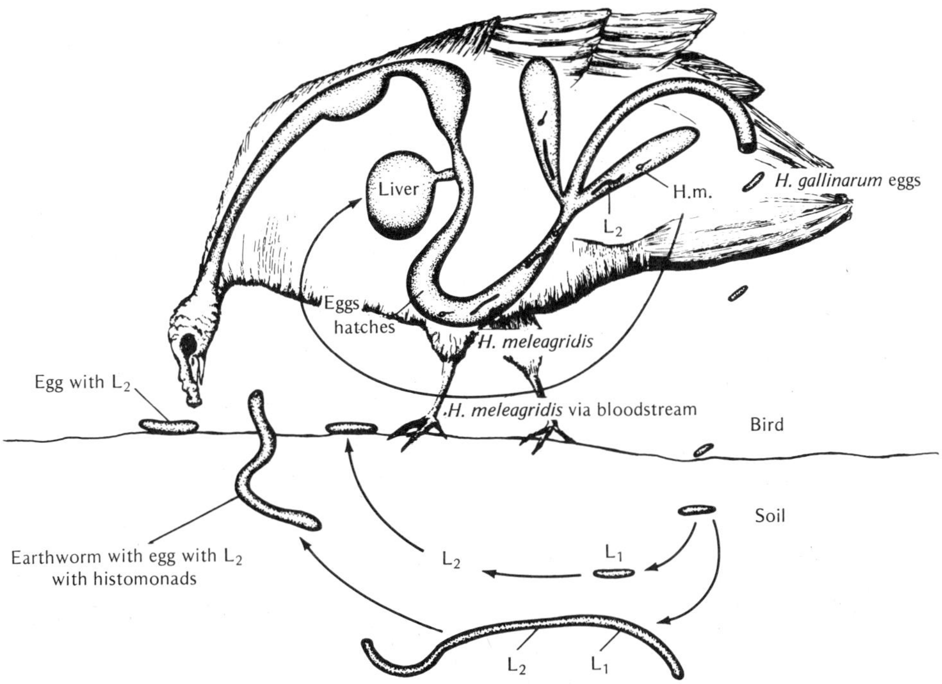


**FIG. 6.7**
Life cycle of *Histomonas meleagridis.*

and kills the protozoa; however, if the gizzard is made alkaline, infections are more likely. Thus, the egg capsules of the worm serve as a means for these delicate protozoa to pass through the hostile environment of the upper digestive tract of the bird.

Once released from the egg capsules, the histomonads multiply in the cecum of the bird. Worms become reinfected by ingesting the trophozoites with other intestinal content.

Depending on the species and age of the host, *H. meleagridis* may remain a relatively harmless commensal, or it may invade the tissues of the gut or liver and cause the death of the host. Although there are sometimes losses in young chickens, histomoniasis is not usually a problem in the domestic fowl. The principal bird in which economic losses occur is the domestic turkey. In young birds, from a month to three months of age, death losses may approach 100% in a flock. Older turkeys are not as susceptible, but death losses may still take a quarter of a flock.

Early in the infection, the histomonads multiply in the cecum, invading the tissues of the ceca and causing severe damage. Some of the histomonads may pass by way of the bloodstream from the ceca to the liver. The liver lesions are usually extensive.

EPIDEMIOLOGY AND CONTROL. Control of histomonosis is based on (1) isolation of turkeys from chickens and (2) chemoprophylaxis.

Where the risk of infection of turkey poults is high, chemoprophylaxis may be practiced, but the side effects of the drugs are such that they should be avoided if possible.

Isolating turkeys from chickens is desirable, but may be difficult. Scrupulous washing of footwear, tools, and machinery is necessary to ensure that eggs of *Heterakis* are not transported from one place to another. The time and effort involved may be prohibitive. Also, the *Heterakis* eggs that contain histomonads survive for an extraordinarily long time. Farr (1956) found that plots contaminated with infected *Heterakis* eggs were still capable of transmitting infections to clean birds 66 weeks later.

Earthworms probably serve as a means of helping both the histomonads and heterakids to survive for long periods outside of the avian host. When earthworms ingest infective eggs of *Heterakis*, the worms hatch and invade the tissues of the earthworm. They remain there for two years or perhaps longer.

Isolation, then, means not only spatial isolation, but also isolation in time. Obviously, the risks of infection are such that only ground not occupied by chickens for years should be used for raising turkeys.

Some additional control methods are to raise turkey poults on wire and to keep age groups of turkeys separated. Also, if turkeys are herded, they should be kept moving and, under ideal conditions, not return to the same pasture during the same growing season.

*Parahistomonas wenrichi* (Lund) is a parasite of the intestinal tract of turkeys. It is closely related to *H. meleagridis*, but sufficiently different structurally to justify separate taxonomic status. *P. wenrichi* is of interest because it has low pathogenicity but protects a bird against infection by *H. meleagridis*. It has been suggested that infection with *P. wenrichi* might provide protection from the effects of *H. meleagridis* and be used as an artificial immunizing agent.

NAME OF ORGANISM. *Dientamoeba fragilis*, Jepps & Dobell, 1918.

HOSTS AND LOCATION. The principal host is humans, but it is also found in some monkeys: baboon, macaques, and chimpanzees. The site of infection is the large intestine.

DISTRIBUTION AND IMPORTANCE. The distribution is cosmopolitan. Estimates of the prevalence of *D. fragilis* in the human population worldwide vary from 4 to 20%. Of 64,544 fecal samples examined in Ontario, 4.1% were found to contain *D. fragilis*. Prevalence in any one area may vary considerably even from these rather wide ranges.

MORPHOLOGY. The organisms have a broad size range, from 3 to 22 μm according to Levine (1973) or from 4 to 18 μm according to Kudo, (1966). They average 9 μm. There are typically two vesicular nuclei, which have a cluster of granules forming the endosome (see Fig. 9.1). The granular cytoplasm contains bacterial inclusions. Cysts are unknown; only trophozoites have ever been found.

The taxonomic status of *D. fragilis* remained clouded until recent years. Since the organism is usually about 9 μm in diameter, it can be appreciated that some structures approach the resolving power of the light microscope and are difficult to see with certainty. Jepps and Dobell originally described the organism as an ameba, but later Dobell (1940) concluded that *D. fragilis* might be an aberrant flagellate closely related to *Histomonas meleagridis*. Studies using silver staining, electron microscopy, and serological analysis of proteins have allowed the conclusion that *D. fragilis* is a trichomonad. The name implies that *Dientamoeba* is an ameba, but the principles of zoological nomenclature require that the original name be retained. *Dientamoeba* is also firmly entrenched in the parasitologic literature as an ameba, so it may be some years before its proper taxonomic location is recognized throughout the biological literature.

LIFE CYCLE. *D. fragilis* multiplies by simple, binary fission; there are no sexual stages. Cysts have never been found. Transmission to a new host takes place by mouth, probably within the egg of the human pinworm, *Enterobius vermicularis* (Yang & Scholten, 1977).

DIAGNOSIS. Organisms may be found in fresh fecal samples, but they die rather quickly outside the host. The living trophozoites are active and move progressively with a clearly defined ectoplasm. In fixed and stained preparations, the organisms may well be missed by inexperienced persons. One needs to look for the typical outline and marginal appearance of an ameba and the two well-stained nuclei. The organism may be cultivated in media such as Balamuth's or a variety of others.

HOST–PARASITE INTERACTIONS. Evidence leads to the conclusion that *D. fragilis* is mildly pathogenic. The principal signs and symptoms are enteritis with diarrhea, abdominal pain, fever, vomiting, flatulence, general weakness, and weight loss. About 25% of infected persons experience symptoms of infection.

TREATMENT. Infections with *D. fragilis* respond to treatment with drugs used against *Entamoeba histolytica* (see Chap. 9).

EPIDEMIOLOGY AND CONTROL. As is true with other trichomonads, *D. fragilis* exists only as a trophozoite, and transmission from one host to the next presents a problem to the organism. Studies such as that of Dobell (1940) indicated that the trophozoites were not infective when administered by mouth; Dobell failed to infect himself even though he drank cultures containing large numbers of trophozoites. Later studies (Burrows & Swerdlow, 1956) presented epidemiologic and cytologic evidence that *D. fragilis* is transmitted in the egg of the human pinworm, *Enterobius vermicularis*. In reviewing a number of aspects of human infections with *D. fragilis*, Yang and Scholten (1977) concluded that there was good, though inconclusive, evidence for transmission by means of pinworm eggs. For example, there is about a 20 times greater association of pinworms and *Dientamoeba* than would be expected by chance.

As a general rule, proper disposal of human feces is recommended as a means of controlling infections with enteric protozoa. With *D. fragilis,* however, the principle does not seem to apply. If we were dealing only with the trophozoites, sanitary disposal of human waste might apply because the organisms live for only hours outside the host; however, the trophozoites seem not to be infective for humans when taken by mouth.

Pinworms are intestinal parasites of humans, but are not generally transmitted through contaminated food or water (Chap. 30). Pinworm eggs containing infective larvae are usually transmitted from one person to another through contaminated clothing and bed linen. An individual may also infect himself or herself directly by scratching the perianal region and then transferring eggs to the mouth on the fingers. Transmission of *E. vermicularis* is so easy that when it is diagnosed in one member of a family, usually all members of the household undergo treatment. It can be appreciated that such ready transmission of the pinworm provides a good vehicle for transmission of *D. fragilis*.

In sum, control is based on treatment of individuals showing symptoms of infection and a confirmed identification of the organisms. It does not appear to be justified to treat persons harboring pinworms alone for infection with *D. fragilis*.

## REFERENCES AND FURTHER READINGS


Burrows, R. B., and M. A. Swerdlow. 1956. *Enterobius vermicularis* as a probable vector of *Dientamoeba fragilis*. *Am. J. Trop. Med. Hyg.* **62**:156–167.

Camp, R. R., C. F. T. Mattern, and B. M. Honigberg. 1974. Study of *Dientamoeba fragilis* Jepps and Dobell. I. Electronmicroscopic observations of the binucleate stages. II. Taxonomic position and revision of the genus. *J. Protozool.* **21**:69–82.

Dobell, C. C. 1940. Researches on the intestinal protozoa of monkeys and man. X. The life history of *Dientamoeba fragilis* – observations, experiments, and speculations. *Parasitology*. **32**:417–461.

Dwyer, D. M. 1974. Analysis of the antigenic relationships among *Trichomonas*, *Histomonas*, *Dientamoeba* and *Entamoeba*. III. Immunoelectrophoresis technics. *J. Protozool.* **21**:139–145.

Farr, M. M. 1956. Survival of the protozoan parasite *Histomonas meleagridis* in the feces of infected birds. *Cornell Vet.* **46**:178–187.

Fulton, J. D. 1969. Metabolism and pathogenic mechanisms of parasitic protozoa. In T.- T. Chen (ed.), *Research in Protozoology*. Vol. 3. Pergamon Press, New York, pp. 389–504.

Hibler, C. P., D. M. Hammond, F. H. Caskey, A. E. Johnson, and P. R. Fitzgerald. 1960. The morphology and incidence of the trichomonads of swine, *Tritrichomonas suis* (Gruby and Delafond), *Tritrichomonas rotunda*, n. sp. and *Trichomonas buttreyi*, n. sp. *J. Protozool.* **7**:159–171.

Honigberg, B. M. 1963. Evolutionary and systematic relationships in the flagellate order Trichomonadida. *J. Protozool.* **10**:20–62.

Honigberg, B. M., and C. J. Bennett. 1971. Light microscope observations on structure and division of *Histomonas meleagridis* (Smith). *J. Protozool.* **18**:687–697.

Honigberg, B. M., W. A. Daniel, and C. F. T. Mattern. 1972. Fine structure of *Trichomitus batrachorum* (Perty). *J. Protozool.* **19**:446–453.

Honigberg, B. M., and J. Kuldova. 1969. Structure of a nonpathogenic histomonad from the cecum of galliform birds and a revision of the trichomonad family Monocercomonadidae. *J. Protozool.* **16**:526–535.

Honigberg, B. M., C. F. T. Mattern, and W. A. Daniel, 1968. Structure of *Pentatrichomonas hominis* (Davaine) as revealed by electron microscopy. *J. Protozool.* **15**:419–430.

Jirovec, O., and M. Petrü. 1968. *Trichomonas vaginalis* and trichomoniasis. *Adv. Parasit.* **6**:117–188.

Kudo, R. R. 1965. *Protozoology*. 5th ed. Thomas, Springfield, Illinois.

Lee, D. L. 1969. The structure and development of *Histomonas meleagridis* (Mastigamoebae:Protozoa) in the female reproductive tract of its intermediate host, *Heterakis gallinarum* (Nematoda). *Parasitol.* 59:877 – 884.

Lee, D. L. 1971. Helminths as vectors of micro-organisms. *In* A. M. Fallis, (ed.), *Ecology and Physiology of Parasites*, University of Toronto Press, Toronto, pp. 104 – 121.

Lee, D. L., P. L. Long, B. J. Millard, and J. Bradley. 1969. The fine structure and method of feeding of the tissue parasitizing stages of *Histomonas meleagridis*. *Parasitol.* 59:171 – 184.

Levine, N. D. 1973. *Protozoan Parasites of Domestic Animals and of Man*. 2nd. ed. Burgess, Minneapolis.

Lund, E. E. 1969. *Histomonas* and histomoniasis. *In* C. A. Brandly and C. E. Cornelius (eds.), *Adv. Vet. Sci. Comp. Med.* **13**:355 – 390.

Mattern, C. F. T., B. M. Honigberg, and W. A. Daniel. 1967. The mastigont system of *Trichomonas gallinae* as revealed by electron micrococoscopy. *J. Protozool.* **14**:320 – 339.

McDougald, L. R. and W. M. Reid. 1978. *Histomonas meleagridis* and its relatives. In J. P. Kreir (ed.), *Parasitic Protozoa*. Vol. III. Academic Press, New York.

McLoughlin, D. K. 1963. A systemic treatment for bovine trichomoniasis. *J. Parasitol.* **49**:959.

McLoughlin, D. K. 1965. Dimetridazole, a systemic treatment for bovine venereal trichomoniasis. I. Oral administration. *J. Parasitol.* **51**:835 – 836.

McLoughlin, D. K. 1968. Dimetridazole, a systemic treatment for bovine venereal trichomoniasis. II. Intravenous administration. *J. Parasitol.* **54**:1038 – 1039.

McLoughlin, D. K. 1970. Dimetridazole, a systemic treatment for bovine venereal trichomoniasis. III. Trials with cows. *J. Parasitol.* **56**:39 – 40.

Neva, F. A. 1974. Protozoal diseases endemic in the United States. The teaching of parasitology as determined by endemic protozoal diseases. *Am. J. Trop. Med. Hyg.* **23**:731 – 735.

Schuster, F. L. 1968. Ultrastructure of *Histomonas meleagridis* (Smith) Tyzzer, a parasitic amoebo-flagellate. *J. Parasitol.* **54**:725 – 737.

Smith, T., and H. W. Graybill. 1920. Blackhead in chickens and its experimental production by feeding embryonated eggs of *Heterakis papillosa*. *J. Exper. Med.* **32**:143 – 152.

Stabler, R. M. 1947. *Trichomonas gallinae*, pathogenic trichomonad of birds. *J. Parasitol.* **33**:207 – 213.

Stabler, R. M. 1948a. Variations in virulence of strains of *Trichomonas gallinae* in pigeons. *J. Parasitol.* **34**:147 – 149.

Stabler, R. M. 1948b. Protection in pigeons against virulent *Trichomonas gallinae* acquired by infection with milder strains. *J. Parasitol.* **34**:150 – 153.

Suriyanon, V., K. E. Nelson, and V. C. NaAyudha. 1975. *Trichomonas vaginalis* in a perinephric abscess. *Am. J. Trop. Med. Hyg.* **24**:776 – 780.

Warton, A., and B. M. Honigberg. 1979. Structure of trichomonads as revealed by scanning electron microscopy. *J. Protozool.* **26**:56 – 62.

Wenrich, D. H. 1943. Observations on the morphology of *Histomonas* from pheasants and chickens. *J. Morphol.* 72:279 – 303.

Yang, J., and T. Scholten. 1977. *Dientamoeba fragilis*: A review with notes on its epidemiology, pathogenicity, mode of transmission and diagnosis. *Am. J. Trop. Med. Hyg.* **26**:16 – 22.

# 7

# *Giardia and Other Diplomonads*

The order Diplomonadida contains the only protozoa that are bilaterally symmetrical. They have two nuclei and six or eight flagella. The organisms form cysts and are transmitted through food or water. Reproduction is by binary fission and there are no sexual stages in the life cycle.

A transition can be seen from the free-living genera *Trepomonas* and *Hexamita* which are both free-living and symbiotic to *Spironucleus, Octomitus* and *Giardia* which are parasitic (Brugerolle, 1975). In this transition, the motility of the organisms is reduced as is their ability to ingest particulate food. *Giardia*, the final stage in the series, takes up nutrient across its external membrane and is adapted for attachment to the intestinal surface by its adhesive disk.

## *GIARDIA* AND GIARDIASIS

In recent years knowledge of *Giardia* has accumulated rapidly, and some of the concepts that have been held for many years require modification. The information presented here is based on studies in progress and on clinical observations. The review by Meyer and Radalescu (1979) is the most recent synthesis of information on *Giardia*. These authors conclude that *G. duodenalis* parasitizes a wide range of mammals, *G. muris* is in rodents and birds, and *G. agilis* is in amphibia.

The consensus had been that *Giardia* spp. have a high degree of host specificity; when a *Giardia* sp. was found in a species of host from which *Giardia* was not known, a new species was described. Thus, we have *G. bovis*, *G. canis*, and *G. muris*, among others, named for the hosts from which they came. The organisms are fairly small, usually about 12 μm in length, and they look much alike in the usual stains such as HIH (Heidenhain iron hematoxylin). Also, until recently it was not possible to cultivate *Giardia* in vitro, and an investigator was faced with trying to study organisms that would survive, at best, for a few hours outside of the living host. Cross-transmission studies could be done only with cysts harvested from feces.

Methods of cultivation (Meyer, 1976), carefully designed cross-transmission experiments, and means of harvesting cysts from large volumes of water have required a reassessment of the concept of a narrow host range in the genus *Giardia*. There is interchangeability of *Giardia* among humans, dogs, and cats. *Giardia* of beaver are infective for humans. Examination of surface waters used for city water supplies where giardiosis has been a problem has shown the presence of *Giardia* cysts.

NAME OF ORGANISM AND DISEASE CAUSED. *Giardia lamblia* Kofoid and Christiansen, 1915, or *G. duodenalis*, the currently accepted name, is referred to by European workers as *Giardia intestinalis* and sometimes in older literature as *Lamblia intestinalis*.

The disease is usually referred to as *giardiasis*, but should more properly be called *giardiosis*. It is sometimes called *lambliosis*. Based on current information, some important synonyms are *G. canis*, *G. bovis*, and *G. cati*.

HOSTS AND LOCATION. In humans and other primates, dogs, cats, rodents, lagomorphs, and probably ruminants, the organisms live in the upper part of the small intestine. They do not usually invade the tissues, but lie on the surface of the intestinal mucosa, sometimes entering the cysts of Lieberkühm. Two studies have presented good evidence at both light microscopic (LM) and electron microscopic (EM) levels that *G. duodenalis* penetrates gut tissues and may be found inside epithelial cells (Brandborg et al., 1967; Morecki & Parker, 1967).

DISTRIBUTION AND IMPORTANCE. Distribution is cosmopolitan. Surveys indicate about a 7% prevalence in temperate North America. Some populations show as many as 67% of persons to be infected.

Giardiosis occurs in humans both as sporadic individual infections and in epidemic form. Epidemics usually are associated with faulty water delivery systems. The incidence of giardiosis over any given period in a locality is not well documented. In Colorado, for example, giardiosis is a reportable disease, and the state department of health generally lists about 400 cases per year. However, unless an outbreak has occurred where a concerted effort is made to determine how many persons actually suffered from the disease, many sporadic infections probably go undiagnosed. Neva (1974) has stated that he is "unaware of any meaningful data on overall or regional prevalence of giardiasis with the exception of unsolicited reporting of cases to the Center for Disease Control in numbers of 500 to 1000 for recent years."

MORPHOLOGY. The trophozoites of *G. duodenalis* average about 13 μm in length with a range of from 9 to 21 μm. They average about 10 μm in width; the organisms are broadly rounded anteriorly and taper to a point posteriorly (Fig. 7.1; see also Fig. 9.1,*E*). When seen in stained fecal smears, the eight flagella appear only as wispy filaments at the margins of the organism. Each organism has two nuclei, which have a delicate outer limiting membrane and a large, darkly staining central endosome. There are two median bodies, clawlike structures that stain darkly with hematoxylin and axostyles or axial rods. In life, the organisms can be seen to be about 3 μm dorsoventrally, with the anterior portion of the body modified into a concave disk.

Under a compound microscope the cysts are ellipsoid in outline and usually about 8 by 10 μm. Sometimes the sides of the cysts are almost straight, but usually they will be rounded (Fig. 7.2; see also Fig. 9.1,*F*). Most cysts have a double set of organelles. There are

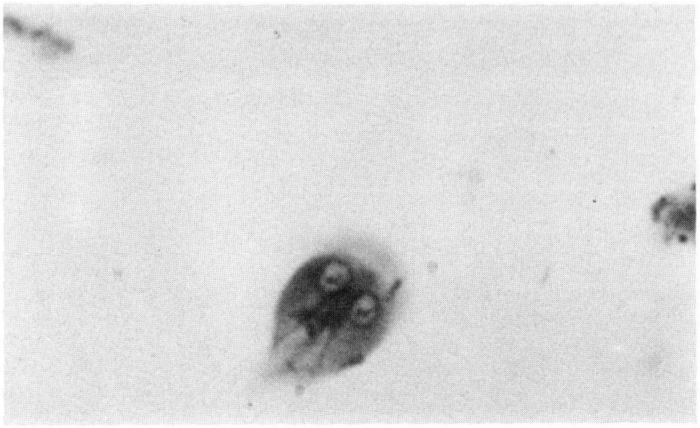

**FIG. 7.1**
A *Giardia duodenalis* trophozoite.

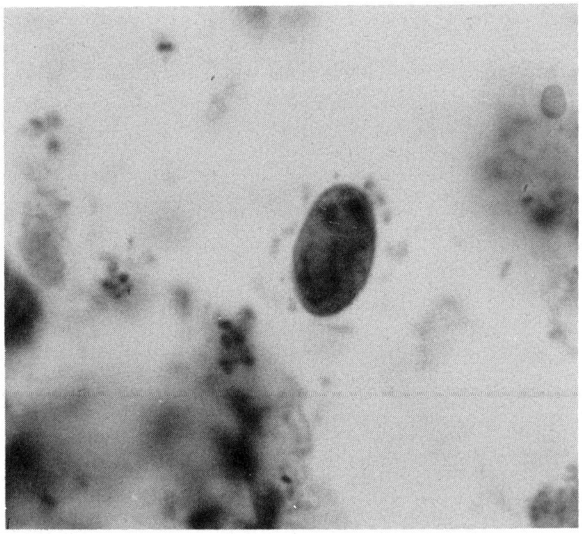

**FIG. 7.2**
Cyst of *Giardia duodenalis* in a fecal smear.

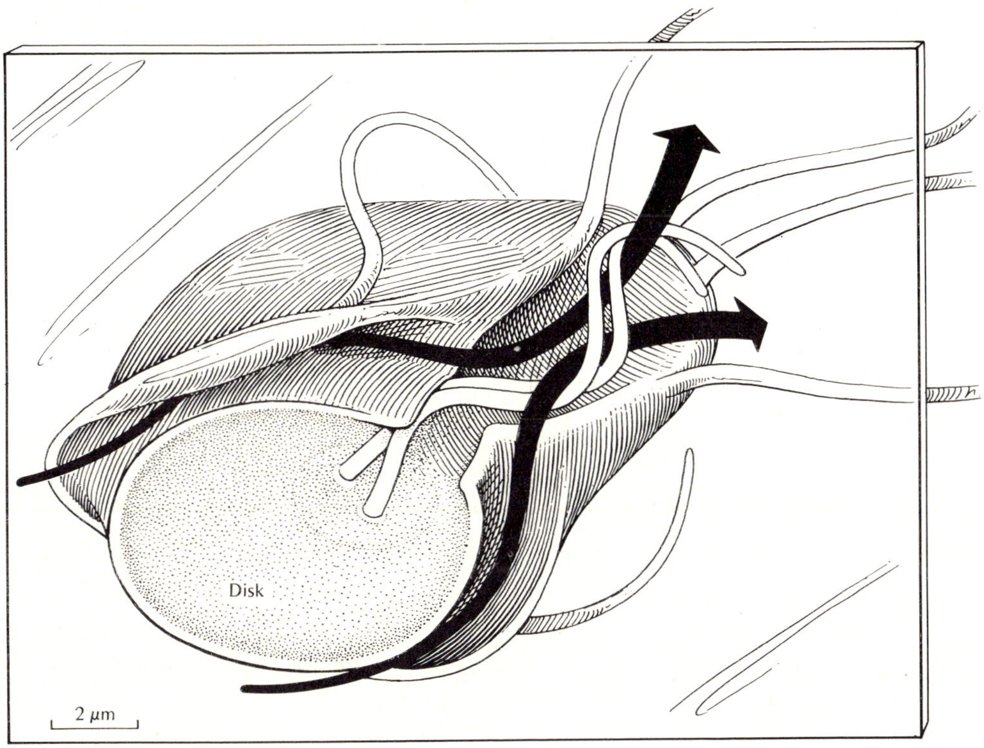


**FIG. 7.3**
Ventral surface of *Giardia* sp. showing the movement of water through the action of the flagella. [Redrawn from Holberton, 1973.]

four nuclei and four median bodies, but other structures cannot be seen clearly. In living cysts, the organisms are often motile and can be seen to roll around with considerable vigor within the cyst wall.

Studies using the electron microscope have confirmed the general features of *Giardia* spp. seen at the LM level. The organisms lie closely applied to the microvilli of the intestinal epithelial cells; lateral to the disk are flanges or extensions of the cytoplasm (Fig. 7.3). The organism is organized around the movement of fluid under the disk from anterior toward the posterior. The flagella are on the ventral surface and serve to move fluid through the channel formed by the concave ventral surface (Holberton, 1973).

The disk is supported by a network of microtubules and the disk itself appears not to be the absorptive surface of the organism. One might think that the movement of fluid across the ventral surface serves to place nutrient in contact with the membrane where uptake occurs; however, ferritin-tagged materials were taken up only on the dorsal surface of the organism (Bockman & Winborn, 1968).

The subject of whether *Giardia* has an axostyle has been discussed at length by Cheissin (1965). Structures seen at the LM level on either side of the midline have been interpreted to be axostyles. At the EM level, it can be seen that the caudal axonemes (the flagella still inside the body of the organism) are surrounded by two semiellipses of microtubules. In Cheissin's terminology, the microtubules form an axial rod, which originates about in the region of the flagellar basal bodies and extends to the posterior margin of the disk. Further posterior, the microtubules remain close to the axonemes but do not surround them. The term *axial rod* is probably more correct than *axostyle* for the structure since the close association with the mastigont apparatus that the axostyle has in the trichomonads is missing. Cheissin interpreted the function of the axial rods as stiffening of the caudal portion of the organism.

LIFE CYCLE. *Giardia* reproduces by binary fission; there are no sexual stages. Trophozoites live in the small intestinal lumen or paramucosal lumen. The trophozoites form cysts in the lower small intestine, usually in asymptomatic infections, and these pass out with the feces. Cysts are probably infective to another host as soon as they are passed. The cysts are usually ingested with contaminated food or water.

DIAGNOSIS. Determining that an individual is clinically affected by *G. duodenalis* is based not only on finding the organism, but also on eliminating other protozoan and microbial agents as the cause of the intestinal problems.

**TABLE 7.1**
Frequency of Symptoms in Persons Infected with *G. duodenalis.*

| SYMPTOMS | PERCENTAGE |
|---|---|
| Bloating, belching, or flatulence | 97 |
| Diarrhea or frequent loose stools | 85 |
| Nausea | 79 |
| Weight loss | 56 |
| Abdominal cramps | 44 |
| Loss of appetite | 41 |
| Vomiting | 24 |
| Fever | 0 |

[From Barbour et al., 1976]

The signs and symptoms most often seen in *Giardia* infections in humans are bloating or flatulence, diarrhea, and weight loss in the face of good appetite (Table 7.1). *Entamoeba histolytica* infection as well as infection with a variety of bacteria may also cause some of these symptoms.

Trophozoites of *G. duodenalis* may be found in a freshly passed fecal specimen, especially if the person is showing diarrhea. In some instances it may be necessary to pass a tube by mouth through the stomach into the lumen of the small intestine and aspirate some of the content. A technique that probably is less distressing to the patient involves passing a weighted string to the duodenum by means of natural peristalsis. The string is retrieved after about 4 hr and the duodenal content adhering to it is examined for trophozoites (Beal et al., 1970). Since *Giardia* trophozoites are delicate creatures, it is essential to examine any specimen immediately upon obtaining it. The technique itself merely involves mixing a tiny portion of the sample with physiological salt solution on a microscope slide, adding a coverslip, and examining it for motile organisms under a compound microscope.

Cysts may be concentrated by coverslip flotation or sedimentation methods such as the formalin–triton–ether (FTE) method and then sought using a compound microscope.

A stained fecal smear should be made when it is not possible to examine a fecal specimen immediately or when a suspected *Giardia* has been found.

HOST–PARASITE INTERACTIONS. It has been clearly established that *G. duodenalis* can be pathogenic in humans, although its pathogenic potential had been a subject of debate. As with other infectious agents, infection results in a spectrum of responses from none at all to a severe, debilitating diarrhea and other symptoms (Table 7.1).

The ways in which *Giardia* damages the host are still uncertain. It is generally agreed that there is interference with absorption of materials from the small intestine. A frequent finding in symptomatic infections is the passage of fat with the feces (steatorrhea), which indicates that the fat is either not emulsified by bile or cannot be absorbed in the small intestine because the epithelial cells are covered with *Giardia.* Micrographs of intestinal histologic sections show that there can be an almost solid mat of organisms on the epithelial surface of the gut. There may, therefore, be a mechanical barrier between the absorptive cells and the lumen of the gut (Yardley & Bayliss, 1967).

Properly chosen material from infected persons has shown little tissue damage. The villi were normal, microvilli were not affected, and there was no inflammatory response (Brandborg et al., 1967). Interestingly, *Giardia* can invade cells of the gut; they were found squeezing between epithelial cells and distributed down to the submucosa. Some were seen within cells, and both the *Giardia* and the host cells appeared to be normal (Brandborg et al., 1967; Morecki & Parker, 1967).

Typically, signs and symptoms of infection occur 9 days after exposure to cysts, but onset may vary from 5 to 20 days or may, in fact, never appear even though the host has a patent infection. Most experimental infections last only a few days, but of 40 men infected in one study, 2 remained positive for 3 months. Only about 10 cysts are needed to establish an infection, but 100 administered to an individual will guarantee infection (Rentdorff, 1954). A person can be reinfected repeatedly, and the results may differ with each infection.

Whether there is immunity to *Giardia* infection is problematic. Reinfection occurs and there seems to be little inflammatory response in the gut. It is still moot what determines whether there will be clinical signs and symptoms in an infection and whether the infection will clear up spontaneously or continue with sporadic evidence of its presence for weeks or months. There are a few reports in the literature of especially severe infections in persons who cannot respond normally to infectious agents, so there may be an immunity that has not yet been clearly defined.

TREATMENT. The drug of choice is quinacrine (Atabrine) taken by mouth in an amount of 100 mg three times a day for 5 to 8 days for adults. Usually, one course of treatment will be sufficient, but sometimes a second course is required. Amodiaquin has been used successfully as a single dose of 600 mg. Metronidazole (Flagyl) has also been used with success but is not recommended because of possible carcinogenesis (see *Trichomonas vaginalis*).

EPIDEMIOLOGY AND CONTROL. Giardiosis occurs both sporadically and epidemically. Although firm evidence is lacking, sporadic infections result principally from contact with other infected persons or animals. On the other hand, epidemic infections seem to occur where there has been a breakdown in a community's water delivery system.

Epidemic outbreaks have occurred in areas where there has been a breakdown in sewerage or where water treatment is inadequate. A classic example of a breakdown occurred in the ski resort of Aspen, Colorado, where old sewer pipes leaked sewage into a well serving a part of the town. At least 25 persons suffered from clinical infections with *G. duodenalis* (Moore et al., 1969). The cysts of *Giardia* are not killed by the levels of chlorine usually employed in municipal water treatment plants. Recent outbreaks have often occurred in water supplies that were chlorinated but not filtered. In these cases, surface waters have been used for potable water, and despite chlorination, persons became infected and showed clinical signs and symptoms of giardiosis. In most instances, contamination of the water supply seemed not to come from human excreta. We then must come back to the concept of host specificity in looking for the source of the infection. It seems likely that some animal or animals other than humans are the source of the contamination. Dogs, cats, and beaver have been implicated, and the evidence is good, but partially circumstantial, that feces of these animals provide the cysts that enter the water supply (Davies & Hibler, 1979).

A number of reports have appeared in the medical literature indicating that travelers in Russia, particularly to the city of Leningrad, run a high risk of contracting giardiosis. In some reports, almost 70% of persons visiting that city become infected; 23% of United States travelers experienced infections between 1969 and 1973 (Jokipi & Jokipi, 1974).

Control of giardiasis involves the following:

1. Treatment of infected persons and companion animals
2. Elimination of the source of cysts
3. Proper treatment of potable water

Chemotherapy is generally successful and the risks or side effects are minimal. There is no question that a person showing symptoms of infection with *Giardia* and having a confirmed diagnosis should be treated. Agreement is not universal whether asymptomatic persons should be treated; however, if one takes into account that signs and symptoms may occur at some later date and that an infected person sheds cysts and is therefore a potential spreader of the infection, treatment is justified.

In addition to treatment of infected persons, especially food handlers, companion animals (dogs and cats) should also be examined and, if infected, treated also. Where there are sporadic cases of giardiosis, the source of the infection may be a companion animal, and treating only one of the infected parties may not be fruitful; as in humans, dogs and cats may or may not be symptomatic.

If an epidemic of giardiosis occurs, the source of the infection must be found. The prime suspect should be the water supply, and it should be determined whether cysts appear in water used for domestic purposes. Fortunately, water can be passed through a millipore filter in large volumes, the filter washed, and the residue examined microscopically for cysts. This technique has been used successfully in a number of instances to determine the source of cysts.

Assuming that there are cysts in the water, a systematic determination of where they are coming from should be made. In a rural area, leakage from pit toilets or septic tank drainage fields into underground waters is a likely possibility. In small municipalities, a broken sewer leaking into a well should be looked for. In both of these instances, fluorescein dye can be used to determine whether it will move from the waste disposal system to the water delivery system. Coliform bacteria counts will also indicate whether there is fecal contamination of the water supply.

In communities that obtain their water from surface impoundments, a search for fecal contamination of the watershed should be made. Contamination from human sources can usually be controlled, but that from animals is difficult or impossible.

When the source of cysts has been found to be the potable water, the first step is to make certain that all water used for drinking, tooth brushing, and so on is boiled or treated with elemental iodine. Water used in cooking need not be boiled.

The next action is to prevent cysts from reaching the water tap. Chlorination at the level used in most municipal water plants is insufficient to kill cysts of *Giardia*, and they can live in water for up to 3 months. Their hardiness is further demonstrated by the fact that 2.5% (0.17 *M*) phenol is required to kill them. The recommended treatment after chlorination is sand filtration. Such treatment virtually ensures that cysts of *Giardia* as well as other protozoan cysts and helminth eggs will be filtered out.

## *HEXAMITA* AND HEXAMITIASIS

These organisms are quite similar to *Giardia* and one might be led to conclude that, because of their name, they have six flagella. The misnomer came about because the original describer thought that they had six flagella, but later investigations showed that they have eight. The rules of zoological nomenclature require that the name be retained even though it may not accurately describe the organism. Brugerolle (1975) has studied diplomonads extensively and suggested that the *Hexamita* species we discuss here are more properly called *Spironucleus*. If there is general acceptance of his proposals, we will speak of spironucleiasis rather than hexamitiasis in the future.

NAME OF ORGANISM AND DISEASE CAUSED. *Hexamita meleagridis*, (McNeil, Hinshaw and Kofoid, 1941) causes hexamitiasis or hexamitosis in turkeys.

HOSTS AND LOCATION. The organism is found in a wide range of gallinaceous birds including turkey, peafowl, and pheasant. It is usually located in the upper part of the small intestine, but may sometimes occur in the large intestine.

DISTRIBUTION AND IMPORTANCE. Distribution is probably cosmopolitan. In the United States, estimates of losses are about $1.8 million.

MORPHOLOGY. These are pyriform, bilaterally symmetrical organisms about 9 μm long and 3 μm wide (Fig. 7.4). There are eight flagella, all arising from basal bodies at the anterior end; six exit anteriorly and two exit posteriorly. Axial rods or axostyles run the length of the organism in association with the posterior flagella. Two nuclei are located at the anterior end, and are round or elliptical and have large endosomes. Cysts are ellipsoid and about 8 μm long.

LIFE CYCLE. Multiplication is by binary fission. Transmission from one bird to the next is through cysts usually ingested with contaminated food or water.

DIAGNOSIS. Clinical signs are suggestive, but finding the organism in large numbers in the intestine is crucial. Birds show a foamy diarrhea, fail to gain weight properly, and become listless.

HOST–PARASITE INTERACTIONS. Infection with *H. meleagridis* is widespread, but disease occurs sporadically, usually only in birds up to 10 weeks of age. The principal site of damage is the small intestine, where an enteritis may be seen at postmortem examination. Although deaths have been recorded in 7 to 80% of birds, death losses are usually low; however, birds that recover from infection may gain poorly.

TREATMENT. Drugs are administered in food or water, but there has been only moderate success in

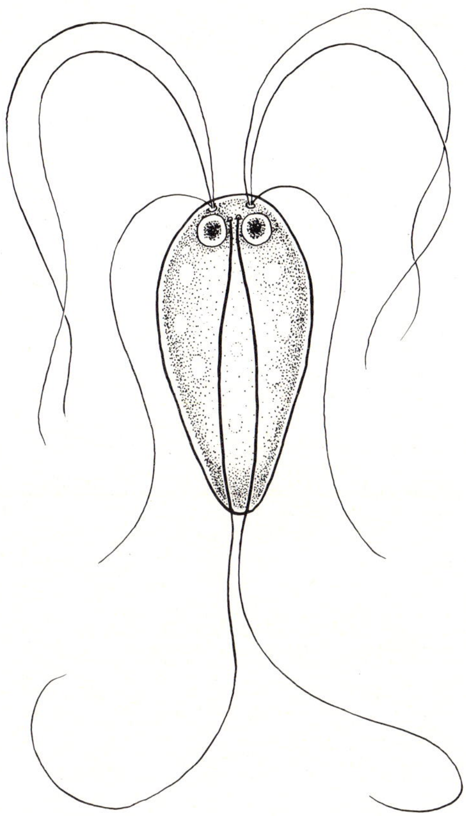

**FIG. 7.4**
*Hexamita meleagridis* trophozoite.

treating hexamitosis. Enheptin, furazolidone, and nithiazide all have some value in reducing the effects of the infection.

EPIDEMIOLOGY AND CONTROL. Prevention of significant infection in young birds is based on the following:

1. Separation of age classes of birds
2. Cleaning premises between batches of birds
3. Allowing time between batches of birds
4. Making sure that workers and tools do not transfer infection between flocks of birds on the premises

It is a general principal, in infections in which older animals shed organisms into the environment and younger animals are particularly susceptible to infection, to keep the age classes physically separate from one another. A variation on this principle is separation in time. If a batch of birds has been raised and then sent to market, a two-week period is recommended before bringing in new poults.

In many infectious diseases, separating susceptible animals from older or infected animals by 50 or 100 m, or keeping them in separate buildings, may be adequate to prevent infection. One must then watch that the infection is not carried on footwear, tools, or machinery from one place to the other. Where risks of infection are high and profit margins low, a clear set of procedures must be established which all workers on the premises adhere to. A breakdown in procedure on any one day may show up as an economic loss within a short period.

Procedures to prevent tracking of infection are, by and large, simple and logical. When feasible, the susceptible animals should be tended first if the same person tends both groups. If duties can be arranged so that the same persons do not tend old and young birds, that is preferable. If vehicles are used to haul feed, the wheels and tires should be hosed off at some intermediate point between the flocks. Workers should not go among birds if it can be avoided, but when it is necessary, at least clean boots should be worn.

NAME OF ORGANISM AND DISEASE CAUSED. *Hexamita salmonis* (Moore, 1922) causes hexamitosis or hexamitiasis.

HOSTS AND LOCATION. Salmonid fish are the host, and the infection develops in the gut, principally the pyloric ceca.

DISTRIBUTION AND IMPORTANCE. Distribution is probably cosmopolitan. Economic losses are uncertain.

MORPHOLOGY. Same as for *H. meleagridis*.

LIFE CYCLE. Same as for *H. meleagridis*.

DIAGNOSIS. Finding large numbers of organisms in the pyloric ceca in fish that fail to grow is indicative of hexamitosis.

CULTIVATION. This organism was cultivated by Uzmann and Hayduk (1963).

HOST–PARASITE INTERACTIONS. There is no consensus on the pathogenicity of *H. salmonis*, but the weight of the evidence indicates that the organism causes weight losses and deaths in salmonid fish in hatcheries. Fulfilling Koch's postulates (see Chap. 6) in their entirety has not been possible because of differences in susceptibility of the hosts or the virulence of the protozoa.

TREATMENT. Treating hatchery fish for infectious disease involves incorporating the drug in the feed. A number of compounds have been found to be efficaceous in treating hexamitosis:

| DRUG | AMOUNT | DAYS GIVEN |
|---|---|---|
| Dimetridazole | 0.15% | 3 |
| Enhepten | 0.2% | 3 |
| Cyzine | 20 ppm | 3 |
| Carbasone oxide | 0.2% | 3 |

EPIDEMIOLOGY AND CONTROL. Outbreaks of hexamitosis probably occur when rearing raceways become contaminated with cysts and the level of contamination becomes high. Since outbreaks are sporadic, it seems likely that factors other than contamination of the environment bring about clinical manifestations of infection.

In the face of an outbreak, chemotherapy should be applied immediately. Between batches of fish, rearing tanks should be cleaned and allowed to dry well.

NAME OF ORGANISM AND DISEASE CAUSED. *Hexamita muris* (Grassi) causes hexamitiasis or hexamitosis.

HOSTS AND LOCATION. *H. muris* has been reported from laboratory rats, mice, and other rodents. Its host range is not known. It is found in the small intestine.

*H. muris* is found frequently in the intestines of laboratory rodents. As a general observation, the organisms bloom when the host has an enteritis. There is also evidence that *H. muris* may cause an enteritis (Matthiesen et al., 1976).

## REFERENCES AND FURTHER READINGS

Barbour, A. G., C. R. Nichols, and T. Fukushima. 1976. An outbreak of giardiasis in a group of campers. *Am. J. Trop. Med. Hyg.* **25**:384–389.

Beal, C. B., P. Viens, R. G. L. Grant, and J. M. Hughes. 1970. A new technique for sampling duodenal contents: Demonstration of upper small-bowel pathogens. *Am. J. Trop. Med. Hyg.* **19**:349–352.

Bockman, D. E., and W. B. Winborn. 1968. Electron microscope localization of exogenous ferritin within vacuoles of *Giardia muris*. *J. Protozool.* **15**:26–30.

Boorman, G. A., P. H. Lina, C. Zrucher, and H. T. Nieuwerkerk. 1973. *Hexamita* and *Giardia* as a cause of mortality in congenitally thymus-less (nude) mice. *Clin. Exp. Immunol.* **15**:623–627.

Brandborg, L. L., C. B. Tankersley, S. Gottleib, M. Barancik, and V. E. Sartor. 1967. Histological demonstration of mucosal invasion by *Giarida lamblia* in man. *Gastroenterology* **52**:143–150.

Brugerolle, G. 1974. Contribution a l'étude cytologique et phylétique des diplozoarires (Zoomastigophora, Diplozoa, Dangéard, 1910). III. Étude ultrastructurale du genre *Hexamita* (Dujardin, 1836). *Protistologica* **10**:83–90.

Brugerolle, G. 1975. Contribution a l'étude cytologique et phylétique des diplozoaires (Zoomastigophorea, Diplozoa, Dangéard, 1910). VI. Caractères généraux des diplozoaires. *Protistologica*. **11**:111–118.

Cheissin, E. M. 1964. Ultrastructure of *Lamblia duodenalis* I. Body surface, sucking disc and median bodies. *J. Protozool.* **11**:91–98.

Cheissin, E. M. 1965. Ultrastructure of *Lamblia duodenalis*. II. The locomotory apparatus, axial rod and other organelles. *Arch. Protistenk.* **108**:8–18.

Davies, R. B., and C. P. Hibler. 1979. Animal reservoirs and cross-species transmission. In W. Jakubowski, and J. C. Hoff. (ed.), *Proc. Sympos.* September 18–20, 1978, Cincinnati, Ohio, National Technical Information Service, Springfield, Va., pp. 104–126.

Ferguson, H. W. 1979. Scanning and transmission electron microscopical observations on *Hexamita salmonis* (Moore, 1922) related to mortalities in rainbow trout fry *Salmo gairdneri* Richardson. *J. Fish Dis.* **2**:57–67.

Filice, F. P. 1952. Studies on the cytology and life history of a *Giardia* from the laboratory rat. *Univ. Calif. Publ. Zool.* 57:53–146.

Friend, D. S. 1966. The fine structure of *Giardia muris*. *J. Cell Biol.* **29**:317–332.

Gangarosa, E. J., and J. A. Donadio. 1970. Surveillance of foodborne disease in the United States. A comparison of data—1968–69. *J. Infect. Dis.* **122**:354–358.

Hoare, C. A. 1955. Life cycle of *Hexamita meleagridis*. *Vet. Rec.* **67**:324.

Hoffman, G. L., and F. P. Meyer. 1967. *Parasites of North American Freshwater Fishes*. University of California Press, Berkeley.

Holberton, D. V. 1973. Fine structure of the ventral disc apparatus and the mechanism of attachment in the flagellate *Giardia muris*. *J. Cell Sci.* **13**:11–41.

Hughes, J. M., M. H. Merson, G. F. Craun, and L. J. McCabe. 1973. Outbreaks of waterborne disease in the United States, 1973. *J. Infect. Dis.* **132**:336–339.

Jakubowski, W., S. L. Chang, T. H. Ericksen, E. C. Lippy, and E. W. Akin. 1978. A large volume sampling of water supplies for microorganisms. *J. Am. Water Works Assoc.* **70**:702–706.

Jakubowski, W., and J. C. Hoff. 1979. Waterborne transmission of giardiasis. *Proc. Sympos.* September 18–20, 1978, Cincinnati, Ohio. National Technichal Information Service, Springfield, Va.

Jokipi, L., and A. M. Jokipi. 1974. Giardiasis in travelers: A prospective study. *J. Infect. Dis.* **130**:295–299.

Khouw, B. T., H. D. McCurdy, Jr., and R. E. Drinnan. 1968. The axenic cultivation of *Hexamita* inflata from the Canadian oyster, *Crassostrea virginica*. *Can. J. Microbiol.* **14**:817–821.

Kofoid, C. A., and O. Swezy. 1922. Mitosis and fission in the active and encysted phases of *Giardia enterica* (Grassi) of man, with a discussion of the method of origin of bilateral symmetry in the polymastigote flagellates. *Univ. Calif. Publ. Zool.* **20**:199–234.

Levine, N. D., P. D. Beamer, and E. McNeil. 1952. *Hexamita* (Protozoa: Mastigophora) from the golden pheasant. *J. Parasitol.* **38**:90.

Matthiesen, T., I. Kunstyo, and K. Tuch. 1976. *Hexamita muris* infection in mice and European hamsters in a laboratory animal colony. *Z. Versuchstierkunde* **18**:113–120.

McElwain, I. B. 1966. Cyzine efficacy for trout hexamitiasis. Unpublished master's thesis. Colorado State University, Fort Collins.

McElwaine, I. B., and G. W. Post. 1968. Efficacy of cyzine for trout hexamitiasis. *Prog. Fish Cult.* **30**:84–91.

McNeil, E. C., W. R. Hinshaw, and C. A. Kofoid. 1941. *Hexamita meleagridis* sp. nov. from the turkey. *Am. J. Hyg.* **34**:71–82.

Meyer, E. A. 1976. *Giardia lamblia*: Isolation and axenic cultivation. *Exp. Parasitol.* **39**:101–105.

Meyer, E. A., and R. Radalescu. 1979. *Giardia* and giardiasis. *Adv. Parasitol.* **17**:1–47.

Moore, G. T., W. M. Cross, C. D. McGuire, C. S. Mollohan, N. N. Gleason, G. R. Healy, and L. H. Newton. 1969. Epidemic giardiasis at a ski resort. *New Engl. J. Med.* **281**:402–407.

Morecki, R., and J. G. Parker. 1967. Ultrastructural studies of the human *Giardia lamblia* and subadjacent jejunal mucosa in a subject with steatorrhea. *Gastroenterology.* **52**:151–164.

Neva, F. A. 1974. Protozoal diseases endemic in the United States. *Am. J. Trop. Med. Hyg.* **23**:731–735.

Padchenko, I. K., and N. G. Stolyarchuk. 1969. On possible circulation of *Lamblia* in nature. *Prog. Protozool.* **3**:311–312.

Plan, B., L. Harter, and F. Frost. 1980. *Bibliography* Giardia *Literature*. Washington State Department of Social and Health Services, Office of Environmental Health Programs, Seattle.

Rentdorff, R. C. 1954. The experimental transmission of human intestinal protozoan parasites. II. *Giardia* cysts given in capsules. *Am. J. Hyg.* **29**:209–220.

Sheffield, H. G., and B. Bjorvatin. 1977. Ultrastructure of the cyst of *Giardia lamblia*. *Am. J. Trop. Med. Hyg.* **26**:23–30.

Takano, J., and J. H. Yardley. 1965. Jejunal lesions in patients with giardiasis and malabsorption. An electron microscope study. *Bull. Johns Hopkins Hosp.* **116**:413–429.

Thompson, R. G., D. S. Karandikar, and J. Leek. 1974. Giardiasis. An unusual cause of epidemic diarrhoea. *Lancet* **7858**:516–616.

Uzmann, J. R., and S. H. Hayduk. 1963. *In vitro* culture of the flagellate protozoan *Hexamita salmonis*. *Science* **140**:290–292.

Wright, R. A. 1975. Giardial infection from water. *Ann. Intern. Med.* **82**:589–590.

Yardley, J. H., and T. M. Bayliss. 1967. Giardiasis. *Gastroenterology.* **52**:301–302.

# 8

# Termite and Wood Roach Symbiotes

Throughout this book we deal almost entirely with those symbiotes that are harmful to their hosts, that is, those considered to be parasites. In the chapter on enteric protozoa of humans (Chap. 9), we emphasize that the array of symbiotes found in any host are usually commensals, which do not harm the host in any measurable way. Earlier (Chap. 1) we alluded to the work of Lincicome (1971), who set forth the hypothesis that the long association between a symbiote and its host will lead not to commensalism but rather to mutualism. In short, both the symbiote and the host would be benefitted if the symbiote produced something that helped the host leave more offspring.

A number of prokaryotes and eukaryotic protists provide their hosts with substances required for normal life. Many insects have bacteria in the cells of their intestinal tracts that are essential for the insect's life. Herbivores such as large ruminants, horses, plant-eating rodents, and rabbits rely on bacteria and protozoa for digesting cellulose.

Even though cellulose is one of the most common organic compounds in our environment, few animals are capable of digesting it. Some mollusks such as snails digest cellulose, and some insects have cellulases that allow them to feed on plant material and digest it directly. No vertebrates can digest cellulose without the aid of microorganisms. In large herbivores, ciliates, bacteria, and yeasts break down the plant material they eat (Chap. 11). In wood-eating insects such as termites (order Isoptera) and wood roaches (order Blattaria), digestion of cellulose is carried out by flagellated protozoa, and this association provides one of the best examples of mutualism to consider. Termites and roaches are both hemimetabolic insects (Chap. 50) that have a common ancestor in an extinct roach.

## TERMITE SYMBIOTES

It is sometimes stated categorically that all termites have flagellated protozoa that digest the cellulose they eat. This is true in the families of lower termites (Kalotermitidae, Mastotermitidae, Hodotermitidae, Rhinotermitidae), but not of the higher termites (Termitidae). Members of the family Termitidae are far more numerous than the other families, including three quarters of the known species (Krishna, 1970). These higher termites feed on plant materials but can digest only cellulose that has first been broken down by fungal or bacterial action. They do have protozoa in their intestines, but there is no evidence that these serve significantly in digesting cellulose.

Much of the work on termite symbiotes was done before the 1960s, and the details of their interrelations have been reviewed by Honigberg (1967, 1970). Kirby, Cleveland, and Hungate undertook a major series of studies defining relations between host and symbiote and aspects of cytogenetics, many of which are cited in Honigberg (1970). Yamin (1979) has cataloged all of

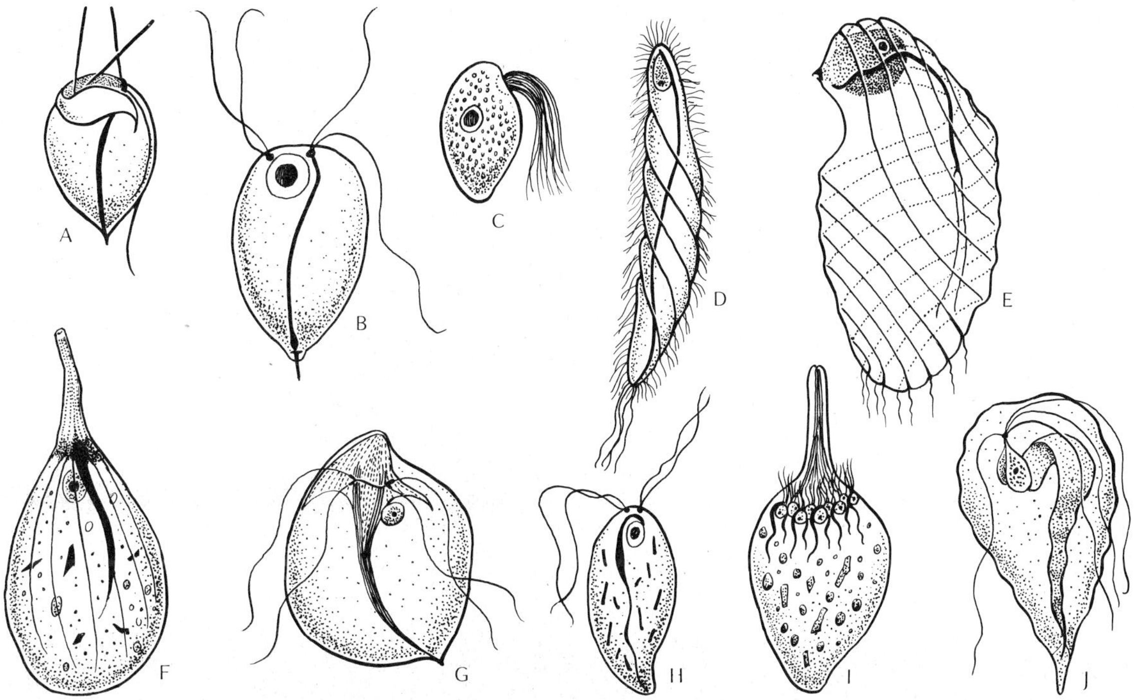


**FIG. 8.1**
Some symbiotic flagellates. (*A*) and (*B*) *Monocercomonoides;* (C) *Callimastix;* (D) *Dinenympha;* (*E*) *Pyrsonympha;* (*F*) and (G) *Oxymonas;* (*H*) *Polymastix;* (*I*) *Proboscidiella;* (*J*) *Saccinobaculus.*

the flagellates from the lower families of termites and *Cryptocercus* in such a way that if either the host or the mutualist is known, one can readily find the pertinent description. Termites have the usual types of protozoa found in the intestines of nearly all higher metazoa: flagellates, amebas, gregarines, coccidia, and microspora. It is the flagellates that are involved in cellulose digestion, and many are structurally complex. The orders containing termite flagellates are Oxymonadida, Trichomonadida, and Hypermastigida. The characteristics of these groups are given at the end of Chapter 2. Figures 8.1, 8.2, and 8.3 illustrate a selection of symbiotic flagellates.

The host range of many of the flagellates is rather narrow, and it appears that the symbiotes and their hosts have evolved together. Insects arose in the Paleozoic era, about 250 million years B.P., and it appears that the Hypermastigida had completed differentiation by the Mesozoic, about 150 million years B.P. Both morphologic and host-range evidence indicate that the Hypermastigida were associated with the extinct roach that gave rise to both the termites and the roaches extant today.

None of the flagellates is known to produce cysts, except when gamonts are formed, and transmission is thought to occur principally by proctodeal feeding, since the protozoa are found in large numbers in the hindgut. In some instances, the lining of the hindgut is retained at molting; the flagellates contained in it are carried over, and then escape and recolonize the gut.

The essential role of the flagellates in digestion of cellulose was shown in studies in which the protozoa were eliminated from the gut of termites by exposing them to high oxygen tensions, high temperature, or starvation. The flagellates were eliminated, but the bacteria and fungi also found in the gut were retained. When defaunated termites were fed wood that had not undergone decomposition or cellulose, they survived only for short periods. Evidence of what carbon sources are provided to the termite by its flagellates is not unequivocal. It is clear that the flagellates have a cellulase, but whether they provide glucose or acetate is not certain; both have been found in various studies. Some success has been achieved in cultivating a few flagellates (Yamin, 1978), and determining their secreted end-products should be feasible.

The flagellates seem not to be a source of protein for the host because individual protozoa live for a long time. In ruminants, the ciliates involved in cellulose digestion pass into the lower part of the digestive tract in large numbers and serve as a significant source of protein for the host. This does not seem to occur in termites, but it is not known what their source of protein is.

## WOOD ROACH FLAGELLATES

The wood roach, *Cryptocercus*, harbors 16 genera of flagellates, some of which are shared with termites. Wood roaches are found in deciduous forests where they live in fallen, decaying trees; because they are wood-eaters, they have the same relationship with their enteric fauna as the termites.

Many of the flagellates of wood roaches undergo sexual reproduction; these phenomena were studied in detail by Cleveland, who abstracted much of his work

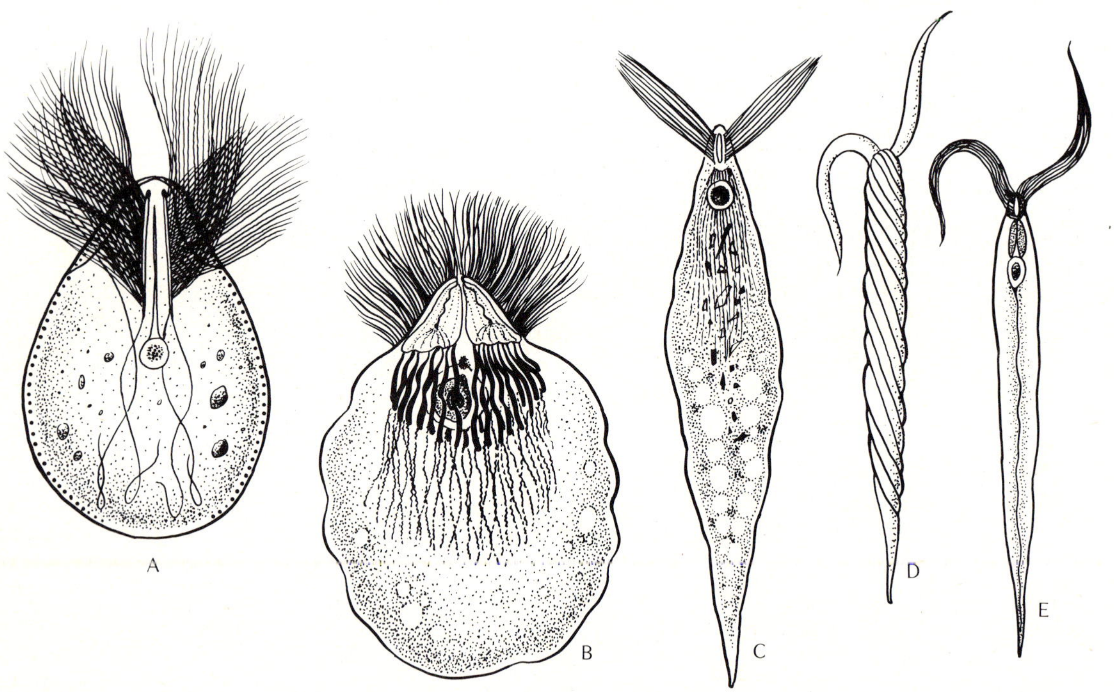


**FIG. 8.2**
Some symbiotic flagellates. (*A*) *Staurojoenina;* (*B*) *Barbulanympha;* (C) *Urinympha;* (*D* and *E*) *Hoplonympha.*

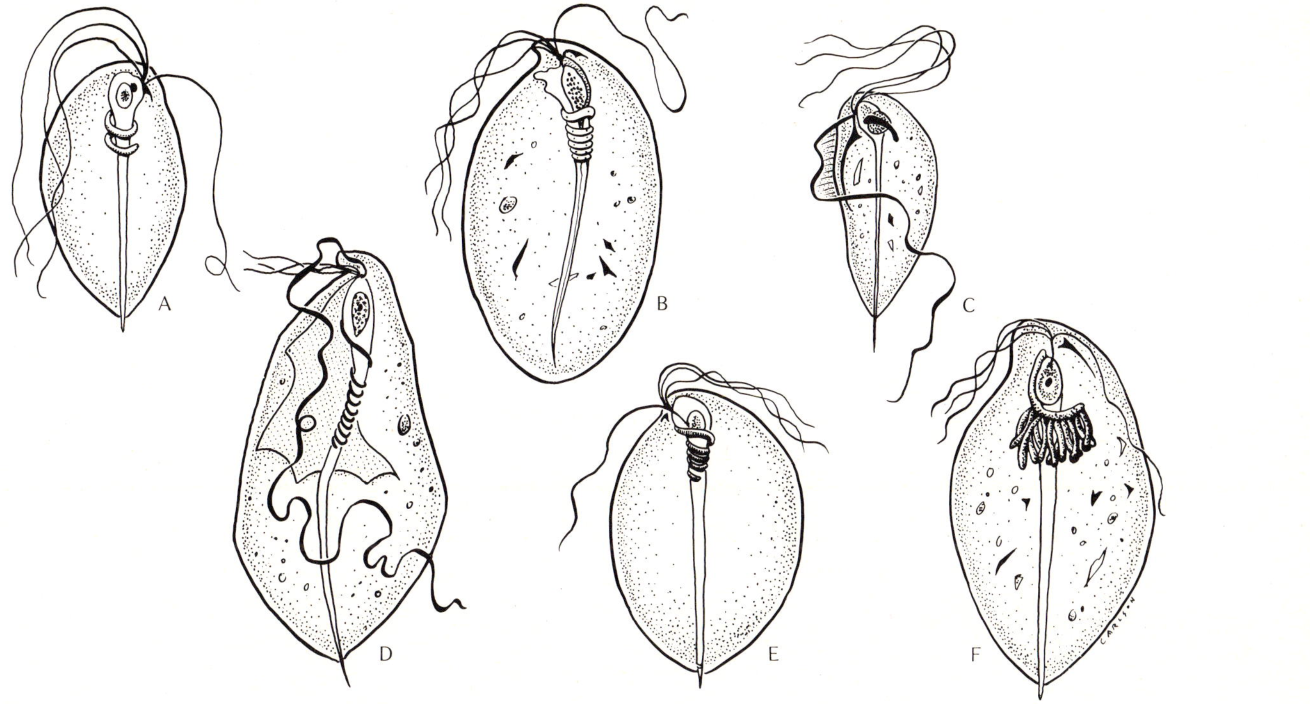


**FIG. 8.3**
Some symbiotic flagellates. (*A*) *Metadvescovina;* (*B*) *Caduceia;* (*C*) *Foaina;* (*D*) *Macrotrichomonas;* (*E*) *Hyperdevescovina;* (*F*) *Pseudodevescovina.*

in a summary paper (Cleveland, 1956). The initiation of sexual reproduction in some ten genera Cleveland studied is controlled by the insect molting hormone ecdysone (Chap. 48). The flagellates in adult termites, or in those that are not in some stage of the molting period, never show sexual activity. There are a number of other instances in which the activities of symbiotes are correlated with the host's: reproduction of rabbit fleas only when the doe host is pregnant (Chap. 53), migration of *Toxocara canis* larvae in the pregnant bitch, and conjugation of the ciliate *Nyctotherus cordiformis* of frogs only during the metamorphosis of the tadpole host.

*Trichonympha* is found in both wood roaches and termites, but it shows sexuality only in wood roaches. The organism spends the greater part of its life cycle in the haploid condition (Fig. 8.4), and when the sexual phase of the life cycle begins, it encysts and loses most of its organelles. The chromosomes are duplicated and then undergo meiotic division, in which the daughter chromosomes can be seen to stain differently from one another; the lighter-staining chromosomes go to one pole and form the female gamont and the darker ones form the male gamont. The gametes excyst and then a male gamete enters a fertilization ring at the posterior end of the female. The male disintegrates, its haploid nucleus migrates to the female nucleus, and fuses with it. Two meiotic divisions occur, and four haploid individuals are produced.

In *Barbulanympha*, much the same cycle is seen (Fig. 8.5) as in *Trichonympha*, but some of the individuals undergo autogamy instead of having complete separation of the gametes. In addition, endomitosis may also occur. There are a number of other

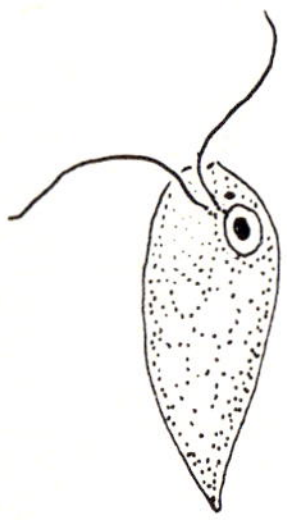

A

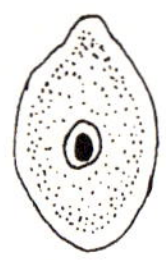

B

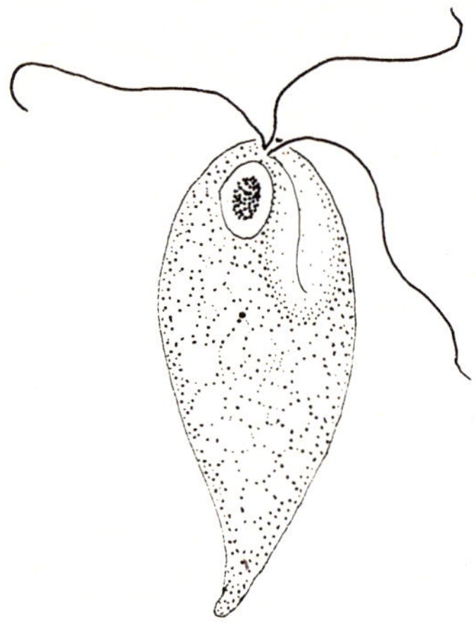

C

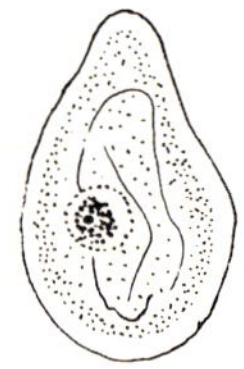

D

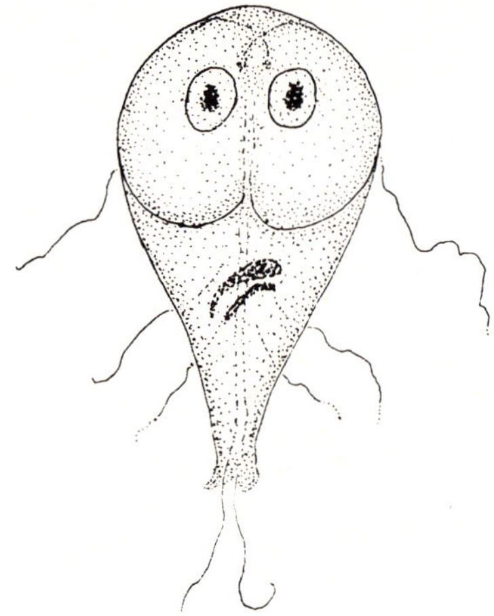

E

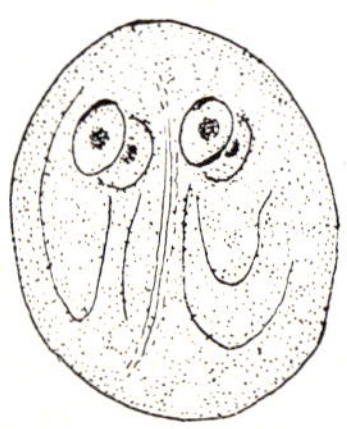

F

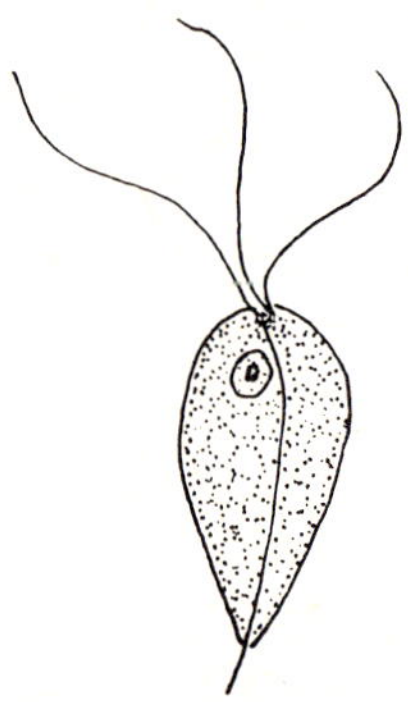

G

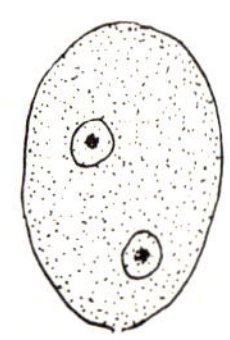

H

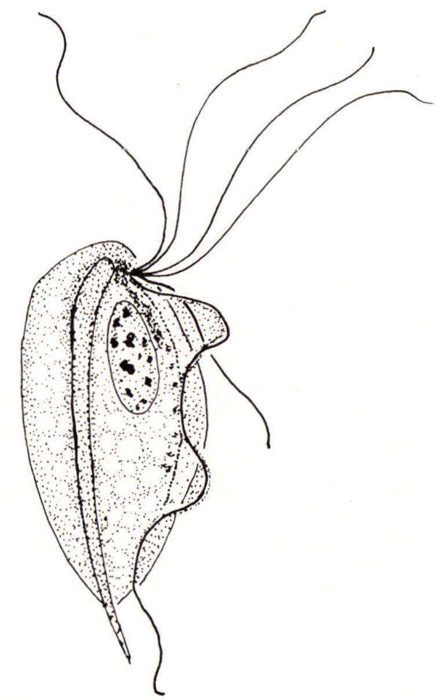

I

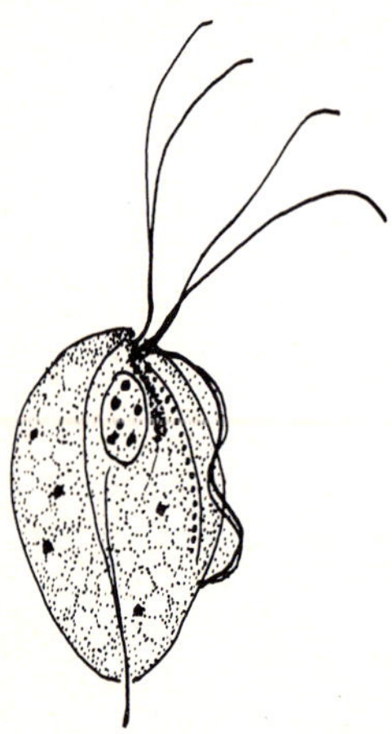

J

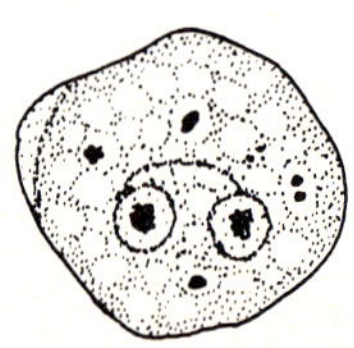

K

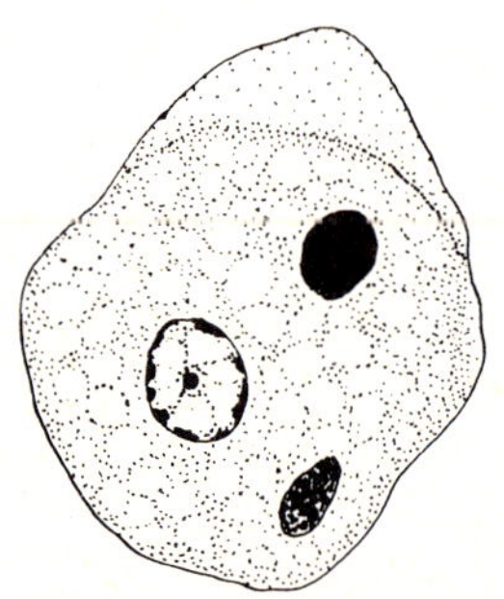

L

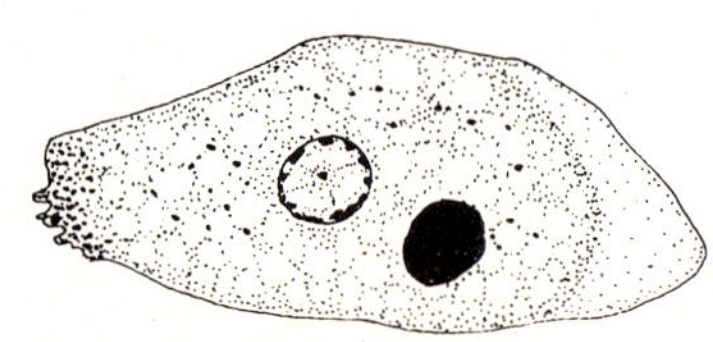

M

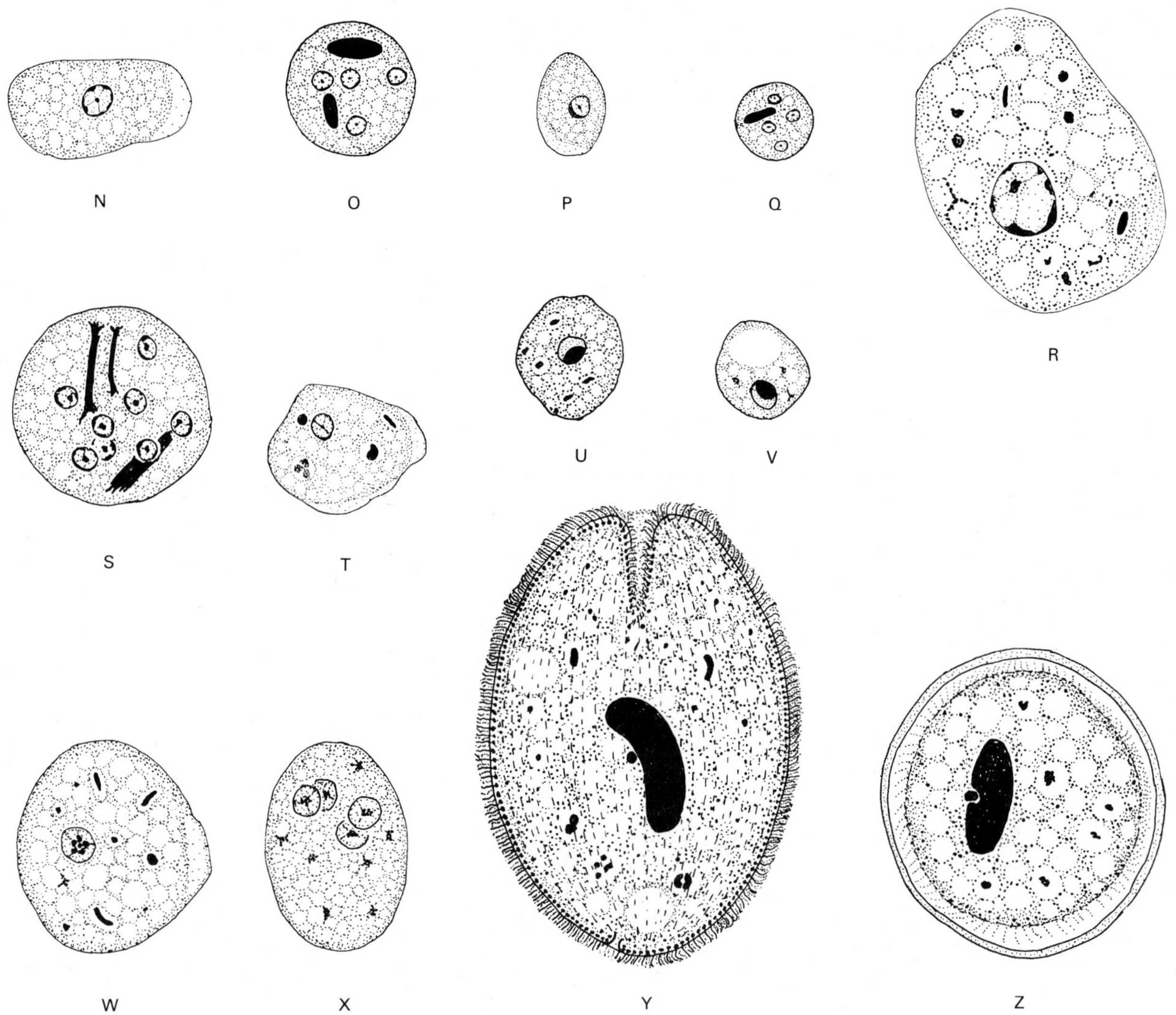


**FIG. 9.1**
The enteric protozoa of humans as seen in smears stained with iron hematoxylin. Some structures are shown that ordinarily do not stain well with hematoxylin. (*A*) *Retortamonas intestinalis*, trophozoite; (*B*) cyst; (*C*) *Chilomastix mesnili*, trophozoite; (*D*) cyst; (*E*) *Giardia duodenalis*, trophozoite; (*F*) cyst; (*G*) *Enteromonas hominis*, trophozoite; (*H*) cyst; (*I*) *Pentatrichomonas hominis*, trophozoite; (*J*) *Trichomonas tenax*, trophozoite; (*K*) *Dientamoeba fragilis*, trophozoite; (*L*) and (*M*) *Entamoeba histolytica*, large race trophozoite; (*N*) *Entamoeba histolytica*, small race trophozoite; (*O*) *Entamoeba histolytica*, small race cyst; (*P*) *Entamoeba hartmanni*, trophozoite; (*Q*) *Entamoeba hartmanni*, cyst; (*R*) *Entamoeba coli*, trophozoite; (*S*) *Entamoeba coli*, cyst; (*T*) *Entamoeba gingivalis*, trophozoite; (*U*) *Iodamoeba buetschlii*, trophozoite; (*V*) *Iodamoeba buetschlii*, cyst; (*W*) *Endolimax nana*, trophozoite; (*X*) *Endolimax nana*, cyst; (*Y*) *Balantidium coli*, trophozoite; (*Z*) *Balantidium coli*, cyst. Magnification: A–K, 2700X; L–X, 1300X; Y and Z, 700X.

the period 1962 to 1972. During 1962 to 1969 an average of 67 deaths was reported (Neva, 1974). Amebiasis is not a reportable infection, so the figures for numbers of infections are undoubtedly low; however, the 67 deaths per year may more nearly represent the real situation for the United States. Although there are continuing sporadic cases of amebiasis in humans, occasional severe local outbreaks may represent an acute problem for the communities involved.

MORPHOLOGY. The trophozoites of large race *E. histolytica* range fom 20 to 30 μm in diameter; some have been reported as large as 60 μm (Fig. 9.1*L*). The small race range from 12 to 15 μm (Fig. 9.1*N*). Other than size difference, the two races are indistinguishable from one another (Table 9.1). There is a thick, clear ectoplasm and a granular endoplasm that may contain ingested red blood cells. The single nucleus is about 4 μm in diameter. The chromatin is located at the nuclear membrane and is generally distributed evenly around the periphery. A small endosome is usually located in the center of the nucleus.

The nucleus is not clearly seen in the living trophozoite, but the motility of *E. histolytica* is distinctive. The trophozoites form "eruptive pseudopodia," in which there is a rapid extrusion of a broad, ectoplasmic pseudopod formed in just a second or two before the granular endoplasm flows into it.

The cysts of the small race are 10 to 20 μm in diameter, averaging 12 μm (Fig. 9.1*O*). They are spherical, with a distinct outer cyst wall. The cytoplasm is evenly granular and contains no food inclusions. When first passed, the cyst is uninucleate, but over about a 24-hr period, the nucleus divides twice and most cysts will have four nuclei. The nuclei are 2 to 3 μm in diameter and have the same structure as the nucleus of the trophozoite. The chromatoid bodies have rounded ends.

LIFE CYCLE. *E. histolytica* divides by simple binary fission in the intestine of its host. Cysts are formed on occasion and pass out with the feces. Another host is infected on ingestion of the cysts, and excystation takes place in the small intestine. The tetranucleate organism divides almost immediately into four small amebas, which are then carried to the large intestine, where they establish.

DIAGNOSIS. The principal means of determining whether a person is infected with *E. histolytica* are the following:

1. Finding the cysts or trophozoites in the feces
2. Cultivating the organisms from feces
3. Performing serological tests
4. Determining clinical signs and symptoms

Determining whether a person or an animal is infected with *E. histolytica* or other intestinal protozoa must be approached with care in laboratory procedures, and requires skill in identifying organisms that may be present. For example, if the host has received medication of any kind, the numbers of protozoa may be reduced to the point where they cannot be found until several days later. If a fecal specimen cannot be examined within 30 min of being passed, it must be properly preserved for examination later. Persons working with patients or in diagnostic laboratories are referred to detailed explanations such as that of Garcia and Ash (1979) for proper procedures and precautions to be taken.

Since the organisms live in the large intestine, it is often possible to find the living amebas in a freshly passed stool. The formation of pseudopodia in the typical explosive fashion is a good indication that *E. histolytica* has been found. However, permanent stained slides must also be made so that the organisms can be measured and nuclei examined for structural details. The slides also provide a permanent record for verification of the identification.

When a person is showing diarrhea, there are normally no cysts in the feces; trophozoites only will be seen at this time. Cysts are seen when the person passes a normal formed stool.

Cysts are found by using either a sedimentation or a flotation procedure to separate them from the bulk of fecal debris.

Culture techniques are usually used when more direct methods fail. A small amount of fecal material is placed in a culture medium, such as Balamuth's egg infusion plus antibiotics, to suppress growth of bacteria. When incubated at 37°C, protozoa usually multiply sufficiently by 48 hr to be found readily in the bottom of the culture tube.

Serological techniques are most useful in detecting extraintestinal amebiasis or in differentiating among species of amebas. Amebas are sometimes carried from the colon to other sites in the body such as the liver. The organisms multiply and form abscesses, which create a serious health problem. Although symptoms are suggestive, a strong serological reaction gives the physician a good indication of what the problem may be.

CULTIVATION. *E. histolytica* was first cultivated in a biphasic medium consisting of a coagulated egg slant overlaid with horse serum – Ringer solution (Boeck & Drbohlav, 1925). The egg yolk infusion medium developed by Balamuth (1946) is easier to prepare and more reliable. Other media have been developed as well, but they all have the same drawback: the amebas are cultivated in the presence of bacteria. Although certain kinds of studies can be done and the media are useful for diagnostic purposes, axenic cultivation is a goal that many investigators have worked toward.

Diamond was the first to cultivate *E. histolytica* axenically and, over the years, the method has been refined (reviewed by Diamond, 1968). It is worth considering Diamond's approach. *E. histolytica* can be grown in the absence of bacteria in a fluid medium together with a kinetoplastid, either *Trypanosoma* or *Crithidia* (Phillips, 1950). Diamond established cultures with *Crithidia* from single bacteria-free cysts of *E. histolytica*. Once they were growing well in monoxenic culture, the trophozoites were transferred to a biphasic medium. The result was that the *Crithidia* died out over a period of several transfers and the amebas were left as the only living agents in the culture. The original medium contained, in addition to balanced salts and commercially available bacterial media, chick embryo extract and horse serum. Since that time the medium has been simplified, and it has been made more reliable for the cultivation of diverse isolates of *E. histolytica*.

HOST – PARASITE INTERACTIONS. *Entamoeba histolytica* is found as two races: a large, pathogenic race and a small, noninvasive, nonpathogenic race (Table 9.1); however, this concept may not be entirely correct (see Elsdon-Dew, 1968). The small race lives as a commensal in the colon of the host and does not cause damage. The large race invades and can cause a great deal of damage to the colon and other body organs of the host.

Invasiveness is sporadic and there are many theories on the number of species represented by *E. histolytica* or related organisms (Table 9.2). Elsdon-Dew (1968) made a persuasive case for the large, invasive race being derived from the small, noninvasive race; we are calling this the *neounicistic hypothesis*. The small race is never found to be invasive and forms cysts. Once having invaded the tissue, the large race loses its ability to form cysts. Invasiveness is considered to be an abnormal activity for *E. histolytica*; once having entered the tissue, it can no longer perpetuate itself because it loses the form essential for transmission to another host. If this is true, invasiveness is a nonadaptive variation for the organism.

When the amebas become invasive, there is a gradual onset of signs and symptoms. Usually abdominal tenderness and diarrhea are experienced at about the same time. Diarrhea may progress to dysentery, in which there is passage of blood and mucus over a period of weeks. Diarrhea may be interspersed with constipation. If the infection progresses so that there is increasing damage to the large intestine, there may be severe loss of blood and electrolytes. Finally, peritonitis may result if the intestinal wall is perforated, allowing bacteria to reach the abdominal cavity.

When amebas begin to invade the colon, they form small ulcers at the surface of the intestinal epithelium. These ulcers become deeper and spread laterally as the colonies of amebas enlarge. Portions of the intestinal mucosa or epithelium lack adequate circulation because of this undermining and sheets of tissue may slough off. There are two effects: bleeding occurs and the gut loses its first line of defense, the intact epithelium. In severe invasion, the amebas reach the submucosal and muscular layers of the gut and cause the formation of large necrotic areas in the gut wall.

The factors that cause the amebas to become invasive have been the subject of considerable study and debate. There are at least two variables in the system, the host and the parasite, but in addition, environmental conditions have an influence on how the partners interact. Let us look at the components separately to find some patterns.

Human susceptibility can be viewed as a continuum ranging from those few in the population who may be completely resistant to those who will be moderately susceptible to those few at the other extreme who will succumb if exposed to the agent. A number of postmortem studies have shown that older individuals (60 years and older) in the population are unlikely to harbor amebas. Although the reasoning is not flawless,

**TABLE 9.2**
Concepts of Pathogenenis in Amebiasis

| HYPOTHESIS | UNICISTIC (1913) | DUALISTIC (1925) | NEODUALISIC (1957) | NEOUNICISTIC (1968) |
|---|---|---|---|---|
| Nonpathogenic amebiasis | *E. histolytica* Small race | *E. hartmanni* | *E. hartmanni* | *E. hartmanni* |
| | Large race, commensal phase ("minuta" form) | *E. dispar* | *E. histolytica* Avirulent, large race | *E. histolytica* Avirulent, cyst-forming small race |
| Pathogenic amebiasis | ↕ | | *E. histolytica* Virulent, large race | ↓ |
| Subclinical | | *E. dysenteriae* | Dormant in lumen of gut ↓ | |
| Clinical | Virulent, tissue phase ("magna" form) | | Invades gut wall | Virulent, non-cyst-forming, large race |

it could be concluded that these older persons represent the resistant portion of the population and have survived longer than the more susceptible individuals. It has also been observed that British military personnel stationed in India suffered more frequently from amebiasis than their native counterparts. It is likely that the British population had not been subjected to the selective pressure of living in an endemic area where the risk of infection with *E. histolytica* was high; the Indian population had been subjected to that pressure and the highly susceptible persons in the population had been removed, leaving the more resistant individuals to propagate and increase genetic resistance in the population. In the experimental infection studies done in humans, the shortest prepatent periods have been a few days, but in some persons, amebas were not found for more than 3 months. Thus, the same isolate reproduced better in some persons than in others.

Bacterial associates play a role in the ability of amebas either to establish in a host or to cause damage once established. Amebas are not generally capable of establishing in a gnotobiotic (germ-free) animal. Likewise, those carried in culture with the protozoon *Trypanosoma cruzi* were incapable of establishing in experimental animals unless they were first returned to cultures containing bacteria.

In culture, amebas tend to lose the ability to invade the tissues of experimental animals, but if they are transferred rapidly from one host to the next, invasiveness is regained. Rapid transfer is a technique used to increase virulence in bacteria that have been maintained in culture for a long period, and evidently the same principle pertains to *E. histolytica*. What change occurs in the amebas is a matter for conjecture.

Some of the other factors that have been pointed to as providing conditions conducive to invasion by amebas are the following:

1. Temperature fluctuations in the host
2. Abnormal secretory function
3. Irritant foods
4. Inadequate diet (i.e., low protein)
5. Inflammatory and erosive processes

It should be borne in mind first that many of these factors have been surmised from observations on individuals who were ill. Further, even if they are valid observations, a mechanism is not obvious, with one exception: in the case of erosive processes such as ulcerative colitis in which the integrity of the epithelium is lost and the amebas can readily gain access to the tissue.

The difficulties in establishing firm principles on invasiveness can be appreciated when one considers the problems encountered in identifying quite similar organisms such as the large and small races of *E. histolytica* and *E. hartmanni*. Experimental studies and good clinical observations indicate different susceptibilities of individuals and populations. There are some tantalizing data on the role of bacteria in promoting the invasiveness of the amebas, but the mechanisms of what happens in the colon of an infected person are not at all clear.

Immunity probably does not play a role in resistance to invasion by *E. histolytica*. An infected person develops antibodies, and these can be revealed by a number of precipitation and fluorescent antibody methods; however, antibodies do not appear to be protective either in clearing an individual of infection or in preventing a subsequent infection.

Meerovitch (1982) has correlated work done by him and his co-workers and that of a number of other investigators, and has presented the hypothesis that *E. histolytica* and other members of the genus were originally parasites of herbivores and are poorly adapted to carnivorous hosts. The amebas also inhabit a tiny niche in the gut, neither in the lumen nor in the paramucosal lumen, but rather in an area where the conditions are microaerophilic. His hypothesis states that when the oxygen tension in the intestinal tissue declines slightly, the amebas move closer to the tissue and tend to become invasive. It is an interesting concept that merits further investigation.

Students may believe that we know everything important about the principal disease agents of humans and domesticated animals. But here we have an agent that infects, conservatively, 5% of the people of the world, thus 150 million persons may be infected at any one time. And at the end of thousands of studies we have some quite general theories about invasiveness that scientists and physicians agree on, but that are nevertheless so general they have only limited usefulness. There is more to be learned about even common disease agents.

Extraintestinal amebiasis usually occurs in persons with severe amebic dysentery. When the colon is damaged to the point where amebas reach the submucosal area of the tissue, the amebas can gain access to blood and lymph vessels, and are carried by the caval and portal circulations to other parts of the body.

The principal site of extraintestinal infection is the liver. The organisms invade the liver tissue and form abscesses that increase in size, progressively damaging the liver. Affected persons usually have moderately severe symptoms which, together with physical signs and laboratory tests for liver function, will usually lead to a proper diagnosis. Because the liver is such an important organ, early diagnosis and treatment is essential for an uneventful recovery.

EPIDEMIOLOGY AND CONTROL. Transmission of *E. histolytica* takes place principally by ingestion of contaminated food or water. In some instances, transmission may occur through close contact between infected and uninfected persons such as members of the same household, or by food handlers. However, the main lines are through faulty sewage disposal systems or direct fecal contamination of food.

The cyst is, for practical purposes, the only important stage in transmission. The trophozoite survives for only a short time outside the host and cannot survive the low pH in the stomach.

Cysts are susceptible to freezing and to temperatures above 50°C (Table 9.3). Cool temperatures are most conducive to long-term survival, and as the temperature exceeds 30°C, the death rate increases. It is clear also from the table that access to oxygen is important for survival; at comparable temperatures, survival is better in water than in feces. Soil is likely to have adequate oxygen, except where there is excessive moisture or organic material at the surface, and survival in soil has been good in the few studies reported.

When human excreta is used as fertilizer for crops, it appears that holding the material in a tank for 2 weeks would be sufficient to kill all of the cysts. All other infectious agents might not be killed, however.

Knowledge of the resistance of *E. histolytica* cysts to chemicals is of importance in the treatment of potable water and disinfection of objects and areas that may be contaminated. The objective of chlorination of water is to kill enteric bacteria; coliform bacteria are killed by concentrations of chlorine of from 0.1 to 1.0 mg/l. On the other hand, *E. histolytica* requires 5 mg/l for an hour to kill the cysts. In another study, cysts remained alive after 30 min exposure to 10 mg/l of chlorine (Becker et al., 1946). Therefore, the usual levels of chlorination of drinking water are insufficient to kill cysts of *E. histolytica.*

**TABLE 9.3**
Survival of Cysts of *Entamoeba Histolytica*

| TEMPERATURE (°C) | MEDIUM | TIME EXPOSED | RESULT |
|---|---|---|---|
| −28 | Water | 7.5 hr | Dead |
| −15 | Water | 24 hr | Dead |
| 0 | Feces | 10 days | Dead |
| 0 | Water | 17 days | Alive |
| 0 | Water | 90 days | Alive |
| 2 – 6 | Water | 40 days | Alive |
| 10 | Water | 30 days | Alive |
| 16 – 20 | Feces | 10 days | Dead |
| 16 – 20 | Water | 10 days | Alive |
| 20 | Water | 10 days | Alive |
| 27 – 30 | Feces | 9 days | Dead |
| 27 – 30 | Feces | 4 days | 50% alive |
| 28 – 34 | Soil | 8 days | Alive |
| 30 | Water | 3 days | Alive |
| 37 | Feces | 3 days | Dead |
| 46 – 47 | Water | 1 hr | Alive |
| 52 | Water | 1 min | Dead |
| 68 | Water | 5 min | Dead |
| 100 | Water | 5 – 10 sec | Dead |

Adapted from several sources.

Certain common disinfectants are active against cysts of *E. histolytica* but the concentrations required reveal the rather high resistance of the cysts to chemical insult. Cresol (isomers of methyl phenol) kills cysts at a concentration of 1:30 in 1 min and at 1:100 in 30 min; the common disinfectant Lysol is a mixture of soap and cresol. Phenol at a dilution of 1:40 kills cysts in 15 min and at 1:100 in 7 hr. A 50% solution of ethyl alcohol kills cysts within an hour, probably by desiccation, and demonstrates the susceptibility of cysts to drying.

Food handlers have been implicated in the transmission of *E. histolytica*, and they are important in some circumstances. A 12-fold reduction in the prevalence of infection in Americans working in an oil camp in Venezuela was obtained when infected food handlers were identified and treated. However, in another study, infected personnel in the U.S. Navy were not implicated in transmission of amebiasis. Likewise, although 52.4% of workers in a factory in South Bend, Indiana became infected through faulty plumbing, only 3.7% of their family contacts were found to be infected, a level no different from that of the population in general in the community. The differences probably lie in the personal habits of the food handlers and in sanitary practices such as hand washing.

Food may be contaminated with cysts of *E. histolytica* through the use of human feces as fertilizer, of irrigation water contaminated with feces or water, or of ice in processing and shipping of fresh fruits and vegetables. Since cysts are killed readily by boiling water (Table 9.3), hazards of infection lie principally in fruits and vegetables normally eaten raw.

Certain domiciliated insects may play a role in the transmission of *E. histolytica* cysts. Both filth flies and flesh flies have been found to harbor cysts for about 2 days in their intestines when fed them experimentally. Also, flies harboring cysts have been caught in houses where infected persons lived. Cockroaches have also been fed cysts, and cultures were obtained as long as 48 hr after feeding.

An outbreak of amebiasis may have aspects that do not fit textbook descriptions. A case in point was an outbreak of amebic dysentery in a wood-working factory in South Bend, Indiana (Brooke et al., 1955). The factory employed more than 1500 persons; over the 2-year period 1951 – 1953, 48 persons working in the factory were diagnosed as having clinical amebiasis. Of 1542 employees examined, 52.4% were found to be infected.

When public health epidemiologists investigated the premises, they found that there were three sources of water: (1) a 40-ft (12-m) well used for a fire sprinkler system; (2) a 118-ft (36-m) well for potable, boiler, and industrial water; and (3) city water as a backup. Neither the sprinkler system nor the city water were considered to be sources of infection. The remaining source of water came from a well situated 80 ft (25 m) from a pumping station and distribution point (Fig. 9.2). The water was pumped from the well to a 1300-gal (4900-l) holding tank. Water was then moved through a 6-in. (15-cm) cast iron pipe by suction pumps to a pressure tank and out through the distribution system. The sewer line roughly paralleled

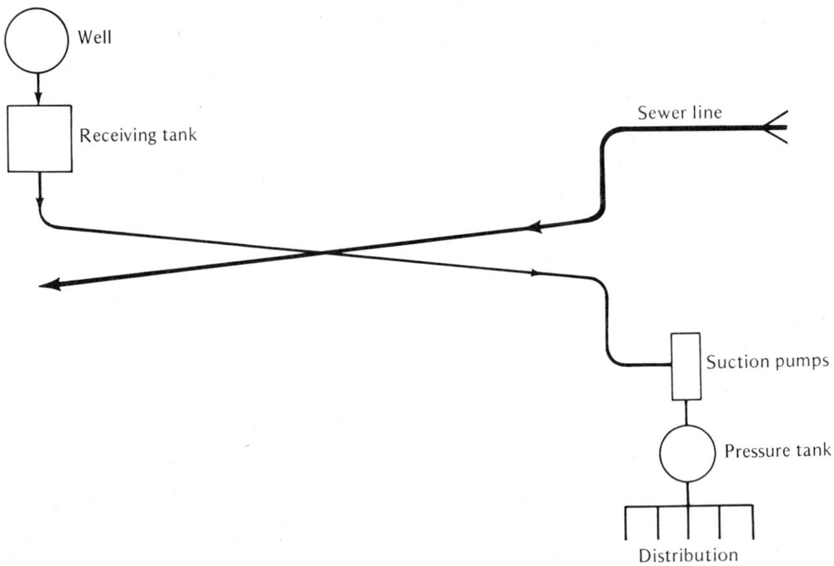


**FIG. 9.2**
Relation of water and sewer lines in a woodworking factory where an outbreak of amebiasis occurred.

the suction line, but was thought to be separated from it by 5 ft (1.5 m). Some problems had been experienced earlier with bacterial contamination of the water, but they had been taken care of by chlorination.

Ultimately, it was found that the water line and the sewer line were no more than 5 in. (12 cm) apart and that the sewer line leaked. Although no break in the water line was found, fluorescent dye moved from the sewer into the water line when the suction pumps were operated.

In this case, the systems were well designed and maintained within acceptable limits. The outbreak occurred because of a slight leak in the sewer, and a negative pressure as great as 15 in. (381 mm) of mercury developed when moving water from the receiving tank at the well to the pumping station. It was generally felt that if the sewer and water lines had been separated by 2 m, the filtering action of the soil would have been sufficient to prevent transfer of cysts from the sewer to the suction line (see Offutt et al., 1955, and Brooke et al., 1955, for details of this outbreak).

During 1933 there was an outbreak of amebic dysentery associated with two adjacent hotels in the city of Chicago. The following summary is based on the meticulous report of Bundesen et al. (1936). During the period from 1 June 1933 to 30 June 1934 there were 1409 confirmed cases of amebic dysentery originating in the city, and ultimately it was determined that 1209 of these persons had had contact with one or both of the hotels. Among these persons there were 62 deaths.

Investigation of the plumbing in the hotels showed that there were two major sources of potential contamination of the potable water: (1) cross-connections between water and sewer pipes and (2) dripping from a faulty joint in a sewer into an uncovered tank of chilled water; minor sources from (3) rusted sewerage, which allowed sewage to drip in food preparation areas; (4) back-siphonage from toilets into the water system; and (5) a sewer that ruptured during a heavy rainstorm also existed.

As a general rule, amebiasis does not follow an epidemic pattern, but in this instance deterioration and changes in the plumbing of the hotels led to contamination of the drinking water and food. The extent of the contamination is exemplified by the fact that 37.8% and 47.4% of the employees in the two hotels, respectively, were found to be infected.

In light of the foregoing discussion, there are a

number of steps that can be taken to prevent the transmission of *E. histolytica.*

1. Proper treatment of potable water. Since cysts are not susceptible to the usual levels of chlorination used for domestic water supplies, water should be passed through a sand filter; filtration will also remove other cysts such as those of *Giardia*. If such treatment is not done, water should be boiled for drinking, washing vegetables, and such things as brushing the teeth.
2. Proper disposal and treatment of human excreta. The best method of disposal is through a well-designed, modern sewage treatment plant. Second, properly constructed pit toilets will retain cysts and prevent their dissemination. If human excrement is used as fertilizer, the material should be held for at least 2 weeks in a tank so that normal bacterial action will hasten the death of the cysts. Promiscuous defecation or the use of fresh night soil on crops should be curtailed.
3. Proper installation and maintenance of water delivery and sewage systems. Water and sewage systems tend to grow and develop in a rather loosely controlled manner. Even in well-run communities, one finds that the plans in the city engineer's office do not match reality or that parts of the system are patched together in odd ways.

   Also, systems or parts of systems do not last forever. Sewers leak at the joints; water mains rust through. But usually nothing is done unless the street caves in or a disease outbreak occurs. The examples given here of amebiasis epidemics originating in a hotel and factory show that location of pipes and maintenance of the systems are of the utmost importance in prevention of disease.
4. Raising and preparing food free of contamination. As indicated, separation of the food and water supplies from human excreta is essential. This holds true at all steps in the chain from raising the crop to processing, cooking, serving, and cleaning utensils after eating. Food handlers have been shown to be sources of infection, especially if their personal habits of cleanliness are less than desirable; food handlers should be examined for infection and treated if necessary. Insects such as flies and cockroaches should be excluded from areas where food is stored or prepared.
5. The only other hosts of *E. histolytica* considered to be of significance are dogs. They are known to be naturally infected with *E. histolytica*, but there is disagreement on whether they play a role in the transmission of infection to humans. Until better data are developed, it is probably best to be on the safe side and consider the dog to be a reservoir.

Despite this list of ideal procedures, what are the practicalities of control? Amebiasis is widespread in all kinds of climates, but the prevalence of infection and disease is higher in tropical than temperate areas. The epidemic occurrence of amebic infections in Chicago and South Bend, as discussed earlier, show how minor malfunctions in a high-technology society can have catastrophic results. Even so, greater public health effects of amebiasis occur in less developed countries and tropical countries. In those areas, the prevalence of infection may exceed 80% and the ratio of persons with clinical signs and symptoms is greater than in temperate-climate, affluent countries. The construction and maintenance of water and sewerage systems require both capital and personnel trained in sanitary engineering. Both are likely to be in short supply in a developing country, and it may be necessary to retreat to a level of appropriate technology.

At a public health level, prevention of indiscriminate defecation, tanking excrement before use as fertilizer, and requiring food handlers in restaurants to wash their hands are all feasible in communities ranging in size from a village of a few hundred persons to a city.

At a personal level, boiling potable water, not eating fresh fruits and vegetables that cannot be peeled, and keeping insects out of the home can all be done by persons who recognize what the problem is and are motivated to carry these measures through. Therein lies the difficulty. Where amebiasis has a high prevalence, people are faced with a multitude of problems, perhaps the least of which is the prevention of disease.

NAME OF ORGANISM AND DISEASE CAUSED. *Entamoeba hartmanni* (synonym: *E. histolytica* in part) is probably nonpathogenic.

HOSTS AND HOST RANGE. Probably the same as for *E. histolytica*.

DISTRIBUTION AND IMPORTANCE. Distribution is cosmopolitan. Its prevalence is rather difficult to determine since *E. hartmanni* was included in *E. histolytica* until recent years, and its small size also may lead to its being overlooked. In an extensive tabulation of prevalence of infections, *E. hartmanni* was generally rated equal to *E. histolytica*: for example, of 14,744 military veterans, 2.0% were infected with *E. histolytica* and 1.6% with *E. hartmanni*, and in Tennessee 6.2% of 2657 persons examined were infected with *E. histolytica* and 6.9% with *E. hartmanni* (Elsdon-Dew, 1968). The organism inhabits the large intestine of the host.

MORPHOLOGY. *E. hartmanni* is a small ameba and its nuclear structure is somewhat different from that of *E. histolytica* (Fig. 9.1*P*). The cysts are 4 to 10 μm (Fig. 9.1*O*); the division between cysts of *E. hartmanni* and *E. histolytica* is at 10 μm. Cytoplasmic inclusions in the trophozoites include bacteria.

LIFE CYCLE. The organism is transmitted as a cyst.

DIAGNOSIS. The principal problem in the diagnosis of this infection is the small size of the organism, which causes it to be overlooked. When found, it must be differentiated from *E. histolytica* (Table 9.1).

HOST–PARASITE INTERACTIONS. *E. hartmanni* is nonpathogenic or, at most, causes only mild symptoms of enteritis.

NAME OF ORGANISM. *Entamoeba coli* (Grassi, 1879) Casagrandi and Barbagallo, 1895 is nonpathogenic.

HOSTS AND HOST RANGE. Humans and other primates serve as hosts. The organisms live in the large intestine.

DISTRIBUTION AND IMPORTANCE. Distribution is cosmopolitan. The prevalence ranges from about 9 to 50%; the U.S. population has a 20 to 30% infection rate.

MORPHOLOGY. This organism is somewhat larger than *E. histolytica*, and the nuclear structure (Fig. 9.1*R*) allows it to be differentiated from *E. histolytica* and other amebas as well. The cysts are from 10 to 33 μm and, when mature, have eight nuclei (Fig. 9.1*S*).

LIFE CYCLE. Transmission from one host to another takes place by mouth through cysts.

DIAGNOSIS. Trophozoites or cysts may be found in the feces, and it is important to differentiate this species from *E. histolytica*.

HOST–PARASITE INTERACTIONS. Nonpathogenic.

THERAPY. Not required.

EPIDEMIOLOGY AND CONTROL. Since this organism is nonpathogenic, control is not required. Prevalence of *E. coli* in a population has sometimes been used to determine the prevalence of *E. histolytica* infections, since *E. coli* is somewhat easier to find.

NAME OF ORGANISM. *Entamoeba moshkovskii* Chalaya, 1941 is not a parasitic organism and has been found only in sewage. The trophozoites are 9 to 29 μm, usually averaging about 12 μm. Its nuclear structure is almost identical to that of *E. histolytica*. Cysts range in size from 7 to 17μm and have four nuclei when mature, but the chromatoid body is rather diffuse. This organism was first found in Moscow and has since been found in other parts of Russia, in Brazil, Britain, Canada, and the United States. The principal significance of *E. moshkovskii* is that it is morphologically similar to *E. histolytica*, and the trophozoites or the cysts isolated from sewage might be misidentified. On cultivation, *E. moshkovskii* grows optimally at 24°C, but grows poorly or not at all at 37°C, the normal temperature for cultivation of *E. histolytica*.

NAME OF ORGANISM. *Endolimax nana* (Wenyon & O'Connor, 1917) Brug, 1918, is a small ameba (Fig. 9.1*W*,*X*) that inhabits the large intestine of humans, where it feeds on bacteria. Transmission from one host to the next is through cysts, which are ingested with contaminated food or water. There is no evidence that *E. nana* is pathogenic for humans. The organism is found in perhaps 30% of the human population and is important because it is essential to differentiate *E. nana* from the other amebas that may be pathogenic to humans.

NAME OF ORGANISM. *Iodamoeba buetschlii* (von Prowazek, 1911) Dobell, 1911 (synonyms: *Iodamoeba williamsi*, *I. bütschlii*).

HOSTS AND LOCATION. Humans and other primates and swine are the hosts. The organism lives in the large intestine.

DISTRIBUTION AND IMPORTANCE. In most human populations, the prevalence of infection is 5% or less but probably is higher in warm climates. This organism can perhaps be invasive, but the number of cases on record is small.

MORPHOLOGY. The trophozoites are 4 to 20 μm and have nuclei with a large, eccentric endosome and a fine nuclear membrane. There are usually a large number of inclusions in the cytoplasm representing ingested food particles (Fig. 9.1*U*). The cyst is 6 to 15 μm and has a typical nucleus and usually a large iodinophilous vacuole; cysts are uninucleate (Fig. 9.1*V*).

LIFE CYCLE. Transmission from one host to the next takes place through cysts.

DIAGNOSIS. Cysts or trophozoites may be found in the feces.

HOST–PARASITE INTERACTIONS. *Iodamoeba* is usually considered to be nonpathogenic, and as long as it remains in the intestine, it seems to be. There are a few cases on record of the organism forming abscesses in organs other than the gut such as the liver, lungs, and brain, but the organisms may actually have been *Naegleria* or *Acanthamoeba* (Chap. 10).

TREATMENT. Not required.

EPIDEMIOLOGY AND CONTROL. Transmission from one host to the next is through food or water contaminated with cysts. There is some evidence that people may become infected from contact with swine. There is little in the literature indicating methods of control that may be used, but the same principles probably apply here as to *E. histolytica*.

## AMEBAS OF ANIMALS OTHER THAN HUMANS

Amebas are widely spread as parasites of vertebrate and invertebrate animals; however, as a general rule, any one host has relatively few species of amebas. A host species that might have 6 to 12 species of flagellates and apicomplexans will probably have only 1 or 2 species of amebas. With a few exceptions, these organisms are not known to be pathogenic.

*Entamoeba invadens* Rodhain, 1934 is a cosmopolitan parasite of reptiles, turtles, snakes, and lizards, whose intestinal tract and liver it inhabits. The trophozoites are 9 to 38 μm, averaging 16 μm, and the cysts range in diameter from 11 to 20 μm, averaging 14 μm. The morphology and formation of eruptive pseudopodia are quite similar to those of *E. histolytica*.

This is a pathogenic organism and may cause the death of the host. It is sometimes reported as a problem in captive reptiles in which it invades the stomach, small and large intestine, and liver. The organism is readily maintained in culture with bacteria. It has been raised axenically and has also been cultivated on a defined medium. *E. invadens* is especially useful in parasitology courses because it is structurally similar to *E. histolytica* but carries no hazard of human infection. *E. terrapinae* is similar to *E. invadens*, but its principal hosts are turtles and it is nonpathogenic.

Some amebas found in domesticated animals are *Entamoeba bovis* of cattle and *E. gallinarum* of gallinaceous birds such as chickens and turkeys. Swine have two species, *E. debliecki* and *Iodamoeba buetschlii*; the latter is also a parasite of humans.

Among companion animals, dogs are naturally infected with *E. histolytica* and perhaps some of the other enteric amebas of humans. Cats do not have amebas that they can call their own but can be experimentally infected with *E. histolytica*.

Amebas of invertebrates are not prominent parasites, but they are present in many species of hosts and some make good material for laboratory study. Cockroaches, for example, frequently have two species of amebas, *Endamoeba blattae* and *Entamoeba thomsoni*, which can be easily studied alive under a microscope.

This list is not exhaustive, and the student is referred to Kudo (1966) or Jahn et al. (1979) for a more extensive treatment.

## REFERENCES AND FURTHER READINGS

Balamuth, W. 1946. Improved egg yolk infusion for cultivation of *Entamoeba histolytica* and other intestinal protozoa. *Am. J. Clin. Path.* **16**:380 – 384.

Becker, E. R., C. Burks, and E. Kaleita. 1946. Cultivation of *Endamoeba histolytica* in artificial media from cysts in drinking water subjected to chlorination. *Am. J. Trop. Med.* **26**:783.

Boeck, W. C., and J. Drbohlav. 1925. The cultivation of *Endamoeba histolytica. Am. J. Hyg.* **5**:371 – 407.

Brooke, M. M., D. M. Melvin, R. Sappenfield, F. Payne, F. R. N. Carter, A. C. Offutt, and W. W. Frye. 1955. Studies of a water-borne outbreak of amebiasis, South Bend, Indiana. III. Investigation of family contacts. *Am. J. Hyg.* **62**:214 – 226.

Bundesen, H. M., J. I. Connoly, I. D. Rawlings, A. E. Gorman, G. W. McCoy, and A. V. Hardy. 1936. Epidemic amebic dysentery. The Chicago outbreak of 1933. *Nat. Inst. Health Bull.* No. 166.

Diamond. L. S. 1968. Techniques of axenic cultivation of *Entamoeba histolytica* Schaudinn, 1903 and *E. histolytica*-like amebae. *J. Parasitol.* **54**:1047 – 1056.

Elsdon-Dew, R. 1968. The epidemiology of amoebiasis. *Adv. Parasitol.* **6**:1 – 62.

Faust, E. C. 1960. The multiple facets of *Entamoeba histolytica* infection. *Intern. Rev. Trop. Med.* **1**:43 – 76.

Feria-Velasco, A., and N. Trevino. 1972. The ultrastructure of trophozoites of *Entamoeba histolytica* with particular reference to spherical arrangements of osmiophilic cylindrical bodies. *J. Protozool.* **19**:200 – 211.

Garcia, L. S., and L. R. Ash. 1979. *Diagnostic Parasitology. Clinical Laboratory Manual*, 2nd ed. C. V. Mosby Co., St. Louis.

Hoare, C.A. 1958. The enigma of host–parasite relations in amoebiasis. *Rice Inst. Pamphlet* **45**: 23 – 35.

Jahn, T. L., E. C. Bovee, and F. Jahn. 1979. *How to Know the Protozoa*, 2nd ed. W. C. Brown, Dubuque, Iowa.

Juniper, C. K., Jr. 1971. Amebiasis in the United States. *Bull. N.Y. Acad. Sci.* **47**:488 – 461.

Kudo, R. R. 1966. *Protozoology*, 5th ed. Charles C Thomas, Springfield, Ill.

Ludvik, J., and A. C. Shipstone. 1970. The ultrastructure of *Entamoeba histolytica. Bull. WHO* **43**:301 – 308.

McConnachie, E. W. 1969. The morphology, formation and development of cysts of *Entamoeba. Parasitology.* **59**:41 – 53.

Meerovitch, E. 1982. The jigsaw puzzle of host–parasite relations in amoebiasis begins to take shape. In E. Meerovitch, (ed.), *Aspects of Parasitology.* Institute of Parasitology, McGill University, Montreal.

Meerovitch, E., R. D. P. Eaton, and J. McLaughlin. 1977. Function aspects of host-parasite relations in amoebiasis. *Proceedings of the International Conference on Amoebiasis*, organized by *Centro de Estudios sobre Amebiasis*, Mexico City, October 27–30, 1975, pp. 628–635. Instituto Mexicano del Seguro Social, Mexico.

Mueller, J. F. 1959. Is *Chilomastix* a pathogen? *J. Parasitol.* **45**:170.

Neal, R. A. 1966. Experimental studies on *Entamoeba* with special reference to speciation. *Adv. Parasitol.* **4**:2 – 51.

Neva, F. A. 1974. Protozoal diseases endemic in the United States. *Am. J. Trop. Med. Hyg.* **23**:731 – 735.

Noble, G. A., and E. R. Noble. 1952. Entamoebae in farm animals. *J. Parasitol.* **38**:571 – 595.

Offutt, A. C., B. A. Poole, and G. G. Fassnacht. 1955. A water-borne outbreak of amebiasis. *Am. J. Publ.Health* **45**:486 – 491.

Padilla y Padilla, C. A., and G. M. Padilla. 1974. *Amebiasis in Man.* Charles C Thomas, Springfield, Ill.

Phillips, B. P. 1950. Cultivation of *Entamoeba histolytica* with *Trypanosoma cruzi. Science* **111**:8 – 9.

Sepulveda, B., and L. S. Diamond. 1976. Amebiasis. *Proceedings of the International Conference* organized by the *Centro de Estudios sobre Amebiasis*, Mexico City, October 27 – 30, 1975. Instituto Mexicano del Seguro Social, Mexico.

# 10

# *Free-Living Amebas as Pathogens*

Free-living aquatic and terrestrial amebas have always been considered merely elements in the microenvironment that cycle energy and prey on one another without any direct importance to humans. In the late 1950s, it became necessary to reconsider whether these small amebas were as innocuous as they had been thought. At first, they were laboratory curiosities that had appeared in monkey kidney tissue cultures; then they were found to produce disease in laboratory rodents inoculated experimentally. Finally, natural infections in human and other hosts were recognized. The question then raised was, how widespread and important are these would-be parasites?

NAME OF ORGANISM AND DISEASE CAUSED. A number of organisms have been found to have the ability to live in the tissues of mammals and to have at least some pathogenic potential. The following list includes those on record as of 1980.

Order Amoebida Ehrenberg, 1830

*Acanthamoeba culbertsoni* (Singh & Das, 1972)

Syn. *Hartmannella culbertsoni* S. & D., 1972

*A. rhysodes* (S. & D., 1972)

Syn. *H. rhysodes* S. & D., 1972

*A. polyphaga* (Puschkarew, 1913)

*A. astromyxis* (Ray & Hayes, 1954)

Syn. *H. astromyxis* R. & H. 1954

*A. hatchetti* Sawyer, Visvesvara, & Harke, 1977

*A. royreba* Willaert Stevens & Tyndall, 1978

Order Schizopyrenida Singh, 1952

*Naegleria fowleri* Carter, 1970

Syn. *N. aerobia* S. & D., 1970

*N. jadini* Willaert & Ray, 1973

*N. lovaniensis* Stevens, DeJonckheere & Willaert, 1980

The principal disease agent is *Naegleria fowleri*, which is the cause of primary amebic meningoencephalitis (PAM), or naegleriosis in humans. *Acanthamoeba polyphaga* causes infections of the cornea of the eye. Some of the other organisms are known to be facultatively parasitic, but do not seem to be highly pathogenic; some cause only mild pathogenesis in cell culture, but can infect laboratory animals without causing any discernible damage.

HOSTS AND LOCATION. Monkeys, humans, dogs, laboratory mice, and ox are the hosts. The organism is found in various body systems, mainly in the central nervous sytem, but also in lungs, liver, and other internal organs, the cornea of the eye, and nasal passages. passages.

DISTRIBUTION AND IMPORTANCE. The possibility that free-living amebas could be pathogenic was not known until 1957, when they were found as contaminants of monkey kidney primary tissue cultures (Jahnes, et al., 1957). Monkey kidney was used as the medium in which to grow polio virus for production of the Salk polio vaccine. The organisms found were ultimately given the name of *Acanthamoeba culbertsoni*, but at the time the significance of the finding was not appreciated.

In studies to determine how the organisms gained access to the kidneys, it was found that instilling amebas into the nasal passages of experimental mice and monkeys allowed them to invade. By 1959 it was reported that experimental animals could be killed by inoculation with amebas. In 1961, 19 isolates of amebas were made from the respiratory tracts of 18 persons by investigators searching for respiratory viruses. At this point, it still was not known whether the occurrence of these free-living protozoa was a curiosity or whether they had some pathogenic significance.

Between 1965 and 1969 the picture was clarified by documented fatalities: 4 in Australia, 3 in Florida, 1 in Texas, and 16 in Czechoslovakia. Two retrospective studies (Callicott, 1968; Symmers, 1969) of persons who had died from unexplained meningoencephalitis revealed amebas in the central nervous system in a small percentage of cases. At about this time, it was the consensus that a number of free-living amebas could live and multiply in the tissues of mammals, but that the pathogenic species was *Naegleria fowleri*, an ameba that sometimes transforms into a free-swimming biflagellate form (Duma et al., 1969).

The public health importance of these infections is still not entirely clear. About 60 human cases are on record in the medical literature; additional cases are reported sporadically in newspapers, but it is not known whether these become a part of published case reports. The isolation of amebas from the respiratory tracts of 18 persons indicates that the association with humans is not unusual. The early case reports indicated that all infections were fatal, but it is now known that some are not. Misdiagnosis probably plays a role in obscuring some of the fatalities from amebic encephalitis because the organisms in the tissue may be mistaken for white blood cells. Also, there is some doubt about the reports of *Iodamoeba buetschlii* invading extraintestinal tissues because this species has a nucleus with a large endosome not unlike that of the free-living water and soil amebas.

Amebas are known to be present in animals other than humans, but their importance is even more obscure. The presence in monkeys, of course, was the first indication of parasitism. There are additional reports of natural infections in a dog and a water buffalo.

Based on the locations where infections have been reported (Europe, Africa, Australia, North America), it seems likely that infections are cosmopolitan.

MORPHOLOGY. These organisms are typical of many small, free-living amebas. There is a vesicular nucleus that has a large, dark endosome or nucleolus and a thin outer limiting membrane (see Figs. 10.1 and 10.2). Trophozoites are 15 to 30 μm; cysts are about 11 μm and have pores in the outer membrane that are plugged. Differentiation among species is based on the size of the trophic forms and on cyst structure and size. Differentiation among genera and families is based on pseudopod form, mode of nuclear division, and the surface of the trophozoite.

In the family Valkampfiidae, there are two genera: *Valkampfia* and *Naegleria*. They are differentiated only by the fact that *Naegleria* transforms into a biflagellated organism if the proper conditions are provided (Fig. 10.1A).

Identification of a free-living ameba requires both skill and care. Two excellent handbooks are those of Page (1976) and Singh (1975).

LIFE CYCLE. Amebas multiply only by binary fission. They form resistant cysts, but the stimuli for cyst formation are not known. *Naegleria* spp. develop flagella, usually when the medium is diluted, but these changes are transitory. Normally all of these species are free-living. Occasionally, they gain entrance into humans and probably other mammals through the nasal passages.

DIAGNOSIS. Isolation of the ameba from an infected host, cultivation free of other living organisms, and reintroduction into experimental animals is the principal means of determining whether the organisms are present. Signs and symptoms of infection are useful but are sufficiently vague that a number of other infectious agents might be the cause. Swimming in warm waters from 3 to 7 days prior to the onset of symptoms is cause for suspicion.

HOST–PARASITE INTERACTIONS. The discussion here concerns *Naegleria fowleri* because it is the principal pathogen. It is most likely that the organisms gain entrance to the body through the nasal passages. They do not seem to be neurotropic, but once they have entered the nasal mucosa, they move along paths of least

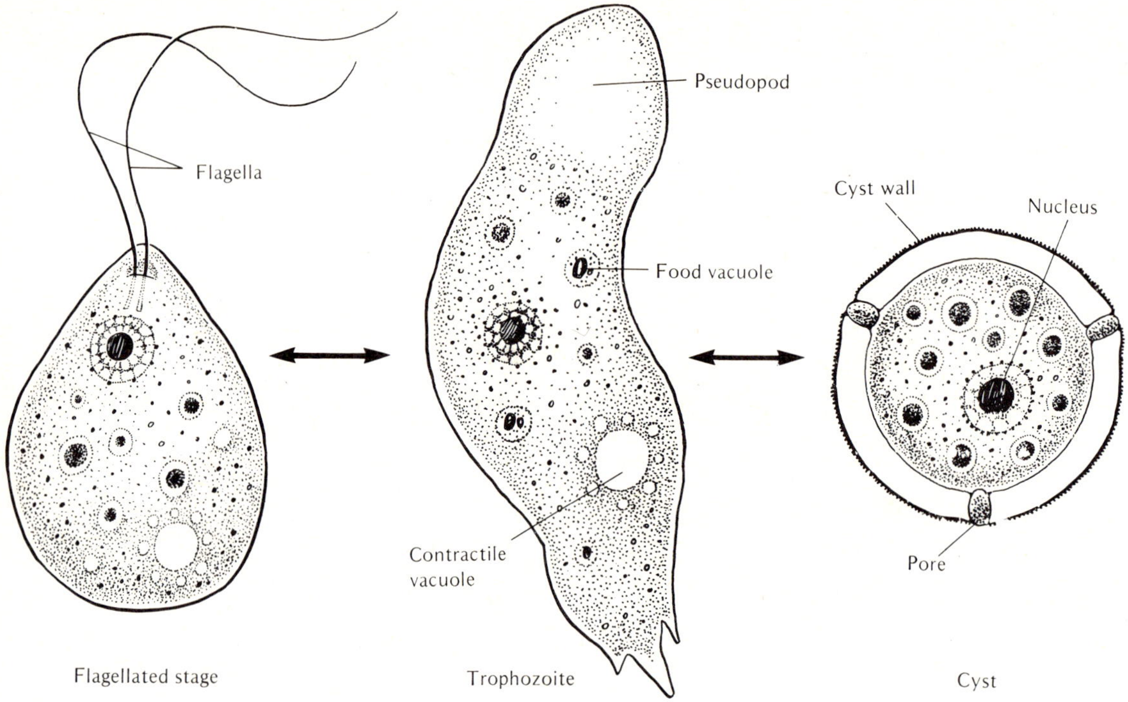


**FIG. 10.1**
Life cycle of *Naegleria fowleri:* flagellated form, trophozoite form, cyst.

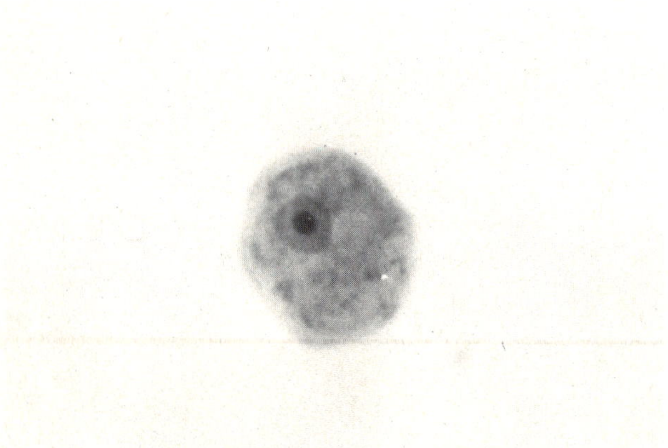

**FIG. 10.2**
Light photomicrograph of *Naegleria fowleri* in a fixed and stained preparation.

resistance, namely, nerve tracts. In the brain they remain associated principally with the membranous coverings of the brain, the meninges. The organisms cause inflammation principally of the meninges, but also of the nerve tissue itself to a degree.

In instances in which exposure could be documented, signs and symptoms of infection appeared in less than a week following the exposure. Death has usually occurred within 10 days of onset. Although there may be some signs of inflammation in the upper respiratory tract early in the infection, usually the first symptoms are lethargy and headache. Within 2 or 3 days the headache becomes more severe and there may be vomiting, stiff neck, and mental confusion. Alteration in normal sensations of taste and smell, double vision, incoordination of motor activity, and bizarre behavior may also occur. The next stage is coma caused by intracranial pressure.

TREATMENT. Among a large number of antibiotics and other antimicrobial drugs that have been tried, amphotericin B has the greatest value as specific treatment for PAM. It is best administered intrathecally and intraventricularly in order to reach the location of infection quickly and with a high concentration of drug. Symptomatic treatment and relief of cranial pressure by craniotomy have also been practiced.

EPIDEMIOLOGY AND CONTROL. Certain patterns of infection in humans with free-living amebas have emerged, but no method for prevention or control has been developed. Except in a few instances, infected people have had a recent history of swimming; in most instances, the weather has been hot, leading to warming of the water, or the waters have originated in warm springs. In some instances, the waters have had some organic pollution.

The general pattern seems to hold in most instances of disease, but there is one case on record of an elderly woman who had no history of recent swimming. Likewise, the apparently normal occurrence of free-living amebas in the respiratory tracts of 18 persons raises a question of the frequency of infection, as does the isolation of amebas from the nasal passages of persons who had recently been swimming.

Free-living potentially parasitic amebas may be found in quite diverse places. The characteristic that is crucial for any organism living as a parasite of mammals is the ability to multiply at 37°C or higher. It was first thought that *Naegleria gruberi* was the principal pathogen in humans, but it was found that, unlike the organism isolated from humans, *N. gruberi* would not grow at 37°C or higher; thus, *N. fowleri* was established as a new species and the principal human pathogen. In searching for potential pathogens, investigators now incubate cultures at temperatures higher than 37°C, even up to 44°C. In Belgium, isolates that were pathogenic for mice could readily be found in waters thermally polluted from power plants and factories. However, *Acanthamoeba hatchetti* was found in brackish water that had a high organic content from sewage pollution but was not considered to be warm. Other isolates of *A. culbertsoni* have been made from areas of sewage disposal. Thus, potentially pathogenic amebas can be found in both warm and cool waters and in areas where there is a great deal of organic pollution or where the waters are clean but warm.

The question then arises of whether it is possible to eliminate the risk of infection in swimming pools. The answer is equivocal at this point. Two reports indicate that chlorination kills the amebas and cysts in swimming pools at 4.0 ppm. On the other hand, another report stated that 10 ppm chlorine was ineffective. One study reported that raising the salt concentration to 7% (weight/volume) would eliminate *Naegleria* for a period of 5 months; however, the finding of potentially pathogenic *A. hatchetti* in brackish water raises the possibility that salination of a pool might be counterproductive.

The few cases on record of infections with free-living amebas in domesticated animals leave no clear pattern or any estimate of their importance. It would be interesting and worthwhile to undertake a retrospective study of tissues taken from animals that had died from unexplained central nervous system disease. If amebic infections were found, it might be possible to discern a pattern of conditions under which animals become infected.

## REFERENCES AND FURTHER READINGS


Ayers, K. M., L. H. Billups, and F. M. Garner. 1972. Acanthamoebiasis in a dog. *Vet. Pathol.* **9**:221–226.

Bos, H. J., H. J. M. Volker-Dieben, and C. C. Kof-van Alphen. 1981. A case of *Acanthamoeba* keratitis in the Netherlands. *Trans. Roy. Soc. Trop. Med. Hyg.* **75**:86–91.

Callicott, J. H. 1968. Amebic meningoencephalitis due to free-living amebas of Hartmanella (Acanthamoebae) Naegleria group. *Am. J. Clin. Pathol.* **49**:84–91.

Carter, R. F. 1972. Primary amoebic meningoencephalitis, an appraisal of present knowledge. *Trans. Roy. Soc. Trop. Med. Hyg.* **66**:193–213.

Duma, R. J., H. W. Ferrell, E. C. Nelson, and M. M. Jones. 1969. Primary amebic meningoencephalitis. *N. Engl. J. Med.* **281**:1315–1323.

Jahnes, W. G., H. M. Fullmer, and C. P. Li. 1957. Free-living amoebae as contaminants in monkey kidney tissue culture. *Proc. Soc. Exp. Med. Biol.* **96**:484–488.

de Jonckhere, J., and H. van de Voorde. 1977. The distribution of *Naegleria fowleri* in man-made thermal waters. *Am. J. Trop. Med. Hyg.* **26**:10–15.

Nerad, T. A., and P. M. Daggett. 1979. Starch gel electrophoresis: An effective method for the separation of pathogenic and nonpathogenic *Naegleria* strains. *J. Protozool.* **26**:613–615.

Neva, F. A. 1970. Amoebic meningoencephalitis—A new disease? *N. Engl. J. Med.* **282**:450–452.

Page, F. C. 1976. *An Illustrated Key to Freshwater and Soil Amoebae.* Freshwater Biol. Assoc. Sci. Publ. No. 34. Titus Wilson & Son, Ltd., Dendall.

Sawyer, T. K., G. S. Visvesvara, and B. A. Harke. 1977. Pathogenic amoebas from brackish and ocean sediments, with a description of *Acanthamoeba hatchetti*, n. sp. *Science* **196**:1324–1325.

Singh, B. N. 1975. *Pathogenic and Non-pathogenic Amoebae.* John Wiley & Sons, New York.

Symmers, W. St. C. 1969. Primary amoebic meningoencephalitis in Britain. *Br. Med. J.* **4**:449–454.

Visvesvara, G. S., and W. Balamuth. 1975. Comparative studies on related free-living and pathogenic amoebae with special reference to *Acanthamoeba. J. Protozool.* **22**:245–256.

Visvesvara, G. S., and G. R. Healey. 1975. Comparative antigenic analysis of pathogenic and free-living *Naegleria* species by gel diffusion and immunoelectrophoresis techniques. *Infect. Immun.* **11**:94–108.

Willaert, E. 1974. Primary amoebic meningoencephalitis. A selected bibliography and tabular survey of cases. *Ann. Soc. Belg. Med. Trop.* **54**:416–440.

# 11

# *Ciliates and Opalinids*

Although this chapter combines the Ciliophora and the Opalinatea, it is purely for convenience and does not imply any close taxonomic relationship of the two groups (Levine et al., 1980; Corliss, 1979).

## PHYLUM CILIOPHORA

The ciliated protozoa comprise a group of eukaryotic protists in which the two most striking characteristics are the infraciliature (silverline system) and the fact that they are heterokaryotic (have two kinds of nuclei in each individual). The other principal characteristic lies in their sexual phenomena, most often conjugation. Most students have studied *Paramecium* in a beginning biology course, and it is a good representative of the group (Fig. 11.1).

The infraciliature is made up of ciliary basal bodies and strands of material connecting them. A so-called kinety consists of the basal body or kinetosome and the strands that run anteriorly from it (Fig. 11.2). When a properly fixed ciliate is treated with silver nitrate, and then exposed to ultraviolet light, the kineties turn black and can be seen readily in the light microscope (Fig. 11.3). The infraciliature is used in the identification and description of species of ciliates, but it has greater significance in working out pathways of evolution in the ciliates and in fundamental studies on cellular development (Jones, 1974; Corliss, 1979).

All ciliates have two kinds of nuclei: a macronucleus, which serves the vegetative functions of the individual, and a micronucleus which serves as a stable repository of genetic information (Fig. 11.1). Conjugation is a phenomenon common among ciliates which involves exchange of micronuclear material.

During division in the ciliates, the plane of division is transverse and oral structures reform in the posterior individual during division (= homothetogenic division). In contrast, division in flagellates is essentially longitudinal or symmetrogenic.

Classification of the ciliates is based on oral structures, ciliation, and modes of reproduction. Recent classifications of the ciliates are contained in Levine et al. (1980) and Corliss (1979). Jones (1974) contains general information on ciliates.

Most ciliates feed on particulate food, some as small as bacteria; other ciliates are predaceous and feed on relatively large organisms. Some species obtain nutrient through membrane uptake (endocytosis). Food is contained in vacuoles and digested in them; then the residue is eliminated into the surrounding medium through the cytopyge (Fig. 11.4).

Ciliates play an important role in the digestion of cellulose by some herbivorous animals. Ruminants and equids have an extraordinary array of ciliated protozoa in their intestinal tracts. Vertebrates cannot digest cellulose, and must rely on symbiotic organisms to break the molecule into smaller fragments. In ruminants and equids, both ciliates and bacteria are involved

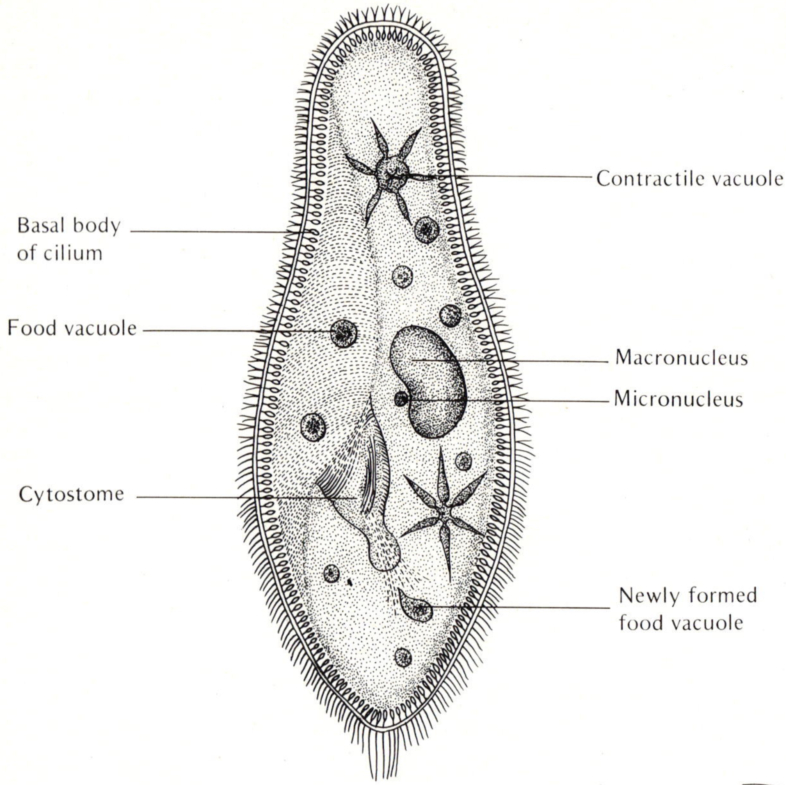


**FIG. 11.1**
*Paramecium.*

in cellulose digestion. In studies done on their role in ruminant nutrition, the results generally show that bacteria alone are sufficient to allow digestion of cellulose but the animal gains weight better when both bacteria and ciliates are present in the rumen. These microorganisms not only digest cellulose but serve as part of the food of the host as well. It has been estimated that 20% of the protein in ruminant nutrition is derived from the ciliates that pass from the rumen into the intestine. Much of the work on these structurally complex organisms has concerned their morphologies and systematics; Levine (1973) is a good source of information on structure and important works.

Only a few ciliates are of economic or medical importance as parasites of other animals; most are free-living organisms. We will consider some of the better-known ciliates that are harmful or beneficial to their hosts.

NAME OF ORGANISM. *Balantidium coli* (Malmsten, 1857) Stein, 1862, has been classified in recent schemes

**FIG. 11.2**
*Tetrahymena piriformis*. Somatic cortex. [From Allen, 1967.]
C = cilium, mb = mucigenic body, pt = postciliary set, bt = basal microtubules, kf = kinetodesmal fiber, M = mytochondrion, er = endoplasmic reticulum, e = amorphous ectoplasmic material, a = alveoli.

in Kinetofragminophorea, Trichostomatida (synonym: *B. suis* McDonald; see Levine, 1940). This organism causes balantidial dysentery or balanditiosis.

HOSTS AND LOCATION. Hosts include pigs, humans, and other primates. Some animals such as rodents may be infected experimentally. The large intestine is the site of the organism.

DISTRIBUTION AND IMPORTANCE. Distribution is cosmopolitan. The organism is common in pigs, being found in from 20 to 100% of individuals in various populations. In humans the prevalence is spotty; some surveys indicate that prevalence is less than 1%, but there are populations in which the infection is common. Not many systematic surveys have been done for infection in humans. In primates other than humans, infections are common.

The normal hosts are pigs, and the organisms seem to cause little, if any, problem in them. In primates *B. coli* is invasive, causing diarrhea and dysentery. There is little information on the overall importance of balantidiasis in primates other than humans.

MORPHOLOGY. The organisms are ciliated uniformly over the body with about 60 kineties; the mouth is at the anterior end and of relatively simple structure (Fig. 9.1*Y*). There is a rather extreme size range running from 30 to 150 μm, but averaging 45 by 60 μm with an ovoid shape (Levine, 1940). There is a large macronucleus and a small micronucleus, usually closely applied to the concave side of the macronucleus, and two contractile vacuoles, one anterior and one posterior. The cytopyge is near the posterior tip (Paulin & Krashinennikow, 1973). Cysts are subspherical and 40 to 60 μm (Fig. 9.1*Z*).

LIFE CYCLE. *B. coli* is transferred from one host to the next in the cyst form. The organism reproduces by binary fission in the large intestine of the host. Conjugation occurs in *B. coli* and has been studied in culture.

DIAGNOSIS. Since the parasite is found in the large intestine and causes diarrhea, the trophozoites can be found by a direct examination of the feces using a wet mount preparation. Clinical signs and symptoms of balantidial dysentery are much like those of amebic dysentery, and it is necessary to differentiate between the two infections.

Cysts and trophozoites may be found in histologic section, most often during postmortem examination.

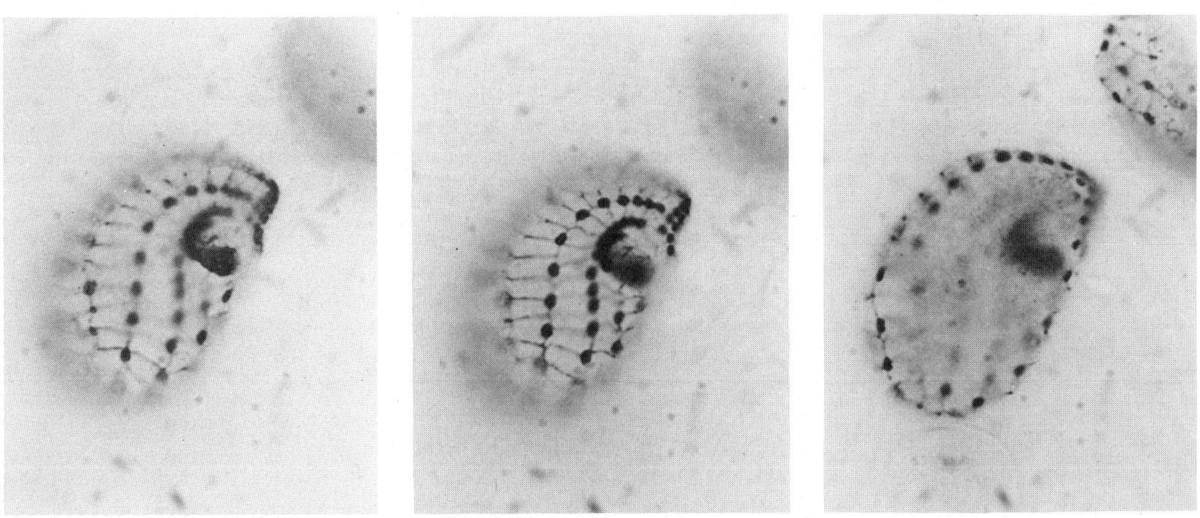

**FIG. 11.3**
Infraciliature or silverline system of *Colpoda steinii*; the organism is suspended in gelatin, and the three views show the basal bodies and kinetodesmata as seen when focusing down through a specimen.

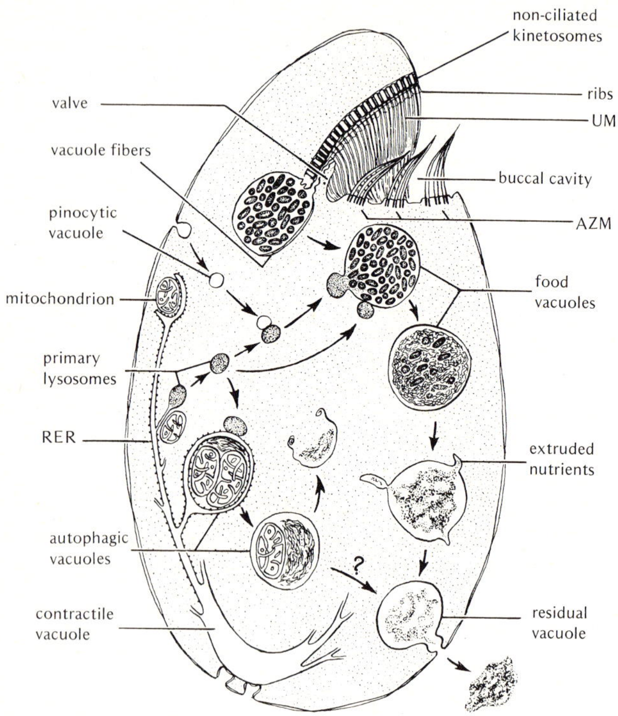


**FIG. 11.4**
Ingestion and digestion of food in *Tetrahymena*. [From Elliot & Clemmons, 1966.] (Abbreviations are UM—undulating membrane; AZM—adoral zone of membranelles; RER—rough endoplasmic reticulum.)

This species is easily recognized by its large size and by the presence of the macro- and micronucleus.

As with other intestinal protozoa, permanent slides should be made so that the identification can be confirmed and a permanent record of the infection exists (Chap. 9).

CULTIVATION. A number of media have been used to cultivate *B. coli* in the presence of bacteria: Balamuth's egg infusion, Ringer–horse serum, and Difco's *Entamoeba histolytica* medium; in all instances, rice starch is provided as a source of particulate carbohydrate (Auerbach, 1953; Lotze, 1939; Levine, 1940).

HOST–PARASITE INTERACTIONS. The normal hosts of *B. coli* are pigs, especially domestic swine. In swine, there is normally little or no damage, but there are a few instances on record in which pigs have been affected by the parasite.

In humans, the organism is invasive (Tempelis & Lysenko, 1957) and damages the large intestine, leading to diarrhea, dysentery, and ulceration. There are a few case reports of *B. coli* in extraintestinal sites. Higher primates other than humans seem to have their own *B. coli* and probably do not become infected from pigs. There seems to be little information on the development of immunity to *B. coli* in humans.

TREATMENT. Broad-spectrum antibiotics have been found to be useful in treating human infection. Carbarsone and diiodoquine have also been used, but not with uniform success.

EPIDEMIOLOGY AND CONTROL. *B. coli* sometimes is a problem in humans, and there seems to be a relationship between human infection and an association with swine. High prevalence of infection in the human population generally correlates with close association with swine. Swine do not usually show any signs of infection, but they may have enormous numbers of *B. coli* trophozoites in the large intestine. It appears that swine shed large numbers of cysts which are ingested by people who may not recognize the hazards of infectious diseases. Person-to-person transfer, if any, is probably not especially important in maintaining and transmitting the infection.

Reducing the incidence of infection presents a moderately difficult problem. Where balantidial dysentery is a public health problem, the economic level is likely to be low, as well as the educational level. Swine are an important part of the diet, and raising them is nearly free of cost because they scavenge on anything edible. Areas where balantidiosis is prevalent in humans also have warm climates, which promote survival of the cysts.

The solution is to erect some sort of barrier between humans and swine. This might be done by fencing pigs out of the village or town, or at least from some areas of it. Persons who tend the animals might be encouraged to wash their hands and feet after being with swine so as not to track cysts home with them or carry the cysts to their mouths on their hands.

If such a program could be established, the incidence of other parasites, namely, *Ascaris suum* and *Trichuris trichiura*, would also be reduced.

NAME OF ORGANISM AND DISEASE CAUSED. *Ichthyophthirius multifiliis* Fouquet, 1876, causes ichthyophthiriasis, white spot, or "ick." In recent classifications the organism has been included as Oligohymenophorea and Hymenostomatida.

HOSTS AND LOCATION. The organism parasitizes a wide array of fish including carp, catfish, and trout in fresh water. It is found under the scales on the body, on gills, and sometimes in the cornea of the eye (Kudo, 1934).

DISTRIBUTION AND IMPORTANCE. Distribution is cosmopolitan, but is most important in fish in small ponds where the water is warm, and in aquaria (Pearson, 1932).

MORPHOLOGY. The trophont or "adult" form on the fish varies from 100 µm up to 1 mm in length; the shape is ovoid, with a mouth at the anterior tip (Fig. 11.5). Tomites or "swarmers" are 30 to 50 µm long and average 20 µm. Theronts, also known as "swarmers," are 10 by 40 µm. Tomites are surrounded by a gelatinous wall. There is a single large macronucleus and a micronucleus, and many contractile vacuoles (Mugard, 1948, 1949).

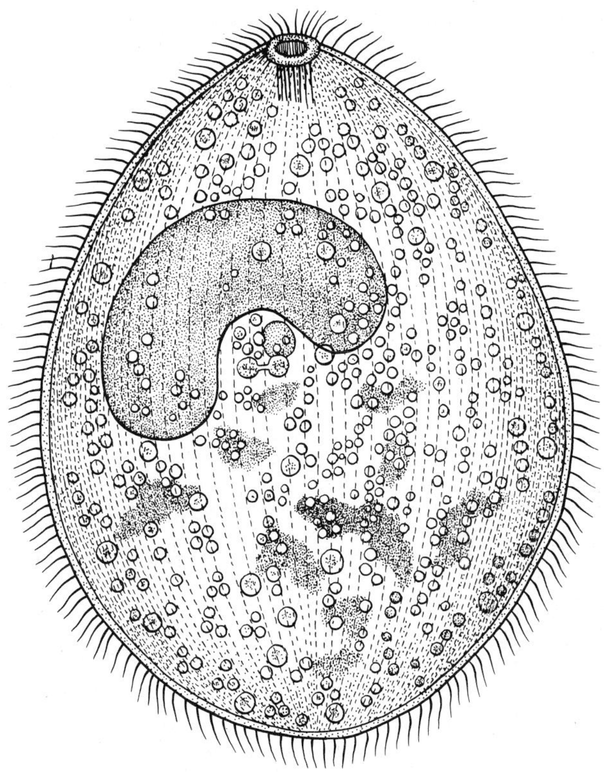

**FIG. 11.5**
Trophont of *Ichthyophthirius multifiliis.*

LIFE CYCLE. The theront is the infective form; it transfers from the free-living phase of the cycle to the fish. The theronts may live as long as 96 hr, but their infectivity decreases after 48 hr. Once having reached a fish, the theront enters its tissue by using a penetrating apparatus at the anterior end; there is no mouth at this stage of the cycle. Once inside the tissues of the fish, the organism is spoken of as a *trophont*; it grows without undergoing any division to produce an enormous uninucleated organism. During this growth the ciliate continually rotates in the cavity of the tissue. When full grown (i.e., approaching 1 mm in length), the trophont drops off of the host, forms a gelatinous envelope around itself, and attaches to the substratum. Division takes place quite rapidly by a process referred to as *palintomy* (Corliss, 1979). Within about 8 hr, up to

Metcalf, M. M. 1923. The opalinid infusorians. *U.S. Nat. Mus. Bull.* **120**:1–147.

——— 1940. Further studies on the opalinid ciliate infusorians and their hosts. *Proc. U.S. Nat. Mus.* **87**:465–634.

Mugard, H. 1948. Régulation du nobre des cineties au cours du cycle de croissance et de division chez un cilié: *Ichthyophthirius multifiliis* Fouquet. *Arch. Anat.* Micros. **37**:204–213.

——— 1949. Contribution à l'étude des infusoires hymenostomes histiophages. *Ann. Sci. Nat. Zool.* (ser. 11) **10**:171–268 (1948 year).

Paulin, J. J., and S. Kraschinennikow. 1973. An electron microscope study of *Balantidium caviae. Acta. Protozool.* **12**:97–103.

Pearson, N. E. 1932. Ichthyopthiriasis among fishes of a pond in Indianapolis. *Proc. Indiana Acad. Sci.* **41**:455–463.

Tempelis, C. H., and M. G. Lysenko. 1957. The production of hyaluronidase by *Balantidium coli. Exp. Parasitol.* **6**:31–36.

Wessenberg, H.S. 1978. Opalinata. In J. Kreier (ed.), *Parasitic Protozoa.* Vol. II. Academic Press, New York, pp. 551-581.

# 12

# Introduction to the Apicomplexa

Members of the Apicomplexa are a diverse group of parasites. What they have in common are sexual development and components of the apical complex (Fig. 12.1). The apical complex remained undescribed until good thin-sectioning techniques for electron microscopy were developed in the early 1960s. As knowledge has developed, organisms without the apical complex that had different modes of development were removed from a rather ill-defined group called the Sporozoa. For example, the Myxozoa (Chap. 18) and the Microspora (Chap. 19), formerly included in the Sporozoa, clearly deserved separate status based on careful transmission electron microscope (TEM) studies.

As the name implies, the apical complex is located in one end of a zoite, called the *anterior end*, since they move with that end forward. The complex consists of one or two polar rings with associated subpellicular tubules, which run most of the length of the zoite (Fig. 12.1), a conoid (Fig. 12.2), rhoptries, and micronemes. Because of their location and appearance in the electron microscope, it has been suggested that these structures are involved in penetrating host cells. The organism illustrated in Figure 12.1 is a zoite of the genus *Eimeria,* and all of the elements are present. In some groups such as the gregarines (Chap. 13), some of the components are seen early in development but modified into holdfast organelles such as the mucron. In members of the Haemosporina, some of which cause malaria in humans (Chap. 16), the conoid is lacking, and in the piroplasms (Chap. 17), the elements are reduced, but the rhoptries and micronemes can usually be seen (see Fig. 17.1).

Life cycles of apicomplexans include both asexual and sexual development. We can look upon the basic life cycle as having these components:

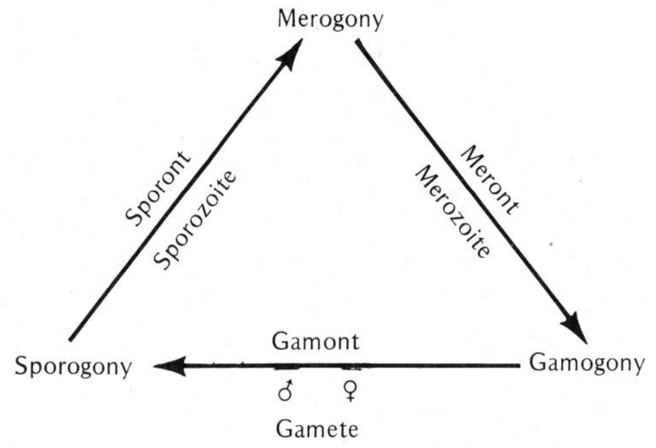


The stages in the diagram represent both sexual and asexual reproduction. Gamonts develop into either macro- (female) or microgamonts (male); when fertilization of the macrogamont takes place, a new individual, the sporont, results.

The sporont, meront, and microgamonts multiply by an asexual process called *schizogony.* In schizogony, nuclear replication occurs, and then daughter cells are formed all at one time. The process is the same in sporogony, merogony, and gamogony; only the product

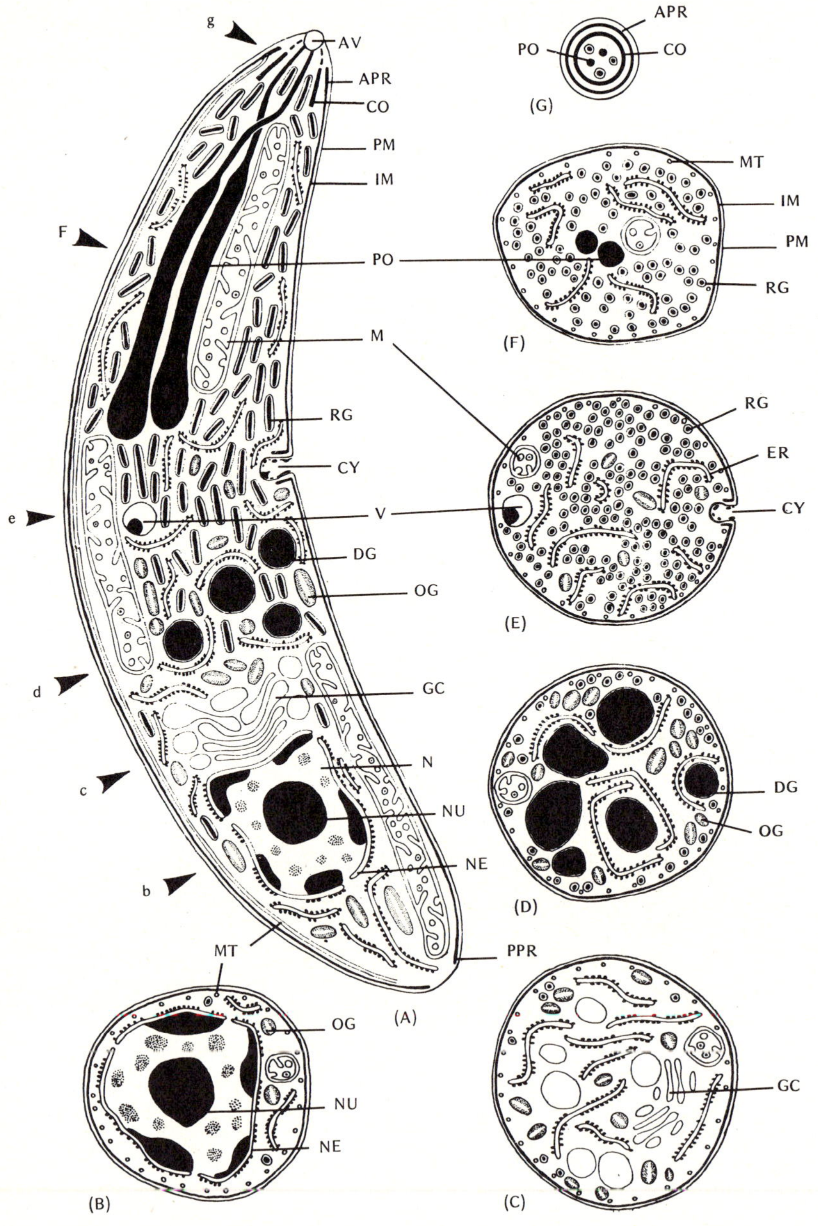


**FIG. 12.1**
A zoite of an apicomplexan showing structures revealed by electron microscopy. (APR = anterior polar ring, AV = anterior vacuole, CO = conoid, PM = outer pellicular membrane, IM = inner pellicular membrane, PO = rhoptries M = mitochondrion, RG = micronemes, CY = cytostome (micropore), V = unidentified body, DG = dark granule, OG = opaque granule, GC = Golgi, N = nucleus, NU = nucleolus, NE = nuclear envelope, PPR = posterior polar ring, MT = subpellicular tubules, ER = endoplasmic reticulum) [From Andreassen & Behnke, 1968.]

of the reproduction is different. When the sporont undergoes sporogony, the product is a large number of sporozoites. The sporozoite is the infective stage of the life cycle; it is always this stage that enters a new host. In the meront, the product of merogony is the merozoite; the merozoite may either go through one or more additional generations of merogony or become a gamont. The cycle always runs in the direction indicated in the diagram; sporogony, for example, never follows merogony.

| CHARACTER | SUBCLASS | |
|---|---|---|
| | GREGARINIA | COCCIDIA |
| Location of development | Extracellular, or only partly within the host cell, in most species | Intracellular |
| Apical complex | Rudiments present that are transformed into holdfast structures | Present in zoites, but not all elements in all groups |
| Merogony | Present in a few species, but usually macroscopic uninuclear gamonts are generally present | Present in all groups; large numbers of merozoites produced |
| Gamogony | Takes place in syzygy, with equal numbers of each gamete produced | Gametes produced independently in most groups |
| Microgametes | No visible organelles of locomotion in most | Flagella present in most groups |
| Sporogony | Principal means of replication in most groups | In vector-borne species replication by sporogony is extensive, but in species with direct transmission, few sporozoites produced, usually fewer than 16 |

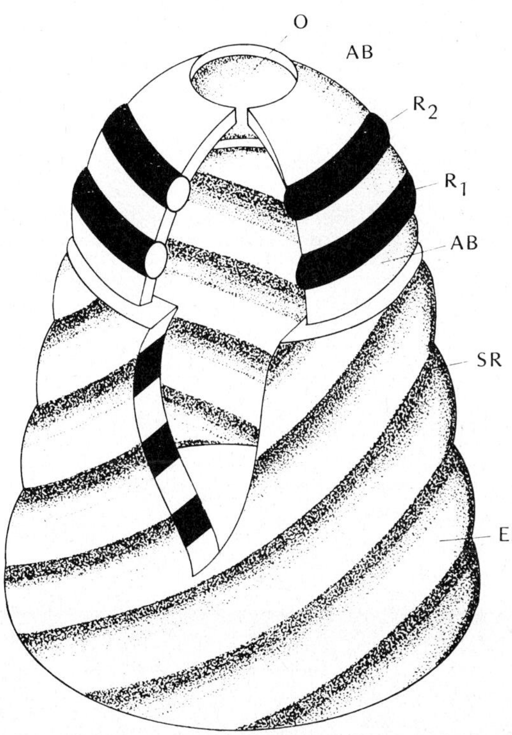


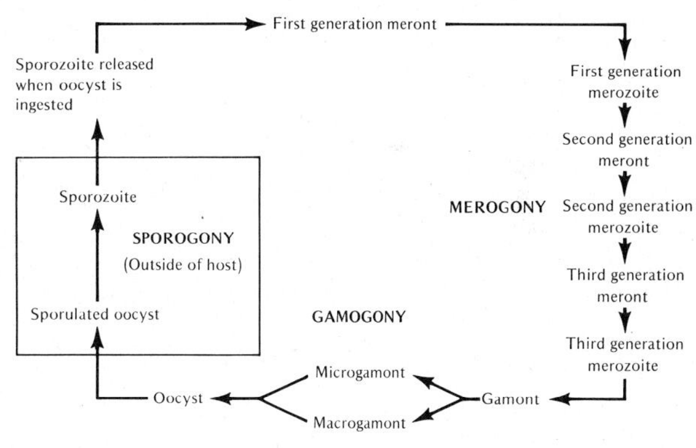


**FIG. 12.2**
Diagram of a conoid (O = orifice or anterior opening, $R_1$ and $R_2$ = conoidal rings, AB = anterior element, SR = spiral elements, ES = Conoidal envelope.) [From Scholtyseck et al., 1970.]

The sporozoan life cycle is not nearly as complex as the life cycle of a digenetic trematode, for example, but students sometimes have difficulty in grasping the basic stages and applying them to organisms in which the theme varies somewhat or more than one host exists. The problem is due largely to an abundance of inconsistent terminology; terminology used here will conform as nearly as possible to that proposed by Levine (1971).

Now, let us take a specific example and look at how the life cycle proceeds. The complete life cycle of *Eimeria tenella*, one of the economically important sporozoa of poultry, is as follows:

Merogony takes place three times in *E. tenella*, but the pattern is the same as the general one; the sporont is contained within a resistant structure called an *oocyst*. When sporogony has been completed, the infective stage (the sporozoite) is formed, and if the oocyst is ingested by a susceptible animal, the life cycle continues.

Classification of the Apicomplexa has remained in a state of flux since the late 1950s. It is likely that the classification presented here will remain the consensus for some time (Levine et al., 1980).

Within the phylum Apicomplexa there are two classes: Perkinsea and Sporozoea. There is only a single species in Perkinsea; it has an incomplete conoid and lacks sexual reproduction. The major group, the Sporozoea, is more difficult to define, but all members have at least some of the elements of the apical complex and probably all have sexual reproduction. The subclasses are the Gregarinia and the Coccidia, which are readily differentiated from one another.

## REFERENCES AND FURTHER READINGS

Andreassen, J., and O. Behnke. 1968. Fine structure of merozoites of a rat coccidian *Eimeria miyairii*, with a comparison of the fine structure of other sporozoa. *J. Parasitol.* **54**:150–163.

Levine, N. D. 1971. Uniform terminology for the protozoan subphylum Apicomplexa. *J. Protozool.* **18**:352–355.

Levine, N. D., et al. 1980. A newly revised classification of the protozoa. *J. Protozool.* **27**:37–58.

Scholtyseck, E. H., H. Mehlhorn, and K. Friedhoff. 1970. The fine structure of the conoid of Sporozoa and related organisms. *Z. Parasitenk.* **34**:68–94.

# 13

# *The Gregarines*

As members of the phylum Apicomplexa, the gregarines have some elements of the apical complex and all are parasitic. They are mostly coelozoic parasites of insects, but they are found widely spread through invertebrates in general and some are intracellular.

At about the time when the phylum Apicomplexa was established, data obtained through transmission electron microscope (TEM) studies showed that a conoid, rhoptries, and a polar ring were present in gregarines (Schrével, 1971). These structures are sometimes converted to holdfast organelles (Fig. 13.1), but their presence in a portion of the life cycle further strengthened the Apicomplexa as a coherent group.

The gregarines seem primitively to have had all three phases of the apicomplexan life cycle: merogony, gamogony, and sporogony. In most species, merogony has dropped out and is replaced by a large (500 μm or greater), uninucleated organism called a trophozoite or a sporadin, but which actually becomes a gamont (Fig. 13.2). The classification within the subclass Gregarinia is based on the life cycle pattern. The order Archigregarinida has merogony, but the Eugregarinida (true gregarines) lack merogony and the Neogregarinida seem to have acquired merogony secondarily (Levine et al., 1980).

The features that are important in both the identification and classification of the gregarines are the following:

**1.** The large trophozoite (Fig. 13.2), which may begin life intracellularly but later becomes coelozoic and attached to the cell by a holdfast (epimerite, mucron) (Fig. 13.1). In the suborder Aseptatina, there is no differentiation within the trophozoite, but in the suborder Septatina there is a septum near the anterior end of the organism dividing it into an anterior protomerite and a posterior deutomerite (Fig. 13.2).

**2.** Gamogony takes place with the gamonts in syzygy, the union of two individuals during gamogony (Fig. 13.3). At the end of gamogony, there may be either isogametes (Fig. 13.3) or, less frequently, macrogametes and motile microgametes. In either event, there is syngamy and then sporogony.

**3.** The oocyst has a resistant outer wall and naked sporozoites within it (Figs. 13.3 and 13.4). The size, shape, and color of the oocyst as well as the number of sporozoites are useful in identification. It should be pointed out that oocysts of gregarines are frequently seen in the feces of animals such as birds that eat insects or earthworms; the unwary can mistake such oocysts for parasites of the vertebrate.

**4.** Host range. The species of hosts in which any one gregarine can complete its life cycle is limited. Usually, they will develop in a few closely related species or genera. Despite the widespread occurrence of gregarines in many invertebrate animal groups, they seem to have only a slightly detrimental effect on their hosts. Evidence of a detrimental effect on

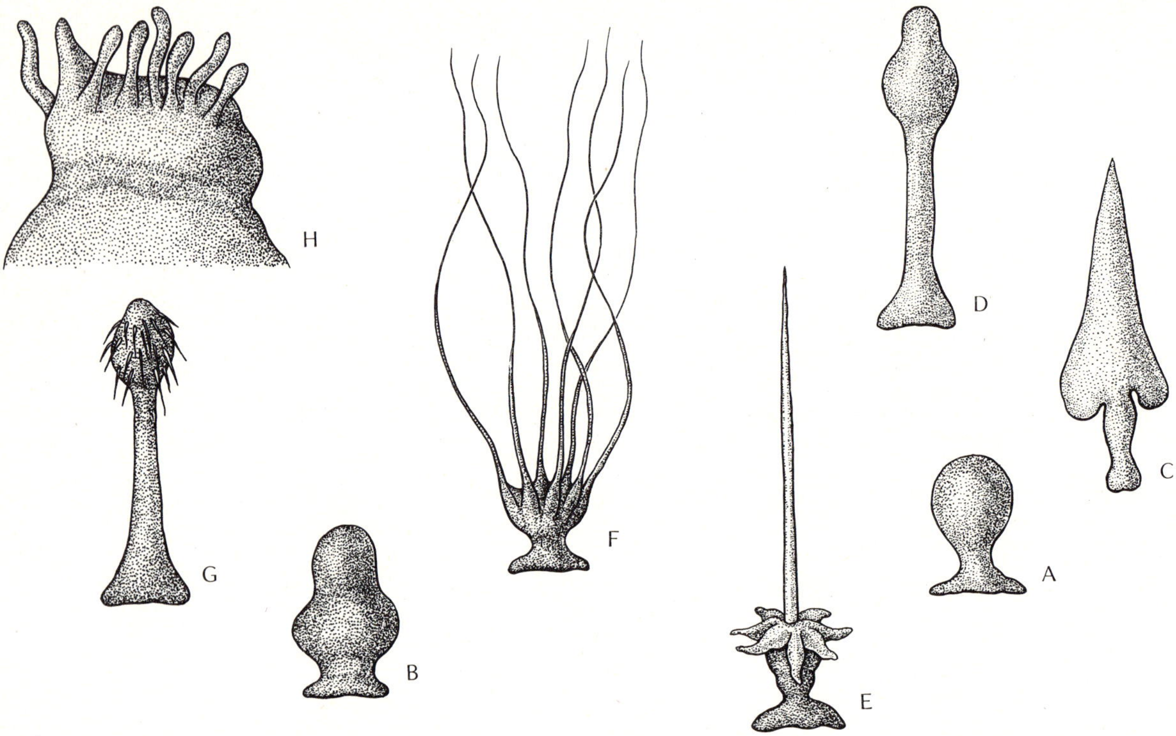


**FIG. 13.1**
Structure of the epimerite of various gregarines. *(A) Gregarine longa, (B) Sycia inopinata, (C) Pileocephalus heeri, (D) Stylocephalus longicollis, (E) Beloides firmus, (F) Cometoides erinitus, (G) Geniorhynchus monnieri, (H) Echinomera hispida.*

populations occurs only in a few instances, such as *Gregarina locustae* of grasshoppers.

In this chapter we will consider only a few gregarines as examples of the hundreds that may be found almost anywhere. Additional information on identification of species may be found in Watson (1916), Kamm (1922), Kudo (1966), and Grassé (1953). For a recent consideration of classification see Levine (1979).

The two species that have been chosen here as examples of gregarine life cycles show the variation in development one can expect. *Lipocystis polyspora* of the scorpion fly has all three phases of the typical apicomplexan life cycle: sporogony, merogony, and gamogony; thus, it is somewhat atypical for a gregarine (Fig. 13.3). The sexual phase of the cycle has isogametes, which develop in syzygy within a gametocyst.

*Monocystis lumbrici* (Fig. 13.4), a parasite in the seminal vesicles of the earthworm, *Lumbricus terrestris*, has a sporozoite, which transforms into a large trophozoite inside earthworm sperm mother cells. Trophozoites transform into gamonts, which secrete a cyst wall; the gamonts produce isogametes. The cycle is completed with the formation of oocysts.

Motility in the gregarines and other apicomplexans has been a matter for study and speculation for some time. Nearly all of the gregarines glide with a stately, progressive movement, but there are no organelles that can be associated with movement.

A variety of speculations have been made on the mechanism of movement. One theory was based on the observation that some trophozoites leave a trail of mucoid material as they move. It was supposed that mucus piled up at the posterior of the organism and it

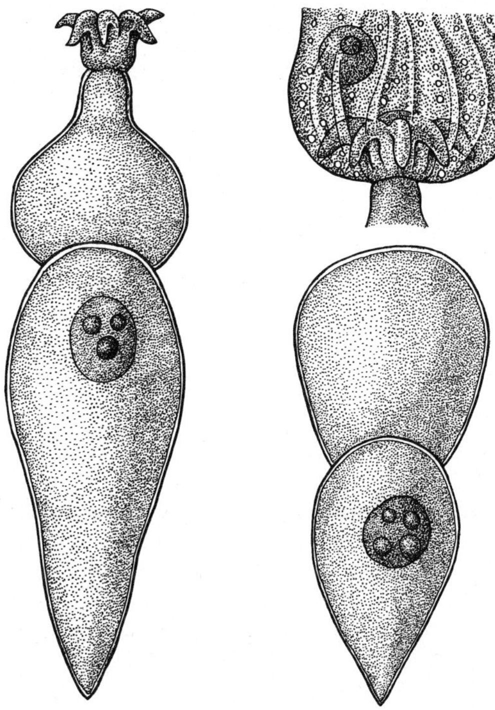

**FIG. 13.2**
Body form of *Corycella armata.*

pushed off from this. Other studies proposed that myofibrils in the body provided the force needed to move forward. A study by Vavra and Small (1969) did much to resolve the problem. By scanning electron microscopy (SEM) it was shown that the folds in the pellicle of the trophozoites undulate in such a way that the waves are out of synchrony. The authors postulated that interaction between mucus in the folds and the rippling of the folds moves the mucus toward the posterior, thereby propelling the organism forward. Attachment to a substrate was considered not to be essential since some trophozoites were seen to move while suspended in fluid. In work on another species, *Gregarina garnhami,* it was found that attachment to the substrate by mucus was essential to progressive movement and that the ridges seemed not to move laterally. Attachment to the substrate was seen to be in patches so that only a small proportion of the organism was in contact with the surface at any one time (Walker et al., 1979; Mackenzie & Walker, 1983). The question of the exact mechanism of movement and whether the same mechanism(s) is used in different species remains open.

The question might also be raised whether work on an arcane protozoan has any relevance to more important problems. In this case, describing and being able to manipulate motility in gregarines has an impact in cell biology and perhaps in control of diseases caused by apicomplexans. Contractility in nonmuscle systems and motility of cells are both significant problems in cell biology. If we can describe a system and manupulate it, we then may be able to reduce or enhance it. Thus, if we can manipulate motility in the gregarines, we may have a way of controlling motility in zoites of the coccidia or malarial parasites (Chaps. 14, 15, 16), thereby preventing infection.

## GREGARINE CLASSIFICATION

### *Subclass Gregarinia Dufour, 1828*

Mature gamonts large, extracellular; mucron or epimerite in mature organism; mucron formed from conoid; generally syzygy of gamonts; gametes usually similar (isogamous) or nearly so, with similar numbers of male and female gametes produced by gamonts; zygotes form oocysts within gametocysts; life cycle characteristically consisting of gamogony and sporogony; in digestive tract or body cavity of invertebrates or lower chordates; generally narrow host range.

**Order Archigregarinida Grassé, 1963.**
Life cycle apparently primitive, characteristically with three schizogonies (merogony, gamogony, sporogony); parasites of annelids, sipunculids, hemichordates, and ascidians.

**Family Selenidioididea Levine, 1971**
Aseptate species, foliaceous or cylindroid, with longitudinal striations (myonemes) and body often curved or crescent-shaped; anterior end contains bacilliform chromatic bodies (which are probably rhoptries; see Schrével, 1971); merogony in an epithelial cell; mucron or epimerite present; gamont develops extracellularly; syzygy generally tail-to-tail (caudal); oocysts spherical or ellipsoidal, containing 4-n sporozoites.
*Selenidioides*

**Family Exoshizonidea Levine, 1971**
Merogony present, extracellular merozoites being formed as a cluster at the anterior end of the meront; gametocysts, gametes, and oocysts unknown.
*Exoschizon*

**Order Eugregarinida Léger, 1900**
Merogony absent; typically parasites of annelids and arthropods, but some species in other invertebrates and lower chordates.

**Suborder Blastogregarinina Chatton & Villneuve, 1936**
Gametocysts absent; gamogony while gamonts are still attached to intestine with anisogamous gametes budding from gamonts; syzygy absent; oocysts with

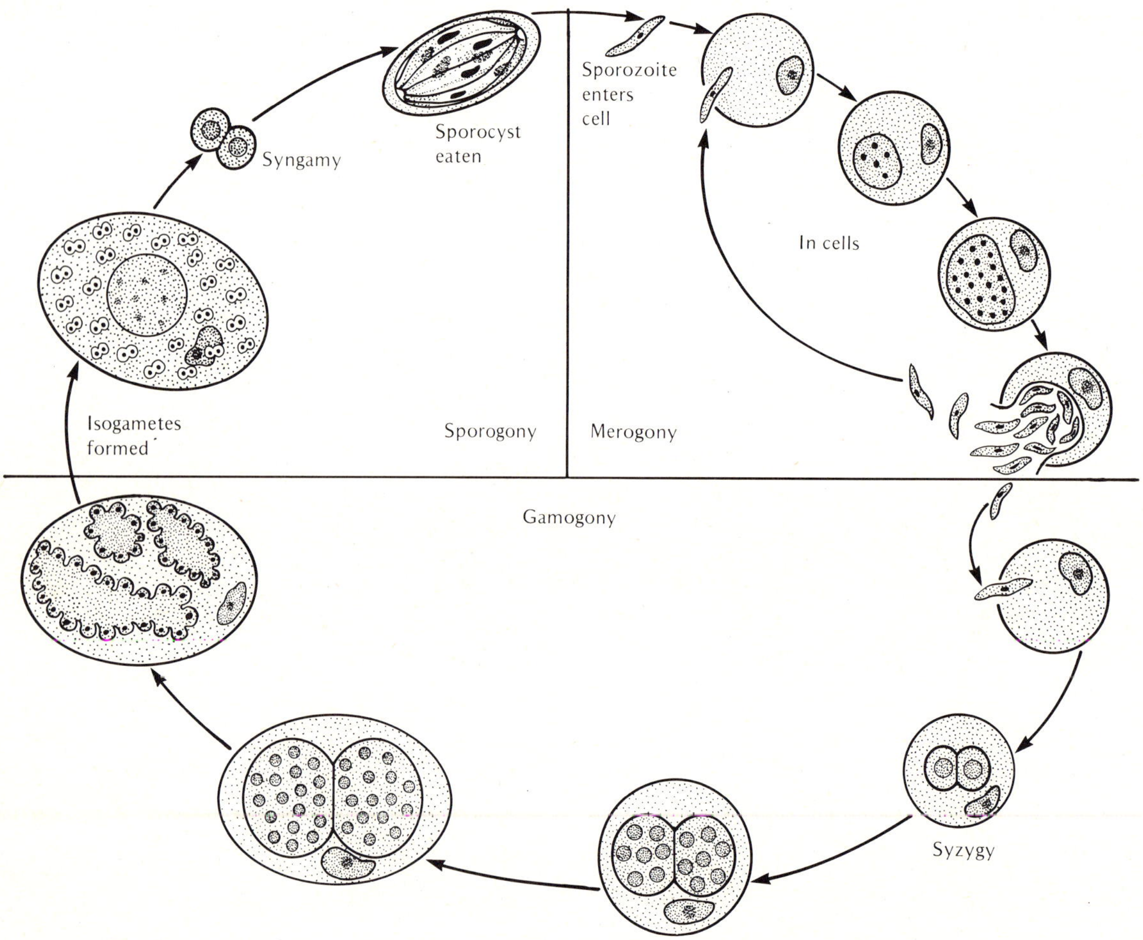


**FIG. 13.3**
Life cycle and structure of *Lipocystis polyspora*.

10 to 16 naked sporozoites; gamont a single compartment with mucron, but without definite proto- and deutomerite; in marine polychaetes. *Siedleckia*

**Suborder Aseptatina Chakravarty, 1960**
Gametocyst present; gamont a single compartment without definite proto- and deutomerite, but with mucron (epimerite?) in some species; syzygy present. *Lecudina, Monocystis, Selenidium, Urospora, Stylocephalus*

**Suborder Septatina Lankester, 1885**
Gametocyst present; gamont divided into proto- and deutomerite by septum; with epimerite; in intestine of invertebrates, mostly arthropods. *Cephaloidophora, Gregarina, Stenphora*

**Order Neogragarinida Grassé, 1953**
Merogony present, presumably acquired secondarily; in Malpighian tubules, intestine, hemocoel, or fat bodies of insects.
*Gigaductus, Mattesia, Lipocystis, Caulleryella*

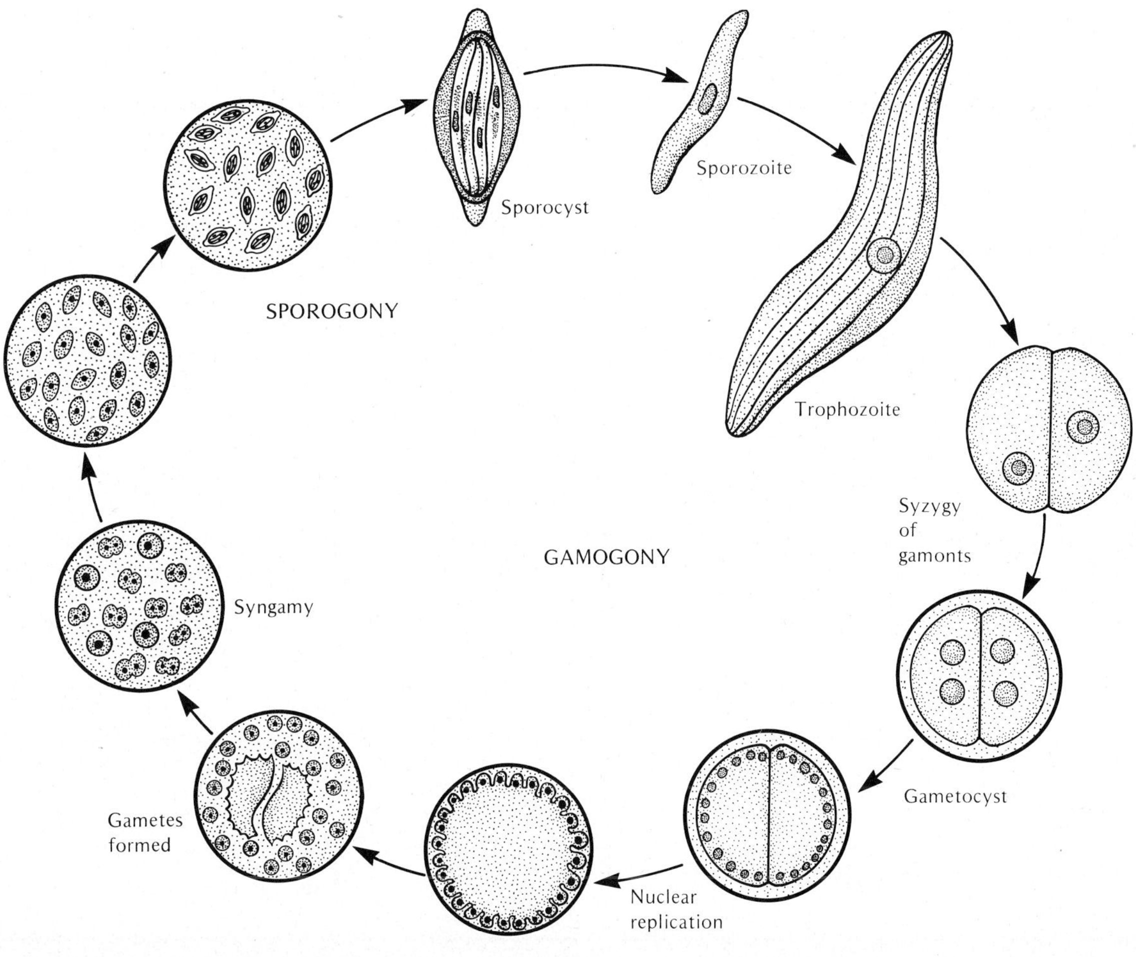


**FIG. 13.4**
Life cycle and structure of *Monocystis lumbrici*.

## REFERENCES AND FURTHER READINGS

Grassé, P. P. 1953. Classe des grégarinomorphes (Grégarinomorpha n. nov. Grégarinae Haeckel, 1866; Grégarinidea Lankester, 1885; gregarines de auteurs). In P.P. Grassé (ed.), *Traité de Zoologie* **1**(2): 550–690.

Kamm, M. W. 1922. Studies on gregarines II. Synopsis of the polycystid gregarines of the world excluding those from Myriopoda, Orthoptera and Coleoptera. *Ill. Biol. Monogr.* 7:1–104.

Kudo, R. R. 1966. *Protozoology*, 5th ed. Charles C Thomas, Springfield, Ill.

Levine, N. D. 1979. New genera and higher taxa of septate gregarines (Protozoa, Apicomplexa) *J. Protozool.* **26**:532–536.

Levine, N.D., et al., 1980. A newly revised classification of the protozoa. *J. Protozool.* 27: 37-58.

Mackenzie, C., and M. H. Walker. 1983. Substrate contact, mucus and eugregarine gliding. J. Protozool. **30**:3–8.

Schrével, J. 1971. Observations biologiques et ultrastructurales sur les Selenidiidae et leurs conséquences sur la systématique des Gréginomorphes. *J. Protozool.* 18:448–470.

Vavra, J., and E. B. Small. 1969. Scanning electron microscopy of gregarines (Protozoa, Sporozoa) and its contribution to the theory of gregarine movement. *J. Protozool.* **16**:745–757.

Walker, M. H., C. Mackenzie, S. P. Bainbridge, and C. Orme. 1979. A study of the structure and gliding movement of *Gregarina garnhami*. *J. Protozool.* **26**:566–574.

Watson, M. E. 1916. Studies on gregarines including descriptions of twenty-one new species and a synopsis of the eugregarine records from the Myriopoda, Coleoptera and Orthoptera of the world. *Ill. Biol. Monogr.* **2**:213–468.

# 14

# The Coccidia

The coccidia are apicomplexans that have all of the elements of the apical complex (Fig. 12.1). They are intracellular parasites principally of vertebrates, but some are found in invertebrates as well. The life cycle is typical of the Apicomplexa, with sporogony, merogony, and gamogony all occurring. Transmission from one host to the next takes place through a resistant stage, the oocyst, and in most instances sporogony or sporulation takes place outside the host.

The coccidia were generally said to have direct life cycles; only a single host is used to complete the development. During the late 1960s, however, evidence began to accumulate indicating that there are some exceptions to this principle, and information, concepts, and nomenclature related to coccidia that have merogonous development in extraintestinal sites (*Toxoplasma*, *Sarcocystis*, *Frenkelia*, etc.) have changed rapidly since 1970. This chapter will therefore first take up those coccidia (principally the genus *Eimeria*) found in the intestinal tract or its adjuncts such as the liver and that have direct life cycles. Then we will consider the coccidia that have some development in extraintestinal sites and may require more than one host for the completion of the life cycle (Chap. 15).

Classification of the coccidia at the generic level is based, first, on the structure of the oocyst (Fig. 14.1; Table 14.1) and then on the characteristics of the life cycle. In the organisms illustrated in Figure 14.1, there is an outer wall, usually double, called the *oocyst wall*. Contained within the oocyst are *sporocysts*, which have a characteristic number and themselves contain a characteristic number of sporozoites in each species (Table 14.1). Levine (1973) has developed for the coccidia a periodic table that is more complete than that given here.

In most species oocysts are passed from the host when the oocyst has a uninucleated sporont. Sporogony is called *sporulation* in the coccidia and takes place outside the host; infectivity of the organism is reached only after sporozoites have been formed.

Sporulation requires moderate temperatures, between 15 and 32°C, and oxygen. Each species of coccidium is characterized by the temperature and time over which its sporulation takes place and its resistance to environmental insult. Many species are extraordinarily resistant to both physical and chemical agents. This resistance is best illustrated by the conditions under which oocysts are allowed to sporulate in the laboratory. Since oxygen is required and oocysts are most often recovered from feces, some way must be found to reduce bacterial growth; 2.5% potassium dichromate, 2% sulfuric acid, or 2% formalin are the chemicals most often added to the suspension of oocysts.

After sporulation, oocysts usually remain alive for at least months and in some species 3 years or more when kept cool. Respiration at this time is at a level so low that it is impossible to measure.

When oocysts are ingested by an appropriate host, the sporozoites become active within a few hours or

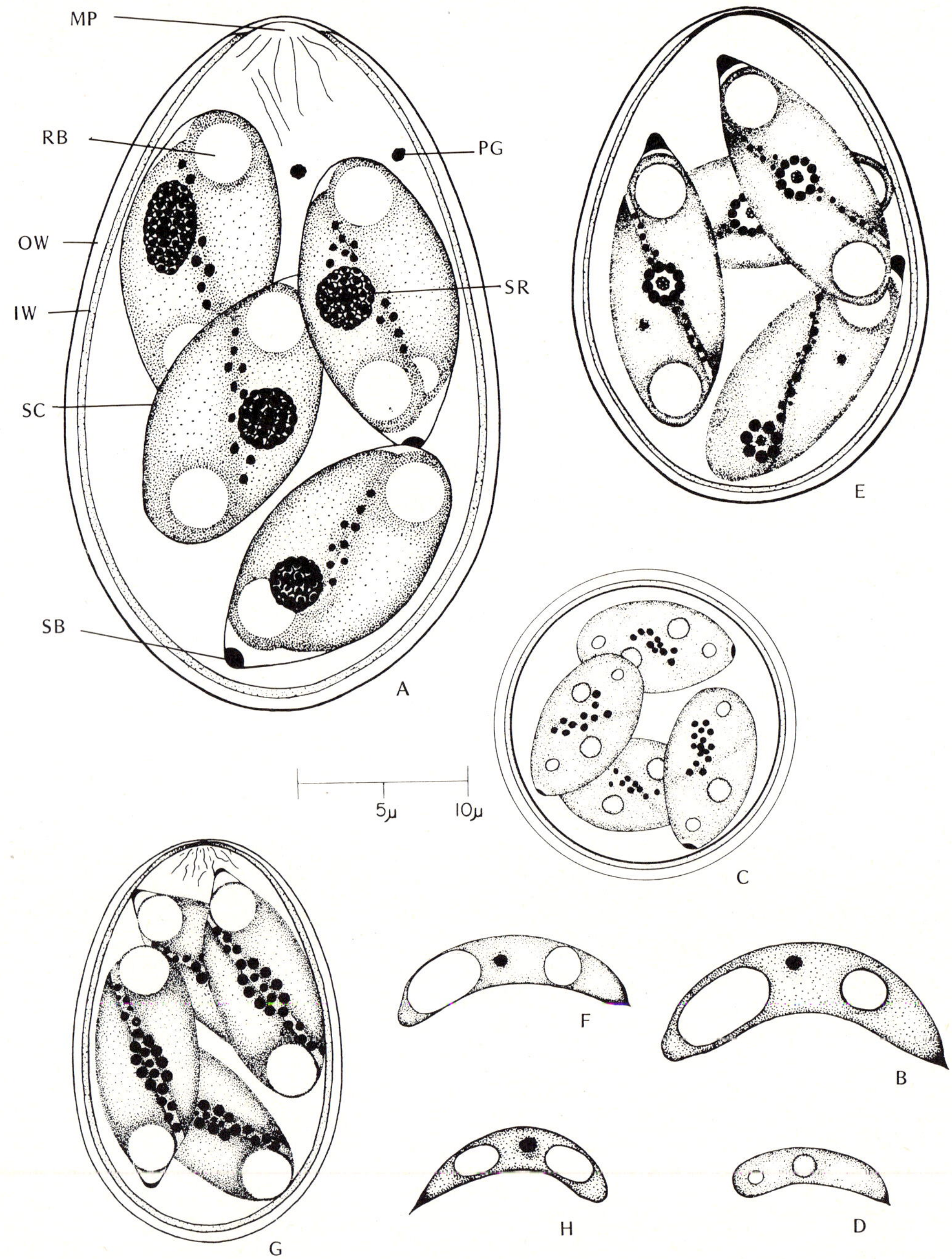
MP
RB
PG
OW
IW
SR
SC
SB
A
E
5μ
10μ
C
G
F
B
H
D

**TABLE 14.1.**
Generic Characteristics of the Oocysts of Some of the Coccidia

| | NUMBER OF | |
|---|---|---|
| GENUS | SPOROCYSTS | SPOROZOITES/OOCYST |
| *Cryptosporidium* | 0 | 4 |
| *Pfeifferinella*<br>*Schellackia*<br>*Tyzzeria* | 0 | 8 |
| *Caryospora* | 1 | 8 |
| *Cyclospora* | 2 | 2 |
| *Isospora*<br>*Toxoplasma*<br>*Besnoitia*<br>*Frenkelia*<br>*Hammondia*<br>*Sarcocystis*<br>*Cystoisospora* | 2 | 8 |
| *Dorisiella* | 2 | 16 |
| *Sivotashella* | 2 | 32 |
| *Eimeria* | 4 | 8 |
| *Wenyonella* | 4 | 16 |
| *Hoarella*<br>*Octosporella* | 8 | 12 |

even a few minutes, leave the sporocyst and oocyst, and invade the tissue of the host. The sporozoites respond to reducing conditions, enzymes, and wetting agents which signal to them that they have reached an environment that will allow the life cycle to continue. Nearly all eukaryotic parasites show a life cycle pattern in which development proceeds to a specific point and then stops. Development will continue only when the organism is stimulated by environmental conditions indicating that a proper host has been reached.

Development within the cells of the host takes place through a series of merogonous generations followed by gamogony and the formation of oocysts. The cycle is then terminated. In most of the coccidia, the number of asexual generations (merogonies) is fixed, but in some species, the number is variable or may continue apparently indefinitely.

Coccidia are widely spread parasites of vertebrates and some invertebrates. Most are intestinal inhabitants, and vertebrates that have been studied extensively are usually found to have from five to ten species of coccidia. The genera most often encountered are *Eimeria* and *Isospora*. There are literally thousands of species of coccidia that have been described from all classes of vertebrates. In all but a small percentage, the only stage of the life cycle known is the oocyst.

The coccidia are important to humans because they are pathogenic for both domesticated animals and humans. Coccidiosis is the cause of losses exceeding $100 million annually in food animals in the United States. *Toxoplasma gondii* infects about one third of the population of North America and causes thousands of human perinatal deaths annually. Most of the species of economic importance are intestinal forms with direct life cycles, but some species of medical importance have not only intestinal stages but also develop in other tissues and have optional or mandatory intermediate hosts in which merogony takes place.

## INTESTINAL COCCIDIA

### *Coccidiosis of Chickens*

The coccidia of poultry have been studied extensively, because coccidiosis is a cause of severe losses in poultry. Since there is a great deal of information on the coccidia of chickens, they will be used here to illustrate many of the general principles in life cycles, development, host–parasite interactions, and control of coccidial infections. In subsequent sections, only unique or specialized aspects of coccidiosis of other host animals will be discussed, and it can be assumed that the generalizations made about chicken coccidia apply.

**FIG. 14.1**
Oocysts of some coccidia of cattle. (*A, B*) Oocyst and sporozoite of *Eimeria auburnensis*; (*C, D*) oocyst and sporozoite of *E. zuernii;* (*E, F*) oocyst and sporozoite of *E. bovis;* (*G, H*) oocyst and sporozoite of *E. ellipsoidalis*. MP = micropyle; RB = refractile globule of sporozoite; OW = outer wall of oocyst; IW = inner wall of oocyst; PG = polar granule; SC = sporocyst; SB = Steida body; SR = sporocyst residuum. [From Nyberg, P. A., and D. M. Hammond. 1965. *J. Parasitol.* 51:669–673.

NAME OF ORGANISM AND DISEASE CAUSED. All of the coccidia of economic importance in chickens are members of the genus *Eimeria*: *E. acervulina*, *E. brunetti*, *E. hagani*, *E. maxima*, *E. mivati*, *E. mitis*, *E. necatrix*, *E. praecox*, and *E. tenella*. These organisms cause coccidiosis and coccy.

HOSTS AND LOCATION. The host of these species is the domestic chicken, *Gallus gallus*. The life cycles of the coccidia of chickens cannot be completed in other hosts such as turkeys, pheasant, or ducks; for all practical purposes, the chicken is the only host for the above *Eimeria* spp.

Each species of coccidium develops in a particular region of the intestine (Fig. 14.2). These regions can be shifted somewhat under experimental conditions, but the preferred sites are characteristic for a species. How this segregation came about is problematic, but having the developmental stages in a particular region of the gut increases the chances of fertilization during the sexual phase of the life cycle.

DISTRIBUTION AND IMPORTANCE. With the distribution of chickens to all parts of the world, their coccidia have been distributed with them, and all species are considered to be cosmopolitan.

Economic losses from coccidiosis come from a number of sources: weight losses, reduced feed efficiency, reduced egg production, and death. Additional costs lie in drugs, veterinary services, and overhead in labor and facilities to prevent clinical infections.

Estimates of economic losses have varied widely. The U.S. Department of Agriculture (USDA) (1965) has estimated that there is a $35 million loss annually in the United States, but others have estimated losses to be as high as $200 million. In two studies at veterinary diagnostic laboratories in poultry-raising areas of the United States, 16 and 19% of birds submitted were found to be suffering from coccidiosis; thus, a fifth to a sixth of clinical diseases of all types were found to be coccidiosis.

MORPHOLOGY. As a general rule, species of coccidia can be identified from the structure of the oocyst. In chickens the oocysts of the various species are similar to one another, but some differentiation is feasible (Fig. 14.2).

Note also that other characteristics of the species are useful in differentiation. The location in the gut has been mentioned. In addition, the sporulation time of the oocysts, the location in epithelial or subepithelial tissue, the type of lesions caused, and the prepatent period all are useful in differentiating the species.

LIFE CYCLE. *Eimeria tenella* is economically the most important coccidium of chickens and has been studied extensively. *E. tenella* is a typical eimerian in that it has merogony and gamogony within the host and sporogony outside it in a direct cycle. There are two or three merogonous generations in the cycle followed by gamogony, which terminates the life cycle (Fig. 14.3).

When sporulated oocysts are ingested, they experience a series of stimuli that trigger excystation of the sporozoites. The gizzard of birds is a specialized structure that grinds food, and the oocysts seem to require this abrasive treatment for activation of the sporozoites. They then require reducing conditions, as provided by cysteine, ascorbic acid, and $CO_2$. Having been thus conditioned, the sporozoites are activated by bile and trypsin. *E. tenella* sporozoites leave the oocyst and sporocysts in the lower part of the small intestine and then invade the tissue of the intestinal ceca.

Sporozoites enter the tissue of the ceca through the tips of the villi. They migrate down to the deeper parts of the tissue via the lacteals in the center of the villi. Having reached nearly the base of the cells forming the intestinal epithelium, the sporozoites enter cells in the crypts of Lieberkuehn and round up to form meronts.

The first merogony takes place in epithelial cells. Each sporozoite may produce up to 900 merozoites and they are released from the host cell late on the second or early in the third day after infection.

The second merogony takes place in epithelial cells, which sink from their surface location deeper into the tissue to a submucosal location. As many as 300 merozoites are produced by each meront, and they are released from the host cell late on the third to early on the fourth day after infection.

In *E. tenella* third-generation merogony takes place in epithelial cells. The consensus is that this third generation is optional; when one studies tissue sections infected with *E. tenella*, these meronts are seldom seen; however, third-generation meronts were present in culture (Doran, 1970)

The second- or third-generation merozoites invade epithelial cells and develop into gamonts. The early

| CHARACTERISTICS | E. ACERVULINA | E. BRUNETTI | E. HAGANI | E. MAXIMA | E. MIVATI | E. MITIS | E. NECATRIX | E. PRAECOX | E. TENELLA |
|---|---|---|---|---|---|---|---|---|---|
| Oocyst size AV(μm) = | 18.3 × 14.6 | 24.6 × 18.8 | 19.1 × 17.6 | **39.5 × 20.7** | **15.6 × 13.4** | 16.2 × 16.0 | 20.4 × 17.2 | 21.3 × 17.1 | 22.0 = 19.0 |
| Length = | 17.7 – 20.2 | 20.7 – 30.3 | 15.8 – 20.9 | **21.5 – 42.5** | 11.1 – 19.9 | 14.3 – 19.6 | 13.2 – 22.7 | 19.8 – 24.7 | 19.5 – 26.0 |
| Width = | 13.7 – 16.3 | 18.1 – 24.2 | 14.3 – 19.5 | **16.5 – 29.8** | 10.5 – 16.2 | **13.0 – 17.0** | 11.3 – 18.3 | 15.7 – 19.8 | 16.5 – 22.8 |
| Zone Parasitized | 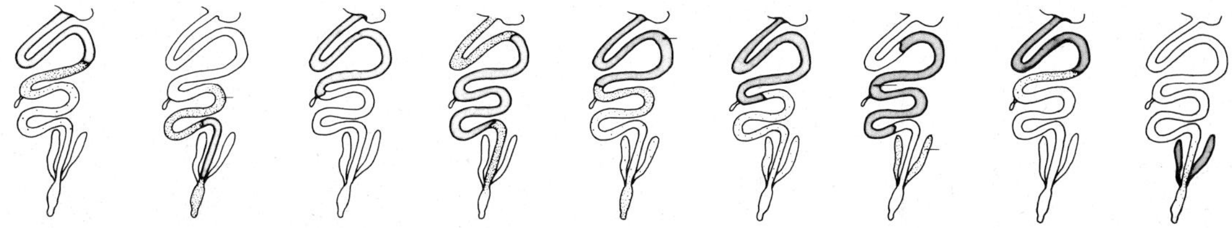 | | | | | | | | |
| Macroscopic Lesions | **White bands,** thick wall, petechiae | **Etechiae, catar. enteritis, blood, coag. necrosis** | Pinhead hemorrhages, petechiae | Petechiae, pink, thick int. walls | **Rounded lesions,** petechiae, congestion | None | Ballooning, white spots, hemorrhages | None | **Hemorrhagic thickening of ceca** |
| **Intestinal Contents** | **Creamy, mucoid** | **Mucoid, bloody** | **Mucoid** | **Mucoid, bloody** | **Creamy, mucoid** | **Mucoid** | **Mucoid, bloody** | **Mucoid casts** | **Bloody cores** |
| **Mortality** | * | ** | * | ** | * | * (Young birds) | *** | 0 | *** |
| **Morbidity** | *** | * | * | *** | ** | * | *** | * | *** |
| **Sporulation Minimum, Hour** | 17 | 18 | 18 | 30 | 12 | 18 | 18 | **12** | 18 |
| Prepatent Period, Hour | 97 | 120 | 99 | 123 | 93 | 99 | 138 | **84** | 138 |
| Immunity, Speed of Development | Prompt | "Sub-immediate" | "Prompt" | Prompt | Delayed | Slow, little | Delayed | Prompt | Rapid |
| Immunity, Duration | Moderate | Moderate | ? | Short | Short | Moderate | Long (?) | Moderate | Long |
| Schizont, Max., in Microns | 10.3 | 30.0 | | 9.4 | 17.3 | 11.3 | **65.9** | | 54.0 |
| Schizont, Tissue Location | Above epithelial nuclei | Second-generation subepithelium | Epithelial | Above epithelial nuclei | Above epithelial nuclei | Below epithelial nuclei | Second-generation subepithelium | Below epithelial nuclei | Second-generation subepithelium |
| Gametocyte Tissue Location | Above or below nuclei | Below epithelial nuclei | ? | **Deep in epithelium** | **Above or below nuclei up to 5 per cell** | Above or below epithelial nuclei | Above or below epithelial nuclei | Above or below epithelial nuclei | Below epithelial nuclei |

* —low.
** —moderate.
*** —high.

**FIG. 14.2**
Characteristics of chicken coccidia. (Items shown in bold face are diagnostic.) [Redrawn with modifications from Reid and Long, 1979.]

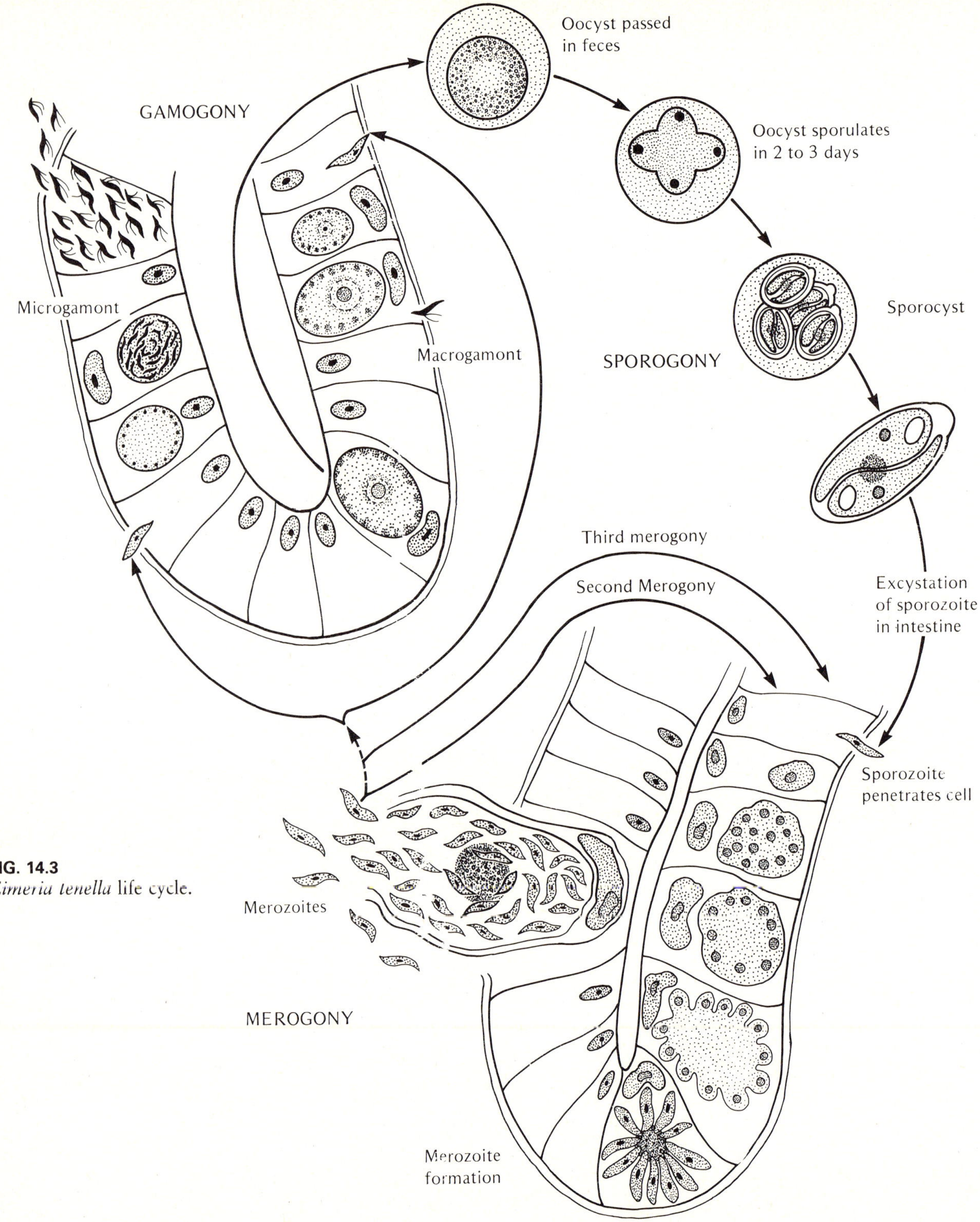


**FIG. 14.3**
*Eimeria tenella* life cycle.

gamonts are not distinguishable from early meronts, nor are the macro- and microgamonts distinguishable from one another at this time. When the microgamonts begin to undergo nuclear replication, the nuclei move to the margins of the gamont and the compact, darkly staining nuclei can be recognized. Late on the fifth day or early on the sixth day after infection the flagellated microgametes have been formed, and they leave their host cells to seek a macrogamont.

The macrogamonts are not motile, and fertilization takes place within the host cell. After fertilization, the oocysts walls are laid down. Two kinds of wall-forming bodies have been identified by electron microscopy, and these bodies move peripherally in the gamont. In histologic sections prepared for examination under a light microscope and stained with H&E, the wall-forming bodies take eosin stain and are recognized as red granules. As the bodies reach the margin of the fertilized gamete, they coalesce and form the outer wall of what is now an oocyst. During this time, the organism retains a single nucleus. Oocysts begin to appear in the feces of the bird late on the sixth day after infection and continue to be shed for about 4 days. Peak numbers of oocysts are seen in the feces during the seventh day after infection.

Regardless of whether there are two or three merogonous generations in *E. tenella*, the cycle is self-limiting. When merogony has been completed, gamogony takes place and the cycle in the host is then finished. Most infectious agents continue to reproduce in the host until immunity somehow limits their life span. In the coccidia, the number of merogonies is apparently programmed and, having completed them, the organisms commence gamogony. Such programming can best be demonstrated by experiments in which merozoites were transferred from infected animals to animals that had never been infected with coccidia. Oocysts were produced at the same time in the recipients as would be expected if the parasites had remained in the same host.

When oocysts have been passed from the host, sporogony or sporulation begins. Temperatures above about 35°C and low oxygen tension inhibit sporulation in most coccidia of birds and mammals, so nothing happens until the oocysts encounter lower temperatures and oxygen outside the host. In *E. tenella*, oocysts can sporulate in as few as 18 hr under the optimal conditions of adequate oxygen and 29°C. However, most oocysts will not reach infectivity at these conditions until about 30 hr.

DIAGNOSIS. Determining whether birds are ill with or have died from coccidiosis is based on the following:

1. Clinical signs of diarrhea, sometimes with blood
2. Weight losses or reduced feed efficiency
3. Finding large numbers of oocysts in the droppings
4. Finding typical lesions at postmortem examination
5. Finding a typical pattern of infection in the flock in which a few birds with signs on a certain day are followed by increasing numbers of birds showing illness during the next few days

A poultry raiser must be sensitive to minor changes in behavior and food and water consumption of the birds in a flock. Chickens that are ill tend to eat and drink less and sit hunched up with ruffled feathers. This is not an indication of coccidiosis, but only that something is not right.

In almost all instances in which there is a disease problem, a few birds are removed from the flock, killed, and examined postmortem. One looks for changes in the gut such as thickening of the intestinal wall and redness or leakage of blood from the tissue into the lumen. Scrapings of the gut examined microscopically will reveal large numbers of developmental stages of coccidia, usually gamonts and oocysts. The content of the lower part of the digestive tract will have large numbers of oocysts in it.

Determining the cause of losses in poultry should be left to a skilled diagnostician. There are a number of infectious agents that cause diarrhea and weight loss in birds, and finding a few oocysts in the feces or a few lesions in the gut indicative of coccidial infection may or may not answer the question of what is causing the disease.

HOST–PARASITE INTERACTIONS. In the coccidia, interaction with the host can be looked upon as having three phases:

1. Establishment
2. Damage to the host
3. Immunity

When a sporulated oocyst is ingested by a host, the sporozoites are activated by factors in the gut (see Life Cycle). The sporozoites escape from the oocyst and invade the tissue of the host; by and large, this is a nonspecific reaction. It was shown early in the studies on the coccidia that sporozoites will leave the oocyst and enter the tissues of nearly any species of host (Andrews, 1927). The sporozoites respond to general conditions that are found in the intestinal tracts of most warm-blooded animals; however, development and multiplication probably are dependent on specific factors within the cells of the host that provide the parasites with their nutrient requirements. We have only fragmentary information on the nutritional requirements and metabolic pathways of the coccidia, but most logically, the high degree of host specificity shown by most members of the group is based on nutrient requirements.

Not only do the coccidia have a high degree of host specificity, they also have a high degree of site preference in the host. Coccidia of chickens have particular locations along the length of the gut that they prefer (Fig. 14.2). In addition, they select specific kinds of cells in which to develop. The *Eimeria* spp. of chickens all develop in epithelial cells, although the cells sometimes move into other locations (see second-generation meronts of *E. tenella*). In *E. bovis* of cattle, the first-generation meronts are located in endothelial cells of the lacteals of the villi in the lower part of the small intestine. The second-generation meronts and the gamonts are in epithelial cells of the crypts of the large intestine.

The organisms show not only a preference for a cell type but a predilection for specific locations within the cell. The first-generation meronts of *E. tenella* lie on the basal side of the epithelial cells and the second-generation meronts locate on the luminal side. Six or so species of *Eimeria*, such as *E. alabamensis* of cattle, develop within the nuclei of their host cells.

Not all species of coccidia are equally pathogenic. In the case of the coccidia of chickens, *E. tenella* is considered to cause the greatest economic loss overall. *E. brunetti*, *E. maxima*, *E. mivati*, and *E. necatrix* cause less, but still severe, economic loss. The three remaining species, *E. hagani*, *E. mitis*, and *E. praecox*, may cause losses under some circumstances, but the damage to the host is mild and short-lived.

The characteristics of the coccidia that determine their degree of pathogenicity are the following:

1. Reproductive potential
2. Location in the tissues
3. Strain of coccidium

Each merogonous generation has a characteristic number of merozoites and finally of oocysts that are produced, and the number of cells potentially destroyed by ingestion of a single oocyst can be calculated. Thus, if one oocyst has eight sporozoites, multiplying this by the number of merozoites in the first generation and then by the number in the second generation gives the number of cells that *can* be destroyed. In *E. tenella*,

$$8 \text{ (sporozoites)} \times 900 \text{ (1st-generation merozoites)} \times 350 \text{ (2nd-generation merozoites)} = 2.52 \times 10^6$$

Often one wants to determine the reproductive index (RI), which is the number of oocysts produced per oocyst fed; this is obtained in the same way except that the product is divided by 2, since only the macrogamonts appear in the feces as oocysts:

$$\frac{8 \times 900 \times 350}{2} = 1.26 \times 10^6$$

These are useful working figures, but they use the largest number of merozoites produced by each merogonous generation. *E. tenella* has an observed RI of about 60,000. Some of the other pathogenic coccidia of chickens have lower RIs: *E. acervulina*—9000; *E. maxima*—11,000; and *E. necatrix*—50,000. Certain of the coccidia that have large numbers of first-generation merozoites are also highly pathogenic; examples are *E. tenella* of chickens and *E. zuernii* of cattle.

The depth in the tissue of the gut seems also to be a factor in the degree of pathogenesis of many species. In *E. tenella*, the second-generation meronts are subepithelial, and at about the time when the second-generation merozoites leave the cell, there is disruption of large areas of epithelium. Large amounts of blood are then lost, the loss of the epithelium removes the first line of defense, and bacteria can invade the gut tissue. In contrast, *E. debliecki* of swine inhabits only the epithelial cells of the intestinal villi; although a slight diarrhea may be seen in infected animals, the damage to the host is minor.

The number of oocysts ingested by an animal also determines the amount of damage that occurs. From what has been said, the more cells infected, the greater the damage to the host. It follows that if thousands or millions of oocysts are ingested, the damage to the tissues of the host will be greater than if tens or hundreds of oocysts are ingested.

As is true of nearly all parasitic and microbial agents of disease, pathogenic potential is not uniform throughout the species. Thus, with the coccidia of chickens, isolates made in different places will cause greater or less damage when introduced into experimental hosts. What has been said here about certain species may be true for most isolates, but some will be more and some less damaging to the host.

If one examines tissues of animals affected with coccidiosis, there are characteristic changes that can be seen grossly at postmortem examination and in tissues prepared for microscopic examination. The wall of the gut is usually thickened, indicating retention of excessive amounts of fluid (edema). There may be blood in the lumen of the gut, indicating blood loss (hemorrhage), or merely retention of excessive amounts of blood in the tissue (hyperemia). There is also infiltration with various kinds of white cells or leukocytes, indicating a foreign body reaction and the development of an immune response. The greater the tissue changes, the larger will be the number of parasites seen, and the greatest changes will be observed in the vicinity of the protozoa. In most instances, the appearance of clinical signs of coccidiosis coincides with the appearance of oocysts in the feces. An exception to this principle is a heavy infection with *E. tenella* in which blood begins to be lost and illness becomes apparent on the fifth day after infection when the second-generation meronts are maturing.

In typical coccidial infections, a few oocysts are seen in the feces when gamogony is occurring, and the number increases rapidly for about 2 days, peaks, and then drops precipitously; the gut tissue then heals. In light infections, the gut will be nearly normal within about a week after oocyst production has ceased. In heavy infections, there are nonreversible changes in the gut, and the consensus seems to be that such animals will not be thrifty.

For the livestock raiser, rapid weight gain and good feed conversion (amount of feed/amount of gain) are important for a satisfactory economic return. It is especially true in the poultry industry, where the profit margin on any one bird may be pennies, that a little more time to reach market weight or a little wasted feed through inefficient conversion may mean the difference between profit and loss on a flock. Most of the studies on weight gain and feed conversion have been done with chickens, and it can be seen that during the peak of the coccidial infection both are poor. However, the infected birds do gain more rapidly after the infection is over and begin to approach the weights of the uninfected birds.

Immunity is an important aspect of coccidial infections in poultry, and of other animals as well. After an animal has passed through even a subclinical infetion with *E. tenella*, it will be immune to infection when challenged by oral inoculation of infective oocysts. Immunity will be evident in two ways: (1) by reduction in the expected numbers of oocysts appearing in the feces and (2) by showing mild or no clinical signs of infection such as reduction in weight gain or diarrhea. Not all species react in the same way; in some species multiple exposures may be necessary to obtain a high level of immunity.

There is little or no cross-immunity between species of *Eimeria* in the chicken. If an animal is exposed to *E. tenella*, for example, it is completely susceptible to infection with *E. necatrix* or *E. acervulina*. The few instances of cross-immunity between species are not generally considered to be of much importance as a means of limiting infections.

TREATMENT. Effective drugs for use against the coccidia of chickens were not developed until 1940, when the sulfa drugs came into use and were found to be effective. During the intervening 40 years, at least ten classes of compounds have been used successfully as anticoccidials for chickens (Fig. 14.4).

Two approaches are used in administering drugs to chickens to prevent economic losses: (1) treatment of the flock after signs of infection are seen and (2) administration of drugs on a continuous basis to prevent the development of significant infections, that is, prophylaxis.

In both instances, drugs are put into the feed or water and birds self-medicate. Treatment or prevention of diseases in poultry flocks is nearly always carried out in such a way that individual birds are not handled. Profit margins in the poultry industry require that labor costs through handling be kept to a minimum. Therefore, anticoccidials are mixed with feed in precise amounts or introduced into the water by devices that monitor the amount of drug distributed throughout the poultry house.

Treatment of birds after signs of coccidiosis are seen has one advantage and several disadvantages. On the plus side, chemicals are not used until necessary, thus having both economic benefits and no environmental

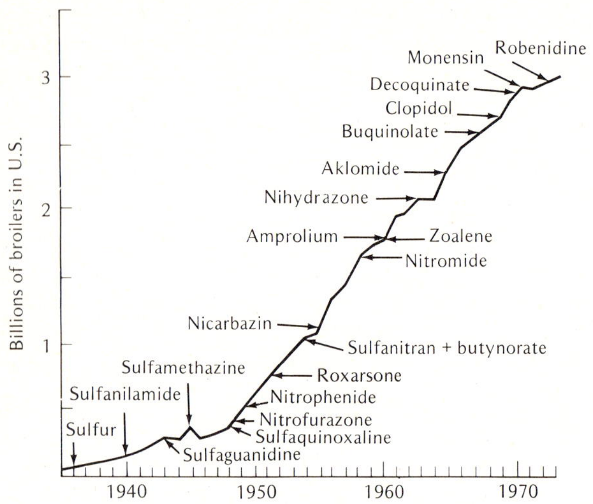


**FIG. 14.4**
Anticoccidials used in the raising of poultry or in laying hens. [Redrawn from Reid & McDougald, 1981.]

contamination. On the negative side, coccidiosis must be diagnosed early; a few hours' delay can usually be tolerated, but a day's delay may mean excessive losses. Second, by the time signs of coccidiosis appear, chemotherapy will have little effect on the course of the disease; birds in the flock not yet showing signs will be helped somewhat. In a large poultry house, coccidiosis may start in one part and spread; in this case, drug treatment will probably be effective. As indicated earlier, signs of infection usually appear at about the time when oocysts are being formed or are appearing in the feces. Few drugs have much effect on the sexual stages of the cycle. Another problem in treating sick birds through self-medication is that one of the first signs of coccidiosis (and other diseases) is reduced feed and water consumption. Obviously if they do not eat and drink, they will not get much drug.

Most of the effort in developing anticoccidials has been directed toward prevention of disease. Chemicals are added to feed with the objective of preventing the development of the merogonous stages of the cycle. The advantages of the prophylactic approach are that (1) losses are prevented, since the drugs act against the early developmental stages of the life cycle and (2) immunity in the birds is allowed to develop through partial development of the parasites. Properly done, losses from coccidiosis do not occur. Disadvantages lie (1) in the cost of administering compounds when problems might not arise, (2) in residue of compounds in meat or eggs, and (3) in the potential resistance to the drugs developed by the parasites, which reduces their efficacy.

Different compounds have different levels of activity and act at different stages in the life cycle. Compounds are spoken of as being potent, intermediate, or weak in their ability to suppress the development of the endogenous stages of a coccidium. Table 14.2 lists some of the drugs and their level of activity. The potent compounds will prevent development to such an extent that almost none of the organisms will complete the cycle and produce oocysts. At the other end of the spectrum, the weak anticoccidials allow production of large numbers of oocysts. One might think that weak drugs would be undesirable, but some contamination of the litter is considered to be good since it is the means by which birds develop a high level of immunity through reinfection. Potent anticoccidials prevent development of the organisms to such an extent that there is relatively little contamination of the litter with oocysts; if the drug is withdrawn a week before the birds go to market, for example, they may be completely susceptible to infection and develop coccidiosis before being shipped.

**TABLE 14.2**
Levels of Activity of Anticoccidials

| POTENT | INTERMEDIATE | WEAK |
|---|---|---|
| 1. Nicarbazin (NICARB, Merck & Co.) | 1. Amprolium plus ethopabate (AMPROL PLUS, Merck & Co.) | 1. Amprolium (AMPROL, Merck & Co.) |
| 2. Buquinolate (BONAID, Morton-Norwich) | 2. Monensin (COBAN, Eli Lilly Co.) | 2. Zoalene (ZOAMIX coccidiostats, Dow) |
| 3. Nequinate (STATYL, Ayerst) | | |
| 4. Decoquinate (DECCOX, Hess & Clark) | | |
| 5. Clopidol (COYDEN—25 coccidiostat, Dow) | | |
| 6. Robenidene (CYCOSTAT, American Cyanamid) | | |

*Source: The Practicing Nutritionist.* Vol. 10, No. 1, 1976.

A continuing problem in using anticoccidials for poultry is the development of drug resistance. As a general rule, an anticoccidial will be in use for about 2 years before its efficacy decreases for this reason. Coincidentally, a new drug has come on the market about every 2 years. Drug resistance comes about as a result of a genetic change in the organisms that allows them somehow to avoid the effects of the drug. In a series of well-designed experiments, Weppleman et al. (1977) showed that *Eimeria tenella* undergoes genetic changes prior to exposure to a drug that allow resistance. Glycarbylamide is known to allow development of drug resistance quite readily. After a single exposure to it, it was calculated that $2.4 \times 10^{-7}$ oocysts were preadapted for resistance. Another drug, amoquinate, had $5.8 \times 10^{-8}$ oocysts with prior resistance, a level more than an order of magnitude greater than that of glycarbylamide. Other drugs tested were amprolium, monensin, robenidine, and nicarbazin, which showed a mutation level less than $7.5 \times 10^{-9}$ per oocyst, another order of magnitude less likely to develop resistance than amoquinate. These findings are generally consistent with observations made in the poultry industry on the length of time a drug has good efficacy.

A further difficulty may arise where resistance to one particular product confers resistance to all members of that class of compound. This was found with the quinolones. Three different quinolones were marketed—buquinolate, decoquinate, and methyl benzoquate. When a strain of coccidium developed resistance to one, it was also resistant to the other two compounds.

The whole question of drug resistance poses problems for both poultry and egg producers and for the pharmaceutical industry. Costs of developing and marketing an anticoccidial exceed several million dollars for any one product. Costs include laboratory screening of chemicals with potential activity, testing of birds in small batteries, testing for residues in eggs and meat, and field trials involving literally millions of birds. A significant portion of the cost in North America and Western Europe comes from obtaining governmental clearance for marketing of the drug. In the United States, the Food and Drug Administration (FDA) has final authority in deciding whether any proposed drug is both safe and effective; other countries have similar agencies. The rewards can be great, but the risks are also great. In the late 1950s, glycarbylamide was placed on the market with great fanfare, but it disappeared within a year because of drug resistance. On the other hand, monensin was introduced in about 1970, and during a 10-year period, resistance has not been a problem. Means of circumventing the effects of drug resistance are discussed in the next section on control of coccidiosis in poultry.

EPIDEMIOLOGY AND CONTROL. Coccidiosis in poultry results from the ingestion of large numbers of sporulated oocysts by nonimmune birds. Reduction in the effects of such infections is brought about by the development of natural immunity and use of drugs to reduce the reproduction of the coccidia.

Knowing the conditions under which oocysts develop to infectivity and then survive is essential to designing control programs. Unsporulated oocysts are no hazard, and the sporozoites must be completely formed before they can excyst and invade the tissues of the host.

After sporulation has taken place, good survival of the oocysts requires some humidity and a little oxygen. A relative humidity of 25% or less is detrimental, and wet conditions, in which there is a high level of bacterial growth, will shorten survival. Freezing is detrimental to survival, especially when conditions allow repeated freezing and thawing or when the temperature sinks below about –10°C.

It is generally agreed that coccidia are found wherever chickens are raised. If one starts a flock from clean, newly hatched chicks, there is almost a written guarantee that they will become infected with at least three species of *Eimeria*. During the first 2 weeks of life, chicks are relatively refractory to infection with coccidia. Then at 2 to 4 weeks of age they become quite susceptible and the effects of coccidial infection may begin to appear. At first a few birds will appear listless and take less than normal amounts of feed and water. With *E. tenella*, this indicates that the second-generation meronts are reaching maturity and damage to the gut is becoming severe. Within half a day, more and more birds show signs of coccidiosis and the following day, weight losses, diarrhea, and deaths begin to occur. At this time, the epizootic is in full swing, and little can be done to retard losses. But within 3 days, weight loss, diarrhea, and deaths taper off and the remaining birds return to nearly normal increases in weight. In most instances, the birds have become immune and the outbreak is over.

What do we know about the parasite, host – parasite

interactions, and drugs that can be used to reduce the effects of coccidiosis?

1. Oocysts are ubiquitous, and the probability of avoiding infection is so low that it is not worth considering.
2. The cycle is self-limiting; after the predetermined number of asexual generations, gamogony occurs, oocysts are produced, and the infection is over.
3. Infection induces immunity. Even relatively small numbers of oocysts, especially if ingested over a period of several days or weeks, give rise to an immunity that protects the birds from the pathogenic effects of the coccidia.
4. Treatment after clinical signs have appeared is generally ineffective in preventing losses.
5. Certain chemicals given in the feed will reduce the reproduction of the coccidia and the pathogenic effects they normally would cause.
6. At proper levels, some of the chemicals will allow sufficient development of the coccidia for immunity to reach a protective level.

As knowledge of anticoccidials developed, it was found that mixing drugs with feed would prevent coccidiosis, but when the drug was withdrawn, birds were often completely susceptible and would succumb to infection. The amount of drug in the feed was then reduced to the point where some development would take place so that immunity would be induced. This worked better, but it could not be guaranteed that birds would ingest sufficient numbers of oocysts to become immune. Dr. Allen Edgar developed a system in which living oocysts of the six major pathogens were mixed with the feed and with an anticoccidial drug to ensure that an immunizing infection would occur. The basic concept in Edgar's method is that controlled infections will result in a reliable immunity. Edgar's system is called *Coccivac* and has been widely accepted in the raising of fryers.

Let us consider further the effects of using a potent, intermediate, or weak anticoccidial in a flock. A potent drug will allow little development and practically no passage of oocysts into the environment. A weak anticoccidial drug will allow considerable contamination of the environment. Note that weight gains are better in the treated than the nonmedicated birds and that feed conversion is about the same in all groups.

The problem arises when the drug is withdrawn. Half of the birds on a potent anticoccidial drug die; development of the endogenous stages is so restricted that immunity does not occur, and there is still sufficient environmental contamination with oocysts that the effects are severe.

After continued administration of a drug in the feed for some period of time, efficacy may decrease as a result of drug resistance on the part of the parasite. Companies raising large numbers of chickens will monitor feed efficiency and usually become aware of the problem when it arises. Two methods are used by raisers to avoid the effects of resistance—switching and shuttling. Switching is using one product for a batch or two of birds, and then using another for a batch or two. Shuttling is a technique in which different drugs are used successively during the growth of the same batch of birds. One drug might be used for a week or two and then a second drug and perhaps a third. In both methods the assumption is the same: different classes of drugs block different metabolic pathways, and the probability of the parasites becoming resistant to two different compounds is extremely low. The assumption is generally true, but strains of coccidia that are resistant to two or three different classes of compounds have been isolated.

In the industrialized countries of the world, and some of the developing ones now, poultry production is almost entirely a high-technology enterprise. Tens of thousands of birds are housed in a single building. Buildings are climate controlled and the day length is set for maximum weight gain or egg production. Feed is blended by computer-controlled machinery to take into account the needs of the animals at their stage of growth and the nutrient quality of the feed. Poultry production thus has become a capital-intensive enterprise as opposed to the labor-intensive enterprise of 50 years ago. When everything goes well, high technology is a culmination of human scientific and engineering endeavors. When something goes awry, disaster results. Poultry raisers plan carefully to prevent disease outbreaks, not only of coccidiosis but of other transmissible diseases as well. A detailed account of all kinds of diseases of birds raised for food is contained in the book entitled *Diseases of Poultry* (6th edition, 1972) edited by M. S. Hofstad and others.

## Coccidosis in Turkeys

Nine species of coccidia infect turkeys—seven *Eimeria*, one *Isospora*, and one *Cryptosporidium*—but only *E. meleagrimitis* and *E. adenoeides* are considered economically important. Coccidiosis in turkeys is a continuing economic threat, but with only two species considered to be economically important, the problem is usually not as complex as in chickens. An early estimate (USDA, 1954) indicated about a \$500,000 annual loss in the United States from coccidiosis in turkeys.

## Coccidiosis in Cattle

There are 12 to 14 species of *Eimeria* that infect domestic cattle, *Bos taurus*; some are illustrated in Figure 14.1. Those of greatest economic importance are *E. bovis* and *E. zuernii*, but *E. auburnensis* and *E. alabamensis* also cause diarrhea on occasion (Levine & Ivens, 1970).

## Coccidiosis in Sheep

There are ten or so species of *Eimeria* that infect domestic sheep. Of these, *E. ovinoidalis* (=*E. ninakohlyakimovae*), *E. ahsata*, *E. ovina* (=*E. arloingi*), and *E. parva* cause losses in sheep, probably in about the order in which they are listed.

Economic losses from coccidiosis in sheep have not been well documented. As early as 1949 estimates for losses in the United States were \$10 million annually, and although the figure has often been repeated, its basis is rather tenuous.

## Coccidiosis in Rabbits

The domesticated rabbit is *Oryctolagus cuniculi*; it is native to Europe. This is the animal raised for food and for use in laboratory experimentation. *Oryctolagus* has perhaps eight or so species of *Eimeria*, and a few of them have been reported from native North American rabbits (*Sylvilagus* spp), the cottontail (Duszynski & Marquardt, 1969). In most instances, good cross-infection experiments have not been done. The following discussion concerns *Oryctolagus*.

Coccidiosis in rabbits usually occurs where the animals are kept under less than ideal conditions of cleanliness. It is a rare rabbit that does not shed oocysts continuously in the feces, and the hazard of contamination of the environment with sporulated oocysts always exists. The species that cause most losses are *E. steidai*, which infects the epithelium of the bile ducts of the liver and causes so-called liver coccidiosis, *E. magna*, *E. media*, *E. perforans*, and *E. residua*, all of which are intestinal forms.

Control of coccidiosis in rabbits is based on two main methods: (1) cages are constructed so that they are easily cleaned and kept clean, and (2) chemoprophylaxis is used in young animals so that infections are suppressed and immunity can develop.

## Cryptosporidiosis

*Cryptosporidium* spp. infect mammals, birds, and reptiles where they live in the brush border of the intestinal cells (Vetterling et al., 1971) and sometimes develop in other organ systems such as the trachea and kidney. *Cryptosporidium muris* was described from a mouse in 1907 by Tyzzer; since then a large number of species have been described. This coccidium has generally been considered another of the many commensals that can be found in almost any host. A few cases were on record of humans or other animals suffering from intractable diarrhea caused by *Cryptosporidium*. In the late 1970s, information began to accumulate that cryptosporidiosis was a disease of significance. The spreading epidemic of acquired immunodeficiency disease syndrome (AIDS) showed that cryptosporidiosis was one of the many infections from which its victims suffered. Domestic animals also were found to be clinically affected by *Cryptosporidium* spp. (Reese et al., 1982).

Cryptosporidiosis has become a concern not only because it has occurred in hosts that were not immunocompetent, but also because it has occurred in normal hosts, and it appears that the host specificity is low. In a number of case reports in which persons have become infected from livestock, the diarrhea was both severe and prolonged. Diagnosis of cryptosporidial infections is generally done by looking for the oocysts in feces following concentration by a flotation technique. The oocysts are small and often present in small numbers, so they are easily overlooked. Serological studies have shown that many species of animals have antibodies to *Cryptosporidium* spp. (Tzipori & Campbell, 1981), but the prevalence of the infection and its importance are not yet clear.

## REFERENCES AND FURTHER READINGS


Andrews, J.M. 1927. Host–parasite specificity in the coccidia of mammals. *J. Parasitol.* **13**:183 – 194

Canning, E. U., and M. Anwar. 1968. Studies on meiotic division in coccidial and malarial parasites. *J. Protozool.* **15**:290 – 298.

Doran, D. J. 1970. *Eimeria tenella*: From sporozoites to oocysts in cell culture. *Proc. Helm. Soc. Wash.* 70:84 – 92.

Duszynski, D. W., and W. C. Marquardt. 1969. *Eimeria* (Protozoa: Eimeriidae) of the cottontail rabbit *Sylvilagus audobonii* in northeastern Colorado, with descriptions of three new species. *J. Protozool.* **16**:128 – 137.

Hammond, D. M., and P. L. Long (eds.). 1973. *The Coccidia.* Eimeria, Isospora, Toxoplasma *and Related Organisms.* University Park Press, Baltimore.

Hofstad, M.S., et al. (eds.). 1972. *Diseases of Poultry*, 6th ed. Iowa State University Press, Ames.

Jensen, J. B., and S. A. Edgar. 1978. Fine structure of penetration of cultured cells by *Isospora canis* sporozoites. *J. Protozool.* **25**:169–173.

Kheysin, Y. M. 1972. *Life Cycles of Coccidia of Domestic Animals.* K.S. Todd (ed.), F.K. Plous (trans). University Park Press, Baltimore.

Levine, N. D. 1973. *Protozoan Parasites of Domestic Animals and of Man*, 2nd ed. Burgess, Minneapolis.

Levine, N. D., and V. Ivens. 1965. *The Coccidian Parasites (Protozoa, Sporozoa) of Rodents.* Ill. Biol. Monogr. No. 33.

———1970. *The Coccidian Parasites (Protozoa, Sporozoa) of Ruminants.* Ill. Biol. Monogr. No. 44.

———1981. *The Coccidian Parasites (Protozoa, Apicomplexa) of Carnivores.* Ill. Biol. Monogr. No. 51.

Long, P. L. (ed.). 1981. *The Biology of the Coccidia.* University Park Press, Baltimore.

Long, P. L., and C. Horton-Smith. 1968. Coccidia and coccidiosis in the domestic fowl. *Adv. Parasit.* **6**:313 – 325.

Long, P. L., and B. J. Millard. 1979. Rejection of *Eimeria* by foreign hosts. *Parasitology.* **78**:239–247.

Marquardt, W. C. 1973. Host and site specificity in the coccidia. In D. M. Hammond and P. L. Long (eds.), *The Coccidia.* Eimeria, Isospora and Toxoplasma. University Park Press, College Station, Pa.

Mayberry, L. F. 1973. Transmission of *Eimeria separata* from the normal host, *Rattus*, to the mouse *Mus musculus. J. Parasitol.* **59**:198 – 199.

McDougald, L. R. 1978. The growth of avian *Eimeria* in vitro. In P. L. Long, K. N. Boorman and B. M. Freeman (eds.), *Avian Coccidiosis.* Proc. 13th Poult. Sci. Sympos., September 14 – 16, 1977, pp. 185 – 223.

Pellèrdy, L. P. 1963. *Catalogue of Eimeriidea (Protozoa: Sporozoa).* Publishing House of Hungarian Academy of Science, Budapest.

———1965. *Coccidia and Coccidiosis.* Publishing House of Hungarian Academy of Science, Budapest.

———1969. *Catalogue of Eimeriidea (Protozoa, Sporozoa).* Supplementum I. Akad. Kiado, Budapest.

Reese, W. C., W. L. Current, J. V. Ernst, and W. S. Bailey. 1982. Human and calf cryptosporidiosis: A case report and results of comparative infections in mice and rats. *Am. J. Trop. Med. Hyg.* **31**:226 – 229.

Reid, W.M., and P.L. Long. 1979. A diagnostic chart for nine species of fowl coccidia. Univ. of Georgia, College of Agriculture, Experiment Station, Research Report.

Reid, W. M., and L. R. McDougald. 1981. New drugs to control coccidiosis: To be or not to be. *Feedstuffs* 53:27 – 28.

Tyzzer, E. E. 1907. A sporozoan found in the peptic glands of the common mouse. *Proc. Soc. Exp. Biol. Med.* 5:12 – 13.

———. 1929. Coccidiosis in gallinaceous birds. *Am. J. Hyg.* **10**:269–383.

Tzipori, A., and I. Campbell. 1981. Prevalence of *Cryptosporidium* antibodies in 10 animal species. *J. Clin. Microbiol.* **14**:455 – 456.

Vetterling, J. M., A. Takeuchi, and P. A. Madden. 1971. Ultrastructure of *Cryptosporidium wrairi* from the guinea pig. *J. Protozool.* **18**:248 – 260.

Wang, C. C. 1976. Some procaryotic characteristics of ribosomes from *Eimeria tenella. Fed. Proc. Am. Soc. Exp. Biol.* 35:786.

———. 1978. Biochemical and nutritional aspects of coccidia. In P.L. Long, K.L. Boorman, and B. M. Freeman (eds.), *Avian Coccidiosis.* Proc. 13th Poult. Sci. Sympos., September 14–16, 1977, pp. 135–184.

Weppelman, R. M., G. Olson, D. A. Smith, T. Tamas, and A. Van Iderstine. 1977. Comparison of anticoccidial efficacy, resistance and tolerance to narasin, monensin and lasalocid in chicken battery trials. *Poul. Sci.* **56**:1550 – 1559.

# 15

# *Extraintestinal Coccidia*

The organisms included in the extraintestinal coccidia have *Isospora*-like oocysts (see Table 14.1) and two hosts in the life cycle, one of which is obligatory and the other either obligatory or optional. In the cycles as currently known, one host is a predator and the other host is a prey animal. The gamogonous phase of the life cycle always takes place in the carnivorous or omnivorous host, and sometimes the merogonous phase of the life cycle does. The merogonous phase of the life cycle always takes place in the prey host.

Two genera of these coccidians have been known for a relatively long time. Meronts of *Sarcocystis miescheriana* (=*S. suihominis*) were first described by Kühn in 1865; meronts of *Toxoplasma gondii* were first described by Nicolle and Manceaux in 1908 from a small North African rodent, the gondi. Although both of these genera were generally recognized to be protozoa, their precise taxonomic position awaited the development of thin-sectioning techniques for the electron microscope. In the early 1960s, a century after *Sarcocystis* was first recognized, the elements of the apical complex were described and found to occur in organisms that caused malaria, coccidiosis, toxoplasmosis, and sarcocystosis. With this information, it became more likely that *Toxoplasma* was a coccidium, and in 1969 workers in three different laboratories simultaneously determined that *T. gondii* is a coccidium whose principal host is the domestic cat and whose merogonous development most often takes place in the mouse. (Fig. 15.1) (Frenkel et al., 1970; Sheffield & Melton, 1970). The work of Fayer (1972) showed that *Sarcocystis* isolated from grackles and allowed to develop in tissue cultures produced what appeared to be oocysts. Almost immediately thereafter workers in several laboratories, especially in Germany, showed that sarcocysts from the muscles of domesticated animals gave rise to isosporoid oocysts when fed to dogs or cats (Mehlhorn & Heydorn, 1978).

Knowledge on the economic and medical importance of some of the tissue coccidia is just now unfolding. Starting in the early 1940s, *T. gondii* came to be recognized as a significant pathogen of humans, and its role has become clearer with the passage of time. The importance of *Sarcocystis* has remained obscure, but recent work indicates that some infections in domestic animals are pathogenic and that some persons may be affected by *Sarcocystis* infections.

## TOXOPLASMA AND TOXOPLASMOSIS

NAME OF ORGANISM AND DISEASE CAUSED. *Toxoplasma gondii* Nicole & Manceaux, 1908, causes toxoplasmosis. It has been the consensus that there is a single species of *Toxoplasma*, but Levine (1977a) admitted seven species to membership. *T. gondii* is the species of importance. Among those included is the controversial *Hammondia hammondi* Frenkel & Dubey, 1975, which Levine calls *T. hammondi* (Frenkel & Dubey, 1975).

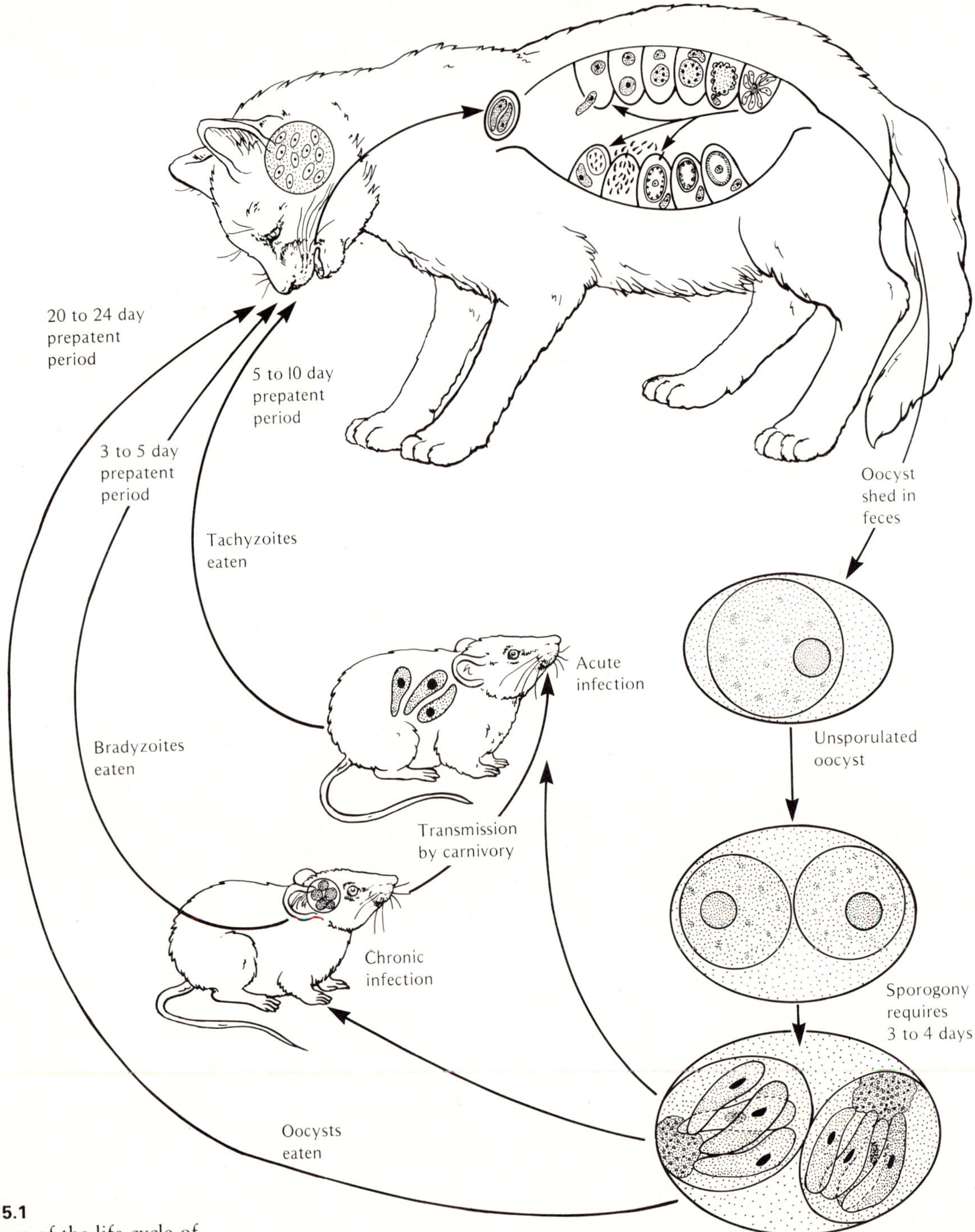


**FIG. 15.1**
Diagram of the life cycle of *Toxoplasma gondii*.

HOSTS AND HOST RANGE. About 200 species of mammals and birds have been found to be infected with meronts of *T. gondii*. The definitive hosts are the domestic cat and other felids. Sexual development (gamogony) occurs only in felids, so far as is known. Intermediate hosts are principally domiciliated rodents and humans. Asexual development can take place in a wide range of mammals and some birds.

DISTRIBUTION AND IMPORTANCE. *T. gondii* is widespread among both human and other animal populations. About one-third of persons living in North America show antibodies to *T. gondii*. In islands in the South Pacific infection rates approach 100%. Although there are exceptions, the prevalence of infection is higher in warm, moist areas than in those which are dry and cold. The percentage of persons who are seropositive rises with age at a fairly regular rate (Fig. 15.2).

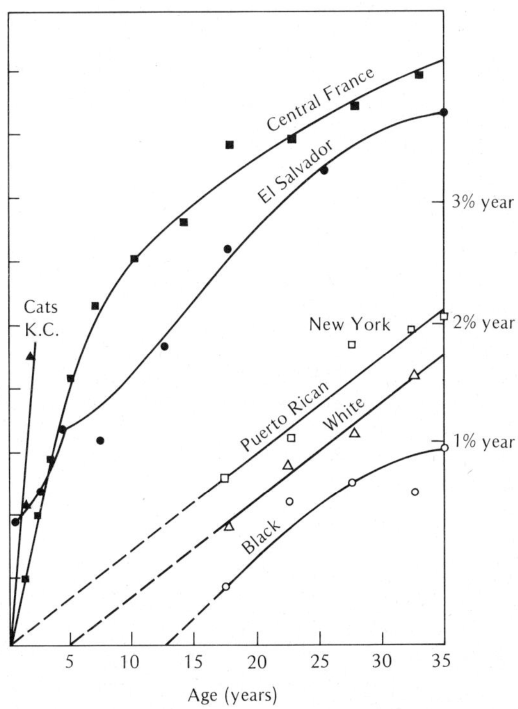


**FIG. 15.2**
Prevalence rate of *Toxoplasma* antibodies in the human population. [Redrawn from Frenkel, J.K. 1973. Toxoplasma in and around us. *Bioscience* **23**: 343-352 © 1973 by the American Institute of Biological Sciences.]

About 19% of cats were found to be infected in four surveys in the United States. In a study in Kansas City, about 50% of cats were found to be seropositive by the time they were 2 years old (Dubey et al., 1970).

Domestic animals that serve as possible sources of infection for humans have fairly high levels of infection: swine—24%; sheep—29%; cattle—16%; chickens—19% of 62 pools of blood or serum.

Levels of infection in a population are interesting, but the real issue is what effect *T. gondii* infections have on humans or other animals either as individuals or as groups. So far as humans are concerned, the greatest effect is on the developing fetus. If a woman becomes infected during pregnancy, there is a 40% chance that the fetus will be infected. It is estimated that about 3000 children born in any one year in the United States will be affected to a greater or lesser extent by congenitally acquired *Toxoplasma* infection. Of these, 150 to 450 will die, 240 to 300 will have severe brain and eye damage, 300 to 390 will have moderate to severe eye damage, and 1750 to 2150 may develop eye problems before maturity. The educational cost is estimated to range between $31 and $40 million (in 1973 dollars) for these children born each year (Frenkel, 1973).

There are a number of case reports of toxoplasmosis causing losses in domesticated animals. The overall significance of infection is not well documented. Toxoplasmosis is considered to be an important cause of perinatal loss of lambs in New Zealand and Britain.

MORPHOLOGY. The sporulated oocysts have a thin, flexible outer wall, and two sporocysts each having four sporozoites (Fig. 15.3). The oocysts average 13 by

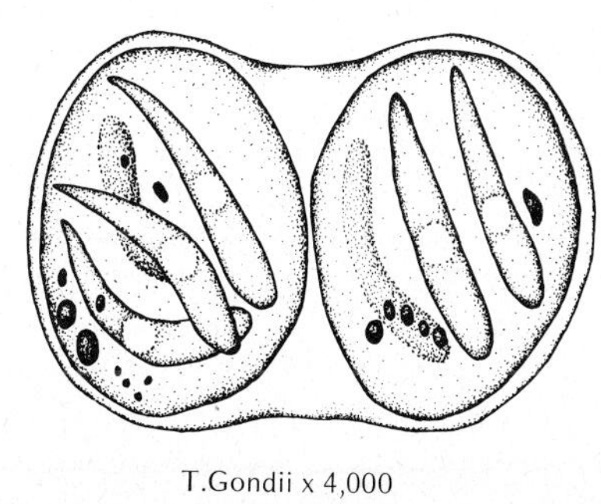


**FIG. 15.3**
Sporulated *Toxoplasma gondii* oocyst.

12 μm with a range of 12 to 15 by 10 to 13 μm. Sporocysts are 9 by 6.5 μm with a range of 8 to 10 by 6 to 8 μm. The sporozoites are 8 by 2 μm and have a refractile body.

In the cat, five different merogonous generations have been described (Frenkel, 1973). The meronts of these generations range from about 2 to 17 μm in diameter, and the merozoites range from 1.5 to 4.5 μm in length. These stages occur principally in epithelial cells of the villi of the jejunum.

In extraintestinal sites both in the cat and other hosts, there are two merogonous generations, (1) tachyzoite-forming groups and (2) bradyzoite-forming pseudocysts.

Tachyzoite-forming groups are found early in infections when they can be seen in vacuoles in the cytoplasm of the host cells. Sixteen or so merozoites are produced by endodyogeny. Endodyogeny (=merogony) is a process of internal budding in which two daughter cells are formed within the body of the mother cell, which dies when the progeny are released. Multiplication takes place rapidly and the organisms are not set off from the host tissue except by the parasitophorous vacuole, a clear area separating the parasite from host cell cytoplasm.

Bradyzoite-forming pseudocysts are seen later in tissue infections and are the chronic form of the infection. Multiplication is by endodyogeny but occurs more slowly than in tachyzoite-forming groups. In addition, the zoites are set off from the host tissue by a tough pseudocyst wall that seems to prevent a significant inflammatory reaction and the development of enough immunity to eliminate them. The pseudocyst wall protects the zoites from the action of some proteases and probably allows the pseudocysts to pass unharmed through the stomach and upper part of the small intestine of an animal that eats the tissues of an infected host.

Bradyzoites may remain within the tissues of an experimental animal for at least 5 years (guinea pigs), and probably for many years in the human host. If the level of immunity declines, the bradyzoites may revert to tachyzoites and multiply rapidly again. In some cases in which persons have received drugs that reduce the immune response, such as may be the case in cancer treatment, *Toxoplasma* tachyzoites have reappeared and their rapid multiplication has led to the early death of the patients.

LIFE CYCLE. Both the asexual and sexual development of *T. gondii* can be completed in the cat, or other species of animals may serve as intermediate hosts in which only asexual development takes place. If an intermediate host is involved, the organism must transfer back to a cat for the sexual phase of the cycle to occur.

When a cat ingests a sporulated oocyst, the sporozoites escape into the small intestine and invade epithelial cells of the gut. There are at least two asexual generations, and likely five, in the epithelial cells of the small intestine; then there is a sexual generation, which produces the oocysts. The prepatent period in the cat is 20 to 24 days (Fig. 15.1).

Although *T. gondii* has a typical coccidian life cycle, and completes development in the intestine of the cat, it can also leave the gut and develop in other tissues, both in the cat and in other hosts.

When oocysts are ingested by a host other than a cat (a mouse, for example), the organisms leave the gut and develop in other tissues. In such a host, only asexual development takes place; gamogony is never seen. Two types of meronts are seen: tachyzoites, or rapidly developing merozoites, early in the infection and bradyzoites, or slowly developing merozoites, later in the infection.

If a cat ingests an animal that has tachyzoites in its tissues, the cat sheds oocysts 5 to 10 days later. If a cat ingests an animal that has bradyzoites in its tissues, it sheds oocysts 3 to 5 days later.

Oocysts are shed in an unsporulated condition. They sporulate and become infective in 24 to 72 hr at 24°C.

DIAGNOSIS. In determining whether a particular host or population is infected with *T. gondii*, the following may be looked for:

1. Oocysts in the feces
2. Antibodies in the serum
3. Meronts or zoites in the tissues
4. Clinical signs and symptoms

In cats, *Toxoplasma gondii* infection may be diagnosed by finding the typical oocysts in the feces during the patent period. The patent period, the time when oocysts can be found in the feces, ranges from a week to 3 weeks during the initial infection. Since immunity develops in the cat after infection, only small numbers of oocysts are likely to be shed thereafter and only sporadically.

The most reliable general technique, and the one most widely used for diagnosis of *Toxoplasma* infection, is based on finding antibodies in the serum of the host. There are a number of serological methods that have been used:

1. Sabin – Feldman dye test (DT)
2. Complement fixation test
3. Hemagglutination test (HA)
4. Intradermal skin test
5. Fluorescent antibody test (FA)

In most instances the DT, HA, and FA tests can be used interchangeably, and the results are both specific and sensitive. It is generally recommended that the dye test not be used, because it requires the use of living *T. gondii* zoites, which may present a hazard to laboratory workers.

For a survey of a population for the prevalence of *T. gondii*, techniques in addition to serology may be used. Normally, the stage looked for is the bradyzoite-forming pseudocyst in the tissues of potential intermediate hosts. After the death of the host, tissues are removed and injected intraperitoneally (IP) into mice. After 5 days, mice are examined by making Giemsa-stained smears of tissues and examining them for zoites under a compound microscope.

In some instances, such as determining postmortem whether a person is infected, tissues may be taken and histologic sections examined for the presence of meronts.

The signs and symptoms of *T. gondii* are essentially the same in humans and other animals. The signs are both varied and nonspecific, and a diagnostician may be hard put to differentiate toxoplasmosis from other infectious diseases. Among the changes that may be seen are fever, swollen and painful lymph nodes, fatigue, skin rash, encephalitis, chorioretinitis, pneumonia, abortion, or premature birth. There will be a preponderance of certain signs and symptoms in any individual or in an outbreak, but aside from ocular involvement or prenatal infections one would not be led to suspect toxoplasmosis. If a bacterial infection is suspected and antibiotic therapy is without effect, toxoplasmosis is a possible alternative. In the end, the presence of antibodies in the blood is crucial for a positive diagnosis.

HOST – PARASITE INTERACTIONS. Infections with *T. gondii* in all species of hosts may range from inapparent, to subclinical, to mild, to acute (sometimes resulting in death), to chronic infections that may last for many years.

The consensus is that the most important infections in humans take place in developing fetuses. The results of these infections can range from quite mild to fatal. In severely infected infants, the infection is disseminated widely in the tissues and as the child gains ability to produce its own antibodies, some of the infection is likely to be controlled; however, the infection may proceed in the central nervous system and eye. There may be blindness, mental retardation, or death.

Acquired infections in humans take place after birth and result from the ingestion of infective oocysts or tissues infected with bradyzoites. Laboratory-acquired infections are included in this category as well. As indicated in the Diagnosis section, many kinds of cells and tissues may be infected; however, early in the infection there is often localization in lymph nodes with concomitant enlargement. Later, the infection may spread to the central nervous system and the eye. In the eye there may be inflammation of the retina and its underlying layer, the choroid, leading to blindness.

In the cat, the infection is both intestinal and systemic. In the intestine there is extensive cellular destruction and loss of epithelium such as takes place in other intestinal coccidioses. The organism may also leave the gut and infect other tissues. In mice and other laboratory rodents, there may also be considerable intestinal damage before other tissues become involved.

Tissue damage in other animals is about the same as in humans. In sheep abortion or perinatal death may occur; these problems have been seen in other animals as well. When the organism invades the central nervous system, encephalitis often results; pneumonia or pneumonitis may occur if the lungs are affected.

The mechanisms of damage of *Toxoplasma* are associated with cellular destruction, inflammation, blockage of blood vessels, and hypersensitivity. Early in the infection, the first three are most likely to cause the damage. Later in the infection, hypersensitivity seems to be the major cause of damage.

Immunity plays a major role in lessening the effects of *Toxoplasma* infection. In infants, some of the systemic damage may be reduced through maternal antibody and later from the development of antibodies as the individual becomes immunocompetent. The same is true in adult hosts; the development of specific immunity reduces the deleterious effects of the infection. Damage to the central nervous system and to the eye results from the inability of the immune response antibodies and cells to reach the site of infection in adequate amounts. When bradyzoite-forming pseudocysts develop, the thick pseudocyst wall seems to isolate the organisms from the host so that there is no significant inflammatory reaction, and over a period of months or years, antibody titers in the blood are observed to decline.

Hypersensitivity plays a role in damage to host tissues during the chronic phase of the infection. When a pseudocyst in the tissues ruptures, the resulting reaction damages tissues in its vicinity and causes a group of cells to die (Frenkel, 1973).

TREATMENT. Sulfonamide drugs potentiated with pyrimethamine have been most effective in treatment of clinical infections in humans. Better therapeutic agents are needed.

EPIDEMIOLOGY AND CONTROL. *T. gondii* is transmitted by the following mechanisms:

1. Ingestion of infective oocysts produced by cats
2. Ingestion of bradyzoites in pseudocysts in the tissues of infected animals
3. Transplacental transfer
4. Injection of zoites by hypodermic needle or splashing of infectious material into the eye (This means of infection is limited to accidental or purposeful transmission in the laboratory or during postmortem examination of infected hosts and will not be considered in the general discussion to follow.)

The domestic cat is the most important animal in the continued transmission of *T. gondii* and the one that is involved in prenatal, transplacental infections. Cats produce oocysts for 1 to 3 weeks. The oocysts sporulate and become infective outside the host in 2 to 3 days at 24°C, 5 to 8 days at 15°C, and 14 to 21 days at 11°C. At temperatures below 4 and above 37°C, sporulation does not take place. It has been documented that the oocysts can survive 2 weeks in "dry" feces. They can probably survive several weeks outside the host under conditions of cool (5 – 15°C) temperatures and in the presence of oxygen.

Oocysts are resistant to 2% sulfuric acid (in which they can be stored), 24 hr in 5% sodium hypochlorite, or 6 hr in 10% formalin. In this regard they are like other coccidian oocysts. Although washing with soap will remove the greater percentage of oocysts from the hands or other surfaces, even strong disinfectants may have little effect on their infectivity.

Small rodents such as house mice serve as the principal alternate or intermediate hosts. Transmission to the mouse takes place through ingestion of infective oocysts or through cannibalism. Completion of the gamogonous phase of the cycle takes place when an infected mouse is eaten by a cat.

In addition to the basic cat – mouse cycle, other animals may be integrated into the cycle as intermediate hosts (Fig. 15.4). These infections may take place through ingestion of infective oocysts or by carnivorism, and a wide array of birds and mammals are important hosts. Transmission by carnivorism can take place indefinitely and undoubtedly is a means of perpetuating infections in the environment.

Transplacental infections take place when pregnant women ingest infective oocysts, produced usually by domestic cats. It is the consensus that women who have not been infected previously, and therefore do not have any immunity, can transmit zoites to the developing fetus through the placenta.

Since the domestic cat is the most important definitive host of *T. gondii*, control programs are directed toward ensuring that people, especially pregnant women, do not ingest infective oocysts. Pregnant women should avoid contact with cats and especially with feces of cats such as in litter boxes. Cats that are kept in the house, or at least do not hunt, are a minor hazard, but a single infected house mouse eaten by a cat could cause a shower of millions of oocysts over a 3-week period.

The role of the cat in transmitting infection both horizontally and vertically in the human population places the cat owner in a difficult position. As is true of nearly all infectous diseases, most *Toxoplasma* infections are inapparent; however, a small percentage of persons may suffer severe debilitating infection, blindness, and even death. About 40% of women who

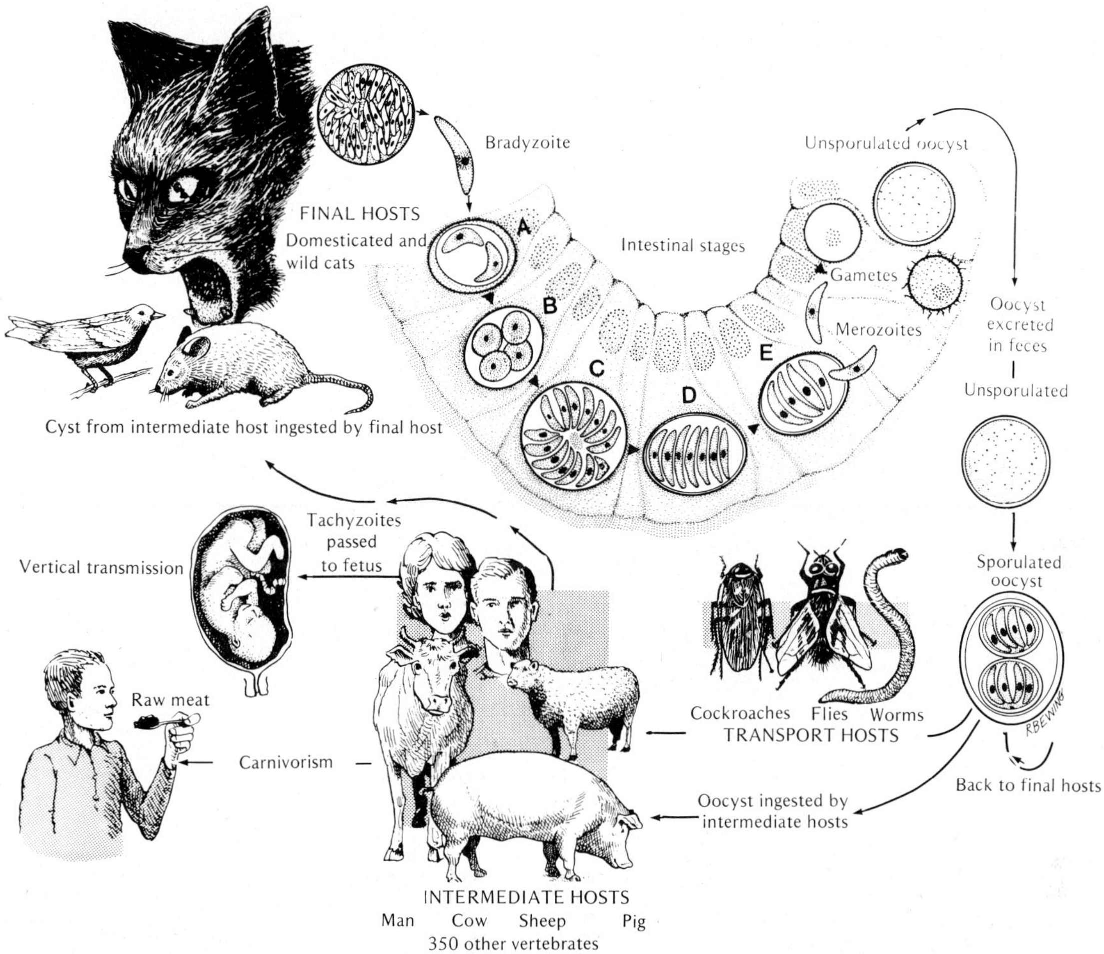


**FIG. 15.4**
Life cycle and modes of transmission of *Toxoplasma gondii*. [Courtesy of R. B. Ewing, Animal Parasitology Institute, ARS, USDA, Beltsville, Maryland.]

become infected while pregnant will transmit the infection to their infants transplacentally. The risk is small, but each person must decide whether the risk is small enough to have the companionship afforded by a cat.

Carnivorism is a means not only of perpetuating infection but also of infecting humans. Eating raw or insufficiently cooked meat leads to infections. In European and North American countries, infection is most likely to be obtained from beef and pork. Some persons eating "steak tartar," a mixture of uncooked ground beef and condiments, have developed severe clinical infections. Other animals, such as chickens and rabbits, are good hosts as well. Meat that has been heated to 60°C (140°F) is noninfective, and freezing to –40°C (–40°F) for 9 days also frees it of infection. Smoking, if properly done at a high temperature, also kills bradyzoites.

Artificial immunizing agents are still in the early stages of development. A safe, effective vaccine that could be given to women in their child-bearing years would be a blessing to children and parents and relieve society of a $30 million annual burden in caring for affected children. Vaccination of cats could be of value, but sterile immunity would have to be maintained for a long peiod, which would be difficult.

## SARCOCYSTIS AND SARCOCYSTOSIS

A number of genera have isosporoid oocysts and two-host life cycles: *Sarcocystis*, *Besnoitia*, and *Frenkelia*. Among the many species in this general category are some that are of medical or economic importance, but their impacts on human and animal health are only now being clarified. For example, in 1961 an outbreak of disease in a dairy herd near Dalmeny, Ontario, resulted in the death of 17 out of a herd of 36 animals; 8 other affected animals survived. The only significant infectious agent found was an organism that looked like *Toxoplasma gondii* but could not be confirmed serologically (Corner et al., 1963). The relationship to *Sarcocystis* was clarified by an investigation of another outbreak in New York, similar to the Dalmeny disease, in which eight yearling heifers died. Typical *Sarcocystis* sporocysts were found in the feces of a dog on the farm (Frelier et al., 1976).

During the years between the description of Dalmeny disease and the conclusion that it was caused by *Sarcocystis*, information had developed on life cycles and potential pathogenesis of a number of species (Dubey, 1976; Leek et al., 1977; Leek & Fayer, 1978; Fayer et al., 1976). In those *Sarcocystis* spp. infecting ruminants, the other host was a carnivore or, in two instances, humans. The pattern then became clear that the intermediate host was a herbivore and the definitive host was a carnivore, or at least a meat eater, since humans are not classified as carnivores. In all instances the two-host cycle is obligatory, unlike *T. gondii*, in which it is optional.

The general pattern of herbivore – carnivore or prey – predator transmission has been demonstrated in a number of other cycles as well. For example, *Isospora buteonis* of hawks has a vole (*Microtus agrestis*) as its intermediate host, and its proper name is *Frenkelia microti*. *Sarcocystis idahoensis* has a cycle that alternates between deer mice (*Peromyscus maniculatus*) and the gopher snake (*Pituophis melanoleucus*) (Bledsoe, 1980). *Sarcocystis rileyi*, known to be widespread in the somatic muscles of ducks, has been found to have skunks as its definitive host.

Because knowledge of life cycles has developed so rapidly and because oocysts and sarcocysts have been associated in unexpected combinations, nomenclature in the group is in a frightful state. The problem lies in (1) the separate descriptions of muscle stages and fecal stages, sometimes when methods and descriptions were only rudimentary, (2) subdivision of species, as it has been determined that similar oocysts comprise more than one species, (3) the problem of determining in each instance which of the original names has priority according to the rules of zoological nomenclature, and (4) differences of opinion among investigators as to what constitute valid generic and specific characters (Levine & Tadros, 1980; Frenkel et al., 1979; Kalyakin & Zasukhin, 1975; Frenkel, 1977; Levine & Ivens, 1981; Levine, 1977b). We will look here at a single species as a representative of the group.

NAME OF ORGANISM AND DISEASE CAUSED. *Sarcocystis cruzi* (Hassleman, 1923); synonyms: *S. bovicanis*, Heydorn et al., 1975; *S. fusiformis*, *Isospora bigemina*, and *I. hominis* cause sarcocytosis, sarcosporidiosis, and Dalmeny disease.

HOSTS AND LOCATION. Definitive hosts are canids in whom the intestine is parasitized. The intermediate host is the domestic ox, in which the parasites are located principally in the somatic muscles.

DISTRIBUTION AND IMPORTANCE. Distribution is cosmopolitan. The prevalence of infection seems to approach 100% in cattle, but cattle are infected by at least two other species of *Sarcocystis* and by *T. gondii*, so we know that they nearly always harbor zoites of some species of tissue coccidium.

A number of outbreaks of disease have been described (Corner et al., 1963; Frelier et al., 1976; Schmitz & Wolf, 1977), and additional outbreaks continue to be reported. Losses to the livestock industry are still unknown.

Cattle are sometimes condemned at slaughter for eosinophilic myositis, a condition in which the muscles of the animals have a gross greenish discoloration. It has been suggested but not proved that eosinophilic myositis is the result of *Sarcocystis* spp. infection.

MORPHOLOGY. Sporocysts in the feces of the definitive host are 10.8 by 16.3 μm. Other closely related species such as *S. fusiformis*, *S. tenella*, *S. miescher-*

iana, and *S. equicanis* have sporocysts much the same size as those of *S. cruzi.*

The cysts in the muscles of the intermediate host are septate, that is, they have compartments; the metrocytes are 7 by 5 μm, and the merozoites are 10 to 14 by 2 μm. (Fig. 15.5).

LIFE CYCLE. The infection is passed from the intermediate host, the ox, to the definitive host, the dog, when raw meat is eaten. The merozoites leave the sarcocyst, after being digested from it, and enter the tissues of the intestine, where they are seen within cells about a day later. The merozoites differentiate into macro- and microgamonts in the same way as *E. tenella* (Chap. 14). After fertilization, the oocysts remain in the cells and sporulate. Sporulation of a coccidium in a warm-blooded host is unusual, and is a distinguishing characteristic of *Sarcocystis* spp. The sporulated oocyst has a thin outer wall and two sporocysts, each of which has four sporozoites. Because the outer wall of the oocyst is so delicate, only sporocysts are seen in the feces. The prepatent period is 9 to 10 days and the patent period is more than 6 weeks.

When sporulated oocysts are eaten by an ox, the sporozoites leave the sporocyst and enter extraintestinal cells. Merogony occurs in various organs, but apparently with high frequency in the endothelial cells of small blood vessels. The number of merogonous generations is not known.

About a month after infection, parasites are seen within the cells of somatic muscles. The first parasites seen are small, rounded forms, the metrocytes. Multiplication occurs by endodyogeny and secondarily by schizogony, and the organisms gradually develop into the longer merozoites. The infective merozoite stage is reached about 3 months after initial infection of the intermediate host.

DIAGNOSIS. The infection in the definitive host is diagnosed by finding sporocysts in the dog's feces. For intermediate hosts, clinical signs are suggestive, but finding the developing sarcocysts on histologic examination is essential. Serological tests showing a rising titer against *Sarcocystis* provide good evidence of the infection as well.

CULTIVATION. Only the sexual stages have been cultivated. Merozoites dissected from infected muscle have been inoculated into cell culture and oocysts have formed.

HOST–PARASITE INTERACTIONS. In the definitive host, infection is pathogenic and leads to diarrhea, which may be severe. Oocyst production may continue

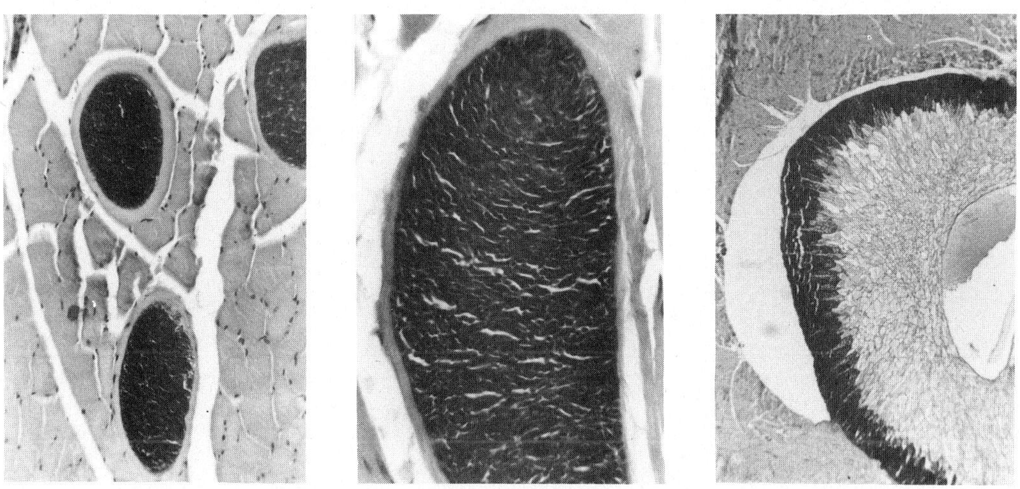

(A) (B) (C)

**FIG. 15.6**
*Sarocystis* pseudocysts. (*A*) Three cysts in infected animal muscle; (*B*) numerous zoites inside a sarocyst; (*C*) cyst older than *A* and *B*. The cyst has become partitioned and zoites are present only at its periphery.

for more than 6 weeks, during which there may be intermittent diarrhea. The reason for the long period of oocyst production is not known.

Infection of the intermediate host leads to a number of signs including fever, failure to eat, abortion, and reduced milk production. There may be ulcers and hemorrhaging from the tissues of the mouth and nose. Early in the infection, the organisms invade and multiply in the blood vessels throughout the body; the liver seems to be a preferred site.

After the organisms reach the somatic muscles, pathogenesis is generally reduced, but there may be muscular weakness. Usually there is little host reaction to the parasites within the muscle cells; the usual large numbers of white cells in the vicinity of the foreign body are not present. There is some indication that a foreign body response occurs when the organisms in the sarcocysts die.

TREATMENT. Symptomatic treatment may help some hosts during the acute phase of the disease.

EPIDEMIOLOGY AND CONTROL. There are currently few useful recommendations for control. Part of the difficulty was indicated in our earlier discussion of taxonomic confusion. We do not know with any degree of certainty what the host range is of the organisms in either the definitive or intermediate hosts; it has been said that the definitive host range is narrow and the intermediate host range can be broad, but the data are still quite limited. It is only a slight exaggeration to state that the environment is saturated with *Sarcocystis*. Nearly every sample (90% or more) of beef or pork picked at random at a grocery store has zoites of something in it. Some, perhaps 5%, will be *T. gondii*, but the greater percentage will be *Sarcocystis* spp.

If we go to the other hosts, the predators or scavengers, the prevalence of infection is likely to be low, perhaps 10% or less. On the other hand we have no information on the incidence of infection; there are no prospective studies on how often dogs are infected and the numbers of sporocysts that they might shed.

Consider the situation in which great numbers of cattle and sheep are raised. The principal cattle- and sheep-raising areas of the world are vast grasslands where 30 to 60 acres of land might be required to support one cow and calf. Most of these grasslands have wild carnivores and scavengers as well as domestic dogs; all are likely to feed on carcasses of animals. With current methods of husbandry, there is no way that the level of infection can be modified.

In some outbreaks of sarcocystosis in cattle, the animals have been confined in relatively small quarters (e.g., dairy cattle). The culprit spreading the disease agent has sometimes been found to be a farm dog. Under these conditions, the owner can control the food that the farm dogs are fed and make certain that they do not eat any uncooked beef, even such things as bones and small trimmings.

## SARCOCYSTOSIS IN HUMANS

Three intestinal coccidia are known to infect humans: *Isospora belli*, which is probably an intestinal parasite only; *S. tenella*, which is probably not pathogenic; and *S. miescheriana*, which is pathogenic for humans (Chap.9). Reviews of sarcocysts in humans (Jeffery, 1974; Beaver et al., 1979) have helped to define the problem somewhat but leave one wishing there were more information. It is believed that all three of these species infect the intestine only. But we know that humans are occasionally infected with parasites in the muscles and therefore they are the intermediate host to something. There are fewer than 50 case reports of sarcocysts in humans. In some instances the persons were clinically affected with muscle pain and weakness, among other symptoms and signs.

## REFERENCES AND FURTHER READINGS

Al-Khalidi, N. W., and J. P. Dubey. 1979. Prevalence of *Toxoplasma gondii* infection in horses. *J. Parasitol.* 65:331–334.

Beaver, P. C., R. K. Gadgil, and P. Morera. 1979. *Sarcocystis* in man: A review and report of five cases. *Am. J. Trop. Med. Hyg.* 28:819–844.

Bledsoe, B. 1980. Transmission studies with *Sarcocystis idahoensis* of deer mice (*Peromyscus maniculatus*) and gopher snakes (*Pituophis melanoleucus*). *J. Wildl. Dis.* 16:195–200.

Corner, A. H., D. Mitchell, E. B. Meads, and P. A. Taylor. 1963. Dalmeny disease. An infection of cattle presumed to be caused by an unidentified protozoon. *Can. Vet. J.* 4:252–264.

Dubey, J. P. 1976. A review of *Sarcocystis* of domestic animals and other coccidia of cats and dogs. *J. Am. Vet. Med. Assoc.* **169**:1061 – 1078.

Dubey, J. P., and J. K. Frenkel. 1972a. Extra-intestinal stages of *Isospora felis* and *I. rivolta* (Protozoa:Eimeridae) in cats. *J. Protozool.* **19**:89 – 92.

———. 1972b. Cyst-induced toxoplasmosis in cats. *J. Protozool.* **19**:155 – 177.

Dubey, J. P., N. L. Miller, and J. K. Frenkel. 1970. Characterization of a new fecal form of *Toxoplasma gondii*. *J. Parasitol.* **56**:447 – 456.

Fayer, R. 1972. Gametogony of *Sarcocystis* sp. in cell culture. *Science* **175**:65 – 67.

Fayer, R., A. O. Heydorn, A. J. Johnson, and R. G. Leek. 1979. Transmission of *Sarcocystis suihominis* from humans to swine to nonhuman primates (*Pan troglodytes*, *Macaca mulata*, *Macaca irus*). *Z. Parasitenk.* **59**:15 – 20.

Fayer, R., A. J. Johnson, and M. Lunde. 1976. Abortion and other signs of disease in cows experimentally infected with *Sarcocystis fusiformis* from dogs. *J. Infect. Dis.* **134**:624 – 628.

Frelier, P., I. G. Mayhew, R. Fayer, and M. N. Lunde. 1976. Sarcocystosis: A clinical outbreak in dairy calves. *Science* **195**:1341 – 1342.

Frenkel, J. K. 1973. *Toxoplasma* in and around us. *Bioscience* **23**:343 – 352.

———1974. Advances in the biology of the Sporozoa. *Zeitschr. Parasitenk.* **45**:25 – 62.

———1977. *Besnoitia wallacei* of cats and rodents with a reclassification of other cyst-forming isosporoid coccidia. *J. Parasitol.* **63**:611 – 628.

Frenkel, J. K., and J. P. Dubey. 1975. *Hammondia hammondi* gen. nov. and sp. nov. from domestic cats, a new coccidian related to *Toxoplasma* and *Sarcocystis*. *Zeitschr. Parasitenk.* **46**:3 – 12.

Frenkel, J. K., J. P. Dubey, and N. L. Miller. 1970. *Toxoplasma gondii*: Fecal stages identified as coccidian oocysts. *Science* **167**:893 – 896.

Frenkel, J. K., A. O. Heydorn, H. Mehlhorn, and M. Rommel. 1979. Sarcocystinae: *Nomina dubia* and available names. *Z. Parasitenk.* **58**:115 – 139.

Heydorn, A. O., and R. Gestrich. 1976. Beiträge zum Lebenszyklus der Sarkosporidien. VIII. Entwicklungsstadien von *Sarocystis ovicanis* in Schaf. *Berl. Muench. Tieraerztl. Wochenschr.* **89**:1 – 5.

Hutchison, W. M., J. F. Dunachie, J. Siim, and K. Work. 1970. Coccidian-like nature of *Toxoplasma gondii*. *Br. Med. J.* **1**:142 – 144.

Jeffery, H. C. 1974. Sarcosporidiosis in man. *Trans. R. Soc. Trop. Med. Hyg.* **68**:17 – 29.

Kalyakin, V. N., and D. N. Zasukhin. 1975. Distribution of *Sarococystis* (Protozoa: Sporozoa) in vertebrates. *Folia Parasitol.* (Praha) **22**:289 – 307.

Leek, R. G., and R. Fayer. 1978. Sheep experimentally infected with *Sarcocystis* from dogs. II. Abortion and disease in ewes. *Cornell Vet.* **68**:108 – 123.

———1979. Survival of *Sarcocystis* sporocysts in various media. *Proc. Helminthol. Soc. Wash.* **46**:151 – 154.

Leek, R. G., R. Fayer, and A. J. Johnson. 1977. Sheep experimentally infected with *Sarcocystis* from dogs. I. Disease in young lambs. *J. Parasitol.* **63**:642 – 650.

Levine, N. D. 1977a. Taxonomy of *Toxoplasma*. *J. Protozool.* **24**:36 – 41.

———1977b. Nomenclature of *Sarcocystis* in the ox and sheep and of fecal coccidia of the dog and cat. *J. Parasitol.* **63**:36 – 51.

Levine, N. D., and W. Tadros. 1980. Named species and hosts of *Sarcocystis* (Protozoa: Apicomplexa: Sarcocystidae). *System. Parasitol.* **2**:41 – 59.

Levine, N. D., and V. Ivens. 1981. *The Coccidian Parasites (Protozoa Apicomplexa) of Carnivores.* Ill. Biol. Monogr. No. 51. University of Illinois Press, Urbana.

Mehlhorn, H., and A. O. Heydorn. 1978. The Sarcosporidia (Protozoa, Sporozoa) life cycle and fine structure. *Adv. Parasitol.* **16**:43 – 92.

Munday, B. L., I. K. Barker, and M. D. Rickard. 1975. The developmental cycle of a species of *Sarcocystis* occurring in dogs and sheep with observations on pathogenicity in the intermediate host. *Z. Parasitenkd* **46**:111 – 123.

Pacheco, N. D., H. G. Sheffield, and R. Fayer. 1978. Fine structure of immature cysts of *Sarcocystis cruzi*. *J. Parasitol.* **64**:320 – 325.

Saxon, S. A., W. Knight, D. W. Reynolds, S. Stagno, and C. A. Alford. 1973. Intellectual deficits in children born with subclinical congenital toxoplasmosis. *J. Pediatrics* **85**:792 – 797.

Schmitz, J. A., and W. W. Wolf. 1977. Spontaneous fatal sarcocystis in a calf. *Vet. Pathol.* **14**:527 – 531.

Sheffield, H. G., and M. L. Melton. 1970. *Toxoplasma gondii*: The oocyst, sporozoite and infection of cultured cells. *Science* **167**:892 – 893.

# 16

# *Malaria in Humans and Related Infections*

Malaria is considered by many to be the most important infectious disease of humans on a worldwide scale. It is most common in tropical and subtropical regions of the world but has established itself in temperate climates as well. During the late 1930s and early 1940s about 300 million persons were infected at any one time in the world.

NAME OF ORGANISM AND DISEASE CAUSED. Four species of *Plasmodium that* normally infect humans are shown in the table below.

HOSTS AND LOCATION. For practical purposes, the species in the table below are found only in humans. Some other primates have been found to be infected with organisms indistinguishable from these, but the importance of these hosts as reservoirs of infection is still an open question. A number of primates other than humans have been infected experimentally with human malaria organisms, and humans have also served as an experimental or accidental host for malarias derived from other primates. Control programs are designed on the premise that the human is the only host.

| NAME | DISEASE | DISTRIBUTION |
|---|---|---|
| *Plasmodium vivax* (Grassi & Felletti, 1890) Labbé, 1899 | Benign tertian malaria; vivax malaria | Cosmopolitan in tropical areas except in parts of tropical Africa; most common species in subtropical and temperate areas. |
| *P. falciparum* (Welch, 1897) Schaudinn, 1902 | Malignant tertian malaria; falciparum malaria, blackwater fever; aestivoautumnal malaria | Tropical and subtropical areas of Africa and Asia; prior eradication in U.S. and Mediterranean |
| *P. malariae* (Laveran, 1881) Grassi & Feletti, 1890 | Quartan or malariae malaria | Principally in southeast Asia; also in African and Indian subcontinent; rare in Western Hemisphere |
| *P. ovale* Stephens, 1922 | Ovale or tertian malaria | Tropical Africa on the west coast and in Ethiopia |

The human malarial parasites are found in the liver during the prepatent period and later in the erythrocytes. In the mosquito vector, the parasites are located first in the wall of the midgut; then, as the sporozoite stage is reached, they are located in the hemocoel and the salivary glands.

DISTRIBUTION AND IMPORTANCE. Malaria is a cosmopolitan disease in tropical and subtropical areas of the world (Fig. 16.1); as indicated in the previous section, certain species are more common in some areas than in others. The importance of malaria can hardly be overstated. Prior to extensive control programs following World War II, at least 300 million persons were infected worldwide at any one time. About 1% of those persons died from the infection. In addition, it greatly increased their susceptibility to other diseases, which also took their toll.

During the time of the Roman Empire, the low-lying lands near Rome were a source of anopheline mosquitoes, which maintained a constant transmission of the organism, and the so-called bad air associated with swamps gave malaria its name. Some evidence has been presented that as the Roman Empire went into decline, mosquito breeding areas increased, leading to greater transmission of malaria, which hastened the further decline of the empire.

During World War II, nearly 500,000 military personnel had malaria. In the conflict in Korea (1950–1953) 6119 cases were reported in military personnel in Korea, but more than 20,000 cases were reported after the personnel had returned to the United States. The war in Viet Nam gave much the same picture, with 16,105 cases encountered in military personnel in the United States after they returned from southeast Asia. Many times more lost man-days in Viet Nam were due to malaria than to combat casualties as such. Many other military campaigns have had equal or greater casualties from malaria, and the outcomes of wars have been influenced more by this disease than by combat wounds.

With the development of control methods starting in the early part of this century and gaining momentum in the late 1930s, malaria was significantly controlled in many localized areas throughout the world. With the development of dichlorodiphenytrichloroethane (DDT) and its relative cheapness, large-scale control of malaria vectors was achieved during the 1950s and early 1960s. In 1955 there were 200 million to 225 million malaria cases worldwide, and more than 2 million deaths annually (Diehl, 1955). Malaria has been eradicated in some areas and nearly eradicated in others; however, there has been a resurgence in many of the developing countries of the world. India, for example, had only 40,000 cases in 1966, but the number had risen to 5.8 million by 1976.

LIFE CYCLE. The pattern of development in members of the genus *Plasmodium* follows the generalized life cycle set forth at the beginning of Chapter 12. In *Plasmodium* as opposed to *Eimeria*, a mosquito vector is required for completion of the cycle. Maturation of the gamonts, fertilization, and sporogony take place in the mosquito, the definitive host. Merogony and gamogony take place in the vertebrate host, the intermediate host (Fig. 16.2). Selected stages are illustrated in photomicrographs in Fig. 16.3.

We will use the life cycle of *Plasmodium vivax*, the cause of benign tertian malaria, as the example. Some differences between *P. vivax* and the other three human species will be discussed later.

When a mosquito takes a blood meal, the long proboscis probes in the skin of the host, searching for a capillary; at the same time, salivary fluid originating in glands in the thorax is injected to keep the blood from clotting. During the injection of salivary fluid, sporozoites are injected from the salivary glands into the skin or directly into the bloodstream. The sporozoites remain in the circulating blood for only about 30 min.

The sporozoites enter liver cells and begin their exoerythrocytic (EE) development. The sporozoites transform into meronts (cryptozoites) and multiply by schizogony. At least one more asexual generation occurs in liver cells, and these generations are called *metacryptozoites*. In *P. vivax*, the number of asexual generations that takes place in the liver is probably limited, and once the merozoites leave the liver, other generations of merozoites do not return to the liver cells for development.

At 8 days after sporozoite infection, merozoites derived from the EE cycle begin to appear in the circulating red blood cells. This is the end of the prepatent period. All further development occurs in the erythrocytes.

When *P. vivax* is first seen in the circulating red cells, there are asexual stages only, and all phases of merogony may be seen at any one time. As the infection progresses during the next week or so, the asexual

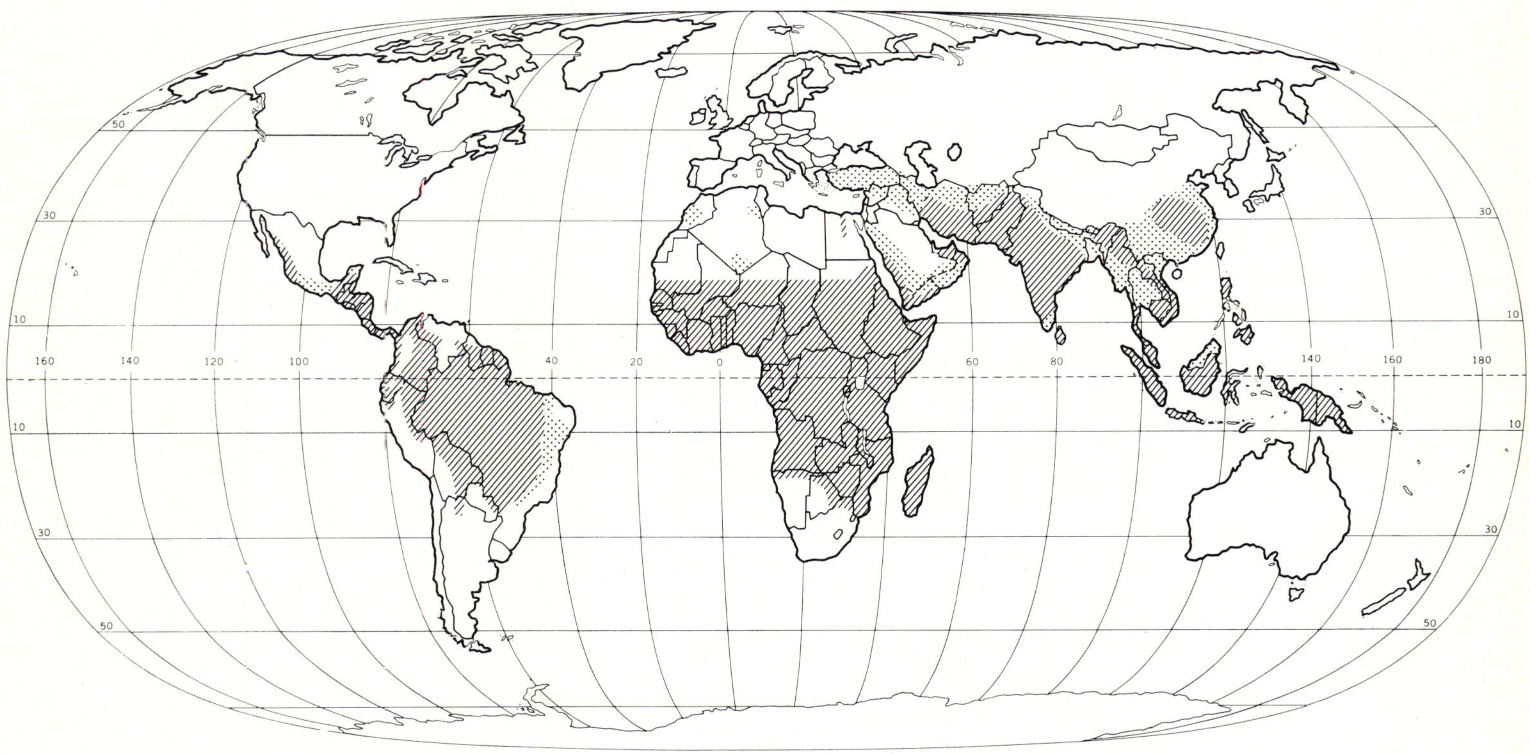

**FIG. 16.1**
Map of the distribution of human malaria. (Diagonal lines represent areas of frequent transmission; stippled areas represent infrequent transmission.)

**FIG. 16.2**
Life cycle of *Plasmodium* of humans.

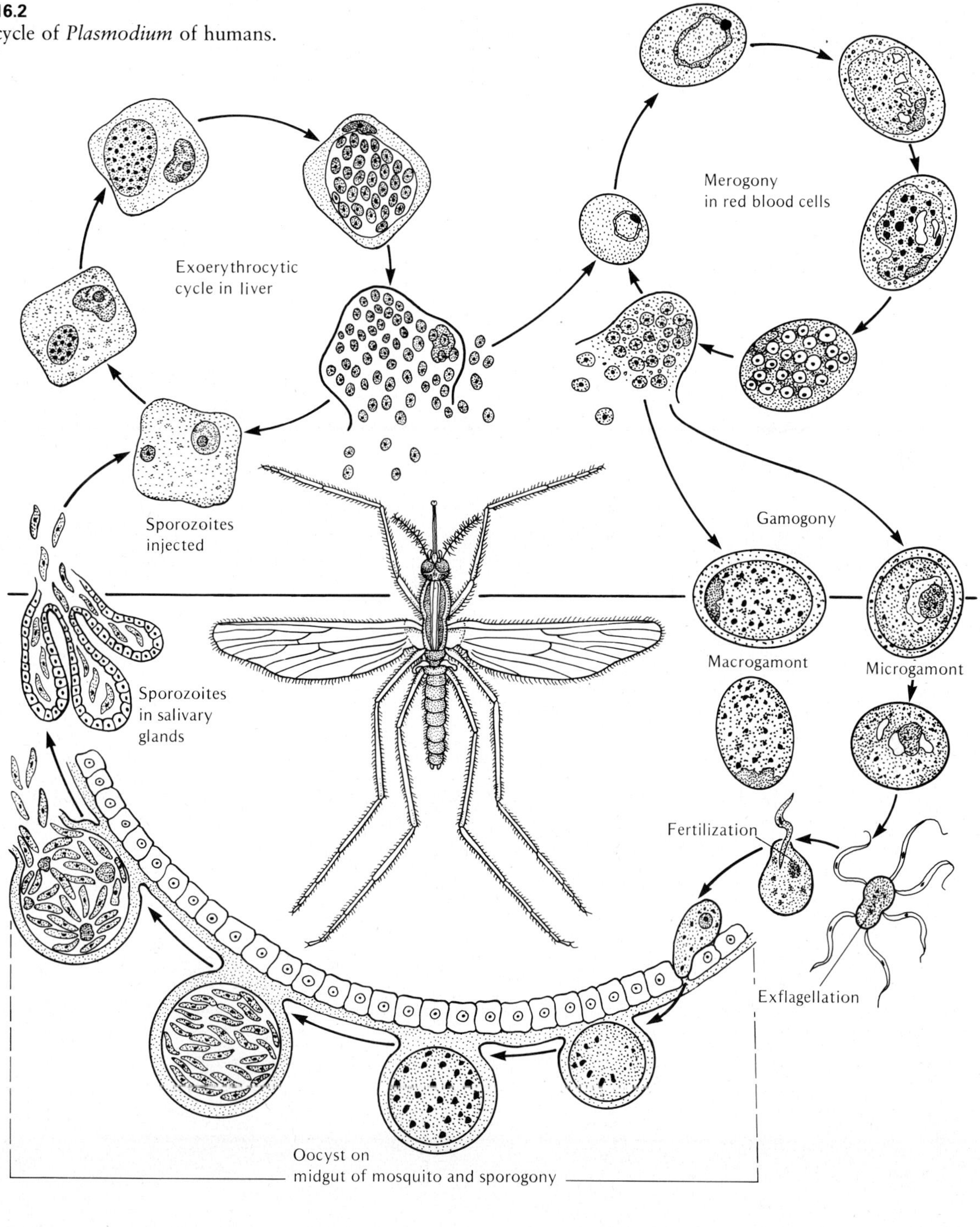

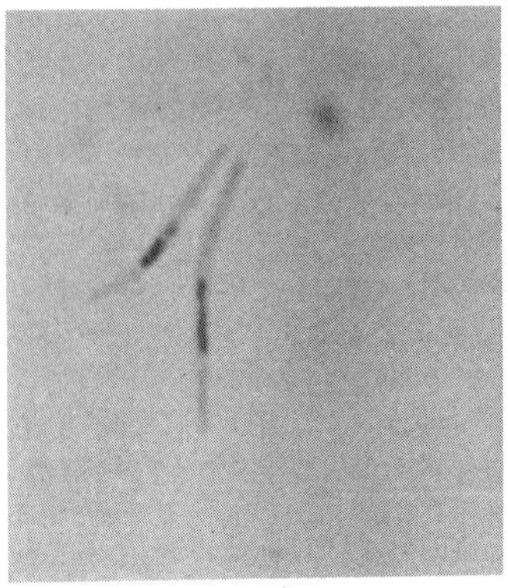

*A*

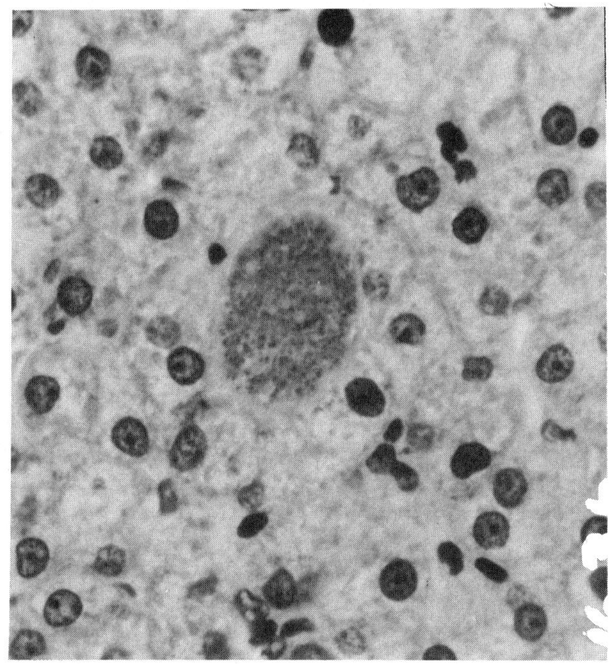

*B*

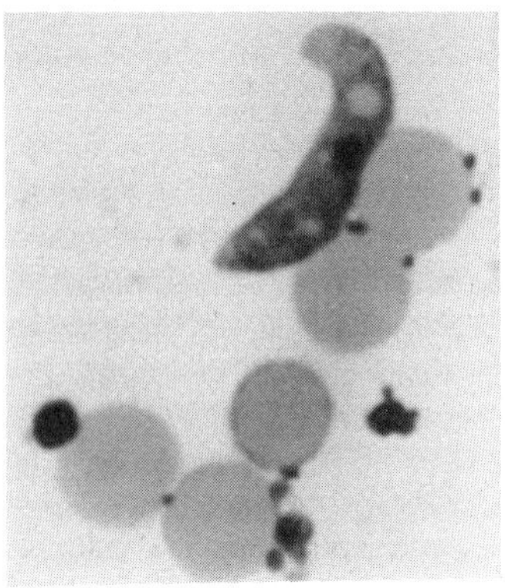

*C*

**FIG. 16.3**
Life cycle stages of *Plasmodium*. (*A*) sporozoite; (*B*) exoerythrocytic schizont; (*C*) ookinete; (*D*) oocysts on mosquito stomach; (*E*) brain capillary clogged by *P. falciparum*.

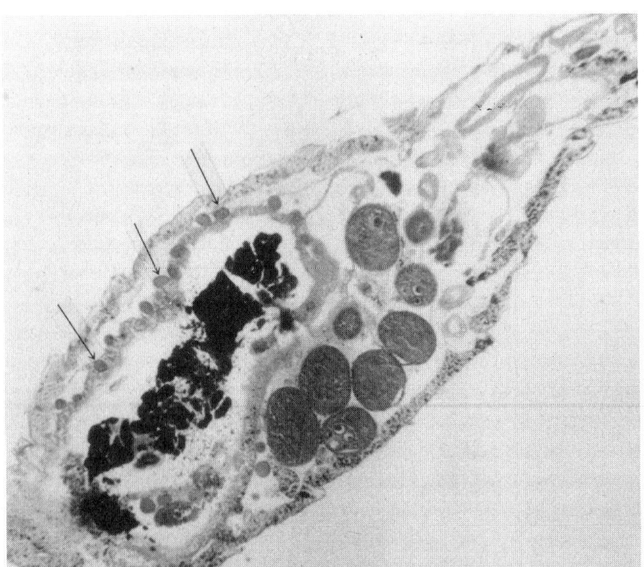

*D*

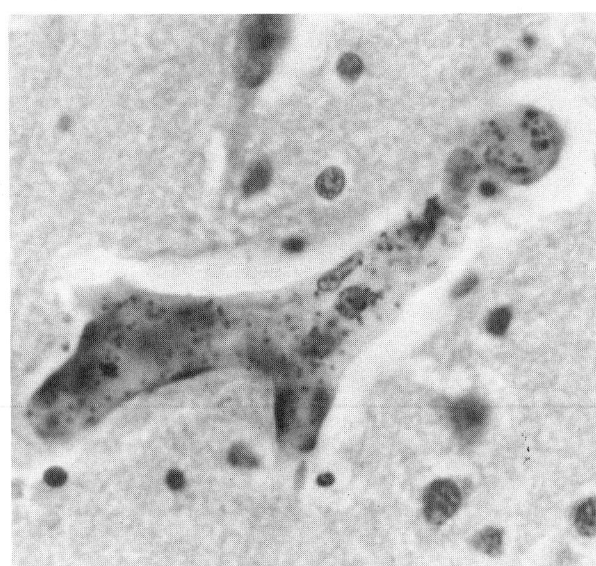

*E*

stages tend to become synchronized, so that a blood film taken at any one time shows a preponderance of organisms in the same stage of merogony. Also, as the infection progresses, a small number of merozoites develop into sexual stages (gamonts, or gametocytes) in the circulating red cells.

Merogony (schizogony) has a 48-hr cycle in *P. vivax*, that is, 48 hr from the time a merozoite enters a red cell until the daughter cells are produced. Merozoites are perhaps 2 μm long when they attach to the surface of the host red cell. They penetrate probably by using the components of the apical complex and pulling in the external membrane of the red cell. Once inside, the merozoite rounds up to form the early meront, a so-called ring stage. The blue ring, 2 to 4 μm in diameter, has what appears to be a clear vacuole in the center and a single round nucleus, which stains red in Romanowsky stains (methylene blue and eosin) (Fig. 16.4).

As the ring begins to grow, it becomes more irregular in shape, the vacuole becomes less definite and pink stippling (Schüffner's dots) form in the cytoplasm of the red cell. The meront (trophozoite) ingests hemoglobin by pinocytosis and by pinching off small bits. The hemoglobin is digested and the undigested portion remains as tiny black granules that sometimes appear greenish because of their refractile quality. The hemoglobin residue is hemozoin and is characteristic of infections with *Plasmodium* and its close relatives.

Nuclear replication takes place through an amitotic splitting of the nuclear material. This is the stage usually called a *presegmenter*, in which nuclear replication occurs, but no cytoplasmic division. In *P. vivax* a so-called mature presegmenter has 12 to 18 (maximum of 24) nuclei (Table 16.1).

Merozoite formation takes place in such a way that each nucleus is encased in an envelope of cytoplasm and plasma membrane. The segmenter stage is reached when the merozoites form.

Forty-eight hours after a merozoite enters a red cell, 12 to 18 merozoites have formed, the red cell breaks down, and the merozoites are released to seek new host red cells. During the breakdown of the red cell, hemozoin, red cell membranes, and metabolic byproducts are released into the bloodstream of the host. Since there is synchrony in development, a large number of red cells break down at nearly the same time; the chills and fever characteristic of clinical malaria are brought about by this massive release of toxic materials. The reproductive potential of *P. vivax* can be appreciated by calculating the possible increase over an 8-day period. If we assume that one merozoite produces 18 daughter cells in 48 hr and that all of the daughter cells invade and reproduce at the same level, one merozoite will have increased to more than 10,000 during 8 days. It is obvious that such increases cannot be supported indefinitely.

Gamonts begin to appear in circulating red cells 3 to 5 days after the infection has become patent. It is not until this time that the infection can be transferred successfully to a mosquito vector.

Gamonts have a relatively short life in the bloodstream; if they are not picked up by a mosquito within about 48 hr of their formation, they will die. They do not undergo any further development as long as they remain in the bloodstream of the vertebrate host.

When an uninfected mosquito takes a blood meal from a person, both sexual and asexual stages are ingested. The meronts die and do not particpate in the transfer of the infection to a new host. The gamonts undergo change quite soon after ingestion by a mosquito. Although the organisms are haploid, nuclear activity takes place in both macrogamonts and microgamonts. As a result of reduced temperature and a change in the blood pH, nuclear division begins in the microgamonts within minutes after the blood has reached the midgut of the mosquito. Within 15 min, exflagellation has been completed with the formation of six to eight motile, threadlike microgametes, each about 10 μm long and with a single nucleus in the center. The microgametes swim off in search of a macrogamete.

The changes in the macrogametocytes or macrogamonts are not as striking as in the microgamonts. The nucleus moves to the periphery of the gamont and protrudes slightly at the surface to form a cone. The microgamete enters the macrogamete at the point where the nucleus protrudes. Fertilization has taken place, a zygote has been formed, and the organism is now diploid.

The zygote is called an *ookinete*, and it moves from the lumen of the mosquito gut to the periphery and squeezes between gut cells. During the next 10 days to 3 weeks sporogony occurs. Within a day or so the ookinete has laid down a membrane and nuclear division has begun; this is the oocyst stage. Meiosis takes place 2 to 3 days after the oocyst has formed. Sporogony is a schizogonic process in which more than 10,000 sporozoites may be formed in each oocyst.

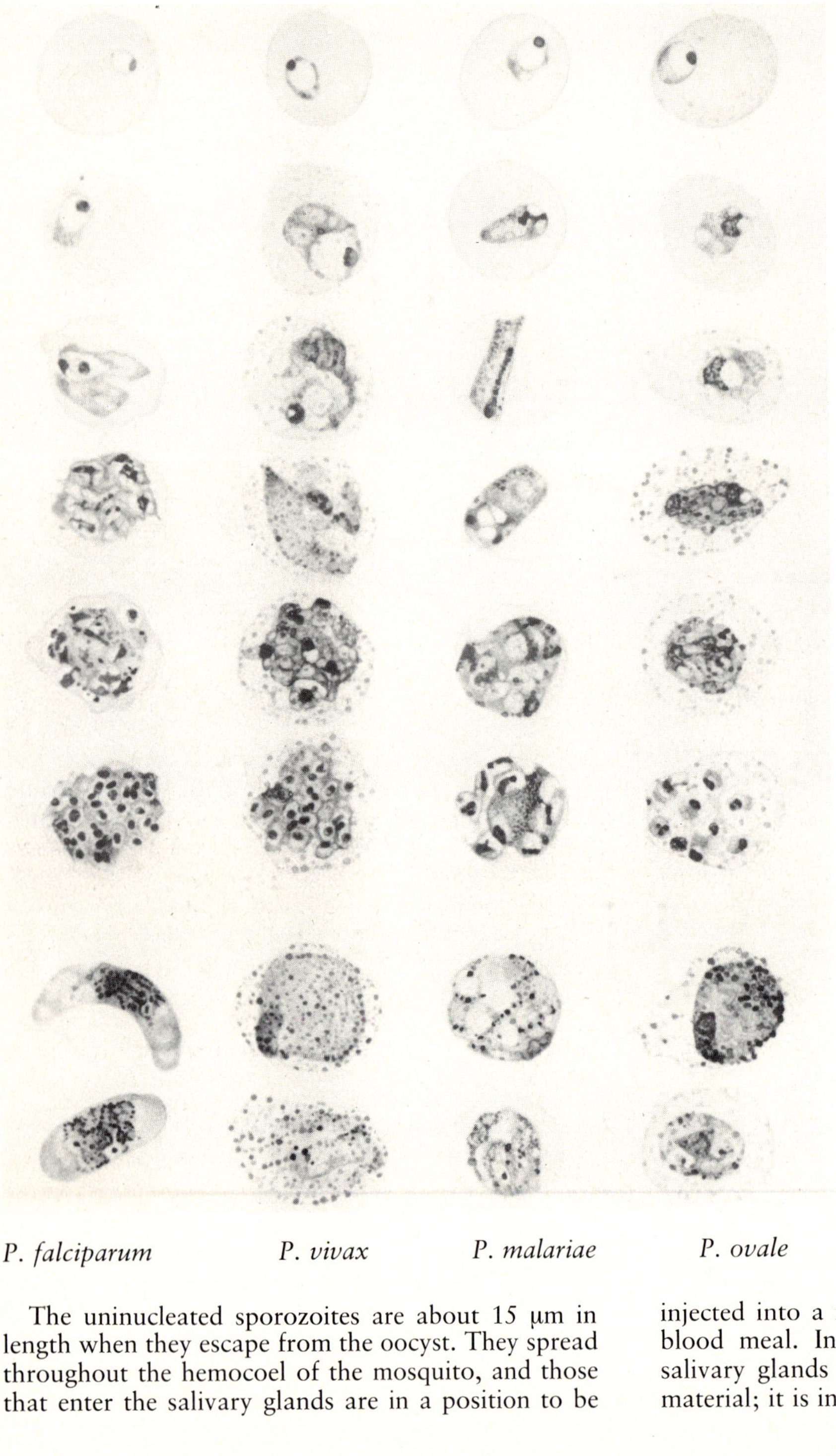

**FIG. 16.4**
Human malarial parasites as they appear in a blood film. [Reproduced by permission from Diggs, L.W., D. Sturm and A. Bell, 1972. *The Morphology of Human Blood Cells,* 4th ed. Abbott Laboratories, North Chicago.]

*P. falciparum* *P. vivax* *P. malariae* *P. ovale*

The uninucleated sporozoites are about 15 μm in length when they escape from the oocyst. They spread throughout the hemocoel of the mosquito, and those that enter the salivary glands are in a position to be injected into a new host when the mosquito takes a blood meal. Interestingly, many sporozoites in the salivary glands have two or three bits of chromatin material; it is inferred that they have undergone some precocious development preparing for entry into hepatic cells of the new host.

The life cycle described above is that of *P. vivax*, but there are some differences among species that are important because of the effect on the host or how they affect transmission.

EXOERYTHROCYTIC CYCLE. All species of *Plasmodium* affecting humans have stages in the liver, and none are able to reenter liver cells once they have entered red cells. The table below lists the principal differences among the species.

Of particular interest is the fact that *P. falciparum* has only a single EE generation and does not retain a liver infection after the erythrocytic phase has begun. The residual liver infection serves as a source of parasites for relapses that may occur some years following infection with *P. vivax* and *P. ovale* (see Relapse).

The prepatent periods given are the minimums for each species, and there may be considerable variation in the time after infection when parasites are first seen or when the clinical attack may occur. In *P. vivax*, the initial attack may not occur until 3 or 4 months after infection with some geographic isolates. Treatment with antimalarials may also delay the patent period.

There are both morphological differences (see Morphology) and differences in time of replication among the species during the blood phase of the cycle (Table 16.1).

RELAPSE AND RECRUDESCENCE. It has been known for a long time that persons who have suffered from an attack of malaria may later suffer another attack without having been reinfected. This recurrence of malaria may be either a relapse or a recrudescence. Coatney (1976) has considered these phenomena and has differentiated between them in the following way:

> *Relapse*: The reappearance of parasitemia in a sporozoite-induced infection following adequate blood schizonticidal (meronticidal) therapy.
>
> *Recrudescence*: Reactivation after a period of relative inactivity.

The principal difference to be noted is that in relapse, the parasites have unequivocally been eliminated from the circulating blood cells, whereas in recrudescence the parasites probably remain in the circulating blood cells. The implication is that the infection is retained somewhere other than the erythrocytes if relapse occurs.

Among the species of *Plasmodium* infecting humans, *P. vivax* and *P. ovale* have true relapse. *P. malariae* has only recrudescence and *P. falciparum* has neither.

In *P. malariae* infections, malaria can recur after long periods of time, sometimes years, but if there has been adequate treatment of the blood stages, the patient will be cleared of infection.

What stages are retained and where are they located in *P. vivax* and *P. ovale*? (Note: most of the work has been done with *P. vivax* since *P. ovale* is rare). (1) It has been found clearly by massive blood transfusions that the parasites do not remain in the peripheral blood. (2) Treatment of experimental animals with drugs such as the 8-aminoquinolines, which kill sporozoites, eliminates relapse. (3) Blood transfusion during the prepatent period fails to produce infections in the recipients, but the donors may later suffer relapse.

| SPECIES OF PLASMODIUM | NO. OF EE GENERATIONS | MEROZOITES/ SPOROZOITE | PREPATENT PERIOD (MINIMUM IN DAYS) | LIVER INFECTION RETAINED |
|---|---|---|---|---|
| *vivax* | 2[b] | 1,000 | 8 | Yes |
| *falciparum* | 1 | 40,000 | 5.5–6 | No |
| *malariae*[a] | 2[b] | 2,000 | 11–12 | No |
| *ovale* | 2 | 15,000 | 9 | Yes |

[a] Some information on *P. malariae* is not available and data have been taken from *P. inui*, a closely related species that infects higher primates.
[b] There are at least two EE generations; there may be more, but the exact number is not known.

**TABLE 16.1**
Comparative Characters of Plasmodia of Humans (Stained thin smears)

| STATE OR PERIOD | PLASMODIUM VIVAX | PLASMODIUM MALARIAE | PLASMODIUM FALCIPARUM | PLASMODIUM OVALE |
|---|---|---|---|---|
| IN MAN | | | | |
| Early trophozoite or ring | Relatively large; usually one chromatin dot, sometimes two; often two rings or more in one cell | Compact; one chromatin dot; double cell infections rare | Small, delicate; sometimes two chromatin dots; multiple red cell infection common; appliqué forms frequent | Compact; one chromatin dot; double infection uncommon |
| Large trophozoite | Large; markedly ameboid; prominent vacuole; pigment in fine rodlets | Small; often band-shaped; not ameboid; vacuole inconspicuous; pigment coarse | Medium size; usually compact, rarely ameboid; vacuole inconspicuous; rare in peripheral blood after half grown; pigment granular | Small; compact, not ameboid, vacuole inconspicuous; pigment coarse |
| Young schizont or presegmenter | Large; somewhat ameboid; chromatin masses numerous; pigment in fine rodlets | Small; compact; chromatin masses few; pigment coarse | Medium size; compact; chromatin masses numerous; pigment granular; rare in peripheral blood | Medium size; compact; chromatin masses few; pigment coarse |
| Mature schizont or segmenter | Larger than normal red cells; may have double rosette | Smaller than normal red cells; single rosette | Smaller than normal red cells; single rosette | Larger than *P. malariae*; irregular rosette |
| Number of merozoites | 8–24, usually 12–18 | 6–12, usually 8 | 8–26, usually 8–18 | 6–16, usually 8 |
| Microgametocytes (usually smaller and less numerous than macrogametocytes) | Spherical; compact; no vacuole; single large nucleus; diffuse coarse pigment; cytoplasm stains light blue | Similar to *P. vivax* but smaller and less numerous | Crescents usually sausage-shaped; chromatin diffuse; pigment scattered, large grains; nucleus rather large; cytoplasm stains darker blue | Similar to *P. vivax* but somewhat smaller; never abundant |
| Macrogametocytes | Spherical; compact; larger than microgametocyte; smaller nucleus; pigment same; cytoplasm stains darker blue | Similar to *P. vivax* but smaller and less numerous | Crescents often longer and more slender; chromatin central; pigment more compact; nucleus compact; cytoplasm stains darker blue | Similar to *P. vivax* but somewhat smaller; never abundant |

**TABLE 16.1 continued**

| STATE OR PERIOD | PLASMODIUM VIVAX | PLASMODIUM MALARIAE | PLASMODIUM FALCIPARUM | PLASMODIUM OVALE |
|---|---|---|---|---|
| IN MAN | | | | |
| Pigment | Short, rather delicate rodlets irregularly scattered; not much tendency to coalesce | Seen in very young rings; granules rather than rods; tendency toward peripheral scatter | Pigment granular; early tendency to coalesce; typical single solid mass in mature trophozoite; coarse scattered "rice grains" in crescents | Similar to but somewhat coarser than *P. vivax* |
| Alterations in the infected red cell | Enlarged and decolorized; Schüffner's dots usually seen | Cell may seem smaller; fine stippling occasionally seen | Normal size but may have "brassy" appearance; Maurer's dots (or clefts) may be seen; host cell of crescent barely seen | Enlarged and decolorized; Schüffner's dots (or James's stippling) early and prominent at all stages; numerous oval-shaped red cells; or crenated margins |
| Length of asexual phase | 48 hr or a little less | 72 hr | 36–48 hr | 48 hr or a little longer |
| Prepatent period: Minimum | Usually 13–17 days<br>8 days | Usually 28–37 days<br>14 days | Usually 8–12 days<br>5 days | Usually 14–16 days<br>8 days |
| Usual incubation period | 8–31 days, average 14 | 28–37 days, average 30 | | 11–16 days, average 14 |
| Interval between parasite patency and gametocyte appearance | 3–5 days | 7–14 days; appearance irregular and numbers few | 7–12 days | 12–14 days; appearance irregular and numbers few |
| IN THE MOSQUITO | | | | |
| Developmental period in mosquito | 16–17 days at 20°C (68°F); 10 days at 25°C (77°F); 8 days at 28–30°C (82.4–86°F) | 30–35 days at 20°C; 25–28 days at 22–24°C (75.2°F) | 22–23 days at 20°C; 10–12 days at 26.7°C (80°F) | 16 days at 25°C |
| Young oocyst (pigment) | Fusiform pigment, often in chain | Rounded or angular masses; golden brown | Square or rectangular blocks; black | Coccobacillary in form, with 15–30 grains |
| Sporozoites (considerable size variations in different mosquitoes) | About 8 μm | Relatively large; about 12 μm; chromatin diffuse | About 9 μm | Larger than *P. malariae* |

We are left with tissue sporozoites or tissue meronts as the source of organisms for relapse. The bulk of the evidence supports the hypothesis that a stage called a *hypnozoite*, which is a somewhat modified sporozoite, is the dormant stage in the liver of the host. The idea is an interesting and tantalizing one. What allows it to remain in an undifferentiated state for years? What stimulus causes it to begin development? If waning immunity is not the major factor in relapse, what is?

MORPHOLOGY AND DIAGNOSIS. The blood stages of *Plasmodium* are looked for and examined in making a positive diagnosis of malaria. Diagnosis of malaria is based on the following:

1. Finding malarial parasites in stained blood films
2. Clinical signs and symptoms
3. Serological tests
4. The patient's prior presence in a malarious area

Of these, finding the organisms is the only certain means of determining that a person is suffering from malaria. The other three are important factors in leading a diagnostician to suspect malaria. (The specific characteristics of each of the species are described in Table 16.1 and shown in Fig. 16.4.)

HOST–PARASITE INTERACTIONS. During the early phase of the infection of humans with *Plasmodium* spp., there is no discernible damage. After the organisms begin to leave the liver and appear in the circulating blood, mild symptoms are experienced. Within about a week after patency, the parasites have increased in number and their replication in the red cells has become synchronized. It is at this point that the malarial paroxysm is seen—the typical chills and fever of a malarious attack.

*Plasmodium vivax*, *P. ovale*, and *P. falciparum* have much the same generation time, 48 hr. On the other hand, *P. malariae* has a 72-hr generation time, and the paroxysms occur every 3 days (Fig. 16.5).

Immunity to infections with *Plasmodium* has beneficial effects but also some adverse effects on the host.

The beneficial effects of immunity to malaria are the following:

1. Prevention of infection, reinfection, or superinfection with the same species or strain
2. Reduction of multiplication of the parasite
3. Destruction of parasites
4. Aid in the repair of tissue

Immunity appears to act in reducing the number of parasites in the blood during a series of paroxysms. Over a period of about 2 weeks in a *P. vivax* clinical infection, the numbers of parasites are reduced and the paroxysms become less severe, ultimately disappearing. Thus, developing immunity during the initial stages of the infection reduces the effect on the host and controls the infection.

Premunition is a condition in which a host must retain an infection with an infectious agent in order to maintain resistance to reinfection. Malaria seems to be a good example of premunition. If an individual is cleared of infection by treatment with drugs or through development of a sterile immunity naturally, the immunity wanes in a few months and the person becomes completely susceptible to reinfection. In areas where there is likely to be year-round exposure to sporozoites of *Plasmodium*, malarious attacks are seldom seen in adults; children suffer from attacks, but as they become older, immunity and continual reinfection hold the clinical symptoms in check.

Development of a vaccine against malaria has been the objective of a number of research groups for a good many years, but a good vaccine for human use has not yet appeared on the market. The problems lie in the following:

1. The short time after a person has been cleared of infection when there is effective immunity
2. Lack of cross-protection between species of *Plasmodium*
3. Lack of cross-protection between strains of the same species

The proteins of *Plasmodium* spp. infecting humans are poor antigens and they do not raise the level of immunity sufficiently to produce an adequate response after the infection has waned. In many of the common bacterial or viral diseases such as measles, smallpox, or diphtheria, clinical disease is followed by sterile immunity, which is retained usually for the life of the person. Such is not the case with *Plasmodium*, and the lack of good, natural immunity has hampered the development of an artificial vaccine.

The rather poor antigenicity of *Plasmodium* spp. not only hampers development of both natural and artificial immunity but may also have detrimental effects on the

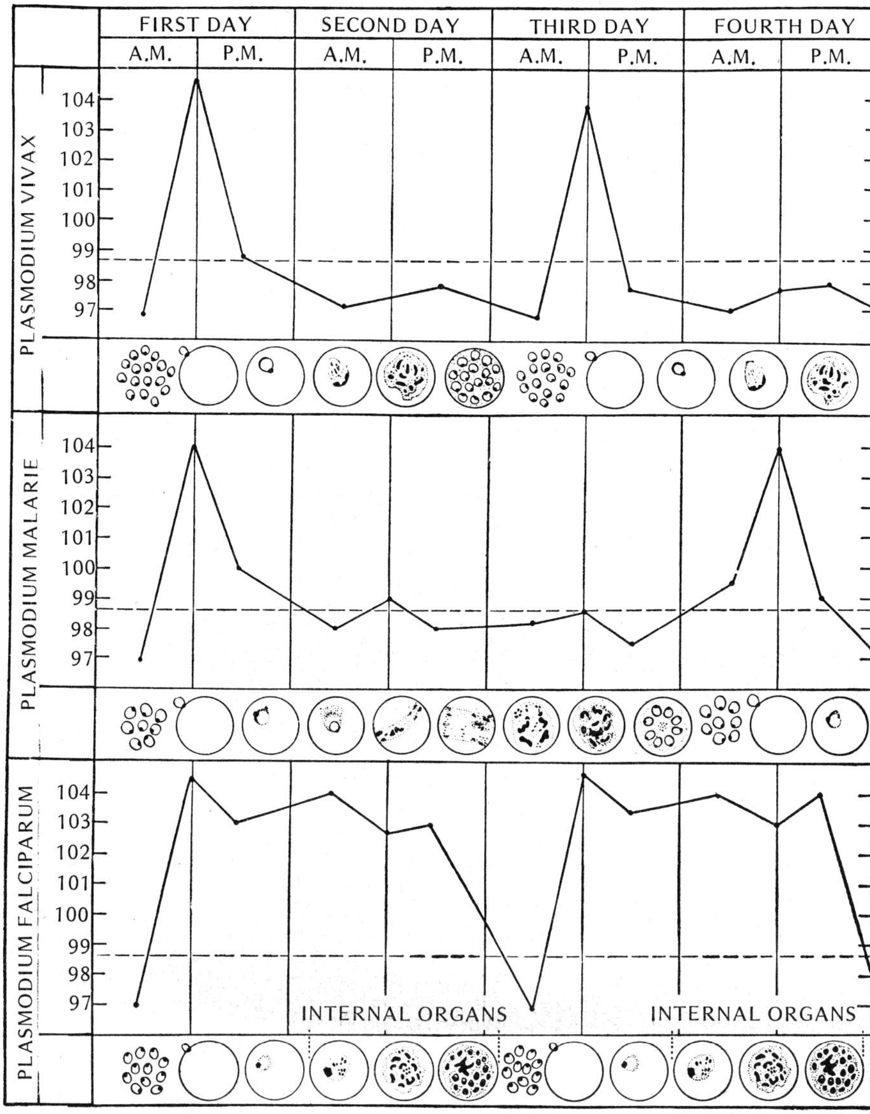


**FIG. 16.5**
Body temperature and development of malarias of humans. [Redrawn from Belding, D. L.. *Textbook of Clinical Parasitology.* 2nd ed. Appleton-Century-Crofts, New York, 1965.]

host. Many eukaryotic parasites that have long developmental times are poorly antigenic. The hypothesis that has been set forth is that parasites which take more than 10 days or 2 weeks to reach the stage when they can be transmitted to a new host must be poorly antigenic or they would be removed by the host defenses.

In an evolutionary context, parasites that are poorly antigenic would be favored for survival, and antigens that the host can readily recognize would drop out over a period of time. One possible result would be that not only would the parasite be poorly antigenic, but it might mimic the host proteins as a means of "hiding" from the immune response (Damian, 1964). *Plasmodium* spp. are clearly poorly antigenic, and it has been found that some species have proteins similar to those of their hosts. The disease called *blackwater fever* is a manifestation principally of *P. falciparum* infections in which there is lysis of red cells that may be sufficiently severe to cause the death of the host. It is theorized that the host becomes hypersensitive (allergic) to its own red cells because of their similarity to the proteins of the parasite. The reaction is thereby turned against the host to its detriment.

Blackwater fever occurs in areas where there is a high prevalence of *P. falciparum* infection and is manifested, in part, by black urine. The dark color represents hemoglobin from lysed red cells. Most often it is seen in persons who reside in highly endemic areas and have long-term, continuous exposure to *P. falciparum*. A person who has an attack of blackwater fever has about a 25% chance of dying if untreated, and once an individual has had an attack, additional attacks are likely.

The interactions of human, *Plasmodium*, and mosquito represent a fascinating aspect of evolution in action. Each of the participants has developed mechanisms for coping with attempts to harm it or remove it completely from the population. Let us look at each separately:

HUMANS. Three genetic changes are seen in humans, which confer greater or lesser degrees of resistance to malaria.

1. Sickle cell. The sickle cell name comes from the observation that persons with the trait has some distorted, elongate red cells. Sickle cell trait is a single-locus mutation that is generally lethal in the homozygous condition but confers some resistance to falciparum malaria in the heterozygous condition. The sickle trait results in a change in the protein portion of the hemoglobin molecule such that glutamic acid is replaced by valine.

   The current view on the mechanisms of action of the sickle cell trait is as follows:

   (a.) Homozygous red blood cells are changed to the sickle shape as a result of the formation of needlelike crystals of hemoglobin when exposed to the low oxygen tension present in tissues. The malarial parasites are mechanically disrupted by the hemoglobin needles and thus are destroyed. Usually the homozygous sickle cell individual dies from anemia, not from malaria.

   (b.) Individuals heterozygous for sickle cell anemia have small knobs on red blood cell surfaces. These knobs are induced by the malarial parasites, and they cause infected red blood cells to stick to narrow peripheral capillary walls. These infected cells, stuck as they are, remain in low oxygen tension longer than uninfected cells. This lowered oxygen tension causes sickling of the red blood cell along with leakage of potassium from the cell. It is thought that the lowered potassium level kills the malarial organisms. Uninfected red blood cells have no knobs on their membranes and thus continue to move throughout the circulatory system. The infected cells are selectively removed, thus providing an individual heterozygous for sickle cell anemia with an effective means of coping with malaria. Unfortunately, progeny from heterozygotes face the prospect of perhaps dying from either malaria or sickle cell anemia, depending on whether they are homozygous against or for sickle cell anemia. Only the heterozygous children will survive in an endemic malarious area (Friedman & Trager, 1981).

   Sickle cell trait is obviously not a "good solution" for genetic resistance to malaria, but it does exemplify the evolutionary pressure exerted in hyperendemic areas where falciparum malaria may cause severe sickness or death may occur in 10% of infants. In central Africa as many as 20% of persons in some populations have the sickle trait, and in some places in the Middle East, as many as 30%; it is not true that only black Africans have the sickle trait.

2. Glucose-6-phosphate dehydrogenase deficiency in red cells. Not a great deal is known about this deficiency, but children with it have lower parasitemias than normal children.

3. Duffy blood group. This factor is probably a single-locus mutation and persons having the Duffy blood group negative phenotype are completely resistant to vivax malaria. The genotype is present in about 90% of west African blacks and in about 70% of American blacks. It has been known for generations that many blacks are resistant to malaria, and the Duffy blood group appears to be the explanation. It should be noted that it relates only to

infection with *P. vivax*. The mechanism appears to reside in receptors on the red cell surface to which the *P. vivax* attaches; in Duffy-negative cells, the receptors are lacking and the parasite cannot attach and therefore cannot enter the red cell. There is no evidence that a person who is Duffy negative has any ill health related to it.

*PLASMODIUM.* The only known change that has been found in the causative agent is drug resistance. In many parts of the world where drugs have been administered to a large proportion of the human population for a long period of time, strains of *Plasmodium* resistant to chloroquine have been found.

MOSQUITO. Since World War II, mosquitoes have been subjected to a great deal of evolutionary pressure from insecticides. With any rapidly reproducing organism with a short life span, one might expect that it could evolve rapidly enough to develop survival techniques. Within a few years after the wholesale use of DDT was begun, resistant populations began to emerge. Since then, populations resistant to all of the other major insecticides have been found, and in some cases the mosquitoes are resistant to more than one compound. This is a physiological resistance in which the insects are either able to metabolize the insecticide rapidly or are unaffected by the usual concentrations used. The other means of resistance is behavioral resistance. Much of malaria control depends on spraying dwellings with insecticides, such as DDT, which have a long residual action. Most of the mosquitoes that are good vectors of malaria enter dwellings, take a blood meal, and then rest on the nearest surface to digest the meal; the DDT on the surface will kill them. Because there is a range of behaviors in the population, some mosquitoes fly out of the house and thereby do not contact the treated surfaces. It is clear that these mosquitoes will be favored for survival, and populations have been selected for flying out of dwellings to find resting surfaces. They survive to return later and transmit *Plasmodium*.

Each of the members of the *Plasmodium-human*-mosquito triangle has been subjected to pressures by the other members and responded with means of surviving and passing their DNA to the next generation. Before the development of technology, humans appeared to be the only member that developed resistant mutations. It is only during the past 40 years that intense pressure has been placed on the vector and the causative agent through the development of various drugs and insecticides. It seems likely that we will see further changes in the not too distant future.

EPIDEMIOLOGY AND CONTROL. The elements in the transmission of malaria are the causative agent, the vector, humans, and the environment. The characteristics of each and the ways in which they interact determine whether malaria will occur in a particular area and what the pattern of transmission will be. The distribution of malaria in the world is determined to a great extent by the distribution of the mosquito vectors. The mosquito family Culicidae contains the subfamilies Culicinae, Anophelinae, and Toxorhynchitinae, but only the Anophelinae serve as vectors of human *Plasmodium*. Mosquitoes are found essentially worldwide and in all kinds of habitats from the tropics to the arctic and from the Dead Sea to almost 5000 m above sea level. The only requirements for them to complete the life cycle are an aquatic habitat for the larvae and a source of blood for the females to develop eggs (Toxorhynchitinae excepted).

Although members of the Anophelinae are widely spread, some even being found in cold northern climates, they are principally found in warm and hot climates. Of something less than 400 species and subspecies of *Anopheles*, about 65 are known to be vectors of *Plasmodium* spp. of humans. In the Western Hemisphere, there are two vectors extending from all of North America south through most of Mexico. For Central America, a much smaller area, there are five proven vectors. In Europe there are four vectors in the northern part, but seven in the Mediterranean area. Thus, in the warmer areas more species can transmit the parasite; more species means that more different ecological niches will be inhabited, thereby increasing the probability that a potential vector will be found in any one locality.

Each species of mosquito not only is structurally distinct but has its own habitat preference, developmental cycle, and host preferences for blood. These factors, together with the physiological characteristics which determine whether the mosquito will be a good host for the reproduction of the parasite, influence whether or to what extent malaria may be transmitted in a particular locality.

Mosquitoes are holometabolic insects and as such have an egg, four larval stages, a pupa, and a winged adult or imago (see Fig. 54.6). Eggs may be laid on land,

but all of the larval stages and the pupae are aquatic; the adults emerge from the pupae at the surface of the water. The time from egg to adult will vary from perhaps 3 or 4 days to several weeks, depending on the species and rearing conditions of temperature and food. Adults may live a month or more, but only a minority do so; attrition is high in these delicate creatures. In order to complete transmission, a female mosquito must take at least two blood meals. Blood is required for the maturation of significant numbers of eggs, and a female takes a blood meal, oviposits within a few days, takes another blood meal, and oviposits a second brood. A long-lived female may lay five or more batches of eggs.

The potential for explosive epidemics in areas where unstable malaria exists is exemplified by both accidental and purposeful transmission. In one experiment a single mosquito infected 40 persons. Late in the Korean War a marine suffered a malarious attack while camping near a girls' camp in the Sierra mountains of California. He was in the vicinity for only a single night during the paroxysm, but 35 girls later came down with malaria (Brunetti et al., 1954). In Syria, a man suffered a relapse with *P. vivax* and there were 52 known cases acquired from this single individual (Macdonald et al., 1968).

Non-mosquito-borne malaria has been a problem at various times. It first came to light in the late 1930s in New York City when an outbreak of malaria occurred; investigation revealed that the infection was passed from a few individuals who had the infection through shared needles by heroin users. There have been about 200 cases in New York, and a few other outbreaks have been reported. During the years 1958 through 1977 there were 56 needle-borne cases of malaria in the United States, 47 of which occurred in 1971, during the Viet Nam War.

Malaria transferred by blood transfusions has not been a serious problem, but there are a few sporadic instances on record. Over the 20 years from 1958 to 1977 there was an average of 2.5 cases of transfusion malaria in the United States. Russell et al. (1963) suggest that persons who have had falciparum malaria may safely donate blood 1 year after an attack and those who have had vivax malaria after 3 years. Those who have had *P. malariae* should not be donors. Most blood banks will not accept blood from a person who has had malaria, and in our opinion that is the proper decision to make unless the need for blood overrides the risk.

Prenatal transmission of malaria can occur, but is not common. It is seemingly a rare occurrence for either sporozoites or meronts to pass from the maternal circulation to the fetus through the placenta.

TREATMENT. In the history of medicine, the treatment of malarial paroxysms with quinine may well be the first instance in which a substance had a specific curative effect for an infectious agent. The myth is often repeated that the Countess of Chinchon was afflicted with malaria in Peru in about 1640. She was given the bark of a local tree by her physician and experienced a miraculous cure. She was so enthusiastic over the results of this treatment that she carried some of the tree bark back to Spain and was instrumental in establishing it as a cure for malaria.

In reviewing the history of malaria, Russell et al. (1963) relate that while the Countess was in Peru with her husband she was in remarkably good health, never suffered from malaria, but died during the return trip before reaching Spain. The traditional story is a myth whose source is unknown. The real story is more interesting, if somewhat less romantic than the myth.

The Amerinds in South America probably used the bark of cinchona and other trees to reduce fever. Since malaria did not exist in the Western Hemisphere until European colonization, the Amerinds did not use it originally against malaria. The first documented use of the bark of *Cinchona ledgeriana* against malaria was in 1600 when a Jesuit missionary, Juan Lopez, was cured of a malarious attack. The first cinchona bark reached Europe in 1632 and was widely used in treating fevers.

During the early years of colonization and exploration of the New World, plants were transported back and forth between the Old World and New World in great profusion. Likewise, there was considerable trade in potential medicinals. One of the tree barks was that of *Myroxylon peruiferum*, the extract of which was used to reduce fever, and a brisk trade grew up by the late 1500s. Demand outstripped supply, and traders began to substitute the bark of *Cinchona ledgeriana*, which looked much like *M. peruiferum*. Physicians thought that they were using a single medicinal, but results were inconsistent and therefore confusing. Ultimately, it was determined that the fake had specific activity against intermittent fevers, and *C. ledgeriana* was established as the source of the active ingredient for treating malaria.

Carolus Linnaeus, the Swedish botanist who established binomial nomenclature for naming plants and

animals, named *Cinchona ledgeriana*. He erred twice, first in perpetuating the myth of the countess' involvement in discovering a cure, and then in misspelling her name.

Extracts of cinchona bark contain more than 20 alkaloids, but only four of them have a significant effect on *Plasmodium*. The one with the highest level of activity is quinine, a quinoline derivative. Quinine was used for 300 years as the only drug for malaria until a concerted search was made for alternatives in the 1920s.

Quinine has a good effect against all species of *Plasmodium*, but it has some serious side effects such as ringing in the ears, deafness, dizziness, and various effects on vision. The side effects can be serious in some individuals, but the real stimulus toward finding substitutes was the interruption of the German supply of quinine in World War I.

The first useful synthetic antimalarial was an acridine, quinacrine (Atabrine, Mepacrine), used extensively in World War II as a suppressant and a chemotherapeutic. Although it is a highly effective drug, quinacrine has side effects such as discoloration of the skin and gastrointestinal disturbances.

Quinacrine was replaced by the quinolines when they were developed in the late 1940s. The quinolines have structures similar to quinine, and there are two groups of substituted quinolines, the 8-aminoquinolines and the 4-aminoquinolines. The 4-aminoquinolines are represented by chloroquine and amodiaquine. The 8-aminoquinolines are represented by pamaquine, pentaquine, and primaquine.

Since the development of these compounds, other groups of chemicals have also been found to have efficiency: pyrimidines (pyrimethamine, sulfadiazine), sufones (dapsone), and biguanides (proguanil).

The side effects seen with quinine were serious in some persons, and there have been some difficulties with each of the synthetic antimalarials. Quinacrine can usually be given prophylactically for long periods, but in therapeutic doses gastrointestinal problems are sometimes seen. Proguanil is a good drug in many respects, but the plasmodia developed resistance to it quite quickly.

Chloroquine rarely causes side effects, but since about 1960, more and more isolates of *P. falciparum* have been found to be resistant to it. Dapsone was originally used against Hansen's disease (leprosy), and has been found to be effective against drug-resistant *P. falciparum*, but regrettably it appears to be carcinogenic. None of the drugs currently cleared for use has a significant effect against sporozoites, that is, none is sporonticidal.

Antimalarials are used as (1) thereapeutics, (2) prophylactics, and (3) suppressives, but no one drug suffices for all three purposes.

Therapeutic administration of a drug takes place during a malarial paroxysm so that the parasites can be cleared from the blood within a day or two and the patient can recover clinically. Prophylactic or suppressive administration of drugs is done on a long-term basis with low dosage of the compound. The purpose is to prevent the appearance of clinical symptoms (suppression) or to kill the parasites in an early stage of development (prophylaxis). Chloroquine can be given both to treat an attack and to suppress the development of malaria; in an attack 3 g are given over a 3-day period, but only 0.25 g/week are given as a suppressive.

Current malaria chemotherapy (Maugh, 1977) uses chloroquine to destroy blood merozoites plus primaquine to destroy liver meronts. Mefloquine is just coming into use to treat chloroquine-resistant falciparum malaria. It has no serious side effects and is effective in a single dose. It has also been used to treat resistant vivax malaria, but must be used with a meronticidal (antischizontal) drug such as primaquine to prevent relapses.

This discussion has taken only a general approach to chemotherapy of malaria. Although recommendations for the drugs to use under different circumstances have not changed radically, they do vary somewhat from year to year. For those living in or planning to visit a malarious area, a physician or local health department should be consulted for current recommendations.

The problem of drug resistance is becoming greater as large numbers of persons receive suppressive as well as therapeutic quantities of drugs. Quinine has been used for over 300 years without the development of drug resistance, although some populations of *Plasmodium* require relatively high dosages. Quinine is exceptional among drugs, however; in nearly all instances where drugs have been used against nearly any kind of parasitic agent, microbial, protozoan, helminthic, or arthropodal, resistance has been seen within a few years of introduction. The agents causing malaria are merely examples of the general principle. *Plasmodium* has quite readily developed resistance to proguanil and pyrimethamine; resistance was first seen in them in the late 1940s and early 1950s, respectively. By 1977 chloroquine-resistant *P. falciparum* had been found in 8 countries of South America and 11 Asian countries; resistant forms were found in East Africa in 1979.

Control of malaria in, its simplest terms, directs efforts at

1. The vector
2. The parasite
3. Personal protection

By far the technologically simplest of these efforts is personal protection, which involves the following:

1. Screening of houses or sleeping under mosquito netting
2. Use of insect repellents
3. Avoiding mosquitoes by either not going where they are abundant or staying indoors when they are most likely to bite (i.e., at dusk and dawn for many species)

In the tropics, many of the houses are not amenable to installing screens; screens may also be prohibitively expensive, and some persons will not install them because they reduce the amount of air moving through the house.

For practical purposes, repellents are most useful in the short term, if one is going into a swampy area or must be out at night where mosquitoes are abundant. An interesting method of repelling insects was developed by the U.S. military by impregnating a mesh jacket with repellent; the jacket can be worn to repel flying insects or placed on the ground to repel crawling arthropods such as ticks.

Efforts directed against the vector mosquito may be aimed at the larva or the adult. Different methods are required for each.

Much of the early effort to reduce the incidence of malaria was directed at the larval stages. In the early 1900s there were no effective insecticides for use against adult mosquitoes. Currently measures to reduce populations of preadult mosquitoes may be categorized as

1. Chemical
2. Environmental alteration
3. Biological control

Chemical control is achieved by the following steps:

1. Oiling sites of larval development with No. 2 diesel oil. The oil is taken into the breathing tubes of the larvae and kills them.
2. Using organochlorine or organophosphate insecticides; they have supplanted oiling almost completely. These compounds may be distributed by a variety of means and either as wettable powders or in granules impregnated with insecticide. DDT was among the first of the synthetic insecticides to be used as a larvicide, but it has been supplanted by organophosphate and carbamate insecticides in many countries because of its environmental effects.

When insecticides were developed that were active against adult mosquitoes, a shift in the emphasis of control programs began. The pyrethrins in the 1930s and then the organochlorine insecticides in the 1940s provided tools not previously available. Most malaria was known to be acquired in and around homes; the important vectors are most often domiciliated species of *Anopheles* such as *An. albimanus* in Mexico or *An. stephensi* in India. It was then possible to direct control against the adult mosquitoes in a relatively limited area.

In *Practical Malariology* (Russell et al., 1963) it is emphasized that mosquito control in most endemic areas is best achieved by directing efforts toward the adults. Since domiciliated mosquitoes are the principal vectors, spraying insecticides in and around houses is the main line of attack. Unskilled workers can be taught in a few hours where and how to apply insecticide. On the other hand, larval control requires a significantly higher level of technical skill.

DDT has been the principal toxicant for control of adult mosquitoes in and around dwellings. It has the advantage of being a contact poison, that is, it can be absorbed through the legs of a resting mosquito, and it has a long residual action. A properly treated house can remain mosquito-free for 3 months or more. In tropical areas where mosquitoes are prevalent year-round spraying may be required only four times a year. If mosquitoes are absent for part of the year, as in a warm-temperate climate, spraying need be done only during the summer.

The persistence of DDT is good and bad. The long residual activity means that houses can be treated infrequently; however, it breaks down slowly and may remain in the ecosystem for years. The well-publicized detrimental effect on reproduction of raptorial birds is one of the long-term hazards of DDT. Alternative insecticides have been sought, and in recent years there has been a shift from the organochlorine to organophosphate and carbamate insecticides.

Although domiciliated mosquitoes are usually the vectors of malaria, woodland- or swamp-inhabiting mosquitoes are sometimes important also. The problem then is how to control anophelines in forests, farmland, and swamps. Starting in the late 1960s the technique of ultralow volume (ULV) application of insecticides was introduced. In ULV application, insecticide of 90% or greater purity is sprayed over a relatively large area from either aircraft or a moving ground vehicle. For example, 95% malathion has been used for control of both culicine and anopheline mosquitoes at the rate of 3 oz/acre (0.22 l/hectare). ULV spraying is a high-technology method and is effective only when (1) droplet size is 30 μm or less, (2) wind speed is low enough to prevent drift and (3) the equipment can be monitored to ensure proper delivery.

Six insecticides were approved for use in the United States as of 1977; most are organophosphates. Experience has shown that there is immediate reduction in the population of adult mosquitoes, but the population will recover within 2 weeks if there is adequate water for development of the larvae. Experience in Haiti showed that transmission of malaria could be interrupted by ULV application of malathion, but transmission resumed when the population of *An. albimanus* rose again. The cost of aerial application of malathion was about twice that of administering prophylactic antimalarials to the people in the test area.

Environmental alteration or source reduction is directed most often toward drainage to reduce swampy areas or toward channeling of small streams. With the advent of large water control projects, strategic lowering and raising of water levels was developed as a means of reducing mosquito populations, both vectors and pests. This technique was even practiced in ancient times when it was recognized that the "bad air" of swamps could be hazardous to health. It was employed as the important method in the early part of this century and must still be used as part of an overall control program. The general principles are to remove standing water by means of ditching or installing subsurface drainage pipe. Running waters also serve as sites of larval development, particularly where the water moves slowly through plant-choked shallow areas. Removal of weeds or straightening the stream bed or placing it in a concrete channel have all been successful.

Biological control has been the dream of investigators for generations. What sort of natural agent can be exploited as an automatic control for unwanted insects? In 1924 Kudo discovered a microsporan, *Nosema anophelis*, that developed in the fat body of *Anopheles quadrimaculatus*. The organism was pathogenic, but it could not be exploited for practical control. More recently, Vavra and Undeen (1970) described *Nosema algerae*, which decimated laboratory colonies of *Anopheles stephensi*. This parasite has great potential for biological control and is being studied in field investigations to determine how it may be used practically.

In some areas, a small mosquito-eating fish, *Gambusia affinis*, is used in mosquito control. In warm climates, it will live year-round, but in temperature climates it may be killed during the winter and must be reintroduced in spring.

A nematode, *Reesimermis* spp., has been used successfully in mosquito control in the southern United States and clearance to use it commercially is being sought.

A planarian (Platyhelminthes, Turbellaria) has been found to feed on mosquito larvae and to reduce mosquito populations.

Two species of *Bacillus*, a bacterium, have been used in mosquito control; each of them shows promise of having qualities that will reduce mosquito populations.

As discussed elsewhere in this book (see Schistosomes), biological control is not an automatic system. The habits of the target organism must be known, and usually a high level of technology and expertise is necessary for the production and dissemination of the control agent. Failures are sometimes experienced for no apparent reason. The principles of mass action pertain to biological control. A relatively large reduction in numbers can be achieved when the population is high; when the population is reduced to a low level, it is much more difficult for the predator or the infectious agent to find the target organism. If the predator or infectious agent dies out, the mosquito population may rebound quite quickly.

ERADICATION OF MALARIA. Eradication of a disease agent means that the organism is completely eliminated from a particular geographic or political area. Eradication is not to be attempted lightly, and success has been achieved with only a few infectious agents and in only limited areas, with the exception of smallpox virus.

Following World War II, it was concluded that malaria was the most serious infectious disease of humans on a worldwide basis, and control efforts had not been sufficiently successful. However, by the late 1940s knowledge and techniques had developed to the point where eradication was being discussed. The factors that allowed consideration of an eradication program were the following:

1. The limited host range of *Plasmodium* spp. Those in humans infect only humans. (Further work showed that this principle was not as true as had been thought at the time.)
2. The anopheline vectors were almost always species that rested in dwellings after feeding and were therefore accessible.
3. DDT had been developed in the late 1930s and tested during World War II, so its characteristics were known. Its high toxicity for insects and its long residual action made it ideal for mosquito control.
4. Good chemotherapeutic and chemoprophylactic agents had been developed and tested during and after the war. There was, therefore, a good armamentarium of drugs that could be used as part of the overall effort.

In 1955 the World Health Organization (WHO) launched a program whose objective was to eradicate malaria on a global scale (Fig. 16.6). In any one area, the phases of program were as follows:

1. Preparation
2. Attack
3. Consolidation
4. Surveillance

The principles involved in these four phases of malaria eradication are the logical progression of actions that anyone would take in setting up a control or eradication program for a transmissible disease.

1. Preparation. Time: 18 months. This is the phase for gathering information, personnel, and facilities. The extent and severity of malaria in the human population must be determined by blood film, splenic index, and public health records. The principal vector(s) of malaria must be determined and their biologies worked out. The general plan is developed, budgets are constructed, personnel are engaged, and facilities of all sorts are found, purchased, or built.
2. Attack. Time: 4 years. The principal thrust in the attack phase is against the anopheline vector through spraying of houses with insecticide. As large a segment as possible of the population is given prophylactic drugs such as chloroquine, and all clinical cases are treated therapeutically. Interruption of transmission should take place during the second year of the attack—that is, there should be no new cases of malaria after this time.
3. Consolidation. Time: 3 years. With transmission interrupted in the major part of the area, pockets of infection can then be located and eliminated. Residual spraying of some dwellings is continued. Seven years from the beginning of the attack phase, the residual or latent infections in the human population should have been almost completely eliminated. Mosquito control should keep the population of vectors, at least those associated with humans, at a low level. Even if a few persons in the population suffer from relapse or recrudescence, their number should be so small that the probability of transmission is insignificant.
4. Surveillance. Time: 2 years or indefinitely. Assuming that a good public health organization is in place, there should be adequate personnel to monitor and treat the occasional case or small outbreak that may occur as a result of importation of malaria.

Eradication of malaria was begun more than 25 years ago by WHO. One might well ask if it has been successful. The truthful answer is, only partly. Eradication has been accomplished in the United States, some islands in the Caribbean Sea, Western Europe, Australia, and a few other countries. Note that where eradication has been successful, the countries are either affluent, on the margins of the endemic areas, or geographically isolated. Success was not experienced in

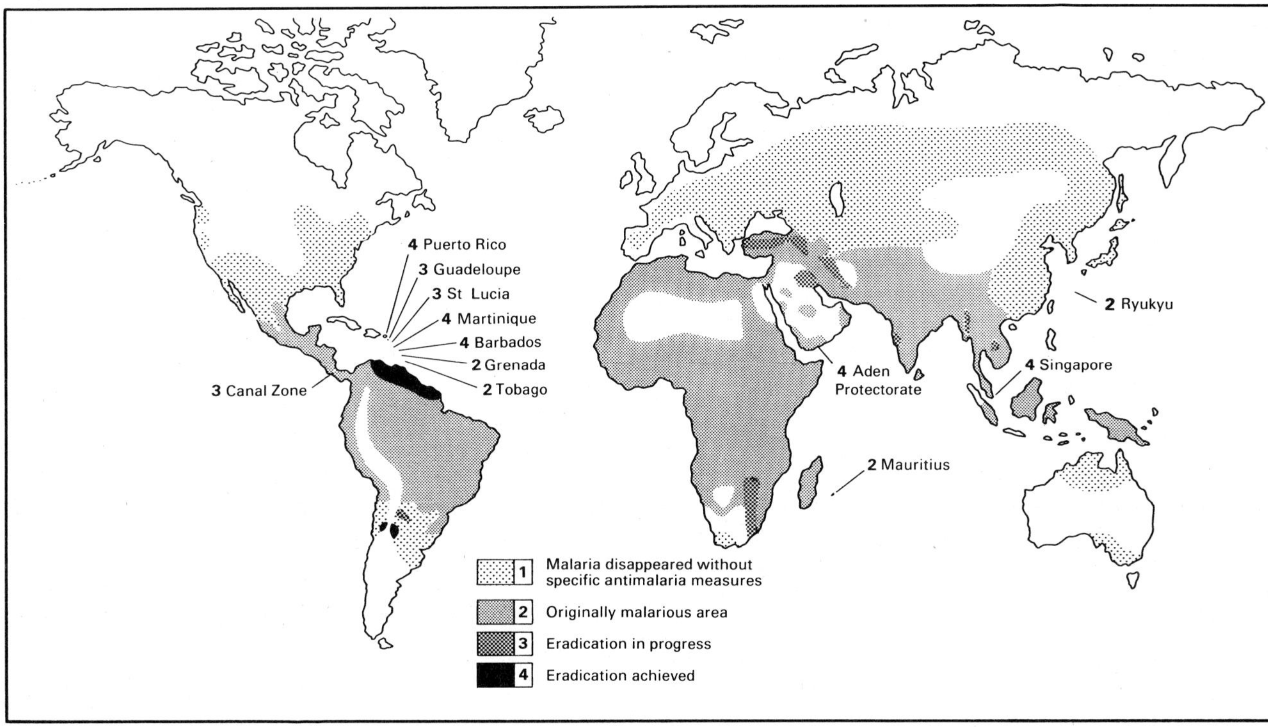


(A)

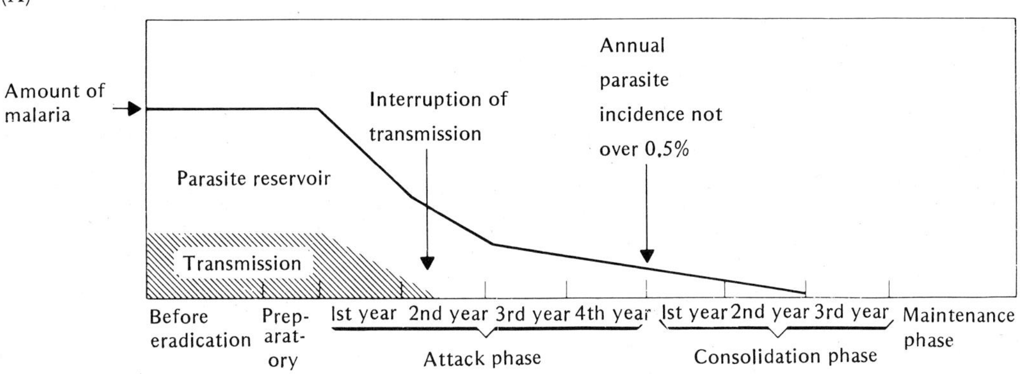


(B)

**FIG. 16.6**
(*A*) Map of Malaria eradication; (*B*) Diagram of malaria eradication. In the attack phase the mosquito population is lowered so that transmission ceases; thereafter, annual parasite incidence shows a slow decline. [Courtesy of World Health Organization.]

the tropical areas of the Americas, Africa, Asia, and the Pacific islands. This is not to say that the impact of malaria has not been lessened, but the original objective has not been reached.

Why has eradication not been achieved when the finest minds in malariology, public health, and entomology reasoned that it could be done? The answer seems to be partly biological, but to a greater extent it is politicoeconomic. It has been pointed out that the following are requisites for malaria eradication in an area:

1. There is a single vector which seeks resting sites within houses
2. There is a relatively high economic level in the country, and a relatively well-educated population.
3. There is a sufficient supply of trained personnel (physicians, entomologists, laboratory technicians, etc.)
4. The government is stable
5. The government is determined to eradicate malaria and has the power to do so

Although the absolute importance of some of these points may be debated, they have generally been found to be relevant to eradication programs. Where formerly endophilic mosquitoes were selected for exophilism through spraying of houses, control programs have become more difficult. Where public health officials have been poorly informed by persons less than expert, malaria has broken out again. Where governments have shifted funds during the surveillance phase to what appeared to be more pressing problems, malaria has surged to the forefront again (Gabaldon, 1972).

Despite the failure to eradicate malaria on a global scale, progress has been made; thus, many people who live in areas where malaria was formerly a severe threat are now at little or no risk. Malaria eradication was begun in the Americas in 1958, and by 1970 there were only 8346 cases of malaria in all countries, only 9% of which were contracted locally (autochthonous). In South America, 55% of the population by 1970 was living in areas where malaria was either in the consolidation or surveillance phases (Garcia-Martin, 1972). The situation in Africa is not as good as in the Americas. Of the 250 million inhabitants of Africa, 230 million originally lived in malarious areas. As of 1974, only 4.43 million were living in areas where malaria had been eradicated and 19.7 million were protected by some measures. About 90% of the population lacked the protection of an organized public health program.

There are a number of examples of eradication programs that almost succeeded. Sri Lanka had about 2.5 million cases of malaria annually prior to eradication programs. By 1963 the number was down to 16; but efforts were reduced, and in 1975 there were more than half a million cases. India's malaria cases were reduced from 75 million to 125,000 by 1965, but in 1975 the number was back up to 4 million.

Since malaria is so widespread and the costs of control are so high, it is important to determine how best to use available resources. Malaria seems to be the first disease in which concerted efforts were made to use quantitative methods for predicting the results of control (Bailey, 1982). As early as 1909, only 11 years after he described the development of *Plasmodium* in the mosquito, Sir Ronald Ross made the first attempt to write a mathematical expression for malaria transmission. He concluded that it was not necessary to eradicate the vector, but only to reduce its numbers below a certain threshold to break the cycle of transmission. Ross further refined his original equation, but nearly 40 years passed before a number of investigators in different institutions, attacking the problem simultaneously, began to refine both the mathematics and the biological elements of malaria. In developing a usable expression, the following ten parameters must be taken into account (Bailey, 1982):

$x$ = proportion of human population infectious to mosquitoes (i.e., the gametocyte rate)
$h$ = rate at which new infections take place
$r$ = recovery rate
$m$ = mosquito density/person
$a$ = average number of persons bitten per day by any one mosquito
$s$ = proportion of mosquitoes with sporozoites
$b$ = proportion of mosquitoes with sporozoites that are actually infectious
$n$ = time required to reach infectivity in the mosquito
$p$ = probability that a given mosquito survives over a given day
$v'$ = mosquito death rate

Nearly all of these parameters must be determined by carefully designed studies in each locality; an educated guess will not do.

When a mathematical model or expression has been developed, it should do the following:

1. Represent the real world in a fairly precise way
2. Be predictive (i.e., if a parameter is changed, the output should tell what will happen)

**3.** Be sufficiently accurate that decision makers can be guided by the model

Any model should allow one to ask the question, "what if?" For example, "what if we vaccinated 90% of the population against malaria?" In one model at least (May, 1983), the answer is, "not enough," since about 99% need to be vaccinated in order to stop transmission. And that is precisely the kind of information, if it is accurate, that a public health administrator needs in order to use resources as efficiently as possible. Going back to the early part of this section, there are a number of possible points in the cycle of transmission of malaria that can be attacked. For ultimate success to be achieved in control, the method used and the intensity of its application must be determined.

## OTHER HAEMOSPORINA

Although our greatest concern here is malaria of humans, related infections occur in other primates, rodents, birds, and reptiles. The bird malarias have been especially important in the development of knowledge on malaria and in searching for potential antimalarial drugs. The mosquito transmission of *Plasmodium* was accomplished by Sir Ronald Ross in 1898, working with *Plasmodium relictum* in sparrows. The exoerythrocytic (EE) cycle of *Plasmodium* was first worked out with bird malaria, and it became clear that such a cycle was probably also present in human malaria. During World War II, bird malaria species (especially *P. gallinaceum*, *P. cathemerium*, and *P. relictum*) were used in screening potential antimalarials.

The bird malarial parasites were good models for laboratory study, but it was felt that a species that infected mammals would be a better model because it would probably be more like the species in humans. A search was begun shortly after World War II and *Plasmodium berghei* was described from an African rodent by Vincke and Lips in 1948. The organism was adapted to laboratory mice and has since become the subject of thousands of studies on all aspects of the infection (Aviado, 1967; Garnham, 1973).

*Haemoproteus* and *Leucocytozoon* are bird parasites whose life cycles are similar to that of *Plasmodium* except that asexual stages do not appear in the circulating blood cells; the asexual stages remain in internal organs such as the spleen and lung. The gamonts (gametocytes) of *Haemoproteus* occur in erythrocytes, whereas those of *Leucocytozoon* are in leukocytes or erythrocytes, depending on the species (Bennett et al., 1975). These cells cannot be recognized with certainty as either WBCs or RBCs, because they are significantly altered by the time the infected cells reach the circulating blood. In any locality, a sample of birds will show perhaps a 10% prevalence of infection with parasites in the genera *Plasmodium*, *Haemoproteus*, and *Leucocytozoon*, and they are generally nonpathogenic. A few species such as *L. smithi* in turkeys and *L. simondi* in ducks are pathogenic and cause losses both in domesticated and wild birds; both species are transmitted by black flies (Diptera, Simuliidae).

## REFERENCES AND FURTHER READINGS


Aikawa, M. 1971. *Plasmodium*: The fine structure of malarial parasites. *Exp. Parasitol.* **30**:284–320

Aviado, D.M. 1967. *An Index to the Bibliography on* Plasmodium berghei. *1948–1966.* Defense Documents Center, Cameron Station, Va.

Bailey, N. T. J. 1982. *The Biomathematics of Malaria.* Macmillan, New York.

Belding, D. L. 1965. *Textbook of Clinical Parasitology*, 2nd. ed. Appleton-Century-Crofts, New York

Bennett, G. F., M. Laird, R. A. Khan, and C. M. Herman. 1975. Remarks on the status of the genus *Leucocytozoon* Sambon, 1908. *J. Protozool.* **22**:24–30.

Bruce-Chwatt, L. J. 1973. Global problems of imported disease. *Adv. Parasitol.* **11**:75–114.

———. 1980. *Essential Malariology*. Springer, New York.

Brunetti, R., R. F. Fritz, and A. C. Hollister, Jr. 1954. An outbreak of malaria in California. *Am. J. Trop. Med. Hyg.* **3**:779–788.

Coatney, G. R. 1971. The simian malarias: Zoonoses, anthroponoses, or both? *Am. J. Trop. Med. Hyg.* **20**:795–803.

———. 1976. Relapse in malaria—An enigma. *J. Parasitol.* **62**:3–9.

Damian, R. T. 1964. Molecular mimicry: Antigen sharing by parasite and host and its consequences. *Am. Natur.* **98**:129–149.

Diehl. H. S. 1955. World Health Assembly – Malaria eradication. *Science* **122**:126.

Fallis, A. M., S. S. Desser, and R. A. Khan. 1974. On species of *Leucocytozoon. Adv. Parasitol.* **12**:4–67.

Fletcher, A., and B. Maegraith. 1972. The metabolism of the malaria parasite and its host. *Adv. Parasitol.* **10**:31–48.

Friedman, M. J., and W. Trager. 1981. The biochemistry of resistance to malaria. *Sci. Am.* **244**:154 – 164.

Gabaldon, A. 1969. Global eradication of malaria: Changes of strategy and future outlook. *Am. J. Trop. Med. Hyg.* **18**:641–656

———. 1972. Difficulties confronting malaria eradication. *Am. J. Trop. Med. Hyg.* **21**:634–639.

Garcia-Martin, G. 1972. Status of malaria eradication in the Americas. *Am. J. Trop. Med. Hyg.* **21**:617–634.

Garnham, P. C. C. 1966. *Malaria Parasites and Other Haemosporidia.* Blackwell, Oxford.

———. 1973. Recent research on malaria in mammals excluding man. *Adv. Parasitol.* **11**:603–628.

Harrison, G. 1978. *Mosquitoes, Malaria and Man: A History of the Hostilities since 1880.* E. P. Dutton, New York.

Huff, C. G. 1963. Experimental research on avian malaria. *Adv. Parasitol.* **1**:1–65.

———. 1968. Recent experimental research on avian malaria. *Adv. Parasitol.* **6**:293–311.

———. 1969. Exoerythrocytic stages of avian malaria and reptilian malarial parasites. *Exp. Parasitol.* **24**:383–421.

Krier, J. P. (ed.). 1980. *Malaria.* 3 Vols. Academic Press, New York.

Macdonald, G., C. B. Cuellar, and C. V. Fall. 1968. The dynamics of malaria. *Bull. WHO* **38**:743–755.

Maegraith, B. 1968. Liver involvement in acute mammalian malaria with special reference of *Plasmodium knowlesi* malaria. *Adv. Parasitol.* **6**:189–231.

Maegraith, B. G., and A. Fletcher. 1972. The pathogenesis of mammalian malaria. *Adv. Parasitol.* **10**:49–75.

Maugh, T. H. 1977. Malaria drugs: New ones are available, but little used. *Science* **196**:415.

May, R. M. 1983. Parasitic infections as regulators of animal populations. *Am. Sci.* **71**:36–43.

Peters, W. 1972. Advances in malariology relating to control and eradication. *Br. Med. J.* **28**:28–33.

———. 1974. Recent advances in antimalarial chemotherapy and drug resistance. *Adv. Parasitol.* **12**:69–114.

Russell, P. F., L. S. West, and R. D. Manwell. 1963. *Practical Malariology*, 2nd. ed. Oxford University Press, London.

Sherman, I.W. 1979. Biochemistry of *Plasmodium* (malarial parasites). *Microbiol. Rev.* **43**:453–495.

Speer, C. A., and P. H. Silverman. 1979. Recent advances in applied malaria immunology. *Z. Parasitenk.* **60**:3–17.

Spencer, H. C., et al. 1978. The Duffy blood group and resistance to *Plasmodium vivax* in Honduras. *Am. J. Trop. Med. Hyg.* **27**:664–670.

Trager, W., and S. M. Krassner. 1967. Growth of parasitic protozoa in tissue cultures. In T.-T. Chen (ed.), *Research in Protozoology.* Pergamon Press, Oxford, pp. 357–382.

Trager, W., and J. B. Jensen. 1976. Human malaria parasites in continuous culture. *Science* **193**:673–675.

Vavra, J., and A. H. Undeen. 1970. *Nosema algerae* n. sp. (Cnidospora, Microsporida) a pathogen in a laboratory colony of *Anopheles stephensi* Liston (Diptera, Culicidae). *J. Protozool.* **17**:240–249.

# 17

# *Piroplasmea*

Members of the subclass Piroplasmea are principally parasites of vertebrate blood cells and are transmitted by ticks. They are tiny organisms, seldom more than 3 or 4 μm long, and are most often seen as parasites of circulating erythrocytes. Most of the species are found in mammals, but they are known also from reptiles and birds.

The importance of the piroplasms on a worldwide basis is difficult to overstate. Significant losses from piroplasms in cattle, sheep, and goats are continually experienced in nearly all parts of the world. All of the major land masses are enzootic areas for piroplasmoses, and livestock losses continue to be severe despite control programs. Infections with members of the genera *Babesia* and *Theileria* present formidable problems in the tropical and subtropical areas of South America, Africa, Asia, the Middle East, and Europe. It is the dryer areas of the world that can best be used by grazing animals where ticks are abundant and where piroplasmoses are also enzootic.

In this chapter we will look at a few of the piroplasms of herbivores and dogs. However, it should be kept in mind that piroplasms are widely spread among vertebrates and ticks.

Although the piroplasms are parasites of animals other than humans, some human infections are on record. This topic will be discussed later in the chapter.

The subclass Piroplasmea contains organisms that are parasites of various kinds of cells of the blood system and are transmitted by ticks; they are small apicomplexans, but the apical complex is reduced usually to a polar ring and rhoptries (Fig. 17.1); those piro-

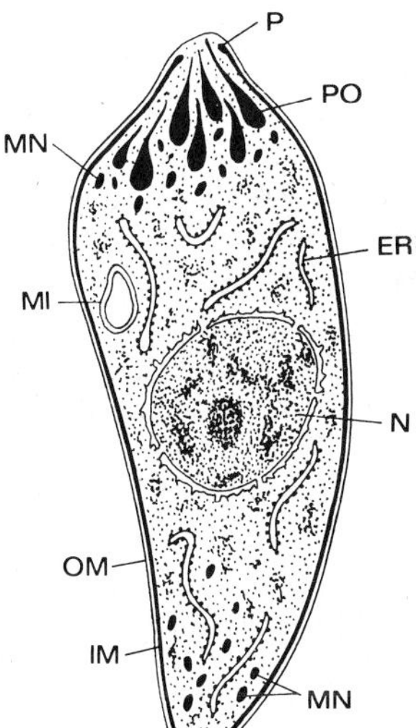


**FIG. 17.1**
Diagram of *Babesia ovis* zoite with reduced apical complex. ER = endoplasmic reticulum; IM = inner membrane; MI = mitochondrion; MN = micronemes; N = nucleus; OM = outer membrane; P = polar ring; PO = rhoptries (paired organelles). [From Friedhoff & Scholtyseck, 1968.]

plasms that parasitize red cells do not form hemozoin, as do most members of the genus *Plasmodium* and other Haemosporina. The classification of the group has been discussed by Levine (1971).

Transmission by ticks has some implications of interest. In all of the cases we know of, ticks are the vectors of piroplasms. In a number of species, such as *Babesia bigemina*, the piroplasm is transmitted from one generation of ticks to the next by infection of the ova (transovarially or TOT) in the female tick. Some piroplasms have been found to be transmitted serially through many generations of ticks without ever passing through a vertebrate host. It has been suggested that the piroplasms are, in fact, parasites of ticks and the vertebrates are merely accidental hosts.

## BABESIOSIS OF DOMESTIC ANIMALS

NAME AND DISEASE CAUSED. *Babesia bigemina* (Smith & Kilbourne, 1893) causes Texas cattle fever, red water fever, splenetic fever, piroplasmosis, and babesiosis.

HOSTS AND LOCATION. The normal vertebrate host of *B. bigemina* is the ox (*Bos taurus*), but some other animals can be infected either transiently or at low levels of parasitemia. In the vertebrate, the piroplasms inhabit erythrocytes only. Ticks of the family Ixodidae (hard ticks) are vectors of *B. bigemina* (Chap. 57). In North America the one-host tick, *Boophilus annulatus*, is the vector, but other ticks such as *B. microplus* are vectors in Australia and other countries. The host range in the mammal is narrow but broad in the vector.

MORPHOLOGY. In the ox, the parasites are seen in red blood cells as piriform or teardrop-shaped organisms about 3 μm long (Fig. 17.2). The nucleus stains red, and the cytoplasm blue with Romanowsky stains. They appear not unlike a ring stage of *Plasmodium*, with the cytoplasm being a wisp of blue surrounding what looks like a vacant area.

In the tick, the parasites vary from forms that are about 1 μm up to the vermicules, which are 11 μm long.

LIFE CYCLE. *Babesia bigemina*, like the other piroplasms, is transmitted from one vertebrate host to the next by means of ticks that serve as biological vectors. In the ox, the parasite develops in erythrocytes and the single trophozoites multiply by binary fission. In a blood film, one most often sees two pear-shaped organisms with their tails near one another (Fig. 17.2). Some time after division, the red cell breaks down and the trophozoites enter uninfected cells.

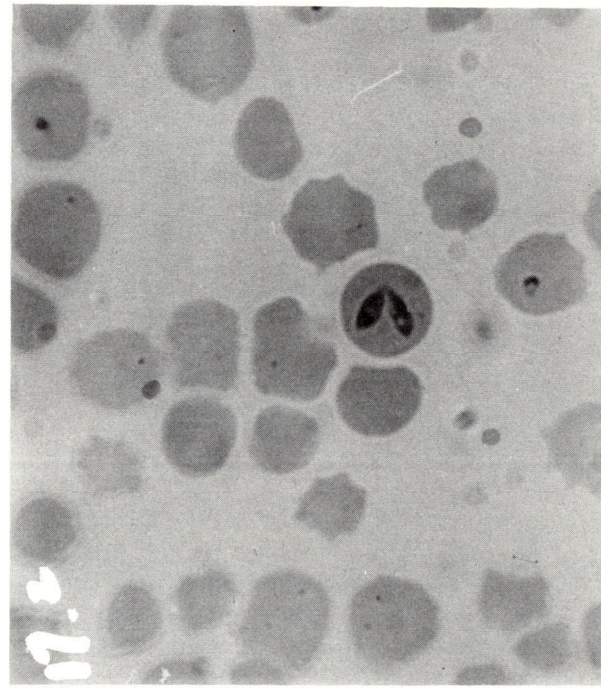

**FIG. 17.2**
Merozoites of *Babesia bigemina* in a red blood cell.

Development in the tick is considerably more complicated than that which takes place in the mammal. Two important points should be recognized about the development of *B. bigemina* in the tick: (1) Transovarial transmission occurs in the tick; that is, a female tick infected as an adult transmits the infection to her offspring and the second generation of ticks then transmits the infection to an uninfected ox. (2) Development in the tick is still not clearly described. Early studies such as that of Dennis (1932) indicated that there was sexual development similar to that of *Plasmodium*, but later investigations consistently reported that development was asexual only. Some recent studies indicate that sexual reproduction occurs in *Theileria* and *Babesia* (Mehlhorn et al., 1975; Schein et al., 1975; Weber & Friedhoff, 1977; Rudzinska et al., 1979).

When a trophozoite is ingested by a tick while taking a blood meal, it escapes from the red cell and enters a cell in the intestine of the tick. It grows into an ameboid form and multiplies by binary fission; the host cell may therefore contain more than 200 daughter cells at 72 hr.

The organisms become elongated, vermiform stages or vermicules about 11 μm long. These vermiform stages enter oocytes, hemolymph cells, and cells of the Malpighian tubules at about 72 hr after infection; as far as transmission is concerned, it is the oocyte invasion that is important for transmission of the parasites.

During embryonation of the egg, further replication of the piroplasm occurs. In this case, multiplication is by schizogony in the intestinal cells of developing and larval ticks.

It is only after the larval tick has fed and molted to the nymphal stage that transmission of the piroplasm is likely to take place. When the nymphal stage has been reached, the parasite then moves to the salivary glands of the host and continues to replicate. Thus, 8 to 10 days after the larva attaches, an ox may be infected, but not before. Having reached an uninfected ox, *B. bigemina* enters erythrocytes and multiplies.

DIAGNOSIS. During the acute phase of the infection, parasites are readily found in a stained film of peripheral blood.

If the animal has gone through the clinical phase and become immune, parasites are seldom seen in blood films. In such a case, there are serological tests such as a complement fixation test, which can be used to determine whether the animal has been infected.

Animals infected with *B. bigemina* become listless, fail to eat properly, and are anemic. If the breakdown of red cells has been massive, hemoglobin passes through the kidneys and the urine is red. Since the signs of infection may be similar to those of a number of other diseases, the search for parasites in the blood cells should be pursued with diligence.

HOST–PARASITE INTERACTIONS. *B. bigemina* lives and reproduces in the red blood cells of the host. As it develops and divides, it consumes some of the content of the erythrocyte and the red cell breaks down, releasing the parasites, which then invade other erythrocytes. Much of the damage to the host revolves around the breakdown of erythrocytes and the consequent anemia; however, effects are seen in other organs as well, such as the spleen, which becomes enlarged, and the liver, which become enlarged and yellowish.

There is considerable difference between reactions of young and mature animals to infection with *B. bigemina*. In most diseases, young animals are more severely affected than older ones, but the reverse is true in babesiosis. Young animals usually show mild clinical signs of infection but recover within a week or so. They are then resistant to further infection and remain free of clinical signs for life. The immunity that develops is not a sterile immunity, however; a low level infection is retained and a state of premunition is achieved. In premunition, the host retains immunity by virtue of residual infection.

Mature animals suffer severe clinical effects when they are infected as adults. Mortality from babesiosis may exceed 75% in some populations.

The only method of preventing infection through immunity is the purposeful infection of young animals with blood transfused from older ones. This method is used especially when animals are to be moved into an endemic area.

TREATMENT. A number of compounds or classes of compounds have been developed for use against babesias. All drugs are administered by injection and have various undesirable side effects. Oddly enough, species larger than 3 μm are susceptible to trypan blue, but those smaller than 3 μm are not.

Trypan blue was first used early in the twentieth century; it must be given IV and stains tissues blue-green. The effect lasts for several months. Acriflavine and acaprin have been in use for more than 40 years. With these drugs parasites are reduced in number but not eliminated. The host, therefore, becomes premunized and resistant to reinfection. Among the most effective drugs are various aromatic diamidines (stilbamidine, pentamidine, berenil), which have a relatively low risk of serious side effects.

Tick control has evolved from the use of dips containing some fairly hazardous materials in the early part of the twentieth century to technologically sophisticated methods of applying systemic acaricides. The livestock dip most used during the Texas cattle fever eradication program was based on arsenic trioxide. During the middle 1930s rotenone was used extensively, but it was not until the 1940s, when the organochlorine insecticides (DDT, lindane) were developed, that highly efficacious and low-risk compounds were available. Instead of completely immersing animals in a dipping vat, it was now possible to treat them by spraying, if care was taken to wet the animals thoroughly. The development of the systemic organophosphate compounds (coumaphos, ruelene, fenthion) allowed wholly new approaches to getting the acaricide to the tick. Animals can be sprayed with an organophosphate compound, but it is not necessary to reach the tick directly because the acaricide is absorbed through the skin and reaches the tick through the bloodstream. A later approach is to apply a concentrated solution or suspension of acaricide to only a limited area of the animal's body. Plastic ear tags impregnated with an organophosphate may also be attached to the ears of cattle to control ticks for as long as 3 months.

EPIDEMIOLOGY AND CONTROL. The control and ultimate eradication of Texas cattle fever from the United States is a saga that began in the nineteenth century when James Law established veterinary medicine as a curriculum at Cornell University. It was here that the early leaders in animal diseases in the United States were trained. Men such as Daniel E. Salmon, John R. Mohler and Cooper Curtice (Fig. 17.3) became the core of the group that was to attack the problem.

Texas cattle fever was a severe problem in the southeastern states and Texas during the middle 1800s. Losses by cattle raisers were considered by some to be intolerable, but there were also losses outside the enzootic area. When cattle were trailed north to markets, losses were observed in animals living along the trails where the disease was not ordinarily seen. The owners of such animals were understandably angry at their animals, dying from a disease contracted from the cattle trailed through the area. Their protests stimulated a study of the cause of the disease.

Some work was undertaken by members of the Bureau of Animal Industry (USDA) and it was Salmon who suggested that Texas cattle fever was an infectious disease transmitted by ticks. The decision was made to

**FIG. 17.3**
Cooper Curtice, USDA veterinarian, examining an animal that died of Texas cattle fever (date about 1892).

attack the problem, and in 1891 Dr. Theobald Smith and Dr. Frederick L. Kilbourne were sent to study the mechanism of transmission and the causative agent.

Smith and Kilbourne not only described the causative agent of splenetic fever but also proved that the tick *Boophilus annulatus* (*Margaropus annulatus*) transmitted the agent.

Further work by Cooper Curtice showed transovarial transmission of *B. bigemina* and set the stage for an avalanche of studies by both USDA investigators and personnel at state experiment stations throughout the Southeast. By the turn of the century, sufficient information was at hand to attack the problem, and in 1903 the Secretary of Agriculture quarantined all cattle south of a line where *B. annulatus* was known to overwinter (Fig. 17.4). Cattle could not be moved north of this area unless they had been treated and found to be free of ticks.

In 1906 a program of eradication of *Boophilus annulatus* from the United States was established. It was concluded that there was sufficient information on the tick to make elimination possible and that the *B. bigemina* infection would thereby die out of its own accord.

The decision was clearly a good one, but not without its difficulties. For example, in 1933 (Fig. 17.4), after nearly 30 years of work, there were still sizable areas of Texas and Florida that still had ticks. Also, not all livestock raisers were in favor of the stringent federal regulations of quarantine, dipping, and inspection. In the papers of Fred C. Bishopp at Colorado State University are documents originating from the Bureau of Animal Industry and from cattlemen's newspapers of the time. In a USDA news release of 1916 the following is stated:

> As a result of ten years' experience in tick eradication, the Federal officials in charge of the campaign are convinced that the hearty cooperation of the people themselves is absolutely essential to success. The Department of

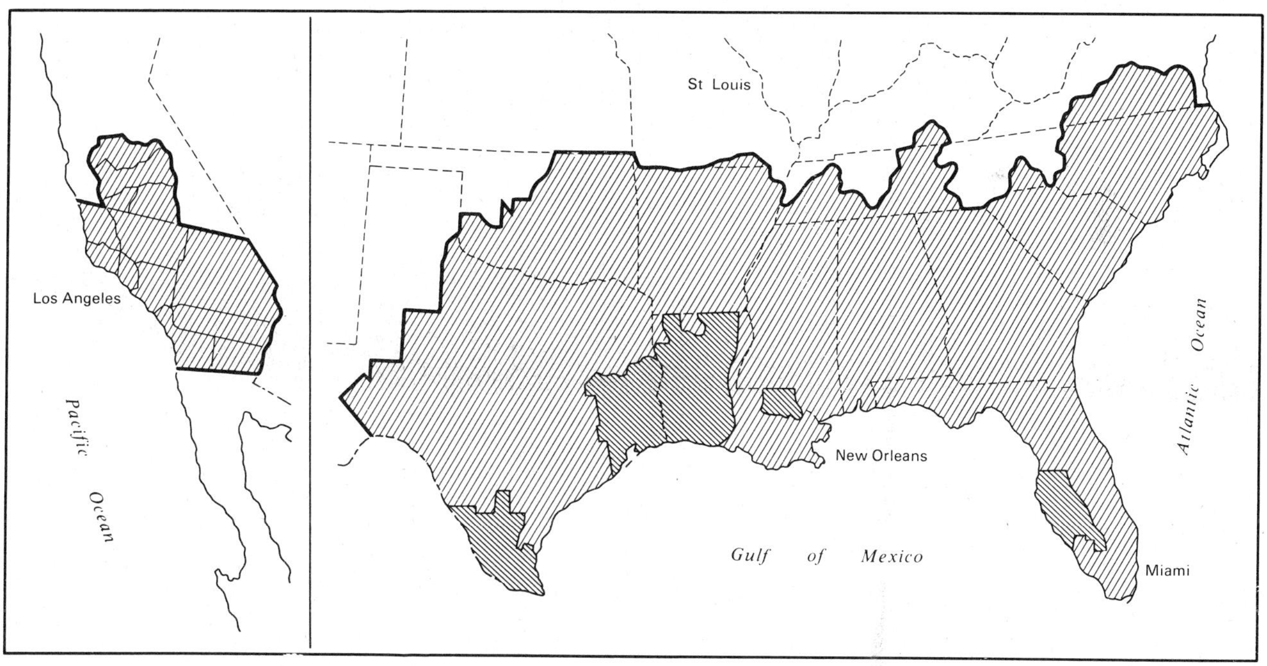


**FIG. 17.4**
Map of the 1933 cattle tick quarantine area in the United States. (Lighter areas represent the limits of the distribution of Texas cattle fever and the darker areas in the south represent the distribution in 1933.)

> Agriculture's representatives in the field, therefore, have been warned of the danger of beginning systematic work in any county before public opinion is ready to lend vigorous support to the enterprise. It frequently happens that a few of the more progressive farmers are earnest advocates of tick eradication, while the rest of the community may be either indifferent or actually hostile. In a number of instances in the past when tick eradication has been begun under these circumstances it has been found necessary, after two or three years, for the Federal government to withdraw its men because the results of their work were, to a great extent, nullified by popular indifference. To make a success of eradication, all cattle in the county must be dipped regularly. In practice, this is only possible when the great majority of the people are so strongly in favor of the work that they will insist upon the enforcement of the necessary regulations.

The term "actually hostile" is used with good reason. In the summer of 1915 three dipping vats in one county of Alabama were dynamited, apparently by a few dissidents who were not in agreement with the vote of 9 to 1 in favor of cooperating in the eradication program. In some other instances, federal inspectors were shot at. However, the program was skillfully managed by the Bureau of Animal Industry by putting the major responsibility for joining the program on the states and counties and then providing the expertise needed to implement eradication.

The program had two crucial parts: quarantine to prevent movement of animals and the lure of increased income if redwater fever were eradicated. There was a continuous stream of hard data from the state experiment station workers and persuasive articles in various farm periodicals advocating eradication. In a bulletin of the Arkansas Agricultural Experiment Station of 1908, W. Lenton discussed tick eradication during the previous year and attempted to persuade livestock raisers as follows: "If every farmer understood this question and was willing to spend a few cents and a few minutes' time each week between now and the first of next November, every tick in Arkansas could be killed by that date." Except during winter, cattle could be sent to market only for slaughter, not for fattening in northern feedlots. Lenton went on to state that "there is practically only one outlet for such cattle: they must go as 'canners.' Other papers comparing untreated and treated cattle said that for an expenditure of about $3, an owner could realize $10.

Eradication of a disease is possible only when certain factors come together. What were they in the case of Texas cattle fever?

1. The distribution of the sole vector in North America was limited. The original habitat of *B. annulatus* was Florida, and the tick had been spread by movement of domesticated animals. Therefore, it was distributed mostly at the margin of its possible range in areas where there was frost in the winter. Some of the methods used in control, such as leaving animals off contaminated pastures for a period of time, as well as the residue of populations in Texas and Florida (Fig. 17.4) in the early 1930s, support this view. In its northern limits, *B. annulatus* died out on pasture during the winter.
2. Relative resistance of young animals. Immunity and premunition could be induced in young animals by transfusing them with blood from older infected animals. Such purposeful infection allowed the introduction of stock into enzootic areas, but presented no risk of transmission to other animals once the tick population had been eradicated.
3. Development of effective technology. Although some of the early topical treatments for ticks such as petroleum were somewhat effective, they had some toxicity. The development of effective dips containing arsenic trioxide was crucial. The other side of the coin was that there was no effective treatment for use against *Babesia.* Thus, there were no alternatives.
4. There is only one host for *B. bigemina*. If an infectious agent has a broad host range, the probability of eradication being successful is nil.

During the eradication phase, all cattle in counties cooperating in the program were required to be dipped in an approved acaricide every 2 weeks during the tick season. They could not be moved unless they had been dipped and inspected. When a county was declared tick-free, cattle could then be moved at will. The motivation for livestock raisers to join the eradication program was that they could then send their cattle to feedlots and obtain more money than if they were sent directly to slaughter. Eradication of Texas cattle fever from cattle in the continental United States was considered to be complete by 1939.

At present, the United States is still free of *B. bigemina* infection in cattle, but Mexico has the disease. Therefore, all cattle entering the United States from Mexico must be dipped in an approved acaricide and inspected before they are allowed to cross the border. Such programs are carried out by personnel of the Animal and Plant Health Inspection Service (APHIS) of the USDA.

NAME AND DISEASE CAUSED. *Babesia caballi* (Nuttall & Strickland, 1910) causes equine piroplasmosis.

HOSTS AND LOCATION. This parasite infects not only horses but other equids as well. The parasite is found in the erythrocytes of the mammalian host.

DISTRIBUTION AND IMPORTANCE. *B. caballi* is distributed in areas with warm climates.

MORPHOLOGY. The erythrocytic forms are 2 to 5 μm long, generally piriform.

LIFE CYCLE. In the horse, the parasites are found in the red blood cells, where they multiply by binary fission. Only one or two trophozoites are found in each red cell. In the tick, the life cycle is similar to that of *B. bigemina*. *B. caballi* can undergo either transtadial (stage-to-stage) or transovarial transmission; on a global basis, many species of ticks serve as vectors of *B. caballi*. In the two- and three-host ticks there is transtadial transmission, and transovarial transmission in the one-host ticks.

DIAGNOSIS. The only certain means of diagnosis is by finding the parasites in the blood of an infected animal. A number of serological tests have been developed that allow an investigator to determine whether an animal has been infected and with what species; however, the best evidence is finding the parasite.

HOST–PARASITE INTERACTIONS. Unlike *B. bigemina* infection, with which mortality exceeds 75%, mortality due to infection with *B. caballi* is less then 50% overall. Also, different isolates have different levels of pathogenesis.

EPIDEMIOLOGY AND CONTROL. On a global basis, there are about ten species of ticks that serve as vectors of *B. caballi*. In the United States, Central America, islands of the Caribbean, and South America, *Dermacentor nitens* is the vector.

*B. caballi* was introduced into the United States around 1960, most likely from Cuba, and was found in 1961. It spread quickly through southern Florida, with up to 200 cases found annually in the United States in the 1960s and early 1970s. Infections have also been found in other states, but these are considered to have either been imported or become infected in Florida and been shipped to other areas.

The establishment of *B. caballi* in the United States illustrates the principle of ready transmission of an agent when imported into an area where a good vector already exists.

When *B. caballi* was found in a racehorse, a program of quarantine, testing, and tick control was instituted by APHIS. The outbreak was contained, but not without considerable cost. It is unlikely that *B. caballi* would establish in parts of the United States other than southern Florida and Texas. The question might legitimately be asked whether the effort at control is economically justifiable. The answer is affirmative because of the continual movement of horses, especially racehorses and show horses. Even though *D. nitens* would not establish permanently in cooler areas of the country, it could spread the infection during the summer months. The economic losses might then be far greater than the costs of control.

NAME OF ORGANISM AND DISEASE CAUSED. *Babesia canis* (Piana & Galli-Valerio, 1895), causes canine babesiosis or piroplasmosis and malignant jaundice.

HOSTS AND LOCATION. Several species of canids, both domestic and wild, are natural hosts. In the mammal the parasites are found in the erythrocytes.

Ixodid ticks are the vectors, and seven species have been implicated. In North America and many other areas of the world *Rhipicephalus sanguineus*, the brown dog tick, is the principal vector (Chap. 57).

DISTRIBUTION AND IMPORTANCE. *B. canis* is a cosmopolitan parasite, principally in warm climates. It has been reported from the United States mostly in southern states but, less frequently, as far north as New Jersey. In many tropical and subtropical areas of the world the prevalence in dogs is high.

MORPHOLOGY. Among piroplasms, *B. canis* is quite

large; when piriform, they are as large as 5 μm and the round ones are about 3 μm in diameter. The shape is variable.

LIFE CYCLE. The development in the red cells of the vertebrate host is by binary fission. In the tick, development is similar to that of *B. bigemina*, with transmission in *R. sanguineus* being both transtadial and transovarial (Fig. 17.5).

DIAGNOSIS. As with other babesioses, finding the parasite in the erythrocytes is the definitive evidence. Clinical signs of infection with *B. canis* are quite variable. Red cell destruction gives rise to anemia and sometimes hemoglobin in the urine; animals are weak and lethargic; fever is usually present. Beyond these rather obvious signs, there may be gastrointestinal ulceration, central nervous system involvement, and bleeding from the ears or muzzle. Both young and mature animals may be infected, but younger animals show less severe clinical effects than older ones on initial infection.

HOST–PARASITE INTERACTIONS. The damage to the dog varies with the strain of parasite and the age of the dog. The breakdown (hemolysis) of the red cells is the most obvious sign, and it has widespread effects. Anemia, splenic enlargement, kidney damage, and impaired circulation in some organs all result from the hemolysis.

If an animal survives the infection, it will become immune and remain refractory to reinfection through premunition.

TREATMENT. A number of drugs have been used against *B. canis*, but acaprin and phenamidine appear to be the most effective.

EPIDEMIOLOGY AND CONTROL. *Rhipicephalus sanguineus*, a three-host tick, is the most important

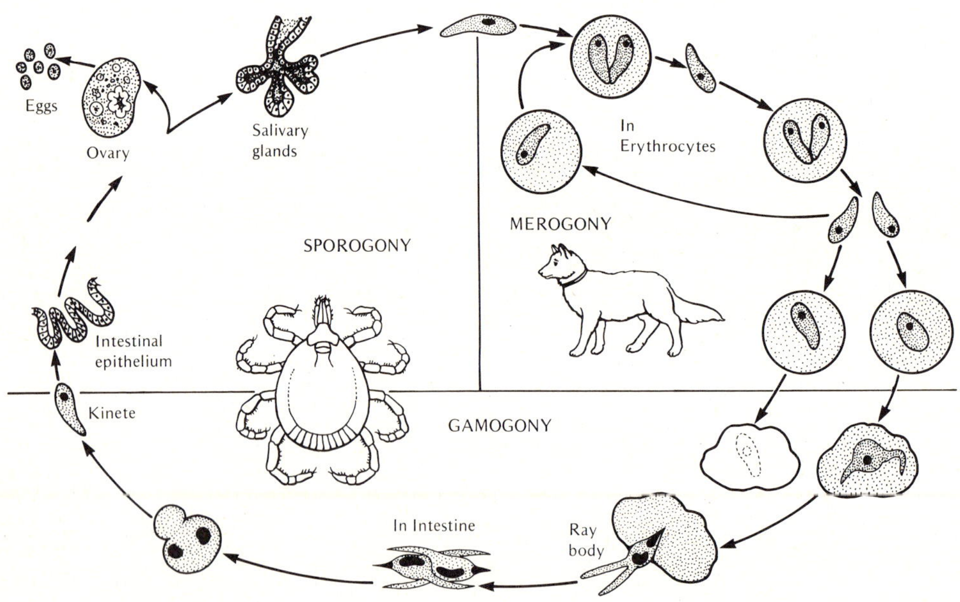


**FIG. 17.5**
Diagram of the life cycle of *Babesia canis*.

vector of *B. canis*. It is found in most parts of the world, although it is probably native to Africa. Although it is a warm-climate tick, it is extremely hardy. The ticks will not establish in colder climates, but they may colonize houses. Thus, in the Rocky Mountain area of the United States, *R. sanguineus* is sometimes found in houses, where it completes its life cycle quite well.

The host range of *R. sanguineus* also aids in its broad distribution. Although it prefers dogs, it has been found on 11 species of large herbivores, three carnivores other than the dog, rodents, lagomorphs, bats, 12 kinds of birds, reptiles, and 2 primates including humans. It is, however, an uncommon parasite of humans.

Throughout much of its range, there appears to be no concerted effort to control *B. canis* infections in dogs. Individual animals may be treated, but the obligation for control ultimately falls on the individual owner. If animals are confined to a fairly small area such as a city house lot, the best approach to preventing infection with *B. canis* is to control ticks by doing the following:

1. Treating animals with an acaricide
2. Using a flea and tick collar on the animal
3. Applying an acaricide to the premises

If an animal is taken into an enzootic area, it should be watched carefully for signs of infection and then treated if found to be infected. The dog should then develop immunity and be resistant to reinfection.

## *BABESIA* INFECTIONS IN HUMANS

Four species of *Babesia* have been found as parasites of humans:

| ORGANISM | NORMAL HOST |
|---|---|
| *B. bovis* | Ox |
| *B. divergens* | Ox |
| *B. equi* | Horse |
| *B. microti* | *Microtus* (rodent) |

Cases of infection in humans have been reported from Yugoslavia, Scotland, Ireland, France, and the United States.

The first three instances of infection with *Babesia* in humans were all in persons who had had their spleens removed. The spleen is a major line of defense against piroplasmosis in domestic animals and also appears to serve the same function in humans. These infections were therefore looked on as oddities in persons who were not able to respond immunologically to the infection. Then, in 1969 a woman on Nantucket Island, Massachusetts, was found to be infected with what ultimately was determined to be *B. microti*; she was clinically ill, but had not had a splenectomy.

An investigation of how the woman became infected brought about a significant change in our viewpoint of babesiosis in humans. Nantucket is a small island a few miles off of Cape Cod. It has a year-round population of a few hundred persons and a large tourist influx during the summer. As of 1976 seven persons had been found on Nantucket to be infected with *B. microti*. Not all had clinical symptoms, but those who did were typical: chills, fever, headache, lethargy, muscle pain.

Of six species of rodents and lagomorphs examined on Nantucket for the presence of piroplasms, only two, the vole (*Microtus pennsylvanicus*) and the white-footed mouse (*Peromyscus leucopus*), were found to be infected with *B. microti*. The prevalence in the rodents was high—5 of 5 voles and 31 of 39 mice. *Ixodes dammini* Spielman et al., 1979, is the vector on Nantucket, where about 5% of ticks have been found to have piroplasms in their salivary glands.

The importance of piroplasmosis in humans is still undetermined. The first individuals found to be infected with *Babesia* died because they could not respond immunologically in a normal manner. None of the persons on Nantucket died as a result of the infection, but some had felt ill for months and were incapacitated by the disease. Also, there was no association with livestock. The association rather was with rodents, but it was peripheral. The known distribution of *B. microti* infections in humans is being extended; in addition to Nantucket Island, infections have been found on Cape Cod, and on Long Island, New York (Teutsch et al., 1980).

Piroplasms are rather common in various animal groups. A study done in Maryland, for example, showed that 29 of 30 raccoons were infected with *B. procyoni* and all of 13 skunks were infected with an organism morphologically similar to *B. caballi* (Frerichs & Holbrook, 1970). Other studies on rodents have confirmed that *Babesia* is common.

DIAGNOSIS. Diagnosis of *Babesia* infection in humans presents some problems. The organisms are small and sometimes present in only a small percentage of red cells. They may be mistaken for ring stages of *Plasmodium*, particularly *P. falciparum*. Except in individuals who are immunologically incompetent, such as those who have been splenectomized or may be taking corticosteroids, the parasitemia may be at a low level. Thus, if a physician requests a differential white blood cell count for a person showing rather vague symptoms, the laboratory technician may well miss the infection. Alternatively, in an area where malaria is endemic, a misdiagnosis of malaria may be made. It is possible to differentiate infections serologically, but there are few laboratories in the world with the antigens to carry out the tests; the Center for Disease Control in Atlanta, Georgia, is the principal one in North America.

Treatment for babesiosis in humans is not satisfactory. Chloroquine has been used with some apparent success clinically, but is not considered to have much effect on the parasites. Pentamidine has been used with some success. Diminiazene aceturate, used in livestock, was administered to a man suffering from babesiosis; he was cured, but suffered neurologic sequelae that may have been caused by the drug (Teutsch et al., 1980).

## EAST COAST FEVER

NAME OF ORGANISM AND DISEASE CAUSED. *Theileria parva* (Theiler, 1904) Bettencourt, Franca, & Borges, 1907, causes East Coast fever, corridor disease, and bovine theileriosis.

HOSTS AND LOCATION. Domestic cattle and both domestic and wild buffalo are hosts. In the mammalian host the organisms replicate in lymphocytes and also parasitize red cells, but do not multiply there.

DISTRIBUTION AND IMPORTANCE. The original distribution of East Coast fever was from South Africa northward to Kenya in the area of Africa whose drainage is toward the Indian Ocean. Control programs have limited the range, particularly in the southern portion of the enzootic area.

*T. parva* has a reservoir in the African buffalo (*Syncerus caffer*), and these animals are a constant source of infection for domestic cattle, both the ox and the zebu.

East Coast fever has been a serious threat to livestock in East Africa for many years; mortality in domestic cattle may exceed 90%.

MORPHOLOGY. The forms that occur in the lymphocytes multiply, and are referred to as *Koch's bodies* or *Koch's blue bodies*. They average 8 μm, are roundish, and are multinucleate.

The forms found in the red cells are nonmultiplying and probably are gamonts. They are tiny organisms, no more than 1 by 2 μm, and with shapes varying from rods to broad ovals. The nucleus is a red dot and the cytoplasm a little wisp of blue in Romanowsky stains; multiple infections of red cells are common.

LIFE CYCLE. *T. parva* and other theilerias are so small that studies of development have been difficult to do; therefore, there are aspects of development that are still in doubt (Fig. 17.6). The following is a brief discussion of the major aspects of development in the ox and the tick.

Upon entry into the mammal, when an infected tick takes a blood meal, the sporozoites are filtered out in lymph nodes, usually in the neck region (e.g., prescapular nodes). The organisms undergo schizogony in the lymphocytes and seemingly cause nuclear and cytoplasmic division of the lymphocytes (DeMartini & Moulton, 1973a and b).

Some of the merozoites enter red cells, but replication does not occur here. However, it is the erythrocytic parasites that can be transferred successfully back to a tick.

When a clean tick feeds on an infected animal, the parasites in the red cells escape and undergo transformation. It was the consensus that *Theileria* does not have sexual development, but recent studies at the electron microscope level have required a reassessment of this view (Mehlhorn et al., 1975; Schein et al., 1975). A gamont and zygote (?) have been found (Fig. 17.7).

Ultimately, the parasites enter the cells of the salivary glands of the tick and are ready to be injected into a clean host.

In *Theileria* spp. there is only stage-to-stage transmission in the tick. Usually the tick is infected as a larva.

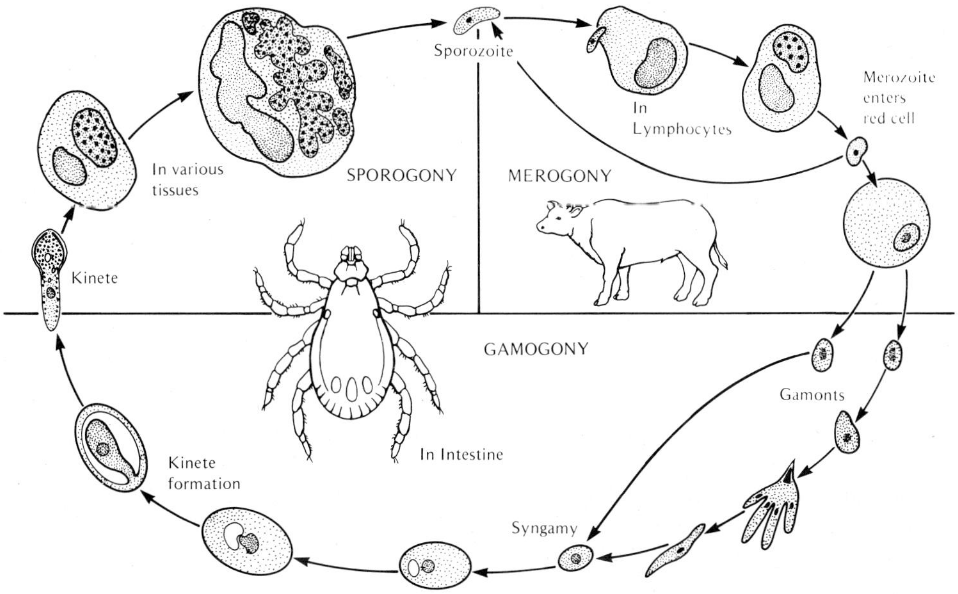


**FIG. 17.6**
Diagram of the life cycle of *Theileria parva*.

DIAGNOSIS. Determining whether an animal is infected is based on the following:

1. Clinical signs
2. Finding the organism in a peripheral blood film
3. Finding the organism in a smear made from a lymph node puncture
4. Serological tests

In the face of suspicious signs of infection such as fever, swollen lymph nodes, and emaciation, it is best to make both a blood film and a lymph node puncture from the animal. A Romanowsky stain will reveal the typical form in the erythrocytes. The forms in the lymphocytes are typically multinucleated schizonts.

The serological tests indicate whether the animal is infected but cannot differentiate among species of piroplasms.

HOST–PARASITE INTERACTIONS. Much of the damage to the mammalian host in *T. parva* infections involves the lymph system: lymph nodes and spleen. However, other organ systems such as the kidney, lungs, and liver are affected also.

A wide range of virulence occurs in different isolates of *T. parva* and there are some differences in the ways in which the host is affected. As with *Babesia* infections, older animals are more severely affected than younger ones.

Immunity develops in animals that survive the infection, and it appears to be a sterile immunity. In some of the other species of *Theileria*, there is premunition, that is, some organisms are retained. Immunization with a mild strain has been attempted, but it does not usually provide adequate protection. Unlike in most *Babesia* infections, the tick vector usually remains infected for a relatively short period of time.

TREATMENT. There is no satisfactory treatment for *Theileria parva* infections. A variety of drugs have been tried for all the species of bovine *Theileria*, and though some, such as tetracyclines, seem to prevent rapid multiplication of the parasites, there is no good treatment for clinically affected animals.

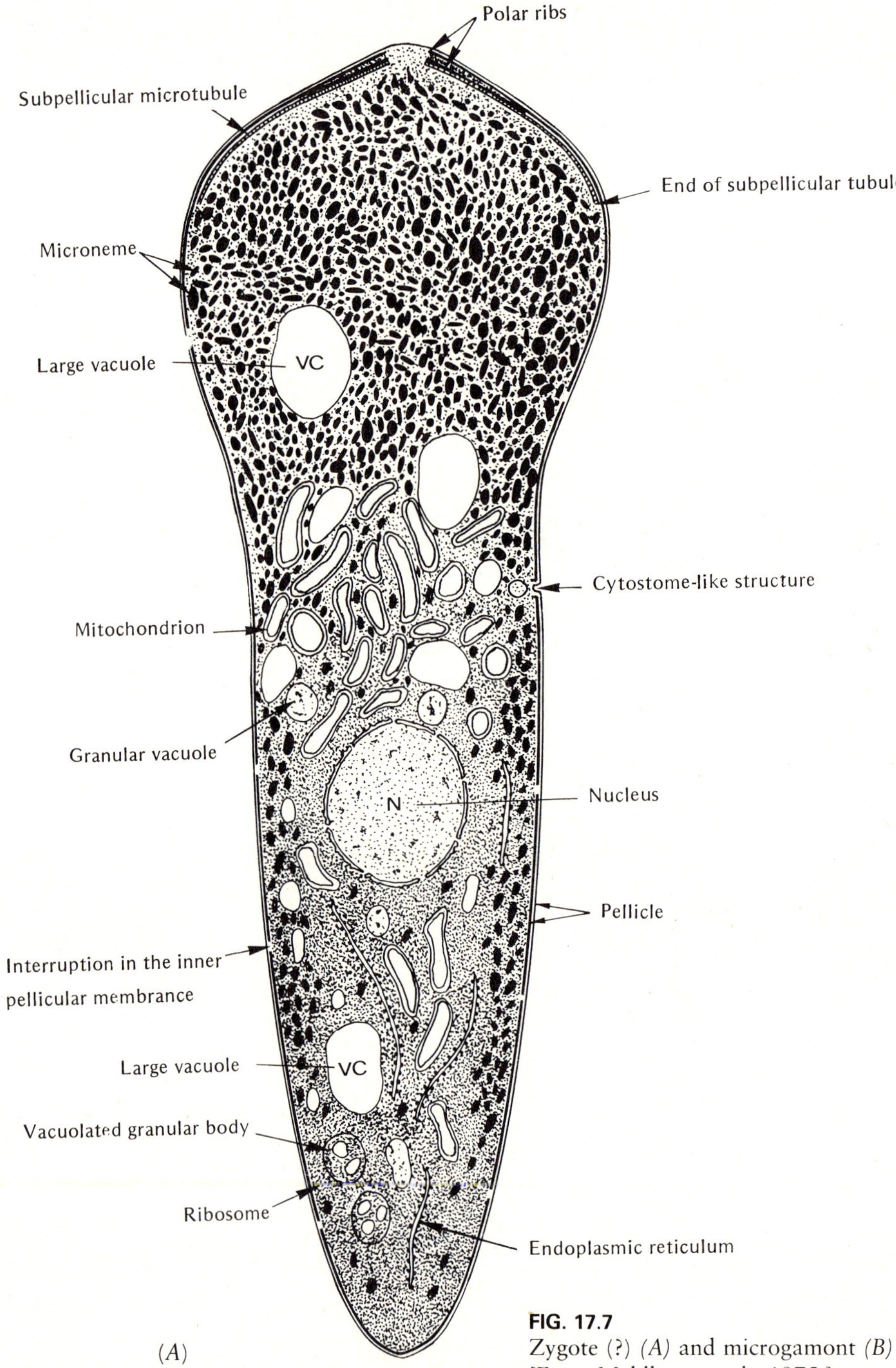


**FIG. 17.7**
Zygote (?) *(A)* and microgamont *(B)* of *Theileria parva*. [From Mehlhorn et al., 1975.]

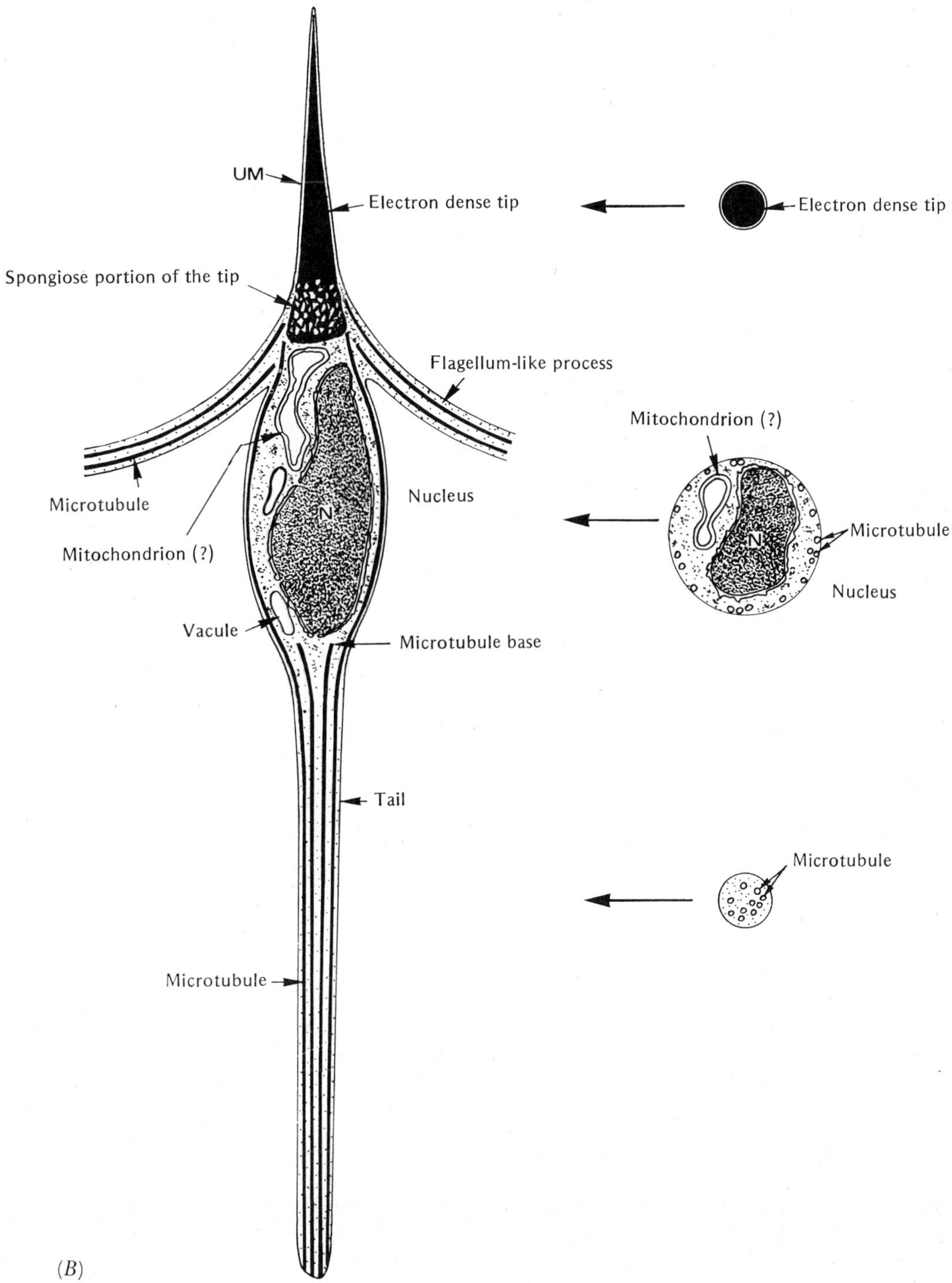
UM
Electron dense tip
Electron dense tip
Spongiose portion of the tip
Flagellum-like process
Mitochondrion (?)
Microtubule
Nucleus
N
N
Microtubule
Mitochondrion (?)
Nucleus
Vacule
Microtubule base
Tail
Microtubule
Microtubule

(B)

EPIDEMIOLOGY AND CONTROL. There are three aspects to treatment that need to be considered:

1. Treating the ox to eliminate the infection with the protozoa
2. Treating the ox to eliminate infestation with ticks
3. Treating the area to reduce populations of ticks while off of the host

The control programs for East Coast fever began in the 1920s and 1930s were based on tick control, quarantine, and slaughter. The principal animals of concern are domestic cattle, but the reservoir in buffalo makes the problem of control exceedingly difficult. The approach to control was similar to that applied in Texas cattle fever: to control ticks and not move any animals that harbor ticks. As the control began to have an effect, and areas were found to be free of infected cattle, a test-and-slaughter program was instituted. It is a drastic step to take and one that requires administrators with the courage of their convictions. Usually the cattle owners are reimbursed for any animals that are killed, and adequate indemnity is essential for the success of a program. It should be emphasized also that even now there is neither satisfactory artificial immunization against nor treatment for *T. parva* infections. The armamentarium of control measures is therefore limited.

The second-generation insecticides (DDT and lindane, for example) allowed tick control to be done much more easily and effectively.

## REFERENCES AND FURTHER READINGS


Arthur, D. R. 1966. The ecology of ticks with reference to the transmission of protozoa. In E.J.L. Soulsby (ed.): *Biology of Parasites*. Academic Press, New York.

Chapman, W. E., and P. A. Ward. 1977. *Babesia rodhaini*: Requirement of complement for penetration of human erythrocytes. *Science* **196**:67–70.

Commonwealth Bureau of Animal Health. 1973. *Bovine Theileriosis. An Annotated Bibliography*. Commonwealth Agricultural Bureau. Farnham Royal, England.

DeMartini, J. C., and J. E. Moulton. 1973a. Responses of the bovine lymphatic system to infection by *Theileria parva*. I. Histology and ultrastructure of lymph nodes in experimentally infected calves. *J. Comp. Path.* **83**:281–298.

———. 1973b. Responses of the bovine lymphatic system to infection by *Theileria parva*. II. Changes in the central lymph in experimentally infected calves. *J. Comp. Path.* **83**:299–306.

Dennis, E. W. 1932. The life cycle of *Babesia bigemina* (Smith & Kilbourne) of Texas cattle-fever in the tick *Margaropus annulatus (Say)* with notes on the embryology of *Margaropus*. *Univ. Calif. Publ. Zool* **36**:363–398.

Frerichs, W. M., and A. A. Holbrook. 1970. *Babesia* spp. and *Haemobartonella* sp. in wild mammals trapped at the Agricultural Research Center, Beltsville, Maryland. *J. Parasitol.* **56**:130.

Friedhoff, K., and E. Scholtyseck. 1968. Feinstrukturen von *Babesia ovis* (Piroplasmidea) in *Rhipicephalus bursa* (Ixodoidea): Transformation spharoider formen zu vermicularformen. *Z. Parasitenk.* **30**:347–359.

Friedhoff, K., E. Scholtyseck, and G. Weber. 1972. Die Feinstructur de differenzierten Merozoiten von *Babesia ovis* in den Speicheldruzen weiblicher Zecken. *Z. Parasitenk.* **38**:132–140.

Friedhoff, K., and G. Weber. 1973. Identification and characterization of certain developmental stages of some *Babesia* species. *Isotopes and Radiation in Parasitology III*. pp. 165–168. Intern. Atomic Energy Agency, Vienna.

Healy, G. R., and N. Gleason. 1976. Human babesiosis: Reservoir of infection on Nantucket Island. *Science* **192**:479–480.

Levine, N. D. 1971. Taxonomy of the piroplasms. *Trans. Am. Micr. Soc.* **90**:2–33.

Mehlhorn, H. 1982. The life cycles of piroplasms and related group. In D. F. Mettrick and S.S. Desser (eds.), *Parasites—Their World and Ours*. Elsevier Biomedical Press, New York.

Mehlhorn, H., and E. Schein. 1977. Electron microscopic studies of the development of kinetes in *Theilera annulata* D. & L., 1904. (Sporozoa, Piroplasmea). *J. Protozool.* **24**:249–257.

Mehlhorn, H., E. Schein, and M. Warnecke. 1979. Electron-microscopic studies on *Theileria ovis* Rodhain, 1916: Development of kinetes in the gut of the vector tick, *Rhipicephalus evertsi* Neumann, 1897, and their transformation within cells of the salivary glands. *J. Protozool.* **26**:377–385.

Mehlhorn, H., G. Weber, E. Schein, and G. Buscher. 1975. Elektronomikorskopische Untersuchung an Entwicklungsstadien von *Theilera annulata* (Dschunkowsky & Luhs, 1904) im Darm und in der Hämolymphe von

*Hyalomma anatolicum excavatum* (Koch, 1844). *Z. Parasitenk.* **48**:137–150.

Piesman, J., and A. Spielman. 1980. Human babesiosis on Nantucket Island: Prevalence of *Babesia microti* in ticks. *Am. J. Trop. Med. Hyg.* **29**:742–746.

Purnell, R. E. 1977. East Coast fever: Some recent research in Africa. *Adv. Parasitol.* **15**:83–132.

Riek, R. F. 1964. The life cycle of *Babesia bigemina* (Smith and Kilbourne, 1893) in the tick vector *Boophilus microplus* (Canestrini). *Austral. J. Agric. Sci.* **15**:802–821.

———. 1966. The development of *Babesia* spp. and *Theileria* spp. in ticks with special reference to those occurring in cattle. In E. J. L. Soulsby (ed.), *Biology of Parasites.* Academic Press, New York, pp. 15– 32.

Ristic, J., J. D. Conroy, S. Siwe, G. R. Healy, A. R. Smith, and D. L. Huxsoll. 1971. *Babesia* species isolated from a woman with clinical babesiosis. *Am. J. Trop. Med. Hyg.* **20**:14–22.

Rudzinska, M. A., A. Spielman, R. F. Riek, S. J. Lewengrub, and J. Piesman. 1979. Intraerythrocytic "gametocytes" of *Babesia microti* and their maturation in ticks. *Can. J. Zool.* **57**:424–434.

Schein, E., G. Buscher, and K. T. Friedhoff. 1975. Lichtmikroskopische Untersuchungen über die Entwicklung von *Theileria annulata* (Dshunkowsky und Luhs, 1904) in *Hyalomma anatolicum excavatum* (Koch, 1844). *Z. Parasitenk.* **48**:123–136.

Shortt, H. E. 1973. *Babesia canis*: The life cycle and laboratory maintenance in its arthropod and mammalian hosts. *Intern. J. Parasitol.* **54**:1095–1098.

Spielman, A., C. M. Clifford, J. Piesman, and M. D. Corwin. 1979. Human babesiosis on Nantucket Island, USA: Description of the vector, *Ixodes dammini*, n. sp. *J. Med. Entomol.* **15**:218–234.

Teutsch, S. M., P. Etkind, E. L. Burwell, K. Sato, M. M. Dana, P. R. Fleishman, and D. D. Juranek. 1980. Babesiosis in post-splenectomy hosts. *Am. J. Trop. Med. Hyg.* **29**:738–741.

Weber, G., and K. T. Friedhoff. 1977. Preliminary observations on the ultrastructure of supposed sexual stages of *Babesia bigemina* (Piroplasmea). *Z. Parasitenk* **53**:83–92.

# 18

# *Myxozoa*

All of the organisms in this phylum of protozoa are parasitic. Their hosts are nearly always fish, though a few species of amphibia and reptiles and a scattering of invertebrates are also infected. In addition to being parasitic, the Myxozoa are characterized by a resistant spore with one to six coiled polar filaments; two to six plates or valves form the outer covering of the spore (see Fig. 19.1, B). Development within the host is asexual; large schizonts or trophozoites reproduce by schizogony or by budding. Spores are multicellular in origin and arise in the plasmodia. Myxozoa are both coelozoic and cytozoic, but in some instances it is not possible to tell whether they are within or between cells.

The Myxozoa and the Microspora were long classified as members of the Cnidosporidia or Cnidospora (see Kudo, 1966), because both have polar capsules containing coiled filaments. In both, the spore is the stage that is transmitted from one host to the next. As knowledge developed, differences in spore structure and in the function of the polar filament, as well as in the manner of spores formation, became evident. In addition, the development of the organisms in the tissues of the host was found to be different. These differences between the two groups are summarized in Table 18.1. Although still included among the protozoa, the Myxozoa have metazoan affinities (Grassé, 1960), and it has been suggested that they are actually highly modified cnidarians (jellyfish, sea anemones) (Vavra, 1971).

Myxozoa are of economic importance because they infect fish caught for food. There are production losses and deaths, and some fish must be discarded because they are unsightly and not considered to be fit for human consumption. No Myxozoa species is known to infect humans or to be of any hazard to human health.

About 300 species of Myxozoa have been described. Most of them are in teleost (bony) fish and their importance is not known. In this chapter, we will look at *Myxosoma cerebralis* as an example of a myxosporan of economic importance and then discuss the group generally.

Identification of Myxozoa is based principally on the structure of the spore (Fig. 18.1). Kudo's text (1966) is a good introduction to the Myxozoa and provides some good illustrations of various genera and species; the same author's monumental monograph (Kudo, 1920) is still a good reference for most species. Some useful descriptions and distributional information are contained in Cone and Anderson (1977); Kudo (1926); Meglitsch (1960); and Sanders et al. (1970)

NAME OF ORGANISM AND DISEASE CAUSED. *Myxosoma cerebralis* (Hofer, 1903) causes whirling disease and twist disease.

HOSTS AND LOCATION. Salmonid fish of many species are the hosts, and the organism resides in the cartilage.

DISTRIBUTION AND IMPORTANCE. Distribution is nearly worldwide. The original location was western

**TABLE 18.1**
Comparison of the Phyla Microspora and Myxozoa

| | MICROSPORA | MYXOZOA |
|---|---|---|
| TROPHIC STAGE | | |
| 1. Level of organization before sporogenesis | A simple plasmodium; all nuclei are equipotential and all become nuclei of sporoplasms | A complex plasmodium: part of the nuclei differentiate with surrounding cytoplasm into generative cells, which later undergo sporogenesis; the rest of the nuclei and plasma provide "food and shelter" until sporogensis is completed; then they degenerate |
| 2. Presence of cytoplasmic organelles | Only ergastoplasmic lamellae and smooth-membraned vesicles; no pinocytosis observed | Mitochondria, typical Golgi, ergastoplasm, fibrils, microtubules, large variety of vesicles and granules; no centriole; pinocytosis observed |
| SPORE | | |
| 1. Degree of complexity | Sporoplasm that has produced protective shell and organelles important for extrusion | Sporoplasm protected by differentiated sister cells, these producing shell and polar capsules |
| 2. Extrusible apparatus | Tubule + polaroplast + polar cap + posterior vacuole; 1 set to each spore | Filament within a polar capsule produced by a capsulogenic cell; 1 to 6 to each spore |
| 3. Function of filament | Injection of sporoplasm into the host cell | Anchorage of spore to the gut epithelium |
| 4. Mechanism of extrusion | Osmotic event; no universal agent for experimental triggering | Prebuilt pressure; triggered easily by urea |
| 5. Composition of shell | Chitinous | Proteinaceous |

[Modified from Lom & Corliss, 1967]

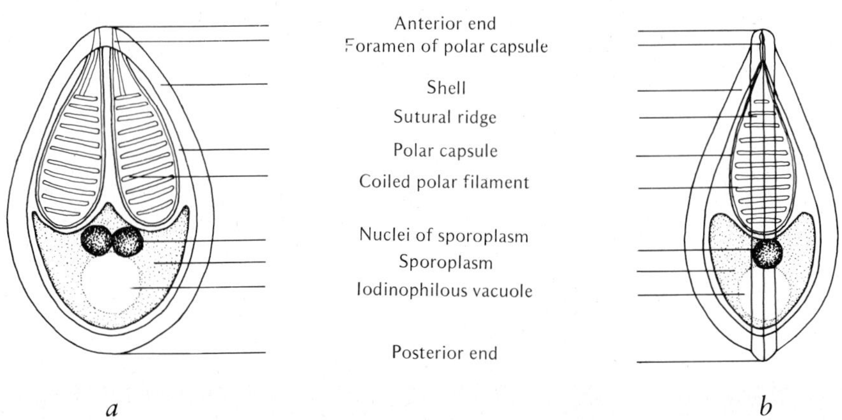


**FIG. 18.1**
Diagram of a myxozoan spore viewed from the front (*a*) and side (*b*). [Redrawn from Kudo, 1920.]

Europe, but the infection has spread to North America and South Africa. The infection is not generally pathogenic to fish in its original western European hosts, but infections in North American trout are pathogenic and have caused extensive losses in hatcheries (Bogdnovoa, 1967; Hoffman et al., 1962).

MORPHOLOGY. Spores are 6 to 10 μm in diameter, oval, and have two valves and two polar capsules. The spores have a mucous envelope; in scanning electron microscopy (SEM) the opening for the extrusion of the polar filament can be seen (Fig. 18.2) (Lom, 1969; Lom & Hoffman, 1971; Lom & de Puytorac, 1965).

LIFE CYCLE. The spores are ingested, and presumably factors in the gut cause extrusion of the polar filaments; the filaments slow down the passage of the spore through the intestine. The sporoplasm escapes from the spore and moves by ameboid movement through the gut wall to the final site of development. *M. cerebralis* develops in cartilage, where large plasmodia (or trophozoites) form. By 4 months after infection, the plasmodia may reach 1 mm in diameter, and it is at this time that the spores begin to form.

Spores are formed within the plasmodium and are multinuclear in origin. They presumably reach the outside when the fish dies. Spores must undergo an aging process before they become infective; epidemiologic evidence has shown that spores are eaten by tubificid worms (Annelida: Oligochaeta) in which they somehow mature and become infective for fish (Markiw and Wolf, 1983).

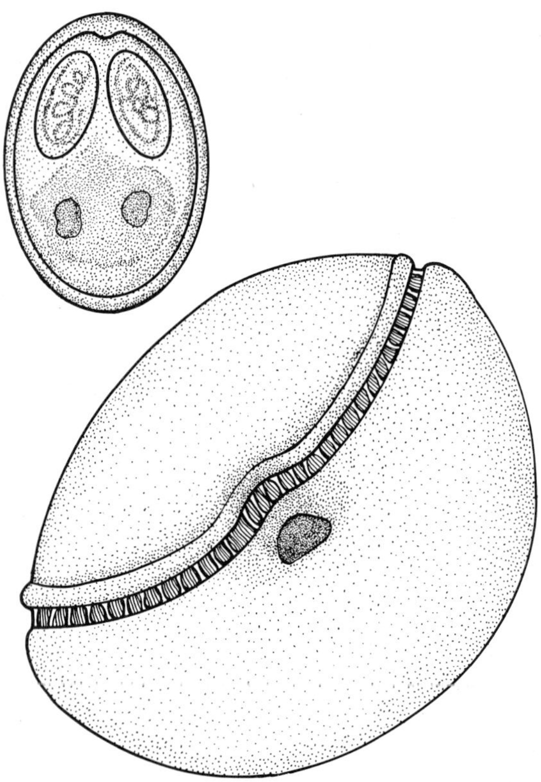

**FIG. 18.2**
Spore of *Myxosoma cerebralis* with an open canal for filament discharge shown on the lower drawing. [Redrawn from Lom & Hoffman, 1971.]

DIAGNOSIS. Finding the typical spores in cysts in cartilage is the surest method of determining whether *M. cerebralis* is the cause of a problem. Skeletal distortions and the typical swimming in circles are cause for suspicion that whirling disease is present.

HOST–PARASITE INTERACTIONS. In the normal host, probably *Salmo trutta*, a central European trout, the effect is minor; however in the rainbow trout, *Salmo gairdneri*, pathogenesis is severe. The parasite develops in cartilage, and the location of the infection determines the effect on the host. The lesions result principally from the breakdown of cartilage and the consequent pressure on nervous tissue or sense organs. When the cartilage surrounding the inner ear is weakened, for example, the balance of the fish is disturbed and typical whirling is observed. If the caudal portion of the vertebral column is affected, there may be pressure on the nerves and the posterior portion of the fish turns dark because control of the nerves controlling the chromatophores is lost. Fish become uncoordinated, various parts of the skeleton become distorted, and they sink to the bottom, exhausted. Predation or death soon occurs.

EPIDEMIOLOGY AND CONTROL. Infection of rainbow trout with *M. cerebralis* is a good example of what may happen when animals or their products are moved around the world. Infection of salmonid fish has been known since the early twentieth century. It was not known in North America until about 1952, when fish in the eastern United States were found to be infected. Whirling disease was a problem all across the United States by the middle 1960s, especially in hatchery fish. It is likely that the organism was transferred to North America with *Salmo trutta* imported from Europe. In addition, rainbow trout are good disseminators of the organism because they become so heavily infected; following World War II, a market developed in Europe for rainbow trout and they were shipped widely for table use (Hoffman, 1970).

The hardiness of the spores is also important in spreading infection. Spores are known to survive freezing, so even frozen fish may be a source of infection for fish in a new locality (Hoffman & Putz, 1969). Apparently frozen fish were the vehicle for spreading the disease in Europe.

Since there is no treatment for whirling disease, and since the spores are so hardy, the best approach to control is prevention. Accordingly, countries now free of infection such as Great Britain, Australia, New Zealand, Japan, and Canada do not allow the importation of salmonid fish or their products.

## OTHER MYXOZOA

As indicated earlier, of more than 300 species of Myxozoa that are known, the pathogenesis of only a few is known. Among fresh water fish, especially those in hatcheries or relatively small lakes, the effect on populations has been documented in a few instances; however, such information on marine fish is sketchy (Lom, 1970; Noble, 1966; Sindermann, 1970).

Some of the same manifestations are seen in myxozoan and microsporan infections (Chap. 19). The bodies are distorted, there may be tumorlike growths, or the muscle may become unsightly. The names of some of the infections indicate the signs of infection: "wormy" halibut, which occurs along the Pacific coast of North America, is caused by *Unicapsula muscularis*; "boil disease" is caused by *Myxobolus pfeifferi* in barbel; "tapioca disease" of salmon is caused by *Kudoa thyristes*.

A good deal is still unknown about the Myxozoa, and the difficulty in obtaining certain types of information has hampered overall knowledge. The first Myxozoa were described about 100 years ago, and new species are continually being described; however, the gaps in our knowledge represent a considerable barrier to manipulation or control of infections (Noble, 1944, 1950; Sprague, 1969).

Upon entry into a host, the polar filament(s) is extruded, but the stimuli for extrusion are poorly understood. In *Myxosoma cartilaginous* of bluegill (Centrarchidae), the filament can be extruded by 1% or stronger KOH, by 0.26% or stronger NaOCl (bleach) or by the quarternary ammonium disinfectant Roccal. Such stimuli relate not at all to what the spores would encounter in the gut of a fish. Urea often triggers the extrusion of the polar filament, but this is not a physiological trigger either.

After being triggered, the sporoplasm leaves the spore capsules and migrates to the tissue, where development occurs. If it is gut, presumably the sporoplasm simply enters nearby tissue. On the other hand, there is little information on how the sporoplasms reach cartilage, muscle kidney, liver, or other organs.

Once in the tissues, development follows, but even then, information is fragmentary and sometimes conflicting. Kudo (1966) illustrated two quite different life cycles of *Sphaeromyxa sabrazesi*, a parasite of the gallbladder of the seahorse and other fish. Much of the difficulty with precise descriptions of the life cycles lies in the inability to produce infections artificially. Fish can sometimes be infected by placing them in tanks with infected fish, but exactly when they become infected remains in doubt. Direct infections of individual fish with spores have usually failed to produce anything. If experimental infections can be achieved with regularity with any species, a variety of techniques would then be available whereby an investigator could track the early stages of infection and begin to fill in some of the important gaps in our knowledge.

## REFERENCES AND FURTHER READINGS

Bogdonova, E. A. 1967. Modern data on the distribution and biology of *Myxosoma cerebralis* (Protozoa, Cnidosporidia) as agent of whirling disease of salmonids. *Bull. Off. Int. Epizoot.* **69**:1499–1506.

Cone, D. K., and R. C. Anderson. 1977. Myxosporidan parasites of pumpkinseed (*Lepomis gibbosus* L.) from Ontario. *J. Parasitol.* **63**:657–666.

Grassé P. P. 1960 Les Myxosporidies son les organismes pluricellulaires. *C. R. Acad. Paris* **251**:2638–2640.

Hoffman, G. L. 1970. Intercontinental and transcontinental dissemination and transfaunation of fish parasites with emphasis on whirling disease (*Myxosoma cerebralis*). In S. F. Snieszko (ed.), *A Symposium on Disease of Fishes and Shellfishes*. American Fisheries Society, Washington, pp. 69–81.

Hoffman, G. L., C. E. Dunbar, and A. Bradford. 1962. Whirling disease of trout caused by *Myxosoma cerebralis* in the United States. U.S. Fish & Wildlife Service Spec. Sci. Rept. No. 427.

Hoffman, G. L., and R. E. Putz. 1969. Host susceptibility and the effect of aging, freezing, heat and chemicals on spores of *Myxosoma cerebralis*. *Prog. Fish Cult.* **31**:35–37.

Hoffman, G. L., R. E. Putz, and C. E. Dunbar. 1965. Studies on *Myxosoma cartilaginis* n. sp. (Protozoa: Myxosporidia) of centrarchid fish and a synopsis of the *Myxosoma of North American freshwater fishes. J. Protozzol.* **12**:319–332.

Kudo, R. R. 1920. Studies on Myxosporidia. A synopsis of genera and species of Myxosporidia. III. *Biol. Mongr.* **5**:239–503.

———. 1926. On *Myxosoma catostomi* Kudo, 1923, a myxosporidian parasite of the sucker, *Catostomus commersonii. Arch. Protistenk.* **53**:191–214.

———. 1966. *Protozoology*, 5th ed. Charles C Thomas, Springfield, Ill.

Lom, J. 1969. On a new taxonomic character in Myxosporida, as demonstrated in description of two new species of *Myxobolus. Folia Parasitol.* (Praha) **16**:97–103.

———. 1970. Protozoa causing diseases in marine fishes. In S. F. Snieszko (ed.), *A Symposium on Diseases of Fishes and Shellfishes*. American Fisheries Society, Washington, pp. 101–123.

Lom, J., and J. O. Corliss. 1967. Ultrastructural observations on the development of the microsporidian sporozoan *Plistophora hyphessobryconis* Schaeperclaus. *J. Protozool.* **14**:141–152.

Lom, J., and P. dePuytorac. 1965. Studies on the myxosporidian ultrastructure and polar capsule development. *Protistol.* **1**:53–65.

Lom, J., and G. L. Hoffman. 1971. Morphology of the spores of *Myxosoma cerebralis* (Hofer, 1903) and *M. cartilaginis* (Hoffman, Putz and Dunbar, 1965). *J. Parasitol.* **57**:1302–1308.

Markiw, M. E., and K. Wolf. 1983. *Myxosoma cerebralis* (Myxozoa: Myxosporea) etiologic agent of salmonid whirling disease requires tubificid worm (Annelida: Oligochaeta) in its life cycle. *J. Protozool.* **30**:561–565.

Meglitsch, P. A. 1960. Some coelozoic Myxosporidia from New Zealand fishes. I. General and family Ceratomyxidae. *Trans. Roy. Soc. N. Z.* **88**:265–365.

Noble, E. R. 1944. Life cycles in the Myxosporidia. *Quart. Rev. Biol.* **19**:213–235.

———. 1950. On a myxosporidian (protozoan) parasite of California trout. *J. Parasitol.* **36**:456–460.

———. 1966. Myxosporidia in deepwater fishes. *J. Parasitol.* **52**:685–690.

Sanders, J. E., J. L. Fryer, and R. W. Gould. 1970. Occurrence of the myxosporidan parasite *Ceratomyxa shasta*, in salmonid fish from the Columbia River basin and Oregon coastal streams. In S. F. Sneiszko (ed.), *A Symposium on Diseases of Fishes and Shellfishes*. American Fisheries Society, Washington, pp. 69–81.

Sindermann, C.J. 1970. *Principal Diseases of Marine Fish and Shellfish*. Academic Press, New York.

Sprague, V. 1969. Need for drastic revision of the classification of subphylum Amoebogena. In A. A., K. Strelkov, M. Sukhanova, and I. B. Raikov (eds.), and *Progress in Protozoology*. Abstracts of papers read at the Third International Congress on Protozoology, Leningrad, July 1961, p. 372.

Vavra, J. 1971. Physiological, morphological, and ecological considerations of some microsporidia and gregarines. In A. M. Fallis (ed.), *Ecology and Physiology of Parasites*. University of Toronto Press, Toronto, pp. 92–103.

# 19

# *Microspora*

The Microspora is a phylum of obscure, but nonetheless important, parasites of a wide array of hosts. All of the Microspora are parasitic principally in insects and fish. They are transmitted from one host to the next as a spore, the form from which the group obtains its name.

The spores are typically quite small (Fig. 19.1). Most of them are ovoid or ellipsoid and are about 2 by 4 μm. A few, such as *Stempelia magna*, have spores that are 14 by 4.5 μm, and some such as *Perezia lankesteriae*, a hyperparasite of the gregarine, *Lankesteria ascidiae*, have spores as small as 2.5 μm.

The spore has a smooth outer wall, and unlike the Myxozoa there are no valves or sutures. The sporoplasm has a single nucleus and a hollow filament coiled inside the spore (Fig. 19.2). This coiled tubule is everted when the spore enters the body of a potential host and actually injects the sporoplasm into a cell. The filament, when everted, may be from 25 to 500 μm, depending on the species, but most are 100 μm or less in length; the diameter is about 0.1 μm.

Spores are usually ingested by a host, but sometimes the sporoplasm is injected through the integument. Whatever the mode of entry, the filament is triggered and the sporoplasm escapes from the spore by moving down the length of the tube. What triggers extrusion of the tubule is not well known. Under experimental conditions, change in osmotic pressure, drying, and mechanical pressure have all been used. The precise conditions in a host have not been determined.

Once inside a host cell, the sporoplasm undergoes merogony (Fig. 19.3). There are one or more merogonous generations, and then sporogony takes place. Depending on the species, a lesser or greater number of spores may be formed. Note that the spores are formed within the parasite, not by the formation of a wall surrounding the whole parasite. The spores reach the outside most often with feces or urine, but may also do so through the integument or when the host dies and breaks down through natural processes.

There are between 500 and 600 described species of microspora and perhaps 200 additional species are known but not adequately described. New ones are being added continually and it is clear that only a small percentage of species has been described.

Although economically important Microspora have been known for more than a hundred years, only recently has their pervasive importance been recognized. We know now that they have a great deal of potential for controlling populations of economically important insect pests and that they are common in fish and mammals among whom they cause disease.

Microspora have an important place in the history of biology and in the development of knowledge of infectious agents. During the early 1860s a disease called pébrine swept through the silkworm-raising industry of Europe and nearly eliminated it. In France there was concern that the silk industry might disappear, and Louis Pasteur was asked to look at the problem. For a period of about 5 years, from 1865 to

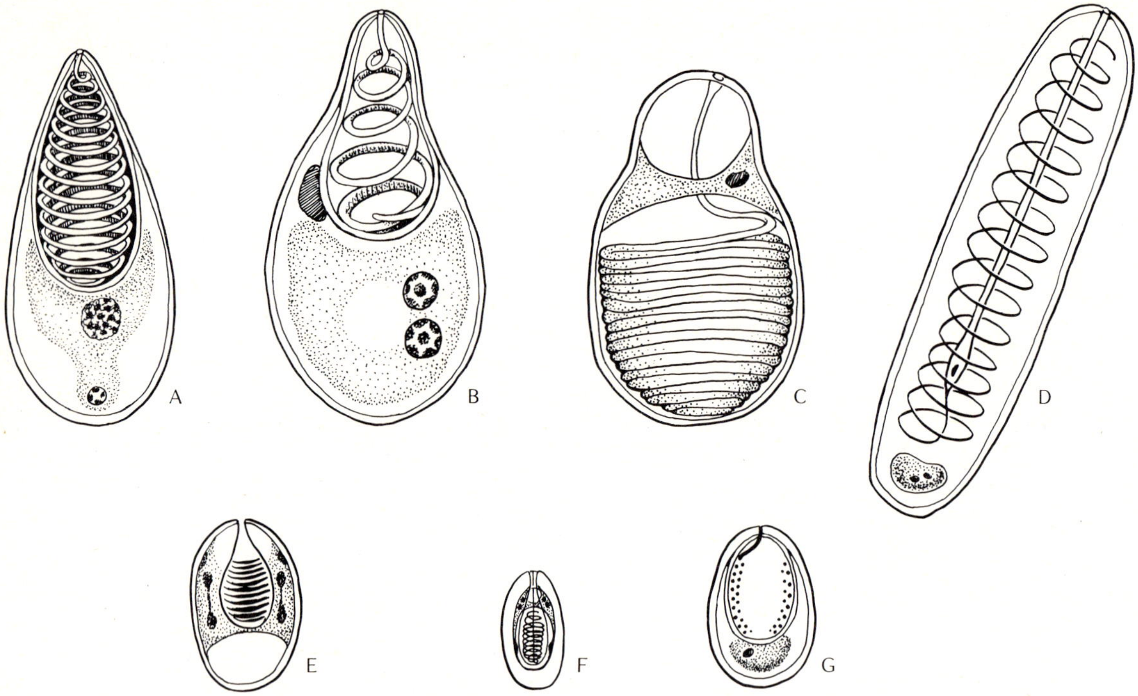


**FIG. 19.1**
A selection of spores of Microspora and Myxozoa. (*A*) *Stempellia magna* from the mosquito, *Culex;* (*B*) *Myxobolus toyamai* from the carp, *Cyprinus carpio;* (C) *Plistophora longifilis* from the fish, *Barbus;* (*D*) *Mrazekia argoisi* from the amphipod, *Asellus;* (*E*) *Thelohania giardi* from the crustacean, *Crangon;* (*F*) *Nosema bombycis* from the silkworm, *Bombyx mori;* (*G*) *Nosema apis* from the honeybee, *Apis mellifera.* [Redrawn with modifications from Kudo, 1924.]

1870, Pasteur studied the transmission of *Nosema bombycis*, the cause of pébrine. Through the experiments and observations of Pasteur and his students, methods were developed to diagnose infected silkworms and to rear them free of infection. Pasteur had previously worked in chemistry and on fermentation of wine, but after his studies on pébrine, he entered the world of infectious diseases and his great discoveries in bacterial and viral diseases followed.

In this chapter we will deal only with a few species of microspora that are important to humans for either their harm or potential benefit.

NAME AND DISEASE CAUSED. *Nosema bombycis* Naegeli, 1857, causes silkworm disease, pébrine, and Flechenkrankheit.

HOSTS AND LOCATION. The normal host is *Bombyx mori*, the silkworm, but infections can be established in other insect species as well. In the silkworm the infection occurs in all life stages, including the egg, and all body tissues are infected.

MORPHOLOGY. The spores are oval and measure 3 to 4 by 1.5 to 2 μm.

DISTRIBUTION. Cosmopolitan.

LIFE CYCLE. The spores are ingested and the polar filament injects the sporont into the cells of the gut. The sporoplasm transforms into a trophozoite and undergoes merogony. There is more than one merogonous generation, and then the merozoites reinvade cells to

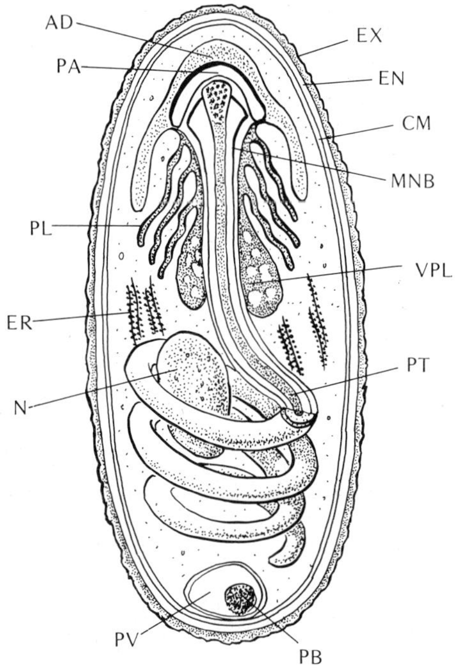


**FIG. 19.2**
A diagram of a microsporan spore as seen in the electron microscope. (AD = anchoring disk of the polar tube; CM = cytoplasmic membrane of the spore content; EN = endospore; EX = exospore; MNB = manubroid part of the filament; N = nucleus; PA = polar aperture; PB = posterior body; PT = polar tube; PL = lamellae of the lamellar polaroplast; PV = posterior vacuole; ER = endoplasmic reticulum densely populated with ribosomes; VPL = vesicular part of the polaroplast).

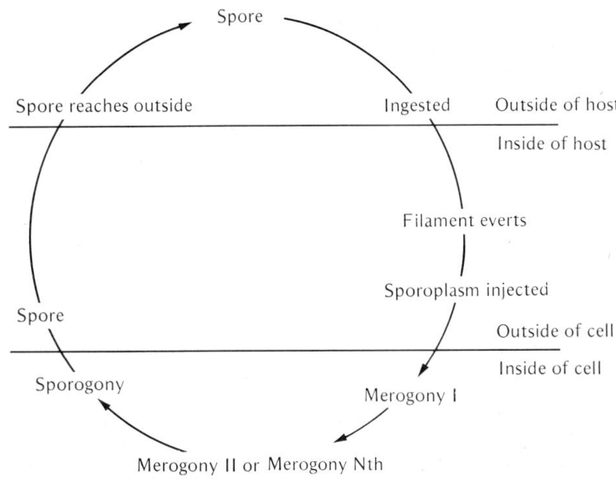


**FIG. 19.3**
Generalized life cycle of a microsporan.

form sporonts. Each sporont gives rise to only a single spore. Spores may be formed as early as 48 hr after infection.

HOST–PARASITE INTERACTIONS. *N. bombycis* is a highly pathogenic parasite that has decimated the silkworm industry at various times and places. All tissues of the developing stages are infected and there may be a range of effects. In heavy infections the larvae die, or they may not be able to spin a coccoon. In lighter infections the life cycle of the insect may be completed, but the organism can be transmitted from mother to offspring through the egg, causing the newly hatched larvae to die.

DIAGNOSIS. The larvae have a speckled surface, as though covered with flakes of pepper (pébré). This sign is good evidence of pébrine, but other conditions such as bacterial infections cause similar changes in the larvae. The most certain method of diagnosis is by finding the characteristic spores in the tissues of the larva or of the adult. The animal is teased apart and the spores are looked for under the higher powers of a compound microscope.

EPIDEMIOLOGY AND CONTROL. Pébrine is spread by contamination of an area with spores which are ingested by silkworms. As a general rule, spores are released to the outside when a host dies and disintegrates. Since the host produces large numbers of spores, the environment becomes contaminated and the disease spreads rapidly. Because of carryover from one generation to another through transovarial transmission, control, and eradication pose difficult problems.

Pasteur's work during 1865–1870 in France developed principles that are still valid today: good diagnosis and use of only clean stock. Pasteur and his students studied the transmission of the disease and determined that the so-called seed or eggs were a principal source of infection. If it were possible to obtain eggs free of infection, these could be used the year following an outbreak to establish a healthy colony. Thus, his recommendations were to use eggs only from adults known to be free of infection by keeping each batch of eggs identified with a particular adult female. If we were dealing with large farm animals, we would call this a "quarantine, test-and-slaughter program." Although the terms may be different, the principles are the same. Fortunately, the spores of *N. bombycis* do not remain alive from one season to the next outside a host. If the area is cleaned and allowed to remain vacant until the next spring, one can begin anew.

NAME AND DISEASE CAUSED. *Nosema apis* Zander causes *Nosema* disease, dwindling diseases, nosematosis, and winter losses.

HOSTS AND LOCATION. The principal host is the honey bee, *Apis mellifera*, but it may infect 13 or more species of insects. The organism infects the midgut and the Malpighian tubules.

MORPHOLOGY. Spores are ellipsoidal, measure 4 to 6 by 2 to 4 μm, and have a long polar filament with up to 44 coils.

LIFE CYCLE. The sporoplasm infects cells of the midgut of the bee. There are two merogonous generations (Gray et al., 1969). The meronts are long chainlike organisms, and the second generation has four double nuclei. Spores are produced about 7 days after infection.

DISTRIBUTION AND IMPORTANCE. Cosmopolitan.

DIAGNOSIS. A number of diseases of bees caused by bacteria, fungi, and viruses may have effects similar to those of *Nosema* disease. It is essential to distinguish which disease is the cause of losses because the methods of treatment and control will vary. Often it is necessary to submit bees to a laboratory that specializes in diagnosis of their diseases.

Nosematosis is a disease of adult bees that is usually seen by the beekeeper as a dwindling in the population of adult bees in the hive. Weak, dying, and dead bees are seen around the hive. The only certain method of diagnosis is to find the typical spores in the tissues of affected bees, although if the midgut of the bee is swollen and pale, one would suspect the disease.

HOST–PARASITE INTERACTIONS. *N. apis* principally affects the intestinal tract of the insect, but it also affects the Malpighian tubules. The normal gut function is disrupted and the bees become weak and cannot fly as strongly as they normally would. Since there is always some turnover in a hive of bees, the losses are not usually recognized until the disease is well established. Thus, the number of bees in an affected hive dwindles.

TREATMENT. Fumagillin (Abbott) can be used to treat an affected hive by offering a sucrose solution containing the drug. Fumagillin seems to provide good control.

EPIDEMIOLOGY AND CONTROL. Because the infection is spread from one bee to the next through fecal contamination, there are four measures that are essential for the minimizing of losses from *Nosema* disease:

1. Purchasing only clean bees
2. Early diagnosis
3. Treatment
4. Cleaning the hives after an outbreak

The bee industry in the United States is controlled, so that in each state there is a system of inspection and certification of the health of a newly purchased colony. There is also periodic inspection of hives to ensure continued freedom from infectious diseases. Such strict regulation may seem excessive until one considers that the bee industry in the United States in 1973 was estimated to have a value of $2.11 billion. Of this total, $2 billion was for use of bees as pollinators of various crops such as fruit trees and alfalfa. Honey accounted for $106 million and the rest was beeswax. The importance of bees has brought the assurance that bees purchased from reliable raisers are free of *Nosema* disease as well as other diseases.

If nosematosis or other disease is suspected, prompt diagnosis is essential. Help in diagnosis may be obtained from the state college of agriculture or the state apiarist, sometimes located in the same place. Once a positive diagnosis has been obtained, fumagillin will control the disease.

Wintertime is especially difficult for bees because they must survive on stored honey for several months. If the colony has been weakened through nosematosis or other infectious disease, winter losses may be excessive. The question of keeping infected bees over the winter must therefore be addressed. Fumagillin usually gives good control of nosematosis, but the hive and surroundings remain contaminated for a considerable period of time. To ensure minimal risk, the hive should either be destroyed or thoroughly fumigated in fall, and then clean bees should be purchased in spring.

NAME AND DISEASE CAUSED. *Encephalitozoon cuniculi* Levaditi, Nicolau and Schoen, 1923, synonym: *Nosema cuniculi*, *E. negri*, cause encephalitozoonosis.

HOSTS AND LOCATION. This organism was originally described from the European rabbit, *Oryctolagus cuniculi*, but it infects, either naturally or experimentally, laboratory mice and rats, guinea pigs, the domestic dog, and perhaps humans. The organism did not develop in three rhesus monkeys (Shadduck et al., 1979) and an apparent *Encephalitozoon* in a squirrel monkey was morphologically different from *E. cuniculi*. At this point, lack of data on the host range and differing structures in different species of hosts leaves in doubt how many species actually infect various mammals.

Infections are widely disseminated in the host. Most often the central nervous system and kidney are infected, but in heavy infections all tissues may have organisms in them.

DISTRIBUTION AND IMPORTANCE. *E. cuniculi* infections are common in laboratory animals, especially rabbits; up to 84% of 51 rabbits have been found to be infected (Malherbe & Munday, 1958).

The effect on individuals and populations of animals is more difficult to determine. However, outbreaks of encephalitozoonosis have been reported in both rabbits and dogs.

The high prevalence in laboratory experimental animals is disturbing. It is essential that no extraneous factors invalidate data when experimental animals are used. It is known that hosts that are not immunologically competent may be affected by *E. cuniculi*, and there are some instances in which tumors have been invaded by Microspora and their size has been reduced (an unexplained observation). *E. cuniculi* is not normally looked for, and what role it may play in affecting the outcome of experiments is unknown at this point.

MORPHOLOGY. The oval spores average 2.5 by 1.5 μm, and have a relatively thick outer wall and a short polar filament of 20 to 25 μm.

The merogonous stages are quite small, ranging from about 1 μm up to 2.5 by 4 μm (Fig. 19.4). *Encephalitozoon* remains as a unikaryon form, as opposed to the diplokaryon, which is an identifying feature of *Nosema*.

LIFE CYCLE. Transmission of the agent takes place by mouth and the polar filament probably everts in the intestine of the host. How the infection is spread to various other organ systems is not known. The number of merogonous generations is not clear; it appears that there might be continuous merogony if the infection is widespread and heavy in a host. Each sporont gives rise to two uninucleated spores. It has been demonstrated that spores may pass with the urine if there is infection

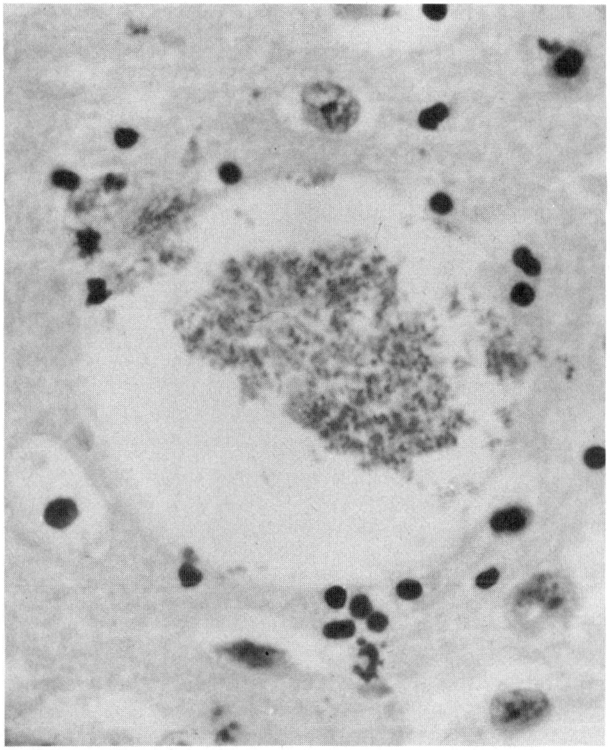

**FIG. 19.4**
*Encephalitozoon cuniculi* meronts in a rabbit brain.

of the kidney; experimental animals have been infected through urine containing spores. Whether spores may reach the outside through other routes is not known. Prenatal transmission seems to take place since rabbits taken by cesarean section and raised under gnotobiotic conditions have been found to be infected.

DIAGNOSIS. Infections with *E. cuniculi* are almost always found through histologic examination of tissues. Since the organisms are so small and their distinguishing features are so difficult to see, a person experienced in identifying members of the group should be consulted. There are staining characteristics that can be used to differentiate the organism from other intracellular organisms (Pakes, 1974).

CULTIVATION. *E. cuniculi* can be propagated in cell culture.

HOST–PARASITE INTERACTIONS. In most instances, *E. cuniculi* infection is inapparent. Sometimes it becomes a fulminating, lethal infection, however. Deaths are mostly in young animals.

The effects of infection are predominantly in the central nervous system and secondarily in the kidney; other tissues may also be affected. The signs of infection are related principally to damage to the brain: incoordination, weakness, convulsions, abnormal behavior. None of these is diagnostic for encephalitozoonosis.

TREATMENT. No treatment is known.

EPIDEMIOLOGY AND CONTROL. The only known means of transmission are through urine containing spores and prenatal transfer. Even laboratory animals derived from supposedly uninfected mothers have been found to be infected. Prevalence of infection varies widely in laboratory animals, and the prevalence in uncaged animals is not known. At this point there are no recommendations for controlling the infection.

## MICROSPORA AS BIOLOGICAL CONTROL AGENTS

Around 1920, a number of Microspora were found that were pathogenic for harmful insects. As an example, among many Microspora described by Kudo was *Nosema anophelis*, which attacked the fat bodies of larval mosquitoes. It reduced stored fat in the larvae so that they were unable to complete development. It appeared that the system might be manipulated to control mosquito populations. (It should be remembered that insecticides such as DDT were still 20 years in the future.) Newspapers, ever alert to a story, declared that the scourge of mosquitoes and the diseases they carry would soon be a thing of the past. It is clear that such a desirable state has not yet arrived despite the passage of more than 50 years. What has delayed achieving control through the use of Microspora?

Probably three principal factors delayed development of this kind of biological control:

1. Unwillingness of administrators to commit public funds to development
2. Relatively primitive technology in biology
3. Development of fairly good insecticides in the 1930s and then the development of organochlorine insecticides (DDT, lindane, dieldrin) in the 1940s

The course of research and development is inevitably influenced by social, political, and economic forces. So it is with biological control in general. The commitment of funds to such a program was probably inconceivable when technology for mass-producing insects and their parasites was still not developed to a high level. Systems for delivery and broadcasting of the parasites were still to be developed. In the years before World War II, economic conditions militated against such programs. During and immediately following the war, development of second-generation insecticides was explosive; it appeared that our insect problems could be taken care of through more and better chemicals. And the control achieved was truly miraculous.

Then, as the 1960s unfolded, it became clear that pesticides were hazardous to many more than the target organisms. As tighter regulations have been imposed and the technological climate has changed, biological control is now an attractive adjunct and alternative.

As of 1980, some organisms are in the final stages of approval for use and others are in the stage of field testing for efficacy and safety. *Nosema locustae* has been approved for control of grasshoppers and is offered to them in a grain bait containing the spores. A number of species have been used with some success in controlling insect pests of coniferous trees; in these cases, the spores are sprayed onto the foliage in liquid.

*Nosema algerae* is undergoing field testing for the control of anopheline mosquito vectors of malaria.

Problems continue in the development of Microspora as biological control agents. Clearance for use requires proof of both efficacy and safety. Efficacy is easier to demonstrate; this can be done in both laboratory and field tests. Safety is difficult to prove since it is necessary to determine the host range of the potential product and to show that there is no harmful acute or chronic effect on any animals important to humans, including humans. In most instances, the expense of obtaining clearance for such a product will be borne only by a large commercial company, and the company must be able to see an adequate return on its investment. The question that managers inevitably ask is, can we predict that there will be a long-term market for a product that continues to multiply in the pests it aims to control? If we sell one batch of spores, is there any need to sell more the next season? The answer is a qualified yes. When a pest is recognized as a potential problem in a certain area, the rapid amplification of the infectious agent can be accomplished most readily by introducing large numbers into the environment (see Tanada in Vavra & Sprague, 1976).

The issue of human safety is difficult to evaluate at this point. There are three reports of microsporan infections in humans (Strano et al., 1976). One of these was in an immunologically deficient infant. The human hazard is present, but because of the difficulty in diagnosing infections, its overall importance is not known.

## MICROSPORA OF FISH

The two principal animal groups that serve as hosts of Microspora are insects and fish. A large number of species of Microspora have been described from fish, and the prevalence of infection is high. The genera most often seen in fish are *Glugea*, *Pleistophora*, and *Nosema*.

The effects on individual hosts vary from inapparent to severe interference with normal function and death. For example, *Nosema branchiale* has been reported from a number of species of cod (Gadidae), but even though there were 60 cysts on a single gill arch in heavily parasitized fish, there did not seem to be any detrimental effect. (Kabata, 1959). *Pleistophora ovariae* infects the ovaries of golden shiner, and causes low-grade losses when these minnows are raised as fish bait (Summerfelt & Warner, 1970). On the other hand, *Glugea anomala* contributed to a mass die-off of stickleback in Russia.

Microspora may infect nearly any organ system of fish including the gut, reproductive system, muscles, gills, brain, and eye. *Glugea stephani* is a common parasite of flounder in temperate waters, and it causes extensive damage to the intestine of the host. The effect is to cause almost complete replacement of the normal tissue of the gut with huge numbers of cysts containing the spores. *Glugea hertwigi* parasitizes not only the gut but also the gonad; fish are then unable to reproduce. A number of species, such as *Pleistophora macrozoarcidis*, cause tumorlike deformities or xenomas (tumors formed of tissue other than that of the animal) on the surface of fish.

The losses from microsporan infections in fish arise from

1. Failure to grow properly
2. Death losses, or weakening so that the animals are readily removed by predation
3. Failure to reproduce
4. Discarding for esthetic reasons fish that might otherwise be eaten

As we have seen with quite a number of diseases, determining the extent of losses, or placing an economic value on them, is difficult. However, there are a number of instances, quite well documented, in which there have been losses. In flounder off Cape Cod, the young of the year were heavily infected, but the second-year fish had little infection; it was concluded that the young fish died in their first year (Stunkard & Lux, 1965). A massive die-off of smelt (*Osmerus*) was observed in Quebec in which 85% of the fish were affected (Legault & Delisle, 1967). Sindermann (1970) has concluded that the "effects [of the Microspora] on the host are among the most severe of any parasite group" on marine and estuarine fish.

## REFERENCES AND FURTHER READINGS

Cali, A. 1970. Morphogenesis in the genus *Nosema*. *Proc. 4th Intern. Colloq. Invert. Pathol.*, pp. 431–438.

Cox, J. C., R. C. Hamilton, and H. D. Attwood. 1979. An

investigation of the route and progession of *Encephalitozoon cuniculi* infection in adult rabbits. *J. Protozool.* **26**:260–265.

Gray, F. H., A. Cali, and J. D. Briggs. 1969. Intracellular stages in the life cycle of the microsporidian *Nosema apis*. *J. Invert. Pathol.* **14**:391–394.

Hazard, E. I., and D. W. Anthony. 1974. A redescription of the genus *Parathelohania* Codreanu 1966 (Microsporida:Protozoa) with a reexamination of previously described species of *Thelohania* Henneguy 1892 and descriptions of two new species of *Parathelohania* from anopheline mosquitoes. USDA, ARS, Tech. Bull. No. 1505. Washington.

Hazard, E. I., and S. W. Oldacre. 1976. Revision of Microsporida (Protozoa) close to *Thelohania*, with descriptions of one new family, eight new genera and thirteen new species. USDA, ARS, Tech. Bull. No. 1530. Washington.

Kabata, Z. 1959. On two little-known Microsporidia of marine fishes. *Parasitol.* **49**:309–315.

Kudo, R. R. 1924. A biologic and taxonomic study of the microsporidia. *Ill. Biol. Monogr.* **9**:83–344.

Kudo, R. R., and E. W. Daniels. 1963. An electron microscope study of the spore of a microsporidian, *Thelohania californica*. *J. Protozool.* **10**:112–120.

Larsson, R. 1980. Ultrastructural study of *Toxoglugea variabilis* n. sp. (Microsporida, Thelohaniidae), a microsporidian parasite of the biting midge *Bezzia* sp. (Diptera:Ceratopogonidae). *Protistolog.* **16**:17–32.

Legault, O. R., and C. Delisle. 1967. Acute infection by *Glugea hertwigi* Wessenberg in young-of-the-year rainbow smelt, *Osmerus eperlanus mordax* (Mitchill). *Can. J. Zool.* **45**:1291–1294.

Lom, J. 1970. *Protozoa Causing Diseases in Marine Fishes.* Spec. Publ. No. 5. American Fisheries Society. Washington, D.C., pp. 142–160.

Malherbe, H., and V. Munday. 1958. *Encephalitozoon cuniculi* infection of laboratory rabbits and mice in South Africa. *S. African Vet. Med. Assn.* **29**:241–246.

Pakes, S. P. 1974. Protozoal diseases. In S. H. Weisbroth, R. E. Flatt and A. L. Kraus (eds.), *The Biology of the Laboratory Rabbit.* Academic Press, New York, pp. 264–286.

Pakes, S. P., J. A. Shadduck, and A. Cali. 1975. Fine structure of *Encephalitozoon cuniculi* from rabbits, mice and hamsters. *J. Protozool.* **22**:481–488.

Shadduck, J. A., and M. B. Polley. 1978. Some factors influencing the *in vitro* infectivity and replication of *Encephalitozoon cuniculi. J. Protozool.* **25**:491–496.

Shadduck, J. A., W. T. Watson, S. P. Pakes, and A. Cali. 1979. Animal infectivity of *Encephalitozoon cuniculi. J. Parasitol.* **65**:123–129.

Sindermann, C. J. 1970. *Principal Diseases of Marine Fish and Shellfish.* Academic Press, New York.

Sprague, V. 1977. Classification and phylogeny of the Microsporidia. In L.A. Bulla, Jr., and T. C. Cheng (eds.), *Comparative Pathobiology*, Vol. 2. Plenum Press, New York, pp. 1–30.

Sprague, V., and J. Vavra. 1977. *Systematics of the* Microsporidia. Plenum Press, New York.

Sprague, V., and S. H. Vernick. 1971. The ultrastructure of *Encephalitozoon cuniculi* (Microsporida, Nosematidae) and its taxonomic significance. *J. Protozool.* **18**:560–569.

Strano, A. J., A. Cali, and R. C. Neafie. 1976. Microsporidiosis. In *Pathology of Tropical and Extraordinary Diseases.* AFIP Press, Washington, D.C. pp, 336–339.

Stunkard, H. W., and F. E. Lux. 1965. A microsporidian infection of the digestive tract of the winter flounder, *Pseudopleuronectes americanus. Biol. Bull.* **129**:371–387.

Summerfelt, R. C., and M. C. Warner. 1970. Incidence and intensity of infection of *Plistophora ovariae*, a microsporidian parasite of the golden shiner, *Notemigonus crysoleucas*. In S. F. Snieszko (ed.), *A Symposium on Diseases of Fishes and Shellfishes.* Spec. Publ. No. 5, American Fisheries Society, Washington D.C., pp. 142–160.

Vavra, J., and V. Sprague (eds.). 1976. *Biology of the Microsporidia.* Plenum Press, New York.

Vavra, J., and A. H. Undeen. 1970. *Nosema algerae* (Cnidospora, Microsporida), a pathogen in a laboratory colony of *Anopheles stephensi* Liston (Diptera, Culicidae). *J. Protozool.* **17**:240–249.

Youseff, N. N., and D. M. Hammond. 1971. The fine structure of the development stages of the microsporidian *Nosema apis* Zander. *Tissue and Cell* **3**:283–294.

# PART 2

# *Platyhelminthes*

20

# Introduction to the Flatworms, Monogenea, and Aspidogastrea

The phylum Platyhelminthes includes a variety of free-living and parasitic flatworms of relatively primitive structure. Within the limitation of being aquatic or living at least in moist environments, they have successfully invaded a variety of ecologic niches. The mostly free-living members, the Turbellaria, seem to be the stem forms for the parasitic classes, the flukes and the tapeworms (Fig. 20.1).

There is a tendency to look on the parasitic flatworms as being degenerate. But a more appropriate viewpoint is that they have become highly specialized in order to survive and reproduce in particular environments. The tapeworms, for example, have lost some sense organs and are often spoken of disparagingly for that reason; however, cave fish have become blind, but no one considers this reprehensible. In both instances, the organisms have evolved so as to be frugal with their resources; they do not allocate energy for structures that give no competitive advantage. Surely this economical use of energy is to be admired rather than condemned.

Flatworms have certain characteristics that are important for classification and identification. The platyhelminths are members of the group referred to as *acoelomate bilateria.* Bilateral symmetry is found in nearly all of the parasitic metazoa, but the fact that they lack a body cavity (acoelomate) allows one to identify them readily in whole mounts or in tissue sections. The organ systems are embedded in a connective tissue network (parenchyma) which, together with the tegument or outer covering, allows their identification as members of the phylum.

The excretory system is based on the flame cell or protonephridium (Figs. 20.2 and 20.3). It is primarily osmoregulatory and functions adequately in small animals. The pattern of distribution of flame cells within an individual is sometimes used for classification and identification.

The nervous system is not usually seen in standard preparations of flatworms; it is composed of ganglia at the anterior end and two longitudinal nerve cords, with cross-commissures, running the length of the body. (Fig. 20.4). In the free-living forms such as planaria, the significance of this arrangement is clear because cephalization occurs first phylogenetically in the flatworms. In the Turbellaria, there is development of a head and associated sense organs such as eyes and tactile cilia (Fig. 20.5). In an evolutionary context, this is the first time that animals meet the environment head-on. Cephalization is somewhat less important in the parasitic flatworms than in the free-living ones, but a dominance of the head is still shown and movement is made with that portion anterior.

The digestive system in the platyhelminths is incomplete. The mouth enters a blind sac, so that the residue of any food ingested must be egested through the mouth (Fig. 20.1).

The greatest number of platyhelminths are hermaphroditic or monoecious. The sexes are separate in a few instances, such as the blood flukes and a small

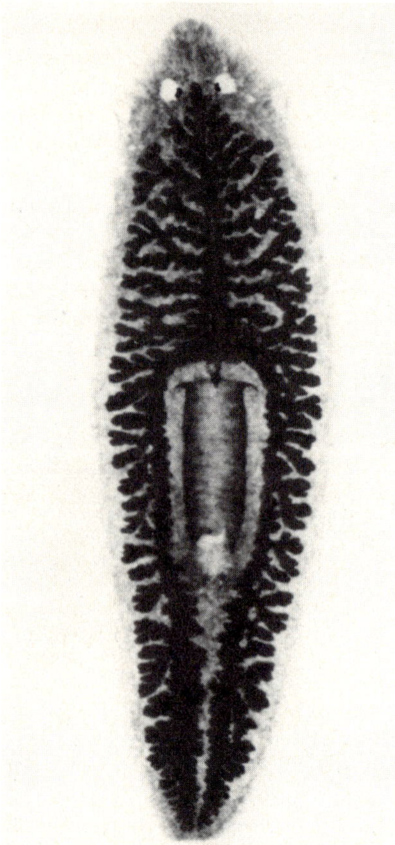

FIG. 20.1
A free-living turbellarian.

number of species of tapeworms. The reproductive systems are used almost more than any other structures for identification and classification of parasitic flatworms.

The class Trematoda includes a diverse group of flatworms all of which are parasitic. They are small to medium-sized with most being from 2 to 30 mm long. Some adults are less than 1 mm in length, and at the other extreme a few flukes exceed 10 cm and one has been reported to be over 12 m long (Noble, 1967). They are generally flat and rather leaf-shaped. All but one or two flukes have incomplete digestive tracts and, with few exceptions (Schistosomatidae, some Didymozoidea), are hermaphroditic.

Within the class Trematoda, there are three subclasses, Didymozoidea, Aspidogastrea, and Digenea. They are differentiated on the basis of the structure of both adults and larvae and on life cycle patterns.

The class Monogenea may be recognized by the small size of the adults, mostly from 5 mm to less than 1 mm in length, a body usually divided into two regions, the body and the opisthaptor, the latter being a posterior holdfast. The oral sucker is not often well developed and internal structures are difficult to see in most stained specimens. As indicated by the name, the Monogenea have direct life cycles, that is, there are no biological vectors. The Monogenea generally are ectoparasites of aquatic, poikilothermic vertebrates, mostly fish. They are found on the gills and, less often, in the mouth, urinary bladder, and ureters. Within the limits of hosts and an aquatic environment, the Monogenea have been highly successful; thousands of species have been described from all parts of the world.

The subclass Aspidogastrea is a rather obscure group found primarily in freshwater clams (Unionidae) but

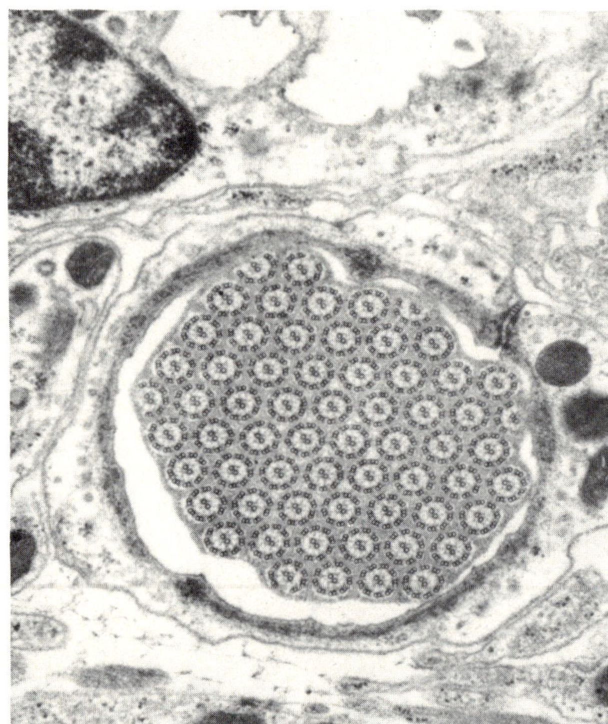

FIG. 20.2
Cross section of a flame cell.

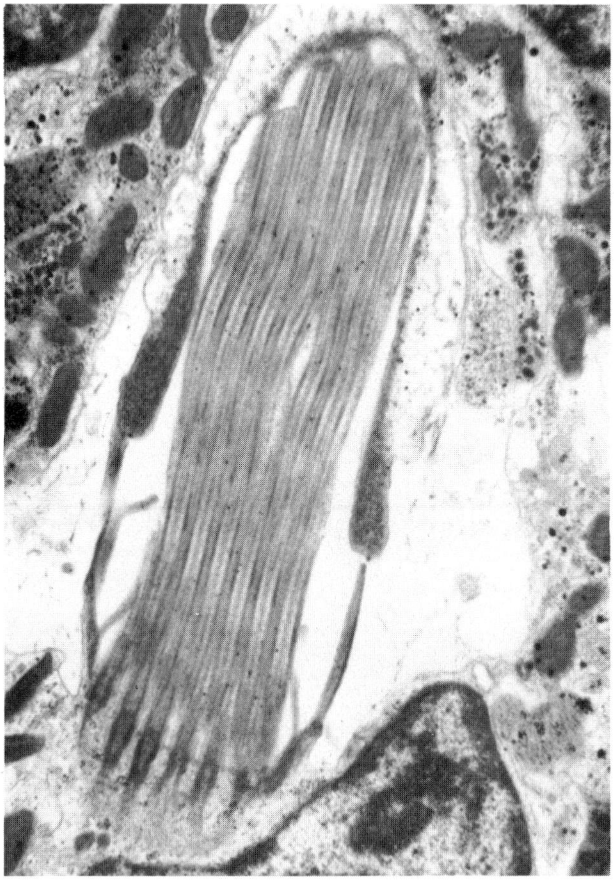

**FIG. 20.3**
Longitudinal section of a fluke flame cell.

occasionally in both freshwater and marine snails and fish and in freshwater turtles. They are endoparasites mostly of the vascular, digestive, or excretory systems of their hosts. They are frequently found by students in preserved clams used for dissection of nonvertebrate animals. The members of the group can be recognized readily by the large ventral adhesive organ, which is compartmentalized. Most of them fall within a size range from 1 to 3 mm in length; some are twice as long as wide and other are nearly as broad as long. Their life cycles are not well known; however, both direct and indirect cycles have been described.

The subclass Digenea is the largest group and the most important medically and economically. The habitats in which its members are found have the greatest range both for hosts and habitats outside of the hosts. With few exceptions, the definitive hosts are vertebrates and the first intermediate hosts are mollusks. Recognition of adult members of the subclass is relatively simple, because they are all endoparasite and both the Monogenea and Aspidobothrea have distinctive holdfasts. For example, a leaflike worm found in the digestive system of a higher vertebrate will nearly always be a digenetic fluke.

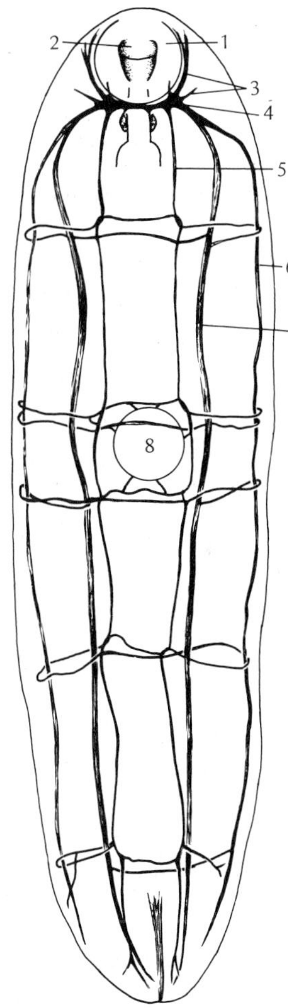


**FIG. 20.4**
Diagram of the nervous system of a distome fluke. (1 = oral sucker; 2 = mouth; 3 = anterior nerves; 4 = "brain" ganglia; 5 = ventral chord; 6 = lateral chord; 7 = dorsal chord; 8 = acetabulum.) [Redrawn from Hyman, 1951.]

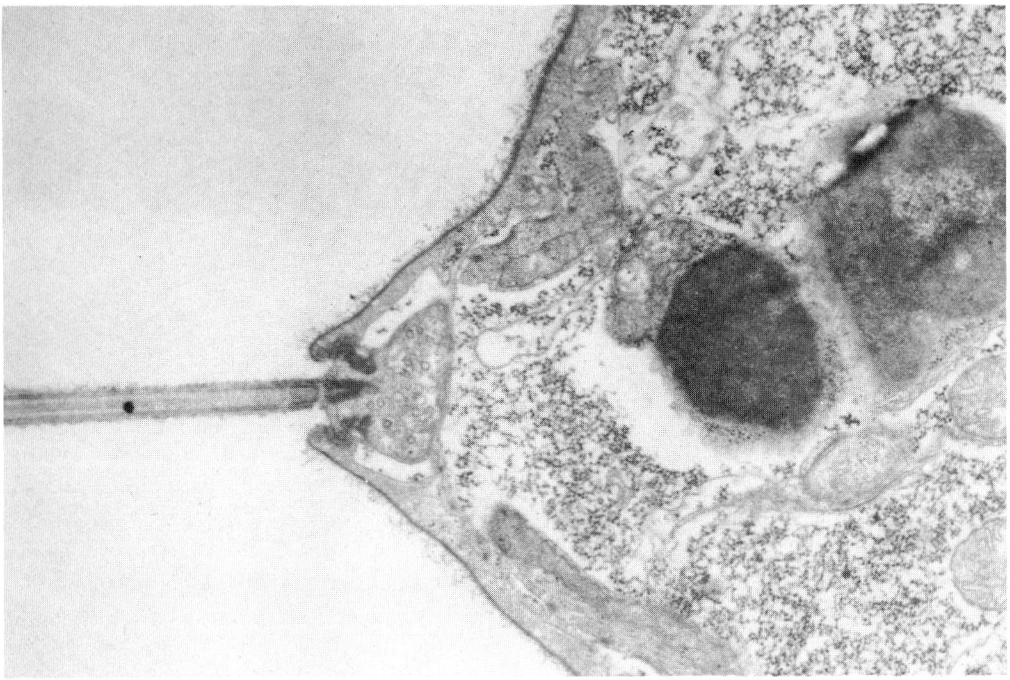

**FIG. 20.5**
Sensory cilium.

## CLASS MONOGENEA

Members of the class Monogenea are all parasitic flatworms, most with simple one-host cycles, hence the designation "one beginning." They are typically ectoparasites of aquatic vertebrates, especially on the skin and gills of fishes. They feed on mucus, epithelial cells, and blood from wounds inflicted by their hooks. Usually they are not pathogenic, but massive infections may occur in artificial situations such as fish hatcheries and ponds, resulting in serious pathology and economic loss. *Dactylogyrus*, *Gyrodactylus*, *Microcotyle*, *Ancyrocephalus*, and *Benedenia* have caused deaths in fish by suffocation and by wounds that permitted fatal bacterial infections to develop. A few species are endoparasitic, living in the mouth, alimentary tract, or urinary tract. Whether ectoparasitic or endoparasitic, most monogeneans are both host- and site-specific.

MORPHOLOGY. Adult monogeneans are oval to circular and range from 1 mm to 3 cm long. The most characteristic structure is the large, complex posterior holdfast or opisthaptor. In the subclass Monopisthocotylea, the opisthaptor is usually a single muscular disk that develops directly from the larval (oncomiracidial) haptor. Usually 1 to 3 pairs of large hooks (anchors) and 2 to 16 small hooks (hooklets) are present (Fig. 20.6). Eyes are usually present, but a genitointestinal canal is absent.

In the subclass Polyopisthocotylea, eyes are usually absent and a genitointestinal canal is present. The opisthaptors in this subclass are quite complex, with multiple suckers, clamps, and hooks (Fig. 20.7). The anterior holdfast, the prohaptor, is far less spectacular. It may simply have glands opening through head organs and marginal ducts, one or more muscular suckers surrounding the mouth, or head lappets.

Monogeneans are hermaphroditic and usually cross-fertilize; male and female reproductive organs are similar to those of the Digenea, and will not be discussed in detail here. A notable difference from the Digenea is the genitointestinal canal present in many Polyopisthocotylea. This is a tube that connects part of the intestine with the oviduct or ootype (Fig. 20.8). The function of this connecting tube remains obscure, but three suggestions have been put forth. First, many free-living Turbellaria also possess this structure, and those who work on Turbellaria regard this as evidence that eggs were once discharged via the intestine or mouth; thus, it is a primitive oviduct - uterus. Second, it may be an intestinal pouch, much like the human appendix (Hyman, 1951). An intriguing third possibility was discussed by Rohde (1975). He noted that when worms were squeezed, surplus material was discharged through the genitointestinal canal into the gut, presumably to be digested and recycled.

LIFE CYCLE. The Monogenea have simple one-host life cycles with an egg, a ciliated larval stage, the oncomiracidium, and an adult. An exception is the family Gyrocotylidae, which has no oncomiracidium stage; members are viviparous and give birth to a subadult. Monogenean eggs are usually fusiform and have filaments at one or both ends for attachment to hosts (Fig. 20.9). The oncomiracidium hatches from the egg, swims to a new host, sheds its ciliated cells, and develops into an adult (Fig. 20.10). The oncomiracidia of *Entobdella soleae* (Fig. 20.11) can find their correct hosts in total darkness with uncanny accuracy using chemical detection of mucus on the scales of the fish (Kearn, 1967). Since the fish can swim much faster than the oncomiracidia, the probability of reaching a fish would seem to be remote; however, hatching behavior in *E. soleae* is regulated to increase the odds in favor of the parasite. Most hatching occurs at dawn when the night-feeding fish (*Solea solea*) return to their daytime resting places (Kearn, 1973).

Members of the family Gyrocotylidae lack the egg and oncomiracidium and instead give birth to subadults (Fig. 20.12). Inside the uterus of the subadult

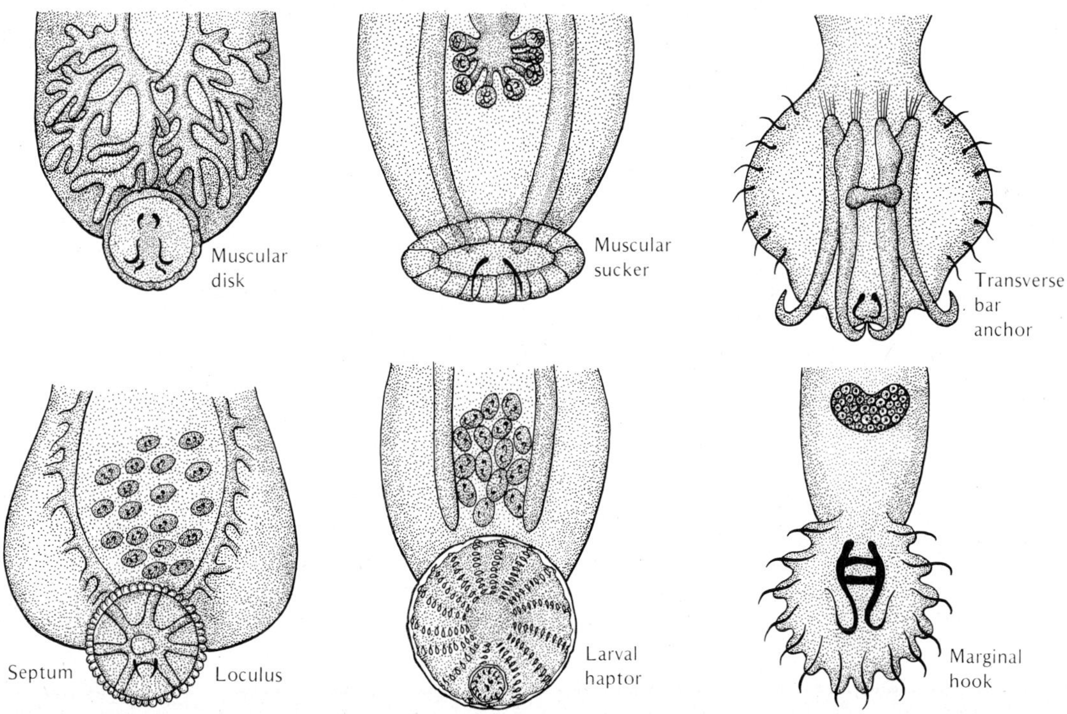


**FIG. 20.6**
Opisthaptors of Monopisthocotylea.

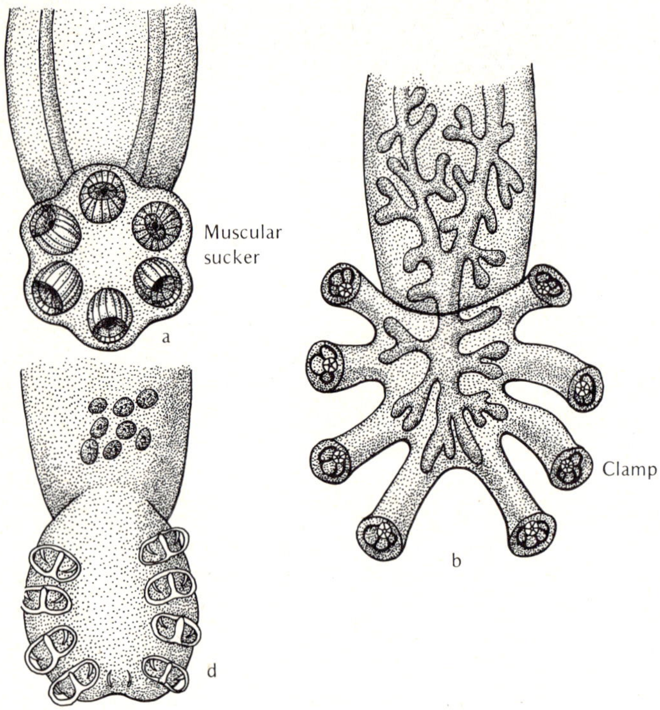


**FIG. 20.7**
Opisthaptors of Polyopisthocotylea.

may be another subadult, the "grandchild"; this unusual type of reproduction is thought to be a form of sequential polyembryony.

EVOLUTION. The Monogenea have been classified as members of the class Trematoda for many years. Most classifications included two subclasses, the Monogenea and the Digenea; however, this concept has changed in recent years. Current evolutionary schemes place weight on larval forms, emphasizing that "ontogeny recapitulates phylogeny," as promulgated by Haeckel in the nineteenth century. Beginning with Bychowsky's (1957) thorough analysis, Llewellyn (1965, 1970) and others developed the concept that monogeneans are more closely related to cestodarian tapeworms than to digenetic trematodes. Some of the evidence supporting this concept is as follows:

1. The hooks of cestodes and monogeneans are stabilized by disulfide bonds and those of the digeneans are not.
2. The oncomiracidium of monogeneans and the lycophore, coracidium, and hexacanth of tapeworms are armed, whereas the digenean miracidium is not.
3. Adult digeneans typically have an infolded tegument, whereas monogeneans and cestoideans have microvilli or microtrichs on their external surfaces (Lyons, 1973).
4. The monogeneans and cestodes are usually host-specific for the vertebrate hosts but the digeneans are host-specific for the invertebrate hosts.

It is the consensus that the free-living rhabdocoel-like ancestors gave rise to three distinct lines, which are

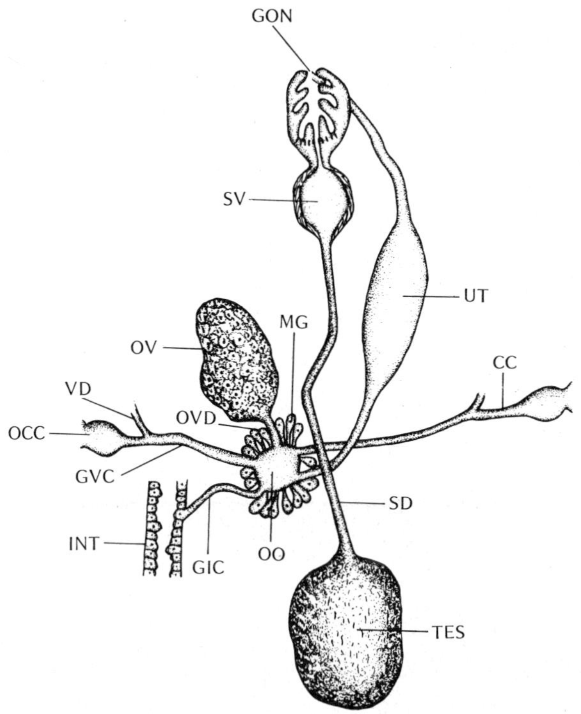


**FIG. 20.8**
Monogenean reproductive system. (CC = copulatory canal; GIC = genitointestinal canal; GON = gonopore; GVC = common genitovitelline canal; INT = intestine; MG = Mehlis' gland; OCC = opening of copulatory canal; OO = ootype; OV = ovary; OVD = oviduct; SD = sperm duct; SV = seminal vesicle; TES = testis; UT = uterus; VD = vitelline duct). [Redrawn with modifications from Cheng, T.C., *General Parasitology*, Academic Press, New York, 1973.]

given separate class status: Monogenea, Cestoidea, and Trematoda.

## REPRESENTATIVE MONOGENEANS

*Gyrodactylus* spp. are viviparous monogeneans that can cause serious skin and gill lesions in trout, bluegill, goldfish, and other fishes if they are crowded together in fish ponds. Newborn subadults have no cilia and closely resemble the adult; they have an opisthaptor with a pair of anchors and 16 marginal hooklets. Typically the subadults attach to the skin or gills of the same host as the parent worm. They have no cilia, and thus are poor swimmers; presumably they transfer to another host when there is contact between the hosts, such as crowding in a fish pond.

*Entobdella soleae* (Fig. 20.10) is a more representative example of the order Monopisthocotylea than the pathogenic *Gyrodactylus* mentioned previously. *E. soleae* has the usual monogenean life cycle with an egg attached to sand grains, a free-swimming oncomiracidium, and an adult on the scales of the common sole, *Solea solea.* In nature, soles have 1 to 9 adult parasites per fish, but up to 200 have been counted on fish kept in aquaria (Kearn, 1971). The parasite erodes only the superficial epidermis of the host; thus the danger of host secondary bacterial infection is minimal. This minimal pathogenesis is the rule for most monogeneans in nature, even those that inhabit the gills and feed on blood.

*Octomacrum lanceatum* (Fig. 20.13), of the order Polyopisthocotylea, has eight clamps on the opisthaptor and two small suckers on the prohaptor. It is found on the gills of suckers (genus *Catostomus*) and, although a blood feeder, it rarely causes serious damage because only a few parasites are on any fish. It has a typical monogenean life cycle with an egg, oncomiracidium, and adult.

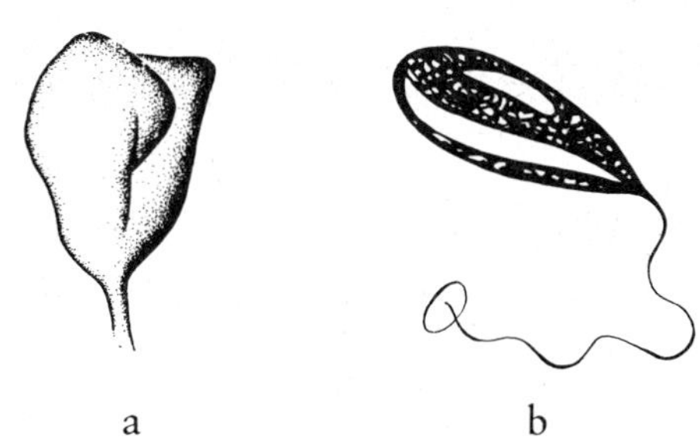


**FIG. 20.9**
Monogenean eggs. (*a*) *Thaumatocotyle dasybatis;* (*b*) *Udonella caligorum.* [Redrawn with modifications from Price, 1938.]

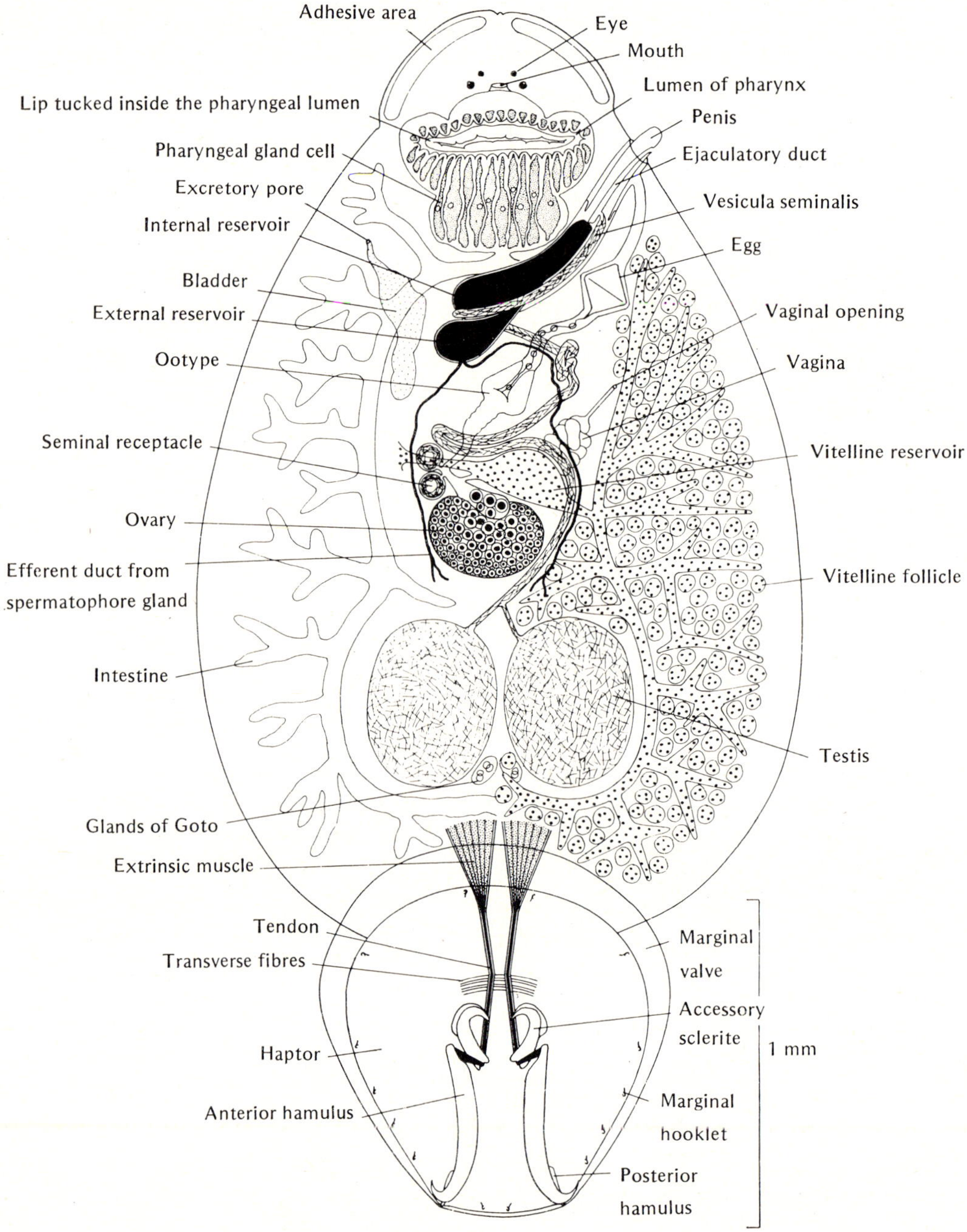


**FIG. 20.10**
Adult monogenean, *Entobdella soleae*. [From Kearn, 1971.]

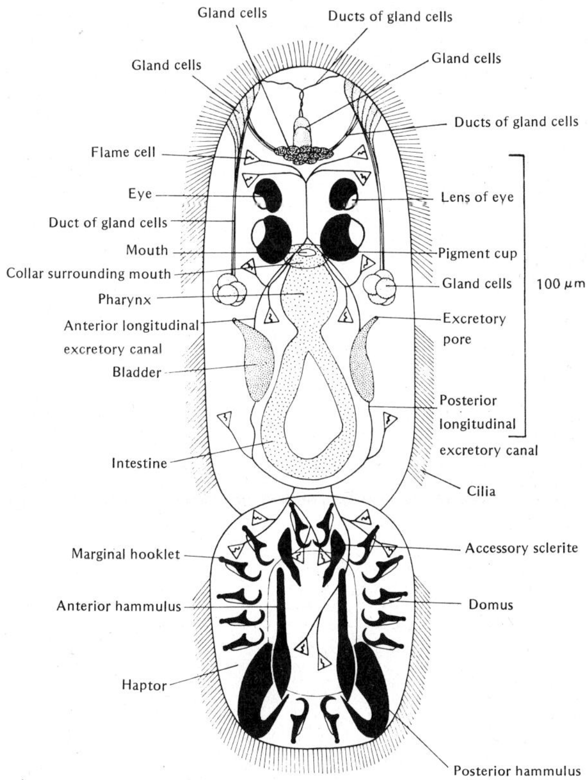


**FIG. 20.11**
Oncomiracidium of *Entobdella soleae.* [From Kearn, 1963b.]

CLASSIFICATION. The characteristics most often used in the class Monogenea are external features such as opisthaptor morphology and variations in male copulatory apparatus. This is in contrast to the class Trematoda, in which detailed examination of internal reproductive organs is required for classification. Good sources of information are Bychowsky (1957), Yamaguti (1963), and Schell (1970).

## SUBCLASS ASPIDOGASTREA

The subclass Aspidogastrea, also called Aspidobothrea and Aspidocotylea, contains a small number of species that have features of digeneans and monogeneans, but the digenean affinities are strongest. Aspidogastreans are primarily endo- and ectoparasites of

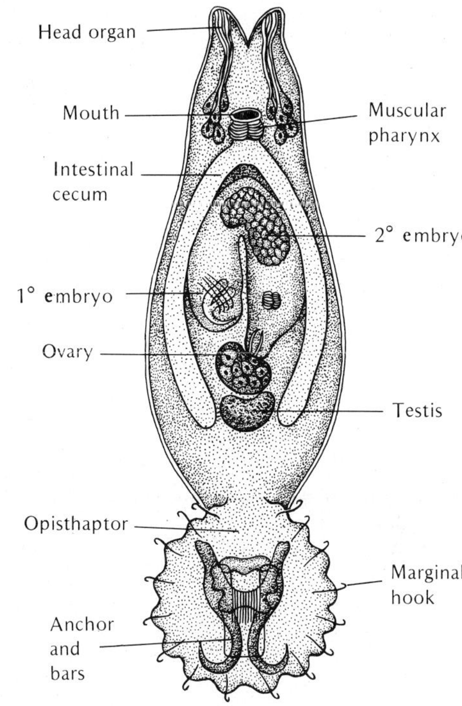


**FIG. 20.12**
Diagram of an adult *Gyrodactylus rhinichthius* containing two generations of developing embryos.

mollusks, but also are parasites of elasmobranchs, teleosts, and turtles. No aspidogastreans are known to be of medical or economic importance.

MORPHOLOGY. Adult aspidogastreans have a characteristic ventral disk with one to four rows of alveoli (Fig. 20.14), or a single row of suckers. Hooks are not present. The ventral disk is isolated from the dorsal body by a longitudinal septum, another characteristic structure that is composed of muscle and connective tissue. Reproductive organs are similar to those of the Digenea, with one exception—the oviduct is divided into compartments by numerous septa; this feature is apparently unique to the aspidogastreans.

LIFE CYCLE. The life cycles of the aspidogastreans are probably direct; one life cycle has been completed in the laboratory (Heuhner & Etges, 1977). Adults (Fig. 20.15) produce eggs; some have opercula and others do not. Eggs may be ingested or hatch, releasing a

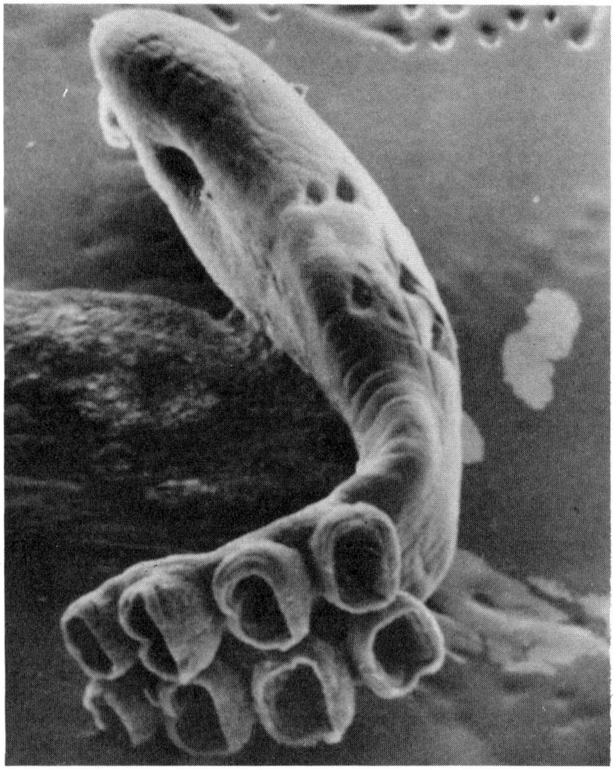

**FIG. 20.13**
Scanning electron micrograph of the polyopisthocotylean, *Octomacrum lanceatum.* [Courtesy of Sharon Haines.]

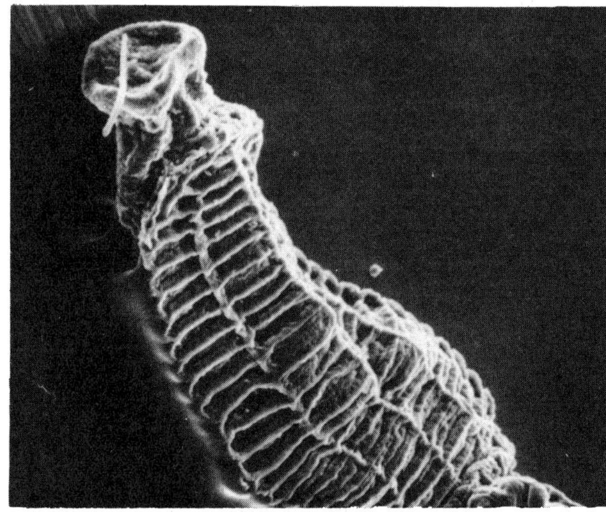

**FIG. 20.14**
Scanning electron micrograph of *Aspidogaster conchicola* adult. [Courtesy of Dr. Tyler A. Woolley.]

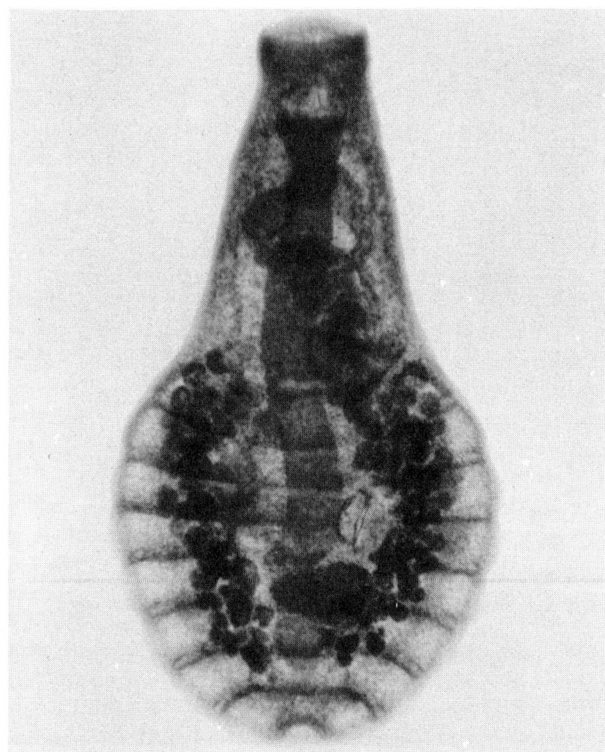

**FIG. 20.15**
Adult *Cotylapsis insignis.*

characteristic larva, the cotylocidium (Wootton, 1966), which has a posterior sucker (Fig. 20.16), and may be ciliated or not. The cotylocidium develops directly into the adult by growth of the posterior sucker into the adult ventral disk. Most aspidogastreans parasitize mollusks. Those in vertebrates may have been ingested while still inside the mollusks; however, *Lobatostoma ringens* remains immature in mollusks and becomes gravid only in the teleost host (Hendrix & Overstreet, 1977), and *Cotylogaster occidentalis* produces more viable eggs in fish than in mussels (Fredericksen, 1978). Thus, the tendency toward a two-host life cycle may be developing in some aspidogastreans.

Aspidogastreans have similarities to monogeneans, including direct life cycles without alternation of generations, and with digeneans, including morphologic similarities and having both mollusk and vertebrate hosts. Distinct aspidogastrean characteristics include the ventral disk, the longitudinal septum, the septate oviduct, and direct development in mollusks. On the basis of this, Rohde (1972) suggested that the Aspidogastrea have many archaic features that merit separate taxonomic status, but have enough features in common with the Digenea to imply that they are closely related and probably diverged from primitive Digenea. He further stated that the aspidogastreans are primitive parasites because they can survive for a long time in simple media outside a host, they have low host and site specificity, they have a complex nervous system, and there are only a few species. All of these characteristics are different in the successful parasitic groups such as the Digenea.

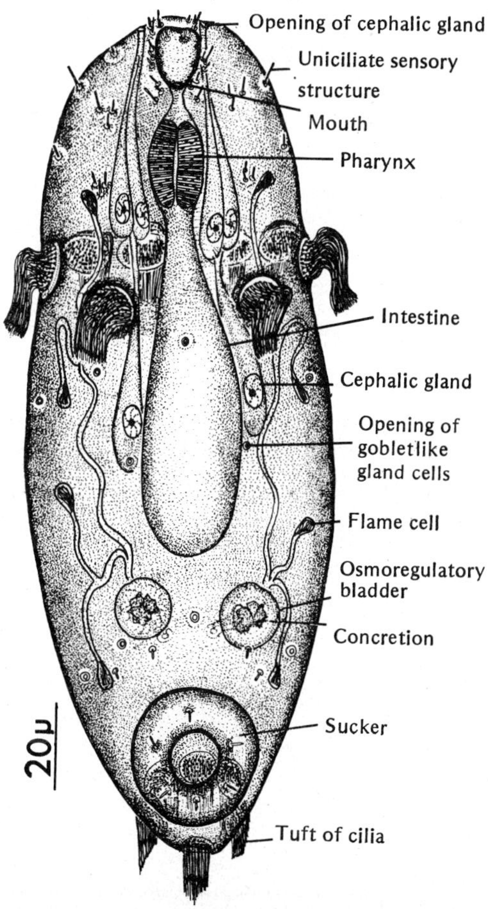


**FIG. 20.16**
Diagram of a *Cotylogaster occidentalis* cotylocidium. [Redrawn from Fredericksen, 1978.]

## GENERAL FEATURES OF DIGENETIC FLUKES

The subclass Digenea includes the flukes that are of greatest importance to humans because of either the medical or economic losses they cause. Included in the group are the blood flukes or schistosomes, which are generally considered to be the most serious helminth human parasite on a worldwide basis. In the case of livestock, the common liver fluke causes untold economic losses in sheep- and cattle-raising areas of the world. Also, a number of flukes (echinostomes, heterophyids) normally found in other animals occur in humans and sometimes cause severe illnesses.

EVOLUTION AND LIFE CYCLE. There is so much diversity in the structures of adults and life cycles of Digenea that students often cannot see the forest for the trees. In the first place, the major dichotomy in the Digenea is based on the way in which the excretory bladder forms in the final larval stage, the cercaria. This division into the superorders Epitheliocystidia and Anepitheliocystidia (LaRue, 1957) has been widely accepted, but except for specialists in fluke development, there are probably not many parasitologists who would recognize an excretory bladder in a developing cercaria. So we start off with an obscurity that most of us take on faith but that is crucial to comprehending the classification of the group. Unfortunately, this system also has serious problems. The cercaria of *Schistosoma mansoni* has an excretory atrium which is lined with epithelium when examined by electron microscopy (Powell, 1973) and is not anepitheliocystid, as proposed by LaRue. Second, the major recognition features of the adults do not always relate to taxonomic affinities (Fig. 20.17). Last, the life cycles include perhaps six larval stages, some of which reproduce asexually and most of which are found in snails.

The crucial step of making sense out of the group lies in the conclusion that gastropod mollusks (snails) were the original hosts of those flatworms that later became the Digenea. Among the many hundreds of life cycles of digenetic flukes that are known, only a small percentage use animals other than snails as their first intermediate hosts. In a few cases, bivalve mollusks, mostly freshwater clams of the family Sphaeriidae, rarely a scaphapod, and in one case a marine annelid, are first intermediate hosts.

The precursors of the Digenea were most likely free-living flatworms of the class Turbellaria

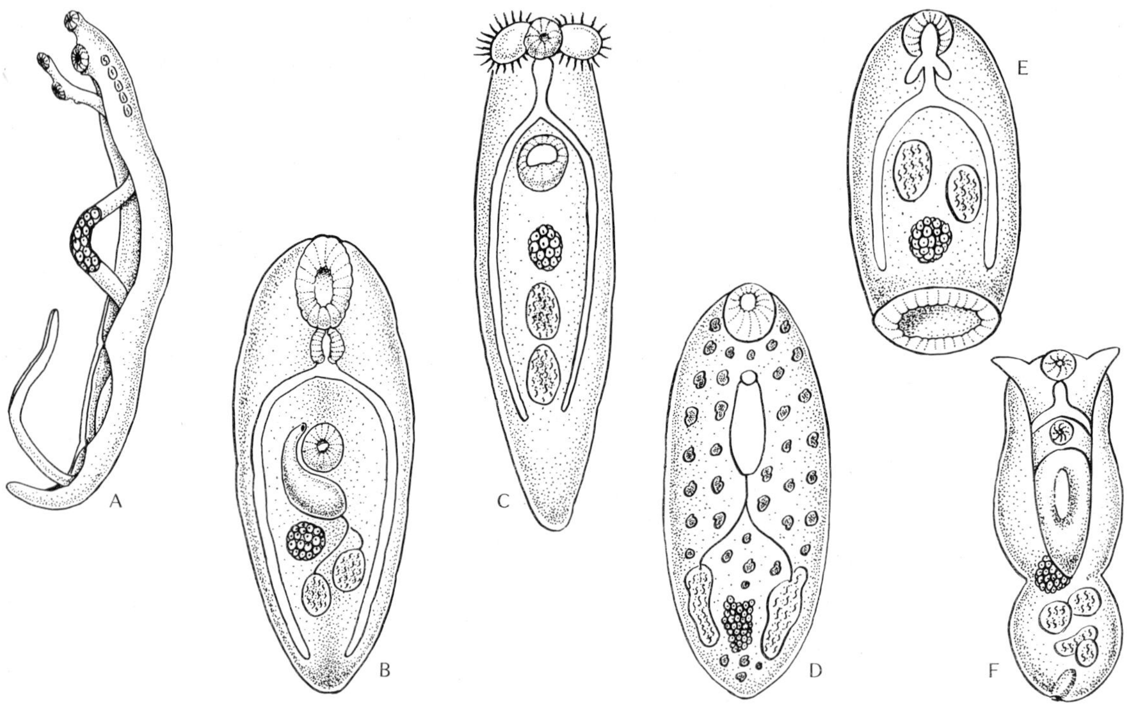


**FIG. 20.17**
Diagram of basic body types of adult digeneans. (*A*) schistosome; (*B*) distome; (*C*) echinostome; (*D*) monostome; (*E*) amphistome; (*F*) holostome.

(Llewellyn, 1965). Although some members of the Turbellaria, such as the Temnocephala, are parasitic, most are free-living and feed on decaying organic matter, both plant and animal. As detritus feeders, they feed on dead snails. It is not much of a step for such a phoretic symbiont to crawl into the mantle cavity or the pulmonary chamber of the snail and to find conditions where predators are few and food is adequate. Once having established a symbiotic relationship, the partnership would have evolved so as to damage the snail minimally or even to benefit it (Cheng, 1971).

The scenario which has been proposed by Pearson (1972) for the evolution of the life cycles of digenetic trematodes is briefly as follows. The forms from which the Digenea arose were most likely free-living rhabdocoel Turbellaria. These predaceous animals were ectocommensals in the mantle cavity of aquatic snails. The earliest parasites penetrated the integument and entered the viscera of their snail hosts.

The presence of the tailed, juvenile stage, the cercaria, in nearly all of the life cycles of the Digenea implies that the adult or preadult forms of these protoparasites escaped from the snail and moved away. Cercariae of present-day trematodes without tails have lost them secondarily. The free-living adults produced eggs that hatched to motile juveniles and these, in turn, sought out snails to serve as hosts (Fig. 20.18).

Thus, the juvenile entered the viscera of snails, became sexually mature, and then escaped to lay its eggs. There is ample precedent among parasites, especially arthropods, but also in some parasitic Turbellaria, for direct life cycles in which the preadults are parasitic and the adults are free-living.

The reasoning entailed in the addition of larval stages (sporocyst, redia) is somewhat more tenuous than the acquisition of a parasitic relationship with snails in the first place. Basically, there developed two generations of adults in a snail, the first parasitic and viviparous and the second free-living and oviparous. The life cycle, then, had established alternate generations—viviparous and oviparous. The viviparous gen-

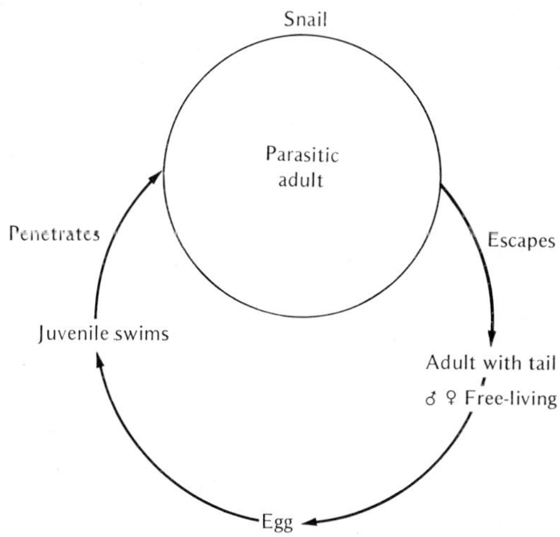


**FIG. 20.18**
Hypothetical one-host life cycle of a protodigenean.

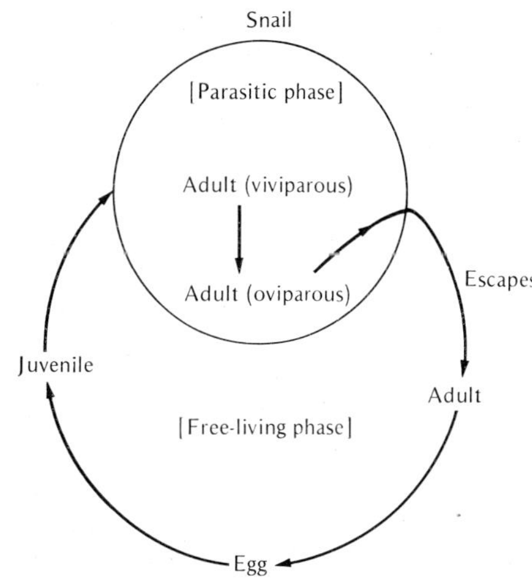


**FIG. 20.19**
Hypothetical origin of alternation of generations in the digenean life cycle.

eration subsequently lost the male system and its female system dedifferentiated into a fragmented ovary, now represented by scattered germ cells (Fig. 20.19). The gut was lost when feeding through the body surface became more efficient. The life cycle stage, which is the end product of the viviparous parasitic generation, is the sporocyst (Fig. 20.20). The means by which further larval generations (daughter sporocyst, redia) were added is highly conjectural and will not be considered here.

The redia is more complex than the sporocyst in that it has a gut, albeit simple, and usually some small tegumentary projections, the ambulatory buds (Fig. 20.21). Although the redial stage obtains nutrient through the body surface, it also ingests host tissues. As the life cycle progresses through its series of larvae, each succeeding stage becomes structurally and functionally more like the adult.

Having established a life cycle with alternation of generations, one parasitic and the other free-living, the next problem is to introduce the second host. It is most likely that the first chordate hosts were bottom-feeding fish such as the cyclostomes (jawless primitive fish). The near-ubiquity of tails on cercariae indicates that the free-living generation escaped from the snail to be eaten by the second host rather than while still in the snail. The swimming or creeping adults were preadapted to parasitism and found the intestinal tract of their cyclostome hosts to be similar to their free-living niche: low in oxygen, rich in organic matter, and dark. Those individuals capable of surviving and continuing to lay eggs in the gut of such hosts would have an adaptive advantage in having their eggs distributed over a wide area and would be protected from predation by other animals. By now the ability to absorb food through the tegument may have been well established. Obligatory parasitism was just a step away. Then came all the trappings of membership in the fraternity of parasites: holdfast structures, close ecologic and physiological dependence on a host, and evolution with the host.

With such an evolutionary perspective, the present-day digenetic trematode life cycle gains a degree of rationality. The basic stages are egg, miracidium, sporocyst, redia, cercaria, usually metacercaria, and adult (Fig. 20.22). Egg, miracidium, cercaria, and adult occur in all life cycles, but there is considerable variation in the sporocyst and redial and metacercarial generations. For example, sporocysts can give rise to daughter sporocysts or rediae or cercariae. In some instances, the sporocyst generation has dropped out and the first stage in the snail is a redia, which may give rise to daughter rediae and then to cercariae. The principal points to remember are that both sporocysts and rediae are reproductive and the reproduction is asexual. For any one miracidium entering a snail, a few hundred to several thousand cercariae will be produced. Additional hosts may be added in the form of second and third intermediates, but reproduction takes place

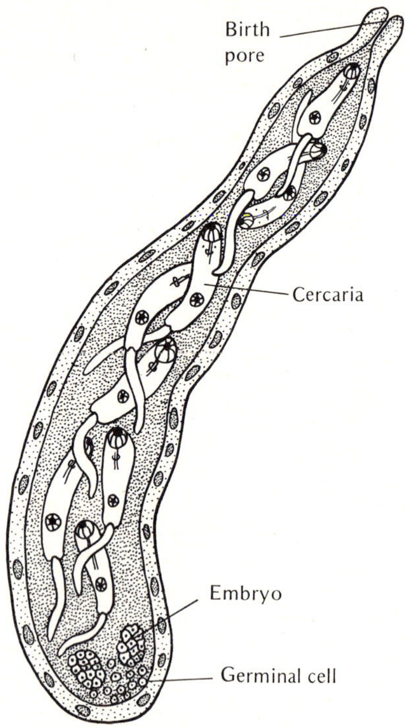


**FIG. 20.20**
Diagram of a sporocyst.

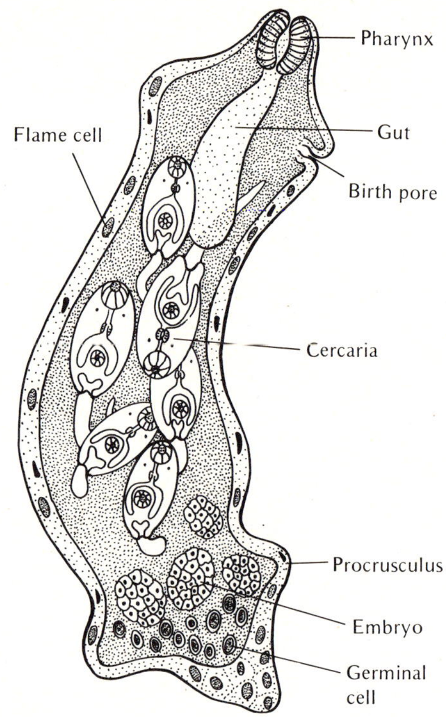


**FIG. 20.21**
Diagram of a daughter redia.

only in the snail and in the final or definitive host. If additional intermediate hosts are added to the cycles, they fit into a food chain, and they either concentrate preadults or move them up to a trophic level where there is a high probability that they will be eaten by the proper definitive host.

ADULT STRUCTURE. Adult flukes are nearly all monoecious, that is, they have both male and female reproductive systems in the same individual. Most are macroscopic, but a few are so small that a dissecting microscope is needed to find them, or at least to see them well enough to distinguish them from extraneous material. In life they are usually flattened, although some, such as the amphistomes (Fig. 20.17), are rotund. Some structures can be seen when the animals are alive, but fixed and stained preparations are essential for accurate identification.

In a stained, sexually mature digenetic trematode, the features that are most striking are the intestine, ventral sucker or acetabulum, and the reproductive systems (Fig. 20.23). The gut is incomplete (no anus); the mouth is most often located at the anterior end and is surrounded by an oral sucker. There may or may not be a muscular pharynx followed by an esophagus. The intestinal ceca are most often thin-walled structures extending through the greater length of the body. In some flukes the ceca are branched (*Fasciola*) (see Fig. 22.1) or unite near the posterior to form a loop (Cyclocoelidae); in a very few instances (*Opecoezina*) there is an anus.

The ventral sucker or acetabulum is an organ of attachment and does not serve in feeding. Whether or not it is present, its size and location are important in the identification of any trematode.

The male reproductive system is relatively simple. It usually consists of two testes, each with a vas efferens which may join to form the vas deferens as they extend anteriorly. The male system terminates at the cirrus, the intromittent organ. Sperm coming from the vas deferens are stored in the seminal vesicle and transferred through the protrusible cirrus to the genital pore of the female system. Variations in the male system used in classification and identification are the number, size,

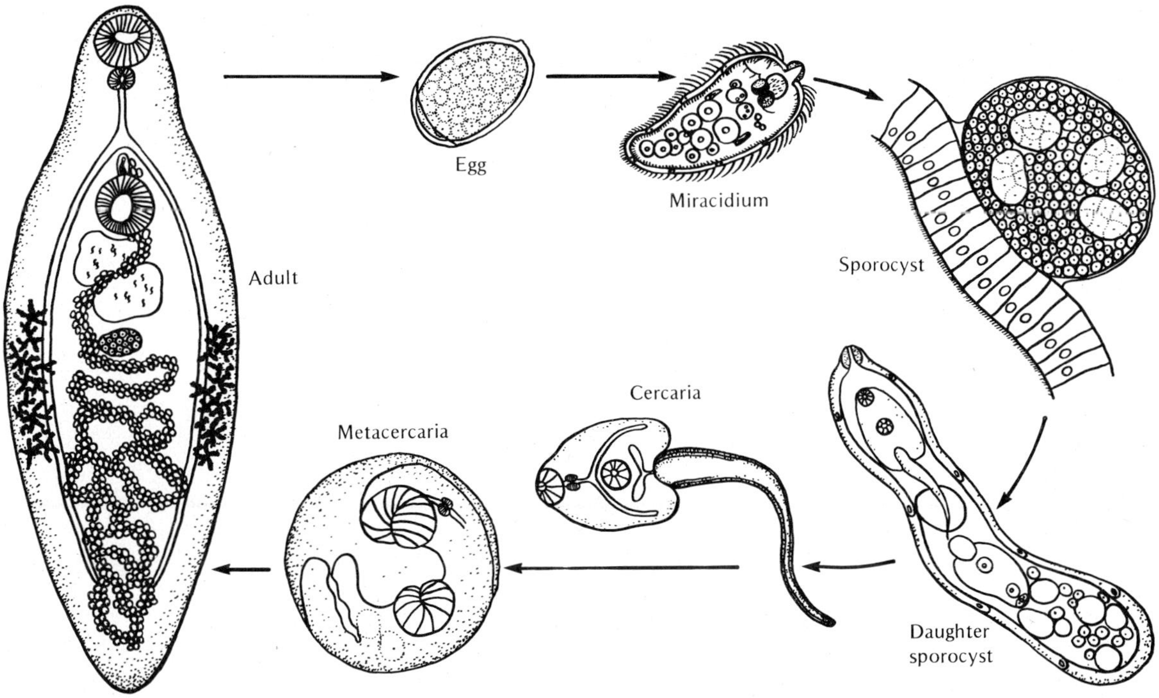


**FIG. 20.22**
Diagram of the life cycle of a digenetic trematode.

shape, and location of the testes, the presence or absence of the cirrus sac, as well as whether the seminal vesicle is external or internal (Fig. 20.23).

The female system has a number of components in which there is considerable variation from group to group. The crux of the whole system is the ootype into which the products of the ovary, vitelline glands, and Mehlis' gland are introduced and where fertilization takes place (Fig. 20.23). Sperm are stored in a seminal receptacle until needed. Laurer's canal extends from the surface of the animal to the oviduct and represents a vestigial vagina. Laurer's canal is no longer used in copulation, but sperm and vitelline secretions might be excreted through it. The ovary is usually single and the ova move from it to the ootype, where they are fertilized.

The vitellaria are located at the lateral margins of the body and their cells produce yolk, which serves as energy storage for the eggs, and a proteinaceous material, which will become one of the egg capsules. The products of the vitelline glands are carried by ducts that join the lateral vitelline ducts at about the level of the ovary. The lateral vitelline ducts either enter the ootype directly or enter a vitelline reservoir where the material is stored. For purposes of identification, the location of the vitelline glands and how they are clustered are both important.

Mehlis' gland is a rather diffuse group of secretory cells surrounding the ootype and with openings into the ootype; its function is not known with certainty.

Following fertilization, the egg shell or capsules are laid down around the zygote. The size and shape of the egg are determined at this time. As the egg moves into the uterus, further changes take place in the eggshell, most likely by quinone tanning of the protein layer (Smythe, 1966).

The uterus is most often a long, winding, looping structure which contains hundreds or sometimes thousands of eggs. Eggs go from the ootype into the uterus and move forward until they are released from the genital pore in the region of the cirrus. Sperm from the male system of the same individual (Fig. 20.24) or from another individual enter the genital pore and pass the full length of the uterus to be stored in the seminal

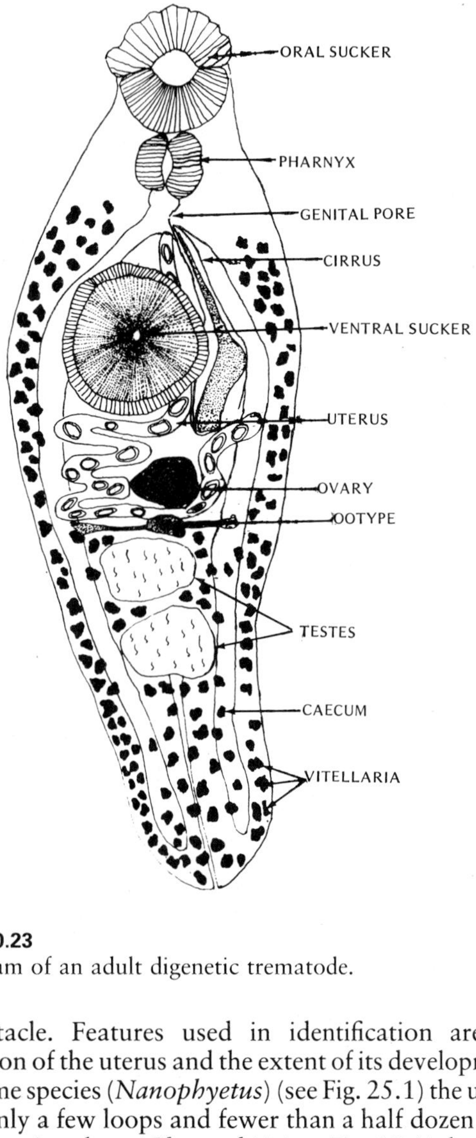


**FIG. 20.23**
Diagram of an adult digenetic trematode.

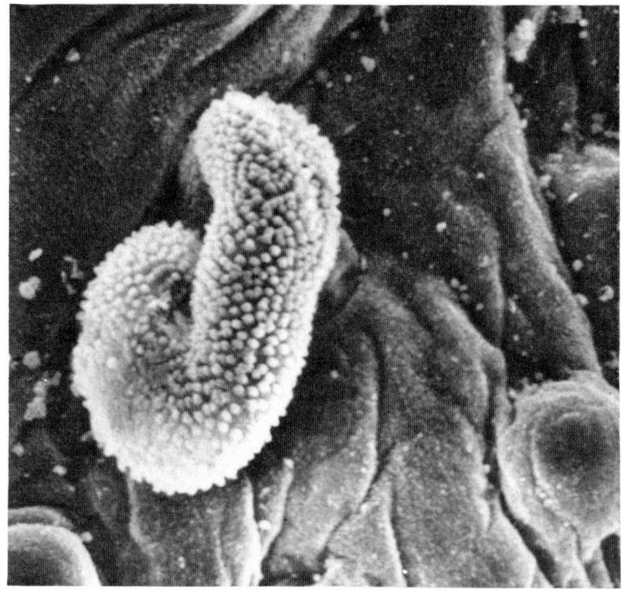

**FIG. 20.24**
A scanning electron micrograph of self-fertilization. The cirrus is protruding from and then reentering the genital pore.

receptacle. Features used in identification are the location of the uterus and the extent of its development. In some species (*Nanophyetus*) (see Fig. 25.1) the uterus has only a few loops and fewer than a half dozen eggs, whereas in others (*Clonorchis*) (see Fig. 23.1) there are many loops and more than a hundred eggs.

In *Dicrocoelium* (see Fig. 22.5) the uterus loops posteriorly (descending loop) before it turns anteriorly (ascending loop), also a constant characteristic. The presence or absence of a muscular structure, the metraterm, near the opening of the uterus is also a constant and identifying characteristic.

About the only portion of the excretory system that can be seen in stained, adult flukes is the excretory bladder near the posterior end. The bladder is usually seen as a vacant or poorly stained area surrounded by parenchyma; its size and shape are used in identification.

Body shape, spination, and the oral and ventral suckers are all important in identification of adult flukes but do not necessarily relate to the taxonomic position of any one individual. Early systems of classification were based on whether there was an oral sucker only (monostomes) or also a ventral sucker (distomes). The location of the oral and ventral suckers determined whether the animal was a gasterostome (mouth in the middle) or an amphistome (suckers at either end). This terminology is not only entrenched but also useful because one can refer to an echinostome and immediately produce a mental picture of a fluke with spines surrounding the mouth (see Fig. 24.3). The echinostomes represent the extreme of spination in the digenea; identifying characteristics are the presence or absence of spines on the body and their location. Sometimes it is necessary to examine the margins of an animal carefully under a compound microscope in order to see the spines.

CERCARIAE Cercariae are easy to collect by simply crushing snails or allowing them to shed cercariae

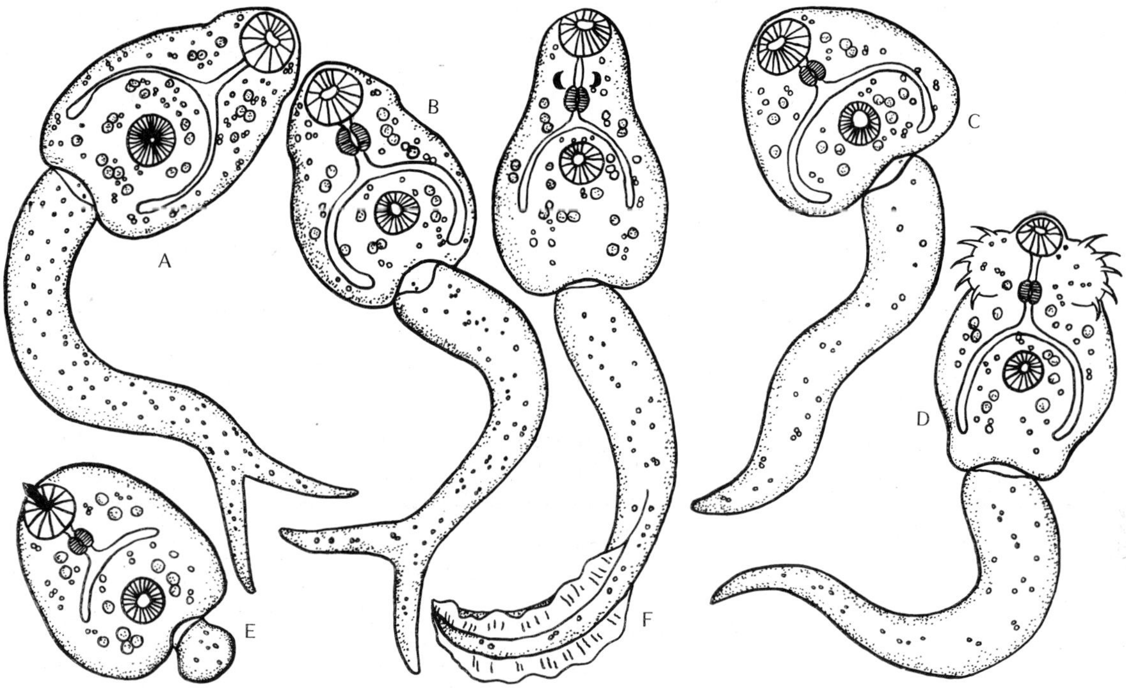


**FIG. 20.25**
Diagram of cercarial types. (*A*) apharyngeate furcocercous; (*B*) pharyngeate furcocercous; (*C*) gymnocephalus; (*D*) echinostome; (*E*) xiphidiomicrocercous; (*F*) opthalmopleurolophocercous.

naturally. Because cercariae have many morphological characteristics similar to those of adult flukes, the cercarial "types" often provide valuable information about the trematode fauna in a certain locality. Representative cercarial types are illustrated in Fig. 20.25.

## CLASSIFICATION OF THE PHYLUM PLATYHELMINTHES

**Phylum Platyhelminthes**
Dorsoventrally flattened Bilateria, without coelom, anus, circulatory, respiratory, or skeletal systems; with flame cell protonephridia, parenchyma filling all spaces between organ systems.

**Class Turbellaria**
Tegument cellular or syncytial, usually provided with rhabdoids, body undivided, intestine present except in Acoela, life cycles simple, most free-living, some ecto- or endocommensals or parasitic.

**Class Monogenea**
Oral sucker absent or weak, anterior end usually with a pair of adhesive structures, posterior end with an adhesive disk usually provided with hooks and/or clamps, gut usually simple, all parasitic on aquatic vertebrates, especially fishes, or rarely on aquatic invertebrates.

**Subclass Monopisthocotylea**
Opisthaptor single but may be subdivided by septa and usually developed directly from larval haptor, one to three pairs of large anchors plus marginal hooklets, oral sucker absent, genitointestinal canal absent, eyes often present.
*Gyrodactylus*, *Entobdella*

**Subclass Polyopisthocotylea**
Opisthaptor with multiple suckers, clamps, or anchor complexes, larval haptor reduced or absent, mouth usually surrounded by suckers, genitointes-

tinal canal present, eyes usually absent.
*Octomacrum*

**Class Trematoda**
Tegument syncytial without rhabdoids, body undivided, one or more suckers, digestive tract with intestine commonly branched, all parasitic, mainly in digestive tract of vertebrates.

**Subclass Digenea**
All endoparasitic, two or more hosts in life cycle, the first host usually a mollusk, larval polyembryony, most with two suckers including an oral sucker, no hooks.

**Superorder Anepitheliocystida**
Excretory bladder not epithelial, stylet absent in cercariae.

**Order Strigeata**
Cercariae usually fork-tailed.
*Schistosoma*

**Order Echinostomata**
Cercariae with single tail.
*Echinostoma*, *Fasciola*

**Superorder Epitheliocystida**
Excretory bladder epithelial, oral stylet common.

**Order Plagiorchiata**
Cercariae without caudal excretory vessels, stylet present or absent.
*Dicrocoelium*, *Nanophyetus*, *Paragonimus*

**Order Opisthorchiata**
Cercariae with caudal excretory vessels during development, stylet absent.
*Clonorchis*

**Subclass Aspidogastrea**
Oral sucker absent, anterior end without paired adhesive structures, with a large ventral sucker divided by septa or one row of suckers, no hooks, ecto- and endoparasites with direct life cycles, without metamorphosis, parasites of mollusks, fishes, and turtles.
*Cotylogaster*

**Subclass Didymozoidea**
Tissue-dwelling parasites of fishes, tubular gonads, no life cycles known.

**Class Cestoidea**
Endoparasites without rhabdites, body usually segmented, adhesive structures limited to anterior end except in Cestodaria, no mouth or digestive tract, embryo with hooks, life cycle usually has two or more hosts, most definitive hosts are vertebrates.

**Subclass Cestodaria**
Body undivided, no scolex, one set of reproductive organs, ten-hooked larva.

**Subclass Eucestoda**
Body long and usually divided, one or more sets of reproductive organs per proglottid, scolex usually present, six-hooked larva.
*Taenia*, *Echinococcus*, *Hymenolepis*

## REFERENCES AND FURTHER READINGS

### Trematoda

Cheng, T. C. (ed.). 1971. *Aspects of the Biology of Symbiosis.* University Park Press, Baltimore.

———. 1973. *General Parasitology.* Academic Press, New York.

Dawes, D. 1956. *The Trematoda.* Cambridge University Press, New York.

Erasmus, D. A. 1972. *The Biology of Trematodes.* Crane, Russak & Co., New York.

LaRue, G. 1957. The classification of digenetic Trematoda: A review and a new system. *Exp. Parasitol.* **6**:306–349.

Llewellyn, J. 1965. The evolution of parasitic platyhelminthes. In A. E. R. Taylor (ed.), *Evolution of Parasites,* 3rd Symp. Soc. Parasitol. Blackwell, Oxford, pp. 47–78.

Noble, G. A. 1967. A forty-foot fluke. *J. Parasitol.* **53**:645.

Pearson, J. C. 1972. A phylogeny of life-cycle patterns of the Digenea. *Adv. Parasitol.* **10**:153–189.

Powell, E. C. 1973. Studies on the excretory "bladder" and caudal ducts of the supposed anepitheliocystid cercaria of *Schistosoma mansoni. Z. Parasitenk.* **43**:43–52.

Schell, S. C. 1970. *How to Know the Trematodes.* Wm. C. Brown, Dubuque, Iowa.

Smyth, J. D. 1966. *The Physiology of Trematodes.* W. H. Freeman & Co., San Francisco.

Yamaguti, S. 1958. *Systema Helminthum.* Vol. 1. Digenetic Trematodes of Vertebrates, Parts 1 & 2. Interscience Press, New York.

———. 1971. *Synopsis of Digentic Trematodes of Vertebrates.* Vols. 1 & 2. Keigaku, Tokyo.

### Monogenea

Bychowsky, B. E. 1957. *Monogenetic trematodes, their Systematics and Phylogeny* (trans. from Russian). W. J.

Hargis, Jr. (ed.). American Institute of Biological Sciences., Washington, D.C.

Hendrix, S. S., and R. M. Overstreet. 1977. Marine aspidogastrids (Trematoda) from fishes in the northern Gulf of Mexico. *J. Parasitol.* **63**:810–817.

Hyman, L. H. 1951. *The Invertebrates: Platyhelminthes and Rhynchocoela, the Acoelomate Bilateria.* Vol. II. McGraw-Hill, New York.

Kearn, G. C. 1963a. The life cycle of the monogenean *Entobdella soleae*, a skin parasite of the common sole. *Parasitology.* 53:253–263.

——— 1963b. The egg, oncomiracidium and larval development of *Entobdella soleae*, a monogenean skin parasite of the common sole. *Parasitology.* 53:435–447.

——— 1967. Experiments in host finding and host specificity in the monogenean skin parasite *Entobdella soleae. Parasitology.* 57:585–605.

——— 1971. The physiology and behavior of the monogenean skin parasite *Entobdella soleae* in relation to its host (*Solea solea*). In A. M. Fallis (ed.), *Ecology and Physiology of Parasites.* University of Toronto Press, Toronto, pp. 161–187.

——— 1973. An endogenous circadian hatching rhythm in the monogenean skin parasite, *Entobdella soleae*, and its relationship to the activity rhythm of the host (*Solea solea*). *Parasitology.* **66**:101–122.

Llewellyn, J. 1963. Larvae and larval development of monogeneans. *Adv. Parasitol.* **1**:287–326.

——— 1965. The evolution of parasitic platyhelminths. In A. E. R. Taylor (ed.), *Evolution of Parasites.* 3rd Symp. Br. Soc. Parasitol. Blackwell, Oxford, pp. 47–78.

——— 1968. Larvae and larval development of monogeneans. *Adv. Parasitol.* **6**:373–383.

——— 1970. Monogenea. *J. Parasitol.* **56**:493–504.

Lyons, K. M. 1973. The epidermis and sense organs of the Monogenea and some related groups. *Adv. Parasitol.* **11**:193–232.

Price, E. W. 1938. North American monogenetic trematodes. II. The families Monocotylidae, Microbothriidae, Acanthocotylidae and Unonellidae (Capsaloidea). *J. Wash. Acad. Sci.* **28**:183–198.

———. 1943. North American monogenetic trematodes. VII. The family Discocotylidae (Diclidophoroidea). *Proc. Helm. Soc. Wash.* **10**:10–15.

Rohde, K. 1975. Fine structure of the Monogenea, especially *Polystomoides* Ward. *Adv. Parasitol.* **13**:1–33.

Yamaguti, S. 1963. *Systema Helminthum.* Vol. IV. *Monogenea and Aspidocotylea.* Interscience Press, New York.

**Aspidogastrea**

Fredericksen, D. W. 1972. Morphology and taxonomy of *Cotylogaster occidentalis* (Trematoda: Aspidogastridae). *J. Parasitol.* **58**:1110–1116.

——— 1978. The fine structure and phylogenetic position of the cotylocidium larva of *Cotylogaster occidentalis* Nickerson 1902 (Trematoda: Aspidogastridae). *J. Parasitol.* **64**:961–976.

Heuhner, M. K., and F. J. Etges. 1977. The life cycle and development of *Aspidogaster conchicola* in the snails *Viviparus malleatus* and *Goniobasis livescens. J. Parasitol.* **63**:669–674.

Rohde, K. 1972. The Aspidogastrea, especially *Multicotyle purvisi* Dawes, 1941. *Adv. Parasitol.* **10**:78–151.

Wootton, D. M. 1966. The cotylocidium larva of *Cotylogaster occidentalis* (Nickerson, 1902) Yamaguti 1963 (Aspidogasteridae, Aspidocotylea, Trematoda). *Proc. 1st Congr. Parasitol.* **1**:547–548.

# 21
# *The Blood Flukes or Schistosomes*

The schistosomes are members of the order Strigeatida, but are different from most members of the order and other digenea in that the sexes are separate. The term *schistosome* or *Schistosoma* means split body and refers to the fact that the males have a ventral groove called a *gynecophoral canal.* Adults of the family Schistosomatidae are found in the blood vessels of vertebrates, both birds and mammals. A related family, Spirorchiidae, has turtles as definitive hosts and is also dioecious.

The blood flukes are important to humans primarily as the cause of medical problems in many tropical and subtropical areas of the world, but also as economically important parasites of cattle and other large domesticated animals.

As parasites of humans or other animals, the blood flukes lie in venules, usually in the lower abdomen, where they copulate, and the females lay eggs which reach the outside in either feces or urine. There is only a single intermediate host, a snail, and the fork-tailed cercariae reach the definitive host by actively penetrating the unbroken skin.

Current estimates, based on the consensus of many authorities, are that 200 million persons are infected with members of the genus *Schistosoma*, the blood flukes. This number has remained fairly constant for some years. There have been local successes in controlling the infections, but at the same time, the prevalence of infection has increased in other areas.

It should always be kept in mind that infection and disease are not the same. There is probably no infectious disease in which infection invariably progresses to severe disease. There is instead a spectrum of responses that ranges from inapparent infection through mild clinical signs and symptoms to death. Schistosomiasis fits this general pattern, and in addition, the infections in humans are frequently present for the greater part of a person's life. The effect on a person is directly related to the number of worms harbored, but the effect may be modified by the species and strain of the worm and the person's age, nutritional status, and previous exposure to the parasites. In most cases, there is repeated exposure over a period of years and long-term, chronic disease results. Therein lies the problem of assessing the importance of schistosomiasis in human populations. There may be periods in an individual's life when there is incapacitation, but by and large the infections result in long-term damage to internal organs: large intestine, bladder, liver. The person probably will continue to work, although productivity may be reduced, and the days lost from work will be greater than for a comparable uninfected individual. Death may not be a direct result of blood fluke infection but rather of other infectious agents attacking a body already debilitated.

The human schistosomes are distributed through some of the countries of the world at the lower end of the economic scale (Fig. 21.1), or at least the parasites are found in relatively poor rural people. Many countries of tropical and subtropical Africa that already have severe economic problems are further handicapped by widespread infection among their farmers. In their review of the problems of assessment of the effects of schistosomiasis, Macdonald and Farooq (1973) quoted Gunnar Myrdal: "Men and women were sick because they were poor; they became poorer because they were sick, and sicker because they were poorer." Macdonald and Farooq further discussed the effects of infection with members of the genus *Schistosoma* and pointed out that in parts of South America, 12% of the deaths in hospitals were due to the consequences of schistosomiasis. In Tanzania 20% of persons in some areas have serious damage to the urinary system and can be expected to live only a few years. These researchers further stated that in Tanzania, Zanzibar, Nigeria, and Egypt the pathological changes in children are likely to have their most serious effects in adolescence and early adulthood, just when they are likely to have completed schooling and are ready to become productive members of society.

In addition to death losses and reduced ability to work, there are also costs of medical care. An example comes from a single irrigated property in Tanzania where there were 2300 employees and 84% of field workers were infected with schistosomes. The cost of treatment of these persons was $12,810 (U.S.) or £4124 (British). In the Philippines, the annual cost of schistosomiasis was estimated at $2 million (U.S.) or 13 million pesos (Philippine).

At current estimates of world population, 1 person in 20 is infected with blood flukes. But in determining the impact on a people, it is necessary to investigate a specific area in detail, to determine not only the prevalence of infection but debility and death losses as well.

MORPHOLOGY. The important members of the genus *Schistosoma* found in humans are *S. haematobium, S. mansoni*, and *S. japonicum*. They are slender worms about 16 mm long (Table 21.1), which can be identified by both male and female characteristics (Fig. 21.2). The females are slender, round in cross section, and 15 to 30% longer than the males. The males have a lateral extension of the body posterior to the ventral sucker called the *gynecophoral canal* in which the female worms lie (Fig. 21.3).

The eggs can be distinguished from one another on the basis of both size and other characteristics (Table 21.1). The most distinctive feature used for identification is the spine (Fig. 21.4). In *S. haematobium* the spine is terminal, in *S. mansoni* it is subterminal, and in *S. japonicum* it is a small, subterminal knob that may be lacking entirely in some populations.

Although there are differences among the species, in cercariae they cannot be distinguished from one another with certainty. Cercariae have a body about 180 μm long and a forked tail (Fig. 21.5). There are oral and ventral suckers, a genital primordium in the posterior portion of the body, and a protonephridial system consisting of four or five pairs of flame cells. The intestinal system consists of the mouth, esophagus, and rudimentary intestinal ceca. Three kinds of glands have been noted in the cercariae: escape glands located in the anterior portion of the body, preacetabular glands, and postacetabular glands. The escape glands allow the cercariae to penetrate the epithelium of the snail host in order to reach the outside. The latter two sets of glands secrete substances that aid in the penetration of the definitive host and are discussed in somewhat more detail in the section on life cycle.

Not all furcocercous or forked-tail cercariae are members of the genus *Schistosoma*. Those in the family Schistosomatidae have both oral and ventral suckers and lack both eyespots and a pharynx.

LIFE CYCLE. The development of the schistosomes of humans follows a pattern that is similar for all three species (Fig. 21.6). The male and female adults in the definitive host move against the blood flow into small venules until they can move no farther. Egg laying begins and may continue for a decade or more. The eggs move toward the lumen of the organ, probably by a combination of muscular action of the organ and release of enzymes through pores in the eggshells (Demaree & Hillyer, 1981) (Fig. 21.7). They are laid undeveloped, but have fully formed miracidia within several days. They reach the lumen of the organ, either gut or bladder, and pass to the outside with feces or urine.

The eggs are ready to hatch immediately on reaching the outside and do so mostly under the influence of lowered osmotic pressure; eggs hatch at 0.1% NaCl, but are almost completely inhibited from hatching at 0.6% NaCl. Eggs hatch well at 28°C, but are inhibited from hatching at 4°C and 37°C. Light also is a stimulus for hatching. It is said that the miracidia crack the outer membranes of the eggs, but there is little information on the mechanism of hatching.

The miracidia that escape from the egg capsules swim in the water and actively seek snails to penetrate. Upon

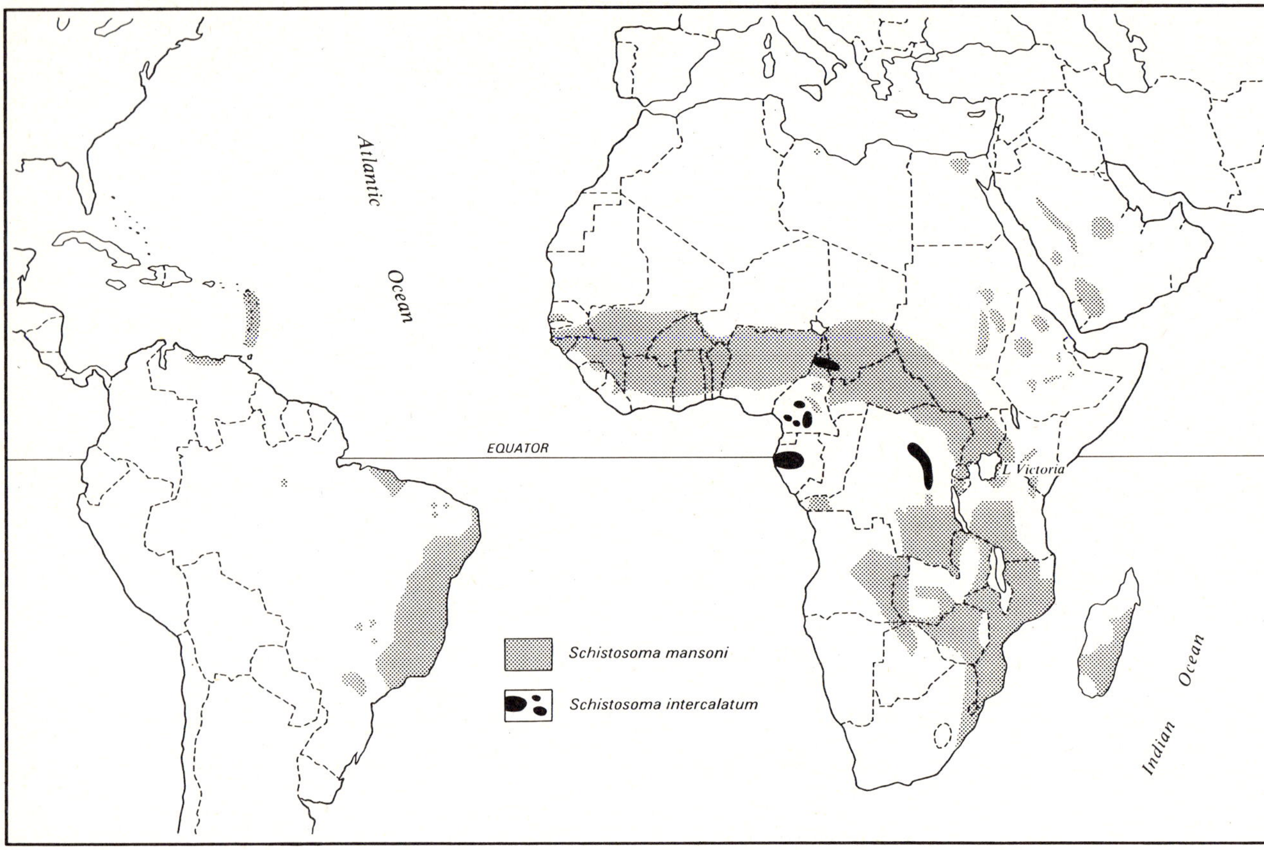


**FIG. 21.1**
Geographical distribution of human schistosomiasis. [Redrawn with modifications from Ansari, 1973.]

finding a suitable snail, the miracidia penetrate the soft tissues of the foot; in the case of the blood flukes, unlike some other digenea, the whole miracidium enters the body of the snail (Fig. 21.8). The mechanism of penetration has not been well worked out. After penetration, the ciliated cells disappear, and transformation into the first generation or mother sporocyst takes place.

The first-generation sporocysts generally remain in the foot of the snail near the point of penetration; they are nonmotile, convoluted sacs less than 1 mm long.

Within 2 to 6 weeks, daughter or second-generation sporocysts form in the central cavities of the mother sporocyst. In *S. mansoni*, 200 to 400 daughter sporocysts form in each mother sporocyst. At this time, the mother sporocyst degenerates, releasing the daughter sporocysts, which migrate into the digestive gland or ovotestis of the snail (Fig. 21.9). The daughter sporocysts are somewhat larger than the mother sporocysts, are motile, and have a spiny snout at the anterior end.

Cercariae form in the central cavity of the sporocyst and leave through a pore at the posterior end. Cercariae may be produced as early as 20 days following infection, but usually in 4 to 7 weeks, and they continue to be produced for more than a year. The normal life span of an infected snail appears to be 1 to 2 months. Production of cercariae depends on the size of the snails; the larger the snail, the greater the production of cercariae. The table on page 229 shows reproduction in

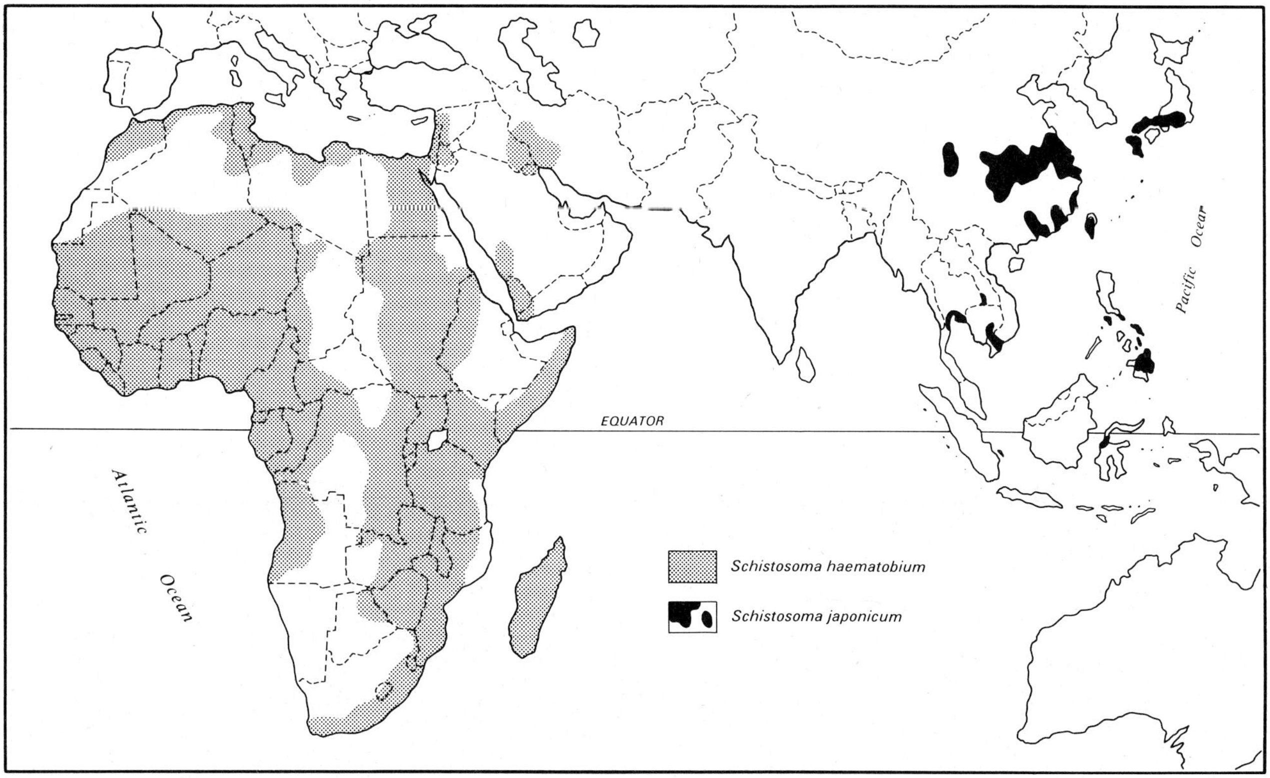


**FIG. 21.1 Continued**

the snail, and also the differences in cercarial production depending on species and location.

In order to give some idea of the contamination of water with cercaria, let us take the first example of *S. mansoni*, and assume that there are 100,000 snails in the vicinity of a community. Then we can see what the contamination of the area might be. A number of surveys on the prevalence of infection in snails indicate that 10% prevalence may be expected in some places. This means that 10,000 snails in the area are infected. If we take 1000 cercaria per day as the low end of production of cercariae, there are 10 million cercariae in the water at any one time. At the high end, 30 million cercariae are ready to infect persons entering the water. Considering the numbers, tipping the balance in favor of the parasite even slightly is likely to bring about an overall increase in the worm burdens of members of the community. We will look at this problem in some detail in the section on epidemiology and control.

The cercariae tend to accumulate in the area of the

| SNAIL | PARASITE | LOCATION | CERCARIA PRODUCE DAILY | CERCARIA PRODUCE TOTAL |
|---|---|---|---|---|
| *Biomphalaria glabrata* | *S. mansoni* | Brazil | 1000 – 3000 | 30,000 – 180,000 |
| *Biomphalaria globosus* | *S. haematobium* | West Africa | 400 | 12,000 – 24,000 |
| *Oncomelania* | *S. japonicum* | Japan | 15 – 150 | 450 – 9 000 |

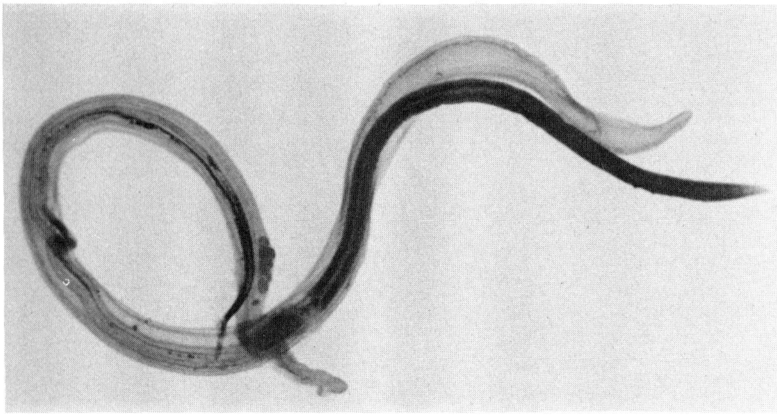

**FIG. 21.2**
A male and female of *Schistosoma mansoni.*

mantle collar of the snail, that is, close to the head of the snail in the region where escape to the outside is simple. The cercariae break out of the epithelium and reach the water. Unicellular escape glands located in the anterior portion of the body are seen only in the unemerged cercariae and presumably are instrumental in the organism's breaking out of the snail tissue. As is true for many trematodes, the cercariae of the blood flukes tend to leave the snail at certain times of the day: *S. mansoni* and *S. haematobium* between 9:30 A.M. and 2:00 P.M. and *S. japonicum* between 10:30 P.M. and 2:00 A.M.

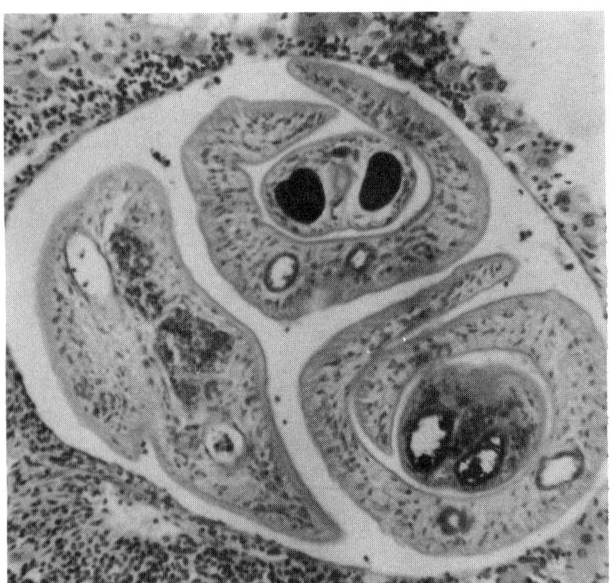

**FIG. 21.3**
Schistosome pairs in cross section. Note the round females in the gynecophoral canals of the males.

The function of the cercaria is to move from the intermediate host through a hazardous environment to the definitive host. The cercariae swim in the water by means of their tails and actively seek a host to penetrate. They can survive about a day, but infectivity declines in a few hours if they fail to find a host. When they contact human skin, cercariae are stimulated to penetrate by warmth, skin surface lipids, and perhaps light. They shed their tails and begin the process.

Penetration takes place using the secretions of the pre- and postacetabular glands. The postacetabular glands secrete a substance that is probably a mucopolysaccharide. This material spreads out from the oral sucker, allows the attachment of the organism to the skin, and may prevent the diffusion away of enzymes. The cercariae often enter the unbroken skin through hair follicles or sebaceous glands, but the exact mechanism of penetration is still unresolved. There is evidence that enzymes are secreted, but Smyth (1966) and Stirewalt (1974) remained equivocal whether these substance actually cause lysis of the skin. The cercariae penetrate the skin at least partly through the muscular action of the oral sucker. Once the organism is in the skin, the preacetabular glands secrete a substance that causes a softening of the skin's keratin layer; Stirewalt (1974) explained the breakdown of keratin as being due to the alkalinity of the secretion.

The time required to begin penetration of the skin is remarkably short. As little as 3 min exposure in water

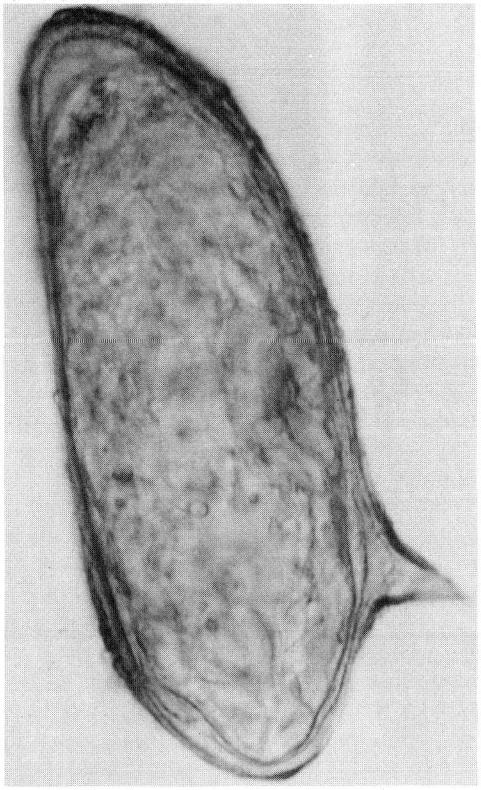

A

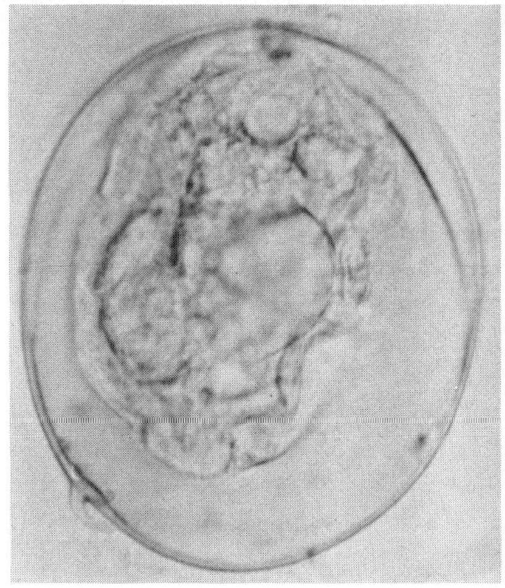

B

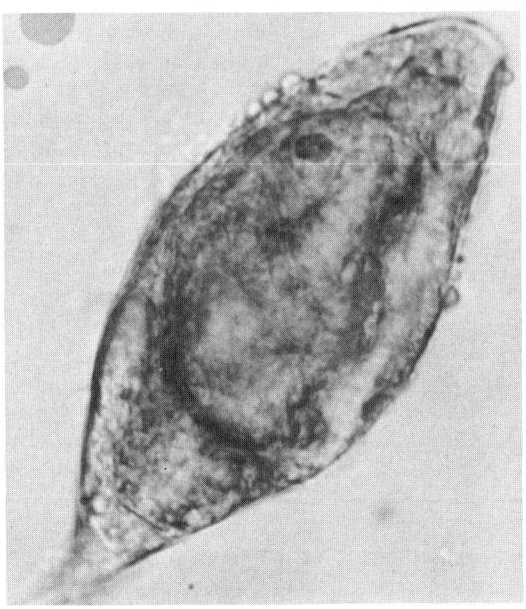

C

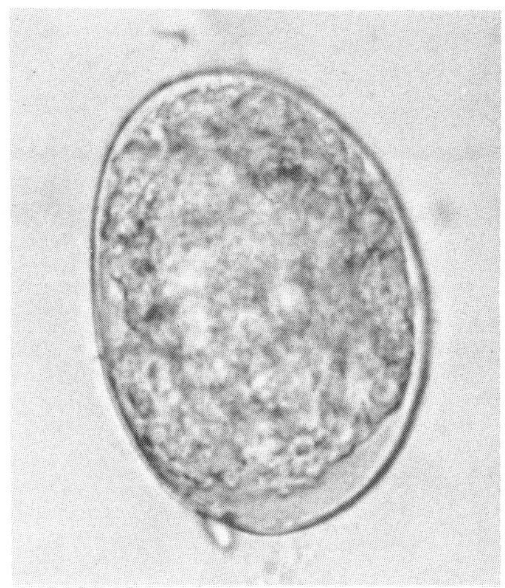

D

**FIG. 21.4**
Schistosome eggs. (*A*) *Schistosoma mansoni;* (*B*) *S. japonicum;* (C) *S. haematobium;* (*D*) *S. makongi.*

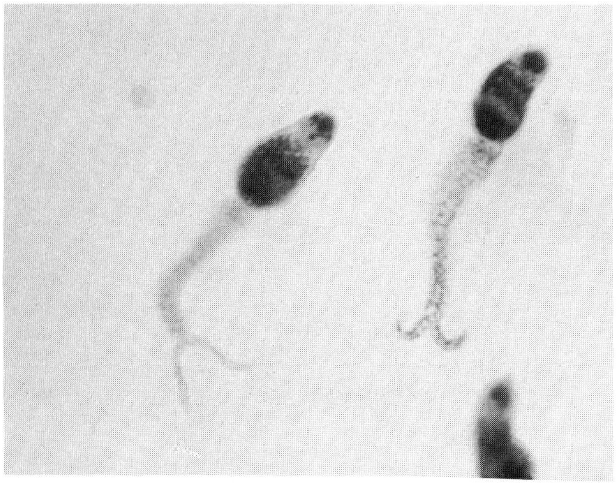

FIG. 21.5
Furcocercous cercariae of *Schistosoma mansoni*.

containing cercaria followed by drying of the skin permits large numbers of cercariae to penetrate the skin of mice. They reach the subepithelial layers within 30 min.

Once the organism penetrates the skin, loses its tail, and secretes the substances from the various glands, it can be considered a schistosomule. Schistosomules are clearly different from cercariae in their ability to tolerate high osmotic pressure, their ability to survive in serum, and their appearance, which is wormlike.

The schistosomules are carried by the circulatory system to the lungs about a day after penetration. They then migrate to the liver, but the exact route is not known. Development takes place in the liver and as the worms approach sexual maturity, they begin to migrate down the hepatic portal vein into the smaller vessels, draining the lower intestine or bladder. Sexual maturity may be reached as early as 7 weeks following infection, depending on the species (Table 21.1).

Studies on egg production in schistosomes have been done in both human and animal infections. The data from human infections are limited in many ways, but they are consistent with animal studies. Egg production per female worm per day ranges from 22 to 3500. The highest figure was found in *S. japonicum* infections in hamsters. An *S. mansoni* female produces perhaps 500 eggs per day, whereas *S. haematobium* probably produces fewer than 100 (Jordan & Webbe, 1969).

There are some problems that the maturing worms need to solve. First, the sexes are separate in the schistosomes. Sex is genetically rather than environmentally determined, so the sex of a worm is fixed when it is in the cercarial stage. Therefore, a worm needs to find one of the opposite sex, and pheromones may be involved. The worms also tend to locate in specific abdominal organs, but there is little information on stimuli that cause them to migrate to one place or another. It has been suggested that *S. mansoni* has a positive tropism for indole and related compounds (Macdonald, 1973). Despite the huge amount of study on the schistosomes, there are many areas that require further study even on the fundamental aspects of the life cycles.

A second problem facing the worms as they mature is how to survive for a long period. Since they have relatively long prepatent periods in the vertebrate host, they must be able to avoid its immune response. The solution they have evolved is an ingenious one; they absorb the ABO blood group antigens and histocompatibility antigens from the host's serum onto their surfaces. The worm antigens are almost completely covered or masked by this subterfuge and the worms are not recognized as nonself. Shedding of the outer portion of the double tegument membrane may also eliminate antigen – antibody complexes. The effectiveness of these techniques is attested to by the fact that worms may live for more than 10 years in a human host. This unique adaptation is called concomitant immunity.

DIAGNOSIS. Diagnosis of infection with members of the genus *Schistosoma* is based on clinical signs and symptoms, a history of living in an endemic area, serological tests, and finding the characteristic eggs.

Signs and symptoms of schistosomiasis are similar to those of a number of other diseases. Persons suffering from schistosomiasis usually are anemic, and have diarrhea and an enlarged spleen and liver. In *S. haematobium* infections, most of the damage is to the urinary system; there is painful urination, sometimes with the passage of blood.

Serological tests are useful during the prepatent period and in chronic cases in which eggs cannot be found. A variety of serological tests have been developed including a slide flocculation test, complement fixation, and a fluorescent antibody test. The circumoval precipitation test, enzyme-linked immunosorbent

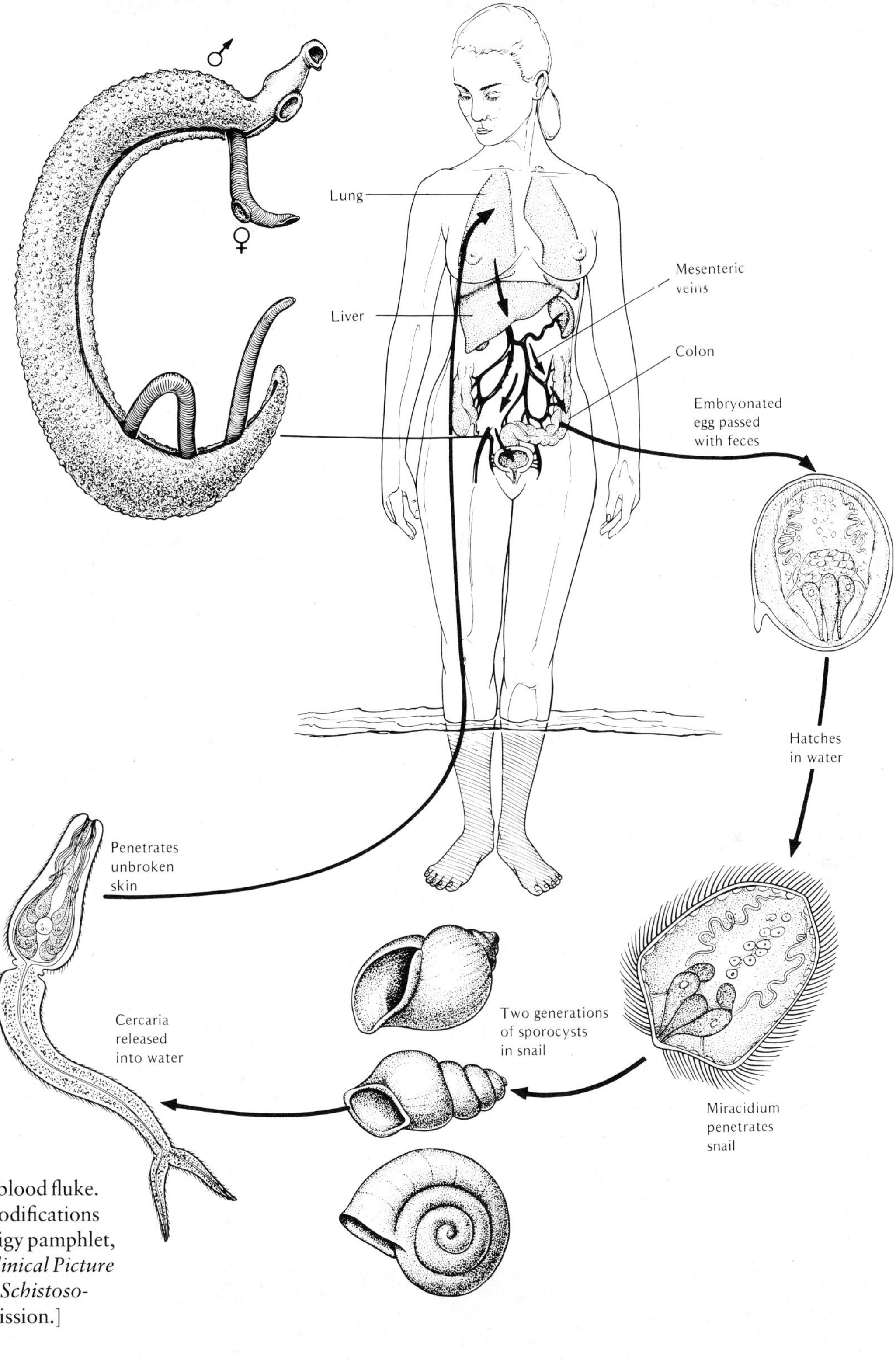


**FIG. 21.6**
The life cycle of a blood fluke. [Redrawn with modifications from the Ciba-Geigy pamphlet, *The Life Cycle, Clinical Picture and Treatment of Schistosomiasis*, with permission.]

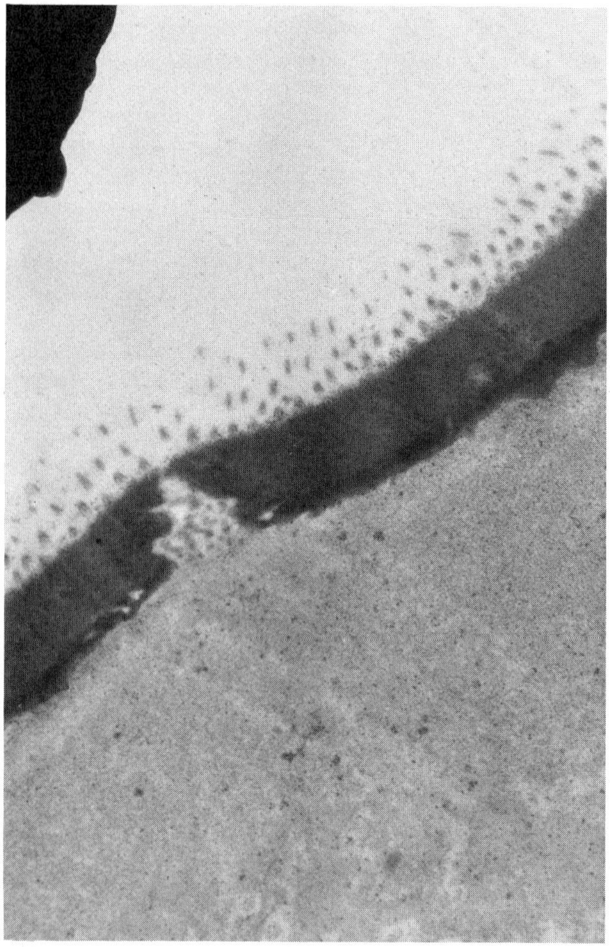

**FIG. 21.7**
Pore in the egg shell of *Schistosoma mansoni.*

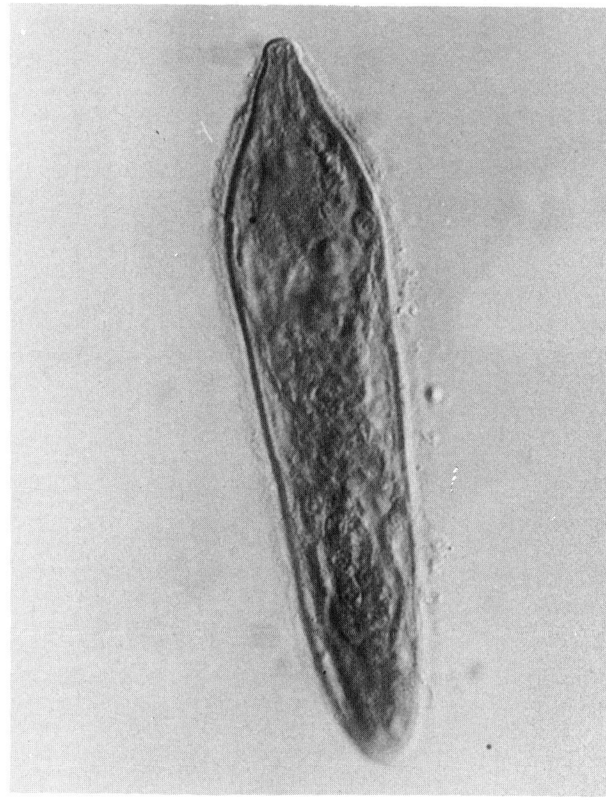

**FIG. 21.8**
Miracidium of *Schistosoma mansoni.*

assay (ELISA), and thin-layer immunoassay have shown promise in diagnosis (Hillyer, 1982).

The most common and conclusive means of diagnosing schistosomiasis is finding the characteristic eggs. Although three species are considered to be the major human parasites, *S. intercalatum* and *S. mekongi* are also found in humans, and species normally found in domestic animals sometimes parasitize humans (Table 21.2). The eggs of these latter worms can usually be distinguished from the usual human parasites. In the opinion of most investigators, finding the characteristic eggs is the only certain sign of infection.

Methods of finding eggs depend on the species and the stage of the infection. If *S. haematobium* is expected, a urine sample should be taken, centrifuged, and the sediment examined microscopically for the characteristic terminally spined eggs. The sample should be taken during early afternoon, a time when the greatest number of eggs per unit volume of urine is passed.

For *S. japonicum* and *S. mansoni*, fecal samples are examined by sedimentation methods designed to remove the greater portion of the fecal debris by screening and/or decanting. Methods most often employed are the filtration staining method, the Kato thick smear, and the modified formol-ether concentration technique (Hillyer, 1982).

In long-standing infections, eggs may not be seen in the feces; the method then used is rectal biopsy. One or two snips of rectal mucosa are taken (the procedure is

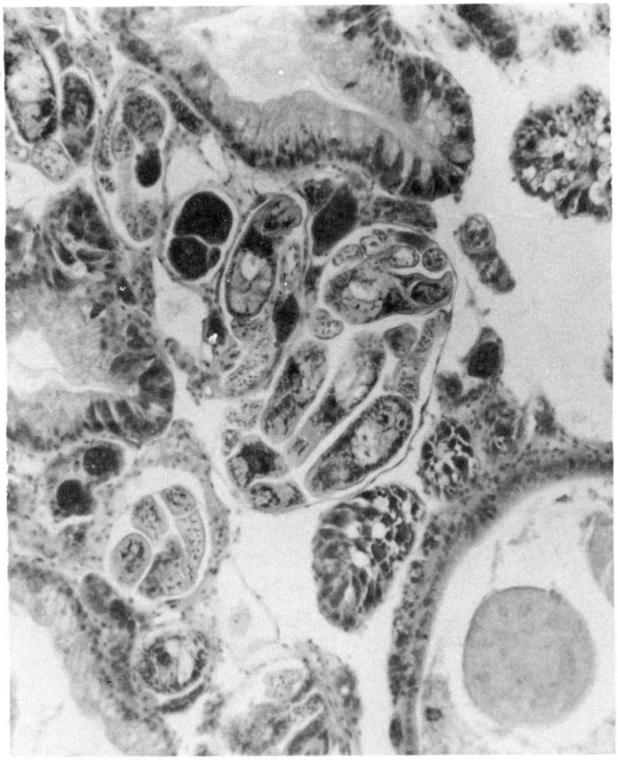

**FIG. 21.9**
Section of daughter sporocysts in a snail digestive gland.

painless if properly done) and the tissue is examined microscopically while pressed between two slides.

Testing viability of eggs is important in determining the stage of infection. In some long-standing infections, dead eggs may be found in feces, urine, or biopsies. Viability may be determined by direct examination of the eggs to look for flame cell action in the miracidia, or eggs may be induced to hatch. As indicated earlier, eggs are passed with a fully formed miracidium, and the eggs hatch when exposed to low osmotic pressure. The standard method involves mixing the feces with physiological salt solution to break it up, then diluting with 10 volumes of water and placing the material in a darkened 1-liter Florence flask which has a short side arm near the top (Fig. 21.10). The flask is left for 2 to 3 hr and then a light is shown on the side arm; the miracidia will collect in the side arm.

Diagnosis not only refers to determining whether an individual is infected, but also to the prevalence of infection in a community. A variety of approaches has been used to determine the prevalence and the incidence of infection in a population. Fecal and urine examinations, autopsy records, serological testing, and skin testing are the major means of screening.

HOST–PARASITE INTERACTIONS. Damage to the definitive host is organized according to the following time sequence:

1. Prepatent period
2. Egg deposition and extrusion
3. Tissue proliferation, repair, and degenerative effects

In the prepatent period, two complexes of signs and symptoms can be seen: invasion of the skin by cercariae and migration of the maturing worms from the liver to the venules in the lower abdomen.

**FIG. 21.10**
Side arm flask used to collect schistosome miracidia.

**TABLE 21.1**
Differential Characteristics of *Schistosoma* Spp. of Humans

| | *haematobium* | *japonicum* | *mansoni* | *intercalatum* | *mekongi* |
|---|---|---|---|---|---|
| **Adults** | | | | | |
| **Male** | | | | | |
| Length (mm) | 10–14 | 12–20 | 6–12 | 11–14 | 10–18 |
| No. of testes | 4 | 7 | 6–9 | 2–7 | 3–8 |
| Posterior union of intestinal ceca | Midbody | Posterior to midbody | Anterior to midbody | ? | Posterior to midbody |
| Tegument | Scattered tubercules and blunt short spines | Minute spines | Numerous tubercules and large, sharp spines | Short, blunt spines | Smooth |
| **Female** | | | | | |
| Length (mm) | 16–20 | 16–28 | 7–17 | 13–24 | 14–20 |
| Ovary | Posterior to midbody | Midbody | Anterior to midbody | Midbody | Midbody |
| No. of eggs in uterus | 20–100 | 50 or more | 1 usually | 10–100+ | 120–130 |
| Eggs produced/female/day | ? | 1400–3500 | 100–500 | ? | ? |
| **Location in host** | | | | | |
| Primary | Bladder and pelvic plexuses | Both superior and inferior mesenteric veins | Veins of small & large intestine near ileocecal junction | Portal and mesenteric veins | Mesenteric veins |
| Secondary | Portal and mesenteric veins | Portal vein | Hemorrhoidal and hepatic veins | ? | Portal vein |
| Prepatent period in weeks | 10–12 | 5–7 | 7–8 | ? | 5 |
| **Eggs** | | | | | |
| Length μm | 112–170 | 70–100 | 114–175 | 140–240 | 58–78 |
| Mean | | 82 | | | 62 |
| Width μm | 40–70 | 50–65 | 45–68 | 50–85 | 43–70 |
| Mean | | 60 | | | 51 |
| Color | Golden brown | Golden brown | Golden brown | | Golden brown |
| Spine | Terminal | Subterminal knob or none | Subterminal spine | Terminal | Subterminal knob |
| Passed in | Urine | Feces | Feces | Feces | Feces |

Penetration of the skin by cercariae is usually without visible effect unless there is massive invasion or the person is immune. In the nonimmune person, there may be small hemorrhages (petechiae), mild itching, and rash at the site of penetration. In the immune person, there is a local reaction by leukocytes, principally eosinophils, indicative of an allergic reaction that develops within 24 hr of infection. Unless the individual has been infected by large numbers of cercariae, the symptoms are mild until the worms begin to migrate into venules. At this time, diarrhea and fever develop. Diarrhea, symptomatic of inflammation of the intestinal tissues, may be accompanied by passage of blood and mucus with the feces when egg laying begins. Metabolic

**TABLE 21.2**
Schistosomes of Domesticated Mammals

| PARASITE | PRINCIPAL DEFINITIVE HOSTS | GEOGRAPHIC DISTRIBUTION |
|---|---|---|
| *Schistosoma japonicum* | Cattle, swine, dogs, humans, sheep, goats | Japan, China, Southeast Asia, Philippines |
| *S. suis* (=*S. incognitum*) | Swine | Orient, India |
| *S. nasalis* | Cattle, sheep, goats, horses | India (one report in the United States) |
| *S. bovis* | Cattle, sheep, goats, humans | Africa, Mediterranean—Italy, Sardinia, Iraq |
| *S. spindale* | Sheep, goats, cattle, buffalo | India, Sumatra |
| *S. indicum* | Sheep, goats, cattle, equines, camel | India |
| *S. mattheei* | Sheep, cattle, equines, baboon, antelope, humans | Africa |
| *S. intercalatum* | Sheep, goats, humans, rat (*Hybomys*) | West Africa—Congo |
| *Ornithobilharzia bomfordi* | Zebu | India |
| *O. turkestanicum* | Cattle, buffalo | USSR, China, Iraq, France |

by-products of the worms probably bring about general toxicosis and fever.

Egg laying and entrapment of the eggs in the tissues cause the greatest long-term damage. The eggs or their products are antigenic; white cells surround the eggs and fibroblasts lay down connective tissue around them. This is a typical foreign-body reaction.

Bear in mind that the adult worms live in a nutritionally rich environment and the females are prolific egg producers. Many of the eggs are trapped in the tissues and never reach the outside, but 250 eggs/ml of urine is common in children infected with *S. haematobium* and 800 eggs/ml of urine has been reported. So, despite the small size of the eggs, many remain in the tissues and their effect is cumulative. Adult worms live as long as 10 years and may live longer than 25. A major long-term effect is the laying down of progressively greater amounts of connective tissue in the infected organs so that they lose elasticity or the blood vessels are occluded and become less able to function normally as time passes. These changes are irreversible.

Not all eggs laid in the venules are trapped in the tissues where they have been laid. Since the blood flows away from the location of the adults toward the heart, eggs may be carried to other parts of the body. Most often eggs are carried to the liver, and the same foreign-body reaction occurs there. Connective tissue is laid down in the blood vessels of the liver, circulation is impeded, and the liver becomes enlarged. Progressive damage to the liver leads to general debility because of impairment to the many metabolic and detoxifying functions the liver normally performs. Other sites where eggs may lodge are the spleen, lungs, reproductive system, and central nervous system, including both the brain and spinal cord. In all instances there is a foreign-body reaction around the eggs.

In *S. haematobium* infections, there may be destruction of the kidney following fibrosis of the bladder and ureters. Lack of the normal peristaltic action of the ureter allows back pressure of the urine to reach the kidney and may cause its destruction. In *S. japonicum* or *S. mansoni* infections in which there is extensive liver damage, an extra load is placed on the heart because of altered blood pressure. The heart may become enlarged and ultimately fail as a result of excessive demands on it.

The evidence that schistosomes cause cancers is circumstantial. It has been noted that persons infected with *S. haematobium* have a higher prevalence of bladder cancers than uninfected persons from the same area. A similar correlation has been made for colonic carcinoma in areas where *S. mansoni* is endemic. There is also evidence for a high prevalence of splenic lymphoma in persons infected with *S. mansoni*. There is currently no way of substantiating these statistical correlations with experimental studies, but the evidence is at least moderately good and should be taken into account in assessing the overall effects of schistosomiasis.

The cellular reaction by the definitive host is both detrimental and beneficial. The reaction is directed primarily against the eggs, and the long-term effects are that the infected organs become progressively less functional. The response ought to be directed against adult worms to wall them off or kill them so that their eggs would cause no damage. But we know that the adults continue to lay eggs for many years.

Over a period of time, infected persons do develop immunity, which results in their failure to show eggs in the feces or urine. In experimental animals, it has been shown that schistosomules are stunted and fail to mature sexually in an immune host. Cercariae may be trapped in the skin of individuals who have been exposed.

Most efforts to develop a vaccine have been directed toward using the antigens of cercariae or schistosomules. Logically, these are the stages against which the immune response should be directed since they have not yet done significant harm, are small enough to be readily surrounded by white cells, and have not yet developed a screen of host proteins. A great deal of work has been done using a variety of antigens: killed cercariae, antigenic extracts, irradiated cercariae, and cercariae from foreign hosts. As yet, no product has been developed that is approved for use against the schistosomes of humans.

TREATMENT. Treatment of schistosomiasis has been hindered by the fact that relatively few drugs are available, their toxic effects are sometimes severe and they are active only against mature worms. The following discussion of drugs (Davis, 1973) gives the compounds roughly in the order in which they were developed and shows that with each new drug side effects have generally been lessened. The purpose of the discussion is to show the kinds of problems, both physiological and social, that may arise when drug treatment is included as a part of a control program.

1. Trivalent antimonials. Tartar emetic (potassium antimony tartrate), sodium antimonial gluconate, and stibocaptate, among others, have been used for years, but are now used infrequently. All are given by either IV or IM injection on schedules that require daily or alternate-day administration for 2 weeks or more. Toxic side effects include nausea, vomiting, liver damage, muscular weakness, cardiovascular irregularities, and skin rash. Complete cure can be expected in 75% or more of individuals who complete the course of treatment. In one carefully controlled study in Africa, only 60% of 160 persons completed the course of treatment. Antimonials are not suitable for mass treatment.

2. Thioxanthones, miracils. Lucanthone and hycanthone are the drugs used. Hycanthone has given cures of from 65 to 95%, with greater success against *S. haematobium* than *S. mansoni*. Administration is by IM injection and a single dose can be used. Immediate side effects are mostly gastrointestinal, but unfortunately, in the long term, hycanthone has been found to be mutagenic, teratogenic, and carcinogenic. Attempts have been made to develop related compounds with less hazard (Reid et al., 1977). These drugs can be used for mass chemotherapy.

3. Niridazole [(5-nitrothiazol-2-y)-imidazolidin-2-one]. This drug is given by mouth and a cure rate of 75 to 95% can be expected. There are some gastrointestinal side effects but, more important, a small percentage of patients exhibit neuropsychiatric changes which may last for weeks. Niridazole can be used for mass chemotherapy.

4. Oxnamiquine, 1,2,3,4-tetrahydro-2[([1-methylethyl]amino)methyl]-7-nitro-6-quinoline methanol, is a tetrahydroquinoline developed during the early 1970s and is highly efficient against *S. mansoni*. A single oral dose gives a complete cure in 85 to 95% of patients. About one third of persons receiving the drug have minor side effects such as dizziness or drowsiness. Oxnamiquine is used in mass treatment programs against *S. mansoni*.

No drug developed to date has all of the characteristics of the ideal drug. A number of workers have taken the view that three types of drug are needed: a prophylactic, a suppressant, and a curative. A prophylactic will prevent infection; usually this kind of drug is taken on a continuous basis (daily or weekly) so that the infective agent is killed or prevented from developing as soon as it enters the body of the host. A suppressant may not have irreversible effects on flukes, but it prevents egg laying. There are three advantages to this kind of drug: it will prevent the damaging effects of eggs being trapped in the tissues, it will serve a public health measure by reducing the numbers of eggs shed into the environment, and it will maintain immunity, thus preventing additional infections. The ideal curative drug kills all of the adult worms present in the host.

How do you find a possible drug among all of the compounds that have been screened? Screening methods are sometimes indirect, but in the case of the antischistosomal drugs, they are nearly all direct. The system is an *S. mansoni*–laboratory mouse model Fortunately, *S. mansoni* can be propagated in the laboratory using snails and laboratory mice as the intermediate and definitive hosts. Thus, there is normal completion of the life cycle and the effects of a potential drug against the flukes can be determined.

A screen consists of infecting animals with a known number of worms, administering a possible drug to the mice, and then evaluating the effect of the chemical on the worms and the mice. The possible combinations of effects are clear, but the one the investigator looks for is the one in which the mouse does not suffer from the toxicity of the drug but the flukes do. In most instances, the system is a blind screen; compounds are chosen almost randomly and activity is determined. In some screens, the investigator may not even know the name or structure of the compounds being tested until one is found that appears to have activity. The attitude is that if it does not work, there is no point in knowing what it is.

Once a compound has shown activity in the rough screen, it is then tested further to determine its level of activity more precisely and to determine both long- and short-term toxicity. Tests are then made in other laboratory rodents such as hamsters and then in monkeys. Further testing is done against different strains of *S. mansoni* and against *S. japonicum*, since it has been well documented that there are differences among them in response to drugs. Finally, the drug will go to clinical trials in people.

It should be emphasized that a drug that has shown preclinical promise may have no effect in humans. In certain antischistosomal drugs (miracils, *p*-aminodiphenoxy alkanes), metabolites formed through the metabolic activities of the vertebrate host are the active compounds. Thus, the peculiarities of the enzyme systems of a particular vertebrate determine whether some of the antischistosomal drugs will have activity. One is also left with the uneasy feeling that because of the differences in the enzyme systems of the mouse, good drugs for humans have been overlooked.

EPIDEMIOLOGY AND CONTROL. The basic pattern of life cycles in the blood flukes was worked out in the first decade of the twentieth century. With that information came the keys to control. It was obvious that sanitary disposal of human excreta and reducing populations of snails were the steps to take. As knowledge increased, additional techniques were developed, so that a list of general methods used in attempting to control schistosomiasis now includes the following:

1. Snail control
2. Sanitary disposal of human excreta
3. Treatment of infected persons
4. Treatment or elimination of reservoir hosts
5. Prevention of penetration by cercariae
6. Immunization of persons at risk of infection
7. Prevention of infections in snails

Each method is directed toward the same end: to break the life cycle or interrupt the chain of infection. If all snails could be removed from an endemic area, the probability of infection would approach zero. If all definitive hosts were treated and maintained free of infection for a year, the cycle would be broken.

Attempts have been made to apply statistical and computer methods in predicting the success of certain control programs. From a theoretical standpoint, it has been shown that the probability of infection is vanishingly small when the prevalence of infected hosts is reduced to a certain level (Fig. 21.11). However, a careful critique of the present state of the art of mathematical epidemiology leads to the conclusion that we are still in an early stage of development of predictive models for schistosomiasis (Thomas, 1973). The results of computer models suggest that the most efficacious approach is a combination of treating the definitive hosts and reducing the number of snails and/or contact with contaminated water. With these combined efforts, eradication can theoretically be achieved in about 4 years.

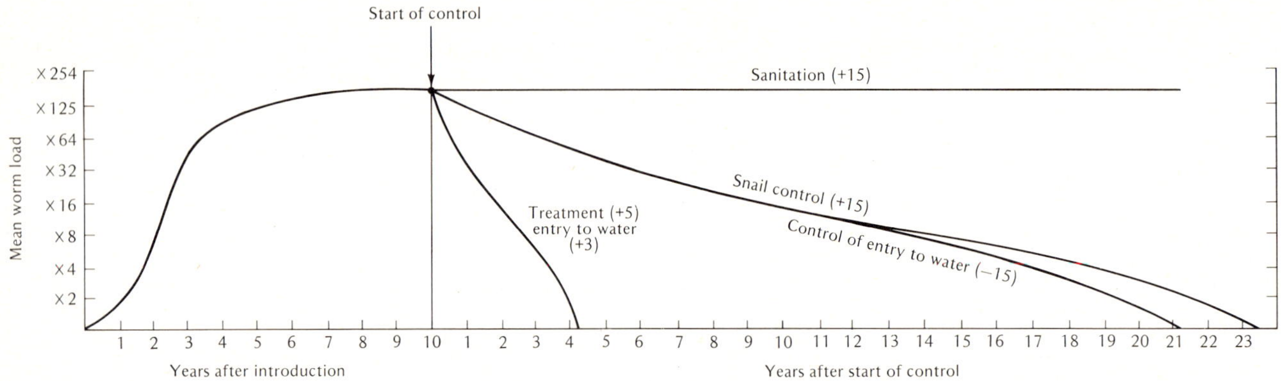


**FIG. 21.11**
Macdonald's computer model of the rise and fall of a helminth infection after four apparently equivalent control programs. Each reduces one of the factors involved to 1/15 its original value. [Redrawn from Macdonald, 1965.]

Snail control is subdivided into

1. Chemical methods
2. Biological methods
3. Ecological methods

Chemical control consists of introducing chemicals such as $CuSO_4$, sodium pentachlorophenate, or niclosamide (*n*-tritylmorpholine,3,5-dibromo-4-hydroxy-4′-nitrazobenzene) into the water.

Since the vectors of schistosomiasis are aquatic snails, chemicals can be introduced into the water and little attention need be given to direct contamination of the terrestrial environment. A variety of methods is used to maintain a proper level of toxic chemical in the water (see Chapter 20 and Malek & Cheng, 1974).

Biological control utilizes diseases and predators of snails such as sciomyzid larvae (Insecta:Diptera), which are predators of snails, or birds that may eat snails or pathogenic bacteria.

A unique biological method of controlling vectors of schistosomes has been the introduction of nonvector snails that compete with the vectors. *Marisa cornuarietis* (Ampullariidae) was introduced into Puerto Rico and was found nearly to eliminate *Biomphalaria glabrata*, the vector of *S. mansoni* (Jobin et al., 1977).

Ecological control entails making the habitat unsuitable for reproduction or survival of the mollusks. The vectors of schistosomiasis are aquatic and prefer to attach to plants in bodies of water where the velocity is fairly slow. Where irrigation is practiced, periodic draining of canals and ditches will reduce populations. Removal of aquatic plants reduces a suitable habitat, and altering streambeds or canal contours to increase water velocity will remove the preferred niche of the snails. As a rule, ecological manipulations require continual effort and monitoring and may therefore bring about only temporary reductions in populations. Lack of efficacy has to do with the great reproductive potential of snails, their ability to disperse from one area to another, and their resistance to drying. When the habitat begins to dry, for example, the snails close up, move down into the mud, and await better conditions. Interestingly, there is evidence that snails may become infected with miracidia even during dry seasons when their populations have reached a low level (Webbe, 1962).

At present, snail control is best accomplished by chemical means. Despite the disadvantages of environmental contamination, the effect on nontarget species, and the possibility of reducing organisms that serve in natural control of snail populations, chemicals are still the most effective. However, instances such as that reported by Jobin et al. (1977) in Puerto Rico, where *Biomphalaria glabrata* was replaced by *Marisa cornuarietis* in 88% of reservoirs, may require reassessment of chemical control in some areas. The cost of control over a 20-year period was $15,000 (1968 U.S. dollars). The cost for chemical control was estimated to be 10,000 times greater and had the disadvantage of contaminating water supplies used for human consumption.

Control of schistosomiasis is not the only motivation for sanitary disposal of human feces and urine. Not only are there esthetic reasons, but there are microbial and parasitic diseases that are transmitted by improper disposal and treatment of human wastes. As indicated in the discussion of Macdonald's 1965 computer model (Fig. 21.11), allowing even 1/15,000 of a community's waste to reach the water may not be a sufficiently high standard of sanitation to affect transmission. Whether proper disposal of feces and urine can be effective in reducing transmission of schistosomes depends on a variety of factors. One of the factors is the host range of a species. *S. mansoni* and *S. haematobium* have been found in hosts other than humans, including rodents and some primates. Despite natural infections in other animals, these definitive hosts of *S. haematobium* seem not to be reservoirs of infection. On the other hand, the host range of *S. japonicum* is considerably broader than that of *S. mansoni* and *S. haematobium*, and the overall effect of sanitary disposal of human excreta may be quite different.

In the case of *S. haemotobium*, hosts other than humans are probably not important, and therefore complete sanitary disposal would have an effect on transmission. Since *S. japonicum* and *S. mansoni* normally infect animals such as cattle and rodents, sanitation campaigns would have little effect. If it were possible to segregate people completely from domestic cattle, it would be possible to control *S. japonicum* by this means, but in endemic areas, cattle are used in rice paddies for plowing and preparation of the seed beds. The farmers have no alternative but to follow behind their beasts in the knee-deep muck to plant their crops. Although machinery is being developed for these purposes by the International Rice Research Institute, immediate replacement of cattle by the internal combustion engine seems unlikely and doubts about the long-term availability of petroleum may jeopardize the eventual switchover.

Host range remains a factor if control is attempted through treatment of infected persons. Let us clarify first that treatment has two objectives: improvement of the health of individuals and the improvement of health of the community. It is the latter with which we are concerned in this discussion. In *S. japonicum*, treatment of everyone in a community is likely to have little effect on rates of transmission even though individuals may be benefited. Infection rates of those worms that are found primarily in humans, *S. haematobium* and *S. mansoni*, can be reduced by treatment of all persons. Unfortunately, it is likely that some human defecation and urination in or near water will always occur, and a small amount of contamination of the environment appears to be sufficient to maintain transmission (Thomas, 1973).

Preventing infections in humans could be accomplished by immunization or by preventing the penetration of cercariae. Although knowledge of immunity has progressed in recent years, there is still no vaccine marketed for developing artificial immunity. Preventing penetration of cercariae can be achieved by applying substances to the skin which act as an impenetrable barrier, or kill or repel them.

One way in which basic biological studies may lead to practical applications has been summarized by Chernin (1974). Even if a small amount of human waste contaminates an area and the number of snails is low, schistosome life cycles can be completed. Part of this success can be attributed to three aspects of snail finding by miracidia:

1. They swim at a velocity of 6.9 m/hr
2. They tend to move to the margins of a body of water where the snails also tend to congregate
3. They sense the presence of a snail and infect one when released 2.7 m away

The third point is of special interest since the snails produce a substance or complex of substances that Chernin called *miraxone*, which attracts snails and acts over a distance; miraxone is probably magnesium ion. Significantly, miraxone is released not only from vectors of schistosomes, but from all snails, and miracidia are attracted to susceptible as well as nonsusceptible snails. In laboratory studies nonsusceptible snails protected susceptible ones. So, the logical areas for investigation are (1) using decoy snails to clear an area of miracidia and (2) trapping miracidia by using miraxone as bait. While research and development have not yet made this method usable in the field, it is both innovative and promising.

It has been known since the work of Cort et al. (1937) that in double infections of snails with trematodes, one of the species may be eliminated. When two species of trematode infected a single snail, one species was likely to be inhibited or sometimes killed. It was not until the middle 1960s, when Kian Joe Lie and his co-workers studied the phenomenon intensively, that the possibility of disease control was considered. Initial studies showed that schistosome infections in snails could be eliminated by infecting them with echinostome miracidia. The key to understanding what happens lies in the structure and food habits of the larval forms: schistosomes form sporocysts only, and echinostomes form rediae. A sporocyst is little more than a germinal sac, but a redia has a mouth, oral sucker, gut, and locomotor appendages. Rediae actively attack sporocysts and eat them. As information accumulated, it became clear that there was another mechanism in addition to direct predation—that of indirect antagonism, in which one species inhibits development through some other mechanism. The mechanism may operate through snail immunity or toxic products produced by one species of larva that acts on the other (Lim & Heyneman, 1972).

Pilot studies have been done to determine whether intertrematode antagonism can be exploited for disease control. Some success has been achieved in these studies, but Lim and Heyneman (1972) have pointed out that whatever use is made of trematode antagonism requires an intimate knowledge of the biology of the organisms that are to be controlled as well as those that will be used in control. If the approach does become practical, it will probably be used for local problems and will require both production of parasites and monitoring of results by trained personnel. The requirements for success that these authors list are as follows:

1. There must be a high probability of infection of a snail by any one miracidium
2. Miracidia must infect all ages of host snail
3. Miracidia must infect snails in all ranges of the natural habitat
4. Once having infected the snail, the larval stages must grow and develop regardless of other trematode infections that may be present

Should one conclude that intertrematode antagonism is not worth pursuing for disease control? Certainly not. If there were simple solutions to the control of schistosomiasis, they would have been found and exploited. This particular method will probably be added to the others and used when circumstances are right. Intertrematode antagonism is the kind of innovative approach that is required for control of this difficult disease complex and can be used as one of the methods to push the probability of transmission toward zero.

Our discussion thus far has looked mostly at biological factors that influence the probability of transmission of schistosomes. Host range, longevity in the definitive host, snail biology, and survival of miracidia, among others, must be considered in a control program, but equally important are the human factors that influence the success of a program. Human behavior, economic level, and attitudes of both politicians and administrators are crucial and must be considered by anyone undertaking a control program.

Although schistosomiasis may have some foci in urban environments, it is primarily an infection of rural or village people at the lower end of the economic scale. Water supplies such as streams, ponds, seeps, and irrigation canals are used for cooking, drinking, and recreation; clothes are washed there and people often must wade through streams to get to market, school, or their fields.

The options for reducing water contact are (1) to change people's behavior or (2) to develop piped, controlled water systems. The probability of changing the way people go about their daily business through an educational campaign is low. The alternative is technological, but there are nevertheless behavioral components.

There have been successes in control campaigns when governmental agencies have provided money to build water supply and disposal systems suitable for the particular community. Merely piping in water to a village where it is stored in a cistern may have little effect on infection rates. Women walk to streams to wash and dry clothing. Areas chosen for washing are likely to have clean rocks where the clothes can be dried and where bundles can be set so that the women do not get their heads muddy when they carry the bundles back home. In planning a washing area, it should approximate the natural facilities at the stream bank and, equally important, be no farther from home. The designer of the laundry must know how the washing is done (most often on the knees), how clothes are dried, and how many women and children need to be accommodated (washing is also a time for visiting).

Swimming is an important recreation for children in areas where schistosomiasis is endemic, and providing uncontaminated water is essential. Teenage children have the highest percentage of infection and most often excrete the highest number of eggs in any population;

they are not only a high-risk group, but they also may contribute greatly to contamination.

As indicated, people sometimes become infected with schistosomes by wading through streams when traveling. It is important to provide footbridges, even simple ones, and to place them where they will be used.

Another part of water supply has to do with providing latrines and bathing facilities. Social customs vary, and it is important to know what they are with respect to modesty and whether a community or family facility is needed. People may not wish to be seen bathing, or even be seen entering a privy. In some areas of Africa an individual's feces are considered to be personal property, and there is fear that they may be stolen by someone and used for witchcraft. In such a circumstance, a person may prefer to defecate in the bush where the feces will be quickly dissipated by rain and invertebrates (Gelfand, 1974).

Providing safe water supplies has been successful in bringing a degree of control of schistosomiasis in Africa and the Philippines. This method is one that should be included in a control program, and benefits will also be seen in the reduction of other waterborne infections.

Even religion plays a role in the epidemiology of schistosomiasis; it probably contributes to the continued transmission of *S. haematobium* and *S. mansoni* in the Middle East and Africa. Islam requires five daily prayers of the faithful, and prior to each prayer the person must undergo ablutions. Ablutions consist of washing the hands, face, and feet. In irrigated areas, ditches and canals may be used at the most immediate source of water, and contamination of the water readily occurs. In Yemen, a cycle of transmission was found in the ablution pools of a mosque where vectors of *Schistosoma* were found in algal mats growing in ablution chambers, phonebooth-sized niches in the wall next to the pool (Kuntz, 1952).

## CERCARIAL DERMATITIS OR SWIMMER'S ITCH

Penetration of schistosome cercariae through the skin of a person or an animal causes some irritation and a mild prickling sensation. In a host that has not previously been exposed to cercariae, the skin reaction is usually mild. However, in a host that has been exposed previously, the reaction is more severe and results in a raised, reddened, itching papule that may persist for 3 days to a week. This reaction is a hypersensitive or allergic response and may be seen in both normal and foreign hosts (Fig. 21.12).

The term *swimmer's itch* is usually reserved for cases of exposure to the cercariae of schistosomes of water-associated birds and mammals (Table 21.3). As indicated, the first infection elicits a mild reaction, but subsequent exposures bring on a more severe reaction. In foreign hosts, cercariae are trapped in the skin where the inflammatory reaction occurs, but there is some evidence that a few cercariae reach the bloodstream in some hosts and undergo a portion of their normal migration.

Cercarial dermatitis has been reported from all parts of the world in tropical, temperate, and arctic climates. It occurs in persons who have been swimming, wading, or working in littoral areas of both marine and freshwater environments. In temperate climates, cases are seen most frequently in the middle and late summer, when the larval stages of the trematodes have had time to develop and shedding of cercariae has reached a high level. Swimmer's itch should be suspected in persons who have an itching, papular eruption and who have been in the water.

In some places, swimmer's itch has been sufficiently severe that beaches have been closed or swimmers

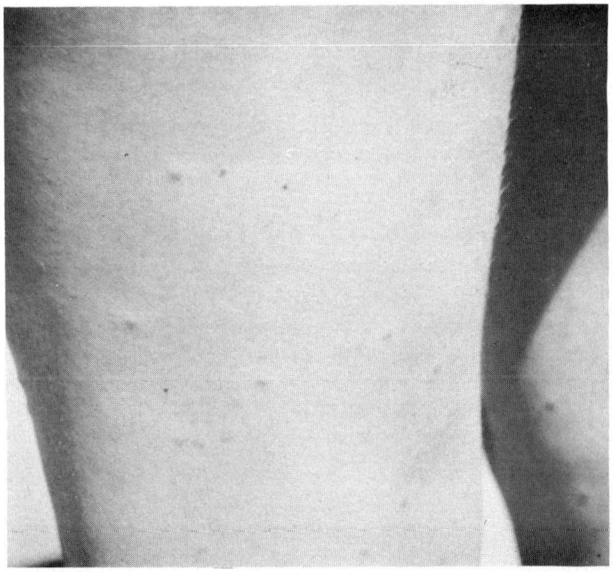

**FIG. 21.12**
Human dermatitis caused by bird schistosomes.

have been warned not to enter the water. Losses in these instances are primarily economic for the operators of businesses catering to vacationers. The affected individuals may suffer intensely for several days, but the effects are not lasting. Affected persons may be treated with lotions or other topically applied preparations designed to reduce itching and inflammation.

Preventing swimmer's itch has two aspects: personal and community. There is some evidence that a brisk toweling upon leaving the water may help prevent penetration of cercariae; as a general rule, cercariae find penetration easier if the skin dries slowly or repeatedly. Penetration may be prevented by applying an oily substance to the skin. At the community level, snail control programs can be effective, but they seem to have been successful only in freshwater areas, not in marine areas.

A control program for cercarial dermatitis should include a determination of the parasite(s) involved and what the definitive and intermediate hosts are. A week spent in a careful epidemiologic investigation may provide the key to control which would not be achieved by hasty application of a molluscicide to a wide area.

## SCHISTOSOMES OF DOMESTIC ANIMALS

The schistosomes of domesticated mammals (Table 21.2) are basically the same in terms of life cycles and host damage as the schistosomes of humans. *S. japonicum* is found in both humans and other animals. Some of the others, such as *S. mattheei, S. intercalatum*, and *S. bovis*, have been found in humans occasionally and have been public health problems in a few instances.

Their locations in the hosts are generally the abdominal veins, and heavy infections cause damage to the intestinal tract and liver. Loss of condition, reduced milk production, and diarrhea are typical signs of infection. Diagnosis is made by finding typical eggs in the feces.

In *S. nasalis* of cattle the adult worms are found in the veins of the nasal passages. In heavy infections, there is ulceration of the nasal epithelium and a copious discharge from the nose. Partial occlusion of the nasal passages gives rise to the term *snoring disease*.

Control of schistosomes of domesticated animals is basically the same as for those of humans. The same drugs are used.

## REFERENCES AND FURTHER READINGS

Ansari, N. (ed.). 1973. *Epidemiology and Control of Schistosomiasis*. University Park Press, Baltimore.

Basch, P. F. 1976. Intermediate host specificity in *Schistosoma mansoni*. *Exp. Parasitol.* 39:150–169.

———. 1981a. Cultivation of *Schistosoma mansoni* in vitro. I. Establishment of cultures from cercariae and development until pairing. *J. Parasitol.* 67:179–185.

———. 1981b. Cultivation of *Schistosoma mansoni* in vitro. II. Production of infertile eggs by worm pairs cultured from cercariae. *J. Parasitol.* 67:186–190.

Basch, P. F., and R. H. Humber. 1981. Cultivation of

**TABLE 21.3**
Some Schistosomes Known to Cause Cercarial Dermatitis (Swimmer's Itch)

| TREMATODE | NORMAL DEFINITIVE HOSTS | INTERMEDIATE HOST |
|---|---|---|
| *Schistosomatium douthitti* | Small rodents | *Lymnaea* spp., *Stagnicolae* spp., *Physa gyrina* |
| *Gigantobilharzia huronensis* | Cardinal and goldfinch | *Physa gyrina* |
| *Trichobilharzia ocellata* (=*Cercaria ocellata*) | Water birds | *Limnaea stagnalis* |
| *T. stagnicolae* | Water birds | *Limnaea stagnalis* |
| *Australobilharzia variglandis* | Terns, ducks | *Nassarius obsoletus* |
| *Heterobilharzia americana* | Dog, nutria, raccoon | *Lymnaea cubensis* |
| *Schistosoma spindale* | Cattle | *Planorbis* spp., *Lymnaea* spp., *Bulineus tropicus* |

*Schistosoma mansoni* in vitro. III. Implantation of cultured worms into mouse mesenteric veins. *J. Parasitol.* **67**:191–195.

Bruce, J. I., and S. Sorman (eds.). 1980. *The Mekong Schistosome.* Malacological Review, Whitmore Lake, Mich.

Chernin, E. 1974. Some host-finding attributes of *Schistosoma mansoni* miracidia. *Am. J. Trop. Med. Hyg.* **23**:320–327.

Cort, W. W., D. B. McMullen, and S. Brackett. 1937. Ecological studies on *Stagnicola emarginata angulata* (Sowerby) in the Douglas Lake region. *J. Parasitol.* **23**:504–532.

Davis, A. 1973. Chemotherapy in control. In Ansari, N. (ed.), *Epidemiology and Control of Schistosomiasis.* University Park Press, Baltimore.

Demaree, R. S., Jr., and G. V. Hillyer. 1981. *Schistosoma* species: Transmission electron microscopy of the circumoval immune precipitin reaction on eggs. *Exper. Parasitol.* **52**:77–85.

Fisher, A. C. 1934. A study of the schistosomiasis of the Stanleyville District of the Belgian Congo. *Trans. Roy. Soc. Trop. Med. Hyg.* **28**:277–306.

Gelfand, K. S. 1974. *A Clinical Study of Intestinal Bilharziasis* (Schistosoma mansoni) *in Africa.* Edward Arnold Publishers, London.

Hillyer, G. V. 1982. Schistosomiasis. In *The Handbook of Parasitic Zoonoses,* Vol. III: *The Parasitic Zoonoses.* CRC Press, Boca Raton, Fla.

Hockley, D. J. 1973. Ultrastructure of the tegument of *Schistosoma. Adv. Parasitol.* **11**:233–305.

Jobin, W. R., R. A. Brown, S. P. Vélez, and F. F. Ferguson. 1977. Biological control of *Biomphalaria glabrata* in major reservoirs in Puerto Rico. *Am. J. Trop. Med. Hyg.* **26**:1018–1024.

Jordan, P., and G. Webbe. 1969. *Human Schistosomiasis.* Charles C. Thomas, Springfield, Ill.

Kuntz, R. E. 1952. *Schistosoma mansoni* and *S. haematobium* in the Yemen, southwest Arabia: With a report of an unusual factor in the epidemiology of schistosomiasis mansoni. *J. Parasitol.* 38:24–28.

Lim, H. K., and D. Heyneman. 1972. Intramolluscan inter-trematode antagonism: A review of factors influencing the host-parasite system and its possible role in biological control. *Adv. Parasitol.* **10**:191–268.

Macdonald, G. 1965. The dynamics of helminth infections, with special reference to schistosomes. *Trans. R. Soc. Trop. Med. Hyg.* **59**:489–506.

———. 1973. Measurement of clinical manifestations of schistosomiasis. In Ansari, N., ed., *Epidemiology and Control of Schistosomiasis.* University Park Press, Baltimore, pp. 354–387.

Macdonald, G., and M. Farooq. 1973. Public health and economic importance of schistosomiasis–its assessment. In Ansari, N., ed., *Epidemiology and Control of Schistosomiasis.* University Park Press, Baltimore., pp. 337–353.

Malek, E. A., and T. C. Cheng. 1974. *Medical and Economic Malacology.* Academic Press, New York.

Mostofi, F. K. (ed.). 1967. *Bilharziasis.* Springer-Verlag, New York.

Pellegrino, J., and N. Katz. 1968. Experimental chemotherapy of Schistosomiasis mansoni. *Adv. Parasitol.* **6**:233–290.

Reid, W. A., P. S. Loizeaux, and M. J. Reardon. 1977. The activity of benzothiopyranoindazole (IA-04 N-OXIDE) against *Schistosoma mansoni* infections in rhesus monkeys. *Am. J. Trop. Med. Hyg.* **26**:937–943.

Smyth, J. D. 1966. *The Physiology of Trematodes.* W. H. Freeman & Co., San Francisco.

Stirewalt, M. A. 1974. *Schistosoma mansoni*: Cercaria to schistosomule. *Adv. Parasitol.* **12**:115–182.

Thomas, J. D. 1973. Schistosomiasis and the control of molluscan hosts of human schistosomes with particular reference to possible self-regulatory mechanisms. *Adv. Parasitol.* **11**:307–394.

Warren, K. S. 1972. The immunopathogenesis of schistosomiasis: A multidisciplinary approach. *Trans. Roy. Soc. Trop. Med. Hyg.* **66**:417–434.

———. 1973. *Schistosomiasis. The Evolution of a Medical Literature, Selected Abstracts and Citations, 1852–1972.* MIT Press, Boston.

———. 1974. Precarious odyssey of an unconquered parasite. *Natural History* **83**(5):47–53.

Warren, K. S., and D. D. Hoffman, Jr. 1976. *Schistosomiasis.* III. *Abstracts of the Complete Literature. 1663–1974.* 2 Vols. Hemisphere., Washington, D.C.

# 22

# *Liver Flukes of Herbivores*

Included in this chapter are some digenetic trematodes of economic and occasional medical importance. They are grouped together principally because of the contrasts that can be seen in epidemiology and control methods. Some occur in the same geographic areas and may be found in the same hosts.

Probably the most important of these flukes, *Fasciola hepatica*, was the first digenetic fluke to have its life cycle uncovered. The work of Thomas in Britain and his publication in 1883 describing the life cycle of *F. hepatica* is interestingly retold by Reinhard (1957).

## FAMILY FASCIOLIDAE

NAME OF ORGANISM AND DISEASES CAUSED. *Fasciola hepatica* (L.), the common liver fluke, causes fascioliasis or liver rot. It may lead to secondary bacterial infections such as black disease in sheep or bacillary icterohemoglobinuria (redwater) in cattle.

HOSTS AND LOCATION. Definitive; hosts include sheep, cattle, and many other ruminants; equids, swine, rabbits, hares, and marsupials are reservoirs; humans are occasionally infected. The fluke is located in the liver. Intermediate hosts include gastropod mollusks of the family Lymnaeidae such as *Lymnaea, Fossaria,* and *Pseudosuccinea.*

DISTRIBUTION AND IMPORTANCE. Worldwide distribution occurs in areas where sheep, cattle and goats are raised and there is a niche for lymnaeid snails. In the United States, organisms are found in the Gulf Coast, the Northwest, Rocky Mountains, Florida, and occasionally elsewhere. They are also found in Great Britain, Ireland, Continental Europe, the Middle East, the Far East, Africa, Australia, and the Caribbean area.

It is generally agreed that *F. hepatica* causes heavy economic losses in livestock worldwide, but putting a value on the global losses has not been possible. Pantelouris (1965) compiled some statistics from the literature of the 1950s and early 1960s (Table 22.1) indicating losses at slaughter, the percentage of animals infected in various parts of the world, and some death losses in specific outbreaks of disease.

Losses are direct and indirect. Direct losses in domestic animals are from chronic and acute fluke infection. Chronic losses are mostly in the form of reduced production of meat, milk, and fiber. Direct losses include condemnation of livers at slaugher; buyers of livestock for packing plants pay less for animals from areas where flukes are endemic. Such flukey livers are used in commercial fish food, but little money is received for them.

In the United States, losses are estimated to be $5.5 million per year from morbidity and mortality, and losses at slaughter are estimated at $2.5 million per year, representing 56 million pounds of liver. In Puerto Rico 32% of cattle were found to be infected at

**TABLE 22.1**
Some Data on the Prevalence of Fascioliasis[a]

| | | Abattoir Statistics (Livers Condemned) (%) | | Estimates of Incidence, Mainly from Detection of Eggs in Feces (%) | | |
|---|---|---|---|---|---|---|
| | | | | Cattle | Sheep | |
| Region | Period | Cattle | Sheep | Infected | Infected | Deaths |
| Anglesey | 1958–59 | | | | | 12.9 |
| Ireland, Northern | | 63.3 | 15.3 | | | |
| Scotland, Central | 1946–54 | | | | 38.6 | |
| Wales | 1958 | | | | | 60 |
| Devon | | 29 | | | | |
| United Kingdom | 1954, June | 17.7 | | | | |
| Poland | | | | 50 | | |
| Lithuania | 1951–52 | | | 91 | 46 | |
| Rumania | 1956 | | | | | 18 |
| Bavaria, Regensburg | 1945 | 8.3 | | | | |
| Bavaria, Regensburg | 1952 | 20.9 | | | | |
| Sweden, Ostergotland | 1953 | 21.5 | | | | |
| Milan, Italy | 1960 | 35.5 | | | | |
| Netherlands | 1957–62 | | 19.1 | | | |
| Rhodesia, Southern | 1956–61 | 10.3 | | | | |
| South Africa | 1951–57 | 5.7 | 2 | | | |
| Japan | 1958 | | | 40 | 15 | |
| Japan, Higashimatsuyama | | 68 | | | | |
| Puerto Rico | 1952 | 7.5 | | | | |
| U.S. Gulf Coast | | 37.5 | | | | |
| United States | | 2.6 | | | | |

[a] [From Pantelouris, 1965.]

slaughter and there was a $1 million annual loss from condemned livers (DeLeon et al., 1981).

Black disease of sheep and redwater of cattle result in indirect death losses from toxins produced by the anaerobic, spore-forming bacilli *Clostridium novyi* (*Cl. oedamatiens*) and *Cl. hemolyticum*, respectively, growing in fluke-damaged livers.

Humans are occasional hosts for *F. hepatica*, and infections are most often obtained through eating watercress (*Nasturtium officinale*) in salads. There are several hundred reports in the literature of human infections, but they are not reportable, so their importance cannot be adequately assessed. The largest outbreak in recent times occurred near Lyon, France, where there were about 500 cases. Signs and symptoms of infection in humans are relatively severe, and diagnosis is made difficult because eggs do not always appear in the feces (Dawes & Hughes, 1964), and eggs of echinostome flukes are similar to *F. hepatica* (Chap. 24).

MORPHOLOGY. Adult flukes reach a size of 30 by 13 mm, but many are smaller. They are brown and flattened, with prominent shoulders and oral cone (Fig. 22.1). The intestine is dendritic and the ventral sucker is located in the anterior one fifth. Most internal organs are obscured by the intestine and vitellaria even in well-stained whole mounts. *F. hepatica* can be distinguished grossly from other liver flukes on the basis of size and shape (Fig. 22.2)

LIFE CYCLE. Eggs (Fig. 22.3) reach the outside by passing down the common bile duct and being voided with the feces. They are undeveloped when passed and require a minimum of 10 days to reach the miracidial stage. The temperatures at which development occurs are 10 to 29°C; below 10°C there is only marginal development of eggs and larval stages, and no emergence of cercariae. Miracidia (Fig. 22.4) hatch in water and penetrate the pulmonary cavity of small lymnaeid snails. One sporocyst and two redial generations give

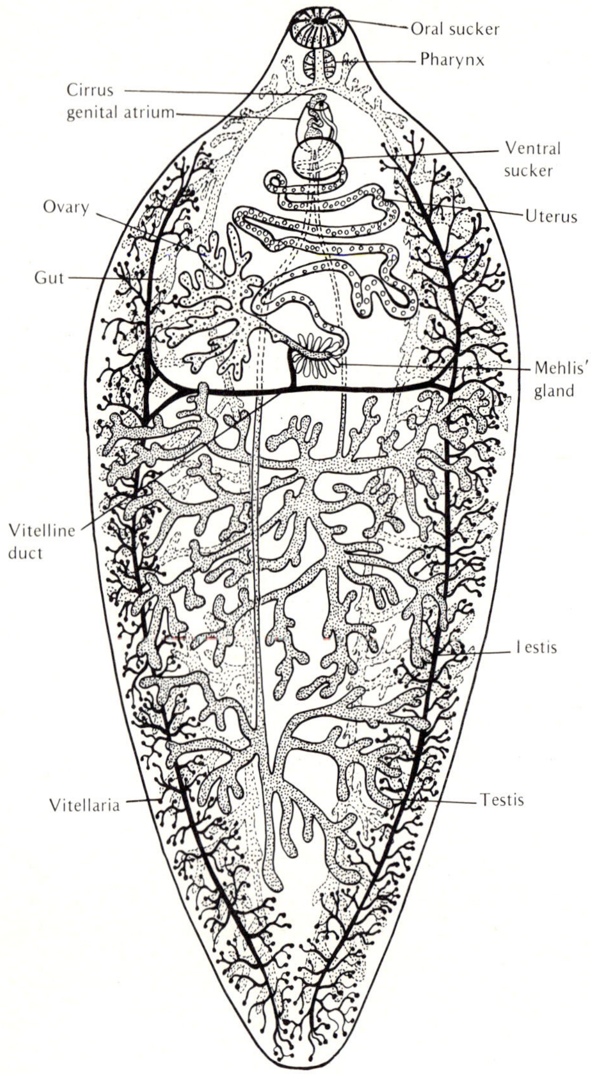


**FIG. 22.1**
Structure of an adult *Fasciola hepatica*.

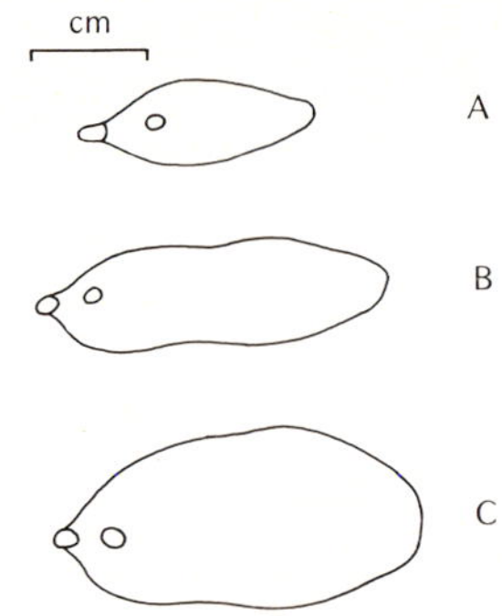


**FIG. 22.2**
Outline drawings of liver flukes of large ruminants. (*A*) *Fasciola hepatica*, (*B*) *Fasciola gigantica*, (*C*) *Fascioloides magna*. [Redrawn from Pantelouris, E.M. 1965. *The Common Liver Fluke*. Pergamon Press, NY.]

rise to gymnocephalous cercariae about 6 weeks later. The cercariae, which are set free in the water, encyst almost immediately on herbage, where they form metacercariae.

Metacercariae, which are eaten with herbage, excyst in the small intestine and migrate through the gut wall to the abdominal cavity. They reach the liver at about day 5 and penetrate its capsule. Juvenile flukes migrate in the liver parenchyma, reaching bile ducts and producing eggs 8 to 12 weeks postinfection (PI). Adult flukes may live longer than 10 years in sheep (Durbin, 1952), but generally less than a year in cattle (DeLeon et al., 1981).

DIAGNOSIS. Finding typical eggs in feces is the method of diagnosis (Fig. 22.3); they are barrel-shaped,

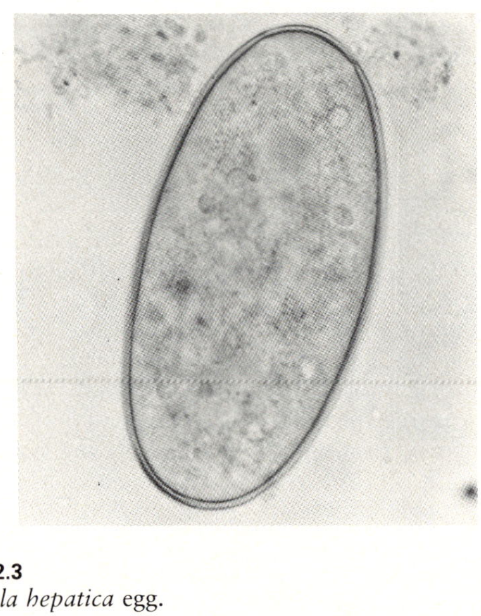

**FIG. 22.3**
*Fasciola hepatica* egg.

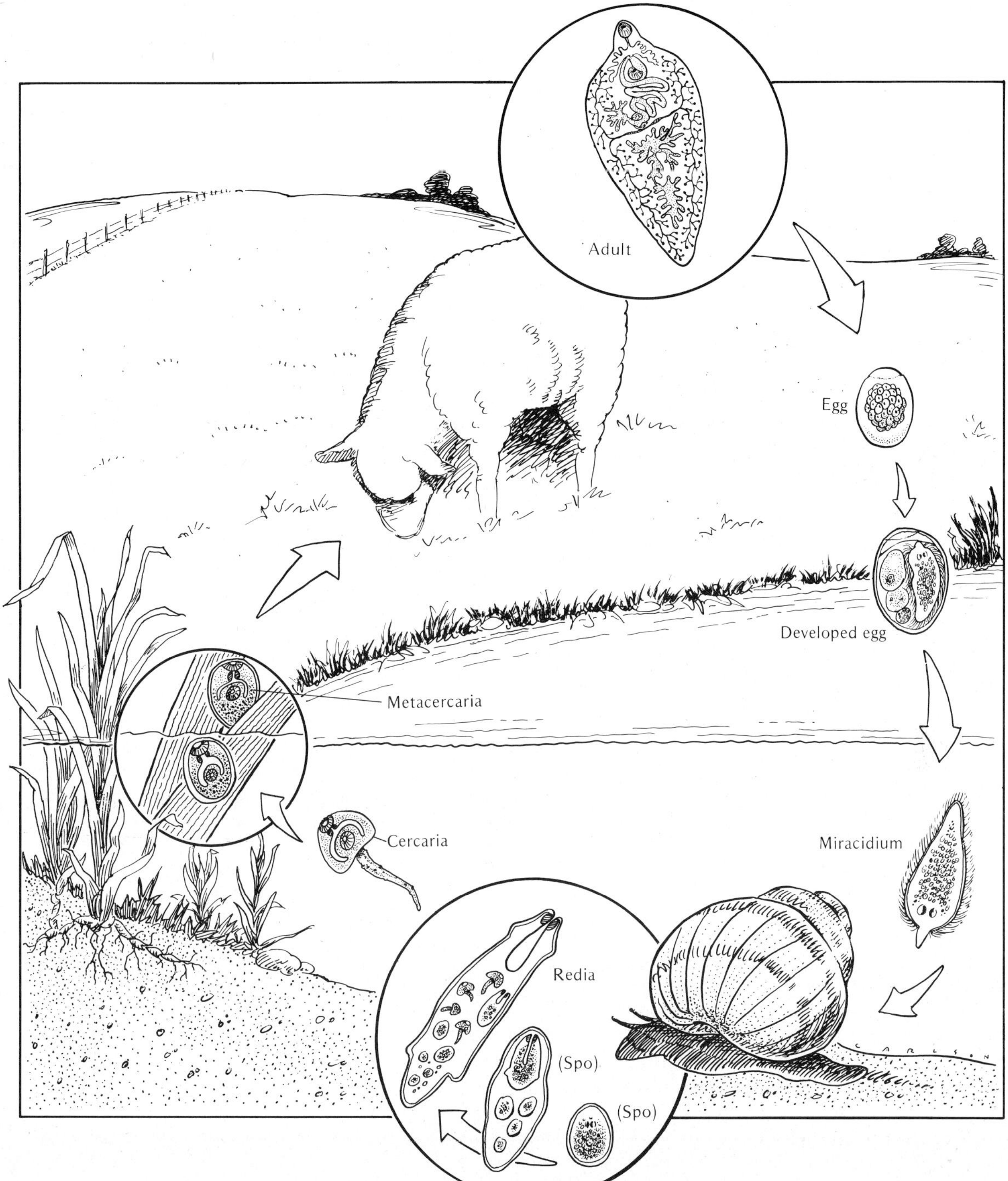


**FIG. 22.4**
Life cycle of *Fasciola hepatica*.

operculate eggs 130 to 150 by 63 to 90 μm. Operculate eggs do not reliably appear in flotation methods, and it is best to use a sedimentation technique. Eggs of *Fasciolopsis* and *Cotylophoron* are indistinguishable from *F. hepatica*, and some others are similar to them.

At postmortem, adult flukes are found in the gallbladder and in larger bile ducts of the liver. In deaths from acute fluke infections, tracks where flukes have migrated may be seen, and it is sometimes possible to find juvenile flukes in the parenchyma of the liver.

In endemic areas, diagnosis may often be made on the basis of clinical signs such as loss of condition, anemia, and failure to gain weight. Knowing the cycle of transmission of *F. hepatica* in a particular area will give a guide to the times of year and conditions under which infection is likely.

HOST–PARASITE INTERACTIONS. Pathogenesis as a direct result of the flukes'. activities may be either chronic or acute. Acute fascioliasis occurs during the preadult migration of the flukes in the parenchyma of the liver, and sometimes other organs, for about 8 weeks. The time of death and severity of the damage are related to the number of metacercariae ingested; deaths occur earlier and the damage is more severe with greater numbers of flukes. In sheep, 1000 or more metacercariae ingested within a short period may cause death. Cattle are more resistant than sheep to the effects of flukes; 10,000 to 20,000 metacercariae were found to cause death (Boray, 1969).

The most striking changes in acute fascioliasis are seen in the liver, where there are hemorrhagic tracts representing the paths of the migrating juvenile flukes. The liver becomes swollen and, in some instances, may become so swollen that the capsule ruptures.

Chronic fascioliasis occurs beyong 12 weeks PI, when the flukes have reached the bile ducts and are maturing sexually. Although production losses are the usual results of chronic infections, deaths also occur. As few as 100 flukes can kill sheep. Although the data are scanty, it appears that more than a thousand adult flukes are required to cause the death of cattle. In chronic fascioliasis, the damage is primarily to the liver, but unlike the acute phase of the disease, the damage is from fibrosis (scar tissue), blockage of the bile ducts, and inflammation of the bile ducts.

The liver performs a vast array of essential functions in vertebrates, and it is axiomatic that liver damage will have a variety of detrimental effects. The adult flukes also cause anemia as a result of blood sucking or from hemorrhages where they graze on the mucosa of the bile ducts. The evidence is equivocal, but seems to favor a hemorrhagic anemia as opposed to a suppression in production of new red cells. The result of liver damage is failure to gain weight in young animals, weight losses in older animals, reduced milk production in cattle, and lowered fleece weights in sheep. Continuous loss of blood puts an extra burden on the bone marrow for production of red blood cells; animals on an otherwise adequate diet are likely to become progressively more anemic with the passage of time. Since more than just red cells are lost, animals become hypoproteinemic, so that the osmotic pressure of the blood is lowered and fluid is retained in the tissue spaces.

No immunity to *F. hepatica* has been demonstrated, although antibodies can be found in the blood of infected animals. Observations in the field indicate that older animals become resistant to infection. Development of vaccines has been attempted, but has not yet been successful. Further efforts to develop practical immunizing agents are continuing.

TREATMENT. The drug of choice for many years for sheep has been carbon tetrachloride at 500 mg/kg. In the 1940s hexachlorethane was developed especially for use in cattle, which may not tolerate carbon tetrachloride well. In using both drugs, precautions must be taken to prevent severe intoxication and even death of animals. In recent years, nitroxynil and clioxanide have been found to be effective against both juvenile and adult flukes. In all drugs, higher doses must be used against juvenile flukes than against the adults in the bile ducts. The drug of choice for use in humans is bithionol.

Strategic treatment should be part of an overall control program; animals are treated after they have become infected but before the flukes have matured. To minimize the losses from black disease or redwater, animals should be vaccinated against these bacterial diseases several months prior to any likelihood of being infected by flukes.

EPIDEMIOLOGY AND CONTROL. Control of liver flukes is effected in three major ways:

1. Snail control
2. Treatment of infected vertebrates

3. Herd management through predicting outbreaks of disease

There is usually a single major snail vector in a geographic area. For example, in Britain and on the European continent, it is *Lymnaea truncatula*. In Australia, *L. tumentosa* is the most important vector, and in North America there are three important vectors, *Fossaria modicella, Stagnicola bulimoides,* and *Pseudosuccinea columella*, depending on the area. Before starting a control program, the vector should be determined and information developed on its reproduction and habitat preferences. For example, lymnaeid snails are air breathers (pulmonates) and are generally thought of as being semiaquatic; they require water for breeding, die when dried out, but are likely to be found in habitats where there is only a film of water such as mudflats near a stream or pond. They are not typically found in fast-moving streams. *L. truncatula* is even found in low-lying pastures where the hoofprint of an animal can provide sufficient water for survival. This type of snail is an opportunist; reproduction is hermaphroditic and takes place all through the growing season. Therefore, populations tend to increase explosively when moisture is abundant and to crash when there is a dry period. The crucial factors in completing the life cycle are adequate moisture for the vector and temperatures above 10°C for development of the larval stages in the vector.

Within the limitations of moisture and temperature, each climatic type shows its own pattern when development, infection, and disease are likely to occur (Table 22.2). Although North America is indicated as having a summer development, it is likely that a subtropical pattern of development takes place in Florida. It should be noted that the broad patterns of development shown in the table are more complex when a specific geographic area is studied; Boray (1969) found five distinct epidemiological patterns in southeastern Australia.

Chemical control of snails is most often attempted by using copper sulfate (blue stone) or sodium pentachlorophenate in the water. A simple means is to immerse a cloth bag containing $CuSO_4$ in a running stream. The problem with this method clearly is that the snails may not be in the stream or ditch, but rather well out of the flow on a mudflat. The solution in Britain has been to broadcast finely granular $CuSO_4$ on pastures where snails are abundant. In Australia, high-pressure spraying of pastures with sodium pentachlorophenate has been found to be effective in irrigated areas.

The semiaquatic nature of lymnaeid snails can be exploited to bring about ecological control of snail populations. Lymnaeids are found in streams where the water flow is relatively slow, in ponds and in lakes near the shore, and on mudflats at the margins of bodies of water. In the Rocky Mountain West, liver fluke infections are associated with irrigation, and extensive irrigation systems distribute water from mountain run-off. Irrigation ditches are constructed so that water flow is slow but constant. As systems age, they tend to erode, so that the sides slope and plants take root in or at the edges of the water. The habitat thereby becomes suitable for lymnaeid snails.

In a study done during the 1930s under WPA auspices, Leonard Swanson (unpublished report) described methods of making the ditch banks vertical and of removing vegetation to reduce the snail habitat. The methods worked and reduced snail numbers. But the study also pointed up that what works in one time and place may be entirely inappropriate in others. Consider that the WPA was instituted when the United States suffered from the worst depression in its history. Some 14 million persons were out of work, with WPA wages of about \$60/month; men were willing to shape ditch banks with shovels and to grub out weeds. During the intervening 40 years the agricultural labor force migrated to urban centers so that a ready supply of labor has gone, daily wages exceed \$25, and agriculture has become heavily dependent on machinery for work requiring almost any earth moving. One could hardly gather the workforce, let alone pay the people living wages for their labors. Machinery, if it can be gotten into the proper areas at all, is expensive and may not be able to do the job required. Thus, this method of control can be used only in areas of the world with low standard of living and an excess of unskilled labor.

Some aspects of habitat aleration are still feasible in a developed country. Snails typically bloom in relatively restricted areas such as mudflats or boggy areas. It is usually possible to fill, drain, or fence boggy areas of a pasture at a modest cost. An owner who could cut death losses by three head of cattle by this means would have perhaps \$300 more beef to market in the first year alone.

Snail control programs are difficult to put into operation under the best of circumstances. Populations bloom when there is even a slight alteration in the

**TABLE 22.2**
Seasonal Occurrence of Fascioliasis

| CHARACTER OF CLIMATE | LOCATION | STAGE | JAN | FEB | MAR | APR | MAY | JUNE | JUL | AUG | SEPT | OCT | NOV | DEC | NUMBER OF GENERATIONS OF CERCARIAE PER YEAR |
|---|---|---|---|---|---|---|---|---|---|---|---|---|---|---|---|
| Continental—Warm summers, cold winters or cold–moist. | Eastern, Northern and Central Europe; hill pastures of Scotland; Northern Asia; N. America | Development | · | · | · | · | + | ++ | ++ | ++ | ++ | · | · | · | |
| | | Infection | · | · | · | · | · | · | · | ++ | ++ | ++ | · | · | ½–1 |
| | | Disease | + | · | · | · | · | · | · | · | · | ++ | ++ | +· | |
| Coastal—wet, cool wet, warm with possible frost | Britain, Ireland; West coast of Europe; subtropics; highlands of tropical Asia | Development | · | · | · | · | + | ++ | ++ | ++ | ++ | ++ | · | · | |
| | | Infection | + | + | · | · | + | + | + | ++ | ++ | ++ | ++ | + | 2 |
| | | Disease | ++ | ++ | ++ | ++ | · | · | · | + | + | ++ | ++ | ++ | |
| Tropical or subtropical—hot, wet, or hot with wet and dry seasons. | Subtropics of Australia Highlands of tropical countries; Southern Hemisphere | Development | ++ | ++ | ++ | ++ | + | · | · | · | ++ | ++ | ++ | ++ | |
| | | Infection | ++ | ++ | ++ | ++ | + | + | + | · | + | ++ | ++ | ++ | 4 |
| | | Disease | ++ | ++ | ++ | ++ | ++ | ++ | ++ | ++ | + | + | + | ++ | |
| Tropical—hot, wet | Tropics | Development | ++ | ++ | ++ | ++ | ++ | ++ | ++ | ++ | ++ | ++ | ++ | ++ | |
| | | Infection | ++ | ++ | ++ | ++ | ++ | ++ | ++ | ++ | ++ | ++ | ++ | ++ | |
| | | Disease | ++ | ++ | ++ | ++ | ++ | ++ | ++ | ++ | ++ | ++ | ++ | ++ | |

[Adapted from Boray 1969.]

ecosystem favoring their survival and reproduction. Chemicals used in control have side effects on other animals in the water such as fish, and may even be toxic for livestock; our concern for environmental contamination limits the use of many toxic substances. Lymnaeid and other snails are readily carried by streams, so that recolonization takes place even if all snails have been killed. Since lymnaeids are self-fertilizing, a single snail left in an area is sufficient to repopulate it within a short time. A control program instituted by a single owner on his own premises is likely to bring little reward for the money and effort spent. The problem, then, is to obtain cooperation of all of the owners in a particular drainage so that an area-wide program can have significant effect.

A different approach to control has been to predict when problems may occur so that animals can be removed from potential problem areas. The work of Ollerenshaw and co-workers (Ollerenshaw & Smith, 1969) provides a basis for predicting when outbreaks of fascioliasis are likely to occur.

Ollerenshaw's studies concern mostly Great Britain. Weather records starting in 1727 and disease outbreak records starting in 1735 were used as a broad frame of reference. More detailed information was obtained from recent outbreaks, snail biology, laboratory experiments on the larval stages of *F. hepatica*, and specific parameters of habitats.

The studies showed that the most important factor is the balance between rainfall and evapotranspiration during the summer and that the month of June in England was especially significant. Evapotranspiration is the total amount of water lost through evaporation from the soil surface and through transpiration from plant leaves. If precipitation exceeds evapotranspiration, the soil gains water and water may be present at the surface.

The critical temperature for larval stages of *F. hepatica* is 10°C (50°F). Below 10°C the eggs will not develop, nor will the larval stages in snails develop or the cercariae emerge. Soil temperatures above 10°C occur in England and Wales from May through October in an average year.

Predicting outbreaks of fascioliasis has been possible by the use of the formula

$$M_t = (R - P + 5)n$$

where

$M_t$ = monthly total
$R$ = rainfall in inches
$P$ = evapotranspiration in inches
$n$ = number of rain days

A figure is calculated for each month, and when the value for a month ($M_t$) approaches 100, pastures tend to be wet during the whole month. With temperatures above 10°C and wet months starting in June, values of 100 (and values over 100 are expressed as 100) for 2 or 3 months give a high level of confidence that acute fluke disease will occur.

The system is a relatively simple one and the calculation for a month having 6 in of rain which fell on 10 days with potential transpiration of 5 in. is:

$$M_t = (6 - 5 + 5)\ 10 = 60$$

It has been pointed out that Ollerenshaw's work has been most directly related to European conditions and may not apply to Australia (Boray, 1969). Control measures there have involved grazing pastures only at certain times of the year coupled with strategic treatment of herds. The basic approach is to utilize possibly contaminated pastures before the cercariae emerge from the snails and generally to treat livestock after they have been infected but before the onset of clinical signs.

Host range, longevity, and egg production of liver flukes in definitive hosts have implications for control. In sheep, they cause relatively little host reaction as encapsulation, and the flukes may live for many years, all the while producing large numbers of eggs. Estimates are that one fluke produces about 20,000 eggs per day. Cattle, on the other hand, are relatively poor hosts in that the infection usually is retained only for months. Egg production by *F. hepatica* is also poorer in cattle than in sheep. While sheep may contaminate pastures with 2 million eggs per day per animal, cattle shed only 50,000 to 100,000 eggs per day per animal. Pasture contamination by cattle therefore increases more slowly and ends sooner than with sheep.

If we make a simple calculation, the difference in pasture contamination by sheep and cattle can be thrown into sharp focus. Stocking rates for sheep on pasture are about eight times higher than for cattle. Taking the figures in the previous paragraph for egg production and assuming equal pasture utilization (one ox or eight sheep) over a 2-month period, the sheep would provide 960 million eggs on the pasture, whereas one ox would deposit 4.5 million, a difference of over 200-fold. In an endemic area, the livestock raiser must watch husbandry practices more carefully with sheep than with cattle.

As indicated in the early part of this chapter, the host range of *F. hepatica* is broad and extends to rabbits, which are common wherever livestock are raised. Although the evidence is somewhat fragmentary, it appears that rabbits cannot by themselves perpetuate fluke infections. Therefore, if a rigorous control program is instituted, it can be directed at domestic animals, only, without seriously affecting the outcome.

### *Other Members of the Fasciolidae*

There are three additional genera whose members are found in the livers or intestines of herbivores or humans. Some are found infrequently or are of little medical or economic importance. The following flukes are briefly considered, because of the losses they may cause.

*Fascioloides magna* is found in North America, where it is a parasite of both wild and domesticated ruminants; the normal hosts are deer. Its life cycle is basically the same as that of *F. hepatica*. In hosts other than deer, the flukes may mature, but eggs do not reach the outside; for example, in cattle and moose there is a strong encapsulating reaction by the host, which walls off the parasite. In sheep, the flukes continue to migrate through the parenchyma of the liver rather than remain in bile ducts. *F. magna* is highly pathogenic for sheep, and pathogenesis seems to result from this continual migration and damage to the animal's liver.

*F. magna* provides an example of interactions that may take place between wild and domestic herbivores. Deer are reservoirs of infection for domestic animals in which the parasites do not mature. The pathogenic effect in sheep is so severe that some workers recommend not raising sheep in areas where *F. magna* is enzootic in deer. *F. magna* and *F. hepatica* may be found in the same geographic area, but mature *F. magna* can be readily distinguished from *F. hepatica* by their large size (as long as 10 cm), and its somewhat different shape (Fig. 22.2).

*Fasciolopsis buski* is a parasite of humans and swine; it is found in the small intestine. Stoll (1947) estimated that there are 10 million human infections in the Orient. The fluke is large, 20 to 75 mm by 8 to 20 mm, and it has a fasciolid life cycle; the snail vectors are members of the Planorbidae. Humans become infected by eating certain aquatic vegetables, or most often by peeling them with the teeth before eating them. Pathogenesis in humans is severe, resulting not only from local damage to the intestine but also from general intoxication. Control methods involve personal protection, snail control, and sanitation. Vegetables should not be peeled with the teeth or they should be immersed for 10 sec in boiling water. Transmission is maintained by fecal contamination of fields where aquatic vegetables are cultivated. Since human excreta or night soil may be essential to agriculture, stopping its use may not be feasible. An alternative is to hold the material for a month in a tight tank before spreading it on the fields. Obviously, snail control in agricultural fields through the use of toxic chemicals should be avoided and other means of reducing snail populations should be attempted (see *Schistosoma* control methods). Changing individuals' food preparation methods may be moderately difficult, and changing agricultural practices is even more difficult. Thus, the broader public health measures such as sanitary disposal of human excreta are clearly the greater challenge in reducing the infection rate from *F. buski*.

## FAMILY DICROCOELIDAE

NAME OF ORGANISM AND DISEASE CAUSED. *Dicrocoelium dendriticum* (syn. *D. lanceolatum*), lancet fluke or little liver fluke, causes dicrocoeliasis.

HOSTS AND LOCATION. Definitive hosts are the sheep, goats, cattle, many wild and domesticated animals, and humans. The fluke is found in the liver. Intermediate hosts include land snails of the genus *Cionella* (North America) and others such as *Limax*, *Helicella*, and some aquatic forms such as *Planorbis*; the parasite first inhabits the digestive gland and later the respiratory chamber. Thirty-eight species of snails in nine families are known to be vectors (Tarry, 1969). Second intermediate hosts are ants of the genus *Formica*; metacercariae are found mostly in the hemocoel but also in the nervous system. Twelve species are known to be vectors in various areas of the world (Tarry, 1969).

DISTRIBUTION AND IMPORTANCE. Distribution is cosmopolitan. Europe, Asia, Africa (Egypt), the Middle East, Australia, South America, and Caribbean islands are included, and, in North America, New York State and eastern Alberta, Canada.

The prevalence in Poland is 80%, and in northern France 37.5% in sheep and 46% in cattle (Tarry, 1969). In the Alma-Ata region of Kazakhstan, mature, coarse-wooled sheep, merino sheep, and goats averaged 1123, 462, and 738 flukes, respectively. The prevalence of infection in goats and mountain goats was 80% and 70%, respectively (Satubaldin, 1958).

The damage to the host is directly proportional to the number of flukes in the liver and the length of time of the infection. Black disease (see *F. hepatica* for description and etiology) sometimes occurs in infected animals. Livers are condemned at slaughter. Losses are seen in the form of reduced production of meat, milk, and fiber.

Although there are documented infections in humans, they are sporadic and probably of no real public health importance.

MORPHOLOGY. Adults are 1.5 to 2.5 mm wide by 10 mm long and pointed toward the anterior, giving the common name of *lancet fluke*; the testes are anterior to the ovary; the uterus has both descending and ascending loops; vitellaria are restricted to the middle quarter of the body (Fig. 22.5).

LIFE CYCLE. Adults lie in the bile ducts or gallbladder; embryonated eggs containing miracidia reach the outside by passing down the common bile duct and are deposited with the feces (Fig. 22.6). Eggs hatch when eaten by a snail. Two generations of sporocysts give rise to cercariae about 3 months after infection. Cercariae are expelled from the branchial chamber of the host in a mucoid envelope called a *slime ball*, which is 1.5 mm in diameter. Cercariae are eaten with the slime ball by ants and develop into metacercariae. Infected ants are eaten by the definitive host and the juvenile flukes reach the liver through either the common bile duct or the circulation. The prepatent period is about 70 days.

DIAGNOSIS. Finding eggs in the feces, embryonated, operculated, and about 26 by 41$\mu$m is the method of diagnosis. In an endemic area, clinical signs and symptoms may be sufficient for an experienced person to make a diagnosis.

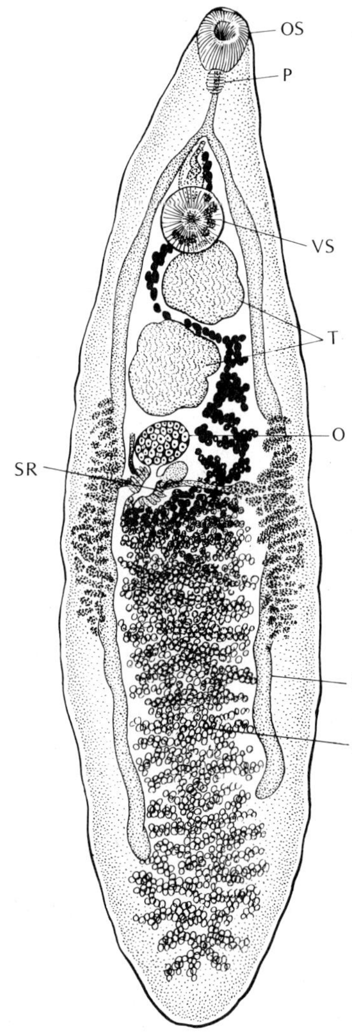


**FIG. 22.5**
Diagram of an adult *Dicrocoelium dendriticum* (IC = intestinal caecum; O = ovary; OS = oral sucker; P = pharynx; SR = seminal receptacle; T = testis; U = uterus; VS = ventral sucker).

Humans serve as the host, but it is wise to be cautious if *Dicrocoelium* eggs are found in feces because erroneous diagnoses have been made where persons have eaten infected liver and passed eggs for a day or two.

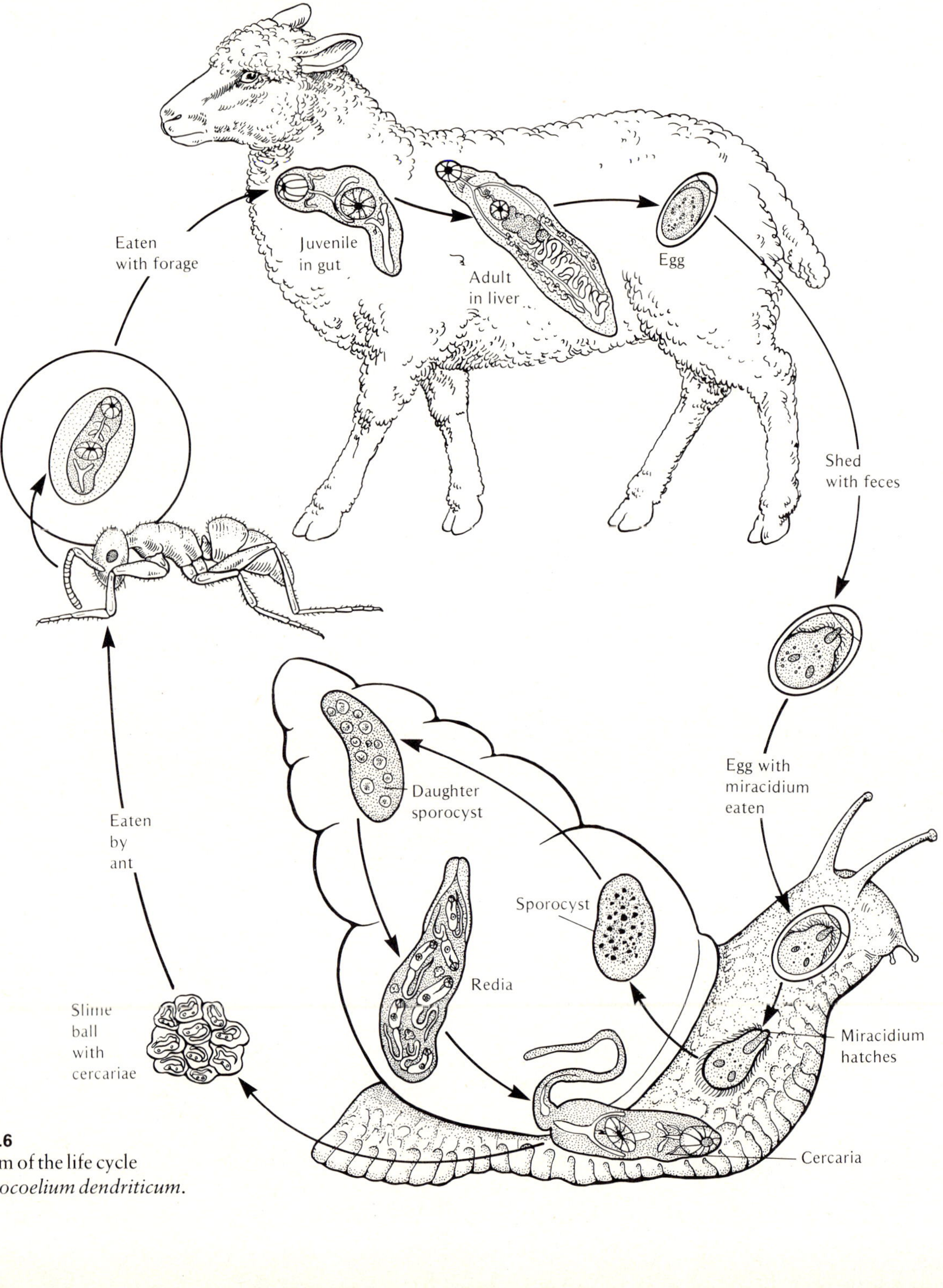


**FIG. 22.6**
Diagram of the life cycle of *Dicrocoelium dendriticum*.

TREATMENT. Hexachlorethane-bentonite at a dosage of 15 g has been recommended for many years. About 20 mg/kg of Hetolin removed 90% of worms in most tests. Thiabendazole (Merck) at 200 or 300 mg/kg is reported to remove more than 90% of adult flukes.

HOST–PARASITE INTERACTIONS. Damage is primarily to the liver, and the changes are directly proportional to the number of flukes inhabiting the bile ducts and the length of time the animal has been infected. As many as 7000 adult flukes have been found in the gallbladder of a sheep (Mapes, 1951), but several hundred seems to be more common. There is inflammation of the bile ducts; progressive development of connective tissue in the parenchyma of the liver leads to cirrhosis. Evidence points to production of a toxin (Mapes, 1951) as the cause of most of the pathology, because the flukes lie free in the bile ducts and do not block the passage of bile. They live at least 6 years in sheep.

In the first intermediate host, the snail, infections are primarily in the digestive gland, but in heavy infections sporocysts are also in the lymph spaces. More than 300 sporocysts have been found in naturally infected snails (Krull & Mapes, 1952c); about 3 months are required for production of cercariae in New York and 4 months in Britain (Tarry, 1969). Cercariae migrate to the pulmonary chamber of the snail and are incorporated into slime balls produced by the snail. Migration of the cercariae and production of slime balls seem to be caused by lowered temperatures or rain.

Successful transfer of infection from the snail through the ant to the definitive host comes about through a combination of food preferences, high infection rates in ants, and behavioral changes of infected ants. Slime balls contain 100 to 400 cercariae and are looked upon as gastronomic prizes by ants, which carry them back to the nest and share them with other members of the colony. In one endemic area, about 30% of ants were found to harbor metacercariae, and 48% were infected in another. Further, infection in an ant sometimes affects the nervous system, so that its mandibles clamp into vegetation when the temperature falls. The ant remains attached to a blade of grass until the temperature rises again, thereby increasing the probability of its being eaten.

EPIDEMIOLOGY AND CONTROL. The lancet fluke is placed in the same chapter as the fasciolids because of the contrast it provides in solving problems of control. Both are liver flukes but they do not generally respond to the same chemotherapeutic agents. Different approaches are also required to break the cycle. The fasciolid life cycles require water at certain stages. The cycle of *D. dentriticum* is entirely terrestrial and requires only damp conditions. Control of the lancet fluke is potentially easier for a single livestock owner than control of the common liver fluke because terrestrial snails migrate more slowly than aquatic snails, which are readily carried by water.

The broad host range of both kinds of flukes presents some difficult problems, but it seems to be of greater significance in the lancet fluke than in the common liver fluke. The consensus is that rodents and rabbits are not major sources of infection for domestic animals with the common liver fluke. However, the wide range of herbivorous animals that serve as normal hosts of the lancet fluke present a different situation. It is usually neither feasible nor desirable to eradicate rodents, rabbits, and large ungulates to eliminate a reservoir of infection for domestic animals. It may be possible to fence out large animals, but rodents and rabbits are almost uncontrollable.

The best approach seems to be a combination of treatment of animals on a herd basis and pasture renovation. Oddly enough, there is little information on control of lancet flukes. In New York, plowing and reseeding of pastures reduced snail populations to about 15% of their original levels. Whether this level is low enough to make transmission unlikely was not stated (Jensen et al., 1955). In Ershov's (1956) book on parasitic diseases of livestock, Antipin claimed that good control can be obtained by hand collecting snails on pastures or by turning poultry out into pastures where snails are found.

In New York Krull and Mapes found that snails occurred most often in raspberry patches or areas of shrubs; usually they were not found in wet areas. Their preferred food was broad-leafed plants; clover and grass were rejected. The snails moved into a variety of habitats during the growing season but congregated in rocky areas toward the fall. Elimination of rocky areas and keeping pastures in grasses might reduce snail populations.

Chemical control of snails on pastures is feasible by broadcasting molluscicides, but toxic effects and contamination of the environment should be weighed carefully.

## REFERENCES AND FURTHER READINGS

Alicata, J. E. 1964. Parasitic infections of man and animals in Hawaii.l *Hawaii Agric. Exp. Stat. Tech. Bull.* No. 61.

Boray, J. C. 1969. Experimental fascioliasis in Australia. *Adv. Parasit.* 7:95–210.

Carney, W. P. 1969. Behavioral and morphological changes in carpenter ants harboring dicrocoelid metacercariae. *Amer. Midland Nat.* **82**:605–611.

Dawes, B., and D. L. Hughes. 1964. Fascioliasis: The invasive stages of *Fasciola hepatica* in mammalian hosts. *Adv. Parasit.* 2:97–168.

DeLeon, D., R. Quinones, and G. V. Hillyer. 1981. The prepatent and patent periods of *Fasciola hepatica* in cattle in Puerto Rico. *J. Parasitol.* 67:734.

Durbin, C. G. 1952. Longevity of the liver fluke, *Fasciola* sp., in sheep. *Proc. Helm. Soc. Wash.* **19**:120.

Ershov, V. S. 1956. *Parasitology and Parasitic Diseases of Livestock* (English trans., 1960).

Foreyt, W. J., and A. C. Todd. 1976. Development of the large American liver fluke, *Fascioloides magna*, in white-tailed deer, cattle and sheep. *J. Parasit.* 62:26–32.

Jensen, P. C., C. R. Mapes, and J. H. Whitlock. 1955. Pasture management and control of the lancet fluke, *Dicrocoelium dendriticum* (Rudolphi, 1819) Looss, 1899. *Cornell Vet.* **45**:526–538.

Kendall, S. B. 1970. Relationships between species of *Fasciola* and their molluscan hosts. *Adv. Parasit.* 8:251–258.

Krull, W. H. 1958. The migratory route of the metacercaria of *Dicrocoelium dendriticum* (Rudolphi, 1819) Looss, 1899 in the definitive host. *Cornell Vet.* **48**:17–24.

Krull, W. H., and C. R. Mapes. 1952a. Studies on the biology of *Dicrocoelium dendriticum* (Rudolphi, 1819) Looss, 1899. (Trematoda: Dicrocoelidae), including its relation to the intermediate host, *Cionella lubrica* (Muller,). III. Observations on the slimeballs of *Dicrocoelium*. *Cornell Vet.* **42**:253–276.

——— 1952b. —— IV. Infection experiments involving definitive hosts. *Cornell Vet.* **42**:277-285

——— 1952c. —— V. Notes on infections of *Dicrocoelium* in *Cionella lubrica*. *Cornell Vet.* **42**:339–351.

——— 1952d. —— VI. Observations on the life cycle and biology of *C. lubrica*. *Cornell Vet.* **42**:464–489.

——— 1952e. —— VII. The second intermediate host of *Dicrocoelium dendriticum*. *Cornell Vet.* **42**:603–604.

——— 1953a. —— VIII. The cottontail rabbit, *Sylvilagus floridanus mearnsi*, as a definitive host. *Cornell Vet.* **43**:199–202.

——— 1953b. —— IX. Notes on the cyst, metacercaria, and infection in the ant, *Formica fusca*. *Cornell Vet.* **43**:389–410.

Mapes, C. R. 1951. Studies on the biology of *Dicrocoelium dendriticum* (Rudolphi, 1819) Looss, 1899 (Trematoda: Dicrocoeliidae), including its relation to the intermediate host, *Cionella lubrica* (Muller). I. A study of *Dicrocoelium* and *Dicrocoelium* infection. *Cornell Vet.* **43**: 382–432.

Mapes, C. R. and W. H. Krull, 1951. II., Collection of the snail, *Cionella lubrica* and its maintenance in the laboratory. *Cornell Vet.* **41**:433–444.

Merck, Sharp & Dohme. 1971. Second International Liver Fluke Colloquium. Wagenigen, The Netherlands, 2–6 October 1967.

Ogambo-Ongoma, A., and J. Goodman. 1976. *Fasciola gigantica* Cobbold, 1856 in the snail. *J. Parasitol.* **62**:33–38.

Ollerenshaw, C. B., and L. P. Smith. 1969. Meteorological factors and forecasts of helminthic diseases. *Adv. Parasit.* 7:283–323.

Olsen, O. W. 1944. Bionomics of the lymnaeid snail, *Stagnicola bulimoides techella*, the intermediate host of the liver fluke in southern Texas. *J. Agr. Res.* **69**:389–403.

——— 1944. Critical tests with a hexachlorethane-bentonite-water suspension as a fasciolicide in cattle. *J. Parasit.* **30** (Suppl.):14.

——— 1945. Ecology of the metacercariae of *Fasciola hepatica* in Southern Texas and its relationship to liver fluke control in cattle. *J. Parasitol.* **31** (Suppl.):20.

——— 1946. Hexachlorethane-bentonite suspension for the removal of the common liver fluke, *Fasciola hepatica*, from sheep. *Am. J. Vet. Res.* 7:358–364.

——— 1947. Longevity of metacercariae of *Fasciola hepatica* on pastures in the upper coastal region of Texas and its relationship to liver fluke control. *J. Parasitol.* **33**:36–42.

——— 1949. White-tailed deer as a reservoir host of the large American liver fluke. *Vet. Med.* **44**:26–30.

Palmieri, J. R., J. T. Sullivan, and C. K. Ow-Yang. 1977. Occurrence of a sporocyst generation of *Fasciola gigantica* in the mollusc, *Lymnaea rubiginosa*. *J. Parasitol* 63:299–300.

Pantelouris, E. M. 1965. *The Common Liver Fluke*. Pergamon Press, New York.

Ratcliffe, L. H. 1968. Hatching of *Dicrocoelium* eggs. *Exp. Parasitol.* *23*:67–68.

Reinhard, E. G. 1957. Landmarks in parasitology. I. The

discovery of the life cycle of the liver fluke. *Exp. Parasit.* 6:208–232.

Satubaldin, Kh. S. 1958. Characteristics of the helminth fauna of domestic and wild sheep and goats in the Alma-Ata region. In S. N. Boev (ed.) *Contributions to Helminthology*. Israel Program for Scientific Translations. Jerusalem, 1968, pp. 332–340.

Stoll, N. R. 1947. This wormy world. *J. Parasitol.* 33:1–18.

Tarry, D. W. 1969. *Dicrocoelium dendriticum*: The life cycle in Britain. *J. Helminthol.* 43:403–416.

# 23

# *Clonorchiasis and Opisthorchiasis*

Several flukes of the family Opisthorchiidae are important agents of human disease. These include the oriental liver fluke of humans, *Clonorchis sinensis*, and two cat liver flukes, *Opisthorchis felineus* and *O. viverrini*. Humans are occasionally infected with another opisthorchid, *Metorchis conjunctus*.

NAME OF ORGANISM AND DISEASE CAUSED. The most important member of the family Opisthorchiidae with respect to human disease is *Clonorchis sinensis* (Cobbold, 1875) Looss, 1907 (= *Opisthorchis sinensis*). The disease is called *clonorchiasis, opisthorchiasis* or *oriental liver fluke disease*.

The generic name *Clonorchis* has been and still is the subject of discussion. *Opisthorchis* was described in 1875 and *Clonorchis* in 1907; the major basis for separating them was the shape of the testes—highly branched in *Clonorchis* and lobed or only slightly branched in *Opisthorchis*. Additional species of *Opisthorchis* have been described where testes range from oval to highly branched, thereby seeming to invalidate the newer *Clonorchis* genus. As Malek (1980a) pointed out, however, there are reasons for retaining *Clonorchis*. *Clonorchis* has been commonly used in medical literature for about 70 years, and it refers to a specific disease caused by a specific parasite. Even though adult morphologies are similar, larval stages (metacercarial and cercarial flame cell patterns and body sizes) are different in each, thus giving additional support for the retention of both genera.

HOSTS AND LOCATION. Humans are the most suitable definitive host for *Clonorchis sinensis*. Most fish-eating mammals and a few fish-eating birds can be infected, but dogs and cats are probably the most important reservoir hosts (Komiya, 1966). Adults are in the liver.

DISTRIBUTION AND IMPORTANCE. *Clonorchis sinensis* is common in Japan, Korea, China, Hong Kong, and Indochina. This distribution coincides with the distribution of the snail first intermediate host, *Parafossarulus manchouricus*, and closely related hydrorbid snails (Malek, 1980a). Millions of people are infected. Nearly 100% of people in certain villages in China (Komiya, 1966) and over 65% of the population of Hong Kong are infected (Hou & Pang, 1964; Belamaric, 1973).

MORPHOLOGY. Adult flukes are 10 to 25 mm long and 3 to 5 mm wide (Fig. 23.1). Their tegument is quite transparent, which makes them ideal for studying internal structures in parasitology classes. The acetabulum is slightly smaller than the oral sucker and is located near the end of the anterior third of the body. The middle third of the body contains the vitellaria and uterus. Two highly branched testes occupy most of the posterior third of the body. A three-lobed ovary and a large seminal receptacle are just anterior to the testes.

Eggs are oval and yellow-brown (Fig. 23.2). Mature

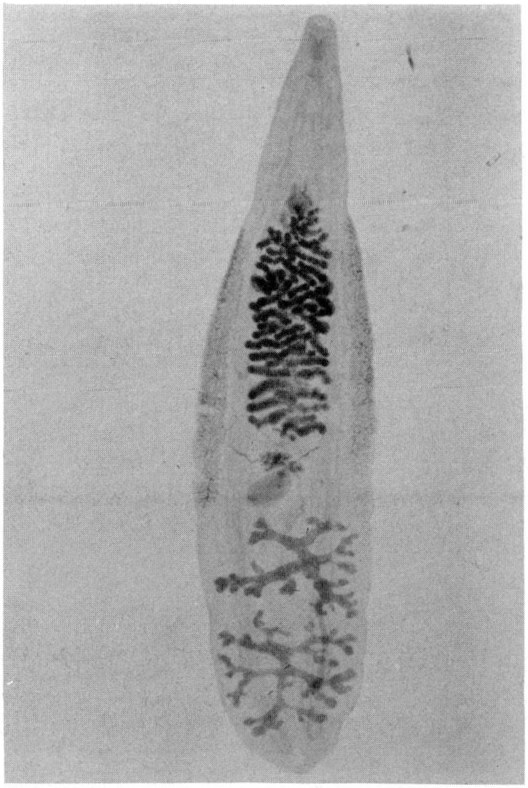

**FIG. 23.1**
*Clonorchis sinensis* adult.

eggs are 26 to 30 μm by 15 to 17 μm and have an etched surface; an operculum fits into a shoulderlike thickening of the shell. There is commonly a small knob on the shell opposite the operculum.

LIFE CYCLE. Fully embryonated eggs are passed in stools of infected individuals (Fig. 23.3). Eggs are ingested by the first intermediate host, *Parafossarulus manchouricus*, and closely related hydrorbid snails. When eggs hatch in the alimentary canal of the snail, miracidia penetrate the intestine and develop into sporocysts. Rediae are produced in sporocysts, leave them, and develop in the digestive gland of the snail. Oculate (one pair of eyes), lophocercous cercariae mature in the digestive gland after leaving the rediae, penetrate the surface of fishes, and develop into metacercariae. Metacercariae infect 80 species of freshwater fishes, but only a dozen of them are responsible for most human infections. In addition to fishes, 3 species of freshwater shrimp harbor infective metacercariae and have been implicated in transmission in China (Yoshimura, 1965). Metacercariae encyst in subcutaneous connective tissue and muscle (Komiya, 1966) and become infective in a month.

Definitive hosts ingest uncooked fish containing metacercariae. Metacercarial cyst walls dissolve in the duodenum or intestine (Gibson & Sun, 1971) and juvenile flukes migrate up the common bile duct into the small bile ducts of the liver (Sun et al., 1968). Egg laying begins 3 to 4 weeks after infection.

DIAGNOSIS. Clinical symptoms, along with a history of ingesting raw fish, help to confirm a diagnosis. Symptoms are classified as mild, moderate, or severe, depending upon the worm burden, length of infection, and health of the host (Komiya, 1966). Light infections may be asymptomatic. Heavy infections are characterized by chills and fever, followed by liver enlargement and tenderness, sometimes with splenomegaly. Chronic heavy infections cause cholecystitis (inflammation of the gallbladder) and hepatitis (inflammation of the liver). Persons with moderate infections of 100 to 1000 flukes show appetite irregularity, diarrhea, edema, and some hepatomegaly (liver enlargement).

A definitive diagnosis depends on recovery of eggs from feces or duodenal aspirates. Ether sedimentation using a Tween 80 detergent solution is recommended, especially for diagnosis of light infection (Komiya, 1966). There is no widely accepted serodiagnostic technique (Hillyer, 1982).

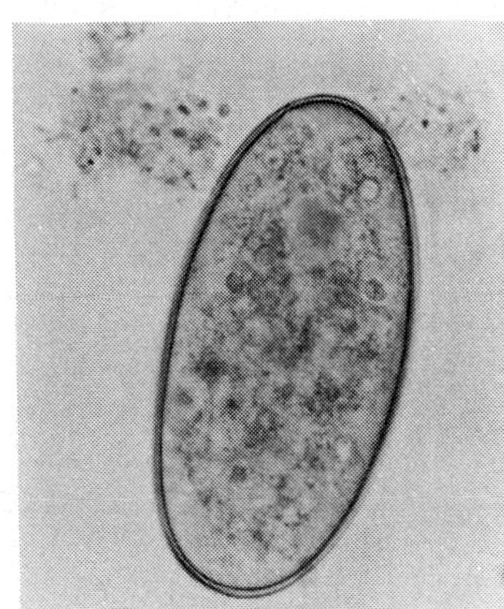

**FIG. 23.2**
*Clonorchis sinensis* egg.

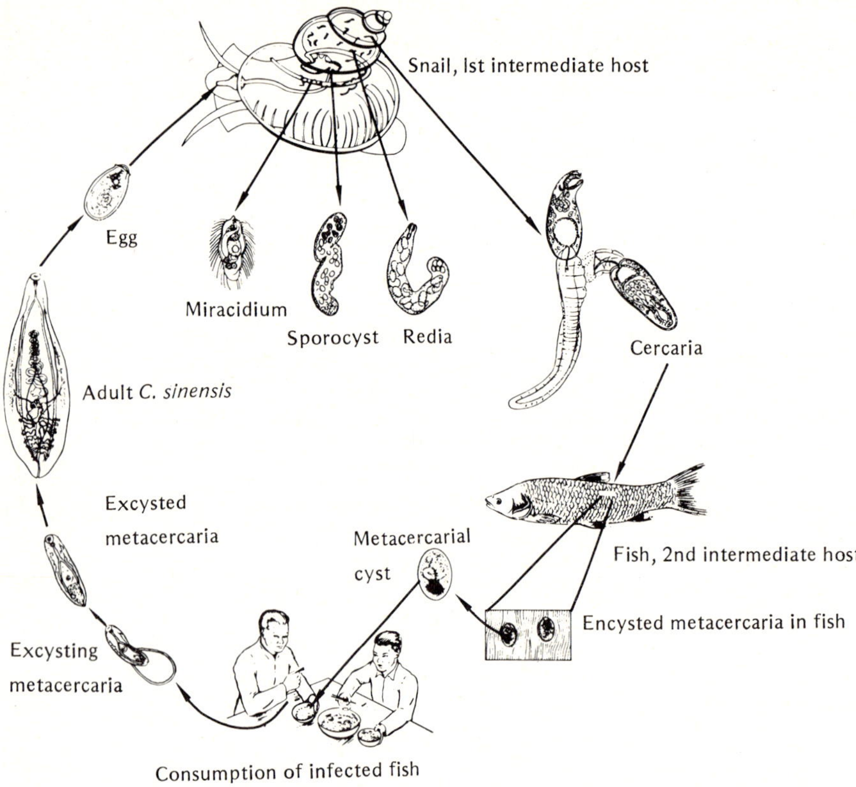


**FIG. 23.3**
Diagram of the life cycle of *Clonorchis sinensis*. [From Yoshimura, 1965.]

HOST–PARASITE INTERACTIONS. *C. sinensis* adults and ova are usually found in bile ducts; the ova apparently evoke no immune response (Sun, 1969). Occasionally an ovum can evoke a nonspecific granulomatous response in the wall of a gallbladder. More typical is bile duct epithelium hyperplasia and subsequent desquamation (shedding). Fibrosis (connective tissue formation) is common in infected bile ducts. Flukes occasionally invade the pancreas and may cause atrophy of the islets of Langerhans in severe infections.

*C. sinensis* has been implicated in bile duct cancer (cholangiocarcinoma) (Belamaric, 1973; Sun, 1980). Eighteen of 19 patients (95%) with intrahepatic bile duct cancer in Hong Kong also had chronic *C. sinensis* infections. Belamaric concluded that severe infections exert a mild but essential carcinogenic effect. He speculated that eradication of *C. sinensis* would eliminate bile duct cancer.

TREATMENT. Hexachloroparaxylol and Hetol [1,4-bis(trichloromethyl)benzene] appear to be useful; Praziquantel is also recommended.

EPIDEMIOLOGY AND CONTROL. Humans must ingest infective metacercariae in raw fish or crayfish to become infected. Raw and pickled fish are delicacies in many parts of the Orient, so millions of people are potential hosts for *C. sinensis* infections. Infection rates correlate with a community's predilection for raw fish (Gibson & Sun, 1971). For example, northern Chinese do not eat raw fish, but southern Chinese do; the prevalence of infection increases from north to south even though animal reservoirs are infected throughout China. Reservoir hosts include dogs, cats, pigs, and rats (Malek, 1980a).

Clonorchiasis can be spread over a wide area because of the longevity of the parasites in the definitive host. A patient acquired the disease in China and lived in Panama for 40 years with no travel to an endemic area (Calero, 1967). This is the longest known duration for a *C. sinensis* infection; it indicates the need to question patients about travel.

The obvious means of control of clonorchiasis is not to eat raw fish. Unfortunately, cultural preferences are difficult if not impossible to change; this appears to be the case with clonorchiasis (Gibson & Sun, 1971). Travelers to the Orient sometimes sample indigenous foods without recognizing the hazards, and persons eating raw fish dishes stand a significant chance of becoming infected. It has been calculated that 10 g of fish contain an average of 2.6 metacercariae and that a single meal at which infected raw fish was eaten might result in the person later having 20 adult flukes (Sun, 1980).

Molluscicides to control the snail first intermediate host are generally not used because they are often harmful to important freshwater fish (Malek, 1980a). Animal and human feces are important fertilizers for fish ponds in the Orient. Feces must be properly composted first, and the process must be rigidly controlled to prevent viable eggs from entering water. A successful composting program is operating now in China (Malek, 1980a). In addition, snail populations have been significantly reduced by removal of water plants when snails lay their eggs (April to September), thereby permitting fish more easily to feed on larval snails (Gibson & Sun, 1971). Thus, even though food habits are difficult to change, progress is being made by breaking the parasite life cycle at the egg and intramolluscan stages.

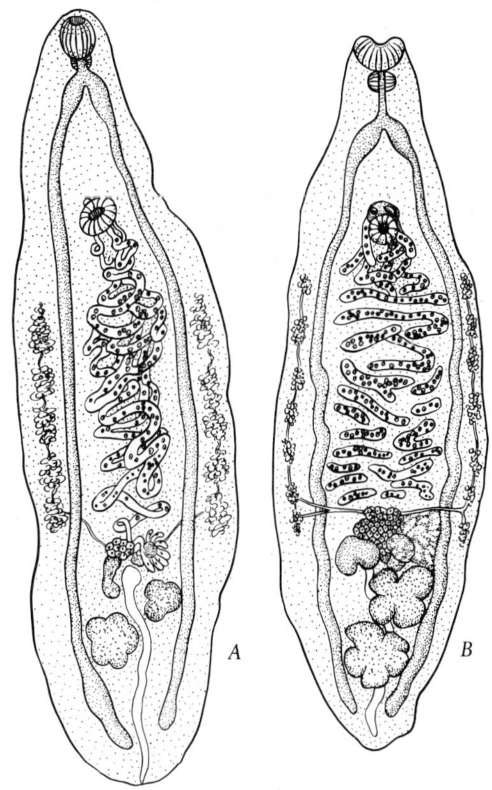


**FIG. 23.4**
*Opisthorchis felineus* (*A*) and *O. viverrini* (*B*) adults.

## OTHER OPISTHORCHIIDAE

Opisthorchiasis is also caused by two flukes, *Opisthorchis felineus* and *O. viverrini* (Fig. 23.4). Both are cat liver flukes that infect other mammals and humans. There are at least 3.5 million human infections of *O. viverrini* in Thailand, but no estimate has been made on the number of human infections throughout the remainder of Southeast Asia (Wykoff et al. 1965). In a review of opisthorchiasis, it was estimated that there are more than 1 million human infections of *O. felineus* in Europe, Russia, North Korea, and probably in Japan, India, and the Philippines (Malek, 1980b).

The life cycles of *Opisthorchis* spp. are similar to that of *C. sinensis*. Bithyneid hydrorbid snails are first intermediate hosts and numerous freshwater fishes are second intermediate hosts. Definitive hosts (including humans) are infected by ingesting metacercariae in raw or undercooked fish.

Wykoff et al. (1965) compared *O. viverrini* and *O. felineus* and concluded that these species could not be distinguished on the basis of the characteristics of adults or eggs, but that they are valid species because of differences in flame cell patterns in larvae. *Opisthorchis* eggs can be distinguished from *Clonorchis sinensis* eggs because they have a smaller diameter (11 μm versus 16 μm) and a less pronounced shoulder at the operculum.

Diagnosis, clinical symptoms, and pathology of opisthorchiasis are similar to those of clonorchiasis.

Like *Clonorchis sinensis* infections, cancer of the liver has also been linked to *Opisthorchis* infections (Sadun, 1955). There is no satisfactory drug to treat opisthorchiasis.

Control measures are similar to those for *C. sinensis*. Because fish are a basic food in endemic areas and because of cultural preferences, raw fish is often eaten. Control must therefore be focused at other stages in the parasite life cycle. Dogs and cats are important reservoirs, so proper sanitation of human feces is not enough. Snail control appears to be the only realistic means of controlling opisthorchiasis. Molluscicides cannot be used because they harm the fish used for human consumption. Therefore biological control of snails ought to be studied for reduction in the number of infections.

*Amphimerus pseudofelineus* ( = *Opisthorchis guayaquilensis*) has been reported in humans and dogs in Ecuador. There is potential for a great number of human infections, because it is found in fish-eating mammals in South, Central, and North America.

*Metorchis conjunctus* is another North American member of the family Opisthorchiidae which occasionally infects humans. It is in the bile ducts of dog, cat, fox, mink, raccoon, and humans (Healy, 1970). The snail *Amnicola limosa porata* is the first intermediate host and the common sucker, *Catostomus commersoni*, the second intermediate host, harbors metacercariae. Fortunately, few people eat suckers, either raw or cooked; therefore, this parasite should not become a serious human health problem.

## REFERENCES AND FURTHER READINGS

Belamaric, J. 1973. Intrahepatic bile duct carcinoma and *C. sinensis* infection in Hong Kong. *Cancer* **31**:468–473.

Calero, C. M. 1967. Clonorchiasis in Chinese residents of Panama. *J. Parasitol.* **53**:1150.

Gibson, J. B., and T. Sun. 1971. Clonorchiasis. In R. A. Marcial-Rojas (ed.), *Pathology of Protozoal and Helminthic Diseases.* Williams & Wilkins, Baltimore, pp. 546–566.

Healy, G. R. 1970. Trematodes transmitted to man by fish, frogs, and crustacea. *J. Wildl. Dis.* **6**:255–261.

Hillyer, G. V. 1981. Immunoserology of fascioliasis, paragonimiasis, and clonorchiasis. In K. Walls (ed.), *Immunoserology of Parasitic Diseases.* Marcel Decker, New York.

Hou, P. C., and L. S. C. Pang. 1964. *Clonorchis sinensis* infestation in man in Hong Kong. *J. Pathol. Bacteriol.* **87**:245–250.

Komiya, Y. 1966. *Clonorchis* and clonorchiasis. *Adv. Parasitol.* **4**:53–106.

Malek, E. A. 1980a. Clonorchiasis. In E. A. Malek (ed), *Snail-Transmitted Parasitic Diseases*, Vol. II. CRC Press, Boca Raton, Fla., pp. 67–89.

———. 1980b. Opisthorchiasis. In E. A. Malek (ed), *Snail-Transmitted Parasitic Diseases*, Vol. II. CRC Press, Boca Raton, Fla., pp. 91–105.

Sadun, E. H. 1955. Studies on *Opisthorchis viverrini* in Thailand. *Am. J. Hyg.* **62**:81–115.

Sun, T. 1969. The non-antigenicity of intact ova of *Clonorchis sinensis. J. Med. Microbiol.* **2**:358–360.

———. 1980. Clonorchiasis: A report of four cases and discussion of clinical manifestations. *Am. J. Trop. Med. Hyg.* **29**:1223–1227.

Sun, T., S. T. Chou, and J. B. Gibson. 1968. Route of entry of *Clonorchis sinensis* to the mammalian liver. *Exp. Parasitol.* **22**:346–351.

Wykoff, D. E., C. Harinasuta, P. Juttijudata, and M. M. Winn. 1965. *Opisthorchis viverrini* in Thailand—the life cycle and comparison with *O. felineus*. *J. Parasitol.* **51**:207–214.

Yoshimura, H. 1965. The life cycle of *Clonorchis sinensis*: A comment on the presentation in the seventh edition of Craig and Faust's *Clinical Parasitology*. *J. Parasitol.* **51**:961–966.

# 24

# *Echinostomatiasis*

NAME OF ORGANISM AND DISEASE CAUSED. Echinostomatiasis is caused by flukes in the family Echinostomatidae. At least 8 genera and 15 species from this family infect humans (Table 24.1), and the potential for additional human infections is great because these flukes typically have low host specificity. A good example of low host specificity is the cosmopolitan parasite, *Echinostoma revolutum*; the adult parasitizes at least 23 species of birds and 9 mammals, including humans. Miracidia infect 10 or more snail species in North America, and metacercariae are found in 16 snail species, 2 species of fingernail clams, 7 species of tadpoles, and 2 species of fish (Olsen, 1974).

HOSTS AND LOCATION. Most echinostomes are incidental intestinal parasites of humans. Natural definitive hosts include many birds and small mammals.

DISTRIBUTION AND IMPORTANCE. The largest number of human echinostome infections are in the Far East, but many echinostomes are cosmopolitan (Table 24.1). Except for an occasional massive human infection, most cause little or no pathology. Some echinostomes in poultry and small mammals cause economic losses (Malek, 1980).

MORPHOLOGY. Echinostomes are medium-sized distome flukes (4–12 mm by 1–3 mm) (Fig. 24.1). The most prominent structural feature is the collar of large spines surrounding the oral sucker, from which the name, *echinostome* is derived. The number, size, and position of spines are of taxonomic value. Collar spines are present in most cercariae (Fig. 24.2), metacercariae, and adult worms (Fig. 24.3). The acetabulum is usually larger than the oral sucker; the genital pore is median and the cirrus pouch is well developed. The uterus is entirely anterior to the ovary and vitelline follicles are abundant; a seminal receptacle is lacking.

LIFE CYCLE. Unembryonated operculate eggs are shed in feces; miracidia are hatched in about 14 days and penetrate snails. A sporocyst generation is followed by two or more generations of rediae. Many echinostome rediae possess lateral body protrusions called *lappets* or *procrusculi* (Fig. 24.4), locomotor appendages that enable rediae to crawl (Dawes, 1968). Echinostome cercariae leave daughter rediae through birth pores located near the anterior end of rediae. Most echinostome cercariae have spines in patterns similar to those of the adults.

Cercariae can encyst (1) in the same snail in which they developed, (2) in other snail species, (3) on snail shells, or (4) on the surfaces of fish, bivalves, and tadpoles (Malek, 1980). Spines can be seen on the juvenile inside bilayered metacercarial cyst. Definitive hosts become infected by ingesting metacercariae. They pass to the intestines, excyst, and develop to adults.

**TABLE 24.1**
Echinostome Parasites of Humans

| | | GEOGRAPHIC DISTRIBUTION | |
|---|---|---|---|
| ORGANISM | SOURCE OF HUMAN INFECTION | ORGANISM | HUMAN INFECTION |
| *Echinostoma revolutum* | Tadpole/frog, snail, clam | Cosmopolitan | Taiwan, Thailand, Indonesia, Mexico |
| *Echinostoma cinetrochis* | Tadpole/frog | Far East | Japan, Taiwan, Indonesia |
| *Echinostoma macrorchis* | Tadpole/frog | Far East | Japan, Taiwan, Indonesia |
| *Echinostoma recurvatum* | Tadpole/frog | Cosmopolitan | Japan, Taiwan, Indonesia |
| *Echinostoma malayanum* | Snail | Far East | Malasia, Singapore, India, Thailand, Sumatra |
| *Echinostoma lindoense* | Clam | Far East, Brazil | Indonesia |
| *Echinostoma hortense* | Fish | Far East | Japan |
| *Echinostoma ilocanum* | Snail | Far East | Java, Philippines |
| *Echinochasmus perfoliatus* | Fish | N. Asia, Europe | Japan |
| *Euparphium melis (=beaveri)* | Tadpole/frog | Asia, Europe, N. America | Romania, China |
| *Echinoparyphium recurvatum* | Tadpole/frog | Cosmopolitan | Indonesia, Japan, Taiwan |
| *Hypoderaeum conoideum* | Snail | Cosmopolitan | Thailand |
| *Paryphostomum sufartyfex* | Clam | India | India |
| *Artyfechinostomum (=P.) mehri* | Snail | India | India |
| *Himasthla muehlensi* | Clam | ? Colombia | New York |

DIAGNOSIS. Initial diagnosis of echinostomatiasis is often difficult. The adult worms cause few lesions and hence few clinical symptoms. In addition, the large operculate eggs (85 μm and larger) are easily confused with the large operculate eggs of *Fasciolopsis* and *Fasciola* (Garcia & Ash, 1979). A definitive diagnosis is made when adult worms are recovered after an anthelmintic is given.

CULTIVATION. Attempts to culture echinostomes have been partially successful. Howell (1968) used a yolk-albumin-saline medium supplemented with yeast extract and obtained eggs from *Echinoparyphium serratum*. Butler and Fried (1977) had limited success culturing metacercariae of *Echinostoma revolutum* in NCTC 135 supplemented with 20% egg yolk.

HOST–PARASITE INTERACTIONS. Most infections with adult echinostomes produce little or no clinical disease, but heavy infections cause diarrhea and abdominal pain.

TREATMENT. Almost any vermifuge will eliminate echinostomes; tetrachloroethylene is the drug of choice (Marcial-Rojas, 1971).

EPIDEMIOLOGY AND CONTROL. The lack of host specificity in all life cycle stages plus the cosmopolitan distribution of many echinostome species suggests that humans are at risk throughout the world. Ingesting uncooked snails, clams, tadpoles, and fish can lead to echinostomatiasis. Fortunately, most second intermediate hosts harboring metacercariae are not eaten raw in most parts of the world, and so the number of human infections is far smaller than the potential risk.

A remarkable story in successful control is illustrated for an echinostome species. As recently as 1965, *Echinostoma lindoense* infected over half of the people living around Lake Lindu in Central Sulawesi (Celebes), Indonesia (Fig. 24.5). Mollusks, especially clams (*Corbicula* spp.), were a major source of animal protein for the people. The mujair fish *Tilapia mossambica* was introduced into the lake and rapidly

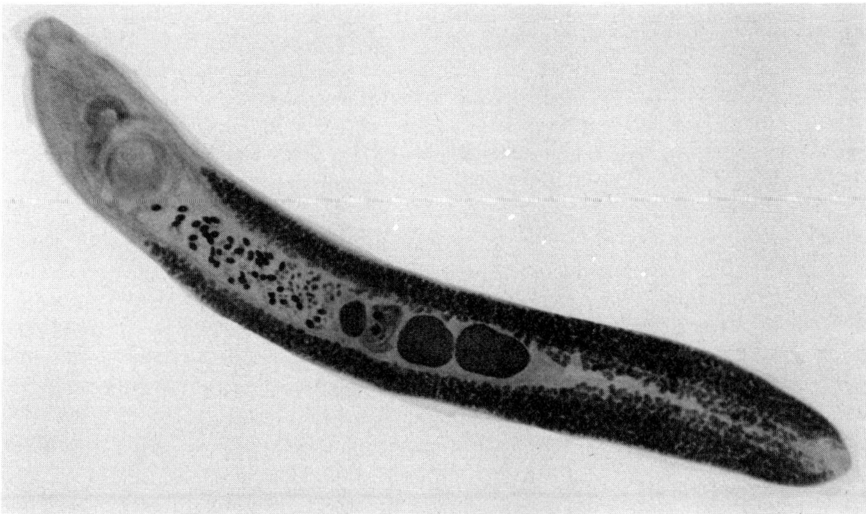

**FIG. 24.1**
Adult *Echinostoma revolutum*.

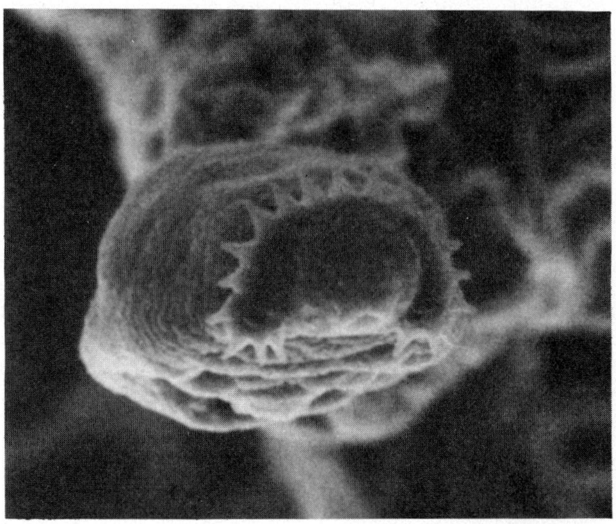

**FIG. 24.2**
"Head-on" view of an echinostome cercaria showing the collar of spines.

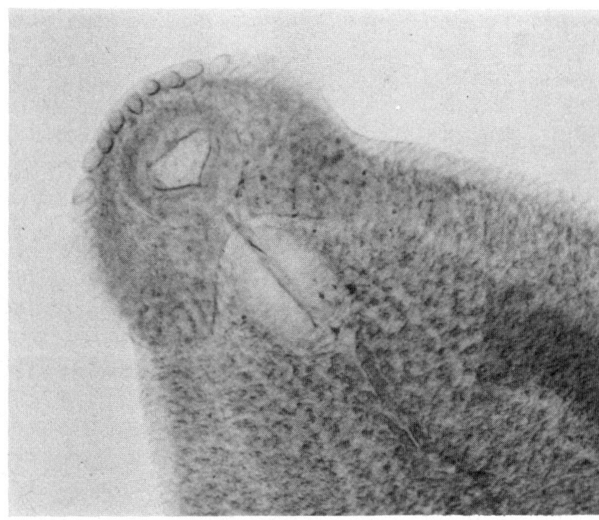

**FIG. 24.3**
Adult echinostome with a collar of spines around the oral sucker.

became abundant. The people changed their diet from clams to fish. This fish is now a daily item in the Lindu diet and is cooked before it is eaten. Consequently, a survey in 1972 of over 1400 people revealed no *E. lindoense* infections (Clarke et al., 1974). Thus, infections with this human parasite were eliminated with just a change in diet. Unfortunately, the more serious parasitic disease caused by *Schistosoma japonicum* (see Chap. 21) is still prevalent in the Lake Lindu population and control efforts have had limited success.

Humans are the only known definitive host for *E.*

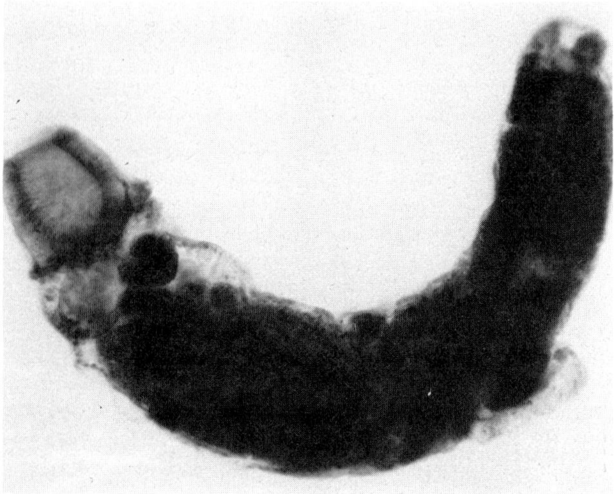

FIG. 24.4
Echinostome redia with a prominent pharynx and one visible procrusculus.

*lindoense*, but there must be animal reservoirs because the parasite was found in *Biomphalaria* snails in Brazil. Since people there do not eat snails at all, there must be a reservoir host maintaining the life cycle, and risk for humans is present (Hillyer, 1981).

Echinostome rediae have been studied extensively as potential predators of other trematode species that do not produce rediae, such as schistosomes (Chap. 21). Echinostome rediae move throughout snails and actively ingest sporocysts. This predatory activity by echinostome rediae on schistosome sporocysts was studied in vitro by Basch and Di Conza (1975) and intramolluscan trematode antagonism was reviewed by Lim and Heyneman (1972). In that review, Lim and Heyneman concluded that echinostomes can be used as biological control agents for other trematode species at the local level if high echinostome infection percentages in snails can be achieved in local susceptible snails. This trematode antagonism combined with other means including sanitary, clinical, and biological control might be used to control some human parasites (see also Chap. 21).

FIG. 24.5
Lake Lindu in Central Sulawesi where human infections of *Echinostoma lindoense* have been eliminated by a change in diet. [Courtesy of Dr. David Kistner.]

## REFERENCES AND FURTHER READINGS

Basch, P. F., and J. J. Di Conza. 1975. Predation by echinostome rediae upon schistosome sporocysts in vitro. *J. Parasitol.* **61**:1044–1047.

Butler, M. S., and B. Fried. 1977. Histochemical and thin layer chromatographic analyses of neutral lipids in *Echinostoma revolutum* metacercariae cultured *in vitro*. *J. Parasitol.* **63**:1041–1045.

Clarke, M. D., W. P. Carney, J. H. Cross, P. Hadidjaja, S. Oemijati, and A. Joesoef. 1974. Schistosomiasis and other human parasitoses of Lake Lindu in Central Sulawesi (Celebes), Indonesia. *Am. J. Trop. Med. Hyg.* **23**:385–392.

Dawes, B. 1968. *The Trematoda.* Cambridge University Press, Cambridge, England.

Garcia, L. S., and L. R. Ash. 1979. *Diagnostic Parasitology. Clinical Laboratory Manual*, 2nd ed. C. V. Mosby Co., St. Louis.

Healy, G. R. 1970. Trematodes transmitted to man by fish, frogs, and crustacea. *J. Wildl. Dis.* **6**:255–261.

Hillyer, G. V. 1981. Introduction to the trematode zoonoses. In Hillyer, G. V. *The Parasitic Zoonoses.* CRC Press, Boca Raton, Fla.

Howell, M. J. 1968. Excystment and *in vitro* cultivation of *Echinoparyphium serratum. Parasitology.* **58**:583–597.

Lim, H., and D. Heyneman. 1972. Intramolluscan inter-trematode antagonism: A review of factors influencing the host-parasite system and its possible role in biological control. *Adv. Parasitol.* **10**:191–268.

Malek, E. A. 1980. Echinostomiasis. In Malek, E. A. (ed.), *Snail Transmitted Diseases,* Vol. II. CRC Press, Boca Raton, Fla., pp. 245–259.

Marcial-Rojas, R. A. 1971. Rare intestinal flukes. In R. A. Marcial-Rojas (ed.), *Pathology of Protozoal and Helminthic Diseases.* Williams & Wilkins, Baltimore, pp. 469–476.

Markell, E. K., and M. Voge. 1981. *Medical Parasitology*, 5th ed. W. B. Saunders Co., Philadelphia.

Olsen, O. W. 1974. *Animal Parasites. Their Life Cycles and Ecology*, 3rd ed. University Park Press, Baltimore.

Ulmer, M. J. 1975. Other trematode infections. In Hubbert, W. T., W. F. McCollough, and P. R. Schnurrenberger (eds.), *Diseases Transmitted from Animals to Man,* 6th ed. Charles C Thomas, Springfield, Ill., pp. 646–677.

Yamaguti, S. 1971. *Synopsis of Digenetic Trematodes of Vertebrates,* Vol. I. Keigaku, Tokyo.

# 25

# Nanophyetus Salmincola *and* "*Salmon Poisoning Disease*"

NAME OF ORGANISM AND DISEASE CAUSED. *Nanophyetus salmincola* is a small trematode which is the vector for a rickettsia, *Neorickettsia helminthoeca*. The infection is usually fatal to dogs, foxes, and coyotes that ingest fishes infected with flukes containg rickettsia. The resultant disease is called *salmon poisoning disease (SPD)*, although salmon are neither poisoned nor do they cause the disease.

HOSTS AND LOCATION. The raccoon (*Procyon lotor*) and spotted skunk (*Spilogale putorius*) are the probable principal definitive hosts (Schlegel et al., 1968). Millemann and Knapp (1970) listed 32 natural and experimental host species for the adult fluke. One human experimental infection has been reported (Phillip, 1958). The adult flukes lie in the paramucosal lumen of the small intestine.

DISTRIBUTION AND IMPORTANCE. Distribution of the fluke is apparently limited by the range of the single first intermediate host to streams in Washington, Oregon, and northern California. Infections in dogs are sporadic but do cause significant losses each year.

MORPHOLOGY. *Nanophyetus salmincola* is 0.8 to 2.5 mm long, with a cirrus sac dorsal to the acetabulum but no cirrus. Vitelline follicles are prominent and uteri typically contain 5 to 16 eggs (Fig. 25.1).

LIFE CYCLE. Eggs are passed unembryonated and take 87 to 200 days to mature and hatch (Bennington & Pratt, 1960) (Fig. 25.1). How miracidia infect snails is unknown. Microcercous xiphidiocercariae develop from rediae and leave the snail with exhalant water currents. Cercariae penetrate the skin of salmonid and other fish and encyst in almost all internal organs, subcutaneous tissue, and fins. Definitive hosts are infected by eating raw or poorly cooked fish containing metacercariae.

Second intermediate hosts are primarily salmonid fishes, but other fish, lamprey, and the giant pacific salamander (Millemann & Knapp, 1970) are also infected. Taxonomy of the snail host currently is uncertain; Millemann and Knapp (1970) called it *Oxytrema silicula*, whereas Malek and Cheng (1974) refer to it as *Goniobasis plicifera* var. *silicula*, and Taylor (1981) called it *Juga silicula shastaensis*.

DIAGNOSIS. Infection with *N. salmincola* is diagnosed by examining feces for eggs. Infection of dogs with *Neorickettsia helminthoeca* is diagnosed by the appearance of clinical signs along with a history of eating raw fish. Prominent signs are fever, lassitude, and diarrhea.

**FIG. 25.1** (***opposite page***) Life cycle and adult structure of *Nanophyetus salmincola*. (*A*) Adult; (*B*) partially embryonated egg; (C) fully developed miracidium within egg capsules; (*D*) free-swimming miracidium; (*E*) redia with cercariae in snail; (*F*) free-living cercaria; (*G*) metacercaria in fish. [From Bennington and Pratt; 1960.]

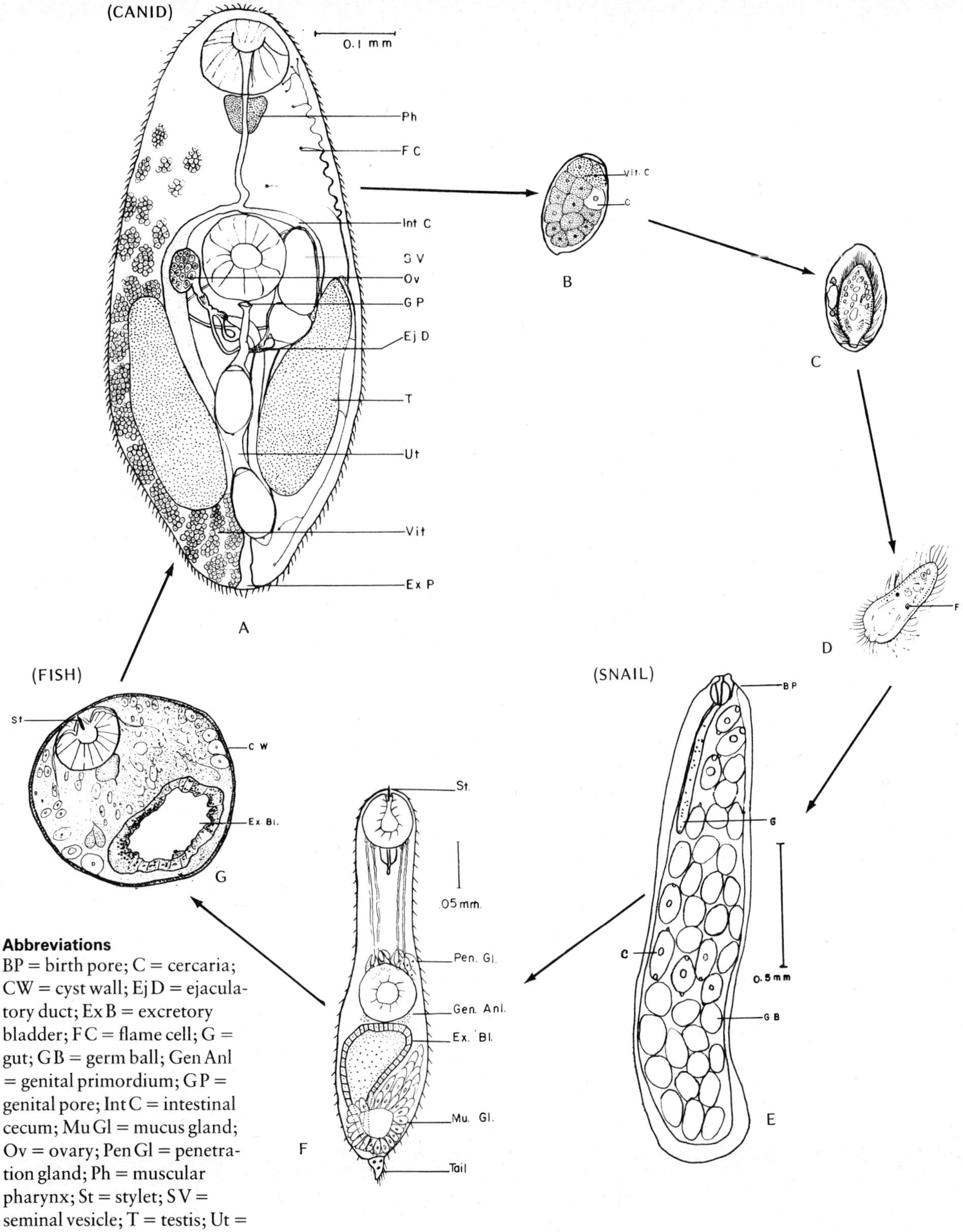


**Abbreviations**

BP = birth pore; C = cercaria; CW = cyst wall; Ej D = ejaculatory duct; Ex B = excretory bladder; FC = flame cell; G = gut; GB = germ ball; Gen Anl = genital primordium; GP = genital pore; Int C = intestinal cecum; Mu Gl = mucus gland; Ov = ovary; Pen Gl = penetration gland; Ph = muscular pharynx; St = stylet; SV = seminal vesicle; T = testis; Ut = uterus; Vit = vitelline gland

HOST–PARASITE INTERACTIONS. By far the greatest effects of SPD are in dogs, foxes, and coyotes. Apparently only canids are seriously affected by SPD. Dogs show sudden onset of fever 5 to 7 days after infection; fever fluctuates for a week. Dehydration and anorexia are common. Lymph nodes enlarge as the disease progresses. About 90% of naturally infected canids die within 10 days to 2 weeks after onset of clinical signs, whereas other mammals such as the raccoon and bear may have only transient fever as an indication of low-grade infection.

The reaction to *Nanophyetus* infections in fish ranges from slight to severe lesions depending on the parasite burden. In heavy infections, almost every organ of the fish is affected, especially muscle, gills, and kidneys. Swimming ability is also affected, at least until the metacercariae encyst (Butler & Millemann, 1971).

TREATMENT. Antibiotics, including chloramphenicol and oxytetracycline, as well as sulfonamides are used to treat SPD (Millemann & Knapp, 1970). Canids that recover from SPD are immune to the rickettsia, but not to *Nanophyetus salmincola* (Milleman & Knapp, 1970).

EPIDEMIOLOGY AND CONTROL. SPD is unique because its vector is a trematode; other rickettsiae such as *Rickettsia* and *Ehrlichia* are transmitted by ticks and lice. Rickettsiae have been reported in all stages of *N. salmincola*, and transovarial transmission was reported (Nyberg et al., 1967). It appears that salmon become infected while still in fresh water; they maintain both the metacercarial and rickettsial infections for 3 to 7 years while in the ocean. The infections are then transmitted to the definitive host when the salmon return to fresh water to spawn.

The only realistic control for SPD is to keep dogs away from streams during salmon migration and not to feed dogs salmon entrails. The swiftly flowing streams make snail control using molluscicides virtually impossible. Treatment of dogs when signs of infection have appeared will help the individual dog, but will not have much effect on the general problem.

OTHER SUBSPECIES. A Siberian subspecies, *Nanophyetus salmincola schikholbalowi*, has a similar life cycle; salmonids and other fishes serve as second intermediate hosts. Some human populations in Siberian villages are reported to have infection percentages of 95 to 98%. These infections result in various gastrointestinal disturbances including diarrhea and gastric pain. Over 500 parasites have been reported from one human host. No SPD organism was found in the Siberian subspecies of fluke (Millemann & Knapp, 1970).

## REFERENCES AND FURTHER READINGS


Bennington, E., and I. Pratt. 1960. The life history of the salmon-poisoning fluke, *Nanophyetus salmincola* (Chapin). *J. Parasitol.* **46**:91–100.

Butler, J. A., and R. E. Millemann. 1971. Effect of the "salmon poisoning" trematode, *Nanophyetus salmincola*, on the swimming ability of juvenile salmonid fishes. *J. Parasitol.* **57**:860–865.

Malek, E. A., and T. C. Cheng. 1974. *Medical and Economic Malacology*. Academic Press, New York, p. 67.

Millemann, R. E., and S. E. Knapp. 1970. Biology of *Nanophyetus salmincola* and "salmon poisoning" disease. *Adv. Parasitol.* **8**:1–41.

Nyberg, P. A., S. E. Knapp, and R. E. Millemann. 1967. "Salmon poisoning" disease. IV. Transmission of the disease to dogs by *Nanophyetus salmincola* eggs. *J. Parasitol.* **53**:694–699.

Phillip, C. B. 1955. There's always something new under the "parasitological" sun (the unique story of the helminth-borne salmon poisoning disease). *J. Parasitol.* **41**:125–148.

Phillip, C. B. 1958. A helminth replaces the usual arthropod as vector of a rickettsia-like disease. *1956 Proc 10th Internat. Congr. Ent.* **3**:651–653.

Schlegel, M. W., S. E. Knapp, and R. E. Millemann. 1968. "Salmon poisoning" disease. V. Definitive hosts of the trematode vector, *Nanophyetus salmincola*. *J. Parasitol.* **54**:770–774.

Taylor, D. W. 1981. Freshwater mollusks of California: A distributional checklist. *Calif. Fish and Game* **67**:140–163.

# 26

# Paragonimus, *the Human Lung Fluke*

A number of mammals, including humans, are infected by lung flukes of the genus *Paragonimus*; Yokogawa (1981) listed over 30 species of flukes, at least 7 of which are implicated in human disease.

NAME OF ORGANISM AND DISEASE CAUSED. *Paragonimus westermani* (Kerbert, 1878) Braun, 1899, is the most important species infecting humans (Nelson, 1975); it causes paragonimiasis, pulmonary distomiasis, and endemic hemoptysis (expectoration of bloody sputum).

HOSTS AND LOCATION. Hosts are carnivores including tigers, lions, leopards, feral cats, pigs, dogs, monkeys, and humans. Adult flukes are found in lungs, but may also be found in almost any body organ or tissue; 30 to 60% of extrapulmonary paragonimiasis were either cerebral or spinal cord infections (Yokogawa, 1969).

DISTRIBUTION AND IMPORTANCE. The organism is distributed in China, Japan, Taiwan, Korea, the Soviet Union, Philippines, Thailand, Indonesia, India, and Sri Lanka. In Korea alone, between 1.0 and 1.5 million people are infected (Yokogawa, 1969).

MORPHOLOGY. Adult flukes are 7 to 12 mm long with a thick, spiny tegument; a small acetabulum is located just anterior to the middle of the body. Often one must remove some tegument and vitellaria to discern the reproductive organ structure. Two posterior lobed testes are present; the ovary is also branched and is located slightly posterior to the acetabulum. There is no cirrus or cirrus pouch. The genital pore is posterior to the acetabulum. Eggs are operculate, irregularly barre-shaped, and are 80 to 120 μm by 48 to 60 μm (Fig. 26.1) in size.

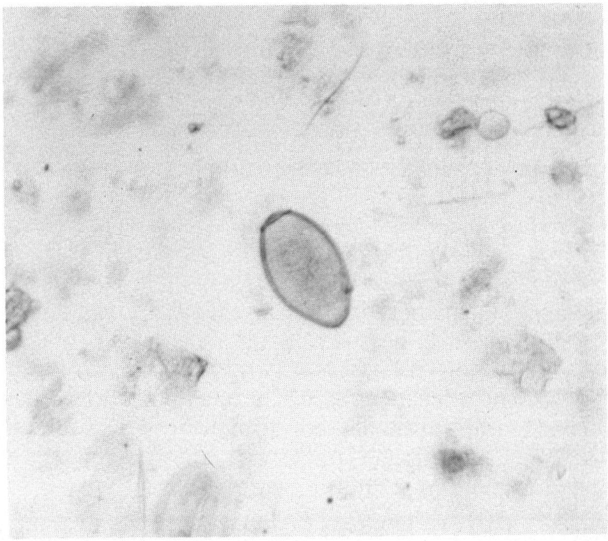

**FIG. 26.1**
*Paragonimus* egg.

LIFE CYCLE. Eggs are shed in sputum, or coughed up, swallowed, and then shed in feces (Fig. 26.2); they hatch in 2 to 3 weeks, and remain alive for several months at low temperatures. Miracidia penetrate the first intermediate host, operculate snails of the family Thiaridae; *Semisulcospira libertina* is the most important first intermediate host (Yokogawa, 1965). Sporocysts develop in the snail and produce mother rediae, which, in turn, produce daughter rediae. Microcercous cercariae emerge from the snails approximately 13 weeks after infection (Chung, 1971).

Crayfish and freshwater crabs are second intermediate hosts for *P. westermani*. Microcercous cercariae crawl along stream beds until they are eaten or penetrate joints of the exoskeleton. Noble (1963) found that if crustacea ingest infected snails they become infected. Metacercariae encyst in the gills, legs, body muscles, and viscera.

Definitive hosts, including humans, are infected by eating raw or poorly cooked crab or crayfish containing metacercariae. Metacercarial cyst walls are dissolved and juvenile flukes are released in the intestines. Juvenile flukes penetrate the intestinal walls and enter abdominal wall muscles, where they remain for 5 to 7 days and then reenter the abdominal cavity. Young flukes penetrate the diaphragm and enter the pleural cavity about 14 days after infection. Seventy days after the infection, the flukes are mature and produce eggs.

DIAGNOSIS. Diagnosis should include microscopic examination of sputum and feces for large operculate eggs (Fig. 26.1). Sputum should be examined following centrifugation and sedimentation in 1 to 2% NaOH. Charcot-Leyden crystals in sputum help to confirm the diagnosis; these crystals are associated with monocytes. Serodiagnostic tests have been developed to identify *Paragonimus* infections, and are especially useful in diagnosing extrapulmonary paragonimiasis and in evaluating chemotherapy (Hillyer, 1981). Intradermal tests distinguish between paragonimiasis and tuberculosis, but not among various *Paragonimus* spp. Nor can they identify a clinical cure. The most reliable serodiagnostic tools available are complement-fixation and precipitation tests. For further information about serodiagnosis, see Hillyer (1981).

HOST–PARASITE INTERACTIONS. There are three characteristic lung lesions in paragonimiasis. First are foreign body granulomata around eggs (Fig. 26.3). This host response is similar to that around schistosome eggs in liver, bladder, and intestines (Chap. 21). Granulomata typically have fibroblasts, lymphocytes, and eosinophils encircling the eggs.

The second lesions are cysts or distoma cysts in which both worms and eggs are encapsulated by fibrous granulation tissue of fibroblasts, lymphocytes, eosinophils, and mononuclear cells. Calcification may occur in chronic paragonimiasis. Apparently, two worms are necessary for formation of a cyst, but surgery usually yields cysts with only one worm. Two worms appear to be present initially, but one of the two dies during chronic infection (Yokogawa, 1965).

The third type of lesion is abscess cavities caused by necrosis of lung tissue and the subsequent liquefication of this tissue, and often of the worms and eggs as well. The pathology of pulmonary paragonimiasis was reviewed by Yokogawa (1965) and Chung (1971).

TREATMENT. Bithionol (Actamer, Monsanto, U.S.A.) is the drug of choice for both pulmonary and subcutaneous paragonimiasis. Because 10 to 20 individual oral doses of the drug must be given on alternate days, other drugs have been sought. Nwokolo and Volkmer (1977) reported 73 to 90% cure rates with a single dose of menichlopholan (Bayer 9015) for pulmonary infections of *P. uterobilateralis* in Nigeria; it is also used in treating fascioliasis and clonorchiasis.

EPIDEMIOLOGY AND CONTROL. Adequate cooking of the crustacean intermediate hosts kills metacercariae and prevents infection. Unfortunately, cultural preferences sometimes interefere with satisfactory control. Some examples discussed by Chung (1971) illustrate the problems. In Japan crabs are usually eaten well cooked, but a common practice is to prepare vegetables using the same chopping block and cutlery immediately after preparation of the crab. Thus, metacercariae are transferred to other foods or to the mouth. In China "drunken crab" is a delicacy. Crabs are immersed in wine for about 12 hr before being eaten, resulting in infection rates approaching 100%. The raw or undercooked crab or crayfish eaten in many other parts of the world leads to paragonimiasis.

Another source of human infection was reported by Miyazaki and Habe (1976). In Japan 136 human cases of paragonimiasis were acquired when people ate raw pork and wild boar which were paratenic hosts for *Paragonimus westermani*; metacercariae were located in the somatic muscles of these hosts.

Control of paragonimiasis through eliminating the snail host has not been successful because their habitat is rapidly flowing water. Elimination of crustacean

**FIG. 26.2**
Diagram of the life cycle of *Paragonimus westermani*.

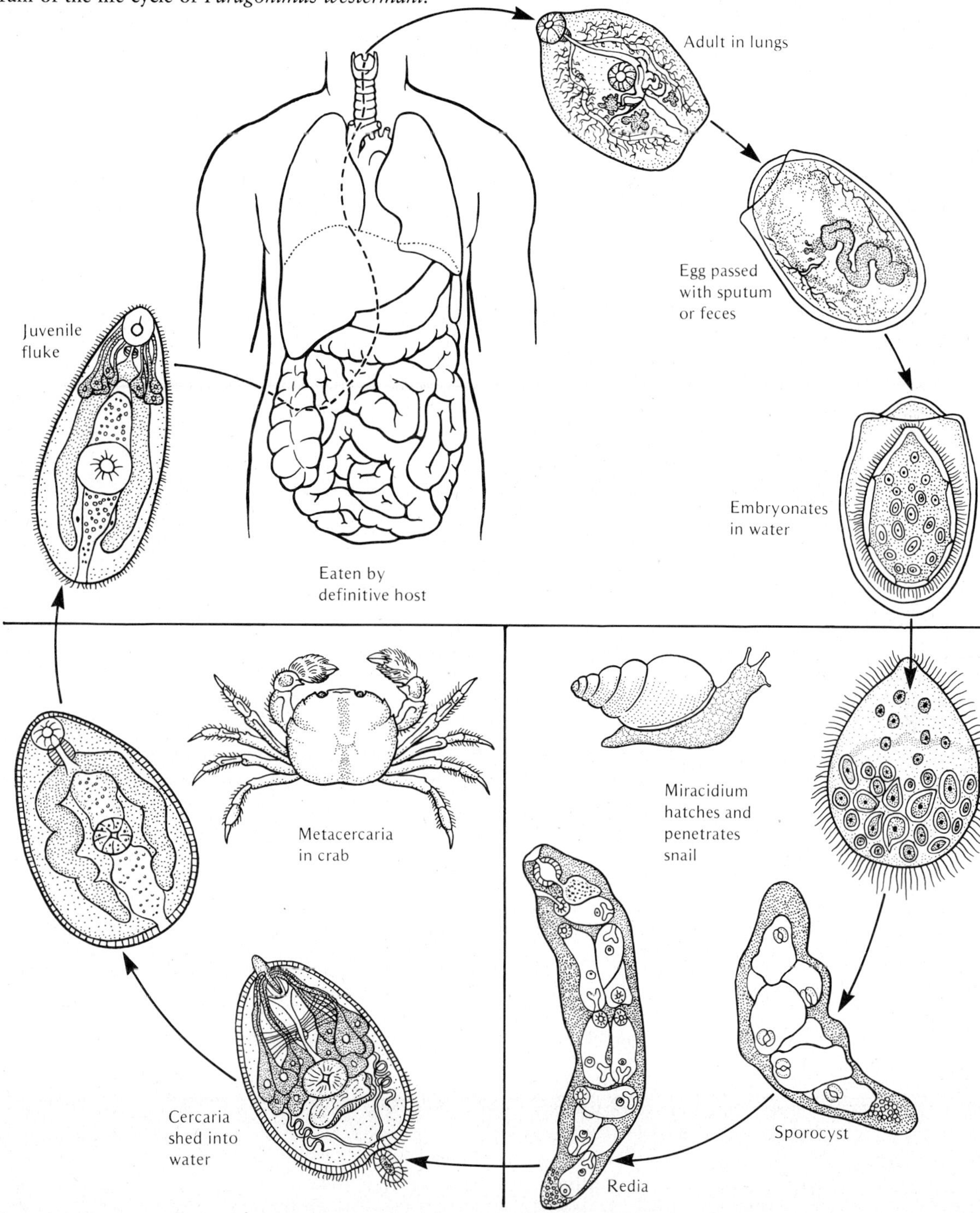

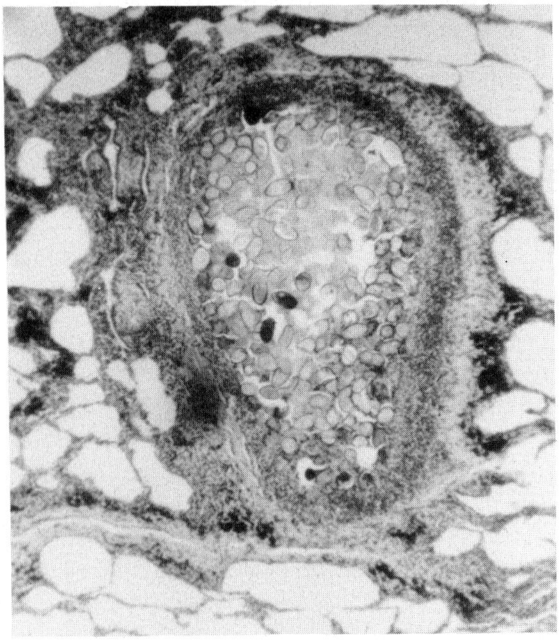

**FIG. 26.3**
Granulomatous response around *Paragonimus* eggs in a lung section.

second intermediate hosts is impractical because they are both a valuable source of protein and a culinary delicacy. Adequate cooking of crabs and crayfish, adequate cleanliness of food preparation utensils, and drug treatment for adult worms offer the most promise for control. Also, other meat from possible transport hosts must be well cooked.

OTHER SPECIES OF *PARAGONIMUS*. Knowledge of the taxonomy and biology of most species of *Paragonimus* is still fragmentary. *P. miyazakii* in Japan, *P. africanus* in the Cameroon and Congo, and *P. uterobilateralis* in Nigeria are also etiological agents of pulmonary paragonimiasis in humans; subcutaneous paragonimiasis is caused by *P. skrjabini* in China and *P. heterotremus* in China, Thailand, and Laos (Yokogawa, 1981).

In the Western Hemisphere human pulmonary paragonimiasis was previously attributed to *P. westermani*. Only *P. peruvianus* (= *P. mexicanus*, *P. ecuadorensis*), found in Ecuador and Peru, is now thought to cause human paragonimiasis (Hillyer, 1981). The status of other Western Hemisphere *Paragonimus* species is currently uncertain (Yokogawa, 1981).

*P. kellicotti* is a lung fluke widely distributed in the eastern United States and Canada. Definitive hosts include cat, dog, pig, goat, mink, muskrat, bobcat, raccoon, opossum, skunk, red fox, coyote, and weasel. *P. kellicotti* can be distinguished from *P. westermani* and other species by its ovarian shape and the thickness of the inner metacercarial wall (Ishii, 1966). Stromberg and Dubey (1978) described its life cycle in cats in detail; *P. kellicotti* does not infect humans.

## REFERENCES AND FURTHER READINGS

Chung, C. H. 1971. Human paragonimiasis (pulmonary distosomiasis, endemic hemoptysis). In R. A. Marcial-Rojas, (ed.), *Pathology of Protozoal and Helminthic Diseases*. Williams & Wilkins, Baltimore, pp. 504–535.

Hillyer, G. V. 1981. Immunoserology of fascioliasis, paragonimiasis, and clonochiasis. In K. Walls (ed.), *Immunoserology of Parasitic Diseases*. Marcel Decker, New York.

Ishii, Y. 1966. Differential morphology of *Paragonimus kellicotti* in North America. *J. Parasitol.* **52**:920–925.

Miyazaki, I., and S. Habe. 1976. A newly recognized mode of human infection with the human lung fluke, *Paragonimus westermani* (Kerbert, 1878). *J. Parasitol.* **62**:646–648.

Nelson, G. S. 1975. Paragonimiasis. In W. T. Hubert, W. F. McCulloch, and P. R. Schnurrenberger (eds.), *Diseases Transmitted from Animals to Man*. Charles C Thomas, Springfield, Ill., pp. 641–645.

Noble, G. A. 1963. Experimental infection of crabs with *Paragonimus*. *J. Parasitol.* **49**:352.

Nwokolo, C., and K. J. Volkmer. 1977. Single dose therapy of paragonimiasis with menichlopholan. *Am. Soc. Trop. Med. Hyg.* **26**:688–695.

Stromberg, P. C., and J. P. Dubey. 1978. The life cycle of *Paragonimus kellicotti*, in cats. *J. Parasitol.* **64**:998–1002.

Yokogawa, M. 1965. *Paragonimus* and paragonimiasis. *Adv. Parasitol.* **3**:99–158.

———. 1969. *Paragonimus* and paragonimiasis. *Adv. Parasitol.* **7**:375–387.

———. 1981. Paragonimiasis. In *The Handbook of Zoonoses*, Vol. III: *The Parasitic Zoonoses*. CRC Press, Boca Raton, Fla.

# 27

# Class Cestoidea: The Tapeworms

## INTRODUCTION AND CESTODARIA

The class Cestoidea of the phylum Platyhelminthes contains the tapeworms. These worms have no digestive tract or mouth, and attachment organs are limited to the anterior end. Because of the lack of mouth and digestive tract, tapeworms have been called the most parasitic of all parasites; our view is rather that they are highly specialized. Cestodes or tapeworms are all parasitic in the digestive tracts and associated ducts of all classes of vertebrates and some freshwater oligochaetes (Chap. 46).

Human and animal tapeworm infections have been known since classical times. Celsus, Pliny, Aristotle, and their contemporaries recognized human intestinal flatworms (*Taenia*). Larval tapeworms were also recognized by early investigators, but they were not associated with their adult forms in humans. The Mosaic law forbids eating meat of the camel, rabbit, hare, or pig, perhaps because of bladderworms (Hoeppli, 1959), but it was not until the nineteenth century that larval and adult worm relationships were described.

MORPHOLOGY. The anterior end of tapeworms is called the *scolex* or *holdfast*; it has various structures for attachment including sucking depressions, hooks, and glandular areas. The sucking depressions are of three types: bothria—a pair of shallow sucking grooves typical of the Pseudophyllidea; bothridia—four leaflike, flexible structures seen in the order Tetraphyllidea; and true suckers or acetabula—four hemispherical mus-

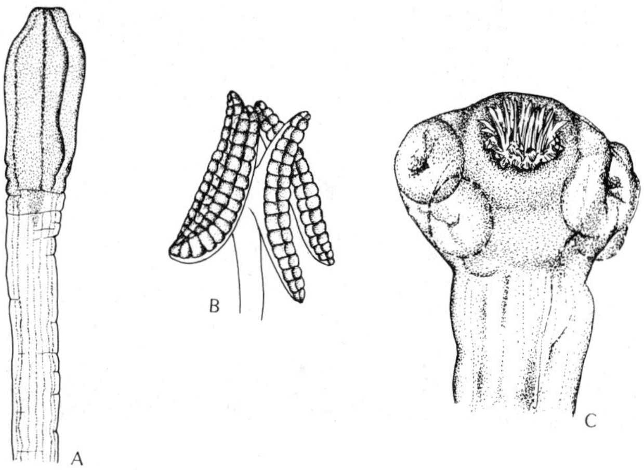


**FIG. 27.1**
Tapeworm scolices. (*A*) Scolex with bothria, *Diphyllobothrium latum*; (*B*) scolex with bothridia, *Echeneibothrium tobijei*; (C) scolex with suckers, *Taenia* sp. [(*A*) redrawn from Rausch, R. A., and D. K. Hilliard. 1970. Studies of the helminth fauna of Alaska. XLIX. The occurrence of *Diphyllobothrium latum* (Linnaeus, 1758) (Cestoda: Diphyllobothriidae) in Alaska, with notes on other species. *Canad. J. Zool.* 48:1201–1219. (*B*) redrawn from Yamaguti, S. 1934. Studies on the helminth fauna of Japan. Part 4. Cestodes of fishes. Japanese J. Zool. 6:1–112.]

cular cups characteristic of several orders including the important Cyclophyllidea (Fig. 27.1). A few tapeworms lack anterior holdfast structures, but most have one of the three types mentioned. In addition, some tapeworms have an apical organ, the rostellum, which may have an additional sucker, glandular area, or hooks (armed).

Posterior to the scolex region there are two body types. The first is a body of a single segment (monozoic) and the other is a body of many segments (polyzoic). Each segment is called a *proglottid*, and the chain of proglottids is called a *strobila*. Usually each proglottid has its own set of reproductive organs. Also, most tapeworms have the immature proglottids closest to the scolex, then a region of sexually mature proglottids, and finally gravid proglottids that have uteri filled with eggs. This form of sequential segmentation is characteristic of the subclass Eucestoda; however, not all tapeworms have a strobila of proglottids produced sequentially. The order Pseudophyllidea, which includes the fish tapeworm of humans, *Diphyllobothrium latum*, has proglottids that are formed "instantaneously"; Wardle et al. (1974) therefore suggested that these should also be called monozoic because they were formed at one time. In addition, these proglottids lack the intervening membranous or muscular partitions charcteristic of polyzoic tapeworms. In many species the proglottids overlap their immediate neighbors and are called *craspedote*; others that do not overlap are called *acraspedote*. Some tapeworms continuously shed terminal proglottids in the feces; these are called *apolytic* and those that do not are called *anapolytic*.

Since tapeworms lack both a mouth and digestive tract, all nutrients must be acquired through the external surface. This external surface is a living, multinucleate (syncytial) tegument and not an inert, secreted cuticle. Lumsden (1975a) reviewed helminth surface structure and function and Smyth (1969, 1972, 1973) and Pappas and Read (1975) reviewed the functions of cestode tegument. The external surface is a chemically complex glycocalyx. Beneath the glycocalyx is a limiting membrane and then an unusual type of microvillar surface projection called *microtriches* (singular, microthrix), which are visible with the electron microscope. Microtriches have characteristic dense caps that point toward the posterior of the worm and are supported by microfilaments. Beneath the microtriches is a vacuolated, syncytial cytoplasm connected to the subtegumental layer by thin protoplasmic extensions. The protoplasmic extensions pass through three muscle layers— circular, longitudinal, and oblique or transverse—before reaching the subtegumental layer. This inner layer contains tegumental cell bodies with nuclei, mitochondria, Golgi apparati, and endoplasmic reticulum cisternae (Fig. 27.2).

The function of microtriches has been the subject of considerable debate. Most workers believe that since the microtriches increase surface area 10 to 50 times (Lumsden, 1975b), there is better nutrient absorption, similar to that of the microvilli in the cells of the vertebrate gut (Lumsden, 1975a); the microtriches may also assist the worm in maintaining its position in the gut because of their posteriorly pointed orientation (Smyth, 1969).

In cross section the tapeworm body is seen to be filled with parenchyma. Muscle layers separate the body into an outer cortex and an inner medulla (Fig. 27.3). The reproductive organs are usually in the medulla. Pairs of dorsal and ventral excretory canals or osmoregulatory canals (Fig. 27.4) traverse the entire strobila from scolex to the last proglottid. The dorsal one is often smaller than the ventral, and this observation can help in orientation when studying stained specimens. Transverse canals can sometimes be seen connecting the longitudinal canals.

Most tapeworms are monoecious, although a few dioecious ones are known. In most tapeworms the male reproductive system matures first (protandry), and may prevent self-fertilization within the same proglottid. The rationale for such a system may seem obscure at first since all proglottids come from the same zygote and are thus genetically identical; however, if sperm and ova develop at different times, the opportunities for genetic change may be increased.

Reproductive systems in members of the class Cestoidea (Fig. 27.5) have the same structures as those of the class Trematoda (Chap. 20) except that some tapeworms lack a uterine pore. A typical mature male system consists of numerous testes which produce sperm; vasa efferentia lead from each testis to a common vas deferens or sperm duct. Some species have a chamber for sperm storage or seminal vesicle. The male copulatory organ, the cirrus, is surrounded by the cirrus pouch. Most often the cirrus pouch and female vagina enter a common chamber, the genital atrium, and share a common opening to the outside, the genital pore.

A typical mature female tapeworm system includes a vagina leading from the genital pore to a seminal receptacle or sperm storage chamber; an ovary which

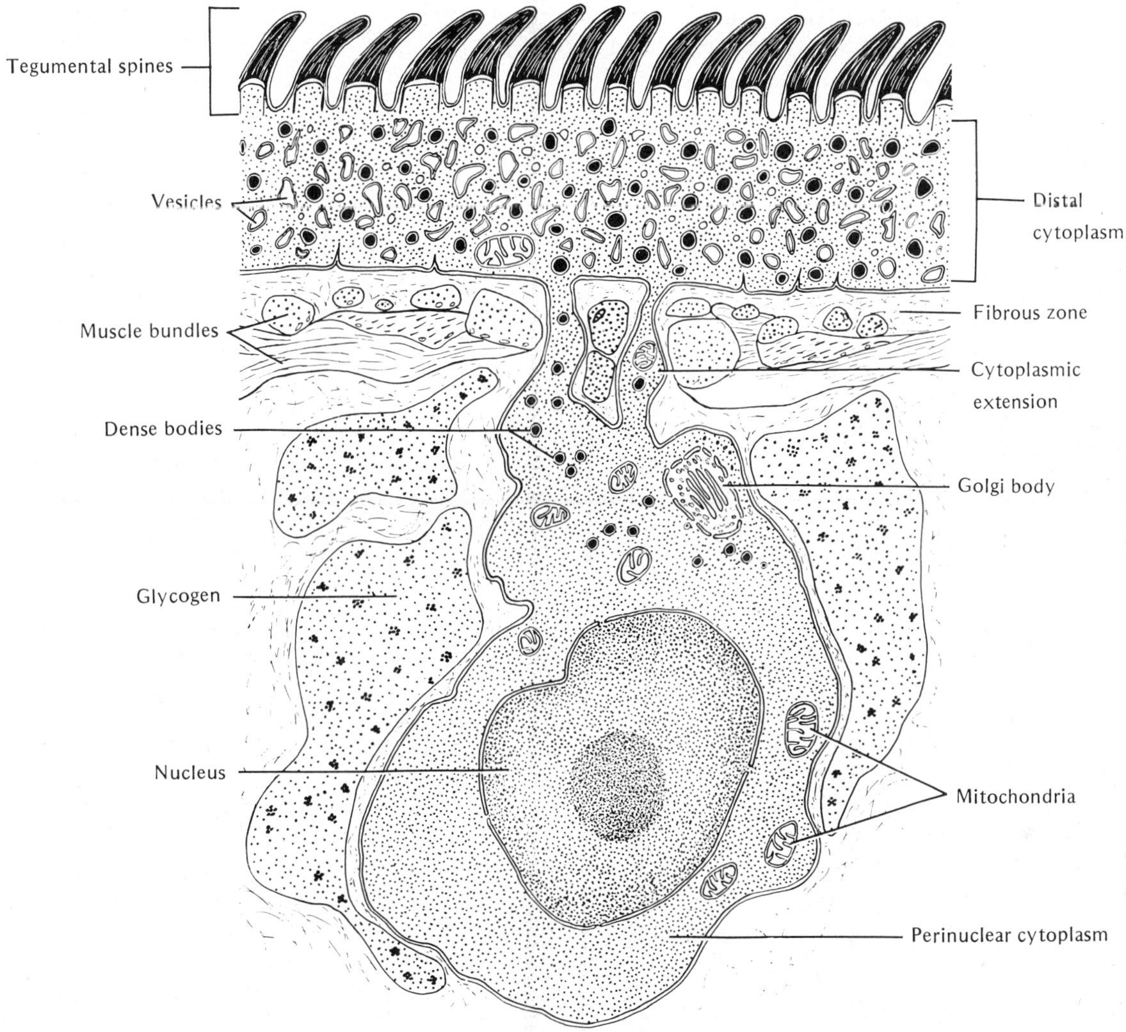


**FIG. 27.2**
Diagram of the tegument of the protoscolex of *Echinococcus granulosus*. [From Morseth, 1967.]

produces eggs or ova; a single oviduct leading from the ovary to conduct eggs; the ootype, which is a chamber joined by the oviduct, the duct from the seminal receptable, and is surrounded by the cells of the Mehlis' gland. The uterus leads from the ootype and may have either an opening or uteropore or end blindly. Proglottids without uteropores detach when gravid (apolysis) and break up to release their eggs. Individual differences will be discussed with each particular organism.

The parenchyma of most cestodes contains structures called *calcareous corpuscles* (Fig. 27.6) because they contain calcium. In addition, inorganic components including magnesium and phosphorus and organic compounds including DNA, RNA, proteins, sugars, and enzymes have also been reported (Smyth, 1969). The function of these corpuscles is unknown, but they are useful in identifying tapeworms in tissue sections (Chitwood & Lichtenfels, 1972). Similar structures are somtimes seen in trematodes but are contained only in the excretory canals.

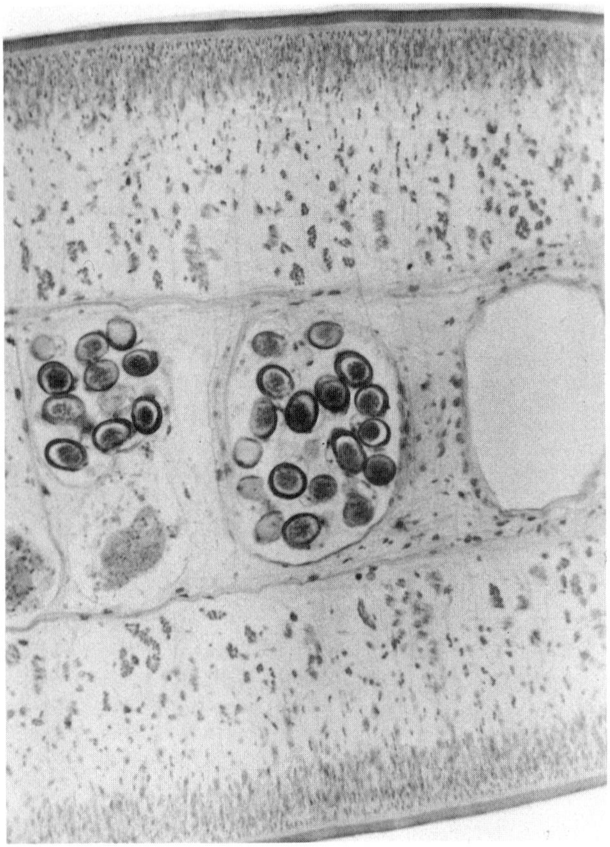

**FIG. 27.3**
A histologic section of *Taenia taeniaeformis* with eggs filling the uterus in the medullary portion.

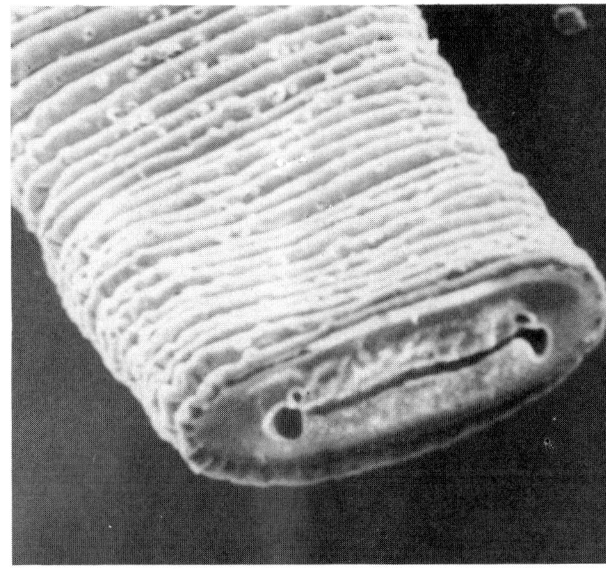

**FIG. 27.4**
Scanning electron micrograph of the cut surface of a mature proglottid of *Hymenolepis diminuta*. Two small dorsal osmoregulatory canals are located above the larger ventral osmoregulatory canals. The transverse canal joining the two ventral canals is also visible.

The nervous system of tapeworms consists of two large nerve trunks that travel from the scolex to the end of the strobila. Lateral branches from the main nerve trunks reach all proglottids in a ladderlike fashion. The area of the genital atrium is richly supplied with nerve endings. The tapeworm "brain" is in the scolex; it consists of dorsal and ventral commissures and accessory nerves connecting with holdfast organs, and so forth. Complexity varies from species to species. Externally, cestodes have a high density of uniciliate sensory cells on the scolex. The proglottids possess both uniciliate and nonciliate sensory cells (Fig. 27.7). These sensory cells are similar to those seen in other invertebrates including the trematodes (Fig. 20.5). The ciliate cells are probably tangosensory (touch) or chemosensory. The nonciliate sensory cells do not reach the tegumental surface and thus probably are not tango- or chemosensors; instead they probably monitor body deformations (stretch sensors) of the tapeworm body wall caused by worm activity or host gut peristalsis or both (Whitfield, 1979).

LIFE CYCLE. The basic life cycle patterns for tapeworms consist of a monoecious (rarely dioecious) adult, which produces an ovum. The ovum is fertilized and the resultant zygote develops into a larva called an *oncosphere*; it usually has six hooks and is called a *hexacanth embryo*. The oncosphere usually is ingested by an invertebrate intermediate host and migrates to a parenteral location such as an insect hemocoel. The larva metamorphoses and grows as a metacestode (i.e., changing cestode) until signs of sexual maturity appear and a scolex with adult structure develops. Usually a vertebrate definitive host ingests the invertebrate containing the metacestode. Final development to a

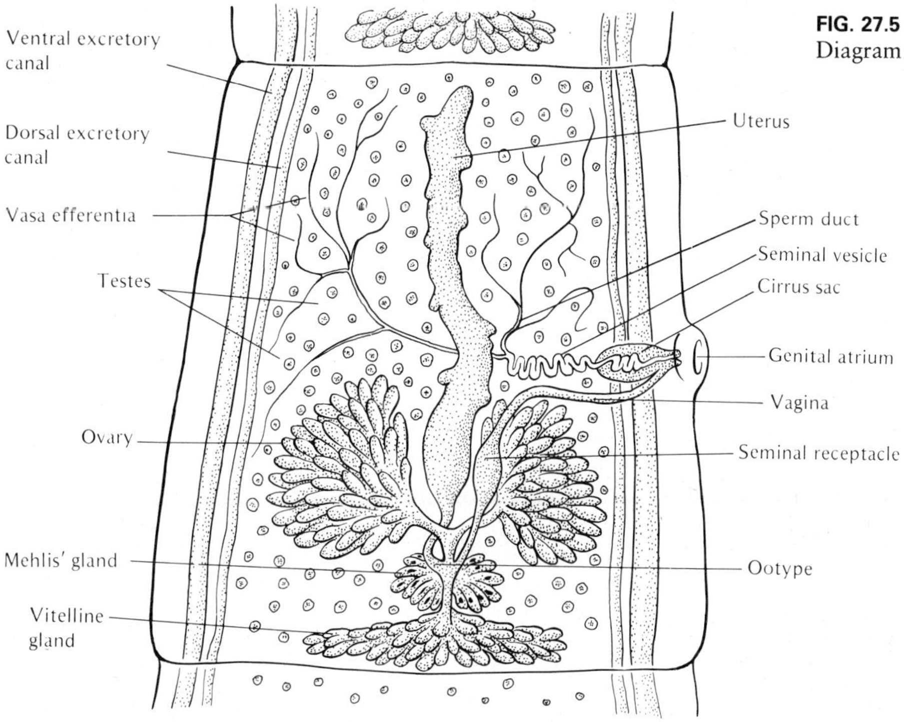


**FIG. 27.5**
Diagram of a mature proglottid of *Taenia* sp.

sexually reproducing adult occurs in the vertebrate digestive tract and associated ducts.

Freeman (1973) proposed terminology that is useful for organizing tapeworm metacestodes. There are two types of metacestode development, primitive and neoteric. Primitive metacestodes develop without a primary cavity or lacuna, whereas neoteric metacestodes do develop one. Primitive or solid metacestodes are subdivided into procercoids, which do not have a scolex similar to the adult, and plerocercoids, which do have a scolex similar to adult worms. Neoteric or hollow metacestodes, all in the order Cyclophyllidea, have two basic patterns of development; the cysticercoid has a lacuna filled with parenchyma and the cysticercus has a fluid-filled lacuna. There are types of metacestodes other than those just mentioned. Most notable are variations of the cysticercus, which include (a) strobilocercous, a cysticercus with some strobilization; (b) coenurus, with multiple scolices inside a nonlaminated lacuna; and (c) hydatid, with multiple scolices inside a laminated lacuna. Some oncospheres and metacestodes are illustrated here (Fig. 27.8). Variations in biology and morphology are discussed with each organism.

CESTODE NUTRITION AND SURFACE ACTIVITIES. Cestodes have been used as models for research on nutrient uptake because they have no gut. Reviews of cestode surfaces and nutrition are in Lumsden (1975b), Pappas and Read (1975), Arme (1975), and Whitfield (1979). The following account was synthesized from these sources.

Most investigators believe that adult cestodes acquire nutrients from the immediate external environment, the host gut, but diffusion and active transport mechanisms and probably endocytosis as well. In addition, the tapeworm tegument must be viewed as a "digestive-absorptive interface" (Whitfield, 1979). The outer surface of tapeworms is studded with enzymes; some are produced by the worm and are called *intrinsic digestive enzymes*. They are membrane-bound and not released into the environment. Other enzymes pro-

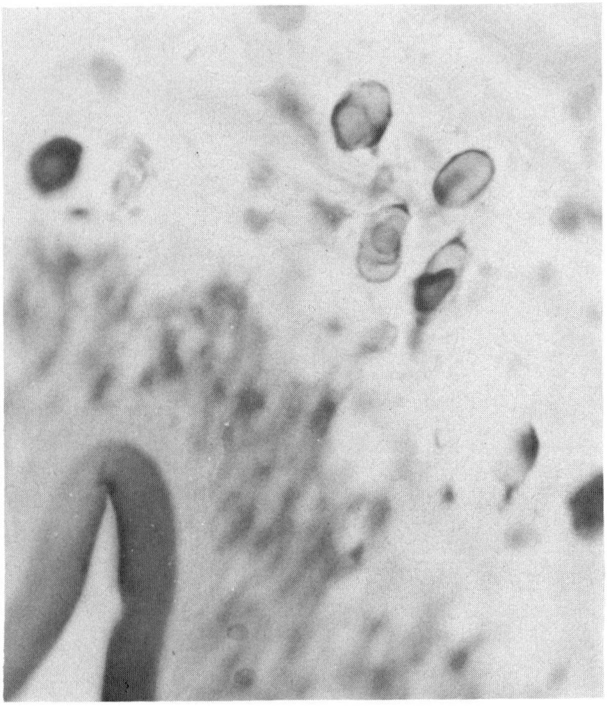

FIG. 27.6
Oval calcareous corpuscles in *Taenia* parenchyma.

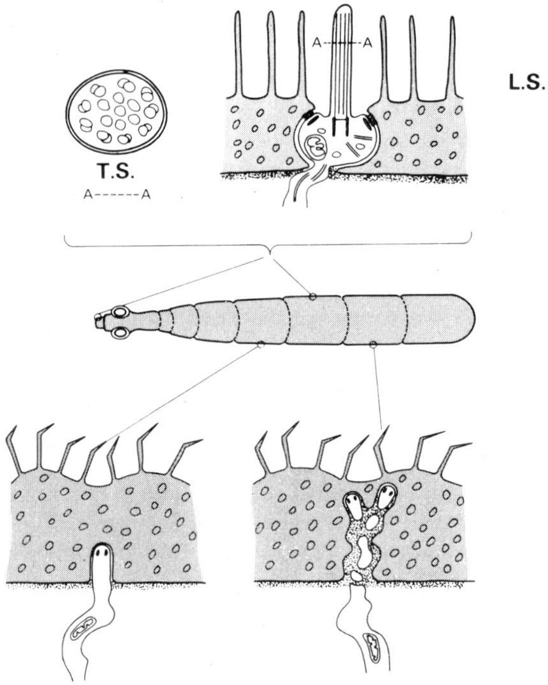


FIG. 27.7
Tegumental sensory endings of tapeworms. The upper tangential section (T.S.) and longitudinal section (L.S.) of a ciliated sensory receptor. The bottom two diagrams are of nonciliated sensory receptors characteristic of proglottid tegument. [Redrawn with modifications from Whitfield, 1979.]

duced by the host and borrowed by the worm for its own uses are called *extrinsic contact digestive enzymes*. Cestode surface activities are illustrated in Fig. 27.9.

Diffusion is the net flow of nutrient from a region of high concentration, in this case the host intestine, to a region of lower concentration, the tapeworm interior. No energy is required for this process. Pyridoxine, a B vitamin, is acquired by simple diffusion.

Most amino acids are acquired by active transport. Active transport moves substances against concentration gradients from a region of low concentration to one of higher concentration. This process occurs only across biological membranes, requires energy, and has many features in common with enzyme–substrate interactions, including uptake kinetics and competitive inhibition. Uptake of glucose by *Hymenolepis* species is an example of active transport.

A number of parasite-produced enzymes are embedded in the tapeworm surface. One example of an intrinsic digestive enzyme is phosphohydrolase, which hydrolyzes glucose-6-phosphate to glucose; this can then be taken into the worm by active transport at specific monosaccharide sites.

Cestodes can also share or borrow enzymes from their hosts. They cannot digest starch at their surface with their own enzymes, so they borrow an extrinsic contact amylase which is reversibly bound to the tegumental glycocalyx. Because this borrowed host enzyme is closer to the worm surface than the host's intestinal cells, the worm has a competitive advantage in acquiring its products. The concept of enzyme sharing may have to be reevaluated, however, in light

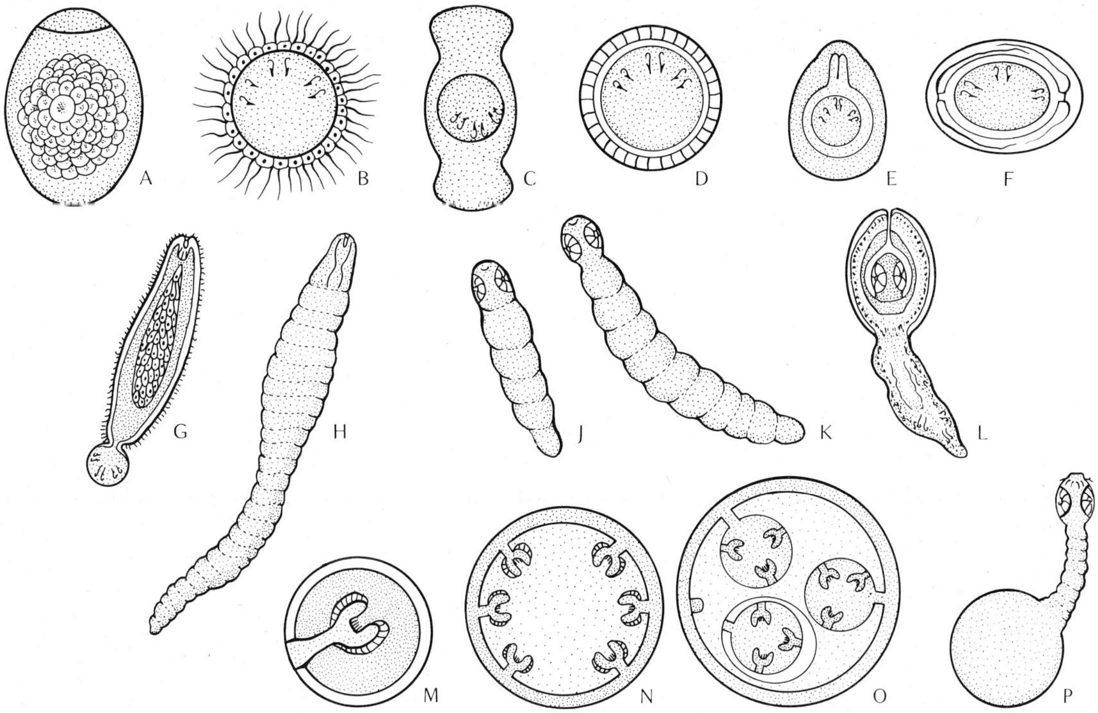


**FIG. 27.8**
Diagram of cestode oncospheres and metacestodes. (*A*) Pseudophyllidean egg; (*B*) coracidium; (*C*) protocephalan egg; (*D*) taenioid egg; (*E*) anoplocephalid egg; (*F*) hymenolepid egg; (*G*) pseudophyllidean procercoid; (*H*) pseudophyllidean plercocercoid; (*J*) protocephalan procercoid; (*K*) protocephalan plerocercoid; (*L*) cysticercoid; (*M*) cysticercus; (*N*) coenurus; (*O*) hydatid; (*P*) strobilocercous.

of Thomas and Turner's (1980) report that protein coming from *Hymenolepis diminuta* interfered with the amylase assay. The extrinsic contact digestive enzyme may be an artifact.

An additional function of the tapeworm's surface that is beneficial for it is the ability to inactivate host digestive enzymes such as trypsin and chymotrypsin. Mammalian intestinal cells do the same thing to prevent their being digested by enzymes in the gut; presumably this is also why the tapeworms do it.

Thus the cestode tegument is a multifunctional surface which processes nutrients and takes them into the body of the worm; at the same time it protects the worm against digestive enzymes of the host.

## CLASSIFICATION OF THE CLASS CESTOIDEA

### SUBCLASS CESTOIDARIA

Monozoic, with a single set of reproductive organs and a ten-hooked larva. No true scolex, but may have an anterior acetabulum. In intestines and body cavities of chimaeroid and siluroid fishes.

The subclass Cestodaria is an admittedly artificial group of aberrant organisms with affinities to monogeneans, because of vestiges of a haptor, and to "true" tapeworms, because of general biology, lack of a gut, and some other features. Usually they

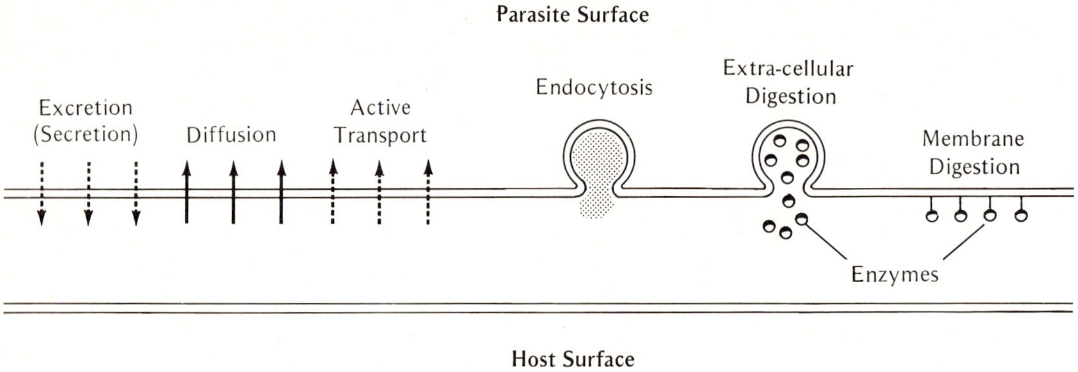


**FIG. 27.9**
Diagram of the activities occurring at a cestode – host interface. [Redrawn with modifications from Smyth, 1973.]

are included with the tapeworms as a separate subclass. They have a characteristic larva, the lycophore or decacanth, which has ten-hooks (Fig. 27.10). These worms are monozoic, that is, they have a single set of reproductive organs. They do not have a true scolex, but may have a single, anterior, cuplike acetabulum (Fig. 27.11). They infect the body cavities and intestines of chimaeroid and siluroid fishes, which are ancient fishes. There are no cestodarians of known economic importance. Information about these unique organisms is in Wardle and Mcleod (1952) and Wardle et al. (1974).

**Order Amphilinidea**
Worms with flattened, monozoic bodies without an anterior holdfast. The anterior end has a boring mechanism, which enables penetration of host body cavities. Worms are found in body cavities of Acepenseridae (sturgeons) in Europe and North America, siluroid fishes in India, and a tortoise in Australia.

**Order Gyrocotyliidea**
Thick, monozoic worms with a single set of reproductive organs and a single, anterior, cuplike, muscular holdfast. They are found in large intestines of chimaeroid fishes in both the Atlantic and Pacific Oceans.

## SUBCLASS EUCESTODA
Monozoic and polyzoic. Single or multiple sets of reproductive organs. Six-hooked larva. True scolex present. Parasites of fishes, amphibians, reptiles, birds, and mammals; one genus matures in the coelom of freshwater oligochaetes.

**Order Caryophyllidea Beneden in Olsson, 1893**
Scolex with shallow grooves or loculi, or fimbriated, weakly developed, lacking true suckers. Strobila

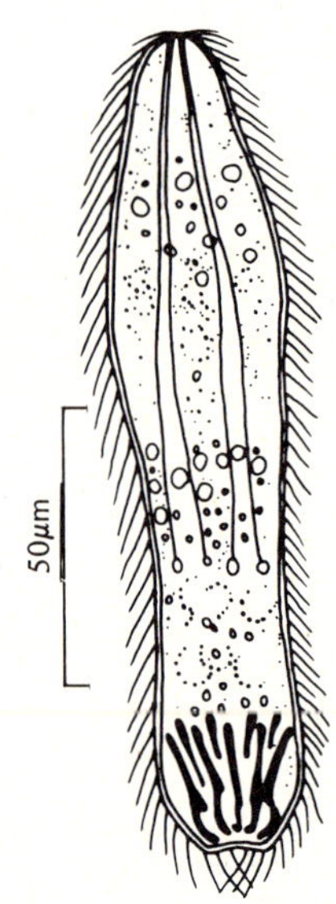


**FIG. 27.10**
Diagram of a lycophore larva of *Gyrocotyle urna*. [Redrawn from Lynch, 1945.]

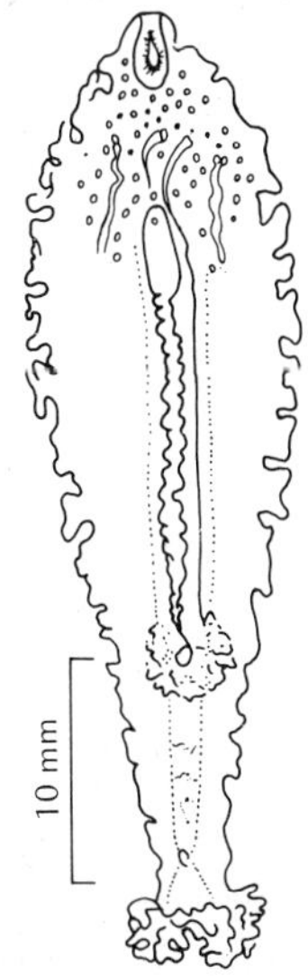


**FIG. 27.11**
Diagram of a *Gyrocotyle urna* adult. [Redrawn from Lynch, 1945.]

monozoic. Genital pores midventral. Testes numerous. Ovary posterior. Vitellaria follicular, scattered or lateral. Uterus a coiled, median-tube opening, often together with vagina, near male pore. Parasites of teleost fishes and aquatic annelids.

**Order Spathebothriidea Wardle and McLeod, 1953**
Scolex feebly developed, undifferentiated, or with funnel-shaped apical organ, or with one or two hollow, cuplike organs. External segmentation absent, internal segmentation present. Genital pores ventral. Testes in two lateral bands. Ovary dendritic. Vitellaria follicular, lateral, or scattered. Uterus coiled, with ventral pore. Parasites of teleost fishes.

**Order Trypanorhynchidea Diesing, 1863**
Scolex elongate, with two or four bothridia, and four eversible (rarely atrophied) tentacles armed with hooks. Each tentacle invaginates into internal sheath provided with muscular bulb. Neck present or absent. Strobila apolytic or anapolytic. Genital pores lateral, rarely ventral. Testes numerous. Ovary posterior. Vitellaria follicular, scattered. Uterine pore present or absent. Parasites of elasmobranchs.

**Order Pseudophyllidea Carus, 1863**
Scolex with two bothria, with or without hooks. Neck present or absent. Strobila variable. Proglottids anapolytic. Genital pores lateral, dorsal, or ventral. Testes numerous. Ovary posterior. Vitellaria follicular, scattered. Uterine pore present, dorsal or ventral. Egg usually operculate, containing coracidium. Parasites of fish, amphibians, reptiles, birds, and mammals.

**Order Lecanicephalidea Baylis, 1920**
Scolex divided into anterior and posterior regions by horizontal groove. Anterior portion cushionlike, or with unarmed tentacles, capable of being withdrawn into posterior portion forming a large suckerlike organ. Posterior portion usually with four suckers. Neck present or absent. Testes numerous. Ovary posterior. Vitellaria follicular, lateral, or encircling proglottid. Uterine pore usually present. Parasites of elasmobranchs.

**Order Aporidea Fuhrmann, 1934**
Scolex with simple suckers or grooves and armed rostellum. No external segmentation, internal segmentation present or not. Genital ducts and pores, cirrus, ootype, and Mehlis' gland absent. Hermaphroditic, rarely dioecious. Vitelline cells mixed with ovary. Parasites of Anseriformes.

**Order Tetraphyllidea Carus, 1863**
Scolex with highly variable bothridia, sometimes also with hooks, spines, or suckers. Myzorhynchus present or absent. Neck present or absent. Proglottides commonly apolytic. Hermaphroditic, rarely dioecious. Genital pores lateral, rarely posterior. Testes numerous. Ovary posterior. Vitellaria follicular (condensed in *Dioecotaenia*), usually lateral. Uterine pore present or not. Parasites of elasmobranchs.

**Order Diphyllidea Beneden in Carus, 1863**
Scolex with armed or unarmed peduncle. Two spoon-shaped bothridia present, lined with minute spines, sometimes divided by median, longitudinal ridge. Apex of scolex with insignificant apical organ or with large rostellum bearing dorsal and ventral groups of T-shaped hooks. Strobila cylindrical, acraspedote. Genital pores posterior, midventral. Testes numerous, anterior. Ovary posterior. Vitellaria follicular, lateral, or surrounding segment. Uterine pore absent. Uterus tubular or saccular. Parasites of elasmobranchs.

**Order Litobothridea Dailey, 1969**
Scolex a single, well-developed apical sucker. Anterior proglottids modified, cruciform in cross section. Neck absent. Strobila dorsoventrally flattened, with numerous proglottids, each with single set of

medullary reproductive organs. Segments laciniated and craspedote; apolytic or anapolytic. Testes numerous, preovarian. Genital pores lateral. Ovary two- or four-lobed, posterior. Vitellaria follicular, encircling medullary parenchyma. Parasites of elasmobranchs.

**Order Nippotaeniidea Yamaguti, 1939**
Scolex with single sucker at apex, otherwise simple. Neck short or absent. Strobila small. Proglottids each with single set of reproductive organs. Genital pores lateral. Testes anterior. Ovary posterior. Vitelline gland compact, single, between testes and ovary. Osmoregulatory canals reticular. Parasites of teleost fishes.

**Order Proteocephalidea Mola, 1928**
Scolex with four suckers, occasionally with apical sucker or armed rostellum. Neck usually present. Genital pores lateral. Testes numerous. Ovary posterior. Vitellaria follicular, usually lateral. Uterine pore present or absent. Parasites of fishes, amphibians, and reptiles.

**Order Cyclophyllidea Beneden *in* Braun, 1900**
Scolex usually with four suckers, rostellum present or not, armed or not. Neck present or absent. Strobila variable, usually with distinct segmentation, hermaphroditic or, rarely, dioecious. Genital pores lateral (ventral in Mesocestoididae). Vitelline gland single, compact, usually posterior to ovary. Uterus variable. Uterine pore absent. Parasites of amphibians, reptiles, birds, and mammals.

## REFERENCES AND FURTHER READINGS


Arai, H. P. 1980. *Biology of the Tapeworm* Hymenolepis diminuta. Academic Press, New York.

Arme, C. 1975. Tapeworm–host interactions. *Symp. Soc. Exper. Biol.* **29**:505–532.

Chitwood, M., and J. R. Lichtenfels. 1972. Identification of parasitic metazoa in tissue sections. *Exper. Parasitol.* *32*:407–519.

Freeman, R. S. 1973. Ontogeny of cestodes and its bearing on their phylogeny and systematics. *Adv. Parasitol.* **11**:481–557.

Hoeppli, R. 1959. *Parasites and Parasitic Infections in Early Medicine and Science.* University of Malaya Press, Singapore.

Lumsden, R. D. 1975a. Surface ultrastructure and cytochemistry of parasitic helminths. *Exper. Parasitol.* **37**:267–339.

———. 1975b. The tapeworm integument: A model system for studies on membrane structure and function in host–parasite relationships. *Trans. Am. Micros. Soc.* **94**:501–507.

Lynch, J. E. 1945. Redescription of the species of *Gyrocotyle* from the ratfish *Hydrolagus coliei* (Lay and Bennett) with notes on the morphology and taxonomy of the genus. *J. Parasitol.* **31**:418–446.

Morseth, D. J. 1967. Fine structure of the hydatid cyst and protoscolex of *Echinococcus granulosus. J. Parasitol.* **53**:312–325.

Olsen, O. W. 1974. *Animal Parasites: Their Life Cycles and Ecology,* 3rd ed. University Park Press, Baltimore.

Pappas, P. W., and C. P. Read. 1975. Membrane transport in helminth parasites: A review. *Exper. Parasitol.* **37**:469–530.

Schmidt, G. D. 1970. *How to Know the Tapeworms.* Wm. C. Brown, Dubuque, Iowa.

Smyth, J. D. 1969. *The Physiology of Cestodes.* W. H. Freeman, San Francisco.

———. 1972. Changes in the digestive-absorptive surface of cestodes during larval/adult differentiation. *Symp. Br. Soc. Parasitol.***10**:41–70.

———. 1973. Some interface phenomena in parasitic protozoa and platyhelminths. *Can. J. Zool.* **51**:367–377.

Thomas, J. N., and S. G. Turner. 1980. A reinterpretation of the evidence for contact digestion in the tapeworm *Hymenolepis diminuta. J. Physiol.* **301**:79P–80P.

Wardle, R. A., and J. A. McLeod. 1952. *The Zoology of Tapeworms.* University of Minnesota Press, Minneapolis.

Wardle, R. A., J. A. McLeod, and S. Radinovsky. 1974. *Advances in the Zoology of Tapeworms, 1950–1970.* University of Minnesota Press, Minneapolis.

Whitfield, P. J. 1979. *The Biology of Parasitism.* University Park Press, Baltimore.

28

# The Pseudophyllidea and Proteocephala

## ORDER PSEUDOPHYLLIDEA

Pseudophyllidean tapeworms are pseudopolyzoic, because they have many proglottids, but they are formed "instantaneously" (see Chapter 27). The scolex has dorsal and ventral grooves called *bothria*. The proglottids are wider than they are long and usually have midventral genital openings. A separate uteropore is also present. The eggs are usually operculate and unembryonated when laid. Testes and vitellaria are scattered throughout most of the proglottid. These are among the largest tapeworms known, some reaching 30 m in length. Of particular importance are the human and animal tapeworms belonging to the genera *Diphyllobothrium* and *Spirometra* (Meyer, 1970).

NAME OF ORGANISM AND DISEASE CAUSED. *Diphyllobothrium latum* (L.) Lühe, 1910 (=*Dibothriocephalus latus*); a number of *Diphyllobothrium* spp. cause diphyllobothriasis or broad fish tapeworm disease in humans; *D. latum* is the most important. Humans are the principal final host, but numerous animal reservoirs exist (von Bonsdorff, 1977).

HOSTS AND LOCATION. The principal definitive host for *D. latum* is the human. Other species of fish-eating mammals are infected; the domestic dog may be an important reservoir. Adult worms reside in the small intestines of definitive hosts.

DISTRIBUTION AND IMPORTANCE. *D. latum* is a cosmopolitan species typically present in subarctic and temperate zones. Infections decrease substantially north of the Arctic Circle and in the tropics. The original focus of infection was northern Europe, from which the worm has spread through migrations of people carrying the infections with them. Human infection reports are common from the Scandinavian countries, the Balkan countries, Europe, Asia, and especially Japan, North America, and South America (Peru and Chili). Additional cases have been reported throughout the world; most apparently result from infections acquired in the previously mentioned areas.

Current estimates of human infections worldwide are about 9 million cases, of which 5 million are in Europe, 4 million in Asia, less than 100,000 in North America, and even fewer in South America (von Bonsdorff, 1977). Serious disease is uncommon, but tapeworm anemia, usually called *tapeworm pernicious anemia*, can result (Vik, 1971).

MORPHOLOGY. *D. latum* is the largest tapeworm that infects humans. Adults 20 to 25 m long and 1.5 to 2.0 cm maximum width have been recorded (von Bonsdorff, 1977). A large worm may have 4000 proglottids (Vik, 1971), and spent portions of the strobila are shed in feces (pseudoapolysis).

The scolex has deep dorsal and ventral bothria (Fig. 28.1), followed by a short neck, and then a strobila of broad proglottids. A mature proglottid has a separate

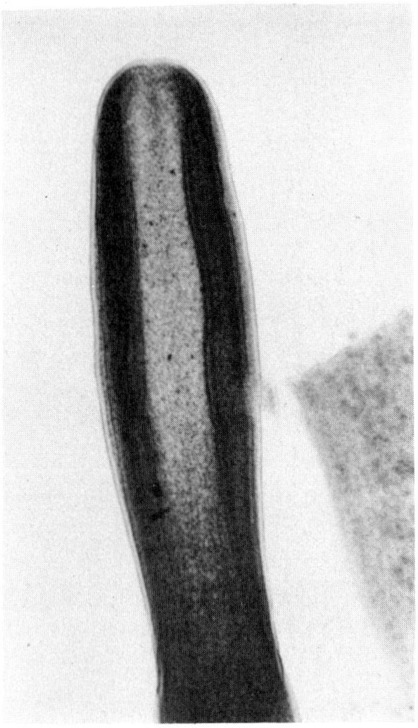

**FIG. 28.1**
Scolex of *Diphyllobothriun latum* with prominent bothrium.

uteropore and a male and a female genital pore, all opening on the midventral surface (Fig. 28.2). Testes and vitellaria are scattered throughout most of the proglottid. The bilobed ovary is located in the posterior part of the proglottid; gravid proglottids appear to the naked eye to be brownish, because the anterior half is filled with uterine loops packed with eggs (Fig. 28.3). The eggs are ovoid and average 65 by 45 μm; they have an operculum at one end and a small knob at the opposite end (Fig. 28.4)

LIFE CYCLE. Excellent reviews of the biology of *D. latum* have been written by Vik (1971), Williams and Jones (1976), von Bonsdorff (1977), and von Bonsdorff and Bylund (1982).

Eggs are passed in the feces and take from 7 to 20 days to develop, depending on temperature and oxygen. Under anaerobic or nearly anaerobic conditions, embryonation proceeds for 5 to 6 days at 20°C, but the embryo then dies. Eggs develop better on a sandy or rocky substrate than on mud, probably because of availability of oxygen.

Hatching takes place under the influence of light after the coracidium is completely formed. Both the amount of light and its wavelength are important. Maximal hatching is seen at 2 wavelengths—in the yellow (500–600 nm) and in the ultraviolet (250 nm)—and only 30 to 60 sec of light at 50 to 100 lux are required to start the process.

The spherical, ciliated coracidium is about 50 μm when it escapes from the egg capsules, but it increases in size to about 100 μm. Under good conditions, a coracidium may live 5 days while swimming in the water. Coracidia can tolerate salinity of 0.4% well but die quickly at 0.9%; therefore, they can live in brackish estuaries where the salinity is 0.2 to 0.4%.

First intermediate hosts are copepods, usually of the genera *Diaptomus* and *Cyclops*. About 40 species of the copepods have been infected with *D. latum*, and although there are differences in susceptibility, the host specificity is rather low. After being eaten by a copepod, the hexacanth embryo escapes from the ciliated covering and penetrates to the hemocoel, probably in the same way as the hexacanth of *T. saginatus* (Silverman & Maneeley, 1955)—by holding on with its hooks and secreting a substance to allow it to squeeze between cells. The hexacanth transforms into a procercoid, which is fully developed in about 14 days at 20°C. The procercoid is 500 to 600 μm long and about 300 μm wide, with a saclike structure at its posterior, the cercomere, which contains all that remains of the hexacanth, its hooks.

The second intermediate hosts are planktivorous fish, which eat the infected copepods. The procercoid migrates into the extraintestinal tissues of the fish and develops into the plerocercoid (Fig. 28.5), which is usually from 6 to 10 mm long. Nearly any tissue of the fish may be infected, but muscle, ovary (hard roe), and liver are especially important for transmission to humans, because these are the parts most often eaten. Different fish are important in different parts of the world (Table 28.1), but pike, perch, and burbot are especially significant. The plerocercoid infection is passed from small planktivorous fish to larger predatory fish without change in the plerocercoid. It moves from the gut to the tissues of its new host and continues its existence there. A large pike or burbot can harbor

A

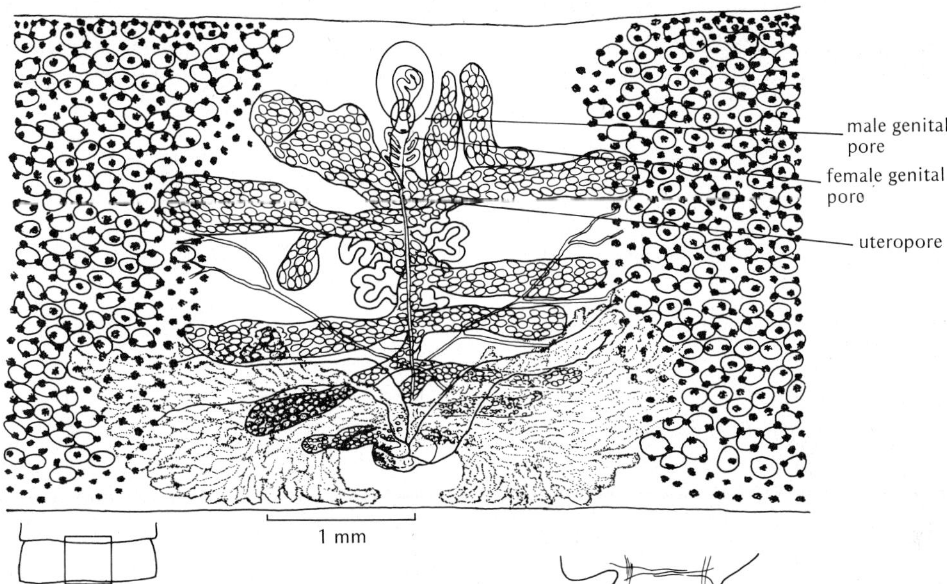


B

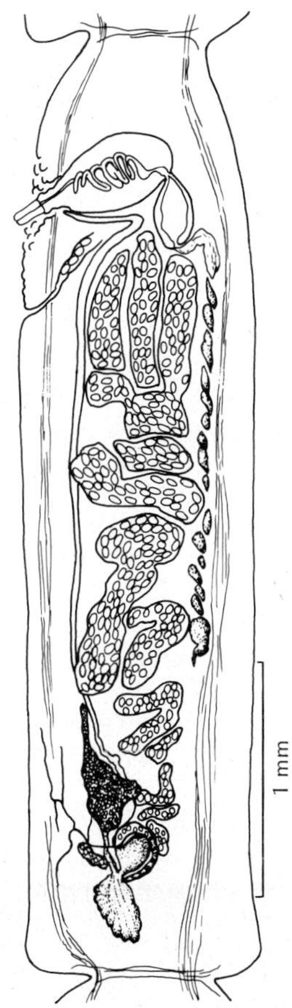


**FIG. 28.2**
Diagram of mature proglottids of *Diphyllobothrium latum*. (*A*) ventral view; (*B*) sagittal section. [Redrawn with modifications from Rausch & Hilliard, 1970.]

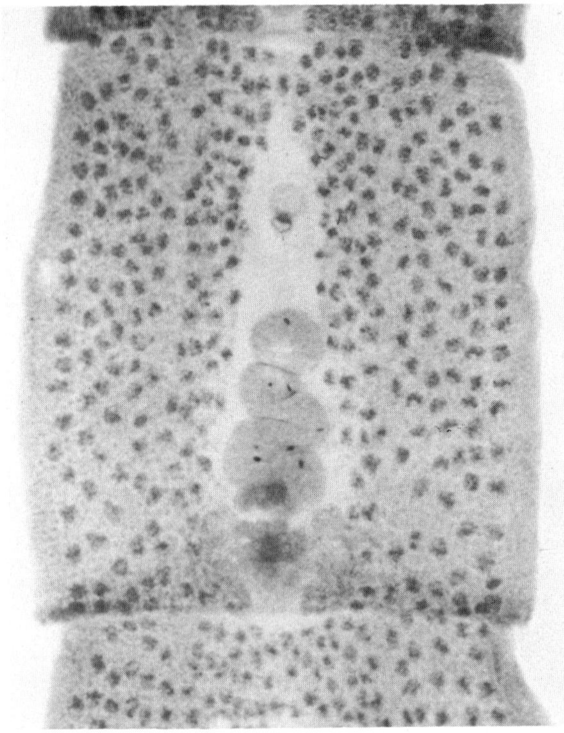

FIG. 28.3
Gravid proglottid of *Diphyllobothrium latum*.

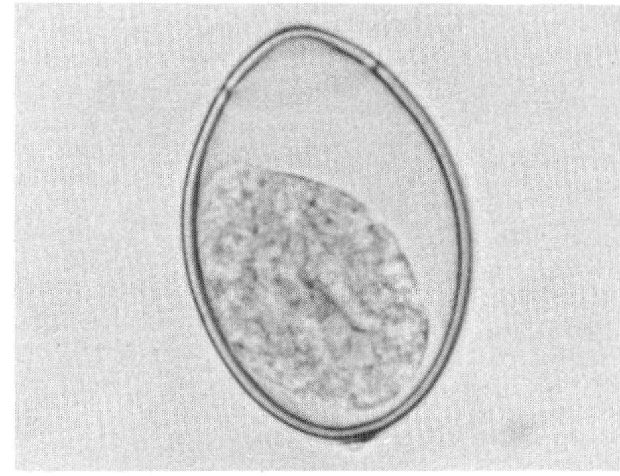

FIG. 28.4
*Diphyllobothrium latum* egg with operculum at one end and knob at the opposite end.

1000 plerocercoids, and in some areas, nearly 100% may be infected.

Humans or other definitive hosts become infected from eating raw or insufficiently cooked fish harboring the plerocercoid. Dogs may be important reservoir hosts in some areas, but the question remains open (Meyer, 1970). The plerocercoid attaches to the intestinal wall with its primitive holdfast and grows to the adult. The worms grow at a rate of 5 cm a day and eggs are produced in from 25 to 30 days after infection. Infections may persist in humans for as long as 30 years.

DIAGNOSIS. Finding eggs in the feces is the most certain diagnostic method. Eggs of *Paragonimus* are much like those of *D. latum* but somewhat larger.

HOST–PARASITE INTERACTIONS. *D. latum* infections in humans usually cause little discomfort. The voracious appetite attributed to tapeworm infections is a myth (von Bonsdorff, 1977), but some vague abdominal discomfort, diarrhea and weakness may occur. *D. latum* rarely causes intestinal obstruction. There is local epithelial atrophy where the scolex is attached to the intestinal villi. When the villi are destroyed, the worm releases and moves to another site.

Tapeworm anemia, called *prenicious anemia*, is the

TABLE 28.1
Second Intermediate Hosts of *Diphyllobothrium latum*

| FISH | GEOGRAPHIC LOCATION EURASIA | NORTH AMERICA | SOUTH AMERICA |
|---|---|---|---|
| *Esox lucius* (pike) | x | x | |
| *Stizostedeon vitreum* (wall-eyed pike) | | x | |
| *S. canadense* (sand pike) | | x | |
| *Perca fluviatilis* (perch) | x | | |
| *Perca flavescens* (perch) | | x | |
| *Lota lota* (burbot) | x | | |
| *Lota maculosa* (burbot) | | x | |
| *Onchorhynchus* spp. (trout) | | x | |
| *Salmo lacustris* (lake trout) | | | x |
| *Salmo irideus* (trout) | | | x |
| *Salmo gairdneri* (rainbow trout) | | | x |

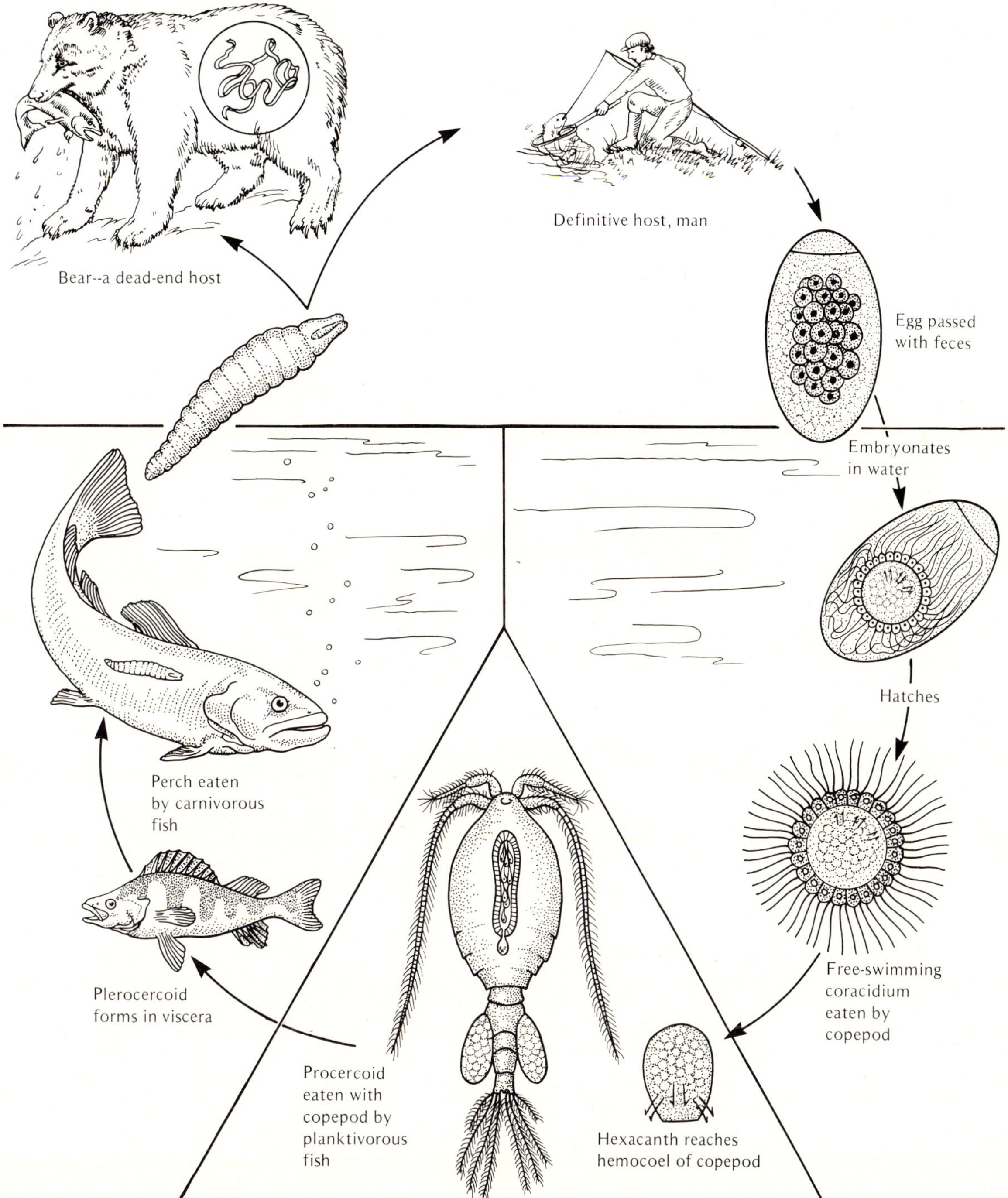


**FIG. 28.5**
Diagram of the life cycle of *Diphyllobothrium latum*.

most significant host–parasite interaction. Tapeworm anemia is rarely fatal, but von Bonsdorff (1977) recommended retaining the term *tapeworm pernicious anemia* because it is established in medical literature. He gave an interesting account of the accidental discovery of tapeworm anemia by a physician name Reyhner. The physician had a patient with "pernicious anemia" who also was passing tapeworm proglottids. When he treated the patient for the tapeworm, the anemia was cured. His accidental bedside observation led to the discovery of a new disease.

Tapeworm anemia occurs in about 1 of 50 infected people, especially in Scandinavia; the disease results primarily from vitamin $B_{12}$ deficiency. The worm absorbs vitamin $B_{12}$ ingested by the host and reduces the amount of the vitamin needed for hematopoiesis (Vik, 1971). Clinically the host has a macrocytic, megaloblastic anemia with leukopenia (von Bonsdorff, 1977). Nyberg et al. (1961) studied 1345 subjects in Finland; of the 366 tapeworm infections (27.2%), 8 (1.9%) persons had megaloblastic anemia and another 1.1% were borderline cases. Of the worm carriers, over 50% had a serum vitamin $B_{12}$ concentration under 100 pg/ml; normal values are 150 to 900 pg/ml, and values below 100 are considered to be pathologic (Vik, 1971). Thus, the potential for more than the observed 2% tapeworm anemia exists. Probably the nonanemic hosts utilized stored vitamin $B_{12}$, and months or even years could pass before anemia developed.

TREATMENT. Numerous folk remedies have been used to expel tapeworms; some are quite effective, but they usually have severe side effects. One example is an extract of the rhizome of a male fern, *Dryopteris felix mas*. The active ingredients are acylphloroglucinoles (von Bonsdorff,1977) and a derivative, Desaspidin is in current use. The most widely used drugs are niclosamide (Yomesan) and praziquantel (Biltricide). In addition to being wormed, anemic patients should be given vitamin $B_{12}$ to replenish fully the body stores that may have been depleted.

EPIDEMIOLOGY AND CONTROL. *D. latum* is primarily an infection of humans, but a number of other species of mammals can also be infected. Among domestic animals, dogs, cats, and swine are readily infected, and wolves, foxes, bears, seals, and other fish-eating wild animals have been found to be infected in nature or have been infected experimentally. Epidemiological studies indicate that there is no feral cycle of *D. latum*, but rather that humans must be present for its maintenance. The role of humans seems to be one of providing the right epidemiological conditions for transmission (Rausch & Hilliard, 1970).

The conditions for continuation of the life cycle are as follows:

1. An aquatic environment suitable for the development of the eggs
2. An aquatic environment suitable for the copepod first intermediate host
3. Fecal contamination of water with eggs (such contamination may come about through inadequate sewage treatment or through people defecating in or near water)
4. Food habits that include eating raw or inadequately cooked fish, especially those species that serve as the second intermediate host

Considering this list, one can appreciate that the conditions are not often satisfied. The aquatic system must be relatively free of organic matter so that the dissolved oxygen does not drop too low for the eggs to develop. Water temperature should be 15 to 20 °C, but it can range from 4 to 25°C. Copepods are ordinarily found in waters that are still or that do not run rapidly. Infection of copepods takes place near the shoreline, so a shallow littoral area is conducive to bringing the coracidia and the copepods near one another. A threshhold density of about 1000 copepods per cubic meter must be met before infection can be maintained in them (von Bonsdorff & Bylund, 1982).

The type of lake is important for infection of the second intermediate hosts as well as the first. Deep oligotrophic lakes have low rates of infection in fish, whereas shallow eutrophic lakes are likely to have high levels of infection even where the prevalence of infection in the human population is low.

The last element is the human, who must be considered from the standpoint of both dietary and sanitary habits. Eating raw or partially cooked fish is associated either with primitive conditions, where fuel for cooking is not available, or with areas where certain fish dishes are a traditional part of the diet. Even the preparation of fish plays a part in transmitting the infection; for example, gefüllte fish is cooked before being served, but it is often tasted by the cook when it is seasoned to make sure that it is properly spiced.

Promiscuous defecation is an obvious enough problem, but even in a well-designed treatment plant, 5% of the *D. latum* eggs may be found in the effluent. Where the system is overloaded or not well run, higher numbers escape treatment (von Bonsdorff & Bylund, 1982). An individual infected with *D. latum* may pass from 2 million to 40 million eggs per day, and that means that a minimum of 100,000 eggs a day would survive treatment from that one person. Since we know that infections can last for 30 years, a minimum of 1 billion eggs would be disseminated into the environment even under conditions of good sewage treatment.

*D. latum* has been distributed from its ancestral home in northern Europe to a number of other places that have all the right conditions. The infections in Alaska were probably carried there by Russians who maintained a colony on the Alaskan coast from about 1741 to 1867, and the endemic are in the midwestern United States was probably established by immigrants from Scandinavia (von Bonsdorff, 1977).

Control of diphyllobothriasis has four possible facets:

1. Eliminating copepods, fish, or both
2. Treating infected persons
3. Preventing contamination of waters with tapeworm eggs
4. Changing dietary habits

All but the first of these are feasible and are used in designing control programs. Although the infection is not especially hazardous to health, infected individuals should be treated in order to reduce possible contamination of the environment. As indicated, egg production of the worms is high, infections last a long time in humans, and humans are the principal definitive host. Reservoir hosts therefore will probably not be able to maintain the life cycle if human infections are eliminated. It should be noted that a few infected persons can transmit the infection to a large number of fish. At a lake in Sweden, 100% of susceptible fish were infected, but only 20 persons in the vicinity were infected (cited in von Bonsdorff & Bylund, 1982).

Modern sewage disposal, or at least efficient pit toilets, is essential for containment of any eggs that might reach the environment. Since eggs are not resistant to either freezing or drying, holding treated sludge for a period of time after treatment further reduces the probability of environmental contamination with eggs.

When campaigns to educate people about the source of infection have been instituted, changing dietary habits has been successful. For example, 34,000 infections recorded in Finland in 1959 were reduced to fewer than 2300 by 1975 following an intensive educational program (von Bonsdorff, 1977). The total infection rate in Finland dropped from 20 to 1.8%. The endemic area surrounding the Great Lakes of the United States seems to have declined; infected fish have not been found (Peters et al., 1978). It seems likely that the habit of eating raw fish has declined. In parts of Alaska and Canada, diphyllobothriasis remains a problem, because sewage disposal is poor and fish are still eaten raw (Rausch & Hilliard, 1970).

Cooking fish is the most effective way of killing the plerocercoids. As with many parasites, heating to 55 to 60°C kills them; this is about the temperature at which protein coagulates and kills plerocercoids. If fish are smoked, therefore, the temperature in the smokehouse needs to reach 55 to 60°C. At the other end of the temperature scale, freezing for 24 hr or more at home freezer temperature (0°C, −18°F) kills plerocercoids. Curing fish for 5 days in 10% salt solution also kills plerocercoids.

## *Other Species*

In addition to *Diphyllobothrium latum*, von Bonsdorff (1977) listed 11 other species of *Diphyllobothrium* , two of *Diplogonoporus* and one of *Schistocephalus* as pseudophyllidean tapeworms that infect humans who ingest plerocercoids in fishes. Rausch and Hilliard (1970) reported that *D. latum, D. dendriticum, D. lanceolatum, D. ursi,* and *D. dalliae* had been recovered from humans in Alaska. In a review of marine helminths and human disease, Williams and Jones (1976) listed *D.pacificum* from Peru as well as *D. latum* and probably two species of *Diplogonoporus* that infect humans in Japan. *Diplogonoporus* is easily identified by its double set of reproductive organs per proglottid.

## *Sparganosis*

In addition to being a definitive host for adult pseudophyllidean tapeworms, humans can also be a transport host for pleroceroids. Unidentified plerocercoids were originally given the generic designation *Sparganum*, and the term *sparganosis* is still used for plerocercoid infections (Tansurat, 1971). Today most spargana are known to be plerocercoids of members of the genus *Spirometra*. Human cases of sparganosis have been reported woldwide, but most are from China, Japan, and Southeast Asia. The taxonomy of this genus is uncertain. The Far Eastern parasite is called *Spirometra mansoni* (Mueller, 1974; Meyer, 1970) or *S. erinacei* (Yamaguti, 1959), and the parasite found in the United States is usually called *S. mansonoides*. To add to the confusion, Mueller (1974) admitted that if the rules of zoological nomenclature were strictly followed, the generic name should be *Luehella*; Schmidt (1970) followed the rules and called these parasites *Luehella*, but the name *Spirometra* is firmly entrenched in both parasitological and medical literature and will probably persist.

The genus *Spirometra* has a typical pseudophyllidean life cycle; the coracidium escapes from an operculate egg and is ingested by a copepod. The copepod harbors the procercoid stage. Second intermediate hosts include amphibians, reptiles, and mammals but not fish (Tansurat, 1971). Definitive hosts are cats, related species, and, rarely, reptiles; they are infected by eating second intermediate hosts infected with plerocercoids.

Humans can become infected in three ways (Tansurat, 1971)

1. A human who drinks water containing copepods infected with procercoids will become a true second intermediate host. The procercoid will penetrate the intestinal wall and enter the subcutaneous tissue, where it will develop into a plerocercoid.

2. Humans can ingest uncooked or poorly cooked frogs, snakes, birds, or mammals containing plerocercoids. They also migrate through the intestinal wall and encyst as spargana; humans are a paratenic host in this event.

3. By pressing the flesh of an infected frog as a poultice to ulcers, wounds, or sore eyes, individuals are infected. Plerocercoids migrate from the frog to the human tissue, and once again the person becomes a paratenic host. Poulticing is a common medical practice for treatment of skin and eye diseases in Viet Nam, Thailand, and other Far Eastern countries.

Spargana are usually encapsulated by fibrous connective tissue and can appear grossly like tumors of the eye, subcutaneous tissues, or almost any visceral organ; they can cause elephantiasislike symptoms if they locate in lymphatic ducts (Mueller, 1938).

Satisfactory treatment is obtained by surgical excision of the nodule containing the parasite. Control methods include ensuring safe drinking water, cooking of frog, snake, and mammal flesh, and avoidance of poultices. All of these are difficult to accomplish with the current socioeconomic situation in the Far East. Since this disease is a zoonosis in which cats are principal definitive hosts, elimination of adult worms is impractical. Fortunately, the number of cases of sparganosis with severe lesions is small.

## ORDER PROTEOCEPHALA

The order Proteocephala is composed of about 30 genera, all endoparasitic in fishes, amphibians, and reptiles as adults (Yamaguti, 1959; Freze, 1965). The adults and the two metacestode stages, procercoids and plerocercoids, all have scolices with four suckers; in some species an additional fifth apical sucker is present as well. A distinctive morphological feature of adult proteocephalans is the lateral location of the vitellaria in mature proglottids. Most proteocephalans belong to the genus *Proteocephalus*, and one of the best known of these is *Proteocephalus ambloplitis*, a parasite of bass and perch.

NAME OF ORGANISM AND DISEASE CAUSED. *Proteocephalus ambloplitis* (Leidy, 1887) Benedict, 1900 is known as the *bass tapeworm*.

HOSTS AND LOCATION. Adults are common intestinal parasites in bass, *Micropterus dolomieui* and *M. salmoides*, but are also found in the yellow perch (*Perca flavescens*), the bowfin (*Amia calva*), and other freshwater fishes.

DISTRIBUTION AND IMPORTANCE. The bass tapeworm infects fish throughout much of North America. Wandering plerocercoids sometimes cause so much

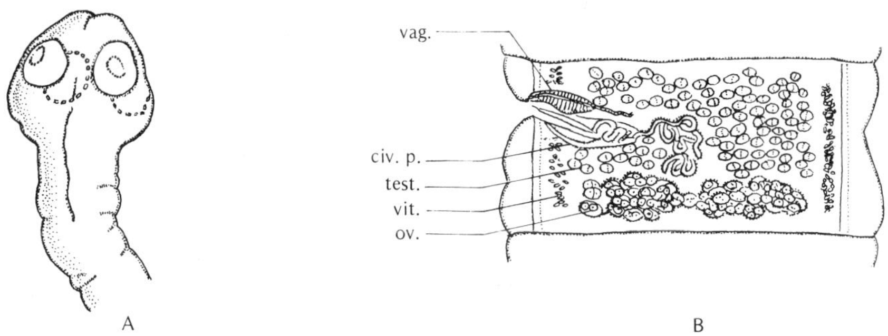


**FIG. 28.6**
Diagram of *Proteocephalus* sp. (*A*) Scolex; (*B*) mature proglottid (civ. p. = cirrus pouch; ov. = ovary; test. = testis; vag. = vagina; vit. = vitellaria) [(*A*) Redrawn from Hoffman, G. L. 1967. *Parasites of North American Freshwater Fishes*. University of California Press, Berkeley. Calif.; (*B*) redrawn from Van Cleave, H. J., and J. P. Mueller. 1934. Parasites of Oneida Lake fishes. Part III. A biological and ecological survey of the worm parasites. *Roosevelt Wild Life Annals* 3:161–334.]

fibrosis of the gonads that bass are rendered sexually sterile (parasitic castration). Because bass are an important sport fish, this disease has received widespread attention (Fischer & Freeman, 1969; Wardle & McLeod, 1952; Becker & Brunson, 1968).

MORPHOLOGY. Adults have lateral genital pores and vitellaria that extend from the anterior to the posterior ends of the mature proglottids (Fig. 28.6). Adults, procercoids, and plerocercoids all bear four suckers.

LIFE CYCLE. Usually the life cycle of *P. ambloplitis* involves three hosts. Gravid proglottids are passed in host feces; they rupture upon contact with water, releasing embryonated eggs. Eggs are ingested by crustaceans including *Cyclops* sp. and *Hyallela* sp., and procercoids develop in the hemocoel. When infected crustaceans are eaten by second intermediate hosts, centrarchid or percid fishes, plerocercoids develop in the viscera. Definitive hosts become infected by eating infected smaller fish. Often, large bass harbor numerous plerocercoids; they were once thought to be a dead end in the life cycle because it was unlikely that an even larger fish would eat them. However, Fischer and Freeman (1969) found that the usual three-host life cycle is often a two-host cycle: plerocercoids in the viscera can penetrate back into the gut of the same fish host and develop into adults.

EPIDEMIOLOGY AND CONTROL. The anthelminthic di-*n*-butyl tin oxide removes all gut tapeworms (Fischer & Freeman, 1969). The treatment must be repeated yearly in hatchery stock because visceral plerocercoids are unaffected. The best method of control is to start with clean fry for stocking purposes.

## REFERENCES AND FURTHER READINGS


Becker, C. D., and W. D. Brunson. 1968. The bass tapeworm: A problem in northwest trout management. *Prog. Fish Cult.* 30:76–83.

Freze, V. I. 1965. *Essentials of Cestodology*. Vol. V. *Proteocephalata in Fish, Amphibians and Reptiles. Izdat. Akad. Nauk.*, Moscow.

Fischer, H., and R. S. Freeman. 1969. Penetration of parenteral plerocercoids of *Proteocephalus ambloplitis* (Leidy) into the gut of smallmouth bass. *J. Parasitol.* 55:766–774.

Meyer, M. C. 1970. Cestode zoonoses of aquatic animals. *J. Wildl. Dis.* 6:249–254.

Mueller, J. F. 1938. Studies of *Sparganum mansonoides* and *Sparganum proliferum*. *Am. J. Trop. Med.* **18**:303–328.

———. 1974. The biology of *Spirometra*. *J. Parasitol.* **60**:3–14.

Nyberg, W., R. Gräsbeck, M. Saarni, and B. von Bonsdorff.

1961. Serum vitamin $B_{12}$ levels and incidence of tapeworm anemia in a population heavily infected with *Diphyllobothrium latum*. *Am. J. Clin. Nutr.* **9**:606–612.

Peters, L., D. Cavis, and J. Robertson. 1978. Is *Diphyllobothrium latum* currently present in northern Michigan? *J. Parasitol.* **64**:947–949.

Rausch, R. L., and D. K. Hilliard. 1970. Studies on the helminth fauna of Alaska. XLIX. The occurrence of *Diphyllobothrium latum* (Linnaeus, 1758) (Cestoda: Diphyllobriidae) in Alaska, with notes on other species. *Can. J. Zool.* **48**:1201–1219.

Schmidt, G. D. 1970. *How to Know the Tapeworms*. Wm. C. Brown, Dubuque, Iowa.

Silverman, P. H., and R. B. Maneeley. 1955. Studies on the biology of some tapeworms of the genus *Taenia*. III. The role of the secreting gland of the hexacanth embryo in the penetration of the intestinal mucosa of the intermediate host, and some histochemical reactions. *Ann. Trop. Med. Parasitol.* **49**:326–330.

Tansurat, P. 1971. Sparganosis. In R. A. Marcial-Rojas (ed.), *Pathology of Protozoal and Helminthic Disease*. Williams & Wilkins, Baltimore, pp. 585–591.

Vik, R. 1971. Diphyllobothriasis. In R. A. Marcial-Rojas (ed.), *Pathology of Protozoal and Helminthic Diseases*. Williams & Wilkins, Baltimore, pp 572–584.

von Bonsdorff, B. 1977. *Diphyllobothriasis in Man*. Academic Press, New York.

von Bonsdorff, B., and G. Bylund. 1982. The ecology of *Diphyllobothrium latum*. *Ecol. Dis.* **1**:21–26.

Wardle, R. A., and J. A. McLeod. 1952. *The Zoology of Tapeworms*. University of Minnesota Press, Minneapolis.

Williams, H. H., and A. Jones. 1976. *Marine Helminths and Human Disease*. Misc. Publ. No. 3. Commonwealth Agricultural Bureaux, Farnham Royal, Bucks, England.

Yamaguti, S. 1959. *Systema Helminthum*. Vol. II. *The Cestodes of Vertebrates* Interscience, New York.

# 29
# *Cyclophyllidea*

The order Cyclophyllidea includes most of the tapeworms of medical and veterinary importance as well as most of those that infect birds and mammals. The scolex with four suckers and a compact vitelline gland are identifying features of the adults of this order. They range in size from a few millimeters long in *Echinococcus* to over 12 m long in *Taenia saginata*. A few, such as *Echinococcus*, infect humans as larvae and in others, such as *Taenia saginata*, the beef tapeworm, humans are the definitive host. *Taenia solium* is a double threat since both larvae and adults can infect humans, occasionally simultaneously.

## FAMILY TAENIIDAE

Taeniids are the most important adult tapeworms of humans, canids, and felids; in addition, humans, cattle, and swine are infected by larval taeniids. Therefore, they are of both medical and veterinary importance.

Traditionally taeniids have been placed in several genera in whose adult forms are quite similar but whose larvae are different (Wardle et al., 1974). *Taenia* has a cysticercus (Fig. 29.2), *Multiceps* has a coenurus (Fig. 29.5), *Hydatigera* has a strobilocercus (Fig. 29.6), and *Echinococcus* has a hydatid (Fig. 29.9). *Taeniarhynchus* is identical to members of the genus *Taenia* except that it lacks rostellar hooks. Wardle and McLeod (1952) retained the various genera, but with the following reservations: "This practice opens up the question as to whether, in the absence of any practicable method of counting chromosomes, the specific and generic units of zoological classification shall be based upon readily recognizable characteristics of *adult* animal, or upon features of the life cycle and biology, a procedure while eminently desirable is rarely attainable." Studies such as those of Verster (1967, 1969) have led to the conclusion that adult structure is more important than that of the larvae, and larval evolution may have taken place secondarily (Moore, 1981). Two genera are now accepted in the Taeniidae: *Taenia* and *Echinococcus*.

NAME OF ORGANISM AND DISEASE CAUSED. *Taenia saginata* Goeze, 1782 (syn. *Taeniarhynchus saginatus*), causes taeniasis, cysticercosis, *Cysticercus bovis*, beef tapeworm, and measly beef.

HOSTS AND LOCATION. Humans are the only definitive host. The main intermediate hosts are domesticated members of the artiodactyl family Bovidae and the reindeer, *Rangifer tarandus* (family Cervidae). Adults are in the small intestine of the definitive host; cysticerci are in the striated muscles of the intermediate host. Whether humans are infected with cysticerci, despite 12 records in the medical literature, is still an open question (Pawloski & Schultz, 1972).

DISTRIBUTION AND IMPORTANCE. Beef tapeworm infections in humans are cosmopolitan. It is the most common taeniid tapeworm of humans. Over 39 million infections are estimated throughout the world, and the prevalence of infection in the United States is 23 per 100,000 (Pawlowski & Schultz, 1972). Infections seldom lead to serious disease, the infected persons usually are in a good nutritional state because they have adequate protein in the diet.

The greatest financial losses are in the beef cattle industry. An annual loss of $500,000 is estimated in the United States due to carcasses that are condemned or must be frozen to comply with meat inspection regulations. In developing countries, the loss is estimated to $25 per animal (Pawlowsky & Schultz, 1972).

MORPHOLOGY. Unlike other taeniids, neither the adult nor the metacestode of *Taenia saginata* has rostellar hooks, although they occur ephemerally during larval development. Adults are usually 4 to 12 m long and 12 to 14 mm wide. They have a typical taeniid reproductive system including a bilobed ovary; a vaginal sphincter is found in this species and *T. hydatigena* only (Verster, 1967). Gravid proglottids have 14 to 32 lateral uterine branches; the number of branches is sometimes used in identification (see Diagnosis).

The spherical eggs with their thick, striated capsules (Fig. 29.1) cannot be distinguished from other taeniid eggs with any degree of confidence.

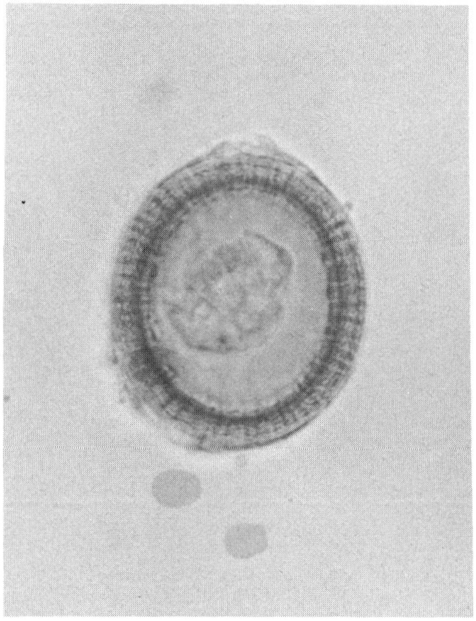

**FIG. 29.1**
Taeniid egg with characteristic striated shell.

The fully developed cysticercus is white, about 10 mm in diameter, and has a single unarmed scolex inverted in a fluid-filled bladder (Fig. 29.2). Before an association was made between the adult and metacestode of *T. saginata*, the name of the metacestrode, *Cysticercus bovis*, was established. Regrettably we still have two so-called scientific names for the same organism. Both names continue to be used in the scientific literature and in meat inspection, as do the names of some other adult and larval tapeworms (Table 29.1).

LIFE CYCLE. Humans become infected by eating cysticerci in raw or partially cooked beef. The scolex evaginates in the small intestine, attaches, and the proglottids form. About 3 to 4 months after ingesting cysticerci, gravid proglottids begin to appear in the feces. Proglottids are shed irregularly and may even crawl out of the anus; they are motile most often in the late afternoon and early evening (Pawłowsky & Schultz, 1972). Cattle become infected by eating eggs while grazing; the cysticerci in the muscles reach infectivity about 60 days after being eaten.

DIAGNOSIS. The symptoms of infection with *T. saginata* do not clearly relate to a tapeworm infection. There may be abdominal discomfort, diarrhea, and frequent hunger pangs, hardly characteristics of anything specific. The most frequent sign is the discharge of proglottids, which can hardly go unnoticed. The proglottids may be examined, and the characteristic bilobed ovary in the mature proglottid and the vaginal sphincter muscle in the gravid proglottid are good indications of the presence of *T. saginata*. The number of uterine branches has been used to distinguish between *T. saginata* (14–32) and *T. solium* (7–16), but it is not a reliable characteristic because of the difficulty in obtaining a good preparation and wide variation in the number of uterine branches (Verster, 1967). If a person is given an anthelmintic and worms are passed, the scolices can be examined, and if they are a taeniid type, but lack hooks, the identification is almost certain.

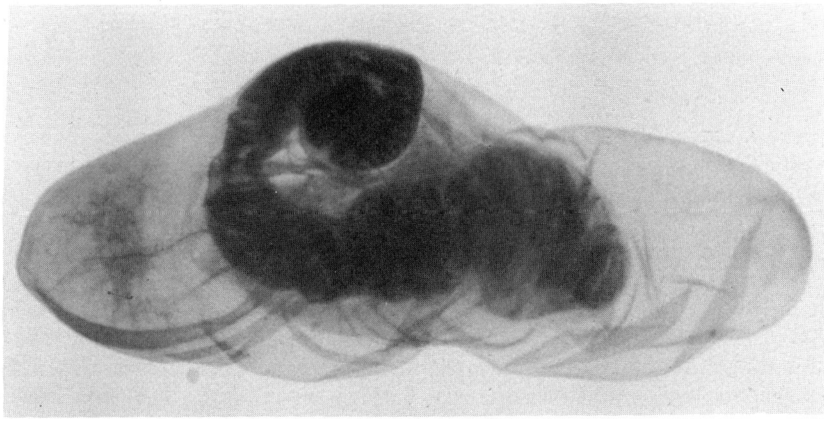

**FIG. 29.2**
*Taenia* cysticercus.

**TABLE 29.1**
Tapeworm Synonyms

| NAME OF ADULT | NAME OF METACESTODE[a] |
|---|---|
| *Taenia hydatigena* | *Cysticercus tenuicollis* |
| *Taenia pisiformis* | *Cysticercus pisiformis* |
| *Taenia ovis* | *Cysticercus ovis* |
| *Taenia taeniaeformis* | *Cysticercus fasciolaris* |
| *(=Hydatigera t.)* | Strobilocercus |
| *Taenia solium* | *Cysticercus cellulosae* |
| | *C. racemosus* |
| *Taenia krabbei* | *Cysticercus tarandi* |
| *Taenia crassiceps* | *Cysticercus longicollis* |
| *Taenia saginata* | *Cysticercus bovis* |
| *(=Taeniarhynchus saginatus)* | |
| *Taenia multiceps* | *Coenurus cerebralis* |
| *(=Multiceps m.)* | |
| *Taenia serialis* | *Coenurus serialis* |
| *(=Multiceps s.)* | *C. glomeratus* |
| *Echinococcus granulolsus* | Hydatid |
| | Unilocular cyst |
| *Echinococcus multilocularis* | Alveolar hydatid |
| | Multilocular hydatid |
| *Spirometra* | *Sparganum* |
| *Spirometra mansonoides* | *Sparganum proliferum* |

[a] "Scientific" names of metacestodes are in italics; the names of the metacestodes are in Roman type; names of adults are the only acceptable scientific names, but the names of the metacestodes are still used.

Finding the eggs is a reliable method of determining that a person has a taeniid infection, but an identification to species cannot be made from the eggs because they are so similar. The principal methods are as follows:

1. Perianal swabs made on three alternate days. The material is examined as a wet mount under a compound microscope.
2. Thick smears of feces may be examined also.
3. Eggs in feces may be concentrated by either flotation or sedimentation techniques.
4. Serological tests may be done to determine whether the person has antibodies to tapeworms in the blood. This is not a highly specific method.

HOST–PARASITE INTERACTIONS. Most *T. saginata* infections cause little damage to the host (see Diagnosis), but occasionally there is sufficient irritation to the gut to cause muscular spasms giving considerable abdominal pain. Occasionally the worms migrate into the appendix, causing severe reactions (Berry & Burrows, 1955) leading to appendicitis.

TREATMENT. The drug of choice is niclosamide; it is highly effective and is usually tolerated well.

EPIDEMIOLOGY AND CONTROL. In their review of taeniid infections, Pawlowski and Schultz (1972) commented on the paradox that "infection is prevalent in developed countries because they are rich and in the developing nations because they are poor!" They were referring to the consumption of raw beef dishes such as steak tartar or rare steaks among well-to-do persons and poor sanitation and lack of fuel that lead to a high prevalence of infection in both humans and cattle among poor peoples.

Many infections in cattle are acquired by indirect

means. Eggs can be spread by flies or passed through the digestive tracts of birds. They can also pass sewage treatment unharmed and appear in the effluent or in sludge drying pits. In the southwestern United States, farm workers may defecate in irrigation ditches and their feces will then spread on fields when the water is turned on; if one of them carries a tapeworm infection, it can be carried some distance.

An outbreak of bovine cysticercosis in a large feedlot in California illustrates the potential losses that an owner can incur. One infected worker infected almost 5% of cattle over several months, causing a large financial loss when the cattle went to slaughter.

Although cysticerci seem not to be pathogenic for cattle, they do live for long periods—at least a year, and perhaps as long as 3 years—so holding cattle until the cysticerci die is not feasible.

Recommendations for control of bovine cysticercosis include the following:

1. Treatment of any infected workers on a ranch or feedlot
2. Education of employees about the mode of transmission and personal hygiene
3. Provision of sanitary facilities that are readily accessible and will contain any infectious agents
4. Periodic examination of employees to ensure that they have not become reinfected
5. Preventing unauthorized persons from entering the premises (Slonka et al., 1978)

In a more general context, the following are important for control of infections in humans:

1. A meat inspection program to find infected carcasses
2. Condemnation of heavily infected carcasses
3. Freezing of lightly infected carcasses
4. Awareness on the part of the consumer of the possibility of infection so that beef is eaten only when cooked sufficiently to kill the cysticerci (60° C, 140° F)

Although meat inspection detects only three fourths of infected carcasses because of the small size of the cysticerci, the probability of infection is reduced by inspection (Dewhirst et al., 1967). Beef should be cooked so that the cysticerci in its center will be killed; adequate cooking also kills other parasites such as *Toxoplasma gondii* and *Sarcocystis* (Chap. 15). In light infections, freezing the carcass kills the cysticerci:

| TIME (HR) | TEMPERATURE (°C) |
|---|---|
| 360 | −5 |
| 216 | −10 |
| 144 | −15 or lower |

Although the risk of infection is small where these precautions are observed, the predilection of some persons for rare beef ensures that the beef tapeworm will be with us for the forseeable future.

NAME OF ORGANISM AND DISEASE CAUSED. *Taenia solium* L. causes taeniasis, cysticercosis, *Cysticercus cellulosae*, *C. racemosis*, and pork tapeworm.

HOSTS AND LOCATION. Humans are the definitive host; adults live in the small intestine. Pigs are the most common intermediate hosts, but dogs, monkeys, camels, deer, sheep, cats, and humans may harbor the cysticercus. Cysticerci tend to localize in striated muscle, but can be found in almost any organ.

DISTRIBUTION AND IMPORTANCE. *T. solium* is a cosmopolitan parasite with a prevalence of infection in the human population of 2 to 3% based on a number of studies; however, in certain localities its prevalence is much higher.

It is rarely found in humans in the United States. Even though the pork tapeworm occurs less often than the beef tapeworm in humans, it is a more significant medical problem. Humans can be a host of the cysticercus as well as the adult; it is unusual, but not unknown, for the same species of vertebrate to serve as both definitive and intermediate host for a tapeworm. *Hymenolepis nana* is another example. Like the beef tapeworm, the pork tapeworm causes little intestinal damage to humans, but when the eggs are ingested, cysticerci develop in striated muscles, liver, or the central nervous system (CNS), often with serious or even fatal consequences. In addition, one person with an adult worm who does not have fastidious personal habits can be a source of infection for other members of a household.

About 2 to 5% of pigs are infected with *T. solium* cysticerci worldwide. Figures for economic losses to "measly pork" are difficult to obtain for most parts of the world, but they must be in the millions of dollars annually.

MORPHOLOGY. Adults are 2 to 8 m long and have a typical taeniid rostellum with a double row of hooks (Fig. 29.3). The ovary in the mature proglottid has three lobes (Fig. 29.4) and the gravid uterus has 7 to 16 lateral branches. Eggs are indistinguishable from other taeniid eggs. The mature cysticerci are about 5 by 9 mm and white; they have an invaginated scolex with hooks and a fluid-filled bladder.

LIFE CYCLE. The life cycle of *T. solium* is the same in many respects as that of *T. saginata*; humans, the definitive host, are infected by ingesting raw or poorly cooked pork containing cysticerci. The scolex evaginates in the small intestine and attaches to the wall of the gut, where it strobilates. Gravid proglottids develop about 5 weeks after infection. Gravid proglottids are shed only during a bowel movement since the proglottids are only weakly motile (unlike *T. saginata*). The eggs are resistant to environmental hazards and remain alive for months. After ingestion by the intermediate host, eggs hatch in the small intestine, penetrate the intestinal mucosa, and enter small mesenteric veins. Having reached the circulatory system, the hexacanth embryos are carried throughout the body and lodge especially in the striated muscles, liver, and brain. The cysticerci reach infectivity in about 2 months.

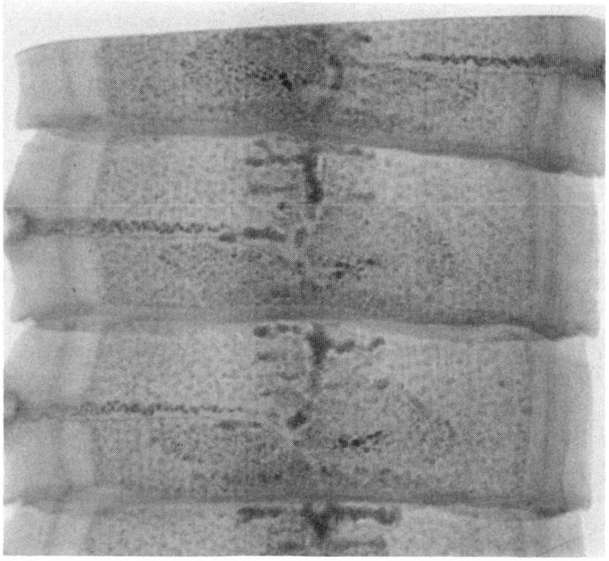

**FIG. 29.4**
Mature *Taenia solium* proglottids with three-lobed ovaries.

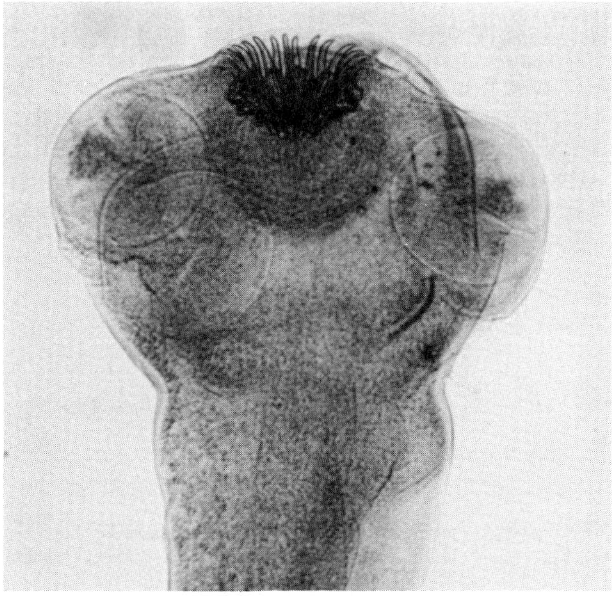

**FIG. 29.3**
*Taenia solium* scolex.

When a human is the intermediate host, the person becomes infected by ingesting eggs in contaminated food or water (heteroinfection) or by carrying them on contaminated fingers to the mouth (autoinfection).

DIAGNOSIS. See *T. saginata* when humans are suspected of being the definitive host. Cysticercosis is diagnosed by serological tests, x-rays, and biopsies. In swine going to slaughter, the cysticerci are looked for in the muscles.

HOST–PARASITE INTERACTIONS. Intestinal infections with *T. solium* have the same clinical manifestations as those of *T. saginata*. Usually only one adult worm is present, hence, the name *solium*; but two or more are occasionally found and some persons harbor other species of tapeworms as well.

Serious effects may result from cysticercosis in humans. Although they may be found in almost any organ, cysticerci are commonly located in striated muscles, liver, the eye, and the CNS, where thy most often are in the meninges and the ventricles of the brain (Marquez-Monter, 1971). About 30% of human infections are in the brain; these cysticerci are usually less than 20 mm in diameter with a volume of only a few milliliters, but in one case report, a cysticercus had a diameter of 50 mm and a volume of 60 ml (Berman et al., 1981). In an aberrant form of *T. solium*, apparently found only in the CNS, a cysticercus has daughter cysticerci budding from the bladder wall and forming an invasive, proliferating parasite. This type of metacestode is referred to as a *racemose cysticercus* and is sometimes given the quasi-scientific name of *Cysticercus racemosis*. No viable scolices have been found in racemose cysticerci (Beaver & Rolon, 1981).

Living cysticerci usually elicit little inflammatory response. They damage the host by forming space-occupying lesions, that is, their growth puts pressure on the adjacent tissue and interferes with its function or causes necrosis due to reduced circulation. In the ventricles of the brain the organisms often cause hydrocephalus because the cerebrospinal fluid cannot be properly drained. When cysticerci die, they often evoke severe inflammatory reactions (Marquez-Monter, 1971) and persons so affected are the cases most often seen by physicians.

TREATMENT. Niclosamide is the drug of choice for intestinal infections; it is highly effective and well tolerated. Cysticercosis is best treated in humans by surgical removal of the cysts, when possible. If surgery is not feasible, or if cysts rupture during surgery, mebendazole, a benzimidazole, can be administered, but the drug is still experimental for this use. Flubendazole, also a benzimidazole, was successful in treating cysticercosis in swine (Tellez-Giron et al., 1981), but human trials have not been done. If chemotherapy is considered, the beneficial effects should be weighed against the foreign body reaction against dead cysticerci.

EPIDEMIOLOGY AND CONTROL. Intestinal infections in humans are acquired by eating inadequately cooked pork. Pigs are infected by eating the gravid proglottids passed with human feces. Since a single proglottid may have hundreds of eggs, a pig may obtain a heavy infection throughout its body by eating only a few proglottids.

Following are the principal means of control:

1. Adequate cooking of pork
2. Freezing pork after slaughter and before it is marketed
3. Sanitary disposal of human waste
4. Meat inspection
5. Prompt treatment of persons found to be infected
6. An educational program aimed at changing habits of personal hygiene so that persons do not reinfect themselves or others they have contact with

These principles are much the same as would be applied in the control of *T. saginata*, but the extra hazard of humans being an intermediate host makes prompt treatment and a program designed to change human behavior essential to prevent the devastating effects of blindness and destruction of the CNS. In many parts of the developing world, pigs are kept in rather informal conditions, that is, allowed to run wild and to eat whatever they can find. They readily eat human feces, and if allowed to do so, a high prevalence of infection can develop. In addition, fuel may be expensive or some peoples prefer their pork less than well done. Serious problems hamper breaking into this cycle and blocking it. In tropical countries, animals are slaughtered and consumed usually on the same day. Freezing of meat is seldom an option. Although there are meat inspection systems for meat that is sold in markets, much is killed for local consumption. Since the definitive host range is limited to humans, emphasis on altering human behavior could have dramatically effective results. Although hosts other than the pig can be infected, they are seldom important in maintaining the cycle. Therefore, sanitary disposal of human waste and efforts to ensure cooking of pork as well as personal cleanliness should result in a reduction in the incidence of infection. If facilities and finances allow it, persons coming to dispensaries and hospitals can be examined for tapeworm infection and treated forthwith if found to be infected.

## OTHER *TAENIA* SPECIES

*Taenia pisiformis* Bloch, 1780, is often used in parasitology classes as a typical taeniid tapeworm. This cosmopolitan worm occurs in dogs and other canids, but rarely in cats. Cysticerci are in the viscera of rabbits and hares, in which they most often can be found attached to the intestinal mesenteries; perhaps 30% of cottontail rabbits are infected. Dogs that are allowed to run loose in areas where there are rabbits frequently become heavily infected. Some intestinal distress is experienced by dogs that are infected, and treatment is justified; a number of highly effective taeniacides are available. Keeping dogs from running loose is the best method of preventing infection. *T. pisiformis* is not a parasite of humans.

*Taenia multiceps* Leske, 1780 (syn. *Multiceps multiceps*), and *T. serialis* Gervais, 1845 (syn. *M. serialis*), have a coenurus metacestode that produces many protoscolices by internal budding (Fig. 29.5). The names given to the larvae are *Coenurus cerebralis* and *C. serialis*, respectively. The metacestode resembles a cysticercus in its nonlaminated outer wall, but it has multiple protoscolices like a hydatid. *T. multiceps* is a cosmopolitan parasite of canids: dogs, foxes, coyote, and jackal. The metacestode occurs in herbivores such as sheep, in which it is located most often in the CNS; because of pressure put on the brain or spinal cord, it produces blindness, incoordination, or circling, which giving rise to the common names of "gid" or "staggers." In the United States giddy sheep have been a cause of significant loss to sheep raisers, but infections have not been reported for some years. A few infections have been reported in humans, especially from sheep-raising areas; presumably human infections have been acquired through close association with infected dogs. Adults of *T. serialis* are in dogs and coyotes, and the metacestodes are in rabbits and sometimes in rodents.

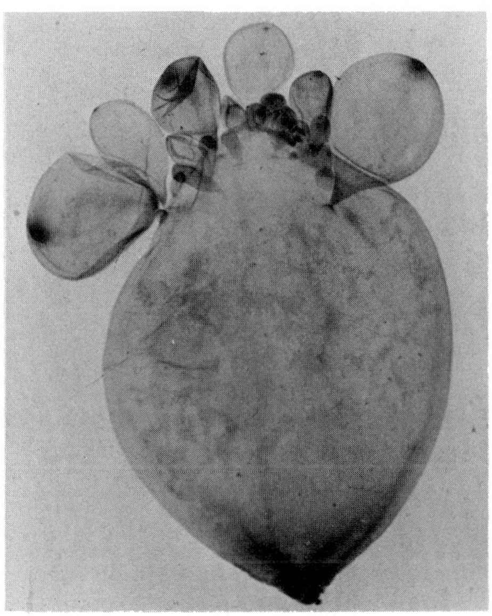

**FIG. 29.5**
*Taenia multiceps* coenurus with several everted protoscolices.

*Taenia taeniaeformis* Batsch, 1786 (syn. *Hydatigera taeniaeformis*), has a metacestode called a *strobilocercus* (Fig. 29.6). The strobilocercus is elongate, with a segmented body and a small, terminal bladder; the scolex is often everted while the metacestode is still in its intermediate host (Wardle et al., 1974). The definitive hosts are cats and the intermediate hosts are rodents; the most common cycle is between domestic cats and mice or rats on which they prey. Interestingly, infection of rats extending over many months results in a small percentage of them developing cancer of the liver. There are a few other instances in which cancers seem to be caused by helminths, such as *Spirocerca lupi*, and they have been used effectively in laboratory studies on carcinogenesis.

## HYDATID DISEASE

Four species of *Echinococcus* ("prickly berry" for the hydatid) cause hydatid disease: *Echinococcus granulosus*, *E. multilocularis*, *E. oligarthus*, and *E vogeli*; the status of a fifth species, *E. cruzi*, is uncertain (D'Alessandro et al., 1979). Both *E. granulosus* and *E. multilocularis* were originally parasites of canids and their prey (Rausch, 1967). *E. granulosus* originally cycled between the wolf and large ungulates such as deer, European elk, and caribou (=reindeer). The northern strain maintains this association, but the

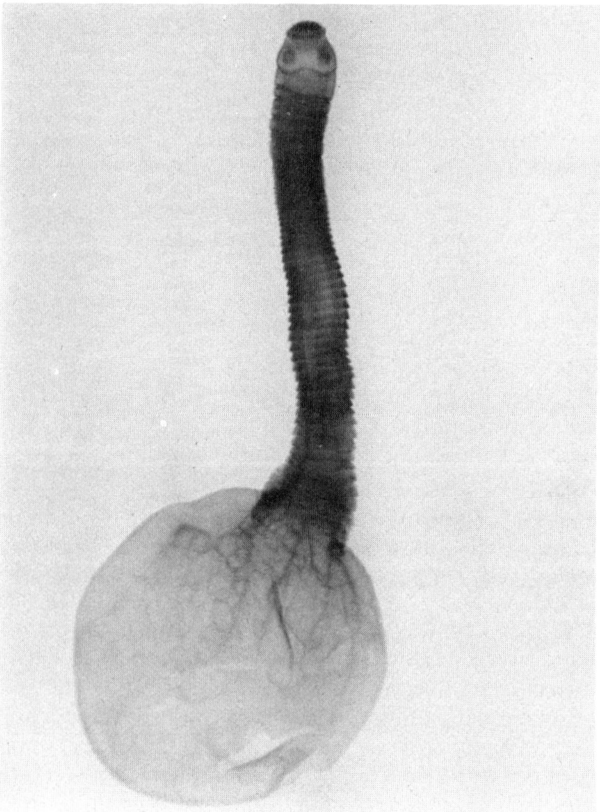

**FIG. 29.6**
Photograph of a *Taenia taeniaeformis* strobilocercus.

European or southern strain has become adapted to domestic animals and cycles between dogs and domestic ruminants.

*E. multilocularis* originally cycled between foxes and microtine rodents such as *Microtus* spp., *Clethrionomys* spp., and the lemming, *Lemmus sibiricus*. Both species have been distributed beyond their original northern location, but *E. multilocularis* to only a limited extent (Fig. 29.7). Some importations such as those into Australia and parts of North America have been well documented, but much of the movement of these parasites is lost in prehistory or is unknown.

NAME OF ORGANISM AND DISEASE CAUSED. *Echinococcus granulosus* (Batsch, 1786) cause echinococcosis, echinococciasis, hydatid disease, chydatidosis, hydatosis, and unicystic or unilocular hydatid disease.

HOSTS AND LOCATION. Adult worms are in the small intestines of dogs and wolves but not of foxes. Hydatids occur in the liver or lungs of deer and domestic ungulates; they occur in humans mostly in lung and liver but occasionally in other organs as well.

DISTRIBUTION AND IMPORTANCE. Two major strains of *E. granulosus* are known: the northern strain found in wolves, deer, and other wild herbivores, the cycles of which is usually referred to as the *sylvatic cycle,* and the European or cosmopolitan strain, which occurs in dogs and domestic ungulates (D'Alessandro et al., 1979). Over 100 human cases of unicystic hydatid disease were reported from Alaska between 1950 and 1966, and about 75% of the Eskimo population is at risk from the northern strain (Wilson et al., 1968). In the continental United States, 123 cases of unicystic hydatid disease were reported between 1900 and 1974. The current major problem areas in the United States are on sheep ranches in California, Utah, New Mexico, and Arizona where humans are sometimes an incidental participant in the dog–sheep cycle (Pappaivonou et al., 1977). About 70% of sheepdogs surveyed in northern California were infected with adults worms (Sawyer et al., 1969), a prevalence comparable to that of the sheep-raising areas of New Zealand and Australia (Warren, 1974), and probably similar to that of other sheep-raising areas of the world.

MORPHOLOGY. Adults are quite small, measuring 3 to 6 mm in length. The scolex has an armed rostellum, a short neck, and commonly one immature, one mature, and one gravid proglottid (Fig. 29.8). Eggs from gravid proglottids are identical to others in the family Taeniidae.

The hydatid, the metacestode stage, has a laminated outer wall that covers a thin, germinal layer (Fig. 29.9). Brood capsules bud from the germ layer into the cyst cavity—so-called endogenous budding. Protoscolices are formed in the brood capsules and remain attached to the surface from which they budded. Occasionally, daughter brood capsules form within the primary ones; these are miniature hydatids including the laminated outer wall. The term *hydatid sand* refers to hydatid contents released from a cyst; there are free-floating protoscolices and daughter brood capsules in the fluid which give it a granular appearance. Since the protoscolices remain attached to the germinal membrane under normal circumstances, it is only in a degenerating or

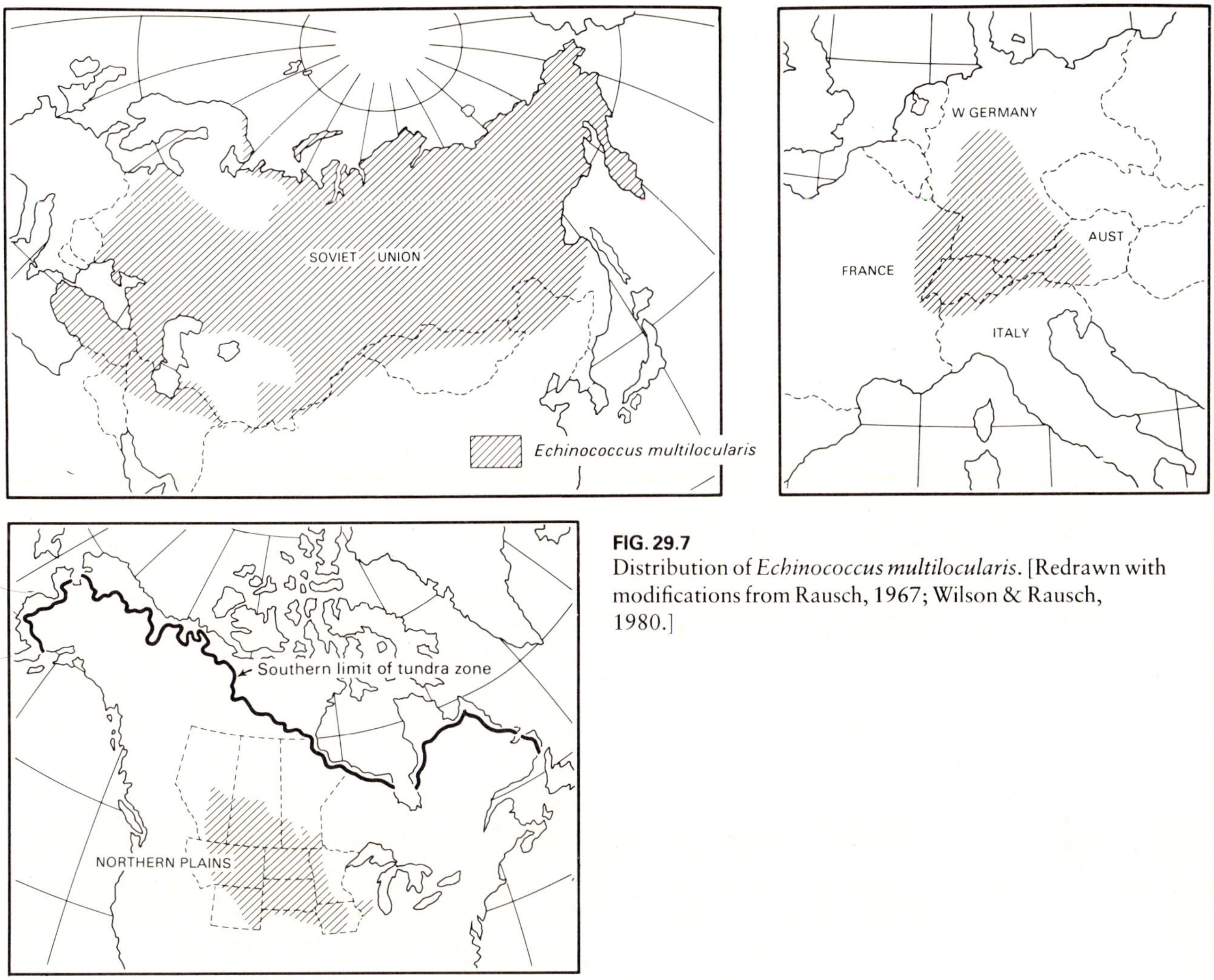


**FIG. 29.7**
Distribution of *Echinococcus multilocularis*. [Redrawn with modifications from Rausch, 1967; Wilson & Rausch, 1980.]

improperly preserved hydatid that the protoscolices are free. After surgical removal, the large and small hooks from the scolices may be measured and the metacestode identified as to species (Table 29.2). *E. granulosus* is the only species with a unicystic hydatid.

LIFE CYCLE. Gravid proglottids and eggs are shed in the feces. Like other taeniid eggs, they are resistant to environmental hazards and to various disinfectants. Intermediate hosts, most often ungulates but occasionally humans, ingest eggs as contaminants of food or water or through close association with canids. Hexacanth embryos hatch in the small intestine and penetrate the intestinal mucosa to reach the circulatory system. The hexacanths are carried by the bloodstream until they lodge in small vessels, commonly in the liver or lung, where they develop into the hydatid. Cysts grow to 4 to 5 mm in about 3 months.

Cysts containing fully developed protoscolices must be ingested by a carnivore of the family Canidae for the adult stage to develop; bile salts of other kinds of animals are lethal to the protoscolices. The protosco-

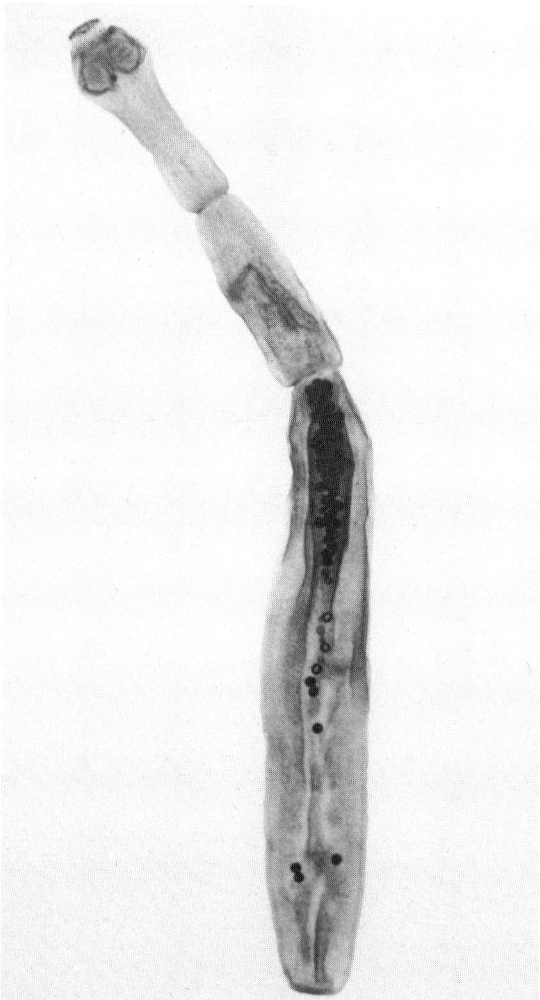

**FIG. 29.8**
Adult *Echinococcus granulosus*.

lices attach to the wall of the intestine and develop to the sexually mature adults in about 8 weeks.

DIAGNOSIS. Determining whether a person is infected with a hydatid is based on the following:

1. An association with dogs or wild canids
2. X-rays of the infected part of the body
3. Serology
4. Finding the hydatid during surgery and identifying the organism

Signs and symptoms of the infection are usually related to pressure on tissues adjacent to the growing hydatid. The effects on normal function of the organ or tissue may be similar to those of a tumor; hydatids often are mistaken for tumors. X-rays show a growing mass and add information for the diagnostician. Specific diagnosis is made through the Casoni skin test and indirect hemagglutination; the latter cross-reacts with *Taenia solium* antibodies, thereby preventing a definitive diagnosis (Schantz et al., 1980). Because the growth of the hydatids is so slow, 10 or more years may elapse between infection and the development of clinical disease; how the infection was obtained is often obscured by a murky history.

CULTIVATION. Early experiments in the cultivation of *E. granulosus* produced only metacestodes, not the adult worms. As work progressed, it was possible to produce adult worms from protoscolices. The worms resembled adults from natural infections, but fertile eggs were not produced (Smyth & Davis, 1972).

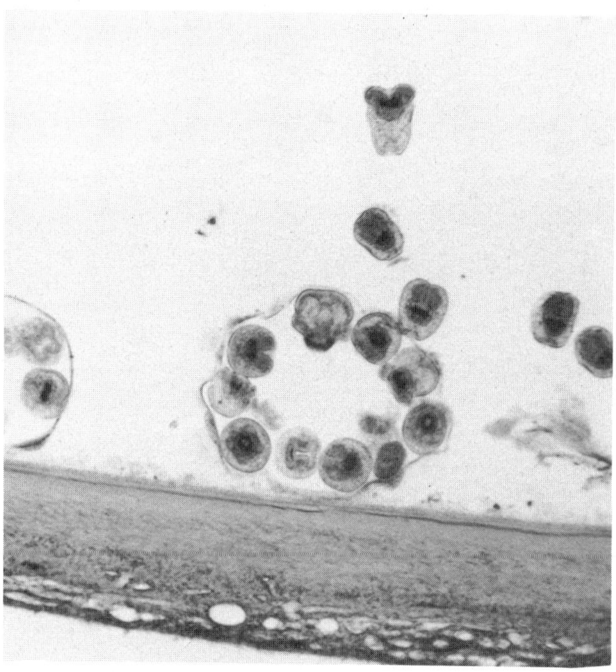

**FIG. 29.9**
A portion of a unilocular hydatid cyst of *Echinococcus granulosus* with endogenous budding.

**TABLE 29.2**
*Echinococcus* Characteristics

| | | NATURAL HOSTS | | | MEAN PROTOSCOLEX HOOK SIZE (MM) | |
|---|---|---|---|---|---|---|
| SPECIES | DISTRIBUTION | DEFINITIVE | INTERMEDIATE | HYDATID TYPE | LARGE | SMALL |
| *granulosus* | N. Hemisphere cosmopolitan | Canidae | ungulates | Unicystic | 22 | 18 |
| *multilocularis* | N. Hemisphere Holarctic | Canidae | rodents | Polycystic | 27 | 23 |
| *oligarthus* | Central & South America | Felidae | agouti | Polycystic | 33 | 26 |
| *vogeli* | Central & South America | Canidae | paca | Polycystic | 42 | 33 |

HOST–PARASITE INTERACTIONS. Hexacanth embryos pass through the intestinal mucosa, reach the hepatic portal circulation, and are carried throughout the body. In the northern strain, the cysts are found almost exclusively in the lungs of deer. In the southern strain, about 60% are trapped in the liver, 20% in the lungs and the remainder in almost any other organ of the body including kidney, somatic muscle, spleen, the cavity of long bones, and the central nervous system (Poole & Marcial-Rojas, 1971).

The damage to the host depends on the location and size of the hydatid. Many years may pass after initial infection before clinical signs and symptoms appear, because the cysts grow slowly. The cyst compresses adjacent tissues and reduces the blood supply, causes pain, or reduces the function of the affected part. If the cyst is untreated and occupies a crucial organ such as the brain, the space-occupying lesion may cause the death of the host.

The host responds by a foreign-body reaction and production of antibodies. The cyst becomes surrounded by connective tissue, and after considerable time it may become calcified. In addition, a host often becomes hypersensitive to the fluid that leaks from the hydatid in small amounts; this fluid contains protein and is allergenic. If a cyst is ruptured either through some normal process or during surgery, the fluid may cause anaphylaxis and immediate death.

*E. granulosus* is one of the few helminths for which we have an explanation of the mechanism of its host specificity. It is well known that various herbivores, swine, and humans can be intermediate but not definitive hosts. Likewise, it is known that certain carnivores (dog, fox, coyote, wolf) can only be definitive hosts. When the eggs are eaten, the outer capsules are removed by slightly alkaline conditions and the hexacanth embryos are then activated by bile or intestinal contents, or both. Activation of the embryos takes place in either human bile or intestinal juice of ruminants; they are not activated in bile or intestinal juice of dogs and cats. When the protoscolices are eaten by a potential definitive host, bile again plays a key role in their survival and establishment of infection. Bile from hare, rabbit, ox, sheep, and human cause lysis and death of protoscolices, whereas that of dog, cat, and fox has no effect on them. The key lies in the specific bile salt that predominates in the bile of a species (Smyth &Haslewood, 1963; Smyth, 1964). Thus, the conditions that the parasite meets upon entering the host are crucial in determining whether the life cycle can continue.

TREATMENT. Surgery is the recommended treatment for hydatid disease, but removal of a hydatid cyst must be approached with care. As indicated earlier, an individual harboring a hydatid often develops hypersensitivity to the fluid in the cyst. If the cyst is broken during surgery and the fluid leaks into the body cavity, the patient may die within minutes from anaphylactic shock, a manifestation of hypersensitivity. The other result of breaking the cyst is that the protoscolices may set up secondary sites of infection in other parts of the body (assuming that the patient survives). In order to prevent this secondary infection, it is sometimes recommended that the cyst be injected with alcohol or iodine to kill the protoscolices before it is removed.

Mebendazole has been used successfully in some individuals who have hydatids in locations that are inoperable (Keystone & Murdoch, 1978). Adult worms in dogs are susceptible to treatment with praziquantel (Wilson & Rausch, 1980).

EPIDEMIOLOGY AND CONTROL. The original cycle of *E. granulosus* was

Wolf ↔ Ungulate

However, with colonization of many kinds of habitat by humans and the domestication of various large ungulates, this tapeworm was carried along and a number of variations on the original natural cycle were established (Rausch, 1967):

Dog ↔ Wild Ungulate

Dog ↔ Domestic Ungulate

Wild Carnivore ↔ Domestic Ungulate

In addition to the dog, some of the definitive hosts that have become integrated into the cycle are foxes in the genera *Alopex* and *Vulpes*, the hunting dog of Africa, *Lycaon pictus*, jackals in Africa and Russia, and the coyote, *Canis latrans*. Unlike most other tapeworms, which do not have metacestode reproduction, a single hydatid may produce literally thouands of protoscolices; this increases the probability that the worm will have sufficient variability to develop in a species of host new to it. Adaptability and rapid evolution seem to be characteristics of *Echinococcus* spp.

Of the preceding cycles, the dog–domestic ungulate contains the greatest hazard to human health. Dogs may eat the entrails of animals that die or are butchered on the ranch or station. The continual shedding of eggs and the close association of humans and dog create a situation in which a person may occasionally ingest a few living hexacanths and eventually develop a damaging infection.

Hydatid disease is a problem in all of the livestock-raising areas of the world, especially where sheep are raised in large numbers. Often sheepdogs are the definitive hosts and sheep the intermediate hosts. This domestic cycle is the one most often seen and, as indicated, the one greatest hazard to humans. Some of the major areas of infection in the United States are California, Arizona, and Utah; other areas are Australia, New Zealand, and the Middle East. It should be noted that the distribution of *E. granulosus* is essentially worldwide and with the movement of livestock it can appear nearly anywhere.

The following are the elements of a control program for hydatid disease:

1. Keeping the viscera of domestic animals from being eaten by dogs
2. Treatment of infected dogs with an effective anthelmintic
3. Education and publicity of the hazards associated with echinococcosis

In certain areas of the world such as Iceland, Australia, and New Zealand, great strides have been made in reducing the prevalence of infection in both dogs and sheep, but only Iceland has achieved eradication. Effective control requires a relatively stable, accessible population of livestock raisers and an infrastructure of parasitologists and public health personnel. If it is not possible to determine who owns what animals or if people are nomadic, various elements of the cycle will slip through the net that one wishes to establish.

In the small focus of infection in Utah, it was possible to examine and treat all dogs by sending teams of people out periodically to hold clinics. An essential part of the program was to reach both livestock and dog owners so that they would recognize the risk to human health and take action to ensure that once the dogs were treated they did not eat the entrails of any sheep that died or were butchered for home consumption. Meetings and printed materials were available to all adults. In order to make certain that the information reached all age groups, a tapeworm coloring book was produced and distributed to children in the grade schools. What parent could ignore a 7- year-old with his or her own tapeworm in hand as a trophy of the day's artistic endeavor?

NAME OF ORGANISM AND DISEASE CAUSED. *Echinococcus multilocularis*, in the adult form, causes the same disease caused by *E. granulosus*; the metacestode causes multilocular or alveolar hydatid disease.

HOSTS AND LOCATION. Natural definitive hosts are foxes, and the intermediate hosts are microtine rodents.

Domesticated animals, especially dogs and cats, are definitive hosts and are sources of eggs that can infect humans; mice harbor metacestodes on farms and in some urban areas (Leiby & Kritsky, 1972).

DISTRIBUTION AND IMPORTANCE. *E. multilocularis* is enzootic in Alaska, Canada, central Europe, Siberia, and northern Japan, and occurs widely in central North America (Fig. 29.7). Since the hydatid grows by external budding, it has often been compared to an invasive neoplasm (cancer); similar to a cancer, if a portion becomes detached, it can be carried by the blood and establish (metastasize) in another location of the body. Untreated alveolar hydatid disease is fatal in about 75% of infections in humans; it is one of the most lethal human helminthic diseases. The primary hepatic lesions resemble cancer clinically and errors in diagnosis are common, probably resulting in a lower incidence of infection than is actually the case. Alveolar hydatid disease is a major problem in Siberia, where, for example, 22 cases per year have been reported in Kamchatka alone (Wilson & Rausch, 1980).

MORPHOLOGY. Adults of *E. multilocularis* and *E. granulosus* are similar, but the former has fewer and smaller testes and the genital pore is slightly different (Wardle et al., 1974). The metacestode forms in the intermediate hosts are different as well; *E. granulosus* has a unilocular cyst and *E. multilocularis* has an alveolar or multilocular cyst (Fig. 29.10). Alveolar cysts in humans are not infectious since they do not contain protoscolices.

LIFE CYCLE. The life cycle pattern of *E. multilocularis* is basically the same as that of *E. granulosus* in that it is a carnivore–herbivore cycle, but the hosts and therefore the epidemiology are different. Microtine rodents such as voles (*Microtus* spp.) and lemmings (*Lemmus* spp.) are the intermediate hosts, an canids such as foxes and the coyote are the definitive hosts in feral cycles. In domiciliated cycles, dogs and cats become infected.

DIAGNOSIS. Since the alveolar cyst grows irregularly in the tissue, usually the liver, the signs and symptoms of infection may relate to impaired liver function or to pressure and interference of the function of the liver or other organ in which it is located. Antibodies are formed and may be shown by the Casoni intradermal test or by serological methods.

HOST–PARASITE INTERACTIONS. The growth form of the alveolar cyst is instrumental to the way in which it damages the host and to the possibility of treatment of the intermediate host. The alveolar cyst tends to grow in an irregular form through the tissue, with buds extending from the surface. The cyst is not a discrete body, as in *E. granulosus*; surgery, therefore, is often difficult, because the parasite must be dissected out of the organ. Only about one quarter of liver infections can be treated surgically. Limited trials have shown that mebendazole (a benzimidazole) is active against alveolar cysts; five persons were treated and were still alive 5 years later (Wilson & Rausch, 1980).

Praziquantel is the drug of choice in the definitive hosts.

EPIDEMIOLOGY AND CONTROL. The feral cycle involves wild canids such as foxes and coyote, and many of the infections with *E. multilocularis* are in fur trappers. In pelting animals, trappers may inhale the eggs thrown into the air from the pelt or get a little intestinal content on their hands. Since the eggs are quite resistant, they might remain for many weeks on a knife or equipment used to stretch hides. Some potential methods for preventing infection in humans are

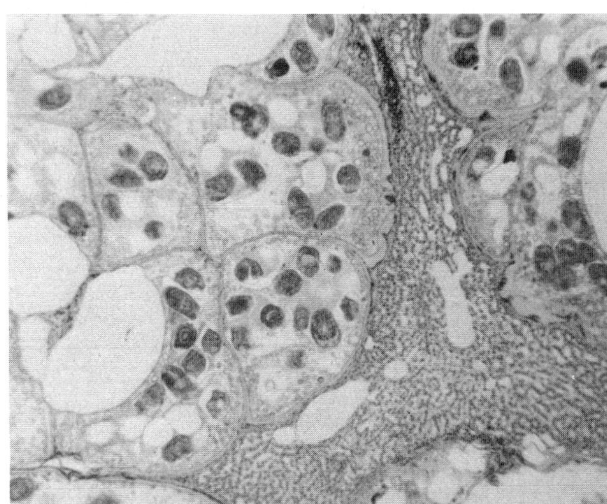

**FIG. 29.10**
Section of a multilocular hydatid cyst of *Echinococcus multilocularis* with exogenous budding.

evident: wearing a mask to keep from inhaling eggs, washing hands, and disinfecting tools. These methods may not be practical considering that trapping of foxes in the arctic is usually done in rural, relatively primitive areas.

If a cycle occurs near human habitation, companion animals should be monitored and treated. Under some conditions, a rodent control program can be instituted near towns or ranches, but with rapidly reproducing rodents such as *Microtus*, success in reducing numbers may be elusive.

## OTHER *ECHINOCOCCUS* SPECIES

The other species currently accepted are *E. multilocularis* Leuckart, 1863, *E. oligarthus* (Diesing, 1863), and *E. vogeli* Rausch & Bernstein, 1972; all have metacestodes with multiple chambers produced by external or exogenous budding. The terms *alveolar, multilocular hydatid*, and *polycystic cyst* have been used for these metacestodes.

*E. vogeli* produces polycystic hydatid disease in humans in Central and South America. Metacestodes differ in several respects from *E. multilocularis*: most notable is the presence of numerous protoscolices in the cysts. The normal definitive host is the bush dog, *Spethos venaticus*, and the natural intermediate host is the paco, *Cuniculus paca* (D'Alessandro et al., 1979).

*Echinococcus oligarthus* also produces polycystic hydatid disease in Central and South America, but no confirmed human infections have been reported (D'Alessandro et al., 1979). The adults are intestinal parasites of cats such as the jaguar (*Panthera onca*), not dogs, and the metacestodes are in rodents such as the agouti (*Dasyprocta punctata*). The habits of cats make it unlikely that humans would commonly become infected by fecal contamination with *E. oligarthus*.

## FAMILY HYMENOLEPIDIDAE

Hymenolepididae comprise small tapeworms of birds and mammals throughout the world. They have long, slender strobilae, armed rostella (usually), and one to four testes per proglottid.

NAME OF ORGANISM AND DISEASE CAUSED. *Hymenolepis nana* Siebold, 1853, the dwarf tapeworm, causes hymenolepiasis.

HOSTS AND LOCATION. This tapeworm is found in the small intestines of rats, mice, and humans.

DISTRIBUTION AND IMPORTANCE. *H. nana* is cosmopolitan, and is the most common tapeworm of humans throughout the world. Relatively high infection rates in children are found in central Europe and Latin America, but range from 0.3 to 3% in the southern United States (Warren, 1974).

MORPHOLOGY. This is the smallest human tapeworm, usually 7 to 100 mm long as an adult. Adult worms have a prominent armed rostellum, a scolex with four suckers, proglottids that are wider than long, and three testes per proglottid. Eggs are 30 to 47 μm and have a thin outer shell. The oncosphere is surrounded by a membrane with two characteristic polar knobs and polar filaments (Fig. 29.11). The metacestode is a cysticercoid.

LIFE CYCLE. Eggs of *Hymenolepis* have the hexacanth embryo formed when they pass from the host; therefore, no period of development outside the host is required. *H. nana* is unique among cyclophyllidean tapeworms in that an intermediate host is optional. Intermediate hosts are arthropods such as grain beetles or larval fleas; the cysticercoid develops in the hemocoels of these hosts when they ingest the oncospheres. When an infected arthropod is eaten by a person, the scolex of the cysticercoid attaches to the small intestine and develops into the adult.

The alternative method of infection of the definitive host is for a person to ingest the egg containing the oncosphere. The embryo escapes from the egg capsules in the small intestine and enters the lacteal of a villus in the small intestine. After developing for 2 weeks or more, the worm returns to the lumen of the intestine and forms a strobila. Eggs from gravid proglottids bring about autoinfection by hatching while still in the host and forming cysticercoids in the intestinal tissue; infections can thus become extraordinarily heavy.

DIAGNOSIS. Both sedimentation and flotation methods concentrate eggs so that they can be found in a fecal sample. The characteristic polar knobs and filaments (Fig. 29.11) are diagnostic.

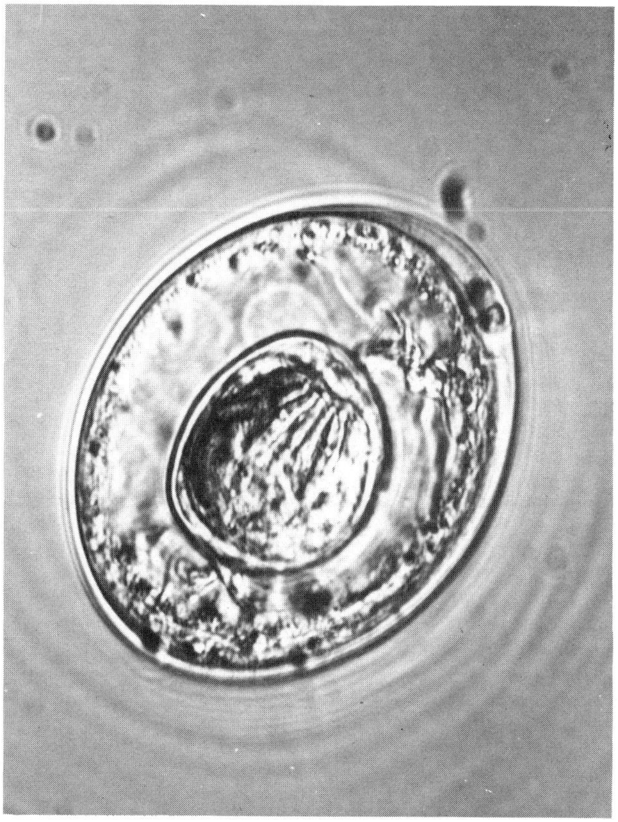

**FIG. 29.11**
Interference micrograph of a *Hymenolepis nana* egg. Characteristic polar knobs and filaments are present. [Courtesy of Dr. Robert Mead.]

CULTIVATION. *Hymenolepis* was the first tapeworm to be cultivated in vitro, and both *H. nana* and *H. diminuta* have produced gravid proglottids from cysticercoids in culture. Also, oncospheres have produced cysticercoids in culture (Berntzen, 1970).

HOST–PARASITE INTERACTIONS. Light infections are asymptomatic, but heavy infections cause abdominal pain, diarrhea, headache and dizziness, among other vague symptoms.

The method of infection and the development of immunity are interrelated. If a cystercoid is ingested, there is little development of immunity, and through autoinfection the number of worms may become quite large. On the other hand, if eggs are ingested, immunity usually develops quickly.

TREATMENT. Although both niclosamide and praziquantel are effective, the former is the drug of choice (Jones, 1979).

EPIDEMIOLOGY AND CONTROL. Infection with the dwarf tapeworm is not usually considered to be a public health problem but rather is a familial one and is approached in that way. If a person in a household is found to be infected, control is undertaken by doing the following:

1. Determining which persons are infected
2. Treating all infected individuals
3. Cleaning the dwelling—clothing, bedclothes, bathrooms
4. Killing insects in the house and making it less attractive for them
5. Teaching measures of personal hygiene, especially to children who might not yet have developed fastidious habits

Those steps are directed at breaking the microcosm of transmission in the household. It is important to rid all persons of their infection and to undertake preventive measures at the same time. Cleanliness and personal hygiene are the keys to ensuring that transmission is stopped. The dwarf tapeworm is often a problem in children who are institutionalized, and under this circumstance, they may not be capable of readily learning procedures of personal hygiene. In addition, publicly supported institutions often operate on minimal amounts of money and may not have the resources to maintain adequate levels of cleanliness despite the desires of their personnel.

*Hymenolepis diminuta* Rudolphi, 1819, is a cosmopolitan parasite of rats and mice and only rarely of humans. Despite its species name, *H. diminuta* is a relatively large tapeworm, sometimes reaching a meter in length, but usually it is about 30 cm long and has an unarmed rostellum (Fig. 29.12). Intermediate hosts are insects, usually beetles, and in the laboratory, it can be propagated readily in the confused flour beetle, *Tribolium confusum*. Eggs resemble those of *H. nana*, but lack polar knobs and filaments (Fig. 29.13). *H. diminuta* has been used extensively in physiological, fine structural, and biochemical studies; it has been studied so much that tapeworm physiology is, in fact, *H. diminuta* physiology. Arai (1980) has reviewed all aspects of studies on this tapeworm.

## FAMILY DILEPIDIDAE

*Dipylidium caninum* L. is a cosmopolitan parasite of dogs and cats. Intermediate hosts are *Ctenocephalides canis* and *C. felis*, the dog and cat flea, respectively, which harbor cysticercoids. Adults are referred to by the common name of the *double-pored tapeworm* because they have two sets of reproductive organs (Fig. 29.14). Proglottids are passed in the feces and the egg packets (Fig. 29.15) are eaten by larval fleas. The cysticercoid develops in the hemocoel of the intermediate host. The infection is retained in the larva when it pupates and emerges as an adult flea; the definitive host is infected when it eats an infected flea. Dogs and cats are mildly affected by *D. caninum*; they usually have a rough hair coat and may have diarrhea. The proglottids are quite motile and may crawl out of the anus of the dog or cat. Since the effects are generally mild, the first sign of infection in a companion animal is likely to occur when a proglottid is found inching its way across the sofa. Diagnosis of the infection entails examining a proglottid for the typical structure of the species or finding the typical eggs in packets (Fig. 29.15).

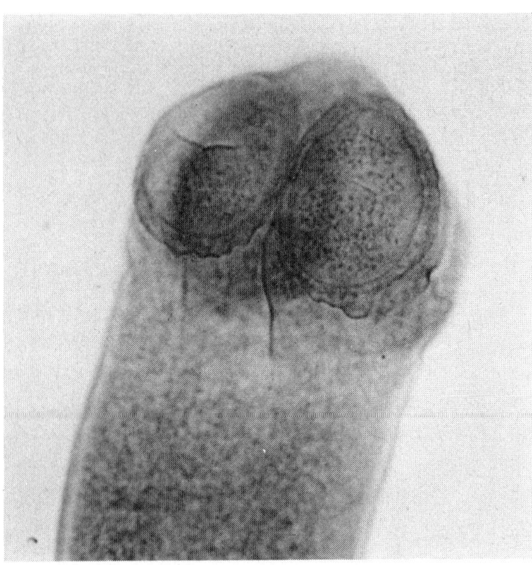

**FIG. 29.12**
Scolex of *Hymenolepis diminuta* with an unarmed rostellum.

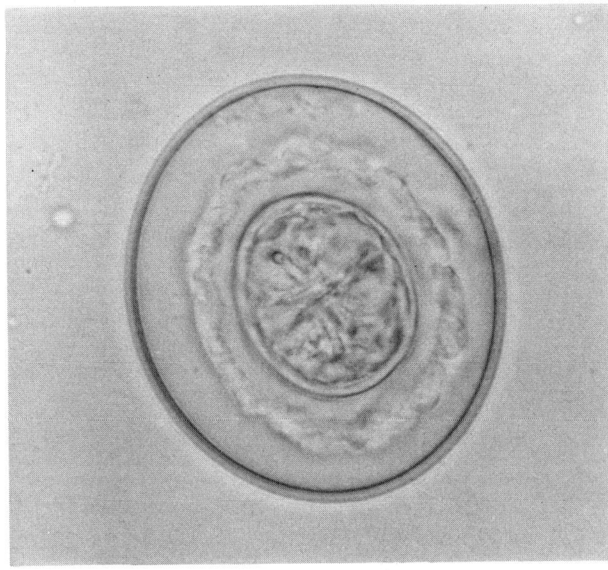

**FIG. 29.13**
*Hymenolepis diminuta* egg without polar knobs and filaments.

When *D. caninum* establishes in a household, three steps should be taken:

1. All companion animals are treated. Niclosamide is effective (Jones, 1979).
2. Flea control is instituted by treating companion animals and by eliminating larvae from the house (see Chap. 53).
3. Periodic examination of companion animals is done to ensure that any reinfections are found early and treated.

In warm climates in which fleas reproduce well out-of-doors and may be active year-round, flea control may not be effective. Even if the inside of the house is cleaned up, the fleas outside are a massive reservoir that may be a continuing source of infection, since they live where the animals defecate.

In addition to infecting companion animals, *D. caninum* occasionally infects children. There are about 100 case reports describing infections in children;

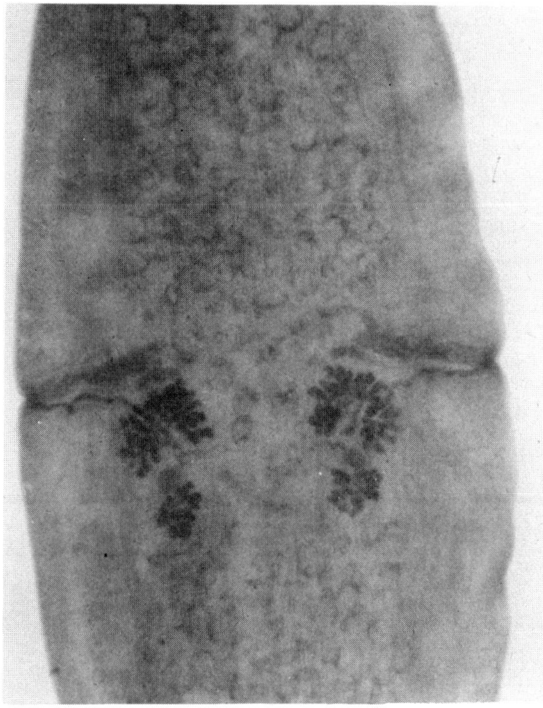

**FIG. 29.14**
Proglottid of *Dipylidium caninum* with characteristic double set of reproductive organs.

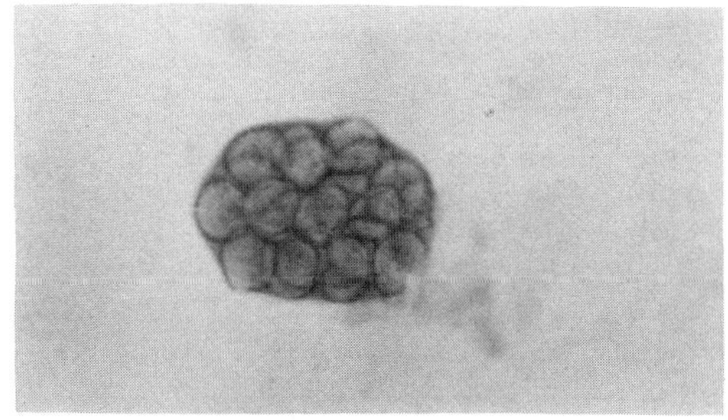

**FIG. 29.15**
*Dipylidium caninum* egg packet.

moderately severe intestinal disturbances occur, such as abdominal cremps and diarrhea. Children become infected if they are closely associated with infected dogs or cats. They ingest infected fleas or parts of fleas just as the normal hosts do.

## FAMILY ANOPLOCEPHALIDAE

The members of this family are mostly large organisms (more than 15 cm long) that have a scolex with large, fleshy suckers lacking hooks and one or two sets of reproductive organs with lateral genital pores. The testes are numerous and scattered in the proglottid. The eggs have a heavy hyaline outer capsule and the hexacanth embryo is encased in a pyriform apparatus, a distinctive structure with two fingerlike projections. Most of the life cycles use grass mites (Acarina, Oribatidae) as intermediate hosts in which the cysticercoid develops. The definitive hosts are herbivorous animals which become infected through eating infected grass mites with the forage (Table 29.3).

Some of the species such as *Cittotaenia* and *Moniezia* are cosmopolitan; others such as *Thysanosoma actinoides* are limited in their distribution, which is only the western United States.

*Moniezia* spp. may reach 4 m in length, but usually they are less than a meter. They may, however, occur in sheep or cattle in large numbers and are quite striking when seen at postmortem examination. When the small intestine is opened, literally handfulls of tapeworms spill out of the gut. It is tempting to lay the cause of death of the animal on something that is as striking as a huge mass of tapeworms, but that is not the case. It is axiomatic that tapeworm tissue is formed of the nutrients that would otherwise have gone to the host, but most studies point toward little weight reduction in

**TABLE 29.3**
Common Herbivorous Animals and Their Anoplocephalid Parasites

| Parasite | Hosts |
|---|---|
| *Cittotaenia* spp. | Rabbits & Hares |
| *Moniezia benedeni* | Cattle & Sheep |
| *Moniezia expansa* | Sheep & cattle |
| *Anoplocephala* spp. | Equids |
| *Thysanosoma actinoides* | Deer & sheep |
| *Monoecocestus* spp. | Porcupine |
| *Avitellina* spp. | Ruminants |
| *Stilesia* spp. | Ruminants |

infected animals. Some well-controlled studies indicate that young animals suffer from infections and gain less well than uninfected animals.

*Thysanosoma actinoides* is found in the high plains and Rocky Mountain area of the western United States as well as sporadically in other parts of the Western Hemisphere, and is the cause of some economic losses in lambs. *T. actinoides* is located in the upper small intestine, and the scolex is frequently attached in the bile ducts of the liver with the strobila extending down the common bile duct into the duodenum. The intermediate hosts of the fringed tapeworm are probably bark lice (Insecta, Psocoptera), in which the cysticercoid develops (Allen, 1959). Studies on weight gains have frequently shown that uninfected animals gain a little better than infected ones, but most losses come from condemnation of livers at slaughter (Allen, 1973).

## REFERENCES AND FURTHER READINGS


Allen, R. W. 1959. Preliminary note on the larval development of the fringed tapeworm of sheep, *Thysanosoma actinoides* Diesing, 1934, in psocids (Psocoptera:Corrondentia). *J. Parasitol.* **45**:537–538.

———. 1973. The biology of *Thysanosoma actinoides* (Cestoda: Anoplocephalidae), a parasite of domestic and wild ruminants. *N. Mex. Ag. Exp. Stat. Bull.* 604.

Arai, H. P. (ed.), 1980. *Biology of the Tapeworm* Hymenolepis diminuta. Academic Press, New York.

Beaver, P.C., and F. A. Rolon. 1981. Proliferating larval cestode in a man in Paraguay. A case report and review. *Am. J. Trop. Med. Hyg.* **30**:625–637.

Berman, J. D., P. C. Beaver, A. W. Cheever, and E. A. Quindlen. 1981. Cysticercus of 60-milliliter volume in a human brain. *Am. J. Trop. Med. Hyg.* **30**:616–619.

Berntzen, A. K. 1970. Continuous axenic culture of the successive stages of *Hymenolepis nana*. *J. Parasitol.* **56**:397.

Berry, L. R., and R. B. Burrows. 1955. Appendicitis with cestodes. *Arch. Path.* **59**:587–593.

D'Alessandro, A., R. L. Rausch, C. Cuello, and N. Arestizabal. 1979. *Echinococcus vogeli* in man, with a review of polycystic hydatid disease in Colombia and neighboring countries. *Am. J. Trop. Med. Hyg.* **28**:303–317.

Dewhirst, L. W., J. D. Cramer, and J. J. Sheldon. 1967. Analysis of current inspection procedures for detecting bovine cysticercosis. *J. Am. Vet. Med. Assn.* **150**:412–417.

Freeman, R. S. 1973. Ontogeny of cestodes and its bearing on their phylogeny and systematics. *Adv. Parasitol.* **11**:481–557.

Gemmel, M. A., and P. D. Johnstone. 1977. Experimental epidemiology of hydatidosis and cysticerosis. *Adv. Parasitol.* **15**:312–369.

Greenberg, A. E., and B. H. Dean. 1958. The beef tapeworm, measly beef and sewage—A review. *Sewage and Industrial Wastes* **30**:262–269.

Jones, A. W. 1979. Niclosamide as treatment for *Hymenolepis diminuta* and *Diplyidium caninum* infection in man. *Am. J. Trop. Med. Hyg.* **28**:300–302.

Keystone, J. S., and J. K. Murdoch. 1978. Mebendazole. *Ann. Intern. Med.* **91**:582–586.

Laws, G. F. 1968. Physical factors influencing survival of taeniid eggs. *Exper. Parasitol.* **22**:227–239.

Leiby, P.D., and D.C. Kritsky. 1972. *Echinococcus multilocularis*: A possible domestic life cycle in central North America and its public health implications. *J. Parasitol.* **58**:1213–1215.

Marquez-Monter, H. 1971. Cysticercosis. In Marcial-Rojas, (ed.), *Pathology of Protozoal and Helminthic Diseases.* Williams & Wilkins, Baltimore, pp. 592–617.

Meyer, M. C., and O. W. Olsen. 1980. *Essentials of Parasitology*, 3rd ed. William C. Brown, Dubuque, Iowa. pp. 266.

Moore, J. 1981. Asexual reproduction and environmental predictability in cestodes. (Cyclophyllidea: Taeniidae). *Evolution* **35**:723–741.

Pappaivanou, M., C. W. Schwabe, and D. M. Sard. 1977. An evolving pattern of human hydatid disease transmission in the United States. *Am. J. Trop. Med. Hyg.* **26**:732–742.

Pawlowski, Z., and M. G. Schultz. 1972. Taeniasis and cysticercosis (*Taenia saginata*). *Adv. Parasitol.* **10**:269–343.

Poole, J. B., and R. A. Marcial-Rojas. 1971. Echinococcosis. In R. A. Marcial-Rojas (ed.), *Pathology of Protozoal and Helminthic Diseases.* Williams & Wilkins, Baltimore, pp. 635–657.

Rausch, R. L. 1967. On the ecology and distribution of *Echinococcus* spp. (Cestoda: Taeniidae) and characteristics of their development in the intermediate host. *Ann. Parasitol.* **42**:19–63.

Rausch, R. L. R. Rausch, and D. D'Alessandro. 1978. Discrimination of the larval stages of *Echinococcus ologarthus* (Diesing, 1863) and *E. vogeli* Rausch and Bernstein, 1972 (Cestoda: Taeniidae). *Am. J. Trop. Med. Hyg.* **27**:1195–1202.

Rybicka, K. 1966. Embryogenesis in cestodes. *Adv. Parasitol.* **4**:107–186.

Sawyer, J. C., P. M. Schantz, C. W. Schwabe, and M. W. Newbold. 1969. Identification of transmission foci of hydatid disease in California. *Public Health Rep.* **84**:531–541.

Schantz, P. M., D. Shanks, and M. Wilson. 1980. Serologic cross-reactions with sera from patients with echinococcosis and cysticercosis. *Am. J. Trop. Med. Hyg.* **29**:609–612.

Slais, J. 1973. Functional morphology of cestode larvae. *Adv. Parasitol.* **11**:396–480.

Slonka, G. F., W. Matulich, E. Morphet, C. M. Miller, and E. V. Bayer. 1978. An outbreak of bovine cysticercosis in California. *Am. Trop. Med. Hyg.* **27**:101–105.

Schmidt, G. D. 1970. *How to Know the Tapeworms.* W. C. Brown, Dubuque, Iowa.

Smyth, J. D. 1964. The biology of the hydatid organisms. *Adv. Parasitol.* **2**:169–219.

——— J. D. 1969. The biology of the hydatid organisms. *Adv. Parasitol.* 7:327–347.

Smyth, J. D., and Z. Davis. 1972. *In vitro* culture of the strobilar stage of *Echinococcus granulosus* (sheep strain): A review of the basic problems and results. *Inter. J. Parasitol.* **1**:631–644.

Smyth, J. D., and G. A. D. Haslewood. 1963. The biochemistry of bile as a factor in determining host specificity in intestinal parasites, with articular reference to *Echinococcus granulosus. Ann. N.Y. Acad. Sci.* **113**:234–260.

Telléz-Girón, E., M. C. Ramos, and M. Montante. 1981. Effect of flubendazole on *Cysticercus cellulosae* in pigs. *Am. J. Trop. Med. Hyg.* **30**:135–138.

Voge, M. 1967. The postembroyonic developmental stages of cestodes. *Adv. Parasitol.* 5:247–297.

———. 1973. The post-embryonic developmental stages of cestodes. *Adv. Parasitol.* **11**:707–730.

Wardle, R. A., and J. A. McLeod. 1952. *The Zoology of Tapeworms.* University of Minnesota Press, Minneapolis.

Wardle, R. A., J. A. McLeod, and S. Radinovsky. 1974. *Advances in the Zoology of Tapeworms, 1950–1970.* University of Minnesota Press, Minneapolis.

Warren, K. S. 1974. Helminthic diseases endemic in the United States. *Am. J. Trop. Med. Hyg.* **23**:723–730.

Wilson, J. F., A. C. Diddams, and R. L. Rausch. 1968. Cystic hydatid disease in Alaska. A review of 101 autochthonous cases of *Echinococcus granulosus* infection. *Ann. Rev. Resp. Dis.* **98**:1–15.

Wilson, J. F., and R. L. Rausch. 1980. Alveolar hydatid disease. A review of clinical features of 33 indigenous cases of *Echinococcus multilocularis* infections in Alaskan Eskimos. *Am. J. Trop. Med. Hyg.* **29**:1340–1355.

Verster, A. 1967. Redescription of *Taenia solium* Linnaeus, 1758 and *Taenia saginata* Goeze, 1782. *Ztschr. Parasitenk.* **29**:313–328.

———. 1969. A taxonomic revision of the genus *Taenia* Linnaeus, 1758. *Onderstepoort J Vet. Res.* **36**:3–58.

# PART 3

# *Nematodes*

# 30

# *Introduction to the Nematodes*

Among multicellular animals, the phyla Arthropoda and Mollusca are considered to be the most successful since they have about 800,000 and 80,000 species, respectively. The phylum Nematoda, or roundworms, must be considered nearly as successful as the arthropods because there are some 500,000 species in the world, though only about 10,000 have been described. Hundreds of new species are described each year.

Success in an animal group is measured by the number of species it has, but the number of individuals must also be taken into account. Although nematodes are both free-living and parasitic, the greater number by far of species and individuals are free-living. Like the English sparrow, nematodes are ubiquitous, as can be readily seen by looking at almost any environment with organic matter in it. For example, if you pull a tuft of grass including some of the roots from the lawn and place it in a Baermann funnel, you will find from 12 to 50 nematodes in the bottom of the funnel after a few hours. Some data from Hyman (1951) are instructive: In soils of Utah and Idaho there are between 800 million and more than a billion nematodes per acre of soil and these include at least 35 species. Farm soils in north China contain 2 billion to 6 billion per acre. Filter beds for water purification have thousands of millions of nematodes in the top layer. In marine areas 4.4 million nematodes were found in surface muck off the Dutch coast; even in the lower reaches of a beach off Massachusetts there were 527 million per acre in the top 3 in. of sand.

Free-living nematodes have been found in almost every conceivable habitat, including aquatic and terrestrial, marine and fresh water, hot and cold water, in deep seas, and floating in the community of plants and animals in the Sargasso Sea. Their only universal requirement is that there be a moderate amount of water for them to move and feed in for a part of the life cycle. Their numbers are greatest where there is a large amount of organic matter and small where the amount of organic matter is low, such as a sandy beach.

Since we are concerned here with parasites, the question might be raised of why we make so much of free-living roundworms. The reason is that the basic structure and life cycle of all nematodes are remarkably homogeneous. The parasites among them inhabit not only other animals, but plants as well. The parasites of plants are of considerable economic importance in citrus fruit, pineapple, root crops such as sugar beets, and tomatoes, among others. The parasites of animals are found in nearly every animal species that has been examined and in one host or another in every organ system and structure that one might think of (excluding hair, feathers, and cornified structures).

The basic structural pattern and the pattern of development of a nematode are admirably suited for invasion of many niches. This is basically what we are emphasizing when we list the free-living and parasitic habits in which nematodes are found. Although the origins and affinities of the nematodes are not clear, it is generally agreed that the early forms were tiny aquatic animals attached to the substrate by their tails, where they trapped particulate food with bristles surrounding the mouth (Fig. 30.1).

Nematodes are elongate worms, round in cross section with a mouth at the anterior tip (Fig. 30.2) and a subterminal anus (Fig. 30.3). The body is covered with a secreted cuticle (Fig. 30.4) which is underlain by a single layer of neuromuscular cells. The nervous system consists of two ganglia dorsal and ventral to the esophagus connected by nerves which make up the circumesophageal nerve ring. Nerves run anteriorly to the sense organs and posteriorly, dorsally, ventrally, and laterally in the hypodermis. A cross section of *Ascaris* shows these nerve trunks clearly, as well as the connections of the neural portion of the neuromuscular cells with the dorsal and ventral nerve cords (Fig. 30.5). The nerves innervate the muscles to control muscular contraction, the amphids and lips at the anterior end and the phasmids at the posterior end of the worm. Both the amphids and phasmids are sensory structures.

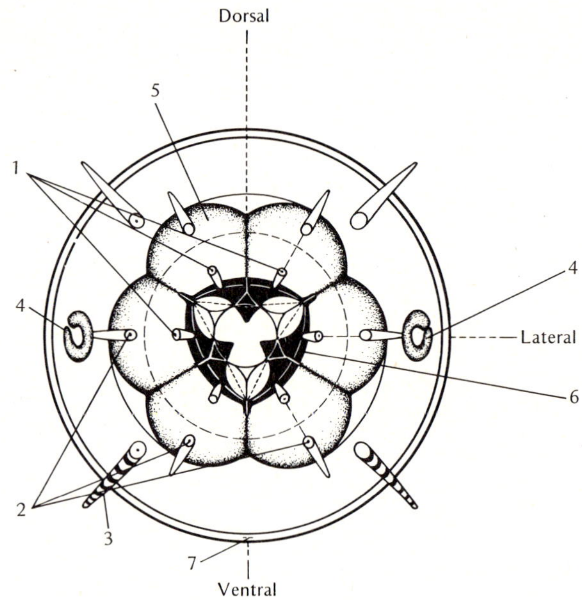


**FIG. 30.2**
A primitive nematode seen *en face*. (1) Cephalic bristles; (2) inner labial bristles; (3) outer labial bristles; (4) amphids; (5) lips; (6) buccal capsule.

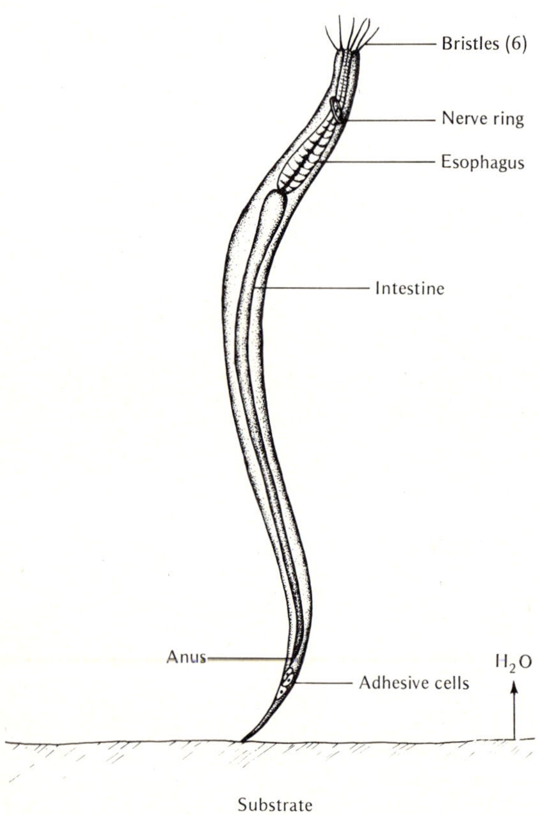


**FIG. 30.1**
A hypothetical primitive nematode attached to the substrate.

The excretory system is usually composed of a gland, pulsatory vesicle, and excretory pore near the posterior end of the esophagus, but there are also longitudinal canals that run the length of the animal.

Sexes are usually separate, although there are a few monoecious species and a few that are parthenogenetic. The female system consists of a uterus, usually with two branches, that connects with an elongate ovary (Fig. 30.3). Ova move from the ovary into the upper part of the uterus, where fertilization takes place and egg capsules are laid down. The eggs reach the outside through the short vagina and the vulva located somewhere along the length of the body. Some species have muscular ovejectors at the vagina to expel eggs. The male system is similar to that of the female in that the testis is an elongate structure in which the spermatids are formed. They mature in the vas deferens, are stored in a seminal vesicle, ejected by a muscular ejaculatory duct, and passed to the female. The

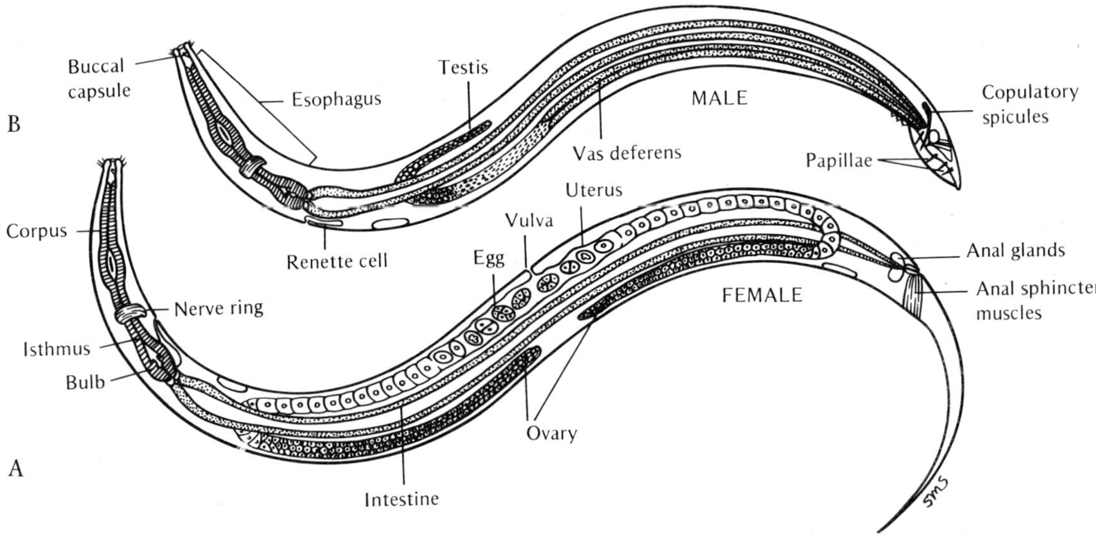


**FIG. 30.3**
Hypothetical (*A*) female and (*B*) male nematodes.

principal differences are that in the male system there is a common opening with the intestine at the cloaca and there are additional structures, such as copulatory spicules and bursae, which facilitate the passing of sperm to the female. Spermatozoa are unique among the animals in that they lack flagella and have ameboid movement.

The basic structure of a nematode is a tube within a tube. The tubes, intestinal tract, and reproductive system lie within a cavity, the pseudocoel. The structure is closely related to a nematode's function.

If you watch a small free-living or parasitic nematode under a microscope you will see that the esophagus is constantly pumping. Small particles in the medium move into the mouth and then pass down the esophagus to the intestine. Occasionally, the muscles that keep the anus closed will open the anus and allow some of the gut content to escape into the medium. In addition to feeding, the esophagus pumps fluid in order to maintain a high hydrostatic pressure in the body of the worm. This hydrostatic pressure not only maintains the round shape of the worm, but is essential for its motility. A study by Harris and Croften (1957) considered the theoretical and practical importance to the nematode of maintaining a high hydrostatic pressure. The study showed that efficient motility is possible in an animal that has a simple nervous system and only a single layer of muscles because of the physical principle that a liquid is incompressible. This means that any shortening or narrowing in the body's diameter will produce an immediate transfer to another part of the body, which will either elongate or increase in diameter.

The cuticle of nematodes is important in their physiology and in relation to the host. It is secreted by the hypodermis and consists of several layers; there is some variation in the layers in various species of nematodes and in the stages of the life cycle, but a general pattern can be seen in most of them. *Ascaris suum* has been studied both at the light microscope (LM) and electron microscope (EM) levels and provides an adequate model for other nematodes (Fig. 30.5). There is a clear outer layer of cortex underlain by a matrix and then three layers of fibers. The basement membrane rests on the hypodermis.

The cortex of the cuticle is at the interface with the host. The proteins in it seem to be poorly antigenic, so that there is often little host response to the worm. The fibers in the fiber layer lie at an angle to one another,

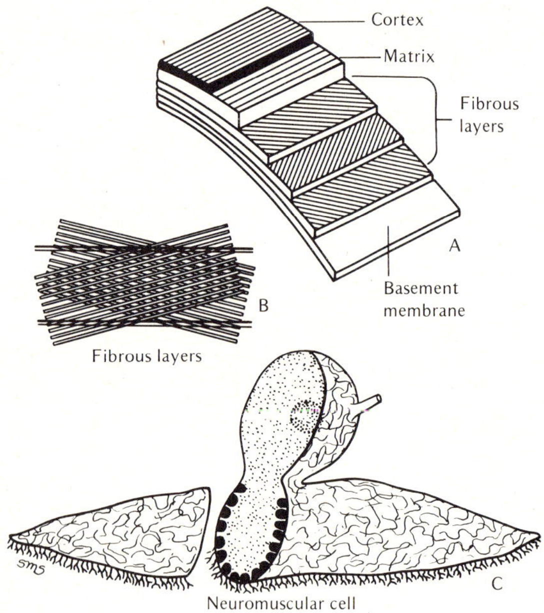


**FIG. 30.4**
Structure of the nematode body wall. (*A*) Cuticle of *Ascaris*, (*B*) fiber lattice showing the oblique orientation of the layers, (*C*) a neuromuscular cell such as is found in *Ascaris*. [Redrawn with modifications from Crofton, M.D. 1966. *Nematodes*. Hutchinson University Library, London.]

usually 70 to 75 degrees (Fig. 30.5*B*). It is this layer which controls the transfer of forces in the movement of the animal, as discussed.

Cuticle and growth in nematodes are interrelated. Growth in nematodes is separated by a series of four molts. Prior to each molt, a new cuticle is formed beneath the old one. This general growth and molting pattern has been found in all nematodes that have been studied with care. After hatching from the egg, the first stage larva ($L_1$) feeds, grows, and then molts to the second stage larva ($L_2$). The process continues to the $L_5$, which then becomes the adult through maturation of the reproductive system (Fig. 30.6).

Molting is a complex process which is under hormonal control, as in insects (Davey & Hominick, 1973; Davey & Sommerville, 1974). It has been said that molting in nematodes serves the same function that it does in insects; the worm can grow only to a certain size and then must molt in order to grow larger. However, the $L_5$ of *Ascaris* grows from about 22 mm to 300 mm without molting, and it therefore appears that molting serves a different function.

Regardless of the origin and function of molting in free-living nematodes, the process has been adapted to the parasitic mode of life. The typical free-living nematode moves continuously from one stage of the life cycle to the next when there is adequate food, moisture, and temperature. On the other hand, the parasitic nematodes have adapted molting to control their development and to increase the probability of completing the life cycle. As an example, the trichostrongyloid nematodes have direct life cycles in which there is a free-living phase. Eggs are shed with the feces of the host, and after they develop for a few days to 2 weeks, the $L_1$ hatches. The $L_1$ feeds, and then molts to the $L_2$, which feeds and molts. The $L_3$ is the infective stage, which is a nonfeeding stage encased or ensheathed in the cast cuticle of the $L_2$. Thus, molting has gone only halfway: the body of the $L_3$ has separated from the $L_2$ cuticle but has not crawled out of it. The $L_3$ remains ensheathed until it has been ingested by the proper host, at which time the sheath is shed and the $L_3$ invades the tissue to become a parasitic form. The same is true of vector-borne nematodes; they develop to a certain stage in the vector and continue development only when they have been transferred to the next host in the cycle. The infective stage may vary depending on the particular group of nematodes, but the same phenomenon is seen:

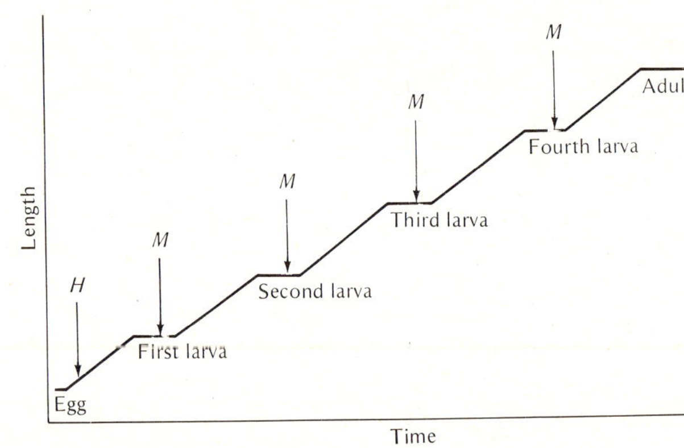


**FIG. 30.6**
Growth and molting in nematodes. (H = egg hatches; M = molt occurs) [Redrawn from Lee, D.L. 1965. *The Physiology of Nematodes*. W.H. Freeman, San Francisco.]

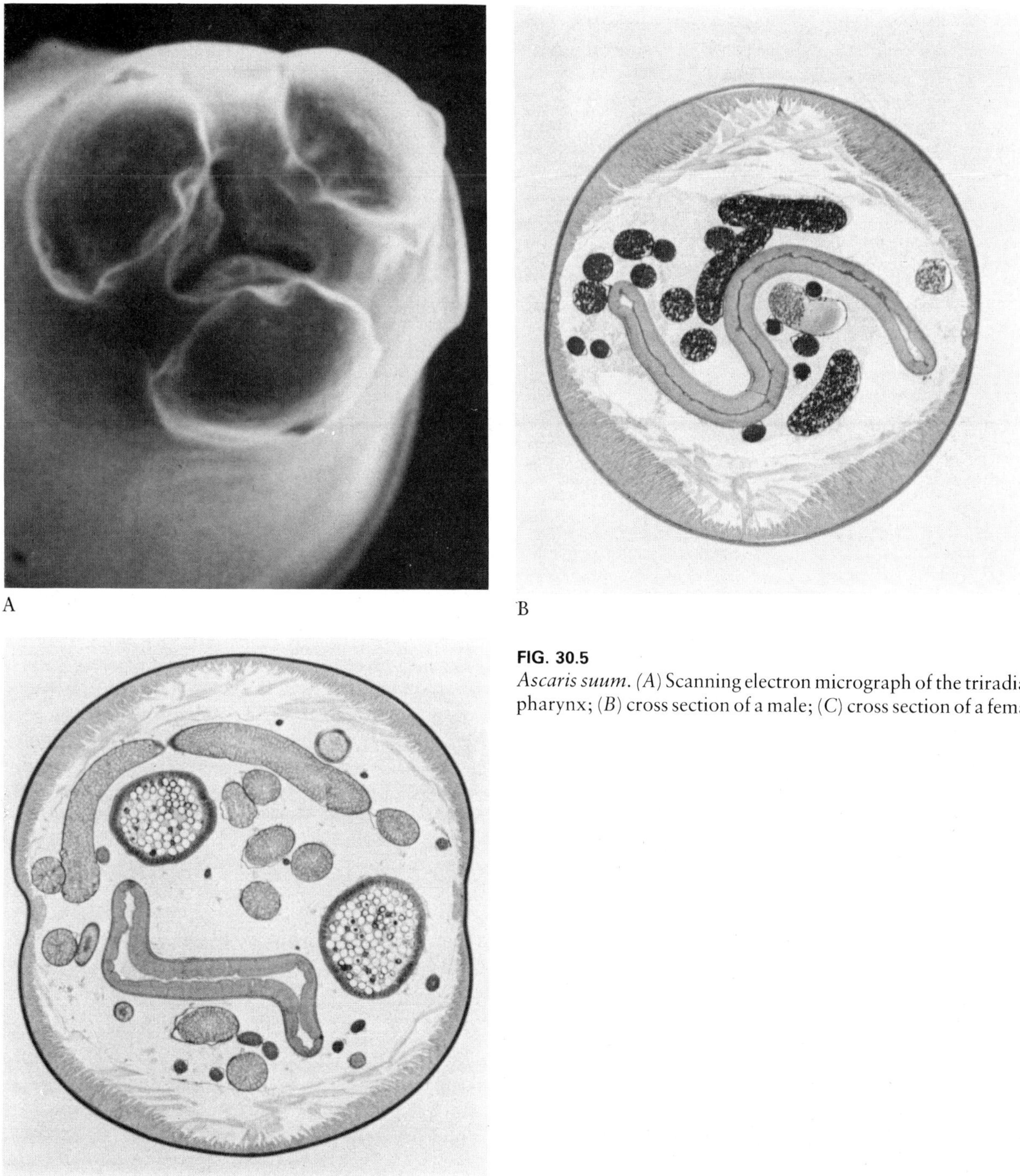

A

B

C

**FIG. 30.5**
*Ascaris suum.* (*A*) Scanning electron micrograph of the triradiate pharynx; (*B*) cross section of a male; (*C*) cross section of a female.

development stops at a certain stage and continues only when the larva has received the signal that it has reached the proper environment and further development is feasible.

Many aspects of the development and structure of nematodes have been treated briefly here. Additional detail may be obtained in books such as Levine (1980), Hyman (1951), Chitwood and Chitwood (1974) and Lee (1965).

Classification and identification of nematodes are based largely on the following features:

1. Phasmids, presence or absence and structure
2. Lateral excretory canals, presence or absence (Fig. 30.7)
3. Esophageal structure (Fig. 30.8)
4. Structure of the buccal capsule and lips or papillae on the anterior end (Fig. 30.9)
5. Structure of the male copulatory organs (Fig. 30.3)
6. Structure of the posterior tip of the female
7. Structure of the eggs

The classification used here is that of Levine (1980), which is based mostly on Chitwood's proposals. The phylum Nematoda has two classes, Secernentea (originally Phasmidia) and Adenophorea (originally Aphasmidia), and the division is based on the fact that the Secernentea has phasmids (structures in the caudal region of the worm of unknown function) and lateral excretory canals and the Adenophorea lacks both. Unfortunately, the phasmids are difficult to see and the lateral excretory canals can be seen only in tissue sections, for practical purposes. In identifying nematodes one ignores both phasmids and excretory canals and uses the other five characteristics to place a specimen in the proper class, order, or lower taxon.

## CLASSIFICATION OF THE PHYLUM NEMATODA

Elongate worms, cylindrical in cross section, and bilaterally symmetrical; covered with a cuticle which is secreted by the hypodermis; segmentation, if present, is only superficial; body cavity a pseudocoelom; digestive tract complete and usually with a subterminal anus; appendages lacking, but there may be bristles or cuticular protuberances; subepidermal musculature of longitudinal fibers only; brain in the form of a circumenteric ring with attached ganglia; muscular pharynx long and triradiate; excretory system composed of one or two gland cells, of canals, or both; mostly dioecious, but a few are parthenogenetic or hermaphroditic; gonads single or double; male duct enters cloaca; female duct has its own opening; growth by a series of four molts.

**Class Secernentea (=Phasmidea)**
Phasmids usually present; lateral excretory canals; caudal glands absent; external amphids porelike, usually small; cervical papillae (deirids) usually present; caudal alae or copulatory bursa common; usually 4 to 6 pseudocoelomocytes.

**Order Tylenchida**
Stylet present in mouth; single excretory canal; esophagus composed of corpus often with bulb, isthmus, and nonvalved glandular region.
*Anguina, Heterotylenchus*

**Order Rhabditida**
Stylet absent; excretory canals paired; esophagus usually composed of corpus, isthmus, and bulb, but the bulb is sometimes absent; transverse vagina, not heavily muscled; caudal alae, if present with papillae rather than muscular rays.
*Caenorhabditis, Pelodera, Rhabiditis, Strongyloides*

**Order Strongylida**
Corona radiata or three or six lips; buccal cavity surrounded by esophageal tissue; larval esophagus consists of pro- and metacorpus, isthmus, and bulb; adult esophagus clavate; excretory system with paired lateral canals and paired subventral glands; female with transverse or short vagina vera and single or double vagina uterina, often heavily muscled (ovejector); male with caudal alae usually containing well-developed muscular rays forming the true or strongylid compulatory bursa; two equal spicules.

**Superfamily Ancylostomatoidea**
Buccal capsule subglobular, never hexagonal in transverse section; lips and corona radiata absent; oral opening unarmed or with teeth and cutting plates; bursa well developed.
*Ancylostoma, Necator, Bunostomum, Uncinaria, Gaigeria, Globocephalus*

**Superfamily Strongyloidea**
Mouth surrounded by corona radiata, sometimes 6

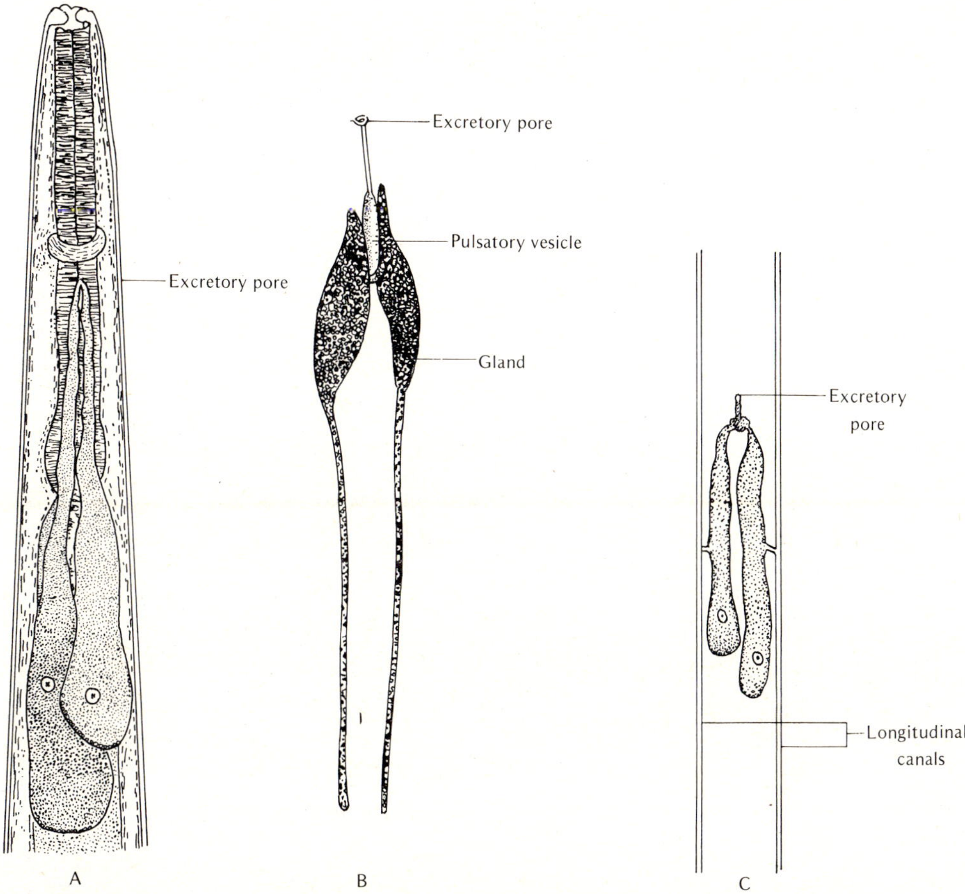


**FIG. 30.7**
Nematode excretory systems. [Redrawn with modifications from Hyman, 1951.]

lips, oral opening sometimes hexagonal, teeth and cutting plates absent; bursa well developed.

*Strongylus, Craterostomum, Chabertia, Oesophagostomum, Stephanurus, Syngamus, Triodontophorus, Oesophagodontus, Mammomonogamus*

**Superfamily Trichostrongyloidea**
Mouth small, lips absent or surrounded by three or six small lips; often with cuticular inflation at mouth region; usually numerous longitudinal ridges; $L_1$ often with valved esophageal bulb and conical tail ending in a simple point; bursa well developed.

*Cooperia, Cooperioides, Haemonchus, Hyostrongylus. Nematospiroides, Heligmosomoides, Nematodirus, Nippostrongylus, Ostertagia, Trichostrongylus*

**Superfamily Metastrongyloidea**
Mouth small, sometimes even absent, surrounded by 6 lips or lips small; longitudinal ridges absent; $L_1$ without esophageal bulb, tail usually asymmetrical; bursa small, rays somewhat fused; parasites of the lungs and blood vessels of vertebrates.

*Angiostrongylus, Dictyocaulus, Elaphostrongylus, Metastrongylus, Muellerius, Paraelaphostrongylus, Protostrongylus*

**Order Ascaridida**
Generally heavy-bodied worms; adults with all

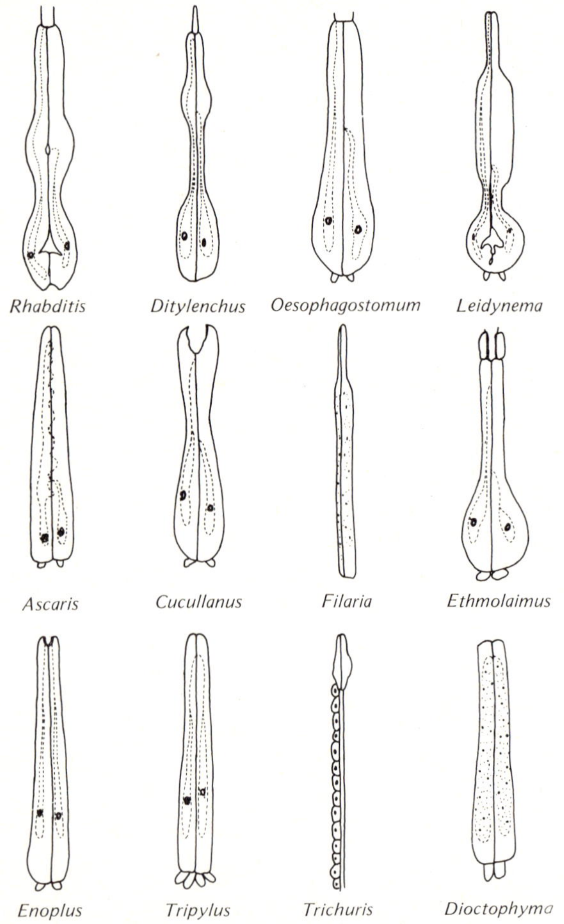


**FIG. 30.8**
Esophageal structures in various nematodes. [Redrawn from Crofton, M.D. 1966. *Nematodes*. Hutchinson University Library, London.]

gradations in esophageal structure from rhabditiform to cylindrical; three, two, or no lips; preanal sucker in male present or absent; caudal alae, if present, with papillae; two, one, or no spicules; generally with direct life cycles, except in Subuluroidea.

*Ascaris, Ascaridia, Heterakis, Toxocara, Toxascaris, Anisakis, Contracaecum, Phocanema, Subulura*

**Order Oxyurida**
Worms with short, stout bodies and often with elongated tails in the females giving rise to the common name of *pinworms*; males with a small number of caudal papillae and a single spicule; preanal sucker absent except in a single genus; esophagus with a large bulb; eggs often asymmetrical and embryonated while still in female; direct life cycles; parasites of the lower part of the digestive tract.

*Enterobius, Oxyuris, Skrjabinema, Syphacia*

**Order Spirurida**
Stout to threadlike worms; anterior extremity bilaterally symmetrical and lacking lateral external labial papillae; caudal papillae ventral or ventrolateral; esophagus with a short, anterior muscular portion and a long, glandular posterior portion; indirect life cycles; parasites of the anterior portion of the gut of vertebrates or of the blood and tissue spaces.

**Suborder Camallanina**
Larva lacking cephalic hook but with long, pointed tail and conspicuous phasmids; esophageal glands uninucleate; parasites of the gut of lower vertebrates or other organs in both lower and higher vertebrates; indirect life cycles with copepods as intermediate hosts.

*Camallanus, Dracunculus, Philometra, Philometroides, Avioserpens*

**Suborder Spirurina**
Larva with cephalic hook or spine and inconspicuous porelike phasmids; esophageal glands multinucleate; parasites of the gut or extraintestinal tissues of all vertebrates; intermediate hosts are arthropods, often insects and ticks; ten superfamilies including the Filarioidea, which is most important, as it contains many parasites of humans and domestic animals.

*Acuaria, Physaloptera, Stephanofilaria, Elaeophora, Wuchereria, Brugia, Thelazia, Habronema, Diplotriaenia, Dipetalonema, Onchocerca, Litosomoides, Dirofilaria, Loa, Mansonella*

**Class Adenophorea (=Aphasmidia)**
Phasmids lacking; lateral excretory canals absent; esophagus cylindrical.

**Order Dorylaimida**
Amphids pocketlike, porelike, or tuboid; excretory system poorly developed or absent; stylet present; esophageal gland orifices posterior to nerve ring and uninucleate; neither caudal alae nor bursa present in male.

**Superfamily Mermithoidea**
Male with 1 or 2 spicules and 2 testes; females

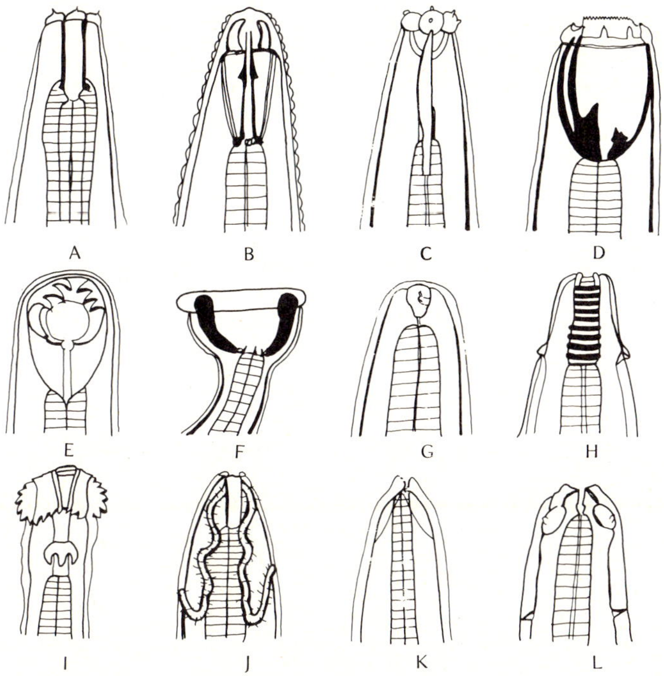


**FIG. 30.9**
Buccal capsules or mouth structures from various nematodes. (*A*) *Rhabditis*; (*B*) *Rotylenchus*; (*C*) *Dorylaimus*; (*D*) *Strongylus*; (*E*) *Ancylostoma*; (*F*) *Syngamus*; (*G*) *Haemonchus*; (*H*) *Physocephalus*; (*I*) *Seuratia*; (*J*) *Acuaria*; (*K*) *Enterobius*; (*L*) *Filaria*. [Redrawn from Crofton, M.D. 1966. *Nematodes*. Hutchinson University Library, London.]

usually with 2 ovaries; larvae parasites of arthropods and adults free-living.

*Mermis, Romanomermis*

**Superfamily Trichinelloidea**
Amphids porelike; esophageal glands free in body cavity forming stichostome; male with one testis and no or one spicule; female with elongate, tubular vagina and one ovary; eggs usually operculate; both adults and larvae parasitic.

*Capillaria, Trichinella, Trichosomoides, Trichuris*

**Order Dioctophymatida**
Amphids pocketlike, porelike, or tuboid; esophageal glands inside wall of cylindrical esophagus; reproductive system single in males and females; male with 1 spicule and caudal sucker; eggs operculate.

*Dioctophyma, Eustrongylides, Hystrichis*

## REFERENCES AND FURTHER READINGS

Anderson, R. C., A. G. Chabaud, and S. Wilmott (eds.). 1974, *Keys to the Nematode Parasites of Vertebrates*. Serial Publ. Commonwealth Inst. Helminthol, England.

Andrássy, I. 1976. *Evolution as a Basis for Systematization of Nematodes*. Pitman Publishing, San Francisco.

Bird, A. F. 1971. *The Structure of Nematodes*. Academic Press, New York.

Bryant, C. 1970. Electron transport in parasitic helminths and protozoa. *Adv. Parasitol.* 8:139–172.

Caveness, F. E. 1964. *A Glossary of Nematological Terms*. Pacific Printers, Ibadan, Nigeria.

Chitwood, B. G., and M. B. Chitwood. 1974. *Introduction to Nematology*. University Park Press, Baltimore.

Chitwood, M.B. 1969. The systematics and biology of some parasitic nematodes. In M. Florkin and B. T. Sheer (eds.), *Chemical Zoology*, Vol. III. Academic Press, New York, pp. 223–244.

Chitwood, M. B., and J. R. Lichtenfels. 1972. Identification of parasitic metazoa in tissue sections. *Exp. Parasitol.* 32:407–519.

Crofton, H. D. 1966. *Nematodes*. Hutchinson University Library, London.

Croll, N. A. 1977. Sensory mechanisms in nematodes. *Ann. Rev. Phytopathol.* **15**:75–89.

Davey, K. G., and W. M. Hominick. 1973. Endocrine relationships between nematodes and their insect hosts.

*Exp. Parasitol.* 33:212–225.

Davey, K. G., and R. I. Sommerville. 1974. Molting in a parasitic nematode, *Phocanema decipiens.* VII. The mode of action of the ecdysial hormone. *Int. J. Parasitol.* 4:241–259.

Faust, E. C., P. R. Russell, and R. C. Jung. 1970. *Clinical Parasitology.* Lea & Febiger, Philadelphia.

Goodey, J. B. 1963. *Soil and Freshwater Nematodes,* 2nd ed. John Wiley & Sons, New York.

Harris, J. E., and D. H. Crofton. 1957. Structure and function of nematodes: Internal pressure and cuticular structure in *Ascaris. J. Exp. Biol.* **34**:116–130.

Hyman, L. 1951. *The Invertebrates.* Vol. III. *Acanthocephala, Aschelminthes, Entoprocta.* McGraw-Hill, New York. York.

Inglis, W. G. 1971. Speciation in nematodes. *Adv. Parasitol.* 9:185–226.

Lee, D. L. 1965. *The Physiology of Nematodes.* W. H. Freeman, San Francisco.

——— 1972. The structure of helminth cuticle. *Adv. Parasitol.* **10**:347–379.

Levine, N. D. 1968. *Nematode Parasites of Domestic Animals and Man.* Burgess Publ. Co., Minneapolis.

——— 1980. *Nematode Parasites of Domestic Animals and Man,* 2nd ed. Burgess Publ. Co., Minneapolis.

McLaren, D. J. 1976. Nematode sense organs. *Adv. Parasitol.* **14**:195–267.

Olsen, O. W. 1974. *Animal Parasites,* 3rd ed. University Park Press, Baltimore.

Poynter, D. 1966. Some tissue reactions to the nematode parasites of animals. *Adv. Parasitol.* **4**:321–384.

Rogers, W. P. 1962. *The Nature of Parasitism.* Academic Press, New York.

Rogers, W. P., and R. I. Sommerville. 1963. The infective stage of nematode parasites and its significance in parasitism. *Adv. Parasitol.* **1**:109–178.

——— 1968. The infectious process and its relation to the development of early parasitic stages of nematodes. *Adv. Parasitol.* **6**:327–348.

Sasser, J. N., and W. R. Jenkins, eds. 1960. *Nematology.* University of North Carolina Press, Chapel Hill.

Silverman, P. H. 1965. *In vitro* cultivation procedures for parasitic helminths. *Adv. Parasitol.* 3:159–222.

Silverman, P. H., and E. L. Hansen. 1971. *In vitro* cultivation procedures for parasitic helminths: Recent advances. *Adv. Parasitol.* 9:227–258.

Soulsby, E. J. L. 1965. *Textbook of Veterinary Clinical Parasitology.* Vol. 1. *Helminths.* E. A. Davis, Philadelphia.

von Brand, T. 1973 *Biochemistry of Parasites,* 2nd ed. Academic Press, New York.

Wallace. H. R.·1971. The movement of nematodes in the external environment. In A. M. Fallis (ed.), *Ecology and Physiology of Parasites.* University of Toronto Press, Toronto, pp. 202–212.

Webster, J. M. 1969. The host–parasite relationship of plant-parasitic nematodes. *Adv. Parasitol.* 7:1–40.

Yamaguti, L. 1961. *Systema Helminthum.* Vol. III. *Nematodes.* Pts. 1 and 2. Interscience., New York.

Zuckerman, B. M. (ed.). 1980. *Nematodes as Biological Models.* Vol. 1. Academic Press., New York.

# 31

# Strongyloides *and Other Rhabditid Nematodes*

The members of the order Rhabditida are singularly unremarkable in their structure; they are small, transparent worms with little development of head or tail structures (Fig. 31.1). Their characteristics are outlined at the end of Chapter 30.

One of the interesting aspects of the rhabditid nematodes is their clear transition from free-living to parasitic modes of existence. We can see in this group species that are completely free-living but can invade hosts and develop in the skin (*Pelodera strongyloides*) or sometimes in internal organs of the host (*Rhabditis axei*). Even those that are obligate parasites (*Strongyloides* spp.) maintain a free-living sexual generation alternating with a parthenogenetic one.

Except for members of the genus *Strongyloides*, the parasitic and facultatively parasitic species in the order are of only minor medical or economic importance. We will consider only one facultative parasite and *Strongyloides*.

Lest we think that nematodes are all harmful, we will do well to consider two species of rhabditids that serve as experimental animals in fundamental biological studies. *Caenorhabditis briggsae* is a free-living species that has been maintained in axenic culture and has served in nutritional studies. Curiously enough, the list of essential nutrients for *C. briggsae* is about the same as those required for growth of parasitic nematodes in vitro (Sayre et al., 1963).

A second member of the genus, *C. elegans*, has become vital as a tool in studying development and genetics. Mutants are readily produced and extensive studies are underway in which changes in the genome are correlated with phenotypic changes. *C. elegans* seems to be filling the same role in eukaryotic genetics

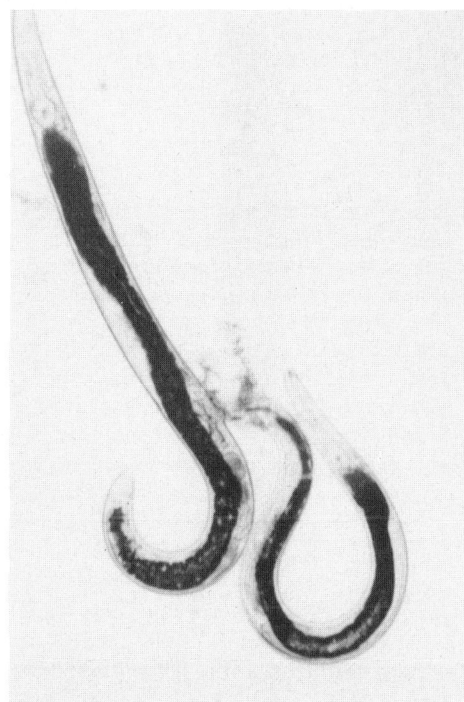

**FIG. 31.1**
*Pelodera strongyloides* male and female *in copula*.

that *Escherichia coli* has filled in prokaryotic genetics (Edgar & Wood, 1977).

NAME OF ORGANISM AND DISEASE. *Pelodera strongyloides* (Schneider, 1860) Schneider, 1866 (syn. *Rhabditis strongyloides*); no name has been given to the infection caused, but it could be referred to as *cutaneous nematodiasis*.

HOSTS AND LOCATION. Dog, ox, sheep, and rodents are hosts. The organism lives in the skin and orbit of the eye.

DISTRIBUTION AND IMPORTANCE. Distribution is cosmopolitan. This species occurs infrequently in domestic animals; it causes the host some discomfort and may lead to secondary bacterial infections.

MORPHOLOGY. Females are 1.3 to 1.5 mm long; males are 0.9 to 1.2 mm long. The vulva of the female is just posterior to the midpoint of the body. The male has a pair of fused spicules 65 to 73 μm long and a rather flat, spindle-shaped gubernaculus 47 μm long. $L_1$s are about 180 μm long and the organism grows through a series of molts, becoming larger at each stage (Fig. 31.2).

LIFE CYCLE. At room temperature (20–22°C) the organism passes from egg to adult in about 4 days. Under crowded cultural conditions or drying, the $L_3$, and perhaps others larval stages, form *dauer* larvae, a general term for a resistant form that can survive for many months.

CULTIVATION. The organism normally lives in areas of decaying organic matter and feeds on bacteria. It can readily be cultivated on petri dishes containing nutrient agar or blood agar. It feeds on bacteria, but also seems to feed on the medium itself since pits form in the surface where the worms are located. *P. strongyloides* is easy to keep in the laboratory and is used in many parasitology courses as a type nematode.

DIAGNOSIS. Nematodes that enter the skin of an animal cause crusty, weeping lesions and some loss of hair. Scrapings of the area reveal the living nematodes when examined as a wet mount under a compound microscope.

Other organisms that may cause similar lesions are chewing lice such as *Bovicola bovis* or mites such as *Chorioptes, Psoroptes,* or *Sarcoptes.* If no nematodes are found, skin scrapings are digested briefly in hot O.1*N* KOH and the fragments of the arthropods are looked for using a compound microscope.

HOST–PARASITE INTERACTIONS. In domesticated animals, the worms are limited to the skin, where they cause principally superficial damage.

TREATMENT. Some symptomatic treatment of the skin to promote healing may be used.

EPIDEMIOLOGY AND CONTROL. *P. strongyloides* is normally a free-living organism. It becomes a parasite under conditions in which animals such as cattle or dogs are housed under unsatisfactory conditions with wet and soiled bedding. If an animal lies on wet bedding and there are bacteria growing in it, the skin becomes soft and its cornified, protective barrier breaks down. The nematodes normally present under such conditions will move from the bedding to the skin and continue their life cycle there.

Control of *P. strongyloides* infections is directed at providing better conditions for the infected animals. When animals have clean, dry bedding, the lesions usually clear up spontaneously.

Our interest in *Pelodera* lies in its transition from a purely free-living organism to a facultative parasite. Its success as a parasite is marginal, but the proper conditions allow it to invade the skin and live there for a while. The percutaneous mode of infection of *Strongyloides* spp. and the maintenance of a free-living generation illustrate the further transition from facultative to obligate parasitism.

## *STRONGYLOIDES* INFECTIONS

*Strongyloides* spp. are tiny worms ranging in length from 2.2 mm for *S. stercoralis* to 9 mm for *S. westeri*; most are 3 to 6 mm long. All of the species have much the same structure and life cycle. They differ from one another in details of structure, life cycle, and pathogenesis. We will use *S. papillosus* of ruminants as a typical member of the genus and then comment on other species as pertinent. Little (1966a & b) has considered taxonomy and identification.

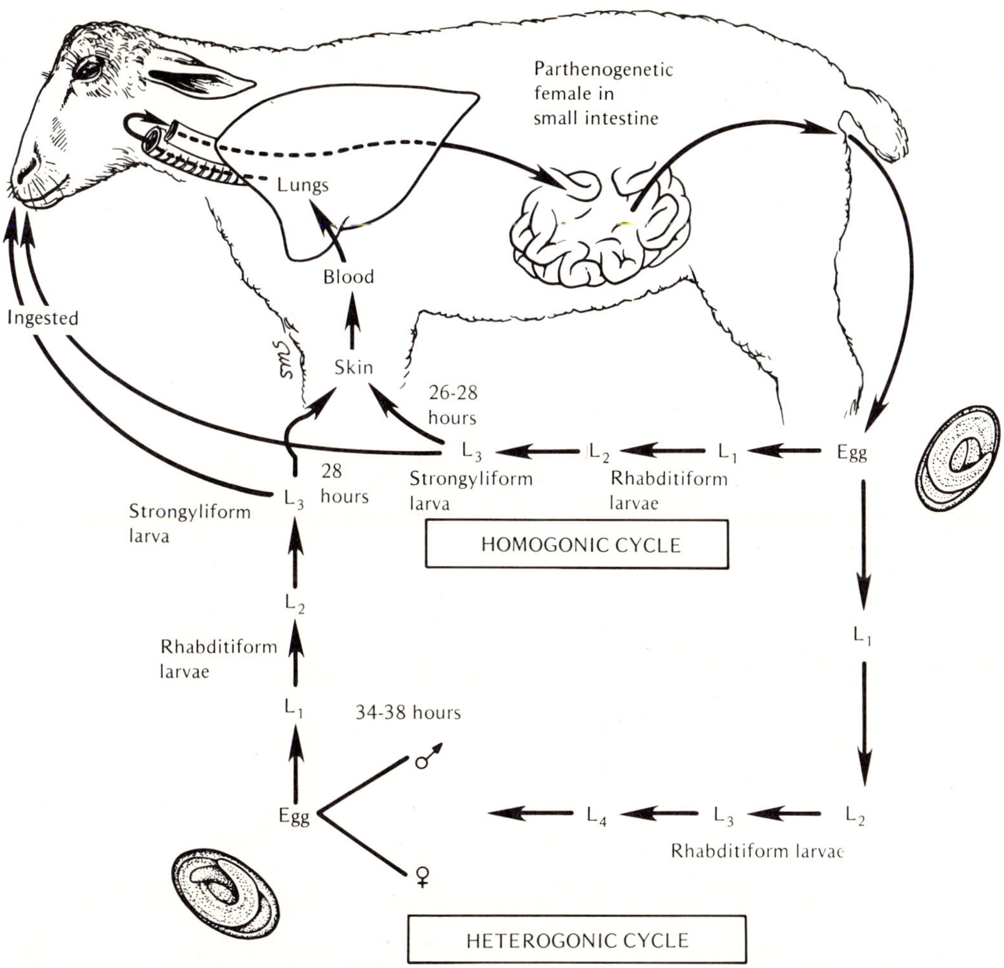


**FIG. 31.2**
Diagram of the life cycle of *Strongyloides* spp.

There are some common features of *Strongyloides* spp. that are important in dealing with them as agents of disease:

1. The small size makes them difficult to find at postmortem examination. More than the usual care is required to look for them using a dissecting microscope in examining intestinal content or scrapings.
2. All of the species have the potential of a homogonic cycle and a heterogonic cycle in the free-living phase (see Fig. 31.2). In the homogonic cycle, the organisms develop directly to the infective $L_3$. In the heterogonic cycle they develop through a free-living sexual generation and then the progeny develop to the $L_3$.
3. The parasitic phase of the cycle consists of parthenogenetic females only. A few reports of finding parasitic males have not been confirmed. Thus, there is an alternation of generation in the life cycle.

4. The free-living stages are short-lived and highly susceptible to environmental conditions except within a narrow range. The $L_3$s typically survive only a few days in the environment. *S. papillosus* $L_3$s were found to survive as long as 40 days at 20°C, and this can be considered to be maximal survival time in the genus. Cold, freezing, drying, anaerobic conditions kill the $L_3$s within a short time.
5. Whether the free-living stages develop directly to the $L_3$ through the homogonic cycle or the heterogonic cycle depends on conditions inside and outside the host. The conditions that shift the cycle through one phase or the other seem to be somewhat different for each species.
6. Any one species has only a moderate host specificity. Usually a species is found naturally in from 3 to 6 species of hosts; development may occur in experimental animals taxonomically far removed from the normal hosts.

NAME OF ORGANISM AND DISEASE CAUSED. *Strongyloides papillosus* (Wedl, 1856) Ransom, 1911, causes strongyloidiasis and threadworm infection.

HOSTS AND LOCATION. Sheep, cattle, goats, other ruminants both domestic and wild, pig, and rabbits are hosts. Adults are in the upper part of the small intestine.

DISTRIBUTION AND IMPORTANCE. The organisms are cosmopolitan. Natural infections in ruminants are nearly always benign, but the prevalence of infection is high in young animals.

MORPHOLOGY. Parasitic females are 3.5 to 6 mm long by 0.05 to 0.065 mm wide and have a filariform esophagus about 0.98 mm long or about one sixth the body length. Eggs in feces are about 55 by 35 μm and contain a fully developed $L_1$ when passed. The infective $L_3$ is 520 to 710 μm long, with a filariform esophagus 200 to 270 μm long; the $L_3$ is not ensheathed. All stages except the parasitic female and the $L_3$ have a rhabditiform esophagus, in which there is an anterior corpus, a narrow isthmus, and a distinct bulb at its posterior end.

LIFE CYCLE. The parasitic females lie in the paramucosal lumen of the small intestine, where they lay eggs with nearly fully developed $L_1$s that pass with the feces; the $L_1$s hatch in 6 hr at 27°C. If the organisms develop directly through the homogonic cycle, they reach the $L_3$ in 26 to 28 hr. If the organisms develop through the heterogonic cycle, the egg-laying adults are reached in 34 to 38 hr after leaving the host. The infective $L_3$s are reached about 28 hr later (Fig. 31.2).

The $L_3$s may be eaten with forage or may penetrate the unbroken skin. If eaten, the larvae probably remain in the gut and develop directly to parthenogenetic females. If they penetrate the skin, the $L_3$s enter the bloodstream and are carried to the lungs, where they are trapped. The larvae then break out of the air sacs, are carried anteriorly, coughed up, swallowed and carried to the small intestine. They molt to $L_4$ either in the lungs or the intestine; the last molt takes place in the intestine. Sexual maturity is reached 7 to 9 days after infection.

The factors that influence the organism to develop directly to the $L_3$ or, alternatively, through the free-living sexual generation have been the subject of investigation by a number of workers. A number of interpretations and theories have been set forth as a result of experiments on the more common species of *Strongyloides*. Current interpretations for *S. papillosus* and *S. ransomi* are that the sex of the offspring of the parasitic females is determined by conditions in the host. The immune status of the host seems to be the most important factor in producing male and female offspring which have the potential of developing into the free-living sexual generation. Males do not appear until several weeks after the parthenogenetic females have established, and although there is a possibility that the worms produce males as they age, most of the evidence points toward immunity as the cause. Once outside the host, environmental conditions cause the female eggs to shift either to free-living adults or to the homogonic $L_3$s. For example, hydrogen ion concentrations below pH 5.9 or above pH 7.2 favors development directly to $L_3$ whereas hydrogen ion concentrations between this range favor the heterogonic cycle. As a general rule, the heterogonic cycle is favored by optimal conditions (Moncol & Triantaphyllou, 1978).

Males seemingly participate in sexual reproduction only as tokens. They produce sperm, copulate with the females, and the sperm enter the unfertilized eggs. The sperm nuclei then degenerate and embryonation takes place. The interpretation is that the sperm initiate fertilization, but do not provide any genetic material for the new individual. This has been termed *obligate pseudofertilization* (Triantaphyllou & Moncol, 1977).

DIAGNOSIS. The most common method of diagnosis is finding the typical embryonated eggs or hatched $L_1$s in the feces. It is said that the $L_1$s fail to float in the usual flotation media used in fecal examinations; an alternative is to use a method such as the Baermann funnel to allow the larvae to migrate out of the feces, where they can then be collected.

HOST–PARASITE INTERACTIONS. *Strongyloides* spp. cause damage to the host at three different points:

1. At the site of penetration of the skin
2. During migration through the tissues and the lungs
3. After reaching the intestine and becoming sexually mature

In the case of *S. papillosus* specifically, little overall economic effect occurs in natural infections. However, penetration of the skin by $L_3$s allows the invasion of bacteria such as *Bacteroides nodosus*, one of the causative agents of foot rot in sheep. For a long time there was doubt that *S. papillosus* was pathogenic at all, but massive inocula have been found to cause considerable damage to both lambs and calves. Inocula of $10^4$ or more $L_3$s were lethal for lambs and $2 \times 10^4$ or more were lethal for calves. Animals are not ordinarily exposed to that many larvae at one time.

If one were to do an epidemiological study of gastrointestinal nematodes of ruminants, eggs of *S. papillosus* would be the first to appear in the feces. In lambs, for example, eggs appear within about the first 6 weeks or 2 months of life, and are seen for the next 5 months and then disappear. Typically, they are not seen in large numbers at any time; in ewes only an occasional egg is seen.

Two factors act to limit the life of infection—immunity and age resistance. As an animal is exposed to $L_3$s, a skin reaction develops that traps many of the larvae in the skin. Circulating antibodies have also been found, which seem to inhibit those that reach the bloodstream. Age resistance probably also acts since much larger inocula are required to produce pathogenesis in larger calves.

*Strongyloides* spp. may not develop in any host, but the $L_3$s may penetrate the skin of almost any potential host. Cutaneous larva migrans has been seen in humans as a result of penetration of the larvae of *S. papillosus*, *S. myopotami* and *S. procyonis*. The larvae sometimes migrate from the skin to internal organs such as the lung and cause some lung damage and eosinophilia (Loeffler's syndrome). Cutaneous larva migrans is discussed further in Chapter 32.

TREATMENT. None is required.

EPIDEMIOLOGY AND CONTROL. The free-living stages of *Strongyloides* do best in climates that are warm and moist. The $L_3$s are particularly susceptible to environmental extremes. As indicated earlier, cold, drying and anaerobic conditions are all detrimental to the $L_3$s. At refrigerator temperatures about half (47%) of the larvae die in 2 days (Turner & Wilson, 1961).

Despite experiments showing the pathogenesis of large inocula given to young animals, *S. papillosus* rarely causes observable damage under conditions in which animals are usually raised.

NAME OF ORGANISM AND DISEASE CAUSED. *Strongyloides stercoralis* (Bavay, 1876) Stiles and Hassall, 1902, causes strongyloidiasis.

HOSTS AND LOCATION. Humans, dogs and other canids, cats, coati mundi, and both higher and lower primates are hosts. The organisms live in the small intestine.

DISTRIBUTION AND IMPORTANCE. The organisms are cosmopolitan, but prevalence is highest in warm, moist climates. In the United States there are about 400,000 human infections at any one time. Under some conditions, infection rates can become quite high; for example, 18% of 1437 inmates in a New York mental hospital were infected (Yoeli et al., 1972). In the rural South, up to 4% of persons may be infected. In dogs in the United States, infections run between 1 and 2%.

MORPHOLOGY. The parthenogenetic parasitic females (Fig. 31.3) are 1.7 to 2.7 by 0.03 to 0.04 mm. Eggs are 55 to 60 by 28 to 32 μm but usually hatch before leaving the host.

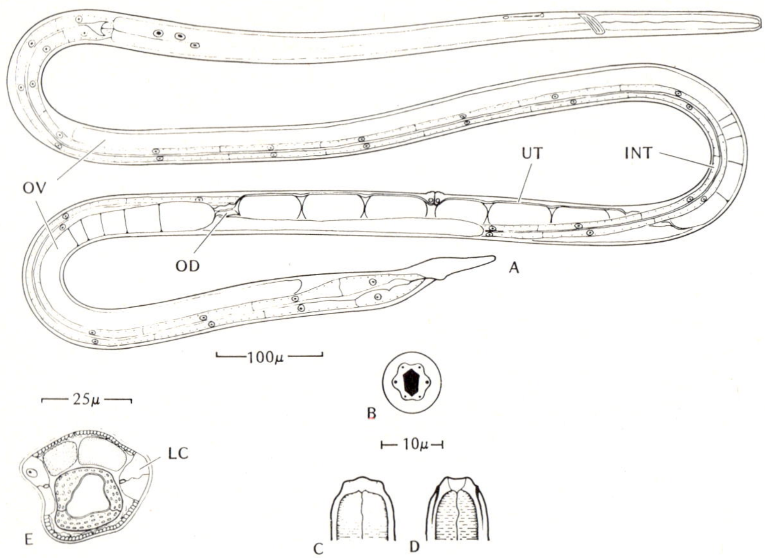


**FIG. 31.3**
Adult parthenogenetic female of *Strongyloides stercoralis*.
(*A–E*), Parasitic female: (*A*) entire worm (INT=intestine; OV=ovary; OD = oviduct; UT = uterus containing eggs); (*B*) en face view; (*C*) head, lateral view; (*D*) head, dorsal view; (*E*) transverse section at level of ovary (LC = lateral chord.) [From Little, 1966a.]

LIFE CYCLE. See *S. papillosus* (Fig. 31.2).

DIAGNOSIS. Finding typical eggs or, more often, larvae in the feces is the method of diagnosis. Clinical signs and symptoms include diarrhea, abdominal pain, weight loss, and eosinophilia, but these are common to a number of intestinal infections, especially helminths.

HOST–PARASITE INTERACTIONS. The host range of *S. stercoralis* is of concern, especially the possibility of cross-transmission between dogs and humans. No definite statement can be made about whether it is likely to occur, because there are geographic races that vary in their host range. Experimental transmission from dog to human appears not to have been done, but there is a report of accidental transmission of *S. stercoralis* to an animal caretaker who had contact with dogs (Georgi & Sprinkle, 1974).

Although there are sometimes skin and pulmonary problems in persons infected with *S. stercoralis*, most of the damage takes place in the intestine when the worms have matured. Since the worms burrow into the mucosa, they cause damage to the intestinal mucosa and may allow subsequent invasion by intestinal bacteria.

In some instances, a peculiarity of the *Strongyloides* life cycle and the level of immunity in the host may combine to cause severe health problems. In *S. stercoralis* in humans, the $L_3$ may be reached before the larvae have left the body. It appears that the worms become infective in the gut and then penetrate either in the lower part of the large intestine or in the perianal region; this type of infection is probably best referred to as *retrofection*, although the terms *hyperinfection* and *autoinfection* have been used for the same process. As discussed for *S. papillosus*, immunity develops in a host upon repeated infection, and retrofection would not ordinarily occur to a significant degree; however, the infection can be overwhelming in a person unable to respond immunologically. A number of instances have been reported in which persons receiving corticosteroids or drugs used against cancer have developed heavy infections with *S. stercoralis* (Georgi & Sprinkle, 1974).

TREATMENT. Thiabendazole is the drug of choice; other drugs have been used, but they have more side effects than thiabendazole.

EPIDEMIOLOGY AND CONTROL. At the public health level, the elements that need to be considered in a control program are the host range, disposal of human waste, contamination of the area with feces of dogs and cats, and the entry of the $L_3$s via both the mouth and the skin.

*S. stercoralis* is likely to be a widespread problem

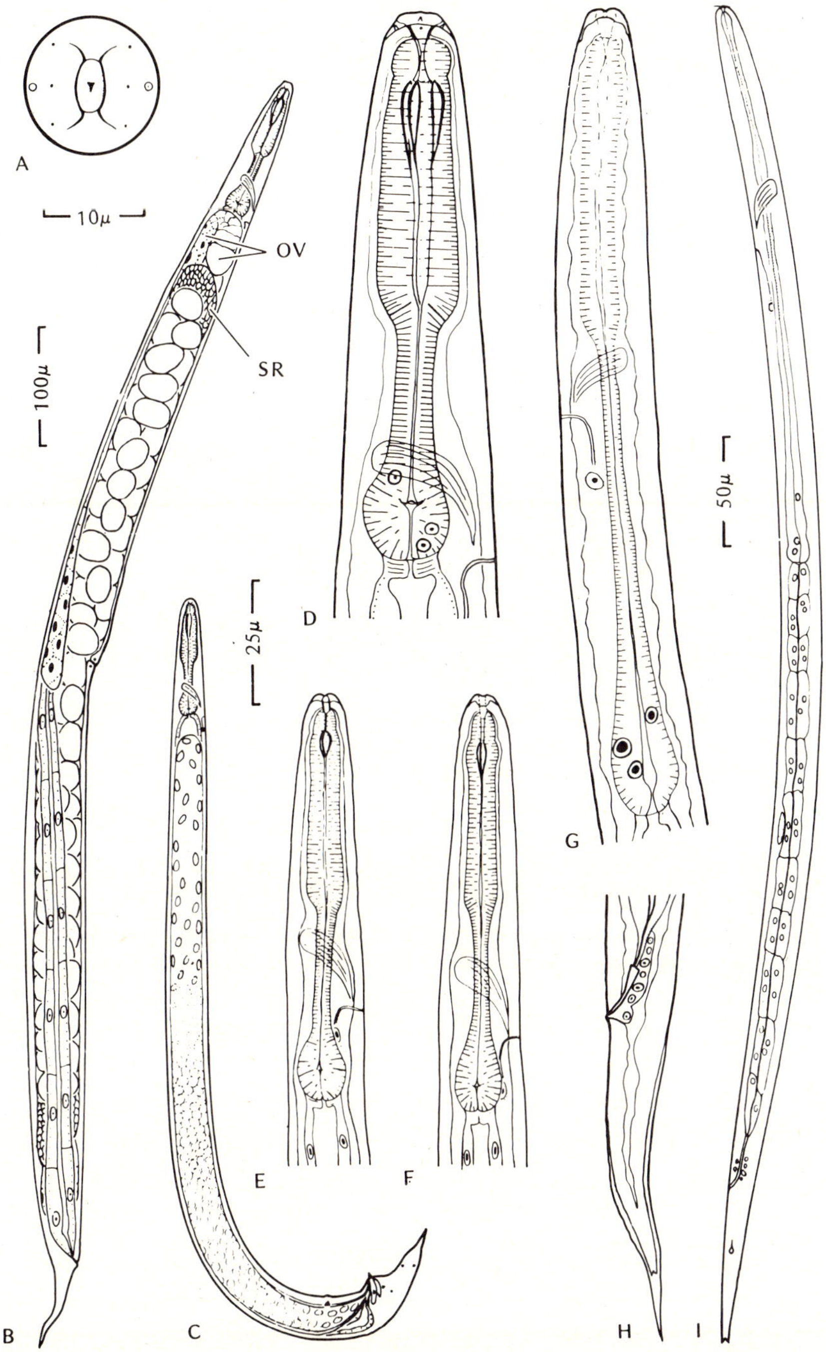


**FIG. 31.4**
*Strongyloides stercoralis* free-living stages. (*A*) Free-living female, en face view; (*B*) free-living female, lateral view (OV=ovary; SR=seminal receptacle containing sperm); (C) free-living male; (*D*) anterior end of free-living female showing details of esophagus; (*E*) newly hatched, first-stage larva obtained by duodenal aspiration from human; (*F*) first-stage larva from freshly passed feces of same patient as was larva in *E*; (*G*) second-stage larva developing to the filariform stage; cuticle is separating at anterior end; (*H*) tail of same larva as shown in *G*; notched tail is developing and cuticle is separating at tip and in rectum; (*I*) filariform larva. [From Little, 1966a.]

only in areas with warm, moist climates. Control involves sanitary disposal of human waste, treatment of infected persons, and elimination of infections in companion animals and/or stringent animal control measures. Since *S. stercoralis* is usually a significant problem in economically poor areas, all the difficulties of implementation of control discussed in the chapters on blood flukes, ascarids, hookworms, and other intestinal parasites pertain here as well.

If a member of a family is diagnosed as having *S. stercoralis* infection, all of the other members and companion animals should be examined and treated if necessary. The source of the infection should be sought in the vicinity; it might even lie outside the family, such as a neighborhood cat that defecates in a sandbox.

Instances such as described by Yoeli et al. (1972) in institutionalized persons present a particularly difficult problem. Mentally ill or retarded people may not be capable of developing fastidious habits of defecation and cleanliness. The rapid development of the free-living stages of *Strongyloides* makes such persons particularly prone to infection. An 18% infection rate in retarded persons is a case in point. The possibility of transmission to attendants should also not be ignored. Under this trying circumstance, mass treatment may be simplest approach to control.

NAME OF ORGANISM AND DISEASE CAUSED. *Strongyloides ransomi* Schwartz and Alicata, 1930, causes strongyloidiasis in swine.

HOSTS AND LOCATION. The host is swine, and the organism lives in small intestine as an adult.

DISTRIBUTION AND IMPORTANCE. The organism is cosmopolitan. Prevalence of infection in swine at slaughter is 10 to 14% but undoubtedly higher in young animals; it is higher in warm, wet climates than in cool and cold climates.

Losses to *S. ransomi* are probably limited to young pigs during the first few months of life. The USDA estimated (1965) that the annual loss in swine is $2.7 million.

LIFE CYCLE. It is the same as that of *S. papillosus*. The principal difference from the other members of the genus is the transmission of larvae from the sow to the pigs either prenatally or in the milk.

DIAGNOSIS. Eggs may be found in the feces by flotation methods, but in young pigs the infection is likely to be pathogenic prior to sexual maturity of the worms. In young pigs that suffer from diarrhea and emaciation, it is best to examine the intestinal tissues of dead or dying animals for worms. Since a number of bacterial and viral agents cause similar signs in young pigs, those should be ruled out by a competent diagnostic laboratory before concluding that *S. ransomi* is the problem.

HOST–PARASITE INTERACTIONS. As is true with the other members of the genus, *S. ransomi* is found in greatest numbers in younger animals. Pathogenesis occurs when large numbers of larvae are transmitted to the pigs prenatally or with milk shortly after birth. Much the same situation pertains here as with *Toxocara canis* and *Uncinaria lucasi*; the pregnant female accumulates larvae in her tissues which then move to the developing offspring before birth or shortly after birth through the milk.

Damage to pigs takes place before the worms are sexually mature. There may be damage to extraintestinal tissues as the larvae migrate through them, and there is considerable damage to the intestine after the worms arrive there (Fig. 31.5).

EPIDEMIOLOGY AND CONTROL. In the southeastern United States, *S. ransomi* has been observed as a problem in young pigs in years when the spring was especially wet. Such conditions apparently allow large numbers of free-living stages to develop, which then accumulate in the tissues of the pregnant sow and are passed to the offspring.

The conditions under which sows are housed during pregnancy are crucial for preventing infection of their pigs. It is essential that bedding be clean and dry and that it be a material that will not hold water. Since the $L_3$ can be reached quickly, and skin penetration takes place readily, it is essential that contamination with feces be kept to a minimum and that it be removed within a day after deposition.

## OTHER *STRONGYLOIDES* SPECIES

*S. westeri* of horses is a cosmopolitan parasite that probably has only a low degree of pathogenicity. *S. ratti* of rodents infects laboratory rats and has been used

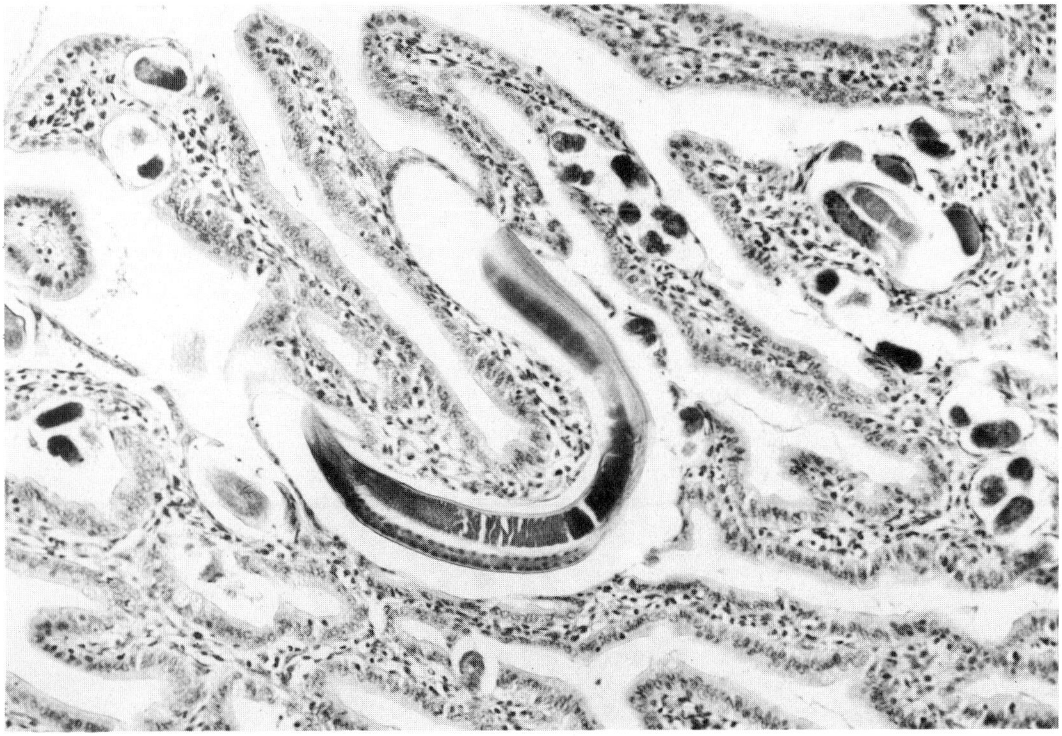

**FIG. 31.5**
*Strongyloides ransomi* larvae in the intestine of a young pig. [Courtesy of Dr. T.B. Stewart.]

extensively as a laboratory model of strongyloidiasis. *S. fuelleborni* is found in higher primates and may occur in humans on occasion.

## REFERENCES AND FURTHER READINGS

Basir, M. A. 1950. The morphology and development of the sheep nematode, *Strongyloides papillosus* (Wedl, 1856). *Can. J. Res. Ser. D. Zool. Sci.* **28**: 173–196.

Chang, P. C. H., and G. L. Graham. 1957. Parasitism, parthenogenesis and polyploidy: The life cycle of *Strongyloides papillosus*. *J. Parasitol.* **43**(Suppl.):13.

Edgar, R. S., and W. B. Wood. 1977. The nematode *Caenorhabditis elegans*: A new organism for intensive biological study. *Science* **198**:1285–1286.

Galliard, H. 1967. Pathogenesis of *Strongyloides*. *Helm. Abst.* **36**:247–260.

Georgi, J. R., and C. L. Sprinkle. 1974. A case of human strongyloidiasis apparently contracted from asymptomatic colony dogs. *Am. J. Trop. Med. Hyg.* **23**:899–901.

Levine, N. D. 1980. *Nematode Parasites of Domestic Animals and Man*, 2nd ed. Burgess Publ., Minneapolis, Chap. 2.

Little, M. D. 1966a. Comparative morphology of six species of *Strongyloides* (Nematoda) and redefinition of the genus. *J. Parasitol.* **52**:69–84.

———. 1966b. Seven new species of *Strongyloides* (Nematoda) from Louisiana. *J. Parasitol.* **52**:85–97.

Moncol, D. J. 1975. Supplement to the life cycle of *Strongyloides ransomi* Schwartz & Alicata, 1930 (Nematoda: Strongyloidinae) of pigs. *Proc. Helm. Soc. Wash.* **42**:86–92.

Moncol, D. J., and A. C. Triantaphyllou. 1978 *Strongyloides ransomi*: Factors influencing the in vitro development of the free-living generation. *J. Parasitol.* **64**:220–225.

Rivera, E., N. Maldonado, E. Velez-Garcia, A. J. Grillo, and

G. Malaret. 1970. Hyperinfection syndrome with *Strongyloides stercoralis*. *Ann. Intern. Med.* **72**:199–204.

Sayre, F. W., E. L. Hansen, and E. A. Yarwood. 1963. Biochemical aspects of the nutrition of *Caenorhabditis briggsae*. *Exp. Parasitol.* **13**:98–107.

Triantaphyllou, A. C., and D. J. Moncol. 1977. Cytology, reproduction and sex determination of *Strongyloides ransomi* and *S. papillosus*. *J. Parasitol.* **63**:961–973.

Turner, J. H. 1959. Experimental strongyloidiasis in sheep and goats. I. Single infections. *Am. J. Vet. Res.* **20**:102–110.

Turner, J. H., and G. I. Wilson. 1961. Experimental strongyloidiasis in sheep and goats. V. The effect of certain environmental conditions and chemicals on the infective larvae of *Strongyloides papillosus*. *J. Parasitol.* **47**:30.

Yoeli, M., H. Most, J. Hammond, and G. P. Schneinesson. 1972. Parasitic infections in a closed community. Results of a 10-year survey in Willowbrook State School. *Trans. Roy. Soc. Trop. Med. Hyg.* **66**:764–776.

# 32

# *Hookworms and Introduction to the Strongyles*

The principal characteristic of the members of the order Strongylida is the copulatory bursa at the posterior end of the male worms; it is composed of three groups of muscular rays. Occasionally reference is made to a bursa in another order of nematodes, but the true copulatory bursa occurs only in the Strongylida. The bursate nematodes are small to medium-sized, ranging from about 2 mm to 20 mm long; they are generally slender. Head structures range from almost none in the Trichostrongyloidea to elegant leaf crowns in the Strongyloidea and cutting plates in the Ancylostomatoidea.

The life cycles are all basically the same. With the exception of some of the Metastrongyloidea, they all have direct life cycles. The eggs are passed with the feces in an early stage of embryonation and they develop on the ground through the $L_1$ and $L_2$ as feeding stages. When the $L_2$s molt to the $L_3$, they become ensheathed, nonfeeding stages. Entrance into a host takes place by mouth with contaminated food or water; the hookworms have the option of entering through the intact skin of the host.

Hookworms form a fairly homogeneous group of strongyles in structure, life cycle, and epidemiology; they have large buccal capsules with cutting plates. The anterior portion of the worm is bent dorsally (Fig. 32.1) so that when one looks at the head, the dorsal teeth are at the bottom and the ventral teeth at the top (Fig. 32.2). The term *hookworm* comes from the general body shape. The copulatory bursa is well developed and the spicules are long and slender.

The life cycles are all alike. $L_1$ and $L_2$ are free-living, feeding stages and $L_3$ is the ensheathed, infective stage. Infection can take place either by mouth or by penetration of the unbroken skin. Worms penetrating the skin migrate through the bloodstream to the lungs, are coughed up, swallowed, and carried to the small intestine, where they mature. $L_3$s ingested with food or water usually remain in the gut and do not migrate through the body.

The distribution of hookworms is generally in warm,

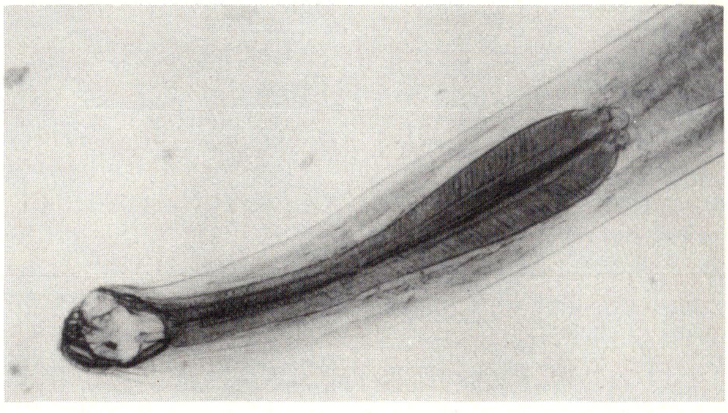

**FIG. 32.1**
Anterior end of *Ancylostoma caninum* with the characteristic dorsal bend of the head.

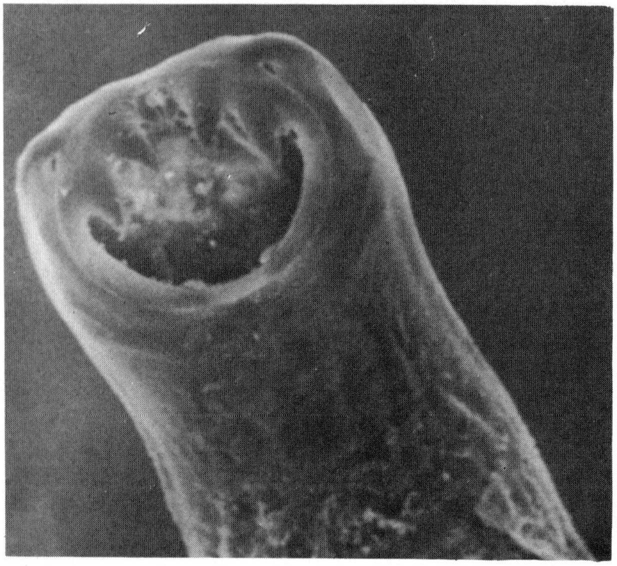

**FIG. 32.2**
Scanning electron micrograph of *Ancylostoma caninum* with the three large pairs of ventral teeth at the top.

moist climates. Many of the medically and economically important hookworms are cosmopolitan in tropical and subtropical areas (Table 32.1); however, some are limited in their distribution and on occasion may be carried outside their limited range by being introduced into special niches such as mines. The distribution and global migration of the hookworms have been discussed in some detail by both Levine (1980) and Miller (1979).

There are perhaps ten hookworms of interest because of their importance to humans or because they demonstrate certain principles in parasitology (Table 32.1). Note, for example, that *Ancylostoma caninum* is found in warm climates, whereas *Uncinaria stenocephala* is found in colder climates. The difference in distribution seems to lie in the preferences of the free-living stages: *A. caninum* has its optimum at about 30° C and *U. stenocephala* has its optimum at about 20° C, Therefore, *U. stenocephala* replaces *A. caninum* in the northern states of the United States and in Canada.

*U. lucasi* of the northern fur seal is a pathogen of young fur seals in the Pribilof Islands of the Bering Sea just north of the Aleutian Islands of Alaska. These isolated, wind-swept islands are an inhospitable locale for a parasite we think of as being found in warm climates. Fur seal pups were found to die of massive hookworm infection within the first few weeks of life. Initial investigation showed that contamination of the beach areas was not important for infection, but how the pups became infected was the puzzle. The landmark work of Olsen and Lyons (1965) revealed that the female seals were infected during one summer and carried the larvae in their tissues for nearly a full year, passing them to their pups through the milk. Transmission in the milk also occurs in *A. caninum* (Stone & Girardeau, 1968), and there is some indication that human hookworms may also be passed in the milk (Miller, 1981).

In this chapter we will consider only a few hookworms, but the principles apply to many other species in the group.

NAME OF ORGANISM AND DISEASE CAUSED. *Ancylostoma caninum* (Ercolani, 1859) Hall, 1913, causes hookworm disease, ancylostomiasis, and hookworm anemia. Additional species of hookworms are found in dogs (Table 32.1) but they will be considered here only incidentally.

HOSTS AND LOCATION. These hookworms are found in a variety of canids and other carnivores; they are the more prevalent hookworm of dogs. Adult worms occur in the small intestine of the definitive host; the larvae migrate from skin through the bloodstream to the lungs before reaching the gut.

DISTRIBUTION AND IMPORTANCE. They are cosmopolitan in warm and temperate climates throughout the world; prevalence is high in warm climates, but infections are found in cold to temperate areas such as the northern United States and southern Canada. If we take some of the data for dogs in North America and arrange them from warm to cooler climates, we can see that the prevalence generally decreases toward the north:

| LOCALITY | PERCENTAGE INFECTION |
|---|---|
| Mexico | 50–98 |
| Hawaii | 71 |
| Georgia | 86 |
| Oklahoma | 72 |
| Kansas | 21 |
| New Jersey | 22–27 |
| Illinois | 7 |
| Quebec | 9 |

It is difficult to assess the importance of the infection of companion animals either in lost days at work or in production losses, as might be done with some other hosts such as humans or domesticated animals. However, animals do become anemic and undoubtedly do not feel well. There is also some hazard of skin infections in humans (see *Cutaneous Larva Migrans*)

MORPHOLOGY. Male worms are 12 wide by 0.36 mm long and have a prominent buccal capsule and bursa. The female worms are 14 to 20 wide by 0.5 mm long and have a pointed tail. The buccal capsule has a pair of ventral teeth that have three points; these are usually spoken of as being three pairs of teeth (Fig. 32.3E); the dorsal teeth are within the buccal capsule. As noted in the introduction, the head of hookworms is bent dorsally and the arrangement of dorsal and ventral is reversed from what one usually expects. The large bursa is typical of the hookworms (Fig. 32.3F). A distinguishing feature is the structure of the dorsal ray; it can often be used in identifying hookworms to species. The spicules are long and slender, about 860 μm in length.

LIFE CYCLE. Eggs are passed in the feces and reach the outside when the embryo is in an 8- or 16-cell stage. The $L_1$ hatches in about a day at 23°C; the $L_3$ is reached in 4 to 5 days at 23°C. Entrance into the host takes place by mouth or by penetration of the unbroken skin. If ingested, most the larvae remain in the gut, but some migrate from gut to blood to lungs to trachea to esophagus and back to the small intestine—the same pattern seen in those that enter the skin. The prepatent period is 2 to 3 weeks (Fig. 32.4). Puppies may be infected prenatally or through the milk.

DIAGNOSIS. The usual method of determining whether an animal is infected is finding typical eggs in the feces (Fig. 32.5). In areas where hookworm is a continuing problem, it should be suspected when animals become listless and have a rough hair coat and pale mucus membranes indicating anemia.

Some of these clinical signs can be related to other parasitic infections such as *Dirofilaria immitis* (Chap. 41) or inadequate diet. Therefore, a fecal examination is important to ensure a proper diagnosis.

HOST–PARASITE INTERACTIONS. Damage to the host involves the following:

1. The site of larval penetration
2. Larval migration
3. The intestine
4. Extraintestinal sites.

The $L_3$s are about 730 by 22 μm and enter the unbroken skin of the host through the hair follicles (Vetter, 1970). The exact means by which penetration of the skin occurs is still a matter for investigation, but the larvae move down the hair follicle and exit near

**TABLE 32.1**
Some Hookworms of Economic and Medical Importance

| NAME | HOSTS | GEOGRAPHIC DISTRIBUTION |
|---|---|---|
| *Ancylostoma caninum* | Canines and other carnivores | Cosmopolitan in warm and warm-temperate climates |
| *A. ceylanicum* | Humans, cats, dogs | Asia, South America |
| *A. braziliense* | Dogs, cats, and other carnivores | Africa, Central and South America, Gulf Coast of United States |
| *A. duodenale* | Humans and some other primates | Subtropical Europe, Asia, South America, Africa |
| *A. tubaeforme* | Cats | Cosmopolitan |
| *Necator americanus* | Humans | Tropical: United States, Central and South America, Africa, Orient, Australia |
| *Uncinaria stenocephala* | Dogs and other canines | Northern United States and Canada |
| *U. lucasi* | Fur seal | Bering Sea |
| *Bunostomum trigonocephalum* | Sheep, goats | Warm, moist climates |
| *B. phlebotomum* | Ox | Warm, moist climates |
| *Gaigeria pachyscelis* | Sheep | India, Africa |

**FIG. 32.3**
Hookworm heads and tails. *Ancylostoma ceylanicum* head (*A*) and bursa (*B*); *Ancylostoma braziliense* head (*C*) and bursa (*D*); *Ancylostoma caninum* head (*E*) and bursa (*F*); *Ancylostoma tubaeforme* head (*G*) and bursa (*H*); *Necator americanus* head (*I*) and bursa (*J*); *Ancylostoma duodenale* head (*K*) and bursa (*L*).

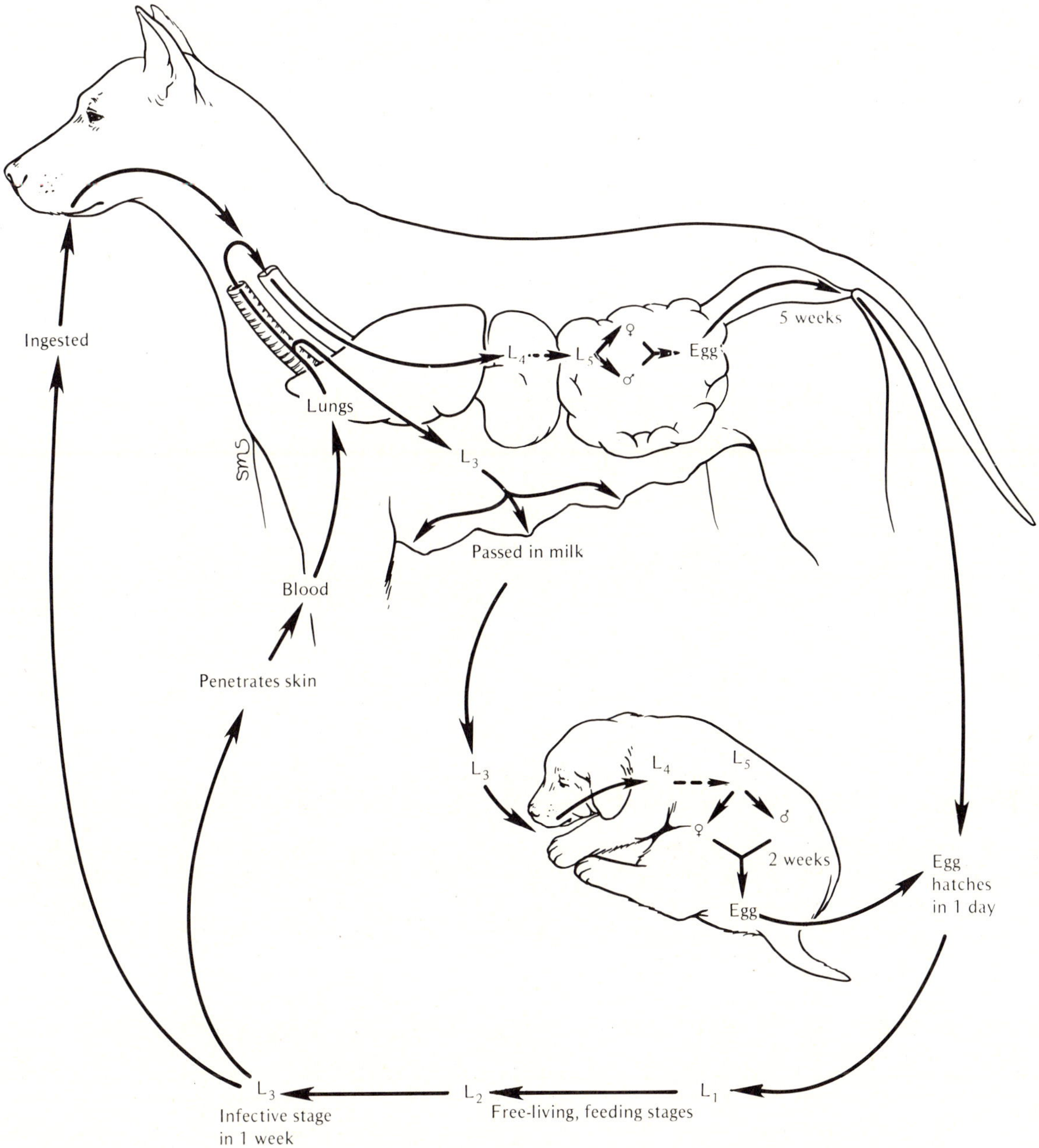


**FIG. 32.4**
Life cycle of *Ancylostoma caninum*.

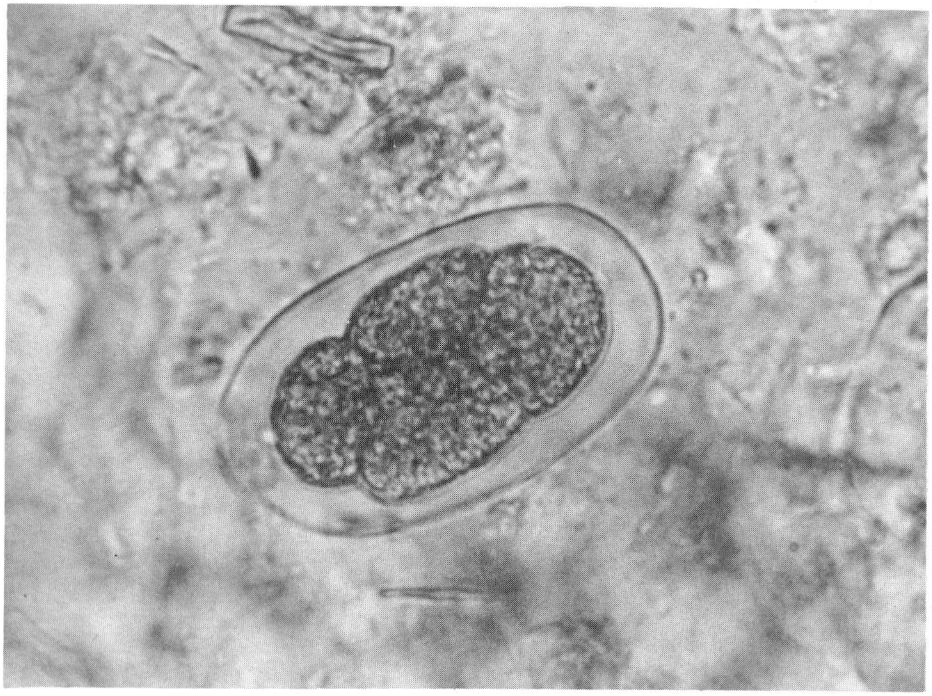

**FIG. 32.5**
*Ancylostoma caninum* egg.

the basement membrane into the dermis. In the normal host, they soon reach the bloodstream and are carried to the lungs; however, in an abnormal host, hookworm larvae may remain in the subcutaneous tissue and migrate there for some weeks (see *Cutaneous Larva Migrans*). There is some slight damage to the skin where each larva enters; the greater the number of larvae, the greater the damage.

During migration, the principal site of damage is the lungs, where larvae break out of the capillaries into the alveoli. There may be pneumonia and coughing at this time.

In the intestine the worms molt from $L_4$ to $L_5$ and begin to attack the intestinal mucosa. It has been the consensus that hookworms suck blood, but Kalkofen (1970, 1974) concluded that the worms feed on intestinal tissue and that ingestion of blood and blood loss are incidental. When worms feed, they pull a plug of intestinal mucosa into the buccal capsule (Fig. 32.6.); the tissue is digested before it passes to the intestine. Digestion requires about 10 min and the worm then takes another bite of tissue. Blood is lost when the capillary bed in the subepithelial layer is ruptured; blood flows out around the worm and some is ingested. A number of observers have pointed to the fact that blood ingested by hookworms seems to pass through the gut unchanged; red cells are unaltered and the blood is still bright red when leaving the anus of the worms.

Kalkofen's (1970) view of feeding is clearly at odds with the views held by many investigators and persuasively set forth by Roche and Layrisse (1966) in their exhaustive study of hookworm anemia. They relied particularly on direct observation of feeding hookworms in the open intestine of animals and on *in vitro* studies. They note that (1) large quantities of blood pass rapidly through the gut of worms, (2) worms spew out large volumes of blood when the body wall is nipped with an instrument, (3) there is continual pumping of the esophagus of worms under observation, and (4) the intestine fails to bleed extensively where worms have detached. Finally, it appears that a factor in the blood is required for glycogenesis in the worms (Fernando & Wong, 1964).

Whichever view of blood loss is correct, a great deal of blood is passed into the intestinal lumen. Such blood will be partially digested and recycled. The important

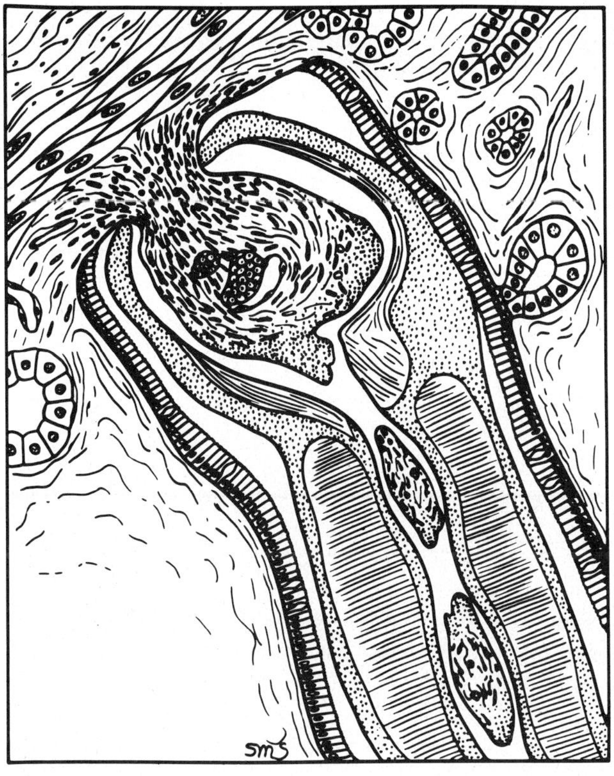

**FIG. 32.6**
Mode of feeding of a hookworm.

point is not how much blood leaks from the intestine into the lumen, but rather how much is ultimately lost. The following table shows some aspects of blood loss in *A. caninum* infections (Miller, 1966):

| NO. OF WORMS/LB OF BODY WEIGHT | TOTAL NO. OF WORMS |
|---|---|
| 8–27 | 240–934 |
| 30–64 | 371–972 |
| MILLILITERS OF BLOOD LOSS/ WORM/DAY | CALCULATED/ LOSS/YEAR/ 100 WORMS |
| 0.07–0.20 | 2.5–7.3 L |
| 0.01–0.09 | 0.36–3.3 L |

Obviously, the calculated blood loss is not realized, or animals could not survive a moderate infection.

Animals that have been repeatedly infected develop immunity; this observation was used to develop an irradiated larval vaccine against *A. caninum* (Miller, 1971). A series of studies growing out of the work on lungworm disease in cattle (Chap. 36) showed that vaccination against hookworms was feasible. A decision was made by Jensen-Salsbery Laboratories to produce and market a hookworm vaccine in the United States. Basic studies done by Miller starting in the 1960s found that it was feasible to vaccinate dogs against hookworm and that they would be well protected against the effects of the worms.

The problems to be overcome for the commercial success of the vaccine were partly biological, partly economic. Miller (1978) reviewed the sequence of events that led to the withdrawal of the vaccine against *A. caninum* after 14 years of basic research and development. A number of hurdles had to be cleared before a product could be made available to veterinarians for use. For example, federal standards for biologics require near sterility, and the fecal culture method of producing $L_3$s led to so much contamination with bacteria that a sterile culture technique had to be developed. The more laborious procedures for sterile culture of the free-living stages had an advantage in there being less damage to the $L_3$s at harvest and a longer shelf life for the vaccine. Once the larvae had been irradiated, storage temperature had to be controlled at between 10 and 15°C for a maximum shelf life of 12 to 14 months. Temperatures as low as 4°C and as high as 20°C could be tolerated only briefly. Thus, special shipping containers had to be developed as well as special incubators to be placed in the veterinarian's refrigerator (normally set at 4–6°C).

After the biological problems of production, storage, and adverse vaccine reactions were overcome, the vaccine did not sell well. Part of the problem was a matter of education and part was economics. The vaccine did not produce a sterile immunity. Hookworm eggs appeared in the feces, and it seemed impossible to educate either veterinarians or owners to the fact that low-level production of eggs did not necessarily indicate that the animal was adversely affected. Perhaps of greater importance to veterinarians was the economic return and the fact that adequate control of hookworm disease could be achieved by routine fecal examination and treatment of dogs. Since the market for the vaccine failed to develop, it was withdrawn in 1975 after being in use for about 2 years.

TREATMENT. A number of drugs are on the market that have a high degree of efficacy, but no treatment should be given until a proper diagnosis has been made. All drugs are toxic to a degree, and the animal may have more than one parasite that requires treatment.

Among the drugs given by mouth are *n*-butyl chloride, toluene, and bephenium. Disophenol can be given by subcutaneous injection. All of these drugs remove 90% or more of mature worms, but do not remove larvae and may vary somewhat in their effect on different hookworm species. Some, such as toluene, remove both hookworms and ascarids.

EPIDEMIOLOGY AND CONTROL. The development, survival, and migration of the infective stage of hookworms are crucial in the epidemiology of the infection. Adding to the complex is egg production of female worms. Hookworms produce relatively high numbers of eggs, considering their size, and they may live for more than a year. Animals may be exposed to large numbers of $L_3$s where there are moderate temperatures, moisture and protection from sunlight. *A. caninum* $L_3$ is reached in the following time periods at various temperatures:

| TEMPERATURE (°C) | TIME (DAYS) |
|---|---|
| 37 | 2 |
| 30 | 2.6 |
| 23 | 4.5 |
| 17 | 9 |
| 15 | 22 |

The $L_3$s can live for many weeks where conditions are moist, the temperature is about 20°C, and there is protection from direct sunlight. Conditions detrimental to the free-living stages of *A. caninum* are temperatures below freezing and above 37°C, drying, and exposure to sunlight. The $L_3$s of *U. stenocephala* survive freezing better than those of *A. caninum*, and their optimal temperature of development is lower, in keeping with its location in the northern United States and Canada.

As indicated (Table 32.1), hookworms are more prevalent in climates that are warm and moist than in those that are cold (*U. stenocephala* excepted), dry, or both. Thus, problems are likely to be more severe in the southern United States than in other areas. Untreated infections can last in a host for more than a year and contamination with $L_3$s can build up rapidly in the summer almost anywhere.

As is generally true of other strongyle $L_3$s, hookworm infective larvae tend to migrate from their site of development in the feces. This will place them in areas where they can readily reach a new host.

Companion animals allowed to roam will almost certainly become infected with hookworms. Likewise, animals in kennels that lack hygienic conditions can become heavily infected.

Control of hookworm disease in companion animals is based on the following:

1. Keeping animals in a clean environment
2. Surveillance by periodic fecal egg counts
3. Treatment

Of these means of control, providing a clean environment is easiest with companion animals that are restricted to a fenced yard or run. Once treated and kept in clean quarters, animals do not usually build dangerously high levels of infection. They can be examined periodically and treated as needed. Animals that are allowed to roam will inevitably become infected. Surveillance and regular treatment are essential in warmer climates.

If dogs are kept in kennels for either boarding or breeding, hookworm disease (as well as other helminthic and infectious diseases) can become severe. Kennel floors should be made of concrete and arranged in such a way that they will dry and can be exposed to the sun for a part of the day. Cleaning by hosing with water should be done daily so that as few as possible of the free-living stages remain in the area. When control has been achieved, any animals introduced should be quarantined, examined, and treated as necessary.

NAME OF ORGANISM AND DISEASE CAUSED. *Ancylostoma duodenale* (Dubini, 1843) Crepling, 1845; *Necator americanus* (Stiles, 1902) Stiles, 1903, and *A. ceylanicum* (Looss, 1910), cause hookworm disease, hookworm anemia and ancyclostomiasis.

HOSTS AND LOCATION. *A. duodenale* and *N. americanus* are the principal hookworms that infect humans. Following early migration, they are found in the small intestinal lumen. Hookworms sometimes mature in abnormal hosts, and hosts other than the principal ones can be infected with human hookworms.

DISTRIBUTION AND IMPORTANCE. Human hookworm is endemic in warm, wet climates (Table 32.1), but each species has some limitation on its distribution. Also, in some locations, such as in coal mines in central Europe, the distribution may either be extended or found in small pockets.

Estimates of the prevalence of infection in the human population range from 450 million to 700 million persons worldwide. In his often-quoted article entitled, "This Wormy World," published in 1947, Norman Stoll compiled data indicating that 457 million people were infected worldwide. Thus, there are about the same number of infected people now as 40 years ago. Since that time, the global population has about doubled, so there is some consolation that the per capita prevalence of infection has declined.

Because hookworm infection occurs in the poorest parts of the world, it interacts with other factors to impose a significant drain on the health of people. Inadequate diet and continual blood loss sap the strength and sense of well being. Other helminthic and protozoan infections are common in warm climates, so that any one person is almost certain to harbor at least one pathogenic parasite, and many have more than one.

In the United States, the area of endemic hookworm infection is the 11 southeastern states. From 1910 to 1935 the numbers of persons infected in that area declined from 4.5 million to 1.75 million and has declined still further in the intervening 50 years. However, there are still pockets of relatively high infection in the United States where 12 to 15% of children have been found to be infected. Current data are rather scattered, but there has been a decline in the Southeast from 42% to perhaps 4% in prevalence from early in this century to the present time.

MORPHOLOGY. In general size, the two principal human hookworms are similar, but can be differentiated statistically:

| | *A. duodenale* | *N. americanus* |
|---|---|---|
| Male | 8–11 × 0.45 μm | 6–9 × 0.3 μm |
| Spicules | 2000 μm | 900–1000 μm |
| Female | 10–18 × 0.6 μm | 9–11 × 0.35 μm |
| Eggs | 40 × 60 μm | 30–40 × 60–76 μm |

The two can readily be identified by an examination of the ventral teeth, (Fig. 32.3 *I, K*), but differentiation from other closely related species is more difficult. *A. duodenale* needs to be differentiated from *A. braziliense* and *A. ceylanicum*, in approximate descending order of their probability of occurring in humans.

*N. americanus* is the only member of this genus that has been reported from humans, but there are some similar species such as *N. suillus* from swine that are found in the same geographic areas.

The principal characteristics used in identifying species of hookworms are dorsal and ventral teeth, size and shape of the buccal capsule, dorsal ray of the bursa, length of the spicules and the structure of their tips, the lateral rays of the bursa, and the posterior end of the female (Fig. 32.3). Because of the similarity between *A. braziliense* and *A. ceylanicum*, the two were not differentiated until Biocca's (1951a) careful study.

LIFE CYCLE. Similar to that of *A. caninum*.

DIAGNOSIS. The principal means of determining whether a person is infected with hookworms is by examining the feces for eggs. The technique most often used is a flotation method, but a sedimentation method can also be used.

In some instances it is important to determine the level of infection in an individual or a community, and quantitative egg counts can provide useful information. Taking a number of studies into account, "light" infections are represented by egg counts up to about 2000 eggs/g of feces. Above 2000 eggs/g infections are considered to be "heavy." There is a direct correlation between the number of worms and the number of eggs per gram of feces; there is also a direct correlation between the number of worms and the severity of anemia. Neither correlation is necessarily linear, but the number of eggs per gram of feces can be used as a guide to possible clinical disease.

The clinical sign of anemia coupled with fatigue are suggestive of hookworm infection in an endemic area.

HOST–PARASITE INTERACTIONS. The ways in which hookworm cause damage to their human hosts may be described as follows:

1. *Primary*, in which the signs and symptoms are directly related to the worms:
   a. Skin penetration
   b. Migration through the lungs and pharyngeal regions
   c. Presence of worms in the small intestine, where they cause direct effects

2. *Secondary*, in which disturbances of the physiology and/or blood of the host result from chronic infection.

Penetration of the skin causes a slight lesion at the site which may be no more than an itching or prickling sensation; however, upon subsequent infection there is likely to be a more severe reaction due to hypersensitivity.

As the worms break into the alveoli of the lungs and migrate anteriorly, there may be some chest pain and inflammation of the pulmonary and pharyngeal tissues.

When the worms reach the small intestine, diarrhea and abdominal pain occur, among other gastrointestinal signs and symptoms.

The most significant aspect of hookworm infection is anemia as a result of long-standing infection. The severity of hookworm anemia is related most importantly to the number of worms and the nutritional status of the person. In general, anemia results from depletion of iron stores and the inability of the person to maintain a normal amount of hemoglobin.

Removal of blood by human hookworm can be calculated as follows:

| SPECIES | BLOOD LOSS | |
|---|---|---|
| | MILLILITERS/WORMS/DAY | MILLILITERS/100 WORMS/YEAR |
| *N. americanus* | 0.01–0.04 | 365– 1,460 |
| *A. duodenale* | 0.05–0.3 | 1,825–10,950 |

One hundred worms is not generally considered to be a severe infection despite what appears to be dramatic loss of blood. This is because much of the blood is recycled. The greater portion of the protein and iron will be digested, absorbed, and resynthesized into plasma and red cells. The more anterior the site of the worms in the small intestine, the greater the recycling; if the worms lie in the lower part of the small intestine, there will be greater real losses.

An inexorable loss of iron depletes iron stores so that eventually there is insufficient iron to make hemoglobin (Fig. 32.7). The normal amount of hemoglobin (15 g/100 ml of blood) is maintained until iron reserves or stores have been depleted; then there is a decline in hemoglobin to a point where the individual stabilizes at about 50% of normal hemoglobin. A person with such anemia would clearly not feel normally vigorous.

There are other factors contributing to hookworm disease, such as impaired liver function and loss of protein from the gut, but they are beyond the scope of this discussion. Further information may be obtained in Roche and Layrisse (1966) or Miller (1979).

As indicated earlier in the section on *A. caninum*, a high level of protective immunity can be developed in the dog; however, there is no conclusive evidence for protective immunity in humans (Miller, 1979). A considerable body of circumstantial evidence points to there being a protective immunity. The evidence suggesting a protective immunity is based on the following:

1. Epidemiologic studies
2. Observations after treatment
3. Comparison to infection in other hosts such as dogs

The epidemiologic pattern of hookworm infection in the human population is an increasing prevalence of infection up to about age 10 years and then a leveling off or decline until age 30 or 40 years; finally, there may be an increase in prevalence in elderly people. One could conclude that the susceptibles are removed from the younger population, leaving the resistant ones, but this seems not to be the case. Following treatment, adults tend to have relatively low worm burdens, but children under 10 years of age reacquire high levels of infection.

The pattern of acquisition and maintenance of worm infection is the same in dogs and people; therefore, humans appear to develop immunity upon repeated or prolonged exposure to infective larvae.

If there is good immunity, the issue of developing a vaccine should be addressed. With the constant drain on the health of most of those 600 million persons infected around the world, a hookworm vaccine would be of great benefit to those countries where the prevalence is high. The development of a vaccine is hampered not by technology, but rather by ethics. There is no way of testing the efficacy and safety of a vaccine for humans in any other host. Developing an irradiated larval vaccine requires that living, perhaps pathogenic, organisms be given to children and then their immunity following challenge observed. Who will assume the risks of such a trial?

The prepatent period in hookworm infections is generally 20 to 50 days, depending on the species, and it

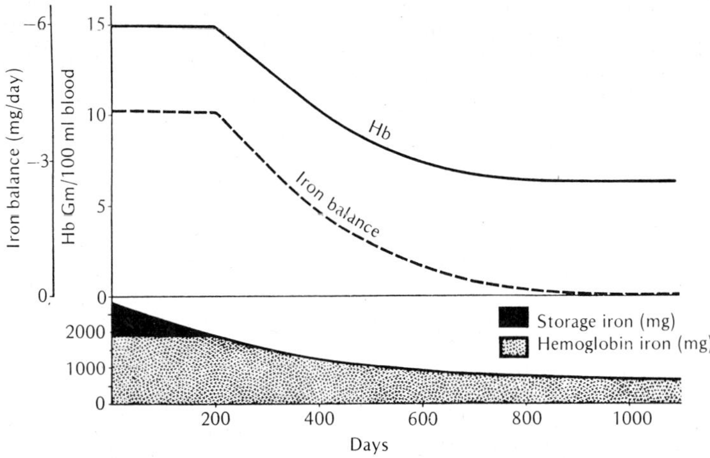


**FIG. 32.7**
Iron depletion and the development of hookworm anemia in humans. (Redrawn from Roche & Layrisse, 1966.)

generally has been held that once they have infected a host, the worms march inexorably toward sexual maturity. However Schad et al. (1973) and Nawalinski and Schad (1974) showed that development might not proceed at the time *A. duodenale* infects a host; sexual maturity can be delayed as long as 40 weeks after infection, nearly six times the usual period. Their work was done in the region of Calcutta, India, where there is a rainy season followed by a long, hot, dry season. Persons who become infected during the wet season harbor preadult worms all through the dry season; egg production commences shortly before the rains begin. It is not known whether the long prepatent period is inherent in the worms, or whether they somehow sense seasonal change from inside the host and then commence reproduction. Whatever the mechanism, the worms have been selected so that they do not jeopardize their resources until there is a high probability that the life cycle will be completed. The phenomenon of delayed or arrested development has long been known in trichostrongyles of domestic ruminants (Chap. 35), but this is the first clear-cut instance of arrested development in a nematode parasite of humans.

TREATMENT. A number of highly effective anthelmintics are available for use in humans, including piperazine and benzimidazoles.

EPIDEMIOLOGY AND CONTROL. Transmission of hookworm in the human population is dependent on a combination of the following biological and climatic factors and human culture:

1. Egg production of the female worms
2. Longevity of adult worms
3. Development and survival of the free-living stages as related to weather and climate
4. Immunity of the definitive hosts
5. Nutritional level of the human population
6. Technological and economic level of the area

The potential contamination with the free-living stages of hookworms is truly great. *A. duodenale* produces 10,000 to 30,000 eggs/female/day and *N. americanus* produces 5000 to 10,000 eggs/female/day. In areas where hookworm is endemic, egg counts of 2000 eggs/g of feces are common. An adult human produces perhaps 100 g of feces in a day, so that one person might contaminate the area with 200,000 eggs every day. In rural Venezuela about 7% of the population had eggs in excess of 2000 eggs/g of feces (Roche & Layrisse, 1966). A few persons in a population may produce 15,000 eggs/g of feces. Since worms may live in the host for more than one year, the potential contamination of soil is horrendous.

Moderate temperatures of about 15 to 35°C, oxygen, and some moisture promote the development and survival of hookworm free-living stages. Sandy or sandy-loam soils allow good survival of infective larvae.

Freezing, high temperatures (above 37°C), drying, and exposure to sunlight are all detrimental to the survival of the free-living stages. When exposed to any one of these conditions, they die within a few days.

Hookworm free-living stages are similar to other

strongyles in their pattern of development and behavior. The $L_1$s and $L_2$s remain in feces, where they feed, grow, and molt. The $L_3$ tends to migrate and will move out of the feces. They move toward subdued light and require a film of moisture in which to move. If there is minimal moisture, a larva may move along with a droplet of water surrounding it and be held there by surface tension. Some early workers reported that hookworm larvae can migrate vertically as much as a meter through sandy soil; although this may not be true, it does emphasize that the $L_3$s may move outward from the place where they have been deposited, and they can reach the surface from below ground.

Disposal of feces deposited on the ground takes place through a combination of weather and microbial and invertebrate animal activity. Invertebrates are the most important in the early dispersal of feces, and in warm climates; beetles seem to be the principal agents that break up and carry them away. In the southern United States, for example, dung beetles (family Scarabeidae) and burying beetles (family Silphidae) can remove the feces of a family of four to six persons during the warm part of the year. It is still not clear whether the beetles are detrimental or beneficial to the development and survival of hookworm free-living stages.

During the middle part of the nineteenth century anemia in coal miners became a problem, or at least came to the attention of the medical community then. Investigation by Italian parasitologists, in the main, showed that people working in mines suffered from hookworm anemia. Working conditions in mines in the nineteenth century left much to be desired, and sanitation was poor. People sometimes worked only partly clothed in cramped, hot, humid, dark mines. If people defecated in the mines, someone was likely to come along later and be exposed to the infective larvae. Thus, special conditions lead to problems with these worms where they would not ordinarily be expected.

Hookworm disease and the prevalence of infection are greatest in the less developed countries of the world, especially those with warm climates. Where there is little sanitary disposal of human waste, subsistence farming, and too little money for the purchase of shoes, hookworms will be prevalent.

A total campaign against hookworm disease involves the following:

1. Mass chemotherapy
2. Sanitary disposal of human waste
3. Wearing shoes
4. Administration of iron salts
5. Ensuring adequate dietary protein

We tend to think of sanitary disposal of human waste as a panacea for control of intestinal infections, but as Scott and Barlow (1938) showed, nearly complete disposal is needed before there will be significant reduction in transmission. This does not mean that the attempt should not be made.

Mass chemotherapy is most beneficial for those who have had sufficient exposure during their lives to have developed immunity. Children merely reacquire infection after treatment.

Oddly enough, help sometimes comes from unexpected sources. Hookworm infection in Egypt has declined because of the availability of inexpensive plastic shoes. The protection that these shoes provide seems to have reduced the prevalence of hookworm infection (Walton, 1981). Merely running an educational campaign on the importance of wearing shoes would probably have had little impact, but the availability of cheap, comfortable shoes has helped to alleviate the problem.

The campaign against hookworm disease in the United States entailed all of the above measures, and was partially successful. By the 1930s, prevalence had been reduced by about two-thirds in the southeastern states from what it had been in the early part of the century. The real gains came with general improvement in the economic level during the 1940s, and hookworm disease is no longer a widespread problem.

## CUTANEOUS LARVA MIGRANS

Human skin may be invaded by a number of species of nematode larvae, with the result that there are skin lesions at the site of entry. The larvae remain in the skin and cause inflammation, which may last for a few weeks. The pattern is typical of a number of helminth parasites that may invade a foreign host: the normal migration is altered.

The principal hookworms causing cutaneous larva migrans are *A. braziliense. Uncinaria stenocephala, and Bunostomum phlebotomum*; *A. caninum* and *A. ceylanicum* play only a minor role. In the rhabditoid nematodes, *Strongyloides papillosus* also causes skin lesions in humans.

In addition to the technical term *cutaneous larva migrans,* the disease is also referred to as *plumber's itch, ground itch,* and *creeping eruption.*

The common names for the disease are descriptive not only of the conditions under which a person may become infected but also of the appearance of the skin. The larvae penetrate the skin, sometimes remaining in place, but sometimes wandering in the skin, causing inflamed, sinuous tracks. Repeated infection may give rise to a more severe reaction at the site of entry indicating the development of hypersensitivity.

Usual treatment is to freeze the skin with a spray of ethylene chloride; the larvae are killed and then cleaned up by phagocytic cells. Thiabendazole ointment is even more effective.

## REFERENCES AND FURTHER READINGS


Ball, P. A. J. 1966. The relationship of host to parasite in human hookworm infection. In A. E. R. Taylor (ed.), *The Pathology of Parasitic Diseases.* Blackwell, Oxford, pp. 41–48.

Biocca, E. 1951a. On *Ancylostoma braziliense* (de Faria, 1910) and its morphological differentiation from *A. ceylanicum* (Looss, 1911). *J. Helminthol.* **25**:1–10.

———. 1951b. On *Ancylostoma braziliense* (de Faria, species from felines, closely related to *A. duodenale. J. Helminthol.* **25**:11–18.

———. 1954. Ridescrizione di *Ancylostoma tubaeforme* (Zeder, 1800) Parassita del gatto, considerato erzoneamente sinonimo di *Ancylostoma caninum* (Ercolani, 1859), parassita del cane. *Riv. Parasit.* **15**:267–278.

Burrows, R. B. 1962. Comparative morphology of *Ancylostoma tubaeforme* (Zeder, 1800) and *Ancylostoma caninum* (Ercolani, 1859). *J. Parasitol.* **48**:715–718.

Faust, E. S., P. F. Russell, and R. C. Jung. 1970. *Clinical Parasitology*, 8th ed. Lea & Febiger, Philadelphia.

Fernando, M. A., and H. A. Wong. 1964. Metabolism of hookworms. II. Glucose metabolism and glycogen synthesis in adult female *Ancylostoma caninum. Exp. Parasitol.* **15**:284–292.

Georgi, J. R. 1968. Estimation of erythrocyte loss in hookworm infection by whole-body $^{59}$Fe counting. *J. Parasitol.* **54**:417–425

Kalkofen, U. P. 1970. Attachment and feeding behaviour of *Ancylostoma canimum. z. Parisitenk.* **33**:339–354.

———. 1974. Intestinal trauma resulting from feeding activities of *Ancylostoma caninum. Am. J. Trop. Med. Hyg.* **23**:1046–1053.

Levine, N. D. 1980. *Nematode Parasites of Domestic Animals and Man* 2nd ed. Burgess, Minneapolis, pp. 75–99.

Miller, G. C. 1981. Helminths and the transmammary route of infection. *Parasitology* **82**:335–342.

Miller, T. A. 1966. Blood loss during hookworm infection determined by erythrocyte labelling with radioactive $^{51}$Cr. I. Infection of dogs with normal and x-irradiated *Ancylostoma caninum. J. Parasitol.* **52**:844–855.

———. 1971. Vaccination against canine hookworm diseases. *Adv. Parasitol.* **9**:153–183.

———. 1978. Industrial development and field use of the canine hookworm vaccine. *Adv. Parasitol.* **16**:333–342.

———. 1979. Hookworm infection in man. *Adv Parasitol.* **17**:315–384.

Nagahana, M., and Y. Yoshido. 1965. Complete development and migratory route of *Anclyostoma duodenale* in young dogs. *J. Parasitol.* **51**(Sect. 2):52.

Nawalinski, T. A., and G. A. Schad. 1974. Arrested development in *Ancylostoma duodenale*: Course of a self-induced infection in man. *Am. J. Trop. Med. Hyg.* **23**:895–898.

Olsen, O. W., and E. T. Lyons. 1965. Life cycle of *Uncinaria lucasi* Stiles, 1901 (Nematoda: Ancylostomatidae) of fur seals, *Callorhinus ursinus* Linn. on the Pribilof Islands. *J. Parasitol.* **51**:689–700.

Roche, M., and M. Layrisse. 1966. The nature and causes of hookworm anemia. *Am. J. Trop. Med. Hyg.* **15**:1031–1102.

Schad, G. A., A. G. Chowdhury, C. G. Dean, V. K. Kochar, T. A. Nawalinski, J. Thomas, and J. A. Tonascia. 1973. Arrested development in human hookworm infections: An adaptation to a seasonally unfavorable external environment. *Science* **180**:502–504.

Scott, J. A., and C. H. Barlow. 1938. Limitations to the control of helminth parasites in Egypt by means of treatment and sanitation. *Am. J. Hyg.* 27:619–648.

Stoll, N. R. 1947. This wormy world. *J. Parasitol.* **33**:1–18.

———. 1962. On endemic hookworm, where do we stand today? *Exp. Parasitol.* **12**:241–252.

Stone, W. M., and M. Girardeau. 1968. Transmammary passage of *Ancylostoma caninum* larvae in dogs. *J. Parasitol.* **54**:426–429.

Vetter, J. C. M. 1970. Skin penetration by hookworm larvae. *J. Parasitol.* **56** (Suppl) (4/2):479.

Walton, S. 1981, U.S.–Egypt Nile Project studies high dam's effects. *BioScience* **31**:9–13.

# 33

# *Roundworms of Horses*

Control of roundworms in horses presents some interesting problems because of peculiarities of pathogenesis in some species, different times of acquisition of other species, and interactions among populations of parasites.

The worms that are of greatest concern in horses are the so-called strongyles, members of the superfamily Strongyloidea, family Strongylidae; they usually are separated into two groups, the large strongyles and the small strongyles. The large strongyles are *Strongylus vulgaris*, *S. equinus*, and *S. edentatus*, which range in size from 15 to 45 mm in length. The small strongyles encompass 16 genera and about twice that many species; most of them fall within a size range of 10 to 20 mm in length. Other nematodes of importance in horses are *Ascaris equorum* (=*Parascaris equorum*) and the pinworm, *Oxyuris equi*. The larval forms of *Gasterophilus intestinalis* and *G. nasalis* (Insecta, Diptera, Gasterophilidae) (Chap. 55), usually called *bots*, are also intestinal parasites and are often treated at the same time as nematodes.

NAME OF ORGANISM AND DISEASE CAUSED. *Strongylus vulgaris* (Looss, 1900); *S. equinus* Mueller, 1780.; and *S. edentatus* (Looss, 1900) are large strongyles or palisade worms that cause helminthosis, strongylosis, and colic.

HOSTS AND LOCATION. Equids in general are the hosts. Although the emphasis here is on horses, these parasites are found in the ass, zebra, mule, and other members of the family as well. The adults are in the lumen of the cecum and colon. Preadults are in the tissues of the gut or in other locations such as the mesenteric arteries, liver, and pancreas, depending on the species (see Life Cycle).

DISTRIBUTION AND IMPORTANCE. The organisms are cosmopolitan. *Strongylus* spp. cause the greatest losses from helminths in horses; *S. vulgaris* is the most important of the three. The prevalence of infection with one or more of these worms approaches 100% in foals, but the pathogenesis seen in any one host is directly related to the intensity of infection.

The dollar losses from large and small strongyles usually quoted are those of the United States Department of Agriculture (USDA). The morbidity and mortality were estimated to be $5.4 million annually in the United States in 1965.

MORPHOLOGY. As members of the Strongylidae, the large strongyles have a large, almost globular buccal capsule and a modification of the lips into a comblike structure called a *leaf crown* or *corona radiata*. These species have both an external leaf crown at the anterior tip of the body and an internal leaf crown just posterior to the external and quite a bit smaller (Fig. 33.1).

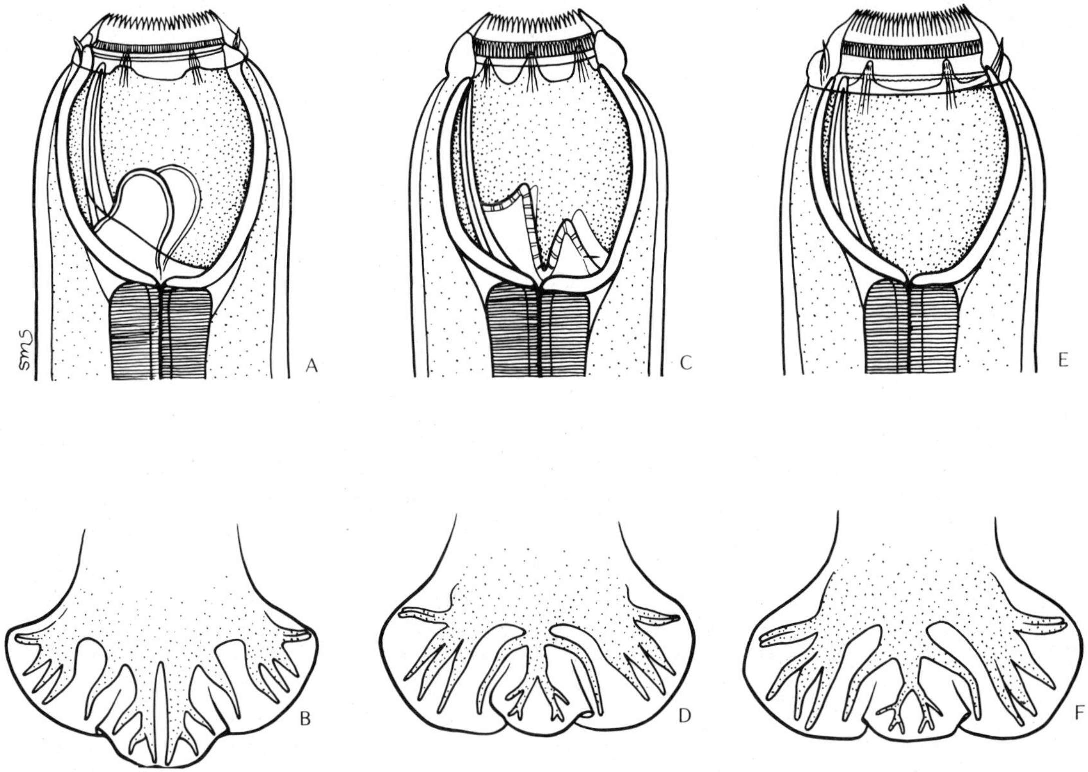

**FIG. 33.1**
Heads and bursae of the large strongyles of horses. (*A* & *B*) *Strongylus vulgaris*. (*C* & *D*) *S. equinus*. (*E* & *F*) *S. edentatus*.

*S. vulgaris*. Males are 0.75 to 0.95 by 11 to 16 mm. Females are 1 to 1.4 by 20 to 25 mm.

*S. equinus*. Males are 1.1 to 1.3 by 26 to 35 mm. Females are 1.8 to 2.3 by 38 to 55 mm.

*S. edentatus*. Males are 1.3 to 1.5 by 23 to 28 mm. Females are 1.6 to 2.2 by 33 to 44 mm.

The worms are usually differentiated from one another by the head structures, but there are differences in the male copulatory structures as well.

LIFE CYCLE. The eggs pass with the feces in about a 16-cell stage and hatch within a day or 2, releasing the $L_1$. The ensheathed $L_3$ can be reached in 2 to 3 days at 35°C, but usually requires a week or 2 at lower temperatures. The $L_3$ is ingested with herbage or water. After being ingested, the $L_3$ enters the wall of the intestine, but each of the three species of large strongyles has a different site of development in the host from this point on.

*S. vulgaris*. The larvae most likely migrate from the intestinal tissue to arterioles in the gut and then move upstream to the mesenteric artery; they attach to the artery wall, remain there for 6 to 7 months, and then move downstream to the colon. They enter the lumen and begin to lay eggs. The prepatent period is about 200 days.

*S. equinus*. The larvae migrate in the wall of the gut until they lie under the serosa or outer layer, where they grow and molt. After about 11 days they leave the modules and migrate to the liver or pancreas or remain in the peritoneal cavity. They return to the gut and mature in the colon and cecum about 118 days after infection.

*S. edentatus*. The larvae pass clear through the gut and come to lie under the peritoneal lining of the abdominal cavity, where they remain for about 3 months. They then move back to the wall of the large intestine, where they form hemorrhagic nodules that ultimately break, releasing the near-adults into the lumen. The prepatent period ranges from 280 to 355 days.

DIAGNOSIS. The following should be taken into account when determining whether an animal or group of animals should be treated for strongyle infections:

1. General physical condition
2. Clinical signs
3. Clinical history of the animal(s)
4. Fecal egg counts

As indicated in the section on epidemiology and control, a program of routine, periodic treatment is usually recommended, but there have been occasions when this has not been done or when the program has broken down for some reason.

A person trained in diagnosis and treatment of diseases or one who works with horses closely will recognize the usual signs of helminthosis: rough hair coat, loss of condition, failure to eat properly, pale mucous membranes indicating anemia, diarrhea, and indications that the animal is in pain.

If one sees some of these signs, and if the fecal egg count is high, treatment is justified. The dilemma is, what is a high egg count? Also, the preadults, which cause much of the problem, do not produce eggs. Some investigators recommend treating horses when egg counts reach 500/g; others recommend it at 1000/g, but egg counts may exceed 2000/g in horses that are clinically normal.

The eggs of the three *Strongylus* spp. and of the small strongyles cannot be distinguished from one another. This further clouds the interpretation of fecal egg counts. However, the eggs of *Ascaris* and *Oxyuris equi* are distinct, as are those of the tapeworms sometimes found in equids.

If it is important to determine which strongyles are present in a host, one can allow the eggs and larvae to develop in the feces and then identify the $L_3$s (see Levine, 1980).

In deciding whether to treat or to change an established program, the history of the animal(s) is important. This will help to determine the probability that strongyles are causing the problem.

To reiterate, only a part of diagnosis is aimed at the health of the individual; the other part is herd health and prevention of infection in young animals, which must be taken into account in determining the severity of infection and the benefits that may accrue.

TREATMENT. The benzimadazoles such as thiabendazole, mebendazole, and cambendazole have been highly successful in recent years, but worms tend to become resistant to them upon continued administration. The organophosphates trichlorfon and dichlorvos are also highly effective.

All of these anthelmintics are given by mouth; this creates problems because it has generally been necessary to pass a stomach tube to ensure that the correct dose reaches the proper place. Except for those who do it routinely and have proper equipment for restraint of the animal, passing a stomach tube in a 500-kg animal, who does not like it, can be hard work. Attempts have therefore been made to develop other means of getting the drug into the intestine of the horse. The drug has been put in the feed for some time; this must be done in such a way that each individual obtains the proper dosage. A paste preparation may be placed on the tongue of a restrained animal. The drug is automatically swallowed when the tongue is released.

Only a single drug is currently being used as an injectable preparation. Ivermectin (Merck) is marketed as Eqvalan for use in horses against both large and small strongyles. It is effective against a broad spectrum of nematodes including the adults of large strongyles, both adults and larvae of small strongyles, *Parascaris equorum, Habronema, Draschia, Oxyuris, Setaria, Trichostrongylus axei*, and microfilariae of *Onchocerca*. Second- and third-stage larvae of *Gasterophilus* are also removed.

Ivermectin is the generic name of one of a group of compounds called *avermectins*, which are fermentation products produced by *Streptomyces avermitilis*. Ivermectin, a macrocyclic lactone, is referred to as 22,23-dihydroavermectin B-1 and is one of 8 closely related chemicals. Its activity against parasites is on the nervous system, in which it stimulates the release of GABA (gamma-aminobutyric acid), a neurotransmitter. The broad spectrum of activity of ivermectin extends to many nematodes and arthropod parasites of livestock and companion animals including lungworms, trichostrongyles, filiarioids, and scab mites.

HOST–PARASITE INTERACTIONS. Although we deal with the large strongyles as a group, they affect the host in different ways and to differing degrees. As adults, all three species attach to the wall of the large intestine, where they ingest intestinal mucosa and suck blood; the host may become anemic or the lesions where intestinal mucosa has been removed may become infected. The damage to the host from the adults is bad enough, but the damage by the preadult stages is greater and more difficult to deal with.

*S. vulgaris* is the most pathogenic of the three species. As the worms migrate and grow in the arteries leading to the gut, including the large mesenteric artery, they cause blood clots to form in the artery. Normally, the slick internal surface of the artery does not stimulate the clotting mechanism; however, if that surface is changed somehow by the worm, a clot forms and may continue to increase in size almost indefinitely. This clot attached to the surface of the blood vessel is called a *thrombus* (pl. *thrombi*). It impedes the normal flow of blood, causing either blockage or ballooning of the artery, or both. The resulting pathological condition is called an *aneurysm*. The larger the artery or the greater number of aneurysms formed, the more tissue has inadequate blood supply. There may be tissue necrosis for lack of adequate blood supply or the gut may telescope on itself (intussusception). The animal feels pain, fails to eat properly, and may roll on the ground, which further aggravates the condition by causing the gut to twist upon itself.

As is true with helminths in general, the greater the number of larvae ingested by an animal, the greater the damage. In experimental infections with from 2500 to 5000 larvae administered in a single dose or doses of 250 larvae given every 4 or 5 days to a total of 6250, the effects on foals were severe. Animals lost weight, had fever, diarrhea, and colic, and moved little. At the postmortem examination there were severe lesions in the intestinal tract, and the blood vessels had damage from thrombi and embolisms (clots carried by the bloodstream) (Drudge et al., 1966).

*S. edentatus* ranks behind *S. vulgaris* as a cause of damage. The nodules produced by its encysted larvae may cause bleeding and interfere with the blood supply to the intestine, depending on their location. The animal may have considerable blood loss into the abdominal cavity, colic, and persistent diarrhea because of interference with normal intestinal function.

*S. equinus* causes small but not negligible damage by encysting under the peritoneal lining. Its major damage is to the liver and pancreas, in which the $L_4$s develop.

Immunity to *Strongylus* spp. probably develops on repeated infection; however, older horses may sometimes be severely affected and even die.

EPIDEMIOLOGY AND CONTROL. A number of problems must be overcome in developing a control program for large strongyles:

1. A relatively small number of infective larvae can have severe effects on foals.
2. The free-living stages quickly reach $L_3$ during the growing season.
3. The free-living stages are quite hardy, living for months and even overwintering in areas where there is only occasional frost.
4. Egg production by adult worms is strikingly high, and an adult horse with only a low-grade infection may contaminate the pasture with more than 13 million eggs in a day. There are seasonal differences in the number of eggs in the feces of animals; it is highest in summer.
5. None of the anthelmintics now recommended (with the possible exception of ivermectin) has a significant effect against preadult parasitic stages.

The approach of most control programs falls into the category of preventive public health. The following program has been modified somewhat from Drudge and Lyons (1978):

1. Treat all horses on the premises.
2. Any horses introduced to a farm should be quarantined and then treated before being allowed in with the residents.
3. Treat mares regularly. In Kentucky, treat mares every 8 weeks; some recommend treating as often as once a month in other areas.
4. Do not treat foals. Treating young animals has no effect on their acquiring infective larvae or the preadult stages, which are most damaging. Treatment of the mares protects the foals.

5. Alternate drugs so that problems with resistance are reduced. Follow the directions on dosage and other procedures so that full efficacy and safety will be achieved.
6. Monitor the efficiency of the program by periodic fecal egg counts.

What about those persons who have only one or two horses? Most of these animals are mature and relatively resistant to infection. However, as noted, animals of any age can be severely affected by strongyles. Those who have only one or a few horses usually do not have the skill or facilities for treating animals or for making fecal egg counts. Their animals may be boarded at a stable with those of a number of other owners, and there may be no control on animals entering the premises. It is therefore important for the owner to determine what treatment, if any, should be given to animals entering a stable for boarding. Also, one should retain the services of a veterinarian who is skilled in diagnosis and treatment not only of helminthoses but of other diseases of horses as well.

NAME OF ORGANISM AND DISEASE CAUSED. Small strongyles include 16 genera and more than twice that number of species in equids. Some of these are *Cylicocercus, Cylicocyclus, Poteriostomum, and Cyathostomum* (=*Trichonema* of earlier times is still retained by some authors). (See Levine, 1980, Soulsby, 1965 or Georgi, 1980, for more complete information on small strongyles.) Strongylosis and helminthosis are the diseases caused.

LOCATION IN THE HOST. The organisms are located in cecum and colon, in gut tissue as larvae and in the lumen as adults. Horses and other equids are the hosts.

DISTRIBUTION AND IMPORTANCE. Distribution is cosmopolitan, though species have been reported only from some localities. These are of lesser importance than the large strongyles.

MORPHOLOGY. Most species are 10 and 20 mm long, but some males may be as short as 5 mm and have stout bodies. Small strongyles are members of the superfamily Strongyloidea but have relatively small buccal capsules. The variations in leaf crowns and male genitalia are the principal means of distinguishing among the genera and species.

LIFE CYCLE. The life cycle is direct see (*Strongylus* spp). The infective larvae do not leave the gut, but enter the tissues and develop; they return later to the lumen of the large intestine. The prepatent period may be as short as 6 weeks in some species or as long as 50 weeks in others. A large population of larvae may be retained perhaps for many years in the tissues of the large intestine, with only a proportion of the larvae reaching sexual maturity at any one time (Gibson, 1953).

DIAGNOSIS. See *Strongylus* spp.

TREATMENT. See *Strongylus* spp.

HOST–PARASITE INTERACTIONS. Damage to the host by the small strongyles is relatively small but is not to be ignored. In some species, the nodules in the tissue of the colon may give rise to colic and interfere with normal intestinal function.

The adult worms such as *Cyathostomum* spp. are said to suck blood, but other genera and species seem to graze on intestinal epithelium or to ingest gut contents.

Gibson's (1953) study of arrested development in small strongyles of horses was important in itself, but also led to studies on other nematodes in which the prepatent period may be extraordinarily long. His work showed that when animals are treated with phenothiazine, adult worms are shed with the feces and fecal egg counts drop to zero. Then, at 4 to 6 weeks after treatment, egg counts increase. If animals are treated repeatedly, the same pattern is seen each time, but fewer worms are shed and the egg counts are not quite as high as the previous time. Since these animals were held under conditions where reinfection was highly unlikely, it is clear that when the adult population was removed, some of the larvae continued development and became adults.

One interpretation of these observations is that immunity keeps the larvae from maturing; when the adults are removed, immunity wanes and some larvae can then mature. In our opinion, it is more logical that there is some sort of chemical communication between adults and larvae that inhibits the development of a portion of the population. The result is a feedback system that retains the population of adults at a moderate level, thereby keeping the harm to the host at a reasonable level and ensuring that there are some worms around to complete the life cycle later. Another interpretation is that there is constant turnover in the population, with some adults being lost continually and replaced by larvae. According to this view, the larvae would have matured anyway.

EPIDEMIOLOGY AND CONTROL. The same principles of development and survival of the free-living stages that were discussed in connection with *Strongylus* spp. pertain to the small strongyles.

Control programs for large strongyles take care of small strongyles at the same time.

## REFERENCES AND FURTHER READINGS

Drudge, J. H. 1972. Metazoal diseases, endoparasitisms. In E. J. Catcott and J. F. Smithcors (eds.), *Equine Medicine and Surgery*, 2nd ed. American Veterinary Publ., Santa Barbara, Calif., pp. 157–179.

Drudge, J. H., and E. T. Lyons. 1978. Equine parasites—Problems and control. *Practicing Vet.* **49**(3):5–9.

Drudge, J. H., E. T. Lyons, and J. Szanto. 1966. Pathogenesis of migrating stages of helminths with special reference to *Strongylus vulgaris*. In E. J. L. Soulsby (ed.), *Biology of Parasites*. Academic Press, New York, pp. 199–214.

Georgi, J. R. 1980. *Parasitology for Veterinarians*, 3rd ed. W. B. Saunders, Philadelphia.

Gibson, T. E. 1953. The effect of repeated anthelmintic treatment with phenothiazine on the faecal egg counts of housed horses, with some observations on the life cycle of *Trichonema* spp. in the horse. *J. Helminthol.* 27:29–40.

Levine, N. D. 1980. *Nematode Parasites of Domestic Animals and Man*, 2nd ed. Burgess, Minneapolis, pp. 100–114.

Soulsby, E. J. L. 1965. *Textbook of Veterinary Clinical Parasitology*. Vol. 2. *Helminths*. E. A. Davis, Philadelphia.

34

# *Kidney Worm of Swine, Nodular Worms, and Other Strongyloidea*

## NODULAR WORMS OF LARGE FARM ANIMALS

NAME OF ORGANISM AND DISEASE CAUSED. *Oesophagostomum* spp. See Table 34.1 for a list of species in sheep, goats, cattle, and swine. Nodular worm, pimply guts, knotty guts, and oesophagostomiasis are caused by the organisms.

HOSTS AND LOCATION. Most of the species have been found in only a few species of hosts that are closely related, but in a few instances they are found in families of closely related hosts (Table 34.1). The large intestine and sometimes the small intestine are the sites of infection.

DISTRIBUTION AND IMPORTANCE. The organisms are cosmopolitan. Some species have been found only in restricted localities, but the species of economic importance seem to be found worldwide. Although these worms are likely to be found almost anywhere, they cause problems in areas that are relatively warm and wet. In the United States, for example, they are a major problem in the area east of the Mississippi River and where the winters are not severe. The worms are a continuing problem in Britain even as far north as Scotland.

Economic losses are due to morbidity, mortality, and inability to use intestines for sausage casing or for "catgut" (*Oe. columbianum*). Annual economic losses from *Oesophagostomum* infection for the United States as of 1965 were estimated as follows: for sheep and goats the loss was $1.7 million; for swine it was $7 million.

MORPHOLOGY. These worms are members of the superfamily Strongyloidea, subfamily Oesophagostominae, and as such have leaf crowns; but unlike other members of the family, they have a relatively small buccal capsule (Fig. 34.1). The vulva of the female is near the posterior end and the males have a large bursa with slender spicules.

LIFE CYCLE. The life cycle is direct. Embryonated eggs hatch to release the $L_1$s; the $L_3$ is the infective stage, which is ingested with the forage. The $L_3$s enter the tissues of the lower part of the small intestine, or more frequently the large intestine, where they encyst. Most species return to the lumen of the gut during the $L_4$, and then molt to become the $L_5$ and adult. In initial infections, the $L_4$s return to the lumen of the gut in about 1 week and the patent period begins at about 3 weeks.

In previously infected animals, the prepatent period is extended and the larvae may remain in the tissues for months.

**TABLE 34.1**
*Oesophagostomum* species in Farm Animals

| SPECIES | NODULE FORMATION | HOSTS |
|---|---|---|
| *Oe. columbianum* | + | Sheep, goat, alpaca, antelope |
| *Oe. venulosum* | 0 | Sheep, bighorn sheep, goat, chamois, argali, camel, alpaca, vicuna |
| *Oe. multifoliatum* | ? | Sheep, goat |
| *Oe. asperum* | ? | Goat |
| *Oe. radiatum* | + | Ox, zebu, water buffalo |
| *Oe. dentatum* | + | Pig, peccary |
| *Oe. brevicaudum* | ? | Pig |
| *Oe. georgianum* | ? | Pig |
| *Oe. quadrispinulatum* | + | Pig |
| *Oe. maplestonei* | ? | Pig |
| *Oe. rousseloti* | ? | Pig |
| *Oe. granatensis* | ? | Pig |
| *Oe. hsungi* | ? | Pig |

+ = Nodules formed; 0 = Nodules not formed; ? = No data on nodules.

DIAGNOSIS. Oesophagostomiasis is generally a herd problem, and determining whether it is the causative agent is based on the following:

1. Herd history
2. Fecal egg counts
3. Clinical signs
4. Postmortem examination

The history of the herd and the signs of disease are helpful. Animals somewhat crowded on moist or wet pastures are likely to pick up $L_3$s of various strongylid nematodes. Diarrhea, loss of condition, and anemia are common signs and symptoms of gastrointestinal helminthoses.

In some species (Table 34.1) nodules form in the wall of the gut, and these can be felt just inside the rectum of an animal. Not all species cause this reaction, however, and it occurs in relatively long-standing infections.

Fecal egg counts give a hint as to the cause of the problem but are not a reliable diagnostic method. There are two caveats:

1. Eggs of *Oesophagostomum* cannot readily be distinguished from those of other strongyles
2. Much of the damage may arise before egg production

The most reliable way of determining the cause of the disease is to perform a postmortem examination. If an investigation is made at a farm or ranch, recently dead or moribund animals may be examined. Sometimes the only alternative is to perform a complete postmortem examination on a few of the sickest animals. If such examinations are done, it must be kept in mind that organisms other than helminths can cause diarrhea and loss of condition, and also that an examination for helminths should be complete and quantitative.

TREATMENT. Treatment for nodular worm is only partly successful. The better drugs such as the organophosphates or benzimidazoles (e.g., thiabendazole) remove only about one third of the larvae but are nearly 100% effective against the adults. Much of the damage to the host is caused by larvae, and more efficient drugs are needed.

HOST–PARASITE INTERACTIONS. There are some generalizations that can be made about the nodular worms, but their damage to the host may vary with the species in question. Much of the damage, whether nodules are formed or not, is caused when the larvae are in the tissues of the gut. There is a foreign-body tissue reaction, and the greater the number of larvae, the greater the gut malfunction.

In instances such as *Oe. columbianum* of sheep, long-term damage arises from nodule formation. In an animal previously unexposed to *Oesophagostomum*, the worms mature and begin to produce eggs about 3 weeks after infection. As the infection continues and as additional $L_3$s are acquired, the larvae are retained in the tissue and a strong response occurs. A granuloma (white cells, fluid, connective tissue) forms around each larva, and this pocket may be infected with bacteria. The nodule may reach 5 to 10 mm in diameter and extends to the muscular layers of the gut; it becomes hard and calcified (Dhar & Singh, 1968). The gut loses its ability to function normally in the vicinity of each nodule. Resorption of water and inorganic balance are disturbed; the obvious sign is diarrhea. Nodule formation is a sign of immunity. However, as is emphasized in a number of instances in this book, the immune

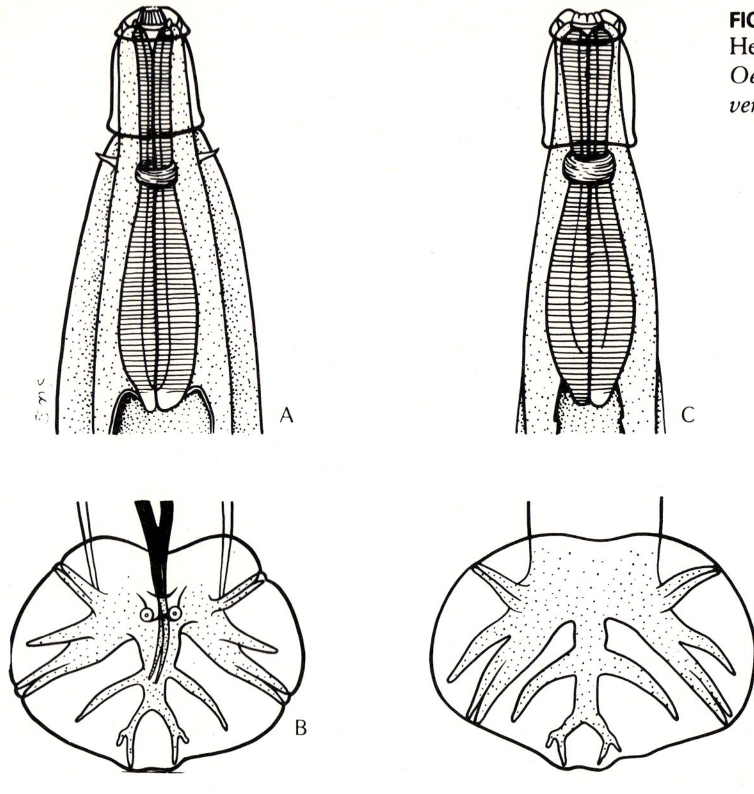


**FIG. 34.1**
Heads and bursae of nodular worms. (*A* & *B*) *Oesophagostomum columbianum;* (*C* & *D*) *Oe. venulosum.*

response is not always beneficial, or the results may be mixed.

The impact of some diseases extends far beyond production losses to owners. In this case, the impact is on the sausage eaters and tennis players of the world. Sturdy sausage casings that have a distinct snap when you bite them are made of intestines, and when the large intestines are full of gritty nodules, they are not esthetically pleasing for casing. In fact, these intestines are condemned for human consumption. Likewise, the so-called catgut used to string tennis rackets is made from the large intestines of sheep, and again, the nodules ruin the intestines for this use.

EPIDEMIOLOGY AND CONTROL. Oesophagostomiasis is a wet or moist climate disease and usually occurs in areas where winters are relatively mild, with infrequent frost or failure of the ground to freeze hard for long periods. Patent infections can be retained for more than 100 days, so animals that go into the winter with an infection can seed pastures with eggs the following spring.

Much of the discussion on control of gastrointestinal nematodes contained in Chapter 35 pertains to nodular worms as well. Prevention is always better than treating a sick animal, and this principle is particularly true with nodular worms, because the larvae cause the greatest damage and there is no good chemotherapy for them.

Young animals are especially susceptible to infection, and the best means of preventing infection is to treat the adults as a herd health measure. As a specific example, let us look at a herd of sheep that is at risk of infection with *Oe. columbianum.* The objective is to prevent seeding pastures, so that ewes are treated before they go out on pasture in the spring. Some worms may mature during the pasture season; they are treated at least once during that time. Monitoring the herd by fecal egg count gives some information on the need to treat. Then, when animals come off of pasture in fall, the whole herd—ewes, lambs, and rams—is treated so that the animals can go into the winter with as little as possible draining them during pregnancy or feeding to market weight.

## KIDNEY WORM OF SWINE

NAME AND DISEASE CAUSED. *Stephanurus dentatus* Diesing, 1839, causes kidney worm disease.

HOSTS AND LOCATION. Swine are the hosts. The organism has been reported occasionally from other hosts such as the ox and ass. The principal location of the adults is in the perirenal fat and around the ureters, but it may be found in nearly any organ system of the body, especially while the preadult stages are migrating.

DISTRIBUTION AND IMPORTANCE. The kidney worm is cosmopolitan in warm, wet climates throughout the world. Losses arise from production losses, mortality, and condemnations of carcasses at slaughter. In 1965 the USDA estimated that losses from kidney worm exceeded $20 million annually in the United States. Of this, about $3.9 million was for condemnations of livers, kidneys, and other parts of the carcass at slaughter. Death losses were less than $1 million. More than 75% of hog livers were condemned in Georgia in the 1970s. The widespread nature of the infection in tropical and subtropical areas suggests that the global losses in swine production are a significant factor in the poor nutrition of people in the less developed countries. In the moist, warm climates of the world, swine are kept almost everywhere and scavenge all kinds of waste materials and plants not otherwise harvestable.

MORPHOLOGY. Males are 1.2 by 20 to 30 mm long and females are 1.8 by 30 to 45 mm long; they have rather irregular mottling of light and dark because the color of the internal organs shows through the cuticle. The broad, cup-shaped buccal capsule has a single leaf crown at the anterior tip and six teeth in the base of the capsule (Fig. 34.2).

LIFE CYCLE. This is a typical strongyloid free-living phase, with the $L_3$ being reached about 3 days after the eggs pass from the host with the urine. Unlike most other strongyloids, *S. dentatus* has three options for reaching the definitive host:

1. It can be eaten with contaminated food or water.
2. It can penetrate the unbroken skin.
3. It can be eaten by earthworms (e.g., *Eisenia foetida*) and use them as transport hosts, which are later eaten by pigs.

If the $L_3$ is eaten directly or with an earthworm, the larva passes from the gut to the mesenteric lymph nodes, where it remains for 10 to 32 days (Lichtenfels & Tromba, 1972), becoming a well-developed $L_4$ before it migrates to the liver via the circulatory system.

If the larva penetrates the skin, it reaches the liver through the circulatory system in 8 to 40 days. The larvae migrate in the liver for a least 2 months, and they have been reported as remaining there for as long as 9 months.

The larva leaves the liver by passing through its capsule into the peritoneal cavity. Although there is some random migration, the larva that completes the life cycle reaches the region of the kidney, where it matures. The female penetrates the wall of the ureter with the posterior part of her body, and the eggs pass out with the urine. Egg laying may begin 4 months after infection, but it usually takes more than 6 months.

DIAGNOSIS. During the patent period, eggs are in the urine. Before egg production begins, one must rely on the history of the animals and the knowledge that kidney worm occurs in the vicinity. If dead or dying animals are available, a postmortem examination may be done; worms should be looked for in the liver, peritoneal cavity, larger blood vessels, and the perirenal fat.

TREATMENT. None is satisfactory, but benzimidazoles have been used with some success.

HOST–PARASITE INTERACTIONS. Damage to the host revolves around three events in the life cycle:

1. Skin penetration
2. Larval migration
3. Egg laying

Skin damage by the penetrating larvae is relatively minor, but there may be swellings at the sites.

The long period of larval migration causes the major damage. The larvae remain during much of the

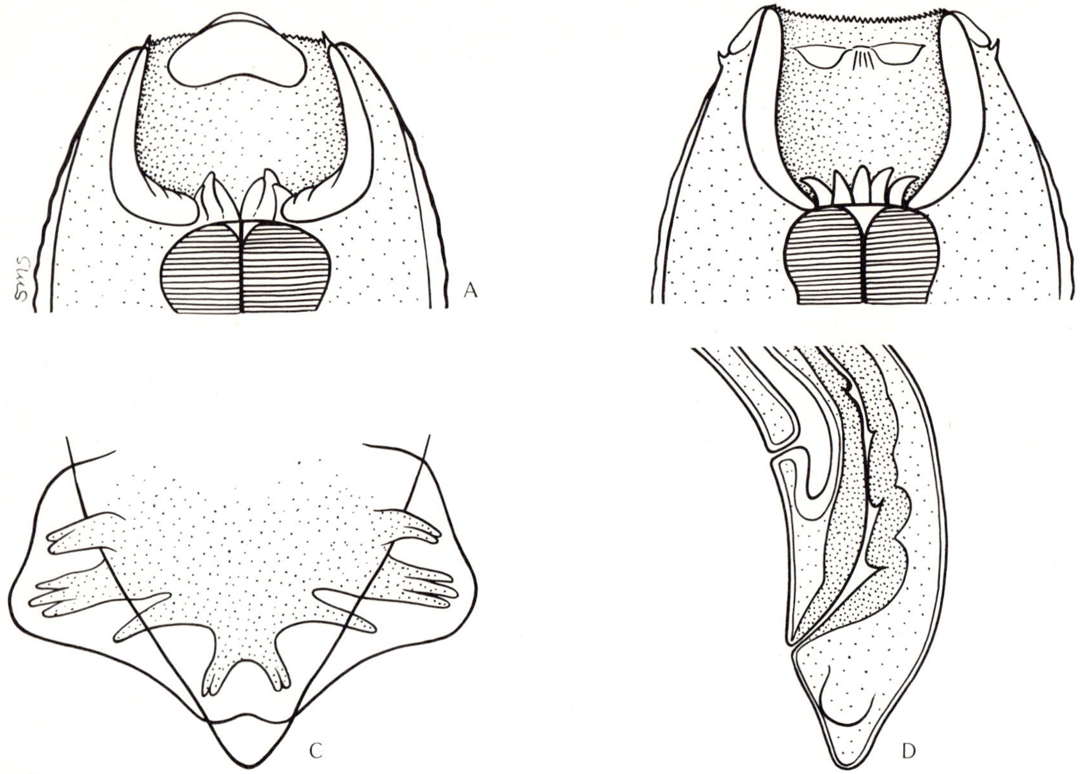


**FIG. 34.2**
*Stephanurus dentatus*, the swine kidney worm. (*A*) Dorsal view of head; (*B*) lateral view of head; (*C*) bursa; (*D*) tail of female.

prepatent period in the liver and literally chew their way through it with the well-developed buccal structures of the $L_4$. Experimental inocula of 10,000 $L_3$s caused so much liver damage that little functional tissue remained (Batte et al., 1960). Extensive liver damage is one of the most life-threatening events that can happen to an animal or a person. The liver has a number of essential functions, such as glycogen storage and detoxification of chemicals, and in severe kidney worm infections, not enough functional tissue remains for the animals to live. The larvae may migrate into the blood vessels, causing thrombi and embolisms; they are also found occasionally in the central nervous system.

When the larvae complete their migration and mature in the vicinity of the kidney, they cause lesions there. Usually a male and female are found near a ureter in a pus-filled nodule. There is damage to the ureter and sometimes the kidney during egg laying since the tissue is penetrated. The lumen of the ureter may be constricted.

Animals that do not succumb to kidney worm infection gain poorly; even under moderate experimental infections of 200 to 1000 $L_3$s, infected pigs gained about ½ lb less per day than uninfected pigs. A 65-lb difference between an infected and an uninfected animal is about $42 (U.S.) at present prices.

EPIDEMIOLOGY AND CONTROL. The limited areas where kidney worm is found tell a good deal about the hardiness of the free-living stages. The climate must be warm and moist; at 11°C and below the larvae are harmed, at 20°C only slight development occurs, and optimal development occurs at 25 to 27°C. Earthworms

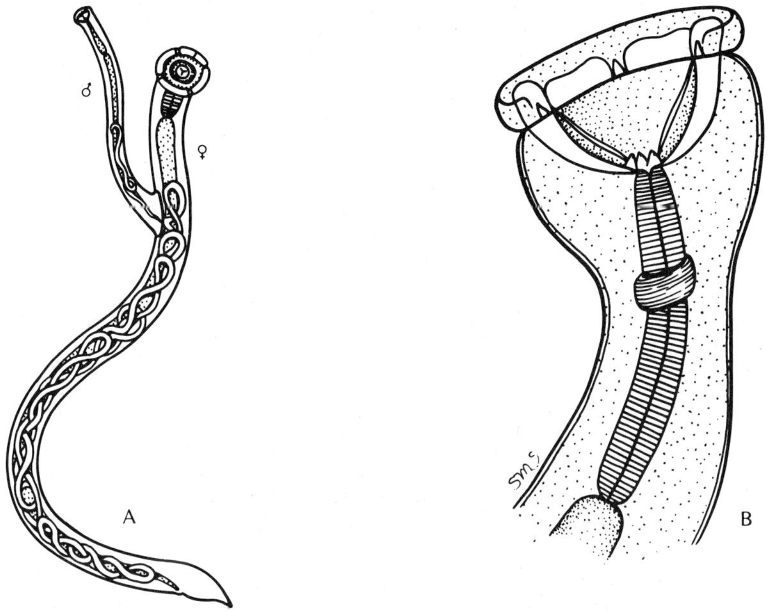


**FIG. 34.3**
Male and female of *Syngamus trachea*. (*A*) Male and female in copula; (*B*) buccal capsule.

are transport hosts, but they also protect the larvae from drying, thereby increasing the length of time that they survive.

Three major problems must be addressed in designing control programs for kidney worm:

1. The prepatent period is about 6 months.
2. The adults may live for 2 years and shed eggs into the environment during that time
3. There is no satisfactory treatment.

On the other hand, some characteristics of the parasite and the host are helpful in designing a program:

1. Narrow host range
2. Short survival time of the free-living stages (not in the transport host)
3. Tendency of domestic pigs, being creatures of habit, to run around the fence line in a pasture

The McLean County system for swine raising discussed in Chapter 38 on *Ascaris* spp., involves farrowing in clean quarters from a clean sow and keeping the animals on clean ground. The problem in applying the McLean County system to kidney worm is obtaining a clean sow; egg production may continue for at least 2 years and there has been no satisfactory chemotherapy. In the southeastern United States, a system was developed in which young, unbred sows (gilts) are purchased from areas where kidney worm does not exist. Six months after the litter is born, the sows and young pigs are shipped out and another group of clean gilts is brought in. Thus, all of the animals are removed from the farm before egg production commences. After three or four farrowings, the residual free-living stages should have died out, the farm will then be clean and conventional hog raising can be started again (Stewart & Tromba, 1957). Although this system involves extra costs, it works.

In a tropical area, it is unlikely that such a system could easily be established. In the United States, the enzootic areas are the southeastern states and Hawaii. In the Southeast, gilts can be purchased from clean areas and shipped in. In the tropics or less developed countries, areas where clean stock can be purchased would not be readily accessible. If such a program were to be instituted, it would need to be done by importing clean swine to a central facility where they could be raised and then provided to farmers as needed. Such a program assumes that feral swine are not present or that owners could fence their own animals and not allow them contact with any others.

## STRONGYLOIDS OF THE PHARYNX AND TRACHEA

The genera *Syngamus, Cyathostoma,* and *Mammomonogamus* contain about 35 species of worms that are found in the pharyngeal region, nasal passages, middle ear, or trachea of birds and mammals. Some of them have been reported from humans, but these seem to be only sporadic infections.

The best known of the group is *Syngamus trachea*, a parasite of the trachea of galliform and other birds; it seems to have a low degree of host specificity. The cycle is both direct and indirect, with a wide array of invertebrates including earthworms, snails, and various insects serving as transport hosts. When large numbers of worms are located in the trachea, passage of air is impeded and the birds gasp for air—thus, the common name of *gapeworm* came about. *S. trachea* is a problem if chickens or other birds are raised as farm flocks; with the development of battery raising of chickens in the United States and other high-technology countries, the prevalence of gapeworm declined. The infection is easily recognized by finding the worms in the trachea at postmortem examination (Fig. 34.3). The male and female remain permanently in copula. Thiabendazole administered in the feed seems to be a good treatment for *S. trachea*.

## REFERENCES AND FURTHER READINGS


Alicata, J. E. 1953. Observations on the lethal action of polyborate on swine kidney worm (*Stephanurus dentatus*) larvae in soil. *Am. J. Vet. Res.* **14**:563–570.

Batte, E. G., R. Harkema, and J. C. Osborne. 1960. Observations on the life cycle and pathogenicity of the swine kidney worm (*Stephanurus dentatus*). *J. Am. Vet. Med. Assoc.* **136**:622–625.

Dhar, D. N., and K. S. Singh. 1968. Mechanism of nodule formation in the nematode *Oesophagostomum columbianum* (Curtice, 1890), Stossich, 1899. *Indian J. Vet. Sci.* **38**:708–725.

Douvres, F. W., F. G. Tromba, and D. J. Doran. 1966. The influence of NCTC 109, serum and swine kidney cell cultures on the morphogenesis of *Stephanurus dentatus* to fourth stage, in vitro. *J. Parasitol.* **52**:875–889.

Lichtenfels, J. R., and F. G. Tromba. 1972. The morphogenesis of *Stephanurus dentatus* (Nematoda:Strongylina) in swine with observations on larval migration. *J. Parasitol.* **58**:757–766.

Ransom, B. H. 1911. The nematodes parasitic in the alimentary tract of cattle, sheep and other ruminants. U.S. Dept. Agric. Bur. Anim. Industry Circ. 116.

Stewart, T. B., and F. G. Tromba. 1957. The control of swine kidney worm, *Stephanurus dentatus*, through management. *J. Parasitol.* **43**(Suppl.):19–20.

Tromba, F. G. 1955. The role of the earthworm, *Eisenia foetida*, in the transmission of *Stephanurus dentatus*. *J. Parasitol.* **41**:157–161.

Waddell, A. H. 1969. The parasitic life cycle of the swine kidney worm *Stephanurus dentatus* Diesing. *Austral. J. Zool.* **17**:607–618.

Yorke, W., and P. A. Maplestone. 1962. *The Nematode Parasites of Vertebrates*. Hafner, New York.

# *Trichostrongyles of Cattle, Sheep, and Goats*

The strongyles discussed here as a group are among the most important helminth parasites of grazing animals. Although there is some variation in the species found in a particular kind of habitat, there is usually a mixture of five or six species in any one locality. It is important to recognize at the outset that any one grazing animal may have a few hundred or a few thousand worms, but these worms often cause no discernible damage to the host. Only under certain conditions are production and death losses observed.

The trichostrongyles are members of the superfamily Trichostrongyloidea and as such have a small buccal capsule and a large copulatory bursa with stout, heavily sclerotized spicules. Their life cycles are direct; $L_3$s are ingested with the forage.

If problems with trichostrongyles arise in grazing animals, there will also usually be problems with lungworms (Chap. 36), nodular worms (Chap. 34), and hookworms (Chap. 32). Overcrowding on moist pastures favors the free-living stages of all of these worms and increases the probability of large numbers of $L_3$s being ingested.

NAME OF ORGANISM AND DISEASE CAUSED. Five principal genera are found in domestic ruminants (Table 35.1); eleven additional genera are of minor importance or are found only in localized areas. The diseases they cause are referred to generally as *helminthosis* or *helminthiasis*, but when one worm predominates, the generic name is often used; thus *Haemonchus* spp. cause haemonchosis and *Trichostrongylus* spp. cause trichostrongylosis. Often a sign of heavy infection is diarrhea, and terms such as *black scours* may be applied to infections with *Trichostrongylus* or *Nematodirus*.

HOSTS AND LOCATION. The hosts listed in Table 35.1 are the principal domesticated ruminants, and it should be noted that most of the species infect all three hosts. There is a moderate degree of host specificity in most species since most will infect many ruminants and sometimes even nonruminants such as horses or humans.

The larval stages of these worms invade the tissues of the intestine, where they lie in the mucosa, but sometimes they invade the submucosal tissues. Adults return to the lumen of the gut but lie closely applied to the epithelium in the paramucosal lumen.

DISTRIBUTION AND IMPORTANCE. Most species are cosmopolitan, but a few, such as *Nematodirus battus*, which is found in Britain, are limited in their distribution.

Although death losses occur from this complex of worms, the main losses are in livestock production, that is, failure to gain properly. The USDA estimated annual losses for the United States in 1965 at $113 million—$92 million in cattle and $21 million in sheep. About one third of the loss is due to deaths and two thirds from production losses (i.e., meat, milk, and fiber).

Two issues remain largely unsolved: whether the annual losses in the United States are accurate and what the global losses are. The USDA figures appear to be large, but the compilations were made conservatively using various kinds of data. Nevertheless, we seldom have reliable local data on production losses that can be used to obtain countrywide estimates. The United States has an extensive monitoring system for agricultural production and quite accurate information for both crop and livestock production, but we really have a difficult time determining what production losses might be. In each state, the state veterinarian's office collects data from practicing veterinarians on the animals that they have treated for various ailments during any month. These figures are difficult to translate into dollars, but there are even further difficulties. Many owners treat their own animals and are not seen by a veterinarian. Low-level production losses are difficult to discern. For example there might be a 1-kg (2-lb) production loss per head in a flock of 1000 lambs when the lambs are weaned. The owner fails to realize perhaps $1200 in weight that the animals never gained.

Another approach to determining losses is to look at "worm production." Just as grain or meat production can be calculated, the amount of nutrient that is diverted into production of worms and worm eggs can also be determined. Boughton (1957) calculated that annual worm production in livestock in the United States amounts to 7000 tons of nematodes and 1000 tons of eggs. Further calculation showed that these worms and eggs had a value of $144 million (1957 dollars). Is this calculated loss merely a trivial exercise? Apparently not. Other lines of evidence point to economic benefits that can be realized when subclinical worm burdens are further reduced through treatment.

MORPHOLOGY. Worms of the superfamily Trichostrongyloidea all have a small buccal capsule and a large bursa. As a general rule, the paired spicules are stout and heavily sclerotized; in addition to the spicules there is a dorsal sclerotization of the cloaca called a *gubernaculum*, which serves as a guide when the spicules are protruded. A good deal of variation occurs in the copulatory apparatus of the males (Fig. 35.1). The females within a genus often cannot be differentiated from one another; specific identification is therefore based on males.

LIFE CYCLE. The general pattern in the life cycle holds for all members of the superfamily: eggs are passed with

**TABLE 35.1**
Principal Trichostrongyloids of Domestic Ruminants

| PARASITE | HOST | | | ADULTS IN | HATCHES AS |
|---|---|---|---|---|---|
| | OX | SHEEP | GOAT | | |
| *Haemonchus contortus* | x | (x) | x | Abomasum | $L_1$ |
| *H. placei* | (x) | x | x | Abomasum | $L_1$ |
| *Ostertagia ostertagi* | (x) | x | x | Abomasum | $L_1$ |
| *O. circumcincta* | (x) | (x) | x | Abomasum | $L_1$ |
| *Trichostrongylus colubriformis* | x | (x) | x | Small intest. | $L_1$ |
| *T. vitrinus* | x | (x) | x | Small intest. | $L_1$ |
| *T. longispicularis* | (x) | x | x | Small intest. | $L_1$ |
| *T. axei* | x | x | x | Abomasum | $L_1$ |
| *Nematodirus spathiger* | x | (x) | x | Small intest. | $L_3$ |
| *N. filicollis* | x | (x) | x | Small intest. | $L_3$ |
| *N. helvetianus* | (x) | x | x | Small intest. | $L_3$ |
| *N. battus* | | (x) | | Small intest. | $L_3$ |
| *Cooperia curticei* | | x | x | Small intest. | $L_1$ |
| *C. oncophora* | (x) | x | | Small intest. | $L_1$ |
| *C. punctata* | (x) | x | | Small intest. | $L_1$ |

[a] (x) = Principal host.

the feces and the $L_3$ is the infective stage. The $L_3$s are eaten with the forage and invade the tissues of the gut.

The times for development to infectivity at any given temperature vary. At constant conditions, *Haemonchus* may reach $L_3$ in 3 days, whereas *Nematodirus* may require 10 days. Also, *Nematodirus*, unlike most of the other genera, remains within the egg capsules until it hatches as an $L_3$; the stages outside the host in this genus rely on stored food for development and survival.

The site of development is somewhat different for each species, and the stimuli for exsheathment differ as well. As discussed in other sections of this book, the host provides a stimulus that triggers the transition from the free-living to the parasitic phase of the cycle. The stimulus for *Haemonchus* is $CO_2$ under reduced conditions in balanced salt solution, whereas *Trichostrongylus* require $CO_2$ and acid pepsin for exsheathment; these worms inhabit the abomasum and small intestine, respectively, and the triggers represent the conditions just anterior to the site where the worm normally establishes (Rogers, 1961).

The prepatent period for most of the worms is 14 to 18 days upon initial infection. With later infection, due to immunity or other undefined factors, the prepatent period may be stretched out for weeks or months.

DIAGNOSIS. Diagnosis is based on a number of factors:

1. History of the herd
2. Clinical signs
3. Postmortem examinations
4. Fecal egg counts

In any one area, there is a high probability that a certain complex of worms cause problems in certain ages of animals at certain times of the year (Table 35.2). Thus, if anemia occurred in animals during the middle of the pasture season, *Haemonchus* would be suspected. If only the lambs in a flock of sheep had severe diarrhea either early or late in the pasture season, *Nematodirus* would be a good bet.

If animals are dead or moribund, a postmortem examination of a few of them is the surest way of determining what the species and number of worms are. Weight loss, diarrhea, and anemia are typical signs of helminthosis in ruminants, but other infectious agents and some intoxications can also cause the same signs. Therefore, it is essential that other causes of disease be kept in mind when performing a postmortem examination.

Fecal egg counts have utility in epidemiologic studies and in diagnosis when the data are conservatively interpreted. The factors that must be taken into account are the following:

1. Egg counts in cattle rarely exceed 300/g, whereas in sheep they are often 1000/g or higher.
2. Different species of worms produce eggs at different rates. A rough rule of thumb is:

| GENUS | EGGS/FEMALE/DAY |
|---|---|
| *Haemonchus* | 5000 |
| *Bunostomum* | 5000 |
| *Trichostrongylus* | 1000 |
| *Nematodirus* | 300 |

(Note: these figures were complied from various sources and are intended only as a general guideline.)

3. Helminthosis is sometimes seen prior to sexual maturity of the worms, such as in *Ostertagia* and *Oesophagostomum* infections.
4. Egg production varies from season to season. Winter time is usually associated with low egg production.
5. Some eggs can be differentiated from one another quite readily. However, a great deal of skill is required to differentiate among *Haemonchus, Trichostrongylus, Bunostomum, Cooperia*, and *Oesophagostomum*. *Nematodirus, Strongyloides*, lungworms, and *Trichuris*, as well as tapeworm eggs and fluke eggs, can be readily identified.

Although we consider the trichostrongyles in this chapter, the gastrointestinal nematodes include hookworms such as *Bunostomum*, *Oesphagostomum* the nodular worm, the threadworm *Strongyloides*, *Trichuris* the whipworm, and members of the Metastrongyloidea, the lungworms. Nearly pure infection with a single species of worm may occur, but this is exceptional. In nearly all helminthoses there will be a mixture of half a dozen species or more. Lungworms, especially in warm, wet climates, are often part of the parasite mix that causes economic losses. Thus, the need for worm counts becomes clearer. In sheep, 1000

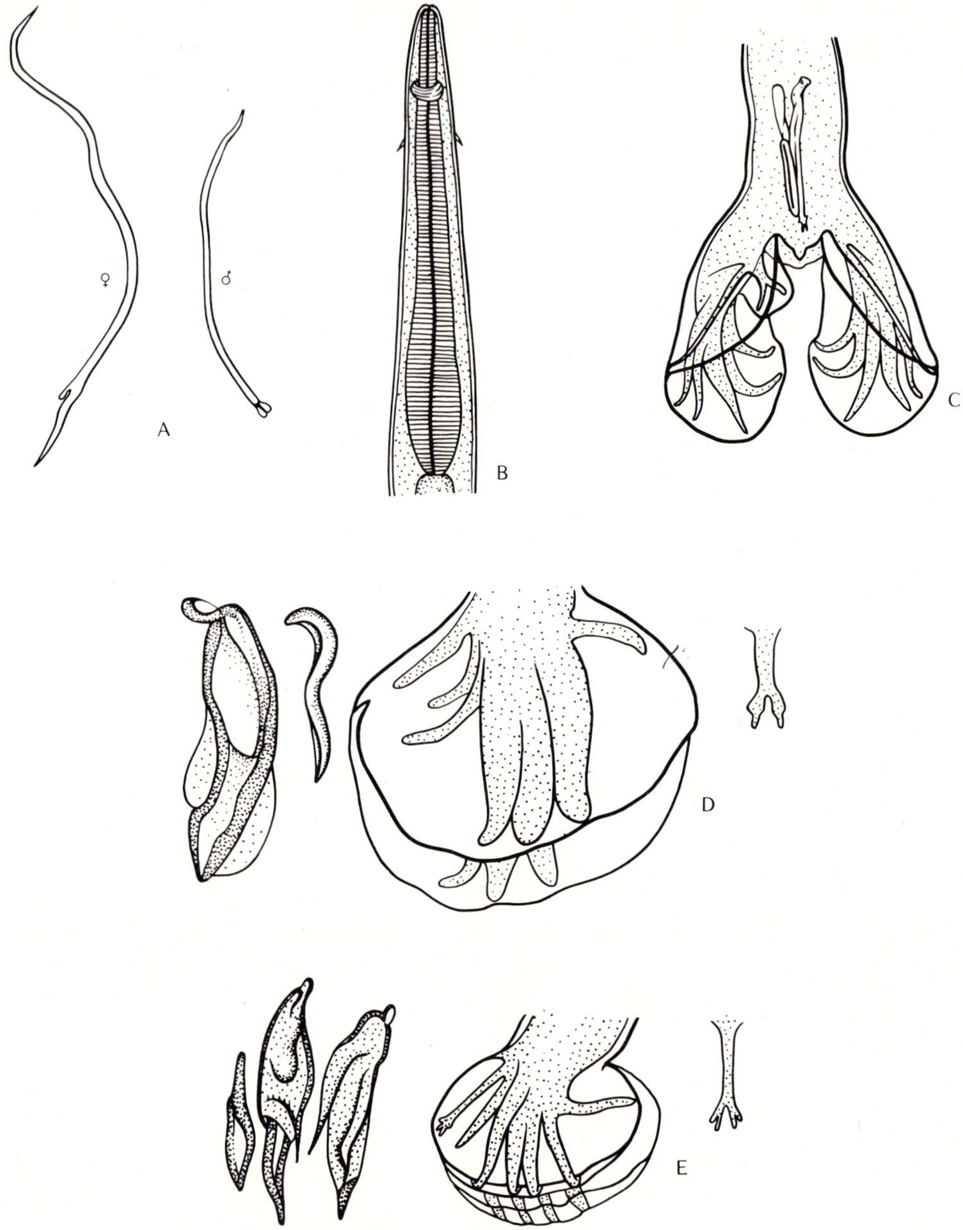


**FIG. 35.1**
*Haemonchus contortus:* (*A*) male and female, three times normal size; (*B*) anterior end of adult worm; (C) bursa of male. (*D*) *Trichostrongylus colubriformis;* (*E*) *Trichostrongylus axei. Ostertagia trifurcata:* (*F*) bursa; (*G*) spicules. *Ostertagia circumcinta:* (*H*) spicules. *Cooperia punctata:* (*I*) bursa. *Cooperia oncophora:* (*J*) spicules. *Nematodirus battus:* (*K*) bursa; (*M*) head. (*L*) spicules of *N. fillicollis (a), N. spathiger (b), N. battus (c).*

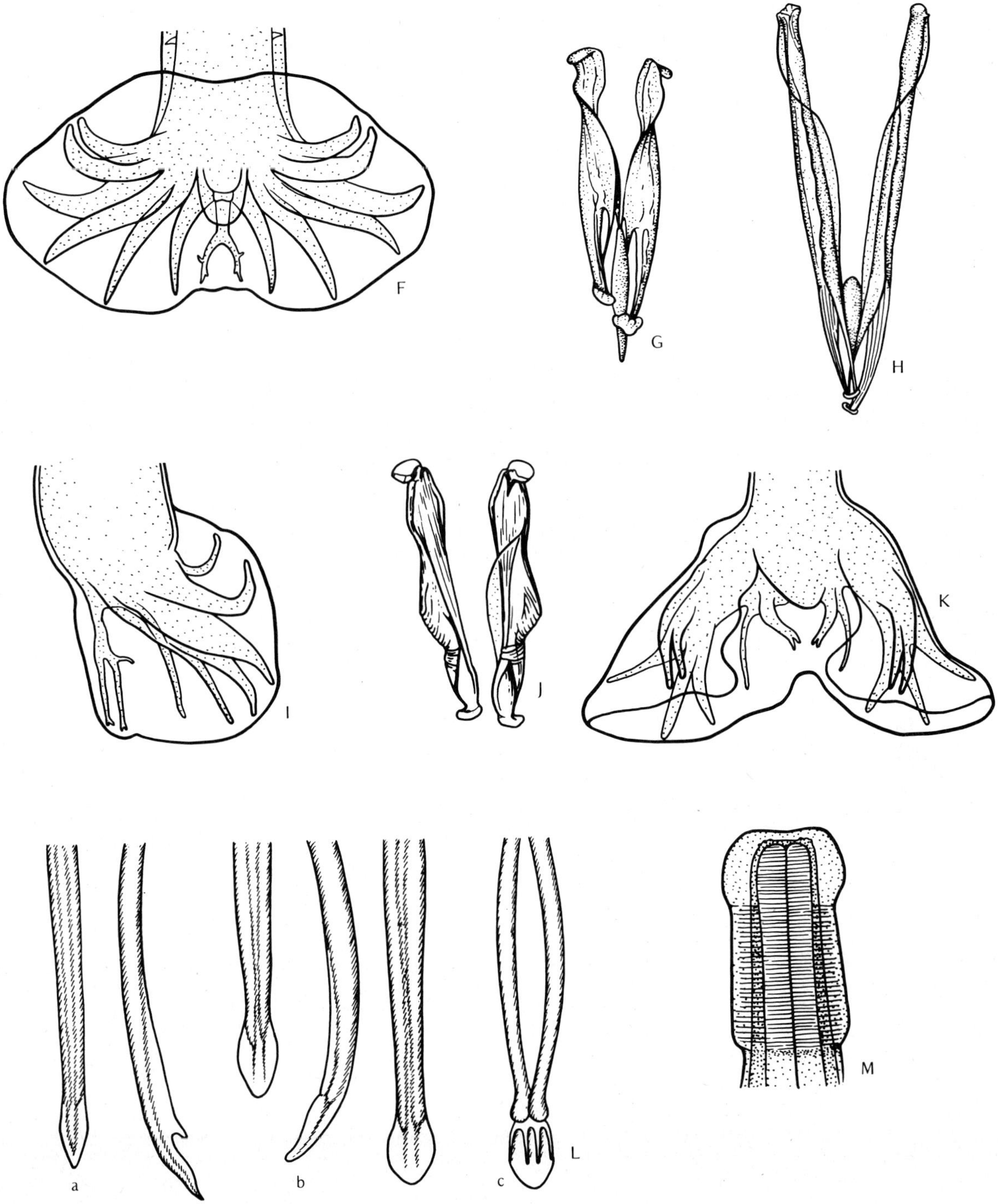
F
G
H
I
J
K
a
b
c
L
M

**TABLE 35.2**
Survival of Sheep Nematode Larvae on Temperate Zone Pastures[a]

| TYPE OF WEATHER | SURVIVAL OPTIMUM[b] | INTERMEDIATE[b] | MINIMUM OR NONE[b] |
|---|---|---|---|
| Warm, moist (summer weather with high maximum temperature and adequate rainfall) | *Cooperia*<br>*Haemonchus*<br>*Oesophagostomum*[c]<br>*Trichostrongylus* | *Nematodirus*<br>*Ostertagia* | |
| Warm, dry (summer weather with high maximum temperature and low rainfall or drought) | | *Cooperia*<br>*Nematodirus*<br>*Trichostrongylus* | *Haemonchus*<br>*Oesophagostomum*[c]<br>*Ostertagia* |
| Cool, moist (early spring or late fall weather with moderate temperatures and adequate rainfall) | *Nematodirus*<br>*Ostertagia*<br>*Trichostrongylus* | *Cooperia*<br>*Haemonchus*<br>*Oesophagostomum*[c] | |
| Cool, dry (early spring or late fall weather with moderate temperatures and low rainfall or drought) | *Nematodirus* | *Ostertagia*<br>*Trichostrongylus* | *Cooperia*<br>*Haemonchus*<br>*Oesophagostomum*[c] |
| Over winter in regions where subfreezing temperatures occur | *Nematodirus*<br>*Ostertagia* | | *Cooperia*<br>*Haemonchus*<br>*Oesophagostomum*[c]<br>*Trichostrongylus* |

[a] From Levine (1963).
[b] Optimum survival = survival of many larvae for 2 or more months during the grazing season, or over winter. Intermediate survival = survival of many larvae longer than 1 month but less than 2 months during the grazing season. Minimum or no survival = survival of few or no larvae after 1 month or less during the grazing season or over winter.
[c] Presumably, *Oe, columbianum*, not *Oe. venulosum*.

to 2000 worms is not usually cause for concern, but when perhaps 5000 worms are present, helminthosis is probable. In cattle, the numbers are higher and worm counts in excess of 5000 are not uncommon in normal animals, but when the number is 10,000 or greater, the worms cause losses.

TREATMENT. Helminthosis in domestic livestock is a herd problem and is approached as such. All of the animals in a herd are treated, though some may not be in immediate need of treatment.

The drugs used against gastrointestinal nematodes of ruminants are the following:

1. Organophosphates: trichlorphon, coumaphos, ruelene, and others
2. Benzimidazoles: thiabendazole, mebendazole, cambendazole
3. Methyridine
4. Thiazoles: tetramisole, levamisole
5. Ivermectin

All are administered by mouth, but tetramisole, levamisole and ivermectin can be administered parenterally. The first two anthelmintics, which have been in use since about 1960, generally have broad spectra of activity and have a little effect against larvae in the tissues. Except for some of the organophosphates, the margin of safety between the therapeutic dose and the toxic dose is broad. Drug-resistant worms have been found where the benzimidazoles have been used repeatedly, and when resistance develops to one compound, it seems to extend to all members of the group (Gibson, 1969). Some of the resistance can be overcome by developing methods that ensure longer contact between the drug and the worms (Prichard et al., 1978).

The technology of administering drugs on a herd

basis has become more sophisticated over the years. In the 1930s, phenothiazine (Ptz) was developed as an anthelmintic; it was a significant improvement over previous drugs such as copper sulfate and nicotine, and concurrently the methods of delivery changed. Ptz was administered principally as a liquid suspension or drench, and the usual method was to fill a 50-ml syringe fitted with a 4-in. nozzle and deliver the liquid into the back of the animal's mouth. A development from the McMaster Laboratory in Australia made the delivery simpler by using an inner-tube reservoir fitted with an automatic syringe; by this method, well over 60 sheep an hour could be treated. A large pill or bolus was also developed, but this has a couple of disadvantages: it might damage the pharyngeal region of the animal's mouth, ruminants tend to spit out such objects, and the material sometimes is packed tightly and it fails to break down quickly enough in the gut.

Handling costs in treating several hundred cattle or 2000 sheep are clearly a major factor. Attempts have been made to administer drugs in the feed or by some route other than the mouth (e.g., parenterally). Drugs given in the feed cut handling costs, but require that the animals be on a feed supplement and be accustomed to it, that the drug be palatable, and that all animals take an equal amount of the supplementary feed. The probability of all these requirements being met is rather low.

Is parenteral injection the answer? Not often. A drug injected intramuscularly (IM), for example, is not where the worms are and is absorbed slowly and diluted through the whole body of the animal. There is the possibility of transmitting some disease of the blood (*Babesia, Anaplasma*) on a contaminated needle. The drug must not be so irritating as to cause tissue damage at the site of injection; an abscess in the middle of the rump roast is not pleasing to the consumer. Ivermectin (Merck) is newly on the market for use in ruminants and may overcome the problems with an injectable preparation.

An alternative is to allow animals to take a small amount of drug daily on a free-choice basis. This was done first with Ptz in a Ptz-salt mixture. Worms were not removed, but egg production was inhibited; there was reduced contamination of pastures and a reduction in transmission as a result. The system has value in a herd health context, but allows drug resistance to develop fairly quickly.

The drugs currently available are generally good. They have, except for some organophosphates, a good margin of safety, have fairly broad spectra of activity, do not cause problems with residues in meat (if the manufacturer's instructions are followed), and have some action against larval stages.

Despite the good quality of the drugs now used, there is room for improvement. We need compounds that have a broader spectrum of activity. Each of the drugs now in use has some genera or species of worms against which it has little activity. One solution has been to mix drugs, such as a combination of trichlorphon and coumaphos (marketed as Neguvon A) to broaden the spectrum. The other problem, activity against larvae, is an important one. In some instances, the major damage is due to larvae (*Ostertagia, Oesophagostomum*), and removing the adults is only a minor palliative. In other instances, a reservoir of larvae may lie dormant for months until the time comes for maturation (see "spring rise" in Host–Parasite Interactions). An ideal drug would be one that would be absorbed through the skin, have a broad spectrum of activity, be lethal to larvae, leave no residues in the meat or other usable parts of the animal, and not allow drug resistance to develop. We have not yet reached that therapeutic utopia.

CULTIVATION. In developing control methods for microbial diseases, the ability to cultivate the organisms was crucial. The manipulation of an organism in the laboratory allows an investigator to determine growth characteristics, to harvest large quantities of material for the production of a vaccine, or to reduce the pathogenicity of the agent (attenuation), among other things. The same needs and objectives apply to helminth infections, and considerable effort has been expended in attempting to cultivate various gastrointestinal and pulmonary nematodes of livestock.

Although axenic cultivation of nematodes goes back to the 1930s with the work of Glaser and his associates with *Neoaplectana glaseri*, it was not until about 1960 that a number of investigators attacked the problem and began to achieve some success with nematodes of livestock. There are a variety of objectives in the cultivation of nematodes:

1. The free-living stages have been cultivated to determine nutritional requirements and to determine how cultural conditions affect the ability of the organisms to reach infectivity.
2. Free-living nematodes and parasitic nematodes

have been compared. In studies on nutritional requirements, it has been found that the free-living nematodes have about the same requirements as parasitic ones.

3. Nematodes have been studied to determine the morphologic changes they go through. In a number of instances, particularly where the nematodes migrate through a variety of tissues, it is difficult to obtain some stages of the parasites, and cultivation has elucidated changes in the structure not seen previously. An example is the $L_4$ mouth structures in the swine kidney worm, *Stephanurus dentatus* (Lichtenfels & Tromba, 1972). Cultivation of *Ascaris* showed the times at which various molts take place, and it was then possible to relate back to stages in swine and determine when the fourth molt occurs (Douvres et al., 1969).
4. Vaccines may be produced. One of the concepts in the development of immunity to nematodes is that secretions and excretions of the worms are antigenic. Axenic cultivation and cultivation in media having only small polypeptides allows the separation of molecules that may be antigenic. The state of the art has been reviewed by Cabrera et al. (1978), and though results are encouraging, only moderate success has been achieved.
5. Drugs may be screened. There is need for better anthelmintics and simple test systems are useful in screening large numbers of potential drugs. *In vitro* systems allow observations to be made on motility, alterations in feeding, and death, among other endpoints.

HOST–PARASITE INTERACTIONS. In looking at damage to the host, we must take into account not only the trichostrongyles, which are the subject of this chapter, but also the nodular worms (Chap. 34), hookworms (Chap. 32), *Strongyloides* (Chap. 31), and lungworms (Chap. 36). The reason is that in any outbreak of helminthosis there usually will be half a dozen or more species of worms from all of these groups contributing to the overall problem. There are instances in which one might see almost pure infections with, say, *Haemonchus* or *Nematodirus*, but these are the exceptions, and usually come about because of some peculiar situation that favors a single species (see Epidemiology and Control).

Because helminthosis is typically caused by a complex of species, the signs and symptoms of infection are usually diarrhea, weight loss, anemia, and dehydration. But the underlying question is, how do the worms damage the host? A part of the answer is that each species among the major pathogens causes damage in a different way. In addition, each species causes more than one type of damage. We can categorize the way that worms damage the host in the following way:

1. Stage of life cycle
   a. Larvae
   b. Adults
2. Type of damage
   a. Anemia
   a. (1) Blood loss: *Haemonchus*, *Ostertagia*, *Bunostomum*
      (2) Suppression of red cell formation: *Trichostrongylus*
   b. Tissue damage
   a. (1) Fluid loss
      (2) Inorganic imbalance
      (3) Reduced efficiency of feed conversion
      (4) Secondary invasion by microbial agents
   c. Immunopathology
   a. (1) Nodule formation: *Oesophagostomum*
      (2) Hypersensitivity

Further complicating the pathologic processes are the following factors:

1. Age of animals
2. Breed of animals
3. Quality of the feed
4. Immune status of the herd
5. Reproductive stage of the herd (pregnant, milking or not)
6. Number of worms in the host(s)

Let us first look at individual species and what we know about their mechanism of damage. Among the simplest is that of anemia caused by blood loss. In *Haemonchus*, *Bunostomum*, and *Ostertagia* the worms suck blood and essentially exsanguinate the host when the intensity of infection is high. In *Haemonchus*, blood sucking begins during the $L_4$. Some of the better studies on blood loss in haemonchosis estimate a blood loss of 0.05 to 0.08 ml/day/worm. It is sometimes said that up to 500 worms in the abomasum of a sheep will not cause significant damage, but using the previous figures, that means there is a loss of from 25 to 40 ml of blood/day and from 3750 to 6000 ml during a 5-month pasture season. If this is scaled up to a clinical level where there may be from 10,000 to 50,000 *Haemonchus* in one animal, the daily blood loss could be from 500 to 4000 ml. Obviously such a loss cannot be sustained for more than a short period. In these cases of uncomplicated haemonchosis, there is a loss of not only red cells but also of serum proteins; the result is anemia, which can readily be seen in the pale mucus membranes and the accumulation of fluid in the tissue spaces. Serum proteins maintain the osmotic pressure of the blood; when proteins are lost, fluid is not returned from the tissue spaces to the blood vessels but instead accumulates. This is best seen in ruminants in a sign called *bottle jaw*, an accumulation of fluid in the submandibular region (Fig. 35.2).

Anemia is often seen in infections with species such as *Ostertagia*, *Trichostrongylus*, and *Cooperia*, which are known to suck little (*Ostertagia*) or no blood. The mechanism leading to anemia seems to lie in damage to the intestine, so that there is leakage of fluids from the tissue into the lumen. Although in most instances the proteins that are lost to the lumen would be digested and recycled, it appears that the diarrhea associated with helminthosis reduces the time for digestion to occur, as well as the ability of the gut to take up nutrients. The result is a heavy loss of protein through the digestive tract. Studies on naturally infected animals indicate that they have adequate stores of everything except protein for production of hemoglobin.

Damage to the intestinal tissue of the host leads not

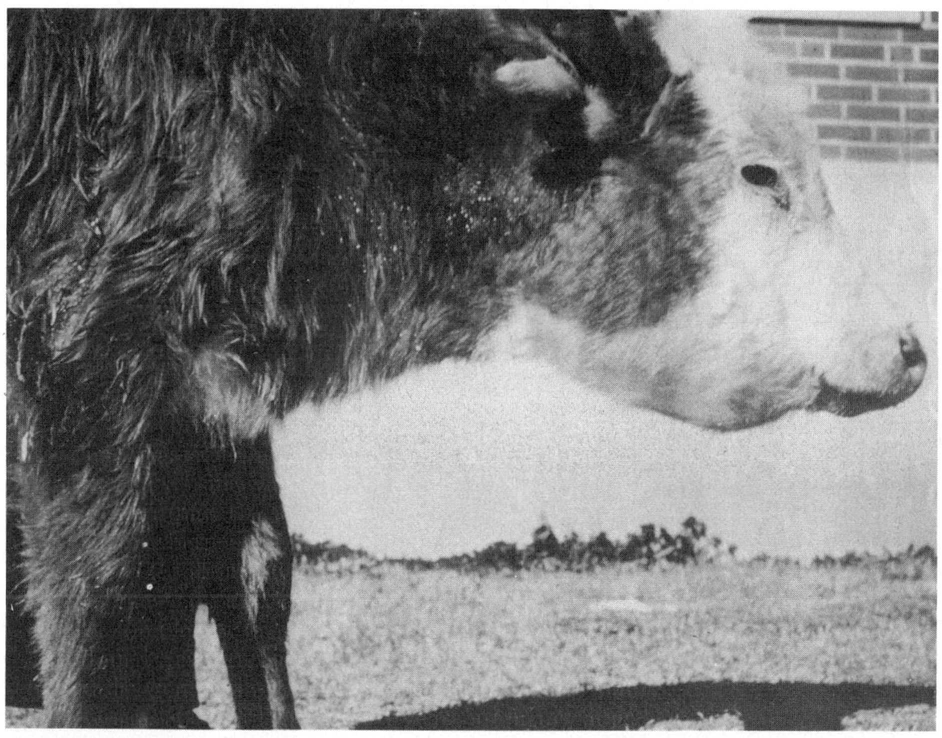

**FIG. 35.2**
Bottle jaw in a calf; this condition is a submandibular accumulation of fluid as a result of hypoproteinemia and is frequently seen in heavy *Haemonchus* infections.

only to anemia, but also to fluid loss, inorganic molecule imbalance, dehydration, and inefficiency in feed conversion. There may also be secondary invasions by bacteria if the intestinal epithelium has been damaged.

Damage to the intestinal tissue by the larval stages and the adults can readily be seen microscopically. Worms invade the glands of the mucosa, or sometimes the submucosal areas, depending on the species. Study of histologic sections reveals that there is cell death in the vicinity of the larvae. Other techniques such as electron microscopy have shown that the tight junctions between epithelial cells are broken and fluid leaks from the tissue to the lumen. The traumatized gut further loses surface by reduction of villi.

Animals treated with a good anthelmintic will be relieved of nearly all of their worms and recover clinically within a matter of several days. However, a common observation is that the ability of animals to grow normally, regain lost weight, and convert feed efficiently has been lost. Some sort of long-term damage to the gut or liver seems to have taken place.

Immunity is usually beneficial to the host (see following), but it can be detrimental also, as has been noted in connection with other eukaryotic parasites: *Schistosoma* spp. and various filarioid nematodes. The nodules that develop around *Oesophagostomum* larvae (Chap. 34) are a case in point. The parasites are successfully walled off, but the gut is left permanently damaged and not able to function normally.

Hypersensitivity (allergy) is a mixed bag of benefit and damage. In the case of certain worms of ruminants, particularly *Haemonchus contortus* of sheep, hypersensitivity is the cause of the so-called self-cure reaction. In a certain percentage of sheep with established infections of *Haemonchus*, the whole adult population may be shed spontaneously. Most often self-cure occurs when sheep receive an inoculum of $L_3$s; the gut becomes highly motile and hyperemic (congested with blood) and the worms are shed 2 to 4 days later. Evidence is good that the $L_3$s trigger an allergic response with release of histamine and other typical reactions. Some animals are resistant, for a time, to reinfection, but others may be reinfected immediately after self-cure.

A good and useful immunity develops to gastrointestinal nematodes on repeated infection. It is not generally manifested as a sterile immunity, such as one typically sees in viral infections, but rather it serves to moderate infection. An animal that has been infected is relatively resistant to challenge, but some of the worms in the challenge inoculum develop. Immunity is manifested in the following ways:

1. Killing some worms
2. Reduction in egg laying of adult females
3. Stunting
4. Inhibition of development of preadults

Most of the immune response is directed against the larvae; in some studies, adults were removed (other than self-cure), but the major action is to kill or prevent development of larvae. The cuticle of nematodes is a relatively poor antigen, and although it is the principal interface with the host tissue, it is not especially important in the immune response. The important factors are excretions and secretions of the larvae; these are fluids released at molting, metabolic end products, and probably substances excreted by the excretory system. In several instances, it has been found that the immune response acts against the $L_4$.

Inhibition of larval development is an important factor in both epidemiology and clinical disease. There is undoubtedly a strong immunologic component in this inhibition, but there are also other factors such as host physiologic state and interaction among the parasites themselves that influence inhibition and its release. There is some discussion of this topic in Chapter 33 on strongyles of horses.

One of the first known instances of larval inhibition was the "spring rise" phenomenon in sheep in Britain. Since the 1930s it had been observed that toward the end of winter or in early spring, preganant ewes not yet on pasture show a striking increase in fecal egg counts. In the 1950s it was finally found that spring rise came about through the maturation of *Haemonchus* larvae, which had lain dormant for many weeks or months, probably since the animals had been taken off pasture in late fall or early winter. Somehow the larvae developed to a certain point and then stopped. Toward the end of pregnancy and at the time when animals would be going out on pasture, development resumed and the worms became sexually mature. The question is whether maturation results from (1) a lowering of the level of immunity, (2) a response to hormonal levels in the ewe, (3) an inherent annual rhythm in the worms, or (4) a change in larval physiology as the weather changes at the end of the pasture season. All except the last of these are still open issues, and in other instances (hookworm and *Ostertagia*) there may be a difference in larvae, depending upon the season.

A second case of inhibited development occurs in *Ostertagia* infections in cattle. Egg laying in *Ostertagia* infections begins 16 to 23 days after administration of $L_3$s. In single infections, clinical signs of diarrhea and weight loss just about coincide with the maturation of the worms. This uncomplicated event is referred to as Type I ostertagosis. In Type II ostertagosis, the appearance of clinical signs is delayed for weeks or months by retention of fourth-stage larvae in the tissue of the abomasum. Typically, animals have been on pasture and have come off at the end of the grazing season to be fed hay or other wintertime feed. The animals may remain normal until later winter or even early spring and then break with clinical ostertagosis coinciding with the maturation of the inhibited $L_4$s. Evidence points toward an immunologic component in releasing the larvae from inhibition.

EPIDEMIOLOGY AND CONTROL. A considerable body of literature exists on the development and survival of the free-living stages of the trichostrongyles, especially *Haemonchus*, *Trichostrongylus colubriformis*, *T. axei*, *Ostertagia circumcincta*, *O. ostertagi*, *Nematodirus spathiger*, and *N. battus*. Other nematodes of ruminants—*Strongyloides*, *Dictyocaulus*, and *Bunostomum*—have been studied extensively (Gibson, 1966; Gordon, 1948, Levine, 1963, Levine 1980; Michel, 1969; Ollerenshaw & Smith, 1966, 1969; Rogers & Sommerville, 1963, 1968; Soulsby, 1965; Thomas, 1974).

The reason for these extensive and intensive studies lies in the crucial role that the free-living stages play in maintaining infection and in allowing clinical infections to develop. The basic questions asked by all of these studies are, under what conditions do animals become infected, and how can the system be manipulated so that animals will not develop harmful infections? Despite many avenues of investigation, these questions are at the core of them all.

Data are gathered in a variety of ways on the free-living stages and on the worm burdens actually found in naturally infected animals.

*Free-Living Stages*. The following studies are done:

1. In the laboratory the time to reach infectivity and the length of survival of infective stages are studied under constant, controlled conditions.
2. In the field eggs and larvae are placed on pasture under the fluctuating conditions found in nature. Materials are placed out at various times and then retrieved periodically, and survival and development are determined. Often pasture clippings are examined for $L_3$s and calculations are made on the number of infective larvae that might be ingested in a day by an ox or a sheep.
3. Purposely contaminated pastures are grazed periodically by sentinel animals. The concept of a sentinel animal is that a susceptible animal is exposed to the infectious agent under natural conditions of transmission for a period and then examined for evidence of infection. A series of groups of, say, five animals are allowed to graze a contaminated pasture each for a week or so and then examined for the presence of infection either by fecal egg count or by postmortem examination.

*Worm Populations*. The kinds of worms that are likely to cause problems in ruminants vary with the area. One way of finding out what worms are present is through the use of "parasite profiles," as suggested by Levine (1963) (Fig. 35.3). These data are for sheep, and one might expect somewhat different results for cattle. Note that *Haemonchus* is the principal parasite in Beltsville, Maryland, *Cooperia* in Auburn, Alabama, and *Ostertagia* in Moscow, Idaho. The situation is somewhat different in Lexington, Kentucky, and in College Station, Texas, where a number of species are more or less evenly distributed in proportion to the population of nematodes.

Parasite profiles are useful as a rough index of what one might expect to be important; however, development of such a profile does not constitute an epidemiologic study. It is necessary to determine not only what species are important, but in what and when, as well. A properly designed epidemiologic study focuses on carefully conducted host examinations of an adequate sample of adults and young of both sexes taken at various seasons of the year. The following discussion of *Nematodirus* disease points up the reasons for such a structured study.

In the broad sense, development to infectivity of the free-living stages of nematodes depends on temperature, moisture, and oxygen. We can generalize rather loosely and state that for trichostrongyles a temperature of from 10 to 35°C (50 to 95°F) allows development.

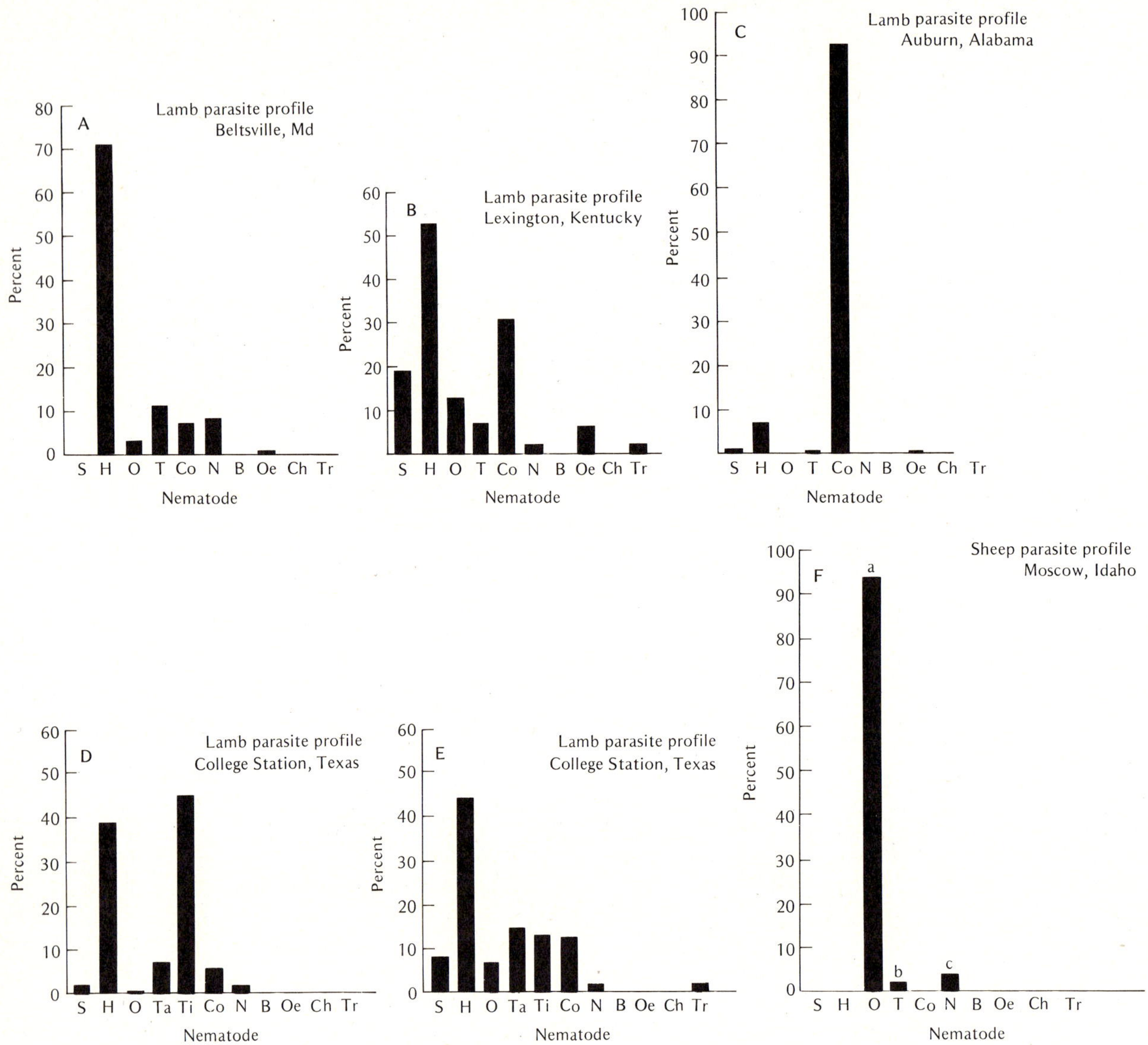


**FIG. 35.3**
Parasite profiles of sheep from different regions. A, Beltsville, Md. B, Lexington, Ky.; C, Auburn, Ala.; D, E, College Station, Tex.; F, Moscow, Idaho. S = *Strongyloides papillosus*, H = *Haemonchus*, T = *Trichostrongylus*, $T_a$ = *Trichostrongylus axei*, Ti = *Trichostrongylus colubriformis*, Co = *Cooperia*, N = *Nematodirus*, B = *Bunostomum*, Oe = *Oesophagostomum*, Ch = *Chabertia*, Tr = *Trichurus*. [Redrawn from Levine, N.D. 1980. *Nematode Parasites of Domestic Animals and Man*, 2nd ed. Burgess Publishing Co., Minneapolis.]

Development is too slow to be significant below 10°C (50°F) and above 35°C (95°F) is harmful to the free-living stages. Moisture needs are not so consistent; some worms require water, others can get along with a saturated atmosphere, and some do well at 50% saturation. Oxygen is required by all worms; if the environment becomes anaerobic through excess water and bacterial action, development is inhibited and eventually the free-living stages are killed.

When we look at individual requirements for development and survival, a pattern emerges of what each species requires and can tolerate. Generalizations made by several researchers were further modified by Levine (1963), Table 35.2, so that it is possible to see the factors that affect populations of various genera of gastrointestinal nematodes. For example, *Cooperia*, *Haemonchus*, *Oesophagostomum*, and *Trichostrongylus* do well where there is a warm, moist summer, but they survive poorly, or not at all, where winters are hard. *Nematodirus* and *Ostertagia*, on the other hand, do well in climates that are more difficult (i.e., cold and dry).

Various approaches have been taken to refine the climatic and meteorologic conditions that affect transmission. In Australia, heavy dependence on the sheep industry made control of infectious diseases a national priority, and the pioneering work of Clunies Ross and H. Mc.L. Gordon laid the groundwork used by parasitologists and livestock raisers around the world. Gordon (1948) used bioclimatographs to describe and predict when transmission would occur. A bioclimatograph (Fig. 35.4) plots mean monthly temperature in a weather shelter against mean monthly precipitation. If we look at the plot for Urbana, Illinois, we see that a dot is placed at the junction of the temperature and precipitation for each month and the months are then connected sequentially with a line. June, July, August, and September have temperatures between 20 and 25°C and 75 to 100 mm precipitation. As the fall progresses, the temperature declines to about 0°C and precipitation to about 50 mm. One can then use a bioclimatograph together with laboratory data on a species of worm to predict when transmission will take place. As noted in the figure legend, *Haemonchus* is transmitted when the precipitation is 50 mm or higher and the temperature is above 15°C.

Bioclimatographs are useful, but they also have some serious drawbacks. Weather shelter data are easy to obtain since records are readily available in nearly any locality. The principal problem is that the temperature is taken 1.5 m above the ground. The real activity is in the plants and at the soil surface; as someone said, very few sheep have become infected in a weather shelter.

What are the factors down where the sheep can become infected? Temperature is stable: seasonal and daily fluctuations are not as great as in the air. Moisture is more constant and may be present when there has been no rain for a long period; even nightly dew can provide water for a portion of the day. Plants provide protection from the harmful effects of ultraviolet irradiation from the sun. In order to deal with these factors, Levine (1963) (Fig. 35.5) looked at available moisture in various climates of North America. In a temperate climate of, say, the midwestern United States, the soil becomes dry toward the end of the growing season; then, with cooler weather it begins to recharge. By the beginning of the growing season, the soil is saturated in a locality such as Urbana, Illinois. In some areas such as Lubbock in west Texas, the soil does not saturate; there is enough moisture for growth of grass and other plants tolerant of arid conditions, but the soil is dry most of the time. Moisture in the soil continually moves out by transpiration from plants and evaporation from soil. If there is water in the soil, there will be moisture near the lower portions of the plants and at the soil surface where the worms are.

Levine carried data on moisture availability and on temperature one step further (Fig. 35.6) to allow prediction of the conditions under which transmission of *Haemonchus* and *Trichostrongylus* might occur in various climates.

On a global basis, *Haemonchus contortus* and *H. placei* probably cause more losses than any other species of nematodes in ruminants. *Haemonchus* is a heavy egg producer; estimates as high as 5000 egg/day/female worm give an idea of the amount of pasture contamination that it can cause when an animal has 1000 to 10,000 female worms in its abomasum. *Haemonchus* is also relatively long-lived; an individual may live for months, and the spring rise phenomenon shows that it can remain in a preadult state for months. Despite the fact that *Haemonchus* prefers warm, moist conditions, it is often found in climates where its presence would be unexpected. For example, *Haemonchus* is a problem in the Red River Valley of North Dakota near Fargo. The bitter winters of this area in great part kill the free-living stages, but the infection can be maintained in ewes over the winter. The author has seen outbreaks of haemonchosis in Montana on irrigated pasture and in animals kept on a low island in a stream. In areas where animals water from troughs, spilled water may allow a lush growth of grass in a small area; animals like this grass and it may have large numbers of infective larvae. Since worms are always present in ruminants, the potential for a rapid increase in the population is also great.

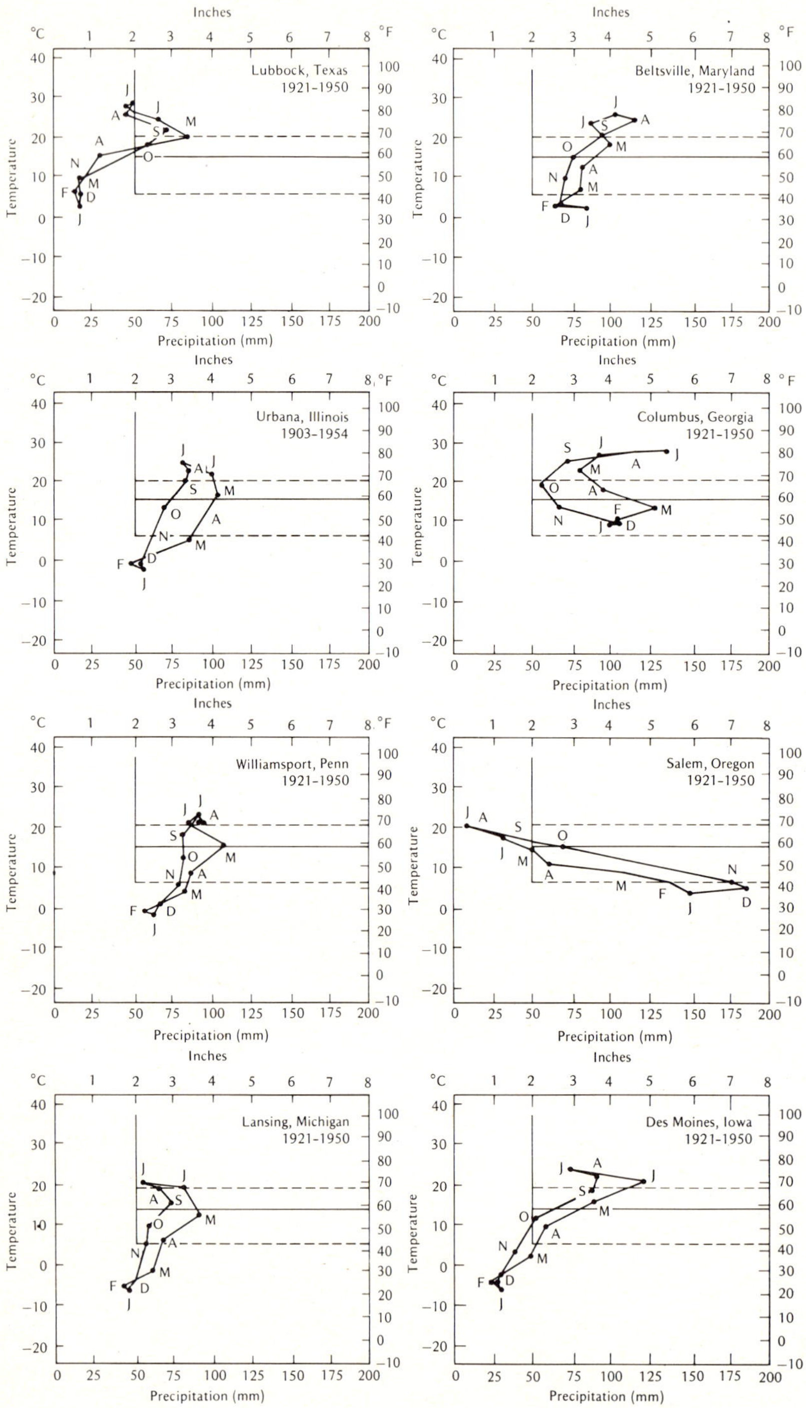


**FIG. 35.4**
Bioclimatographs from various localities in the United States. Transmission of *Haemonchus* takes place under conditions of temperature and rainfall represented between the dotted lines. [Redrawn from Levine, 1963.]

With *Haemonchus* the adult animals maintain a moderate population of worms and pass the infection to younger animals during the pasture season. The situation with *Nematodirus* exemplifies a different epidemiologic picture. *Nematodirus* is found in significant numbers only in lambs. If host examinations are done on ewes, few or none are found; this prevalence mainly in younger animals is one reason for carefully designing an epidemiologic study.

During the mid-1950s it was found that outbreaks of diarrheal disease in lambs in Great Britain were caused by *Nematodirus battus*. Few worms were found in ewes, but lambs were subject to overwhelming infections. The question was, how did the lambs acquire such massive infections? The answer revolves around the slow development of the free-living stages and of their remarkable resistance to environmental insult. *Nematodirus* does not hatch until the $L_3$ has been reached, so it is somewhat different from most of the other trichostrongyles that have free-living, feeding $L_1$ and $L_2$. Even under optimal conditions, *Nematodirus* $L_3$s do not hatch for 2 weeks, whereas *Haemonchus* can reach $L_3$ in as few as 4 days. Careful studies by British workers showed that eggs deposited on pasture during late summer and fall do not reach infectivity until late winter or early spring, at which time they appear in massive numbers. Sheep husbandry calls for lambs to be born just about the time that ewes and lambs go on pasture in the spring. Thus the most susceptible members of the population were being exposed to massive numbers of infective larvae. The concept developed from these investigations was that there is a lamb-to-lamb transfer in *Nematodirus battus*; the ewes are not important. The infection in last year's lamb crop is the source of infection for this year's lamb crop.

With *Haemonchus*, other trichostrongyles, and *Nematodirus*, clinical disease comes about through the exposure of susceptible animals to large numbers of infective larvae over a short period of time. In *Haemonchus*, *Trichostrongylus*, and *Ostertagia*, the infection comes from the ewes (Fig. 35.7); on the other hand, in nematodiriasis, the infection comes from the long development and simultaneous hatching of the $L_3$s as the lambs begin to take solid food (Fig. 35.8). In both cases there is a steep rise in the number of $L_3$s on the herbage. Much has been said about the buildup on pasture of free-living stages through successive generations of worms; calculations have been made showing how such an increase occurs. The problem is that it does not happen that way at all in climates with definite seasonal changes. What the graphs show is the sudden appearance of $L_3$s at a particular time of year. Depending on the species, the time of year varies, but the general principal is that the preinfective stages accumulate on pasture during hard times and develop slowly toward $L_3$. Then, when moisture or temperature reaches the proper level, the development continues to infectivity like a thunderclap.

The following are approaches that have been taken to control harmful populations of nematodes in ruminants:

1. Periodic treatment
2. Stock rotation, also known as *pasture rotation*
3. Grazing of pastures by nonsusceptible species
4. Killing the free-living stages on pasture
5. Maintaining a high level of nutrition
6. Immunization

In a review of the control of nematodes of domestic animals, Michel (1969) discussed these topics. It is clear, when considering the various recommendations, that much of the dogma that has influenced the design of control programs has been without adequate foundation. When a livestock owner has a problem, we attempt to make a set of suggestions that are clear-cut and feasible within available resources of workforce or money. Many of the topics discussed in this chapter muddy the waters. Species of worms differ in development and survival, as do the nutrition, immunity of the hosts, and turnover or inhibition of populations in the hosts. The objective of a control program is to prevent financial loss; eradication of worms of grazing animals is not a feasible goal. Therefore, a livestock owner lives with the problem, recognizes that there is a continual threat of loss, and tries to minimize it.

Among the methods that have been used, stock rotation (pasture rotation, hurdling, pasture spelling) allows animals to graze for a period less than that required for development to the $L_3$ and then animals are moved to clean pasture. Animals are moved through a series of pastures and then returned to the first one, presumably when the free-living stages have died out. The principal problem is that animals cannot be left off of pasture long enough during the growing season for the free-living stages to die. In fact, if animals are returned to the original pasture a month later, the

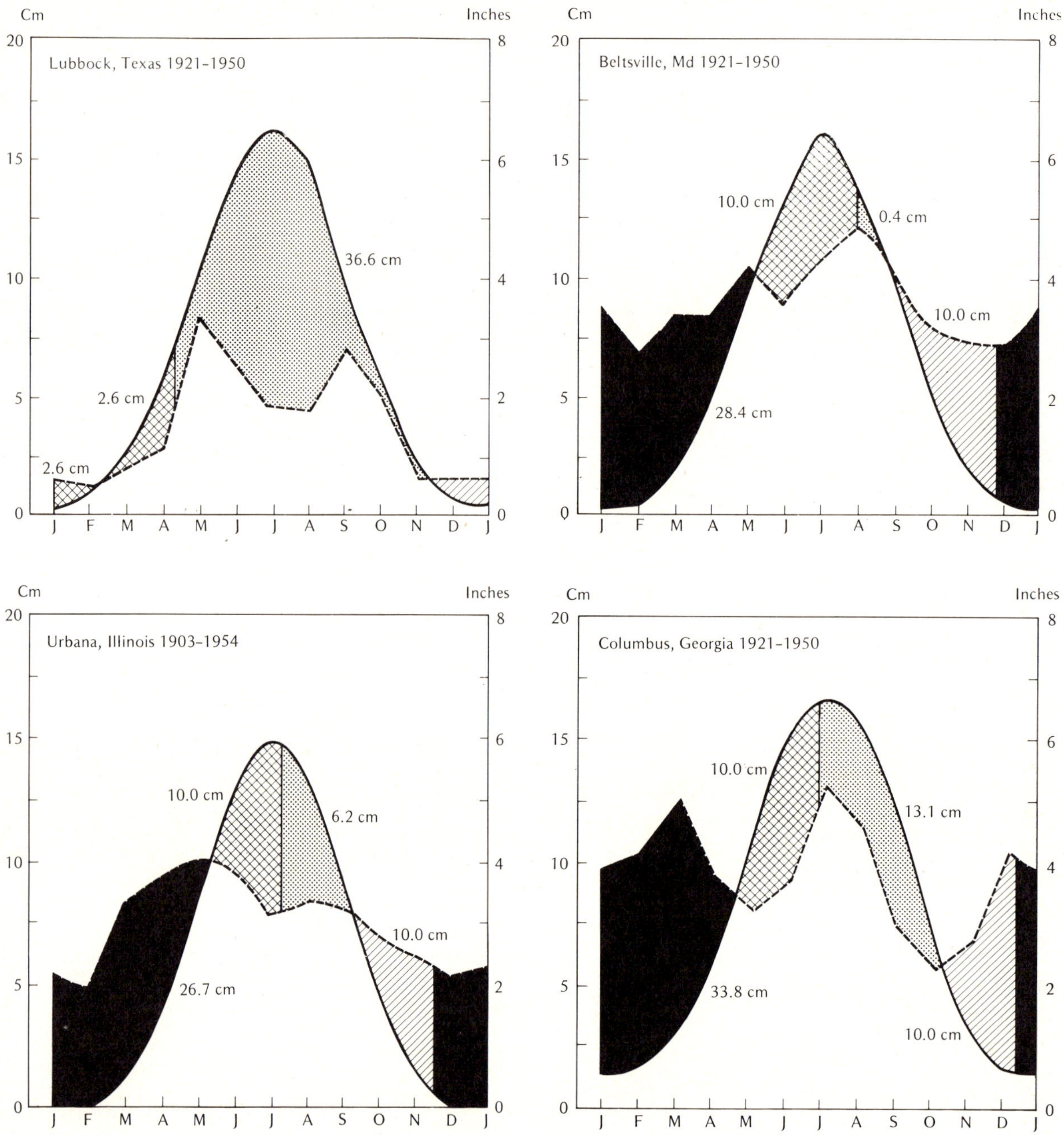


**FIG. 35.5**
Potential evapotranspiration and available moisture for various localities in the United States. [Redrawn from Levine, 1963.]

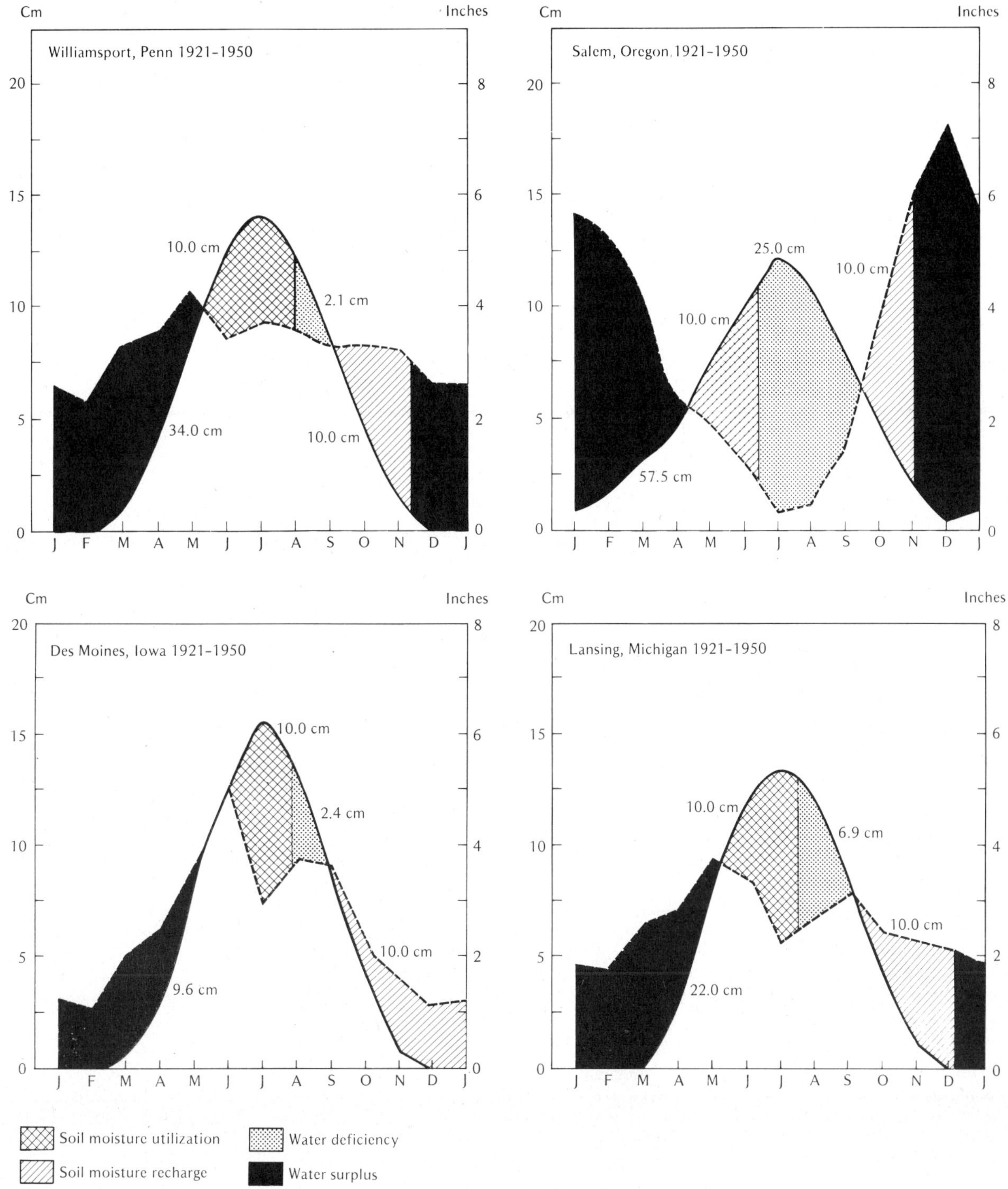
Williamsport, Penn 1921–1950
Cm
Inches
10.0 cm
2.1 cm
34.0 cm
10.0 cm
Salem, Oregon 1921–1950
25.0 cm
10.0 cm
10.0 cm
57.5 cm
Des Moines, Iowa 1921–1950
10.0 cm
2.4 cm
10.0 cm
9.6 cm
Lansing, Michigan 1921–1950
10.0 cm
6.9 cm
10.0 cm
22.0 cm
J F M A M J J A S O N D J
Soil moisture utilization
Water deficiency
Soil moisture recharge
Water surplus

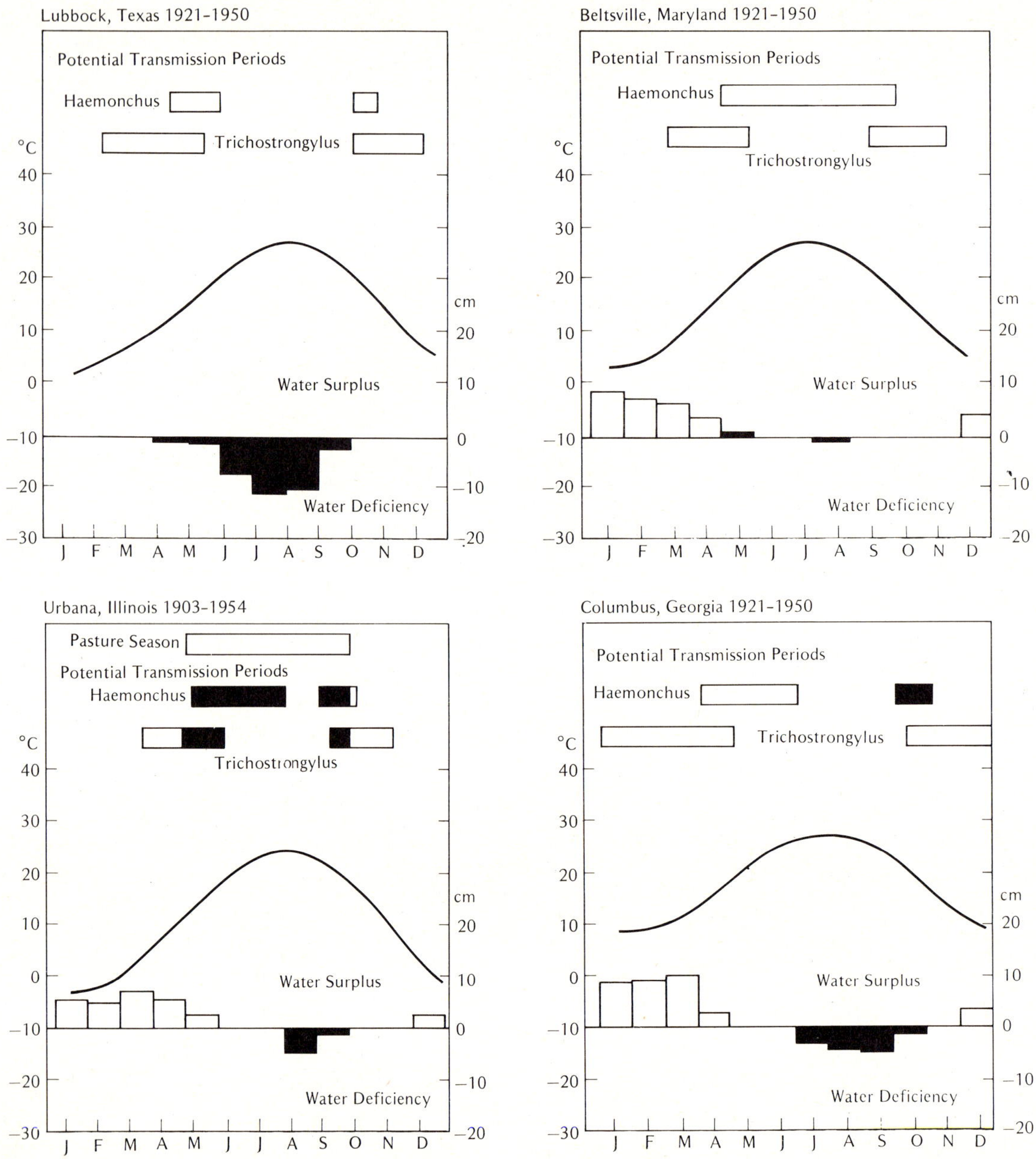


**FIG. 35.6**
Times of the year when transmission of gastrointestinal nematodes can occur in various localities. [Redrawn from Levine, 1963.]

Williamsport, Penn 1921–1950

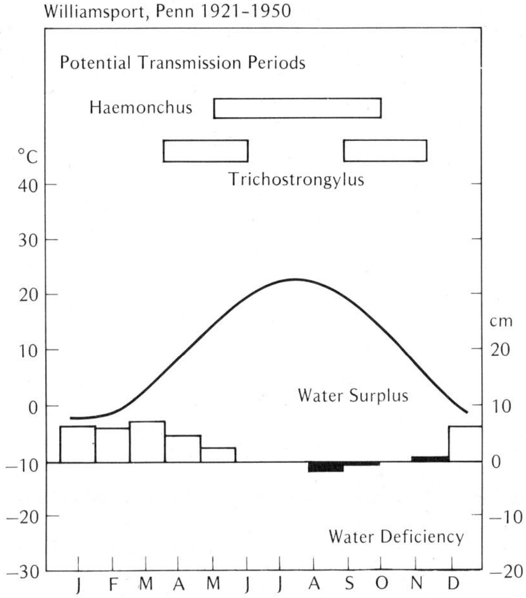


Salem, Oregon 1921–1950

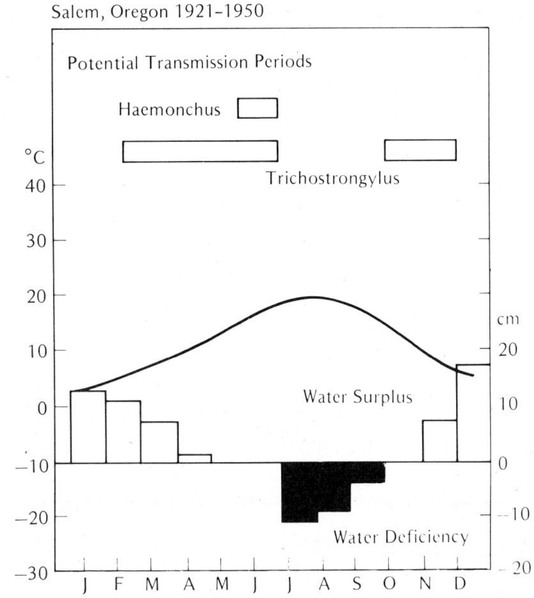


Des Moines, Iowa 1921–1950

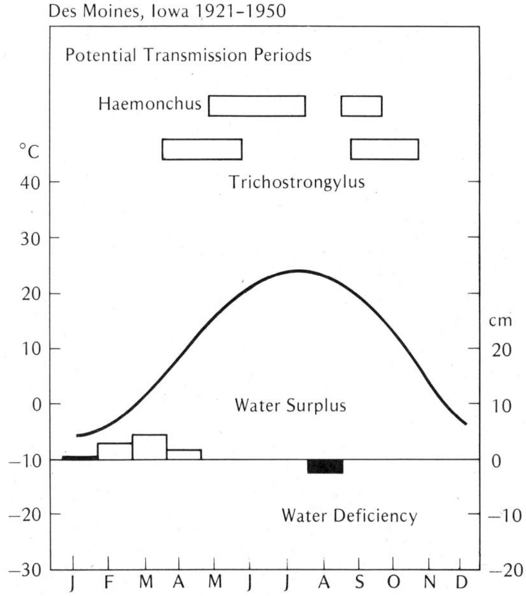


Lansing, Michigan 1921–1950

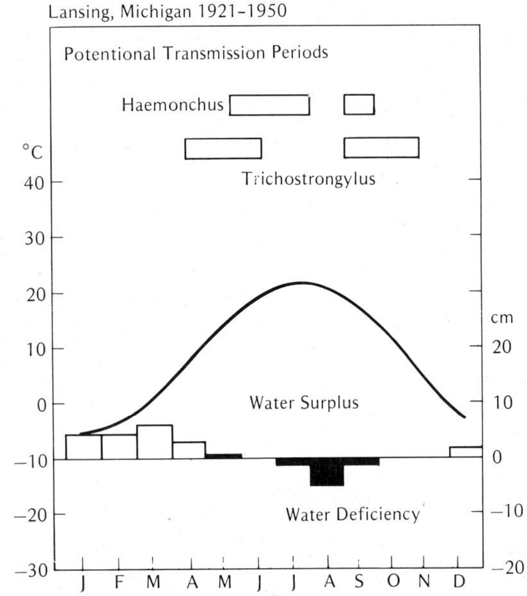

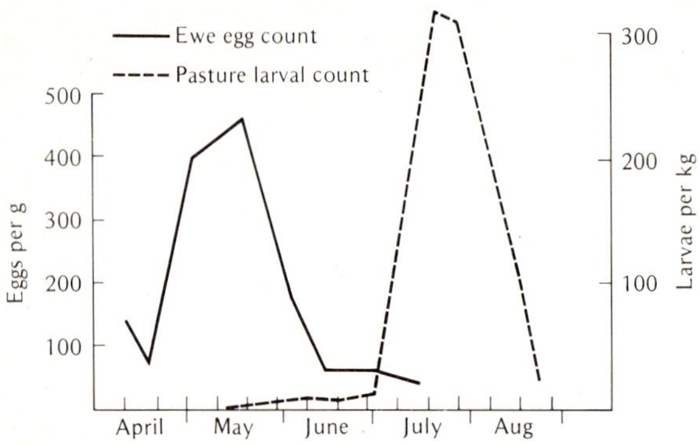


**FIG. 35.7**
Gastrointestinal nematode infection in sheep: egg counts in ewes and contamination of pastures with larvae. [Redrawn from Thomas, 1974.]

population of $L_3$s is likely to be at its peak. Stock rotation, at the end of a number of careful investigations, does not serve to control nematodes, but does allow good utilization of pasture forage.

What about periodic treatment? It is frequently recommended that animals on pasture be treated once a month. The idea is that one never lets the worms build up to the point where they will cause a problem. The anthelmintic used must remove not only adults, but the larvae as well. If the larvae are not removed, they will mature shortly after the adults have been removed and continue to contaminate the pasture and cause harm to the host.

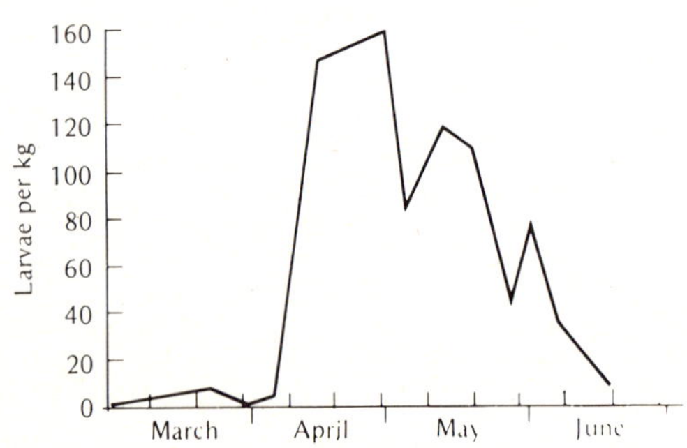


**FIG. 35.8**
Occurrence of third-stage larvae of *Nematodirus battus* on pasture in Britain. [Redrawn from Thomas, 1974.]

The best approach to minimizing losses is to know what worms are likely to cause problems and when it is likely to happen (Ollerenshaw & Smith, 1969). An epidemiologic investigation is done to determine what worms are present and then to monitor both the health of the animals and worm populations. Animals are then treated before there are significant losses. Theoretically, the animals will have developed immunity through subclinical infections and if they are watched for signs of helminthosis, the nutritional level can be maintained so as to minimize the effects of the worms.

A key to control is the development of immunity. In haemonchosis and nematodirus disease, lambs are subjected to numbers of infective larvae before they can develop effective immunity or they are exposed to such large numbers that they are overwhelmed. The goal of many research groups has been to facilitate the development of immunity so that animals can withstand infection. There has been success with lungworms of ruminants (Chap. 36) and with canine hookworms (Chap. 32), but not yet with the gastrointestinal nematodes of ruminants. A good product will no doubt be developed, but it will most likely be one that aids the acquisition of immunity through natural exposure.

The proposal has sometimes been made that rather than have animals graze on pasture, it would be better to bring the pasture to them. The so-called zero grazing or feeding green chop consists of cutting forage and bringing it to animals kept in corrals or buildings where the possibility of their becoming infected with worms is exceedingly low. Additional benefits are that the pasture would be well utilized because it would not be trampled or selectively grazed. Such a system might work where there is a long growing season with abundant rainfall and where both energy and labor costs are low; however, most ruminants are raised on the vast grasslands, savannahs, and steppes that comprise the largest biomes in the world. The zero-grazing proposal ignores the facts of life for the greater part of animal husbandry on earth.

As knowledge has progressed on infectious diseases, attempts have been made to develop mathematical relationships among the factors involved in disease transmission. In this book, some attention has been given to schistosomes of humans and fascioliasis of ruminants. Ollerenshaw and Smith (1969) reviewed the subject regarding gastrointestinal nematodes of ruminants, and they evaluated modeling as a means of controlling nematodiases. Success has been moderate with *Nematodirus* spp.; the beginnings of methods exist for the other trichostrongyles, but little has been done for lungworms of ruminants. The interested student is referred to this excellent review for further information.

## REFERENCES AND FURTHER READINGS


Boughton, D. C. 1957. Helminth production in livestock: Tonnage, carrying capacity, costs and control. *Am. J. Trop. Med. Hyg.* **6**:455–461.

Cabrera, E. J., P. H. Silverman, and J. Guerrero. 1978. Immunoprophylaxis of parasitic infections. *Practicing Vet.* **49**(3):24–26

Douvres, F. W., F. G. Tromba, and G. M. Malakatis. 1969. Morphogenesis and migration of *Ascaris suum* larvae developing to fourth stage in swine. *J. Parasitol.* **55**: 689–712.

Gibson, T. E. 1966. The ecology of the infective larvae of *Trichostrongylus colubriformis.* In E. J. L. Soulsby (ed.), *Biology of Parasites.* Academic Press, New York, pp. 2–13.

——— 1969. Advances in veterinary anthelmintic medication. *Adv. Parasitol.* 7:349–373.

Gordon, H. McL. 1948. The epidemiology of parasitic diseases with special reference to studies with nematode parasites of sheep. *Austral. Vet. J.* **24**:17–44.

Levine, N. D. 1963. Weather, climate and the bionomics of ruminant nematode larvae. *Adv. Vet. Sci.* 8:215–261.

——— 1980. *Nematode Parasites of Domestic Animals and Man*, 2nd ed. Burgess Minneapolis, Chaps. 1 and 6.

Lichtenfels, J. R., and F. G. Tromba. 1972. The morphogenesis of *Stephanurus dentatus* (Nematoda: Strongylina) in swine with observations on larval migration. *J. Parasitol.* **58**:757–766.

Michel, J. F. 1969. The epidemiology and control of some nematode infections of grazing animals. *Adv. Parasitol.* 7:211–282.

Ollerenshaw, C. B., and L. P. Smith. 1966. An empirical approach to forecasting the incidence of nematodiriasis over England and Wales. *Vet. Rec.* **79**:536–540.

——— 1969. Meteorological factors and forecasts of helminthic disease. *Adv. Parasitol.* 7:283–323.

Prichard, R. K., D. R. Hennessy, and J. W. Steel. 1978. Prolonged administration: A new concept for increasing the spectrum and effectiveness of anthelmintics. *CSIRO* **4**:309–315.

Rogers, W. P. 1961. *The Nature of Parasitism: The Relationship of Some Metazoan Parasites to Their Hosts.* Academic Press, New York.

Rogers, W. P., and R. I. Sommerville. 1963. The infective stage of nematode parasites and its significance in parasitism. *Adv. Parasitol.* **1**:109–177.

——— 1968. The infectious process, and its relations to the development of early parasitic stages of nematodes. *Adv. Parasitol.* **6**:327–348.

Soulsby, E. J. L. 1965. *Textbook of Veterinary Clinical Parasitology.* Vol. I. *Helminths.* F. A. Davis, Philadelphia.

Thomas, R. T. 1974. The role of climate in the epidemiology of nematode parasitism in ruminants. In A. E. Taylor and R. Muller, (eds.), *The Effects of Meteorological Factors upon Parasites.* Sympos. Brit. Soc. Parasitol., Vol. 12, pp. 13–32. Blackwell, London.

# 36

# *Lungworms*

Among the bursate nematodes, the superfamily Metastrongyloidea contains the lungworms and some worms of blood vessels. They can be recognized superficially by their small buccal capsule similar to that of the trichostrongyles, a small bursa, and relatively short, uncomplicated spicules. There are lungworms in other groups of nematodes, such as the Strongyloidea and some of the spirurids, but nearly all of the metastrongyles are lungworms.

Most of the lungworms of economic importance are in ruminants or swine. Their life cycles are completed by ingestion of the infective larvae, but some have direct life cycles and some have invertebrate vectors. The effects on the host are often by direct damage, but there are some interesting interactions with other infectious agents that we will look at in some detail.

Since the $L_3$s are ingested, the larvae must migrate to reach the lungs. As is true in a number of other instances, if larvae are ingested by a foreign host, migration may be abnormal. An example is eosinophilic meningitis in humans caused by *Angiostrongylus cantonensis*.

The lungworms also provide an example of efforts expended by investigators to find new solutions to problems in the face of parasites refractory to the usual chemotherapeutic approaches.

## LUNGWORMS WITH DIRECT CYCLES

NAME OF ORGANISM AND DISEASE CAUSED. *Dictyocaulus filaria* (Rudolphi, 1908) Railliet and Henry, 1907; *D. viviparus* (Bloch, 1782) R. and H., produce lungworm disease, husk, hoose, and lungworm pneumonia.

HOSTS AND LOCATION. *D. filaria* is found principally in sheep and goats among domestic ruminants but also occurs in others such as camel and alpaca and some wild ruminants. *D. viviparus* is found principally in the ox, but is also in other ruminants as well, both domestic and wild.

The adults are in the trachea and bronchioles.

DISTRIBUTION AND IMPORTANCE. The organism is cosmopolitan but usually causes problems in mild, wet climates.

In small numbers, lungworms cause no discernible damage to their hosts; however, in large numbers, or in connection with heavy infections with gastrointestinal nematodes, they can cause considerable economic loss.

MORPHOLOGY. *D. filaria* females are from 43 to 112 mm long, with a conical posterior end and the vulva just posterior to the middle of the body. Males are from 25 to 80 mm long, with moderately heavy, bent spicules 400 to 500 μm long (Fig. 36.1).

*D. viviparus* females are from 23 to 80 mm long and have a short, pointed tail; the vulva is near the posterior of the body. Males are from 17 to 50 mm long, with moderately heavy, bent spicules 195 to 215 μm long.

LIFE CYCLE. The cycles of both species are essentially the same. Adults lie in the air passages from the bronchioles to the trachea where the females lay eggs. Eggs are coughed up, swallowed, and normally hatch in transit through the digestive tract. The free-living stages do not feed; the $L_3$ is encased in the cast cuticles of the $L_1$ and $L_2$, and is reached in 4 to 6 days under optimal conditions of temperature and moisture.

The $L_3$s are ingested with the forage, cast the cuticles in the small intestine, and penetrate the gut to enter mesenteric lymph nodes. The $L_4$ is reached about 4 days after infection, and the larvae then leave the lymph nodes to be carried by the vascular system to the lungs. They break out of the alveoli of the lungs and reach sexual maturity about 18 days after infection.

DIAGNOSIS. Clinical signs of infection include rapid, labored breathing, pneumonia, and failure to gain properly. In an enzootic lungworm area, these signs are suggestive. The signs together with finding large numbers of $L_1$s in the feces give good evidence of clinical lungworm infection.

Larvae may often be found in feces examined following one of the standard fecal flotation methods; however, the usual method is to use the Baermann apparatus. Additional methods that can be used are examination of mucus from the respiratory tract for $L_1$s and finding $L_3$s on the herbage where animals are pastured.

HOST–PARASITE INTERACTIONS. A considerable part of the damage to the host occurs during the preadult stage of infection. Larvae are trapped in the capillaries of the air sacs (alveoli) of the lungs and then break out of the tissue into the alveoli themselves. Clinical signs are seen at this time, and the damage to the lung is both direct and indirect. Pneumonia is caused directly by the lungworms, and other infectious agents may also invade the lung tissue because of the rupture of the endothelial surfaces. The adults cause production of excessive amounts of mucus from the lungs, leading to chronic coughing. Animals that die from lungworm infection have large, tangled masses of worms in the air passages as well as a large quantity of mucus. Portions of the lungs are purple and lack the usual spongy texture and pink color of normal lung.

Animals that have been exposed to moderate numbers of larvae develop immunity against the preadult stages in the lymph nodes and lungs. Immunity affects not only the numbers of worms that may reach maturity but also the rate at which the worms develop. The prepatent period of about 3 weeks can be lengthened considerably in immune animals. Although some of the larvae are killed by the immunity, some survive, and these tend to accumulate in the lungs, causing long-term, low-grade damage.

TREATMENT. Some of the benzimidazoles are effective against lungworms. Earlier, cyanacethydrazide (Dictycide) was found to be effective against the adults, but not against the preadults. As noted, much of the damage caused by lungworms is in the preadult stage, and although there may be some beneficial effect from removing adults, it is a second-best solution. Drugs that are highly effective against the preadults are needed.

EPIDEMIOLOGY AND CONTROL. The distribution of clinical lungworm infections indicates the effect of external conditions on the free-living stages. Although lungworms are found nearly everywhere, they are problems in moist, usually warm climates. The larvae do not survive freezing or drying, and under similar conditions survive a shorter time than the free-living stages of the trichostrongyles. Although larvae of *D. viviparus* have been found to survive for nearly 11 months on pasture in Britain, this is quite unusual, and they almost always die within a month after being deposited. Interestingly the cast cuticle in the trichostrongyles is said to provide protection for the $L_3$; however, in the lungworms the $L_3$ is ensheathed in two cuticles and survives rather poorly.

As is true with gastrointestinal nematodes of ruminants, lungworm disease usually follows exposure to large numbers of $L_3$s over a short period. Thus, environmental conditions (temperature and moisture) allow the accumulation of preinfective stages that then become infective within a short time. If the animals on pasture are not immune, clinical disease results.

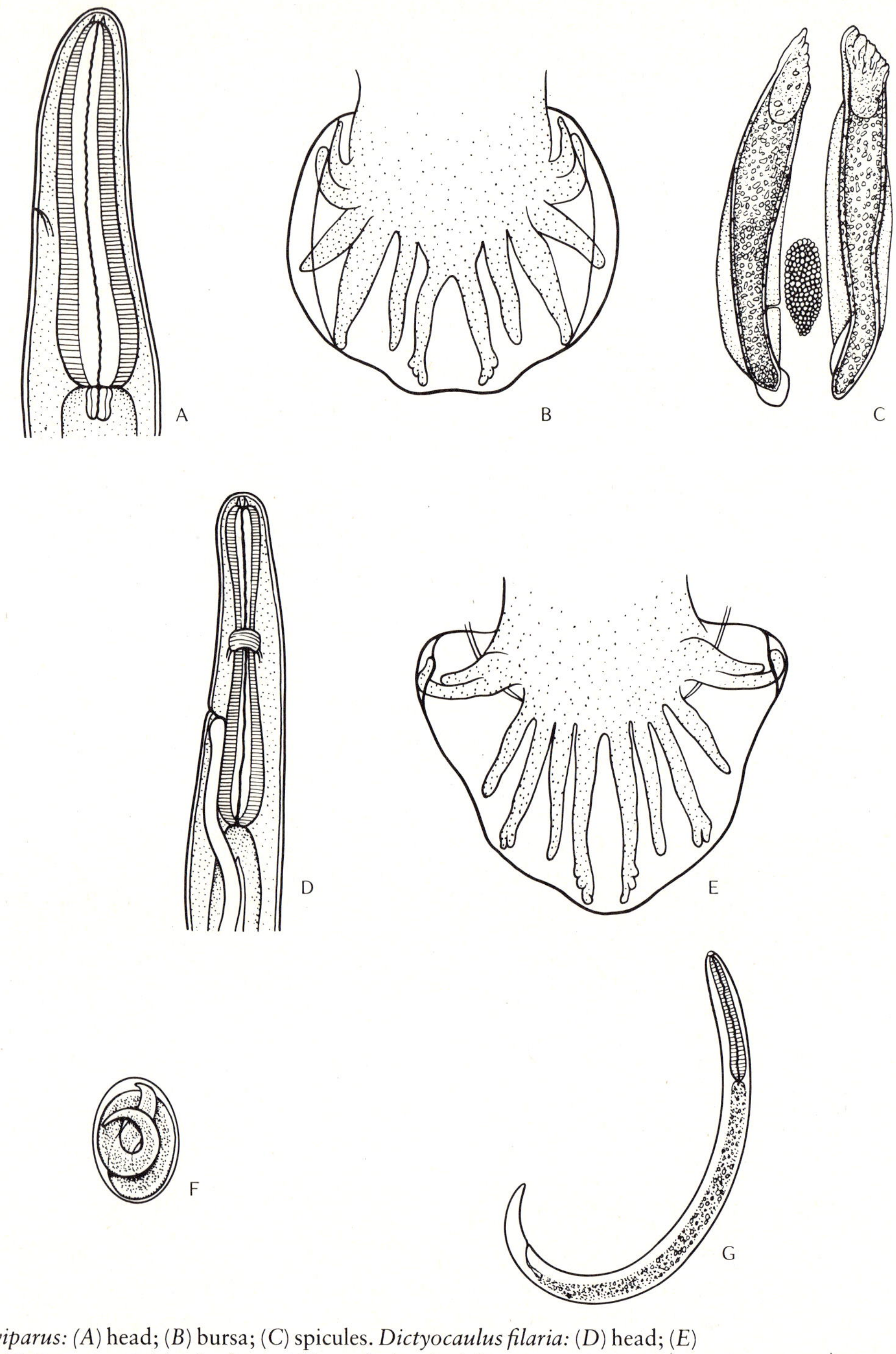


**FIG. 36.1**
*Dictyocaulus viviparus:* (*A*) head; (*B*) bursa; (*C*) spicules. *Dictyocaulus filaria:* (*D*) head; (*E*) bursa; (*F*) egg with $L_1$ as passed in the feces; (*G*) newly hatched $L_1$.

Work in Britain on lungworm disease of cattle brought to light a number of important aspects of transmission. For example, lush, tender grass tends to grow around the margins of a fecal deposit. One might think that this grass would be attractive to cattle, but they tend to avoid it. Without belaboring the point, it is well to remember that ruminants are not four-legged mowing machines. They select particular plants at particular stages of growth, and in this case somehow avoid the place where contamination with lungworm larvae may be greatest. In opposition to this grazing behavior of the larvae and their association with the fungus *Pilobolus*. This fungus is commonly found in manure and it has a sporangium with a cap that is explosively ejected when stimulated by dim light. Infective lungworm larvae curl up on the cap of the sporangium and are carried up to 2 m when the cap is ejected (Robinson, 1962). It appears that this worm–fungus association is a means of distributing the worms (Fig. 36.2).

Since cattle are most seriously affected by lungworms, and since there are no satisfactory drugs for the preadult stages, control of lungworm disease is a difficult problem. A number of approaches are taken:

1. Separation of age classes of animals
2. Treatment of patent infections to reduce pasture contamination
3. Avoidance of contaminated pastures
4. Promoting of immunity artificially

The general principle of separating age classes of animals is a useful one to control various parasitic diseases. In a dairy operation, say, replacement animals are likely to range in age from a few weeks to yearling heifers ready for breeding. If all of these animals are pastured together, the younger animals may be exposed to various parasites before they have the capacity to respond immunologically. However, if young animals are raised using the portable pen system described by Davis et al. (1952), there is.light exposure to infectious agents and time for the immune system to develop.

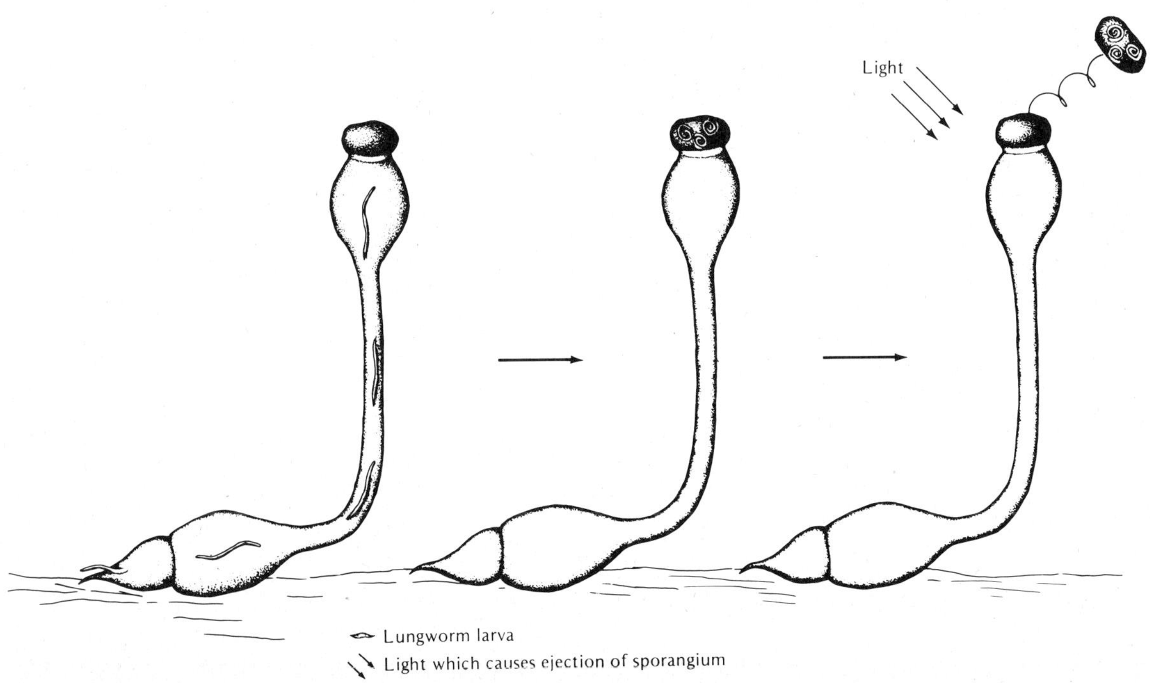


**FIG. 36.2**
Sporangium of the fungus *Pilobolus,* with *Dictyocaulus* larvae on it being ejected into the air.

Points 2 and 3 attempt to do the same thing, —that is, reduce exposure of animals to large numbers of larvae. Despite the relative sensitivity of the free-living stages, it is rare for a pasture to be avoided long enough for all the larvae to die (see Chap. 35).

Immunity has been exploited more successfully than other control measures, but it is not ideal. Lungworm disease in Britain had been a long-standing problem, and lack of adequate control measures stimulated an investigation on immunity during the 1950s at the School of Veterinary Medicine, University of Glasgow, Scotland. Older animals developed resistance of some sort, and investigations found development of a strong, specific immunity. Immunity was expressed against the preadults when they migrated to lymph nodes and lungs. The question to answer was how to produce immunity against the larvae without their damage. The answer lay in x-irradiation of the larvae. Larvae "attenuated" by x-irradiation developed partially in the lungs and then died. The system finally developed calls for administering 1000 irradiated larvae by mouth on two occasions at an interval of 1 month. The immunity produced reduced worm numbers by about 95% on challenge with 10,000 larvae shortly after immunization. Even after 4 months, there is still 89% immunity, as measured by the number of worms that will mature upon challenge. This vaccine is marketed in Europe as Dictol (Allan & Hanbury, Ltd.) and has been useful in controlling lungworm disease.

The vaccine has two drawbacks: shelf life and unexplained failures. It can be readily appreciated that x-irradiated larvae are not perfectly healthy, and one must use such a product soon after preparation. Conditions of shipment and storage are crucial for successful vaccination. In large countries such as Canada and the United States, shipping time may be a problem. In the tropics, maintenance of temperature can be difficult. Some of the unexplained failures may be due to death or the moribund nature of the larvae; whatever the reason, a livestock owner expects all cattle to be protected by a vaccine and may be reluctant to use a product that has a moderate chance of failure.

## REFERENCES AND FURTHER READINGS

Davis, L. R., G. W. Bowman, and D. A. Porter. 1952. Portable pens compared with other enclosures for control of diseases of dairy calves. *Vet. Med.* **47**:485–490.

Michel, J. F. 1969. The epidemiology and control of some nematode infections of grazing animals. *Adv. Parasitol.* **7**:211–282.

———. 1976. The epidemiology and control of some nematode infections in grazing animals. *Adv. Parasitol.* **14**:355–397.

Robinson, J. 1962. *Pilobolus* and the translation of the infective larvae of *Dictyocaulus viviparus* from faeces to pasture. *Nature* **193**:353–354.

Rose, J. H. 1973. Lungworms of the domestic pig and sheep. *Adv. Parasitol.* **11**:559–599.

Soliman, K. N. 1953. Migration route of *Dictyocaulus viviparus* and *D. filaria* infective larvae to the lungs. *J. Comp. Path. Therap.* **63**:75–84.

———. 1953. Studies on the bionomics of preparasitic stages of *Dictyocaulus viviparus* with reference to the same allied species in sheep, *D. filaria*. *Br. Vet. J.* **109**:361–381.

# 37

# *Lungworms with Vectors*

## LUNGWORMS OF BIGHORN SHEEP

The disease complex in Rocky Mountain bighorn sheep (*Ovis canadensis*) called *lungworm pneumonia* is of interest because its etiology involves both a number of biotic factors and human incursion into bighorn habitat. What impact lungworm pneumonia had on populations of bighorn sheep prior to the colonization of western North America is unknown. We do know that the disease is now a controlling factor in populations of bighorn and that further interference by humans is essential to the continued survival of herds in the Rocky Mountains. Having generated the problem, we now have to manage bighorn herds so as to minimize the effect of lungworm infections.

Protostrongyline lungworms are common in ruminants and rabbits in all parts of the world. They have indirect life cycles; terrestrial snails are their intermediate hosts.

NAME OF ORGANISM AND DISEASE CAUSED. *Protostrongylus stilesi* Dikmans, 1931, causes lungworm disease or verminous pneumonia. A synergistic relationship exists between three organisms in bringing about clinical disease: the lungworms, at least one virus, and a bacterium. *P. frosti* Honess, 1942, and *P. rushi* Dikmans, 1937, are also common in bighorn sheep, but they contribute less to the lungworm pneumonia complex than *P. stilesi.*

HOSTS AND LOCATION. Adults are located in the parenchyma of the lung. The principal host is the Rocky Mountain bighorn sheep (*Ovis candensis*), but domestic sheep (*Ovis aries*) are infected occasionally.

DISTRIBUTION AND IMPORTANCE. They are found in North America in the area contiguous with the distribution of bighorn sheep throughout the Rocky Mountain chain. Prevalence of infection ranges from about 30% to nearly 100% in different herds.

In association with at least one virus and bacterium, lungworms cause debilitating and fatal pneumonia, principally in younger animals.

Before 1800 there were about 2 million bighorn sheep in North America. The current population is about 18,000, 1% of the original number. Individual herds have sometimes declined gradually over a period of years, but in other instances there have been spectacular die-offs over a short period (Table 37.1). Whether all of these die-offs were the result of lungworm pneumonia is not known; however, recent investigations have shown that lungworms and associated microorganisms are the principal cause of mortality.

MORPHOLOGY. Since *P. stilesi* adults are embedded in the parenchyma of the lung, the length of the adults is not known. The bursa and spicules of the male are used in identification (Fig. 37.1).

**TABLE 37.1**
Recorded Die-offs of Bighorn Sheep, Some Presumably from Lungworm Pneumonia.

| DATE | HERD LOCATION | NO. IN HERD | PERCENTAGE MORTALITY |
|---|---|---|---|
| 1905–06 | Salmon River, Idaho | ? | Mass |
| 1923–24 | Tarryall Mtns. Colo. | 350 | 96 |
| 1924–25 | Sun River, Mont. | 250 | 74 |
| 1926–27 | Glacier Nat. Park, Mont. | 150 | 17 |
| 1927–29 | Yellowstone Nat. Park, Wyo. | 346 | 78 |
| 1930 | Nat. Bison Range, Mont. | 100 | 66 |
| 1934–36 | Jackson Hole, Wyo. | 1200 | 50 |
| 1936–37 | Glacier Nat. Park, Mont. | 84 | 39 |
| 1940 | Nat. Bison Range, Mont. | 65 | 54 |
| 1952–53 | Pikes Peak, Tarryall and Kenosha Ranges, Colo. | 1500 | 66 |
| 1955–56 | Salmon River, Idaho | 150 | 14 |
| 1955–56 | Glacier Nat. Park, Mont. | 50 | 20 |
| 1939–59 | Rocky Mountain Nat. Park, Colo. | 329 | 36 |

LIFE CYCLE. Despite their lying in the tissue of the lung, eggs laid by the adults reach the alveoli, pass forward, hatch, and are carried out with the feces. $L_1$s enter the snail by penetrating the integument. The intermediate hosts are tiny terrestrial snails in the genera *Pupilla, Vertigo,* and *Euconulus*, but many terrestrial snails can be infected. The larvae grow and molt twice to reach the infective $L_3$. Infectivity for sheep is reached about 2 weeks after infection depending on the temperature. When snails are ingested with the herbage, the larvae leave the sheep intestine and migrate through the bloodstream to the lungs (Fig. 37.2).

Infection can be transferred transplacentally from the pregnant ewe to the lamb. This seems to be an important mode of infection since a high proportion of lambs are infected and thereby eliminated from the population (Hibler et al., 1972; Hibler et al., 1974).

DIAGNOSIS. Larvae are readily found in the droppings of infected sheep, and this technique can be used for determining both whether an individual is infected and the extent of infection in a herd. At postmortem examination, lungworm lesions can usually be seen on the dorsal surface of the diaphragmatic lobes of the lungs. Portions of worms can be dissected out of the tissue for identification. The question of whether lungworm pneumonia may be the cause of illness and death is determined by herd history, examination of vectors, and isolation of the microbial agent(s) in the laboratory.

HOST–PARASITE INTERACTIONS. The lungworms themselves cause damage to a relatively small proportion of the lung. However, their association with viruses and bacteria bring about pneumonia, which results in extensive damage to the lungs followed by debilitation and death.

Young animals are especially susceptible to infection; data indicate that a large proportion of lambs are infected before birth. It is these young animals that are removed from the population; in addition, animals that are not otherwise in good nutritional condition are likely to suffer from the effects of the combined agents and succumb to them.

TREATMENT. Benzimidazoles are highly effective.

EPIDEMIOLOGY AND CONTROL. Lungworm pneumonia in bighorn sheep is the result of interaction between two or three infectious agents combined with the general physical condition of the host. It should be recognized that a die-off of bighorn is merely the culmination of a series of other events that have worked indirectly to the detriment of the sheep.

Bighorn have been hunted, but they were not a staple food such as the buffalo and deer were for early settlers.

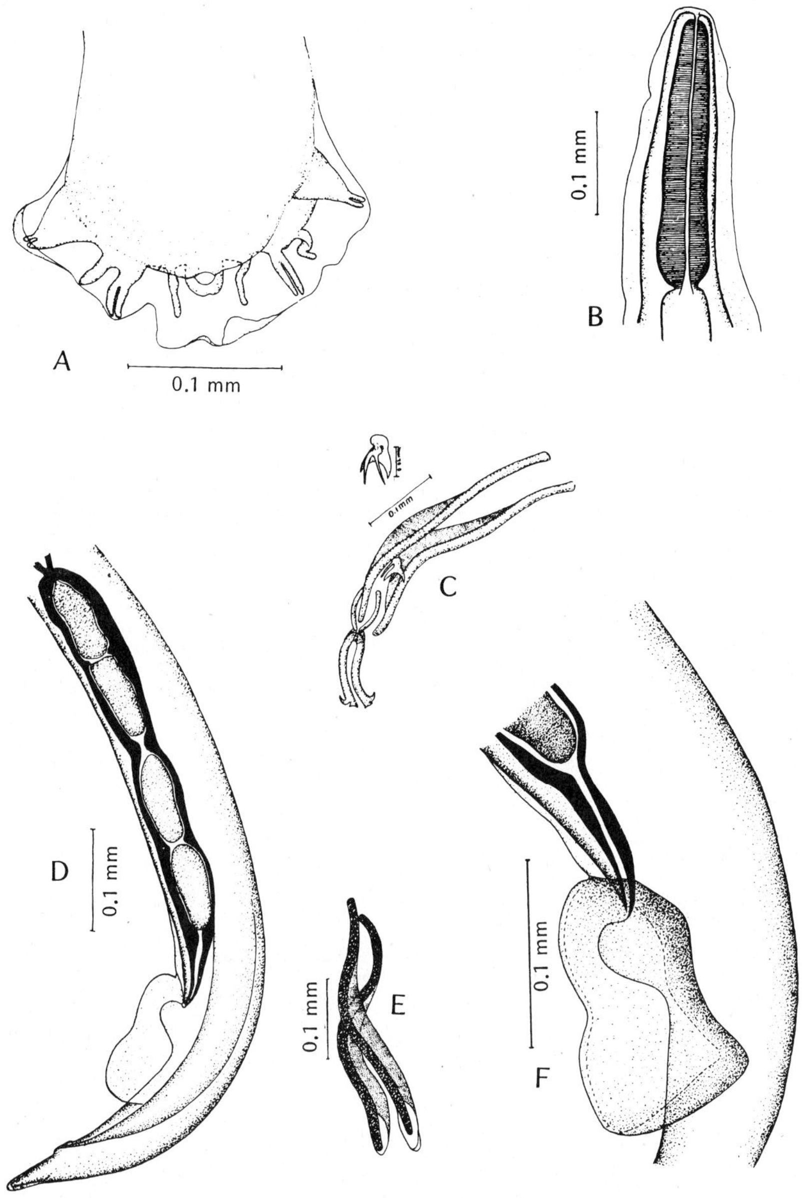


**FIG. 37.1**
Diagram of *Protostrongylus stilesi*. (*A*) Bursa of male; (*B*) esophagus; (*C*) spicules, gubernaculum, and telamon; (*D*) tail end of female; (*E*) spicules; (*F*) vulva and provagina of female. [From Dikmans, 1931.]

Bighorn occupy high mountain areas that are harsh and inaccessible; this niche was where they were seen by early explorers and settlers, except for the population along the breaks of the Missouri River in Montana and the Dakotas. They are wary animals that tend to drift away at the sight of humans. They also do not mix well with other species of ruminants such as domestic cattle and sheep.

Bighorn tend to remain in herds; they migrate seasonally to lower elevations during the winter, but still inhabit a harsh ecologic niche. They also are creatures of habit, using certain bedgrounds that are

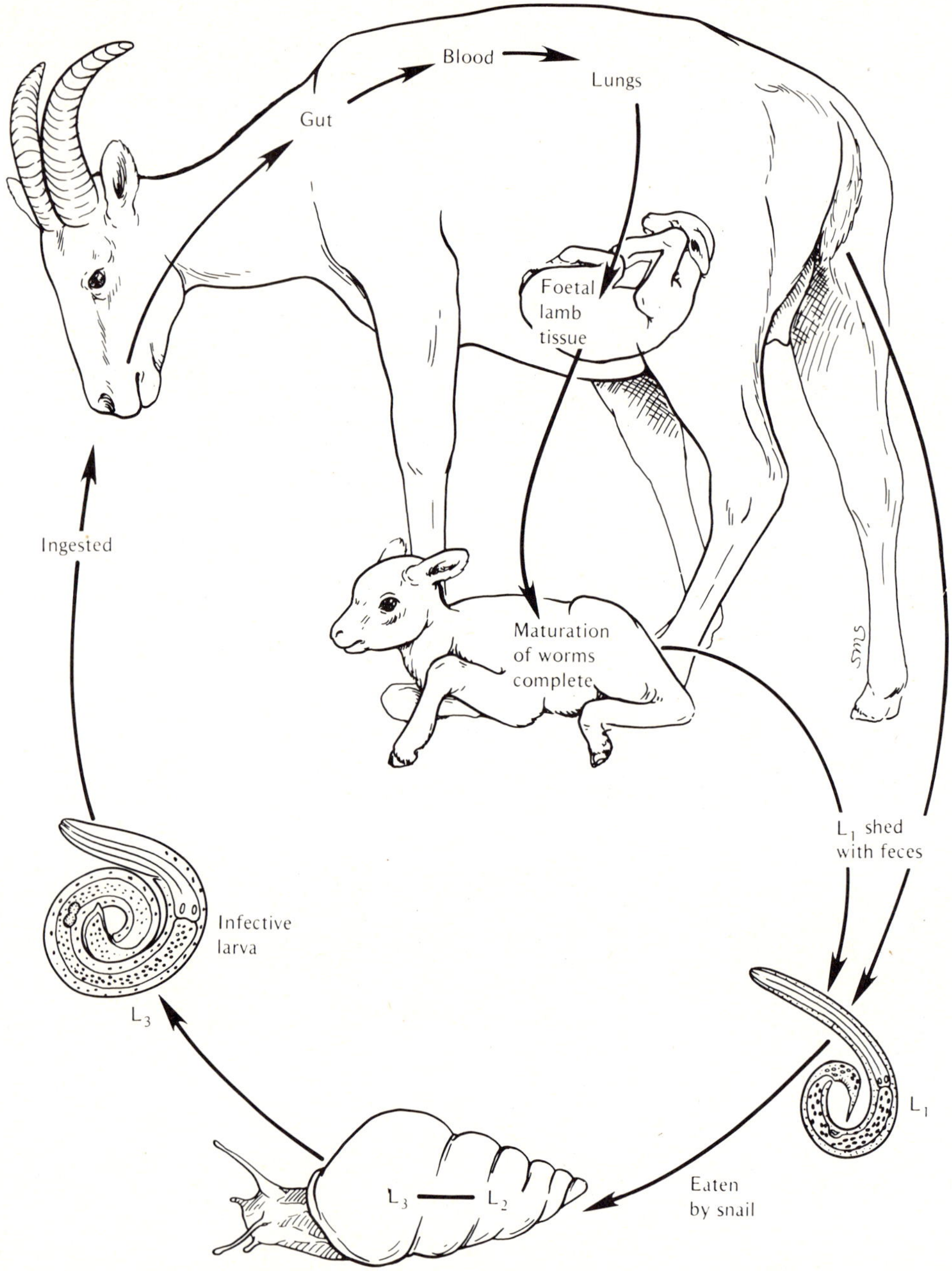


**FIG. 37.2**
Diagram of the life cycle of bighorn sheep lungworms.

free of snow and are somewhat protected from the wind.

With the incursion of humans into the mountains for travel, mining, settlement, and grazing of domestic animals, the habitat and the habits of the bighorn were altered. They drifted away to more restricted areas. Their traditional grazing areas were more and more occupied by sheep and cattle and the forage deteriorated. Bedgrounds became a solid mat of feces, and snails grazed on the feces as a source of food.

It is hard to conceive of these animals being crowded and subject to the pressure of parasitic diseases we associate with the intensive grazing practices used with domestic animals. But that is exactly what happened. Human activities in the mountains restricted the habitat and movement of the sheep and reduced their level of nutrition. Lungworms were present and precipitated the pathologic condition.

Many human modifications of the world are irreversible, and the uses that are now made of high mountain areas for grazing of livestock, recreation, and other activities are not going to be stopped. The problem posed is how to maintain healthy populations of bighorn in the face of heavy pressure from humans. There have been partial solutions to the dilemma. For example, it has been found feasible to administer drugs to animals in fermented apples. Sheep can be lured in to bait stations by providing fermented apples for a period of time. Then, once they are habituated, a drug such as cambendazole can be added to the mash. In herds of central Colorado, this technique has reduced the level of infection so that young lambs survive through their first summer and winter. Unfortunately, the crux of the problem is not whether animals are infected with lungworms, but rather their general health and nutritional level. It is not feasible to return to the eighteenth century and abandon the high mountains. We are left with stopgap measures that buy time until better solutions can be devised. What solutions, if any, will be developed will be revealed with the passage of time.

## LUNGWORMS IN SWINE

Among the most interesting of the complex epidemiologic patterns of helminthic diseases is that of lungworm infection in domestic swine and the associated swine influenza. There are so many factors that come together in this disease complex that the probability of its occurring is absurd. Not only are there the lungworm and the definitive host, which we have seen in *Dictyocaulus*, but an earthworm intermediate host and then the virus of swine influenza carried from one animal to the next by the lungworm. Finally, a precipitating factor of the weather is crucial.

NAME OF ORGANISM AND DISEASE CAUSED. *Metastrongylus apri* (Gmelin, 1790) Vostokov, 1905; *M. pudendotectus* Vostokov, 1905 (=*Choerostrongylus p.*), and *M. salmi* (Godoelst, 1923) cause lungworm disease and are associated with swine influenza.

HOSTS AND LOCATION. The adult worms are in the air passages of the lungs of swine and other members of the family Suidae. Occasionally they are found in ruminants, dogs, and humans.

DISTRIBUTION AND IMPORTANCE. The three species are cosmopolitan in swine, but their prevalences differ. *M. apri* is the most common; its prevalence in domestic swine in moist climates is 50 to 80%. *M. pudendotectus* is the next most frequent, and *M. salmi* has a prevalence of perhaps 4%.

LIFE CYCLE. The development of all three species of lungworms in pigs is essentially the same. The adults lie in the air passages of the lungs, where copulation and egg laying occur. The eggs are passed up the respiratory tree by ciliary action, and when they reach the upper part, they are coughed up and swallowed. The eggs have $L_1$s in them when they are passed with the feces. They hatch when eaten by earthworms (*Lumbricus terrestris, Eisenia foetida,* among others), releasing the $L_1$, which penetrates the gut and develops in the coelom of the earthworm. In 2 to 3 weeks the $L_3$ stage is reached and it encysts in the gut.

The $L_3$s do not develop further until eaten by a susceptible animal; they can remain alive in the earthworm for as long as 4 years. If the earthworm dies, the exposed larvae probably live only for 2 weeks or so.

Pigs are infected usually by eating infected earthworms, but secondarily by eating the free $L_3$s. The $L_3$ is released in the small intestine and passes through the wall of the gut to the lymph system. The $L_3$ or $L_4$ moves by the circulatory system to the lungs, where it is filtered out in the capillaries. The $L_4$ breaks out of the alveoli of the lung, molts once more, and becomes mature about 4 weeks after infection. Egg production may continue for several months.

HOST–PARASITE INTERACTIONS. Swine lungworms are not generally considered to be highly pathogenic; however, they can cause significant damage on their own. During migration of the preadult worms, the greatest damage is to the alveoli of the lungs. Each $L_4$ causes a small hemorrhage (petechia) where it leaves the capillary to enter the air sac; the greater the number of worms, the greater the damage. As the worms mature, they irritate the air passages and may block some of the smaller ones. Pigs breathe hard, cough, and may suffer from chronic pneumonia.

Greater economic losses are caused by swine influenza than directly by lungworm infection. The association of the virus of swine influenza is not an obligate one, but a close relationship has developed so that the virus is protected in the earthworm. Swine influenza is not usually a fatal disease; deaths are usually less than 5% but may reach 25% in some herds. Most of the losses are production losses: Animals fail to gain well, and may be affected for a long period after the acute stage of the infection is over.

It is likely that the disease in swine arose from the organism that caused the pandemic of influenza in humans during 1918. In that pandemic 21 million people died, and probably 25 times more were affected. The disease was first recognized in swine during the summer and fall of 1918. The clinical manifestations were similar to those appearing in humans, and they were different from any disease then known in swine.

The disease has an incubation period of about 2 days, at which time there is onset of fever. Prostration, failure to eat, and respiratory symptoms are manifested as labored breathing and coughing. Death may occur from 3 to 6 days after onset. Recovery is usually seen by the sixth day. After introduction of the disease into a herd, it rapidly transfers from animal to animal. Although animals are severely affected during the disease, the death rate ranges from 1 to 4%. Recovered animals are immune.

Outbreaks occur in fall and early winter, and generally follow cold, wet weather. When winter sets in, outbreaks disappear. The virus overwinters in association with swine lungworms. If a pig is infected with adult lungworms at the time when the disease is in the acute phase, lungworm eggs containing masked virus are shed from the host. The embryonated eggs are eaten by the earthworms. If the earthworm is eaten by a pig, the $L_3$ is digested out, migrates through the wall of the intestine of the pig, reaches the bloodstream, and travels by way of the blood to the lungs.

A pig so infected is, in Shope's words, "in a precarious situation." The virus is still "masked," but influenza is precipitated in 48 hr by any upset. Under farm conditions, it is inclement weather. In the labortory, other stimuli such as injections with bacteria bring on the disease. Interestingly, the virus requires the presence of a bacterium, *Haemophilus influenzae suis*, to make its full effect felt. This bacterium is frequently a normal, nonpathogenic inhabitant of the respiratory tract of swine. The two agents act together to bring about the clinical disease.

The lungworm is a reservoir only during the periods between outbreaks—about 9 months of any one year. It has been found that the virus can persist in lungworm larvae outside the definitive host for as long as 32 months. Further, it can persist in swine lungs for at least 3 months without giving rise to clinical disease. The total time between disease outbreaks can then be nearly 3 years.

EPIDEMIOLOGY AND CONTROL. It is not surprising that the prevalence of lungworm infection in swine exceeds 70% in many areas. The adult stages may persist for 8 months in the pig, and the stages may live for years in the intermediate host.

Kates (1941) studied the survival of eggs by exposing them to various environmental conditions and then determining infectivity for earthworms. An example of the hardiness of the eggs is contained in his study on the effect of freezing where temperatures between −8 and −20 ° C had little effect on viability until more than 49 days had passed. On the other hand, desiccation was detrimental to survival, especially at 37 to 39° C, at which fewer than 1% survived 10 days. At 17 to 22° C, survival was better, but none lived longer than 25 days. At 1 to 2° C, about 10% lived 38 days.

Once they gain the earthworm, the larvae seem to be almost completely protected. In a study made on an abandoned pasture near the Beltsville Agricultural Research Center in Maryland, it was found that earthworms were still infected 4 years after any hogs had been on land. Seventy-five sexually mature earthworms were found to have a prevalence of infection of 46.6% with a level of infection up to 100 larvae. These larvae were found to be infective for a pig. Immature earthworms from the same pasture were uninfected (Spindler, 1938). Interestingly, although lumbricid earthworms have a potential longevity of 4 to 8 years, the life span is considered generally to be no more than a few months. Under protected cultural conditions, however, the following have been found to be maximal survival times: for *Allolobophora longa*, 10¼ years; for *Lumbricus terrestris*, 6 years; for *Eisenia foetida*, 4½ years (Edwards & Lofty, 1972). If one assumes that most earthworms live only a few months and that the maximal life span is 6 years, only 5 to 10% of the original population of the worms would be alive at 4 years. Spindler's (1938) finding that nearly half of the worms were still infected at 4 years is startling.

The recommendations usually made for control of lungworms are that the animals be kept on areas that are dry or well drained and that temporary pastures be used and contaminated pastures be left for more than 4 years before allowing pigs on them. Since pigs will root in the soil if allowed, ringing their noses reduces that proclivity; they then are less likely to eat earthworms. In areas of high technology, raising swine on concrete eliminates the possibility of lungworm infection.

Treating animals with one of the better anthelmintics benefits both the individual and the herd. Animals so heavily infected that they cough and have difficulty breathing benefit from removal of the worms. Since the adult worms continue to lay eggs for months, they also continue to contaminate the area; reduction in the amount of contamination is beneficial to the herd; also, since contamination can persist for years, any measure to desaturate the environment is desirable. Lung damage in the air-passage-dwelling lungworms takes place mainly when they break out of the alveoli. Onset of influenza is also likely at this time. Therefore, much of the damage may be done to the individual by the time treatment is administered.

Control of swine influenza is based principally on vaccination. Protection is good, and in enzootic areas vaccination is given before animals might be stressed by inclement weather.

The relation of swine influenza to human influenza is circumstantial, but nevertheless strong. The disease was not known in swine until the pandemic at the close of World War I, but it spread rapidly. The clinical signs in swine are essentially the same as in humans. New strains of influenza virus may arise from infections in swine, probably by recombination of viruses. In the late 1970s so-called swine flu became of concern to public health officials. A few persons having contact with domestic pigs fell ill and the question arose of whether we were on the verge of a serious epidemic in humans. The virus was of a new antigenic type, and it had a fairly high degree of virulence. The decision was ultimately made to produce enough vaccine to vaccinate all of the persons in the United States who were at high risk—the young and the old. The decision was made in a highly political atmosphere and in such a way that the federal government became legally responsible for the economic and health impacts of the program. The decision and its consequences are an interesting example of biopolitics.

## METASTRONGYLE INFECTIONS IN HUMANS

No metastrongyle is considered to be a normal inhabitant of humans. However, two species found in rodents cause problems: *Angiostrongylus cantonensis* and *A. costaricensis*, found in the Eastern and Western Hemispheres, respectively. We will consider *A. cantonensis* here, but the interested student can obtain information on *A. costaricensis* from Loria-Cortes and Lobo-Sanahuja (1980). Keys to members of the genus are found in Kinsella (1971).

Both of these nematodes are parasites of the arterial system, rather than the lungs, and the problems of pathogenesis for humans result to an extent from their being in an abnormal host.

NAME OF ORGANISM AND DISEASE CAUSED. *Angiostrongylus cantonensis* (Chen, 1935) Dougherty, 1946, causes angiostrongylosis, eosinophilic meningitis, meningitis or tropical eosinophilia in humans. Alicata and Jindrak (1970) reviewed in detail all aspects of angiostrongylosis and the reader is referred to this work as well as to Alicata (1965, 1969a) for additional information.

HOSTS AND LOCATION. *Rattus* spp., including the Norway and black rat, seem to be the principal hosts, but bandicoot rats (*Bandicota* spp.) are also included among the important hosts. The final site of the adult worms in the normal definitive host is the pulmonary arteries; in humans the worms are most often in the central nervous system but may also be found in other organ systems including the eye.

DISTRIBUTION AND IMPORTANCE. Infections in rats are widespread in tropical areas of the Eastern Hemisphere such as the Malay Peninsula, Japan, the islands of both Melanesia and Micronesia, northern Australia, Hawaii, and East Africa including Madagascar. According to Alicata and Jindrak (1970), the original enzootic area of *A. cantonensis* was Madagascar, the large island off the southeastern coast of Africa, from which it spread northward to India and Sri Lanka, thence to the Malay Peninsula. Migration then took place in a number of directions: northward to China, eastward through the islands of the South Pacific, and southward to New Guinea and Australia (Fig. 37.3).

Infections in humans have been found in Southeast Asia, Japan, many islands of the South Pacific, and Hawaii. All well-documented infections have been found to be contiguous with the distribution of the infection in rats.

Although infections are widespread in rats, infections in humans are sporadic in the present enzootic area and they are related to types of food eaten and their method of preparation (see Epidemiology and Control). Infections in humans seldom result in death; it is therefore difficult to determine with certainty whether *A. cantonensis* was actually the cause of eosinophilic meningitis in any specific instance. Most often, only a few persons suffer from the infection at a particular location and time.

MORPHOLOGY. The female worms are 18 to 33 by 0.28 to 0.5 mm long with the vulva about 0.25 mm from the posterior end. Males are 15 to 22 by 0.25 to 0.35 mm and have subequal spicules that are 490 to 1200 μm long. Both sexes feed on blood and therefore have a red intestine around which the reproductive system winds; this gives a barber-pole appearance also noted in such worms as *Haemonchus* (Fig. 37.4).

LIFE CYCLE. The life cycle described here is that which occurs in the normal hosts, rats, and in the invertebrate hosts. As is true with many other metastrongyles, the $L_3$s are ensheathed in the cast cuticles of the two previous stages. The $L_3$s are eaten by rats and within a day reach the central nervous system through

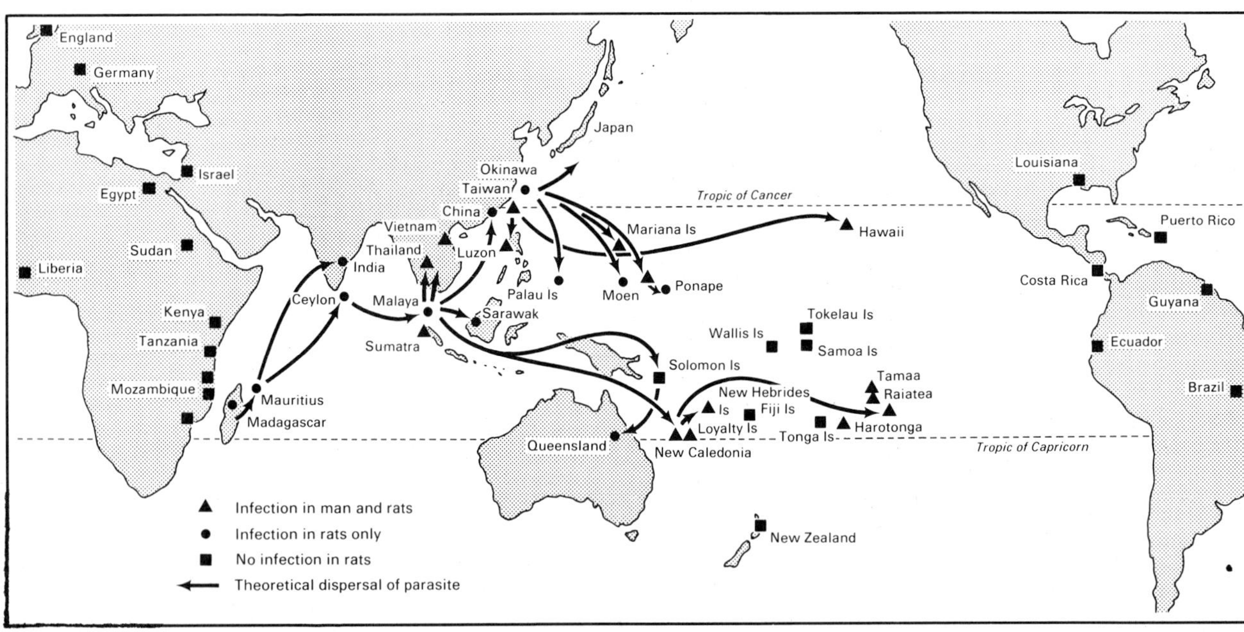


**FIG. 37.3**
Theoretical dispersal of the rat lungworm, *Angiostrongylus cantonensis*. [Redrawn from Alicata, J.E., and K. Jindrak, 1970. Angiostrongylus *in the Pacific and Southeast Asia*. Charles C Thomas, Springfield, Ill.]

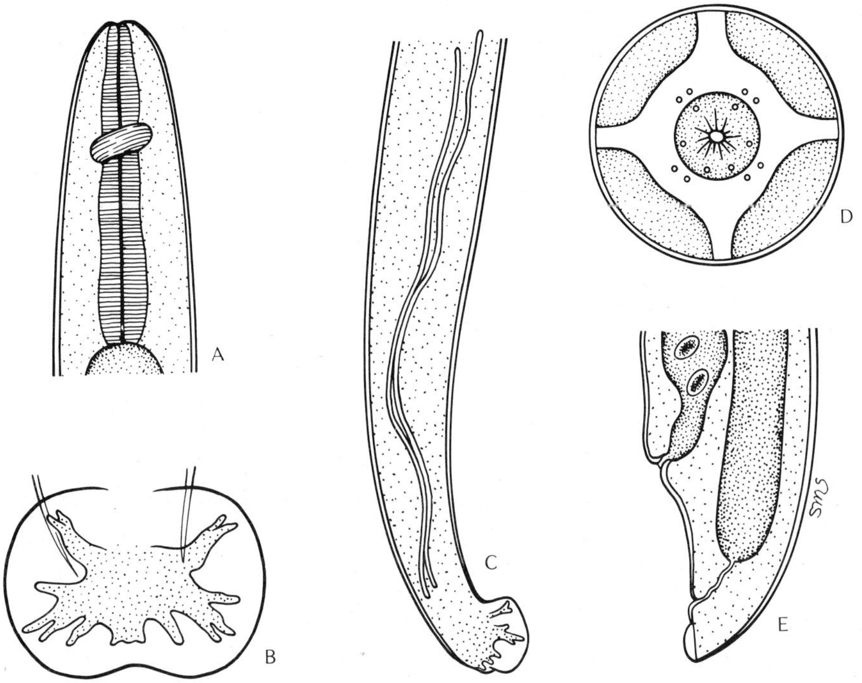


**FIG. 37.4**
Adult of *Angiostrongylus cantonensis*. (*A*) Anterior end; (*B*) bursa of male; (*C*) spicules of male; (*D*) head; (*E*) posterior of female.

the circulatory system. The early adult stage is reached in 11 to 13 days after infection, and migration then begins. The worms move to the meninges covering the brain and remain between the layers (subarachnoid space) for about 2 weeks. They then move to the venous system and are carried through the heart to the pulmonary arteries, where they remain and begin laying eggs. The unembryonated eggs are carried with the arterial blood to the capillaries in the alveoli of the lung, where they lodge. The $L_1$s break out of the alveoli and pass up the respiratory tree and are swallowed; the prepatent period is 42 to 45 days, with larvae appearing in the feces at that time (Fig. 37.5).

The first intermediate host is a snail or slug, usually a terrestrial one. The host range is quite broad, but development takes place more readily in some species than in others (Alicata, 1965). The $L_1$s penetrate the foot of the mollusk and reach the $L_3$ in about 18 days at 21 to 25° C.

Having reached the infective stage, the $L_3$ organism can reach the definitive host by one of the following means:

1. The gastropod can be eaten.
2. The $L_3$s can escape from the gastropod and be eaten with other food or with water.
3. A transport host can be infected by either of these routes and then it, in turn, may be eaten by the definitive host. The most important transport hosts seem to be freshwater crustacea such as prawns and crab, and land planaria. The most frequent mode of infection in wild rats is not known.

The life cycle in humans seems to be the same as in the rat except for the final site. In humans the worms remain in the brain, lying either between the meninges or migrating through the brain tissue itself.

DIAGNOSIS. Diagnosis is difficult in humans; in the other hosts, an examination may be done to find the worms in the pulmonary arteries. Unless the person has died, the diagnosis of *A. cantonensis* infection is indirect and is based on the following:

1. Clinical signs and symptoms (Table 37.2)
2. Food habits and a history of having eaten a likely source of larvae about 2 weeks previously

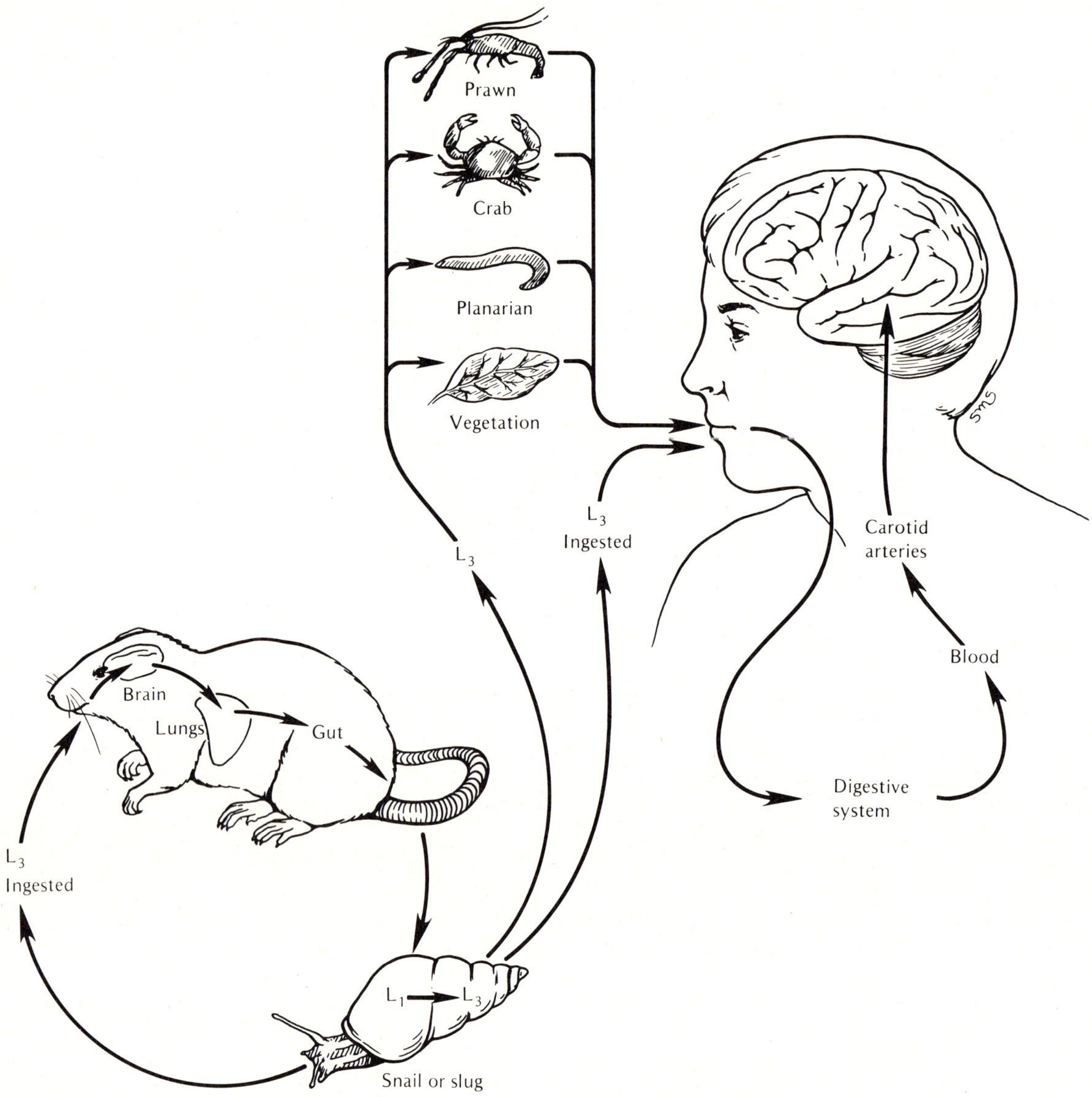


**FIG. 37.5**
Diagram of the life cycle of the rat lungworm, *Angiostrongylus cantonensis,* and avenues of human infection.

3. The presence of eosinophils in the cerebrospinal fluid (CSF)
4. Evidence of antibody production
5. Finding worms in the CSF (rare; only two instances are on record)

There are many causes of meningitis; therefore both a skilled diagnostician and a laboratory capable of isolating and identifying various infectious agents are required to determine whether bacterial, viral, or other parasitic agents may be the cause of the disease. Skin and serological testing can indicate whether a nematode may be involved in producing antibody. A further narrowing of the possible causes of meningitis comes from questioning of the patient. In an enzootic area, the typical symptoms coupled with suggestive laboratory tests and food habits consistent with becoming infected provide a good indication of infection with *A. cantonensis*.

Other infections with which *A. cantonensis* can be confused are *Taenia solium* cysticerci, *Paragonimus westermani*, and *Gnathostoma spinigerum*. Noninfectious processes such as cancer and multiple sclerosis may also cause an increase in eosinophils (Kurberski & Wallace, 1979).

TREATMENT. Thiabendazole has been used with some success in treating laboratory animals. This drug has been suggested as appropriate for treatment of some *A. cantonensis* infections in humans. Since thiabendazole has a low degree of toxicity, harming the patient by the drug is unlikely.

Some reservations have been expressed (Alicata & Jindrak, 1970) on the desirability of treatment in some instances and on the level of drug that might be administered to a patient. The concern is that a worm migrating through the brain tissue clearly causes damage, but a dead worm becomes a foreign body against which a strong leukocytic reaction is directed. The result is a large granuloma, which itself causes damage. Roughly the same hesitation is expressed on the protocol of treatment; if a high level of drug is used, all of the worms may be killed at once, producing a toxic reaction that could be detrimental to the patient. Although some cases are on record in which persons have died from the infection, most recover completely without specific therapy. The physician's cardinal principle pertains: first, do no harm.

**TABLE 37.2**
Clinical Signs and Symptoms of Eosinophilic Meningitis and Presumably Infection with *A. cantonensis*

| SIGN OR SYMPTOM | PERCENTAGE HAVING |
|---|---|
| Headache | 90 |
| Stiff neck | 56 |
| Vomiting | 56 |
| Paresthesia (abnormal sensation) | 54 |
| Fever | 41 |
| Paralysis or tremors | 9 |

[Modified from Kuberski & Wallace, 1979]

HOST–PARASITE INTERACTIONS. Frequently, but not invariably, *A. cantonensis* has a low degree of pathogenesis in the rat. On the other hand, in humans there is a full range of effects including severe damage to the central nervous system leading to death.

In humans it is the consensus that there are many inapparent infections and others in which there may be only mild symptoms of meningitis but with eosinophils in the cerebrospinal fluid. Lesions are related to both the location of the worms and the number in the nervous tissue. Eosinophilia indicates an allergic response to the worms and the meningitis indicates that there is interference with normal nerve impulse propagation and probably swelling of the CNS. If damage is sufficient, paralysis of particular parts of the body or of areas served by the cranial nerves may result.

EPIDEMIOLOGY AND CONTROL. The life cycle is probably maintained in rats by their eating infected snails, but freshwater crustacea are a second source.

The food habits of people in various localities or their ethnic derivation determine whether infection is likely to occur. This is another example of a parasite that cycles silently in the nonhuman population and can be transmitted when humans come into contact with the cycle. In this case, food preferences are the key.

Among certain peoples of the Pacific islands, foods such as tiaro on Tahiti or keloguen on Guam consist principally of grated coconut mixed with the juice from ground or crushed freshwater prawns. On some of the Pacific islands, 5 to 10% of freshwater prawns have been found to be infected, mostly with two or three larvae of *A. cantonensis* but in one instance with over 3000.

The predeliction of Europeans or North Americans for green salads leads to infection. Terrestrial gastropods, and planaria secondarily, shed infective larvae onto plants in the field and the larvae are then eaten with the salad. In some enzootic areas, eosinophilic meningitis occurs in Caucasians and not in the natives because the latter do not usually eat leafy salads.

In a case report (Kojima et al., 1979) a man suffered typical *A. cantonensis* signs and symptoms (Table 37.2) following ingestion of live slugs. The 59-year-old farmer had back pain, and a Chinese physician had recommended eating raw slugs, which he had done for about 4 years, eating from five to ten daily. It is not stated whether the diet of slugs alleviated his back pain.

Since eosinophilic meningitis is usually a sporadic infection in small numbers of persons at any one time, control programs have not been developed. At the public health level, there are two approaches that might be considered:

1. Rat control
2. Education to alter food habits

Since the definitive host range is relatively narrow, rat control might be worthwhile, but in general rats are not restricted to areas of human habititation. Control is therefore likely to be difficult.

The other alternative, altering food habits, may be equally difficult. When it is explained to people what might happen if they eat tiaro or a green salad, the response is likely to be, "yes, what you say is probably true"; then they will continue to eat what they always have.

Drinking untreated surface water is also a hazard since larvae escape from the intermediate host or from planaria and may be transmitted if this happens in small streams.

At the public health level, probably the best that can be done is to monitor the extent of the infection in the intermediate and transport hosts and make available information on the consequences and likelihood of infection. The responsibility then falls on the individual on whether to avoid infection or not.

## REFERENCES AND FURTHER READINGS

Alicata, J. E. 1964. Parasitic infections of man and animals in Hawaii. *Hawaii Ag. Exper. Sta. Tech. Bull.* No. 61.

———. 1965. Biology and distribution of the rat lungworm. *Angiostrongylus cantonensis*, and its relationship to eosinophilic meningitis and other neurological disorders of mand and animals. *Adv. Parasitol.* 3:223–248.

———. 1969a. The present status of *Angiostrongylus cantonensis* infection in man and animals in the tropics. *J. Trop. Med. Hyg.* 72:53–63.

———. 1969b. On the identity of *Angiostrongylus* from rats in Brisbane, Australia, and on the life history of *Angiostrongylus cantonensis*. *J. Parasitol.* 55:323.

Alicata, J. E., and K. Jindrak. 1970. Angiostrongylus *in the Pacific and Southeast Asia*. Charles C Thomas, Springfield, Ill.

Buechner, H. K. 1960. The bighorn sheep in the United States, its past, present and future. *Wildl. Monogr.* 4:1–174.

Dikmans, G. 1931. Two new lungworms from North American ruminants and a note on the lungworms of sheep in the United States. *Proc. U.S. Nat. Mus.* 79 (article 18).

Edwards, G. A., and J. R. Lofty. 1972. *Biology of Earthworms*. Chapman and Hall Publishers, London, p. 72.

Forrester, D. J. 1971. Bighorn sheep lungworm pneumonia complex. In J. W. Davis and R. C. Anderson (eds.), *Parasitic Diseases of Wild Mammals*. Iowa State University Press, Ames, Iowa, pp. 158–173.

Hibler, C. P., R. E. Lange, and C. J. Metzger. 1972. Transplacental transmission of *Protostrongylus stilesi* in bighorn sheep. *J. Wildl. Dis. Assoc.* 8:389.

Hibler, C. P., C. J. Metzger, T. R. Spraker, and R. E. Lange. 1974. Further observations on transplacental transmission of *Protostrongylus stilesi* in bighorn sheep. *J. Wildl. Dis. Assoc.* 10:39–41.

Honess, R. F. 1942. *Lungworms of Domestic Sheep and Bighorn Sheep in Wyoming*. University of Wyoming Ag. Exp. Stat. Bull. No. 225.

Honess, R. F., and N. Frost. 1942. *A Wyoming Bighorn Sheep Study*. Wyoming Fish and Game Dept. Bull. No. 1.

Kates, K. C. 1941. Observations on the viability of eggs of lungworms of swine. *J. Parasitol.* 27:265–272.

Kinsella, J. M. 1971. *Angiostrongylus schmidti* sp. n. (Nematoda: Metastrongyloidea) from the rice rat, *Oryzomys palustris*, in Florida with a key to the species of *Angiostrongylus* Kamensky, 1905. *J. Parasitol.* 57:494–497.

Kojima, S., H. Hata, M. Kobayashi, M. Yokogawa, N. Takahashi, T. Takaso, and J. Kaneda. 1979. Eosinophilic meningitis: A suspected case of angiostrongylosis found in Shizuoka Prefecture, Honshu, Japan. *Am. J. Trop. Med. Hyg.* 28:36–41.

Kuberski, T., and G. D. Wallace. 1979. Clinical manifestations of eosinophilic meningitis due to *Angiostrongylus cantonensis*. *Neurology* **29**:1566–1570.

Loria-Cortes, R., and J. F. Lobo-Sanahuja. 1980. Clinical abdominal angiostrongylosis. A study of 116 children with intestinal eosinophilic granuloma caused by *Angiostrongylus costaricensis*. *Am. J. Trop. Med. Hyg.* **29**:538–544.

Marsh, H. 1938. Pneumonia in Rocky Mountain bighorn sheep. *J. Mammal.* **19**:214–219.

Rose, J. H. 1973. The lungworms of domestic pigs and sheep. *Adv. Parasitol.* **11**:559–570.

Schwartz, B., and J. E. Alicata. 1934. *Life History of Lungworms Parasitic in Swine*. USDA Tech. Bull. No. 456. U.S. Department of Agriculture, Washington, D. C.

Shope, R. E. 1958. Swine influenza. In H. W. Dunne, (ed.), *Diseases of Swine*. Iowa State College Press, Ames, pp. 81–98. (This chapter reviews the work of Shope published over many years, mostly in *Am. J. Hgy.*)

Spindler, L. A. 1938. Persistence of swine lungworm larvae in earthworms. *Proc. Helm. Soc. Wash.* **5**:63.

Uhazy, L. E., and J. C. Holmes. 1973. Lungworms in Rocky Mountain bighorn sheep of western Canada. *Can. J. Zool.* **51**:817–824.

# 38

# Ascaris *of Humans and Swine*

Members of the order Ascaridida are stout-bodied worms that are easily seen with the naked eye. Most of them have three prominent lips surrounding the mouth, but some have none and some have six (see Chap. 30 for a complete description). The posterior ends of the males are often curled ventrally, and most have two stout spicules.

All the members of the order are parasitic and many are of economic and medical importance. The adults are intestinal inhabitants of a wide array of vertebrates, both terrestrial and aquatic.

Students often have problems with the life cycle of an organism such as *Ascaris*, which enters the definitive host by mouth and then migrates through the body to reach the lungs from which the larvae are coughed up and swallowed only to end up in the small intestine where they started. This does not seem to be an entirely logical system, but Sprent (1954, 1962, 1963, for example) has set the life cycles of some of the ascarids in a context that has an evolutionary basis.

*Ascaris lumbricoides* and *A. suum* both have the type of life cycle just described. It appears that in *Ascaris* a multihost cycle is compressed into a single host; thus, the same individual is both the intermediate and definitive host. Sprent (1954) set forth the following hypotheses:

1. The primitive ascarids were free-living marine worms.
2. The terrestrial forms arose secondarily.
3. The indirect (heteroxenous) life cycle pattern is a primitive and widespread feature of both marine and terrestrial species.

The rationale for these conclusions lies in the analysis of the life cycles of various members of the group, and is reasonable and logical:

1. The ancestors of the ascarids were marine bottom dwellers living on decaying organic matter.
2. Some completed the sexual portions of their life cycles in the bodies of decaying invertebrates.
3. The first stage of parasitism arose when these nematodes were ingested by carrion-feeding animals, perhaps shrimplike animals. The nematodes remained in the tissues of this first host.
4. The stage was then set for passing the incipient parasite up the food chain to small fish and then to large fish.
5. The gut of the large fish provides a good source of food for the parasites, and the movement of the fish in seeking food would disseminate eggs of the worms over a large area, thereby increasing the probability of completion of the cycle.

Sprent's theory best explains the transitions that have taken place in those ascarids that migrate in the host.

Thus, one can see that the worms are not as irrational as they appear.

NAME OF ORGANISM AND DISEASE CAUSED. Ascariasis in humans is caused by *Ascaris lumbricoides* Linnaeus, 1758, and in swine by *A. suum* Goeze, 1782.

HOSTS AND LOCATION. Adult *Ascaris lumbricoides* lives in the small intestine of humans and in the chimpanzee, rhesus monkey, gorilla, at least one great ape, and pigs. *Ascaris suum* is the common pig ascarid, but has also been reported from goats, cattle, sheep, and humans.

DISTRIBUTION AND IMPORTANCE. Human ascariasis is cosmopolitan. Because of the number of infections plus the large size of the worms, numerous reports of human infections appeared in early literature. According to Heoppli (1959) *Ascaris lumbricoides* was the first human helminth recorded in Chinese medical literature (about 300–200 B.C.). Frequent reports also appear in early literature from India, Greece, and Rome.

In more modern times, Warren (1974) estimated that there were about 4 million human ascarid infections in the United States in 1972, infecting about 20% of the rural southern population. Estimates of worldwide human ascarid infections run as high as 1 billion (Micozzi & Eveland, 1980), making *Ascaris* one of the most common human parasites.

Usually only mild lung and intestinal lesions occur with ascariasis; however, fatalities also occur.

*A. suum* is considered by many to be the single greatest cause of economic loss in the swine industry. The USDA estimated in 1965 that it caused $34 million loss annually in the United States. In North America, swine husbandry is generally a high-technology enterprise in which attempts are made to keep disease losses to a minimum. In many developing countries pigs are an essential source of protein, but little or nothing may be done to maintain their health and productivity.

MORPHOLOGY. Adult *Ascaris* are large; adult males are up to 30 cm long and 3 mm in diameter, and adult females reach 35 cm in length and 5 mm in diameter. Compare these sizes with a standard lead pencil, which is about 18.5 cm long by 5 mm in diameter (Fig. 38.1). Males have a characteristic curved tail, two stout spicules about 2 mm long, but no gubernaculum. The vulva of the female is near the end of the anterior third of the body.

The mouth is surrounded by three large, fleshy lips, each with two basal papillae (Fig. 38.2). There are dentigerous ridges on each lip in the vicinity of the mouth opening. These ridges were used to distinguish between *Ascaris lumbricoides* and *A. suum* (Sprent,

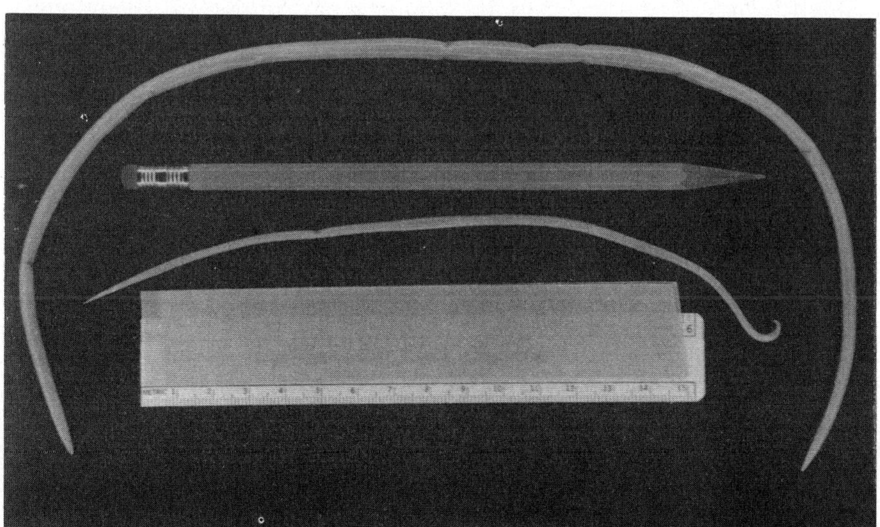

**FIG. 38.1**
Comparison of a lead pencil (*middle*) with an *Ascaris* female (*upper*) and male (*lower*).

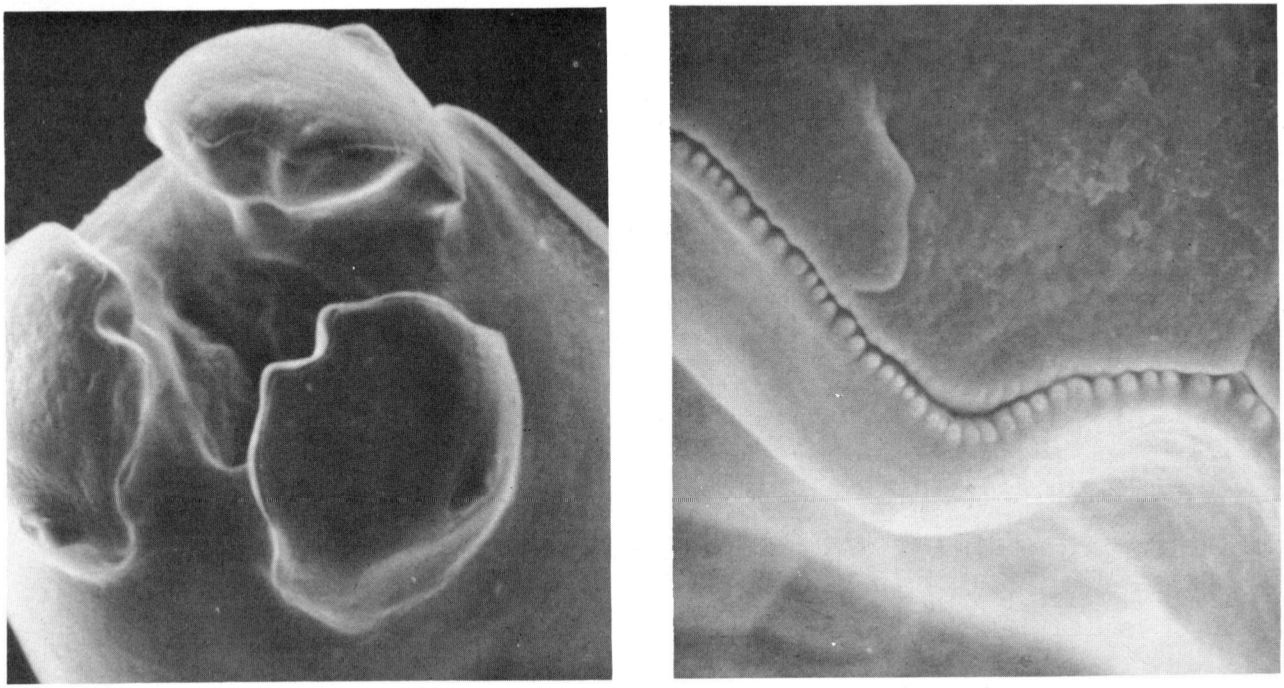

A B

**FIG. 38.2**
*Ascaris suum.* (*A*) head; (*B*) dentigerous ridges.

1952; Ubelaker & Allison, 1972). Madden and Tromba (1976) studied these ridges in *A. suum* of known ages and cautioned against their use because they vary as the worms aged. Thus, it appears that no single characteristic can be used to distinguish between *A. lumbricoides* and *A. suum*. There are, however, physiological differences (discussed in Host–Parasite Interactions), so these two organisms are considered to be separate species.

Fertilized *Ascaris* eggs are 45 to 75 by 35 to 50 μm and have a thick, mammilated shell (Fig. 38.3). They are passed in feces in the one-cell stage. Unfertilized eggs are larger, 85 to 90 by 43 to 47 μm, and have thinner shells (Fig. 38.3). Occasionally eggs lack the characteristic mammilations (Fig. 38.3), which may cause problems in diagnosis.

LIFE CYCLE. The life cycle of *Ascaris* is illustrated in Fig. 38.4. The life cycles of *A. lumbricoides* and *A. suum* are similar. Eggs are shed in the feces in a one-celled stage. They develop within the egg capsules to the $L_2$ in 10 to 20 days; under less than optimal conditions, it may take much longer (reviewed by Levine, 1980). The $L_2$ is the infective stage, but a report of a second molt to $L_3$ within the egg capsules of both *A. suum* and *A. lumbricoides* (Maung, 1978) raised the possibility that the $L_3$ is the infective stage.

We follow the life cycle pattern given for *A. suum* by Douvres et al., (1969). Eggs containing infective $L_2$s hatch in the stomach and small intestine. Larvae penetrate the wall of the small intestine and travel through the hepatic portal circulation, and occasionally the lymphatics, to the liver. The molt from $L_2$ to $L_3$ takes place within 18 to 96 hr after ingestion. A large number of larvae are in the liver 2 days after infection and they leave it between 4 and 9 days. The $L_3$s migrate to the lungs by way of the circulatory system and are caught in the capillaries of the alveoli. They break out of the alveoli into the air spaces, are coughed up, swallowed, and reach the small intestine again. Molt to

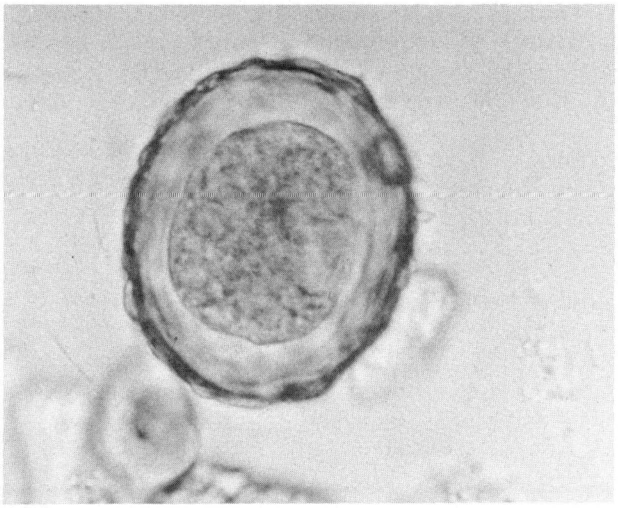

A

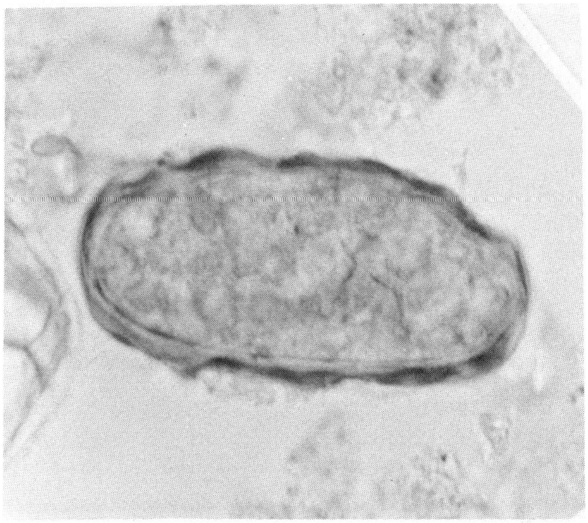

B

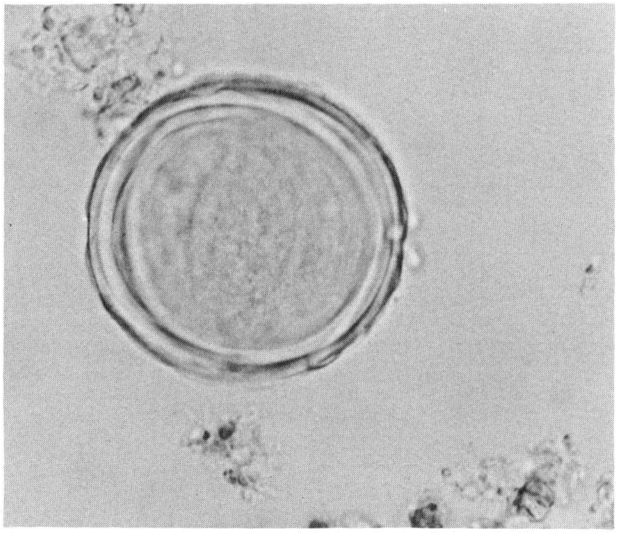

C

**FIG. 38.3**
*Ascaris* eggs. (*A*) Fertilized with mammilated shell; (*B*) unfertilized; (*C*) fertilized but without characteristic mammilations.

$L_4$ occurs at about 23 days; a month after ingestion, the final molt to the adult occurs. Copulation takes place and eggs are shed in the feces beginning 50 to 60 days after infection.

DIAGNOSIS. In both humans and swine, finding eggs in the feces is the best evidence of infection. There is a correlation between the number of worms in the intestine and the number of eggs passed with the feces; since *Ascaris* is such a heavy egg producer, eggs are easily found by a flotation technique.

Before the worms mature, pigs have pneumonia and show labored breathing when the worms reach the lungs. There are few other signs of infection before the worms reach the intestine, but after that weight gains may be less than expected.

There may also be respiratory problems in humans if a large number of eggs is ingested, but usually the signs and symptoms of infection occur after the worms reach the small intestine.

CULTIVATION. A number of investigators have attempted to cultivate *Ascaris* in vitro. Douvres and Tromba (1970) have been the most successful. They began with infective eggs and obtained many $L_3$s and some $L_4$s using culture fluids plus swine kidney cell cultures. Development was slower in vitro than in vivo.

HOST–PARASITE INTERACTIONS. Excellent reviews of *Ascaris* host–parasite interactions were written by Arean and Crandall (1971) and Levine (1980). Symptoms and effects are directly related to worm burden. Light infections are often asymptomatic and cause little lasting damage, but fatalities are known from even light infections.

Humans can be infected by both *Ascaris lumbricoides* and *A. suum*. Similar lesions result; however, the physiology of the two species is markedly different in humans and pigs. Galvin (1968) compared the development of *A. suum* and *A. lumbricoides* in specific pathogen-free pigs as well as rabbits and concluded that in the pig, *A. suum* "completes development in the lungs and migrates to the intestine earlier, it requires less time to reach the egg-laying stage, and when taken into man it develops to maturity less frequently and remains in the intestines for relatively short periods."

At 5 to 6 days after infection, larvae in lungs cause fever, chills, and paroxysmal cough. If enough larvae are present, Loeffler's syndrome, pneumonia associated with eosinophilic infiltration of the lungs, may result. These symptoms last 7 to 10 days but improve progressively (Arean & Crandall, 1971). During the initial phase of the lung pathology, the sputum often contains abundant eosinophils, Charcot-Leyden crys-

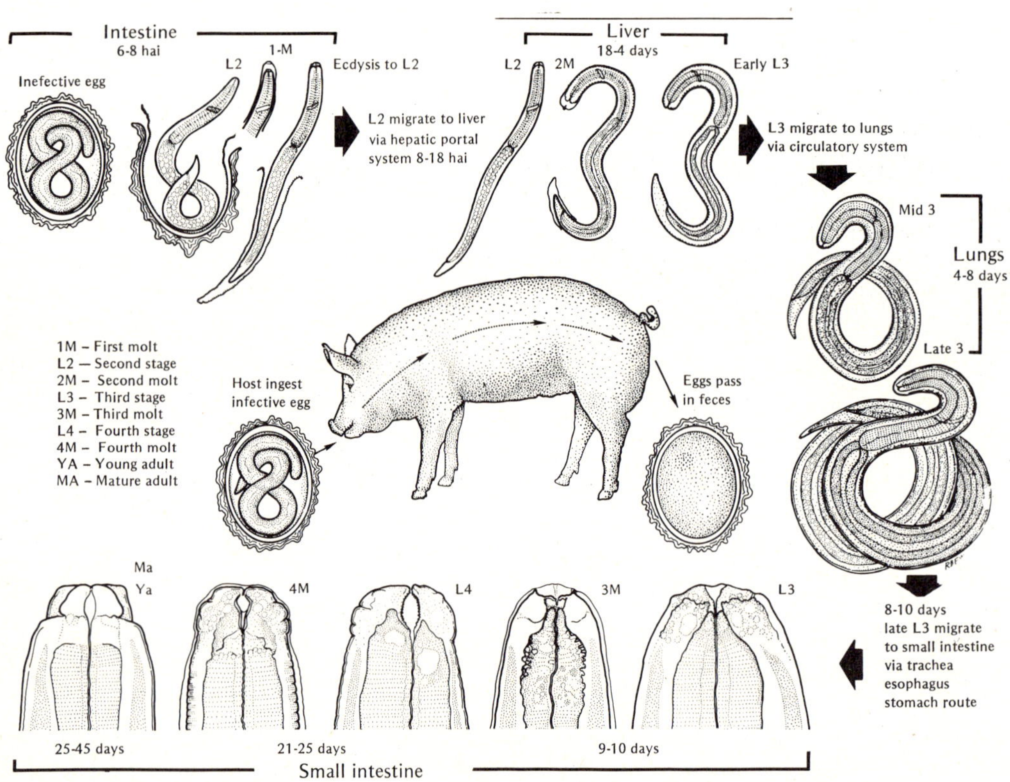


**FIG. 38.4**
Life cycle of *Ascaris suum*. [Courtesy of Dr. F.W. Douvres, Animal Parasitology Institute, USDA, Beltsville, Md.]

tals, and, rarely, larvae. Larvae are sometimes obtained by pleural lavage.

A few adult worms usually cause no clinical disease, but severe disease results from heavy infections. The most severe complications result from parasites wrapping themselves around one another, leading to intestinal obstruction, abdominal pain, loss of weight, nausea, and several other symptoms.

On rare occasions, ascarids may perforate the intestines and cause peritonitis. Wandering ascarids occasionally enter bile ducts, pancreatic ducts, kidneys, or pass from the mouth, nose, anus, or umbilicus, causing physical as well as mental trauma. Female worms apparently seek the curved tail of a male and may pass through openings similar to those described above, or even a shoe eyelet if a male is not present (Beaver, 1964).

If we consider *A. suum* only, losses arise from the following:

1. Migration of the larvae through the lungs, where they may cause pneumonia. Hog raisers refer to the effects of this stage as "thumps," because of the rapid breathing shown by young swine.
2. Exacerbation of other infections during migration. Viral pneumonias may be triggered or made worse by the migration of the larvae through the lungs.
3. Scarring from larvae migrating in the liver. Such livers are condemned at slaughter because of "milk spot."
4. Reduced feed efficiency caused by adult worms. Adults feed on intestinal contents and thereby remove nutrients from the small intestine that would otherwise be absorbed.
5. Plugging of the intestine. Worms sometimes occur in such large numbers that movement of the intestinal contents is impeded.
6. Migration into extraintestinal sites. Trucking of animals to market or fever, among other upsetting factors, may cause worms to migrate. They can move through the intestinal wall into the peritoneal cavity, anteriorly so that they are coughed up, or move into the common bile duct and thence to the smaller bile ducts of the liver. When the worms move into the bile ducts, they may become so tightly packed that the flow of bile is reduced. The tissues of the animals then become yellow and may be condemned at slaughter because they are jaundiced or icteric. Movement into the peritoneal cavity can give rise to bacterial peritonitis, which may be fatal.

Information on immunity to *Ascaris* in humans is sketchy; however, there is a large body of knowledge on immunity in swine. There seems to be little effective immunity against the adult worms but good immunity against the migrating larvae. Most likely, the immunity acts at the molt from the second to the third stages and is both cellular and humoral. This knowledge has been used in attempts to develop a vaccine to prevent ascariasis in pigs, but no product has yet reached the market.

A high percentage of laboratory workers who have studied *Ascaris* have become allergic to it. Their hypersensitivity is severe enough that some investigators have switched their interests to other less hazardous parasites. *Ascaris* in the teaching laboratory carries the hazard of hypersensitivity in either students or instructors as well as the risk of living eggs remaining in the worms even though they have been preserved in formalin. Better nematodes for anatomical study are some of the small ones such as *Pelodera, Rhabditis,* or *Anguillula.*

TREATMENT. Antiminth (pyrantel pamoate) is the drug of choice to treat intestinal ascariasis in humans. Other drugs that have been used successfully include Antepar (piperazine citrate), Mintezol (thiabendazole), and Levamisole (tetramisole). If other helminths are present, the ascarids should be treated first so that they will not be stimulated to wander by ineffective drugs. There is no successful treatment for larval ascariasis.

Swine are treated for helminths by adding drugs to the feed or the water; individual handling of pigs is rather time-consuming. Effective drugs for *A. suum* were not available until the early 1940s, when sodium fluoride (NaF) and then cadmium-containing compounds (CdO and cadmium anthranilate) were developed. Both compounds have problems with toxicity or with residues in the tissues. Piperazines came into use in the middle 1950s, and a number of different formulations are 100% effective against adult worms; their toxicity is low.

Hygromycin-B added a new dimension to treatment of ascarids in swine. Hygromycin is a fermentation product of a strain of *Streptomyces hygroscopicus* and is therefore called an *antibiotic.* It is added to the feed at a rate of 6000 units per pound and effectively removes worms as they mature; it has no effect on the migrating stages. Hygromycin increases feed efficiency under some conditions, so it has other economic benefits.

Pyrantel affects the migrating stages and may have use in preventing lung and liver damage in swine.

EPIDEMIOLOGY AND CONTROL. *Ascaris* infections are more common in humans in the tropics and subtropics than in the temperate zones. But with an estimated 4 million human infections in the United States (Warren, 1974), these numbers are relative. Eggs are extremely resistant and can survive up to 6 years in nature, up to 6 months in sewage sludge, and even on paper money (Levine, 1980). Most parasitology texts state that each female *Ascaris lumbricoides* produces about 200,000 eggs per day for 1 year. However, attempts to quantify the number of female worms using the number of eggs per gram of feces may not be valid because Mello (1974) showed that the number of eggs per gram of feces did not indicate the number of worms in a patient. Nalin and McLaughlin (1976) reported that ascarid eggs can carry viable cholera bacteria (*Vibrio cholerae*).

Adequate sanitation plus education should prevent ascariasis. Lack of adequate fertilizer in many parts of the undeveloped world means that night soil must be used to fertilize crops. Due to the resistance of *Ascaris* eggs to standard disinfectants, the situation would appear to be hopeless. There is a realistic solution though; Thitasut (1961) reported that aqueous solutions containing 120 ppm free iodine killed *Ascaris lumbricoides* and *Toxocara canis* eggs in 5 min or less and did not affect the flavor of leafy vegetables.

One of the problems in control is preventing children from ingesting eggs with contaminated soil. Given the habits of children throughout the world, including pica (dirt eating), playing in contaminated soil and then placing their fingers or other contaminated objects in their mouths, ascariasis will unfortunately be with us until adequate sanitation is accomplished. That prospect is a long way off for many undeveloped nations.

Control of ascarid infections in domestic swine is a particularly difficult problem. High egg production coupled with the long survival of the embryonated eggs outside the host means that a farm can become heavily contaminated in a short time and remain contaminated for years. The systems that have been developed are directed toward reducing infections in and on sows and protecting the young pigs from being exposed to significant infections.

One of the early systems of disease control in swine is the McLean County system of swine sanitation; although we are concerned here with *Ascaris*, other soil- and feces-borne infections are also controlled. The method was first implemented by B. H. Ranson of the USDA in 1921 in McLean County in central Illinois. Where it was adopted and followed faithfully, swine production benefited. The objective of the system is to allow young pigs no contact with anything contaminated with ascarid eggs. Even though the original trials were done more than 60 years ago, the principles are still valid:

1. Prior to the birth of the pigs (farrowing), the quarters are cleaned. The floor should be concrete, but if not, all feces should be removed. The whole farrowing pen should be scrubbed with a brush and a strong detergent. The original recommendation called for boiling lye water.
2. Just prior to farrowing, the sow is washed so that all mud and adhering feces are removed from all parts of the body.
3. The sow and pigs are kept in the clean quarters until they are ready to be moved to pasture. When they are moved, they are carried by truck so that they have no opportunity of contacting eggs on the ground.
4. The pasture should not have been used by pigs for several years, and the water supply should be free of possible contamination with ascarid eggs.

With the introduction of highly effective anthelmintics in the 1940s, it became possible to treat the sows before farrowing, and this additional step is taken to keep contamination of the area to a minimum.

## REFERENCES AND FURTHER READINGS

Arean, V. M., and C. A. Crandall. 1971. Ascariasis. In P. A. Marcial-Rojas (ed.), *Pathology of Protozoal and Helminthic Diseases*. Williams & Wilkins, Baltimore, pp. 769–807.

Beaver, P. C. 1964. *Ascaris* strangled in a shoe eyelet. *Am. J. Trop. Med. Hyg.* **13**:295–296.

Douvres, F. W., and F. G. Tromba. 1970. Influence of pH, serum, and cell cultures on development of *Ascaris suum* to fourth stage *in vitro*. *J. Parasitol.* **56**:238–248.

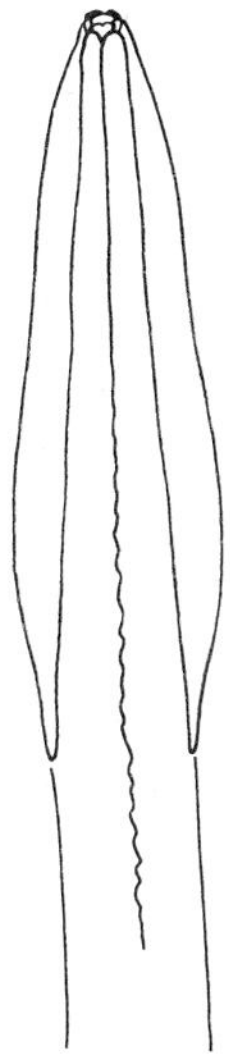

**FIG. 39.1**
Diagram of the anterior end of *Toxocara canis* with prominent cervical alae. [Redrawn from Morgan & Hawkins, 1949.]

fine dentigerous ridges. Prominent long, slender cervical alae (Fig. 39.1) are present. The esophagus has a distinct muscular bulb and a prominent ventriculus (Fig. 39.2).

Males are 4 to 10 cm by 2 to 2.5 mm, have 20 to 30 preanal papillae, five postanal papillae, a narrow terminal appendage, and subequal spicules. Females are 5 to 18 cm by 2.5 to 3.0 mm, and have a vulva in the anterior fourth of the body. Eggs are subspherical, pitted, and are 85 to 90 by 75 μm (Fig. 39.3).

LIFE CYCLE. In the life cycle of *Toxocara canis* (Fig. 39.4) unembryonated eggs are shed in feces. Larvae molt once to the infective $L_2$ in 9 to 15 days under optimal conditions (Schacher, 1957). Dogs ingest ova containing infective $L_2$s. The second-stage larvae are released in the small intestine and penetrate the intestinal mucosa. At this point the life cycle can follow one of two paths, depending on the age of the dog. This larval development and migration was studied by Schacher (1957), Webster (1958), Douglas and Baker (1966), and summarized by Levine (1980).

If a dog less than 3 months old ingests infective ova, larvae hatch in the duodenum, penetrate the intestinal walls, and enter lymph vessels. They go to the liver through the hepatic portal system, arriving there 1 to 2 days after infection. No molt occurs in the liver. Larvae then migrate to the heart through the hepatic vein or vena cava. Peak numbers of larvae in lungs are reached 3 to 5 days after infection. From the lungs, two paths are followed by larvae. Most larvae in young dogs break out of the alveoli, travel up the trachea, are swallowed, and pass to the stomach by 10 days postinfection. The second molt to $L_3$ occurs in lungs, trachea, or esophagus. The molt to $L_4$ occurs in the stomach and small intestines. The fourth molt to adult occurs in the intestines 2 to 3 weeks after infection. Adult worms mate and eggs are passed in feces 4 to 5 weeks after infection.

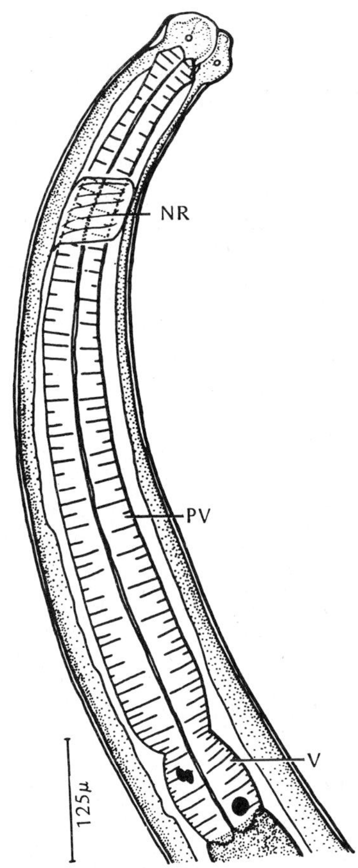


**FIG. 39.2**
Diagram of the esophagus of *Toxocara canis*. NR=nerve ring; PV=proventriculus; V=ventriculus. [Redrawn from Schacher, 1957.]

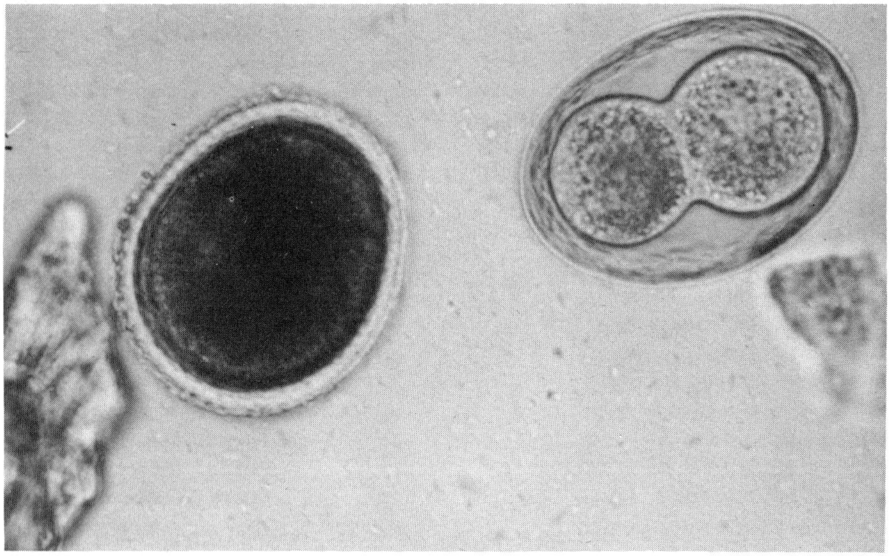

**FIG. 39.3**
Eggs of *Toxocara canis* (*right*) and *Toxascaris leonina* (*left*).

The second route followed by larvae in lungs (especially larvae in older dogs 6 months and older) is to travel by the pulmonary vein to the heart and then throughout the body through the arterial circulation; they leave small blood vessels and enter practically all tissues of both natural and transport hosts. This somatic tissue migration gives rise to visceral larva migrans in humans (discussed in Host–Parasite Interactions). Host defenses destroy some larvae, but many are encapsulated and may remain alive and infective for years.

A third life cycle pattern is the most important method of infection in dogs—transplacental infection. Somatic $L_2$s in pregnant bitches somehow become activated, enter the circulatory system, and eventually pass through the placenta to go to the liver of developing pups. Larvae molt there to the $L_3$. The usual migratory route in young dogs follows after birth, and pups carry adult worms by 1 week after birth and pass eggs when they are 3 weeks old (Sprent & English, 1958). Transplacental infection can occur by activation of somatic $L_2$s or by ingestion of infective ova during gestation. The former is probably the source of most infections. Bitches cleaning puppy feces may be reinfected with additional somatic $L_2$s, ensuring infections in future generations of puppies.

*T. canis* can infect a wide variety of transport hosts including rodents and humans; somatic larvae remain as $L_2$s.

DIAGNOSIS. Toxocariasis is diagnosed by finding eggs (Fig. 39.3) or worms in fecal samples. Puppies that have been infected prenatally typically become potbellied and have a rough hair coat when the worms mature. Most dog raisers need only these signs of infection to know when to institute treatment.

Visceral larva migrans in humans can be confirmed by finding larvae in biopsies, especially liver, but larvae are only rarely found. Glickman et al. (1978) developed an enzyme-linked immunosorbent assay which is the standard diagnostic test used by the Centers for Disease Control to identify both visceral larvae migrans and ocular larva migrans.

HOST–PARASITE INTERACTIONS. *Toxocara canis* produces various responses in dogs, depending on the worm burden and the age of the dog. Infected puppies and kittens may have digestive disturbances, dull coat, and vomit often. Pneumonia is often present in infected young animals. Heavily infected animals may die 2 to 3 weeks after birth.

Adult dogs have granulomatous lesions in almost any tissue or organ caused by migrating larvae. Dogs are apparently afflicted with ocular larva migrans similar to the human disease.

Beaver et al. (1952) proposed the term *visceral larva migrans (VLM)*. The typical patient is a child 1 to 5 years old. Fatalities have occurred, but normally the disease is self-limiting (Glickman et al., 1979).

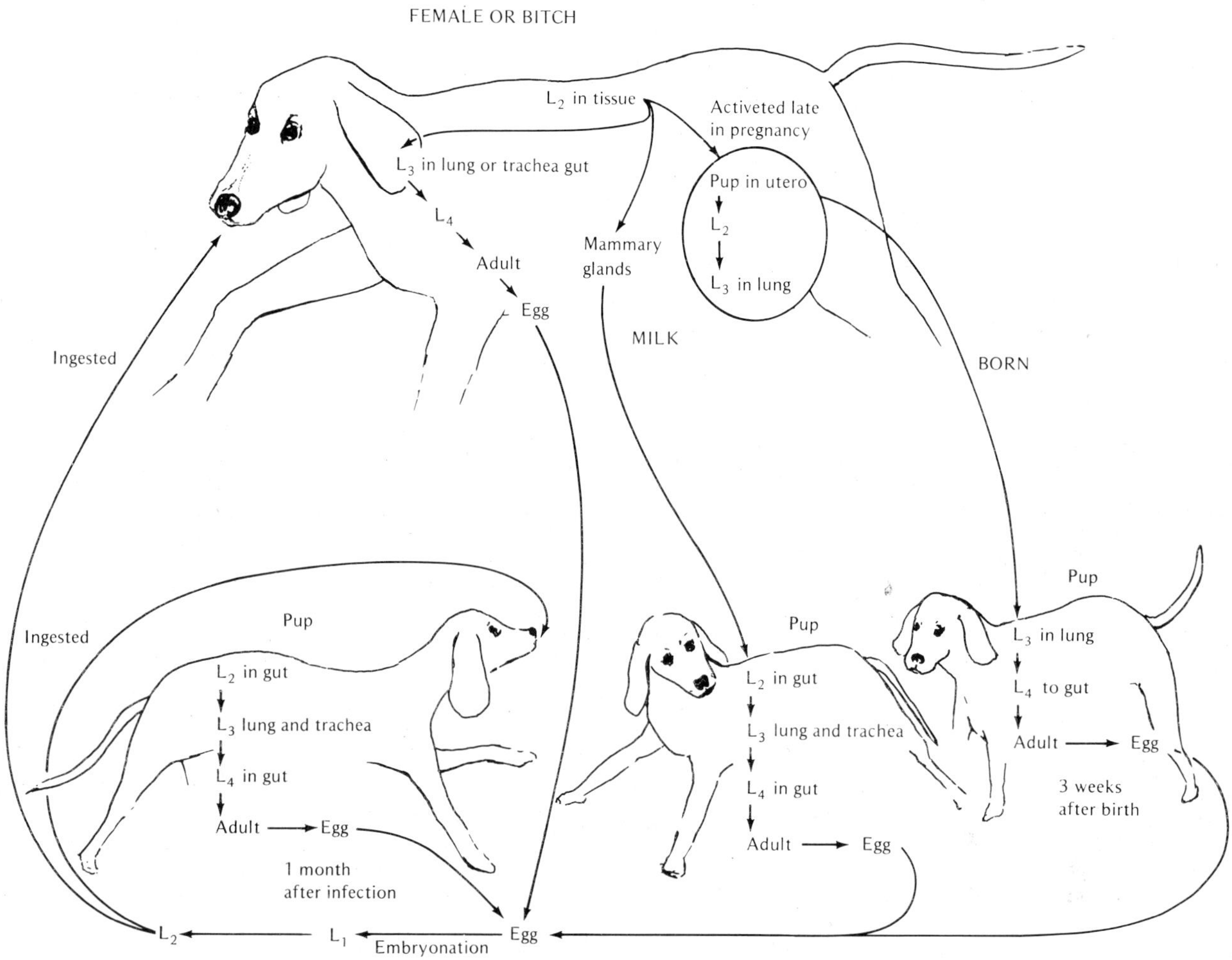


**FIG. 39.4**
Life cycle and transmission of *Toxocara canis*.

Pathogenesis in adult humans results from tissue damage (eosinophilic granulomata) caused by migrating larvae and a strong immune response. Early migration is often accompanied by abdominal pain, diarrhea, and hepatomegaly. Larvae cause pneumonia accompanied by eosinophilia during the lung migration. VLM sometimes occurs in the brain and has tentatively been linked to epilepsy, however, Glickman et al. (1979) reported no evidence to support the hypothesis even though serious damage to the central nervous system can result.

Ocular larva migrans (Wilder, 1950) or OLM is the result of migrating *T. canis* larvae in the eyes. Concurrent ocular and systemic disease are rare (Glickman et al., 1979). OLM lesions look much like those of retinoblastoma, the most common eye tumor in children; this has resulted in the needless removal of eyes from children. The ELISA serodiagnostic test may help to prevent such tragic losses. Apparently, children with OLM have ingested only a few eggs, and hence have only a mild immune response, which allows larvae to reach the eyes.

TREATMENT. Intestinal worms in dogs are treated with piperazine compounds, pyrantel pamoate, di-

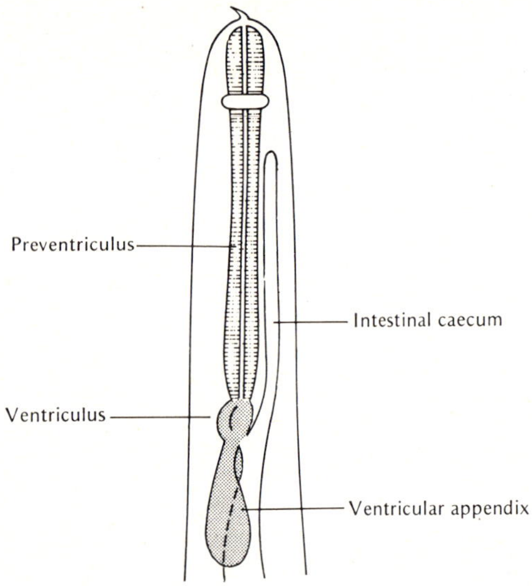


**FIG. 39.5**
Diagram of the anterior portion of the digestive tract of anisakine nematodes. [Redrawn with modifications from Myers, 1976.]

chlorovos, or mebendazole (Schantz & Glickman, 1979). None of these prevents prenatal transmission or is effective against migrating larvae.

Corticosteroids are useful in humans for severe pulmonary and myocardial involvement as well as for eye inflammation. Clinical improvement in both OLM and VLM was reported after the use of thiabendazole, but the effect may be antiinflammatory and analgesic rather than anthelmintic.

EPIDEMIOLOGY AND CONTROL. Infections in dogs throughout the world approach 100% in puppies, but older dogs have fewer than 20% patent infections. Schantz and Glickman (1979) and Glickman et al. (1979) said that the infection percentages are high in puppies because of the efficiency of transplacental infection. With 60 million to 80 million dogs in the United States, it is obvious that many households, parks, and playgrounds are contaminated with *Toxocara* eggs, which potentially can cause either VLM or OLM in children. The problem is compounded by the fact that *Toxocara* eggs are almost as resistant as *Ascaris* eggs and may survive for years.

Because eggs are shed unembryonated, prompt removal and proper disposal of dog feces should significantly reduce human risk. However, a visit to almost any community park illustrates the potential severity of the problem. The results of serodiagnostic tests suggest that VLM is usually a benign, self-limiting disease, but can be serious. On the other hand, OLM frequently results in visual impairment.

Prevention should include controlling dogs, excluding dogs from children's play areas, enforcement of leash laws, worming of dogs, and education of people.

The timing of worming is important because of the parasite's biology. Schantz and Glickman (1979) suggested that young pups and nursing bitches be treated 2, 4, 6, and 8, weeks after parturition because pups can shed eggs at 3 weeks. Weanling pups can be wormed twice 10 to 14 days apart. Older dogs should have fecal examinations once a year and be treated as needed. In addition, puppy and bitch feces should be disposed of often so that eggs do not have a chance to embryonate and infect either dogs or humans. With adequate veterinary care and responsible owner behavior, toxocariasis (VLM and OLM) can be significantly reduced.

## *Other Ascarids*

*Toxocara cati* (syn., *T. mystax*) occurs only in cats and has a migration similar to that of *T. canis* except that it lacks transplacental infection (Sprent, 1956). Its cervical alae are short and broad. Eggs have a pitted surface and are 65 to 75μm in diameter. Eggshell morphology in SEM is different for *T. canis* and *T. cati* (Ubelaker & Allison, 1975). Up to 75% of cats are infected with *T. cati* (Styles & Evans, 1971), but this parasite rarely is implicated in VLM (Glickman et al., 1979). Sprent (1956) said that *T. cati* must be regarded as an occasional human parasite.

The third ascarid found in both cats and dogs is *Toxascaris leonina*. *Toxascaris* has no esophageal ventriculus and its three lips differ slightly from those of *Toxocara*. *Toxascaris leonina* ova are smooth (Fig. 39.3) and measure 75 to 85 by 60 to 75μm. Adults have long, slender cervical alae resembling those of *Toxocara canis*. *T. leonina* has no somatic migration and no prenatal infection. Mice can be transport hosts, but humans are not infected.

## ASCARIDS OF DOMESTICATED ANIMALS

Ascarids of economic importance are found in horses, cattle, and poultry. *Neoascaris vitulorum* (=*Ascaris vitulorum*) is passed prenatally to calves and is pathogenic in young animals. *Parascaris equorum* (=*A. equorum, A. equi, A. megalocephala*) of horses is also pathogenic in young animals but the animals do not become infected before birth. Ascarid infections are not usually seen in adult cattle or horses, and it is therefore necessary to look for the signs of infection in young animals and for eggs in the feces. Prompt treatment to remove the adult worms is called for.

Chickens that are raised on the ground, as opposed to broiler or laying houses, are subject to infection with *Ascaridia galli* and *Heterakis gallinarum*. *A. galli* may affect growth rate in chicks when they average as few as 4.3 worms per bird. After about 3 months of age, birds become resistant to infection with *A. galli*; both age resistance and immunity develop on exposure to the worms.

*H. gallinarum*, the cecal worm, is usally found in small numbers. Its importance lies in its being a vector of *Histomonas meleagridis*, a flagellated protozoan that is especially pathogenic for turkeys (Chapter 6).

## ANISAKIASIS

The disease in humans called *anisakiasis* is caused by infections with nematodes of the subfamily Anisakinae. Smith and Wootten (1978) said that three genera caused human disease—*Anisakis, Phocanema* (=*Terranova*=*Porrocaecum*), and *Contracaecum*.

Adult anisakids resemble *Ascaris*. Three lips are present, usually with dentigerous ridges. The anterior diverticulum of the digestive tract is characteristic of the subfamily (Fig. 39.5). The esophagus is divided into a proventriculus and a ventriculus. Some genera have a ventricular appendix, and some also have an anterior, blind extension of the intestine called the *intestinal cecum*.

Definitive hosts of the Anisakidae are marine mammals and some fish, birds, reptiles, and amphibians (Myers, 1976). Life cycles are usually indirect. Eggs are passed in feces and the first molt occurs in the eggs. The second molt probably occurs in crustaceans such as the shrimp *Euphausia* spp. Smith and Wootten (1978) found a low infection rate in crustacea, but fish consume enormous numbers. Fish and squid harboring third-stage larvae are transport hosts. The $L_3$s are infective for marine mammals and development to the adult occurs in their intestines.

Anisakiasis has been recognized only recently as a human disease because of two factors. First, the disease is usually diagnosed only after surgical removal of parasitized stomach or intestine. And second, recent changes in fishing methods probably have increased the incidence of infections. With modern factory ships, which process fish at sea, hours or even days may elapse before fish are eviscerated. Intestinal larvae ($L_3$) move from intestines into other organs and muscles as the fish die. When humans then ingest uncooked fish, they are at risk (Smith & Wootten, 1978). Identification of the larvae infecting humans is difficult; Myers (1976) recognized five larval types based on anterior digestive tract structure and the position of the excretory pore.

Nematodes in the family Anisakidae are found throughout the world. Smith and Wootten (1978) said that *Anisakis* is the most important genus causing human disease. Human anisakiasis has been reported in Japan, the Netherlands, Korea, Denmark, France, Belguim, and the United States—countries where marine fish and shellfish are eaten raw, pickled, or lightly cooked (Williams & Jones, 1976).

Anisakiasis in humans can produce either gastric or intestinal disturbances (Smith & Wootten, 1978). Symptoms of gastric infection include sudden stomach pains, nausea, and vomiting 4 to 6 hr after eating raw fish. Stomach lesions are typical foreign-body responses with granulomata around the larvae. Intestinal infections cause severe abdominal pain, vomiting, fever, and blood in the stool within 7 days after eating raw fish. Intestinal damage is an Arthus-type reactions in acute cases, characterized by thickening of the intestinal wall along with eosinophilic infiltration. Later, eosinophilic granulomata form around the larvae.

Anisakiasis will probably be seen more often with the change in fishing mentioned earlier and the trend toward "natural" foods, which are often uncooked or undercooked. Two surveys illustrate the potential problem. Bier et al. (1980) and Jensen and Dailey (1980) sampled commercially caught fish in Boston, San Francisco, and Los Angeles. Most fish in both surveys contained potential pathogens, especially on the West Coast, where 41.6% of the fish harbored *Anisakis* larvae.

Control includes not eating smoked, marinated, or raw fresh fish or squid. Heating fish to 60° C or freezing to –20°C for 24 hr kills the larvae (Jensen & Dailey, 1980). All of these methods, unfortunately, are contrary to the natural foods movement.

## REFERENCES AND FURTHER READINGS

Anvik, J. O., A. E. Hague, and A. Rahmann. 1974. A method of estimating urban dog populations and its application to the assessment of canine fecal pollution and endoparasitism in Saskatchewan. *Can. Vet. J.* **15**:219–223.

Beaver, P. C., C. H. Snyder, G. M. Carrera, J. Dent, and J. Lafferty. 1952. Chronic eosinophilia due to visceral larva migrans. *Pediatrics* **9**:7–19.

Bier, J. W., G. J. Jackson, and W. L. Payne. 1980. Recovery of nematodes from Pacific and Atlantic pleuronectid fish, compared by sedimentation and digestion. *J. Parasitol.* **66** (Suppl.):66(Abstr.).

Douglas, J. R., and N. F. Baker. 1966. Some host–parasite relationships of canine helminths. In J. E. McCauley (ed.), *Host–Parasite Relationships.* Proc. 26th Ann. Biol. Colloq., April 23–24, 1965. Oregon State University Press, Corvallis, pp. 97–115.

Glickman, L. T., P. M. Schantz, and R. H. Cypess. 1979. Canine and human toxocariasis review of transmission, pathogenesis, and clinical disease. *J. Am. Vet. Med. Assn.* **175**:1265–1269.

Glickman, L., P. M. Schantz, R. Dombroske, and R. Cypess. 1978. Evaluation of serodiagnostic tests for visceral larva migrans. *Am. J. Trop. Med. Hyg.* **27**:492–498.

Jensen, L. A., and M. Dailey. 1980. Larval anisakine roundworms of marine fishes from Southern and Central California with comments on public health significance. *J. Parasitol.* **66**(Suppl):67(Abstr.).

Levine, N. D. 1980. *Nematode Parasites of Domestic Animals and of Man,* 2nd ed. Burgess, Minneapolis.

Morgan, B. B., and P. A. Hawkins. 1949. *Veterinary Helminthology.* Burgess, Minneapolis.

Myers, B. J. 1976. The nematodes that cause anisakiasis. *J. Milk Food Technol.* **38**:774–782.

Schacher, J. F. 1957. A contribution to the life history and larval morphology of *Toxocara canis. J. Parasitol.* **43**:599–612.

Schantz, P. M., and L. T. Glickman. 1979. Canine and human toxocariasis: The public health problem and veterinarian's role in prevention. *J. Am. Vet. Med. Assn.* **175**:1270–1273.

Smith, J. W., and R. Wootten. 1978. *Anisakis* and anisakiasis. *Adv. Parasitol.* **16**:93–163.

Sprent, J. F. A. 1956. The life history and development of *Toxocara cati* (Schrank, 1788) in the domestic cat. *Parasitology* **46**:54–78.

Sprent, J. F. A., and P. B. English. 1958. The large roundworms of dogs and cats—A public health problem. *Aust. Vet. J.* **41**:161–171.

Styles, T. J., and D. S. Evans. 1971. Intestinal parasites of dogs and cats in Schenectady County. *N.Y. St. J. Med.* **71**:2755–2757.

Ubelaker, J. E., and V. F. Allison. 1975. Scanning electron microscopy of the eggs of *Ascaris lumbricoides, A. suum, Toxocara canis,* and *T. mystax. J. Parasitol.* **61**:802–807.

Webster, G. A. 1958. On prenatal infection and the migration of *Toxocara canis,* Werner 1782 in dogs. *Can. J. Zool.* **36**:435–440.

Wilder, H. C. 1950. Nematode endophthalmitis. *Trans. Am. Acad. Opthalmol. Otolaryngol.* **55**:99–109.

Williams, H. H., and A. Jones. 1976. *Marine Helminths and Human Health.* Misc. Publ. No. 3, Commonwealth Agricultural Bureaux, Farnham Royal, Bucks, England.

# 40

# *Pinworms*

The pinworms (order Oxyurida) received their name because of the long, pointed tail of the female. Their other striking characteristic is the large bulb at the posterior of the esophagus (see Figs. 40.2, 40.3). The females are stout-bodied, and the males, which have only a single spicule, are usually about half the size of the females.

Pinworms are widely spread in the animal kingdom, being found in both vertebrates and invertebrates such as millipedes. They are often found in large numbers in the intestines of their hosts, but only a few species are of medical or economic importance. There are a number of species of *Skrjabinema* in ruminants (Schad, 1959), and laboratory rodents are frequently infected with *Aspiculuris tetraptera* or *Syphacia obvelata*. Striking species found in the cecum of the porcupine are *Evaginurus evaginata* and *Wellcomia evaginata*, in which the vulva of the female grows into an elegant, elongated structure (Fig. 40.1).

General principles of ecology and niche diversification can sometimes be elucidated with unlikely material, as was done by Schad (1963) with the pinworm, *Tachygonetra* of turtles. Eight species of *Tachygonetra* occur in the large intestine of the turtle *Testudo graeca*, and they have different distributions not only along the length of the gut but radially as well. The result is that each species has its own niche and feeds on different food, as shown by different mouth structures.

NAME OF ORGANISM AND DISEASE CAUSED. *Oxyuris equi* (Schrank, 1788) Rudolphi, 1803, causes pinworm infection, oxyuriasis.

HOSTS AND LOCATION. The organism is cosmopolitan in horses. This is a minor infection that can have secondary consequences of some importance in individual animals.

MORPHOLOGY. The females are 40 to 150 mm long with pointed tails that vary considerably in length. Males are 9 to 12 mm long and have a single spicule about 140 μm long (Fig. 40.2). The operculate eggs are somewhat flattened on one side and are 85 to 95 by 40 to 45 μm.

LIFE CYCLE. Eggs containing the infective larvae are ingested with contaminated food or water or with the hair of an infected animal. The preadult stages develop in the paramucosal lumen and they return to the lumen when they become sexually mature. The prepatent period is about 5 months. Eggs are shed containing an $L_1$ and they reach infectivity, probably as an $L_2$, within 3 days of leaving the host.

DIAGNOSIS. The principal signs of infection are patches where hair has been rubbed off at the tailhead. Affected animals rub against a barn wall, fence post or even barbed wire to relieve itching. Finding the adults or

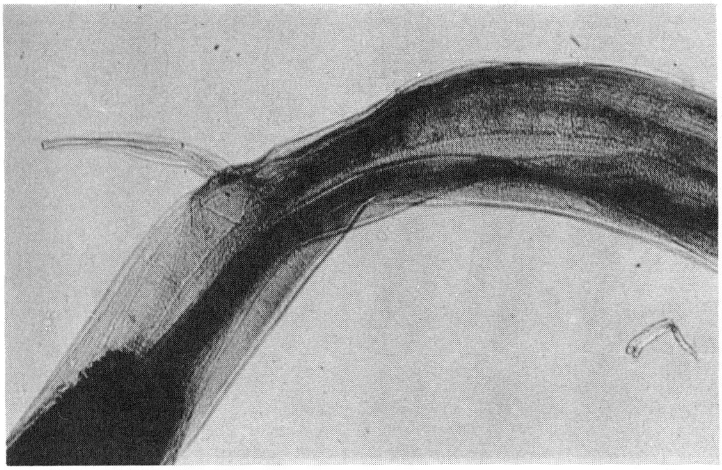

**FIG. 40.1**
A pinworm of the porcupine, *Evaginurus* sp.; the large lateral extension is the vulva. Photographed in interference microscopy.

eggs in the perianal region or sometimes in the feces provides the best evidence of infection.

HOST–PARASITE INTERACTIONS. The preadult stages do not generally cause any damage. When mature, the females migrate posteriorly to lay eggs in the perianal region. Material on the surface of the eggs is irritating and the anal region evidently itches, since animals then begin to scratch it and become nervous.

Scratching can become so vigorous that the horses' hair in the tail and tailhead area is pulled out and the skin becomes raw. There may be secondary invasion by bacteria, and some of the flesh flies such as *Phormia* or *Cochliomyia*, may oviposit in the wounds (Chap. 56).

TREATMENT. Piperazines and benzimidazoles are effective.

EPIDEMIOLOGY AND CONTROL. Pinworm infection is not generally considered to be a herd problem; however, if a routine program of strongyle control has been established (Chap. 33), pinworms will automatically be taken care of by anthelmintic treatment.

NAME OF ORGANISM AND DISEASE CAUSED. *Enterobius vermicularis* (L.) Leach 1853 causes enterobiasis, oxyuriasis, pinworm, or seatworm infection.

DISTRIBUTION AND IMPORTANCE. The organism is cosmopolitan, with greater prevalences of infection in temperate and cold than in tropical climates. In North America and Western Europe, the prevalence of infection is about 30% overall and perhaps 50 to 60% in children. In the tropics, the prevalence is about 10%.

HOSTS AND LOCATION. Humans are the only host, for practical purposes; some other primates and rodents have been found to be infected, but these are rare infections that have no epidemiologic importance. Dogs and cats are never infected. The adult worms are in the large intestine and perianal region.

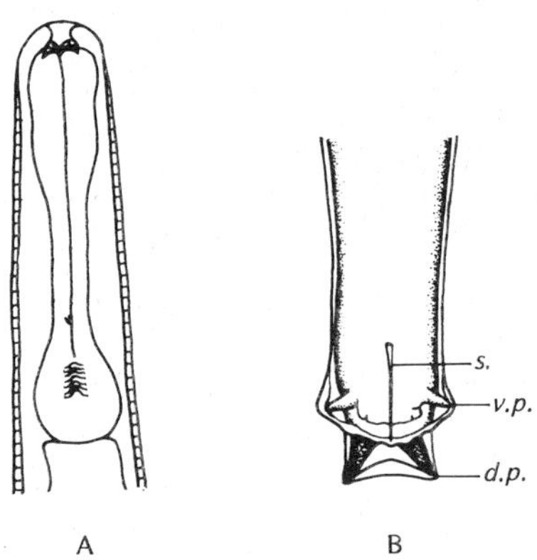


**FIG. 40.2**
*Oxyuris equi.* (*A*) Anterior of female; (*B*) posterior of male: *s* = spicule; *v.p.* = subventral papilla; *d.p.* = dorsal papilla. [Redrawn with modifications from Yorke, W., & P. A. Maplestone, 1962. *The Nematode Parasites of Vertebrates.* Hafner Publishing Co., New York.]

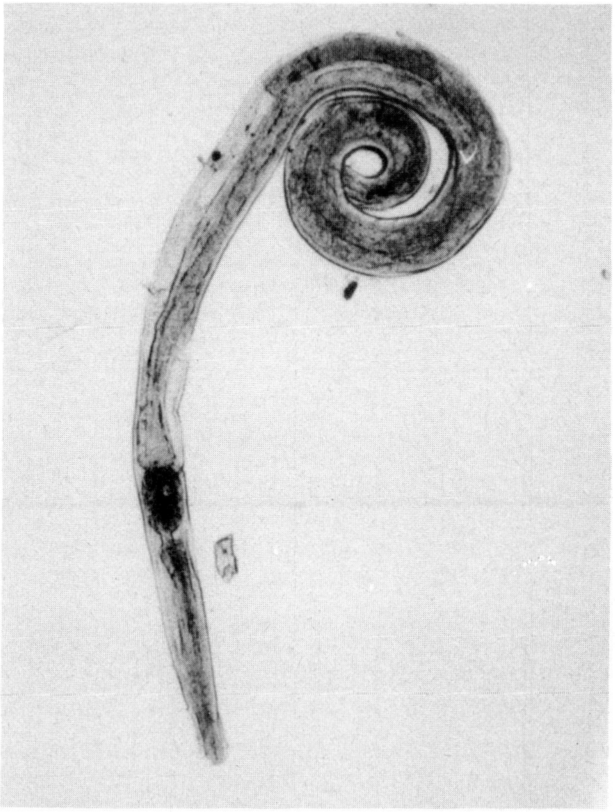

**FIG. 40.3**
Adult male *Enterobius vermicularis.*

MORPHOLOGY. The females are 8 to 13 by 0.3 to 0.5 mm, and they have a pointed tail, which may reach 2 mm. The males have a curved tail with a single spicule and are 2 to 5 by 0.15 mm. Both sexes have an esophagus with a slender anterior portion and a large posterior bulb; there are small cephalic alae at the anterior tip (Fig. 40.3).

The eggs are 50 to 60 by 20 to 32 µm, with a smooth, clear outer capsule; like many oxyurid eggs, they are flattened on one side, giving an asymmetric appearance (Fig. 40.4).

LIFE CYCLE. The females oviposit or actually burst in the perianal region; eggs have infective larvae in them within a day or two after release from the female. Eggs are carried to the mouth of a person on the fingers or are inhaled with dust. Most of the development takes place in the tissue of the large intestine; worms may be found in the paramucosal lumen or as deep in the tissue as the serosa (the external covering of the intestine). The worms return to the lumen of the gut and move posteriorly as they mature. The prepatent period is 36 to 53 days. Retrofection may occur when eggs become infective in the perianal area and the hatched larvae reenter the anus.

DIAGNOSIS. Since the females oviposit outside the host, eggs are seldom found in the feces. The eggs are collected at the perianal region by swab when the person first gets up in the morning. Two different kinds of swabs have been developed, but the simplest and most effective is the Scotch tape swab. A piece of cellophane tape is placed sticky side out over the end of a tongue depressor. The tape is pressed into the folds around the anus to pick up the eggs or pieces of worm; it is then placed sticky side down on a microscope slide and a drop or 2 of toluene is allowed to run under it. The slide is examined under the 10x objective of a compound microscope for the typical asymmetric eggs (Fig. 40.4).

The eggs and fluid from the body of the female worm are irritating and cause intense itching in the anal region. Nervousness, irritability, and scratching are signs often shown by infected children.

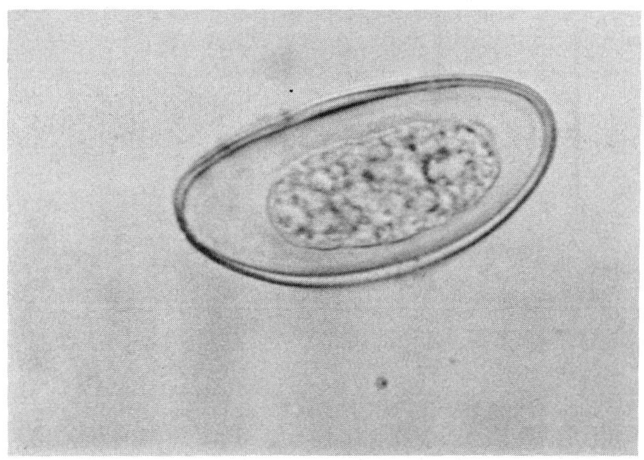

**FIG. 40.4**
*Enterobius vermicularis* egg with characteristic flattening on one side.

HOST–PARASITE INTERACTIONS. The preadult stages of *Enterobius* seldom cause significant damage to the intestine, but a few cases are on record in which they have. The signs and symptoms of infection are a result of the females ovipositing or breaking up in the perianal region. Although they primarily cause itching, there may also be skin eruptions and secondary invasion by bacteria. Vaginitis may occur in some girls from worms crawling from the anus into the genitalia.

Children are twice as likely as adults to be infected with *E. vermicularis*. It is uncertain whether the lower prevalence in adults is due to age resistance or to acquired immunity. The latter seems likely because the preadult worms are in close contact with the intestinal tissue, but there is little direct evidence.

TREATMENT. Two good, safe drugs for use against pinworms are piperazine hexahydrate (Antepar) and pyrvinium pamoate (Povan). Although both are highly effective, it is usually necessary to give a second dose a week or 2 after the first. Povan is a red dye that can stain the teeth and will cause the feces to be red, a circumstance leading to excited telephone calls from parents who have not been forewarned.

EPIDEMIOLOGY AND CONTROL. The high prevalence of pinworm infection indicates both its ease of transmission and the hardiness of the free-living stages. Where the humidity is high, infective eggs can live for 3 weeks or more; low humidity is detrimental to survival. Transmission is principally from the fingers, contaminated clothing, or by inhalation of eggs with dust. Some eggs may be transmitted through contaminated food and water, but this mode is probably of minor importance. Females produce an average of about 11,000 eggs and the contamination of a household can become heavy.

Pinworm infection is a household disease; if one person in a family is found to be infected, usually other children and adults will also be infected. Therefore, everyone in the family should be treated regardless of whether they have evidence of infection. A drug such as Povan has little risk and is highly effective; the only persons exempted from treatment are pregnant women.

Children are usually the first to show signs of infection in a family. Because of their lack of fastidious habits, they are likely to have relatively high levels of infection. Perhaps an outbreak of oxyuriasis in a family provides an occasion for teaching children not to put things in their mouths and to wash their hands after going to the toilet.

In addition to treating the people, scrupulous cleaning of the house and frequent washing of bed linens and underclothing are essential to ridding the house of infective eggs.

Nearly every parasitologist or veterinarian has heard the family dog or cat blamed for bringing pinworms into the house. Veterinarians are sometimes asked to euthanize a dog in order to eliminate the reservoir of infection. The host range of *E. vermicularis* is quite narrow and dogs and cats do not become infected.

*E. vermicularis* is implicated as the vector of the intestinal protozoan, *Dientamoeba fragilis* (Chap. 6).

## REFERENCES AND FURTHER READINGS

Cram, E. B. 1941. Studies on oxyuriasis. IX. The familial nature of pinworm infestation. *Med. Ann. Dist. Columbia* 10:39–48.

Hulinska, D. 1968. The development of the female *Enterobius vermicularis* and the morphogenesis of its sexual organ. *Folia Parasitol.* 15:15–27.

Inglis, W. G. 1961. The oxyurid parasites (Nematoda) of primates. *Proc. Zool. Soc. London* **136**:103–122.

Schad, G. A. 1959. A revision of the North American species of the genus *Skrjabinema* (Nematoda: Oxyuroidea). *Proc. Helm. Soc. Wash.* **26**:138–147.

———. 1963. Niche diversification in a parasite species flock. *Nature* **198**:404–406.

Schüffner, W., and N. H. Swellengrebel. 1943. Retrofection in oxyuriasis. A newly discovered mode of infection with *Enterobius vermicularis. J. Parasitol.* **35**:138–146.

Yorke, W., and P. A. Maplestone. 1962. *The Nematode Parasites of Vertebrates*. Hafner Publishing, New York.

# 41

# *Filarioid Nematodes*

A diverse array of nematodes, all of which have vectors in their life cycles, is included in the order Spirurida. Their distinguishing features are an esophagus with an anterior muscular portion and a posterior glandular portion and unequal spicules in the males.

Only some members of the superfamily Filarioidea will be discussed at length, but there are other spirurids of economic and medical importance that deserve attention. The genus *Physaloptera* contains a large number of species that have elaborate head structures; grasshoppers or beetles are intermediate hosts and definitive hosts are insect eaters. The adults burrow their heads into the stomach or intestine and cause considerable damage to the tissue.

*Spirocerca lupi* is among the few carcinogenic helminth parasites. The main definitive host of *S. lupi* is the dog, which becomes infected by eating infected beetles. The adult worms are located in the lower end of the esophagus, where they lie deep in the tissue. The worms cause a moderate amount of damage directly, but over a long period somehow cause the tissue in their vicinity to become cancerous. Most dogs that develop malignant tumors are 3 to 7 years old, so probably a long-term association is necessary before the inflammation at the site of the worms turns neoplastic (Bailey, 1971).

Humans occasionally become infected with spirurids such as *Gongylonema pulchrum, Gnathostoma spinigerum,* and *Thelazia californensis. G. spinigerum* is obtained by humans by accidentally ingesting infected insects. *G. pulchrum* has a more complex life cycle involving two intermediate hosts, a copepod, and a cold-blooded vertebrate. *Thelazia* has yet another life cycle variation in that it is transferred to the human eye from the mouthparts of muscoid flies such as *Fannia* spp. There are a few hundred case reports of spirurid infections in humans, but their prevalence is not well documented; all are zoonoses, with their normal hosts being ruminants, suids, and carnivores.

Olsen (1974) included a number of different kinds of spirurids and discussed their structure and biology for those interested in obtaining additional information on these food-chain types of life cycles.

Although the order includes a great number of food chain-passed worms, we will be concerned here only with those that are transmitted through bloodsucking vectors—mosquitoes and blackflies in most of the examples. These are in the superfamily Filarioidea and include the spirurids, which are of greatest medical or economic importance, and they are examples of the kinds of helminthoses that are difficult to ameliorate.

The transition from a food-chain type of life cycle to one in which bloodsucking arthropods are intermediate hosts is exemplified by members of the family Diplotriaenidae, whose adults are found in the air sacs, lungs, and subcutaneous tissues of birds and some reptiles. *Diplotriaena bargusinica* is an example of an organism that lies in the air sacs of birds. The eggs pass from the lungs up the trachea and leave the host with the feces. Eggs are eaten by grasshoppers and develop to the infective $L_3$; birds eat infected grasshoppers and thereby obtain the infection. The preadult stages most likely migrate and develop in the vessels of the circulatory system before coming to lie in the lungs; thus, there is a transition from those worms that remain in the gut to those that lie deep in the tissues and, then, to those that use the circulatory system as a migratory route; loss of the food chain would come when some of the larvae remaining in the tissues of blood vessels were ingested by a bloodsucking insect. The evolution of spirurid life cycles has been discussed in papers such as Anderson's (1959).

## *WUCHERERIA* AND *BRUGIA* INFECTIONS IN HUMANS

This section discusses those filarioid nematodes that cause the human disease known, as *elephantiasis*. The important organisms are *Wuchereria bancrofti* and *Brugia malayi*. They are discussed together because they have some common areas of distribution, but also because their differences point up the need for different approaches in control. Sasa's (1976) monograph on filariases in humans is an excellent source of detailed information.

NAME OF ORGANISMS AND DISEASE CAUSED. *Wuchereria bancrofti* (Cobbold, 1877) Silva Araujo, 1877 (syn. *W. pacifica* for the subperiodic form), and *Brugia malayi (Brug, 1927) Buckley, 1958 (syn. Wuchereria malayi*), cause Bancroftian filariasis, Brugian filariasis, periodic filariasis, subperiodic filariases, and elephantiasis.

HOSTS AND LOCATION. Adults are found principally in lymphatic vessels of the extremities. Microfilariae (mf) are in the circulating blood.

DISTRIBUTION AND IMPORTANCE. *W. bancrofti* is the more widespread of the two worms; it is found in both the Eastern and Western Hemispheres, principally in the tropics, but with some extension into subtropical climates and small pockets in temperate areas. The areas of greatest prevalence are in Southeast Asia, the islands of the South Pacific, Africa, the southern portion of the Arabian Peninsula, and India. Areas of lesser importance are in Japan and Korea, West Indies, South America, and areas fringing the Mediterranean Sea. (Fig. 41.1).

*B. malayi* has a limited distribution, principally in Southeast Asia, with more scattered distribution on the Indian subcontinent and on mainland China and Korea (Fig. 41.2).

Stoll's (1947) estimate of filarial infections (*Wuchereria* and *Brugia*) in humans was 189 million persons. A more recent estimate by the World Health Organization (WHO) in 1974 was 250 million infections. In 30 years or so, little or no progress has been made in reducing the prevalence of infection.

Nearly everyone is familiar with the gross deformities that may be seen in long-standing infections with filaria in which the legs or genitalia are swollen and the skin has undergone irreversible changes. Such extreme changes occur only in persons living in endemic areas and who have been exposed and reexposed over many years. The prevalence of infection in endemic areas may exceed 50%, as measured by microfilaremia, but the prevalence of elephantiasis is usually much lower. In some localities 10 to 20% of the population may be affected; however, in Puerto Rico, 24% of males were found to be infected, but the infection is nearly asymptomatic. Needless to say, severe deformity is a great burden to bear.

MORPHOLOGY. *W. bancrofti* males are 28 to 40 long by 0.1 mm, with spicules of unequal length; females are 80 to 100 by 0.27 mm long. The head has eight small papillae. Mf average 280 ± 8μm long (Table 41.1; Fig. 41.3).

*B. malayi* males are 13.5 to 20.5 by 0.075 mm long; females are 43 to 52 long by 0.15 mm. The male spicules are unequal in length. The head has six small papillae. Mf average 220 μm long (Table 41.1; Fig. 41.3).

LIFE CYCLE. The life cycles of *W. bancrofti* and *B. malayi* are similar; that of *W. bancrofti* is given here. The mf circulate in the peripheral blood. The greatest numbers are seen between 10 p.m. and 2 a.m; during the daytime, they will seldom be seen in the peripheral blood. The mf are picked up by mosquitoes during feeding. More than 60 species of mosquitoes in the genera *Culex, Aedes, Anopheles,* and *Mansonia* can support development (Chap. 54).

The mf migrate from the gut of the mosquito into the hemocoel and penetrate muscle cells in the thorax, where they are found within a day after ingestion. They

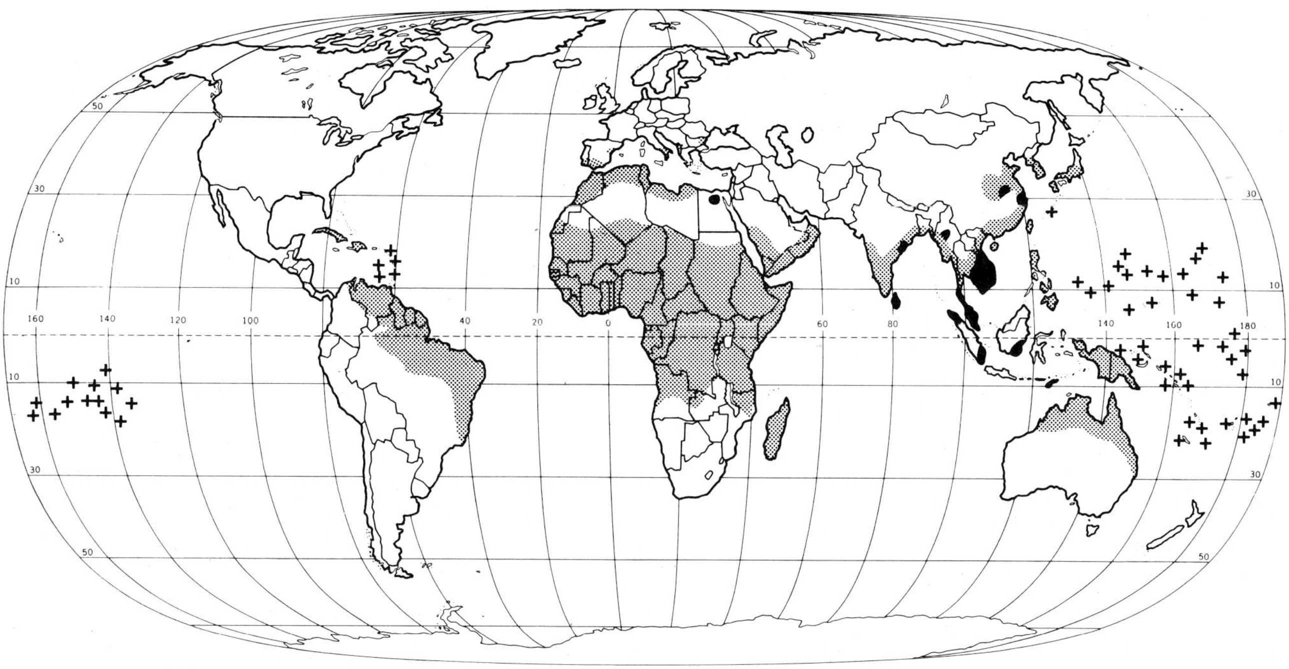

**FIG. 41.1**
Geographic distribution of *Wuchereria bancrofti*. (+ = island infections; solid areas = concurrent *Brugia malayi* infections.)

transform into sausage-shaped forms of about 190 by 14 μm and then begin to grow and molt. The $L_3$ is reached in 6 to 30 days, depending on temperature, and they are about 1700 long by 20 μm.

Transmission to humans takes place when the mosquito returns to take another blood meal. The $L_3$s escape from the mouthparts of the mosquito, probably by sensing the bending of the labium when the mosquito has its fasicle inserted deeply in the skin. The $L_3$s are not injected into the host, but rather penetrate the wound made by the feeding mosquito after the mouthparts have been removed. There is usaully a small pool of fluid at the site of feeding and the $L_3$s are able to survive for as long as 4 min until this pool of hemolymph dries. When they penetrate, they reach the tissue spaces and are carried by the flow of lymph to lymph nodes.

The worms remain in the lymphatic vessels and lymph nodes as they grow and reach sexual maturity. The prepatent period is about a year, but hard data on the time of initial infection are understandably difficult to obtain. Adults may produce mf for 15 years or more.

DIAGNOSIS. Determining whether a person is infected with *W. bancrofti* or *B. malayi* depends on the following:

1. Clinical signs and symptoms
2. Living in or having visited an endemic area
3. Finding mf in the circulating blood

Early in the infection there may be swelling of lymph nodes, skin rash, fever, and urine cloudy with lymph (chyluria). Such signs occur in a variety of infectious diseases and are not specific for filariasis.

Finding the mf in the circulating blood is good evidence when combined with clinical aspects and a high probability of having been exposed. Since there are a number of mf that may be found in humans, it is important to examine them with care and to identify them (Fig. 41.3 and Table 41.1).

Failure to find mf does not mean that the person is not infected with a filarioid (Beaver, 1970); in periodic filariasis, mf are not seen in the peripheral blood during

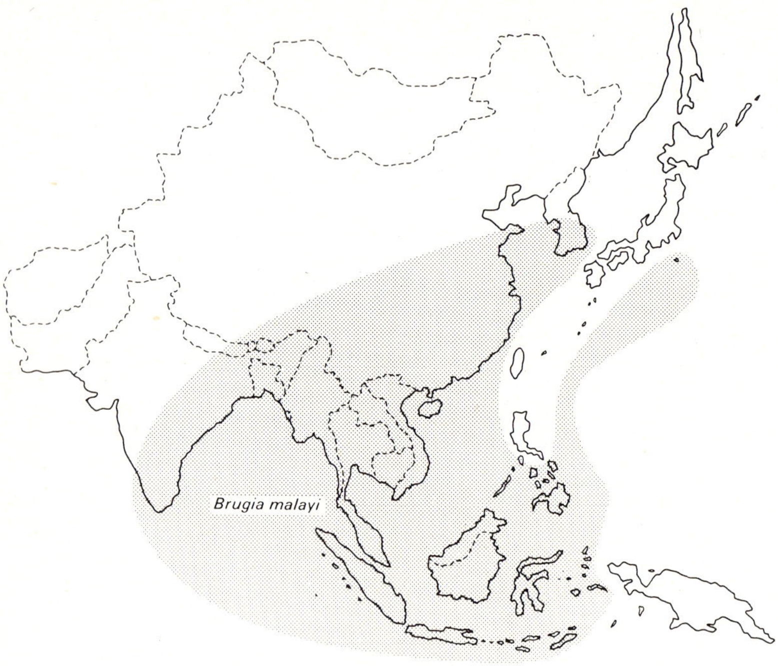


**FIG. 41.2**
Geographic distribution of *Brugia malayi.*

the daytime. As indicated previously, the peak of microfilaremia takes place between 10 p.m. and 2 a.m., and it is important to take the blood sample during that peak time, or at least close to it. If a patient is hospitalized, taking a blood sample in the middle of the night is feasible, but appearing at a person's home at midnight with a syringe and needle at the ready is not likely to elicit enthusiastic cooperation. Epidemiologic studies as well are hampered by these factors. The DEC Provacative Test causes mf to appear in the blood 30 to 60 min after a patient has been given 100 mg of diethylcarbamazine (DEC) (see Treatment); this eliminates the need for night sampling.

Microfilariae are found by using a compound

**TABLE 41.1**
Characteristics of Microfilariae

| | *W. bancrofti* | *B. malayi* | *L. loa* | *O.volvulus* | *D. perstans* | *D. streptocerca* | *M. ozzardi* |
|---|---|---|---|---|---|---|---|
| Length | 200–300 μm | 177–260 μm | 250–300 μm | 285–368 μm or 150–287μm | 150–200 μm | 180–240 μm | 150–200 μm |
| Sheath | Present | Present | Present | Absent | Absent | Absent | Absent |
| Location | Blood | Blood | Blood | Skin | Blood | Skin | Blood & skin |
| Vector | Mosquitoes | Mosquitoes | *Chrysops* | *Simulium* | *Culicoides* | *Culicoides* | *Culicoides* |
| Periodicity | Nocturnal or subperiodic | Nocturnal | Diurnal | Present | None | None | None |
| Tail end | No nuclei Pointed tip | 2 widely spaced nuclei, blunt tip | Nuclei present, rounded tip | No nuclei, pointed tip | Nuclei present, rounded tip | Nuclei present, curved tip | No nuclei, pointed tip |

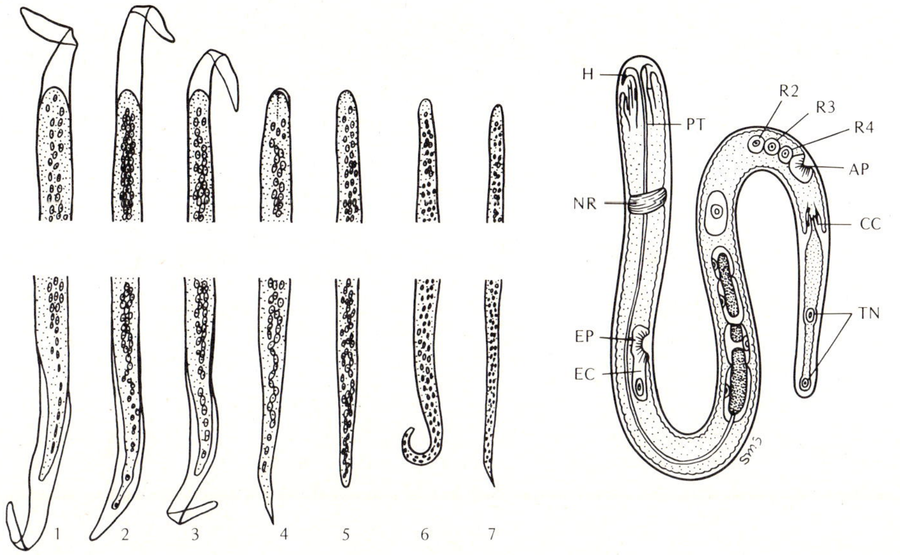


**FIG. 41.3**
Morphology of microfilariae found in humans. (1) *Wuchereria bancrofti;* (2) *Brugia malayi;* (3) *Loa loa;* (4) *Onchocerca volvulus;* (5) *Dipetalonema perstans;* (6) *D. streptocerca;* (7) *Mansonella ozzardi;* AP = anal pore; CC = caudal channel; EC = excretory cell; EP = excretory pore; H = hook; NR = nerve ring; PT = pharyngeal thread; R2–R4 = rectal cells; TN = terminal nuclei.

microscope and one of the following techniques:

1. Stained blood film
2. Counting chamber
3. Membrane filtration (Denham, 1975)

The stained blood film has been the traditional method of finding mf; a thick film is best so that a relatively large volume of blood can be examined easily. The counting chamber method allows one to examine up to 200 $mm^3$ of blood but does not allow identification of the mf species. In the membrane filter method, blood drawn from a vein is passed through a filter, which catches the mf. The membrane filter technique is generally considered to be the most sensitive (Southgate, 1974).

HOST–PARASITE INTERACTIONS. The damage that either *W. bancrofti* or *B. malayi* causes is dependent on the length and intensity of exposure to infective larvae and the response of the individual person. For example, military personnel were exposed to infection on some islands in the south Pacific during World War II for a period of a few months to 2 years. Many had signs and symptoms of infection, but the disease did not progress to the elephantoid stage.

For persons remaining in an endemic area, the disease progresses through four stages:

1. Incubation
2. Symptomless, but patent
3. Acute
4. Chronic

As noted, the $L_3$s are about 1 mm long when they enter the definitive host. They move through the lymph system and are trapped in lymph nodes or in the tissue spaces. There is a reaction around the growing worms giving rise to inflammation (pain, swelling, and tenderness) and usually fever. Seemingly at about the time the worms mature, the symptoms subside and mf are seen in the blood (stage 2). This stage may persist for more than a year or may never progress further. The acute phase (stage 3) is caused both by significant blockage of lymph vessels and by the reaction of the host to the worms. There is swelling due to obstruction of lymph vessels, and a complex of symptoms and signs indicate that hypersensitivity has developed.

After some years, the swelling and obstruction of vessels leads to the irreversible changes in the extremities and genitalia that characterize the disease in the minds of most people. The sequence of events is basically blockage of the lymph vessels, causing retention of fluid distal to the blockage, a foreign body response, and an allergic reaction in the vicinity of the adult worms leading to additional swelling. The skin becomes progressively stretched and loses its normal structure (Fig. 41.4).

The role of protective immunity in altering the course of filariasis is not at all clear. Antibodies are produced in the infected person; they can be detected by a variety of serological techniques, but there is not much evidence that the host is protected by the immune response. Also, the adult worms can live for 15 years or more, which indicates that they either are poorly antigenic or somehow escape the immune response.

Circadian periodicity is the typical pattern of microfilaremia in the definitive host. During the daytime, the microfilaremia is at such a low level that it is unlikely that any will be found in a sample of peripheral blood. During the evening, the mf begin to appear and peak usually between 10 p.m. and 2 a.m.; then they begin to decline and disappear around dawn. Two questions that investigators have tried to answer are, where do they go and why do they cycle?

The answer to the first question is that the mf concentrate in the lungs during the daytime. The second question has two answers. There is a nocturnal periodicity, so that the mf appear in the peripheral blood at the time when there is greatest likelihood of their being ingested by mosquitoes that normally feed from twilight till dawn. How they know it is nighttime is another issue, and most of the evidence points toward oxygen levels in the blood, which tend to decline when a person is asleep. Experiments have shown that the microfilaremia can be reduced at night if a person breathes pure oxygen (Denham & McGreevy, 1977).

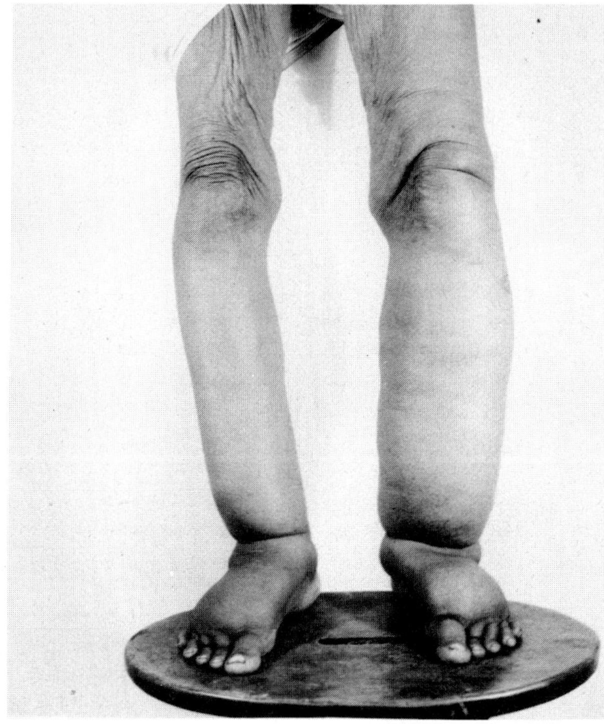

**FIG. 41.4**
Elephantiasis of human legs. [Courtesy AFIP, Negative No. 44430-1.]

In some geographic areas, the nocturnal periodicity of microfilaremia is not as striking; originally this condition was referred to as *nonperiodic filariasis* but is more correctly called *subperiodic filariasis*, because the day-night alteration is damped so that there is about a twofold increase at night rather than complete absence of periodicity. Subperiodic filariasis occurs with both *W. bancrofti* and *B. malayi* in localities where the principal vectors are day-biting mosquitoes. Subperiodic Bancroftian filariasis occurs in the Philippines, Fiji, Samoa, Tahiti, Tokelau, Wallis, and Ellice Islands. The localities of subperiodic Brugian filariasis are unclear, but seem to be based on the host in which it is principally found. The strains of *B. malayi* that cycle mostly in the human population are periodic, whereas those that cycle in feral animals with humans as an occasional host are subperiodic (Denhan & McGreevy, 1977).

The question of why microfilarial periodicity occurs has been addressed (Worms, 1972) but not answered. We can see that adaptation to the time when mosquitoes tend to feed is advantageous. But why not be there all the time? It would not seem to cost the mf anything to circulate continuously, and they would be in an area to be ingested should a mosquito take a blood meal at lunchtime rather than at midnight. Evolutionary dogma states that every characteristic of an organism has adaptive advantage; it is not obvious what advantage is conferred by periodicity.

TREATMENT. Two different aspects of filariasis need to be considered: alleviating the effects of the infection for the individual and reducing the probability of transmission for the general well-being of the population.

The only drug recommended for treatment of either Bancroftian or Brugian filariasis is diethylcarbamazine (hetrazan, DEC). It is active against microfilariae, but there is still some doubt as to its efficacy against adults. It is taken by mouth three times a day for 10 to 30 days. In some instances there are side effects resulting from the death of the mf, and it is therefore difficult to ensure that people will continue to take the drug for the full course. At the end of the course of treatment, the microfilaremia will have been reduced completely, thereby eliminating the probability of transmission. The female worms will at least have been sterilized, and both males and females may have been killed. DEC was developed by Redginal Hewitt at the Lederle Laboratories in about 1945; no better drug has appeared in the intervening 35 years or so.

The gross deformities seen in some cases of filariasis can be prevented by early drug treatment and by bandaging the legs so that fluid cannot accumulate. When the part has been badly affected, the changes are irreversible and surgery helps to reduce elephantoid deformities.

EPIDEMIOLOGY AND CONTROL. In any control program aimed at infectious disease, it is essential to determine exactly what agents and vectors are involved, and filariasis in humans exemplifies that principle as well as any. One needs to know the following:

1. Whether the causative agent is *W. bancrofti* or *B. malayi*
2. Whether the filariasis is periodic or subperiodic
3. What the vectors are

Host range affects the design of a control program and determines whether control can be achieved. Either as the periodic or subperiodic form, *W. bancrofti* has only a single known host—humans. *B. malayi*, on the other hand, has humans as the principal host in the periodic form, but other animals in the subperiodic form. Therefore, after an identification has been made of the species of mf, it is necessary to determine the character of the periodicity. The investigator then knows whether it will be necessary to try to take into account the reservoirs of infection. In subperiodic Brugian filariases, the worms probably circulate almost equally among humans, monkeys, and domestic dogs and cats. Whether it is possible to do something about the infection in hosts other than humans is problematic, but at least the infection there is known.

In any one locality there is usually a single principal vector, although development is possible in other vectors. In filariasis, a wide range of hosts is possible; Bancroftian filariasis can be transmitted by more than 60 species of mosquitoes in the genera *Culex, Aedes, Anopheles,* and *Mansonia*. *B. malayi* has not been studied as extensively, nor does it occur over as wide an area, but about a dozen species in the genera *Aedes, Anopheles,* and *Mansonia* are known to be vectors.

Sites of larval development differ with the species of mosquito; therefore, identifying the mosquito and knowing its biology are important. *Anopheles* vectors of *B. malayi* are usually found in clean water such as that of ditches and wells, but *Aedes togoi* larvae are found in water in small containers such as tin cans or rock holes. *Mansonia* larvae obtain oxygen by piercing plant tissues with their specially adapted air tubes; they are found in waters that have herbaceous plants growing in them.

Taking vector preferences one step farther, the periodic *B. malayi* is found in agricultural areas where *Anopheles* and *Aedes* spp. are associated with human artifacts, whereas the subperiodic forms are usually passed by *Mansonia* spp. in wet, forested areas.

Much of filariases control revolves around vector control, and the discussion in the chapter on malaria in humans pertains here; however, the role of chemotherapy is important in reducing the probability of transmission and some illustrative data are worth considering. In one small drainage, the Reeve Delta, on the island of Fiji, all 1500 residents received treatment with DEC. Before treatment, the prevalence of infection was 79% and the mf density was 91%; 5 months later these were reduced by 92% and 99%, respectively. No mature infections could be found in mosquitoes, and it was considered that transmission had been completely interrupted (Burnett & Mataika, 1961). A great deal of the fundamental work on the control of filariases was done by Dr. John F. Kessel and co-workers shortly after World War II (e.g. Kessel et al., 1958).

## ONCHOCERCIASIS IN HUMANS

NAME OF ORGANISM AND DISEASE CAUSED. *Onchocerca volvulus* (Leuckart, 1893) Railliet & Henry, 1910, causes onchocerciasis, onchocercosis, river blindness, blinding filariasis, and coastal erysipelas.

HOSTS AND LOCATION. For practical purposes, humans are the only host; a few infections have been found in some higher primates, but they probably are not important in maintaining the infection. Adult worms are located in subcutaneous nodules on the trunk, head, and limbs.

DISTRIBUTION AND IMPORTANCE. This parasite is distributed most heavily in subsaharan Africa from the west coast across the continent to Kenya and Tanzania. A few reports of infection have come from Saudi Arabia and Yemen. There are foci of infection in the Western Hemisphere in Guatemala, Mexico, Colombia, Venezuela, and Brazil (Fig. 41.5). Proper habitat for the blackfly vector is the limiting element in distribution.

Stoll (1947) estimated that there were 19.8 million infections worldwide, 19 million of them in tropical Africa. Current estimates of infection are as high as 30 million or 40 million; up to 500,000 persons have been blinded by the infection.

The figures for prevalence of infection are not indicative of the catastrophic impact that onchocerciasis has on the population in an endemic area. The term *river blindness* indicates where people become infected: along rivers and small streams where people live and where good farmland lies. In some areas one-fifth of the people in a village may be blind as a result of *O. volvulus* infection; the effect on the individual and the community in a subsistence farming economy can easily be appreciated. In many instances in Africa, the areas near rivers have been abandoned and people have tried to farm on less desirable land.

MORPHOLOGY. Males are 19 to 42 long by 0.17 mm and have a ventrally curved tail; females are 335 to 500 long by 0.35 mm. The anterior end has eight small papillae and two large lateral ones. A male and a female are usually coiled together in a nodule. Microfilariae are found in two sizes—285 to 368 by 6 to 9 μm or 150 to 287 by 5 to 7 μm (Fig. 41.3 and Table 41.1).

LIFE CYCLE. The mf do not circulate in the blood but are in the skin, where they are picked up when blackflies (Diptera, Simuliidae) (see Fig. 54.6) feed on humans. The cycle is much like that of *W. bancrofti*, with development to the $L_3$ taking place in the thoracic muscles. Infectivity for humans is reached in a week or so after ingestion of the mf. When the blackfly takes another blood meal, the $L_3$s escape from mouthparts and invade the host through the wound made by the vector. The route of migration in the definitive host is not known with certainty, but the adults come to lie in the subcutaneous tissues and begin to produce mf within 6 months to a year.

DIAGNOSIS. The presence of nodules under the skin of persons living in endemic areas is good evidence of infection, especially if a nodule is opened surgically and the adult worms are found.

Since the mf do not circulate in the peripheral blood, skin snips are usually examined for evidence of infection. Small snips of skin are shaved or nipped from the skin surface in such a way that little or no bleeding occurs, and when placed in physiological salt solution the mf migrate out and can be found microscopically. In the absence of nodules, or where surveys for infection are done, the skin snip method is used. In moderate to severe infections, 40 to 245 mf/mg of skin have been found.

HOST–PARASITE INTERACTIONS. The effects on the host are in nodule formation, in which adult worms are encapsulated, in fibrous tissue, in long-term skin changes, in damage to lymph nodes, and in damage to the eye eventually resulting in blindness. The causes of

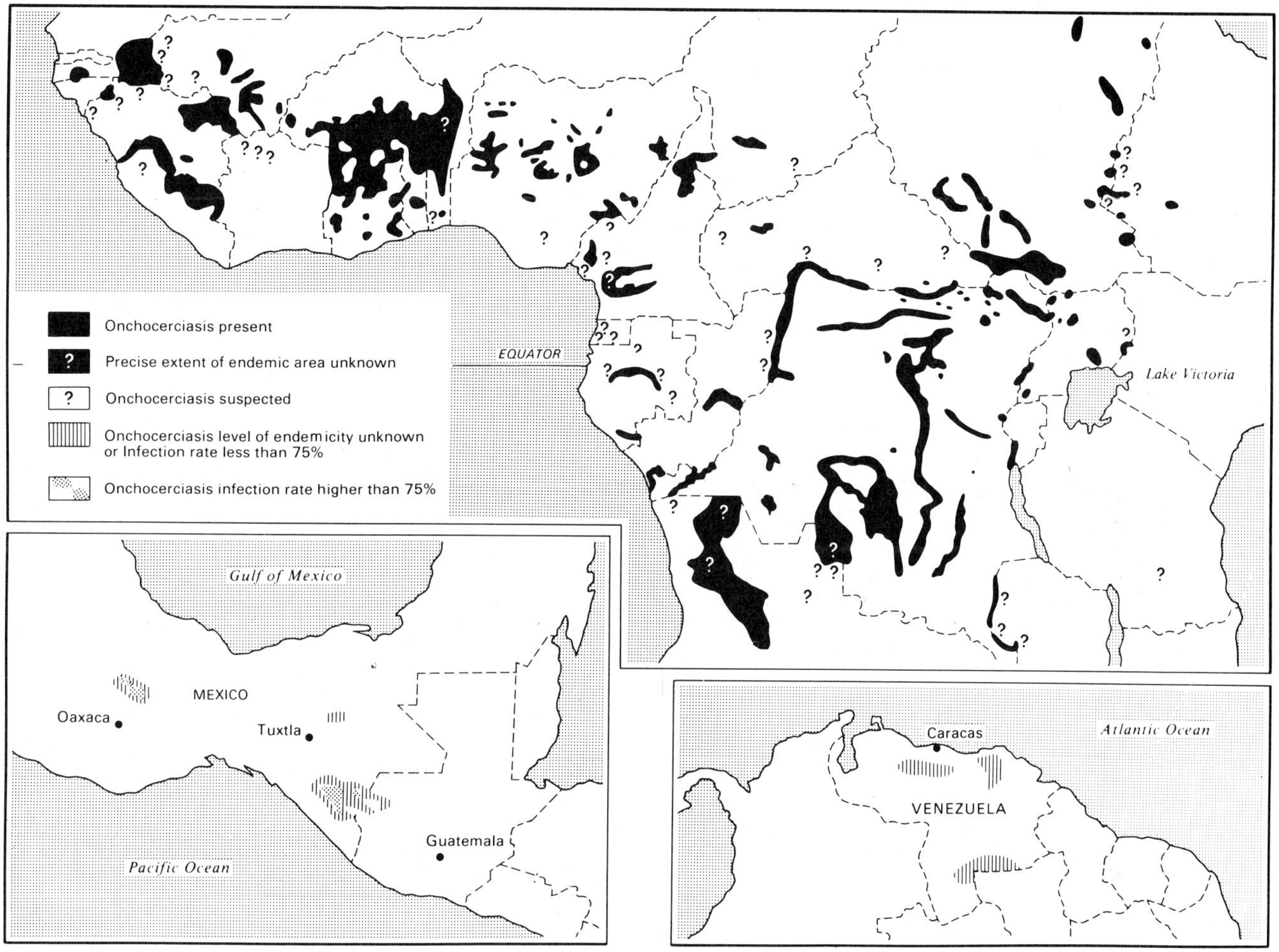


**FIG. 41.5**
Geographic distribution of *Onchocerca volvulus*. [Redrawn from Sasa, 1976.]

the damage are related to immunity, hypersensitivity, and intoxication from worm products or from the worms themselves.

Early in infections the adult worms seem to move about under the skin, but about the time the infection becomes patent, the adults become encapsulated in fibrous connective tissue. The consensus is that these nodules are of no great consequence in themselves but that the adults are responsible for continuing microfilaria production.

The damage to the eye is related to the microfilariae, most likely to direct damage by the mf and to an allergic response to them. Greatest eye damage occurs when most mf are present. Damage is in the cornea, iris, retina, and optic nerve. The latter two areas are probably most important in the complete loss of sight. Blindness is not inevitable; for example, in a survey of 1015 persons in West Africa, 52% had nodules and of them, 9% were blind, 10% had lesions in the retina, 19% had lesions of the cornea, and 4% had involvement of the iris.

Skin changes are related to the breakdown and disappearance of connective tissue in the epidermal and dermal layers; the skin sags and looks prematurely aged. In a few instances, elephantoid changes reminiscent of *W. bancrofti* infections have been seen.

Immunopathology may be important in causing some of the harmful effects of onchocerciasis. Deposition of immune complexes in lymph nodes may trigger inflammation, leading to irreversible changes in the nodes and other organs (Gibson & Connor, 1978).

The role of immunity in protecting a person against reinfection seems to be unimportant. It is known that infections can last for 11 years or more and that the encapsulation of the adult worms does not lead to their demise, at least not within a short period.

Although mf are found in the skin, they are probably normally distributed in a number of other organ systems. They are often found in urine and other secretions and excretions. Some investigators have tried to determine whether there is a periodicity in the mf, as in *W. bancrofti* and *B. malayi*. The investigations have given equivocal answers: in some instances there were peak numbers in skin snips at dusk and dawn, but in other studies there was no statistical significance to the variation. It is important to note that mf are not merely trapped in the skin, passively awaiting a blackfly to arrive and feed, but that they do move about.

In the preceding discussion, some generalizations have been made that may not hold in all localities. There are variations in the place on the body where the nodules are found, for example. In the Western Hemisphere, nodules are most common on the head, whereas in Africa, nodules are more common on the trunk. It is said that nodules occur at the height above the ground where the local vector tends to feed. If this were the case, the nodules should be found on the heads of short people in Africa and on the trunks of tall people in Guatemala. This relationship appears not to have been investigated. Additionally, the pathologic aspects of onchocercosis vary geographically; there seems to be a greater incidence of blindness in the New World than in Africa. In Yemen a different pathologic picture is seen in the lymph nodes from that seen in Africa.

TREATMENT. The objectives of treatment for *O. volvulus* infection relate to both the individual and the health of the community, and both surgery and chemotherapy are used. Obvious nodules under the skin of a person can be opened under local anesthesia and the adult worms removed. If the adults cannot be removed by surgery, only then should chemotherapy be started, because the toxic or allergic reaction to the dead adults can be severe. DEC is recommended to kill the mf, but there may be side effects from their death. Suramin can be used against the adult worms.

EPIDEMIOLOGY AND CONTROL. Control of *O. volvulus* is directed to reduction in the following:

1. Skin nodules containing adult worms
2. Microfilarial numbers
3. The blackfly vectors

Removal of the parasites from the human host benefits the individual but is equally important in reducing transmission. Since humans are the only definitive host, as far as we know, reducing the prevalence and the level of infection in a population ultimately has an effect by itself. Nodulectomy has had its greatest success in the Guatemalan focus of infection; the number of nodules on a person is usually small, and therefore it is possible to remove them from a significant proportion of the population. In Africa the number of nodules on a person is likely to exceed ten and a significant proportion of them have only dead, calcified worms; removal is not feasible and may accomplish little.

Chemotherapy to kill microfilariae with DEC and adult worms with Suramin has the same effect as nodulectomy, but the hazard of intoxication from dead worms is high. Good nursing care and drugs to control the side effects of the anthelmintics are essential for the survival of a heavily infected person. Even if good care can be given, some workers feel that the risk of treatment with Suramin is too great for its routine use.

With the foregoing as a prelude, we are obviously left with blackfly control as the principal means of controlling onchocerciasis. The larvae of these tiny insects, usually no more than 2 mm in length, are associated with running water. They are found in well-aerated water, where they feed on small bits of organic matter; they are among many small stream invertebrates that process leaf litter. Most often larvae are found in cascading streams or shallow riffles where the rocks are partly out of the water. Some species adhere to rooted aquatic plants, and *Simulium neavei* attaches to freshwater crabs (Chap. 54).

Despite their small size, adult simuliids are strong fliers and may migrate several kilometers from their site of larval development. Interestingly, those insects that have relatively heavy infections with *O. volvulus* tend to migrate only short distances.

Relatively few species of blackflies on a global basis are good vectors of *O. volvulus*. In Africa there are two principal species: *Simulium damnosum* and *S. neavei*. In Central and South America there are three species: *S. ochraceum* (the most important vector), *S. metallicum*, and *S. callidum*. With relatively few vectors, the problem of control would seem to be simple because the biology of any one species is fairly circumscribed. Success in reducing fly populations has been difficult to achieve, however, because the habitats of the larvae are usually inaccessible and some species of flies have long flight ranges; *S. damnosum* has a flight range of at least 100 km (60 miles).

Reducing blackfly numbers has involved the following:

1. Source reduction or habitat alteration
2. Killing larvae by putting insecticides in streams
3. Aerial spraying of insecticides to kill adults

Source reduction has been applied in two ways:

1. Cleaning streams of rooted plants
2. Water control

Removing plants eliminates the substrate to which larvae attach. Such a system was used in East Africa, but two problems presented themselves: the cost was too high, and the larvae of *S. neavei* are phoretic on freshwater crabs, so this approach has pretty much been abandoned.

Water control has come about more or less inadvertently with the construction of dams for irrigation and power production. Upstream from a dam, the cascades and riffles that provide substrate and maintain well-oxygenated water are covered up and the habitat of the larvae is gone; however, downstream from the dam there may be a better habitat than before.

Water control allows year-round availability of water for irrigation and other uses, so streams do not dry out seasonally. In addition, dam outlets are constructed of concrete and usually engineered in such a way that the energy is taken out of the water with drops and baffles. Such man-made structures may provide a perfect larval blackfly habitat and exacerbate the fly problem.

Most control programs rely on insecticides, principally DDT (Nelson, 1970). Early larviciding programs in Kenya starting in 1947 relied on large concentrations of DDT applied frequently; by this method and by meticulous monitoring, *S. neavei* was eradicated from an area of 15,000 square miles. Later, aerial spraying also reduced fly populations significantly, but it turned out that insecticide supposedly directed against the adults was actually directed against larvae, because of the extreme sensitivity of the larvae to DDT. For example, larvae are killed by DDT concentrations of 0.03 ppm in water; this is $3 \times 10^{-5}$ g of DDT/liter of water, hardly enough to see if one were to weigh it on an analytical balance.

Success of a control program requires:

1. Monitoring fly populations
2. Monitoring transmission, that is, the incidence of infection during a given period of time
3. Continuing the program for many years
4. Placing most of the available resources in areas of highest transmission rates

These points may seem to be only common sense, but let us consider some of their ramifications. Resources are inevitably limited and a judgment must be made where to put them. In the Kenya eradication program, success was achieved because it was possible to reach the areas of larval development. In other instances, accessible areas may not be where the highest rates of transmission occur. The practicalities of the situation present a moral dilemma: do you go where the problem is greatest or where you have the probability of success? It is essential to continue a program of surveillance until the focus has died out completely. Adult worms may live longer than 11 years, and as long as mf are present, the possibility of transmission exists. Given the long time required for surveillance and control, what is the probability of success in developing countries? At this point the answer must be equivocal. Good success has been achieved in the Volta River drainage of West Africa and in Kenya. In Central America, the inaccessibility of sites of larva development has hampered control. But even if initial success is realized, will hard-pressed governments allow the public health personnel to continue to fight a battle that seems to have been won? If malaria eradication programs are an indicator, the most optimistic answer is, probably not.

We have discussed some undesirable results of control programs against various diseases. It is worth quoting Nelson (1970) in this regard:

Onchocerciasis is an economically important disease, but not a killing disease. It incapacitates large sections of the community and makes them an economic burden. Control of the parasites can do nothing but good; it will not increase the birth rate and produce more mouths to feed, but it will release whole communities from an intolerable burden of misery. Also, by ridding the countryside of biting simuliids, vast fertile areas will be freed for agricultural development. Total eradication is still remote but even with the application of existing knowledge and methods of control the intensity of transmission can be so restricted that many areas can be completely freed of this scourge.

## OTHER FILARIAL INFECTIONS IN HUMANS

In addition to the three filarial infections of humans discussed earlier in this chapter, there are four others that commonly occur in humans and several others that occur sporadically.

*Loa loa* is the eye worm; it occurs in subsaharan Africa principally in rain forest areas, where it is transmitted by *Chrysops* spp. (Diptera, Tabanidae). Although found first in the eye of a West African in 1770, the adults lie in subcutaneous tissues as well, where they cause inflammation and swelling. *Dipetalonema perstans* occurs in North Africa and widely in subsaharan Africa, but is of only minor medical importance. Its vectors are *Culicoides* spp. (Diptera, Ceratopogonidae). *D. streptocerca* and *D. semiclarum* are also found in humans. *Mansonella ozzardi* is a New World parasite that occurs in tropical areas, where it is transmitted by *Culicoides* and *Similium* spp. Although the prevalence of infection in some Amerinds is as high as 96%, the damage to the host is minor or limited to pain in the joints.

Sporadic infection in humans have been seen with *Dirofilaria* spp., which are normally found in various carnivores. Other species of filarioid nematodes probably infect humans, because of the occurrence of so-called occult filariasis in which there is no microfilaremia (Beaver, 1970).

## FILARIASIS IN DOGS

NAME OF ORGANISM AND DISEASE CAUSED. *Dirofilaria immitis* (Leidy, 1856) Railliet & Henry, 1911, causes dirofilariasis, filariasis, and heartworm disease.

HOSTS AND LOCATION. The principal host is the domestic dog; other canids also are good hosts. The domestic cat is seldom infected. There are a number of reports in higher primates, but humans are infrequently infected. Some animals such as the ferret can serve as experimental hosts. The organisms are located as adults in the vena cava, right heart, and pulmonary artery. They are occasionally found in other locations in the dog and especially in foreign hosts.

DISTRIBUTION AND IMPORTANCE. The organism is cosmopolitan. This worm was originally thought to be found in warm, moist climates such as the southeastern United States, but in the late 1950s data began to accumulate showing that *D. immitis* also occurs in temperate and cold climates (Schlotthauer & Griffiths, 1964). There are enzootic areas near Chicago, Illinois, and Minneapolis, Minnesota; the infection can even become prevalent in semidesert locales such as Grand Junction, Colorado, where irrigation is practiced. It is likely that the organism has extended its range in the United States by increased travel and changing of residence, which Americans seem compelled to do. Surveys in the southeastern United States usually show from 10 to 70% prevalence, whereas that in the northern part of the country is from 1 to 8%. Cats generally have a prevalence of infection less than a tenth that of dogs.

Dirofilariasis is considered to be a major disease of dogs in the southeastern United States and other warm climates.

MORPHOLOGY. The adult males are from 12 to 200 by 0.8 mm and the females are 250 to 310 by 1 mm, the females are therefore about 1 ft long (Fig. 41.6). Head structures are minimal, but six papillae surround the mouth. The male has 11 pairs of papillae in the anal region and a pair of unequal spicules 300 and 200 μm long, respectively.

LIFE CYCLE. The infective stage, the $L_3$, develops in mosquitoes, usually in the genus *Aedes*, but other

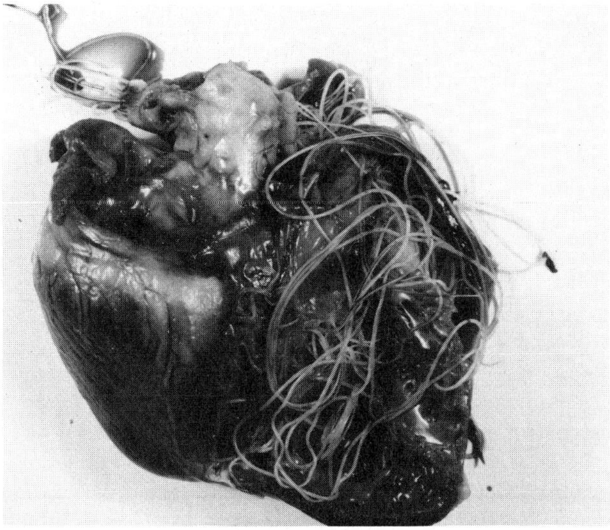

**FIG. 41.6**
*Dirofilaria immitis* adults in a dog heart.

genera are satisfactory vectors in some localities. The $L_3$s escape from the mouthparts of the mosquito during feeding and enter the wound made by the mosquito. Early in the infection the developing worms occur in subcutaneous, adipose, and muscular tissue. When the worms reach early $L_5$, at about 90 days postinfection (PI), they begin to locate in the heart. The prepatent period is about 180 days before microfilariae appear in the blood.

Mf are picked up by mosquitoes during blood feeding. They migrate from the gut to the hemocoel and enter cells of the Malpighian tubules, where they become short, sausage-shaped organisms. They then begin to grow and molt twice to reach the $L_3$ 10 to 16 days PI. The $L_3$s migrate to the head of the mosquito and readily escape when another blood meal is taken (Kume & Itagaki, 1955).

DIAGNOSIS. The principal means of determining whether an animal is infected is by finding the mf in the circulating blood, but clinical signs are also important in enzootic areas.

Finding the mf is done by drawing a small amount of blood into a weak formalin solution and then looking for them in the sediment after centrifugation. As indicated in the section on *Wuchereria* and *Brugia*, mf can be differentiated from one another, and it is sometimes important to do so. *Dipetalonema reconditum* of the raccoon is most often the filariid that needs to be differentiated from *D. immitis*:

| | LENGTH | WIDTH | TAIL |
|---|---|---|---|
| *Dirofilaria immitis* | 286–340 μm | 5–7 μm | Straight |
| *Dipetalonema reconditum* | 269–283 μm | 4.3–4.8 μm | Hooked |

Other mf landmarks can also be used, but the overall dimensions are most important for these two species (Newton & Wright, 1956).

Signs of infection begin to appear more than 6 months after infection. Dogs lack stamina and may become exhausted or breathe hard after only a short run.

HOST–PARASITE INTERACTIONS. Clinical signs appear in dogs as the worms become sexually mature. Given the size, location, and potential number of adult worms, the principal clinical signs involve interference with blood flow. The worms locate in the right heart; when mature they may extend from the vena cava into the right auricle, down into the right ventricle, and out into the pulmonary artery. They impede blood flow and prevent proper closing of the heart valves. There is backpressure from the heart to the liver because the blood does not move through the right heart at a normal rate. The heart works harder in an attempt to empty and also to meet the need to move blood from the heart to all organs of the body.

The results of this study in fluid mechanics are that the liver becomes enlarged as a result of engorgement with blood, the right heart becomes enlarged, and there is insufficient oxygenated blood for the animal to maintain strenuous activity. A frequent pattern is that a dog becomes infected during the summer; it develops a patent (but undetected) infection during the winter at a time when it may be relatively inactive. When spring comes and the exhilaration we all feel wells up, the dog may be turned loose for a romp and suffer from cardiac failure in midgambol. Death or permanent cardiac damage may be the outcome.

There are effects on the host in addition to these

mechanical ones. Lesions occur in the blood vessels such as the pulmonary artery, where there may be inflammation and erosions (Adcock, 1961). The mf may also be pathogenic.

There is nocturnal periodicity in the numbers of mf in the blood of the dog. Peak microfilaremia is from about 8 p.m. to 2 a.m. but periodicity is not as marked as in *Wuchereria bancrofti.*

TREATMENT. Three aspects to therapy of heartworm infection are as follows:

1. Preventing development of worms in the dog
2. Killing the microfilariae
3. Killing or removing the adult worms

In endemic areas, dogs are given DEC before the start of the mosquito season and for 2 months after it is over. This drug prevents the development of the worms that may infect the dog.

If a patent infection develops, the adult worms may be killed with arsenamide or they can be removed surgically in special cases (Horne, 1962). The drug is administered in divided dosages so that only some of the adults are killed at any one time. If a single large dose of drug were to be given, all of the worms would die at the same time; they would flow downstream and damage the lung so severely that the dog might die.

Surgical removal involves chest surgery in which the worms are extracted by cutting into the major vessels. Special circumstances are required because surgery is both risky and expensive.

The mf may be killed by administering dithiazinine iodine, but care must be taken to ensure that all the mf are not killed at once or a toxic reaction may ensue.

EPIDEMIOLOGY AND CONTROL. Although dirofilariasis is more prevalent and more of a problem in warm than in cool climates, it can become a problem in any area where mosquitoes are abundant. The host range is quite broad in the vector (Kartman, 1953), so the probability of its being able to establish in a given area is high. This is confirmed by the cosmopolitan distribution in United States (Marquardt & Fabian, 1966a) and other countries such as Australia (Bemrick & Moorhouse, 1968).

In some instances, the incidence of infection has increased dramatically in a locality such as Minneapolis (Schlotthauer & Griffiths, 1964) or Cook County, Illinois, north of Chicago (Marquardt & Fabian, 1966a). The increased prevalence of infection in various parts of the United States is probably a result of the high mobility of the U.S. population. People move from one part of the country readily and they travel long distances on vacation. Dogs are taken along and carry their infections back home or to their new homes.

Once established in an area, the infection seems to remain enzootic. The reasons seem to be the long life of the adults, 2 to 3 years, and the high microfilaremia in even moderate infections. Hundreds of mf/mm$^3$ of blood are common in dirofilariasis. Dogs are seldom protected from mosquitoes and may sleep in unscreened areas near the house. The role of wild canids in perpetuating the infection has not been systematically investigated, but both foxes and coyotes have been found to be infected in enzootic localities.

No public health measures have been taken to reduce the incidence of *D. immitis* infection in dogs, but community mosquito control programs should reduce the probability of infection. The responsibility for control lies with the individual owner. The steps the owner can take are as follows:

1. Protection of the dog from mosquitoes by providing screened quarters or by keeping the dog confined at times of day when mosquitoes most actively seek blood meals, usually at dusk and dawn.
2. Administration of DEC during and after the mosquito season. In areas where mosquitoes are active year round, DEC must be given continuously.
3. Periodic examination of the dog to determine whether it has a patent infection and to provide treatment if necessary.
4. Periodic treatment, usually twice a year, regardless of whether the animal shows signs of infection. Some veterinarians recommend such periodic treatment in areas where the probability of infection is high; the rationale is that the dog probably is infected and the drug will eliminate any infection that might be there. Arsenamide is not without side effects and the risks of treatment should be balanced against its benefits.
5. Clearance of the microfilaremia by dithiazinine. Mf can survive a long time in the blood of an

animal after the adult filarioids have been removed. These mf provide a pool of infectious agents to be passed on to another animal, and they may also be somewhat pathogenic to the host. Their removal is therefore a public health measure and gives relief to the individual dog.

The importance of *D. immitis* infections in humans is still poorly known. There have been about 35 case reports of persons infected with what is most likely *D. immitis*. In two there were worms in the heart, but most often they have been found in the lungs and subcutaneous tissues (Beaver & Orihel, 1965). Dirofilariasis in humans is clearly not a major problem, but effects can be moderately severe in infected persons. Prevention of infections in humans can be taken care of by control programs for dogs.

## FILARIAL INFECTION OF RUMINANTS AND HORSES

Five genera of filarioid nematodes have been found in domestic herbivores in temperate and warm climates throughout the world. Where systematic studies have been done, the prevalence rate sometimes exceeds 75%. Nearly all species cause at least minor damage to the host and a few cause serious losses in some populations of animals. Significant control programs do not seem to have been put into practice. Some species that are representative or are of interest because of their impact on populations are the following:

*Stephanofilaria stilesi* occurs in the skin of the belly of cattle in the United States; it seems to be most common in cattle in the western United States, where it occurs in up to 90% of animals at slaughter. The skin where the adults and microfilaria are found becomes thickened, inflamed, and leaks blood and serum; such lesions may reach 15 cm (6 in) in diameter. Horn flies, *Haematobia irritans* (Diptera:Muscoidea), are the intermediate hosts (Hibler, 1966). Losses from this parasite lie principally in trimming of small parts of the hide at slaughter, not a major loss.

A number of species of *Onchocerca* lie in ligaments, where they cause inflammation and significant discomfort to the host. For example, *Onchocerca cervicalis* occurs in the ligamentum nuchae, the large ligament that inserts at the back of the head and supports it. About 50% of horses in the United States are infected, and up to 80% of horses in both temperate and warm climates in other parts of the world have been found to be infected. The parasite causes inflammation of the ligament and then necrosis of the connective tissue fibers and calcification later in the infection. Vectors are biting midges of the genus *Culicoides* (Diptera, Ceratopogonidae).

The normal hosts of *Elaeophora schneideri* are mule deer (*Odocoileus hemionis*) and white-tailed deer (*Dama virgiana*) in the United States. In the normal host the parasite causes little damage, but there can be considerable damage in domestic sheep and in elk (*Cervus canadensis*). We have discussed some instances in which migratory patterns are altered in a foreign host and a few in which the infection is more severe than in the normal host. *E. schneideri* is a good example of the latter: sheep suffer from a severe allergic dermatitis and elk are often blinded. The infection in sheep was first described by Dr. H. E. Kemper of the USDA in 1938 in sheep in southern Colorado and northern New Mexico that had grazed on pastures above 2000 m (6000 ft) during the summer. It was later found that elk in the Gila National Forest of north central New Mexico were blinded because of infection. An excellent series of studies was done under difficult conditions on the pathogenesis and the life cycle and impact on populations of elk (Adcock & Hibler, 1969; Hibler & Adcock, 1968; Hibler & Metzger, 1974).

## LABORATORY MODELS OF FILARIOID INFECTIONS

It is evident that there is a good deal of conjecture and partial knowledge of nearly all aspects of filarioid infections in humans and other animals. Although filarioids may infect many hosts, some have a narrow host range or develop abnormally in other hosts. *W. bancrofti*, for example, develops only in humans; on the other hand, the subperiodic form of *B. malayi* develops fairly readily in primates other than humans and in cats. Thus, we have better descriptions of the course of the disease in *B. malayi* than in *W. bancrofti*. An ideal situation is one in which a parasite can be propagated in its normal hosts in the laboratory and studied in a controlled way. *Dirofilaria immitis* has been studied in the dog (Adcock, 1961), but its long prepatent period and the expense of keeping dogs preclude its use as a widespread laboratory model. The best model found to date is *Litosomoides carinii* of the cotton rat; it has been used for a variety of studies on transmission, immunity, and other aspects of host-parasite interactions. A great deal has been learned with *L. carinii*, but it is of limited usefulness; for example, it can be used as a drug-screening system, but with reservations (Bertram, 1966). Better models would allow knowledge to advance more rapidly.

## REFERENCES AND FURTHER READINGS


Adcock, J. L. 1961. Pulmonary arterial lesions in canine dirofilariasis. *Am. J. Vet. Res.* **22**:655–662.

Adcock, J. L., and C. P. Hibler. 1969. Vascular and neuro-ophthalmic pathology of elaeophorosis in elk. *Path. Vet.* **6**:185–213.

Anderson, R. C. 1959. Possible steps in the evolution of filarial life cycles. *Proc. 6th Intern. Congress Trop. Med. and Malariol.* **2**:444–449.

Bailey, W. S. 1971. *Spirocerca lupi:* A continuing enquiry. *J. Parasitol.* **58**:3–22.

Beaver, P. C. 1970. Filariasis without microfilaremia. *Am. J. Trop. Med. Hyg.* **19**:181–189.

Beaver, P. C., and T. C. Orihel. 1965. Human infection with filariae of animals in the United States. *Am. J. Trop. Med. Hyg.* **14**:1010–1029.

Bemrick, W. J., and D. E. Moorhouse. 1968. Potential vectors of *Dirofilaria immitis* in the Brisbane area of Queensland, Australia. *J. Med. Entomol.* **5**:269–272.

Bertram, D. S. 1966. Dynamics of parasitic equilibrium in cotton rat filariasis. *Adv. Parasitol.* **4**:255–319.

Buck, A. A. (ed.). 1974. *Onchocerciasis.* World Health Organization, Geneva.

Burnett, G. F., and J. U. Mataika. 1961. Mass administration of diethylcarbamazine citrate in preventing transmission of aperiodic human filariasis. *Trans. Roy. Soc. Trop. Med. Hyg.* **55**:178–187.

Connor, D. H. 1978. Current concepts in parasitology and onchocerciasis. *N. Engl. J. Med.* **298**:379–381.

Denham, D. A. 1975. The diagnosis of filariasis. *Ann. Soc. Belg. Med. Trop.* **66**:517–524.

Denham, D. A., and P. B. McGreevy. 1977. Brugian filariasis: Epidemiological and experimental studies. *Adv. Parasitol.* **15**:243–309.

Duke, B. O. L. 1968. Onchocerciasis. *Brit. Med. J.* **1968**(4):301–304.

———. 1971. The ecology of onchocerciasis in man and animals. In A. M. Fallis (ed.), *Ecology and Physiology of Parasites.* University of Toronto Press, Toronto, pp. 213–222.

———. 1971. Onchocerciasis. *Brit. Med. J.* **28**:66–71.

Eberhard, M. L. 1979. Studies on the *Onchocerca* (Nematoda: Filarioidea) found in cattle in the United States. I. Systematics of *O. gutterosa* and *O. lienalis* with a description of *O. stilesi* sp. n. *J. Parasitol.* **65**:379–388.

Gibson, C. L. 1952. Comparative morphology of skin-inhabiting microfilariae of man, cattle and equines in Guatemala. *Am. J. Trop. Med. Hyg.* **1**:250–262.

Gibson, D. W., and D. H. Connor. 1978. Onchocercal lymphadenitis: Clinopathologic study of 34 patients. *Trans. Roy. Soc. Trop. Med. Hyg.* **72**:137–154.

Hibler, C. P. 1966. Development of *Stephanofilaria stilesi* in the horn fly. *J. Parasitol.* **52**:890–898.

Hibler, C. P., and J. L. Adcock. 1968. Redescription of *Elaeophora schneideri* Wehr & Dikmans, 1935 (Nematoda: Filarioidea). *J. Parasitol.* **54**:1095–1098.

Hibler, C. P., and C. J. Metzger. 1974. Morphology of the larval stages of *Elaeophora schneideri* in the intermediate and definitive hosts with some observations on their pathogenesis in abnormal definitive hosts. *J. Wildl. Dis.* **10**:361–369.

Horne, R. D. 1962. Evaluation of pulmonary arteriotomy for the removal of *Dirofilaria immitis. JAVMA* **141**:248–255.

Kartman, L. 1953. Factors influencing infection of the mosquito with *Dirofilaria immitis* (Leidy, 1856). *Exp. Parasitol.* **2**:27–78.

Kessel, J. F., J. Laigret, H. N. March, et al. 1958. Epidemiology and control of filariasis with special reference to French Polynesia. *Proc. 6th Intern. Cong. Trop. Med. and Malaria.* **2**:326–338.

Kume, S., and S. Itagaki. 1955. On the life-cycle of *Dirofilaria immitis* in the dog as the final host. *Brit. Vet. J.* **3**:16–24.

Marquardt, W. C., and W. E. Fabian. 1966a. Heartworms in dogs in Illinois. *Ill. Vet.* **9**(2):11–14.

———. 1966b. The distribution in Illinois of filariids of dogs. *J. Parasitol.* **52**:318–322.

Nelson, G. S. 1970. Onchocerciasis. *Adv. Parasitol.* **8**:173–224.

Newton, W. L., and W. H. Wright, 1956. The occurrence of a dog filariid other than *Dirofilaria immitis* in the United States. *J. Parasitol.* **42**:246–258.

Olsen, O. W. 1974. *Animal Parasites. Their Life Cycles and*

*Ecology,* 3rd ed. University Park Press, Baltimore, pp. 466–496.

Rodger, F. C. 1959. *Blindness in West Africa.* H. K. Lewis & Co. Ltd., London.

Sasa, M. 1976. *Human Filariasis: A Global Survey of Epidemiology and Control.* University Park Press, Baltimore.

Schlotthauer, J. C., and H. J. Griffiths. 1964. Canine filariasis in Minnesota. *JAVMA* 144:991–993.

Southgate, B. A. 1974. A quantitative approach to parasitological techniques in Bancroftian filariasis and its effect on epidemiological understanding. *Trans. Roy. Soc. Trop. Med. Hyg.* 68:177–186.

Spencer, H. (ed.). 1973. *Tropical Pathology.* Springer-Verlag, Berlin.

Stoll, N. R. 1947. This wormy world. *J. Parasitol.* 33:1–18.

World Health Organization. 1966. Expert Committee on Onchocerciasis. Second Rept. Tech. Rept. Ser. No. 335. Geneva, Switzerland.

Worms, M. J. 1972. Circadian and seasonal rhythms in blood parasites. In E. U. Canning and C. A. Wright (eds.), *Behavioral Aspects of Parasite Transmission.* Academic Press, New York, pp. 53–67.

Zaman, V. 1979. *Atlas of Medical Parasitology.* Lea & Febiger, Philadelphia.

# 42

# *Guinea Worm*

NAME OF ORGANISM AND DISEASE CAUSED. *Dracunculus medinensis* (L.) Reichard, 1759, is the guinea worm that causes a disease variously called *dracunculiasis, dracunculosis*, or *dracontiasis*—the fiery serpent of the Old Testament.

HOSTS AND LOCATION. Humans and a wide range of other hosts including wild and domestic carnivores, herbivores, and primates exist. Adults lie in the subcutaneous tissues of the definitive host; however, other sites may also be infected.

DISTRIBUTION AND IMPORTANCE. The organism is distributed in the Sudan area of Africa, New Guinea, India, southern Asia, Middle East, some Caribbean islands, and parts of South America including Brazil.

There are about 20 million infections in Africa, the Middle East, and the Indian subcontinent. Infected persons are incapacitated for about 3 months when the worms are under the skin.

MORPHOLOGY. Female worms are from 1.5 to 70 cm by 1 mm, but they may reach a length of 120 cm (47 in.) Males are 12 to 29 by 0.4 mm. The head and tail structures are rather poorly developed (Fig. 42.1).

LIFE CYCLE. Adult female worms lie in subcutaneous tissues and release $L_1$ through a break in the skin. The $L_1$ is eaten by a copepod such as *Cyclops viridis*, in which it develops to the $L_3$ in about 2 weeks. Infection is passed to the host by drinking water contaminated with infected copepods. The prepatent period is nearly a year.

DIAGNOSIS. Finding a portion of the body of the female worm protruding through a blisterlike break in the skin is the means of diagnosis.

HOST–PARASITE INTERACTIONS. The developing stages in the definitive host do not cause any known symptoms, but when the mature female reaches the subcutaneous area, both local and systemic reactions occur. There is an allergiclike reaction in the area where the worm lies, and the person may also suffer from intestinal upset, difficulty in breathing, dizziness, and fainting. Worms sometimes lodge in tissues other than the skin and may cause paralysis if in the CNS. The worms cause an opening to form in the skin, and thereby allow the entrance of bacteria, which may cause a variety of problems if left untreated. Little or no immunity develops from infection.

TREATMENT. The treatment used for thousands of years has been to pull out a portion of the worm through the hole in the skin and to wind it up on a stick. Care must be taken not to break the worm or it will retract, die, and cause a considerable reaction. This treatment is memorialized in the staff of Asclepius with a serpent wound around it.

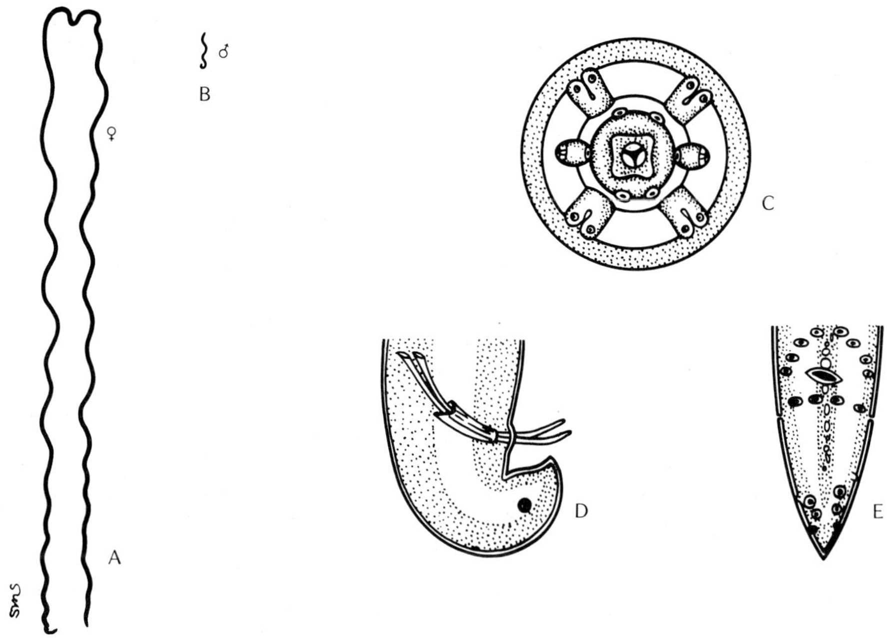


**FIG. 42.1**
Adults of *Dracunculus medinensis*. (*A and B*) Relative sizes of adult male and famale; (*C*) anterior end of male: (*D*) tail of male; (*E*) tail of male showing anus and papillae. [Redrawn from Moorthy, 1937.]

Although some drugs have been used to kill or immobilize the worms, they should still be extracted by surgical or other means.

EPIDEMIOLOGY AND CONTROL. The 1980s have been designated as the International Drinking Water Supply and Sanitation Decade, and the success of the program will have enormous impact on the health of millions of persons in areas where guinea worm is endemic. Programs of control show the simplicity of some solutions and the difficulty in achieving them. Although guinea worm has a broad host range, the mode of infection and the habitat of the intermediate hosts are such that safe drinking water will nearly eradicate the infection from a human population in a year.

Infections with *D. medinensis* are associated with standing water where copepods live. In India, step wells are typically the source of infection; a step well is a community water supply in which stone steps allow people to climb down easily to wherever the seasonal water level is. Particularly at low levels, the water will have a high population of copepods and will be readily contaminated with $L_1$s from persons having contact with the water by carrying it, bathing in it, or washing clothes. As many as 95% of persons have been found to be infected in such communities.

In Africa, where water is often taken from streams, water levels drop during the dry season and water will often be confined to pools before the rains start again. The same factors pertain here as in India.

Recommendations for prevention of guinea worm infection have been to keep people from drinking potentially contaminated water and to prevent those who are infected from having contact with the water supply. It is generally admitted that these recommendations have been distinct failures. However, if it were possible to provide piped clean water that people neither would nor could get into, this infection would almost automatically be prevented. The universal recognition of the importance of clean, safe water for the control of a variety of microbial and parasitic diseases has been an idea too long in incubation. Implementation of safe water supplies has become a high priority of the United Nations, of individual countries, and of the U.S. Agency for International Development (AID). In addition to guinea worm, it will significantly reduce other infections such as blood flukes, amebiasis, and giardiasis.

## REFERENCES AND FURTHER READINGS

Anonymous. 1981. Guinea worm targeted for elimination as part of international clean water decade. *Trop. Med. Hyg. News* 30(3):14–15.

Moorthy, V. N. 1937. A redescription of *Dracunculus medinensis*. *J. Parasitol.* 23:220–224.

Muller, R. 1971. *Dracunculus* and dracunculiasis. *Adv. Parasitol.*9:73–151.

Onamabiro, S. D. 1956. The early stages of the development of *Dracunculus medinensis* (L) in the mammalian host. *Ann. Trop. Med. Parasitol.* 50:157–166.

# 43

# Trichinella, Trichuris, *and* Capillaria

From its discovery in the early part of the nineteenth century until the present time, *Trichinella spiralis* has been the pivotal character in a series of dramas. The discovery of the worm by a medical student, James Paget, in the muscles of a cadaver in 1835 was overshadowed by the published description of his material by the eminent zoologist Richard Owen (later a virulent critic of Charles Darwin's theory of evolution). Infected meat shipped from the United States to continental Europe brought about strained international relations and an embargo on pork from the United States. Epidemiologic studies in the Arctic done under exceedingly difficult conditions led to a clarification on modes of transmission and the probable spread of the infection throughout the world. The efforts to have older ideas on transmission supplanted by more accurate ones have been met with a cultural inertia that demonstrates human unwillingness to shift viewpoints in the face of data to the contrary. All in all, the interaction of *T. spiralis* with humankind is a fascinating intertwining of the biology of the parasite with food habits, politics, and economics.

NAME OF ORGANISM AND DISEASE CAUSED. *Trichinella spiralis* (Owen, 1835) Railliet, 1895, causes trichinellosis; this term is now being used more widely, especially in Europe, but the less correct term, *trichinosis* is embedded in the literature and may never be eradicated. *Trichinella pseudospiralis* Garkavi, 1972, was described from the raccoon (*Procyon lotor*) in Russia and differs from *T. spiralis* principally in that *T. pseudospiralis* larvae do not form a capsule in the muscle, and they continue to move in the muscle (Al Karmi & Flaubert, 1981).

HOSTS AND LOCATION. Humans, pig, bear, seals, walrus, and rodents are hosts. With few exceptions, nearly any mammal can be infected experimentally, and a wide array of carnivores have been found to be naturally infected, as well as about 75 species of wild animals. Although the domestic chicken has been infected experimentally, birds are probably resistant.

Adults lie in the paramucosal lumen of the small intestine. Larvae are in the striated muscles, but may also be found in nearly any other organ including the brain.

DISTRIBUTION AND IMPORTANCE. The organism is cosmopolitan. *T. spiralis* was probably originally prevalent in the Arctic and has spread nearly throughout the world (Fig. 43.1). Its importance lies exclusively in infections of humans; other animals do not seem to be significantly affected. Outbreaks of infection in humans result from eating poorly cooked pork, bear meat, aquatic mammals, or carnivore meat.

During the 1930s and 1940s the prevalence of infection in the United States was 16%, as measured by trichinae in the diaphragms of persons examined postmortem. A more recent study (1966–1970) showed only 4.7% of diaphragms containing trichinae. From

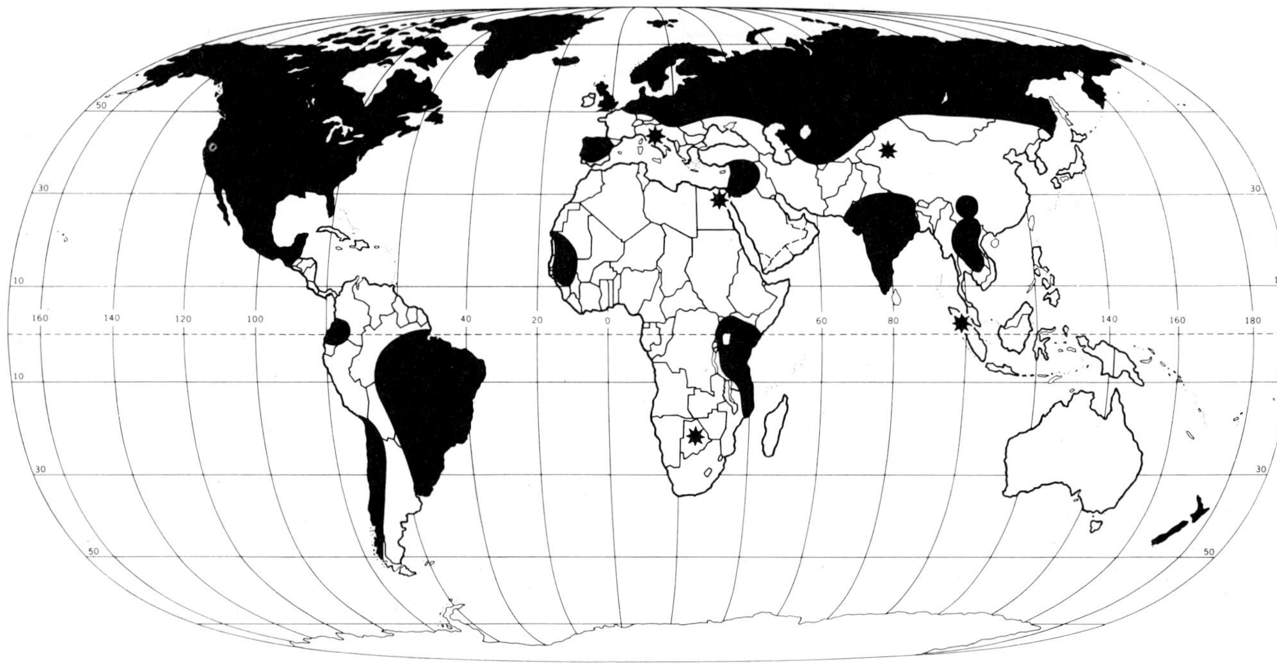

**FIG. 43.1**
Geographic distribution of *Trichinella spiralis.*

1947 to 1971 an average of 244 infections per year were reported in the United States. From 1947 to 1953 there was an average of 12 deaths per year, but from 1954 to 1971 there were only 2.8 per year. Among 10,204 infections, the death rate was 1.25%. Instances of clinical disease are sporadic at present and usually are associated with home-butchered pork or eating inadequately cooked meat of wild animals, most often bear.

Human infection is rare in continental Europe. For example, no human infection has been seen in the Netherlands since 1926. In Romania, pork is preferred well cooked and human infections are rare.

MORPHOLOGY. Adults are tiny worms (Fig. 43.2); the females are 3 to 4 by 0.06 mm and the males 1.5 by 0.04 mm. Members of the superfamily Trichinelloidea have a stichosome, a row of glandular cells forming a portion of the esophagus (see classification of nematodes Chap. 30). In the male, the stichosome extends nearly half the length of the body, but only about one-sixth in the female. The male has a pair of lobed structures at the posterior tip; spicules are lacking.

The larvae in the tissues are coiled up in cysts 250 to 500 μm long (Fig. 43.3); the larva itself is about 1 mm long.

LIFE CYCLE. The larvae in the muscles of a mammal are in an advanced state of development, but are probably still $L_1$s. Transmission to a new host takes place when muscle of an infected animal is eaten by a susceptible host. Development to adult males and females takes place quite rapidly in the intestine of the host; sexual maturity is reached in about 40 hr. After copulation, the females burrow into the tissue of the small intestine and larviposit in the mucosal or even submucosal areas of the gut. Larvae enter the lymph vessels, from which they are carried throughout the body. Females live about 6 weeks and give birth to perhaps 1500 larvae; males live a considerably shorter time. The larvae develop intracellularly, especially in striated muscles, and they reach infectivity about 17 days after reaching their final site.

DIAGNOSIS. Determining whether a person has been infected is seldom done during the intestinal phase of the life cycle. Signs and symptoms of infection at that

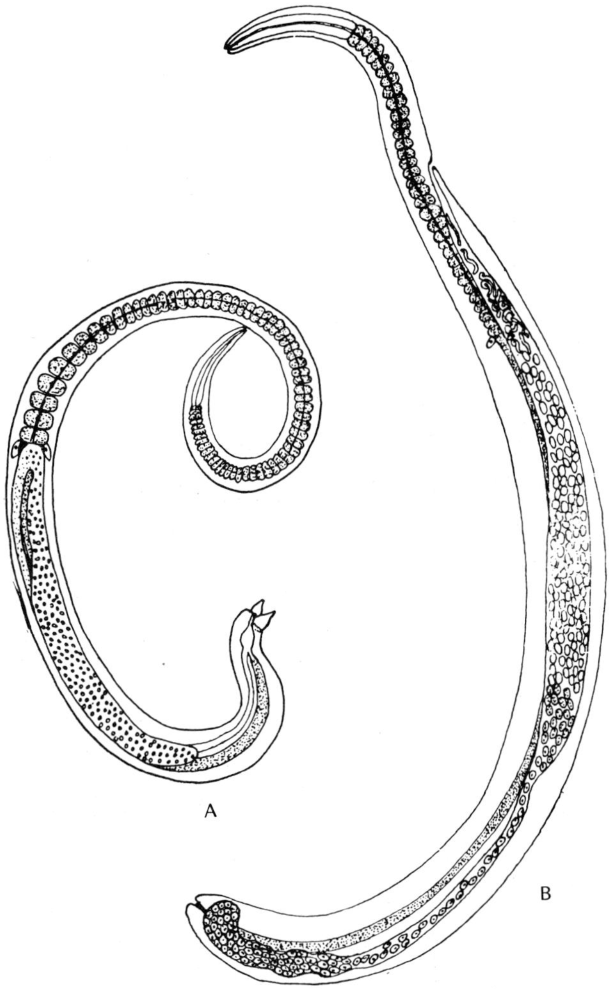


**FIG. 43.2**
Male (*A*) and female (*B*) of *Trichinella spiralis*. [Redrawn from Yorke, W., & P.A. Maplestone, 1962. *The Nematode Parasites of Vertebrates*. Hafner, New York.]

time include abdominal pain, diarrhea, and vomiting; any number of infectious agents can cause the same signs and symptoms.

Trichinosis is usually suspected during the second phase of the infection and is usually diagnosed by examining muscle biopsies.

In epidemiologic surveys, small portions of somatic, diaphragm, or tongue muscle are examined for trichinae. The tissue may be pressed between glass plates and examined under the low power of a compound microscope; alternatively it may be digested with pepsin-HCI for 8 hr and the sediment examined for the freed larvae.

A wide array of serologic tests have been developed, and some of the more recent ones have a high degree of specificity and sensitivity. Such tests can be useful in surveys and in determining who might be infected in an outbreak of trichinosis.

HOST–PARASITE INTERACTIONS. As a general rule, hosts other than humans are not much affected by *T. spiralis*. In experimental studies in pigs, inocula of 100,000 larvae have not generally produced clinical disease, but in one study pigs showed decreased weight gains and probably muscle pain since they were reluctant to move about at 3 weeks after infection.

In humans, the infection has been divided into three phases:

1. Intestinal. Starting a few days after infection, there may be diarrhea, abdominal pain, and vomiting (Table 43.1).
2. Migration. About 9 days after infection large numbers of larvae enter the circulatory system

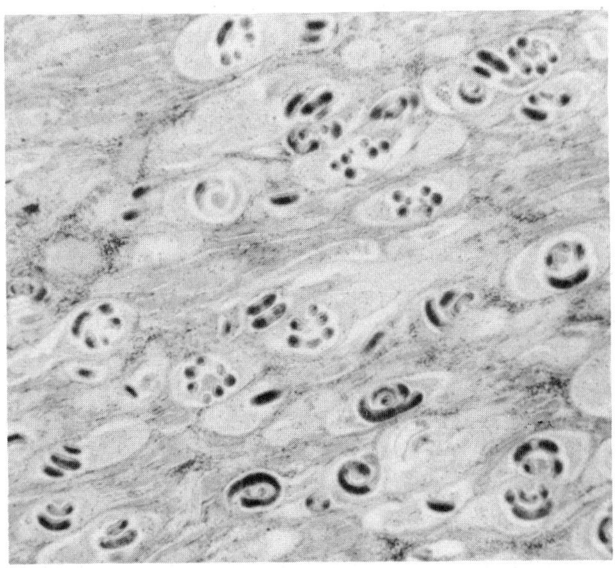

**FIG. 43.3**
Section of muscle heavily infected with *Trichinella spiralis* larvae.

and begin to reach various organ systems. There is muscle pain, and difficulty in breathing, chewing, and swallowing. If death occurs, it usually takes place at this time.

3. Encystment. About 6 weeks after infection, there is a strong reaction to the larvae in the tissues. There is pain and swelling of the tissues and fever.

All of these signs and symptoms vary with the size of the inoculum. Data on the size of the inoculum in human infections are hard to come by. We know that there can be a thousandfold increase over the number of worms ingested and that 100,000 larvae ingested by a pig might give 500 larvae per gram of pork loin. A poorly cooked quarter-pound pork tenderloin would thereby constitute an inoculum of more than 56,000 larvae. This inoculum could produce 56 times $10^6$ larvae in a person or an average of $1.2 \times 10^3$ larvae/g of tissue in an average 68-kg (150-lb) person.

Larvae are not distributed randomly in the host and have a preferred site in each species of host. In humans, diaphragm and tongue have the largest numbers. In swine, using the diaphragm as 100%, other sites are tongue, 53%; shoulder, 44%; ham, 22%; loin, 14%. In cats, the intercostal muscles are preferred; in rodents, the masseter muscles.

Infection with only moderate numbers of larvae confers immunity to reinfection. The response is directed toward the intestinal phase (adults) and is most likely T-cell dependent. When the larvae reach their final sites in the muscles, they become encapsulated by a collagenous capsule elaborated by the muscle fiber. It is generally agreed that larvae may live for 5 to 10 years in humans. This would confirm the lack of significant response against larvae.

Larvae live within the muscle fiber; such infected cells are referred to as *nurse cells* since the larvae apparently elicit production of substances that they require (Purkenson & Despommier, in Kim, 1974). The larvae also take up nutrient across the body surface. Such a phenomenon is rather unusual in nematodes because the cuticle serves as a relatively impermeable barrier; this provides evidence that the larvae newly arrived in the muscle are pre-$L_1$s.

**Table 43.1**
Clinical Features of Patients Who Had Eaten Bear Meat Containing *T. spiralis*

| SIGNS & SYMPTOMS | PERCENTAGE |
|---|---|
| Myalgia (muscle pain) | 100 |
| Fever | 82 |
| Diarrhea | 76 |
| Fatigue | 65 |
| Periorbital edema | 47 |
| Abdominal cramps | 35 |
| Visual disturbances | 35 |
| Nausea and vomiting | 29 |
| Skin lesions | 18 |
| Dyspnea (difficulty in breathing) | 12 |

TREATMENT. Unless an astute diagnosis has been made, treatment of trichinosis is not usually undertaken until late in the second or third phase of infection. At this time most of the adult worms have disappeared. If the infection has been recognized early, in the first or early second phase, thiabendazole has some efficacy. The action of thiabendazole during the time when adult worms are in the intestine is temporarily to sterilize the female worms. The effect is to reduce the number of larvae invading the tissues in the late second and third phases of the disease. Other benzimidazoles in addition to thiabendazole are effective.

After the worms reach the tissues, the benzimidazoles have some effect, but usually symptomatic treatment is also given. For example, cortisone reduces inflammation of the tissues, and it also seems to have some direct effect on the larvae.

EPIDEMIOLOGY AND CONTROL. *T. spiralis* originally occurred in the Arctic region of the Northern Hemisphere. It was passed among various carnivore and carrion-eating hosts such as bear, seal, walrus, fox, and wolf. We tend to think of the infection as being passed directly from one host to the next, but as Madsen (in Kim, 1974) pointed out, some isolates of *T. spiralis* can survive for long periods outside the host. At or near the freezing point, encysted larvae have survived for 3 months up to a year. Thus, the original cycle in the Arctic involved both mammalian predators and scavengers.

The infection seems to have been spread from the Arctic to other regions of the world by increased travel and trade. For example, infected animals and people have been found in Kenya only since the early 1960s; the infection was previously unknown there. As a result of the dissemination of the worm, we now recognize three more or less distinct cycles:

1. Feral. This is the primeval cycle in the Arctic.
2. Semidomestic. This cycle occurs in temperate zones. Both wild and domesticated carnivores as well as some rodents are infected. Infection may spread from infected pork scraps in garbage dumps, or from pigs dying in areas that make them accessible to scavengers. Once established, the cycle can continue through predation and scavenging.
3. Domestic. Pigs are the principal hosts in this cycle, and the cycle is perpetuated by garbage and carrion feeding. Rats have been implicated in the cycle, because of their feeding habits, but Madsen (in Kim, 1974) presented evidence that rats (or other rodents and lagomorphs) are merely indicators of infection in an area and do not contribute significantly to its maintenance.

In all three cycles, humans are only an incidental host and rarely contribute to maintenance of the infection.

The pattern of infection in *T. spiralis* is related to trade and technology. The original focus of infection in boreal regions spread into temperate and finally tropical climates. As the infection spread, it became established as a domestic cycle in swine. At this point, the infection became a more widespread public health problem and was undoubtedly promoted by more sophisticated methods of swine husbandry. At one point, many piggeries were located at the fringes of cities, and the industry was based on feeding garbage. As a result, 5.7% of garbage-fed pigs were infected, whereas only 0.95% of farm-raised pigs were infected, a sixfold difference. At this same time, 16% of the U.S. population was infected.

The relatively high prevalence of infection in swine in the United States led to outbreaks of trichinosis in Germany from pork imported from the United States. Germans (and seemingly their descendants in the United States) prefer rare pork and are therefore at some risk of infection, but pork is inspected at slaughter in Germany for trichinae, so the probability of infection is low. Because trichinosis in Germany resulted from U.S. pork, an embargo was put on pork from the United States.

In the United States, the main reliance for prevention of trichinosis is placed on education of the populace to cook pork well. Inspection of pigs at slaughter is considered to be too expensive and not to give an adequate margin of safety. Given a prevalence of 16% infection of the human population, the campaign could not be considered to be any more than a modest success.

The prevalence of infection dropped to its present low level in the United States for reasons unrelated to trichinosis. The crucial events were an outbreak of vesicular exanthema in swine and the campaign to eradicate hog cholera; both infections are caused by viruses that can be transmitted by ingestion of infected meat. Because of the seriousness of these infections in the swine industry, a federal law was passed in the early 1950s prohibiting feeding of raw garbage to pigs. All garbage had to be cooked to the point where the viruses were killed; fortunately, the trichinae were also killed. Recent studies in humans show a prevalence of infection of less than 5%; thus, human health has benefited from the need of swine raisers to maintain the health of their herds. It should be noted also that high-technology swine husbandry has had a significant impact on infection levels. Keeping swine on concrete and feeding them prepared food for maximal growth and feed conversion clearly make *T. spiralis* infection unlikely. Thus, the prevalence of infection in swine in the United States is now well below 1%; in farm-raised swine, infection is probably about 0.1%.

In North America, most outbreaks of trichinosis are now related to eating home-slaughtered pork from which uncooked sausage is made or from eating wild meat such as bear. Both types of outbreaks emphasize the importance of cooking meat well if there is even a minimal risk of infection.

Control programs for the protection of the human population from trichinosis have several aspects. Some of them are feasible in some times and places; others are not.

1. Cooking potentially infected meat well. The educational campaigns in the United States started early in this century were partially successful. In parts of Europe, there probably was less success; among primitive northern peoples there was probably no impact.
2. Inspecting pork for infection. Modest laboratory facilities and a microscope with a magnification of about 60x are required. The cost per animal for examination is also important; inspection by trichinoscope in Europe has been used to determine which animals are infected and to remove them from the market. Using the trichinoscope in the United States has been

considered to be cost-prohibitive, but recent digestion techniques have reduced the cost to about 9 cents per animal (1970 dollars).

3. Freezing pork products. If pork is frozen, the larvae can be killed. Regulations call for:

| TEMPERATURE | TIME IN DAYS | |
|---|---|---|
| | 6″ THICKNESS OR LESS | MORE THAN 6″ THICKNESS BUT LESS THAN 27″ |
| −30°C or −20°F | 20 | 30 |
| −12°C or −10°F | 10 | 20 |
| −30°C or −20°F | 6 | 12 |

This technique can be used for infected pork only, or for all pork regardless of infection. It should be noted that there are some differences in the sensitivity of different isolates to freezing.

4. Cooking garbage. This method is clearly based on moderate affluence and low fuel costs, but as indicated earlier, it has been successful in the United States in controlling trichinosis and other infections.
5. Eliminating wild reservoirs of infection. Although there may be special local circumstances that will allow this method to be used, it is generally impractical.

## *TRICHURIS* SPP.: THE WHIPWORMS

The members of the genus *Trichuris* are called *whipworms* because they have relatively short, stout bodies and an extremely long esophageal area. The name means "hairlike tail"; it has been pointed out that the name should more correctly be *Trichocephalus*, that is "hairlike head," but the rules of zoological nomenclature call for retention of the original name even if it refers to the wrong end of the worm.

Whipworms live in the large intestine, where they are found almost exclusively in the cecum. The long esophageal region is threaded through the mucosa of the gut with only the body in the lumen (Fig. 43.4). Their life cycles are direct, with eggs (Fig. 43.5) developing to the infective $L_1$ stage in about 2 weeks. The larvae hatch when the eggs are eaten and the larvae enter the mucosa of the intestine. After 10 days to 2 weeks they return to the lumen and seek the cecum, where they mature. The prepatent period may be as short as 30 days in some species or as long as 90 days in others.

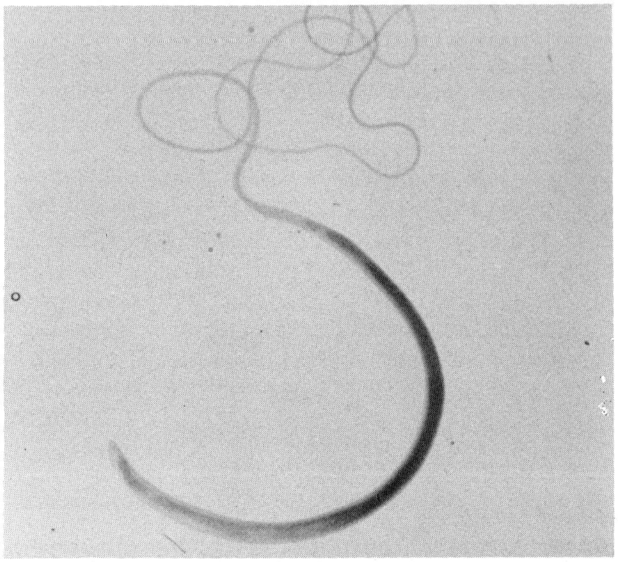

**FIG. 43.4**
Adult *Trichuris*.

Although about 70 species are known from a wide range of mammals, only a few are of economic or medical importance. *Trichuris trichiura* is a cosmopolitan parasite of humans that has mild pathogenicity. *T. suis* is identical anatomically to *T. trichiura*, but they apparently do not cross-transmit. *T. suis* has generally been considered to be benign, but enteric disorders have been found in both naturally and experimentally infected pigs. *T. vulpis* is a cosmopolitan parasite of dogs and other canids, and it can cause severe damage in dogs. The whipworms of ruminants, such as *T. ovis* of sheep, cattle, and other ruminants, seem to be only slightly pathogenic, but they have not been studied extensively.

Removal of whipworms by anthelmintics has been erratic, at best, in the past. Mebendazole is generally effective against *T. trichiura* in humans. At one time, cecectomy was the only effective treatment for *T. vulpis* in dogs, but there are now at least four good drugs: phthalofyne, dichlorvos, mebendazole, and glycobiarsol.

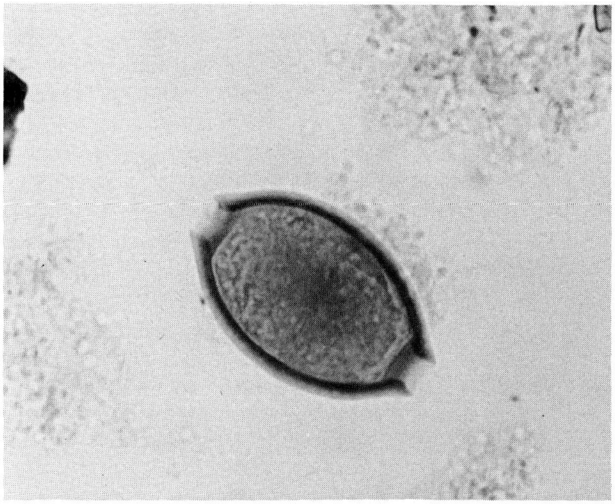

**FIG. 43.5**
Double-plugged *Trichuris* egg.

*Trichuris* has many of the same developmental characteristics as *Ascaris:* slow development outside the host, retention of the infective larva within the egg capsules, and extraordinary hardiness of the embryonated eggs. Therefore, control programs that are effective for ascarids are also effective for whipworms.

## CAPILLARIDS

A number of members of the superfamily Trichinelloidea are referred to as *capillarids*, although they are in the subfamily Capillarinae. They are widespread in the digestive tracts of higher vertebrates and a few of them are either of economic or medical importance.

They are generally found threaded in the mucosa of the intestinal tract: in the esophagus or small intestine. The adults are 0.5 to 4 cm long and quite slender, so that special care is needed to find the worms in the tissue and then to dissect them out. The eggs are similar to those of whipworms; they are double-plugged and yellow or brownish.

*Capillaria philippinensis* is the cause of severe intestinal disease in humans in the Philippines and parts of Southeast Asia. Humans are the only known definitive host, but evidence indicates that infection comes from eating insufficiently cooked fish. Pathogenesis can be severe, and a number of people have died from the infection.

In domestic birds, *C. caudinflata* and *C. contorta* require earthworms as an intermediate host, whereas *C. obsignata* has a direct life cycle. All three species are pathogenic for their definitive hosts and cause losses in free-ranging poultry because of damage to the intestinal tract.

*C. hepatica (Hepaticola hepatica)* is found in the liver of rodents and sometimes in humans in many parts of the world. It is peculiar that in this species the eggs become infective only after passing through the intestinal tract of a bird or mammal.

## REFERENCES AND FURTHER READINGS

Al Karmi, T. O., and G. M. Flaubert. 1981. Comparative analysis of mobility and ultrastructure of intramuscular larvae of *Trichinella spiralis* and *Trichinella pseudospiralis. J. Parasitol.* 67:685–691.

Garkavi, B. L. 1972. *Trichinella* species recovered from wild animals. *Veterinariya Moscow* 10:90–91.

Gould, S. E. (ed.). 1970. *Trichinosis.* Charles C Thomas Publishing Co., Springfield, Ill.

Kim, C. W. (ed). 1974. *Trichinellosis. Proceedings of the 3rd International Conference on Trichinellosis.* Intext, New York.

Larsh, J. E., Jr. 1963. Experimental trichiniasis. *Adv. Parasitol.* 1:213–268.

——. 1968. Experimental trichiniasis. *Adv. Parasitol.* 6:361–372.

Steele, J. H., and P. V. Arumbolo, III. 1975. Trichinosis: A world problem with extensive sylvatic reservoirs. *Intern. J. Zoon.* 2:55–75.

# PART 4

# *Acanthocephala, Nematomorpha, Mollusca, Annelida, and Pentastomida*

44

# *Acanthocephala: Spiny-Headed Worms*

The phylum Acanthocephala (spiny-headed worms) is a small group of pseudocoelomate endoparasites with a number of unique structural and biological features. A few members are important economically and medically; however, they are commonly found as parasites of the intestinal tracts of fish, reptiles, birds, and, less often, mammals in the wild. There are only scattered reports of the effects of Acanthocephala on populations of nondomesticated animals. Moore (1984) has shown how behavior of intermediate hosts of acanthocephalans is altered to make them more likely to be eaten by the definitive host.

A prominent characteristic of the phylum, from which it receives its name, is a protrusible proboscis which is armed with recurved hooks (Fig. 44.1) by which the worm attaches to the wall of the intestine of the definitive host (Fig. 44.2). Hooks and spines are also present on other parts of the body: around the anterior part and the genital openings. The anterior part of the body is divided into the presoma, comprising the proboscis and the unarmed neck; the trunk or body contains the rest of the organs. Sexes are separate; there is no larval reproduction, such as occurs in the digenetic trematodes. There is no digestive tract, and all food is absorbed through the surface of the body. The internal organs lie in a body cavity, the pseudocoelom. Because of their many peculiarities, the Acanthocephala have been a puzzle for taxonomists since the first descriptions were given early in the eighteenth century. Hyman's (1951) review of the various taxa in which they have been placed is instructive and shows why there have been disagreements. Hyman had originally placed the Acanthocephala in the Aschelminthes in the first volume of her series on the invertebrates, but later

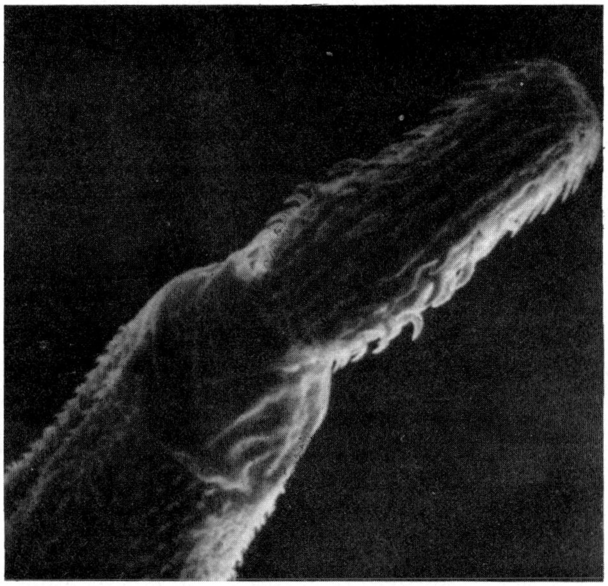

**FIG. 44.1**
Scanning electron micrograph of an acanthocephalan proboscis.

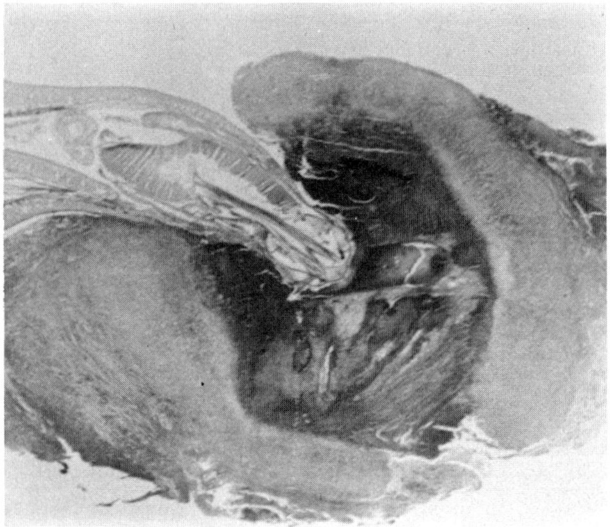

**FIG. 44.2**
Section of *Macracanthorhynchus hirudinaceus* proboscis embedded in a pig intestine. Note the large, dark area of hemorrhage.

reconsidered and gave them phylum status because both their embryogenesis and adult structures did not allow them to be allied with any other group. It is said that she had planned to include the Acanthocephala in the Aschelminthes rather by default, but that Dr. Harley J. Van Cleave dissuaded her because he felt that phylum status was justified for the animals he had studied for so many years.

The fact that the Acanthocephala have no digestive tract made the tegument of interest to many investigators. Light microscopic (LM) studies gave some information, but the advent of thin-sectioning techniques for transmission electron microscopy (TEM) revealed structures that were only vaguely described. As in the tapeworms and digenetic trematodes, the tegument is a syncytium, that is, a multinucleate structure in which there are no discrete cell membranes; however, unlike the flatworms, which have microvilli and microtriches, the Acanthocephala have pores that extend into the tegument (Fig. 44.3). It has been calculated that the pores increase the surface area 44 times over that of a flat surface (Byram & Fischer, 1973). The pores extend into the hypodermis, branch repeatedly, and end blindly (Wright & Lumsden, 1969). The tegument has a secreted glycocalyx, sometimes called an *epicuticle*; then deeper are the striped, felt, and radial layers. Also in the tegument is the fluid-filled lacunar system (Fig. 44.4), which apparently

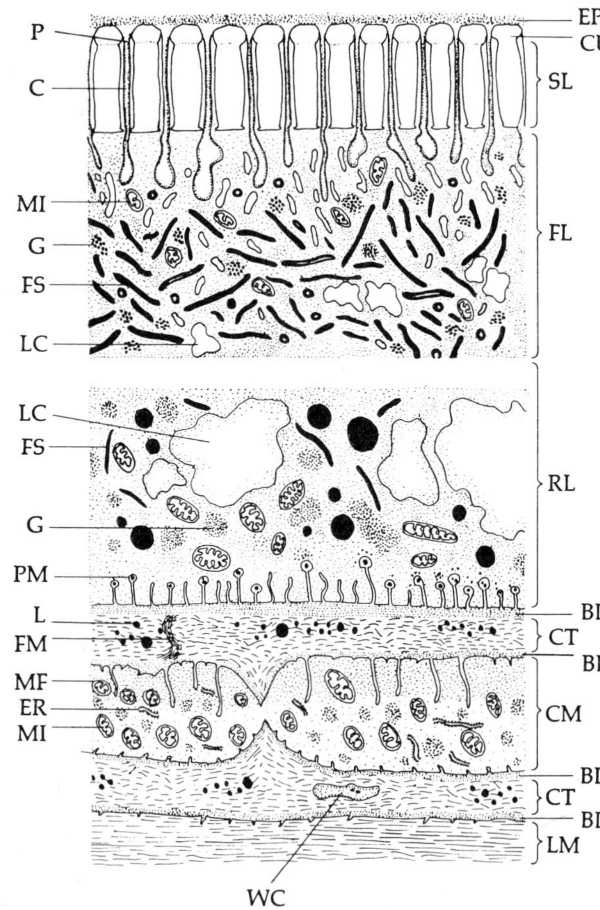


**FIG. 44.3**
Diagram of an acanthocephalan body wall. (BL = basal lamella; C = canal; CM = circular muscle; CT = connective tissue; CU = cuticle; EPI = epicuticle; ER = endoplasmic reticulum; FL = felt layer; FM = fibers attaching muscles to body wall; FS = fibrous strands; G = glycogen; L = lipid; LC = lacunar canal; LM = longitudinal muscle; MI = mitochondrion; MF = myofilaments; P = pore; PM = plasma membrane; RL = radial layer; SL = striped layer; WC = wandering cell.) [From Cheng, T.C., 1973.]

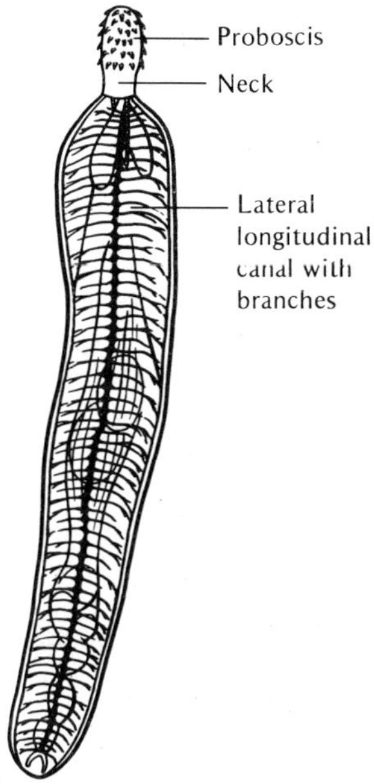


**FIG. 44.4**
Diagram of the lacunar system of *Acanthocephalus ranae*.

serves as a circulatory system as well as a hydraulic system; the pattern of the lacunar system is used to separate the members of the phylum into classes.

The sexes are separate; usually females are larger than males. Other than the tegument and proboscis, acanthocephalans are largely given over to the reproductive system. In both males and females the reproductive organs are attached to a ligament that extends from the base of the proboscis receptacle to the posterior end of the body. Usually the ligament is composed of two ligament sacs; in some males only one sac develops; in females both sacs often rupture as ova develop (Nicholas, 1967). The males have two testes, sperm ducts, and cement glands (Fig. 44.5). The cement glands produce a proteinaceous secretion that forms a fertilization or copulatory cap in the vulva of the female; the cap probably prevents additional fertilization (Parshad & Crompton, 1981). Males have a posterior, eversible copulatory bursa, with which they grasp the female, and a penis to facilitate transfer of sperm.

Initially, the ovary of the female is attached to the ligament, but as she matures, the ovary breaks up into a number of ovarian balls that float free in the ligament sacs. The uterine bell, which serves as an egg-sorting mechanism, is located at the posterior end of the body (Fig. 44.6); immature eggs pass into the bell and then back to the body cavity, and mature eggs pass out of the genital pore (Asaolu, 1980; Whitfield, 1970). Females that have been mated frequently have a characteristic constriction where they were clasped by a male; the copulatory cap in the vulva of the female dissolves before the release of ova (Fig. 44.7).

Early in embryonation, between the 4- and 32-cell stages, an acanthocephalan becomes syncytial; this pattern is similar to that of members of the Aschelminthes. Nuclei continue to divide and sometimes become so large that they are referred to as *giant nuclei*.

The ova passed from the female contain a developing larva called an *acanthor* (Fig. 44.8). When fully developed, the acanthor has bladelike hooks at its anterior end, the aclid organ or rostellum. Ova containing acanthors develop into acanthellae when ingested by the proper intermediate host. The acanthor penetrates the gut of the host and develops to the acanthella in the hemocoel. The features of the adult worm are apparent in the acanthella, but the reproductive system is immature. The acanthella becomes

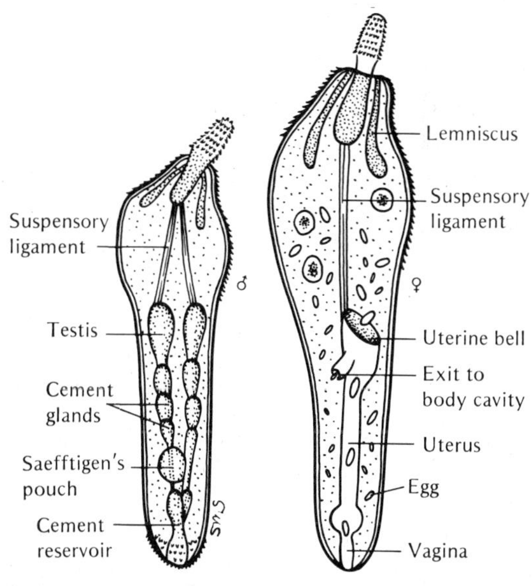


**FIG. 44.5**
Diagram of the reproductive systems of a male and female *Corynosoma*.

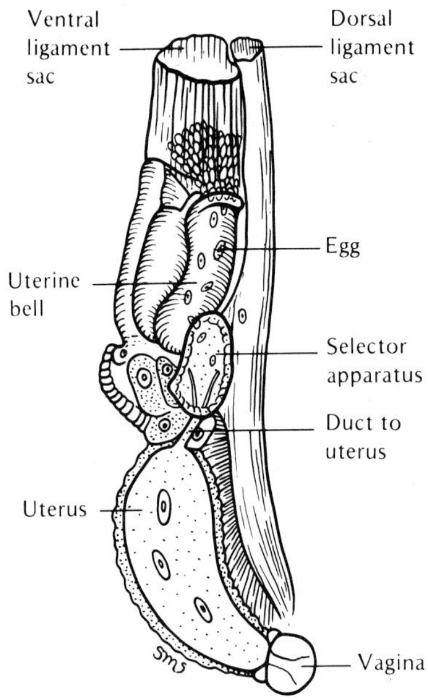


**FIG. 44.6**
Diagram of an acanthocephalan uterine bell.

infective for the definitive host when it has been encapsulated as a cystacanth, that is, surrounded by a hyaline envelope produced by host cells. Cystacanths can be ingested and remain alive in transport or paratenic hosts. The cystacanth is released in the gut of the definitive host, attaches to the gut wall, and matures sexually.

The physiology of the Acanthocephala has not been studied extensively, but some studies give information on the way in which they interact with the host. For example, if the intestine of a host is infected with an acanthocephalan, it appears at first that the worms embed in the gut and remain attached at that spot; however, they do move periodically, one of the influences being other parasites in the gut (Holmes, 1973). The proboscis and the neck are hollow and fluid-filled. The proboscis can be both everted and withdrawn into the body by a combination of muscular movement and hydraulics (Hammond, 1966, 1967). The proboscis is pushed into the wall of the intestine so that the hooks catch; then the neck retractor muscles force fluid out of two lateral sacs, the lemnisci, into the lacunar system of the proboscis, causing it to expand, thus firmly anchoring the worm (Fig. 44.9).

As is true of the tapeworms, acanthocephalans obtain all of their nutrient by absorption across the body surface; the digenetic flukes also obtain a significant portion of their food in the same means despite having an intestine. Substances are picked up at specific sites on the surface of the worm and transported to the interior. Amino acids are absorbed at the surface of the body after being enzymatically released from protein (Uglem et al., 1973). Sugars are absorbed at two or more sites and then energy is obtained by glycolysis. The biology of the Acanthocephala has been reviewed by Nicholas (1967 and 1973). Crompton (1970) took a novel approach to acanthocephalan ecology and physiology. Taxonomy is considered by Yamaguti (1963), Schmidt (1972), and Hyman (1951).

NAME OF ORGANISM AND DISEASE CAUSED. *Macracanthorhynchus hirudinaceus* (Pallas, 1781).

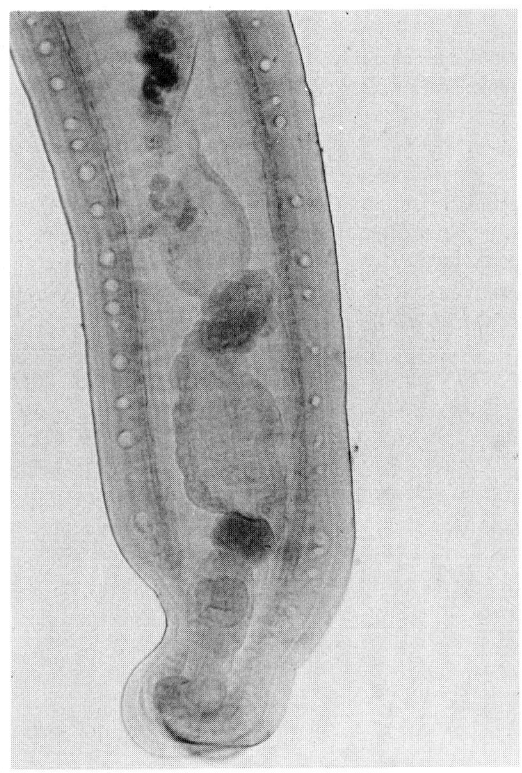

**FIG. 44.7**
Mated female acanthocephalan with characteristic constriction where she was clasped by the male.

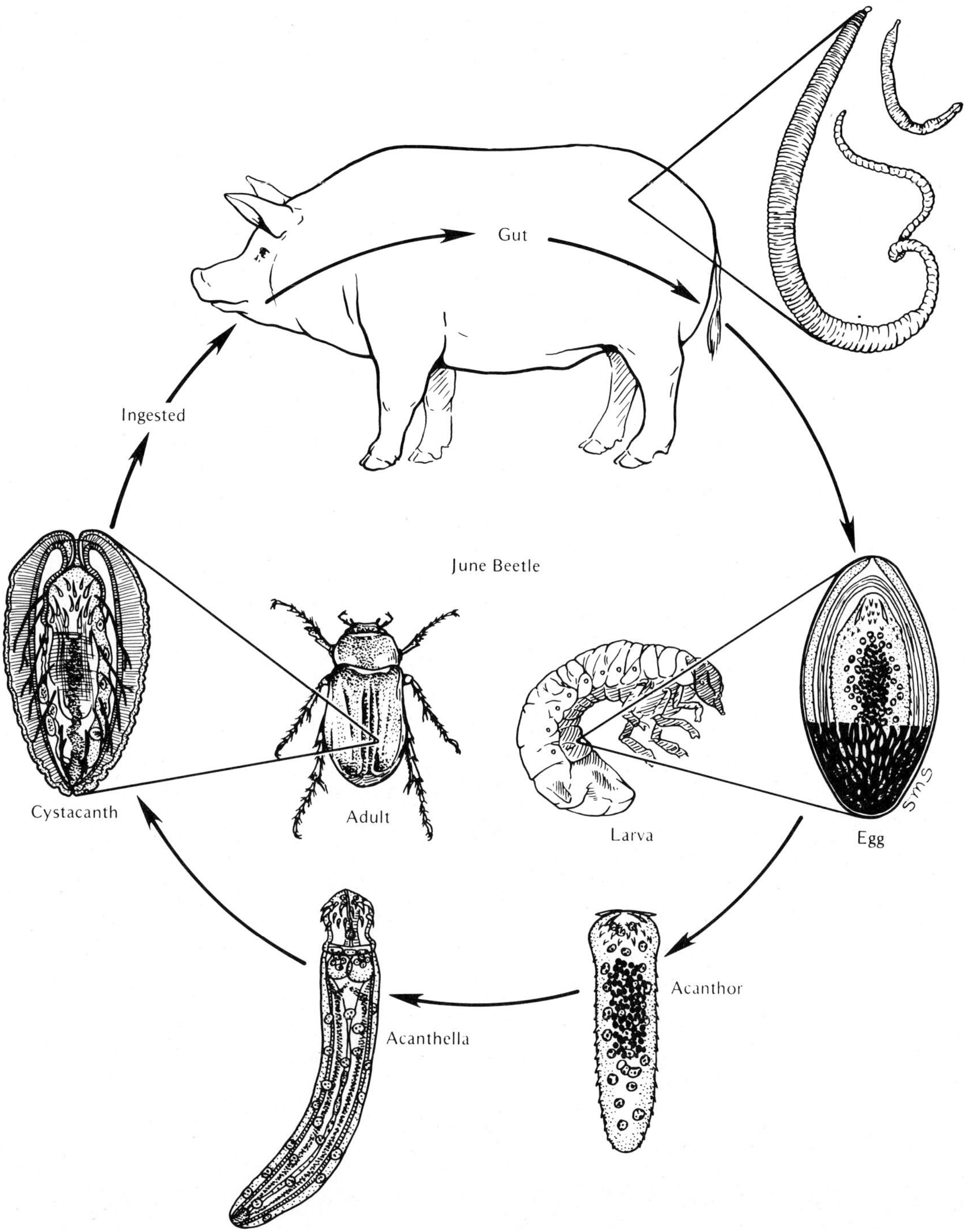


**FIG. 44.8**
Diagram of the life cycle of *Macracanthorhynchus hirudinaceus.*

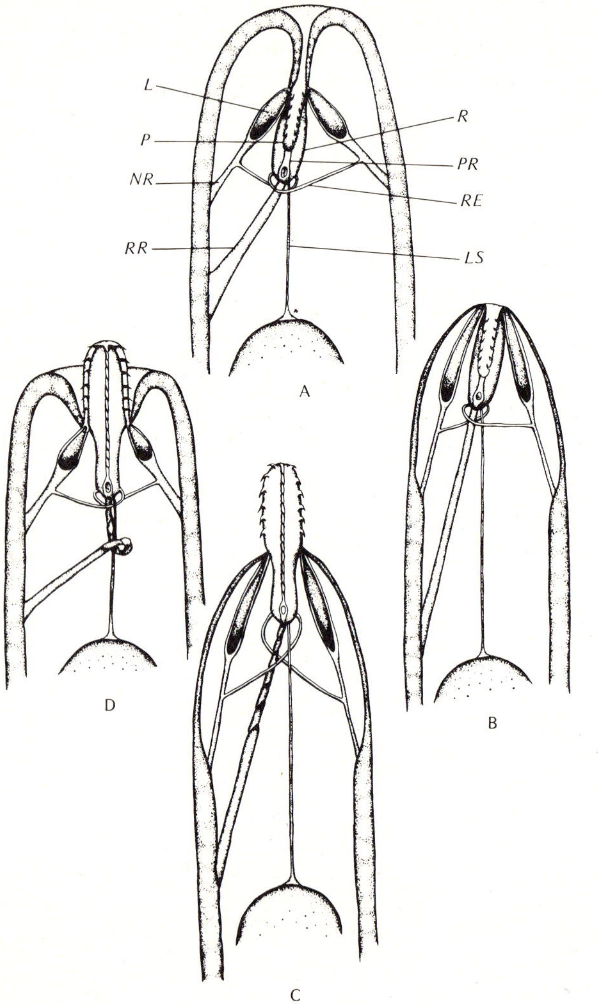


**FIG. 44.9**
Diagram of the eversion of an acanthocephalan proboscis. (*A*) Proboscis withdrawn; (*B*) trunk extended and proboscis withdrawn; (C) trunk extended, proboscis beginning to be everted; (*D*) proboscis everted, neck muscles contracted. (L = lemniscus; LS = ligament sac; NR = neck retractor muscles; P = proboscis; PR = proboscis retractor muscles; R = receptacle; RR = receptacle retractor muscles.) [Redrawn with modifications from Hammond, 1966.]

HOSTS AND LOCATION. Swine, occasionally humans and a variety of other animals such as wild pigs, cattle monkeys, dogs, rodents, and moles are the hosts (Nicholas, 1967).

DISTRIBUTION AND IMPORTANCE. The organism is cosmopolitan in pigs. Although serious intestinal pathology and even death may occur, mild infections causing few lesions and little or no economic loss are typical.

MORPHOLOGY. Females may exceed 35 cm (13 in.) in length; males are about 10 cm (4 in.) long. In life, the worms are usually rather flattened and wrinkled; their large size makes them appear, at first glance, like the large hog ascarid. The proboscis is small and bears only five or six rows of hooks. The eggs are brownish with a sculptured shell and 80 to 100 by 45 to 60 μm (Fig. 44.10).

LIFE CYCLE. The adult worms are attached to the wall of the small intestine; eggs containing acanthors are passed in the feces and are infective when passed. Intermediate hosts are a number of species of beetle larvae including "June bugs," *Phyllophaga* spp. The

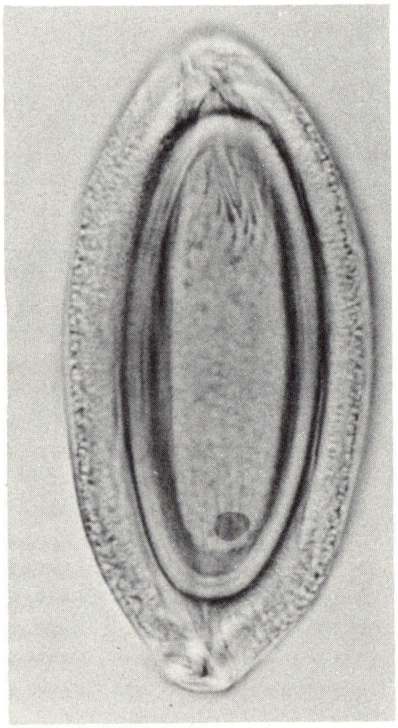

**FIG. 44.10**
*Macracanthorhynchus hirudinaceus* egg containing an acanthor.

acanthor migrates through the gut wall into the hemocoel and develops into an acanthella. By 60 to 95 days after infection, the acanthella develops into a cystacanth. It is released in the small intestine, attaches, and develops to sexual maturity in 2 to 3 months.

DIAGNOSIS. Finding the characteristic eggs in the feces is the means of diagnosis; worms may also be found in the feces after successful drug treatment.

HOST–PARASITE INTERACTIONS. Most infections produce no clinical signs; however, weight loss, diarrhea, and abdominal pain may occur in heavy infections. Usually, there is an inflammatory response where the proboscis is embedded in the gut (Fig. 44.2). If the gut wall is perforated, bacteria may leak into the abdominal cavity and cause peritonitis.

TREATMENT. Benzimidazoles are effective: levamisole has been used in swine, and mebandazole seems to be effective in humans (Goldschmidt et al., 1974).

EPIDEMIOLOGY AND CONTROL. The hardiness of the eggs and the fact that there is an intermediate host in the life cycle make control of infection in swine rather difficult. The eggs remain infective in soil for up to 3½ years in Maryland; they tolerate both desiccation and temperatures below –18°C (0°F) (Kates, 1942). Female worms may produce more than 250,000 eggs per day, so even a moderate infection can lead to heavy contamination of pastures or hog lots. Beetle control has not been successful in controlling infections.

Humans become infected by eating infected beetle grubs. One would think that avoiding infection would be simple.

## OTHER ACANTHOCEPHALAN INFECTIONS

Humans have been reported to be infected with *Acanthocephalus bufonis, A rauschi, Cornosoma stromosum*, and *Moniliformis moniliformis* (Beck & Davies, 1981), in addition to *Macracanthorhynchus hirudinaceus*. Intermediate hosts for *Acanthocephalus* and *Corynosoma* are crustacea, and cockroaches serve for *Moniliformis* (Yamaguti, 1963).

In Texas, dogs and coyotes are infected with *Oncicola canis*, a species that can produce serious disease. Infection probably occurs by eating armadillo (*Dasypus novencinctus*), the transport host; the intermediate host is not known.

*Prosthenorchis elegans* sometimes becomes a problem in caged primates; the intermediate hosts are cockroaches. A cycle can be established under conditions in which cockroach control is difficult. Infections build up in monkeys; sometimes the proboscis penetrates the gut wall, allowing bacterial peritonitis to develop.

## ACANTHOCEPHALA CLASSIFICATION

**Class Archiacanthocephala**
Main longitudinal lacunar canals dorsal and ventral or only dorsal; two ligament sacs persist in females; cement glands separate. Parasites of birds and mammals.
*Moniliformis moniliformis, Macracanthorhynchus hirudinaceus*

**Class Palaeacanthocephala**
Main lacunar canals lateral; single ligament sac in females degenerates; cement glands separate. Parasites of fish, amphibia, reptiles, birds, and mammals.
*Corynosoma stromosum, Acanthocephalus*

**Class Eoacanthocephala**
Main longitudinal lacunar canals both dorsal and ventral; two persistent ligament sacs in females; cement gland single. Parasites of fish, amphibia, and reptiles.
*Neoechinorhynchus emydis*

## REFERENCES AND FURTHER READINGS

Asaolu, S. O. 1980. Morphology of the reproductive system of *Moniliformis dubius* (Acanthocephala). *Parasitol.* 81:433–436

Beck, J. W., and J. E. Davis. 1981. *Medical Parasitology*, 3rd ed. C. V. Mosby Co., St. Louis.

Byram, J., and F. Fischer, Jr. 1973. The absorptive surface of *Moniliformis dubius* (Acanthocephala). I. Fine structure. *Tissue and Cell* 5:553–579.

Cheng, T. C. 1973. *General Parasitology*. Academic Press, New York.

Crompton, D. W. T. 1970. *An Ecological Approach to Acanthocephalan Physiology*. Cambridge University Press, Cambridge.

Goldschmidt, J. M., M. E. Smith, and F. Fleming. 1974. Human infections with *Moniliformis* sp. in Rhodesia. *Ann. Trop. Med. Parasitol.* **68**:363–364.

Hammond, R. A. 1966. The proboscis mechanism of *Acanthocephala ranae*. *J. Exp. Biol.* **45**:203–213.

——— 1967. The mode of attachment within the host of *Acanthocephala ranae*. (Shrank, 1788) Luehe, 1911. *J. Helminthol.* **41**:321–328.

Holmes, J. C. 1973. Site selection by parasitic helminths: Interspecific interactions, site segregation and their importance to the development of helminth communities. *Can. J. Zool.* **51**:333–347.

Hyman, L. H. 1951. *The Invertebrates* Vol. III. *Acanthocephala, Aschelminthes and Ectoprocta*. McGraw-Hill, New York, pp. 1–52.

Kates, K. C. 1942. Viability of the eggs of the swine thorn-headed worm (*Macracanthorhynchus hirudinaceus*). *J. Agric. Res.* **64**:93–100.

Moore, J. 1984. Parasites that change the behavior of their hosts. *Sci. Am.* **250**(5):108–115.

Nicholas, W. L. 1967. The biology of the Acanthocephala. *Adv. Parasitol.* **5**:205–246.

——— 1973. The biology of the Acanthocephala. *Adv. Parasitol.* **11**:671–706.

Parshad, V. R., and D. W. T. Crompton. 1981. Aspects of acanthocephalan reproduction. *Adv. Parasitol.* **19**:73–138.

Schmidt, G. D. 1972. Revision of the class Archiacanthocephala Meyer, 1931 (Phylum Acanthocephala), with emphasis on Oligoacanthocephala Southwell and Macfie. 1925. *J. Parasitol.* **58**:290–297.

Uglem, G. L., P. W. Pappas, and C. P. Read. 1973. Surface amino-peptidase in *Moniliformis dubius* and its relation to amino acid uptake. *Parasitology.* **67**:185–195.

Whitfield, P. J. 1970. The egg sorting function of the uterine bell of *Polymorphus minutus* (Acanthocephala). *Parasitology.* **61**:111–126.

Wright, R. D., and R. D. Lumsden. 1969. Ultrastructure of the tegumentary pore-canal system of the acanthocephalan, *Moniliformis dubius*. *J. Parasitol.* **55**:993–1003.

Yamaguti, S. 1963. *Systema Helminthum*. Vol. 5. *Acanthocephala*. Wiley Interscience, New York.

# 45

# *Phylum Nematomorpha: Horsehair Worms*

The phylum Nematomorpha (=thread form) is a small group of about 230 species of long, slender worms that superficially resemble nematodes. The long-believed myth was that these worms had transformed from horse hairs which had fallen into water, hence the name (Hoeppli, 1959). Also, because the free-living adults often become entangled with one another, they are sometimes called *Gordiacea*, *gordian worms* or *gor diids* after the mythical Gordian knot. References to classification are in Poinar and Doelman (1974).

The adults are free-living in fresh water or moist soil; one genus, *Nectonema*, is pelagic in marine environments as an adult and parasitic in hermit crabs and true crabs as a larva. The larvae of all other nematomorphs are endoparasites of arthropods, especially grasshoppers, crickets, and terrestrial beetles, and sometimes of leeches.

DISTRIBUTION AND IMPORTANCE. Horsehair worms are found worldwide in temperate and tropical freshwater habitats; they have been found recently as far north as the Arctic Circle (Redlich, 1980). Except for rare, accidental infections in humans or other mammals, they are of no medical or economic importance.

MORPHOLOGY. The adult worms are quite slender, with diameters from 0.3 to 2.5 mm and lengths of sometimes 1 m, although most are perhaps 20 cm long. The sexes are separate; the males are shorter than females, except in the genus *Nectonema*, in which the males are longer. Their colors range from gray to black. The anterior end or *calotte* is usually white with a dark ring at its posterior limit. There is a terminal or subterminal mouth. Both sexes have a cloaca, and the posterior end is lobed in some species. Horsehair worms have a cuticle similar to that of the nematodes (Fig. 45.1) (Zapatosky, 1971) and have characteristic *areoles* that are contiguous, rounded, or polygonal areas often projecting above the cuticular surface as papillae with bristles or pores (Fig. 45.2); their function is unknown.

Nematomorphs have no excretory, circulatory, or respiratory structures. They are pseudocoelomate, but the body cavity is filled with parenchyma in most freshwater species, with only limited space remaining. Only the marine form, *Nectonema*, has a complete pseudocoelom (Hyman, 1951). The digestive tract is degenerate and apparently without function. The mode of nutrition is unknown; some authors say that food is not ingested at any time during the life cycle (Pennak, 1978), and others state that nutrients must be absorbed through the tegument (Baer, 1951). The nervous system consists of an anterior nerve ring and a ventral nerve cord; little is known about specialized sensory structures.

The gonads are paired tubular organs that fill most of the pseudocoelom in mature worms. The ovaries are unique in that they transform into uteri by developing lateral diverticuli after producing the eggs. Sperm are rod-shaped and apparently nonmotile (Donin & Cotelli, 1977).

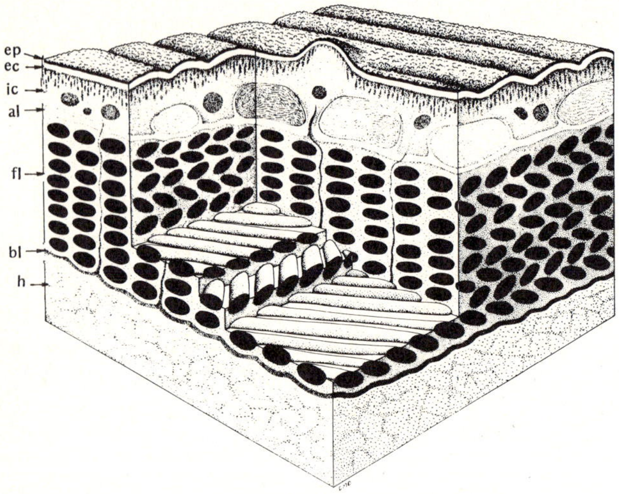


**FIG. 45.1**
Diagram of the cuticle of *Paragordius varius*. (al = areolar layer; bl = basal layer; ec = external cortical layer; ep = ependymal layer; fl = fibrillar layer; h = hypodermis; ic = internal cortical layer.) [From Zapotosky, 1971.]

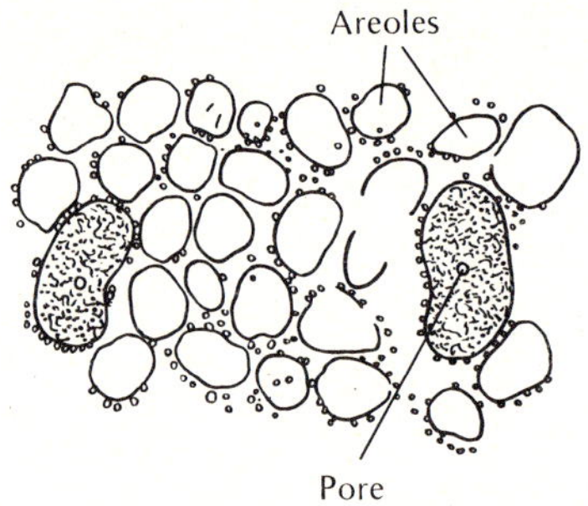


**FIG. 45.2**
Diagram of areoles of *Parachordodes*. [Redrawn with modifications from Hyman, 1951.]

LIFE CYCLE. In temperate climates, the worms escape from the host in the fall and remain near water until the following spring. A male swims to a female, coils the posterior end of its body around the female, and deposits sperm in her cloaca. Fertilization apparently takes place inside the body of the female just before she extrudes the eggs in long, gelatinous strings. Embryonation requires 15 to 80 days. After hatching, the larvae (Fig. 45.3) either penetrate an arthropod host or, more likely, encyst on a substrate such as vegetation near the shoreline. Arthropods ingest the larvae along with the vegetation. The larvae encyst in the gut and then later burrow into the hemocoel and develop into adults. In transport hosts, including certain insects, snails, and fish, cyst walls are digested in the gut, but the freed larvae encyst again until ingested by a suitable host (Baer, 1951). After developing for 2 weeks to several months, young adults break through the exoskeleton of the arthropod (Fig. 45.4) and become free-living. They survive only if they reach moist soil or water. It is said that when the worms have reached their maximum size, the host becomes thirsty and dies near water, thereby releasing the worms in the environment they require. The worms typically emerge in late summer or fall.

HUMAN INFECTIONS. Occasional human infections with horsehair worms have been reported. Most have been found when the worms were vomited or passed in the feces, but no serious illness has been associated with

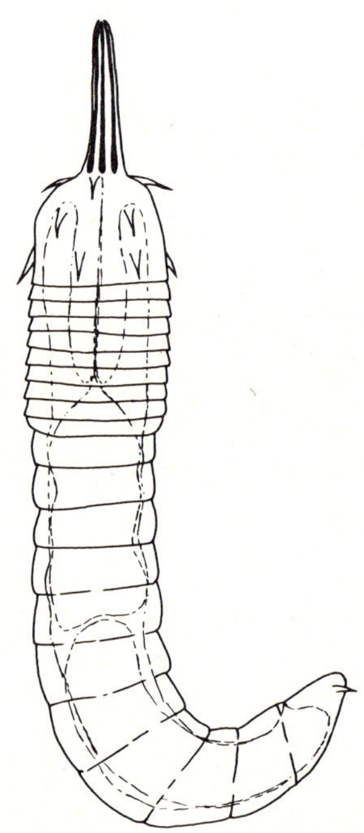

**FIG. 45.3**
Diagram of a nematomorphan larva. [Redrawn from Pennak, R.W. 1978. *Freshwater Invertebrates of the United States*, 2nd ed. John Wiley & Sons, New York.]

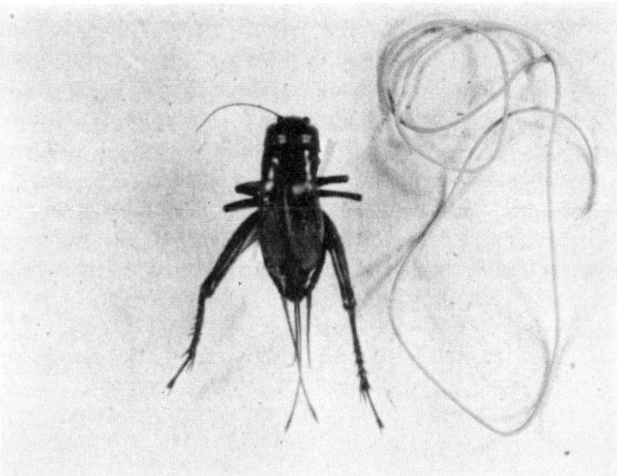

**FIG. 45.4**
Young adult nematomorphans breaking out of a cricket.

them. In one case, an abscess in the orbit of the eye was associated with an infection with *Gordius robustus* (Sayad et al., 1936). In another instance, *Paragorodius esavianus* was expelled from the urethra of a 23-year-old woman with a history of urinary tract problems extending over 3 years (Burger, 1972). In both of these cases, it seems likely that infection was acquired while swimming. The rarity of human infection indicates that it is unlikely to become a widespread problem.

At one time or another, nearly every parasitologist has had to deal with an excited parent carrying a bottle of aftershave lotion containing a horsehair worm and stating that his or her child had passed it. The conversation usually reveals that the worm was found in the toilet. In these cases, the source is most likely a cricket which sought water and died about the time it released the worm.

## CLASSIFICATION

**Phylum Nematomorpha**
**Order Nectonematida**
Pseudocoel open; both dorsal and ventral epidermal chords present; gonads single; marine. A single genus,
*Nectonema*

**Order Gordiida**
Pseudocoel filled with parenchyma; only ventral epidermal chord present; gonads paired; freshwater and terrestrial.
*Gordius, Paragordius*

## REFERENCES AND FURTHER READINGS


Baer, J. G. 1951. *Ecology of Animal Parasites*. University of Illinois Press, Urbana, pp. 43–45.

Burger, R. 1972. *Paragordius esavianus* passed per urethra *J. Urol.* **108**:469.

Donin, C. L. L., and F. Cotelli. 1977. The rod-shaped sperm of Gordioidea (Aschelminthes, Nematomorpha). *J. Ultrastruct. Res.* **61**:193–200.

Hoeppli, R. 1959. *Parasites and Parasitic Infections in Early Medicine and Science*. University of Malaya Press, Singapore.

Hyman, L. H. 1951. *The Invertebrates: Acanthocephala, Aschelminthes, and Entoprocta. The Pseudocoelomate Bilateria*. McGraw-Hill, New York.

Pennak, R. W. 1978. *Freshwater Invertebrates of the United States*, 2nd. ed. John Wiley & Sons, New York. pp. 231–238.

Poinar, G. O., and J. J. Doelman. 1974. A reexamination of *Neochordodes occidentalis* (Mont.) comb. n. (Chordididae: Gordioidea): Larval penetration and defense reaction in *Culex pipiens* L. *J. Parasitol.* **60**:327–335.

Redlich, A. 1980. Description of *Gordius attoni* sp. n. (Nematomorpha, Gordiidae) from northern Canada. *Can J. Zool.* **58**:382–385.

Sayad, W. Y., V. M. Johnson, and E. C. Faust. 1936. Human parasitization with *Gordius robustus*. *J. Am. Med. Assn.* **106**:461–462.

Zapotosky, J. E. 1971. The cuticular ultrastructure of *Paragordius varius* (Leidy, 1851) (Gordioidea:Chordodidae). *Proc. Helminthol. Soc. Wash.* **38**:228–236.

# 46

# *Mollusca and Annelida*

The two phyla discussed in this chapter, Mollusca and Annelida, are not often included in parasitology courses; if they are, only the class Hirudinea, or leeches, is covered. Probably the reason for the omission is that many of these are marine parasites. Yet these organisms have fascinating adaptations for the parasitic mode of existence ranging from slight differences from their free-living counterparts to complete loss of alimentary tracts and increased dependency on their hosts for nutrition. Many mollusks are important because they are intermediate hosts for many trematode and nematode parasites.

## PHYLUM MOLLUSCA

The name Mollusca comes from the Latin *molluscus*, which means soft; mollusks have a characteristic soft body. This phylum contains over 100,000 described species, second in numbers only to the phylum Arthropoda. Mollusks are well known, partly because of the widespread interest in seashell collecting.

The mollusks have a true coelom, specialized respiratory organs (gills or lungs), an open circulatory system, and usually paired metanephridia. Also characteristic are the foot, a muscular portion of the ventral body wall; the pallium or mantle, a pair of folds in the dorsal body wall enclosing the internal organs, forming gills or lungs, and often secreting the shell; and a complete digestive tract which has a rasping organ, the radula. The first larval stage is the trochophore, which is absent in some groups; the veliger is a later larval stage characteristic of many mollusks of interest here.

Parasitism in mollusks usually occurs in larvae in the class Bivalva and in adults in the class Gastropoda. These are the same two classes of mollusks that are important as intermediate hosts of many parasites.

### *Class Gastropoda*

The class Gastropoda includes the familiar snails and slugs plus numerous other organisms such as limpets and conchs. Marine, freshwater, and even terrestrial snails are intermediate hosts for trematodes and nematodes. In addition, many species of marine gastropods are true parasites.

MORPHOLOGY. The freshwater snail, *Goniobasis livescens*, is a typical gastropod. The shell is a hard, coiled structure that protects the soft body parts; it is secreted by the mantle. The opening or aperture of the shell can be covered by the moveable operculum so that the soft body is completely protected. The shell is coiled in a dextral or clockwise direction. The foot and head–trunk regions are united (Fig. 46.1); the latter has two tentacles (organs of touch) and two eyes. Extending from the head–trunk region is the visceral mass, which includes gonads and most of the digestive system. The

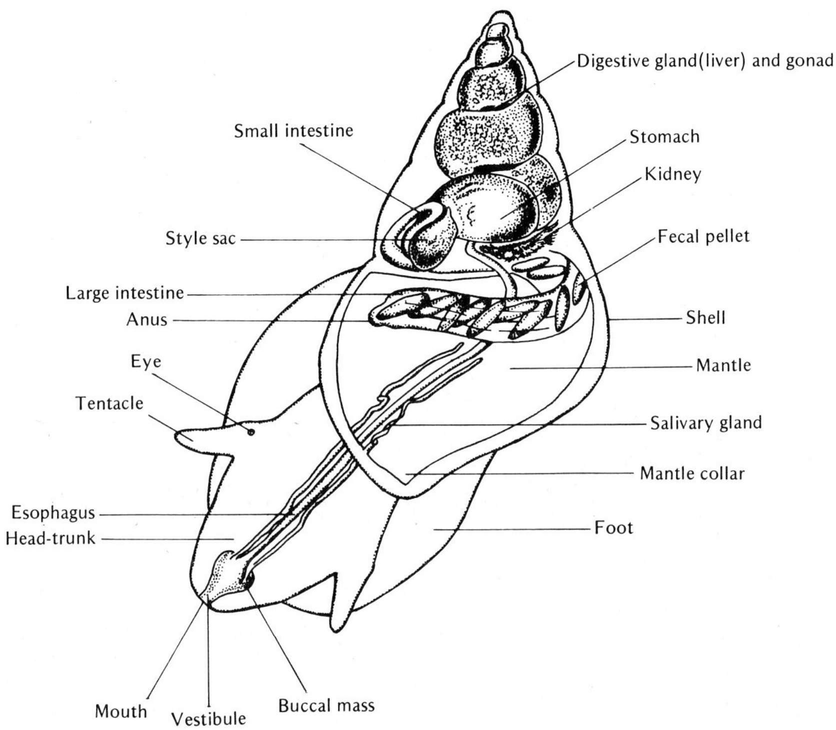


**FIG. 46.1**
Diagram of *Goniobasis livescens*. [Redrawn with modifications from Dazo. 1965.]

visceral mass is located in the upper part of the conical shell.

The digestive system has an anterior mouth opening to the buccal mass, which contains the rasping organ, the radula. The esophagus leads from the buccal mass to the stomach. The style sac containing a crystalline style is located at the distal end of the stomach. The head of the style rotates against a thickened portion of the stomach, drawing food into the stomach and then grinding it (Dazo, 1965). The style sac may also release digestive enzymes into the stomach (Winkler & Wagner, 1959). Leading from the stomach is the small intestine, followed by the large intestine, which ends at the anus. The liver or digestive gland is located in the upper whorls of the shell. This gland has multiple functions, including digestion, secretion, absorption, and storage of nutrients (Owen, 1966; Hyman, 1967).

*Goniobasis* has gills for respiration. They arise as folds from the mantle. The excretory organ, the kidney, is a large, flat organ adjacent to the stomach and intestine. The circulatory system has a 2-chambered heart, a pericardial cavity, six major blood sinuses, and numerous small blood vessels.

The sexes are separate in *Goniobasis*. The males lack external genitalia, making it difficult to distinguish them from females. A large testis or ovary is coiled around the digestive gland in the upper whorls of the shell. Gonopores are adjacent to the anus.

## *Gastropods as Intermediate Hosts*

Miracidia of digenetic trematodes infect snails by penetrating the tentacles, head, trunk, mantle, or foot, or by being ingested and then penetrating the esophagus (Loker, 1978). Eggs containing miracidia can also be ingested and then hatching occurs in the gut (Smyth, 1966). Mother sporocysts usually develop near the site of miracidial penetration—the mantle, foot, or tentacles. Daughter sporocysts or rediae leave mother sporocysts, and most commonly infect the digestive gland (Malek & Cheng, 1974), but may also invade gonads or other snail tissues.

In addition to digenetic trematodes, gastropods also harbor a number of species of larval nematodes, including *Mullerius capillaris*, a sheep parasite; *Aleurostrongylus abstrusus* and *Anafilaroides rostratus*, both cat lung parasites; and *Angiostrongylus cantonensis*, a rat lungworm (Malek & Cheng, 1974). The last is a human parasite, which causes meningoencephalitis in the Far East and Pacific islands, including Hawaii (see Chap. 37).

**FIG. 46.2**
Diagram of *Odostomia scalaris* inserting proboscides between the valves of a clam. [Redrawn from Baer, 1951.]

## *Gastropods as Parasites*

NAME OF ORGANISM. *Odostomia* spp.

HOSTS AND IMPORTANCE. Species of *Odostomia* are found free in the sea, under stones, and associated with bivalve mollusks (Baer, 1951). Two known pathogenic species are *O. bisutularis* Say, 1922, parasitic on the oyster, *Crassostrea virginica*, and *O. eulimoides* Hanley, 1844, parasitic on oysters and mussels (Cheng, 1967).

MORPHOLOGY. Shells are 2 to 7 mm long and a dull white. These snails have a long proboscis, which is inserted between the valves of their mollusk host (Fig. 46.2). They have a "false foot" surrounding the proboscis, plus the other usual gastropod features.

HOST–PARASITE INTERACTIONS. *O. bistularis* pierces the mantle of its host and feeds on blood and fluids, damaging the mantle. Since the mantle secretes the shell, parasitized oysters often have deformed shells. Boss and Merill (1965) found that *O. bisutularis* attaches to a number of species of mollusks, often near the original host, the oyster. *Odostomia* attaches to the host shell by an elastic thread produced by the posterior end of the foot. Hoffman (1979) found that *Odostomia* can detach from the original thread, and then secrete a new thread for reattachment, thus enabling it to move over the mantle surface. *O. eulimoides* lives in small pockets in the mantle of its host, irritating the mantle so much that it continually withdraws until it reaches the adductor muscle. Irritation of this muscle causes the host to open its valves and may lead to its death (Cole & Hancock, 1955).

NAME OF ORGANISM. *Stilifer* spp.

HOSTS AND LOCATION. Members of the genus *Stilifer* are embedded in starfishes and brittle stars in Asian and Pacific waters.

MORPHOLOGY. These gastropods still retain their shells, but the feet are vestigial, and the proboscis is modified for parasitism. There is a well-developed pseudopallium in most species, which may almost cover the entire organism, including the shell (Fig. 46.3).

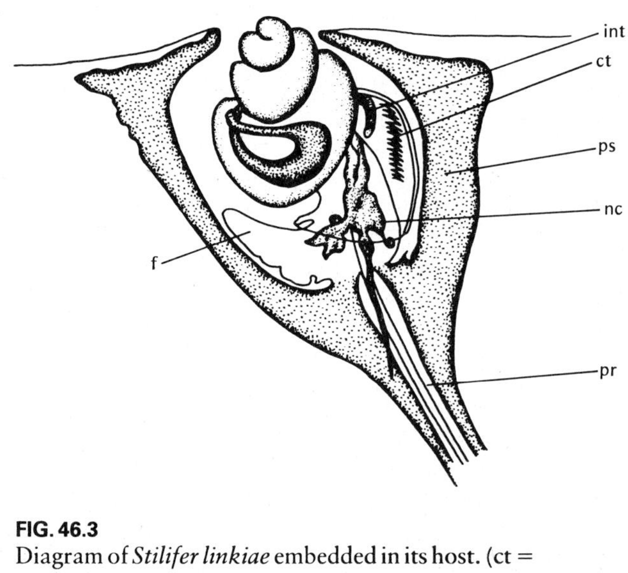


**FIG. 46.3**
Diagram of *Stilifer linkiae* embedded in its host. (ct = ctenidium; f = foot; int = intestine; nc = brain; pr = proboscis; ps = pseudopallium.) [Redrawn from Baer, 1951.]

Contraction of the pseudopallial walls creates a water current in the pseudopallial cavity which supplies water to the gills and also evacuates feces and sex products (Baer, 1951). Both hermaphroditic and dioecious species have been described; Baer thought that some, if not all species, were protandrous hermaphrodites.

LIFE CYCLE. The eggs are enveloped in a gelatinous mass and are laid by females through an opening in the host. They hatch into veliger larvae, develop in about 10 days, and then enter another echinoderm through the ambulacral groove, where they mature.

HOST–PARASITE INTERACTIONS. Members of the genus *Stilifer* are true endoparasites, deriving their nutrition from host fluids. Starfish have a defense mechanism called *autotomy* in which an arm is broken off; the presence of *S. linchias* inhibits autotomy (Davis, 1967).

NAME OF ORGANISM. *Thyonicola mortenseni.*

HOSTS AND LOCATION. *T. mortenseni* adults are embedded in the gut walls of the sea cucumbers, *Thyone* spp. and *Cucumaria* spp.

MORPHOLOGY. The veliger larva is typical, but the adult has become so modified that it is not recognizable as a gastropod (Fig. 46.4*A*). The mouth and all traces of the digestive system have disappeared, as have the shell and mantle. The adult is a sac of gonads in the host's coelom, attached to the gut. Baer (1951) reviewed the Russian literature on the development of this unusual gastropod.

LIFE CYCLE. Eggs within cocoons are voided in the host's feces. The cocoons burst on contact with seawater, releasing the veliger larvae (Fig. 46.4*B*). The larvae are ingested by sea cucumbers, penetrate the gut, and undergo extensive metamorphosis into adults. Males complete their metamorphosis embedded in an epithelial petiole, or male receptacle, in the female's central cavity. The complete life cycle of closely related species was reported by Lutzen (1979).

## *Class Bivalva*

The class Bivalva includes the familiar oysters and clams. They have two shells or valves, hinged dorsally, which completely enclose the body. Larval parasitism, sometimes called *protelian parasitism*, is characteristic of three families of bivalves: the Unionidae, Mutelidae and Mycetopodidae. The Unionidae are cosmopolitan and have a special veliger larva called a *glochidium* (Fig. 46.5); the Mutelidae are found in Africa and have a haustorium veliger; and the Mycetopodidae, from South America, have a lasidium veliger (Barnes, 1980).

The following account of glochidial biology is summarized primarily from the account of Baer (1951). Glochidia are expelled in large numbers from clams and settle to the bottom of the stream. They are so light that the slightest current, such as the passage of a fish, throws up clouds of glochidia. Many enter the fish through the respiratory currents and clamp themselves to the gills. Other glochidia, especially those with spines, may pierce fish tissues with their spines, fold back, and clamp onto their host. The initial attachment in *Anodonta* is with a long sticky thread secreted by the rudimentary foot. The thread attachment is through specific molecules in fish mucus, which explains the reported host specificity for glochidia (Wood, 1974); however, not all glochidia are host specific.

Both the glochidia on the gills and those on other surfaces are quickly encapsulated by host tissues. The larvae liquefy the surrounding host tissues for their own nourishment. They then begin to digest their own bodies, beginning with the adductor muscle. This is not total self-destruction; at the same time new adult organs develop. Eventually the young clams fall free from their hosts and begin the free-living existence. Apparently adult fish are not harmed by glochidial infections, but fish fry have died as a result of secondary infections.

Parasitism by adult bivalves is quite rare. The best known example is the marine genus *Entovalva*, which lives in the gut of sea cucumbers. Other closely related parasitic genera are *Cycladoconcha*, *Montacuta* and *Scintillona* (Boss, 1965).

# PHYLUM ANNELIDA

The phylum Annelida is composed of segmented worms, including the familiar earthworms or nightcrawlers. We see a few annelids in garden soil or on our lawns after a rain, but vast numbers are present as marine benthos. About 8700 species have been de-

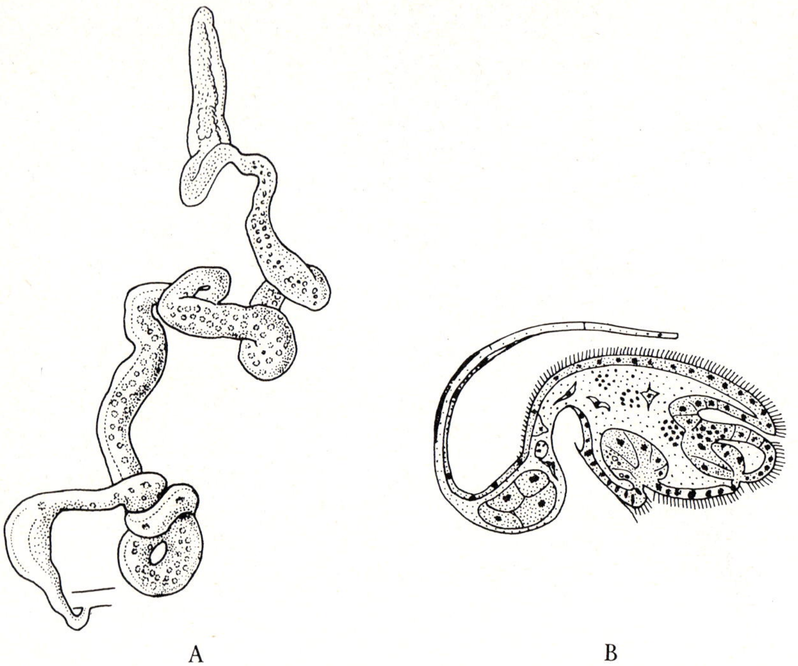


**FIG. 46.4**
Diagram of *Thyonicola mortenseni.* (*A*) Adult; (*B*) veliger larva. [Redrawn with modifications from Baer, 1951.]

scribed; most are free-living, but there are examples of all types of symbioses, including parasitism.

Metamerism, the division of the body into similar segments or metameres, is characteristic of this phylum. Organs such as nephridia are present in most metameres. Annelids have a true coelom, closed circulatory system, complete digestive system, and chitinous setae often present on fleshy appendages called *parapodia.* They respire through the integument or gills.

Most taxonomists accept three classes in the phylum Annelida: Polychaeta, Oligochaeta, and Hirudinea. In addition, three other groups have uncertain affinities. The enigmatic Archiannelida are sometimes placed in a separate phylum or class or are included in the class Polychaeta. No archiannelids are known to be parasitic. The branchiobdellids and acanthobdellids, which have parasitic species, are regarded by most systematists as primitive leeches and are included in the class Hirudinea (Clark, 1978).

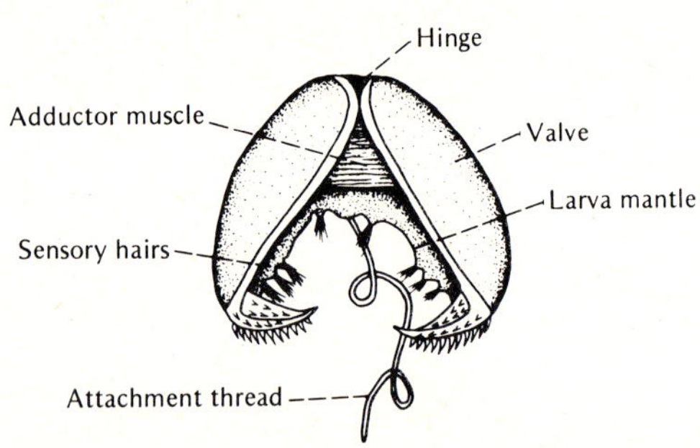


**FIG. 46.5**
Diagram of a glochidium of *Anodonta.* [Redrawn with modifications from Pennak, R.W. 1978. *Freshwater Invertebrates of the United States,* 2nd ed. John Wiley & Sons, Inc., New York.]

## *Class Polychaeta*

Polychaetes have paired appendages, parapods, which bear setae; they are used for locomotion and often for respiration. Parasitism is not common in the class Polychaeta, but there are examples of both ecto- and endoparasites (Clark, 1956).

The ectoparasites are more modified for the parasitic life-style than are the endoparasites. One example is

*Ichthyotomus sanguinarius*, which attaches to the fins of eels by means of protrusible stylets (Fig. 46.6). It has modifications for blood feeding including piercing stylets, an anticoagulant secretion, a sucking pharynx, and an enlarged gut extending into the parapods for storing large quantities of host blood (Clark, 1956). *Ichthyotomus* is a neotenic worm; except for the mature gonads, it resembles annelid larvae more than adults.

Endoparasitic polychaetes have surprisingly few modifications for parasitism, perhaps because they have only recently become parasites. Little is known about these organisms, most from single reports. Clark (1956) reviewed the scattered reports on commensal and parasitic polychaetes; he commented that it is highly unusual for large, predatory polychaetes to become parasites of their own prey, especially without any special modifications or reduction in size.

## Class Oligochaeta

Oligochaetes have no parapodia, but have well-developed metamerism. Like the polychaetes, very few oligochaetes are true parasites, and none are of medical or economic importance except for the few that are intermediate hosts for lungworms (Chap. 37).

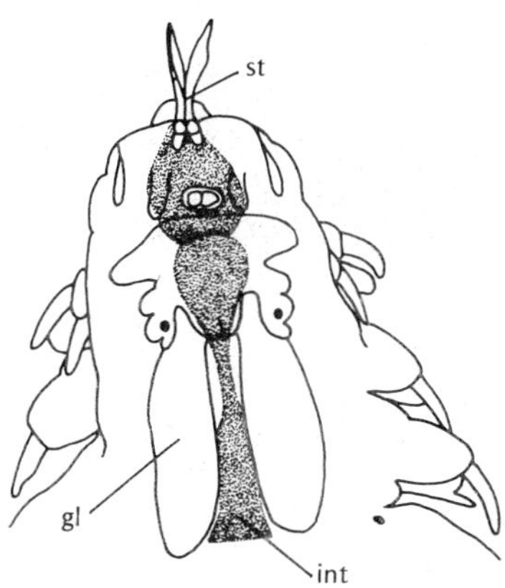


**FIG. 46.6**
Diagram of *Ichthyotomus sanguinarius* anterior end. (gl = glands; int = intestine; st = stylets.) [Redrawn from Baer, 1951.]

Members of the genus *Chaetogaster* (Fig. 46.7) are usually epicommensals or epiphoronts on freshwater mollusks, but some are true parasites. It is interesting to note that *Chaetogaster* eats both miracidia and cercaria (Khalil, 1961). This should be detrimental to the reproductive success of the parasitic digenetic trematodes infecting the snail and thus advantageous to both the annelid and its host snail. This annelid is an example of the complete gamut of symbiotic relationships: (1) commensalism—the annelid eats detritus generated by the snail's feeding activities; (2) mutualism—the snail provides shelter and the annelid reduces the number of digenetic trematodes infecting the snail; and (3) true parasitism—the annelid feeds on snail tissues (Lankester, 1869).

## Class Hirudinea

The class Hirudinea contains about 500 species of marine, freshwater, and terrestrial leeches. Only about

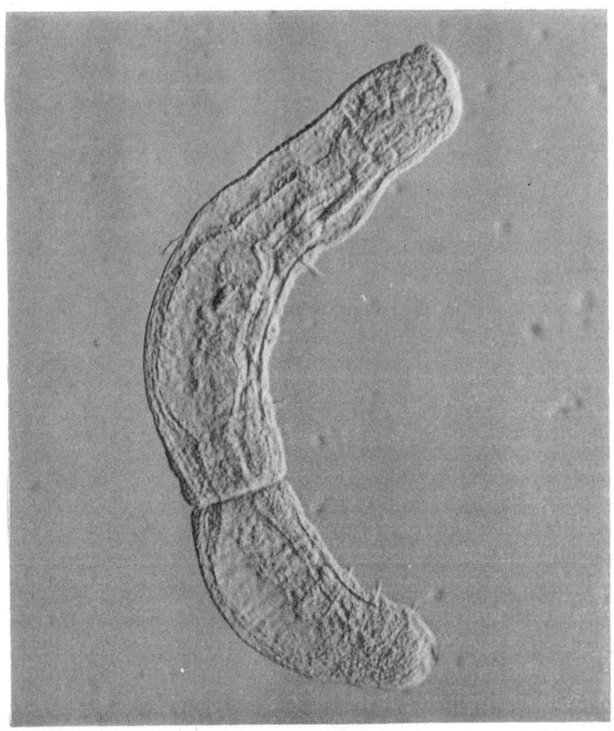

**FIG. 46.7**
Interference contrast micrograph of *Chaetogaster*.

25% are parasitic, despite the notion that all leeches are bloodsuckers (Baer, 1951). Most leeches are predators or scavengers. Those that are true parasites are usually temporary; they abandon their hosts after engorgement with blood or tissue fluids.

In Europe, North Africa, and the Near East, leeches were used for blood letting to cure headaches and a number of diseases. The early Greek and Roman medical writers provide detailed instructions on the techniques of blood letting using leeches. In early Chinese medicine, ground-up leeches were used as a treatment for internal hemorrhaging. Probably this was done because it was thought that if living leeches could remove blood as ectoparasites, their powder should be helpful in removing internal blood; there is no scientific basis for this assumption. See Hoeppli (1959) for a fascinating discussion of the uses of leeches in early medicine.

Leeches differ from other annelids in having (1) a well-developed posterior sucker and an anterior sucker or suckerlike depression, (2) a constant number of body segments, 34 in most leeches, but 14 or 15 in branchiobdellids, (3) no setae or parapodia except in acanthobdellids, and (4) a reduced coelom largely filled with muscle and connective tissue.

NAME OF ORGANISM. *Hirudo medicinalis* L is the medicinal leech.

HOSTS AND LOCATIONS. The medicinal leech is an ectoparasite that feeds on the blood of mammals, frogs, tadpoles, and small fish.

DISTRIBUTION AND IMPORTANCE. This leech is native to Europe and Asia and has been imported into the United States. It was raised extensively for medicinal purposes in Europe and the United States, but that use has drastically declined in the past hundred years.

MORPHOLOGY. A mature medicinal leech can extend to 12 cm long by 1.5 cm wide. The body has 102 annuli or external folds, not counting the posterior sucker. These annuli do not correspond to the 34 segments or metameres characteristic of most leeches; there may be one to five annuli per segment (Fig. 46.8*A*). The 34 segments are present in embryonic development; in the adult they are recognized as paired nerve ganglia, except for the seven that have fused into the large posterior sucker. The anterior sucker is a depression of segments I through IV on the ventral surface. Even though each annulus has sense organs, they are difficult to see, as are the five pairs of eyes on segments I through V.

The medicinal leech has a complete digestive tract. Inside the buccal cavity are three jaws (Fig. 46.8*B*) which make a Y-shaped incision in the host skin. Posterior to the buccal cavity is the muscular pharynx that pumps blood into the largest part of the digestive tract, the crop. It has 11 pairs of lateral branches and can store large quantities of blood. Following the crop is a thin-walled intestine and then a rectum leading to the terminal anus.

*Hirudo* is a hermaphrodite. The gonads are in the anterior part of the body (Fig. 46.8*C*). There are usually ten pairs of testes and a single pair of ovaries (Fig. 46.8*C*). Eggs are fertilized by sperm from another leech, coated with albumin in the oviduct, and passed through the female gonopore into a cocoon secreted by the clitellum. The leech then slips the cocoon containing developing eggs over its head.

The excretory system is composed of 17 pairs of nephridia. The circulatory system has no distinct blood vessels; instead, there are six blood sinuses. The two lateral sinuses have muscular walls and function as hearts to circulate the blood.

LIFE CYCLE. There is an annual life cycle. Leeches mate in the spring; eggs in cocoons are deposited in damp soil. Young leeches develop and mate in the spring of the following year.

## *Other Human and Animal Leeches*

In North America, *Macrobdella* and *Philobdella* are the only common leeches that regularly take human blood (Pennak, 1978). Like the medicinal leech, they firmly attach by their oral suckers and move their jaws back and forth to make painless incisions in the host skin. They then secrete a small amount of an anticoagulant, *hirudin*, which remains in host tissues. Bloodsucking leeches feed infrequently, but take an enormous amount of blood when they do feed. They normally take two to ten times their own body weight and need feed only twice a year (Mann, 1962). They secrete only exopeptidases and need the help of gut bacteria such as *Pseudomonas hirudinicola* to aid digestion (Wilde, 1975).

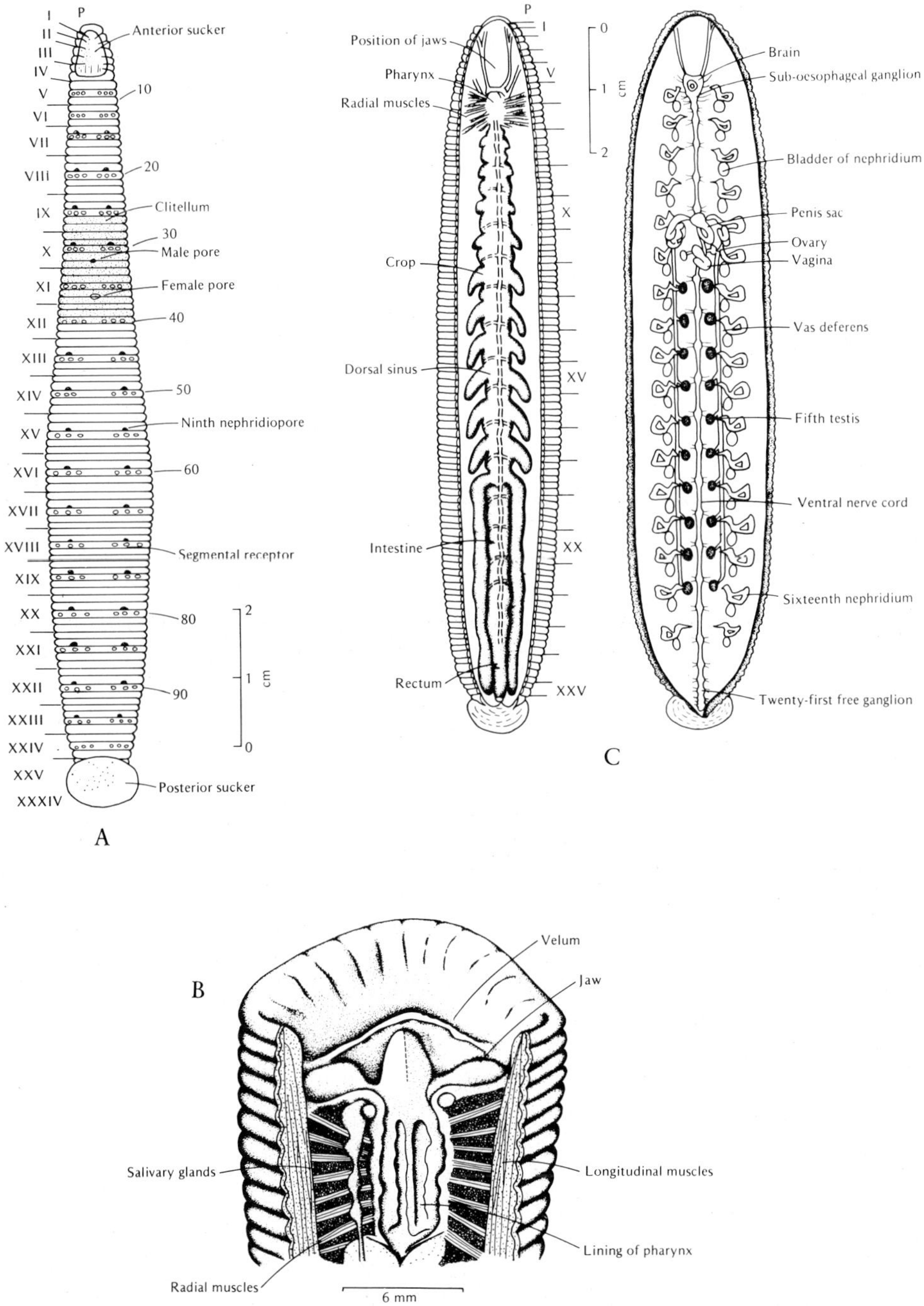


**FIG. 46.8**
Diagrams of the medicinal leech, *Hirudo medicinalis*. (*A*) Ventral view; (*B*) ventral dissection of head; (*C*) internal organs. [Redrawn with modifications from Mann, K.H. 1962. *Leeches (Hirudinea): Their Structure, Physiology, Ecology and Embryology*. Pergamon Press, New York.]

Members of the terrestrial genus *Haemadipsa* are important parasites of humans and domestic animals throughout tropical and temperate regions in Asia and the Far East. A number of other aquatic and terrestrial leeches attach to almost any mammal.

A potentially serious problem is internal hirudiniasis, caused by the accidental ingestion of aquatic leeches of the genus *Limnatis*, commonly known as the *horse leech*. They are common in fresh water in Europe, northern Africa, and Asia. There are numerous case reports of horse and human upper respiratory tract infections and an occasional upper digestive tract infection. Many infections are self-cured, but occasionally severe blood loss and secondary infections result, sometimes leading to death (Cheng, 1973).

Fishes are often parasitized by leeches belonging to the family Piscicolidae, but species of *Actinobdella* and *Placobdella* are also found on fishes. The amount of injury is related to the number of parasites present and the amount of blood removed (Hoffman, 1967).

Leeches are also vectors for protozoa such as *Trypanosoma*, *Cryptobia*, and *Haemogregarina* and are intermediate or final hosts for some digenetic trematodes (Fig. 46.9) as well as cestode cysticerci (Mann, 1962).

## CLASSIFICATION

**Phylum Mollusca**
**Class Monoplacophora**
Bilateral symmetry; broad, flat foot; single shell; internal segmentation only.

**Class Polyplacophora**
Bilaterally symmetrical; elongated body and reduced head; usually a shell of eight dorsal plates; large ventral foot; trochophore larva. Chitons.

**Class Aplacophora**
Worm-shaped body; no typical shell, mantle or foot; calcareous spicules in tegument; radula present; Marine. Solenogasters.

**Class Scaphopoda**
Body enclosed in one-piece, tubular shell; mouth with tentacles; head absent; trochophore larva. Tooth or tusk shells.

**Class Gastropoda**
Body asymmetrical; head well developed; mantle modified into a gill or lung; foot large and flat; shell usually coiled. Snails.

**Class Bivalva (Pelecypoda or Lamellibranchia)**
Bilateral symmetry; body enclosed in a two-lobed mantle; shell of two lateral valves; no head or radula; trochophore or glochidia larva. Clams, oysters, and mussels.

**Class Cephalopoda**
Bilateral symmetry; shell often reduced; foot modified into arms or tentacles; siphon present. Squids and octopods.

**Phylum Anellida**
**Class Polychaeta**
Body of numerous segments with lateral parapodia

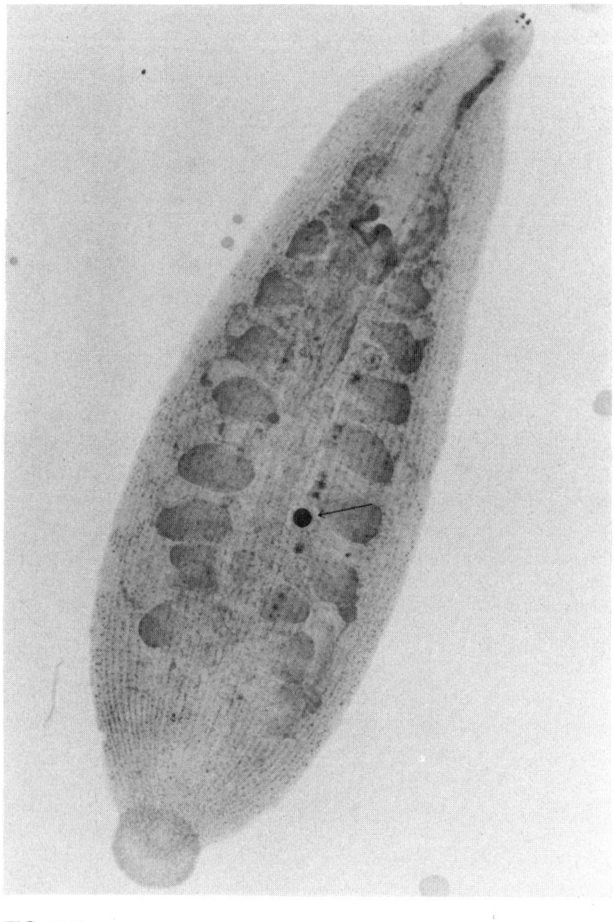

**FIG. 46.9**
Leech as an intermediate host for a fluke. Note the dark, circular metacercaria.

bearing setae; head distinct; no clitellum; trochophore larva usually; mostly marine.
*Nereis*, *Ichthyotomus*

**Class Oligochaeta**
Body of numerous segments with few setae per metamere; head absent; clitellum present; no larva; mostly terrestrial and freshwater.
*Lumbricus*, *Eisenia*.

**Class Hirudinea**
Body usually with 33 or 34 segments; setae usually absent; clitellum present; anterior and posterior sucker usually present; terrestrial, freshwater and marine.
*Hirudo*, *Macrobdella*, *Placobdella*.

## REFERENCES AND FURTHER READINGS

### Mollusca

Baer, J. G. 1951. *Ecology of Animal Parasites*. University of Illinois Press, Urbana.

Barnes, R. D. 1980. *Invertebrate Zoology*, 4th ed. Saunders, Philadelphia.

Boss, K. J. 1965. Symbiotic erycinacean bivalves. *Malacologia* 3:183–195.

Boss, K. J., and A. S. Merill. 1965. Degree of host specificity in two species of *Odostomia* (Pyramidellidae: Gastropoda). *Proc. Malac. Soc. London* **36**:349–355.

Cheng, T. C. 1967. Marine mollusks as hosts for symbiotes. In F.S. Russell (ed.), *Advances in Marine Biology*, Vol. 5. Academic Press, New York.

Cole, H. A., and D. A. Hancock. 1955. *Odostomia* as a pest of oysters and mussels. *J. Mar. Biol. Assoc. U. K.* **34**:25–31.

Davis, L. V. 1967. The suppression of autotomy in *Linckia multifora* (Lamarck) by a parasitic gastropod, *Stylifer linckiae* Sarasin. *Veliger* 9:343–346.

Dazo, B. C. 1965. The morphology and natural history of *Pleurocerca acuta* and *Goniobasis livescens* (Gastropoda:Cerithiacea:Pleuroceridae). *Malacologia* 3:1–80.

Hoffman, D. L. 1979. An attachment structure in an ectoparasitic gastropod (Opisthobranchia). *Veliger* 22:75–77.

Hyman, L. H. 1967. *The Invertebrates: Mollusca*. Vols. V, VI. McGraw-Hill, New York.

Loker, E. S. 1978. Normal development of *Schistosomatium douthitti* in the snail *Lymnaea catascopium*. *J. Parasitol.* **64**:977–985.

Lutzen, J. 1979. Studies on the life history of *Enteroxenous* Bonnevie, a gastropod endoparasitic in aspidochirote holothurians. *Ophelia* **18**:1–52.

Malek, E. A., and T. C. Cheng. 1974. *Medical and Economic Malacology*. Academic Press, New York.

Owen, G. 1966. Digestion. In K. M. Wilbur and C. M. Yonge, (eds.), *Physiology of Mollusca*, Vol 2. Academic Press, New York.

Pennak, R. W. 1978. *Freshwater Invertebrates of the United States*, 2nd ed. John Wiley & Sons, New York.

Smyth, J. D. 1966. *The Physiology of Trematodes*. W. H. Freeman, San Francisco.

Winkler, L. R., and E. D. Wagner. 1959. Filter paper digestion by the crystalline style in *Oncomelania*. *Trans. Am. Micros. Soc.* 78:262–268.

Wood, E. M. 1974. Some mechanisms involved in host recognition and attachment of the glochidium larva of *Anodonta cygnea*. *J. Zool* **173**:15–30.

### Annelida

Cheng, T. C. 1973. *General Parasitology*. Academic Press, New York.

Clark, R. B. 1956. *Capitella capitata* as a commensal, with a bibliography of parasitism and commensalism in polychaetes. *Ann. Mag. Nat. Hist.* 9:433–448.

——. 1978. Composition and relationships. In P. J. Mill (ed.), *Physiology of Annelids*. Academic Press, New York, pp. 1–61.

Hoeppli, R. 1959. *Parasites and Parasitic Infections in Early Medicine and Science*. University of Malaya Press, Singapore.

Hoffman, G. L. 1967. *Parasites of North American Freshwater Fishes*. University of California Press, Berkeley.

Khalil, L. F. 1961. On the capture and destruction of miracidia by *Chaetogaster limnaei* (Oligochaeta). *J. Helminthol.* 35:269–274.

Lankester, E. R. 1869. The sexual form of *Chaetogaster limnaei*. *Quart. J. Micros. Sci.* 9:272–285.

Mann, K. H. 1962. *Leeches (Hirudinea). Their Structure, Physiology, Ecology and Embryology*. Pergamon Press, New York.

Wilde, V. 1975. Investigations on the symbiotic relationship between *Hirudo officinalis* and bacteria. *Zool. Anz.* **195**:289–306.

47

# *Phylum Pentastomida*

The phylum Pentastomida consists of about 90 species of endoparasites of the respiratory tract of vertebrates. The name *pentastome* or "five mouth" is derived from the four anterior leglike protuberances, plus a fifth median projection that actually bears the mouth (Fig. 47.1). Often the "legs" are reduced to little more than the terminal hooks used for clinging to the host tissues. Pentastomes are also called *linguatulids* or *tongue worms* because some, such as *Linguatula serrata*, are found in the nasopharynx of the host (Self, 1969). Most adults are found in reptiles such as snakes and crocodiles, but a few are in birds and mammals.

These unusual parasites have been particularly puzzling to biologists. Most parasites have adults in more highly evolved animals and larvae in less highly evolved animals, suggesting that they evolved along with their hosts. However, in many pentastomids this situation is reversed (Self & Kuntz, 1967); for example, fish are intermediate hosts for crocodile pentastomids and mammals are intermediate hosts for snake pentastomids.

The unusual structure of pentastomids has made determination of evolutionary affinities difficult. They have no circulatory or excretory systems, nor do they have a distinct respiratory system. Similarities to annelids and arthropods have been suggested (Self, 1969), but the evidence more strongly supports an arthropod affinity. The outer covering is a chitinous cuticle (Fig. 47.2) similar to that of arthropods (Demaree, 1973; Trainer et al., 1975), muscles are

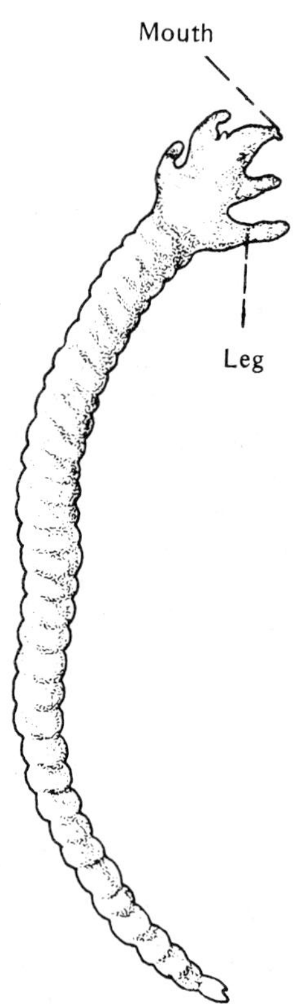


**FIG. 47.1**
Diagram of an adult *Cephalobaenia tetrapoda* from the lung of a snake. [Redrawn from Barnes, 1980.]

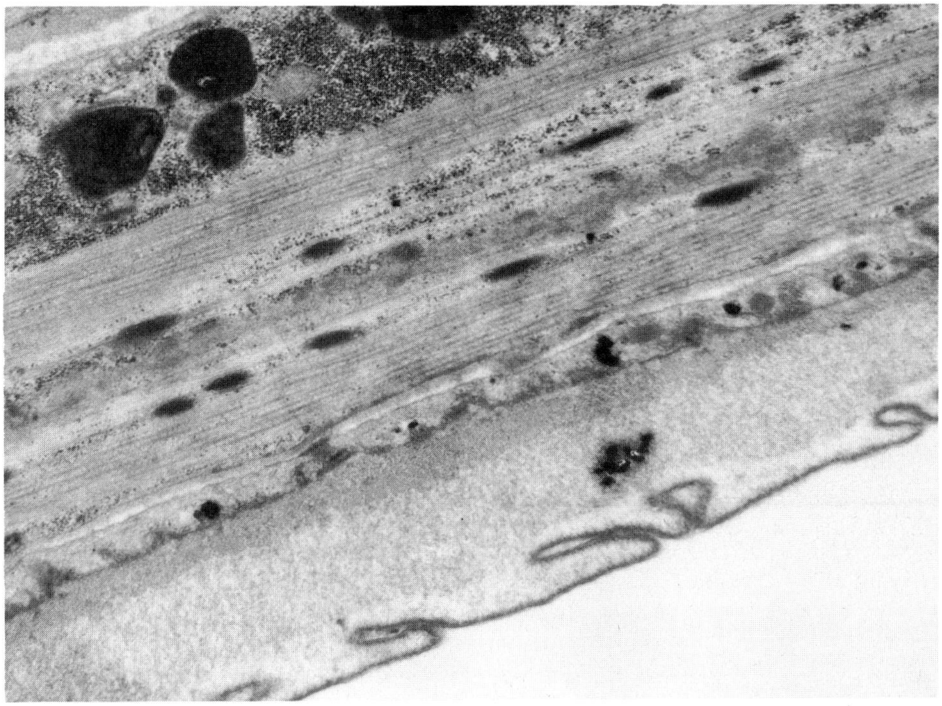

**FIG. 47.2** Transmission electron micrograph of *Porocephalus* nymph cuticle and muscle layer.

striated like those of arthropods (Mill & Riley, 1972); the molting sequence from primary larva to adult is similar (Riley, 1981); embryological development and especially spermatozoan characteristics resemble those of crustacea (Wingstrand, 1972; Riley et al., 1978). Von Haffner (1977) placed the pentastomids in a special class of the Articulata between tardigrades and myriapod progenitors, whereas Riley et al. (1978) recommended that the pentastomes be placed in a subclass of the class Crustacea. Recognizing the arthropod affinities, but realizing that the evolutionary and taxonomic position of the pentastomes is still uncertain, we take a conservative approach and retain them in their own phylum, immediately adjacent to the phylum Arthropoda.

The pentastome digestive tract is a straight tube from a subterminal mouth to a terminal anus. The mouth is surrounded by a chitinous ring called a *cadre*; the muscular pharynx and frontal or head glands have secretions that apparently liquefy host tissue and prevent coagulation of blood.

The nervous system has paired, metamerically arranged ganglia along a ventral nerve cord much like those of annelids and arthropods. There are also specialized sense organs in the tegument.

Sexes are separate; males are usually smaller than females. Males have a single testis connected to a seminal vesicle which is joined by lateral ducts to a pair of ejaculatory organs. The latter connect to a muscular cirrus, which fits into the groove of a dilator organ that may serve as a guide for the cirrus, just as the nematode gubernaculum does, or may actually be an intromittent organ in some species.

Females (Fig. 47.3) have a single ovary that sometimes branches into two lobes with joined oviducts. The single or double oviducts connect with a single, highly coiled uterus leading to the vagina and gonopore. The gonopore is in the anterior end of the abdomen in the order Cephalobaenida and at the posterior end in the order Porocephalida.

Fertilization is internal; embryonated eggs leave the definitive host through nasal and mouth secretions, or are swallowed and passed in feces. Eggs hatch, releasing primary larvae which have four claw-bearing, stumpy appendages (Fig. 47.4). Primary larvae molt several times, becoming nymphs (Fig. 47.5). Infective nymphs

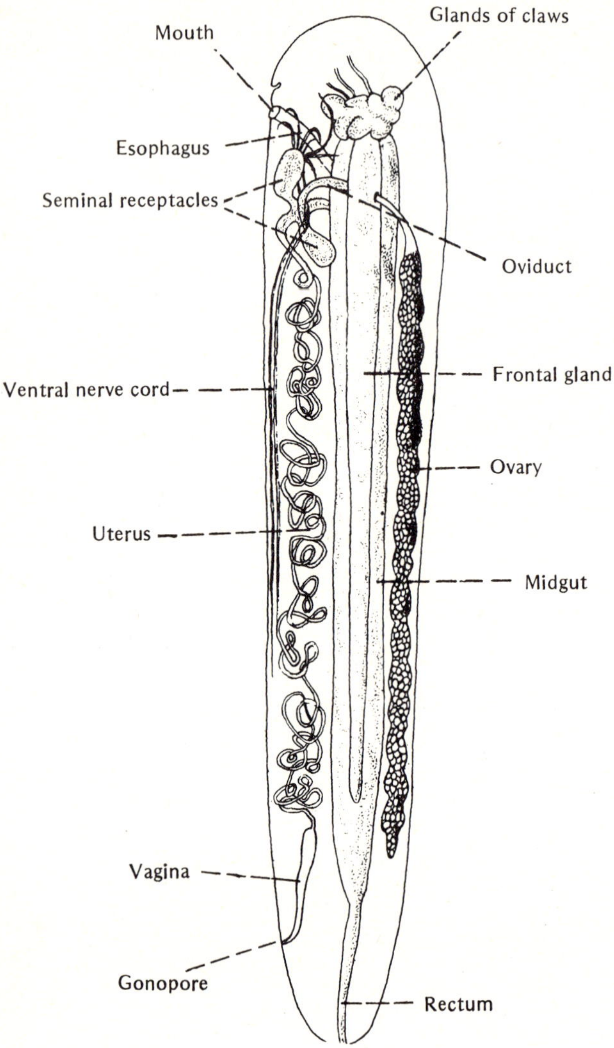


**FIG. 47.3**
Diagram of the internal organs of a female *Waddycephalus teretiusculus*. [Redrawn from Barnes, 1980.]

in intermediate hosts are ingested by definitive hosts, molt several times, and become adults (Fig. 47.1). Complete life cycles are known only for a few species.

NAME OF ORGANISM AND DISEASE CAUSED. *Porocephalus* spp. cause pentastomiasis, porocephalosis, and visceral pentastomiasis.

HOSTS AND LOCATION. Adults commonly occur in the lungs of various snakes, whereas nymphs are found in the mesenteries and internal organs in many mammals, including monkeys and humans.

DISTRIBUTION AND IMPORTANCE. The genus *Porocephalus* occurs worldwide in tropical and temperate zones. Adults usually cause little damage in snakes (Self & Kuntz, 1967), and visceral infections in humans with nymphs are often incidental findings at autopsy. Human infections are common in some parts of Africa and have been reported worldwide, including the United States (Prathap et al., 1969; Fain, 1975; Self et al., 1975; Mapp et al., 1976)

LIFE CYCLE. The best-known life cycle is that of *P. crotali* of rattlesnakes (Esslinger, 1962a, 1962b; Self et al., 1972; Riley, 1981). Eggs are ingested by mammals such as mice; they hatch, releasing primary larvae, which penetrate intestines. The larvae wander about in the peritoneal cavity and molt six times to become infective nymphs. The nymphs then become encapsulated in the host tissues (Fig. 47.5). When a rattlesnake ingests infective nymphs, they penetrate the intestine and adjacent lung, molt three more times and become adults. After mating, eggs are passed in nasal secretions, or are swallowed and pass in feces.

HOST–PARASITE INTERACTIONS. As mentioned, little damage occurs in either the intermediate or definitive hosts. Prathap et al. (1969) described three types of

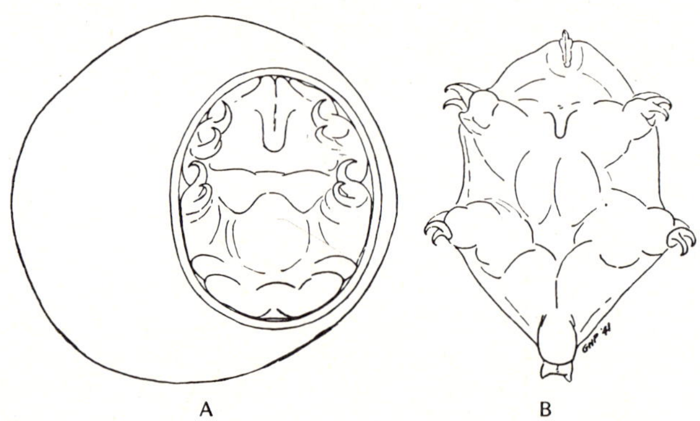


**FIG. 47.4**
Diagram of a larva of *Porocephalus crotali*. [Redrawn from Penn, 1942.]

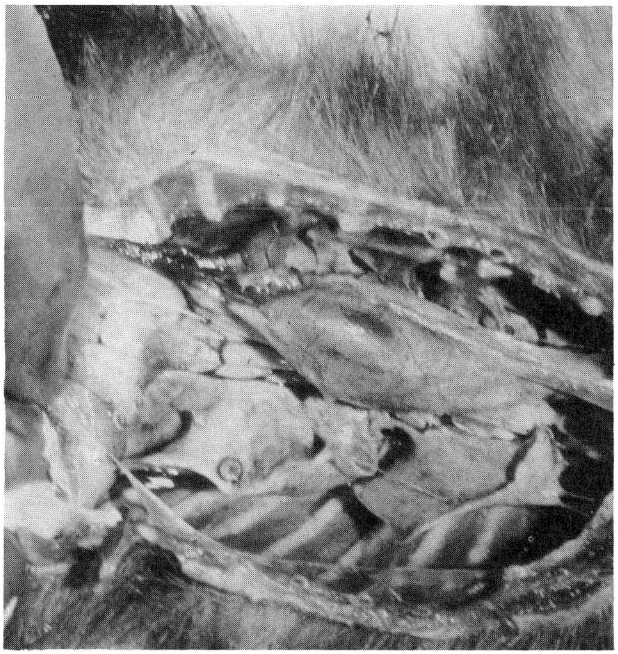

**FIG. 47.5**
*Porocephalus* nymph encapsulated in monkey viscera.

human lesions: encysted pentastomid nymphs, necrotic pentastomid nymphs, and "cuticle" granulomas. The first type is shown in Fig. 47.6; it is a nymph encysted in the mesenteries of a capuchin monkey, *Cebus apella*, from Colombia. Sometimes, however, visceral larva migrans reactions develop in the reptilian and mammalian hosts (Self, 1972).

Halzoun or marrara syndrome is a human disease caused by pentastomids found in the Middle East, Turkey, Greece, Morocco, India, and Sudan. The onset of the disease is seen within minutes to half an hour after the ingestion of raw liver or lymph nodes principally of sheep and goats. The signs and symptoms are usually restricted to the head and neck region, where there is swelling of the tissues of the throat, larynx, tonsils, nasal passages, and lips. The swelling can be so severe that the affected person may die of asphyxiation during the first few hours after onset. Later complications may include facial paralysis and bacterial abscesses in the throat. Not all persons ingesting infected tissues of sheep or goats are affected, but the same person may suffer repeated attacks (Khalil & Schacher, 1965; Schacher et al., 1969). The sudden onset and dramatic swelling of the throat tissues imply that halzoun is a hypersensitivity to the pentastomids.

A dish eaten in the Sudan and associated with marrara syndrome is made from raw stomach, liver, lung, and trachea of sheep, goats, cattle, or camels, seasoned with bile, lemon, spices, and sometimes onion and urine (described in Khalil & Schacher, 1965).

Humans are both the intermediate and definitive host of *L. serrata*, in addition to some other parasitic infections, such as *Taenia solium* and *Trichinella spiralis*.

EPIDEMIOLOGY AND CONTROL. Human infections are acquired in several ways. In Africa, where infections may be common, snakes are often handled, and undercooked python meat is considered to be a delicacy. Water contaminated with eggs or vegetables washed with contaminated water may be sources of infection. In the United States the human visceral infections probably were acquired by eating inadequately cooked rattlesnake meat. Adequate cooking of meat and drinking filtered water prevents human infections.

Other pentastomes of interest include *Reighardia*,

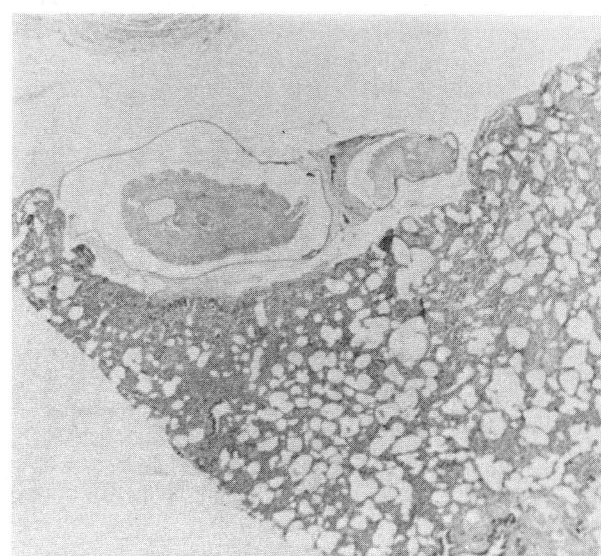

**FIG. 47.6**
Section of a *Porocephalus* nymph encysted on a monkey lung.

which is found in the air sacs of gulls and terns, *Armillifer* spp., which are similar to *Porocephalus* and can cause human infections, and *Linguatula*. The latter is found in nasal passages and frontal sinuses of canids and felids throughout the world. Intermediate hosts, including cattle, sheep, rabbits, and humans, ingest eggs from contaminated water or vegetation, leading to visceral pentastomiasis.

## CLASSIFICATION OF THE PHYLUM PENTASTOMIDA

**Order Cephalobaenida**
Mouth anterior to hooks; hooks without fulcrum; female gonopore at anterior end of abdomen.
*Cephalobaena, Reighardia*

**Order Porocephalida**
Mouth between hooks; hooks with fulcrum; female gonopore near posterior end of body.
*Armillifer, Linguatula, Porocephalus*

## REFERENCES AND FURTHER READINGS


Barnes, R. D. 1980. *Invertebrate Zoology*, 4th ed. W. B. Saunders, Philadelphia.

Demaree, R. S., Jr. 1973. Ultrastructure of pentastome nymph integument. *Proc. 31st Ann. Mtg. Electron Micr. Soc. Am.*, pp., 502–503.

Esslinger, J. H. 1962a. Development of *Porocephalus crotali* (Humboldt, 1908) (Pentastomida) in experimental intermediate hosts. *J. Parasitol.* **48**:452–456.

——. 1962b. Morphology of the egg and larva of *Porocephalus crotali* (Pentastomida). *J. Parasitol.* **48**:457–462.

Fain, A. 1975. The Pentastomida parasitic in man. *Ann. Soc. Belg. Med. Trop.* **55**:59–64.

Heymons, R. 1935. Pentastomida. In *Dr. H. G. Bronns Klassen und Ordmangen des Tierreichs*, Vol. 5. Akademische Verlagsgellschaft M. B. H., Leipzig.

Khalil, G., and J. F. Schacher. 1965. *Linguatula serrata* in relation to halzoun and the marrara syndrome. *Am. J. Trop. Med Hyg.* **14**:736–746.

Mapp, E. M., H. M. Pollack, and L. H. Goldman. 1976. Roentgen diagnosis of *Armillifer armillatus* infestation (porocephalosis) in man. *J. Nat. Med. Assoc.* **68**:198–200.

Mill, P. J., and J. Riley. 1972. An ultrastructural study of the body wall muscles of the pentastomid *Reighardia sternae* Dies. 1864 parasitic on the herring gull *Larus argentatus* Pontopp. *Ztschr. Zellforsch.* **130**:12–17.

Penn, G. H., Jr. 1942. The life history of *Porocephalus crotali*, a parasite of the Louisiana muskrat. *J. Parasitol.* **28**:277–283.

Prathap, K., K. S. Laws, and J. M. Bolton. 1969. Pentastomiasis: A common finding at autopsy among Malaysian aborigines. *Am. J. Trop. Med. Hyg.* **18**:20 27.

Riley, J. 1981. An experimental investigation of the development of *Porocephalus crotali* (Pentastomida:Porocephalidae in the western diamondback rattlesnake (*Crotalus atrox) Int. J. Parasitol.* **11**:127–132

Riley, J., A. A. Banaja, and J. L. James. 1978. The phylogenetic relationships of the Pentastomida: The case for their inclusion within the Crustacea. *Int. J. Parasitol.* **8**:245–254.

Schacher, J. F., S. Saab, R. Germanos, and N. Boustany. 1969. The aetiology of halzoun in Lebanon: Recovery of *Linguatula serrata* nymphs from two patients. *Trans. Roy. Soc. Trop. Med.* **63**:854–858.

Self, J. T. 1969. Biological relationships of the Pentastomida. A bibliography on the Pentastomida. *Exp. Parasitol.* **24**:63–119.

——. 1972. Pentastomiasis: Host responses to larval and nymphal infections. *Trans. Am. Micros. Soc.* **91**:2–8.

Self, J. T., H. C. Hopps, and A. O. Williams, 1972. Porocephaliasis in man and experimental mice. *Exp. Parasitol.* **32**:117–126.

——. 1975. Pentastomiasis in Africans. *Trans. Geo. Med.* **27**:1–13.

Self, J. T., and R. E. Kuntz. 1967. Host–parasite relations in some pentastomida. *J. Parasitol.* **53**:202–206.

Spencer, W. B. 1892. The anatomy of *Pentastomum teretiusculum* (Baird). *Quart. J. Micros. Sci.* **34**:1–73.

Trainer, J. E., Jr., J. T. Self, and K. M. Richter. 1975. Ultrastructure of *Porocephalus crotali* (Pentastomida) cuticle with phylogenetic implications. *J. Parasitol.* **61**:753–758.

von Haffner, K. 1977. Systematic position and the evolution of the Pentastomida based on new comparative research. *Zool. Anz.* **199**:353–370.

Wingstrand, K. G. 1972. Comparative spermatology of a pentastomid, *Raillietella hemidactyli,* and a branchiuran crustacean, *Argulus foliaceus,* with a discussion of pentastomid relationships. *Kong. Danske Vidensk. Selsk. Biol. Skrift.* **19**:1–72.

# PART 5

# *Arthropods*

48

# Introduction to Arthropods

The arthropods are so ubiquitous that they surround us and impinge on nearly every aspect of our lives. When we consider that there are about 800,000 described species in water, on land, and in the air, it can be appreciated that we live in a cloud of insects, mites, crustacea, and spiders. We depend on these creatures for food, pollination of crops, and control of unwanted pests, as well as for disposing of various kinds of dejecta in our environment. In this book, we are concerned with those arthropods that are harmful to humans either directly or indirectly, or with those that demonstrate some principles of symbiosis. Despite our limited view of arthropods here, most are not harmful, and many serve beneficial roles in recycling, controlling populations of other organisms, or as food. As we have discussed in connection with other organisms, arthropods often become problems when the environment is somehow altered by human activities.

The organisms in the phylum Arthropoda belong to diverse groups, but they have some features in common:

1. Bilateral symmetry.
2. Chitinous exoskeleton with jointed legs.
3. Growth by molting, which is controlled by hormones.
4. A complete digestive tract extending from an anterior mouth to a posterior anus.
5. A nervous system consisting of an anterior set of ganglia and commissures, which extend around the esophagus and pass posteriorly as two fused chains of ventral ganglia.
6. True segmentation or metamerism. Primitively, each segment had a pair of legs, neural ganglia, probably an excretory unit, and a set of muscles. In primitive arthropods there is little difference among the segments along the length of the animal (homonymous metamerism), but in more advanced forms, there is movement toward specialized changes in segments (heteronomous metamerism) or the merging of segments into distinct body parts (tagmatosis).
7. The body cavity is a hemocoel and the circulatory system is open. In an open circulatory system, blood moves into the heart through openings or ostia, and is pumped out to various parts of the body, where it leaves the vessels and bathes the tissues directly.

Let us consider the consequences of some of the characteristics of the arthropods. The exoskeleton and mode of growth are fundamental influences on what arthropods are and do. Chitin (an acetylated glucosamine) occurs principally in arthropods, but also occurs in various other animal groups and fungi. It is a tough, flexible compound that is remarkably resistant to bacterial action but is eventually recycled. The hard exoskeleton associated with animals such as crab and lobster is chitin impregnated with calcium salts added for rigidity. In many small arthropods or insect larvae, the exoskeleton is tough but not rigid because of the presence of tanned proteins.

There are two areas in which problems are associated with an exoskeleton: size and growth. The largest arthropods, such as the king crab, are marine. The terrestrial arthropods do not have seawater to support the body and are generally rather small; some tropical insects may approach 15 cm in length, but they are exceptional.

The exoskeleton is a complex structure; exquisite control is required in the formation of a new one at molting. All of the new external structures (Fig. 48.1 *A,B,C*) must be formed below the old exoskeleton and produced in such a way that the next stage can be larger (Fig. 48.2). The molt must be controlled so that it is coordinated and occurs relatively quickly. During the molt, the animal struggles to shed the old exoskeleton; because the new exoskeleton is soft for a short time, the animal is more susceptible to predation than at other times.

Hormones control molting and also differentiation of the body as it matures sexually. Hormonal systems differ in various groups of arthropods; insects are representative of the way control is effected. One hormonal system in insects controls molting and another controls sexual maturity. The stimuli to molt and to mature originate in the brain of the insect (Fig. 48.3). Brain hormone has the prothoracic gland as its target; this gland secretes ecdysone, which causes the new exoskeleton to form and the old one to be shed. Juvenile hormone (JH), secreted by the corpora allata, maintains the insect in a preadult stage. As the amount of JH decreases, the insect moves to the next stage of its life cycle. In the example given of a moth, the caterpillar moves to the pupal stage; when JH is almost gone, the pupa becomes an adult moth.

Much of the basic work on the control of molting and maturation of insects was done by V. B. Wigglesworth in Britain with *Rhodnius prolixus,* a vector of *Trypanosoma cruzi,* and by Carroll Williams at Harvard in the United States with the cecropia moth. The puzzle was fitted together by these two investigators and their students in a series of experiments done mostly in the 1940s. Their aim was to describe the mechanisms by which these extraordinarily complex processes of molting and maturation were controlled. They succeeded in what appeared to be solely an exercise in academic research, but the practical values are now being realized. Some of the third-generation insecticides (Chap. 50) are based on JH mimics. JH is a terpene with a chain of 15 carbon atoms; its synthesis led to the synthesis of JH mimics, compounds sufficiently similar to JH that they have some hormonal activity. A JH mimic (methoprene) is now used to control some insects; its effect is to prevent development in the insect beyond a preadult stage and to cause maldevelopment.

Metamerism is associated with the exoskeleton and with most of the internal organs; it is fundamental to arthropod structure. The arthropods probably arose from wormlike forms intermediate between the arthropods and the annelids (earthworms, polychaetes, leeches) such as *Peripatus* (class Onychophora). Considerable modification has occurred in structure and development; however, segmentation is retained in a number of systems:

1. The nervous system has a pair of ganglia in each segment.
2. Muscles are associated with each segment.
3. Openings to the respiratory system are often present in each segment.
4. The exoskeleton usually retains its segmentation, which can be seen in its primitive condition during embryogenesis.

The digestive tract of arthropods is complete and has regional differentiation. There are various specialized structures in the gut, depending on the type of food the animal takes, but the basic plan is the same in all.

If we go beyond these basic similarities, there is a great deal of diversity in the structures and modes of development in various arthropod groups. Excretory systems are associated with the gut (insects), the antennae (crustacea), or the legs (arachnids). Respiration, that is, gaseous exchange, takes place through the whole body surface in many small species through a tracheal system (insects), wet gills (crustacea) or book gills or a dry surface (arachnids), among others. Light receptors also vary from simple eyes in spiders to compound eyes in the crustacea, and both in the insects. Other sense organs allow the reception of air movement (such as hairs in insects), orientation (statocysts in crustacea), or carbon dioxide (ticks and some insects).

Development is even more diverse and its patterns bear little relationship to each other. Copulatory and reproductive structures aside, embryogenesis and preadult patterns of development in the various groups are quite different. The crustacea have seven different

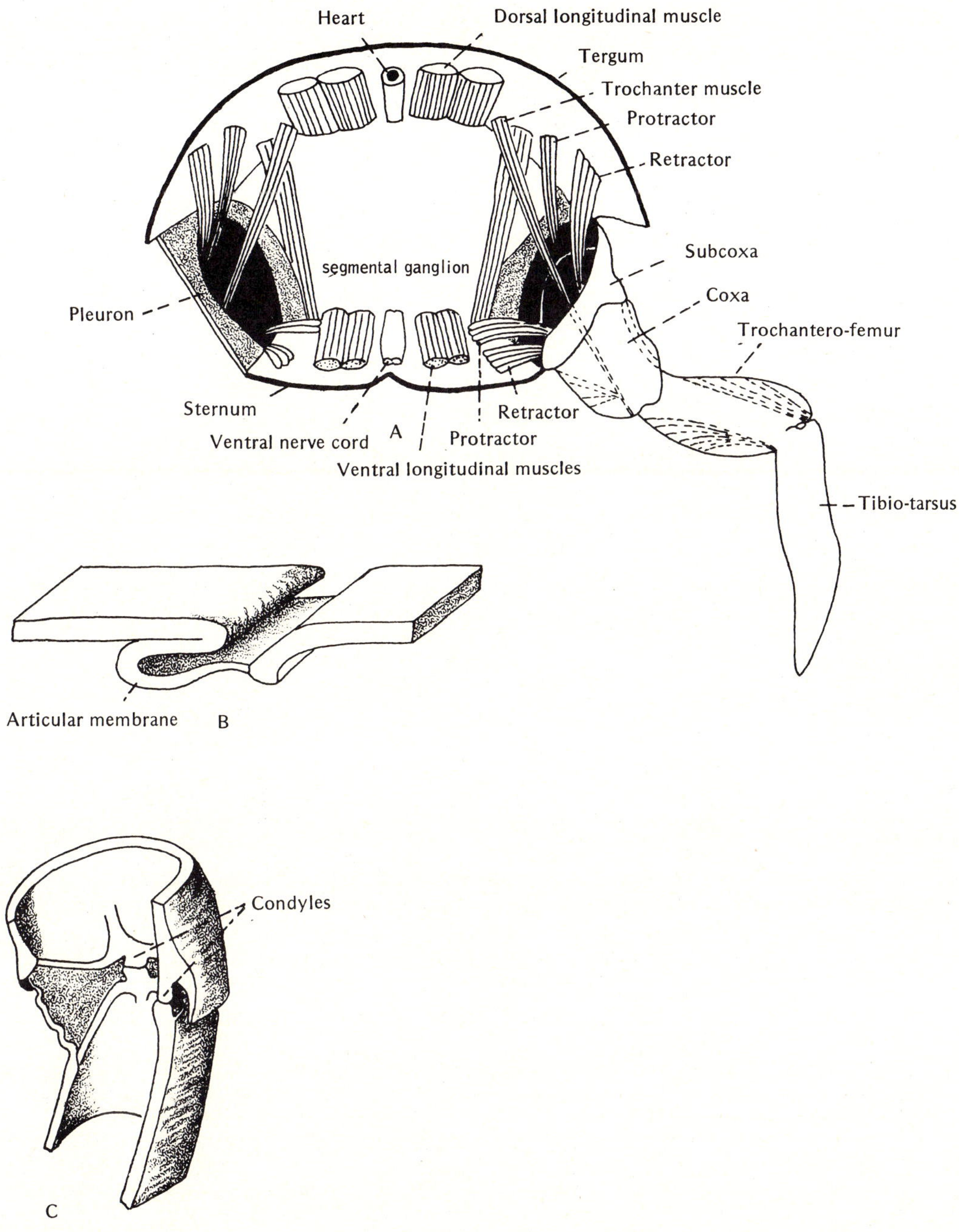


**FIG. 48.1**
General body structure of arthropods. (*A*) Transverse section; (*B*) intersegmental articulation; (*C*) appendicular articulation. [Redrawn with modifications from Barnes, R.D., 1963. *Invertebrate Zoology.* W.B. Saunders, Philadelphia.]

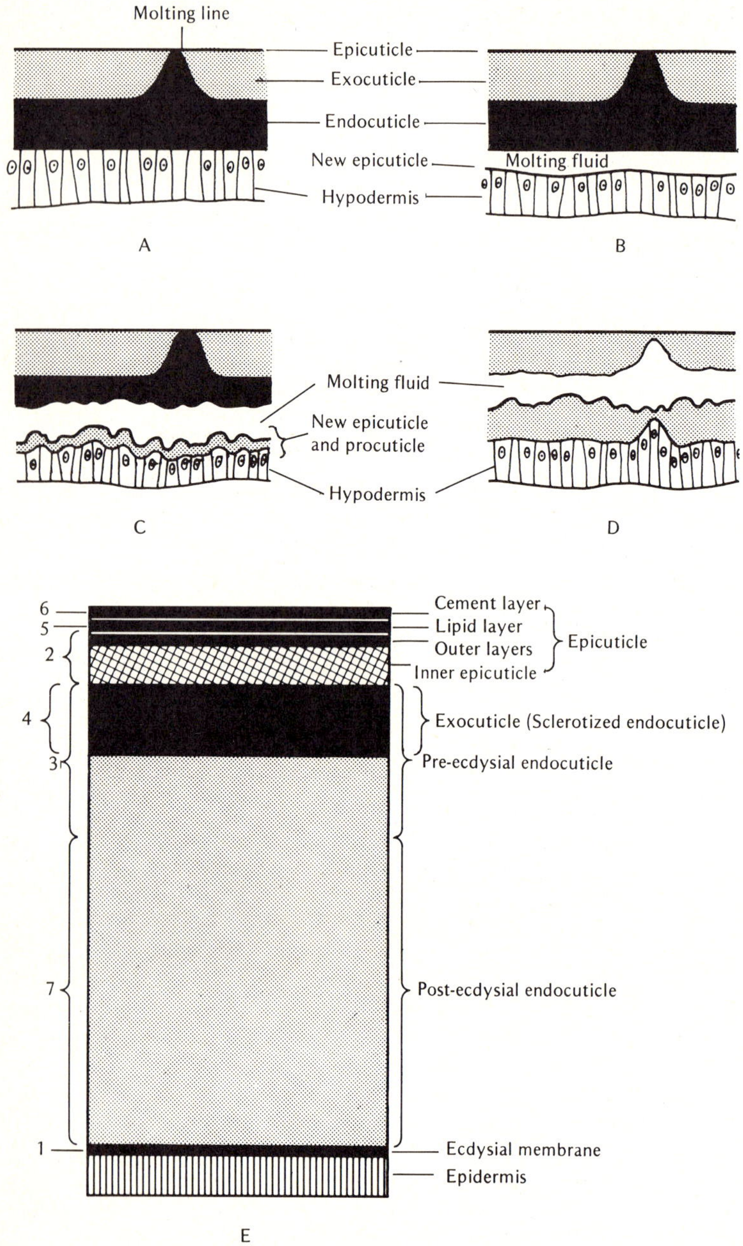


**FIG. 48.2**
Structure of the exoskeleton and molting in arthropods. (*A*) The fully formed exoskeleton and underlying epidermis between molts; (*B*) separation of the epidermis and secretion of molting fluid and the new epicuticle; (*C*) digestion of the old endocuticle and secretion of the new procuticle; (*D*) the animal just before molting, encased within both new and old skelton; (*E*) sequence in which different parts of the cuticle are deposited. [Adapted from Barnes, 1980.]

larval forms (Fig. 49.3), but these do not occur in all life cycles and a few additional forms occur in some groups. The general pattern is for the organism to become more complex and more like the adult at each molt. However, in spiders the body form in the early molts is about the same as in the adults, and the adults continue to molt and to grow larger. The insects, on the other hand, reach the winged adult stage and do not molt again (see

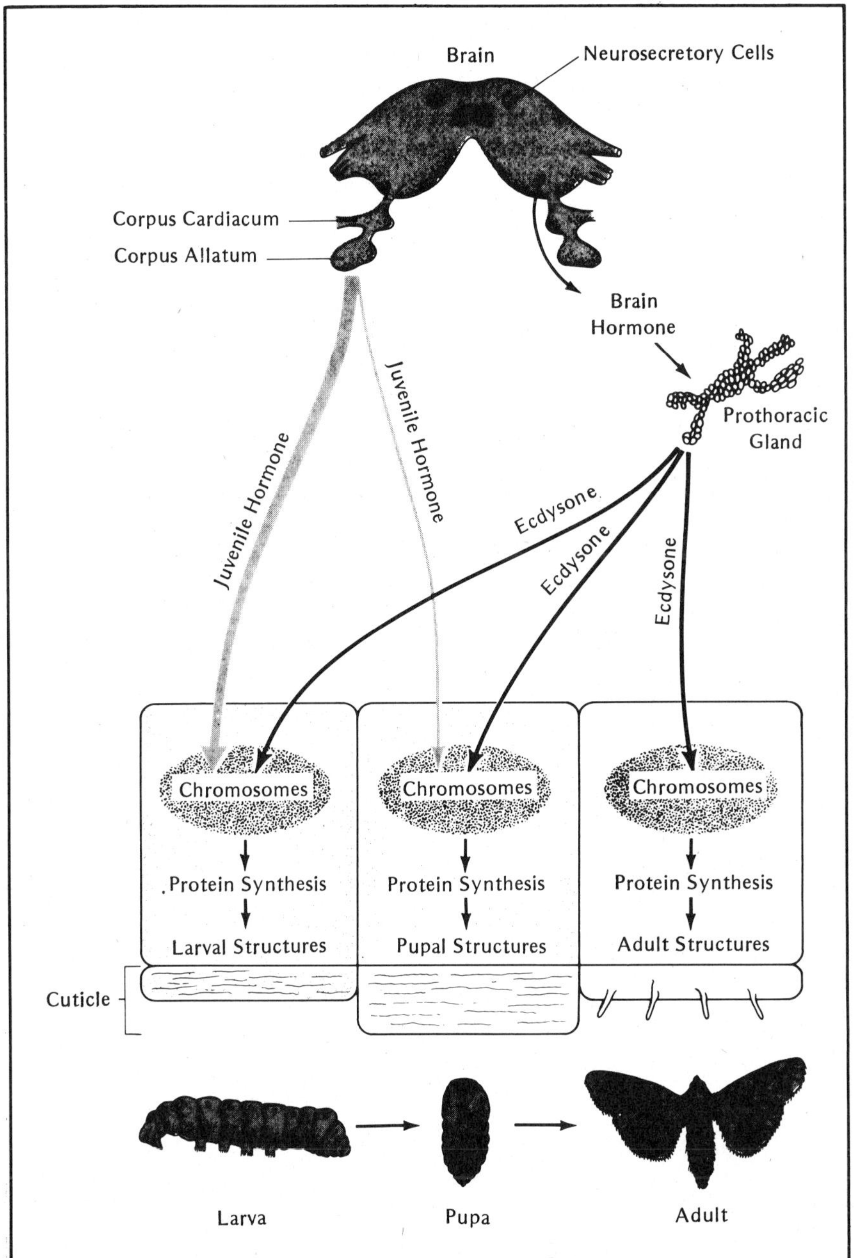


**FIG. 48.3**
Control of molting and maturation in insects.

Fig. 50.10). Taking into account both adult structures and the patterns of development, the consensus is that the arthropods are polyphyletic, that is, they arose from more than a single stem form, most likely as many as four (Manton, 1973, 1977). The classification generally used (see end of this chapter) contains four subphyla, representing the concept of four different origins of the phylum.

## VENOMOUS ARTHROPODS AND ALLERGIC REACTIONS TO ARTHROPODS

A wide array of arthropods causes toxic reactions in vertebrates. The cause of intoxication may be direct (bites, stings, defensive secretions) or indirect because of hypersensitivity (allergy). The effect of such contact with arthropods may be a mild wheal such as results from a mosquito bite, or a moderately severe reaction such as occurs following the bite of a black widow spider, or death due to an anaphylactic reaction, such as in a person hypersensitive to bee stings. Although we are principally concerned with the responses in humans or in animals important to humans, the same effects occur in other vertebrates or (except for hypersensitivity) invertebrates.

Arthropods use various secretions in the process of obtaining food or in defending themselves from predators. Toxicants are found all through the phylum, but those of greatest importance are in the insects, arachnids, and centipedes. Local reactions often occur in animals bitten by bloodsucking arthropods such as insects and ticks. A bloodsucking arthropod must have mouthparts capable of penetrating or lacerating the skin. The clotting mechanism would immediately close the wound and stop the flow of blood unless there were some mechanism to prevent it. Organisms such as mosquitoes, blackflies, fleas, ticks, and mites inject salivary fluid into the wound in order to prevent clotting or to liquefy the tissues. These arthropods' mouthparts differ in origin and function (Harwood & James, 1979), but all of them inject fluid to keep the blood from clotting or to liquefy the tissue. Such secretions are mixtures of substances, but they contain proteins, some of which are enzymatic. They produce an inflammatory response and an itching papule on the skin. If there are only a few bites, the host may be uncomfortable but not significantly affected. If there is massive attack by, say, blackflies (Chap. 54), there may be severe intoxication from the injection of salivary fluid causing prostration or even death. Such occurrences fortunately are rare.

Predaceous arthropods have a number of mechanisms for capturing and immobilizing prey. Beetles often have strong mandibles, which they use to crush their prey; the effect is essentially mechanical. Alternatively, piercing mouthparts (Hemiptera, spiders, centipedes) or stingers at the posterior end (Hymenoptera, scorpions) are specifically constructed to pierce the integument of the prey and to inject an immobilizing toxin. These same structures and toxins are used defensively against people who handle or inadvertently come in contact with these animals. Usually, when a mammal is bitten or stung, the damage and the pain remain localized. This is true when a person is bitten by an insect such as the wheel bug (*Arilus*), which is common in gardens in temperate and warm climates; the pain is likely to be severe, but there is no long-term effect. Occasionally, plant-feeding insects such as thrips, leafhoppers, or boxelder bugs will bite humans, and considerable discomfort together with skin irritation may result, but serious effects are lacking. On the other hand, severe skin lesions can be seen following the bites of some bloodsuckers such as soft ticks or kissing bugs, but most dramatically in the case of the brown recluse spider, *Loxosceles reclusa*, and several related species (Chap. 56). The bite of the brown recluse may be so minor that it is not noticed initially, but within a few hours the site of the bite blisters; there may be systemic reactions causing the person to become prostrated. Death is unlikely except in small children. The long-lasting effects result from extensive tissue destruction in the vicinity of the bite; necrosis of cutaneous and subcutaneous tissue causes extensive scarring, which may require skin grafting.

The black widow spider, *Lactrodectus mactans*, and several related species produce severe systemic reactions caused by injected neurotoxins. There is swelling at the site of the bite, but more severe effects are generalized: muscular pain, nausea, difficulty in speech and breathing, and sweating. The kidneys, liver, and spleen may be damaged. In adults, the effects may last 2 or 3 days; death, if it occurs, is usually in children, aged persons, or those who have cardiovascular disease. The probability of death resulting from a black widow bite is less than 4% (Harwood & James, 1979).

A peculiar type of systemic intoxication is tick paralysis, discussed in detail in Chapter 57. Generally associated with hard ticks, tick paralysis affects mammals by causing a flaccid paralysis that may lead to death if the breathing center in the brain is involved. A few ticks may paralyze even a fully grown animal such as a domestic ox.

Many arthropods have defensive secretions that serve to fend off would-be predators; often these chemicals are exuded from one of the body openings, from between the joints of the skeleton, or from the tips of body hairs. In a few instances, such as in the millipedes or whipscorpions and apparently a few Lepidoptera, there is forcible ejection of secretions. *Rhinocrichus latespargor*, a millipede found in Haiti, is said to eject fluid up to 82 cm (32 in.). Many Hymenoptera use their stings defensively in addition to capturing prey. Defensive secretions may have a bad odor, which serves to drive off the predator; they may cause local irritation of the skin, mucous membranes, or eyes, or in a few instances they may have systemic effects.

Some common insects that have defensive secretions are Lepidoptera larvae and beetles. The hairs or setae on the bodies of lepidopteran larvae are frequently defensive; they may be irritating in themselves, or they may be hollow and secrete a chemical from a gland at the base of the hair. Blister beetles (family Meloidae) exude a toxicant from between the joints on the legs when they are disturbed. This secretion produces a severe burning sensation on the skin and on the mucous membranes of the eye, mouth, or intestine; not only severe discomfort but blistering of the epithelial surface occurs. Large animals such as horses may be killed by ingesting blister beetles baled with hay; they die from severe ulceration of the digestive tract.

Some persons who have been stung by a bee or wasp may become allergic or hypersensitive to the venom. When such a hypersensitive individual is stung again, a violent allergic reaction or anaphylaxis may result. The person may lose consciousness within a few minutes of being stung and, in a very strong reaction, may die within minutes unless epinephrine is administered. About 200 deaths from anaphylactic reactions to stings of bees or wasps are reported annually in the United States.

Hypersensitivity is usually manifested by upper respiratory symptoms or by stronger than normal skin reactions. Children particularly may become allergic to mosquito bites and have excessive swelling of the skin at the site of bites. The housedust mite, *Dermatophagoides farinae*, is one cause of housedust allergy usually manifested as asthma. Insect scales from lepidopterans may also cause hay fever or asthma.

## INSECTICIDES AND ACARICIDES

Populations of parasitic agents become sufficiently high to cause economic losses or medical problems for a variety of reasons (Chap. 1). Not infrequently high populations result from some sort of interference by humans with natural controlling factors. The solution is to interfere additionally to reduce populations to tolerable levels again. All kinds of arthropods may become problems to humans; some examples are barnacles on ships, spiders or scorpions in houses, and sowbugs in the garden. The greatest problems are associated with insects and mites, and most of the principles and techniques of control have been developed for those groups of pests and vectors. Eight principal methods have been developed for arthropod control:

- **A.** Toxicants
  - **1.** Direct
  - **2.** Chemosterilants
- **B.** Biological agents
  - **1.** Predators (invertebrates and vertebrates)
  - **2.** Parasites (viruses, prokaryotes, eukaryotes)
  - **3.** Parasitoids (mostly Hymenoptera)
- **C.** Genetic
  - **1.** Selection for resistant hosts
  - **2.** Autocidal methods
    - a. Release of genetically altered individuals
    - b. Sterile male (and/or female) technique
- **D.** Behavioral modification
  - **1.** Chemical
  - **2.** Physical
- **E.** Environmental modification
- **F.** Immunization
- **G.** Quarantine
- **H.** Test and slaughter

All of these methods are discussed in this text with the various arthropods used as examples, but the general method of applying toxicants recurs as a theme with nearly all of the insects and mites. Literally thousands of toxicants have been and are in use for controlling insects and mites. They are generally spoken of as *insecticides* and *acaricides*, but the term *insecticide* is often used to cover both. The names, formulations and characteristics of the products are so numerous that one is easily overwhelmed. On the other hand, if we take a historical approach, a pattern of development and characteristics can be seen.

## FIRST-GENERATION INSECTICIDES

First-generation insecticides came into use through long-term trial and error in dealing with insect invasions. Although some are still in use, most were developed before 1940. Heavy metals, smothering agents such as petroleum products, and chemicals obtained from plants are included among the early products.

Arsenic was used early in insect control; a mixture known as *Paris green* was in use for a long time against larval mosquitoes. Arsenic has the disadvantage of being toxic and is generally avoided now.

Petroleum products—either petroleum itself or refined fractions—were used early in this century, sometimes being applied directly to animals infected with mites. A major use of diesel oil has been in oiling sites of larval development of mosquitoes. Petroleum can be severely toxic when ingested and is no longer used on animals.

Pyrethrins and rotenone are chemicals obtained from plants. Pyrethrins are refined from a chrysanthemum grown principally in the Middle East and the highlands of Kenya. They have a low mammalian toxicity and immediately knock down insects; their disadvantage is that many insects that are knocked down later get up. Synthetic pyrethrins were developed in the 1970s; these chemicals have the same low mammalian toxicity, but insect kill is better. Rotenone, obtained from the root of the derris plant, has not been used much to control insects but has its greatest use in killing fish.

## SECOND-GENERATION INSECTICIDES

In the 1930s, phenothiazine was found to be an efficient insecticide, and it was used fairly widely. It has the disadvantage of causing skin problems in some humans and animals. The watershed in insecticide technology came with the development of DDT in the late 1930s.

### *Organochlorine or Chlorinated Hydrocarbon Insecticides.*

The organochlorine compounds are mostly cyclic compounds with three or more chlorines. DDT was the first to be developed; it was developed by Geigy in Switzerland and patented in 1940. It has a broad spectrum of activity, that is, it acts against many different insects and it has long residual action. Its first significant use in medical entomology was against mosquitoes and then human lice during World War II. Later, it came into widespread use in controlling not only insects of medical and veterinary importance, but also insect pests of crops. Problems with DDT began surfacing in the late 1940s and early 1950s when both insecticide resistance and food-chain residues became apparent. By about 1960, especially with the publication of Rachel Carson's *Silent Spring*, efforts were begun to limit the use of DDT. Other organochlorines are methoxychlor, toxaphene, lindane (the gamma isomer of benzene hexachloride), and a group of closely related compounds including aldrin, dieldrin, and heptachlor. All of the organochlorines have long residual action and therein lies the dilemma. Their long residual action made them invaluable additions to our armamentarium of insecticides, but it allowed contamination of the environment that was finally considered to be intolerable. Organochlorines are now used in the United States only for limited applications.

### *Organophosphate Insecticides (OPs)*

The organophosphates were an outgrowth of nerve gases developed during World War II. They have the following general structure:

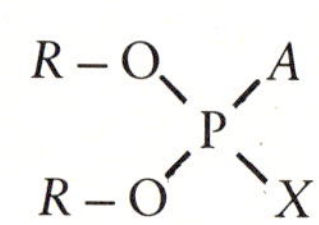


in which $A = \text{O or S}$
$X = \text{variable}$
$R = \text{Me or Et}$

These compounds act by blocking cholinesterase; the result is an accumulation of acetylcholine at synapses and ultimately failure of the nervous system to operate. The OPs have relatively high mammalian toxicity, but they break down rapidly in the environment, particularly under alkaline conditions. Their especially useful characteristic is that they can be absorbed by vertebrates and act against arthropod parasites through the blood; accordingly, they have been called systemic insecticides. Their use against arthropod and helminth parasites of livestock has been widespread; no systemics are used against parasites of humans. Examples of OPs are parathion, malathion, TEPP, coumaphos, ronnel, and trichlorphon.

### *Carbamates*

These compounds have the following general structure:

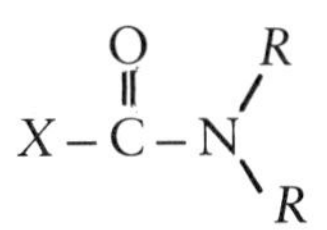

in which $X$ = variable
$R$ = Me or H

The carbamates act in the same way as the organophosphates, that is, by inhibiting cholinesterase. Although their action is the same as that of the OPs, they do not generally act systemically; however, aldicarb is absorbed and has some use against arthropod parasites of domestic animals. The spectrum of activity of individual carbamates is usually narrow, or at least not as broad as that of other second-generation insecticides. Carbamates break down readily in the environment and do not leave harmful residues.

## THIRD-GENERATION INSECTICIDES OR INSECT GROWTH REGULATORS (IGRs)

Earlier in the chapter, in the discussion of molting in arthropods, we alluded to third-generation insecticides that had grown out of the work by Wigglesworth in Britain and Williams in the United States. Juvenile hormone acts to maintain the animal in a preadult stage; when JH drops below a certain level, the adult stage emerges at the next molt, or pupation takes place. It was reasoned that it would be possible to affect development adversely by exposing preadult insects to compounds similar, but not identical, to JH. Thus, the JH mimics were born, and the first product to be cleared for use against insects was methoprene. In the general area of medical entomology, methoprene can be used for mosquito control where there is no contact with food crops, or it can be fed to cattle to control flies in manure (Chap. 55). This latter use is being reconsidered because breakdown of feces is slowed.

Methoprene acts principally by preventing the maturation of insects; its effect may not be seen for at least a week, because insects in late development are not affected. Good control of some insects has been achieved with methoprene; for example, midges (Diptera, Chironomidae) arising from some lakes have been successfully controlled with it.

The other product now available is diflubenzuron, which prevents synthesis of chitin. Because chitin is found principally in arthropods, a chemical that interferes with chitin synthesis will prevent their normal development; insects cannot molt properly, or they die of starvation. Diflubenzuron is marketed as Dimilin, Difluron, and Th-604; it has been cleared for use against mosquitoes where there is no contact with food crops.

With the third-generation insecticides, we have moved into an era when some rationality can be applied to searching for active compounds. The first-generation insecticides were found by trial and error and the second-generation ones by screening large numbers of chemicals against standard test animals in the laboratory. The approach was methodical but blind in the sense that chemicals were (and are) screened for activity without any rational basis. It is hoped that third-generation insecticides will target narrowly on a single or few species, be nonpolluting, and have long useful lives, since resistance probably will not develop.

## CLASSIFICATION OF ARTHROPODS

**Phylum Arthropoda—arthropods**
- Subphylum **Trilobita**—trilobites (fossils only)
- Subphylum **Chelicerata**
  - Class **Merostomata**
  - Subclass **Xiphosura**—horseshoe crabs

Subclass **Eurypterida**—eurypterids (fossils only)
Class **Arachnida—arachnids**
Class **Pycnogonida**—sea spiders
Subphylum **Crustacea** (Mandibulata in part)—crustaceans
Class Branchiopoda
Class Copepoda
Class Ostracoda
Class Cirrepedia
Class Malacostraca
Subphylum **Uniramia**
Class **Onychophora**—onychophorans
Class **Diplopoda** (Myriapoda in part)—millipedes
Class **Chilopoda** (Myriapoda in part)—centipedes
Class **Pauropoda** (Myriapoda in part)—pauropods
Class **Symphyla** (Myriapoda in part)—symphylans
Class (Insecta) **Hexapoda**—insects

## REFERENCES AND FURTHER READINGS

Barnes, R. D. 1980. *Invertebrate Zoology*, 4th ed. Saunders/Holt, Rinehart & Winston, Philadelphia.

Carson, R. 1962. *Silent Spring*. Houghton Mifflin, Boston.

Harwood, R. F., & M. T. James. 1979. *Entomology in Human and Animal Health*, 7th ed. Macmillan, New York.

Herried, C. F. 1977. *Biology*. Macmillan, New York.

Manton, S. M. 1973. Arthropod phylogeny—A modern synthesis. *J. Zool.* **171**:111–130.

———. 1977. *The Arthropoda: Habits, Functional Morphology and Evolution*. Clarendon Press, Oxford.

Romoser, W. S. 1981. *The Science of Entomology*, 2nd ed. Macmillan, New York.

Smith, K. G. V. (ed.). 1973. *Insects and Other Arthropods of Medical Importance*. British Museum of Natural History, London.

# 49

# *Crustacea*

Almost everyone has seen movies or still pictures of "cleaner fishes" busily removing parasites from various other fish species, including normally highly predatory ones. Many of the parasites being removed belong to the subphylum Crustacea. Because this cleaning behavior is so widespread, one can appreciate how important the removal of these pathogenic parasites must be to the infected hosts. As the world's population increases, aquaculture will play a greater role in meeting the need for more protein. Therefore, it is a good idea to gain some understanding of crustacean parasites that can have a significant impact on the aquacultural industry.

Over 30,000 species are known in the subphylum Crustacea, including the well-known and delicious lobster, crab, and shrimp, as well as the lesser known smaller species that form a significant part of the freshwater and marine food chains and the parasitic species. Indeed, many parasitic species have free-living larval stages that are part of the food chain.

The subphylum Crustacea (hard shell) has a number of features distinctive from those of the other arthropods; it is the only major subphylum that is primarily aquatic, has gills for breathing, two pairs of antennae, one pair of mandibles, two pairs of maxillae, and lacks Malpighian tubules. The subphylum is usually divided into the class Malacostraca, which contains the larger organisms such as lobsters and crabs, and the Entomostraca, a group of seven classes, but without formal taxonomic status, containing the smaller organisms.

Crustaceans typically have three tagmata, or body divisions: head, thorax, and abdomen. Often these regions are difficult to distinguish because of fusion of various parts. The head typically bears five pairs of appendages: two pairs of antennae, one pair of mandibles, and two pairs of maxillae. In addition, two types of eyes may be present; the median or nauplius eye is a central, relatively simple structure probably used primarily for orientation, and the much more complex compound or lateral eyes have complex visual discrimination abilities.

Appendages on the thorax are for swimming, and some, such as the maxillipeds, may be modified for feeding. In the primitive state, crustacean appendages were biramous (two-branched), but numerous modifications to this basic plan are seen, especially in the parasites.

## CLASS COPEPODA

The Copepoda (oar foot) is the largest class of Entomostraca, with about 7500 species; they form an important part of aquatic food chains. In addition, both free-living and parasitic copepods are of interest to the parasitological world. Free-living copepods (Fig. 49.1) are intermediate hosts for tapeworms (*Diaptomus* and *Cyclops* for *Diphyllobothrium* spp.), nematodes (*Cyclops* for *Dracunculus medinensis*), and acanthocepha-

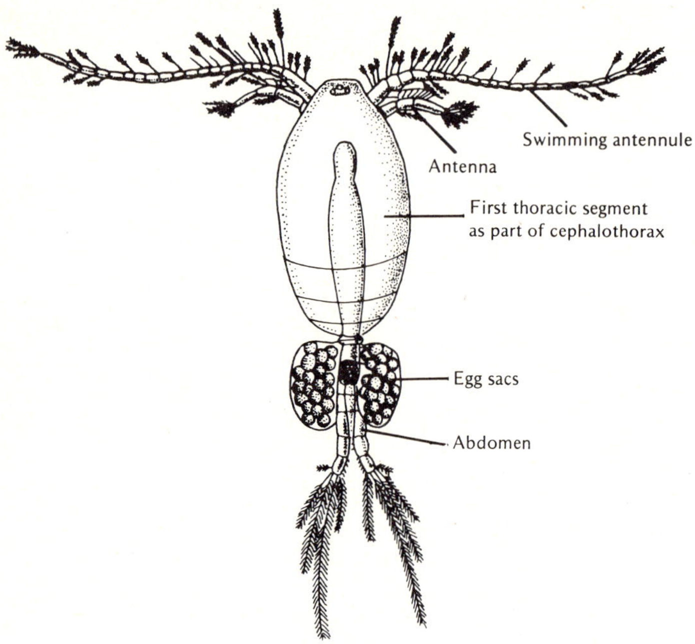


**FIG. 49.1**
Female of *Cyclops*. [Victor Schechter, *Invertebrate Zoology*, © 1959, p. 373. Reprinted by permission of Prentice-Hall, Englewood Cliffs, N.J.]

lans. Several parasitic crustaceans are serious economic problems in commercial fish hatcheries (Kabata, 1970).

The parasitic crustaceans are fascinating organisms with all sorts of modifications for the parasitic way of life. Little is known about their biology, which presents numerous challenges to parasitologists of the future. An excellent discussion of the parasitic copepods, their biology, and what still remains to be done is given by Kabata (1981). We will discuss only a few of the better known species to illustrate basic biological features typical of this diverse group.

NAME OF ORGANISM AND DISEASE CAUSED. *Ergasilus* spp. causes ergasilosis.

HOSTS AND LOCATION. *Ergasilus* spp. are found worldwide; they are serious problems in fish culture and sometimes in nature as well (Hoffman, 1967).

MORPHOLOGY. Parasitic female adult *Ergasilus* spp. have few modifications for a parasitic existence and greatly resemble free-living copepods. Their most obvious modification for parasitism is having greatly enlarged second antennae, which terminate in large claws (Fig. 49.2*A*); these antennae are used to grasp securely gill filaments. The mouthparts of ergasilids resemble those of free-swimming copepods rather than the highly modified mouthparts of many other parasitic copepods (Fig. 49.2*B*). Swimming legs are biramous (Fig. 49.2*C*); some species have stout spines on the first pair of swimming legs, presumably to dislodge tissue and move it toward the mouthparts.

LIFE CYCLE. Complete life cycles have been worked out for only a few species. Wilson's (1911) monograph is still one of the best sources of information on ergasilids. Probably the best known life cycle is that of *Ergasilus sieboldi* (Russian literature reviewed by Kabata, 1981). Copulation occurs while the males and females are free-swimming; the males then die and the females attach to gills and embryonic development occurs in external egg sacs. There are three naupliar (Fig. 49.2*D*) and five copepodids stages, a single preadult stage, and then the adults. All except the fertilized adult are free-swimming.

DIAGNOISIS. Diagnosis is based on finding a parasitic adult female with greatly enlarged second antennae attached to the gills of fish.

HOST–PARASITE INTERACTIONS. Female ergasilids feed on gill tissue and mucus. Kabata (1981) reviewed

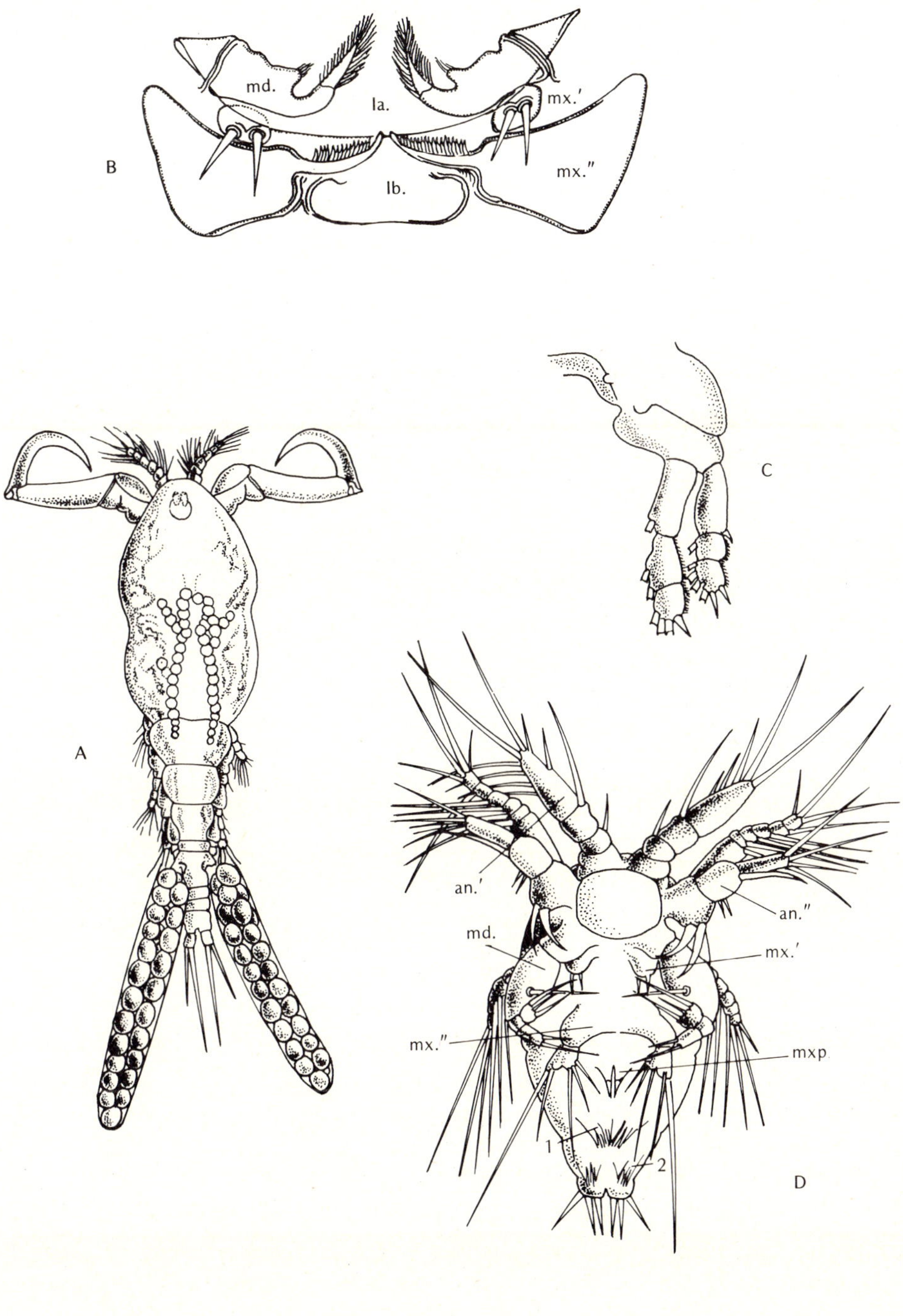


**FIG. 49.2**
*Ergasilus*. *(A)* adult female; *(B)* mouthparts; *(C)* swimming leg; *(D)* nauplius.

**Abbreviations:**

an' = first antenna
an'' = second antenna
la = labium
lb = labrum
md = mandible
mx' = first maxilla
mx'' = second maxilla
mxp = maxillary process

[Redrawn from Wilson, 1911.]

the Polish literature on the feeding of *E. sieboldi*, based on histological examination of parasite gut contents, which were gill epithelium, mucus, erythrocytes, and white blood cells, in that order. Massive infections lead to emaciation, growth retardation, and even death (Bauer, 1962), but infections with a single parasite may have no obvious detrimental effect.

EPIDEMIOLOGY AND CONTROL. Like most other parasitic copepods, *Ergasilus* spp. are problems only when crowding occurs, such as in fish farming. The best control method is not to introduce infected fish into lakes and ponds. Bauer (1962) speculated that the negatively phototropic larvae stay in the vegetation at the bottom; dispersal is reduced in vegetation-rich environments, resulting in only small numbers of copepod infections. Chemical treatments for *Ergasilus* spp. were reviewed by Hoffman and Meyer (1974).

NAME OF ORGANISM AND DISEASE CAUSED. *Lernaea cyprinacea* Linnaeus, 1761, causes anchor worm disease.

HOSTS AND LOCATION. This copepod can probably infect any freshwater fish, as well as frog tadpoles and salamanders (Hoffman, 1967). There is some disagreement about the host specificity of *L. cyprinacea* (Kabata, 1981). Most adult females are embedded near fins, but they can be found almost anywhere on a fish where the copepod can penetrate (Demaree, 1967).

DISTRIBUTION AND IMPORTANCE. This parasite apparently has a worldwide distribution. *L. cyprinacea* may have been introduced into North America with the goldfish, *Carassius auratus* (Bauer, 1962). This parasite is a serious problem in closed environments such as ponds and hatcheries and has been responsible for massive fish kills; in one outbreak 18 tons of carp and 2 tons of goldfish were killed in a 2-week period (Tidd, 1934).

MORPHOLOGY. All stages except the parasitic females have typical copepod structure (Fig. 49.3*A*) and are free-living. The parasitic female embeds her anterior end in fish flesh and grows from about 1.5 mm to over 10 mm in length in about a week. Mature females (Fig. 49.3*B*) lose virtually all resemblance to copepods because growth does not involve mouthparts or swimming legs. The first thoracic segment enlarges greatly, forming the "anchor," which permanently secures the parasite in the host tissues. The anchor consists of two ventral processes and two branched dorsal processes (Fig. 49.3*C*). All thoracic segments enlarge greatly, but the swimming legs do not, so that the legs become difficult to see and the female takes on a wormlike appearance. Mature females have two egg sacs immediately posterior to the pregenital prominence.

LIFE CYCLE. Grabda (1973) reported that there are three naupliar stages, five copepods stages, a preadult, and an adult stage in the life cycle (Fig. 49.3). Earlier literature (Wilson, 1911) suggested that two hosts were needed to complete the life cycle, but in more recent studies both copepodid and adult females were found on the same host, suggesting that at least some can complete the life cycle on one host (Nakai, 1927; Demaree, 1967; Kabata, 1981). In cooler climates there may be only a single generation per year (Russian literature reviewed by Bauer, 1962), whereas in warmer climates the life cycle can be completed in 12 to 14 days (Yashouv, 1959).

DIAGNOSIS. Anchor worms can be seen protruding from various sites on fishes with the naked eye.

HOST–PARASITE INTERACTIONS. Anchor worm infections produce lesions in fish and amphibia ranging from slight to fatal. The head and anchor of adult females are surrounded by host connective tissue. Hemorrhage and ulceration are common; major injury is caused by loss of blood and by secondary infection with bacteria, fungi, and other organisms (Putz & Bowen, 1964). The degree of injury depends on the number of parasites and their attachment sites. Amphibian infections elicit similar responses, except that there is no integumental encapsulation (Tidd & Shields, 1963).

TREATMENT, EPIDEMIOLOGY, AND CONTROL. Anchor worm infections usually result in a single parasite per host fish in flowing rivers and streams and there is little damage (Demaree, 1967), but in closed environments severe infections often result. A number of chemicals kill larvae, but there is no successful treatment for infections with adult females. Thus, prevention is most important; breeding stock should be examined carefully before introducing them into ponds

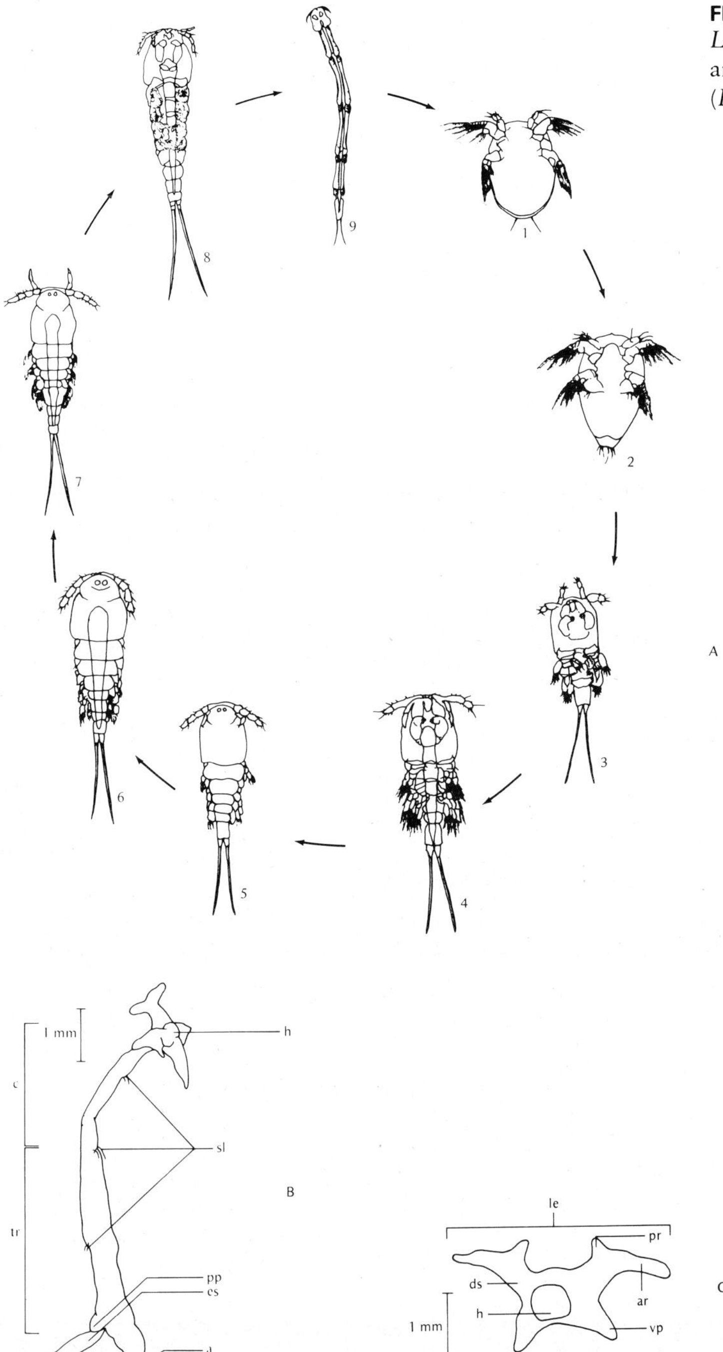


**FIG. 49.3**
*Lernaea cyprinacea.* (*A*) larvae; (*B*) adult female; (C) anchor of adult female. [(*A*) Redrawn from Bauer, 1962; (*B*, C) from Demaree, 1967.]

or hatcheries, and stock fish should be isolated from contaminated waters. Once infections are present, waterways must be drained, chemically treated, and all parasitized fishes destroyed.

NAME OF ORGANISM AND DISEASE CAUSED. *Salmincola edwardsi* (Olsson, 1869) causes gill maggot.

HOSTS AND LOCATION. The principal host in North America is the brook trout, *Salvelinus fontinalis*; infections of many other salmonid fishes are reported worldwide. Like other members of the genus, *S. edwardsi* is a parasite of freshwater fishes, but has been recovered from anadromous fishes when they migrate to salt water. The parasite is found on gills, in the gill cavity, and rarely in other sites.

DISTRIBUTION AND IMPORTANCE. Most members of the genus *Salmincola* have little economic significance, but *S. edwardsi* causes serious damage and has forced the closure of fish hatcheries in the United States (Fasten, 1921).

MORPHOLOGY. Like *Lernaea*, the parasitic adult female of *Salmincola* bears little resemblance to a crustacean. She, too, is permanently anchored to her host. Her anchor, called a *bulla*, is secreted from glands in her head and maxillae. The enlarged maxillae then fuse with the bulla, creating the permanent attachment (Fig. 49.4). Only the bulla is inside the host; the remainder of the parasite is external, including the movable cephalothorax used for "grazing" on host tissues. There is no obvious external segmentation. The males are much smaller than the females, about 0.5 mm, and are not attached with bullae. The unique larva, called a *chalimus*, is described in the life cycle section.

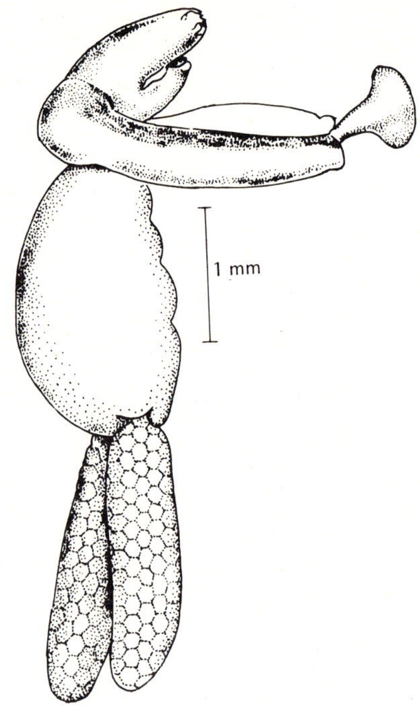


**FIG. 49.4**
Female of *Salmincola* with bulla. [Redrawn from Wilson, 1915.]

LIFE CYCLE. The life cycle of *Salmincola edwardsi* consists of a nauplius, which develops inside the egg, several copepodid stages (chalimus), and the adult. After hatching, the nauplius molts to the free-swimming copepodid stage. This larva spends the daylight hours near the water surface and nighttime near the bottom, a behavior which coincides with that of the brook trout. Upon contact with a fish, the parasite crawls into the gill cavity, rasps a hole in the host tissues with its mouthparts, and then forces the end of the frontal or attachment filament into the cavity. Gluelike secretions of the filament together with the regenerating tissues of the host soon anchor the parasite (Fasten, 1921). The attached larva, the chalimus larva (Fig. 49.5), undergoes several molts while attached by the frontal filament; it loses its segmentation and plumose swimming feet. It takes about 10 days to reach sexual maturity, at which time the males detach from their frontal filaments. A male grasps a female with second maxillae and maxillipeds, and transfers packets of sperm (spermatophores) to her (Wilson, 1915). The male then drops off and dies, while the female breaks free from the frontal filament, secretes a bulla, places it in gill tissue, and attaches her second maxillae to it. The female then develops two egg sacs, which produce nauplii. The complete life cycle of *S. californiensis* was described by Kabata and Cousens (1973).

TREATMENT. There is no satisfactory treatment for adult females, but a number of chemicals including formalin and copper sulfate kill larvae (Hoffman & Meyer, 1974).

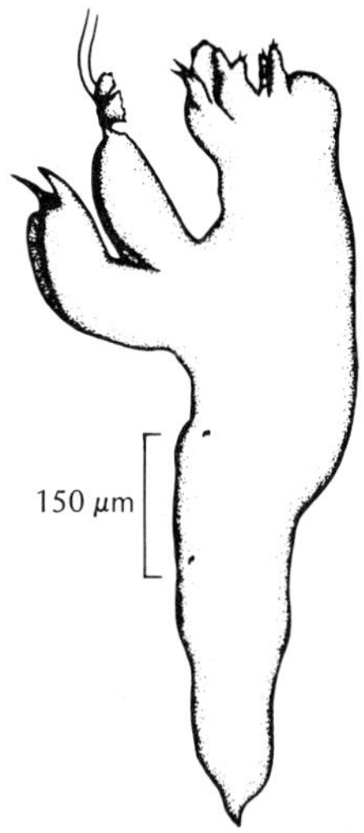


**FIG. 49.5**
Chalimus larva attached to host. [Redrawn with modifications from Kabata & Cousens, 1973.]

EPIDEMIOLOGY AND CONTROL. *S. edwardsi* usually presents problems only in restricted environments such as fish ponds and hatcheries. Fasten's (1921) account of control is still the classic reference. He recommended using sand filters to catch free copepods as the water enters the empoundment, washing the fish fry in salt solutions, destroying infected adult fish, and introducing small fish such as *Gambusia* and *Notropis* to feed on the free-swimming copepods.

There are many other parasitic copepods, often with bizarre shapes, but the same principles apply to them as well. Good sources of information include Yamaguti (1963) and Kabata (1970).

## CLASS BRANCHIURA

The class Branchiura is a group of about 75 species of bloodsucking ectoparasites that infect the skin and gills of freshwater and marine fishes as well as amphibians. Like the parasitic copepods just mentioned, some branchiurans, such as the cosmopolitan *Argulus* spp., cause serious disease and often death to fishes in hatcheries and ponds.

There are a number of differences between the class Copepoda and the class Branchiura. Branchiurans have compound eyes and a large, shieldlike carapace covering the head and thorax; they are not permanently attached to their host, and continue to molt after sexual maturity.

The fish louse, *Argulus* spp., holds onto a host with two sucking disks which are modifications of the second maxillae (Fig. 49.6). The proboscis penetrates the skin, and the parasite ingests host blood and apparently injects toxic substances from glands near the stylet or *sting* (Fig. 49.6) (Bauer, 1962). After copulation the *Argulus* female leaves the host and lays eggs on any submerged surface. Embryonic development lasts from 15 to 55 days, depending on the temperature. The newly hatched larva must find a host within 3 days or it dies.

Certain features of *Argulus* biology make control possible. The parasite cannot tolerate high oxygen tensions (Bauer, 1962); thus a high rate of water flow through a rearing pond is a useful control measure. Placing removable boards as substrates for *Argulus* egg clutches also helps. Hoffman and Meyer (1974) list the many antiparasitic chemicals that have been employed successfully against *Argulus* spp.

## CLASS CIRRIPEDIA

The class Cirripedia is the only group of sessile crustaceans, other than the various parasitic species in other classes. Best known are the barnacles that live attached to rocks and the undersides of ships. With their attached mode of existence, it is not surprising that a number of these organisms have evolved parasitic life-styles.

The cirripeds are attached to the substrate by their antennules; they literally stand on their heads. There are no compound eyes. The carapace is a fleshy covering or mantle that may have calcareous plates, as in the barnacles. The body usually has six pairs of biramous, jointed appendages, the cirri. Most are hermaphroditic.

Some parasitic cirripeds such as *Sacculina*, a parasite of the shorecrab, lose all appendages and segmentation as adults. The larva, called a *cypris* (Fig. 49.7), attaches to a host using its first antennae, the host tegument is perforated, and a simple cell mass, the kentrogen, is injected into the host. This cell mass sends nutrient-absorbing rhizoids throughout the crab's abdomen (Fig. 49.7). Then a male cyprid larva attaches and injects cells into the cell mass, thereby converting

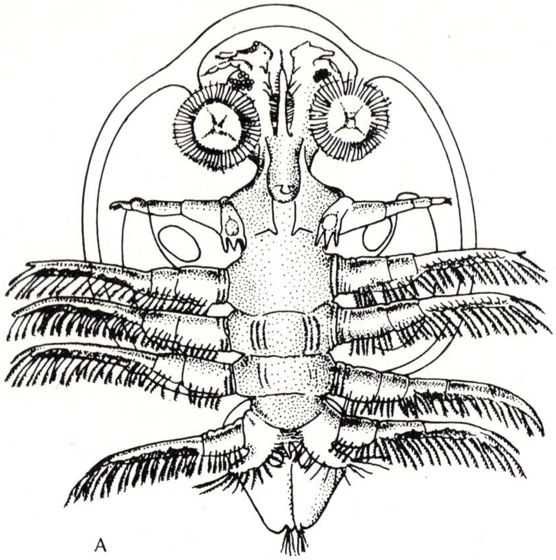


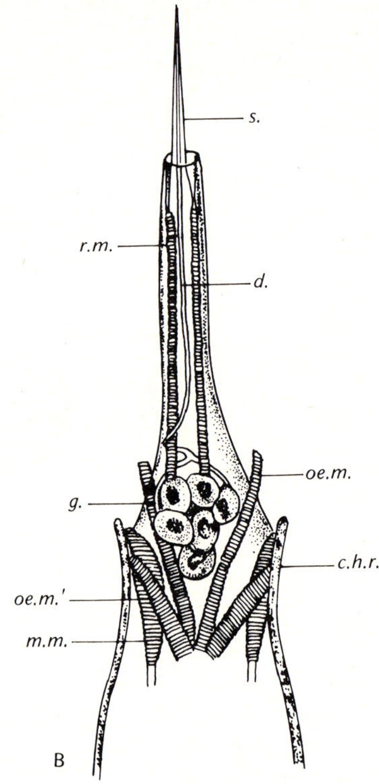


**FIG. 49.6**
*Argulus versicolor. (A)* Female; *(B)* sting.
**Abbreviations**: c.h.r. = chitin rib of proboscis, d. = ducts of poison gland, g. = poison gland, m.m. = mandible muscle, oe.m. = esophagus muscle, oe.m' = side esophageal muscle, r.m. = retractor muscle, s. = sting.
[Redrawn from Wilson, 1904.]

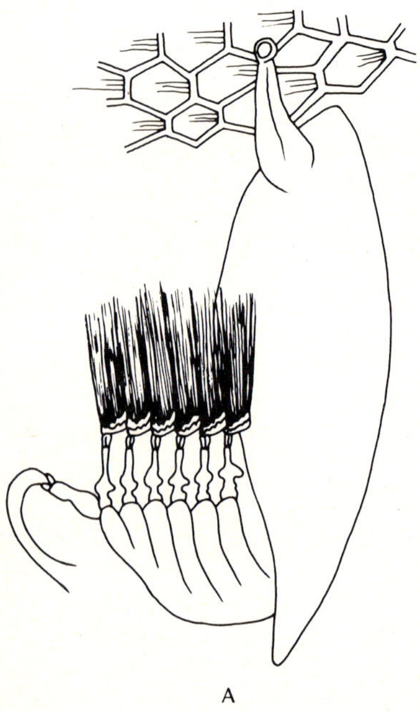


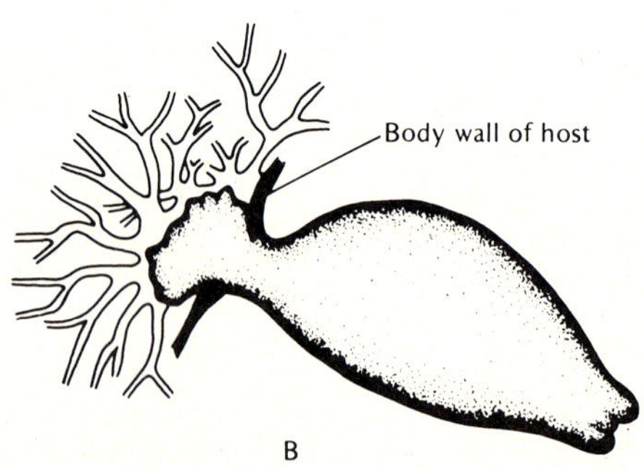


**FIG. 49.7**
*Sacculina.* (*A*) Cypris larva; (*B*) brood chamber. [Victor Schechter, *Invertebrate Zoology,* © 1959, p. 378. Reprinted by permission of Prentice-Hall, Inc., Englewood Cliffs, N.J.]

the female into a hermaphrodite (Day, 1935). Because of the spread of parasite rhizoids, infected crabs often are unable to molt and their gonads are usually destroyed (Reinhard, 1956).

## CLASS MALACOSTRACA

The large crustaceans of the class Malacostraca have only a few parasitic species. Most are isopods; a few are amphipods and decapods. Some cause economic losses to commercial fisheries, but as a group little is known about their biology and effects on their hosts (Kabata, 1970).

## REFERENCES AND FURTHER READINGS


Bauer, O. N. 1962. *The Ecology of Parasites of Freshwater Fish.* Office of Tech. Services, U.S. Dept. Commerce, P.S.T. Cat. No. 622. Translated from Russian (Bull. State Sci. Res. Inst. Lake River Fish.). Vol. XLIX, 1959, pp. 3–216.

Day, J. H. 1935. The life history of *Sacculina. Quart. J. Micros. Sci* 77:549–583.

Demaree, R. S., Jr. 1967. Ecology and external morphology of *Lernaea cyprinacea. Amer. Mid. Nat.* **78**:416–427.

Fasten, N. 1921. Studies on the parasitic copepods of the genus *Salmincola. Am. Natur.* **55**:449–456.

Grabda, J. 1963. Life cycle and morphogenesis of *Lernaea cyprinacea.* L. *Acta Parasitol. Polon.* **11**:169–199.

Hoffman, G. L. 1967. *Parasites of North American Freshwater Fishes.* University of California Press, Berkeley.

Hoffman, G. L., and F. P. Meyer. 1974. *Parasites of Freshwater Fishes.* T. F. H. Publications, Inc., Neptune City, N.J.

Kabata, Z. 1970. *Diseases of Fishes.* Book 1: *Crustacea as Enemies of Fishes.* T. F. H. Publications, Jersey City, N.J.

——. 1981. Copepoda (Crustacea) parasitic on fishes: Problems and perspectives. *Adv. Parasitol.* **19**:1–71.

Kabata, Z., and B. Cousens. 1973. Life cycle of *Salmincola californiensis* (Dana, 1852) (Copepoda: Lernaeopodidae). *J. Fish. Res. Bd. Can.* **30**:881–903.

Nakai, N. 1927. On the development of a parasitic copepod, *Lernaea elegans* Leigh-Sharpe, infesting *Cyprinus carpio* L. *J. Imp. Fish Inst. Tokyo* **23**:39–59.

Putz, R. E., and J. T. Bowen. 1964. *Parasites of Freshwater Fishes.* IV. *Miscellaneous. The Anchor Worm* (Lernaea cyprinacea) *and Related Species.* U.S. Department of the Interior. Bureau of Sport Fish and Wildlife. Fishery Leaflet No. 575.

Reinhard, E. G. 1956. Parasitic castration of crustacea. *Exp. Parasitol.* 5:79–107.

Schechter, V. 1959. *Invertebrate Zoology.* Prentice-Hall, Englewood Cliffs, N.J.

Tidd, W. M. 1934. Recent infestations of goldfish and carp by the "anchor parasite," *Lernaea carassii. Am. Fish. Soc. Trans.* **64**:176–180.

Tidd, W. M., and R. J. Shields. 1963. Tissue damage inflicted by *Lernaea cyprinacea*, Linnaeus, a copepod parasitic on tadpoles. *J. Parasitol.* **49**:693–696.

Wilson, C. B. 1904. A new species of *Argulus*, with a more complete account of two species already described. *Proc. U.S. Nat. Mus.* **27**:635–742.

——. 1911. North American parasitic copepods belonging to the family Ergasilidae. *Proc. U.S. Nat. Mus.* **39**:263–438.

——. 1915. North American parasitic Copepoda belonging to the Lernaeopodidae with a revision of the entire family. *Proc. U.S. Nat. Mus.* **47**:565–729.

Yamaguti, S. 1963. *Parasitic Copepoda and Branchiura of Fishes.* Interscience Publications, New York.

Yashouv, A. 1959. On the biology of *Lernaea* in fish ponds. *Bamidgeh* **11**:80–89.

# 50

# *Insects: An Introduction*

No matter where one goes, insects are an important part of the biological world. The tropics are sometimes thought of as typical of biotopes with insects that are large in both numbers and size. Although this is true, insects also play an important role in the biology of the area at the other extreme of habitat, the Arctic. Insects are opportunists in such a harsh environment, and their populations are likely to explode dramatically during the short summer. For example, mosquitoes rise in hordes from the tundra, making life for humans uncomfortable at best and unbearable at other times.

In the context of ways in which animals may use their resources in reproduction, insects are r-selected (Chap. 1); some exceptions will be pointed out when the specific group is discussed. Animals that are r-selected place their resources of energy and protein into a large number of rapidly growing, independent offspring, as opposed to the K-selected animals that produce a small number of large offspring, which may require extended care by their parents. Mosquitoes, mentioned above, are a good example of r-selection because large numbers of eggs hatch to produce rapidly growing larvae and adults that become sexually mature within a few days of emergence but soon die. As a dramatic example (Herms & James, 1961), if one pair of house flies were to reproduce and all of their offspring were to survive and reproduce for one summer in southern California, the earth would be 47 feet deep in house flies. The upshot of this potential is a group of organisms in a state of rapid evolution which has invaded almost every conceivable ecological niche. It is estimated that there are from 600,000 to more than a million species of insects.

Adaptable, opportunistic, ubiquitous, the insects are both beneficial and harmful to human activities. We depend on them for pollination of many crops such as fruit trees and alfalfa grown for seed. Insects prey on other harmful insects, they serve as recycling centers for organic materials, and they are the principal food for many fish.

Regrettably, these beautiful creatures are generally thought of only for the direct or indirect harm that they do to humans. They destroy crops, bear disease agents, and are pests that suck blood and invade households. In this book we look principally at those insects that are parasitic or are vectors of disease agents; however, whatever their importance, these harmful insects are only a small part of the biota.

The insects share with the other members of the phylum a segmented exoskeleton with jointed legs, an open circulatory system with a dorsal heart, and paired ventral nerve chords. They are differentiated from other members of the phylum by having three distinct body segments: the head, thorax, and abdomen; they have a single pair of antennae and three pairs of legs. Wings are present in most adults and they arise as extensions of the body wall on the meso- and metathorax. The legs all have the same parts: coxa, trochanter, femur, tibia, and tarsi going distally from the body.

The eyes of insects consist of a pair of compound eyes and up to three simple eyes. The compound eyes are usually conspicuous and consist of a cluster of individual light-sensing units, the ommatidia. A pair of simple eyes are located near the medial margin of the compound eyes and a single one usually is between the compound eyes on the midline of the head. Other sense organs that may be present are sound receptors, the tympanum on the thorax of the grasshopper; the halteres, modified hindwings in the Diptera; antennae; and portions of the mouthparts. Insects rely not only on sight but on air movement and the ability to detect small quantities of chemicals in the air and on the surface where they feed. Their ability to find one another for mating also depends on the ability to detect minute quantities of pheromones.

Most of the more highly evolved insects have two pairs of wings, and insects are the most successful of all animal groups in taking to the air. Wings of all insects arise on the meso- and metathorax (Fig. 50.1) and are extensions of the body wall (Fig. 50.2). Wings probably arose in stream insects such as mayflies (order Ephemeroptera) or the stoneflies (order Plectoptera) in which rudimentary extensions of the body wall served as stabilizers in the current. In extant species, functional wings occur only in the adult stage and once an insect has developed functional wings, it does not molt again (with the exception of the Ephemeroptera).

Feeding structures in insects take many forms. A generalized type of mouthparts can be seen in the grasshopper (order Orthoptera) (Fig. 50.3). In this type of chewing mouthpart, there are strong laterally cutting mandibles, and the principal food is plant material, which is cut into small pieces and then passed to the mouth opening. This rather basic type has been modified into a variety of other types depending on the

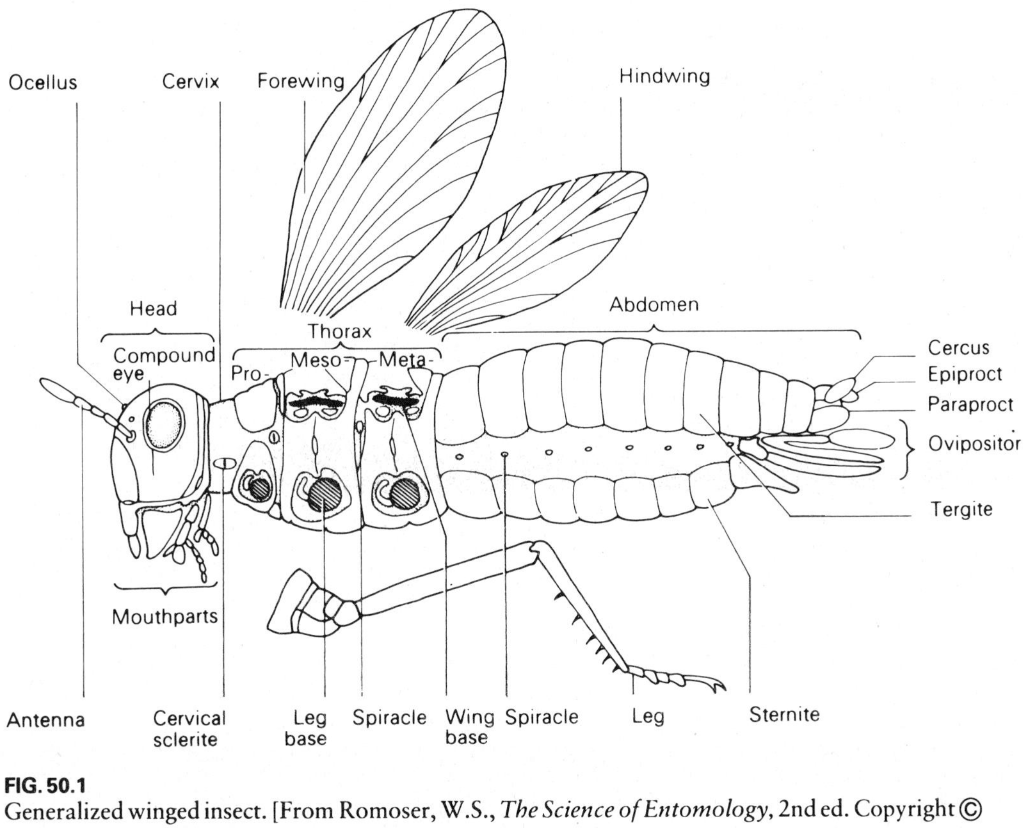


**FIG. 50.1**
Generalized winged insect. [From Romoser, W.S., *The Science of Entomology*, 2nd ed. Copyright © 1981, by Macmillan Publishing Company, Inc., New York.]

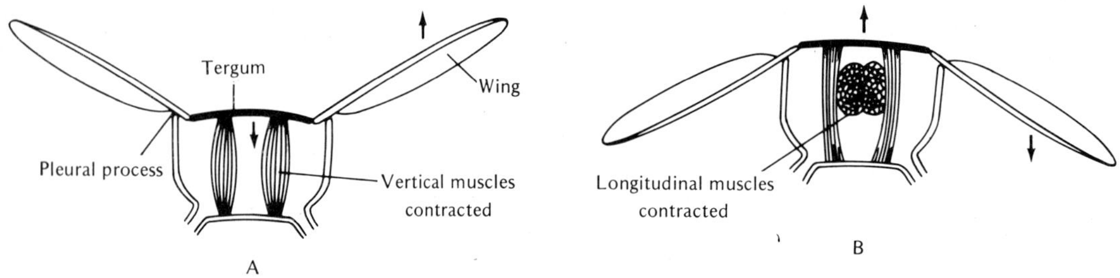


**FIG. 50.2**
Musculature and wing movement in insects. [Redrawn with modifications from Ross, 1956.]

food of the insect. In insects with sucking mouthparts such as the sucking lice, Hemiptera, and mosquitoes, the mouthparts become elongated so as to pierce the integument of the plant or animal on which they feed; however, there is elaboration of different elements in different groups. For example, the mandibles and the maxillae are designed for piercing and cutting in hemiptera and female mosquitoes, but the mandibles are vestigial in the sucking lice and male mosquitoes (Fig. 50.4). House flies have a further modification for obtaining liquid food in which they have a large labellum (Fig. 50.5); the grooved lower surface of the labellum serves to move liquid to the mouth by capillary action. A more complete discussion of mouthparts can be found in Harwood & James (1979). Two main points, however, should be kept in mind: first, that mouthparts and the type of food are inextricably associated and, second, that mouthparts have evolved in different ways to achieve the same end.

The intestinal tracts of insects are all basically the same; the gut has three sections (Fig. 50.6):

1. Foregut, which includes the mouth, pharynx, esophagus, crop, and proventriculus
2. Midgut or stomach
3. Hindgut, which includes the ileum, colon, rectum, and anus

The foregut serves principally to store food for a short period. The salivary glands lie in the thorax and empty into the mouth area; certain functions of the salivary glands are fairly clear, such as production of anticoagulants by bloodsucking insects but others, such as lubrication of the mouthparts, are uncertain.

Digestion and absorption of food take place in the midgut. When food reaches the midgut of many insects, it is surrounded by a peritrophic membrane. This distinct structure arises from the midgut and effectively contains the meal. The peritrophic membrane is important in bloodsucking insects that are vectors of disease agents. In most instances, the infectious agent must escape from the peritrophic membrane in order to reach a location where it can develop further. If the membrane forms quickly, is impenetrable, and lasts a long time, the insect is likely to be a poor vector.

The hindgut, like the foregut, is lined with cuticle and holds undigested food until it is evacuated; it functions with the Malpighian tubules to maintain water and inorganic ion balance.

The threadlike Malpighian tubules are at the junction of the midgut and hindgut. Although they are associated with the gut, they are excretory organs and function not only in the excretion of the end products of nitrogen metabolism but also in water and inorganic ion balance. The major end product of nitrogen metabolism in most insects is uric acid, but other nitrogenous compounds are sometimes produced.

Gaseous exchange in insects takes place through the tracheal system, which opens through pores along the sides of the abdomen (Fig. 50.7). The tracheal system extends throughout the body of the insect, ramifying into smaller and smaller tubules. In some larval forms, the tracheal openings may be located at the posterior end, as in house fly larvae. In some preadult insects such as naiads of mayflies or aquatic larvae of mosquitoes or dobsonflies, gills serve in gaseous exchange.

Insects are dioecious, although some, such as aphids, reproduce parthenogenetically during a portion of the life cycle. Sexual maturity is reached after the final molt. Some insects copulate, oviposit, and die within a day or

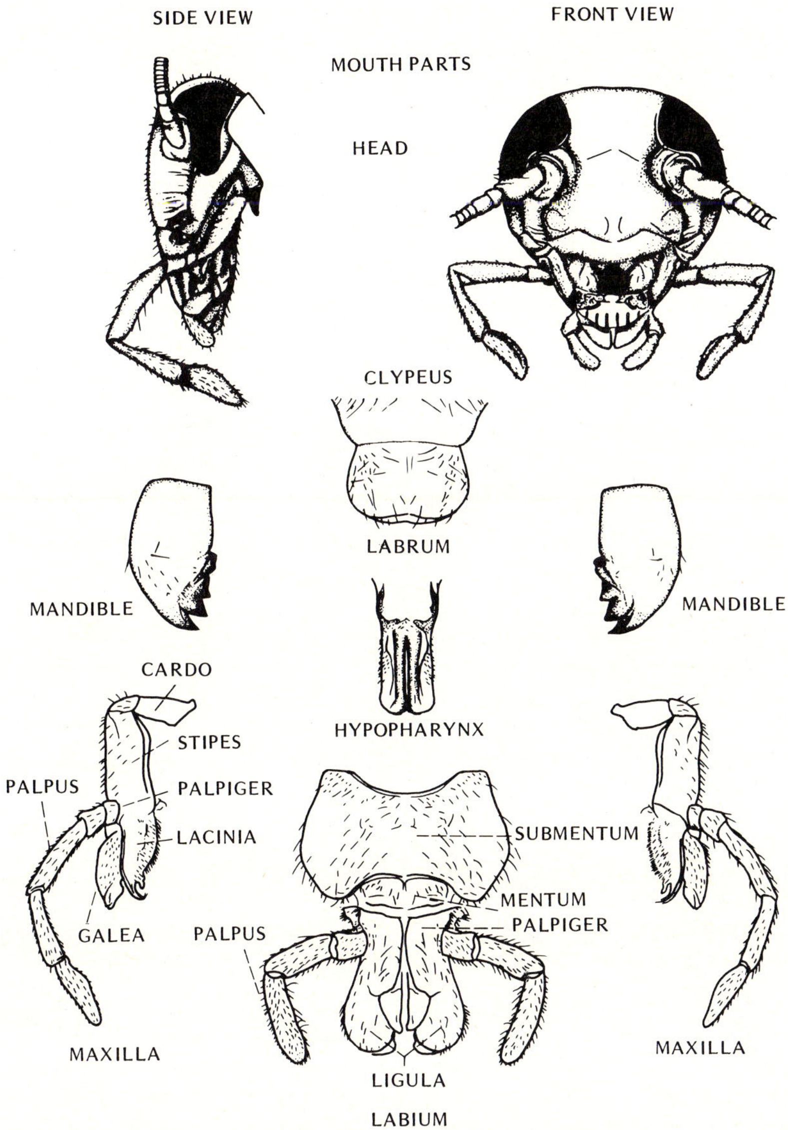


**FIG. 50.3**
Mouthparts of a cockroach. [From Harwood, R. F., and M.T. James, *Entomology in Human and Animal Health,* 7th ed. Copyright © 1979, by Macmillan Publishing Company, Inc., New York.]

two of emerging as adults, but others such as tsetse, the vectors of African sleeping sickness, may produce offspring for many months. The female reproductive system is branched, with an ovary lying at the tip of each of the two branches (Fig. 50.8). Eggs pass down the oviduct and reach the outside through the vagina. The spermatheca stores sperm, and the female may not need to be inseminated each time a batch of eggs is laid; some insects, such as the primary screwworm, are inseminated only once. The other structures are the spermathecal gland and accessory glands. The internal structures of the male are also bilaterally symmetrical, with a pair of testes that produce tailed sperm. The sperm pass down the vasa deferentia and are stored temporarily in the seminal vesicles. Sperm are transferred to the female through the ejaculatory duct and the external genitalia.

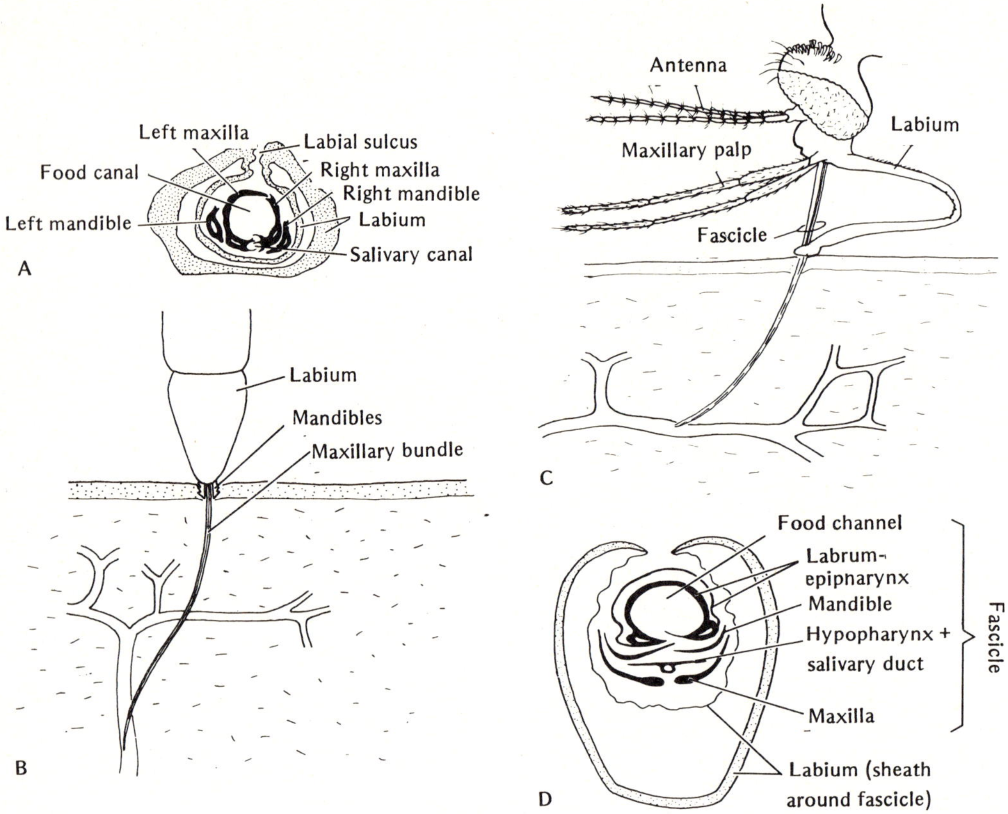


**FIG. 50.4**
Sucking mouthparts of *Rhodnius* (*A* and *B*) and mosquitoes (*C* and *D*). [From Harwood & James, 1979.]

The male genitalia are varied but have paired claspers to grasp the female and an intromittent organ to transfer the sperm to the female. Each group of insects has distinctive structures which are sometimes unique to a species. For example, in the *Culux pipiens* complex, the subspecies can be differentiated only on the basis of the male external genitalia. In the fleas, the males have long, elegant penile rods that can be seen coiled in the body in cleared specimens.

Insects have three main patterns of growth and they are used to establish the highest taxonomic divisions in the class:

1. Ametabola. The early instars are indistinguishable from the later ones except that the organism grows slightly larger with each molt (Fig. 50.9). Examples are the orders Thysanoptera and Protura.
2. Hemimetabola. There is gradual change from the early instars to the final one in which sexual maturity is reached and the wings (when present) are functional (Fig. 50.9). Body proportions change with each molt, and the wing buds appear some time before the final molt. This series is now often referred to as the Exopterygota because the wings develop externally. Included are the orders Ephemeroptera, Hemiptera, Mallophaga, Anoplura, and Dermaptera.
3. Holometabola. There is a striking change from the last larval instar to the adult, and reorganization of the body takes place in a pupa (Fig. 50.9).

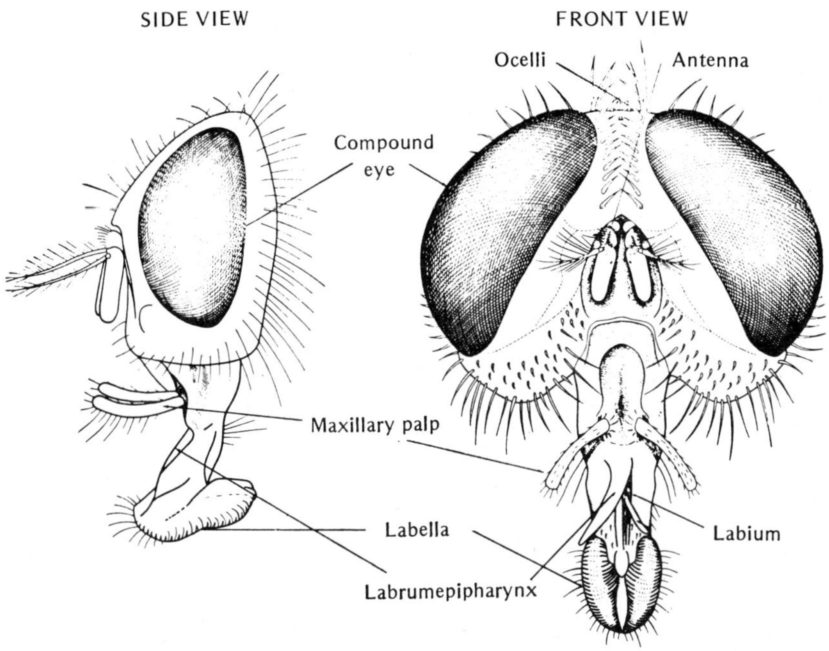


**FIG. 50.5**
Head and mouthparts of a house fly. [From Harwood & James, 1979.]

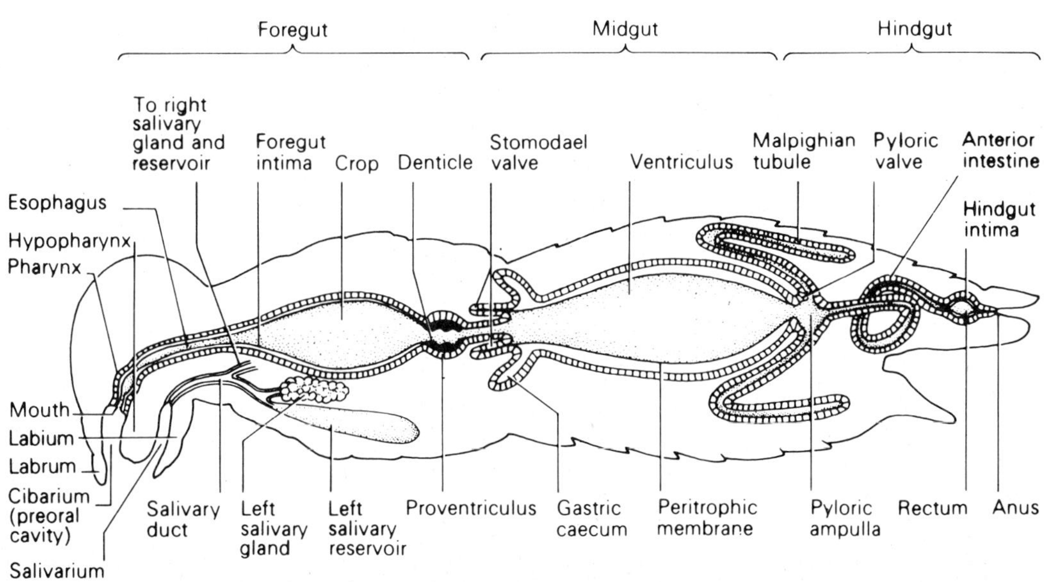


**FIG. 50.6**
Intestinal tract of a generalized insect. [From Romoser, 1981.]

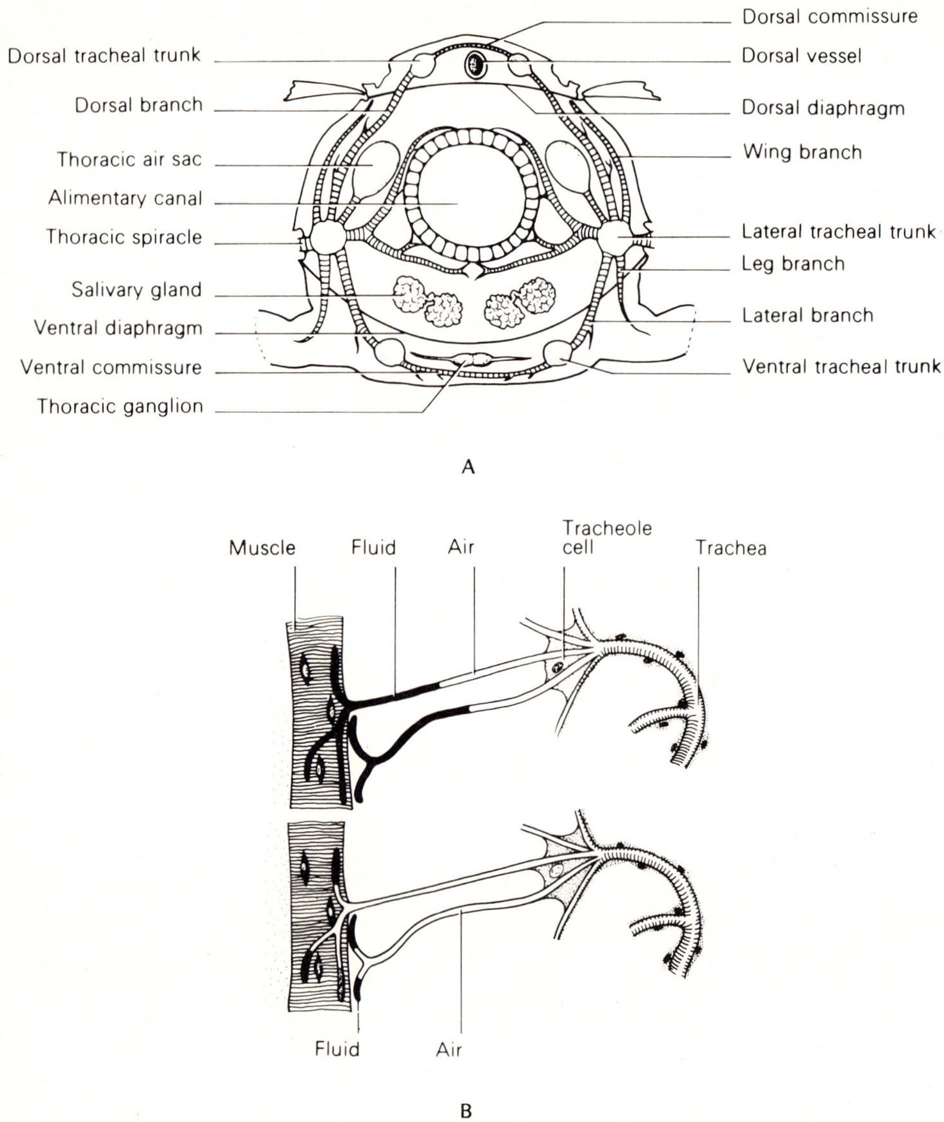


**FIG. 50.7**
Tracheal system of a generalized insect. [From Romoser, 1981.]

The best known example is a butterfly, which has a series of larval instars during which it grows to the final size. It then forms a pupa (sometimes covered by a cocoon), in which it undergoes dramatic reorganization and emerges as the winged, sexually mature adult. This series is now often referred to as the Endopterygota because the wings develop internally. Examples are the orders Diptera, Coleoptera, Lepidoptera, Hymenoptera, and Siphonaptera.

Further subdivision at the ordinal level is based on external characteristics:

1. Wings. The number, 0, 1, or 2, pairs, and the wing form. Primitive wings have a large number of

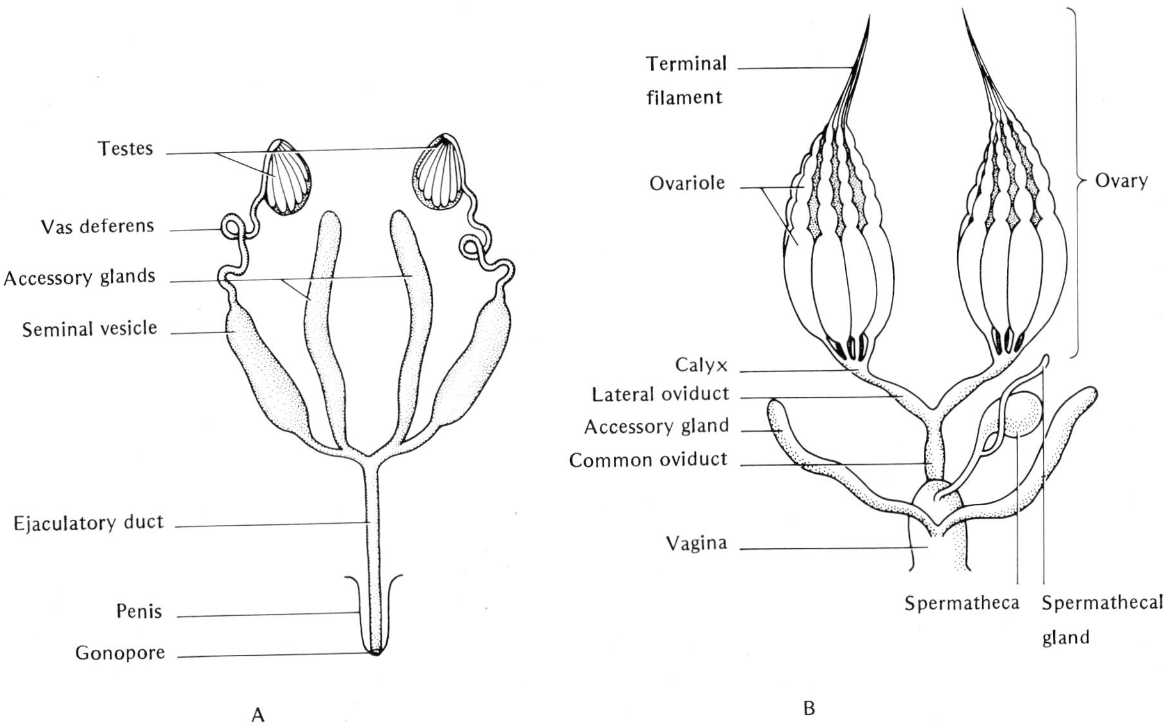


**FIG. 50.8**
Reproductive systems of male (*A*) and female (*B*) insects. [Snodgrass, R. E., 1935. *Principles of Insect Morphology*. McGraw-Hill, New York.]

small veins (e.g. Neuroptera), whereas the more highly evolved wings have a relatively small number of stout veins (e.g., Diptera). Within any one order, the wing form is relatively homogeneous.

2. Mouthparts—chewing or sucking. The embryogenesis of the various parts is the major characteristic. There may be considerable variation in some groups such as the Diptera, which have chewing, sucking, or sponging mouthparts, but generally the mouthparts are much the same within an order.

Other characteristics are flattening of the body, either laterally or dorsoventrally, body proportions, and hairs, or setae, on the body, but these are of lesser importance.

## CLASSIFICATION OF INSECTS

The taxonomy used here is taken principally from Romoser (1981), with some modifications for the sake of simplicity. Not all 31 orders have been included. Those that are described are either related to human and animal health in some way or are the most common orders.

**Class Insecta**
**Series Ametabola or Apterygota**

**Order Thysanura.**
Small, elongate, covered with scales; chewing mouthparts; wings lacking; elongate cerci. Bristle tails, silverfish, firebrats (Fig. 50.10).

These organisms are not of any health significance, but they often become domiciliated and may be nuisances in the house. They move through

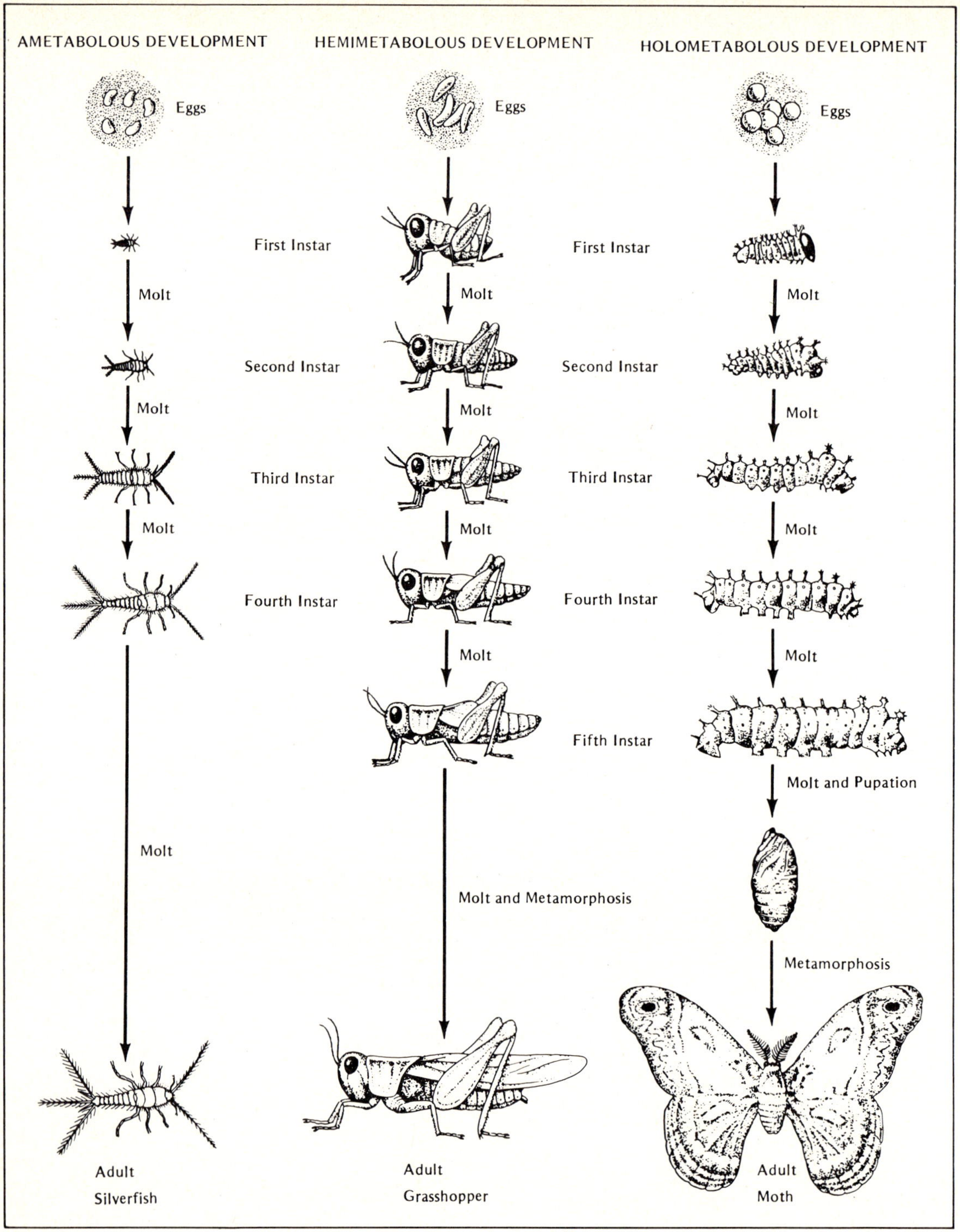


**FIG. 50.9**
Patterns of development in insects. [From Herried, C.F., *Biology*. Copyright © 1977, by Macmillan Publishing Company, Inc. New York.]

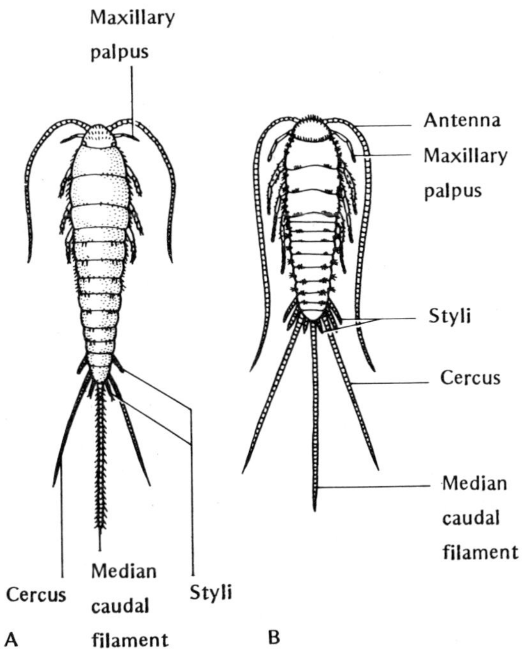


**FIG. 50.10**
Thysanura. The common silverfish, *Lepisma saccharina (A)*, and a firebrat, *Thermobia domestica (B)*. [From Romoser, 1981.]

apartment buildings via pipe chases and are seen in bathrooms; they also eat paper, cloth, and the glue in book bindings.

### Series Hemimetabola or Exopterygota

### Order Odonata

Medium to large with slender abdomen in the adult stage; chewing mouthparts, naiads with prehensile labium; large eyes, short antennae; two pairs of membranous wings with netlike venation; legs adapted for holding prey; naiads aquatic. Dragonflies (wings held at right angle to the body), damselflies (wings held over the abdomen) (Fig. 50.11).

Some odonatan naiads are intermediate hosts of digenetic trematodes such as *Haematoloechus medioplexus*, a common lung fluke of frogs, and *Prosthogonimus macrorchis* the oviduct fluke of many birds.

### Order Ephemeroptera

Small, fragile, and soft-bodied; chewing mouthparts in the nymphs, vestigial in the adults; two pairs of membranous wings with forewings larger than hindwings; wings held vertically at rest; adults have a pair of long caudal cerci. Mayflies (Fig. 50.12).

The mayflies are important because of their role as food for fish. They are among the most common benthic stream invertebrates. They tend to emerge from streams and lakes as swarms in spring and early summer. Their direct importance to humans lies in the huge number of insects, both living and dead, that accumulate in some urban areas; some persons become hypersensitive to them. Mayflies are the only order in which there is more than a single winged stage; the subadult or subimaginal stage has functional wings.

### Order Plecoptera

Small to medium-sized and soft-bodied; chewing mouthparts in the nymphs, but vestigial in the adults; two pairs of membranous wings with the hindwings folded under the forewings when at rest; long cerci at the tip of the abdomen. Stoneflies (Fig. 50.13).

Stoneflies appear quite similar to mayflies as nymphs; like the mayflies, the nymphs are an important food for fish.

### Order Orthoptera

Small to very large, usually with sturdy bodies; chewing mouthparts; well-developed compound eyes; usually three ocelli; filiform antennae, which are sometimes as long as the body; two pairs of wings with netlike venation, forewings parchment-like, hindwings membranous and folded beneath the forewings when at rest; hind legs adapted for jumping. Grasshoppers, crickets, katydids, locusts (Fig. 50.14).

Orthopterans are most important as consumers of crops; in some parts of the world, particularly grasslands, they can cause agricultural disaster. Crickets sometimes move into houses, but cause little other than annoyance with their song, which was once thought to bring the household good luck. Some members of the order are intermediate hosts of spirurid nematodes (Chap. 41) such as *Physaloptera rara* of carnivores.

### Order Dermaptera

Small to medium-sized with narrow bodies; chewing mouthparts, small eyes, lacking in some species; antennae long and slender; two pairs of wings; forewings short and leathery, hindwings membranous and semicircular, folded under the forewings at rest; wingless species common, and abdomen not covered by wings in those species that are alate;

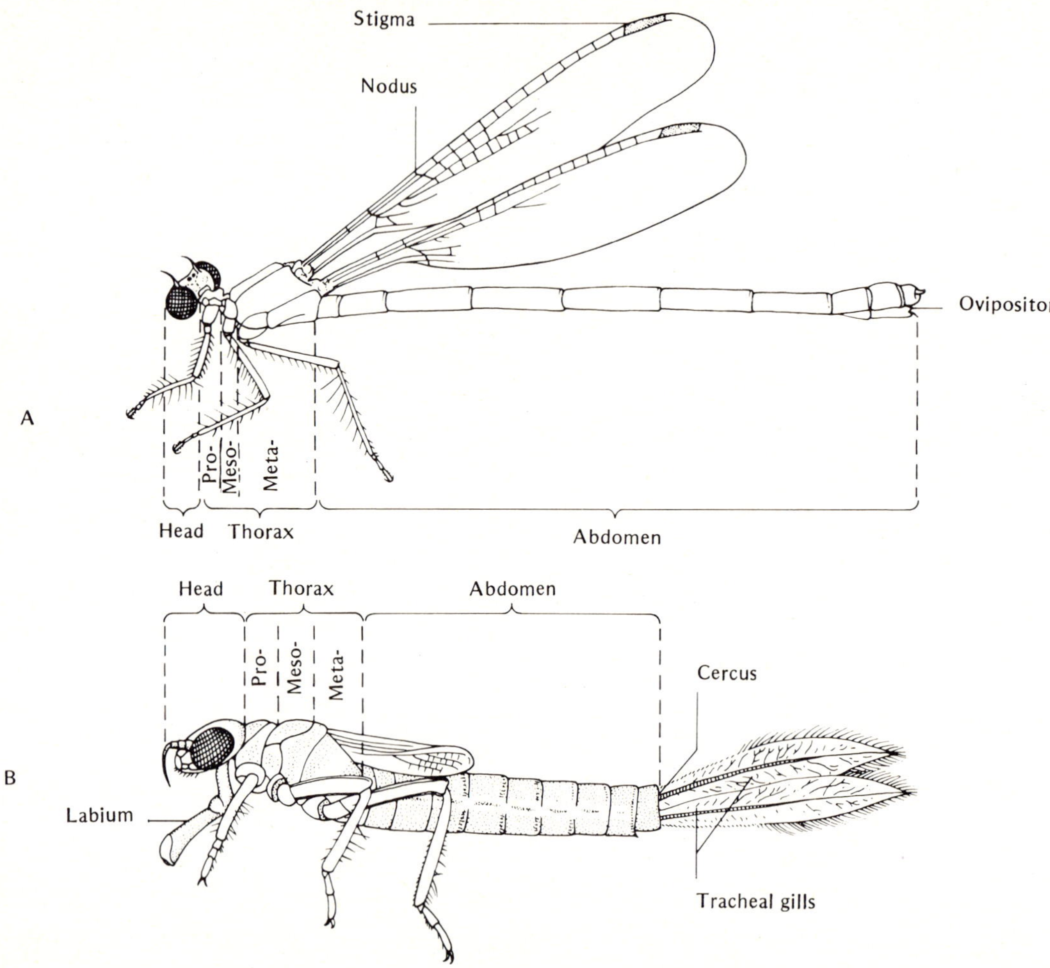


**FIG. 50.11**
Odonata. A dragonfly (*A*) and a damselfly (*B*). [From Romoser, 1981.]

caudal cerci pincerlike and apparently used to capture prey. Earwigs (Fig. 50.15).

Earwigs live in damp places where they feed on organic matter and sometimes on living plants; they sometimes become domiciliated, but do not have any medical importance as vectors of disease agents of humans. Some are intermediate hosts of spirurid nematodes, and a few are symbiotes of bats.

**Order Isoptera**
Small to large with elongate bodies; chewing mouthparts; compound eyes in winged forms, but may be lacking in other forms; antennae short, either moniliform or filiform; wings membranous, both pairs identical. Termites (Fig. 50.16).

Termites are social and have castes, which serve specific functions in the colony. Termites can be confused with ants, but a constriction occurs at the junction between the thorax and abdomen in the Hymenoptera. Termites are important economically in wooden structures (Chap. 8).

**Order Blattaria**
Small to very large, soft bodies with distinct dorsoventral flattening; chewing mouthparts; compound eyes well developed; antennae long and

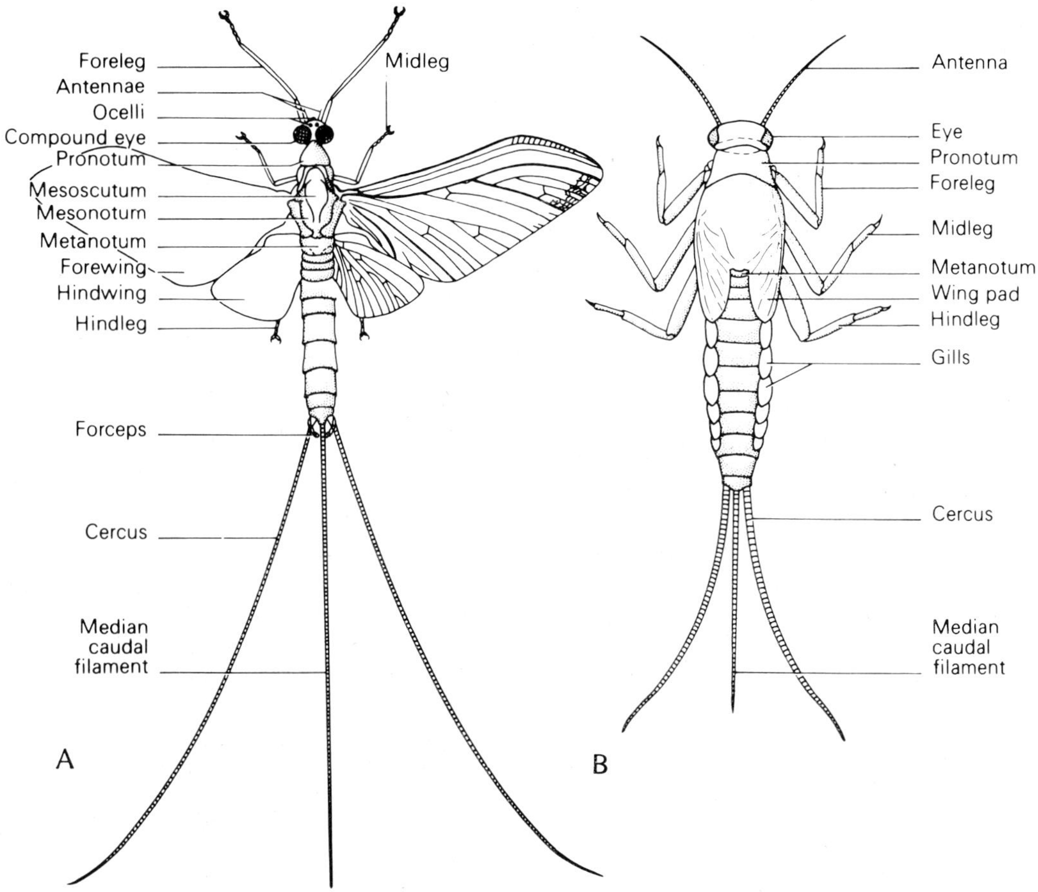


**FIG. 50.12**
Ephemeroptera. An adult (*A*) and a nymph (*B*). [From Romoser, 1981.]

filiform; forewings rather leathery, hindwing membranous; pronotum large and covering the head. Cockroaches, woodroaches (Fig. 50.17).

Cockroaches are highly adaptable insects that often become domiciliated; they are vectors of enteric bacterial diseases and some helminths. They feed on any organic matter and are likely to feed on human or animal feces and then move into a house and defecate on or near food. Cockroaches are nocturnal and are sometimes not seen until the kitchen light is turned on unexpectedly. Control of cockroaches is dependent on insecticides and absolute cleanliness in disposal of food. Insecticides such as carbamates and silicones are effective in reducing populations only when all food is properly disposed of and kitchens do not have any bits of food lying around or containers of food that the insects can enter. Additional information is found in Harwood & James (1979).

Woodroaches live in moist areas of deciduous forests where they remain in decaying logs; they are part of the recycling mechanism in the forest biome. Woodroaches have protozoan symbionts that break down the cellulose the insects eat (Chap. 8).

### Order Mallophaga

Minute, but large enough to see with the unaided eye; broad head; chewing mouthparts; small eyes, sometimes lacking; antennae small with three to five segments, sometimes concealed; wingless; legs mod-

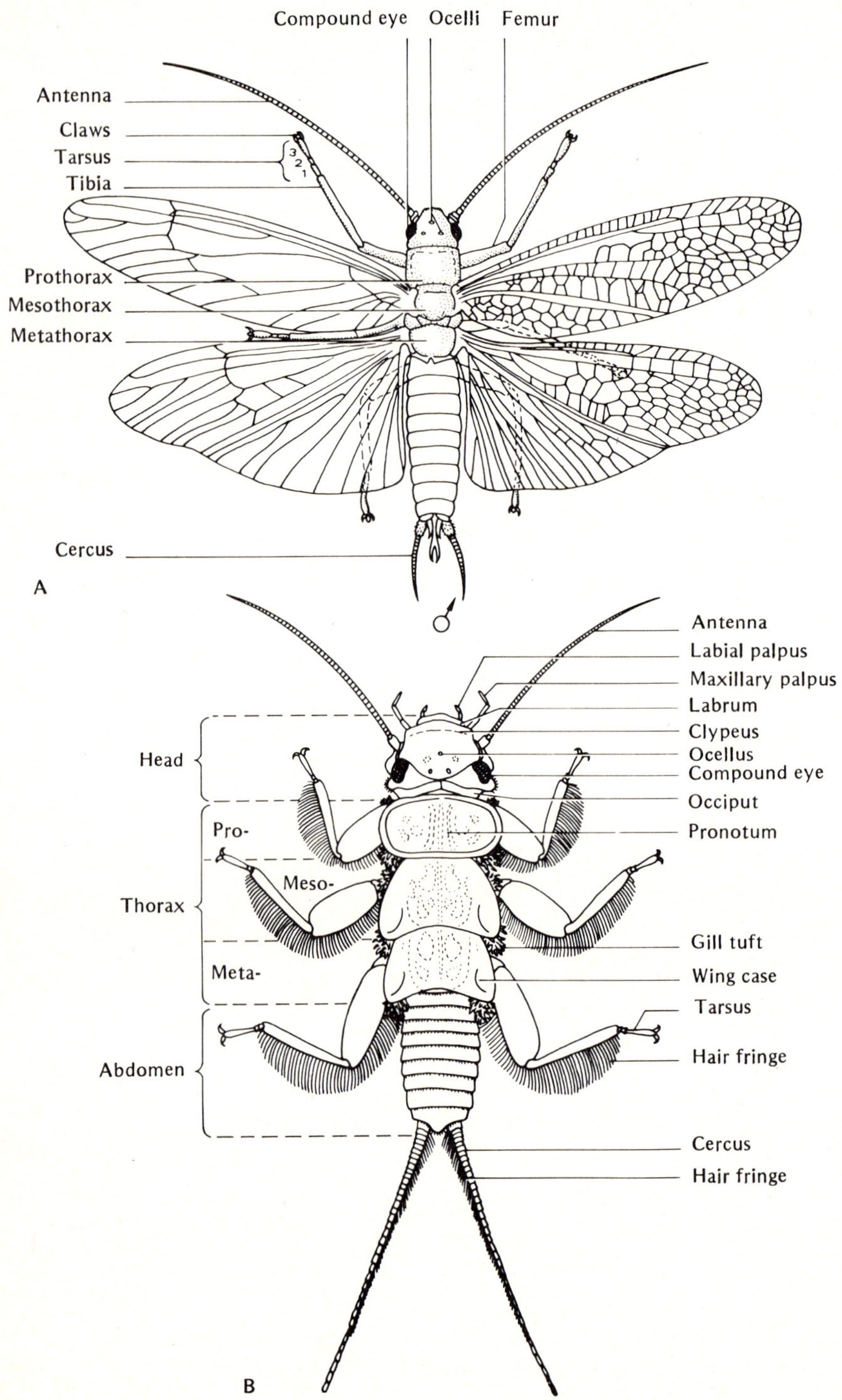


**FIG. 50.13**
Plecoptera. An adult (*A*) and a nymph (*B*). [From Romoser, 1981.]

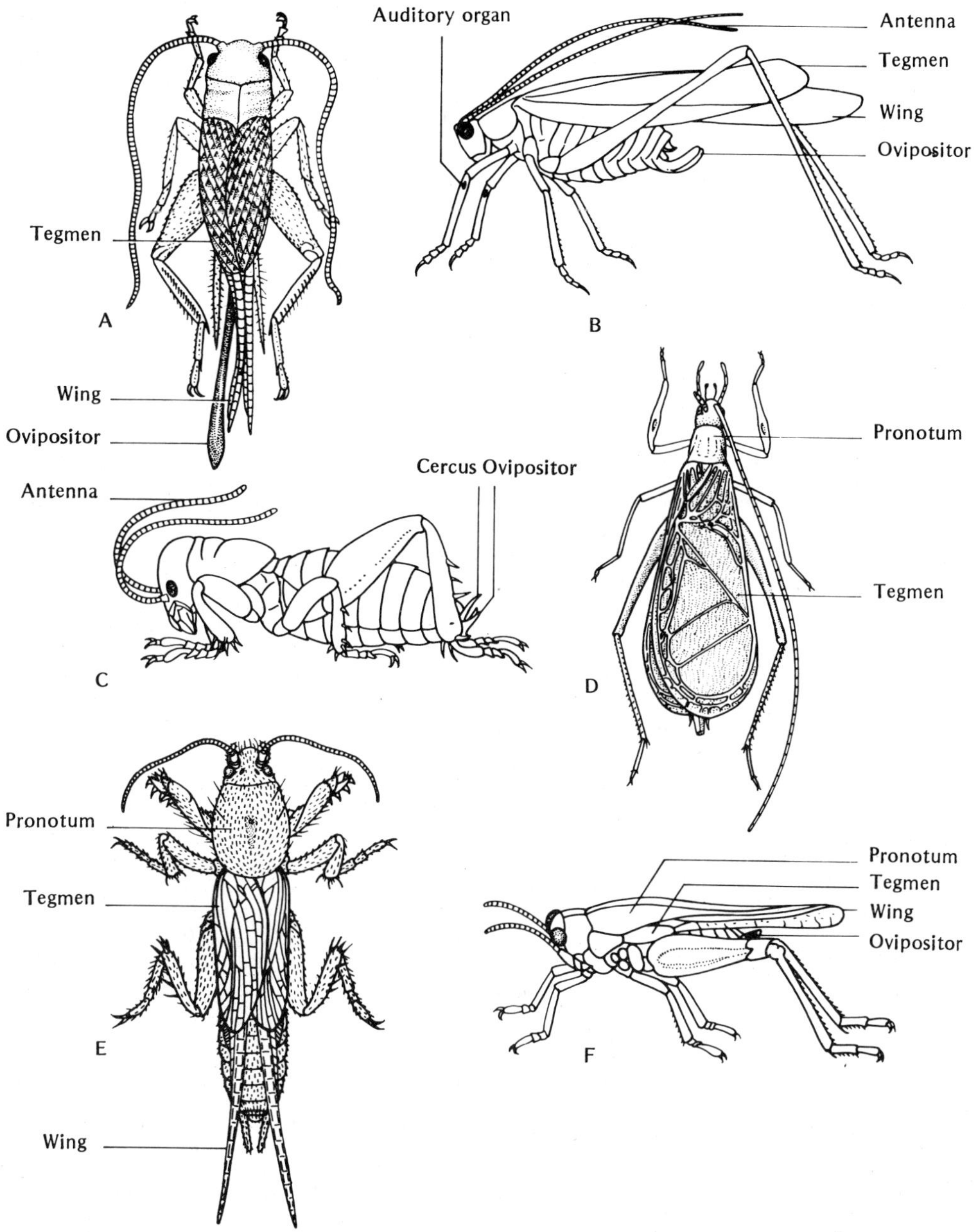


**FIG. 50.14**
Orthoptera, representative members. (*A*) Field cricket, *Gryllus assimilis;* (*B*) bush katydid, *Scudderia furcata;* (*C*) Jerusalem cricket, *Stenopelmatus longispina;* (*D*) tree cricket, *Oecanthus niveus;* (*E*) mole cricket, *Gryllotalpa hexadactyla;* (*F*) pygmy grasshopper, *Telmatettix hesperus.* [From Romoser, 1981.]

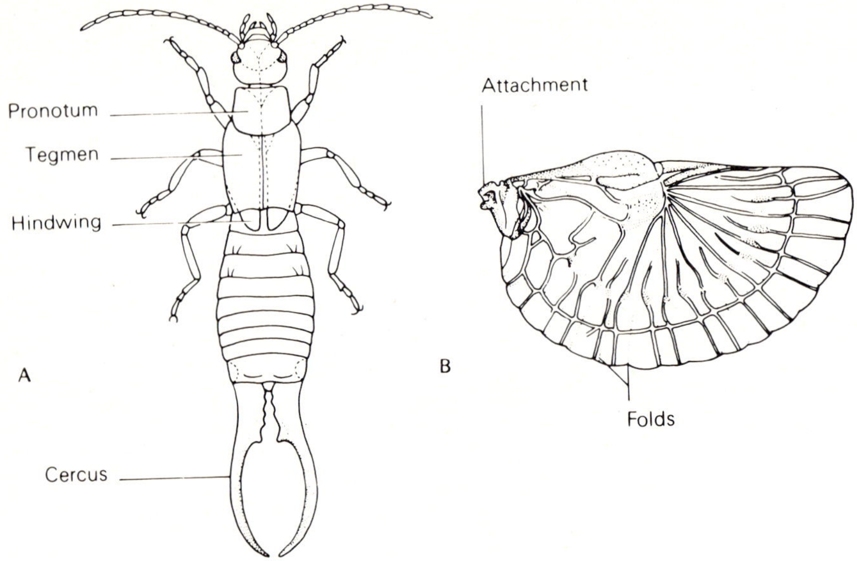


**FIG. 50.15**
Dermaptera, earwig. (*A*) Male; (*B*) right hindwing. [From Romoser, 1981.]

ified for grasping; abdomen usually rather rotund; body flattened dorsoventrally. Bird lice, chewing lice (Chap. 51).

**Order Anoplura**
Minute, but large enough to see with the unaided eye; narrow head; sucking mouthparts; compound eyes small or lacking; short antennae with three to five segments; wingless; legs adapted for grasping. Sucking lice (Chap. 51).

**Order Thysanoptera**
Minute to small, delicate-bodied; rasping-sucking mouthparts; small compound eyes; short antennae; two pairs of narrow wings with fringe hairs and hooked together; some species wingless. Thrips (Fig. 50.18).

Thrips are principally pests on plants and cause considerable economic loss in fruits and truck garden crops. On occasion they cause skin irritation to humans when they attempt to pierce the skin with their mouthparts.

**Order Hemiptera**
Minute to large (more than 3 cm); sucking mouthparts; two pairs of wings, but some forms wingless; Heteroptera have forewings with an anterior leathery portion and a posterior membranous portion; Homoptera have membranous wings with reduced venation; eyes large to absent. True bugs, leaf hoppers, cicadas, aphids, assassin bugs, kissing bugs, bedbugs, scale insects, back swimmers, spittle bugs (Fig. 50.19 and Chap. 52).

This diverse order is often divided into the orders Hemiptera and Homoptera, consisting of the suborders previously referred to in the description of wings. The members range from plant feeders to predators to bloodsuckers; the plant feeders are economically important both for their role in removing plant juices and as vectors of disease agents. The bloodsucking forms in the family Reduviidae are vectors of the causative agent of Chagas' disease (Chap. 4). Some species such as the wheel bug bite humans if handled and cause considerable pain (Chap. 48); aphids and box elder bugs, among others, sometimes attempt to feed on humans, but they do not present important medical problems.

**Series Holometabola or Endopterygota**

**Order Coleoptera**
Minute to very large (greater than 10 cm), usually heavy-bodied; chewing mouthparts; two pairs of wings with the forewing as sclerotized elytra and the hindwings membranous and folded under the forewings at rest; larva is usually grublike, with a distinct head capsule and three pairs of rudimentary legs. Beetles (Fig. 50.20).

There are about 300,000 species of beetles, which makes this the largest of the orders of insects. There

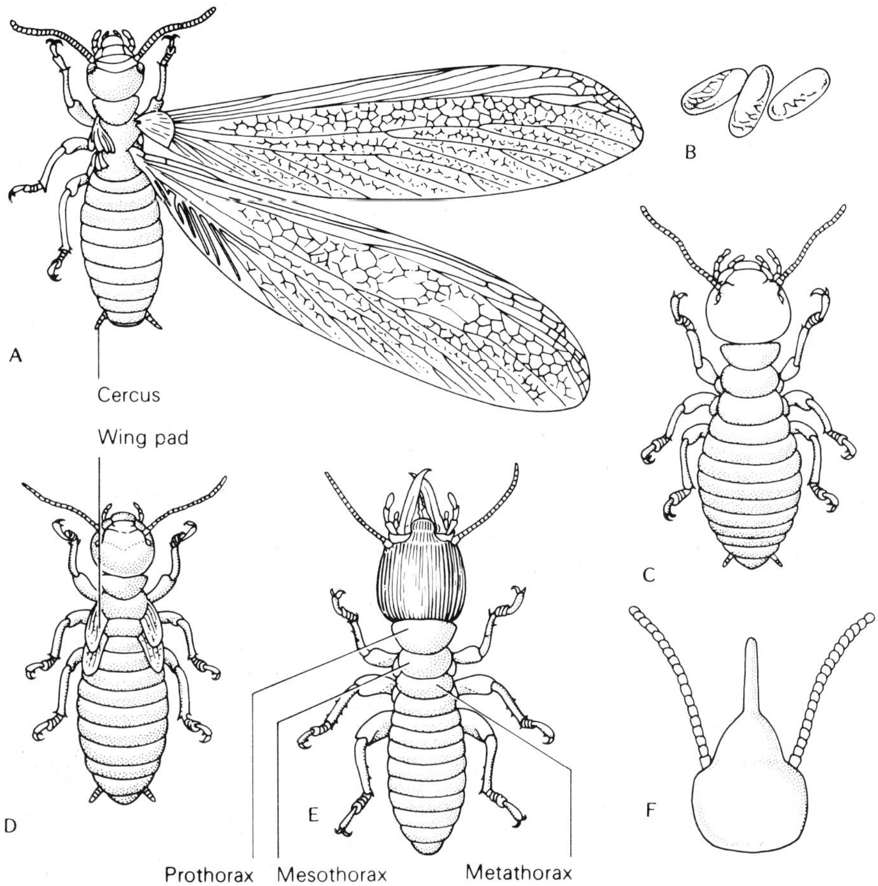


**FIG. 50.16**
Isoptera. (*A*) Winged form, *Zootermopsis angusticollis;* (*B*) eggs; (*C*) third instar nymph; (*D*) last instar nymph; (*E*) soldier; (*F*) head of *Nasutitermes* sp. [From Romoser, 1981.]

is a good deal of diversity in the group in structure, life histories, and food habits. Many beetles are economically important because of the losses they cause in stored food products and crops and as vectors of disease agents of plants. Only a few are parasitic, such as *Platypsyllus castoris* (Fig. 50.21) of the beaver (*Castor canadensis*); these organisms look superficially much like Mallophaga and require careful examination for differentiation.

Beetles have some importance as vectors of helminth parasites. A variety of tapeworms, among them *Hymenolepis diminuta*, *H. carioca*, and *Raillietina* spp., have beetles as intermediate hosts. Some nematodes such as *Gongylonema pulchrum* and *Spirocerca lupi* also use beetles as intermediate hosts. Beetles are not usually associated with the transmission of bacterial diseases in the way cockroaches are, but dung beetles probably harbor enteric bacteria picked up from human or animal feces; fortunately, they do not often associate with humans. Dung beetles serve as important agents in the breakdown and dissemination of feces of mammals (Mohr, 1943; Miller, 1954; Fincher, 1973).

### Order Strepsiptera

Minute, soft-bodied parasites; chewing mouthparts; compound eyes in males and those females that are

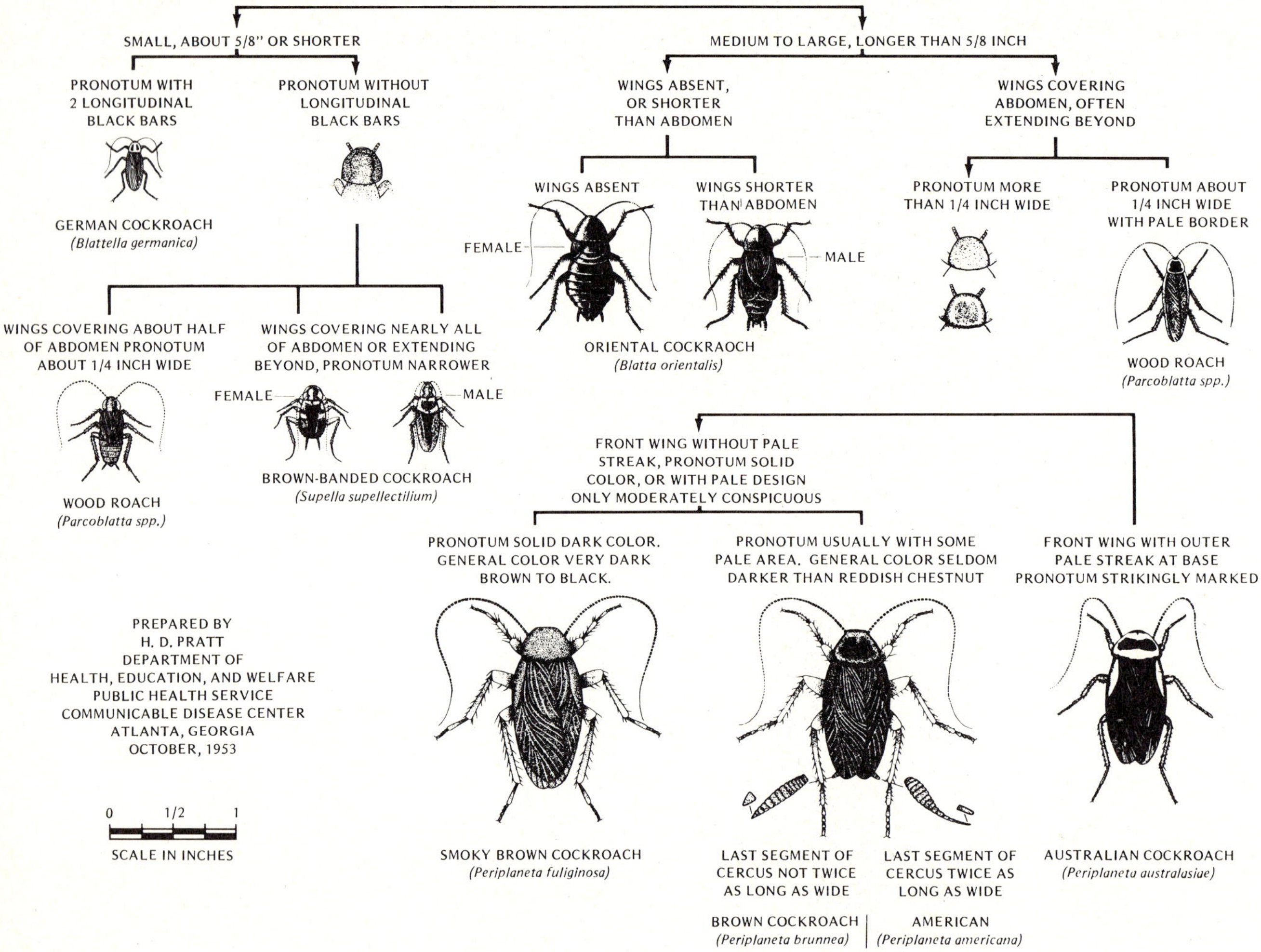


**FIG. 50.17**
Pictorial key to some common adult cockroaches (Blattodea).

free-living; antennae flabellulate in males; two pairs of membranous wings in males; forewings clublike and hindwings fan-shaped; females wingless; larvae and females legless; three to five to genital pores in females. Twisted-wings, stylopids (Fig. 50.22).

The Strepsiptera are endoparasites of other insects, mostly Hemiptera and Hymenoptera; only the males and the first instar larvae are free-living. This is a rather small, obscure order.

**Order Neuroptera**
Small to large, soft-bodied; chewing mouthparts in adults and grasping-sucking in larvae; compound eyes moderately developed; filiform antennae; two pairs of membranous wings about the same size and with many small veins; fluttering type of flight. Lacewings, snakeflies, dobsonflies, hellgrammites, ant lions (Fig. 50.23).

The larvae of the Neuroptera are mostly predaceous and make up an important segment of freshwater biota. Dobsonfly larvae or hellgrammites are especially large aquatic predators which are often mimicked in fishing lures; this indicates the role of many of these organisms both as predators

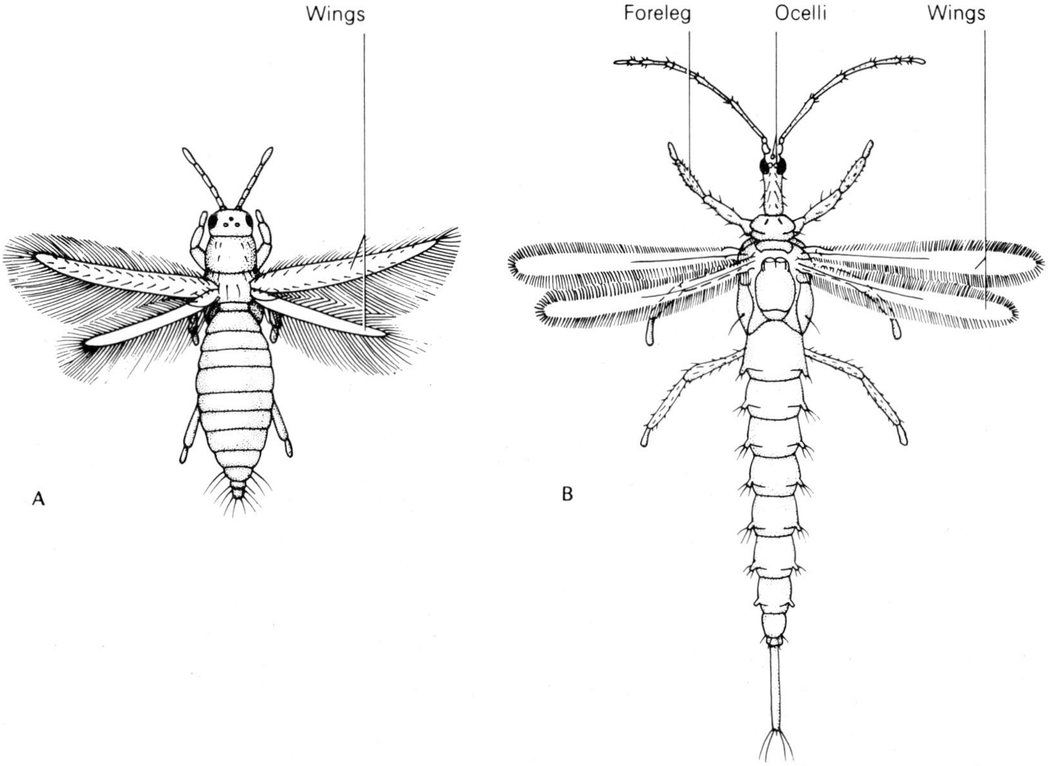


**FIG. 50.18**
Thysanoptera. (*A*) Member of Terebrantia; (*B*) female member of Tubulifera. [From Romoser, 1981.]

and as food for fish. Ant lions are often recognized by the conical pits they make in loose or sandy soil; they kick out little puffs of soil as a means of maintaining the slope of its ant trap.

**Order Diptera**
Minute to large (up to 4 cm); mouthparts vary from piercing-sucking to lacerating to sponging in adults—; larvae with prehensile, trapping, rasping mouthparts; compound eyes may be well developed to absent; antennae from filiform to clublike; one pair of wings; hindwings modified into balancing organs, the halteres; a few species wingless (Chaps. 54 and 55).

The order Diptera is the fourth largest; it has about 80,000 members. They are important as pests of crops, as vectors of disease agents, and as parasites both in the larval and adult stages.

**Order Siphonaptera**
Small (up to about 2 mm), hard-bodied, laterally compressed with many posteriorly pointed setae; piercing-sucking mouthparts in adults and chewing in the larvae; compound eyes absent; two lateral ocelli present in most forms, some lack eyes; short antennae that lie in grooves on the head; legs have exceptionally long coxae adapted for jumping; elaborate copulatory structures in male. Fleas (Chap. 53).

**Order Trichoptera**
Small to medium, soft-bodied; mouthparts usually poorly developed in adults, chewing in larvae; compound eyes moderately developed in adults; two pairs of membranous wings, forewings somewhat larger than hindwings; fluttering flight. Caddis flies, caddisworms, salmon flies (Fig. 50.24).

The caddis flies comprise an important part of freshwater biota as benthic organisms and as food for fish as both larvae and adults. They often emerge from streams in large numbers at a particular time of

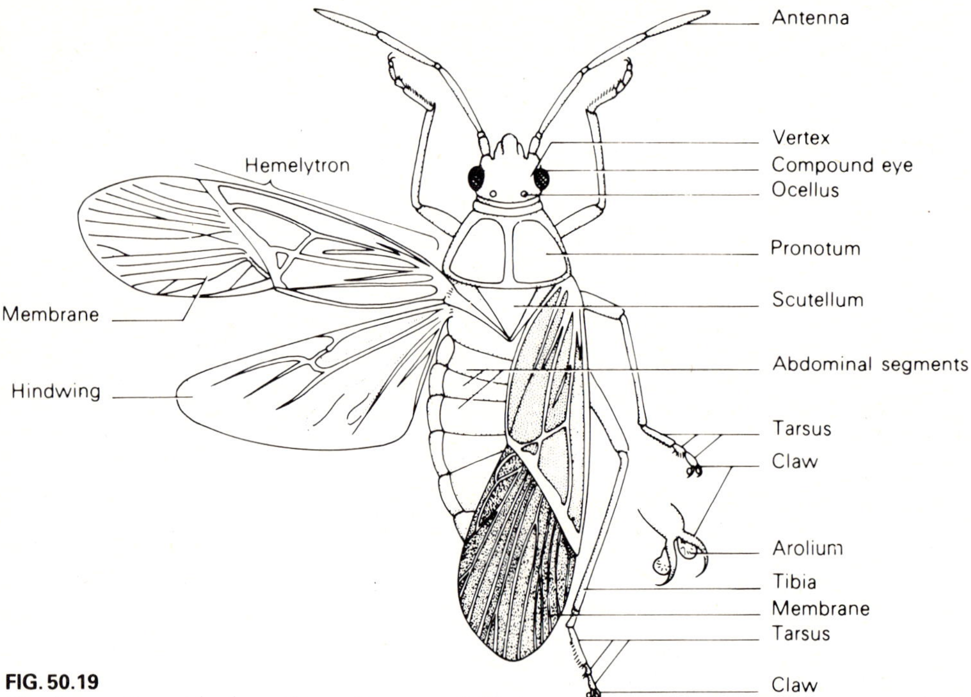


**FIG. 50.19**
Hemiptera. Box elder bug. [From Romoser, 1981.]

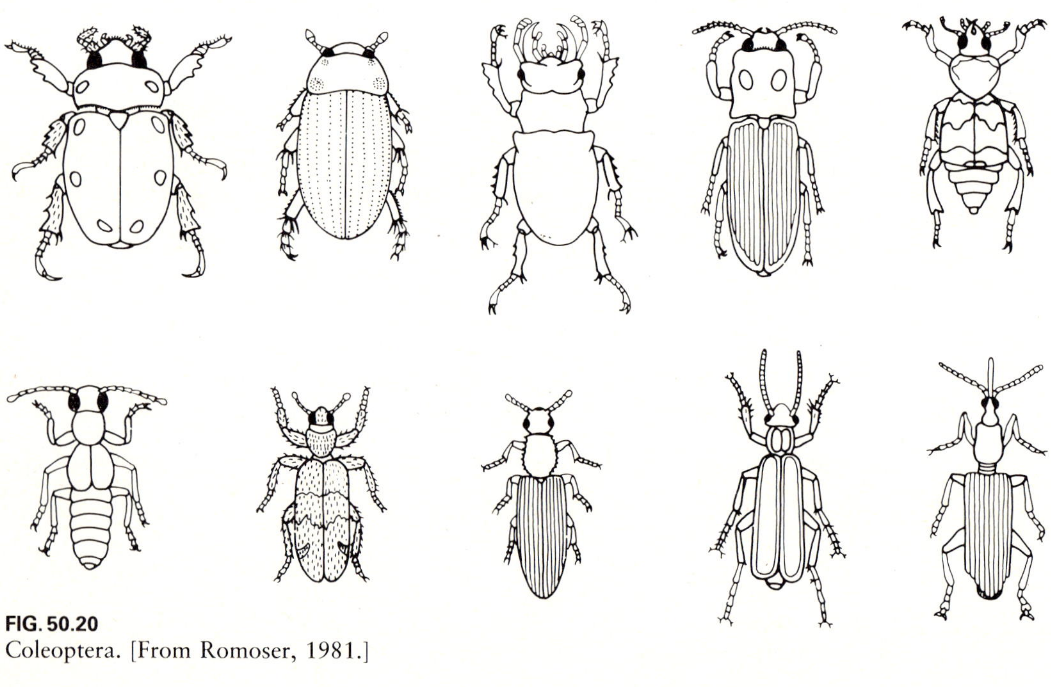

**FIG. 50.20**
Coleoptera. [From Romoser, 1981.]

**FIG. 50.21**
Coleoptera. *Platypsyllus castoris,* a beetle parasitic on beaver. [Microscope slide courtesy of Dr. Richard Eads.]

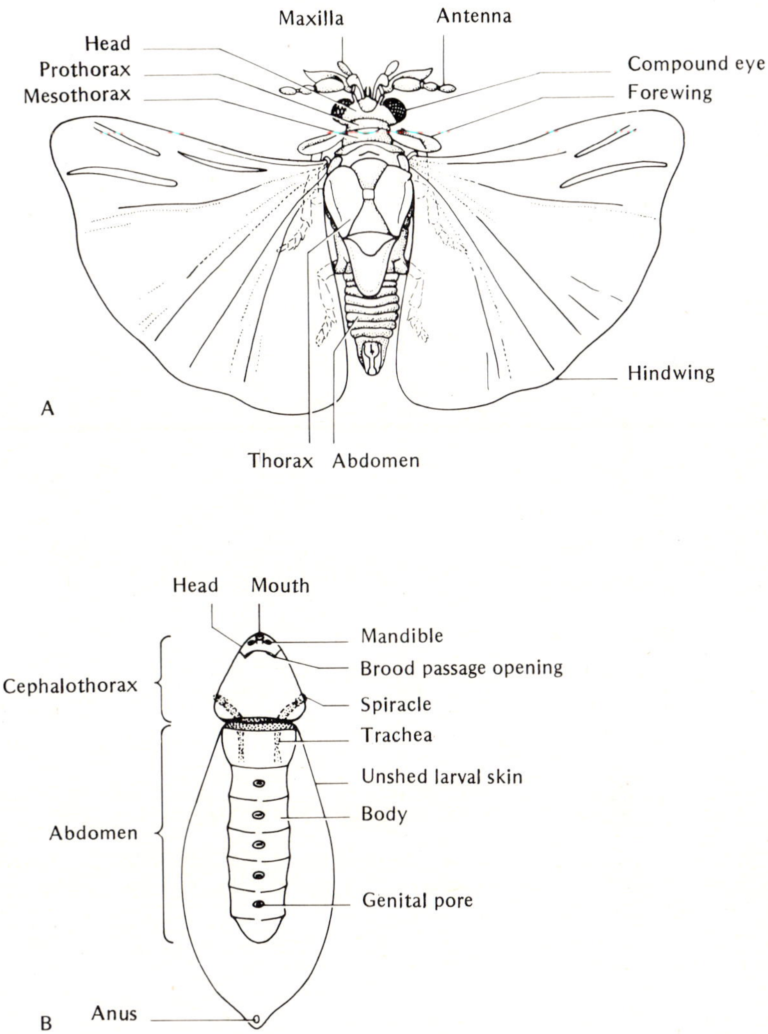


**FIG. 50.22**
Strepsiptera. (*A*) male; (*B*) female. [From Romoser, 1981.]

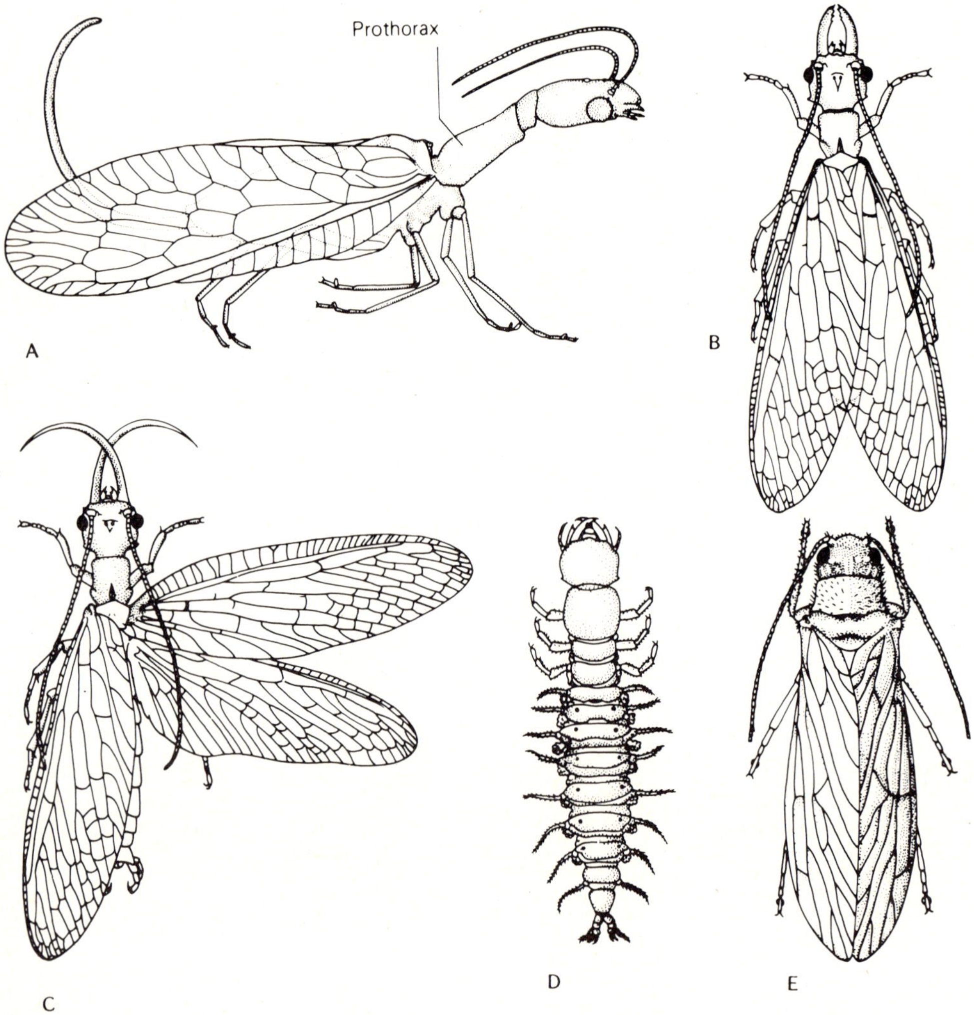


**FIG. 50.23**
Neuroptera. (*A*) Snakefly, *Agulla bractea* (Raphidiodea; Raphidiidae); (*B*), (*C*), and (*D*) dobsonfly, *Corydalis cornutus*, female, male, and larva, respectively (Megaloptera; Corydalidae); (*E*) alderfly, *Sialis mohri* (Megaloptera; Sialidae). [From Romoser, 1981.]

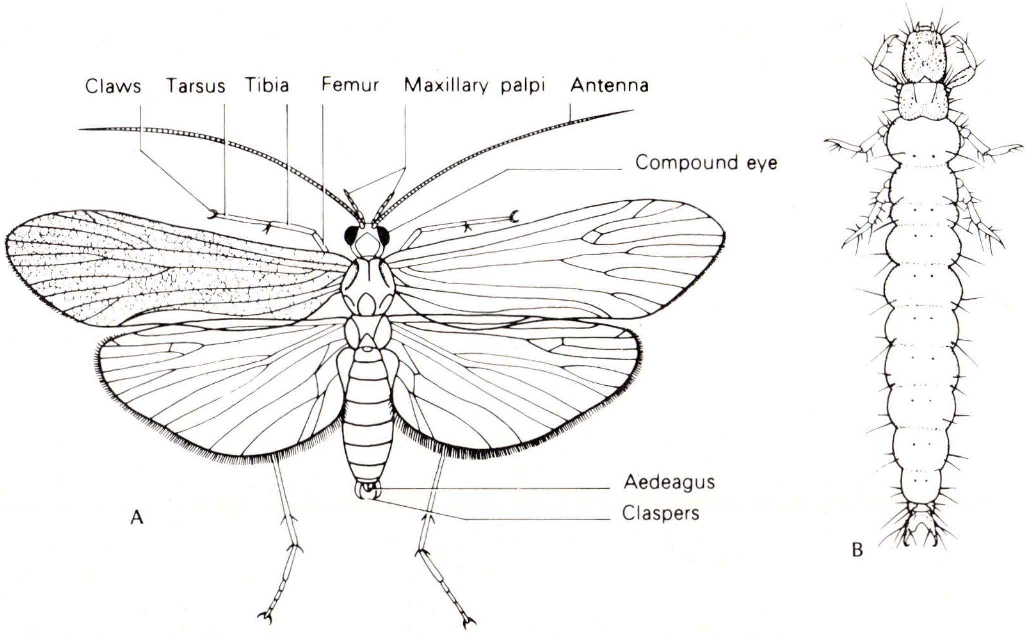


**FIG. 50.24**
Trichoptera, caddisflies. (*A*) Adult; (*B*) larva without larval case. [From Romoser, 1981.]

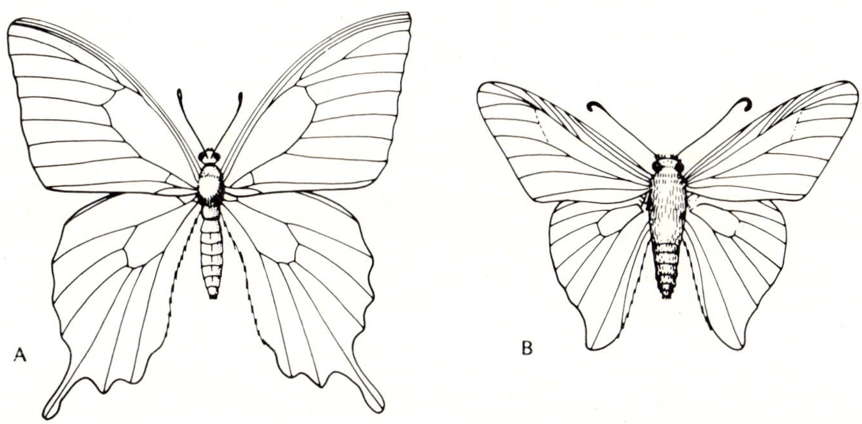


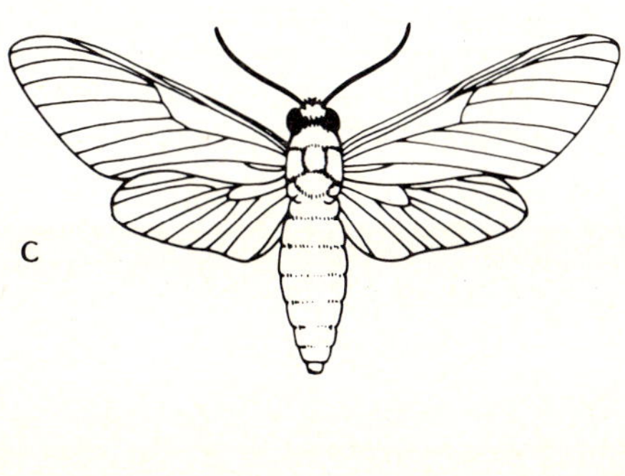


**FIG. 50.25**
Lepidoptera. (*A*) Butterfly; (*B*) skipper; (C) moth. [From Romoser, 1981.]

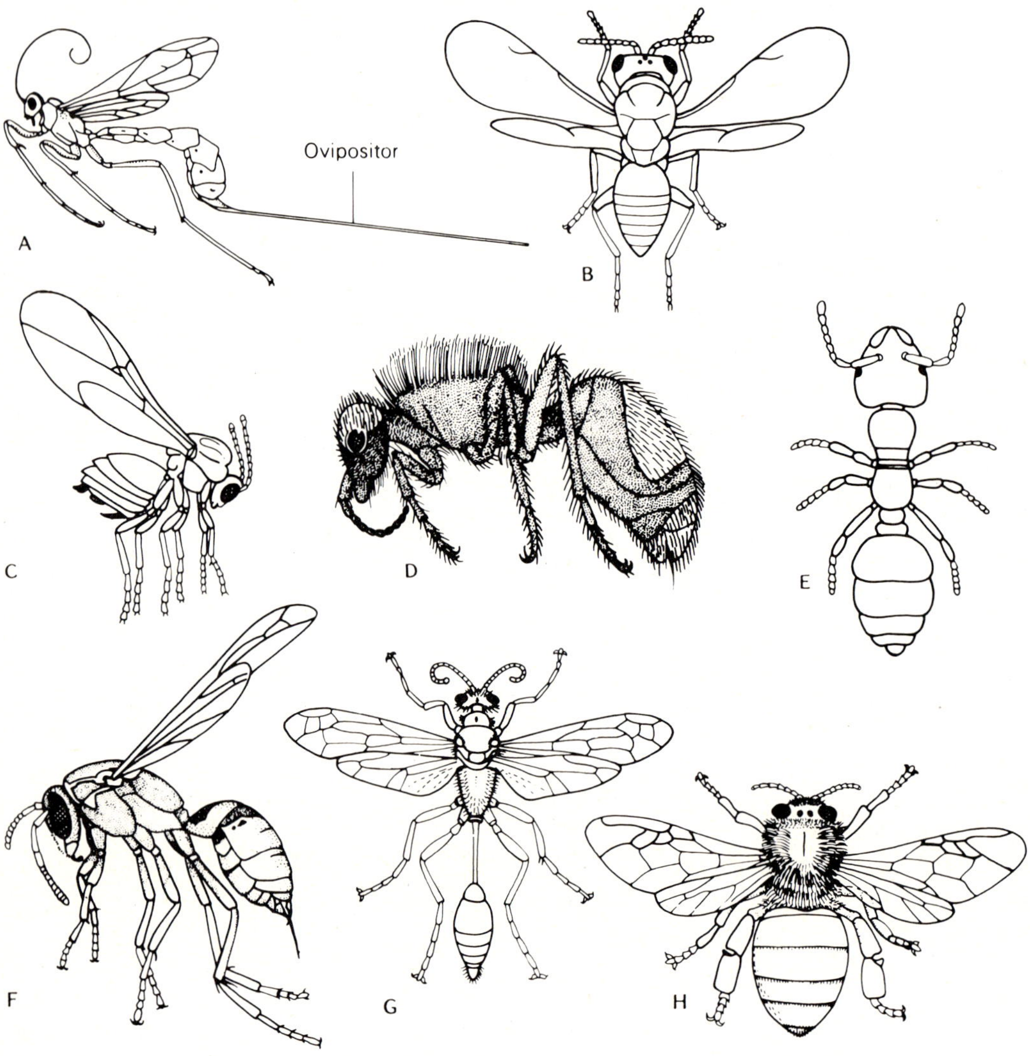


**FIG. 50.26**
Hymenoptera, Apocrita. (*A*) Ichneumon wasp; (*B*) chalcid wasp; (*C*) California gallfly, *Andricus californicus;* (*D*) velvet ant; (*E*) thief ant, *Solenopsis molesta;* (*F*) golden polistes, *Polistes aurifer;* (*G*) thread-waisted wasp; (*H*) honey bee, *Apis mellifera.* [From Romoser, 1981.]

year. Most of the larvae form cases by gluing bits and pieces of plant material or sand into a tubelike structure; these cases are distinctive for a species.

**Order Lepidoptera**

Small to large (up to 15 cm), body covered with scales and often with highly colored wings; mouthparts in adults with a long, sucking proboscis, absent in some, larvae with chewing mouthparts; large compound eyes in adults, larvae with ocelli; antennae variable; two pairs of membranous wings with reduced venation; larvae with prolegs on segments 3 to 6 and 10. Butterflies, moths, caterpillars (Fig. 50.25).

There are about 170,000 species of lepidopterans; many are economically important as destroyers of field crops, truck garden crops, and trees. Some of the names represent the plants that the larvae attack: corn earworm, beet armyworm, sod webworm, grape root borer, spruce budworm. The gypsy moth was inadvertently released on the East Coast of the United States late in the nineteenth century; it has since spread through much of the mid-Atlantic states and annually denudes millions of acres of deciduous forest. At this point there is little effective control, and the spread of the gypsy moth through the rest of the United States seems to be inexorable. In contrast, the silkworm is a beneficial insect, and we nearly all feel pleasure at seeing a brightly colored butterfly feeding on flowers around the house.

Only a few lepidopterans are important in human or animal health. Some Noctuidae feed on secretions around the eyes of domestic animals and are accused of transmitting the causative agent of pinkeye of cattle. In Malaya, there are a few species that suck blood. The scales and urticating hairs of caterpillars may be quite irritating and even cause blistering of the skin.

**Order Hymenoptera**

Small to large (up to 15 cm); mouthparts variable; compound eyes usually well developed in the adults; antennae usually long; two pairs of membranous wings with somewhat reduced venation, forewings larger than hindwings and often hooked together with hamuli; larva legless with distinct head; ovipositor of female often modified into a stinger or a cutting or boring structure. Bees, wasps, hornets, sawflies, ants (Fig. 50.26).

Members of the Hymenoptera are mostly beneficial, serving as pollinators of plants, as natural biological control agents of harmful insects, and as makers of honey and beeswax. The primary harmful species are sawflies, which attack various crops of economic importance such as small grains, grasses, trees, and some small fruits. The stings of bees and wasps are generally painful but not of great importance unless the person has become hypersensitive to the toxin (Chap. 48).

## REFERENCES AND FURTHER READINGS

Borror, D. J., and R. E. White. 1970. *A Field Guide to the Insects of America North of Mexico*. Houghton Mifflin Co., Boston. (An excellent small volume for identification of insects).

Fincher, G. T. 1973. Dung beetles as biological control agents for gastrointestinal parasites of cattle. *J. Parasitol.* 59:396–399.

Harwood, R. F., and M. T. James. 1979. *Entomology in Human and Animal Health*, 7th ed. Macmillan Publishing, Co., New York.

Herried, C. F. 1977. *Biology*. Macmillan Publishing Co., New York.

Herms, W. B., and M. T. James. 1961. *Medical Entomology*, 5th ed. Macmillan, New York.

Knipling, E. F. 1979. *The Basic Principles of Insect Population Suppression and Management*. U.S. Department of Agriculture. Agricultural Handbook No. 512.

Miller, A. 1954. Dung beetles (Coleoptera, Scarabeidae) and other insects in relation to human feces in a hookworm area of southern Georgia. *Am. J. Trop. Med. Hyg.* 3:372–389.

Mohr, C. O. 1943. Cattle droppings as ecological units. *Ecol. Monogr.* 13:275–298.

Romoser, W. S. 1981. *The Science of Entomology*, 2nd ed. Macmillan Publishing Co., New York.

Ross, H. H. 1956. *A Textbook of Entomology*, 2nd ed. John Wiley & Sons, New York.

# 51

# *The Lice: Mallophaga and Anoplura*

Although the Mallophaga and Anoplura are usually placed in separate orders, they are much alike in both structure and life cycle. They are small, wingless insects somewhat flattened dorsoventrally and with moderate setation. They are all parasitic on birds and mammals, where they feed and spend all of their lives; they are hemimetabolic in development and have three nymphal instars. The operculated eggs are attached to hairs or feathers and hatch in about a week to release the first instar nymph. The nymphal stages look much like the adults except for size. Members of both orders have a high degree of host specificity, with each species being found on only a single species of host. They sometimes feed on other hosts for a short period but will not usually remain for more than a day or two.

## MALLOPHAGA

The Mallophaga are the chewing or bird lice. There are about 3000 species in the order; most of them are found on birds, but many mammals have at least one species of biting louse. Humans have no biting lice, but these lice are important on domestic animals, especially chickens, where they cause significant economic loss. The biting lice have opposing mandibles by which they feed on skin scurf or feathers, or cut into the skin of the host and feed on oozing fluids and blood. A few species have mouthparts adapted for piercing and sucking. None is known to transmit infectious agents of humans, but they are hosts for two filarioid nematodes of canids and anatids. *Dipetalonema reconditum* seems to develop in the chewing louse (*Heterodoxus spiniger*) and in the biting louse (*Linognathus setulosus*). *Sarconema eurycera* of geese and swans develops in the chewing louse, *Trinoton anserinum*.

### *Biting Lice of Domestic Poultry*

Among domestic animals, the greatest losses from Mallophaga probably occur in chickens. Chickens are raised under crowded conditions, or are at least housed at night in coops where they are crowded and the humidity is high. The following are common biting lice of chickens (Fig. 51.1):

*Menacanthus stramineus*—the body louse, little yellow louse

*Menopon gallinae*—the shaft louse

*Cuclutogaster heterographus*—the head louse

*Goniocotes gallinae*—the brown chicken louse

*Lipeurus caponis*—the wing louse

Not only are these lice limited to chickens, but each species has its preferred location on the host. The threat of economic problems from lice is always present when chickens are raised. Most Mallophaga feed on feathers and skin scurf, but the body louse may also feed on

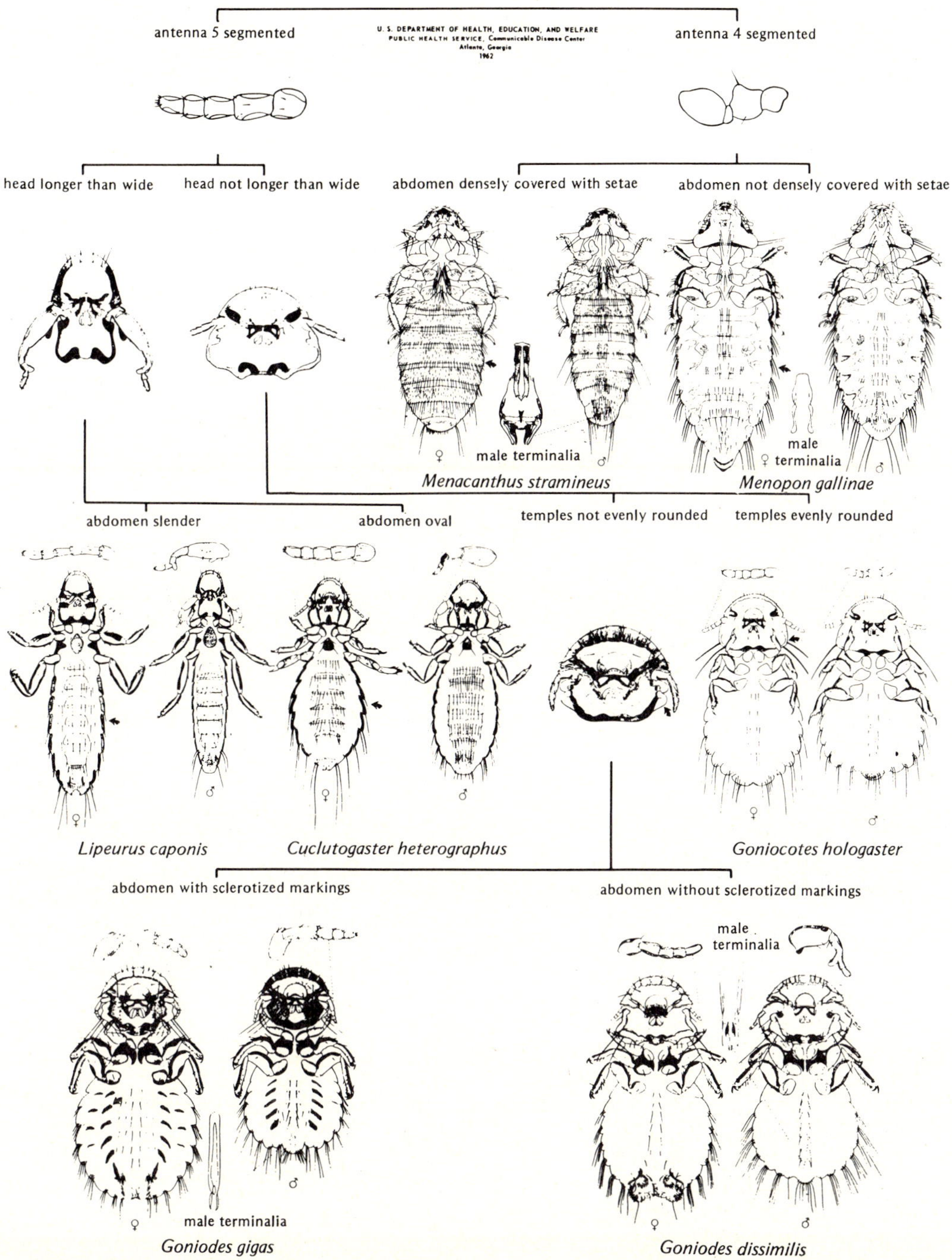


**FIG. 51.1**
Biting lice of chickens.

feather shafts, causing them to rupture; it may also cause the skin to bleed. Birds become restless merely because the lice move about on the skin; they are more bothered when the skin is bitten, and they fail to feed properly, which reduces weight gains as well as egg production.

Typical of lice in general, chicken lice remain on the host nearly all the time; they transfer from one animal to another, usually by direct contact, since they do not live for long off the host. Populations build up rapidly under moderate temperatures and high relative humidity such as might be found in chicken houses. When animals become restless, it can be determined whether lice are the problem by simply ruffling the feathers of a bird the wrong way and looking for the lice on the skin. The head, vent, and undersides of the wings should be looked at with care.

Chicken lice are typically controlled by insecticidal dusts. Care must be taken to ensure that the dust penetrates the feathers so that the toxicant will get down to the level where the lice are. Dusts containing malathion, coumaphos, or sodium fluoride are currently in use. If chickens have roosts, the roost can be painted with a solution of nicotine sulfate.

## *Other Mallophaga on Birds*

Turkeys have two lice: *Chelopistes meleagridis*, the large turkey louse, and *Oxylipeurus polytrapezius*, the slender turkey louse. *Columbicola columbae* is a parasite of pigeons and is infrequently found on other birds.

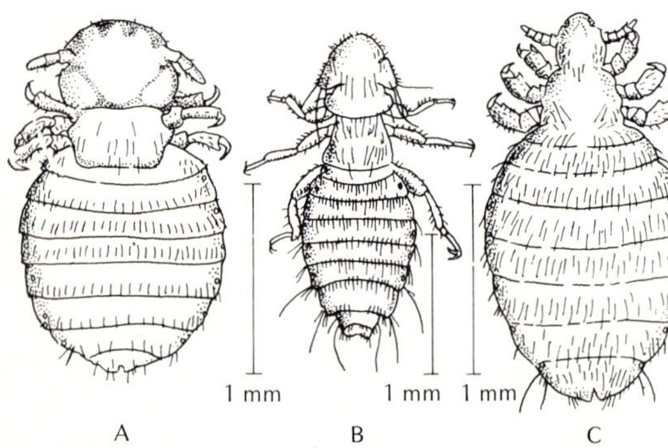


**FIG. 51.2**
Lice of the dog. (*A*) *Trichodectes canis;* (*B*) *Heterodoxus longitarsus;* (C) *Linognathus setosus.*

An interesting study was done by Foster (1969) on some museum specimens of tropical birds. She found that the numbers of Mallophaga fluctuated with the time of year and the breeding cycle of the bird. It appears that lice reproduction is timed to coincide with the reproduction of the hosts to maximize the probability of transfer to new hosts.

Some taxonomic works on Mallophaga are those of Ferris (1951) and Kim & Ludwig (1978). Clay (1970) proposed that the lice are polyphyletic and that the Mallophaga and Anoplura should be a single order with four suborders.

## *Mallophaga of Mammals*

Domestic and companion animals have one or two species of biting louse each. For example, dogs have *Trichodectes canis* and *Heterodoxus longitarsus* (Fig. 51.2) and cats have *Felicola subrostrata* (Fig. 51.3).

The biting lice of dogs and cats may become quite irritating. Unlike fleas, lice remain on their hosts and once the infestation is eliminated, recolonization takes place only if there is contact with another lousy animal. Insecticidal dusts and tick and flea collars are the usual mode of treatment.

Large domestic animals usually have only one or two biting lice: *Bovicola bovis* on cattle, *B. pilosa* on horses, and *B. ovis* on sheep. Studies on *B. bovis* in New York showed that the numbers of lice fluctuated depending on the season (Matthysse, 1944, 1946). When popula-

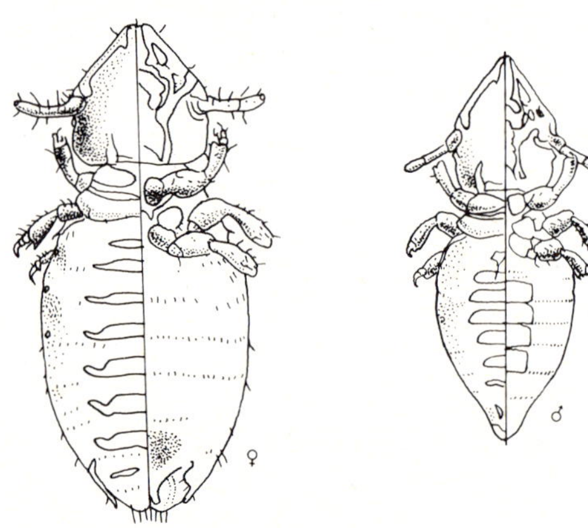


**FIG. 51.3**
*Felicola subrostrata,* biting louse of the cat.

tions were at a low level, the lice were clustered on the head and at the base of the tail. As numbers increased, they moved out along the back and then downward to cover the whole body. Males constituted less than 10% of the populations and in some instances less than 2%. It was possible to rear two generations without any males, and it was concluded that reproduction took place principally by parthenogenesis. The life cycle required about 29 days from egg to egg.

## ANOPLURA

Sucking lice are generally somewhat larger than biting lice and they have narrow heads; another distinguishing feature is that the antennae of the sucking lice are extended out from the head rather than pointed toward the posterior, as in many mallophagans. There are 400-odd species on placental mammals. Members are vectors of important disease agents of humans and of domestic animals.

### *Lice and Louse-Borne Diseases of Humans*

There are three sucking lice found on humans: *Pediculus humanus* L., the body louse, *P. capitis* DeGeer, the head louse, and *Pthirus pubis (L.)*, the crab louse (Fig. 51.4). *P. humanus* and *P. capitis* have been considered to be varieties of *P. humanus*; however, there are slight structural differences between them, and more important, the two will not interbreed (Busvine, 1978).

The prevalence of persons with lice and the numbers of lice per person are related to crowding, the quality of living conditions, and personal hygiene. In Ethiopia, those persons who washed themselves and their clothing frequently were usually free of lice, or at least had low populations of them (Sholdt et al., 1979).

Head lice are a recurring problem among school children, and in the 1970s there seemed to be an increasing incidence.

Body lice spend most of their time in clothing, and oviposit there rather than on the body; wearing apparel is therefore a source of infestation and where one looks for eggs (nits). When conditions are good, head and body louse populations can increase rapidly; a female head louse lays 50 to 150 eggs and a body louse 275 to 300 during a lifetime. The time from egg to egg is about 3 weeks in both species; it has been estimated there can be a 4000- to 5000-fold increase in the population in 3 months (Harwood & James, 1979).

Head and body lice tend to feed frequently and to cause a good deal of irritation of the skin. There is some indication that persons become tolerant to lice over a period of time. Whether immunity develops is still problematic, but circumstantial evidence indicates that it does. (Sholdt et al., 1979). In feeding, lice pump blood through the intestinal tract and soil the skin and clothing of the host with their feces.

Crab lice (papillon d'amour) are often thought of as being venereally transmitted, and this is sometimes the case, but as with other lice, it is mostly a matter of close personal contact between individuals. This species tends to feed almost continuously and to cause considerable itching and skin irritation. Its life cycle requires about a month from egg to egg, but the females are not as prolific as the head and body lice, producing only about 30 eggs in a lifetime.

Lice do not live long off a host; unfed lice can survive about 10 days. The nits are quite resistant, however, and are a residue of infection when control measures are instituted. When lice are recognized as being present, both the person and the clothing must be treated. Reduction in populations can be achieved by dusting inside the clothing with an organochlorine insecticide. A better approach is to wash and heat-treat all clothing and to treat the person with an insecticide. Clothing can be fumigated with methyl bromide or heated to about 90°C (200°F). Most often a single treatment of the person and clothing is not sufficient to eliminate the lice; a second treatment is usually given routinely a week after the first one.

Body lice are vectors of some important disease agents of humans; head lice and crab lice are not usually vectors but can be infected under experimental conditions. The principal diseases are louse-borne or epidemic typhus, caused by *Rickettsia prowazeki*; trench fever, also caused by a rickettsia (*Rochalimaea quintana*); and epidemic relapsing fever, caused by the spirochete *Borrelia recurrentis*.

Of these three diseases, typhus is the most important and has had a decisive effect on human welfare and activities throughout history; Hans Zinsser's (1935) history of the impact of infectious diseases on humans is both readable and informative, and he discusses typhus in some detail. Typhus follows poverty, war, and

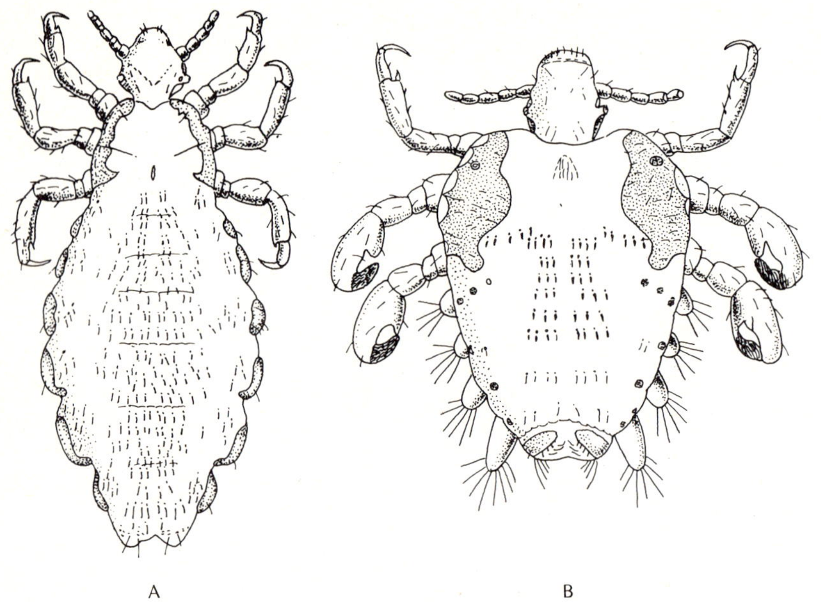

**FIG. 51.4**
Lice of humans. (*A*) *Pediculus humanus* and *P. capitis;* (*B*) *Pthirus pubis.*

natural disaster; wherever there is crowding, interruption of sanitary services, or displacement of people, typhus appears. Humans are the only host of the rickettsia, and lice do not normally live on any host other than humans, so we have a situation in which a low-grade infection in a population erupts when conditions are favorable for it. Infection apparently is maintained in a few individuals for years; when they are placed under difficult living conditions, the infection becomes active and transmission is possible. When a person has a rickettsemia, a louse can be infected and the organisms then multiply in the cells of its gut. The infected cells rupture, releasing large numbers of rickettsiae; transmission takes place through the feces of the louse or when a louse is crushed on the skin. The agent enters the body either through a break in the skin or through the unbroken mucous membranes of the eye, nose, or mouth. The rickettsiae can remain alive for up to 60 days in louse feces, so an area can be contaminated and infection can remain a hazard for some time.

Mortality from epidemic typhus ranges from 10 to 100% in any one outbreak. There is a generalized infection in the body, with the organisms multiplying in the endothelial cells of the smaller blood vessels. When the blood vessels break down, blood leaks into the skin and a skin rash of small hemorrhages appears. There is a good chance of recovery if broad-spectrum antibiotics are given. The impact of typhus can be appreciated when it is considered that in World War I there were 3 million deaths in Europe and 100,000 deaths in North Africa in World War II. A serious epidemic was averted in Naples, Italy, in 1943–44 by the use of DDT dust applied to the clothing of every person who could be reached.

Typhus is peculiar among infectious diseases of humans in that humans are the only known host for both the infectious agent and the vector. It would therefore seem to be an easy disease to eradicate, but it smolders somewhere, awaiting the sort of disruption that will allow it to recur.

Epidemic relapsing fever, caused by *Borrelia recurrentis*, is not as severe a disease as typhus, but it occurs under much the same circumstances. Death rates are less than 5%; currently the known focus of infection is in Ethiopia, but an epidemic occurred in southeast Asia in the 1970s. Transmission from the louse to humans is through the feces of a louse or due to a louse being crushed, releasing organisms onto the skin or mucous membranes. Relapsing fever tends to run a long course with recurring episodes of fever and prostration. Antibiotics are effective. Tick-borne relapsing fever (Chap. 57) is related, but the agent is adapted to transmission by soft ticks.

Trench fever is a short-lived, acute disease with a low

mortality. The individual is prostrated and in a great deal of physical discomfort for about 48 hours, recovers, and then is clinically affected about 5 days later; additional attacks ensue with diminished severity.

## *Lice of Domestic Herbivores*

Lice on cattle and sheep are herd problems, and although there are several species on each (Table 51.1, Figs. 51.5, 51.6), they can usually become problems for the same reasons and are controlled by the same measures. Louse populations fluctuate with the seasons and are more severe in temperate than in warm climates. Populations are generally low during the summer when skin temperatures are high; lice usually cannot survive long at a temperature above 38°C (100° F). When the temperature drops in the fall and animals are gathered together, transfer among animals occurs and populations increase. The USDA estimates that between $50 million and $100 million is lost annually from lice in domestic animals, mostly cattle and sheep. Losses are due to decreased weight gains and damage to fleeces; animals become anemic due to the blood sucking and probably are susceptible to other infections.

Treatment should be given before lice become a problem for two reasons; first, losses will not occur, and, second, treating animals during the cold winter months is hard on them. The standard treatment for lice has been to dip the animals in a contact insecticide. In attempting to reduce the cost of dipping, other means of applying insecticides have been developed and much of the technology is now based on organophosphate insecticides, which may be applied as a spray or a pour-on or injected. There are many reports, such as that of Campbell et al.(1978) in Nebraska, which show that louse populations can be reduced about 90% by all three methods of application.

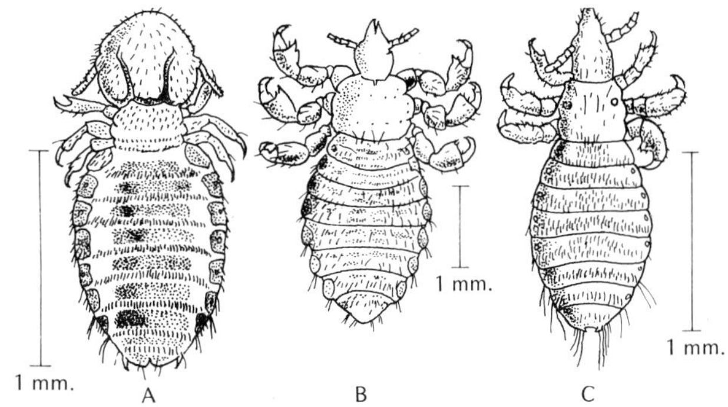


**FIG. 51.5**
Lice of cattle. (*A*) *Bovicola bovis;* (*B*) *Haematopinus eurysternus;* (*C*) *Linognathus vituli.*

In any cattle herd there are always a few animals that are carriers, that is, animals especially susceptible to

**TABLE 51.1**
Lice of Domestic and Companion Animals

| HOST | ORDER | SCIENTIFIC NAME | COMMON NAME |
|---|---|---|---|
| Cattle | Mal | *Bovicola bovis* | Little red cattle louse |
| | Ano | *Haematopinus eurysternus* | Short-nosed cattle louse |
| | Ano | *H. tuberculatus* | Buffalo louse |
| | Ano | *H. quadripertussis* | Tail louse |
| | Ano | *Linognathus vituli* | Long-nosed cattle louse |
| | Ano | *Solenopotes capillatus* | Little blue cattle louse |
| Sheep | Mal | *Bovicola ovis* | Biting louse of sheep |
| | Ano | *Linognathus ovillus* | Body or blue louse |
| | Ano | *L. pedalis* | Foot louse |
| Swine | Ano | *Haematopinus suis* | Hog louse |
| Dog | Mal | *Trichodectes canis* | Biting louse of dogs |
| | Mal | *Heterodoxus longitarsus* | Biting louse of dogs |
| | Ano | *Linognathus setosus* | Sucking louse of dogs |
| Cat | Mal | *Felicola subrostrata* | Biting louse of cats |
| Horse | Mal | *Bovicola equi* | Biting louse of horses |
| | Ano | *Haematopinus asini* | Sucking louse of horses |

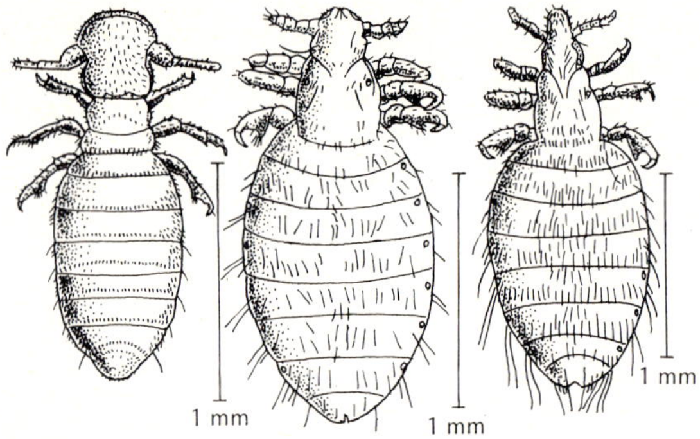


**FIG. 51.6**
Lice of sheep. (*A*) *Bovicola ovis;* (*B*) *Linognathus pedalis;* (C) *Linognathus africanus.*

lice. They maintain infections at a higher level than other animals in the herd and serve as a reservoir of lice when their numbers are likely to increase. There have been a number of explanations for the high susceptibility of carrier animals (reviewed by Sholdt et al., 1979), but none has survived experimental scrutiny.

## REFERENCES AND FURTHER READINGS


Busvine, J. R. 1978. Evidence from double infestations for the specific status of head lice and body lice (Anoplura). *Syst. Entomol.* 3:1–8.

Campbell, J. B., R. G. White, J. B. Baxter, and D. J. Boxler. 1978. *Cattle Louse Control Trials in Nebraska.* Report No. 5. *Ag. Exp. Stat.,* University of Nebraska, Lincoln.

Clay, T. 1970. The Amblycera (Phthiraptera: Insecta). *Bull. Br. Mus. Nat. Hist. (Entomol.)* 25:73–98.

Ferris, G. F. 1951. *The Sucking Lice.* Pacific Coast Entomology Society Memoir. 1.

Foster, M. S. 1969. Synchronized life cycles in the orange-crowned warbler and its mallophagan parasites. *Ecology* 50:315–323.

Harwood, R. F., and M. T. James. 1979. *Entomology in Human and Animal Health,* 7th ed. Macmillan Publishing Co., New York.

Kim, K. C., and H. W. Ludwig. 1978. The family classification of the Anoplura. *Syst. Entomol.* 3:249–284.

Matthysse, J. G. 1944. The biology of the cattle biting louse and notes on cattle sucking lice. *J. Econ. Entomol.* 37:436.

——. 1946. *Cattle Lice, Their Biology and Control. Ag. Exp. Stat. Bull. No. 832.* Cornell University, Ithaca, N.Y.

Sholdt, L. L., M. L. Holloway, and W. D. Fronk. 1979. *The Epidemiology of Human Pediculosis in Ethiopia.* Spec. Pub. Navy Dis. Vect. Ecol. & Control Center, NAS, Jacksonville, Fla.

Zinsser, H. 1935. *Rats, Lice and History.* Little, Brown & Co., Boston.

# 52

# *Hemiptera: Bedbugs and Kissing Bugs*

As indicated earlier (Chap. 50), the Hemiptera, or true bugs, are often divided into two orders—the Hemiptera, which have forewings that are leathery anteriorly and membranous posteriorly, and the Homoptera, which have either membranous or thickened forewings. All have sucking mouthparts and are hemimetabolic in development. The suborder Heteroptera has forewings that are leathery at the anterior and membranous at the posterior. It is this group that is of importance in medical entomology. The order is a large one, with about 23,000 species of Heteroptera and 32,000 species of Homoptera worldwide. Because they have sucking mouthparts, the members feed on juices; most are either plant feeders or are predaceous, but some feed on blood of vertebrates. Many species cause losses in agriculture. Among those that do are scale insects on citrus fruits and horticultural plants, chinch bugs, which feed on small grains, and aphids, which feed on many kinds of plants and transmit a number of viral diseases of plants.

Not all hemipterans are detrimental. The dye cochineal used originally by southwestern Indians is obtained from members of the family Dactylopiidae. Members of the family Lacciferidae provide lac, the basis of shellac. Probably the greatest impact of the hemipterans is in natural biological control; they are among the many animals that modulate populations of other species by preying on them. For example, during the warmer months water striders (family Gerridae) may be found on nearly any pond, where they feed on other small insects; a few water striders kept in an aquarium will consume large numbers of house flies if they are provided in a stunned state. Such a voracious appetite must mean that each water strider disposes of two or three insects daily.

A few bugs bite humans occasionally and may cause temporary discomfort. Some of these are giant water bugs (family Belostomatidae), water scorpions (family Nepidae), backswimmers (family Notonectidae), the wheel bug (family Reduviidae), and the masked hunter (family Reduviidae). One rather obscure family (Polyctenidae) is parasitic on bats, but the two families whose members are important bloodsuckers are Cimicidae and Reduviidae, subfamily Triatominae. It is these which we will consider in detail here.

## FAMILY CIMICIDAE: THE BEDBUGS

Bedbugs are flattened, brownish insects that lack functional wings as adults. The adults have wing pads, but the earlier instars do not. They are usually 5 mm or less in length and have many small hairs on the body. The three-segmented beak is held back under the head in a groove. Some of the genera are *Cimex, Leptocimex, Haematosiphon, Oeciacus,* and *Ornithocoris*. Birds quite commonly have bedbugs, and large numbers are likely to be found in their nests. Usinger's monograph (1966) is particularly useful for taxonomy, biology, and host–parasite interaction in cimicids.

The life cycle is typically hemimetabolic, with four nymphal stages. All stages feed on blood, and they have little host preference; although each species tends to be found on a few host species, they feed readily on other species.

The question whether bedbugs are vectors of disease agents continues to recur. They have all the characteristics one associates with vectors: blood as the sole source of food, close association with the host, frequent feeding, ready dissemination by various means. But they are not known to serve as the primary vectors of any disease agent, although a number of agents have been isolated from them and a number of agents can be propagated in bedbugs in the laboratory. Thus, the hazard of bedbugs serving as vectors seems to be minimal (Burton, 1963).

NAME OF ORGANISM. *Cimex lectularius,* the human bedbug, and *C. hemiterus,* the tropical bedbug.

HOSTS AND LOCATION. About seven species of bedbugs are associated with humans in various parts of the world, but the two preceding species are the most common; in the following discussion *C. lectularius* is the organism principally referred to. They are located on the skin.

DISTRIBUTION AND IMPORTANCE. The organisms are cosmopolitan. The importance of bedbugs is minor; they cause skin irritations, principally through the development of hypersensitivity. Heavy infestation of a home can bring about anemia, loss of sleep, and psychological trauma in its inhabitants, not to mention social disapproval.

MORPHOLOGY. Adults are about 5 mm long and 3 mm broad, with many short hairs (Fig. 52.1). The nymphs are proportionately smaller.

LIFE CYCLE. The female deposits yellowish eggs off of the host and the eggs hatch in about 10 days (range: 4–21). One to three blood meals are taken by each instar and about 8 days are required to develop from one instar to the next. Under laboratory conditions, the life cycle requires 37 to 128 days from egg to egg.

DIAGNOSIS. The presence of bedbugs in a dwelling is usually recognized when a person finds the bites in the morning. There will often be a row of three or four small wheals; this is a good indicator of bedbugs. Further searching under the bed, behind wallpaper, and in cracks of the walls usually reveals various stages including eggs. Bedbugs have scent glands that open on the ventral surface of the thorax, and the odor emitted is said to give a room a characteristic aroma if the population is fairly high.

HOST–PARASITE INTERACTIONS. Bedbugs sensitize the host to the salivary secretions they inject while feeding. There is usually no sensation while the bug is feeding, indicating that there is an anesthetic in the saliva. Over a period of time, many but not all persons become hypersensitive to the saliva of the bugs. The sites of feeding develop as raised, red, itching papules. Interestingly, some persons are not bothered at all by bedbugs and do not develop skin reactions.

TREATMENT. When a person becomes hypersensitive, local application of an ointment containing a corticosteroid reduces the symptoms, as will other topical preparations for insect bites.

EPIDEMIOLOGY AND CONTROL. Bedbugs feed on the host and then return to a hiding place where they spend most of their time. They can survive for a long time without feeding; for example, the larger instars can survive for 80 to 140 days without feeding. Temperatures below 13 to 15°C cause cessation in development and activity, and above about 37°C, *C. lectularius* populations begin to decline. Freezing kills embryonating eggs in about 2 weeks. Organisms can be readily disseminated by contaminated clothing or furniture. A person might stay in a hotel for a day or two and then leave, carrying back home a few bugs in his or her clothing, thereby setting up another colony. Likewise, a bed might be purchased at a garage sale and found later to be the home of some unwanted nocturnal raiders. Populations will build up rapidly in a home and cause considerable discomfort because of the bugs feeding every night.

When bugs are detected in a home, control involves the following:

1. Laundering all clothing that might have eggs or bugs in it
2. Eliminating hiding places of bugs
3. Treatment of the dwelling and furniture with insecticides

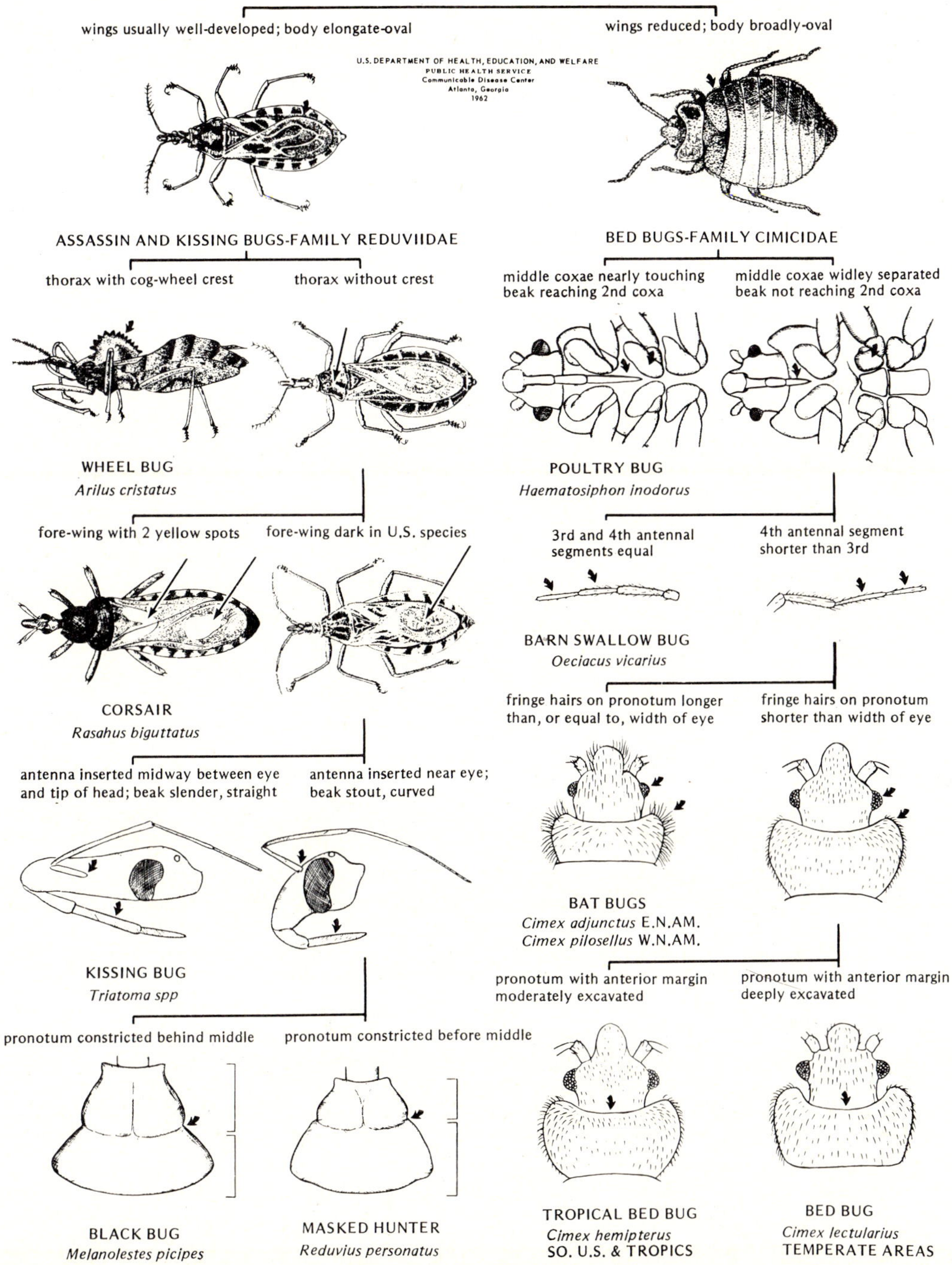


**FIG. 52.1**
Pictorial key to some adult bugs that may bite humans.

It is essential to undertake all of these steps, or residual populations will merely increase again. Clothes washing should be done with a strong detergent in hot water. Eliminating hiding places is time-consuming and involves patching woodwork, caulking cracks, and eliminating pathways by which bugs can crawl from one room or apartment to another. Bugs are often found behind wallpaper or loose baseboards, and these must be glued or nailed tightly. In some instances it is possible to paint and thereby cover up some cracks that serve as a refuge for the bugs. Insecticides currently recommended are organophosphates (malathion, ronnel, dichlorvos). These are sprayed in kerosene or water emulsion on the underside of the bed, mattress seams, woodwork, and so on. Care is needed to avoid the toxic effects of these chemicals; children are especially susceptible. In the past, fumigation of buildings was recommended as the principal means of control, but it is not necessary with the newer insecticides.

## TRIATOMINS—KISSING BUGS

The members of this group have a variety of common names: *kissing bugs, cone-nosed bugs, giant bedbugs, chinche de monte, benchuca,* and *vinchuca,* among others. Members of the subfamily Triatominae, they are bloodsuckers. The common name *kissing bug* comes from its habit of feeding on a person where the skin is thin, near the mouth or eye. Triatomins are important because they suck blood, but, more important, they are vectors of *Trypanosoma cruzi*, the cause of Chagas' disease (Chap. 4)

NAME OF ORGANISM. There are about 35 species of kissing bugs that are vectors of Chagas' disease, including *Triatoma protracta, T. rubida, T. sanguisuga, T. gerstaekeri, T. recurva, T. lectularius, T. neotomae, Rhodnius prolixus*, and *Panstrongylus megistus* (Fig. 52.2).

HOSTS AND LOCATION. Kissing bugs have a low host specificity; certain species tend to be associated with a narrow range of hosts, but they generally feed on whatever animals are available.

DISTRIBUTION AND IMPORTANCE. The vectors of *T. cruzi* are located in the Western Hemisphere from warm-temperate to tropical climates. In the United States, triatomins are found as far north as Maryland, southern Illinois, and northern California, as well as the southern states. The main vectors are in Central and South America. The bites alone of triatomins may cause fairly serious illness.

MORPHOLOGY. The adults may exceed 20 mm (3/4 in.) in length, and the nymphs range upwards from 2 to 3 mm. The heads are long and narrow, with the antennae inserting anterior to the prominent eyes. The thorax has a lateral constriction at about the middle, and the three-segmented beak is held under the head. The abdomen flares laterally and the patterned margins turn upward so that the wings of the adults nestle into this concave area.

LIFE CYCLE. Eggs are laid in animal burrows or crevices in houses where bugs spend the daylight hours. They hatch in about 2 weeks at moderate temperatures. The tiny first instar nymphs feed on either blood or hemolymph of larger instars by inserting the beak between the joints of the exoskeleton. Development is slow, requiring about a year for the five molts to take place. The adult females lay from 100 to 300 eggs in small clutches and then die.

Because the nymphs are wingless, they must walk, but the adults of most species are capable of strong flight. In the laboratory, males were capable of flying for 2 to 3 hr on the average, and up to 6 hr in one instance (Gringorten & Friend, 1979). Dissemination can therefore take place readily when the adult stage has been reached.

DIAGNOSIS. Because triatomins are nocturnal, they are most often seen at night. In addition, persons have reactions to bites received while sleeping.

CULTIVATION. It is sometimes necessary to keep bugs in the laboratory for xenodiagnosis (Chap. 4); they feed readily on small laboratory rodents such as mice.

HOST–PARASITE INTERACTIONS. Triatomins have host preferences but feed on nearly any vertebrate, sometimes even cold-blooded ones. They are attracted to a host by warmth and $CO_2$. They feed for about 10 min if undisturbed and then return to their hiding places off the host. They apparently have an anesthetic in

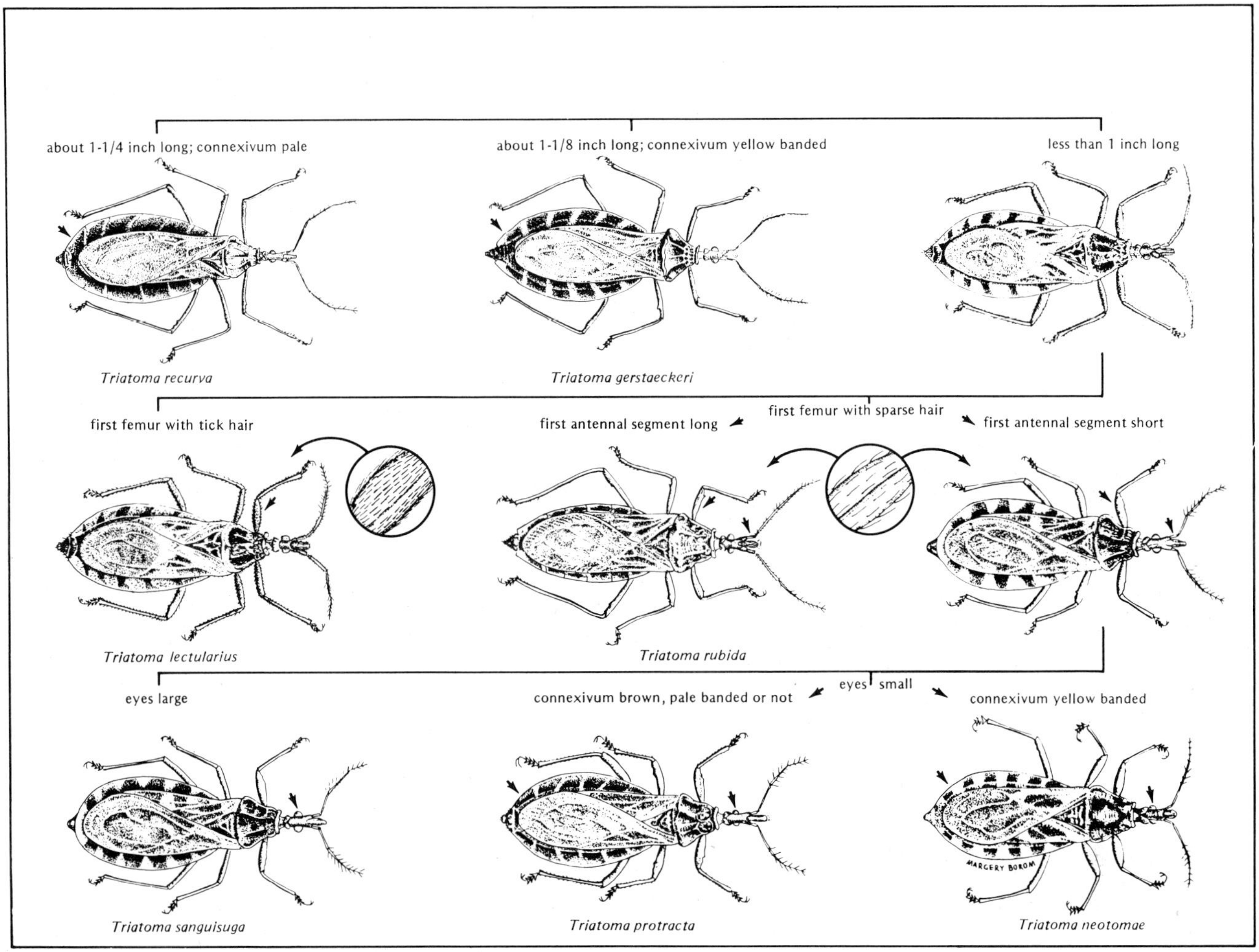


**FIG. 52.2**
Pictorial key to some common kissing bugs in the United States.

their salivary fluid since they cause no pain when feeding; nonhematophagous reduviids usually cause considerable pain.

Despite the lack of pain when feeding, triatomins' bites are likely to give rise at least to an uncomfortable lesion at the site of feeding, often to hypersensitivity, and in some cases they may cause serious illness. *Triatoma protracta* has been found to cause nausea, heartbeat irregularity, shortness of breath, and severe itching (Wood, 1953). Some persons develop strong psychological reactions to having bugs in the house and coming out to feed on them at night; these reactions are referred to as *neuroses*, but they are not entirely irrational.

The principal concern about triatomins is that they are vectors of *Trypanosoma cruzi*, the cause of Chagas' disease (Chap. 4). The important vectors of Chagas' disease become domiciliated; agents that have cycled silently in the wild move into houses and humans are at risk of becoming infected. We tend to think of triatomins as living in primitive and substandard houses, but if conditions are right, they will move into good houses and do well there. Not all species are equally good vectors (Chap. 4).

TREATMENT, EPIDEMIOLOGY, AND CONTROL. These topics are covered in Chapter 4 in connection with Chagas' disease.

## REFERENCES AND FURTHER READINGS

Burton, G. J. 1963. Bedbugs in relation to transmission of human disease. *Publ. Health Rep.* 78:13– 24.

Gringorten, J. L., and W. G. Friend. 1979. Wing-beat pattern in *Rhodnius prolixus* Stal (Heteroptera:Reduviidae) during exhaustive flight. *Can. J. Zool.* 57:391–395.

Usinger, R. L. 1944. *The Triatominae of North and Central America and the West Indies and Their Public Health Significance.* U.S. Public Health Bull. No. 288.

——. 1966. *Monograph of Cimicidae (Hemiptera- Heteroptera).* Vol. 7. Thomas Say Foundation, College Park, Md.

Usinger, R. L., P. Wygodzinsky, and R. E. Ryckman. 1966. The biosystematics of Triatominae. *Ann. Rev. Entomol.* 11:309–330.

Wood, S. F. 1953. Conenose bug annoyance and *Trypanosoma cruzi* Chagas in Griffith Park, Los Angeles, California. *Bull. Southern Calif. Acad. Sci.* 52:105–109.

# 53

# *Fleas and Flea-Borne Diseases*

The fleas (order Siphonaptera) are highly specialized holometabolic insects that are adapted structurally and physiologically for a symbiotic life. Their phylogenetic relations to other insect groups are not known, but during formation of the adult in the pupa, wing buds appear and then are resorbed. The adults are wingless, but the members of the order are related to some winged forms, probably the Mecoptera. Adults are small, seldom reaching more than 4 mm in length; they are laterally compressed and the body is well supplied with bristles, which project posteriorly (Fig. 53.1). The legs are stout and adapted for jumping; the mouthparts are of the piercing-sucking type adapted for capillary feeding (Fig. 53.2). The larvae are detritus feeders and have chewing mouthparts. All fleas are parasitic as adults; birds and mammals are their hosts, and about 1500 species are known.

As holometabolic insects, fleas have these life cycle stages: egg, three larval instars, pupa, and adult (Fig. 53.3). Under optimal conditions, the life cycle may be completed in as short a time as 3 weeks, but it may extend over 2 years or more under adverse environmental conditions or when there are no hosts on which to feed.

The characteristics of fleas that make them successful parasites and vectors of disease agents are these:

1. Bristles on the body, which keep them from being easily dislodged.
2. Mouthparts adapted for sucking blood.
3. Jumping ability. Despite their lack of wings, fleas can leap as much as 45 cm (18 in.) (Rothschild, 1965; Rothschild et al. 1973). If a host appears in its vicinity, a flea can reach it quickly by leaping onto it.
4. Flexibility in the rate of development. When conditions are optimal, the life cycle of *Pulex irritans* may take 60 days, but when a host is not available, it may require 125 days. When fed, it has been observed to live for 513 days.
5. Response to the presence of a host. Fleas may remain in the protective pupal cocoon for long periods before emerging. They are sensitive to vibrations, and when stimulated, they emerge quickly and leap to reach a host. The leaping stimulus is not known for all fleas, but when the light is suddenly reduced, *Ceratophyllus gallinae* responds as if there might be a chicken passing by.
6. Low host specificity. Fleas readily feed on many species of hosts, although they prefer one or a few species.
7. Reproduction is cued to the reproductive state of the host (Rothschild et al., 1970).

The classification and identification of fleas is usually fairly straightforward for a geographic area that has been well studied. Some illustrated keys (Fig. 53.4) have been included for North America. For additional

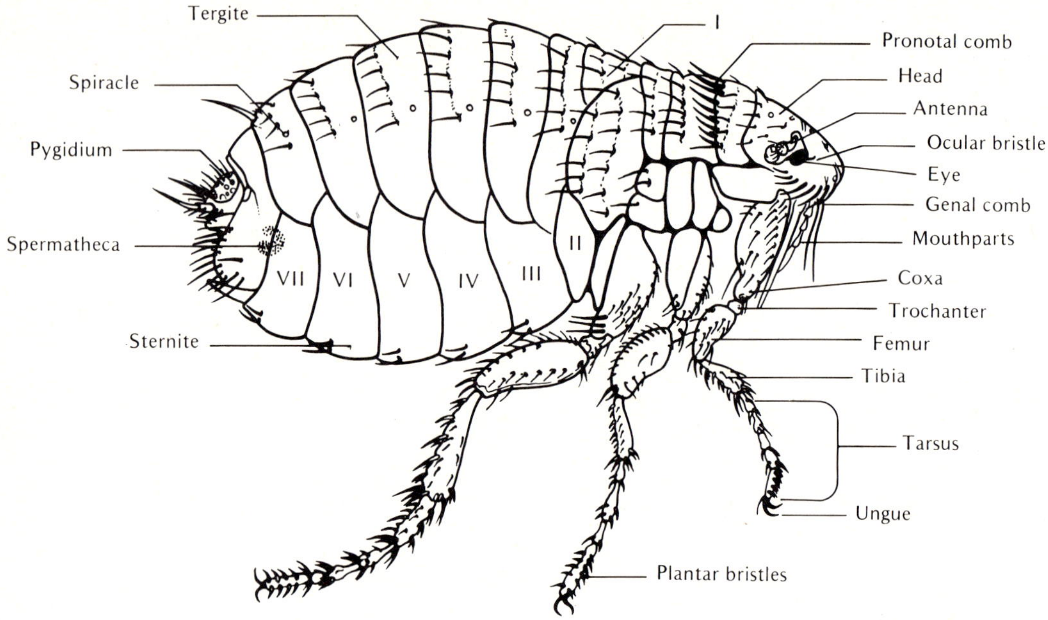


**FIG. 53.1**
An adult flea, *Ctenocephalides felis.*

information, the following are useful: Holland (1949) for North America; Holland (1964) for some general aspects of classification and evolution; Hubbard (1947) and Jellison et al. (1953) for North America; Hopkins and Rothschild (1953, 1956, 1962, 1966, 1971) for a worldwide treatment. Traub and Starcke (1980) edited a book that gives considerable information on fleas.

NAME OF ORGANISM. *Pulex irritans*, also known as the *human flea.*

HOSTS AND LOCATION. Humans, swine, dogs, and other domestic animals are hosts. Adults are occasional feeders on the host; larvae are free-living.

DISTRIBUTION AND IMPORTANCE. The organism is cosmopolitan, usually in warm, moist climates. Adults irritate and sometimes cause hypersensitivity of the hosts on which they feed. They are vectors of plague and perhaps murine typhus (Table 53.1).

MORPHOLOGY Adults are up to about 2 mm long. Females are slightly larger than the males. They lack both genal and pronotal combs (Fig. 53.4).

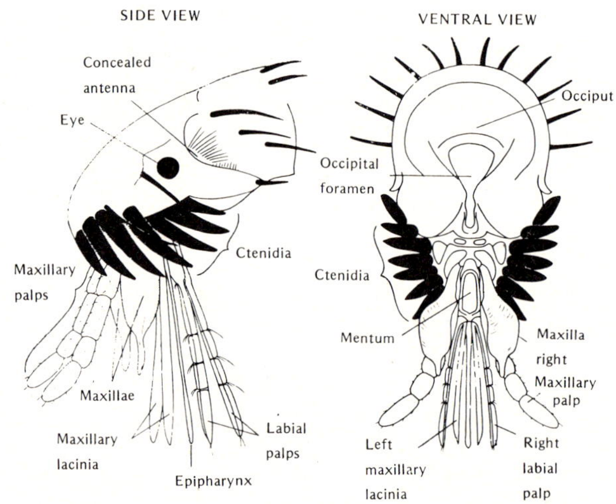


**FIG. 53.2**
Mouthparts of a flea. [From Harwood & James, 1979.]

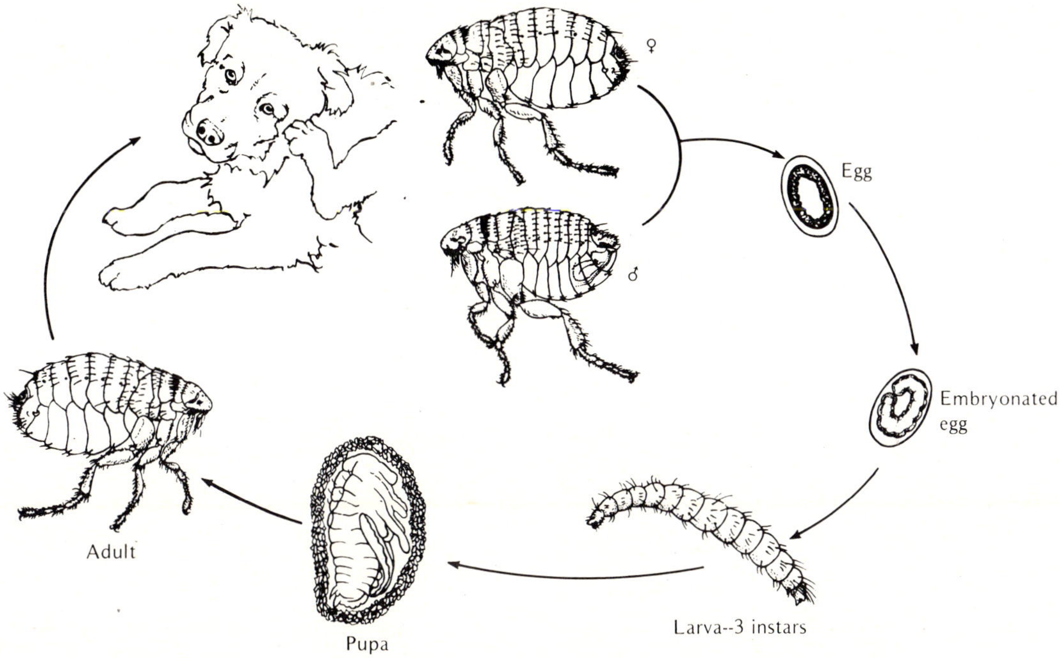


**FIG. 53.3**
Life cycle of a flea.

LIFE CYCLE. The glistening, white eggs are usually laid off the host; a female may lay as many as 448 eggs over a 196-day period. Eggs hatch in 7 to 10 days at an optimal temperature of 25°C. At 35 to 37°C or at 11 to 15°C, development is slower. A relative humidity of 65 to 95% is required for development and survival.

The larvae are elongate, legless, and have sparse bristles. There are three larval stages that may be completed in as few as 9 to 15 days under optimal conditions but may extend over as many as 200 days when food is scarce or the temperatures are low. The larvae feed on organic matter that has a high protein content: dried blood passed as feces from adult fleas and animal feces. The third larval instar spins a loose cocoon and the adult may emerge in as few as 7 days or as many as 300 days, depending on the environmental temperature. The time from egg to adult is usually 63 to 77 days.

DIAGNOSIS. Most often, persons do not see fleas, but rather have small, reddened areas on the skin where the fleas have bitten them. A careful search will usually turn up larvae, pupae, and adults in the furniture, under the bed, or in cracks in the woodwork.

HOST–PARASITE INTERACTIONS. *P. irritans*, an occasional feeder, takes a blood meal and then returns to some hiding place. The biting causes skin irritation in the form of a raised, itching papule, and hypersensitivity may occur in some persons. Persons who live in an area where they are prevalent are often unaffected by fleas, indicating the development of immunity.

TREATMENT. Symptomatic treatment of the skin is adequate in uncomplicated bites.

EPIDEMIOLOGY AND CONTROL. Control of *P. irritans* is directed at those fleas that are in houses and attack humans. Vacuuming rugs, furniture, and cracks where the fleas hide should be done first. If the dog has a place in the house where it lies, the bedding should be thrown out and new bedding provided. Insecticides such as an organophosphate or carbamate powder can be sprinkled in areas where the fleas are likely to lurk. A flea collar for the dog helps to reduce the numbers further.

NAME OF ORGANISM. *Ctenocephalides felis*, the cat flea, and *C. canis*, the dog flea.

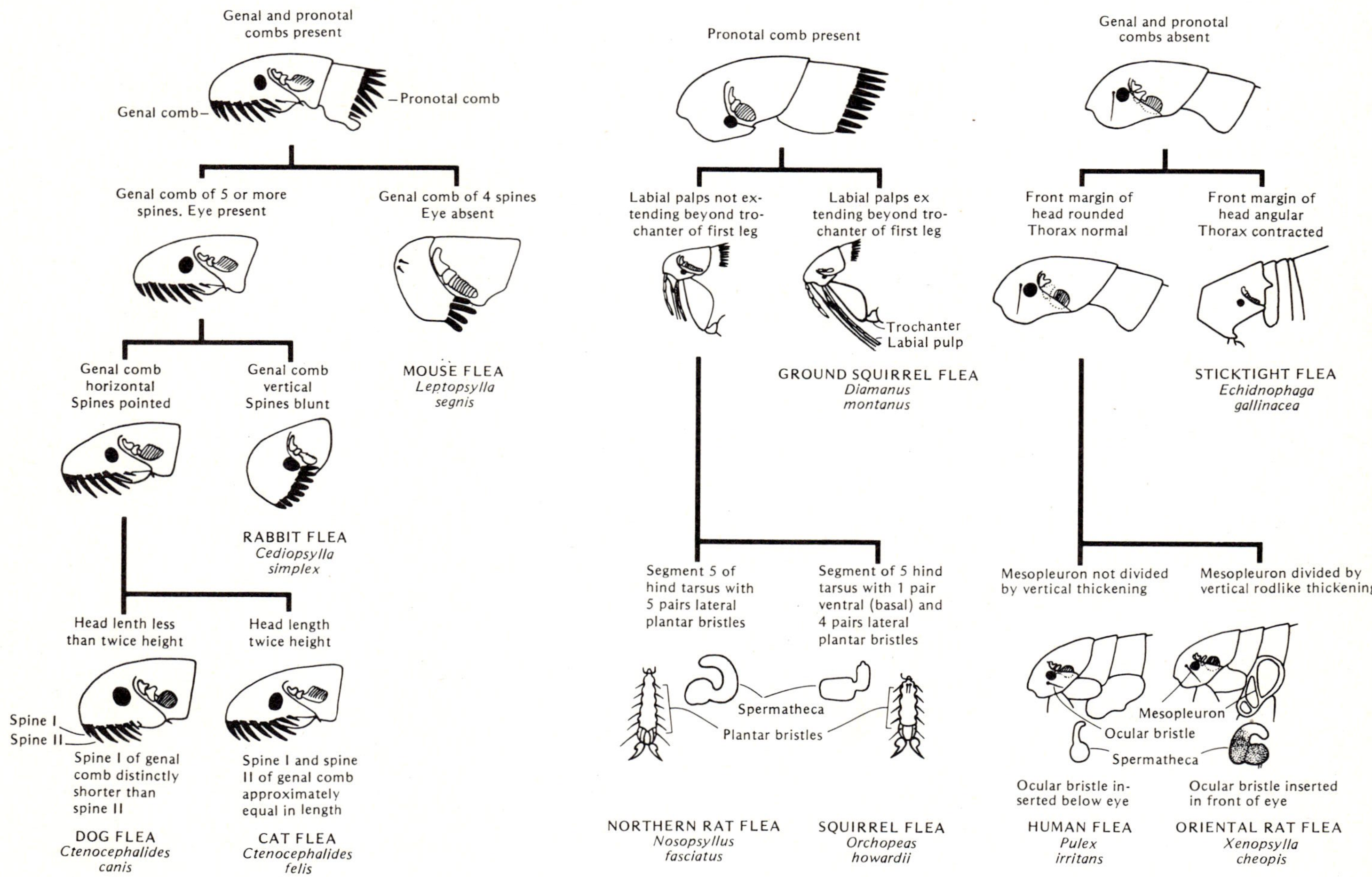


**FIG. 53.4**
Pictorial key to some common fleas of the United States.

**TABLE 53.1**
Disease Agents Transmitted by Fleas

| DISEASE | CAUSATIVE AGENT | NORMAL HOST | VECTORS | HUMAN OR DOMESTIC ANIMALS INFECTED |
|---|---|---|---|---|
| Plague | *Yersinia pestis* (*Pasteurella pestis*) | Rodents | *Xenopsylla choepis* *Diamanus montanus,* *Nosopsylla fasciatus,* *Opischrostis labis* and many others | Humans, dogs and cats occasionally |
| Murine typhus, endemic typhus | *Rickettsia typhi* (*R. mooseri*) | Rats | *Xenopsylla cheopis,* *Nosophylla fasciatus,* Louse: *Polyplax spinulosa,* Mite: *Ornithonyssus bacoti,* | Humans |
| Myxomatosis | Myxoma virus | Rabbit: *Oryctolagus cuniculi* | *Spilopsyllus cuniculi* | No |
| *Dipylidium* infection | *Dipylidium caninum* | Dog and cat | *Ctenocephalides felis and C. canis* | Humans |
| Dwarf tape-worm infection | *Hymenolepis nana* (*Vampirolepis nana*) | Humans, rodents(?) | *Xenopsylla cheopis,* *Nosopsylla fasciatus,* *Pulex irritans,* and others | Humans |
| *Hymenolepis* infection | *Hymenolepis diminuta* | Rodents | *Nosopsyllus fasciatus,* *Xenopsylla cheopis,* Other insects: beetles | Humans |
| *Dipetalonema* infection | *Dipetalonema reconditum* | Raccoon | *Ctenocephalides felis,* *C. canis,* and others | Dog |

HOSTS AND LOCATION. Both species are found on dogs and cats, as well as on humans occasionally. They are located on the skin.

DISTRIBUTION AND IMPORTANCE. Distribution is cosmopolitan except in climates with low relative humidities. The organisms cause skin irritation, usually of a minor nature, and serve as vectors of the double-pored tapeworm, *Dipylidium caninum* (Chap. 29).

MORPHOLOGY. Their length is from 1.5 to 2 mm. The head of the dog flea is proportionally shorter than that of the cat flea and genal spine I is shorter than genal spine II (Fig. 53.4).

LIFE CYCLE. The life cycle follows the same pattern as *P. irritans*. Under optimal conditions, the time from egg to adult is 24 days.

DIAGNOSIS. Finding fleas on a host or sometimes in the bedding of an animal is the means of diagnosis. When fleas are on a dog or cat, the fur can be ruffled the wrong way to see the fleas moving through the fur or on the skin.

HOST–PARASITE INTERACTIONS. Adults feed intermittently but may remain on the host between feedings. The bites of these fleas cause both animals and people considerable irritation.

TREATMENT. The dog or cat may be dusted with a powdered insecticide. Tick and flea collars are somewhat effective. Cats are especially susceptible to the toxic effect of organophosphate insecticides and should be treated with care.

EPIDEMIOLOGY AND CONTROL. As with *P. irritans*, areas where adults and larvae can hide should be vacuumed and animal bedding replaced. In mild climates, significant numbers of fleas are found out of doors and they may be brought into the house continually. This makes for an unending flea problem on dogs and cats, but no attempt is generally made to reduce flea populations outside the house.

NAME OF ORGANISM. *Tunga penetrans* (Linnaeus, 1758) Jarocki, 1838, the chigoe.

HOSTS AND LOCATION. The female embeds in the skin of both birds and mammals. It is found in a variety of burrowing animals such as rodents and some owls. Swine, dogs, and humans are also important hosts.

DISTRIBUTION AND IMPORTANCE. Distribution was originally in South America, but the organism was carried to Africa during the nineteenth century. Most concern is with humans in whom there may be secondary bacterial infections of the skin.

MORPHOLOGY. The female has a sharply angled head and a globular abdomen (Fig. 53.5); the engorged, ovigerous female may reach 5 mm or more in diameter.

LIFE CYCLE. The female enters the skin by slashing it with the mouthparts and then burrows in until only the last two segments of the abdomen are at the surface. Oviposition then takes place; eggs are sometimes retained in the skin. The mouthparts extend as deeply as the capillaries from which the female feeds. Both females and males feed on blood. The rest of the cycle takes place off the host.

DIAGNOSIS. This infection is well known in tropical America and swelling of the skin as a watery blister or a localized inflammation is usually recognized as a chigoe. Successful treatment is evidence that the infection has been correctly diagnosed.

HOST–PARASITE INTERACTIONS. Once in the skin, considerable irritation takes place as the female feeds and engorges; the lesion becomes secondarily infected with bacteria and painful inflammation results. Invasion of the wound with anaerobic spore-forming bacteria may cause gangrene. Although any part of the body may be attacked, the feet and especially the area under the toenails are favored sites. Even in uncomplicated infections, a flea 3 to 5 mm in diameter under the toenail would be quite painful.

TREATMENT. Once it is suspected that a skin lesion is caused by *T. penetrans*, the treatment is quite simple: the flea is dug out with a clean needle and the wound treated with a bactericidal solution. If the bacterial infection has progressed, more extensive chemotherapy is required.

EPIDEMIOLOGY AND CONTROL. The original distribution of the chigoe was from about the northern extremity of South America to about central Brazil (10° South), but it has been reported as far south as Uruguay (32° South). During the nineteenth century, this organism was imported into western Africa, probably the Congo drainage, as a by-product of the slave trade. It was probably carried in sand used as ballast in slave ships sailing from the New World back to Africa. The migration of the flea through tropical Africa was rapid;

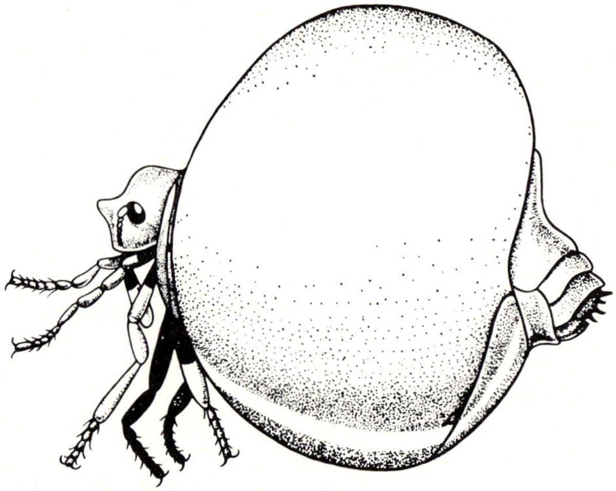

**FIG. 53.5**
*Tunga penetrans*, the chigoe.

within a short time of its being taken to Africa in the 1870s or 1880s, it had been found 200 miles from the coast at Stanley Pool. By the 1890s, it had extended its range to 1500 miles, to the western Rift Valley and Lake Victoria. Unfortunately, the local people did not know the cause of their foot infections and therefore did not know how to treat them. There were instances in which so many people were disabled by the infections that whole villages starved, because the crops were not tended or harvested.

Wearing shoes is usually recommended to prevent infection, but although this is good advice, people living in poor countries may not be able to comply. Persons visiting tropical endemic areas should be aware of the likelihood of infection and either wear shoes or treat themselves when infection is first seen.

## PLAGUE

Plague, caused by the bacterium *Yersinia pestis (=Pasteurella pestis)*, is also referred to as *bubonic plague* and the *black death*. Plague is one of the important diseases of human history. It has followed trade routes, decided the course of wars, and killed untold millions of people during historic times.

Plague is described in the Old Testament; it seems likely that it had its original focus in Africa. Its spread from North Africa and the Middle East to Europe did not take place until the sixth century. Four outbreaks were described by St. Bede in England in the seventh century. In the fourteenth century, Europe was struck with recurring epidemics of plague in which three-fourths of the people in some areas were killed. It has been estimated that 25 million persons died before the epidemics subsided. The impact of these epidemics brought enormous economic and social change, and played a crucial role in the decline of the feudal system in central Europe (Tuchman, 1978; Langer, 1964). The horror of one of the plague epidemics was graphically described by Samuel Pepys in 1664–1665 when the Great Plague of London killed 75,000 persons out of a population of 450,000.

Recent outbreaks of plague have become pandemics, worldwide epidemics, because of rapid transport along trade routes. Over the years 1894 to 1922, plague was spread from Chinese seaports, most likely with pelts of Siberian marmots (Fig. 53.6).

Plague is still a disease of worldwide concern. Where there is disruption of the social system, natural disaster, or poor living conditions, plague is likely to occur. A focus of infection of an especially pathogenic plague organism, which has not been satisfactorily controlled, exists in southeast Asia. In the United States, foci of plague occur from the high plains through the Rocky Mountains to the West Coast (Fig. 53.7).

*Yersinia pestis* cycles in rodent populations in the wild or in urban areas. In most instances, plague is a disease of wild rodents such as prairie dogs and ground squirrels. On occasion it spills over into urban rodents and then into the human population. Rodents not only become infected, but also suffer from clinical disease and may die from plague. Many rodents such as *Rattus norvegicus*, the Norway rat, become infected and maintain a bacteremia for weeks. An infected animal ultimately may die, but recovery is likely. During the bacteremia, fleas feeding on the animal become infected and the bacteria multiply in their intestines.

Transmission takes place from the flea back to another rodent primarily by so-called blocked fleas. In some species of fleas, the bacteria multiply well and there are so many organisms that they tend to move into the anterior portion of the intestinal tract and to block the proventriculus. The bacteria make a sticky mass here, forming a sticky plug that prevents the esophageal pump from taking in blood. The flea attempts to suck blood, fails to do so and, when it relaxes, ejects bacteria back into the wound. Transmission may also take place through fecal droplets contaminated with plague bacilli, which may be rubbed into the skin or enter through a break in the skin such as the wound made by a feeding flea. Handling infected rodents or rabbits is sometimes a means of transmitting infection. Inoculation is the primary mode of infection, however.

Fleas differ in their capacity to transmit plague, and the concept of vector efficiency (VE) has been developed to give a quantitative estimate of how well a species may transmit the agent. VE is the percentage of fleas that become infected after feeding on a host with a bacteremia; however, other factors enter into the efficacy of any species as a vector:

1. Blocking potential: the percentage with a blocked proventriculus
2. Length of life of the flea
3. Population size

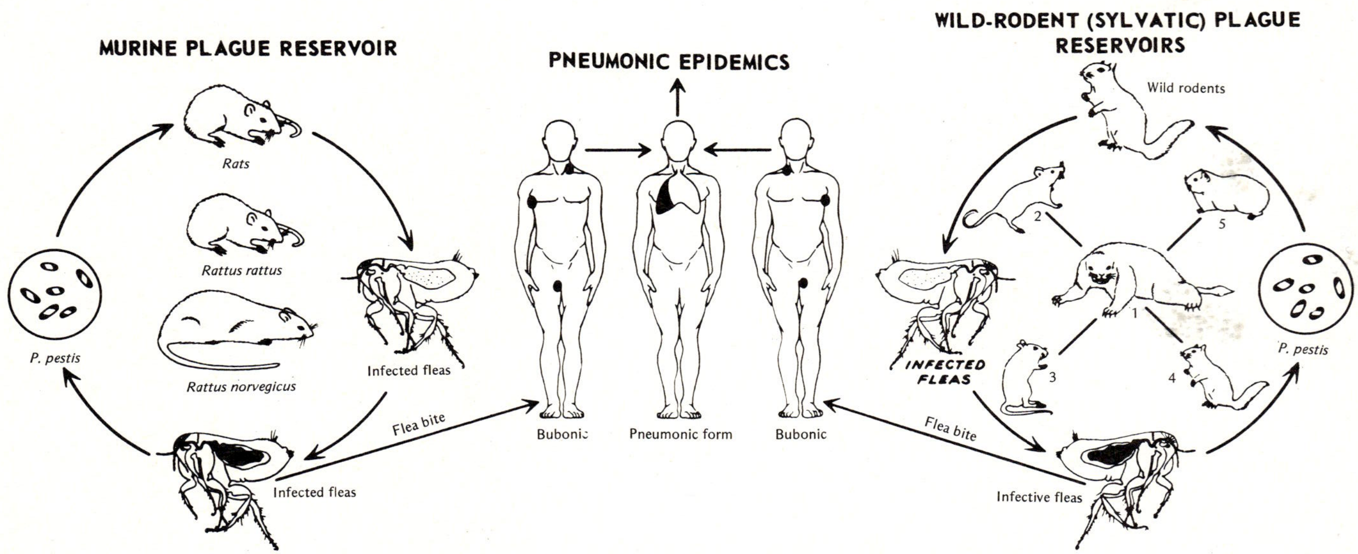


Principal reservoirs of murine plague:
(1) *Rattus rattus*
(2) *Rattus norvegicus*

Forms of human plague

A. Infection through flea bites leads usually to bubonic plague, less frequently to "septicemic" plague without apparent bubos.

B. Direct spread of the infection through bubonic patients with secondary lung involvement is apt to lead to manifestations of primary pneumonic plague.

Map showing incidence of plague in 1964.

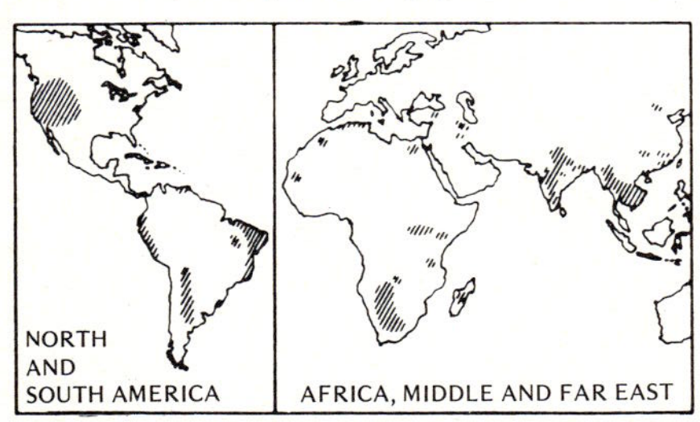


Slightly modified after Meyer, K. F. 1955. The zoonoses and their relation to rural health. *Univ. Calif. Publications in Public Health.* 3:1–50.

Principal reservoirs of sylvatic plague:
(1) Marmota (Central Asia)
(2) Meriones (Western Asia)
(3) Tatera (South Africa)
(4) Citellus (South-East Russia, West of North America)
(5) Caviinae (South America)

Human plague of sylvatic origin is quite frequently contracted through direct contact with infected rodents and not through flea bites.

In the United States the important reservoirs of sylvatic plague include ground squirrels, prairie dogs, wood rats, sage brush voles, meadow mice, deer mice, rabbits, and hares.

**FIG. 53.6**
Epidemiology of plague.

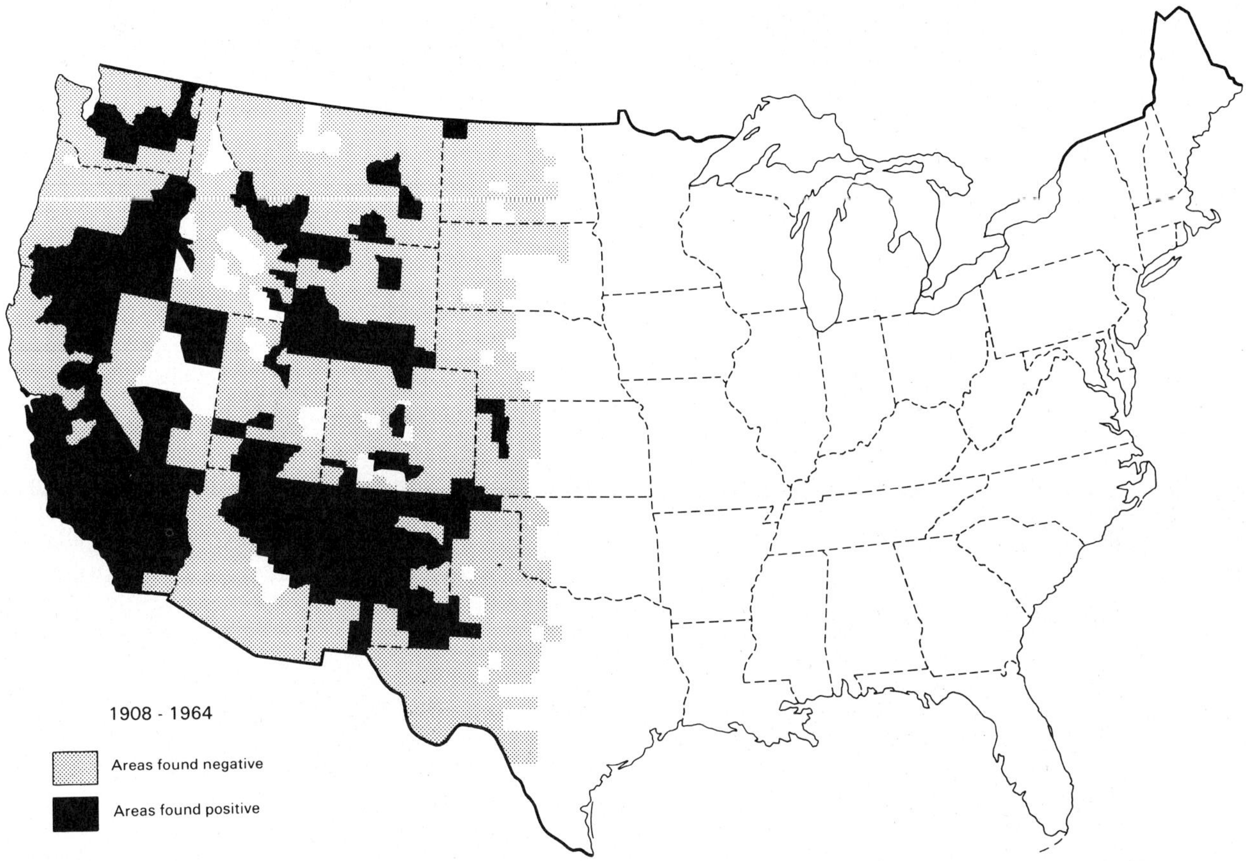


**FIG. 53.7**
Distribution in the United States of *Yersinia pestis,* the cause of plague.

4. Host specificity, that is, the willingness of the flea to transfer to other species of host and to feed
5. Length of time the flea may carry an infection

Of about 100 species of fleas known to transmit the plague bacillus, *Xenopsylla cheopis*, the Oriental rat flea, is the most important vector on a worldwide basis. It blocks well, and the infection remains in it for a long period. Other species of flea are also good vectors, but with the exception of the human flea, they are mostly found in particular geographic areas.

The problem of maintenance of plague in the wild between epidemics has not been satisfactorily explained. It seems likely that there is a relatively low rate of transmission among rodents through fleas that remain infected for many months. Some fleas become infected, but remain unblocked for a long period, perhaps as long as 6 months. When they do become blocked, they are able to transmit the agent. This explains the continuing residue of infection in some areas, particularly in light of the observation that blocked fleas usually do not live longer than 30 days. The following factors come together in plague epidemiology:

1 Susceptibility of the rodents
2 The ability of fleas to serve as vectors (VE)
3 The life cycle and behavior of the fleas

Campestral plague, sometimes called *sylvatic plague,*

refers to the maintenance of plague in wild populations, usually in grasslands. On occasion, plague moves from these centers into the human population through infected fleas. When a rodent dies of plague, its fleas tend to leave for nearly any warm-blooded host that may be nearby. This may be another wild rodent, an urban rodent, or a human.

Urban outbreaks take place primarily through urban rats, *Rattus norvegicus* or *R. rattus*. In the major epidemics in humans or pandemics spread across trade routes, rats have been the principal reservoirs. Subsidence of plague outbreaks probably occurs through dilution of susceptible individuals to the point where transmission is unlikely or by the development of immunity among the survivors.

The manifestations of plague in humans range from mild, almost inapparent infections to severe but nonfatal disease to rapid, fulminating infection resulting in death. There are three types:

1. *Bubonic.* Infection is transmitted by fleas into the skin. Fever, headache, body pain, and prostration are some of the signs and symptoms; swollen lymph nodes are usually seen with the onset of symptoms, which is sudden. The groin and axilla are the most common sites of the so-called bubo. The lymph nodes become so swollen and engorged with bacilli that they may break through the skin to the outside. Recovery from bubonic plague is likely.
2. *Septicemic.* If a flea injects bacilli directly into a capillary, the infection becomes general rather than localized in the lymph system. Recovery is unlikely.
3. *Pneumonic.* In some persons, the infection reaches the lungs and is then spread to other persons through aerosols or sputum. The rapid spread from person to person in a plague epidemic takes place through pneumonic plague rather than through fleas. Recovery is unlikely.

Treatment of plague is done by giving sulfonamides or broad-spectrum antibiotics such as the aureomycin or streptomycin groups. It is important to ensure that the patient is not overtreated. The bacteria contain an endotoxin, and it is likely that many of the symptoms of plague result from release of the toxin. If a high level of antibiotic is given immediately, the patient may die as a result of the release of large quantities of endotoxin within a short time.

The control of plague outbreaks requires a careful approach to ensure that the proper sequence of steps is carried out:

1. Quarantine, so that persons are prevented from entering the area or leaving it without some record of their whereabouts
2. Flea control
3. Rodent control
4. Isolation of human cases, especially of pneumonic plague
5. Vaccination of persons at high risk

The primary objective of this sequence is to contain the outbreak. Flea control must thus precede rodent control, because fleas leave the bodies of dead rodents and readily spread infection. When the flea population has been reduced, the rodents can be controlled by trapping or poisoning.

## REFERENCES AND FURTHER READINGS

Bahmanyar, M., and D. C. Cavanaugh. 1976. *Plague Manual.* World Health Organization, Geneva.

Barnes, A. M. 1982. The surveillance and control of bubonic plague in the United States. *Symp. Zool. Soc. Lond.* No. 50, 237–270.

Fox, I. 1940. *Fleas of Eastern United States.* Iowa State College Press, Ames, Iowa. Reprinted 1968, Hafner Publishing Co.

Harwood, R. F., and M. T. James. 1979. *Entomology in Human and Animal Health*, 7th ed. Macmillan, New York.

Holland, G. P. 1949. *The Siphonaptera of Canada.* Canadian Dept. of Agric. Tech. Bull. No. 70.

——. 1964. Evolution, classification and host relationships of Siphonaptera. *Ann. Rev. Entomol.* **9**:123–146.

Hopkins, G. H. E., and M. Rothschild. 1953, 1956, 1962, 1966, 1971. *An Illustrated Catalog of the Rothschild Collection of Fleas (Siphonaptera) in the British Museum (Natural History)*, Vol. I, Vol. II, Vol. III, Vol. IV, Vol. V. British Museum, London.

Hubbard, C. A. 1947. *Fleas of Western North America.* Iowa State College Press, Ames, Iowa.

Jellison, W. L., B. Locker, and R. Bacon. 1953. *Index to the*

*Literature of Siphonaptera of North America*. Supp. I. 1939–50. Mimeo. Rocky Mountain Lab, USPHS, Hamilton, Mont.

Langer, W. L. 1964. The black death. *Sci. Am.* **210** (2):114–121.

Lewis, R. E. 1972–1975. Notes on the geographic distribution and host preferences in the order Siphonaptera. Parts I–VI. *J. Med. Ent.* **9**:511, **10**:255, **11**:147, **11**:403, **11**:525, **11**:658.

Marshall, A. G. 1981. *The Ecology of Ectoparasitic Insects*. Academic Press, New York.

Pratt, H. D., and H. E. Stark. 1973. *Fleas of Public Health Significance and Their Control*. Pub. No. 75-8267. USDHEW, Center for Disease Control, Washington D.C.

Romoser, W. S. 1981. *The Science of Entomology*, 2nd ed. Macmillan, New York.

Rothschild, M. 1965. Fleas. *Sci. Am.* **213**(6):44–53.

——. 1975. Recent advances in our knowledge of the order Siphonaptera. *Ann. Rev. Entomol.* **20**:241–259.

Rothschild, M., B. Ford, and M. Hughes. 1970. Maturation of the male rabbit flea (*Spilopsyllus cuniculi*) and the oriental rat flea (*Xenopsylla cheopus*): Some effects of mammalian hormones on development and impregnation. *Trans. Zool. Soc. London* **32**:105–188.

Rothschild, M., Y. Schein, K. Parker, C. Neville, and S. Sternberg. 1973. The flying leap of the flea. *Sci. Am.* **229**(5):92–100.

Traub, R., and H. Starcke (eds.). 1980. *Fleas*. Proceedings of a Conference, Ashton Wold, Peterborough, United Kingdom, June 1977. Balkema, Rotterdam.

Tuchman, B. W. 1978. *A Distant Mirror: The Calamitous 14th Century*. Knopf, New York.

54

# *Diptera: Nematocera and Brachycera*

Members of the order Diptera are a diverse group in both structure and development. Beyond the fact that they have a single pair of wings (the hindwings reduced to halteres) and are all holometabolic, they apparently have little else in common. There are about 80,000 species in 140 families, making it the third largest order of insects, exceeded only by the beetles and lepidopterans; diversity might well be expected. Many important members of the order are familiar: mosquitoes, house flies, flesh flies, deer flies, midges, blackflies. Selected members of each group will be examined with respect to structure, development, and biology (Fig. 54.1).

Dipterans are important to humans for a variety of reasons. Many flies, such as the house fly and face fly, are at least bothersome. In addition, many serve as either mechanical or biological vectors of infectious agents. Tsetse transmit African sleeping sickness; mosquitoes transmit malaria, filariasis, and hundreds of viruses; biting midges transmit nematodes and viruses such as blue tongue virus; tabanids transmit tularemia. Since these flies all suck blood, they can be serious pests regardless of whether they are vectors of infectious agents. Many flies are parasitic as larvae; they can be serious medical and economic problems. Examples are the tumbu fly, which often infects the skin of humans in tropical areas; cattle grubs; stomach bots of horses; and the primary screwworm, which infects a wide array of mammals and causes enormous economic loss in the Western Hemisphere.

The suborders Nematocera, Brachycera, and Cyclorrapha are quite well-defined groups with characteristics that set them apart.

NEMATOCERA (THREAD HORN). Adult antennae with at least seven segments that are not fused; no arista; no ptilinum; larva with well-developed head; mandibles work horizontally; pupa not enclosed in the last larval skin. Mosquitoes, midges, blackflies, sand flies.

BRACHYCERA (SHORT HORN). Adult antennae with three segments, although the third segment may appear to have additional segments, so that there may appear to be as many as seven segments; no ptilinum; larval head incomplete, retractile; mandibles of larvae hooklike or knifelike and working vertically; pupa seldom enclosed in last larval skin; adults emerge from pupa through a straight or T-shaped slit. Deer flies, horseflies, snipe flies.

CYCLORRAPHA (CIRCULAR SUTURE). Adult antennae with three segments and an arista; ptilinum present; larva with vestigial head; larval mouthparts reduced to simple hooks, which work vertically; pupa enclosed in last larval skin (coarctate); the adult emerges through a circular opening (ptilinum expands to push off the operculum of the pupal case). House flies, flesh flies, bots, pigeon flies, tsetse.

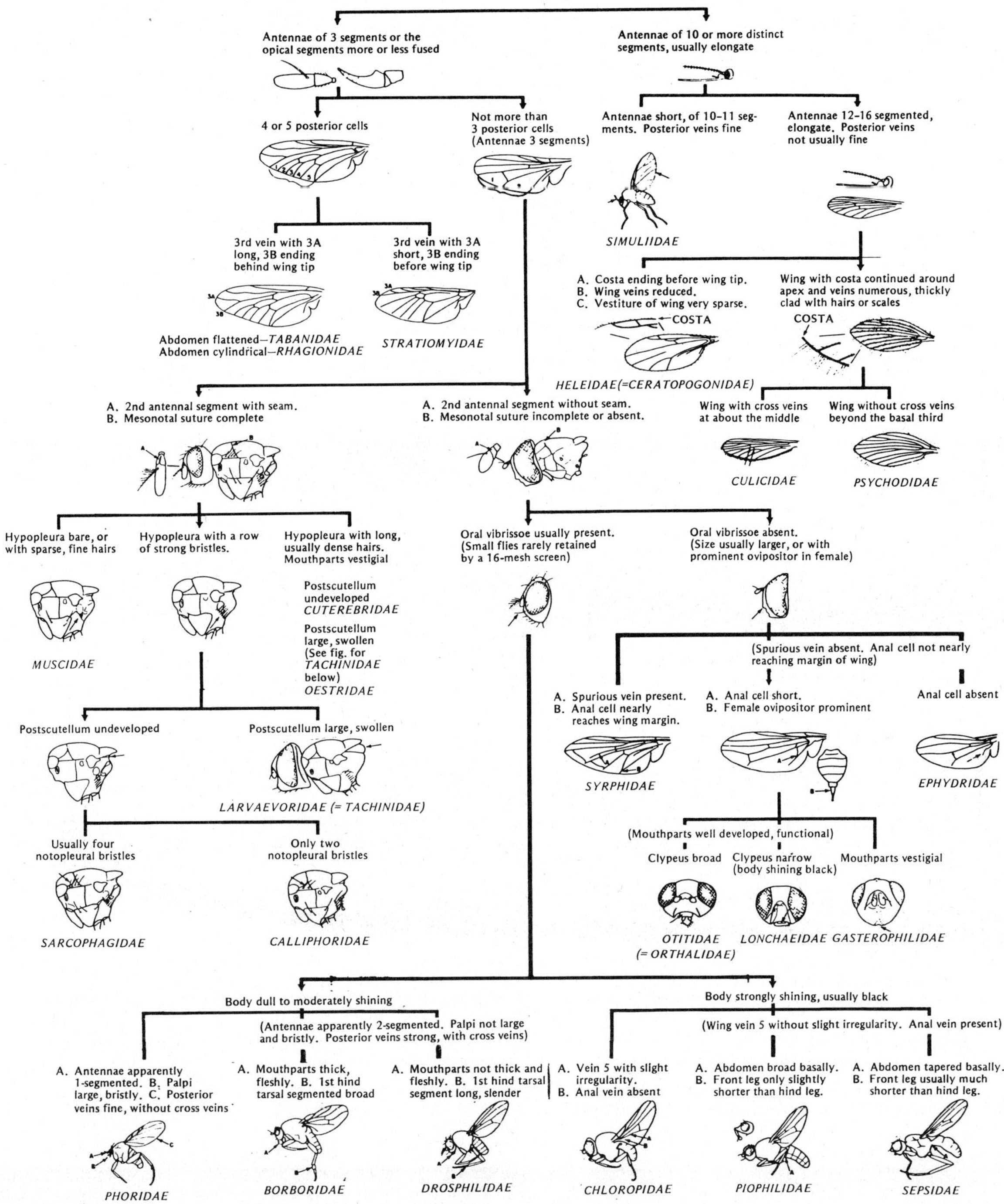


**FIG. 54.1**
Pictorial key to the principal families of Diptera of public health importance.

## SUBORDER NEMATOCERA

### *Family Culicidae—Mosquitoes*

MORPHOLOGY. Mosquitoes are generally 3 to 6 mm long, but a few exceed 9 mm; the wings have flat scales on the veins and the margins; they have an unforked vein between two forked veins on the distal margin; mouthparts are a long proboscis adapted for sucking blood and plant juices; the thorax is broader than the head; antennae are long and plumose in the males, and females have only a few short hairs; larvae are aquatic and have a well-defined head, thorax, and abdomen; an air tube arises on abdominal segment 8. There are about 3000 species in three subfamilies.

Some general works on mosquitoes are by Horsfall (1955), Carpenter & LaCasse, (1955), Bates (1949). Christophers' (1960) treatise on *Aedes aegypti* is a monumental work on a single species. There are a number of regional studies on mosquitoes such Harmston and Lawson (1967) for Colorado, Gjullin and Eddy (1972) for the Pacific Northwest, or Diaz Najera and Vargas (1973) for Mexico. McLintock (1978) has reviewed some aspects of mosquitoes and viruses. The Centers for Disease Control (CDC) publishes a number of specialized works on mosquitoes such as the mosquitoes of the United States (1977); other publications take up individual viral encephalitides in detail and are quite useful.

Mosquitoes are important to humans because they suck blood, but also because they transmit a great number of disease agents. The Toxorhynchitinae are large mosquitoes with a long, curved proboscis (Fig. 54.2); both males and females are plant feeders only. The larvae are predaceous on other mosquito larvae and are used in biological control. In all other species of mosquito, females are blood feeders and secondarily plant feeders; the males feed on plants only.

Identification of mosquitoes, either larvae (Figs. 54.3 and 54.4) or adult females (Figs. 54.2 and 54.5), requires knowledge of elements of anatomy and some care in examining the details of structure. Some, such as members of the *Culex pipiens* complex, are best left to experts, but many can be identified to species with some initial instruction and practice.

LIFE CYCLE. The life cycles of mosquitoes have some common features, but some variations are of interest in developing control programs or alleviating the effects of mosquito-borne disease agents. For example, all mosquitoes require water for larval development. Eggs are laid in or near water, but never in open water. Females respond to a number of environmental stimuli in choosing areas on which to oviposit. Eggs are laid singly or in rafts. *Aedes* and *Psorophora* lay their eggs in damp or dry places, whereas *Anopheles* lays its eggs in the water. *Culex* and *Culiseta* lay their eggs in rafts and *Mansonia* lays its eggs on the underside of leaves of submerged vegetation.

Eggs develop quickly and hatch (Fig. 54.6) in 24 to 48 hr depending on the temperature. Some aedine mosquitoes develop partially and then require a stimulus such as cooling or a particular photoperiod to resume development. Floodwater mosquitoes, some members of the genera *Aedes* and *Psorophora*, lay their eggs in areas that will flood periodically, and in temperate climates eggs laid during one summer do not hatch until the following spring when the area is flooded. They typically come out in hordes at that time.

All mosquito larvae are aquatic; most are filter feeders, but some, such as members of the Toxorhynchitinae, are predaceous. The larval stage may last for as short a time as three days to more than a month, depending on the temperature and species. For example, *Psorophora confinnis* goes from egg to adult in 3 days at 46°C (115°F) in the California desert. *Culiseta inornata* requires 28 to 30 days to complete development at 14°C (60°F). The development of *Aedes aegypti*, the yellow fever mosquito, is almost stopped below about 26°C (80°F). Most culicine mosquitoes complete development in about 7 days. Oxygen is required for larval development, and most larvae obtain it from the air by coming to the surface where they apply their breathing tubes; however, in *Mansonia* oxygen is obtained from plants by puncturing them with a posterior stylet.

The fourth larval stage is the last before the pupa. The pupa is free-swimming but nonfeeding; the pupal stage usually lasts only 2 to 4 days. Pupae require oxygen and have breathing tubes on the thorax and at the posterior end.

Adults emerge from the pupa at the surface of the water surface; 5 to 10 min are required for emergence, drying of the body, and extension of the wings. During this time they are quite susceptible to predation.

Both males and females feed on plant juices and nectar, but only the females take blood. Females

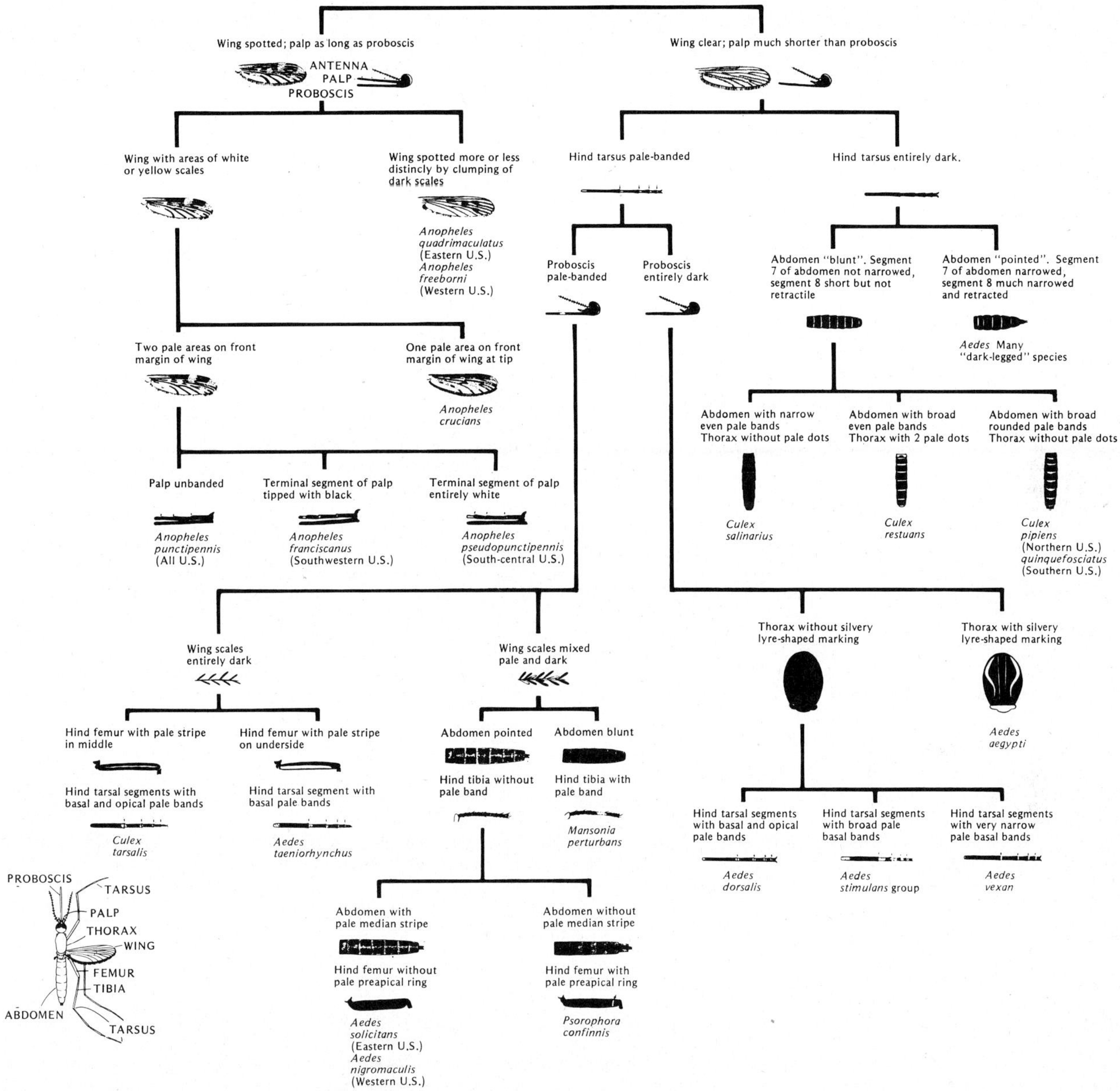


**FIG. 54.2**
Pictorial key to the genera of female mosquitoes in the United States.

ordinarily require a blood meal for maturation of the eggs; the exceptions are those populations that have autogeny, that is, the ability to lay eggs without having taken a blood meal. Even autogenous individuals prefer blood, however, and lay only small numbers of eggs without a blood meal. Females are not ready to take a

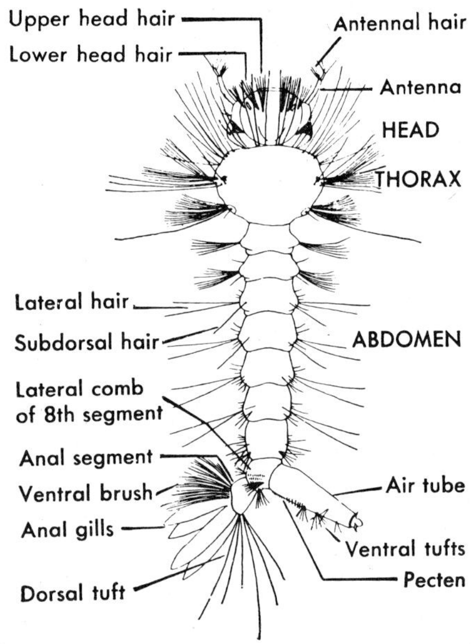


**FIG. 54.3**
Mosquito larva.

blood meal as soon as they emerge, but have a prefeeding phase of 1 to 3 days while the ovarian follicles develop. The next phase, hostseeking, lasts for 3 to 10 days, during which time they seek hosts, take a blood meal, and then rest somewhere while the eggs develop. Last, they seek oviposition sites and lay eggs over about a 3-day period. This cycle, called the *gonotrophic cycle*, is then repeated. One female may have as many as five cycles. Mating of male and female mosquitoes takes place in the air, and only once, as far as is known, despite the fact that the female lays several clutches of eggs.

The flight range of mosquitoes is quite variable; they have mostly short, hopping flight patterns. *Anopheles, Culex*, and *Culiseta* fly perhaps 2.5 km (1.5 miles) in a lifetime. Some mosquitoes have been found as far as 70 to 115 km (50 to 70 miles) from their larval development sites, but they undoubtedly had been carried most of the way by wind.

HOST–PARASITE INTERACTIONS. Certain biological and behavioral characteristics of mosquitoes are important for their ability to serve as vectors of infectious agents. The categories are used principally in relation to humans, but apply to domesticated animals as well:

1. A few domiciliated species that have sites of larval development in or near houses, readily enter them and feed primarily on human blood; usually they fly only a few hundred meters from the site of emergence. *Aedes aegypti* and *Culex pipiens* are examples of mosquitoes that are vectors of viruses. Several species of *Anopheles* that are vectors of malaria fall into this group, and *An. stephensi* is a truly urban species.
2. Species that regularly come in contact with humans outdoors and may come into houses, but do not rest there. They breed in wet fields, irrigation ditches, swamps, and standing water, either natural or man-made: tree holes, banana trees, tin cans, old tires. *Culex tarsalis, An. albipictus*, and *Mansonia* are examples.
3. Saltwater or freshwater marsh, temporary rain pool breeders, or floodwater breeders. Eggs are laid in dry or damp areas and adults emerge 5 days after a rain or flooding. There may be up to 20 million larvae/hectare (50 million/acre). *Ae. sollicitans* and *An. taeniorhynchus* are examples; they may migrate 70 km (40 miles) from the site of emergence. *Psorophora* spp. are also included since they are single-brood per year mosquitoes in the Arctic.
4. Forest canopy mosquitoes breed in treeholes or water-holding portions of leaves. These species come into contact with humans only under special circumstances, that is, during lumbering operations when the mosquitoes are carried from the canopy down to the ground or when trees are planted for shade near dwellings. *Aedes* spp. are examples.

It is important to note that feeding preference is crucial in the ability of mosquitoes to serve as vectors of disease agents. The anopheline vectors of malaria must have humans as the preferred host. The reservoir of St. Louis encephalitis virus is in birds, and the mosquito that transmits the agent to humans must therefore feed readily on both birds and mammals; in the western United States, *Culex tarsalis* plays the role well. Most mosquitoes have rather strict preferences for the hosts on which they will feed—either reptiles, birds, or mammals—but some that feed on mammals prefer domestic animals to humans.

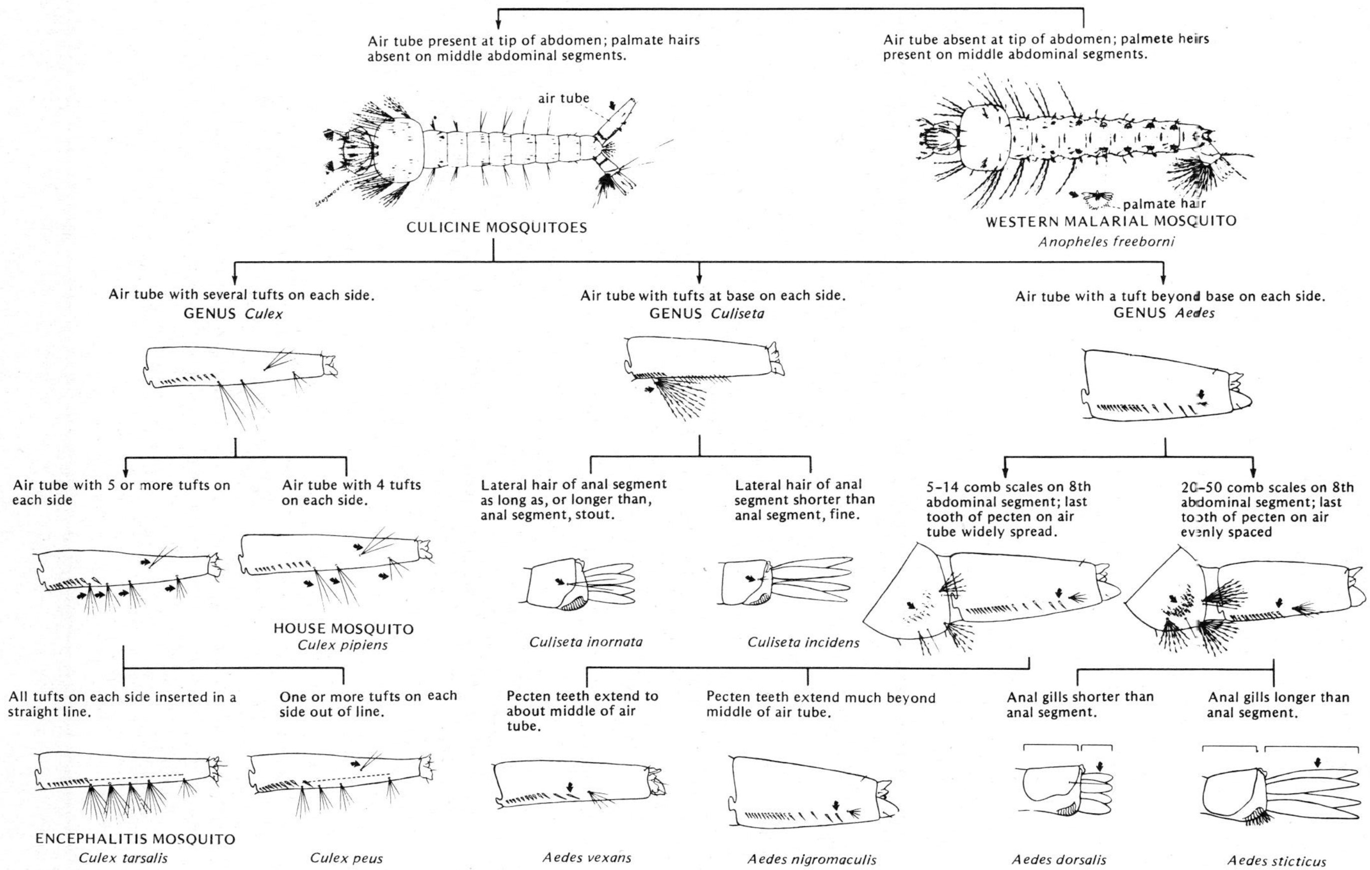


**FIG. 54.4**
Pictorial key to some common mosquito larvae of the United States.

**TABLE 54.1**
Some Disease Agents Transmitted by Mosquitoes

| AGENT OR DISEASE | MOSQUITO | RESERVOIR | HOSTS: HUMANS OR DOMESTIC ANIMALS |
|---|---|---|---|
| Malaria (*Plasmodium*) | *Anopheles* spp. | Human | Human |
| Filariasis (*Wuchereria*, *Brugia*) | *Culex*, *Mansonia* | Human, cat | Human<br>Canids |
| Heartworm (*Dirofilaria*) | *Culex*, *Aedes* | Canids | Dogs, human |
| St. Louis encephalitis virus (SLE) | *Culex tarsalis*<br>*C. pipiens* | Birds | Human |
| Eastern equine encephalitis (EEE) | *Culex* spp. | Birds | Human |
| Western equine encephalitis (WEE) | *Culex tarsalis* | Birds | Human, horse |
| California encephalitis (CE) | *Culiseta*, *Culex* | Rabbits | Human |
| Venezuelan equine encephalitis (VEE) | *Culex* | Mammals | Horse, human |
| Dengue | *Aedes* | Humans, monkeys (?) | Human |
| Yellow fever | *Aedes aegypti* | Monkeys | Human |

CONTROL. Mosquito control is far more complex than we can deal with here in detail. As indicated in Table 54.1, mosquitoes are vectors of many important disease agents throughout the world. The general principles set forth in the following section pertain to all of them, but in each geographic area, the target mosquito must be determined and its habits known before effective control can be achieved. Note also that, because the larvae are aquatic and the adults are terrestrial, entirely different techniques are required for the two stages.

## METHODS OF MOSQUITO CONTROL

### *Larvae*

**1.** No. 2 diesel oil on the surface of water. Larvae surface to breathe and are killed by becoming covered with oil; the oil also has a component that is directly toxic. This method does not work against *Mansonia* because they obtain oxygen from plant tissues.

**2.** Insecticide placed in water. The materials most often used now are organophosphates and carbamates; organochlorines have been phased out because of environmental damage. The principal drawback of this method is that mosquitoes are not the only organisms killed. For example, the preadults of insects, which are food for fish, are also killed.

**3.** Biological control.

a. Predators: (1) fish such as *Gambusia*, (2) *Toxorhynchites* larvae, (3) nematodes such as *Romanomermis petersoni* (= *Reesimermis p.*)

b. Disease agents: (1) microspora, such as *Nosema algerae*, (2) bacteria, such as *Bacillus thuringiensis* and *B. sphaericus*

In general, disease agents that are approved for use are highly specific so that nontarget organisms are not harmed.

**4.** Source reduction or habitat alteration. This method refers specifically to the techniques that do not use chemicals directly to kill insects and are not included under biological control; it can be thought of as physically altering the environment.

a. Fill or drain swampy areas

b. Remove aquatic plants that impede water flow or provide sites of oviposition

c. Remove man-made sources of standing water, such as water, cans, tires, flower pots

d. Use irrigation water properly so that ditches can be flushed or waste water does not accumulate

**5.** Insect growth regulators (IGRs). Methoprene is a juvenile hormone mimic, and dimilin is an inhibitor of chitin formation. Both are placed in the water and cause malformation of larvae.

## *Adults*

**1.** Insecticides.

a. Direct action. Insecticide is sprayed into the air and kills mosquitoes that fly into the spray. Compounds used are carbamates, organophosphates, and pyrethrins.

b. Residual. The insecticide is sprayed onto a surface or released from a carrier into a closed space. DDT has residual action for 4 to 6 months when sprayed onto surfaces in homes at a rate of 2 $g/m_2$. Organophosphates such as dichlorvos have only a short residual action, but can be released slowly from a plastic carrier.

**2.** Personal protection.

a. Repellents or barriers: (1) Deet and ethyl hexanediol repel insects and may be placed on the skin or impregnated into clothing, as has been done for military personnel. (2) Screening of houses and use of mosquito netting at night. (3) Use of barrier animals. Since mosquitoes have feeding preferences, some prefer to feed on animals other than humans. Although possible to use, this method is not ordinarily employed.

b. Traps. Various kinds of devices that emit ultraviolet light lure insects so that they are either electrocuted or drowned in a detergent-water solution.

c. Sound generators are sometimes marketed with the claim that certain frequencies will drive away mosquitoes. These claims have not been substantiated.

d. Birds. Purple martins are known to eat large numbers of insects, and houses for them to use in nesting are sometimes provided. Whether they reduce mosquitoe numbers is unknown.

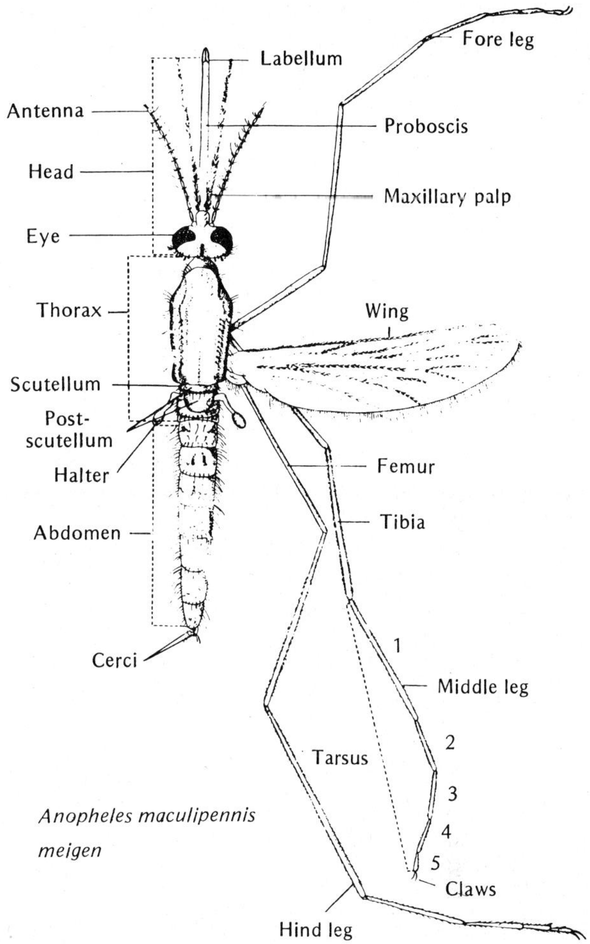


**FIG. 54.5**
Adult mosquito.

## *Family Simuliidae*

NAME OF ORGANISMS AND MORPHOLOGY. The members of the Simuliidae are small flies ranging in size from 1 to 5 mm. They usually have dark bodies from which one of the common names of the group comes—*blackflies*. The thorax is quite convex dorsally, giving rise to the other principal common name—*buffalo gnats*. They range from the tropics to the Arctic. The larvae live in well-oxygenated running water. Males and females feed on plant juices, but only the females feed on blood. Unlike mosquitoes, which are

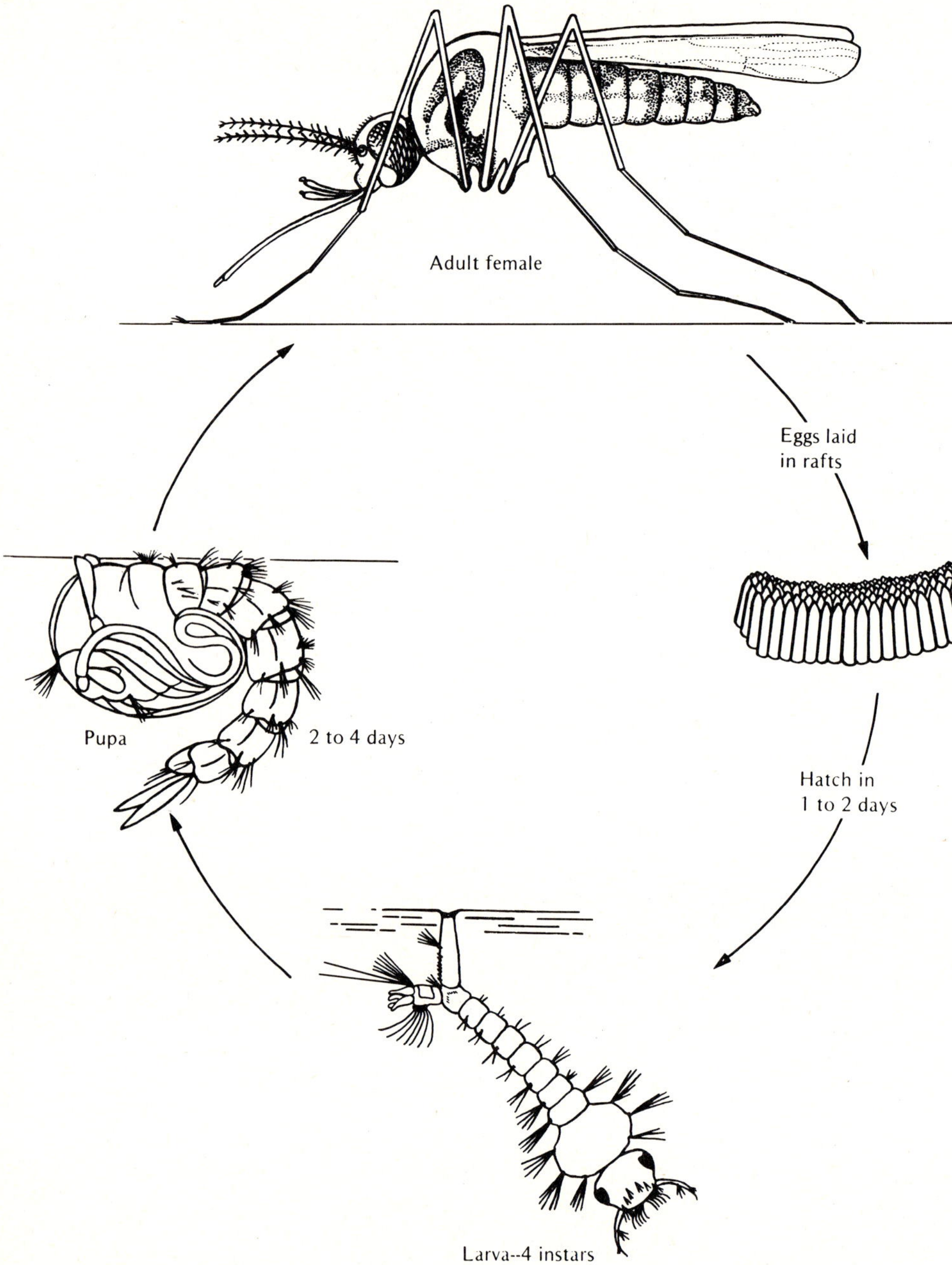


**FIG. 54.6**
Life cycle of a culicine mosquito.

capillary feeders, blackflies are pool feeders; their mouthparts are short (Fig. 54.7), and they lacerate the skin so that the blood wells up to the surface.

LIFE CYCLE. As with mosquitoes, blackfly larvae are always associated with water; eggs are laid by the females in flight on or near water, especially in sheltered

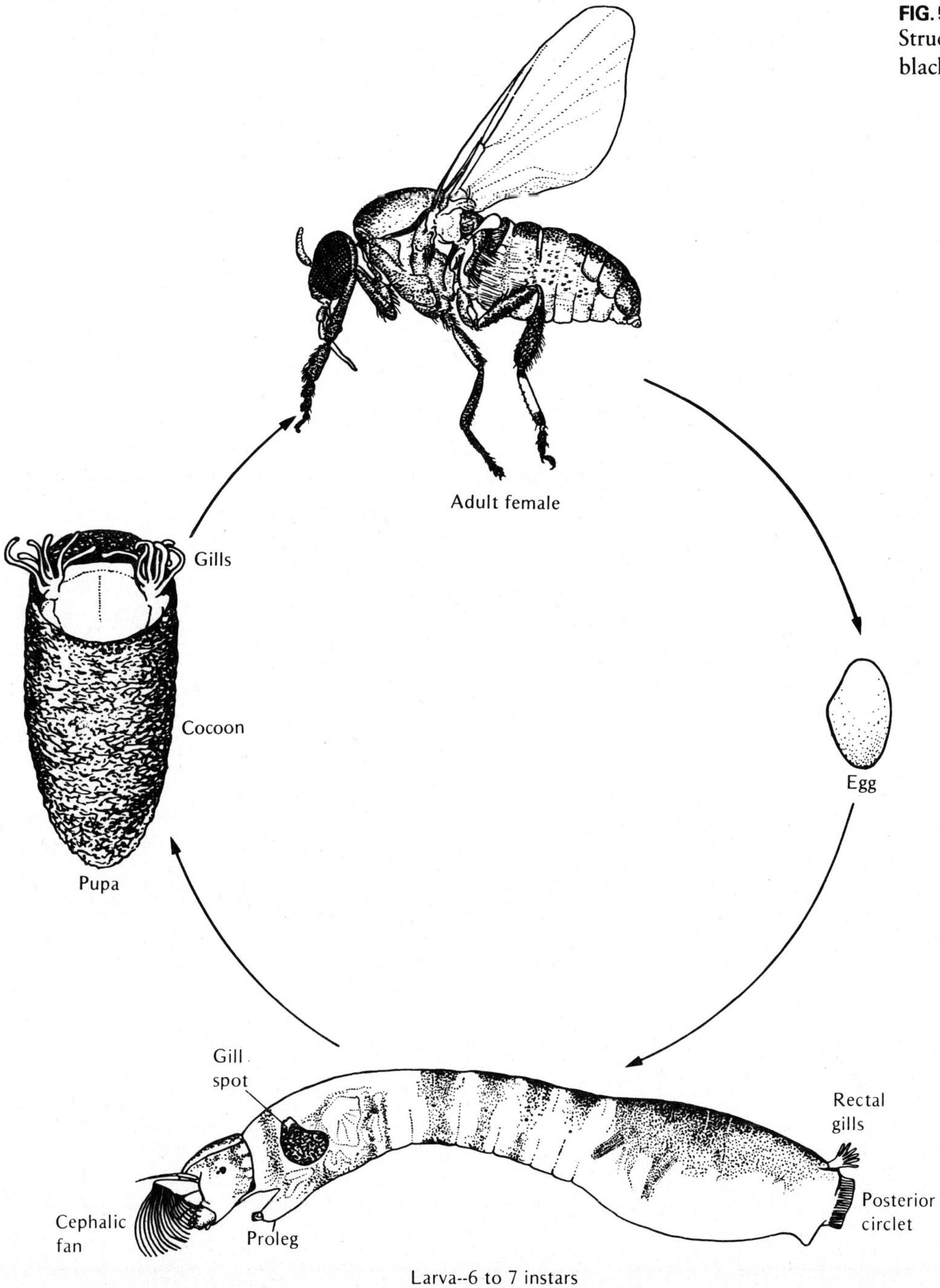


**FIG. 54.7**
Structure and life cycle of a blackfly.

areas. Eggs laid during the warm part of the year hatch within about a week, except for those species that have only a single brood per year; then the development is retarded until the following spring. The larvae have well-developed heads, rather barrel-shaped bodies, and a somewhat swollen posterior abdomen (Fig. 54.6). They attach to the substrate by a sucker at the posterior end. The six or seven larval instars may be completed in as short a time as 10 to 24 days, but may take nearly a year in those species that have a single generation per year (univoltine species). The multivoltine species (those with multiple generations per year) develop rapidly; the pupal stage is usually short, lasting only 2 days in some species but up to 4 weeks in others. The adults emerge from the pupa under water and rise to the surface, where they dry and extend their wings. Copulation takes place in the air in swarms of males; the females lay from 200 to 500 eggs, but some are known to lay up to 800 eggs. The time from egg to egg in multivoltine species ranges from 60 to 100 days.

Blackfly larvae live in well-oxygenated water where they either feed on particulate matter by trapping it with mouth brushes or graze on material on the substrate. The larvae are common in cascading streams and in shallow riffles, where they can be found on the downstream side of objects. In some cascades, the whole downstream surface of a rock may be covered with a dark brown mat of larvae. The predilection for this type of habitat leads blackfly larvae to colonize man-made structures such as concrete drops in irrigation systems (Edmunds, 1954). Some species, such as *Simulium neavei*, attach to freshwater crabs.

HOST–PARASITE INTERACTIONS. In temperate and cold climates there is a single brood of adults per year, and they tend to emerge all at once. The result is that swarms of blackflies appear suddenly and the females look for blood. A number of instances have been reported in which hundreds or even thousands of cattle have been killed by blackfly attacks; death occurs from exsanguination, shock, and suffocation from inhaling the flies. Even under less dramatic conditions, economic losses are experienced in domesticated animals such as chickens, in which there was nearly a threefold decrease in egg production during attacks by blackflies. In areas where blackflies are prevalent, it becomes uncomfortable, at the very least, for humans to be outside. It should be noted that the biting midges (family Ceratopogonidae) are quite similar to blackflies and the two groups are often mistaken for one another. Both groups are small enough to pass through ordinary fly screen, so there may be no place to hide during fly season.

Blackflies are important not only because of their bloodsucking, but also because they transmit some disease agents pathogenic for humans and domesticated animals. The most important agent is *Onchocerca volvulus* of humans, for which some of the principal vectors are *Simulium damnosum*, *S. neavei*, *S. ochraceum*, and *S. metallicum*. *Onchocerca* spp. of domestic animals (Chap. 41) are also vectored by blackflies as well as some biting midges. Blackflies also transmit *Leucocytozoon* spp., some of which are pathogenic for birds.

CONTROL. Blackflies are usually controlled by placing a granular insecticide in the water; where irrigation is practiced, the water level can be lowered so that the stream bed dries out. Screens can be painted with insecticide to keep flies from passing through or to kill them if they do. See Laird (1981) for methods of controlling blackflies.

## *Family Ceratopogonidae*

NAME OF ORGANISMS AND MORPHOLOGY. Adult biting midges are similar in size and appearance to blackflies; they have the same hump-backed appearance and are about 1 to 3 mm long (Fig. 54.8). They are usually brownish, but there is some overlap between the darker midges and the lighter blackflies. The term *no-see-ums* refers to biting midges and gives some idea of their insidious nature; the term may also be used for simuliids in some localities. Atchley et al. (1975) compiled a complete bibliography on ceratopogonids. Keys to the species are found in Blanton and Wirth (1979); despite the title of their work, it covers an area much greater than Florida.

LIFE CYCLE. Ceratopogonids have aquatic larvae, but they inhabit a quite different niche from that of blackflies. Midges have wormlike larvae that are found usually where there is a high content of organic matter. There are four larval instars, a short pupal period, and then the adults emerge. Several broods emerge each year except in climates with short growing seasons; eggs laid in the fall develop slowly until the following spring when the larvae all pupate quickly.

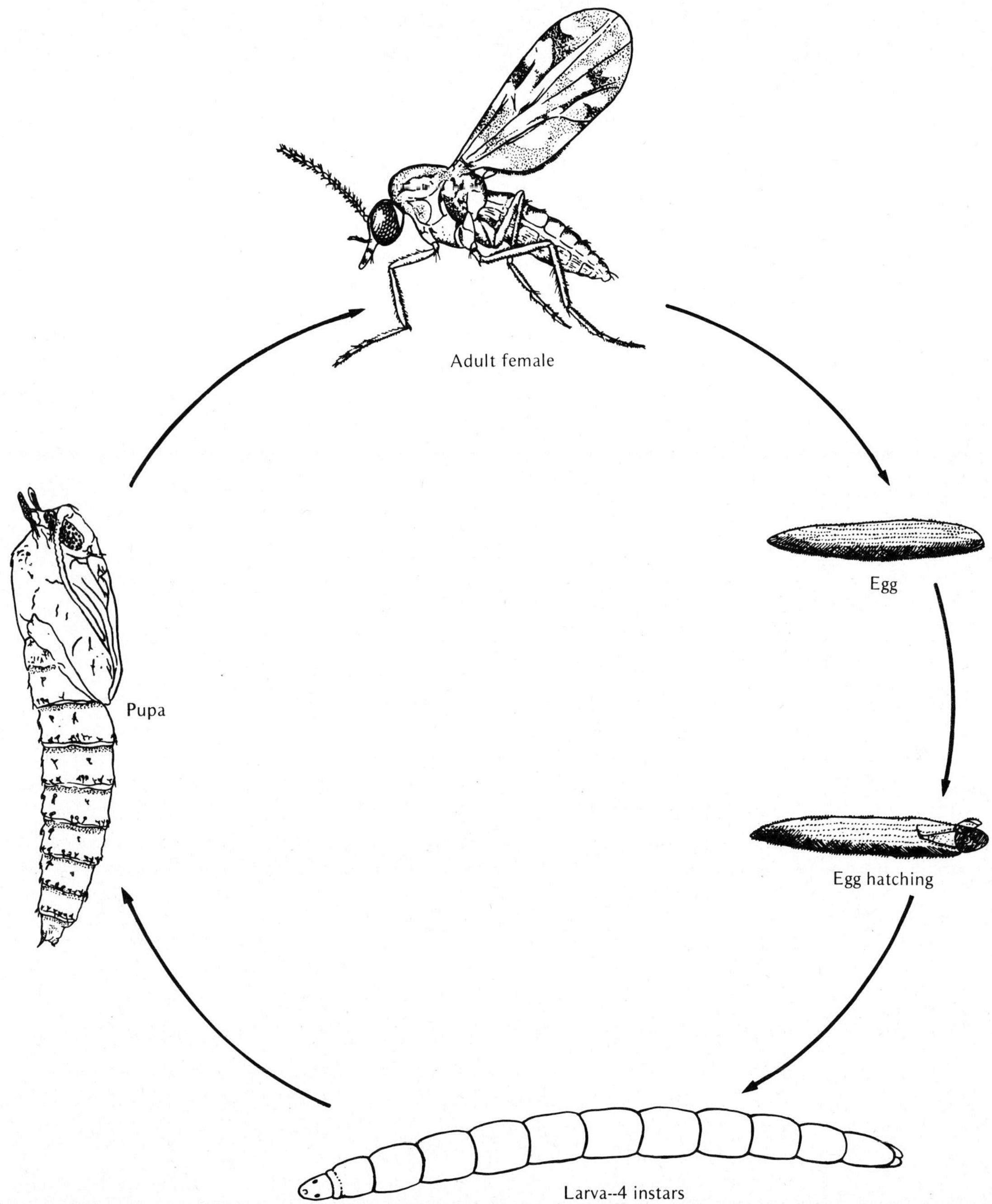


**FIG. 54.8**
Structure and life cycle of a biting midge.

Larval stages are about 0.5 to 3.0 mm long, are slender, and have well-developed head capsules; they are usually grayish. Adults are distinguished from other small flies by their antennae (12 to 16 segments), patterned wings with scales, and their hump-backed appearance.

CULTIVATION. *Culicoides variipennis* has been maintained in continuous laboratory culture for many years (Jones, 1957), and studies on the biology and vector capabilities of biting midges have therefore been possible.

HOST–PARASITE INTERACTIONS. Biting midges are important because they suck blood and they are vectors of a number of disease agents. Like the blackflies, biting midges are pool feeders because their mouthparts are short. Only females suck blood, and they have preferences for certain species of hosts, but they seem not to be as restricted as some species of mosquitoes are. Biting midges feed mostly at dusk and dawn (crepuscular feeding). The bites of ceratopogonids are painful, and because they are pool feeders, they may leave a bleeding wound. They harass both humans and livestock. In cattle, repeated feedings give rise to hypersensitivity, manifested as a strong skin reaction. The skin swells at the site of feeding, oozes fluid, and prevents the midges from feeding. Queensland itch, a disease described in Australia, occurs in horses that have been fed upon repeatedly by ceratopogonids; intense irritation of the skin occurs, and histologic changes such as hyperkeratosis and fibrosis take place.

In addition to their direct damage, midges also transmit both viruses and filarioid nematodes. The viruses that cause blue tongue disease of ruminants and African horse sickness are transmitted by *Culicoides* spp. In North America, *C. variipennis* is the vector of blue tongue virus. Blue tongue disease has great impact on the sheep and cattle industries, in which it causes both acute and chronic disease and is also transmitted from the infected cows to the developing calf (vertical transmission). The filarioid nematodes known to be transmitted by ceratopogonids are *Dipetalonema perstans, Mansonella ozzardi* of humans, *Onchocerca* spp. of large herbivores, and several species of filarioids of birds such as *Splendidofilaria* and *Chandlerella.*

CONTROL. Ceratopogonid larvae develop in a wide variety of aquatic habitats; they are found most often in the substrate of bodies of water that do not move or move rather slowly. One species has been found breeding in rotting cactus in Australia, so only moist conditions may be needed for development of some species. Each species is likely to prefer a particular type of habitat (Jones, 1961), and these range from fresh to brackish water with a full range of organic matter in the water. To further complicate the matter, a widespread species such as *Culicoides variipennis* has different preferences in different geographic areas and climates. Control programs depend on determining the species of most importance and where they develop.

Control of ceratopogonids has been only partially successful; it is based on source reduction and larviciding. Wet, organic-laden areas along streams or the margins of bodies of water may be eliminated by draining or stabilizing a stream bank. Insecticide can be broadcast on remaining areas that are likely to harbor larvae. Control of adults usually involves spraying barns and sheds where animals are kept and screening windows of houses. Painting of screening with insecticide has been partially successful in keeping populations down in houses.

## *Family Psychodidae—Sand Flies*

NAME OF ORGANISMS AND MORPHOLOGY. Adult sand flies are small, dark flies that have hairy wings; the adults (Fig. 54.9) are small enough to pass through standard fly screen. There are two subfamilies: the Psychodinae, which are not bloodsuckers, and the Phlebotominae, which do suck blood, as indicated by their name. All members of the Phlebotominae were formerly placed in the genus *Phlebotomus*, but in recent years, this name and *Sergentomyia* have been applied to the Old World species and *Lutzomyia* and *Psychodopygus* to the Western Hemisphere species (Lewis, 1974). There are about 300 species in the family.

LIFE CYCLE. The larval stages of sand flies are terrestrial and the females oviposit in damp sites where moisture is likely to be retained for a long period. The eggs hatch within 1 to 2 weeks, releasing the larvae, which have distinct heads and a body with a thorax and ten abdominal segments; there are long caudal bristles on the ninth segment. There are four larval stages followed by the naked pupa. Adults emerge 7 to 10 weeks after oviposition.

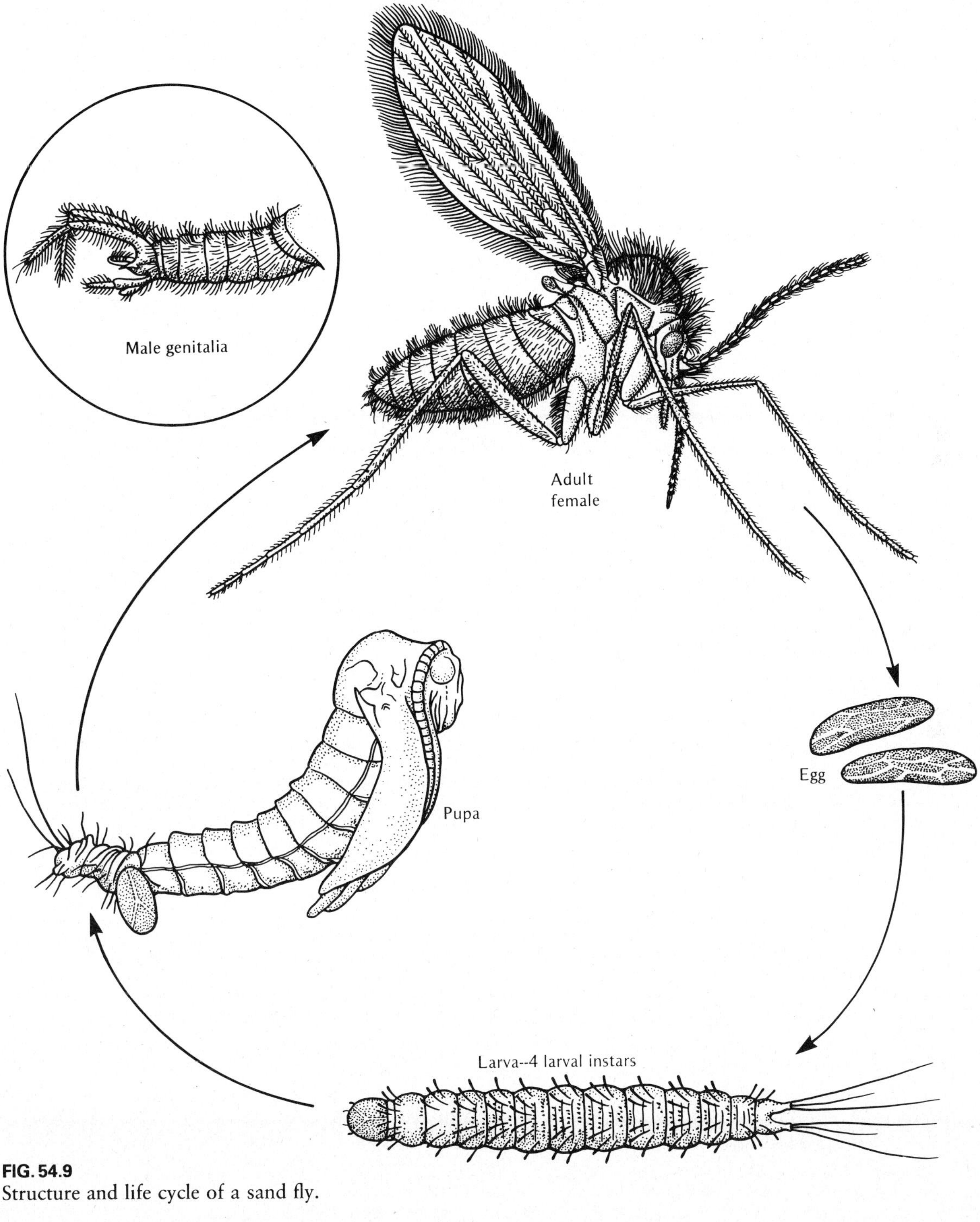


**FIG. 54.9**
Structure and life cycle of a sand fly.

Larvae feed on decaying organic matter of all sorts—plant material, garbage, feces. Only the adult females are blood feeders. These delicate flies do not fly far, but rather take short, hopping flights when there is little wind.

HOST–PARASITE INTERACTIONS. Although phlebotomines suck blood, their importance lies principally in their being vectors of infectious agents. The various agents causing leishmaniasis (Chap. 5), the bacterial disease Carrion's disease, and the viral disease papatasi fever are all transmitted by these flies.

CONTROL. As discussed in Chapter 5, reduction in phlebotomine populations is usually possible only when there are domiciliated species serving as the principal vectors of a disease agent. Personal protection through repellents and screening is in order if one is at risk of contracting a phlebotomine-borne agent.

### *Other Nematoceran Families*

A few other families in this suborder may be encountered when one is looking for a bloodsucker or vector, but they are of little or no medical or veterinary importance.

The crane flies (family Tipulidae) are elegant creatures that look like giant mosquitoes; however, they are not blood feeders. They have grublike aquatic larvae, usually in streams with low organic content.

Members of the family Chaoboridae at one time were included in the family Culicidae as the subfamily Chaoborinae, but have now been separated. The majority do not suck blood but are important primarily because they can be terrible nuisances when they emerge by the billions from lakes.

The dixids (family Dixidae) are similar in appearance to anopheline mosquitoes, but are of no known economic or medical importance. They do not suck blood.

Chironomids (family Chironomidae) have aquatic larvae similar to those of dixids, and the adult midges superficially resemble mosquitoes. These flies are common around streams and ponds and occasionally occur in large enough numbers to become nuisances, but they do not suck blood (Oliver, 1971).

## SUBORDER BRACHYCERA

Two families are of interest in the Brachycera—Tabanidae and Rhagionidae. The tabanids are more prevalent and more important as vectors of disease agents than the rhagionids, not all of which are bloodsuckers. Members of the suborder have larvae with poorly developed heads and mouthparts that are reduced to knifelike or hooklike mandibles that work vertically. The pupae are not enclosed and the adults emerge from a straight or T-shaped slit. The adult antennae have fewer than seven segments and an arista is seldom present. There is no ptilinum.

### *Family Tabanidae—Horseflies, Deer Flies*

NAME OF ORGANISM AND MORPHOLOGY. Tabanid adults are large, heavy-bodied flies that are strong fliers. They have large heads and large eyes, some so large in males that they are spoken of as being holoptic; the entire front of the head consists of what appears to be a single eye. The antennae are short, being composed of three segments, of which the terminal one sometimes is subdivided (Fig. 54.10). Wing venation is characteristic; veins R4–5 diverge widely. Some of the common tabanids are *Chrysops (deer flies), Tabanus* (horseflies), *Haematopota*, and *Hybomitra*.

LIFE CYCLE. The adult females oviposit on vegetation near water, and the eggs hatch in 2 to 3 days. The grublike larvae (Fig. 54.10) are aquatic or at least live in muddy areas; some larvae such as that of *Tabanus* are predaceous, whereas those of *Chrysops* are plant feeders. There are four to nine larval instars, after which the pupa forms; the pupae ordinarily leave the water for drier areas where they complete pupation. The adults emerge in 5 days to 3 weeks. The life cycle is rather slow; in temperate climates there is usually a single generation per year and sometimes 2 to 3 years are required to complete development.

HOST–PARASITE INTERACTIONS. Only the females are blood feeders; the males have poorly developed mouthparts and probably die after mating. Being bitten by a tabanid is a memorable experience. The bite is painful and there may be considerable swelling at the site of the wound. They are mostly pool feeders and they

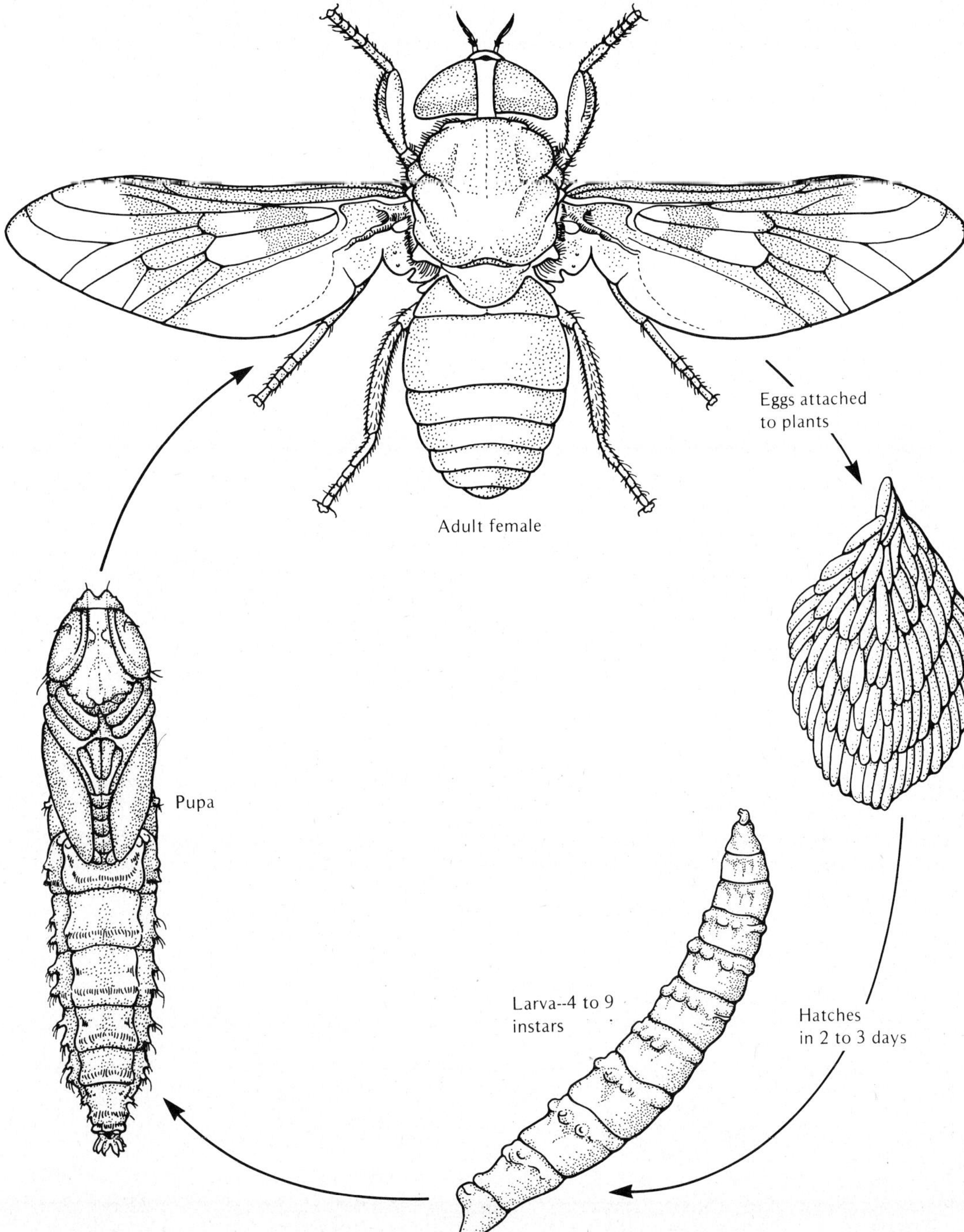


**FIG. 54.10**
Structure and life cycle of a tabanid.

lacerate the skin with their sharp mandibles so that the blood wells up to the surface. Tabanids are usually restless feeders in that they feed briefly and then move to another site or another host, but they can take up to 0.34 ml of blood compared to a biting midge, which may take only about 0.00023 ml.

Tabanids are mostly daytime feeders and harass both humans and other animals. The black horsefly, *Tabanus atratus*, is about 20 mm (3/4 in.) long and makes quite a loud buzz when flying. It is difficult not to cringe when one approaches. Deer flies, *Chrysops* spp., come out of swampy areas, typically cattails, to feed on humans and livestock. Tabanids cause economic losses in the livestock industry resulting from poor weight gains and reduced milk production.

In addition to being pests, the tabanids are either biological or mechanical vectors of both prokaryotic and eukaryotic parasites. The best-known example is the eye worm of humans, *Loa loa*, which occurs in tropical Africa and is transmitted by *Chrysops silicea* and *C. dimidiata*. *Hybomitra* is a biological vector for *Elaeophora schneideri*, a filarioid nematode of large ruminants (Chap. 41). Tabanids also serve as mechanical vectors for *Trypanosoma evansi* (Chap. 3) and for the causative agents of anthrax and tularemia. Krinksy (1976) reviewed tabanids as vectors.

CONTROL. Control of tabanids is usually only partially successful. Reducing the source by draining wet or boggy areas is sometimes feasible, but often the areas where the larvae develop are quite large; a few species require only muddy areas and not much can be done to change those. Along the East Coast of the United States, some success was experienced in treating salt marshes from the air with organochlorine insecticides, but the problem of nontarget organisms being affected was a drawback. Livestock can be protected by self- treatment with treadle sprayers or back rubbers containing insecticide, but animals must have access to such equipment at least once a day. Daily treatment is possible for dairy cattle or where pastures are relatively small, but in grasslands, animals may not be seen for days at a time and they may not come to treatment centers very often. Malaise, traps consisting of a large black ball and $CO_2$ as attractants, have been used to trap tabanids, but they seem not to reduce populations sufficiently to make any difference; the situation is the same as with tsetse, of which Harris traps take large numbers but do not reduce the probability of transmission of trypanosomes (Chap. 3). Deet may repel some tabanids, but apparently it is not effective against all species.

### *Family Rhagionidae—Snipe Flies*

The members of this family are small to medium-sized and have small head and large eyes. The antennae are short and some of them have an arista. Only a few are bloodsuckers, and they are not known to serve as vectors of disease agents. They are vicious biters and feed repeatedly, so it is possible that they at least can be mechanical vectors of disease agents. Some genera are *Symphoromyia, Spaniopsis, Suragina*, and *Atrichops*.

## REFERENCES AND FURTHER READINGS

Atchley, W. R., W. W. Wirth, and C. T. Gaskins. 1975. *A Bibliography and a Keyword-in-Context Index of the Ceratopogonidae (Diptera) from 1758 to 1973*. Texas Tech Press, Lubbock, Texas.

Bates, M. 1949. *The Natural History of Mosquitoes*. Macmillan Publishing Co., New York.

Blanton, F. S., and W. W. Wirth. 1979. *The Sand Flies* (Culicoides) *of Florida (Diptera: Ceratopogonidae). Arthropods of Florida and Neighboring Land Areas*, Vol. 10. Florida Department of Agriculture and Consumer Services, Gainesville, Florida.

Carpenter, S. J., and W. J. LaCasse. 1955. *Mosquitoes of North America (North of Mexico)*. University of California Press, Berkeley.

Centers for Disease Control. 1977. *Mosquitoes of Public Health Importance and Their Control*. CDC Pub. No. 77–8140. USDHEW, Public Health Services, Washington, D.C.

Christophers, S. E. 1960. Aedes aegypti *(L) The Yellow Fever Mosquito; Its Life History, Biology and Structure*. Cambridge University Press, London.

Diaz Najera, A., and L. Vargas. 1973. Mexican mosquitoes. Present geographical distribution (in Spanish). *Rev. Invest. Salud Publica* 33:111–125.

Edmunds, L. E. 1954. A note on irrigation drop structures as breeding sites of black flies in western Nebraska (Diptera: Simuliidae). *Mosq. News*. **14**:65–66.

Gjullin, C. M., and G. W. Eddy. 1972. *The Mosquitoes of the Northwestern United States*. Tech. Bull. ARS, USDA

No. 1447. U.S. Department of Agriculture, Washington, D.C.

Harmston, F. C., and F. A. Lawson. 1967. *The Mosquitoes of Colorado*. USDHEW, PHS, Centers for Disease Control, Washington, D.C.

Harwood, R. F., and M. T. James. 1979. *Entomology in Human and Animal Health,* 7th ed. Macmillan, New York.

Hill, M. A. 1947. Life cycle and habits of *Culicoides*. *Ann. Trop. Med. Parasitol.* **41**:55–115.

Horsfall, W. R. 1955. *Mosquitoes: Their Behavior and Relation to Disease*. Ronald Press, New York.

Jones, R. H. 1957. The laboratory colonization of *Culicoides variipennis* (Coq.). *J. Econ. Entomol.* **50**:107–108.

——. 1961. Observations on the larval habitats of some North American species of *Culicoides* (Diptera: Ceratopogonidae). *Ann. Ent. Soc. Am.* **54**:702–710.

Krinsky, W. L. 1976. Animal disease agents transmitted by horse flies and deer flies (Diptera: Tabanidae). *J. Med. Entomol.* **13**:225–275.

Laird, M. (ed.). 1981. *Blackflies: The Future for Biological Methods in Integrated Control*. Academic Press, New York.

Lewis, D. J. 1974. The biology of Phlebotominae in relation to leishmaniasis. *Ann. Rev. Entomol.* **19**:363–384.

McLintock, J. 1978. Mosquito-virus relationships of American encephalitides. *Ann. Rev. Entomol.* **23**:17–37.

Oliver, D. R. 1971. Life history of the Chironomidae. *Ann. Rev. Entomol.* **16**:211–230.

Smith, K. G. V. (ed.). 1973. *Insects and Other Arthropods of Medical Importance*. London, British Museum (Natural History).

Weinburg, H. B., and H. D. Pratt. 1967. Culicoides. *Public Health Importance, Biology, Survey and Control*. USDHEW, PHS, Centers for Disease Control, Washington, D.C.

55

# *Diptera: Cyclorrapha*

This suborder contains those flies in which the pupa is enclosed in a case composed of the last larval skin. The adults are generally strong fliers and mostly rather heavy-bodied, such as house flies. The mouthparts range from the sponging type of house flies to the piercing type of tsetse to a combination of the two types, as seen in the stable fly, *Stomoxys calcitrans*. Some flies such as *Hypoderma* spp., the cattle grubs, have vestigial mouthparts and do not feed as adults. Larvae are maggots. Their anterior ends have only the remnants of a head and the mouthparts are hooklike structures referred to as the *cephalopharyngeal skeleton*; the spiracles by which they obtain oxygen are at the posterior end.

The classification and identification of cyclorraphan flies is based on the presence or absence of the ptilinum, by which most species break out of the pupal case, and on the wing structure, antennal structure, and setae, especially on the head and thorax. Since there are about 140 families in the order, most of them in the Cyclorrapha, it can be appreciated that identification is difficult except for some common species.

Flies are important because of the following:

1. They are pests that often occur in large numbers and crawl over everything and everyone. Examples are house flies, eye gnats, and fruit flies.
2. They suck blood from humans and other animals, causing considerable discomfort and blood loss. Examples are louse flies and horn flies.
3. They are mechanical vectors of disease agents. Examples are house flies, which transmit the agents causing amebic dysentery and typhoid fever, and eye gnats, which transmit the agents causing pinkeye and yaws.
4. They are biological vectors of disease agents. Examples are tsetse, which transmit the various African trypanosomes (Chap. 3), and muscoid flies such as *Stomoxys calcitrans* (the stable fly), which transmits the nematode causing summer sore in horses.
5. They are parasitic as larvae and may cause extensive tissue damage in their hosts. Examples are blowflies of domestic animals such as *Phormia regina*, cattle grubs, *Dermatobia hominis*, the human bot fly, or torsalo.

LIFE CYCLE AND MORPHOLOGY. The life cycles of other cyclorraphan flies follow the pattern seen in the common blue bottle fly, *Calliphora vicina* (Fig. 55.1). Eggs are laid in decaying organic matter and usually hatch within about a day. The three larval instars develop quite rapidly and reach the pupal stage in about a week. The last larval stage typically leaves the medium in which it has grown and burrows into the soil, where it forms the pupa. During warm weather, the pupal stage

**FIG. 55.1**
Life cycle of *Calliphora vicina*, the blue bottle fly. [Redrawn with modifications from Smart, 1948.]

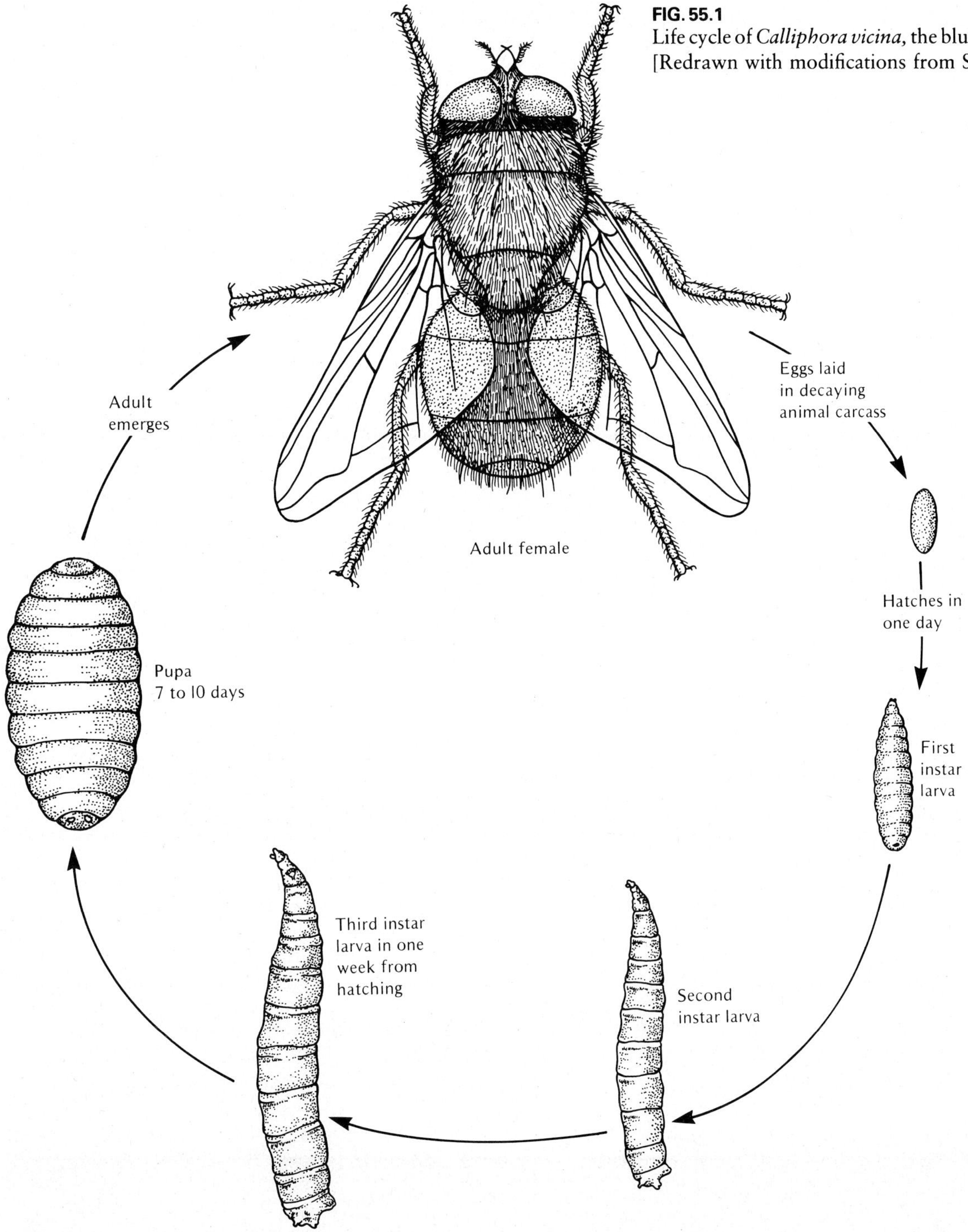

lasts a week to 10 days and then the adult emerges by expanding the ptilinum on the midline of the head, pushing off the round operculum of the pupal case. The adults feed on various kinds of organic matter but are not bloodsuckers. After copulation, the females lay up to 180 eggs at a time and may lay as many as 720 in a lifetime. Overwintering probably takes place in the pupal stage. *C. vicina* is a blowfly that ordinarily oviposits in decaying flesh, but it can become a problem as one of the blowflies or wool maggots of sheep. The blue bottle fly is readily recognized by its relatively large size of 9 to 11 mm (3/8 in.) and the shiny metallic blue abdomen.

There are variations on the pattern seen in *C. vicina*, but the stages remain the same. Some flies do not lay eggs, but give birth to larvae (larviparity); some examples are tsetse and hippoboscids. The larvae are retained in the body of the mother and often are nourished by secretions from her reproductive tract. In a number of instances, larvae are obligate parasites such as stomach bots of horses or the Congo floor maggot, *Auchmeromyia luteola*.

Some general taxonomic works on flies are those of Cole (1969) for adults and Johannsen (1969) for preadult stages (this is a reprint of monographs published in the 1930s). Smart's (1948) book is still useful, although it has been replaced by Smith (1973) as a general work in the area of medical entomology. James's (1947) monograph on myiasis should be available in most libraries; Zumpt (1965) covers much the same information for the Eastern Hemisphere.

## FAMILY MUSCIDAE—HOUSE FLIES AND THEIR RELATIVES

NAME OF ORGANISM AND DISTRIBUTION. The house fly is generally considered to be *Musca domestica*, a grayish fly 6 to 9 mm (5/16 in.) long (Fig. 55.2). A number of other flies that are closely related, and some not so closely related, also come into houses and may bring some of the same health hazards as *M. domestica*. It occurs nearly worldwide in human habitations, but is replaced by a closely related species, *M. sorbens*, in the Orient, Africa, and some Pacific islands. Other flies likely to be found in association with humans (synanthropic) are the cluster fly (*Pollenia rudis*), the lesser house fly (*Fannia* spp.), and various flesh flies such as *Calliphora vicina*, discussed previously as a typical cyclorraphan.

LIFE CYCLE. *M. domestica* has a life cycle nearly identical to that of *C. vicina* but its larvae develop in decaying organic matter of all sorts: feces, garbage, decaying plant material. At summer temperatures, the life cycle is completed in about 10 days. A female lays up to 150 eggs during her lifetime. Under conditions of rapid development and good sites for larval development, house flies can become a serious nuisance.

HOST–PARASITE INTERACTIONS. In addition to being nuisances, house flies can transmit a number of infectious agents such as those that cause amebic dysentery, typhoid fever, and related intestinal infections caused by Gram-negative bacteria. Since the larvae develop in feces and the adults feed on feces as well as ovipositing there, they can readily carry such infectious agents to the house and deposit them from their feet or by vomiting and defecating on food.

CONTROL. Source reduction is the principal technique used in reducing populations of house flies: removing those bacteria-laden materials that draw the adults to oviposit. Garbage must be disposed of frequently, especially in the warm months; garbage cans need to be cleaned and to have tight-fitting lids. Human excrement must be disposed of properly; if there is no sanitary sewer system, privies that are well constructed, tight to the ground, and screened are essential to keep flies out. Animal feces are a site of larval development and feces of companion animals should be disposed of. Feces of large farm animals or chickens present a particularly difficult problem because of the sheer volume of material that is likely to accumulate in a short period with intensive animal husbandry methods. In some instances, it is possible to scatter feces so that they dry rapidly, break down, and do not ferment. A number of natural biological agents such as mites, parasitoid wasps, and some nematodes have been used with some success in reducing fly numbers. In buildings, houses, barns, and milking parlors, screening to keep flies out is the first line of defense. In addition, insecticides can be used to kill those flies that do enter. Methoprene can be fed to livestock to control fly larvae in their feces.

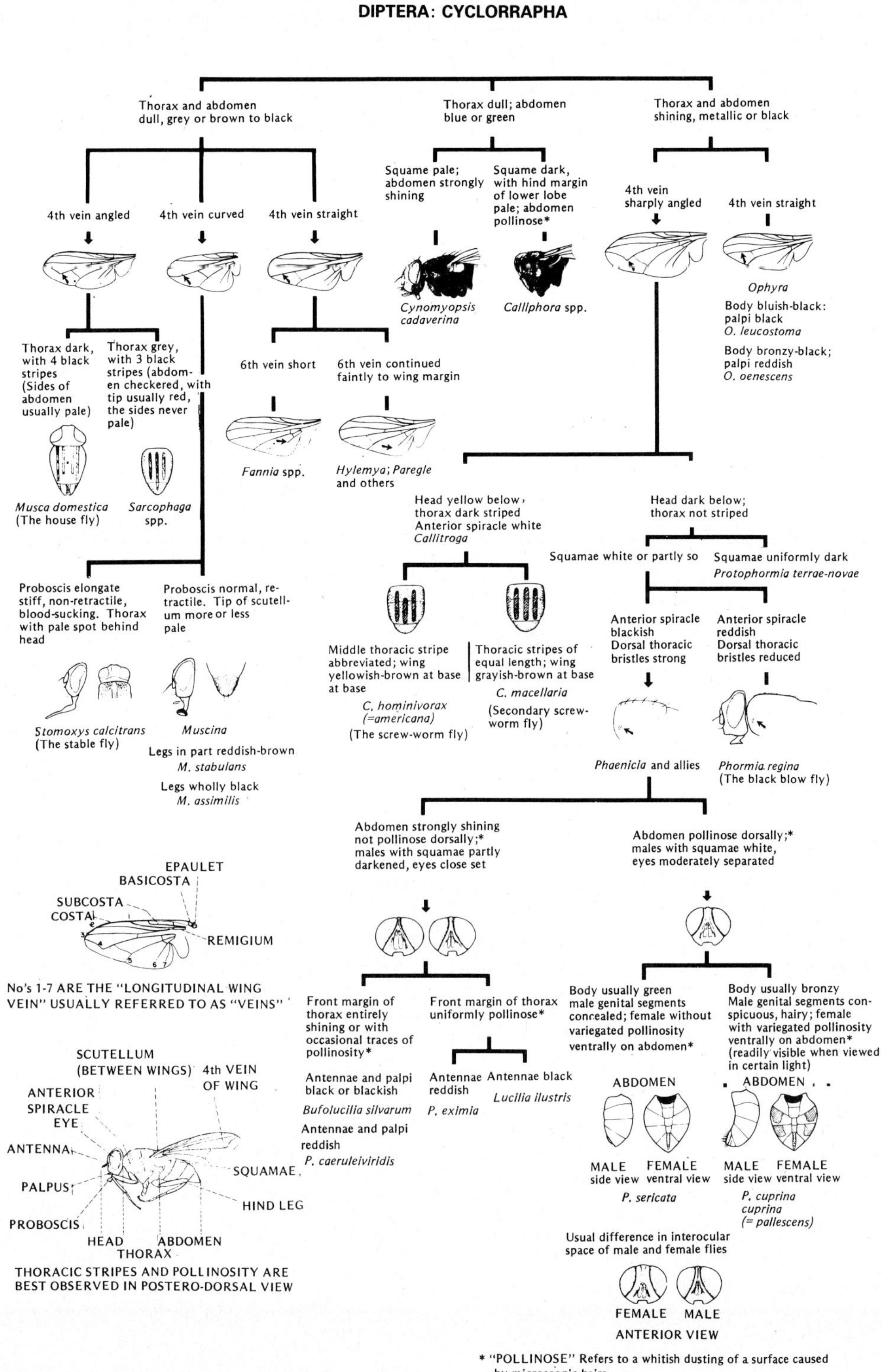


**FIG. 55.2**
Pictorial key to some common domestic flies.

NAME OF ORGANISM, MORPHOLOGY, AND IMPORTANCE. The face fly of cattle, *M. autumnalis*, is closely related to *M. domestica*. The face fly causes severe problems during the summer; estimated annual losses in the United States are $168 million. *M. autumnalis* is a European import that arrived in North America from Europe in about 1950 and has since spread through much of the United States and Canada. Adult face flies are slightly larger than house flies and look very much like them. One of the structural differences is that the white bands between the eyes of the female are broader than those of the house fly. There are other details of structure used for identification.

LIFE CYCLE AND HOST–PARASITE INTERACTIONS. Face flies develop in cattle feces and have about a 10-day cycle in summer. They probably overwinter as adults.

Face fly adult females have sponging mouthparts and dab at secretions from the eyes, nose, and mouth of cattle; they will also cluster around any wounds. Cattle sometimes have their heads covered with hundreds of face flies and are greatly annoyed by them. Face flies also mechanically transmit the causative agent of pinkeye (keratoconjunctivitis) of cattle. Disease transmission aside, animals are so bothered by face flies that they fail to gain weight properly and milk production falls off.

CONTROL. Scattering cattle feces may be feasible as a control measure under some rather restricted conditions, but it is usually not feasible if animals are on large pastures. Most often animals are allowed access to self-treatment devices or their heads can be painted with an insecticide solution.

NAME OF ORGANISM, DISTRIBUTION, AND IMPORTANCE. Another muscoid fly that is a worldwide pest is the stable fly, *Stomoxys calcitrans*. The adults (Fig. 55.2) have about the same size and appearance as house flies; however, they have biting mouthparts and feed on blood. They are the cause of serious economic loss in the livestock industry and are also such pests to humans during the middle and late summer in some areas that they make life on the beach unbearable. They prefer to breed in wet and fermenting plant material but also develop in feces.

NAME OF ORGANISM AND LIFE CYCLE. The horn fly, *Haematobia irritans*, is similar in appearance to the stable fly and, in keeping with its name, tends to cluster around the base of the horns of cattle. It is smaller than the stable fly, but is a vicious biter nevertheless. It prefers to breed in cow manure and develops a little slower than the house fly. A horn fly feeds on nearly any animal, but requires bovine blood for ovarian maturation. It is a vector of *Stephanofilaria stilesi* (Chap. 41). It causes considerable annoyance and blood loss in domestic animals, thereby affecting weight gains and milk production; annual losses in the United States are estimated to be $178 million.

The flies discussed here are similar in appearance, but they can be differentiated by careful study of the adults or the larvae. Since they have somewhat different breeding sites, feeding habits, and potential for disease transmission, it is sometimes important to determine which fly is causing a problem.

Tsetse, members of the genus *Glossina*, are currently found only in Africa, where they are vectors of, *Trypanosoma* spp.; they are discussed in Chapter 3.

## FAMILIES CAUSING MYIASIS

Myiasis is defined as infestation of tissues of vertebrate animals by the larvae of Diptera, which feed on those tissues for a period of time. Occasionally, members of other insect orders have larvae in the tissues of vertebrates. Infection by Coleoptera is called *canthariasis* and that by Lepidoptera *scolechiasis*, but these conditions are rare.

Even the thought of fly maggots burrowing through the tissues is revolting to many people, but the phenomenon is widespread and nearly any warm-blooded animal may be a host. In fact, so many of these associations exist that two classifications have been proposed: (1) Patton's, in which the requirements of the fly are foremost, and (2) Bishopp's, in which the organ system parasitized is the principal frame of reference.

Patton's classification is based on the nature of the association:

1. Obligate parasitism. The fly larvae require a living host for development. Examples are stomach bots of horses, cattle grubs, human bot flies.
2. Facultative parasitism. Adults are attracted to foul aromas such as those caused by wounds or

areas soiled by excrement; they oviposit and the larvae feed on necrotic tissue. In some instances, once invasion of the necrotic tissue has taken place, the larvae may continue deeper and feed on healthy tissue. Examples are blowflies, which normally oviposit in decaying carcasses, and various members of the Sarcophagidae.

3. Accidental parasitism. Larvae find their way into the body and live for a period of time, but are poorly adapted to the habitat. Sometimes the larvae may be located in the intestinal tract, having been eaten with contaminated food. Some of the flies in this category are fruit flies (family Drosophilidae), a number of house flies (family Muscidae), cluster flies (family Calliphoridae), crane flies (family Tipulidae), and rat-tailed maggots (family Syrphidae).

Bishopp's classification (slightly modified) relates to sites of myiasis:

1. Skin
   a. Cutaneous: larvae remain in the superficial skin
   b. Dermal: larvae may penetrate to the connective tissue layer of the skin
   c. Subdermal: larvae may penetrate deeply into the tissues
   d. Blood: the Congo floor maggot, *Auchmeromyia luteola*, of humans, and some related larvae feed on nestling birds and take blood
2. Intestinal: larvae are in the lumen or superficial layers of the gut
3. Urogenital: larvae enter the openings of the reproductive or urinary systems
4. Cephalic
   a. Ear
   b. Eye
   c. Nasopharyngeal region

In both classifications, there are problems with which species fits where, a common failing of any classification. In Patton, it is difficult to determine whether a particular species is facultative or accidental in its relationship to the host. In Bishopp, any one species may be found in different sites, and certain species such as cattle grubs have a long, circuitous migration in which they enter the skin and ultimately reappear in the skin. However, both classifications allow one to deal with the subject in a reasonable way. We will consider some examples from several families of myiasis-producing flies.

## FAMILY CALLIPHORIDAE

NAME OF ORGANISMS AND DISEASE CAUSED. The calliphorids include about a dozen genera of flies that are either obligate or facultative parasites, mostly the latter. There are a large number of species including *Phormia regina, Lucilia caesar, Phaenicia cuprina*, and *Calliphora vicina* (Fig. 55.1), which are called *blowflies*. They ordinarily oviposit in carcasses of animals, including meat for human consumption, but are attracted to anything that has a foul aroma. They thereby become parasites of various mammals including sheep, cattle, and humans.

LIFE CYCLE AND HOST–PARASITE INTERACTIONS. Females typically oviposit near wounds or areas on animals that are soiled with urine and feces. Sheep are particularly susceptible to this attack and are spoken of as being fly-struck; when animals have diarrhea, the hind quarters become soiled and blowflies are attracted. At first, they are only at the surface of the skin, but if the skin is soiled and wet, they begin to move into the skin and subcutaneous tissues, causing necrosis and attracting more ovigerous females.

CONTROL. Prevention of fly strike is based on the following steps:

1. Docking of the tails of lambs. In ewe lambs, the tail is left long enough to extend slightly beyond the vulva so as to prevent sunburn.
2. Selective breeding for minimal wool on the hindquarters.
3. Treating wounds, such as the navel of newborn lambs, with a repellent.
4. Applying a long-residue insecticide on the hindquarters of all animals.
5. Clipping excess wool from the hindquarters.
6. Performing the "Mules" operation, in which some of the skin on either side of the back legs is removed to tighten the skin.

If none of these is sufficiently effective and sheep become fly-struck, it is necessary to handle each sheep for inspection, clipping the wool and treating with a jelly-insecticide product where necessary. The importance of fly strike can be seen in the measures that are taken to prevent it. In Australia, where the sheep industry is a major source of income, millions of dollars lost annually stimulated research programs that allowed considerable reduction in the impact of fly strike. Under any circumstances, eternal vigilance is the price of freedom from fly strike.

NAME OF ORGANISMS AND DISEASE CAUSED. Obligate myiasis is best represented by *Cochliomyia hominivorax (=Callitroga hominivorax)*, the primary screwworm (Fig. 55.3), which has been a scrouge of the livestock industry and wildlife in the Western Hemisphere since prehistoric times. The secondary screwworm, *Cochliomyia macellaria (=Callitroga macellaria)*, is not as invasive as the primary screwworm, and they can be differentiated from one another on the basis of adult head structures and the larval cephalopharyngeal skeleton and anterior spiracles (Figs. 55.4 and 55.5). The Old World screwworm is *Chrysomyia bezziana*, which is found widely spread in the Orient, Africa, and the islands of the Pacific; it is comparable to *C. hominivorax* in its biology and pathogenic effects.

DISTRIBUTION AND IMPORTANCE. *C. hominivorax* caused severe economic losses in the livestock industry all through North America, but especially in Texas and the southeastern states. Some estimates of economic losses are from $70 million to $120 million annually in the United States; in the summer of 1935 in Texas there were an estimated 180,000 deaths in livestock from screwworms. It is also likely that they were a controlling factor in white-tailed deer populations in Texas and the Southeast. But these figures are dull compared to the personal experiences of those who had to contend with screwworms on a daily basis. Some quotations from Richardson (1978) give moving descriptions not usually seen in scientific writings:

> Every spring, as sure as the seasons, and for generations unknown, screwworms began their annual march northward from their overwintering sanctuaries in Mexico and South Texas. Pushed by an unknown force as inexorable as gravity, screwworms move north—ever moving, ever spreading, ever multiplying, ever destroying. No army ever advanced any more surely or methodically. No army was ever more destructive. Attacking, killing, maiming, and destroying, screwworms literally ate their way north. Reaching upper South Texas, they fanned east and west—all the while moving north—dotting the countryside with the dead carcasses of hapless wildlife, cattle, sheep and goats . . . .
>
> There is no way to estimate the man-hours Southwesterners spent in searching for and doctoring screwworm-infested animals. It undoubtedly reached into the billions of hours annually . . . .
>
> Screwworm smear was shipped into Texas by the trainload. Most producers bought it by the case . . . .
>
> In effect, the screwworm ruled the livestock industry of the Southwest for a century and a half. Producers would not brand, dehorn or castrate animals during that season of the year when screwworms were severe.
>
> People would not leave home for more than a day for fear of finding their animals had been eaten alive while they were away.

CONTROL. The following were the methods used for control at that time:

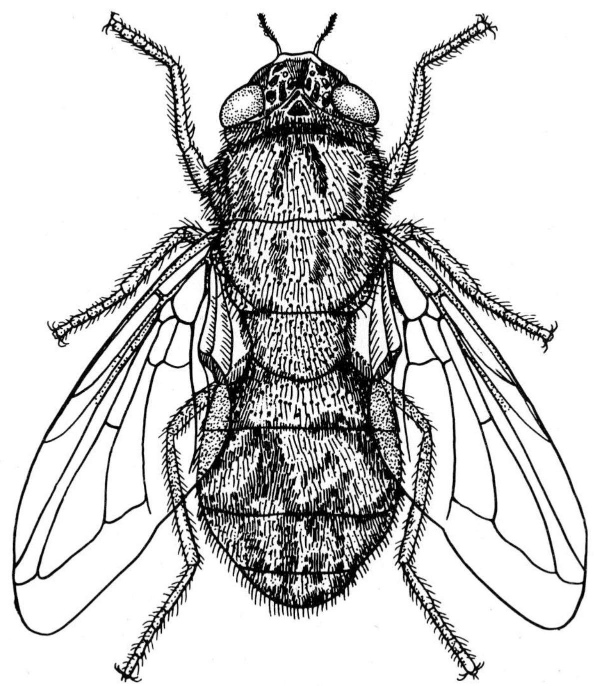

**FIG. 55.3**
The primary screwworm, *Cochliomyia hominivorax*. [From James, 1947.]

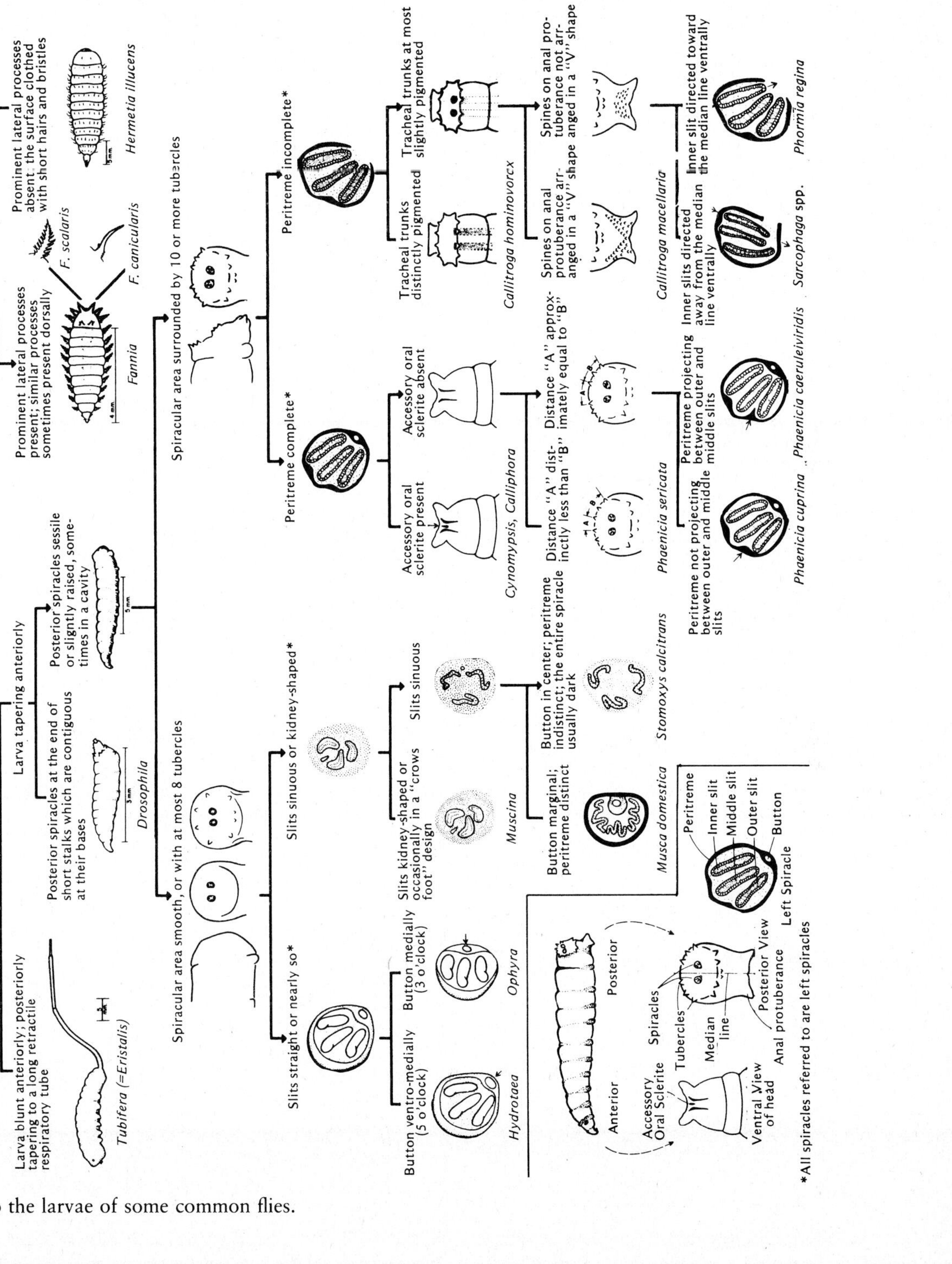


**FIG. 55.4**
Pictorial key to the larvae of some common flies.

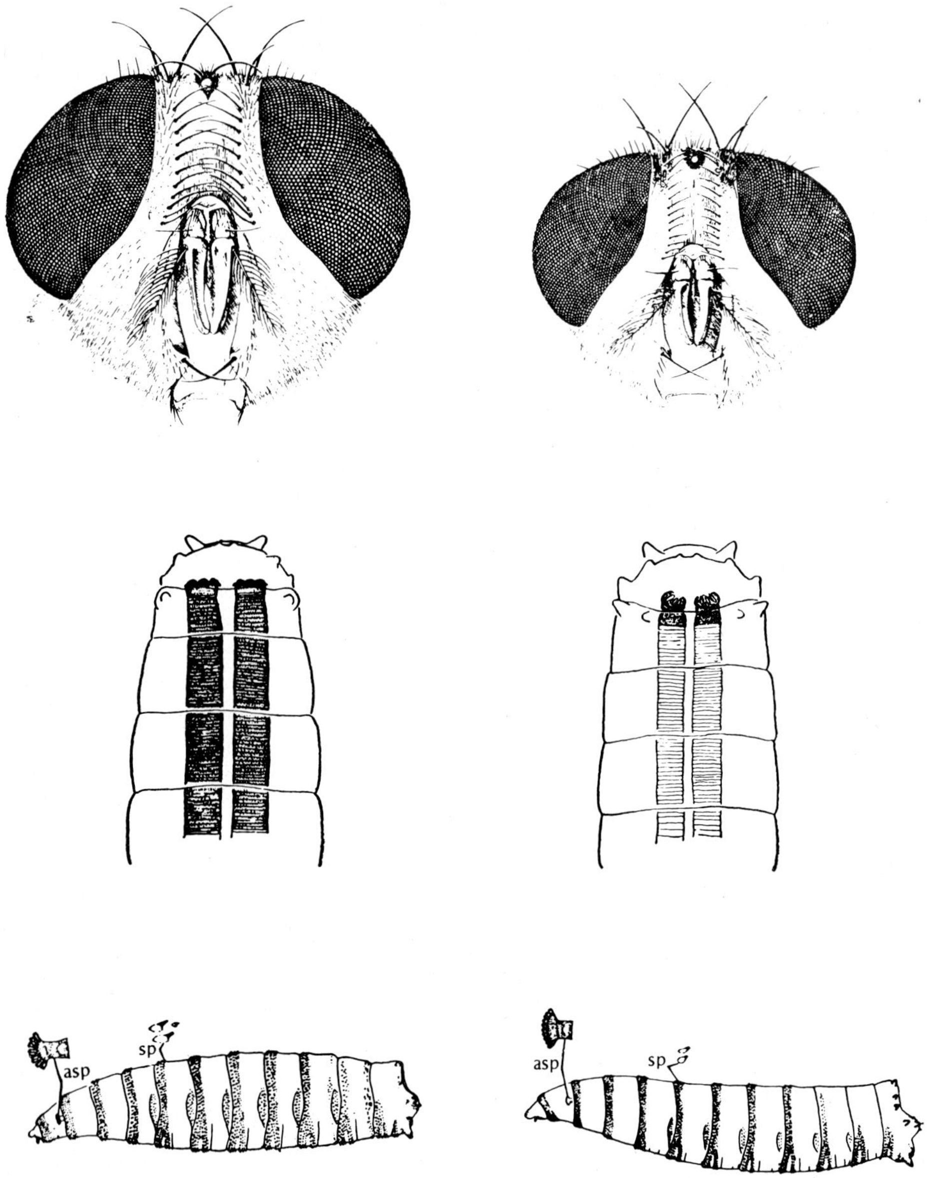


**FIG. 55.5**
Differences between *Cochliomyia hominivorax* (left) and *C. macellaria* (right). Heads; trachea of larvae; side of larvae.

1. Calving and any procedures such as branding and castrating were done in the winter or when it was hot and dry.
2. Barbed wire was replaced with smooth wire to reduce the probability of wire cuts.
3. Animals were inspected regularly and treated. In fact, in the Southwest, cowboys continually rode among the cattle looking for those that had been fly-struck; those that had been were roped and treated immediately.

Where there was concerted effort on the part of all livestock raisers, the impact of screwworms was reduced, but it was often a losing battle. The key to effective control lay in studies done in the middle 1930s by Drs. Knipling and Bushland of the USDA on the basic biology of screwworms. They found that a female screwworm mated only a single time during her life and they reasoned that if she could be mated by a sterile male, she would produce only infertile eggs. It was finally possible to implement the theory in the period following World War II.

The theoretical base was secure, but the following technical problems had to be solved:

1. Raising millions, or even billions, of flies
2. Sterilizing such large numbers of flies reliably by ionizing radiation
3. Distributing the sterilized flies—How many? How often? By what means?

Flies were induced to oviposit in a medium consisting principally of horse meat. They were then irradiated at the proper time by $^{60}$Co and distributed by airplane. The first extensive field test was undertaken in 1952 on the island of Curaçao in the Caribbean. In 1954 as many as 68,000 sterile males were released each week on the island. By the end of 1954, the island was declared to be screwworm-free. The technique had been established and was ready for further exploitation (Knipling, 1960).

A screwworm factory was established at Sebring, Florida, and eradication was planned for the southeastern states. Between July 1958 and July 1959, 2.75 billion flies were released in the Southeast and screwworm infestations dropped from 40,000/year to 0. In 1962 the screwworm factory was moved to Texas and the campaign was begun to push the screwworms southward. From 1963 to 1976 more than 140 million sterile flies were released each week in Texas and northern Mexico. The total number from this one operation over the 14-year period was 1.44 x $10^{12}$, or about 300 sterile males for every person on earth.

The screwworm control program came about just at a time when the economics of cattle raising was changing. The costs of all aspects of cattle raising increased rapidly during the 1960s and 1970s, especially for land and labor, but calf prices remained static. It has been estimated that screwworm treatment required about $10/calf weaned in 1963, but this cost would have risen to about $30 in 1977 if there had not been screwworm control. Thus, the implementation of the sterile male technique has allowed many cattle raisers to remain in business (Richardson & Averhoff in Richardson, 1978).

The total number of screwworm cases in the United States between 1962 and 1976 was:

| YEAR | NUMBER OF CASES |
|---|---|
| 1962 | 51,592 |
| 1963 | 7,111 |
| 1964 | 395 |
| 1965 | 1,060 |
| 1966 | 1,875 |
| 1967 | 875 |
| 1968 | 9,878 |
| 1969 | 219 |
| 1970 | 153 |
| 1971 | 473 |
| 1972 | 95,626 |
| 1973 | 15,916 |
| 1974 | 7,273 |
| 1975 | 17,568 |
| 1976 | 29,671 |

The success that was achieved in the late 1960s and early 1970s obviously had broken down somehow by 1972. In response, the number of flies per unit area was increased, but the number of cases was not reduced. The rearing system and its effects were analyzed (Bush et al., 1976) and it was found that flies had been selected for rapid development at a relatively high temperature and for short flight. The result was establishment of a flight muscle enzyme (α-glycerol phosphate dehydrogenase) allele in such a high proportion of the population that the males flew little and were unable to compete with the wild-type flies.

The solution was to reisolate wild-type flies and establish them in culture. In addition, the flies are constantly monitored to ensure that noncompetitive alleles or mutants do not become established.

It had originally been thought that the sterile male technique (also known as *autocidal control*) would not have the same problems as the various organic insecticides. It is nonpolluting and resistance would not be a problem. Although the prediction has generally been true, this is a highly complex process that requires careful monitoring of all stages in the program, and there are ways in which it can break down, as we have seen.

The question might be raised of whether the war is over. Not yet. The location has shifted to Mexico, with the area of eradication being moved continually southward. The probability of success will be revealed during the next few years.

NAME OF ORGANISM AND HOST–PARASITE INTERACTIONS. The Congo floor maggot, *Auchmeromyia luteola,* is the only bloodsucking larva that attacks humans. The fly occurs throughout much of subsaharan Africa to the Cape of Good Hope and is well adapted to feeding on people who sleep on mats on the ground. The adult flies, which are 9 to 13 mm long, oviposit in dry soil or sand such as might be found in a hut. The larvae are active at night and poke through sleeping mats to pierce the skin and suck the blood of the sleeper. They seem to cause little discomfort and are considered to be more of a nuisance than a serious health problem. Persons who sleep only a few centimeters off the ground are protected from the attack of this maggot.

NAME OF ORGANISM AND LIFE CYCLE. A fly that resembles the house fly and often enters houses is the cluster fly, *Pollenia rudis*. This fly is 7 to 9 mm (1/4 to 3/8 in.), grayish, and dull. The larvae are parasitic in earthworms, in which they grow and which they ultimately devour. They are not usually considered to be a health hazard because their site of development is not likely to have microbial pathogens. Records of their causing enteric myiasis in humans are doubtful. Cluster flies readily enter houses, especially in late summer and fall, and they are likely to be mistaken for house flies. They crawl in through cracks around doors and windows as the weather becomes cool and remain there until a warm day occurs when they will buzz lazily around the house.

## FAMILY SARCOPHAGIDAE

Members of this family have the full range of myiases from obligate to occasional. Among the important species are *Wohlfahrtia vigil, W. opaca, W. vigil, Sarcophaga haemorrhoidalis*, and *S. bullata*. The adult flies are usually 10 to 15 mm (3/8 to 5/8 in.) in length and gray, with hairy bodies. The fully grown larvae sometimes exceed 25 mm (1 in.) and cause considerable damage in small animals. Females larviposit in the vicinity of wounds and the larvae immediately enter them and begin to feed. Their two molts are often completed in 3 to 4 days and they then drop out and pupate.

## FAMILY CUTEREBRIDAE

These flies are usually 20 mm (3/4 in.) or more long and are often covered with short hairs; mouthparts are small or rudimentary. The larvae are often parasites of newborn animals. The genus *Cephenemyia* includes a number of species that are parasites of the nasal passages and upper respiratory tract of large ruminants such as deer and reindeer.

NAME OF ORGANISM AND LIFE CYCLE. *Dermatobia hominis* is called the *human bot*, but is is found in a wide range of other hosts and is a particularly severe problem in cattle in tropical areas. It is widespread in tropical America. Its mode of transmission is unique; the female fly captures a bloodsucking insect such as a mosquito (Fig. 55.6) or stable fly, oviposits on it, and then turns it loose. The eggs embryonate in 5 to 15 days and the larvae leave the egg when the host insect takes a blood meal. The larvae penetrate the skin and feed head down in the lesion. The larvae remain in the skin for 5 to 10 weeks but have been known to remain as long as 3 months. The fully grown third-stage larva is from 18 to 24 mm (1 in.) in length.

## FAMILY OESTRIDAE—CATTLE GRUBS AND NOSE BOTS

NAME OF ORGANISMS. Two species are of principal interest in cattle: *Hypoderma lineatum*, the common cattle grub, and *H. bovis*, the northern cattle grub. The larvae are mainly parasites of the domestic ox, in which they cause great economic loss throughout the world. They are frequently found in the backs of horses, making them impossible to saddle and ride at that time. The adults are hairy, beelike flies (Fig. 55.7) that have only rudimentary mouthparts. They are strong fliers that buzz loudly.

LIFE CYCLE. In temperate climates, the life cycle is governed by the seasons and extends over a full year.

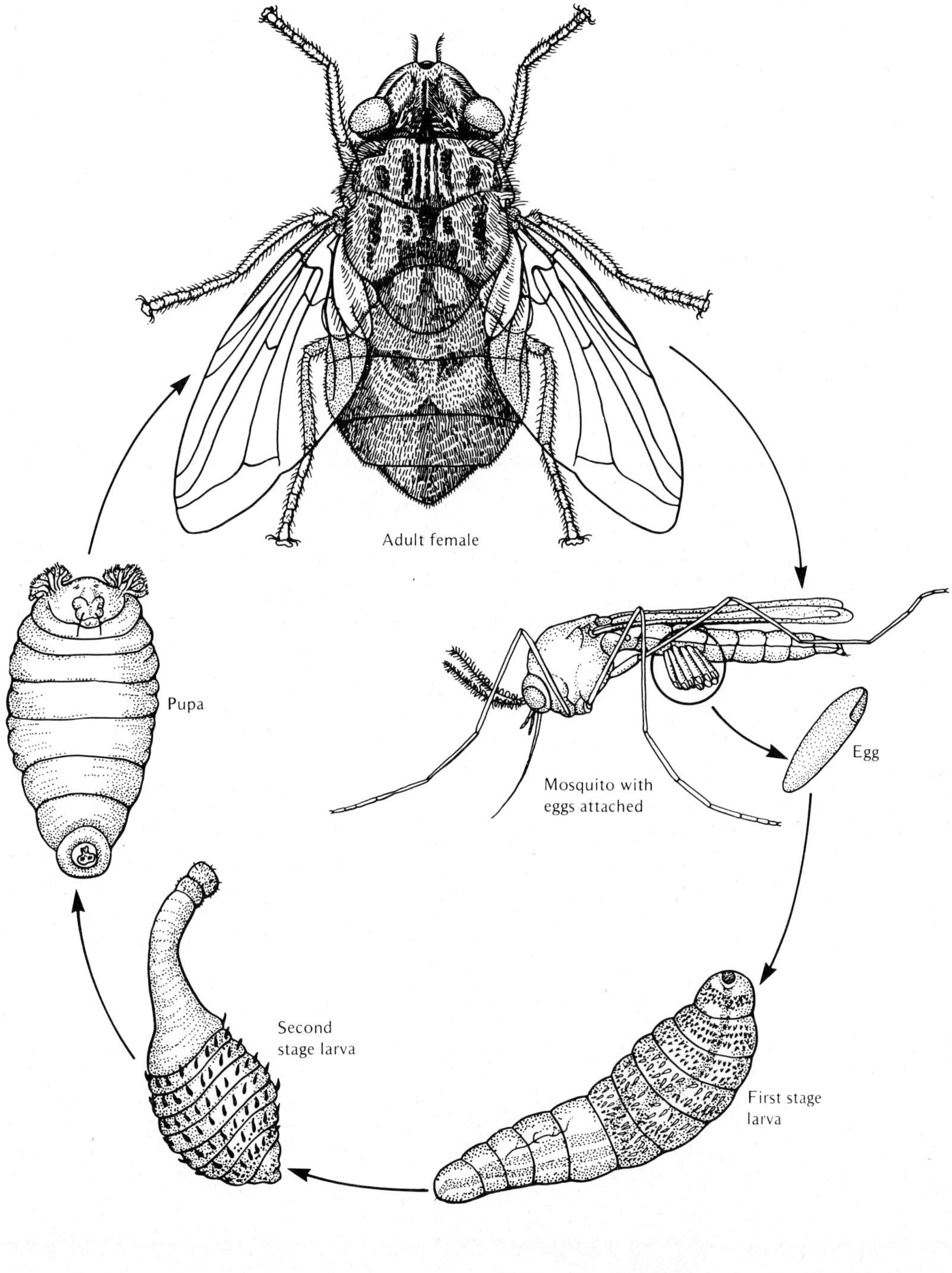


**FIG. 55.6**
Life cycle of *Dermatobia hominis,* the human bot fly. [From James, 1947.]

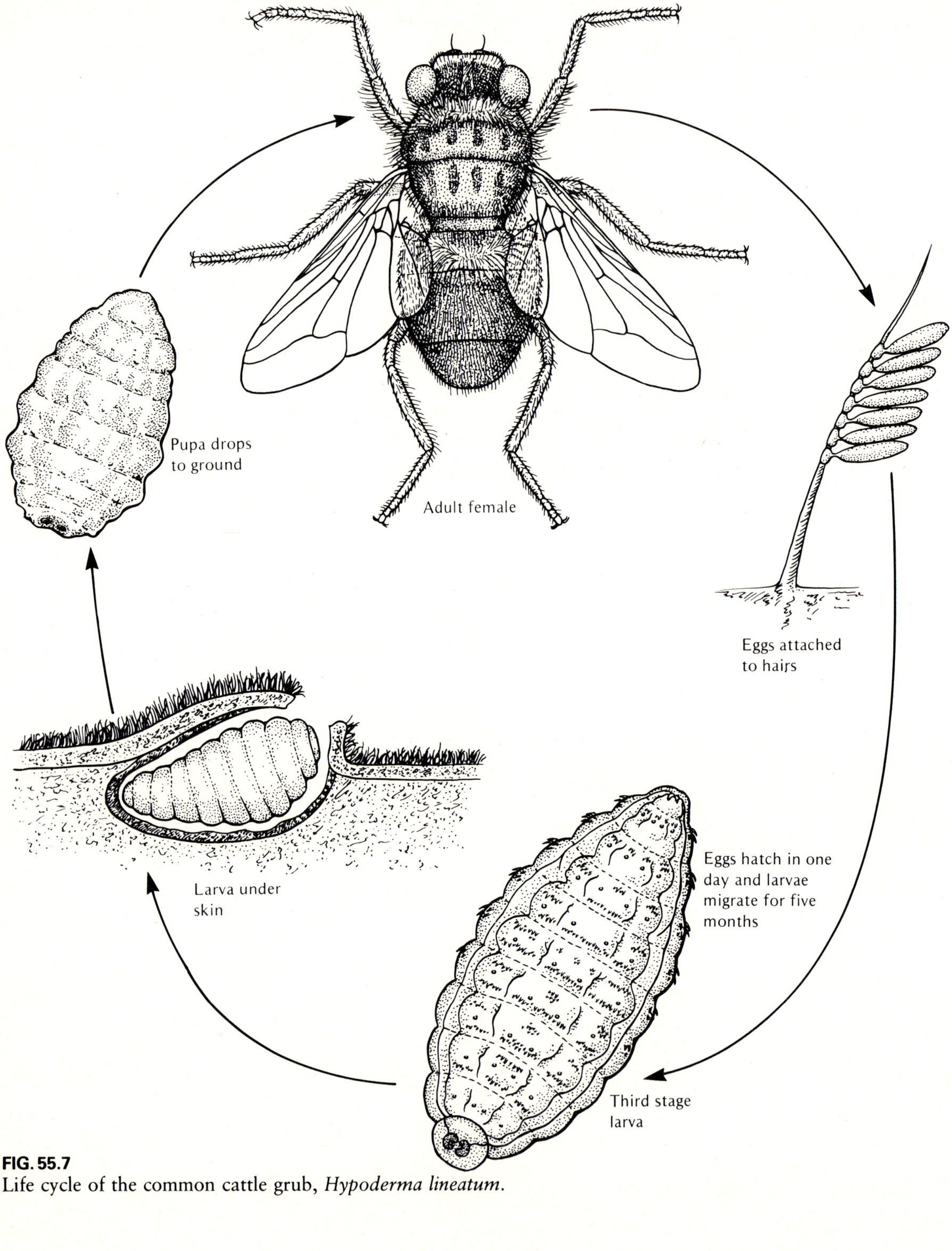


**FIG. 55.7**
Life cycle of the common cattle grub, *Hypoderma lineatum*.

Flies first appear in early spring and are seen until about midsummer. After emergence, the adults copulate; the females oviposit and then die. The adult *H. lineatum* oviposits on the hind legs of cattle by darting in quickly and attaching individual eggs to hairs on the legs. Eggs hatch in 3 to 7 days and the first-stage larvae penetrate the skin, probably through hair follicles. At this time, the larvae are only about 0.7 mm long and are rather smooth. They migrate during the summer by passing through connective tissue planes, and reach the connective tissues surrounding the esophagus in September or October. They are now about 10 mm (½ in.) long and look like a large rice grain—smooth and rather transparent. Starting in about November, the larvae begin to migrate again, move upward through connective tissue planes in the muscles, and come to lie under the skin along the midline of the back. They molt for the first time and begin to grow rapidly. The general view is that they require oxygen and therefore cut a hole in the skin to which they apply their spiracles. The second-stage larvae are about 15 mm (⅝ in.) long; they molt for the second time about 25 days later, and become the spiny, brown grub that may reach a length of 30 mm (1¼ in.). After 50 to 100 days in the back, the larvae shrink somewhat and then crawl out of the hole and drop to the ground to pupate. Depending on the temperature, from 18 to 77 days are required for the adults to emerge.

The life cycle of the northern cattle grub is much like that of *H. lineatum* except that its stages are shifted about a month later, and there are some differences in its migratory pattern. Whereas the common cattle grub lies in the tissue surrounding the esophagus during the fall, the northern species lies in the subdural fat in the spinal canal. In both species, the life cycle is tuned to the climate, and it shifts later in the year in the more northern states and Canada (Fig. 55.8). In Texas the cycle of the common grub is not as synchronized as it is farther north, because the flies are active for a great part of the year.

DISTRIBUTION AND IMPORTANCE. The common cattle grub is found from the northern Great Plains to some tropical areas of Central America; it is distributed all through northern Europe and Asia. The northern cattle grub is found in the northern three or four tiers of states and occasionally in more southerly locations such as the Carolinas. In the Eastern Hemisphere it is more prevalent in northern Europe but also occurs in some warm areas such as North Africa. One of the limiting factors on the distribution of cattle grubs relates to their requirements for pupation; they require relatively dry sites for the survival and development of the pupa. If a locality has a spring climate in which the soil becomes saturated, cattle grubs are absent.

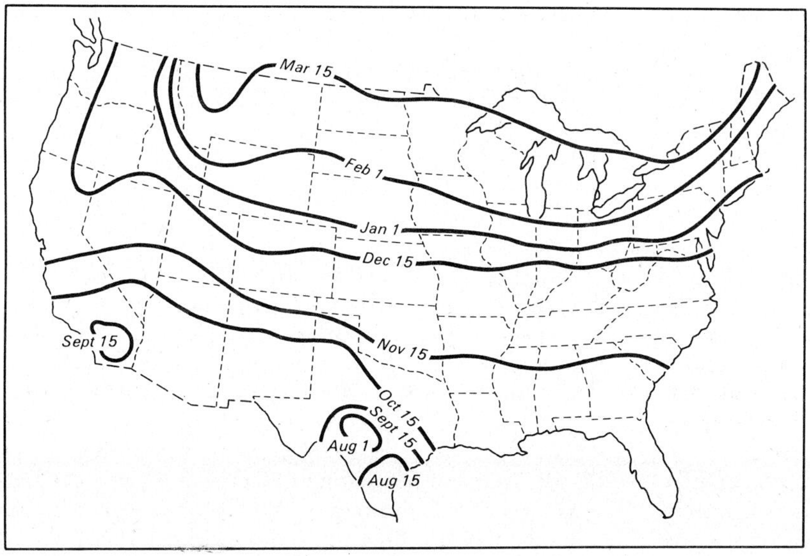


**FIG. 55.8**
Time of first appearance of grubs in the backs of cattle in the United States.

Economic losses from cattle grubs come about through the following:

1. Activity of the flies during egg laying
2. Secondary bacterial infections when the hide is pierced by the second-stage larva
3. Trimming losses when animals go to slaughter with grubs under the skin
4. Leather losses because of holes in the hide or scarring
5. Deaths through anaphylaxis from crushed grubs

During the time when adult flies are active, cattle stand in the shade or in ponds to keep the flies away; flies buzz loudly when flying and when coming in to oviposit, and cattle seem to be afraid of them despite the fact that the adults do not bite and ovipositing causes them no pain. They run to keep away from flies, and there are instances of cattle having been stampeded by flies. One of the common names, *heel fly*, comes from the observation that the flies hover about the lower legs of cattle when ovipositing.

During migration of the first-stage larvae there is some damage to the tissues, but no economic losses have been shown; however, when the grubs reach the back and cut holes in the skin, there is both encapsulation of the larvae and bacterial infection of the area surrounding them. The area around each grub becomes swollen and leaks pus out onto the skin; these so-called warbles reach 20 to 30 mm (1 in.) in diameter. When animals are slaughtered with grubs in the back, the infected areas must be trimmed out, causing some loss of meat and requiring extra labor.

Hides that have six or more grub holes in the back are downgraded and buyers pay less for them. At one time, livestock raisers were not concerned with downgrading of hides since they were such a small proportion of what they received for their animals. In the 1970s hide and leather prices began to rise because of a worldwide shortage of leather. For example, between 1975 and 1978 hides went from $46.50/100 pounds to $78.75/100 pounds, nearly a 60% increase. About 10% of the wholesale price of shoes is for leather. Hides are now a significant factor in the economic return for producers. Total annual losses in the United States from cattle grubs are estimated to be $192 million.

Some protective immunity against cattle grubs probably results from natural infection. Yearling cattle have fewer grubs than calves. However, allergy is an aspect of immunity that is not beneficial, and some animals become hypersensitive. One of the means of eliminating infections is to squeeze grubs out of the back by hand; dairy farmers often do this, because they see animals at least twice a day and they are readily handled. The difficulty comes if a grub is crushed during the process; a great deal of material is released suddenly, and anaphylactic shock may kill the animal.

It might be thought that significant production losses would occur in cattle that can have 50 or more grubs in the back. Interestingly, studies on weight gains and grub infestations have seldom shown that they adversely affect growth. An example of a study in which weight gains were seen is that of Khan and Kozub (1981), in which two groups of steers were 29.9 and 13.4 kg heavier than untreated animals 6 months after treat-

CONTROL. Until about the middle 1950s, cattle grub control was rather primitive. Animals were treated with insecticides after the grubs had reached the backs of animals and had cut holes in the skin. Insecticide was applied as a wash and scrubbed into the holes, or a high pressure sprayer was used to force the material into the holes. The procedure was not satisfactory, because the damage had already been done. The organophosphate insecticides, which had grown out of the development of nerve gases during World War II, were efficient insecticides. Some of them had low mammalian toxicity and were absorbed into the system when administered by mouth, by injection, or applied to the skin. A number of products reached the market at about the same time, but ronnel was the first to show a high level of efficacy and low mammalian toxicity. A number of studies showed that it gave 90 to 100% control of grubs when administered in fall after the fly season. Among the organophosphates currently approved for cattle grub control are ronnel, fenthion, coumaphos, famphur, and crufomate.

As emphasized in other chapters on livestock parasites, economic return governs whether to treat animals and how the treatment is administered. The first systemic insecticides were given by mouth; this method was good for a start, but was rather time-consuming. Later approaches involved giving the insecticide by injection or in a feed supplement. Injection has the disadvantage of depositing toxicant in the muscle, and feeding it ad libitum has the disadvantage that all animals may not take the same amount. The mode of administration in wide use now is a pour-on in which a liquid containing 4 to 13% insecticide is ladled onto the midline of the backs of animals; it is absorbed and kills the migrating larvae. Animals merely need to be lined up in a chute and the material is applied to their backs about as quickly as it can be described. A further refinement is a spot-on preparation, which contains 20% insecticide and a blue dye; animals are herded into a corral and "shot" at a distance with the insecticide, thereby treating and marking them at the same time.

Eradication programs for cattle grubs are underway in some countries, but have not yet succeeded. Part of the problem is in the host range; despite their name, cattle grubs are found in other hosts including other ruminants and horses. Residual populations are therefore difficult to eliminate. Occasional infections occur in humans, but they do not contribute to the maintenance of the populations.

NAME OF ORGANISM AND MORPHOLOGY. The nasal bot of sheep, *Oestrus ovis*, is also called the *head maggot* and is a cosmopolitan parasite of sheep. The beelike adults are 10 to 12 mm (½ in.) long (Fig. 55.9) and do not feed.

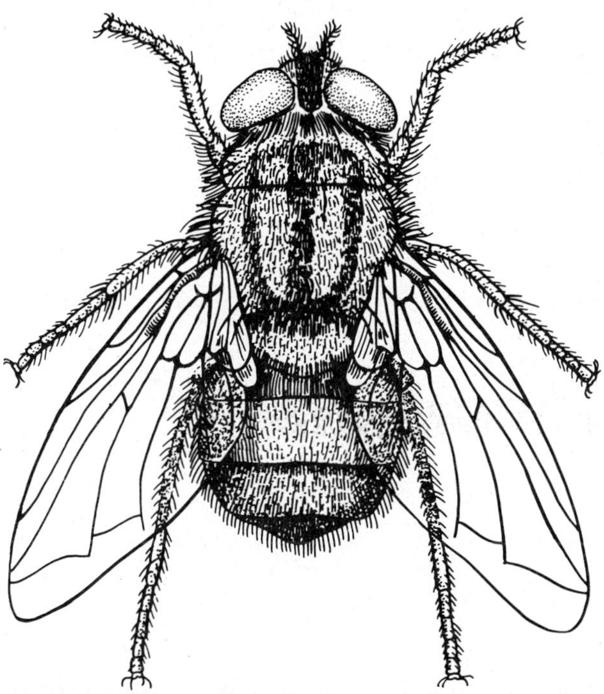

**FIG. 55.9**
Adult of the nasal bot of sheep, *Oestrus ovis*. [From James, 1947.]

LIFE CYCLE. The females larviposit near the nostrils of sheep. The larvae enter the nasal passages, where they remain and develop through the second instar. The third-instar larvae enter the head sinuses and remain for 8 to 10 months. They are about 30 mm (1¼ in.) long when they leave the head in late winter or early spring and pupate in the soil.

HOST–PARASITE INTERACTIONS. The feeding of the larvae in the nasal passages and sinuses causes direct irritation and secondary infections. Infected sheep usually have excessive nasal discharge; occasionally, larvae bring about bacterial encephalitis, causing the death of the host. Humans are sometimes infected; most cases on record have concerned persons associated with sheep, such as shepherds.

CONTROL. Infection can be partially prevented by applying a jellied insecticide on the noses of sheep. Treatment with crufomate by mouth eliminates nearly all of the larvae.

## FAMILY GASTEROPHILIDAE—STOMACH BOTS OF HORSES

NAME OF ORGANISMS AND MORPHOLOGY. On a global basis, six species are found in equids, but only three occur in North America: *Gasterophilus intestinalis*, *G. nasalis*, and *G. haemorrhoidalis*. The adults are from 10 to 17 mm (½ to ¾ in.) long, hairy, and with rudimentary mouthparts; the flies resemble bees.

LIFE CYCLE. The life cycles of the various horse bots are similar and are tuned to the weather in temperate climates. The females oviposit on hairs around the muzzle of the horse (*G. nasalis, G. haemorrhoidalis*) or on the legs (*G. intestinalis*); they hatch in from 2 to 7 days when the horse licks them. The larvae seem normally to penetrate the oral mucosa and migrate to the stomach through the body tissues. They attach in the stomach and probably remain in the same place for 8 to 11 months. They then detach and are carried to the outside where they pupate.

HOST–PARASITE INTERACTIONS. Although horses are annoyed when the adults lay eggs, the principal damage is to the lining of the stomach where the bots have their heads buried deeply in the tissue. Colonies of larvae are often seen that are 10 cm (4 in.) in diameter, and they sometimes block the passage of food from the stomach to the small intestine.

TREATMENT AND CONTROL. Treatment for stomach bots is related to the life cycle and is a type of strategic treatment. Treatment is not given until all fly activity has ceased at the end of summer or early fall; then it is given along with the routine treatment for intestinal nematodes (Chap. 33). Prevention of infection of horses is done most often by currying; the eggs are nearly 1 mm long and yellowish, so they can readily be seen on the legs or around the muzzle of the animal. The minimal time for embryonation is 2 days, so daily currying to remove the eggs is recommended.

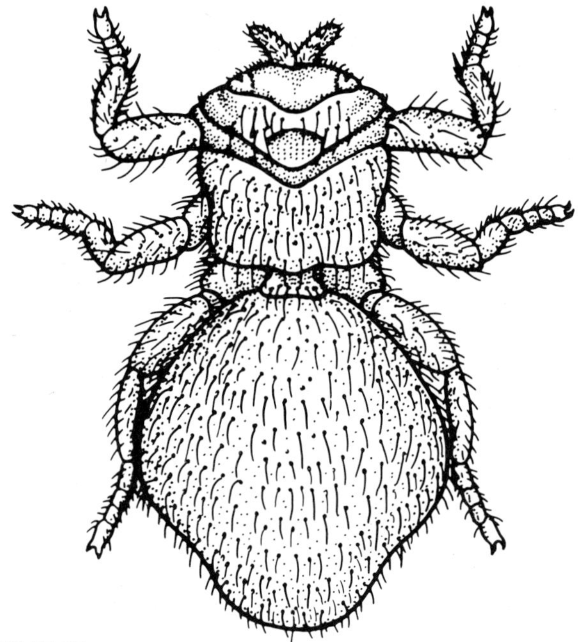

**FIG. 55.10**
Adult of the sheep ked or "tick," *Melophagus ovinus*.

## FAMILY HIPPOBOSCIDAE—LOUSE FLIES AND KED

Flies of the family Hippoboscidae are all bloodsucking parasites of vertebrates. They are flattened, rather hairy, and have inconspicuous antennae. Wings vary from well developed to shortened to absent, depending on the species. The females are larviparous with the larvae being nourished in the body of the female through the third instar. When the larva is born, it develops into a pupa almost immediately. Detailed information on hippoboscids may be found in Bequaert (1942, 1953, 1954, 1955, 1957).

Most hippoboscids are associated with birds and are referred to as *pigeon flies* or *louse flies*. Some, such as *Pseudolynchia canariensis*, are vectors of protozoan hematozoa; this species is a vector of *Haemoproteus columbae* of columbiform birds. *Lynchia hirsuta* is a vector of *H. lophortyx* of quail.

The sheep ked or sheep "tick," *Melophagus ovinus*, is commonly found sheep in cool-temperate (but not warm) climates, where it causes significant economic loss. The whole life cycle takes place on the sheep, and because the ked is wingless (Fig. 55.10), transmission from one animal to another is by close contact. The adults are bloodsuckers, and sheep become anemic when the numbers of ked are high. Control of sheep ked is by contact insecticides, which may be applied by spray or dust when the sheep have been shorn, or by dipping at other times. Related species are found on other mammals such as camel, deer, and antelope.

## REFERENCES AND FURTHER READINGS


Bequaert, J. 1942. A monograph on the Melophaginae, or ked-flies, of sheep, goats, deer and antelopes (Diptera: Hippoboscidae). *Entomol. Am.* **22**:1–210.

——. 1952–1957. The Hippoboscidae or louse-flies (Diptera) of mammals and birds. Part I. Structure, physiology and natural history. *Entomol. Am.* (1952) **32**:1–209; (1953) **33**:211–242. Part II. Taxonomy, evolution and revision of American genera and species. *Entomol. Am.* (1954) **34**:1–232; (1955) **35**:233–416; (1957) **36**:417– 611.

Bishop, F. C., and E. W. Laake. 1949. *Cattle Grubs or Heel Flies with Suggestions for Their Control.* USDA Farmers' Bulletin No. 1596. Washington, D.C.

Bush, G. L., R. W. Neck, and G. B. Kitto. 1976. Screwworm eradication: Inadvertent selection for noncompetitive ecotypes during mass rearing. *Science* **193**:491–493.

Cole, F. R. 1969. *The Flies of Western North America.* University of California Press, Berkeley.

James, M. T. 1947. *The Flies That Cause Myiasis in Man.* USDA Misc. Pub. No. 631. Washington, D.C.

Johannsen, O.A. 1969. *Aquatic Diptera: Eggs, Larvae and Pupae of Aquatic Flies.* Entomol. Reprint Specialists, Los Angeles. (Four Memoirs of Cornell University Agric. Exp. Stat., 1934, 1935, 1936, 1937).

Khan, M. A., and G. C. Kozub. 1981. Systemic control of cattle grubs (*Hypoderma* spp.) in steers treated with Warbex and weight gains associated with grub control. *Can. J. Comp. Med.* **45**:15–19.

Knipling, E. G. 1960. The eradication of the screwworm fly. *Sci. Am.* **203**(4):54–61.

Pratt, H. D., K. S. Littig, and H. G. Scott. 1975. *Flies of Public Health Importance and Their Control.* USDHEW, CDC Pub. No. 77–8140. Centers for Disease Control, Washington, D.C.

Richardson, R. H. (ed.). 1978. *The Screwworm Problem: Evolution of Resistance to Biological Control.* University of Texas Press, Austin.

Smart, J. 1948. *Insects of Medical Importance*, 2nd ed. British Museum, London.

Smith, K. G. V. 1973. *Insects and Other Arthropods of Medical Importance.* British Museum, London.

Zumpt, F. 1965. *Myiasis in Man and Animals in the Old World.* Butterworths, London.

# 56

# *Introduction to Arachnida and Mites*

The subphylum Chelicerata includes those organisms that have piercing-cutting mouthparts called *chelicerae* in opposition to the other principal groups that have mandibles (Crustacea, Insecta). Associated with the chelicerae are the pedipalps, accessory mouthparts that manipulate the food or have sensory functions, depending on the group. There is no head as such, and the segments are fused so that one sees only major body regions; antennae are lacking, and the eyes are simple. They have four pairs of legs and most members are terrestrial. Development patterns are such that young forms appear to be much like the adults. Members are spiders, scorpions, daddy longlegs, sun spiders, pseudoscorpions, and ticks and mites (Fig. 56.1). Savory (1977) provides general treatment of arachnids.

Chelicerates are predaceous (spiders, scorpions, some mites), detritus feeders (some mites), or parasitic (ticks, some mites). They feed mostly on liquid food, but some (scorpions) tear their prey apart with their pedipalps and then liquefy it near the mouth. The only group that sucks blood from vertebrates and serves as vectors of infectious agents are the ticks and mites.

## SUBCLASS SCORPIONINA

Scorpions are readily recognized, because their pedipalps are modified into pincers by which they grasp their prey. The telson or tail is modified into a stinger. Their usual upper limit in length is about 10 cm (4 in.), but one of the African forms, *Pandinus*, reaches 20 cm (8 in.). Their habitats are generally hot, dry climates, where they prey on small invertebrates. Their importance lies in the fact that they occasionally sting humans. Most of the New World forms are only mildly toxic, but some of the African species are highly dangerous to humans, especially to children. They do not seek humans to sting; humans are stung when they accidentally come in contact with scorpions. Keegan (1980) is an excellent source of information on the biology, identification, and hazards of scorpions.

## SUBCLASS ARANEAE

Spiders are ubiquitous predators on other small arthropods. Most are harmless to humans, and many are distinctly beneficial by their feeding on insects. Some especially hazardous spiders are discussed in Chapter 48. Further information may be found in Gertsch (1949) for general information, Kaston (1972) for identification, and Fabre (1919) for some interesting aspects of natural history; Comstock (1940) provides an enormous fund of information on all aspects of spiders; Cloudsley-Thompson (1958) has good information on the natural history and biology of spiders and some other groups.

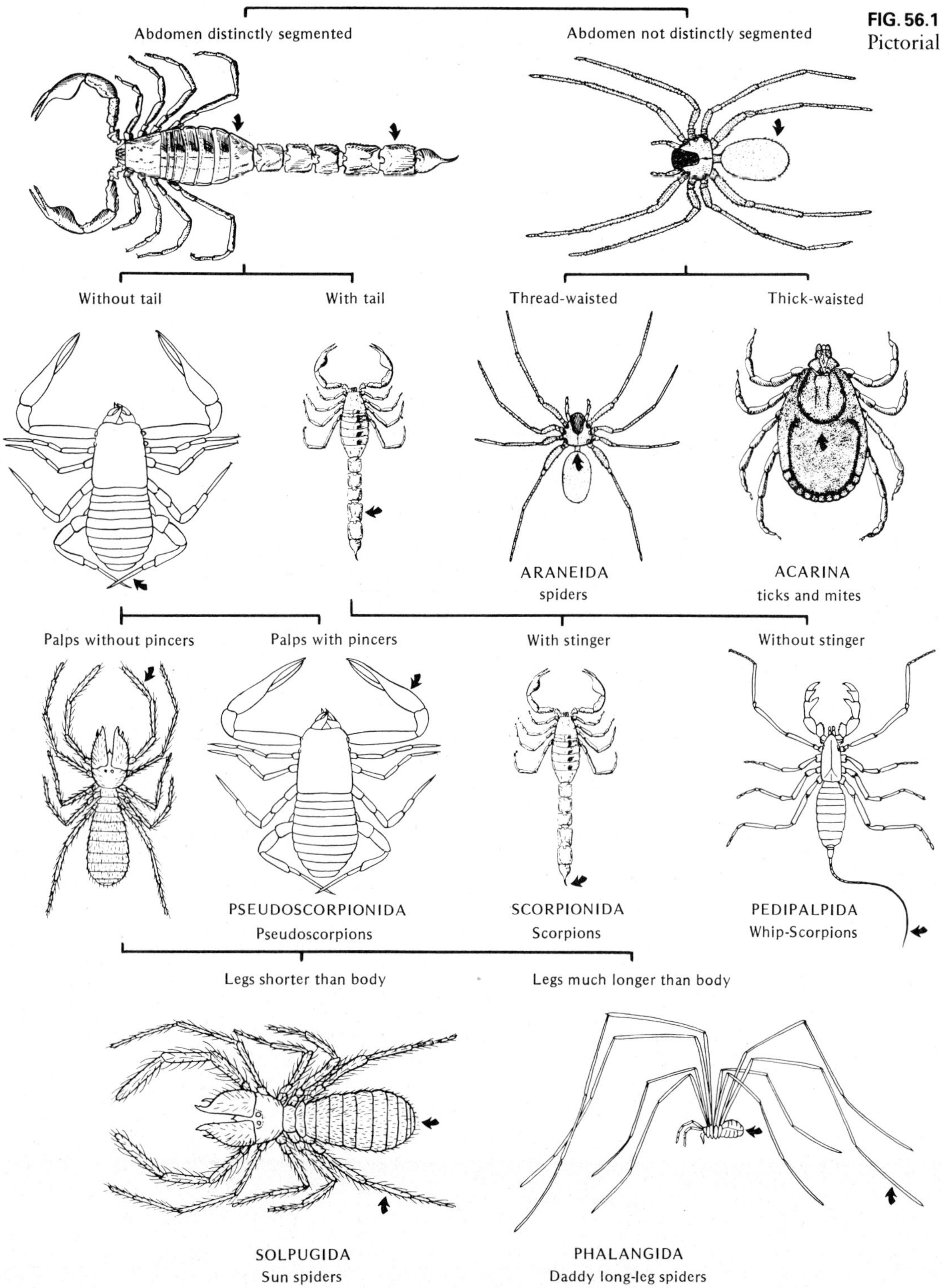


**FIG. 56.1**
Pictorial key to Arachnida.

## SUBCLASS ACARI—THE MITES AND TICKS

Acari, whose members are commonly called *mites* or *acarines*, includes both mites and ticks; ticks are actually giant mites and are spoken of separately only because they are much larger. Mites are diverse both in structure and biology. The greatest number of species are free-living and they may be detritus feeding or predaceous. The parasitic forms feed on plants and animals by puncturing the tissues with their chelicerae and taking juices or blood. The parasites are important as vectors of disease agents of plants and animals as well as for the damage they produce directly through feeding. Mites are found in all sorts of habitats, terrestrial, aquatic, and parasitic. About 20,000 species are known, and additional species are being described as investigators look for them with care. For example, there is a group that is symbiotic in the balancing organs of noctuid moths, and others are found in many other unique habitats as well.

General references on mites and ticks are rather sparse. Baker and Wharton (1952) give a good general introduction to mites. Baker et al. (1956) cover identification of mites only; McDaniel (1979) is a fairly detailed work on identification of both mites and ticks; Arthur's (1962) book gives a great deal of information on ticks as they relate to transmission of disease agents; Woolley (1961) reviews the phylogeny of mites.

Terminology for the body parts of mites and other arachnids takes into account the fusion of the segments. There is no head, but the gnathosoma or capitulum bears the mouthparts (Fig. 56.2). The chelicerae are derived from the first appendages and the pedipalps from the second. The other major body region is the idiosoma, which comprises everything except the gnathosoma. Usually the terms employed for the subdivisions of the idiosoma are the *podosoma*, which bears the legs, and the *opisthosoma*, which is the region posterior to the legs. Leg terminology is similar to that of the insects, but since the legs have more segments, some additional terms have been included. Although there is some variation, the *anus* is usually subterminal and the *genital opening* is at the level of the coxae of the fourth legs.

Classification of mites is based principally on adult structure, and the first characteristic used at the ordinal level is the location and the number of stigmata or spiracles, the openings to the respiratory system. In many classifications (e.g., McDaniel, 1979), the ordinal names state either the number or the location of the stigmata, or both.

The developmental pattern is basically the same in all mites:

$$\text{egg} \rightarrow \text{larva} \rightarrow \text{nymph} \rightarrow \text{adult}$$

The stages usually are similar to one another in general form, but the larva has three pairs of legs and the nymph and the adult have four pairs. The life cycle of the scab mite of humans is an example (Fig. 56.3). The life cycles vary in that there may be more than one nymphal stage and there may be quiescent stages, such as the nymphochrysalis in the chiggers (Trombiculidae). In most cases, females die after laying eggs, but in a few instances, such as the soft ticks (Argasidae), females may continue to lay clutches of eggs for months.

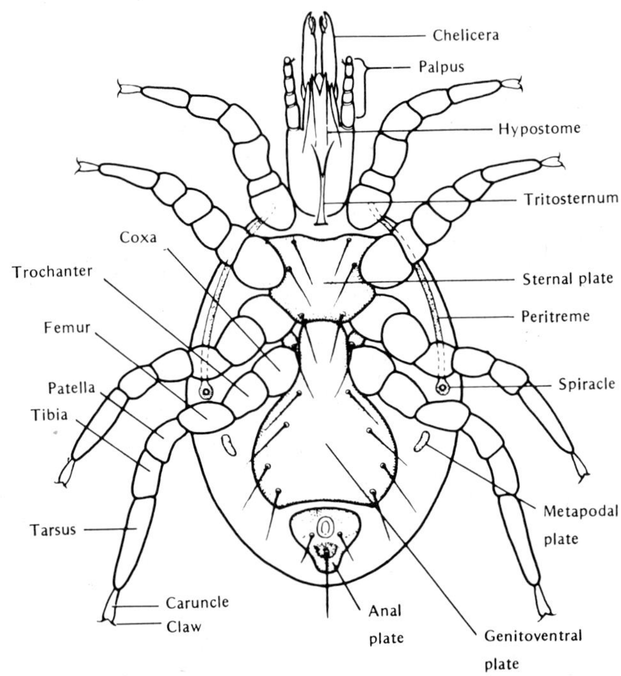


**FIG. 56.2**
Diagram of a mite.

## ORDER PARASITIFORMES

In this major group of mites, a single pair of stigmata is located near the lateral margin of the hysterosoma. Included are the ticks (discussed in detail in Chapter 57) and two other suborders, the Mesostigmata and the Tetrastigmata. The members of the Mesostigmata are common free-living mites found in many habitats including decaying organic matter, nests, and stored foods; they are both free-living and parasitic. The Tetrastigmata have been found in New Zealand, Australia, New Guinea, Sri Lanka, Mauritius, and the Seychelles Islands, and they seem to be predaceous. They will not be considered further. The Cryptostigmata are of little interest in the context of this book except for the Oribatidae, the beetle grass mites, which are intermediate hosts of some anoplocephalid tapeworms (Chap. 29).

### *Suborder Mesostigmata*

*Family Dermanyssidae.* Members of this family are relatively long legged and have piercing chelicerae. A number of species are symbiotes of birds and mammals. Their parasitic habitats are on the skin, in the respiratory tract, or in the outer or inner ear; some are harmful to the host and some serve as vectors of disease agents.

NAME OF ORGANISM. *Ornithonyssus sylviarum,* the northern fowl mite; *O. bacoti,* the tropical rat mite; *O. bursa,* the tropical fowl mite; *Dermanyssus gallinae,* the chicken mite.

HOSTS AND LOCATION. These mites have host preferences, but may be found feeding on a wide array of species. For example, *O. sylviarum* occcurs on chickens, but has been found on more than 10 other species of birds, and feeds on mammals such as humans for short periods. Because of their fairly broad host range and their occasional feeding on humans or companion animals, they are considered here as a group.

DISTRIBUTION AND IMPORTANCE. The organisms are cosmopolitan. All of these mites suck blood from their hosts. They cause anemia when the numbers are high and skin irritation both from feeding and from crawling on the skin. *O. bacoti* is found in warmer climates and may replace the northern fowl mite in those regions. It is the intermediate host of a filarial nematode of cotton rats, *Litomosoides carinii* (Chap.

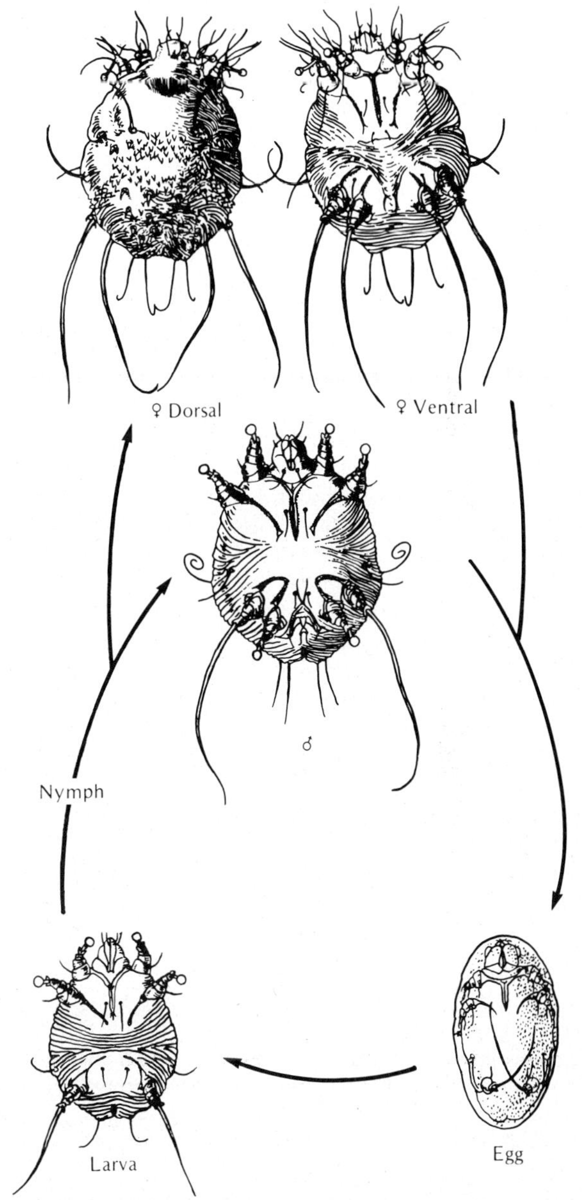


**FIG. 56.3**
Life cycle of the scab mite, *Sarcoptes scabiei.* [Redrawn with modifications from Smart, 1948.]

41). These mites have been implicated as vectors of various viral and bacterial disease agents including St. Louis and Western equine encephalitis, avian spirochetosis, and fowl pox virus, among others.

MORPHOLOGY. All of these mites are about the same size; the females are about 0.5 to 1.0 mm long. The bodies have many bristles, the legs are fairly long, and the chelicerae are stylet-like (Fig. 56.4). It is important to differentiate among the species so that a proper control program can be devised when an infestation is found.

LIFE CYCLE. The pattern of development in these species is:

egg → larva → protonymph → deutonymph → adult

Eggs are deposited in nests in crevices in the house or building or in litter; they are sometimes laid on the host. Hatching may occur within a day, releasing the larva; the larva does not feed but molts to the protonymph, which does feed. After feeding, the protonymph molts to the deutonymph, which may or may not feed, depending on the species. The deutonymph molts to give rise to the adult. Adults copulate within about 2 days after emergence. In general, females feed, and then lay a clutch of eggs and feed again. Their life cycles can usually be completed in as short a time a 7 days, but ordinarily require about 2 weeks

DIAGNOSIS. Infested poultry become fidgety, egg production drops, and, in heavy infestation, the feathers may become discolored because of the large amount of feces deposited by the mites. On any host, the skin irritation may be severe and there may be a rash if hypersensitivity develops. Poultry should be examined individually, looking carefully at the skin, particularly in the region of the vent. The person examining birds may well find mites on his or her own skin. It should be determined whether the infestation is due to biting lice (Chap. 51) or to mites, because the approach to treatment is somewhat different. If the mites are on people, their identification to species will usually point to the source of the mites, whether birds or mammals.

CULTIVATION. All of these mites have been carried in the laboratory, usually by feeding them on chickens or laboratory rats.

HOST–PARASITE INTERACTIONS. Damage to the host involves blood feeding, skin irritation, and development of hypersensitivity. Animals may be killed as a result of extreme anemia. Production losses in poultry result from decreases in egg production or failure to gain weight well.

TREATMENT. Organochlorine and organophosphate acaricides have been used successfully. They are usually applied to birds as a dust; roosts of chickens may be painted with a liquid preparation.

EPIDEMIOLOGY AND CONTROL. Though mites spend most of their time on a host, they may also leave the host for fairly long periods. Some, such as *D. gallinae,* may survive 4 to 5 months off a host without feeding. In considering a control program for poultry, it is essential not only to treat the affected birds but also to clean out poultry houses and to apply insecticide to the structure.

## *Dermanyssids of Wild and Captive Animals*

A number of dermanyssids are found in the respiratory tracts of reptiles and mammals. For example, *Orthohalarachne attenuata* is found in the upper respiratory tract of the Alaskan fur seal (*Callorhinus ursinus*) and Stellar's sea lion (*Eumatopia jubata*) on the West Coast of the United States. *Pneumonyssus simicola* is commonly found in the nasal passages of rhesus monkeys. *Pneumonyssoides caninum* is in dogs and probably is widespread. *Ophionyssus natricis* is commonly found on snakes in zoos, where it can cause serious blood loss. It serves as a vector of a haemogregarine of snakes as well as of some bacterial disease agents.

*Allodermanyssus sanguineus* (Fig. 56.4) is an ectoparasite of the house mouse, *Mus musculus.* Both the host and the parasite are ubiquitous, and the mite is a vector of *Rickettsia akari,* which causes rickettsial pox in humans. Rickettsial pox is a mild, febrile disease that causes a rash similar to that of tick-borne typhus; the disease is not life threatenting and usually does not warrant treatment, although broad-sprectrum antibiotics are sometimes given to hasten recovery.

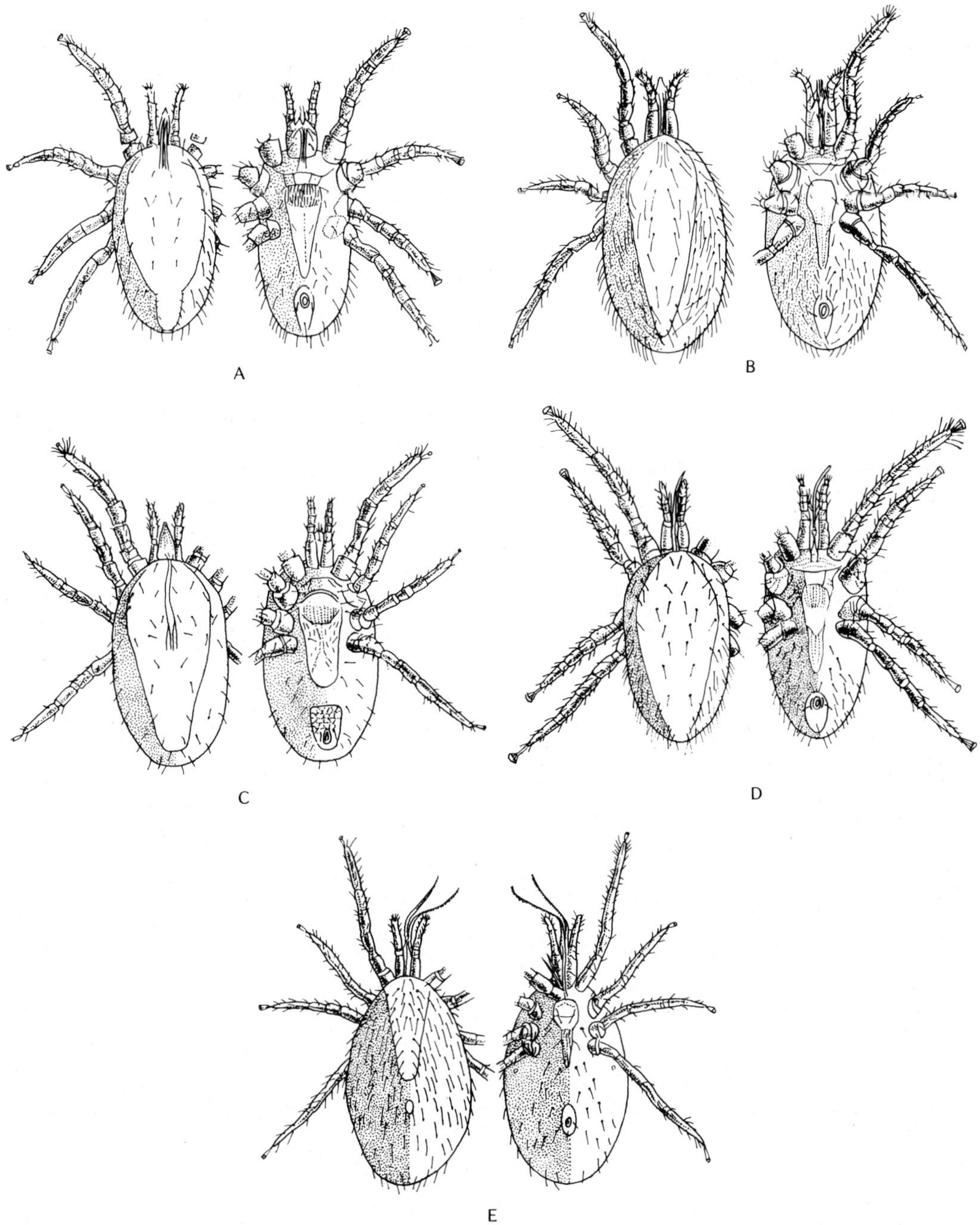


**FIG. 56.4**
Some members of the family Dermanyssidae. (*A*) *Ornithonyssus sylviarum;* (*B*) *Ornithonyssus bacoti;* (*C*) *Dermanyssus gallinae;* (*D*) *Ornithonyssus bursa;* (*E*) *Allodermanyssus sanguineus.* [Redrawn from Baker et al., 1956.]

### *Family Laelapidae*

These mites bear a superficial resemblance to the dermanyssids; they have long legs and considerable setation on the body. There are both free-living and parasitic species in the family. The parasitic forms are important as ectoparasites and as vectors of various disease agents, mostly from rodents to humans.

*Echinolaelaps echidninus* (Fig. 56.5*A*), the spiny rat mite, is representative of a number of closely related species; it is found principally on rodents, especially *Rattus norvegicus,* but feeds on other rodents as well. Its life cycle has

egg → larva → protonymph → deutonymph → adult

The female gives birth to a larva that does not feed. This mite is a vector of *Hepatozoon muris,* a common haemogregarine of rats. *Haemogamasus pontiger* has been implicated in causing skin rashes in soldiers who used straw as bedding. Other species, such as *Eulaelaps stabularis* (Fig. 56.5*B*), may serve as vectors of the agents causing plague and tularemia.

## ORDER ACARIFORMES

The stigmata are of first importance in defining Acariformes, as they are in the Parasitiformes. In the Acariformes, the stigmata are located anterior to the hysterosoma or they are absent. In the suborder Prostigmata, the stigmata are on the anterior portion of the body or absent. The Cryptostigmata have pseudostigmata, which are sensory organs, on the propodosoma; the tracheal system, when present, opens on legs I and III, near the legs, or in the pseudostigmata. The Astigmata lack a tracheal system. The Acariformes also have specialized sensory organs of various types. McDaniel (1979) has keys and further detail on characteristics.

### *Suborder Astigmata*

The Astigmata includes both parasitic and free-living mites. Some of the free-living species, such as *Glycyphagus domesticus* (family Glycyphagidae), are found in food products, and this species causes a skin rash referred to as *grocer's itch. Dermatophagoides farinae* and *D. pteronyssinus* are free-living mites that feed on various kinds of detritus and are often associated with house-dust allergies (Lang et al., 1976). The supercohort Psoroptidia of McDaniel (1979) contains most of the families of medical and economic importance.

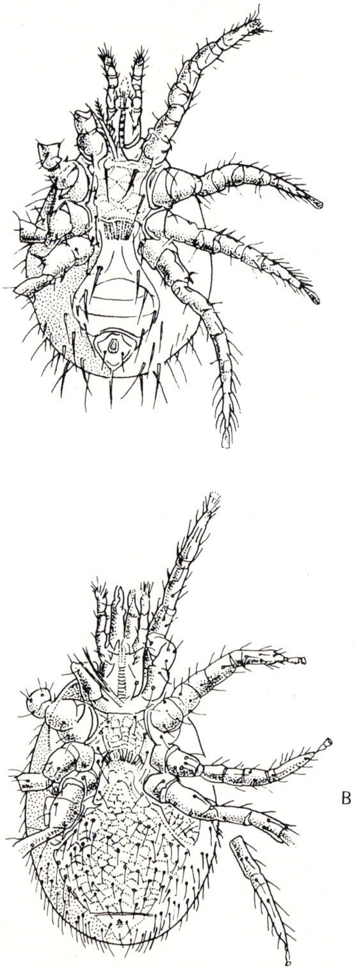


**FIG. 56.5**
(*A*) *Echinolaelaps echidninus;* (*B*) *Eulaelaps stabularis.* [From Baker et al., 1956.]

NAME OF ORGANISM AND DISEASE CAUSED. *Sarcoptes scabiei* is a tiny mite; the females average 0.2 to 0.4 mm long and the males are somewhat smaller (Fig.

56.3). There is a single species with a number of varieties named for the hosts on which they occur. *S. scabiei* var. *humani* is on humans and *S. scabiei* var. *suis* is on swine. They cause sarcoptic mange or scabies; the latter term is also used for other mites.

DISTRIBUTION AND IMPORTANCE. The scab mites are cosmopolitan; although they are of some medical importance, they can cause severe economic losses in livestock and considerable misery in companion animals if left untreated. Mites have a place in the history of modern medicine because the mite that causes scabies was the first infectious agent to be related to a specific disease as a result of investigations done by Bonomo and Cestoni in Italy in 1687 (Graham & Hourrigan, 1977).

LIFE CYCLE. Transmission from one host to another takes place through close contact, and the fertilized female is probably the stage transferred (Fig. 53.3). The female burrows into the skin and lays eggs in a sinuous tunnel, which she forms as she oviposits. The females have two nymphal stages but the males may have only one. The female life cycle is completed in 14 to 17 days and that of the males in 9 to 11 days.

DIAGNOSIS. Skin scrapings at the suspected site of infection are made deeply into the skin and then heated in 10% KOH to dissolve the protein of the skin and hair. The sediment in the tube is examined under a compound microscope for mites and parts of mites. Another recommended procedure is to make the scraping into mineral oil and then examine it under a microscope for mites, eggs, and fecal pellets.

HOST–PARASITE INTERACTIONS. In humans, relatively few mites can cause severe skin reactions. Some persons seem to tolerate the mites with little response, but most others have severe itching and allergic reactions. The irritation and hypersensitivity seem to result from excretions the females deposit in the skin as they burrow. Bacterial skin infections may also occur, probably as a result of scratching. Most of the mites live in places where the skin is thin, such as between the webbing of the fingers or in the folds at the wrist, but other sites are not uncommon.

Domesticated animals often infected with sarcoptic mites are horses, swine, and dogs. In these hosts, the lesions are crusty, weeping patches that are extremely irritating. As indicated from the names, the varieties are found on different species of hosts, but there is mostly host preference, not strict host specificity. The mites from swine, for example, can transfer to humans and may live in the skin for a time.

As is true of a number of transmissible diseases, scabies is usually seen when its hosts are crowded. In humans, crowded, disruptive conditions such as those during a war, natural disaster, or generally poor living conditions are conducive to scabies outbreaks.

TREATMENT. Regardless of the host, the elimination of scab mites is based on the use of acaricides, usually organochlorine compounds. In humans, acaricides are applied to the skin after a hot, soapy bath. All clothing and bedding should also be laundered. The World Heath Organization (WHO) recommends an emulsion containing 6% DDT, 68% benzyl benzoate, 12% benzocaine, and 14% polysorbate 80.

Horses, dogs, and swine, in which most sarcoptic mange is seen, are treated individually by hand as opposed to herd treatment.

EPIDEMIOLOGY AND CONTROL. Scabies is usually a family (in humans) or herd (in other animals) problem. If one host is infected, the others in contact with it also become infected. Because the mites appear not to live long off the host, the host is the principal target of treatment. Before an animal, such as a horse or dog, is introduced into a paddock or kennel, quarantine and inspection are in order. Such measures are not ordinarily taken for humans, but might be considered under some conditions.

Nearly any infectious disease has a cyclic incidence in a population. Scabies is no exception, and epidemiologic data indicate that a cycle about 45 years long occurs in the human population. The peak of infection lasts about 15 years and then it wanes. In the 1970s the prevalence of scabies began to increase and some writers said that there was a pandemic. The cause was laid to various changes in human behavior, such as sexual permissiveness, but such conclusions may have been the result of the observers' biases. There seems to be some immunity resulting from infection, since a person cleared of infection is reinfected only with some difficulty. Again, immunity may play a role in the number of susceptibles in the population and therefore the spread of the infection. However, the length of the cycle, 45 years, indicates that many factors are somehow intertwined in the waxing and waning of scabies.

Two genera of sarcoptic mites that are of importance on birds or smaller animals and sometimes humans are *Notoedres* and *Knemidocoptes* (=*Cnemidocoptes*). *Notoedres cati* has been reported to cause mange in cats, rats, rabbits, squirrels, dogs, and humans. *N. muris* is sometimes a problem in laboratory rats and has been found on other rodents in the wild such as voles.

*K. mutans*, the scaly leg mite, also has a broad host range having been reported from pheasants, turkeys, partridges, finches, parakeets (budgerigars), and various passeriform birds. They burrow into the skin beneath the scales on the legs and feet of birds and cause considerable damage. Birds may become lame, crippled, and even die from the infection. It may be necessary to differentiate this mite from other members of the genus. *K. laevis* is the depluming mite; it burrows into the skin at the base of feathers so that feathers may be lost over a large portion of the body. *K. fossor*, on the other hand, is found at the base of the host's bill.

NAME OF ORGANISM AND DISEASE CAUSED. *Psoroptes* spp. and *Chorioptes* spp. cause psoroptic and chorioptic mange, respectively, scabies or scab.

HOSTS AND LOCATION. These mites are most important as parasites of large farm animals: cattle, sheep, goats, camels, and horses. They are found almost anywhere on the body, but often congregate in the ears, at the base of the tail, or on the withers. The genus *Psoroptes* has a single species, *P. equi*, and many varieties whose names are based on the host on which the mite is found. Thus *P. e. bovis* is found on cattle, *P. e. cuniculi* on rabbits, *P. e. ovis* on sheep, and so forth. Other species designations have been proposed (Sweatman, 1958), but we have taken a somewhat simpler approach by retaining the varietal designations. The same terminology is applied to *Chorioptes bovis*, which has varietal designations depending on the host from which it is taken.

DISTRIBUTION AND IMPORTANCE. Both of these genera are cosmopolitan on domesticated animals. Of the two, *Psoroptes* is the more important economically. In North America, psoroptic mange causes millions of dollars of losses annually.

MORPHOLOGY. These are tiny mites, 0.5 mm or less in length. Their legs extend beyond the margins of the body and they have long setae on the body and some of the hind legs (Fig. 56.6). Differentiation between the two species is usually based on the pedicels of the anterior two pairs of legs; the pedicels are at the distal tips of the legs and terminate at cuplike structures called *suckers*. The pedicels are long and jointed in *Psoroptes* and short and unjointed in *Chorioptes*.

LIFE CYCLE. There are egg, larva, nymph, and adult stages in the life cycle (Fig. 56.7). The whole life cycle is spent on a host.

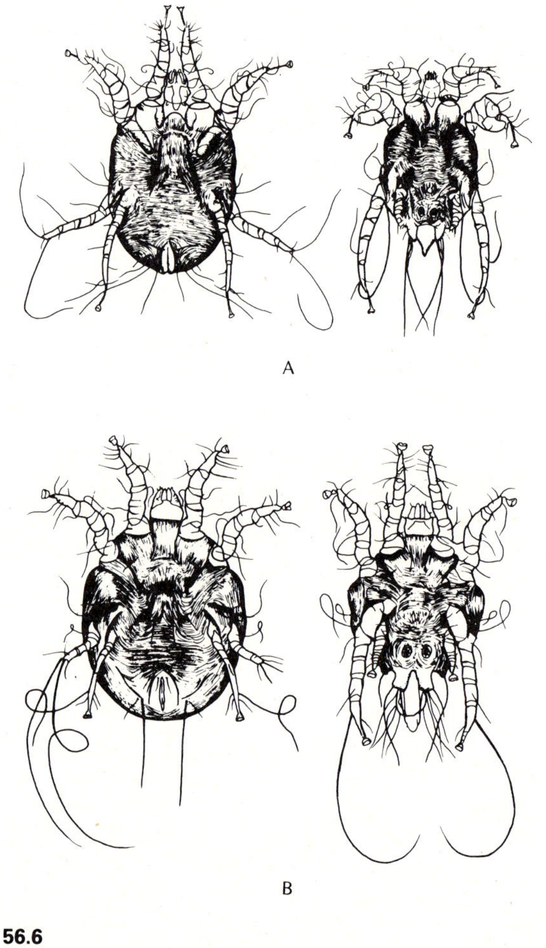


**FIG. 56.6**
*Psoroptes equi* (*A*); *Chorioptes bovis* (*B*). [From Baker et al. 1956.]

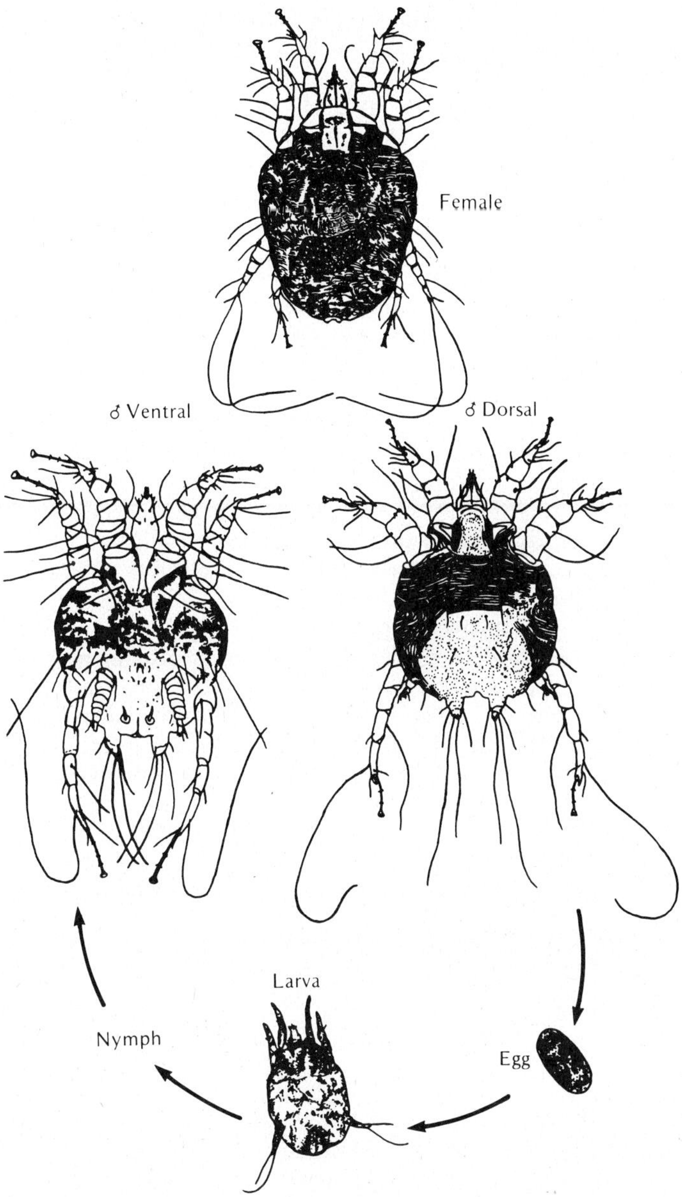


**FIG. 56.7**
Life cycle of *Psoroptes equi.*

DIAGNOSIS. Mites are generally at the margins of skin lesions. Scrapings are made as for sarcoptic mange and examined under a compound microscope for mites and parts of mites. Crusty, weeping lesions are an indication of scab mites, but other infecious agents need to be considered. In sheep, the fleece becomes ragged with long tags of wool hanging down, but this leads one only to suspect mites, not to conclude that they are present.

HOST–PARASITE INTERACTIONS. Chorioptic mite infection is less serious than psoroptic, because the mites remain at the surface of the skin and do not cause either serious damage or much irritation. *Psoroptes,* on the other hand, cuts more deeply into the skin with its chelicerae and causes not only oozing lesions but a great deal of irritation. Infected animals scratch against posts or buildings, further damaging the skin. Sheep pull at the fleece in an attempt to reduce the itching and make the fleece unmarketable.

TREATMENT. Treatment for scabies in large farm animals nearly always requires dipping in a contact insecticide. The organochlorine toxaphene and the organophosphates coumaphos and phosmet are currently used in the United States. Interestingly, lime-sulfur, which has been in use for more than a century, is still an approved acaricide for domestic animals. Ivermectin administered by injection gives good control, but is not yet approved for this use (Chap. 33).

EPIDEMIOLOGY AND CONTROL. Mites spread readily from one animal to another by bodily contact. Therefore, nearly all animals in a herd become infected in a short period. Control is based on the following steps:

1. Proper treatment
2. Inspection after treatment to ensure that all animals are clean
3. Inspection of animals to be introduced into a herd
4. Quarantine of infected animals, which should not be moved until they have been certified as being clean

Laws enabling the control of sheep scab have been on the books for more than 125 years. In the Australian state of New South Wales, laws were enacted in the 1850s to allow the destruction of scabby sheep; no other means of control were available at the time. By the turn of the century, scabies had been eradicated in Australia, New Zealand, and some European countries. During the early part of the twentieth century, sheep raisers in the western United States experienced severe losses from psoroptic scabies, and there seemed to be no way that they could control the infections individually. As a result, they requested help in setting up a control program, and both state and federal agencies cooperated in establishing and implementing a scabies eradication program. By 1943, eradication in the western states had been accomplished and the sheep industry in that section of the country was relieved of at least one economic burden.

Eradication in the rest of the country was not accomplished until 1970, because sheep flocks outside the West are typically small. For example, 8000 flocks of sheep were found in Wisconsin during the eradication program. All of these sheep had to be quarantined until they were inspected and declared to be scabies-free or treated and reinspected until free of mites. About 300 state and federal employees were assigned temporarily to the eradication program and all sheep in the state were inspected in less than a month (Graham & Hourrigan, 1977). The cost of eradication was high but was considered to be worthwhile since the mites could be completely eliminated because of their narrow host range.

A side benefit of extensive treatment for sheep scabies was that lice in sheep were almost eradicated as well. Lice (Chap. 51) are not nearly as severe a problem as mites, but their reduction further helped to produce clean, intact fleeces.

A scabies eradication program is also in place for cattle, but it has not worked as well as for sheep. In the first place, cattle do not have a fleece that is worth money, and they may not be so obviously affected as sheep. Although many cattle raisers recognize the benefits of eradication, a significant number do not want any interference with their right to ship their animals. Without the full cooperation of the cattle raisers, eradication remains a hope for the future.

## *Suborder Prostigmata*

The suborder Prostigmata includes both free-living and parasitic mites with quite diverse life cycles and structures. The stigmata, when present, are on the anterior part of the body, often at the base of the chelicerae. We will consider a few families of importance to humans or domesticated animals.

## *Family Pyemotidae*

Many pyemotids are pests of stored food products and can cause significant losses. A secondary aspect of their being in foods is that they may cause skin irritations in persons coming into contact with them.

Members of the genus *Pyemotes* (Fig. 56.8*A*) as well as other genera, such as *Haemogamasus* (Fig. 56.8*B*), may be found in large numbers in foods, hay, or straw, and people handling such material may suffer from severe dermatitis from having the mites on the skin. The mites puncture the skin, and a wheal arises several hours later at the site of the puncture. In especially heavy attacks, persons may suffer from systemic reactions such as headache or intestinal symptoms and joint pain. Species in other families such as the Tetranychidae may also cause skin rashes.

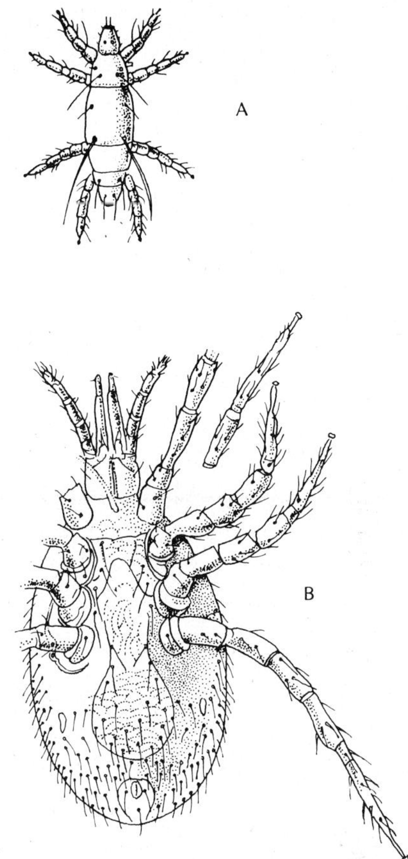


**FIG. 56.8**
(*A*) *Pyemotes ventricosus;* (*B*) *Haemagamasus pontiger.* [From Baker et al., 1956.]

## *Family Demodecidae*

NAME OF ORGANISM. *Demodex* spp. are all parasitic mites of mammals; they live in the skin.

HOSTS AND LOCATION. The members of *Demodex* have a high degree of both host and location specificity. Humans have two species—*D. folliculorum,* which lives in hair follicles, and *D. brevis,* which is found in sebaceous glands. Other species are typically named after the hosts in which they are found: *D. canis* in dogs, *D. bovis* in cattle, and so forth.

DISTRIBUTION AND IMPORTANCE. The organisms are cosmopolitan. There is often a high prevalence of infection with demodecids in both humans and other animals, but the damage to the host is inconsistent. In dogs, bitches may remain without clinical signs except toward the end of pregnancy, when there may be severe skin eruptions. Cattle may have such severe lesions on the skin that the hide has a reduced value for leather, but under most circumstances, up to 90% of hides may be infected without significant damage.

MORPHOLOGY. These mites are elongate with four pairs of stubby legs (Fig. 56.9); they are tiny, 100 to 400 μm long.

LIFE CYCLE. The whole life cycle takes place on the host. Transmission seems to take place through close body contact.

DIAGNOSIS. A number of different methods are used depending on the host and the lesions seen. In humans, expressing the content of pinpoint pustules will often produce mites that can be seen under a compound microscope. Skin snips can be taken and macerated to look for whole mites or fixed to look for them in stained sections. In any event, having the mite intact is important for making an accurate identification. Fecal flotations are done in dogs, since the mites are seen frequently in the feces.

HOST–PARASITE INTERACTIONS. *Demodex* spp. have often been blamed for causing acne in humans, especially adolescents. Although the prevalence of infection in humans is especially high, probably exceeding 90%, and persons with infections have acne, the percentage of persons affected by the mites is low.

In dogs, the situation is different from that in humans, and demodectic mange is quite common. Skin lesions most often become severe in bitches toward the end of pregnancy, when lesions break out on many parts of the body. Immunity seems to play a role in controlling populations of mites, since lesions are seen in older dogs as well as pregnant bitches at times when the ability to respond to infectious agents wanes. When there are skin lesions on the host, it seems probable that the principal cause is skin bacteria more than the mites themselves.

TREATMENT. Organochlorine acaricides applied to the skin are successful in most instances. Success has been seen with systemic organophosphates in animals other than humans.

EPIDEMIOLOGY AND CONTROL So little is known about demodecid mites that the only control currently recommended is treatment of affected individuals.

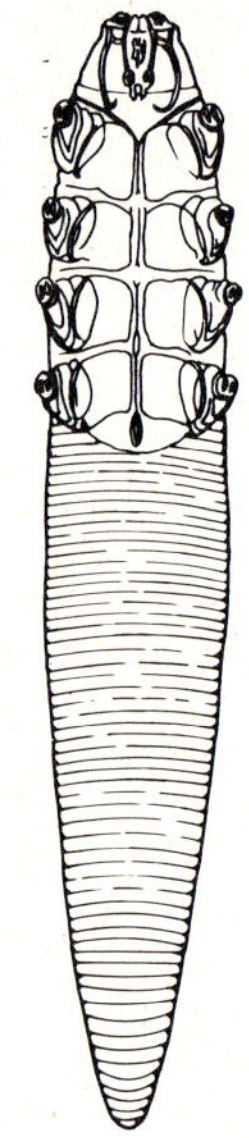

**FIG. 56.9**
The follicle mite, *Demodex canis,* female. [From Baker et al., 1956.]

## *Family Trombiculidae*

Chiggers or trombiculids are widespread and important because they directly damage their hosts and transmit disease agents. Some of their other common names are *red bugs* and *harvest mites.*

NAME OF ORGANISM AND DISEASE CAUSED. Among the common genera in this family are *Leptotrombidium, Euschoengastia, Trombicula,* and *Eutrombicula.* These organisms cause chigger dermatitis and trombidiosis.

HOSTS AND LOCATION. The parasitic stage has low host specificity; the organisms feed on mammals, birds, reptiles, and sometimes amphibia. Not all have such a broad host range, but they generally have catholic tastes in sources of food. They are found on the skin.

DISTRIBUTION AND IMPORTANCE. The organisms are cosmopolitan; chiggers are found in all climates, but are probably more abundant in mild and warm climates. Losses are the result of direct damage to the host by feeding of chiggers and indirect damage through transmission of scrub typhus or tsutsugamushi disease in the Orient and Australasia.

MORPHOLOGY. Most keys for identification of chiggers are based on the larvae (Fig. 56.10). They are usually reddish or orange and barely big enough to see—100 to 300 μm in length before feeding. They have three pairs of long legs, which are well supplied with setae and five-segmented palps. The bodies have long setae.

LIFE CYCLE. The stages in the life cycle of trombiculids are

egg → deutovum → larva → nymphochrysalis → nymph → imagochrysalis → adult

The larva is the only parasitic stage (Fig. 56.10); it usually feeds on a wide range of hosts. The nymphochrysalis is a quiescent stage and is followed by the nymph, which is a free-living, feeding stage. Between the nymph and the adult is another quiescent stage, the imagochrysalis; adults are free-living. The nymphs and adults probably feed on small invertebrates in the litter at or near the soil surface. Males deposit stalked spermatophores, which the females crawl over and take into the genital opening.

Life cycles of chiggers are dependent on weather. In cooler climates they may have three generations each year, but in tropical and semitropical climates, develop-

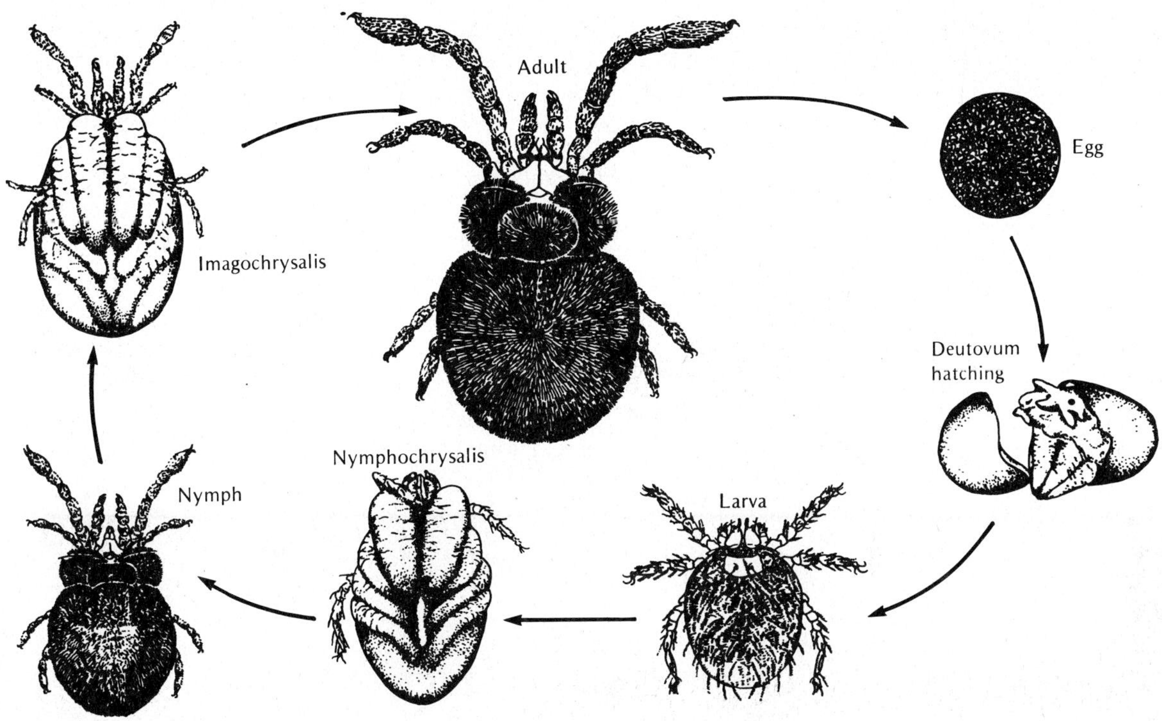


**FIG. 56.10**
Life cycle and morphology of the chigger, *Leptotrombidium akamushi*. [Redrawn from Neal & Barnett, 1961.]

ment takes place year-round. Forty days is about an average time for completion of a life cycle.

DIAGNOSIS. Raised, itching papules and a history of having recently visited grassy or forest-edge areas are usually adequate to determine that a person has been infected. Livestock and poultry are often attacked during warm months, and diagnosis is usually based on recognizing the skin lesions that occur at certain times of the year.

CULTIVATION. Chiggers can be maintained in the laboratory (Lipovsky, 1953).

HOST–PARASITE INTERACTIONS. Chiggers remain on the surface of the skin to feed. A few species such as *Hannemania* embed themselves in the skin of amphibia (Duszynski & Jones, 1973), but generally the larvae insert their mouthparts into the skin and the body remains at the surface. The chigger pierces the skin with its chelicerae, applies its mouth to the skin, and seals the margin of the wound. During feeding, the chigger alternately ejects a salivary fluid, which liquefies the tissue, and then sucks it up. A structure called a *stylostome* forms in response to the salivary fluid; there is proliferation of epidermal cells and inflammation resulting in a structure that has a well-defined morphology (Hase et al., 1978).

The inflammatory response of the host gives rise to the raised, reddened wheal with a depressed center characteristic of chigger bites. The initial response is usually seen within a day after the larva begins to feed and may continue as an intensely itching wheal for a week or more. The inflammatory response need only be initiated for there to be a strong response; even if the chigger begins to feed and is then removed, the inflammation and itching are memorable. It is likely that there is an immune response on the part of the host after the initial attack.

TREATMENT. If one has been in an area where larval chiggers may be active, the first line of defense is a hot, soapy shower or bath followed by a brisk toweling to get the mites off the skin. If the chiggers have been on long enough to inject their salivary fluid, at least moderate inflammation will occur, and symptomatic treatment is all that can be done. The itching can be reduced with proprietary preparations from a pharmacy; calomine lotion and salves containing corticosteroids are examples of preparations used with some success. The victim of an attack should not expect a medical miracle from any concoction; a week's suffering will be followed by a spontaneous and blessed cure.

EPIDEMIOLOGY AND CONTROL. Chiggers are unique mites in that the larva is the only parasitic stage. The larvae typically feed on rodents and birds, and the largest numbers of larvae that may attack humans are therefore found in areas where there are rodents and nesting birds. Each species of trombiculid has its own preferred habitat, but in general, shrubby areas, forest edges, grassy areas, or those that have been cut for timber are most likely to have large numbers of larvae.

In temperate climates, chiggers are seasonal and occur in greatest numbers in spring. Thus, it is possible to avoid areas where they are likely to be. If it is necessary to be in such an area, tick repellents such as deet should be used and clothing tied tightly at the ankles and wrists. Area chigger control has been successful with chlordane or toxaphene dust applied at a rate of 1 lb per 1000 sq ft.

Scrub typhus, or tsutsugamushi disease, is caused by *Rickettsia akamushi* and has a distribution in the Eastern Hemisphere in warm climates: Southeast Asia, India, the Pacific islands, and Australia. The vectors are most often *Leptotrombidium akamushi* and *L. deliense*. The agent is transmitted transovarially in mites, since only the larvae feed on vertebrates. The signs and symptoms of scrub typhus include prostration, headache, fever, a body rash, and central nervous system abnormalities. The individual usually recovers in about a month. Mortality in most outbreaks is 6 to 35%, but in a few instances, it has reached 60%. Phillip (1948) estimated that there were nearly 18,000 cases in the Pacific theater during World War II, and that 150,000 man-days were lost in two outbreaks alone. Broad-spectrum antibiotics are effective in treating scrub typhus.

The agent causing epidemic hemorrhagic fever, probably a virus, is also suspected of being transmitted by chiggers. This disease occurs in Korea and was the cause of great losses among American ground troops during the Korean War.

## *Family Cheyletidae*

NAME OF ORGANISMS AND HOSTS. *Cheyletiella parasitivorax, C. yasguri,* and *C. blakei* seem to have a fairly broad host range, with *C. parasitivorax* usually being found on rabbits, *C. yasguri* on dogs, and *C. blakei* on cats.

DISTRIBUTION AND IMPORTANCE. The mites are cosmopolitan. Most of the members of the family are predaceous, but a few species are found in association with small animals such as rabbits, dogs, and cats, in which they cause dermatitis. They are also sometimes found temporarily on humans.

MORPHOLOGY. *C. parasitivorax* averages 266 by 386 μm and has a few long setae on the body. The palpal claw is strongly developed. The legs have a double row of hairs at their tips rather than claws.

LIFE CYCLE. Poorly known.

DIAGNOSIS. Finding the mites in the fur of an animal is the means of diagnosis.

HOST–PARASITE INTERACTIONS. Mites are found in the fur of animals, where they cause dermatitis; animals scratch a great deal. Curiously, dogs with long, soft hair seem to be especially susceptible to infestion with cheyletids; the cocker spaniel, for example, is the breed that most often has these mites.

TREATMENT. Contact acaricides will usually rid a companion animal of mites. Further cleaning of the animal's bedding and use of an acaricide in the area where the animal sleeps are also in order.

EPIDEMIOLOGY AND CONTROL. The information on this group of mites as parasites is rather sketchy, and treatment of individual animals seems to be the only step recommended.

## REFERENCES AND FURTHER READINGS

Arthur, D. 1962. *Ticks and Disease.* Row, Peterson & Co., Evanston, Ill. (U.S. edition of 1961 Pergamon edition).

Baker, E. W., T. M. Evans, and D. J. Gould. 1956. *A Manual of Parasitic Mites of Medical or Economic Importance.* National Pest Control Association, New York.

Baker, E. W., and G. W. Wharton. 1952. *An Introduction to Acarology.* Macmillan Publishing Co., New York.

Cloudsley-Thompson, J. L. 1958. *Spiders, Scorpions, Centipedes and Mites.* Pergamon Press, New York.

Comstock, J. H. 1940. *The Spider Book,* rev. ed. W. J. Gertsch (ed.). Comstock Publishing Assoc., Ithaca, N.Y.

Duszynski, D. W., and K. L. Jones. 1973. The occurrence of intradermal mites, *Hannemania* spp. (Acarina: Trombiculidae), in anurans of New Mexico with a histological description of the tissue capsule. *J. Parasitol.* **59**:531–538.

Fabre, J. H. 1919. *The Life of the Spider.* Dodd, Mead & Co., New York.

Gertsch, W. J. 1949. *American Spiders.* Van Nostrand, New York.

Graham, O. H., and J. L. Hourrigan. 1977. Eradication programs for the arthropod parasites of livestock. *J. Med. Entomol.* **13**:629–658.

Hase, T., L. W. Roberts, P. K. Hildebrandt, and D. C. Cavanaugh. 1978. Stylostome formation by *Leptotrombidium* mites (Acari:Trombiculidae). *J. Parasitol.* **64**:712–718.

Kaston, B. J. 1972. *How to Know the Spiders,* 2nd ed. Wm. C. Brown Co., Dubuque, Iowa.

Keegan, H. L. 1980. *Scorpions of Medical Importance.* University Press of Mississippi, Jackson, Miss.

Lang, J. D., L. D. Charlet, and M. S. Mulla. 1976. Bibliography (1864–1974) of house dust mites, *Dermatophagoides* spp. (Acarina: Pyroglyphidae), and human allergy. *Sci. Biol. J.* **2**:62–83.

Lipovsky, L. J. 1953. Improved technique for rearing chigger mites (Acarina: Trombiculidae). *Ent. News* **64**:4–7.

McDaniel, B. 1979. *How to Know the Mites and Ticks.* Wm. C. Brown Co., Dubuque, Iowa.

Phillip, C. B. 1948. Tsutsugamushi disease (scrub typhus) in World War II. *J. Parasitol.* **34**:169–191.

Savory, T. 1977. *Arachnida,* 2nd ed. Academic Press, New York.

Smart, J. 1948. *Insects of Medical Importance,* 2nd ed. British Museum, London.

Sweatman, G. K. 1958. On the life history and validity of the species in *Psoroptes,* a genus of mange mites. *Can. J. Zool.* **36**:905–925.

Woolley, T. A. 1961. A review of the phylogeny of mites. *Ann. Rev. Entomol.* **6**:263–284.

# 57

# *Ticks and Tick-Borne Diseases*

The ticks are often looked on as a group apart from the mites, but they are merely large mites and have the same body (Fig. 56.1). As with other acarines, the location of the stigmata is the basis of classification, and in ticks there is a distinct stigmatal or spiracular plate posterior to the hind leg and on the lateral margin of the hysterosoma (Fig. 57.1). Ticks, unlike mites, often have a well-developed hypostome by which they attach firmly to the host (Fig. 57.2), and Haller's organ, a humidity receptor located on the dorsal surface of tarsus I.

Ticks are important because they feed on blood and tissue fluids of humans and other animals, and also because they transmit a large number of disease agents including viruses, rickettsiae, bacteria, protozoa, and nematodes. They can cause a great deal of damage during feeding by removing blood, damaging the skin at the site of feeding, and causing tick paralysis in humans or other animals. They are known to transmit more than 50 pathogenic agents and are suspected of being involved in the transmission of others. Among the most serious diseases transmitted by ticks are the piroplasmoses, which continue to cause serious losses in large ruminants on a global basis (Chap. 17).

Ticks are divided into two families, the Ixodidae or hard ticks and the Argasidae or soft ticks. The division is based on the following characteristics:

Ixodidae. A dorsal scutum is present covering the dorsum of the male but only the anterior part of the female; mouthparts are terminal and visible from above; stigmata are located posterior to coxae IV.

Argasidae. A dorsal scutum is lacking; mouthparts are ventral and not visible from above; stigmata are usually located between coxae III and IV.

Among useful works on ticks are those of Arthur (1962), for a general treatment of ticks, especially as vectors of disease agents; a U.S. Department of Agriculture manual on livestock ticks (Strickland et al., 1976), which contains a great deal of information on the biology of ticks and provides excellent keys for the North American species; and the Centers for Disease Control's short but useful pamphlet on ticks and control of ticks (1978). Harwood and James (1979) provide a more extensive discussion of ticks than is possible here.

## FAMILY IXODIDAE—THE HARD TICKS

Adult ticks range in size from 3 to 10 mm long prior to feeding; the males do not increase much in size on feeding, but the females may be so engorged with blood that their legs are nearly useless. Engorged females may exceed 20 mm in length. The larvae are usually 1 mm or so in length and the nymphs are about twice as large.

The life cycles of hard ticks fit the usual pattern seen in mites:

$$\text{egg} \rightarrow \text{larva} \rightarrow \text{nymph} \rightarrow \text{adult}$$

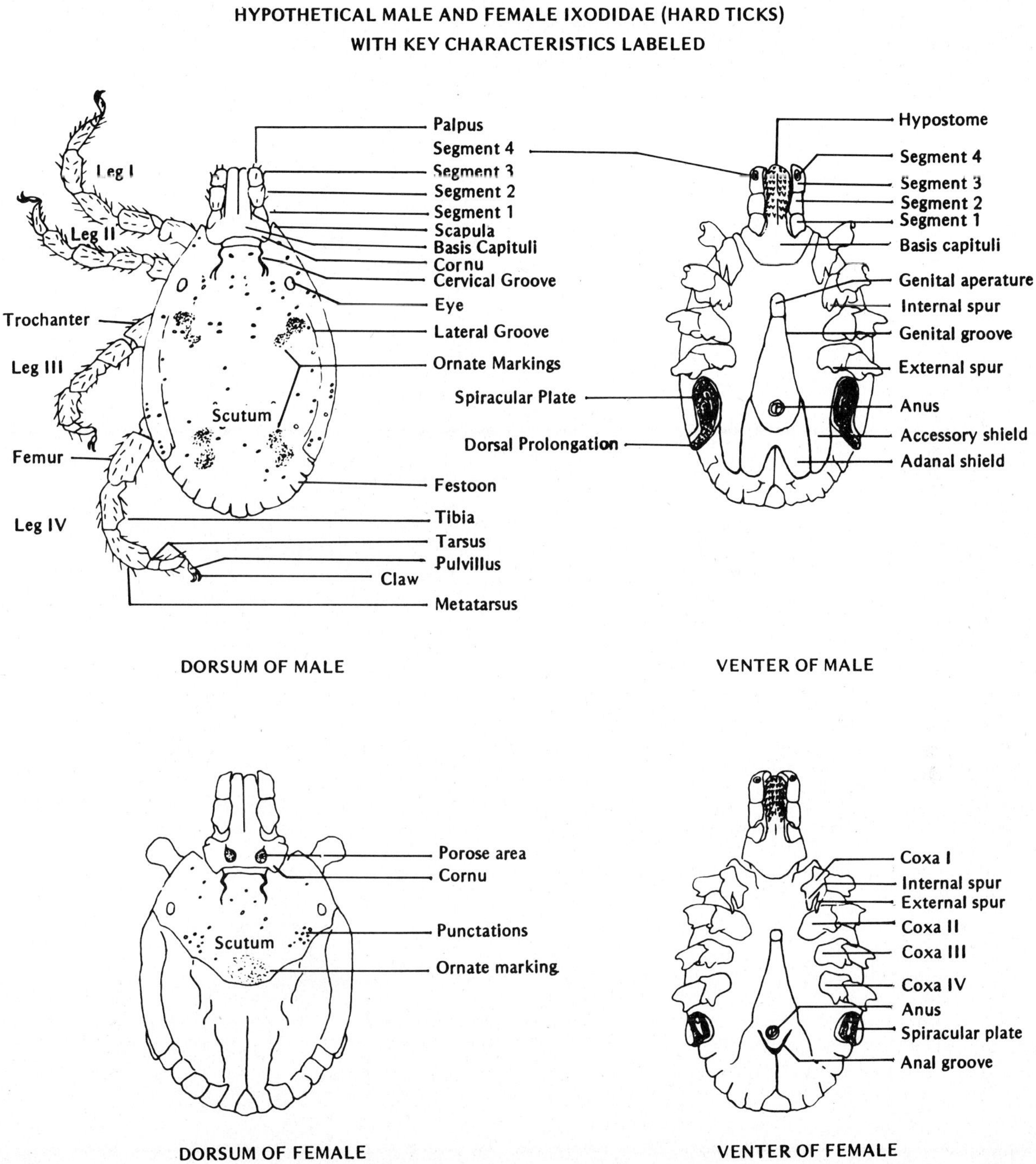


**FIG. 57.1**
External anatomy of male and female ticks.

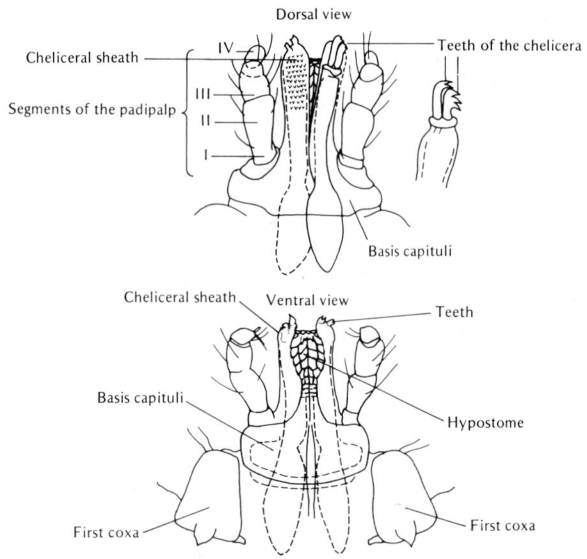


**FIG. 57.2**
Capitulum and mouthparts of a tick.

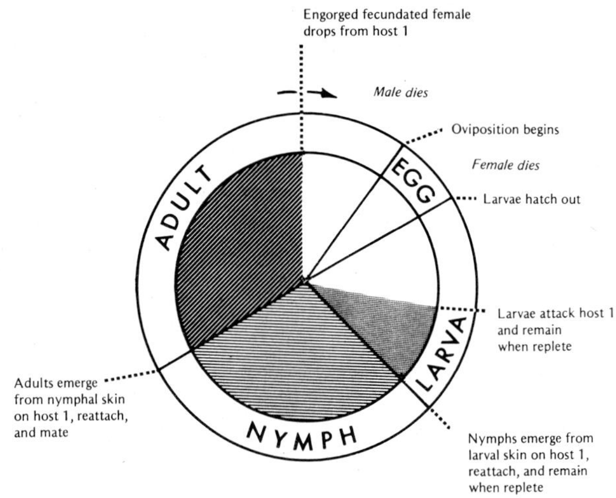


**FIG. 57.3**
One-host type of tick life cycle.

The pattern of development is one of feeding and molting; the larva feeds and then molts to the nymph, which feeds and molts to the adult. The adults copulate on the host; the females drop off, lay eggs, and die. Although this pattern occurs throughout the hard ticks, the number of hosts used by a tick in completing its life cycle is important both for planning control programs and for understanding the epidemiology of a disease agent that may be transmitted by the tick.

We speak of *one-host*, *two-host*, or *three-host* ticks. These terms refer to the number of different individual animals that the tick feeds on during its life cycle. The simplest cycle is that of the one-host tick, in which the tick reaches a host as a larva and remains on it until it reaches sexual maturity (Fig. 57.3).

*Boophilus annulatus,* the cattle fever tick, and *B. microplus,* the tropical cattle tick, are both one-host ticks, as are other members of the genus. *Dermacentor andersoni* is a three-host tick and *D. albipictus,* the winter tick, is a one-host tick, which exemplifies the fact that any genus may have ticks with different patterns of development. Although the cycle is completed on a single animal, it should not be concluded that no transfer of stages between animals takes place. Close contact between cattle, for example, will allow the transfer of *Boophilus annulatus.* In instances where a one-host tick is the vector of an infectious agent, vertical or transovarial transmission (TOT) of the agent usually occurs.

In ticks that have two-host cycles, two different individuals are required for the completion of the life cycle (Fig. 57.4).

Examples of a two-host cycle are those of *Rhipicephalus evertsi, Dermacentor variabilis,* and *Hyalomma aegypticum.* In most instances, the larva

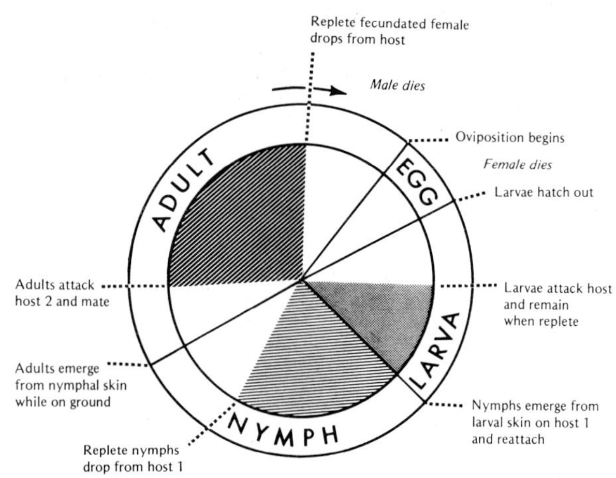


**FIG. 57.4**
Two-host type of tick life cycle.

and the nymph feed on the same host, but in some species the nymph and adult feed on the same host. The hosts of the larvae and nymphs tend to be small rodents, and those of the adults are larger mammals such as canids or herbivores. If a two-host tick is the vector of a disease agent, there can be either transovarial or stage-to-stage transmission, but the former predominates.

In ticks that have three-host cycles, three different individuals are required for completion of the life cycle (Fig. 57.5).

Examples of three-host cycles are those of *Dermacentor andersoni, D. occidentalis, Ixodes ricinus,* and *Rhipicephalus sanguineus*. In three-host cycles, hosts tend to be larger as the ticks move from larva toward the adult stage. Larvae are usually on small rodents, nymphs on larger rodents, and adults on carnivores or larger herbivores. If a three-host tick serves as a vector of a disease agent, there is usually stage-to-stage or horizontal transmission.

NAME OF ORGANISM. *Dermacentor andersoni* (=*D. venustus*) is also known as the *Rocky Mountain wood tick* or *spotted fever tick*.

HOSTS AND LOCATION. Mammals such as small rodents, lagomorphs, porcupine, large herbivores, canids and humans are the hosts. The ticks are found on the skin.

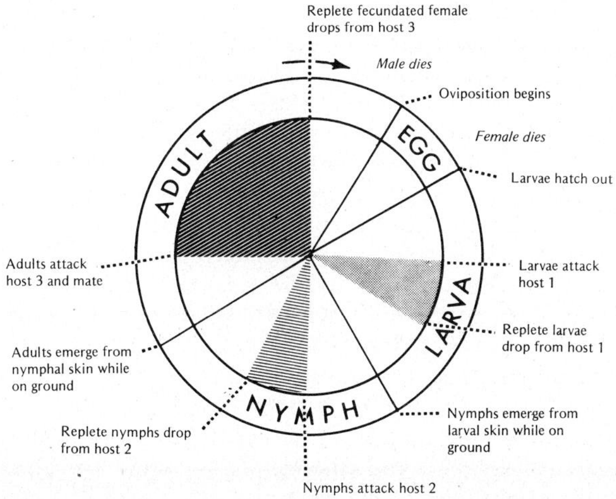


**FIG. 57.5**
Three-host type of tick life cycle.

DISTRIBUTION AND IMPORTANCE. Western North America from northern New Mexico to British Columbia and nearly to the West Coast are the areas of distribution; usually the ticks are found at elevations above 1800 m (6000 ft), but they are also at lower elevations as one goes from the southern to the northern parts of its range (Fig. 57.6). Although the wood tick can be a pest of both humans and other animals, it is of greatest importance as a vector of the agents causing Rocky Mountain spotted fever and Colorado tick fever.

MORPHOLOGY. This is an *ornate tick*, a term which refers to the dark brown pattern on the scutum. It also has festoons, box-like markings on the posterior margin of the opisthosoma (Fig. 57.7).

LIFE CYCLE. The life cycle can be said to start in late winter or early spring when the adults seek hosts such as canids, large herbivores, porcupine, or rabbits. They attach and begin to feed. Copulation takes place on the host; after the female has engorged for about a week, she drops off and begins to oviposit in about another week. The female oviposits for about 3 weeks. Females produce 6400 eggs, on the average. The larva hatches in about 35 days and seeks a small rodent such as the deer mouse (*Peromyscus maniculatus*) to feed on. The six-legged larva feeds for 3 to 5 days, drops off, molts, and remains in the soil until the next spring. The nymphs become active in late winter and seek small rodents such as the golden mantled ground squirrel, *Citellus richardsoni* (=*Spermophilus richardsoni*), on which to feed. They feed for about a week, drop off, and molt to the adult stage.

The cycle of *D. andersoni* is completed in 1 year in the southern and middle parts of its range. In the northern part of its range, or at high altitude, it may take 2 or even 3 years. In cold climates, the larval stage feeds and molts to the nymph in 1 year, and then the nymphal stage may not feed until late summer of the next year, thus putting off the adult stage until 2 years after the egg hatched. All of the stages can survive a long time without feeding, allowing the life cycle to be further stretched out. Larvae and nymphs can survive for more than 500 days without feeding and the adults for more than 1000.

HOST–PARASITE INTERACTIONS. Some ticks have narrow host ranges, but the wood tick exemplifies many three-host ticks in that it has host preferences. It tends

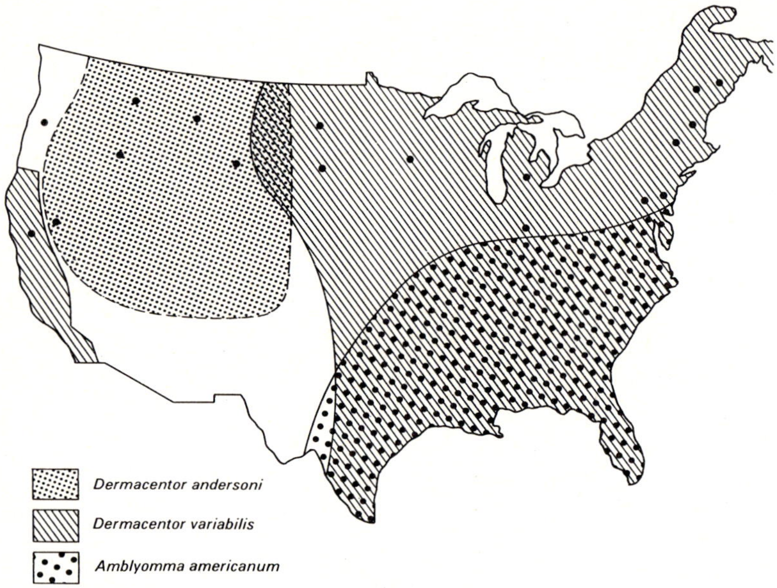


**FIG. 57.6**
Distribution of *Dermacentor andersoni, D. variabilis* and *Amblyomma americanum.* [Redrawn with modifications from Arthur, D.R. 1962. *Ticks and disease.* Harper & Row, New York.]

to feed on larger animals as it moves from the larva to the nymph to the adult. The early stages are in rodent burrows and probably find the host close at hand. The adults seek out hosts by questing; they climb on vegetation and hang head down with the legs extended so as to catch a passing mammal. Ticks also sense $CO_2$ and will crawl toward its source.

TREATMENT. The wood tick is not usually found on livestock in large numbers and cattle are not usually treated for ticks. Owners sometimes treat horses individually with an insecticide such as an organophosphate as a spray or wash. Wood ticks found on humans or companion animals can usually be removed by pulling them off. It is important to grasp the tick mouthparts at the surface of the skin so that nothing is left in the skin to serve as a site for bacterial infection. Removal is most efficiently done by using forceps or tweezers and pulling steadily and firmly until the tick comes off. Many remedies have been suggested for tick removal, including burning them with a cigarette and painting them with fingernail polish, but the treatment of choice is to pull them off, because these other methods do not work well. For humans, it is best to avoid allowing ticks to embed by using a repellent such as deet, tying the clothing tightly at the ankles and wrists, and searching for ticks in the clothing and on the body after a day out of doors. The wood tick usually does not embed for about 3 hr, so one has a little time to procrastinate while the ticks wander about searching for a likely spot to dine.

EPIDEMIOLOGY AND CONTROL. *D. andersoni* is the vector of *Rickettsia rickettsi*, the cause of Rocky Mountain spotted fever (RMSF) or tick-borne typhus and of Colorado tick fever virus (CTF). It also is a vector of the causative agent of canine piroplasmosis (Chap. 17), the causative agent of Q fever (*Coxiellia burnetti*), and the cause of anaplasmosis of cattle (*Anaplasma marginale*). Knowledge of the habits and life cycle of the wood tick can provide the means for avoiding infection with one of the preceding agents. Wood ticks are generally found in areas that are not continuously inhabited by people, and they may be in inaccessible areas so that control by treating the area with an acaricide is not feasible. Wood ticks are typically found in hilly or mountainous areas on south-facing slopes where there may be scattered coniferous trees, but they need low brush, grasses, and an abundance of exposed rocks. This habitat also describes that of small rodents such as deer mice and chipmunks, on which the larvae and nymphs feed. The adult ticks, which are the vectors for RMSF and CTF for humans, are active for a limited time. In Colorado, for example, adult ticks appear in

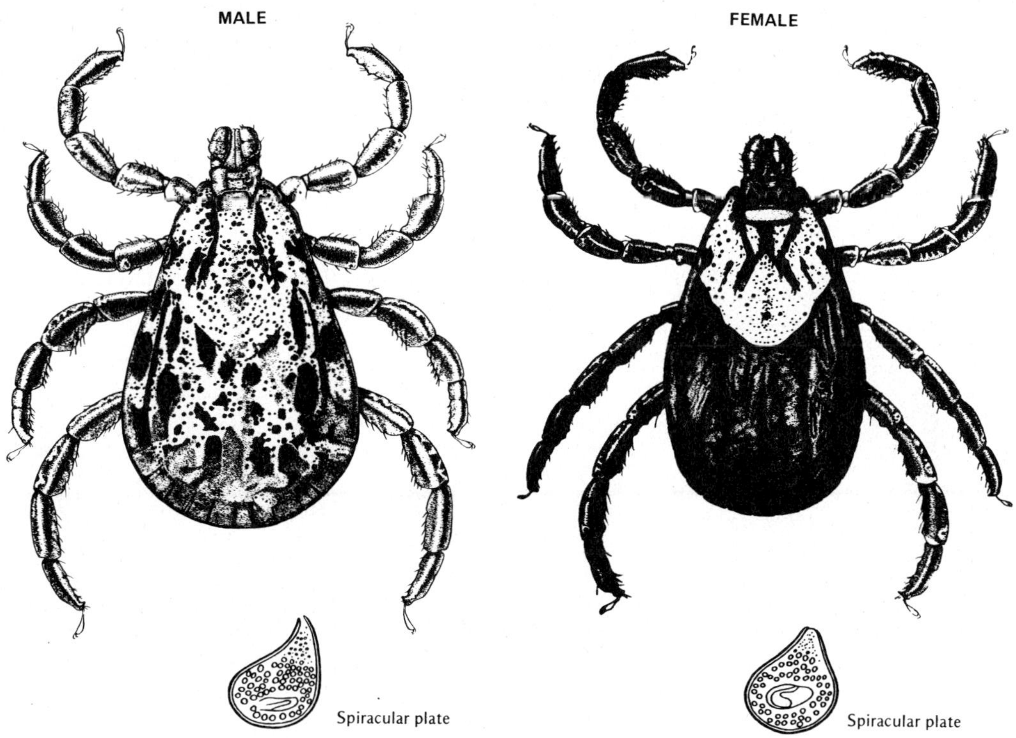


**FIG. 57.7**
Male and female of *Dermacentor andersoni*, dorsal surface.

mid-March and can be collected until mid-June; an occasional tick may be found into August, but they are rare. Thus, where there is risk of infection with RMSF or CTF, we know that the vectors are limited in both time and space. The simplest tactic is to avoid those areas at the time when adult ticks are active. When planning picnic or camping areas, those sites that are likely to have a high risk can be avoided. The habitat type is a good indication that ticks will be present and a simple census will show whether they are there. Locating a campground either in a wooded area or a valley bottom thereby decreases the risk of infection among campers.

Other species of *Dermacentor* include *D. albipictus* (Fig. 57.8), the winter tick, which is a one-host tick with a discontinuous distribution in the northern United States and Canada. It is found on large ruminants such as moose, elk, deer, and bighorn sheep, as well as on horses, bear, and canids. Larvae seek hosts in late fall and the whole life cycle is completed on one animal; the ticks reach maturity in late winter or early spring when they drop off to oviposit. This species can be especially bothersome to cattle and horses during the winter.

*D. nitens*, the tropical horse tick, is found in Florida and southern Texas and is spread throughout the islands of the Caribbean, Central America, and the northern part of South America. Although its name indicates that it is found on horses, it is also a common parasite of many other large herbivores. The horse tick is a one-host tick that prefers to attach in the ears of its host. They may remain attached in the ears for as long as 99 days while feeding and molting. They may clog the auditory canal with their bodies, excrement, and cast skeletons, leading to ear infections. *D. nitens* is the vector of equine piroplasmosis, caused by *Babesia caballi* (Chap. 17).

*D. variabilis*, the American dog tick (Fig. 57.9), is a three-host tick that is distributed widely over the eastern two-thirds of North America from Labrador to Mexico (Fig. 57.6), Although its preferred host is the

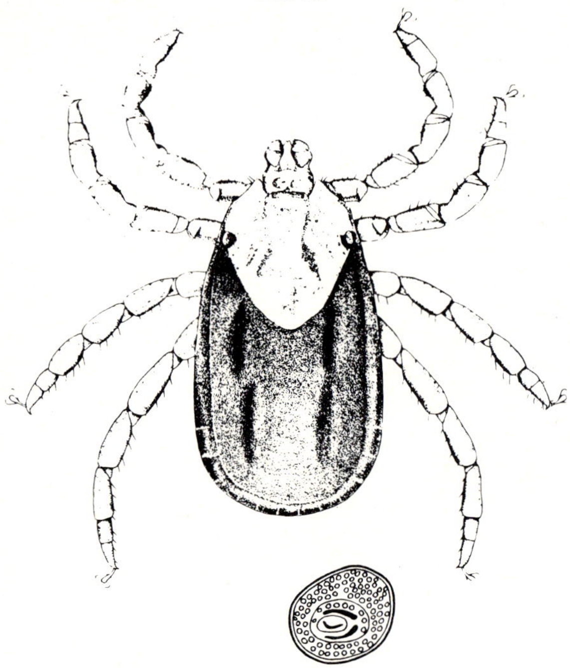


**FIG. 57.8**
Male and female of *Dermacentor albipictus*, dorsal surface.

dog, it is found on other carnivores, suids, some herbivores, rodents, lagomorphs, mustelids, and humans. In temperate climates it is found on dogs during the warmer months, but in warm climates it is active year-round. The dog tick is the principal vector of the causative agent of Rocky Mountain spotted fever in the eastern United States. RMSF was originally found in the West and it was a significant cause of disease in humans; however, in the 1960s, with a decline in incidence in the West, there was an increase in the southeastern United States.

NAME OF ORGANISM. *Haemaphysalis leporispalustris* is known as the *rabbit tick*.

HOSTS AND LOCATION. The rabbit tick prefers to feed on lagomorphs but is found on a wide range of carnivores, rodents, opossum, and birds. About 60 species of birds are known to be hosts of the immature stages.

DISTRIBUTION AND IMPORTANCE. Distribution is throughout North America from Alaska to South America. Heavy infections are known to kill rabbits and birds. For example, the snowshoe hare in Ontario was found to have as many as 2700 ticks in the summer. The rabbit tick also transmits the causative agent of tularemia, *Francisella tularensis*.

MORPHOLOGY. This is a small, inornate brown tick that has festoons at the posterior margin of the opisthosoma (Fig. 57.10).

LIFE CYCLE. The larvae and nymphs are often found on birds, usually ground-feeding species such as grouse, quail, and meadowlark. The preferred hosts for the adults are rabbits. The life cycle is completed in about 110 days; the adults remain on the host for about 17 days while engorging (Campbell & Glines, 1979).

HOST–PARASITE INTERACTIONS. The rabbit tick is adapted for remaining in the vicinity of possible hosts by adjusting the time when it drops from the rabbit. Studies have shown that the ticks drop from the host during late afternoon, a time when rabbits are usually in their resting places.

*H. leporispalustris* is used widely in laboratory studies because it is relatively easy to cultivate in the laboratory. Ticks elicit immunity in the hosts on which they feed, and the rabbit tick has been used to study the phenomenon (McGowan et al., 1979).

NAME OF ORGANISM. *Boophilus annulatus*, the cattle fever tick.

HOSTS AND LOCATION. Cattle are the principal hosts, but the cattle fever tick has been taken from deer and other large ruminants, equids, and rarely dogs and humans. The consensus is that it was originally a parasite of deer and made the transition to cattle. It is located on the skin: legs, belly, and dewlap.

DISTRIBUTION AND IMPORTANCE. This tick is found in tropical and subtropical climates; it may be carried into temperate climates but does not overwinter there. It occurs primarily in the Western Hemisphere south of the United States; distribution was originally through the southern United States, but it has been eradicated from there. It is also in tropical Africa, the Middle East, and the Mediterranean basin. The cattle

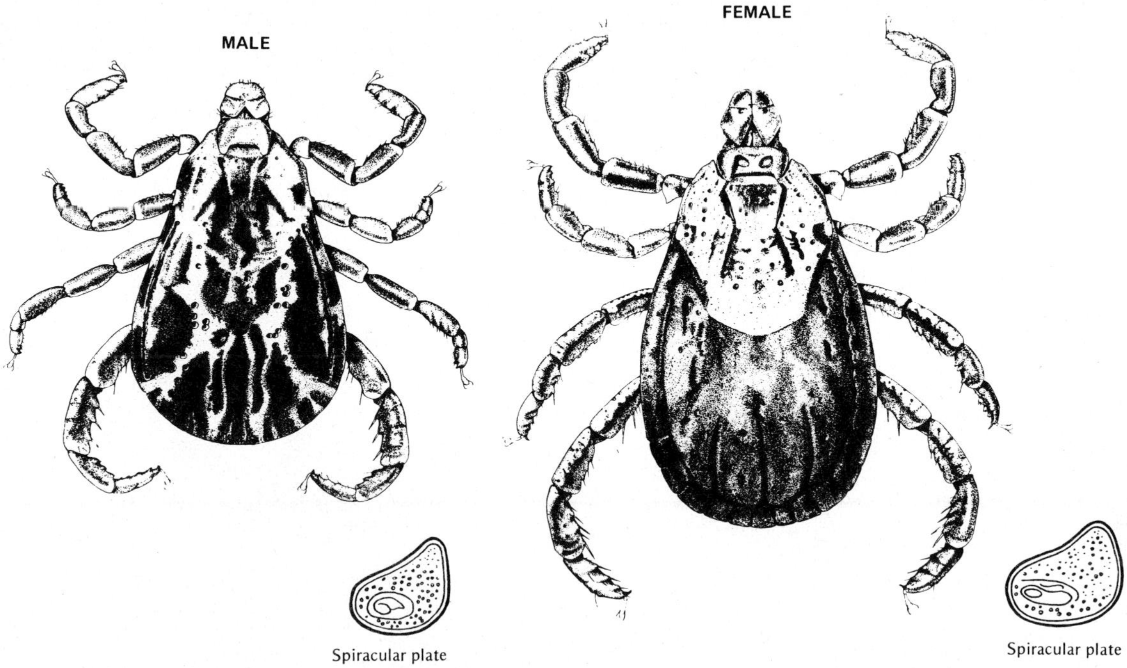


**FIG. 57.9**
Male and female of *Dermacentor variabilis,* dorsal surface.

fever tick is a vector of the causative agent of Texas cattle fever. (Chap. 17).

MORPHOLOGY. *Boophilus annulatus* is a relatively large, inornate brown tick that lacks festoons; as is true in the genus, the pedipalps are short (Fig. 57.11). Eyes are located on the margins of the scutum and the spiracular plate is nearly round. The females swell to an enormous size when engorged.

LIFE CYCLE. *Boophilus annulatus* is a one-host tick; it may complete development from the time of larval attachment until the female drops off in as short a time as 20 days, but usually takes longer and may extend to 66 days. Portions of the life cycle may be extraordinarily drawn out; for example, the females may begin to lay eggs as early as the day after dropping from a host but may not begin for 98 days. Larvae may hatch in 17 days at 30° C, but may require 202 days at low temperatures. Larvae may survive unfed for nearly 8 months (246 days) while awaiting a suitable host.

TREATMENT, EPIDEMIOLOGY, AND CONTROL. See Chapter 17.

Other *Boophilus* species include *B. microplus*, the tropical cattle tick, which is distributed widely in the warmer climates of the Western Hemisphere, Australia, Africa, and the Orient. Cattle are the principal hosts, but it is found on other herbivores, some carnivores, marsupials, and pigs (Fig. 57.12). This is a one-host tick; the adult drops from the host about 23 days after the larva attaches. The larvae feed mostly in the ears of the host, and the nymphs and adults are found widely distributed on the body. *B. microplus* is a vector of *Babesia bigemina, B. ovis, B. caballi, Theileria mutans*, and *Anaplasma marginale*. In addition to being a vector of a variety of disease agents, it causes significant production loss, especially in cattle. In tropical areas, ticks are on cattle year-round and therefore are a continual drain on weight gain and milk production. For example, in a study done in Australia, reductions in weight gain showed the following:

| RETARDATION | NUMBER OF TICKS | | | | |
|---|---|---|---|---|---|
| | 40 | 50 | 60 | 70 | 80 |
| In pounds/year | 56 ± 19.8 | 82 ± 18.7 | 108 ± 19.5 | 134 ± 22.1 | 160 ± 25.9 |
| In kilograms/year | 25 ± 8.9 | 37 ± 8.5 | 49 ± 8.8 | 61 ± 10.0 | 73 ± 11.6 |

The extrapolation of the data in this study showed that there was a reduction of 0.76 kg per tick per year (1.67 lb) on a growing animal (Little, 1963). This is not to say that a single tick would have such an effect, but the figure gives a way of calculating gains when considering whether tick control is warranted. Since *Boophilus microplus* is so prevalent over much of the warm climates of the world where livestock are raised, the losses in production of needed animal protein are truly staggering. For example, losses in Australia are estimated to be $67 million (Australian) annually (Anon., 1975).

NAME OF ORGANISM. *Ixodes ricinus* is known as the *sheep tick* or *castor bean tick*.

HOSTS AND LOCATION. The principal hosts are sheep, but like most ticks it is found on quite a number of other species as well, including cattle, deer, carnivores, mustelids, rodents, rabbits, birds, reptiles, and occasionally humans. On sheep, the tick tends to feed on areas of the body that are only lightly wooled: around the mouth, ears, axilla, and groin.

DISTRIBUTION AND IMPORTANCE. The distribution includes Great Britain, continental Europe, parts of the Middle East, Africa, and Japan. This is generally a cool climate tick; for example, in Iran it is found north of about 35 degrees north latitude, and may be found in other countries as far as 65° north latitude, not far from the Arctic Circle. *I. ricinus* removes significant amounts of blood from its host, and, perhaps more important, serves as a vector of the disease agents causing louping ill in sheep, buotonneuse fever, piroplasmosis, and spring and summer encephalitis.

MORPHOLOGY. The pedipalps are long, as are those of the other members of the genus; the body is reddish brown to black, inornate, and lacks festoons. A distinguishing characteristic of the genus is the anal groove, which curves around the anterior margin of the anus. The male is 2.5 by 1.3 mm long and has scattered hairs on its body. The engorged female is 11 by 7 mm, and the body has hairs, but they are shorter than those of the male (Fig. 57.13 shows an *I. scapularis*, a closely

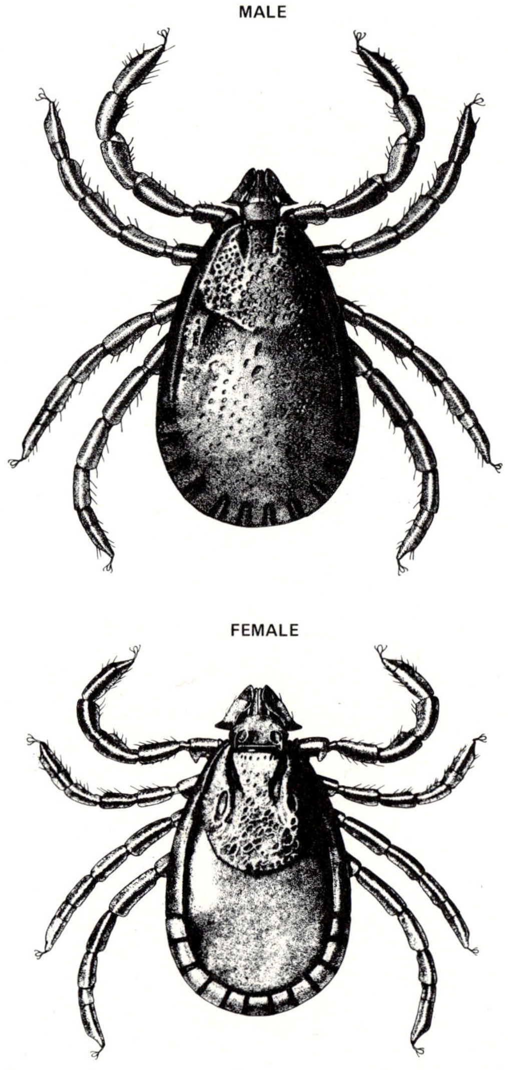


**FIG. 57.10**
Male and female of *Haemaphysalis leporispalustris,* dorsal surface.

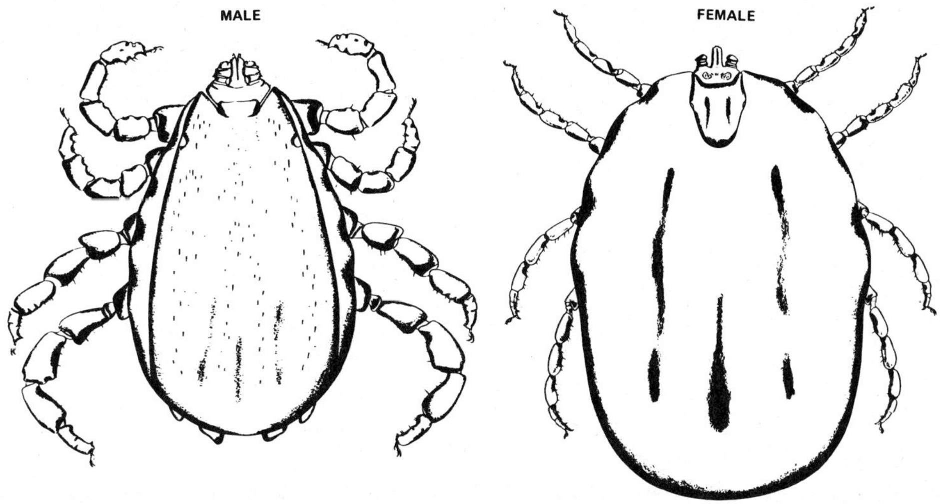


**FIG. 57.11**
Male and female of *Boophilus annulatus,* dorsal surface.

related species). The hexapod larva is 0.6 to 1.5 mm long (Fig. 57.14).

LIFE CYCLE. *I. ricinus* is a three-host tick that often uses the same species of host throughout the life cycle; larvae and nymphs are often found on birds, but sheep may be hosts for all three stages of the cycle. The cycle in Britain requires 3 years for completion, with each stage actively seeking a host during the spring of the year. Adults seek hosts in the spring and females engorge for 5 to 14 days. The males inseminate the females while they are engorging and may remain on the host to fertilize additional females. The females do not lay eggs until 4 to 8 weeks after dropping from the host, and they continue to lay eggs over the next 4 to 6 weeks until early July. The females lay 500 to 2000 eggs and then

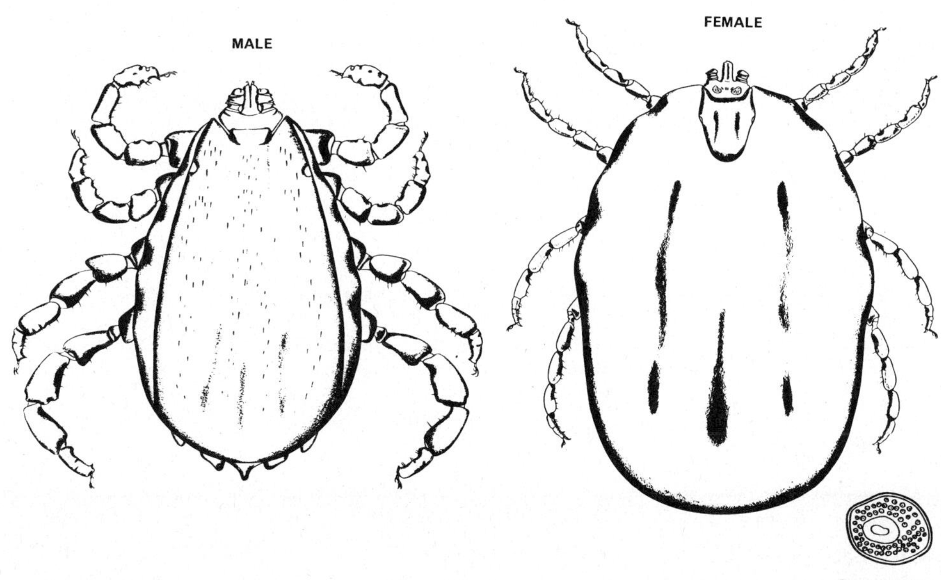


**FIG. 57.12**
Male and female of *Boophilus microplus,* dorsal surface.

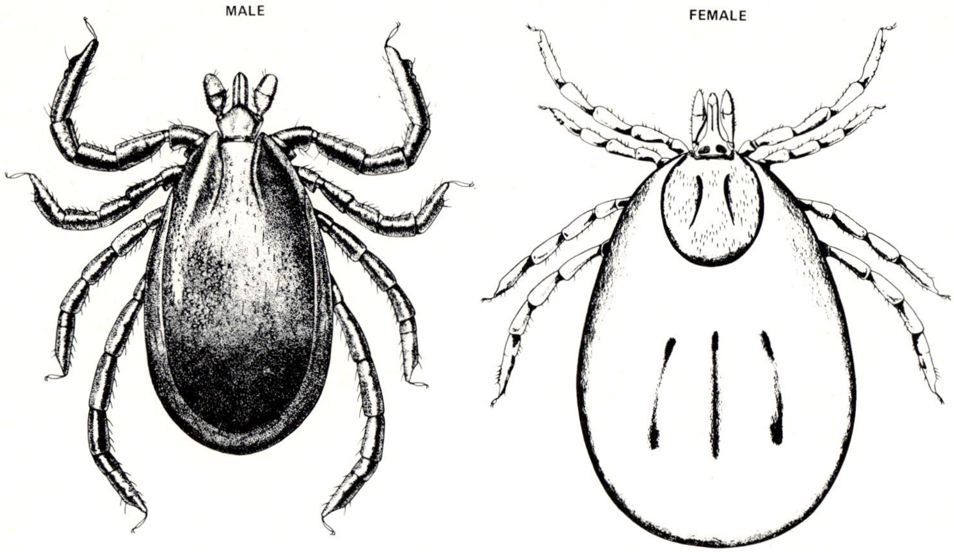


**FIG. 57.13**
Male and female of *Ixodes scapularis,* a species closely related to *Ixodes ricinus,* dorsal surface.

die. Larval development proceeds slowly; the larvae hatch in late summer or fall, but they normally do not feed until the following spring.

In the spring, larvae climb up on plants and wait for a host. The larvae feed for 3 to 6 days, drop off, and molt during the summer. The nymphs also wait until the following spring to seek hosts; they feed for 3 to 5 days, drop off, and molt to the adult. The molt to the adult stage may require as long as 7 months if the nymph has fed late in the season and cold weather has set in. There appears to be a diapause that must be broken so that development can proceed for each of the stages, because although development may have been completed in a particular summer, feeding does not occur until the following spring. In some parts of the British Isles, there are populations of *I. ricinus* that feed in the fall, and this appears to be a stable characteristic.

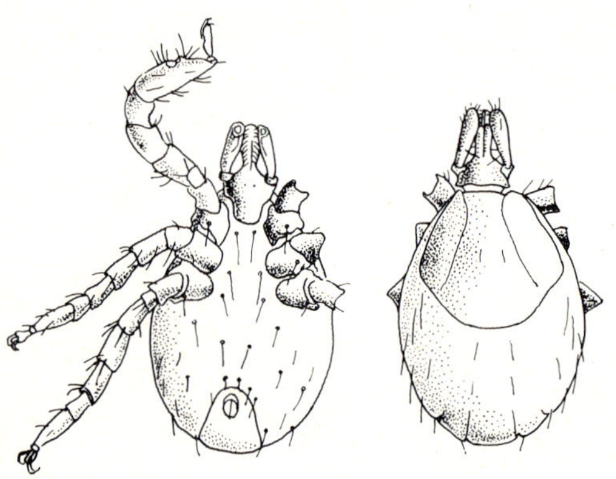

**FIG. 57.14**
Larvae of *Ixodes ricinus.* (*Left*) Ventral surface; (*right*) dorsal surface.

HOST–PARASITE INTERACTIONS. *I. ricinus* has been studied extensively in Britain and many of the studies on it point up the adaptations of ticks as parasites. For example, despite poor eyesight, ticks can sense hosts and either prepare to be swept up by them or move toward a potential host. They are sensitive to the following:

1. Vibrations of the plants they climb to quest. The swaying of a plant in the wind will not activate them, but more rapid vibration causes them to extend their legs so that they can catch on the fur of a passing animal.
2. A reduction in light intensity also causes them to turn toward the shadow and extend their legs.
3. Warm air, as from the body or breath of a mammal, causes them to extend their legs in a questing attitude.
4. Aromas from the hair or wool of mammals also activate them.

**5.** Carbon dioxide is an attractant and may cause ticks to migrate some distance on the ground toward its source. $CO_2$ is used in traps designed to capture ticks.

Ticks that begin to quest have about 30 days in which to find a host and to feed, or they will use up their food reserves and die. Even so, if we consider the life cycle of *I. ricinus*, the long time between feedings, nearly a year, demonstrates marvelous adaptation to survival while awaiting another meal. This ability of ticks to wait for long periods is seen in nearly all species, and is even more remarkable in those found in hot, dry climates where retention of water is likely to be difficult. Water retention is aided by the leathery cuticle and waxes in the exoskeleton.

Blood sucking by *I. ricinus* is usually not considered to be a cause of reduced production in sheep. In a study in which it was assumed that a sheep would have 265 female ticks, 2252 nymphs, and 254,400 larvae on it during a 4-month period, the animal would lose 1249 ml of blood or about 10 ml/day. For an animal on a good plane of nutrition, losing 1250 ml of blood may be tolerable, but if pasture is poor or other disease agents are present, there may be significant loss. In Britain, for example, the nematodes *Haemonchus contortus, Trichostrongylus colubriformis, Oesophagostomum* spp., the digenetic trematode, *Fasciola hepatica*, and the sheep ked, *Melophagus ovinus*, all cause anemia directly or indirectly and are likely to be present in sheep at least in small numbers. If animals are on marginal pasture and have a combination of parasites, as many animals do, another liter of blood loss from ticks could well mean the difference between profit and loss for the sheep raiser.

TREATMENT. As is true of most ectoparasites, a contact acaricide applied by dipping or spraying usually results in an immediate reduction of ticks. Sheep should be in short fleece when a toxicant is applied, because the fleece serves either as a barrier or a blotter for the chemical.

EPIDEMIOLOGY AND CONTROL. The feeding habits of *I. ricinus* allow some measure of control; a tick feeds on a host for a maximum of 25 days during its 3-year life cycle. Although the population of ticks will be active for 3 to 4 months in the spring and early summer, and some demes feed in fall, knowing the short time that they are active can be used to avoid areas where they are present. A tick has about 30 days to find a host or it will die. Partial success in reducing tick populations has been achieved by burning, cultivating, or draining pastures in order to reduce habitat. *Ixodes* is one of the most widespread genera of hard ticks; there are 34 species in the United States and 9 species in Britain.

## FAMILY ARGASIDAE—THE SOFT TICKS

The argasids are leathery and lack the scutum seen in the hard ticks; the mouthparts cannot be seen from above. There are some differences in life cycles compared to those of the hard ticks. The hard ticks have egg, larva, nymph, and adult, whereas the soft ticks have a number of nymphal stages (usually three but sometimes more). There are only four genera: *Argas, Ornithodoros, Otobius*, and *Antricola*; about 140 species are known worldwide.

NAME OF ORGANISM. *Argas persicus*, the fowl tick or blue bug.

HOSTS AND LOCATION. Chickens are the principal host, but the tick has been taken from all kinds of birds including galliform, anseriform, passeriform, and falconiform. It occasionally feeds on rabbits, rodents, and humans. It usually feeds on areas where feathers are sparse, such as beneath the wings.

DISTRIBUTION AND IMPORTANCE. The tick is cosmopolitan, in both temperate and tropical climates. It causes production losses and sometimes death in poultry, as well as being the vector of the organism causing fowl spirochetosis.

MORPHOLOGY. These ticks are yellowish brown when unfed and slate blue when engorged with blood. A distinct sutural line separates the dorsal from the ventral parts of the body; this line can be seen even when the tick is engorged (Fig. 57.15). The males are 3 by 4.5 mm and the females 8.5 by 5.5 mm, although both are occasionally larger.

LIFE CYCLE. The larvae seek hosts and remain attached for 2 to 10 days while feeding. They drop off and molt in 4 to 16 days, and the nymphs seek hosts and

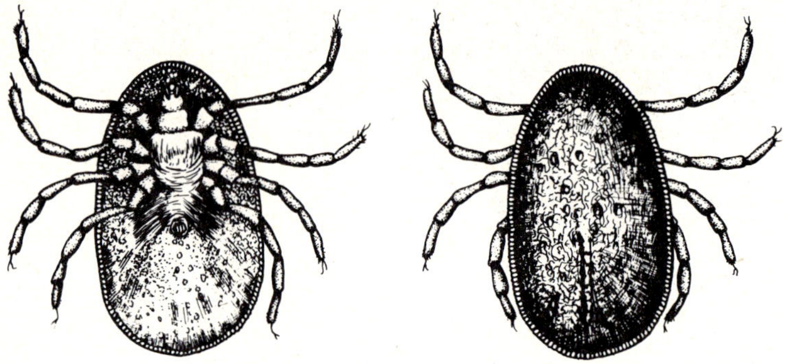

**FIG. 57.15**
Adult of *Argas persicus;* ventral view (left), dorsal view (right). [Redrawn from Cooley, R. A., and G. M. Kohls. 1944. *The Argasidae of North America, Central America and Cuba.* Am. Midland Naturalist, Monograph Series (1) June, 152 pages.]

9, 10.5 Sabon w/ Italic
9 Univers #65

feed rapidly: they usually engorge within 30 min. There are three or more nymphal stages that feed and molt. The adults feed at night and usually engorge in 20 to 45 min. The pattern in the females is to feed and lay eggs and then feed again. A female may feed daily and deposit eggs every day. The number of eggs laid is high early and tends to be reduced as the female ages; as many as 274 eggs can be laid in the first clutches and as few as 25 in the later ones.

DIAGNOSIS. Since the larvae remain attached for several days, they may be found under the wings or around the vent. The nymphs and adults may be found feeding on chickens at night.

HOST–PARASITE INTERACTIONS. The fowl tick takes a large amount of blood and causes production losses because of decreased weight gains and lowered egg production. Heavy infections cause the death of the host through anemia. Tick paralysis (Chap. 48) sometimes occurs in chickens, ducks, and geese infected with *Argas persicus*.

*Argas persicus* is the vector of *Borrelia anserina*, the cause of fowl spirochetosis.

TREATMENT. Most treatments are based on treating birds and their houses with a second-generation acaricide such as an organophosphate. If birds have roosts in a building where they spend the night, the legs of the perch can be placed in cups of motor oil to prevent the nymphs and adults from reaching their objective.

EPIDEMIOLOGY AND CONTROL. *A. persicus* represents almost the ultimate in patience and survival. Even the larvae have been found to survive as long as 246 days without feeding, and an unfed female was reported to survive for 5 years. With such a long survival time, active steps must be taken to kill the stages in the environment. Not much can be done to control stages outside of poultry houses, and surveillance of birds must be continuous so that ticks picked up out-of-doors do not have time to establish in houses after treatment has been administered.

NAME OF ORGANISM. *Ornithodoros moubata* has generally been considered to be a single species, but it has been divided into four species (Walton, 1964). The species of greatest importance on domestic animals is known as *O. moubata*. There seems to be little host specificity; ticks are found on various domesticated animals, especially swine, and they feed readily on humans when in houses.

HOSTS AND LOCATION. The original hosts probably were warthogs, *Phacochoerus*, and the ticks later transferred to domestic swine and then to humans.

DISTRIBUTION AND IMPORTANCE. *O. moubata* (in the narrow sense) is found in arid or semiarid areas of Africa from southern Africa northward to Kenya. Not only is this tick a cause of significant blood loss, but it also is a vector of *Borrelia recurrentis*, the cause of relapsing fever.

MORPHOLOGY. This is a leathery, wrinkled tick with a mammilated exoskeleton (Fig. 57.16). It lacks eyes, unlike *O. savignyi*, a closely related species which has two pairs of eyes. Adults average 6.5 by 8 mm, with engorged individuals reaching a length of 11 mm.

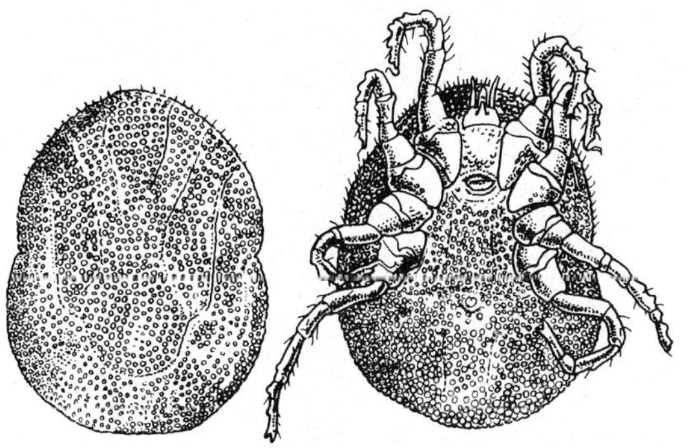

**FIG. 57.16**
*Ornithodoros moubata.* (*Left*) Dorsal surface; (*right*) ventral surface.

LIFE CYCLE. The female oviposits the 0.5-mm eggs in batches averaging 247, with a total of 1926 laid in 7.8 batches. Activity that has been interpreted as "brooding" of eggs has been observed in *O. moubata* since females remain in the vicinity of the eggs and sometimes have been seen with nymphs clinging to them. Eggs hatch in 8 days at 30°C, releasing the hexapod larvae. The larvae do not feed, but molt within a short period of time to the nymphal stage. There are three or more nymphal stages, which feed between molts.

HOSTS–PARASITE INTERACTIONS. Swine can be killed by infections with *O. moubata*, and these ticks may be significant pests of humans when they move into huts and feed on people. This tick is probably of greater importance in its role as a vector of *Borrelia recurrentis*, the cause of tick-borne relapsing fever.

EPIDEMIOLOGY AND CONTROL. Control of *O. moubata* is directed mostly toward preventing the transmission of the causative agent of relapsing fever. In central Africa, the ticks move into houses and feed on people at night. Houses in villages and small towns are largely made of mud. The ticks can crawl into cracks in the walls to spend the day. Control involves decreasing resting sites by plastering the insides of the dwellings to make them smooth and crack-free, and then treating the house with a contact acaricide such as an organophosphate or organochlorine compound.

TICK-BORNE RELAPSING FEVER. Relapsing fever, caused by the spirillar bacterium, *Borrelia recurrentis*, occurs widely throughout the world, and strains of the organism, or organisms, are either tick-adapted or louse-adapted (Chap. 51). In North America, tick-borne relapsing fever is most often contracted when a person enters a nidus where the agent is cycling in rodents and soft ticks. Individual cases often occur in campers who sleep on the ground near rodent burrows. A classic outbreak occurred in 1973 in Grand Canyon National Park when 62 people contracted relapsing fever at a lodge (Boyer et al., 1977). The disease occurred in those persons who stayed in rustic log cabins as opposed to those who stayed in a large central building. Investigation showed that the cabins had large amounts of rodent nesting materials both under the floors and in the attics. The cabins were rather old and not constructed to keep rodents out, so they had moved in and their ectoparasites came with them. It was also observed that rodent populations were at a low point during that summer. The vector, *O. hermsi*, apparently did not have its usual rodent hosts to feed on and moved to the human inhabitants of the cabins instead.

Control of the outbreak consisted of removing the nesting materials from the cabins (a total of 3000 kg were taken out) and then spraying the cabins with a carbamate acaricide. Rodent-proofing of the cabins was also undertaken to prevent recolonization in subsequent years; inquiries during the following year revealed that no additional cases of relapsing fever occurred.

NAME OF ORGANISM. *Otobius megnini* is known as the *spinose ear tick*.

HOSTS AND LOCATION. The ticks are found principally on cattle, but they have only a low degree of host specificity, and are found on other domestic animals, wild ruminants, carnivores, and humans. They feed in the auditory canal.

DISTRIBUTION AND IMPORTANCE. These ticks are cosmopolitan in warm climates. Sometimes they are found in cold climates, but they probably do not perpetuate themselves there.

LIFE CYCLE. Only the larvae and nymphs of *O. megnini* are parasitic (Fig. 57.17). The larvae seek hosts and move to the ears, where they feed in the ear canal. They molt in the ear and the nymphs remain on the same host; once having reached a host, the larvae and

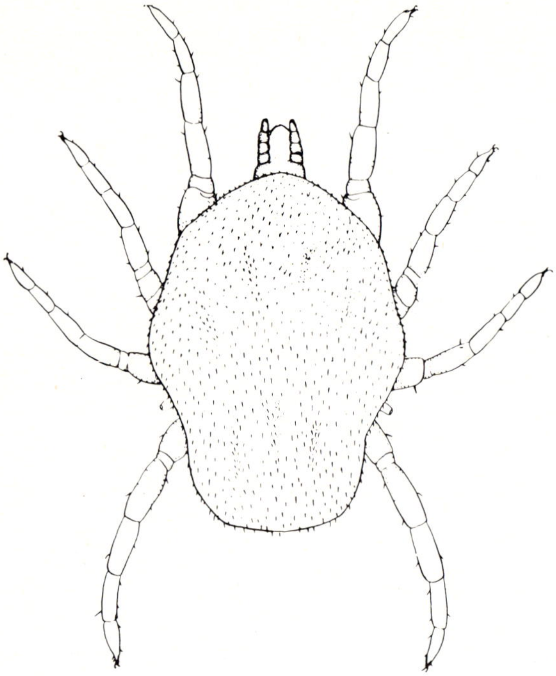

**FIG. 57.17**
Nymph of *Otobius megnini.*

nymphs remain on it for up to 7 months or sometimes more (Fig. 57.18). The nymphs drop from the host and molt to the adult stage; copulation takes place off the host. The adults are free-living.

HOST–PARASITE INTERACTIONS. When large numbers of ticks are present in the ear canal, it becomes plugged with the ticks, their shed skeletons, ear wax, and other debris. This obviously causes cattle considerable distress, and may lead to secondary bacterial infections of the ears.

It is generally stated that the spinose ear tick is occasionally found in the ears of humans; however, infection of humans may be more common than published case reports indicate. We were brought a tick that had been removed from the ear of a woman in northern Colorado by a physician. It had apparently been in the woman's auditory canal for about a year. Dr. Richard Eads of the Centers for Disease Control confirmed the identification and then provided data showing recoveries from seven individuals in 1978 in New Mexico. In some individuals there may be penetration of the tympanic membrane, and in others facial paralysis on one side.

TREATMENT. Generally, a liquid preparation of a contact acaricide is run into the auditory canal. A variety of ways of applying the chemical have been used, but one of the quickest is to use a pump-type oil can so that one shot is easily applied to each ear.

In humans, the ticks can be pulled out with forceps or killed with a solvent such as acetone and then removed.

EPIDEMIOLOGY AND CONTROL. In warm climates, the spinose ear tick can be a continuing problem in livestock, and frequent inspection and treatment is the best method of control.

## TICK PARALYSIS

Tick paralysis is an acute ascending paralysis of vertebrates that results when some ticks attach to engorge. Mammals are most frequently affected, and the disease is usually seen when one or more ticks embed at the base of the skull. Within a day or two of attachment, the person or animal will begin to lose coordination and sensation in the hands and feet. The

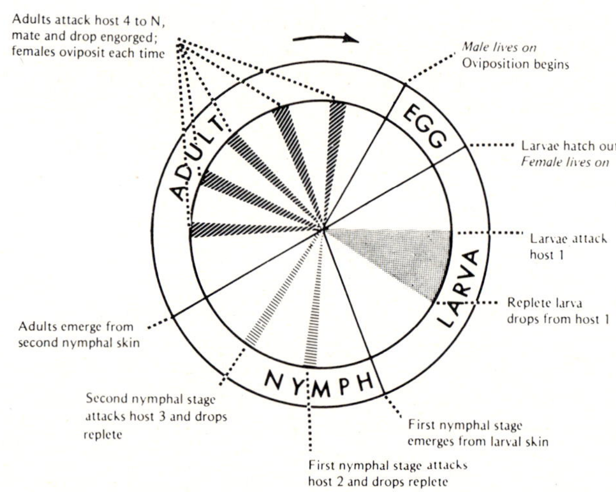


**FIG. 57.18**
Life cycle and feeding pattern of a soft tick.

paralysis progresses centrally and the legs and arms become useless; if the ticks remain attached, the host is likely to die when the paralysis reaches the breathing center of the brain. If the ticks are found and removed, recovery begins almost immediately, and a person or animal that had been helpless will be up and around in a day. Within a few days to a week, all signs of paralysis are usually gone.

Not all ticks are likely to cause tick paralysis, but both hard and soft ticks have been implicated at one time or another.

The following are some of the ticks that are involved:

| TICK | LOCALITY |
|---|---|
| *Dermacentor andersoni* | North America |
| *Dermacentor variabilis* | United States |
| *Amblymomma maculatum* | South America |
| *Ixodes homocyclus* | Australia |
| *Ixodes ricinus* | Central Europe |
| *Ixodes hexagonus* | Great Britain and Continental Europe |
| *Ixodes rubicundus* | South Africa |
| *Rhipicephalus evertsi* | South Africa |
| *Argas persicus* | Cosmopolitan |

The incidence of tick paralysis is erratic. Not all ticks cause it and not all potential victims in a population seem to be susceptible. *Dermacentor andersoni* may cause paralysis in humans or domestic animals all through its range, but the more northerly populations, such as in British Columbia, are more likely to cause the disease. Smaller or younger animals are more likely to show paralysis; thus, children are more susceptible than adults. However, large ruminants such as cattle and bison have been killed by tick paralysis.

The toxin causing tick paralysis is a part of the salivary fluid which the tick injects while feeding. It acts at the myoneural junction, preventing normal control of muscles and resulting in incoordination. The observations that tick paralysis is usually associated with ticks attached near the base of the skull, and the rapid recovery when ticks are removed, lead to the conclusion that the toxin acts locally and that it breaks down rapidly or is rapidly detoxified by the body.

Since tick paralysis is so sporadic, the principal hurdle to overcome is the awareness that it might be a possibility. Children, dogs, cats, livestock, and birds are all likely candidates, and the vague signs or symptoms are likely to be insufficient to elicit a search for ticks. Once that decision has been made and ticks have been found, removal of them is generally all that is needed.

## REFERENCES AND FURTHER READINGS


Anonymous. 1975. *Cattle Tick in Australia. Inquiry by the Cattle Tick Control Commission.* Australian Government Publishing Service, Canberra.

Anonymous. 1978. *Ticks of Public Health Importance and Their Control.* USDHEW, Center for Disease Control, Washington, D.C.

Arthur, D. R. 1962. *Ticks and Disease.* Row, Peterson & Co., Evanston, Ill. (U.S. edition of 1961 Pergamon Press edition).

Boyer, K. M., R. S. Munford, G. O. Maupin, et al., 1977. Tick-borne relapsing fever: An interstate outbreak originating at Grand Canyon National Park. *Am. J. Epidemiol.* **105**:469–479.

Campbell, A., and M. V. Glines. 1979. Development, survival and oviposition of the rabbit tick, *Haemaphysalis leporispalustris* (Packard) (Acari: Ixodidae), at constant temperatures. *J. Parasitol.* **65**:777–782.

Harwood, R. F., and M. T. James. 1979. *Entomology in Human and Animal Health*, 7th ed. Macmillan Publishing Co., New York.

Little, D. A. 1963. The effect of cattle tick infestation on the growth rate of cattle. *Austral. Vet. J.* **39**:6–10.

McGowan, M. J., J. H. Camin, and R. W. McNews. 1979. Field study of the relationship between skin-sensitizing antibody production in the cottontail rabbit, *Sylvilagus floridanus*, and infestation by the rabbit tick, *Haemaphysalis leporispalustris* (Acari: Ixodidae). *J. Parasitol.* **65**:692–699.

Strickland, R. K., R. B. Gerrish, J. L. Hourrigan, and G. O. Schubert. 1976. *Ticks of Veterinary Importance*, rev. ed. APHIS, USDA Agric. Handbook No. 485, U.S. Department of Agriculture, Washington, D.C.

Walton, G. A. 1964. The *Ornithodoros* "moubata" group of ticks in Africa. Control problems and implications. *J. Med. Ent.* **1**:53–64.

# Glossary

**acanthella.** Acanthocephalan larva following the acanthor and prior to the cystacanth; reproductive system is not fully developed.

**acanthor.** Acanthocephalan larva with bladelike hooks; develops inside the egg capsules.

**acetabulum.** Ventral sucker or holdfast of digenetic trematodes; a sucker on a tapeworm scolex.

**acoelomate.** A condition in which a body cavity is lacking, as in the members of the phylum Platyhelminthes, where the organs lie embedded in parenchyma.

**ala (-ae).** Winglike projection such as the cuticular expansions in certain nematodes.

**allergy.** See hypersensitivity.

**alveolus (-i).** The air sac in the lung where gaseous exchange takes place.

**ametabolous.** A type of metamorphosis in insects in which there is no external change as they proceed through a series of molts to the adult; *ametabola* refers to the taxonomic group.

**amphid.** Anterior sensory structures of nematodes (see also phasmid).

**anaphylaxis.** An extraordinarily strong hypersensitive reaction in which the individual may collapse, stop breathing, and die; each species of animal has a peculiar syndrome of signs and symptoms.

**anapolysis.** The process in which terminal, gravid proglottids are not shed in certain tapeworms (ant. apolysis).

**anisogamete.** Morphologically different male and female gametes.

**anterior station.** Protozoan development in the anterior part of an insect vector, for example, Salivaria of the genus *Trypanosoma*; transmission takes place by biting.

**anthelmintic.** A chemical used to remove worms, usually from the intestinal tract of a vertebrate host; a vermifuge.

**antibody.** Serum protein (immunoglobulin) synthesized by lymphoid cells in response to an antigenic stimulus.

**antigen.** Any substance that can stimulate an immune response.

**antigenic mimicry.** Acquisition of or production of host antigens by a parasite so that it is not recognized as nonself, as in *Schistosoma.*

**apical complex.** Organelles characteristic of members of the phylum Apicomplexa: polar rings, subpellicular microtubules, conoid, rhoptries, and micronemes.

**apolysis.** The process in which terminal, gravid proglottids are detached and shed by certain tapeworms (ant. anapolysis).

**areole.** Round or polygonal areas on nematomorphan cuticle; function unknown.

**autogeny.** The ability of some bloodsucking arthropods to lay eggs without having had a blood meal (adj. autogenous).

**autoinfection.** A process in which the progeny of a parasite reinfect the host without passing out of it, for example, *Hymenolepis nana.*

**axostyle.** A longitudinal rodlike or tubelike structure in members of the protozoan order Trichomonadida; probably serves as a cytoskeleton.

**biramous.** Divided into two branches; typical of the terminal segments of the legs of Crustacea.

**bothridium (-ia).** Muscular, leaflike sucking structure on the scolex of tapeworms of the order Tetraphyllidea.

**bothrium (-ia).** Shallow, sucking groove on the scolex of tapeworms of the order Psuedophyllidea.

**bradyzoite.** Slow-growing zoite or meront of the pseudocyst of *Toxoplasma* and related protozoa.

**bug.** An insect of the order Hemiptera; something that has legs and cannot be identified as anything else.

**bulla.** A structure seen in some female symbiotic copepods that serves to anchor the maxillae; elaborated by head and maxillary glands.

**bursa (copulatory bursa).** A muscular copulatory structure at the posterior end of males of the order Strongylida; useful in taxonomy and identification.

**cadre.** Chitinous ring surrounding the mouth of members of the phylum Pentastomida.

**calotte.** Light-colored anterior end of members of the phylum Nematomorpha.

**campestral.** An adjective referring to a plain (see sylvatic).

**cellular immunity.** A specific response to an antigen in which lymphoid cells are the primary effectors.

**cercomer.** Taillike appendage, which often retains the hooks of the hexacanth embryo, on procercoid and cysticercoid larvae of certain tapeworms.

**chalimus.** Specialized copepodid larva of certain symbiotic members of the class Copepoda; larvae attach to their hosts by a filament secreted by the frontal gland.

**chemotherapy.** Use of a chemical that has a specific action in removing or killing a parasitic agent.

**coenurus.** Fluid-filled metacestode of the tapeworm family Taeniidae; has a nonlaminated wall that produces protoscolices; no brood chambers.

**commensalism.** Symbiosis in which there is no discernible damage to the host.

**concomitant immunity.** A form of acquired immunity in which an established infection persists long after resistance has developed against a challenge infection.

**conjugation.** A temporary union of two ciliated protozoans for the exchange of nuclear material.

**conoid.** Spirally coiled filaments in the anterior tip of the zoite of certain apicomplexan protozoa.

**coracidium.** Free-swimming, ciliated embryophore of tapeworms of the order Pseudophyllidea.

**costa.** A striated, rodlike structure that lies just under the recurrent flagellum of certain protozoa of the order Trichomonadida; composed of carbohydrate and protein.

**cuticle.** A secreted surface covering, generally considered to be nonliving.

**cypris.** A parasitic, postnaupliar larva of symbiotic members of the order Copepoda.

**cyst.** A general term used when an organism has a membrane surrounding it, whether the covering is of its making or of host origin, a resistant stage in the protozoa.

**cystacanth.** A juvenile acanthocephalan that is surrounded by a capsule of host origin.

**cysticercoid.** A tapeworm larva in which the scolex develops in its normal position but is surrounded by a fluid-filled bladder; one of the bladder worms or hollow (as opposed to solid) larvae; characteristic of many cyclophyllidean families such as Hymenolepididae, Dilepididae, or Anoplocephalidae.

**cysticercus.** A tapeworm larva in which the scolex develops in an inverted form and which has a fluid-filled bladder surrounding it; one of the bladder worms or hollow (as opposed to solid) larvae; characteristic of the cyclophyllidean family Taeniidae.

**definitive host.** Host in which the sexually mature form of a symbiote is found.

**diarrhea.** Frequent passage of fecal material containing an abnormally high proportion of water (see dysentery).

**didelphic.** Having two ovaries and hence a uterus with two horns; usually applied to nematodes.

**dioecious.** Separate sexes (ant. monoecious).

**direct life cycle.** A life cycle in which only a single host is required for its completion.

**disease.** Abnormal performance of physiologic functions as a result of injury to cells performing those functions.

**domiciliated (domiciliation).** The condition in which an unwanted animal associates with humans, usually by moving into houses; cockroaches, house flies, and house mice are examples.

**dysentery.** A form of diarrhea in which blood and mucus originating in the intestinal tract are passed (see diarrhea).

**ecdysis.** See molt.

**-ectomy.** A suffix meaning to remove, usually by surgery.

**ectoparasite.** A parasite that lives on the external surface or in the integument of a host.

**edema (Br. oedema).** Abnormal accumulation of fluid in cells, tissues, or tissue spaces resulting in swelling.

**egg.** The germ cell of a female, an ovum; the term egg is usually reserved for the ovum plus its outer coverings such as a shell or capsule.

**-emia (Br. -aemia).** A root referring to blood.

**endemic.** A disease or disease agent that occurs in a human community at all times (see also enzootic).

**endodyogeny.** A special form of merogony in which two daughter cells are formed, each with its own pellicle, while still in the mother cell; occurs in certain members of the phylum Apicomplexa such as *Toxoplasma*.

**endopodite.** The inner or medial branch of a biramous appendage.

**entero-.** A root that means cavity but usually is used in reference to the digestive tract.

**entomo-.** A root that refers to insects, as in *entomology*, the study of insects, or *entomophagous*, eating of insects.

**enzootic.** A disease or disease agent that occurs in an animal population at all times (see also endemic).

**epidemic.** A disease or disease agent that spreads rapidly through a human population (see also epizootic).

**epidemiology.** The study of the causes of disease; the complex of factors that lead to disease outbreaks; usually reserved for diseases in humans, *epizootiology* is used for diseases in other animals.

**epizootic.** A disease or disease agent that spreads rapidly through an animal population (see also epidemic).

**erratic parasite.** Parasite normal to a particular host, but in an abnormal location.

**euryxenous.** Having a broad host range.

**eutely.** Having a constant number of cells and usually nuclei.

**exflagellation.** Process of formation of microgametes in certain apicomplexan protozoans such as *Plasmodium*.

**exopodite.** The outer branch of a biramous appendage.

**facultative parasite.** An organism that is normally free-living but can become symbiotic when the proper conditions are presented.

**falx.** Densely ciliated margin on the anterior of certain members of the protozoan subphylum Opalinata.

**feral.** Wild; a feral cycle of a parasitic agent is one that takes place in the wild as opposed to an urban site.

**flame cell (=protonephridium).** The terminal cell of the excretory system in members of the phylum Platyhelminthes; a tuft of cilia in the cavity of the cell beats to move fluid along the tubule; the beating of the cilia give the impression of a flickering flame.

**flotation.** In parasitology, the term is used for techniques of fecal examination in which eggs or cysts are concentrated by suspending them in a medium of high specific gravity so that they will rise to the top of the liquid.

**fomite.** An inanimate object that is involved in the transmission of an infectious agent.

**gamogony (=gametogony).** Formation of gametes.

**glochidium (-ia).** Larva of members of freshwater mussels of the family Unionidae.

**gonotyl.** A muscular sucker, often with spines, surrounding the gential pore in certain digenetic trematodes such as *Heterophyes*.

**granuloma (-ata).** A swelling composed of leukocytes, fluid, and connective tissue; often a foreign-body reaction.

**gubernaculum.** A sclerotized plate located on the dorsal surface of the cloaca of certain male nematodes; serves as a guide when the copulatory spicules are protruded.

**gynecophoral canal.** The longitudinal groove on the ventral surface of male schistosomes in which the female worm lies.

**helminth.** A worm.

**hema- (Br. haema-).** A prefix meaning blood.

**hematophagous (Br. haematophagous).** Bloodsucking; usually refers to the feeding habits of various insects and acarines such as mosquitoes and ticks.

**hematuria.** Blood in the urine; a condition seen in some persons infected with *Plasmodium falciparum*.

**hemimetabolous.** A type of development in insects in which there is a gradual change in the external structure as development proceeds to the adult stage; Hemimetabola is the name of the taxonomic group.

**hemorrhage (Br. haemorrhage).** Escape of blood from vessels.

**hepato-.** A root referring to the liver.

**heterogonic.** Reproduction in which sexual and asexual generations alternate, as in the nematode *Strongyloides.*

**heteroxenous.** Having more than one host required to complete a life cycle, such as in the digenetic trematodes (see homoxenous).

**hexacanth embryo.** See oncosphere.

**holometabolous.** A type of development in insects in which there is a distinct change in the morphology of the stages; the cycle usually includes several larval stages, a pupa, and the adult; Holometabola is the name of the taxonomic group.

**homoxenous.** An adjective referring to a parasite that has a direct life cycle or one in which only a single host is required for its completion.

**horizontal transmission.** Transmission of a parasitic agent among members of a group (see vertical transmission).

**host range.** The number of species of hosts in which a parasitic agent can develop (see euryxenous and stenoxenous).

**humoral immunity.** A specific response to an antigen in which the principal effectors are antibodies that circulate in the blood.

**hyper-.** A prefix meaning more than normal, over or above (ant. hypo-).

**hyperemia (Br. hyperaemia).** An abnormally large amount of blood in a tissue.

**hyperplasia.** An abnormally high number of cells in a tissue.

**hypersensitivity (=allergy).** A condition in which a mammal is sensitized to a particular substance and has an abnormally strong reaction when the substance is contacted again; hay fever is an example.

**hypertrophy.** An abnormal increase in the size of a tissue.

**hypo-.** A prefix meaning less than normal, under or below (ant. hyper-).

**-iasis.** A suffix meaning to be infected with, as in helminthiasis (see -osis).

**imago.** The adult stage of an insect.

**immunity.** A specific response in vertebrates to a foreign protein in which cells respond by producing humoral and/or cellular antibodies.

**immunopathology.** An immune response that is damaging in itself.

**incidence (of infection).** The proportion of a population infected or showing disease over a particular time period; in public health, the incidence of measles, for example, would be given as 23/100,000 persons per year.

**incidental parasite.** Parasite found in a host other than its usual one.

**incubation period.** The period of time between infection and the appearance of clinical signs or symptoms (see prepatent period).

**indirect life cycle.** A life cycle in which more than one host is required for its completion (see heteroxenous).

**inflammation.** A response by a vertebrate to physical or chemical insult in which there is pain, reddening, increased temperature, and swelling at the site (see -itis).

**instar.** A stage in the life cycle of an insect, such as a larval or nymphal instar.

**intermediate host.** Host in which a parasite is not sexually mature.

**-itis.** A suffix meaning inflammation.

**juvenile.** An organism similar to the adult of the species but is sexually immature.

**kentrogen.** The undifferentiated cell mass injected into a host by larvae of certain parasitic barnacles (class Cirripedia).

**kinetoplast.** An organelle characteristic of protozoa of the order Kinetoplastida; actually a mitochondrial compartment packed with minicircles and maxicircles of DNA.

**kinety.** A longitudinal row of ciliary basal bodies and their connecting fibrils which extend anteriorly from each basal body.

**lacunar system.** A system of fluid-filled canals in the tegument of members of the phylum Acanthocephala; apparently serves as a circulatory and hydraulic system.

**larva.** An embryo that becomes self-sustaining and independent before it has developed the characteristic features of the adult form; a larva is anatomically different from the adult.

**larviparous.** Giving birth to larvae.

**Laurer's canal.** A tubule, usually blind-ended, which arises near the junction of the oviduct and vitelline duct in many digenetic trematodes; probably represents a vestigial vagina.

**lemniscus (-i).** One of a pair of lateral, fluid-filled sacs at the base of the proboscis in members of the phylum Acanthocephala.

**lumen (-ina).** The central cavity of an organ such as the lumen of the intestine.

**lycophore.** The ten-hooked or decacanth larva that develops within the egg capsules of tapeworms of the subclass Cestodaria.

**mammillated.** Having many small nipplelike protrusions.

**mastigont apparatus.** A group of structures in members of the protozoan order Trichomonadida: flagella, parabasal body, costa, axostyle, pelta.

**-megaly.** A suffix meaning to be enlarged beyond normal size.

**Mehlis' gland.** Secretory cells surrounding the ootype of certain symbiotic Platyhelminthes; serous and mucous cells are present, but their function is uncertain.

**merogony.** A type of asexual reproduction in which there is nuclear replication without plasmotomy and then two to many merozoites or daughter cells are produced simultaneously; a type of schizogony in which merozoites are formed; examples are found in many members of the Apicomplexa such as *Eimeria* and *Plasmodium* (see schizogony).

**merozoite.** Product of merogony; usually an elongate organism that infects another host cell to undergo either merogony again or gamogony.

**meta-.** A prefix meaning many, after, changed, or behind.

**metacestode.** Tapeworm stage following the oncosphere, but one not yet sexually mature.

**metamerism.** Segmental repetition of homologous parts; in each metamere or segment there are identical structures such as muscles, neural ganglia, and nephridia.

**metraterm.** Muscular, terminal portion of the uterus of certain digenetic trematodes; a diagnostic feature not present in all species.

**microneme.** Slender, chordlike bodies in the anterior of zoites of certain members of the phylum Apicomplexa.

**microtriches (sing. microthrix).** Surface projections from the tegument of tapeworms; resemble microvilli of vertebrate intestinal epithelial cells, but have characteristic dense caps; function to increase surface area.

**molt(=ecdysis).** Shedding of an external covering such as integument or exoskeleton; in arthropods and nematodes, shedding the external covering is integral to growth.

**monoecious.** Both male and female sex organs in one individual; hermaphroditic (ant. dioecious).

**monoxenous.** Having a single host in the life cycle (see heteroxenous).

**monozoic.** Refers to a condition in members of the tapeworm subclass Cestodaria in which there is only a single set of reproductive organs.

**mucous membrane.** Any of several moist surfaces in the body of vertebrates in which there are mucus-secreting or goblet cells; examples are the orbit of the eye, nasal passages, inside the mouth.

**mucron.** An organelle of certain gregarines which anchors the organism to its host cell; formed from elements of the apical complex.

**multivoltine.** Having a number of generations in a year.

**mutualism.** Symbiosis in which both partners benefit.

**naiad.** The preadult stage of an insect, which has hemimetabolic development in an aquatic environment (see nymph).

**neoteny.** A condition in which an individual that still has larval characteristics becomes sexually mature.

**nephro-.** A root referring to the kidney.

**nymph.** The preadult stage of an insect, which has hemimetabolic development in a terrestrial environment (see naiad).

**obligate parasite.** A parasite that requires a host for the completion of its life cycle.

**-oma.** A root meaning tumor or swelling.

**oncomiracidium.** A free-swimming, ciliated larva of members of the class Monogenea.

**oncosphere.** The six-hooked embryo that is contained in the egg membranes of members of the subclass Cestoda; hexacanth embryo.

**oocyst.** A stage in the life cycle of certain members of the phylum Apicomplexa in which the zygote secretes a wall around itself; often highly resistant to environmental conditions.

**ookinete.** The zygote in the life cycle of certain members of the phylum Apicomplexa following syngamy of macro– and microgametes; the term most often refers to the motile stage of *Plasmodium* which is seen in the midgut of the mosquito shortly after syngamy.

**opisthaptor.** The highly specialized, posterior holdfast of members of the class Monogenea; sometimes also refers to the ventral adhesive disk of members of the class Aspidobothrea.

**-osis.** A suffix meaning diseased, as in helminthosis.

**-otomy.** A suffix meaning to cut.

**oviparous.** Egg laying; producing eggs that hatch after leaving the body of the mother.

**ovoviviparous.** Producing eggs with persistent membranes through which the young escape while still within the body of the mother.

**ovum (-a).** The female germ cell.

**papilla (-ae).** A nipplelike structure.

**parabasal body.** An organelle that is part of the mastigont apparatus of certain members of the protozoan order Trichomonadida; has the function of the Golgi of metazoan cells.

**parameter.** A quantity whose value varies with the circumstances in which it is applied.

**parasitism.** An association between the populations of two species in which the smaller (parasite) is physiologically dependent on the larger (host), the prevalence of the parasites and the intensity of the infection are nonrandom and the parasite species has a higher reproductive potential than the host species; the parasite has the potential of harming the host.

**parasitophorous vacuole.** A clear space between an intracellular parasite and the host cell cytoplasm.

**paratenic host (=transport host).** A host in which a parasite resides but does not develop and which is not physiologically essential for the completion of the life cycle.

**parenchyma.** A reticulum of cells between the organs of an animal; also, the cells that perform the principal functions of an organ.

**paroxysm.** A sudden intensification or recurrence of symptoms of a disease.

**parthenogenesis.** Development of an organism from an unfertilized egg; common in insects such as aphids and in some nematodes such as *Strongyloides*.

**patent period.** The period during an infection when the causative agent can be found by some means.

**pellicle.** A double external membrane.

**peritoneum.** A thin membrane of mesodermal origin that lines the body cavity of vertebrates (and some other higher metazoans) and supports the organs of the body cavity.

**peritrophic membrane.** A covering that forms around the blood meal of a hematophagous insect; digestion usually takes place within the membrane; the membrane sometimes serves as a barrier to a parasitic agent moving from the blood to the tissues of the host.

**petechia (-iae).** Pinpoint hemorrhage in a tissue or organ.

**phago-.** A root meaning to eat.

**phasmid.** Sensory pit located on the posterior part of nematodes of the class Secernentea (see amphid).

**-philia.** A root meaning affinity for or an excess of.

**plasma.** The fluid portion of the blood (see serum).

**plasmodium (-ia).** An organism that is multinucleate but has a single outer limiting membrane.

**plasmotomy.** Fission of a multinucleated protist into two or more multinucleated daughter cells without direct relationship to nuclear division.

**plerocercoid.** A larva or metacestode developing from a procercoid in the life cycle of tapeworms of the orders Proteocephala and Pseudophyllidea; this type of tapeworm larva is solid (as opposed to hollow) and has a rudimentary holdfast at the anterior end.

**pleuro-.** A root meaning at the rib; refers to the lining of the thorax.

**polar filament.** A coiled filament or tubule in the spore of protozoa of the phyla Myxozoa and Microspora; the filament is extruded when the spore is ingested by a host and serves either to slow down the passage of the spore through the gut or to inject the sporoplasm into the tissue of the host.

**polyembryony.** A process in which a zygote gives rise to more than one embryo.

**polyzoic.** Consisting of more than one zooid or animal or proglottid.

**premunition.** A type of immunity in which the continued presence of the parasite in the body of the host is necessary for the maintenance of effective immunity.

**prepatent period.** That time after infection but before the causative agent can be found by usual diagnostic techniques.

**procercoid.** A larva or metacestode that does not have a scolex similar to that of the adult; commonly found in members of the tapeworm orders Proteocephala and Pseudophyllidea; this type of larva is solid (as opposed to hollow)

**proglottid.** A body segment of a tapeworm containing a complete set of reproductive organs.

**prohaptor.** The anterior holdfast of members of the class Monogenea.

**prophylaxis.** Prevention; procedures that are carried out to prevent the transmission of the parasitic agent or the occurrence of disease.

**protandry.** The condition in which the male matures before the female; especially refers to a gonad that first produces sperm and then ova.

**protoscolex (protoscolices).** A holdfast of tapeworms of the order Cyclophyllidea which forms from a germinal epithelium in a coenurus, hydatid, or alveolar cyst.

**pseudocoelom.** A body cavity of a metazoan that is not completely lined with mesoderm.

**pter-.** A root meaning wing.

**quarantine.** Limitation in the freedom of movement of humans or animals in order to contain the spread of a parasitic disease; the length of time is slightly longer than the longest known incubation period of the disease in question.

**radula.** Rasping structure inside the mouth of mollusks of the classes Gastropoda, Cephalopoda, Amphineura, and Scaphopoda; the pattern of teeth on the radula is characteristic of a species.

**recrudescence.** The recurrence of signs or symptoms of a disease after an abatement of days or weeks (see relapse).

**relapse.** The recurrence of signs or symptoms of a disease after an abatement of weeks or months; in malaria, the term is used for the reappearance of clinical disease after successful meronticidal treatment.

**renal.** Of or near the kidney.

**reportable disease.** A disease that, by law, must be reported to a health authority such as a state department of health or the office of the state veterinarian; in general, such diseases are of special concern to the health of the human or animal population; examples in the United States are malaria, rabies, and tuberculosis.

**reservoir host.** Any organism or inanimate substrate in which a parasitic agent normally lives and multiplies and from which those hosts important to humans can become infected.

**retrofection.** A process of infection in which the parasite leaves the body of the host and then almost immediately penetrates the skin; an example is *Strongyloides,* a common nematode of humans and dogs.

**rhoptry (-ies).** Saclike, electron-dense structure in the anterior portion of a zoite of a member of the phylum Apicomplexa; perhaps involved in the penetration of host cells.

**Romaña's sign.** Periorbital swelling characteristic of the early stages of infection with *Trypanosoma cruzi,* the cause of Chagas' disease.

**Romanowsky stain.** Any of a number of blood or tissue stains that contain eosin and methylene blue; examples are Wright and Giemsa stains.

**rostellum.** A prominence on the anterior end of the scolex, usually fitted with hooks, of certain tapeworms of the order Cyclophyllidea.

**Saefftigen's pouch.** A muscular, fluid-filled sac in the posterior end of male acanthocephalans that helps to evert the copulatory bursa.

**schizogony.** A type of asexual reproduction in which there are multiple nuclear divisions and then plasmotomy takes place, giving rise to a large number of daughter cells; occurs often in members of the phylum Apicomplexa; merogony, sporogony, and microgametogony are types of schizogony.

**scolex (scolices).** The holdfast or organ by which a tapeworm attaches to the intestine of its host.

**SEM.** Scanning electron microscope.

**sequela (-ae).** A diseased condition resulting from a previous disease.

**serum.** The fluid part of vertebrate blood after the fibrin has been removed.

**sign.** Any objective evidence of disease; fever, diarrhea, and skin rash are examples (see symptom).

**sparganum.** A plerocercoid or larval tapeworm, usually of the order Pseudophyllidea, of unknown specific identity.

**spicule (=copulatory spicule).** An elongate, sclerotized structure of male nematodes used in holding open the vulva of the female during copulation and transfer of sperm.

**spleno-.** A root referring to the spleen.

**spore.** A resistant stage that is formed internally by the mother cell (see cyst).

**sporogony.** Schizogony in which the product is the sporozoite.

**sporozoite.** The infective or transfer stage in members of the phylum Apicomplexa

**stenoxenous.** Having a narrow host range (see euryxenous).

**stichosome.** A column of large, rectangular cells (stichocytes) posterior to the short, muscular esophagus of members of the nematode order Trichurida; function unknown.

**strobila.** A chain of tapeworm proglottids or segments.

**Swimmer's itch (=cercarial dermatitis).** A skin reaction to the penetration of cercariae of certain members of the trematode family Schistosomatidae; actually a hypersensitivity to the cercariae and their secretions.

**sylvatic.** Refers to forest or a wooded area; used as an adjective to describe the location of a disease cycle in the wild, but often used imprecisely, such as in referring to sylvatic plague when the more correct term is *campestral plague*.

**symbiosis.** An intimate association of two organisms of different species.

**symptom.** usually reserved for subjective indicators of a disease condition such as pain, dizziness, nausea (see sign).

**symptomatic treatment.** Nonspecific therapy of a disease that is designed to reduce the symptoms or the effects.

**syzygy.** End-to-end joining of two or more gamonts of members of the sublcasses Gregarina and Coccidia.

**tachyzoite.** Rapidly growing meront or zoites characteristic of the early stage of infection with *Toxoplasma* and related organisms of the phylum Apicomplexa.

**TEM.** Transmission electron microscope.

**temporary parasite.** A parasite that visits a host at intervals and only for relatively short period; examples are mosquitoes and ticks.

**transport host.** See paratenic host.

**trochophore larva.** A free-swimming larval form of some marine annelids and mollusks.

**trophozoite.** The growing, feeding stage of a protozoan, also called the *vegetative stage* (see cyst).

**uniramous.** Having one branch.

**univoltine.** Having one generation per year.

**uterine bell.** A funnel-shaped structure in female acanthocephalans that serves as a sorting device for eggs.

**vector.** An arthropod, mollusk, or other invertebrate that serves to transmit an infective agent to a vertebrate host; a mechanical vector is one in which the parasite neither multiplies nor develops; a biological vector is one in which the parasite either multiplies or develops; some vectors are intermediate hosts, such as snails in the life cycles of digenetic trematodes; some vectors are definitive hosts, such as mosquitoes in transmitting the causative agent of malaria.

**veliger.** Second-stage larva of marine mollusks; the velum or ciliated swimming organ is characteristic.

**vertical transmission.** Transmission of a parasite from one generation to the next through the egg or in utero.

**virulence.** The ability of a parasite to produce pathogenic effects or to invade.

**Winterbottom's sign.** Enlargement of the cervical lymph nodes early in the course of African trypanosomiasis in humans.

**worm.** An organism without legs that creeps or crawls and is not obviously anything else.

**xenodiagnosis.** A diagnostic method in which the natural vector of the suspected organism is allowed to feed on the host and later the organism is looked for in the tissues of the vector; used almost entirely for the diagnosis for Chagas' disease in humans.

**zoonosis (-es).** A disease common to humans and other animals.

**zoonotic agent.** An organism that causes a zoonosis.

# Index